Developments in Plant and Soil Sciences

VOLUME 68

The titles published in this series are listed at the end of this volume.

Progress in Nitrogen Cycling Studies

Proceedings of the 8th Nitrogen Workshop held at the University of Ghent, 5–8 September, 1994

Edited by

O. Van Cleemput, G. Hofman and A. Vermoesen

Partly reprinted from *Plant and Soil*, Volume 181, No. 1 (1996)

SPRINGER SCIENCE+BUSINESS MEDIA, B.V.

Library of Congress Cataloging-in-Publication Data

ISBN 978-94-010-6292-3   ISBN 978-94-011-5450-5 (eBook)
DOI 10.1007/978-94-011-5450-5

Printed on acid-free paper

Contents

---

* Chapters indicated with an asterisk are reprinted from *Plant and Soil* Volume 181, No. 1 (1996). In these chapters the book pagination is printed at the bottom of the pages.

SECTION 2: ASPECTS RELATED TO FERTILIZATION: UPTAKE, AVAILABILITY, EFFICIENCY, TECHNIQUES

SECTION 6: ASPECTS RELATED TO AMMONIA VOLATILIZATION

SECTION 7: ASPECTS RELATED TO NITRIFICATION AND DENITRIFICATION

SECTION 8: ASPECTS RELATED TO LEACHING

SECTION 9: ASPECTS RELATED TO METHODOLOGY

Plant and Soil **68**: xiii, 1996.

Preface

Although nitrogen dynamics in different ecosystems have been studied during several decades, new orientations and other emphases came up. At times, nitrogen was considered as an essential element mostly in terms of productivity. Now, nitrogen is especially seen within the context of productivity in combination with environmental consequences.

More than 100 contributions in this book tackle recent developments within the field of nitrogen advice systems, plant response to fertilization, immobilization and mobilization, nitrification, denitrification, leaching, ammonia volatilization as well as biological nitrogen fixation. A high number of papers is devoted to the formation of gaseous nitrogen compounds, while organic matter turnover is another topic of important interest.

This book contains the proceedings of the "8th Nitrogen Workshop", which was held at the University of Ghent, Belgium, from 5 to 8 September 1994. This workshop was the first one outside the UK, and it was attended by 300 scientists. More than 200 papers and posters were presented: 26% on nitrification/denitrification aspects, 22% on fertilization (uptake, availability and efficiency), 21% on organic matter turnover, 10% on modelling, 8% on leaching, 6% on fertilizer advice systems and nitrogen balances, 2% on biological nitrogen fixation and 1% on ammonia volatilization.

We hope that this book brings a number of ideas for new developments in nitrogen research.

Oswald Van Cleemput   April 1996
Georges Hofman
Annick Vermoesen

Plant and Soil **181**: 47–56, 1996.

Future trends in nitrogen research

S.C. Jarvis
Institute of Grassland and Environmental Research, North Wyke, Okehampton, Devon, EX20 2SB, UK*

Key words: denitrification, gaseous losses, interactions, leaching, mineralization, nitrification, systems

Abstract

N research effort has undergone major changes over recent decades with changing emphasis because of environmental problems and issues. This driving force, coupled with a universal desire to improve N-use efficiency, appreciation of the importance of maintaining soil resource quality and a need to provide integrated landscape managements, will continue to prompt new research areas and issues for study. Already, much information has been provided and new approaches and needs defined. It will be essential in future research to take full note of the many interactions that occur and to provide a mechanistic basis so that scaling of effects can be undertaken with the appropriate simplification without being superficial. Examples of interactions, as well as fundamental gaps in the basic processes are discussed and needs for future research identified.

Introduction

For the efficient functioning of all ecosystems, whether under agricultural management or in natural/semi-natural environments, the importance of nitrogen (N) cycling is profound. This has been appreciated for decades and researchers have therefore investigated many of the components of the cycle. However, despite the longevity of interest and the intensity of research activity, and despite a good appreciation of the basic processes and structure of the cycle, many components are still poorly understood and/or quantified. Further, as demands for new information have arisen, so it has become apparent that results from all situations do not necessarily fit well with existing general concepts and model descriptions.

In large part, the stimulus to research various components of the cycle afresh has come about through environmental pressures. Whilst it is true that many research projects in many parts of the world are still driven simply by the need to achieve crop yields that are closer to the potential for a given site, many others have arisen because of problems of pollution and the needs of policy makers to make attempts to predict pollution effects and/or meet legislative demands. The overall result of this has been, so far as the N-cycle is concerned, (i) a refocusing of effort to meet specific demands, often to provide quantification (e.g. how much NO_3^- leaches from particular managements), (ii) development of novel approaches to tackle hitherto neglected areas (e.g. determination of gaseous N fluxes), and (iii) a much wider appreciation of the complexity and interactions that occur and the need to develop models of varying degrees of sophistication to aid prediction.

At least in the foreseeable future it seems likely that problems of excess, mobile N leaking in one form or another to the wider environment will continue to dictate research demands. Nitrate in waters remains, because of current legislation, a major issue, but volatilization of ammonia (NH_3) and the release of nitrous oxide (N_2O) and the other oxides of N (NO, NO_x) into the atmosphere and waters are all of concern and interest and will require greater understanding. This must also be seen in the context of other driving forces for research on N:

(1) The N cycle is scientifically challenging: its complexity and interactions attract research initiatives, and the availability and applicability of new tools and approaches provide stimuli to researchers.

(2) There is still a need to provide crop (plant) response information. Many areas require increased agricultural production and N, in many circumstances, is a major limiting factor. In all systems, there is a

---

* FAX No.: +44183782139
Plant and Soil is the original source of publication of this article. It is recommended that this article is cited as: *Plant and Soil* **181**: 47–56, 1996.

demand for increased efficiency of utilization and this requires new data for many existing managements.

(3) Novel cropping regimes, crops and management practices that become available have consequences for N flows and an awareness of these is essential. Organic farming, use of land for 'set-aside' and other new managements, put particular demands on nutrient flows which are not yet understood and will require new knowledge.

(4) Increasingly, there will be demands to integrate agriculture more closely with other components of the countryside. Nutrient flows, and that of N in particular, are important features of this integration and will require control and manipulation and, therefore, understanding.

(5) There is growing appreciation of soil as an important resource and of the need to 'protect' that resource. Soil quality is not easily defined (it depends on 'quality for what?') but N status is an essential component and will require characterization and definition in relation to the demands of the particular system.

The present paper is not an attempt to review the overall current status of N research and pinpoint detailed gaps in our knowledge: this would have been difficult because of the breadth and depth of the subject. Instead, it is rather to identify and illustrate, using recent information based in large part on grassland N research from the Institute of Grassland and Environmental Research, general areas and approaches which will be required to meet the demands of the issues already noted. It will concentrate particularly on soil N processes and losses and their impact over a range of spatial and temporal scales.

Soil transformations: current status and research needs

An emphasis on environmental issues with respect to N has brought about changes in research activities over the last 10 years or so. In the first instance there was a need for quantification of losses under various defined circumstances so that some preliminary assessments of impact could be made. Much information of this nature has accumulated, especially with regard to NO_3^- leaching and there is now a substantial literature on the rates of loss of N by leaching and by other means from various managements. Although the emphasis may change between particular processes, there will still be a requirement to provide direct quantification of losses of this nature.

However, numerous measurements have demonstrated the complexity of the various systems, and that, in order to make confident model predictions for other circumstances, a much better understanding of the transformation processes is required. Future research will not only have to probe the processes further but also recognize the very high degree of interaction that occurs and the need to integrate information so that effective, reliable extrapolation can be made over a range of spatial and temporal scales. In the following, by no means exhaustive, examples some of the problems and issues are discussed.

Mineralization/immobilization

The balance between these two microbially based processes is central to the flows and availability of mobile forms of N in the soil. Much research effort has been undertaken to determine rates in the field and under controlled conditions and much information already exists. This has been reviewed extensively recently (Powlson et al., 1994). However, it is not yet possible to provide adequate reliability for prediction which is of immediate relevance to, and needed for, decisions on fertilizer requirements and recommendations, NO_3^- leaching, recycling of N from crop residues and animal manures, maintenance of soil organic matter contents, emissions of trace greenhouse gases and the future management of changing agricultural practices.

Much is known on a qualitative basis about many of the soil factors which influence mineralization but the extent of data is limited and does not necessarily allow a practical interpretation of impact on N cycling in other than a few restricted circumstances. There is little doubt that the practical effects of mineralization can be substantial. Recent studies have attempted, using a field based incubation technique with regular, frequent sampling, to produce an annual net mineralization rate for a range of different grassland soils (Table 1). The results showed a wide range of maximum daily rates from 1.01 to 3.19 kg N ha^{-1} and net annual rates ranged up to 370 kg N ha^{-1} in an intensively managed grazed pasture. Even where there had been no fertilizer N applied for many years 135 kg N ha^{-1} were released. Addition of fertilizer to a previously unfertilized sward significantly increased the net release of N, but withholding N had no effect on mineralization during the year of measurements. Changes in temperature accounted for 35% of the variability in daily rates but there was little significant effect of soil moisture. It is of immediate relevance to the problems of pollution to note that, on average, 27% of the annual release occurred during the period November - January, i.e. when any NO_3^- in the system would have been particularly vulnerable to losses in winter drainage.

The amounts of net release were large and must have considerable impact on the overall efficiency of N in these grassland soils. Preliminary studies of the same soils have indicated significant differences between the distribution of organic materials between different physical fractions of the soils and between the different treatments and differential mineralization/immobilization rates of those materials when measured under controlled conditions. Soil texture apparently exerts control over mineralization by influencing aerobicity, affecting the physical distribution of organic materials and conferring some degree of "protection" through an association of organic matter with clay particles (Hassink et al., 1993). However, there is not yet enough appropriate information to allow the development of effective 'broad-brush' predictive models. This will require better definition of soil organic matter, its location over a fine spatial scale and its interaction with soil biology and environmental status. An important component of this will be to derive/link relationships between soil texture, structure and pore size and to determine whether protected soil organic matter is of practical relevance. We will also need to define the quality of soil organic materials and added residues in terms of N (and other nutrient) supplying potential rather than by chemical composition.

Another important aim in studies of mineralization, and indeed of all soil N cycling processes, will be to consider the architecture of the soil (Dexter, 1988) and the location of the internal N cycle and its relationship to soil organisms. A knowledge of the diffusional constraints to NH_4^+ and NO_3^- movement (i.e. between different pools) will also be required. This would help to provide the mechanistic basis for understanding the controls over mineralization at a fundamental level: the concepts of microsites and different 'pools' of potentially mineralizable N and of mineral N are implicit in this. With this as a basis it should be possible to integrate information sequentially into a series of larger scales.

Soil microbial biomass

Soil microbial biomass has a profound effect on the net release of mineral N but we cannot yet assess the possibility of being able to manipulate N flows though

the complex food webs that this forms part of in soils to the advantage of agricultural and other systems. An understanding of this is essential to aid the development of all management systems. Microbial biomass is important as an agent of change, decomposition and release of N from fresh organic residues and native soil organic matter into more labile forms, as a major sink for 'active' soil N and as a potential source of labile N. One description has been as "the eye of the needle though which all organic matter must pass" (Jenkinson, 1990).

In grassland soils the amounts of soil microbial biomass present were substantial, but were not apparently influenced to a very marked degree by background management (Bristow and Jarvis, 1991) . In other studies summarized in Table 2, there were few differences in trends of biomass N (or C) with time, drainage, withdrawal of fertilizer N from a previously fertilized soil, or addition of fertilizer N to a previously unfertilized soil. There was, however, substantially more biomass in the previously unfertilized than fertilized treatments. In contrast, measurements of ATP, enzyme activity and respiration showed significant effects of both short and long term treatments. It is also apparent that the patterns and effects demonstrated with soil microbial biomass (Table 2) do not conform with those found in the same soils for net mineralization rates (Table 1). There is therefore a need, firstly, to understand the differences in microbial activities which occur in response to short term management changes and to identify and quantify differences in microbial community structure. Secondly, if the aim of making progress in increasing the efficiency of N utilization from all sources is to be realized, the integration of this information with better understanding of the soil organic matter and added residues is essential and will require research at a fundamental level.

Nitrification

As far as the fate of any N released through interaction of mineralization/immobilization processes is concerned, nitrification is an important step. The transformation from a relatively immobile (NH_4^+) to a highly mobile (NO_3^-) stage via NO_2^- provides opportunities for escape of a number of materials (NO_3^-, NO_2^-, NO_x, N_2O, N_2) with the obvious consequences. Despite its importance as a rate limiting process and a reasonable knowledge of the ecology and environmental demands of the organisms involved, it is a poorly defined pro-

Table 1. Mineralization rates in soils under long-term pastures (from Gill et al., 1995)

N treatment (kg N ha^{-1})		Drainage	Total net annual mineralization (kg N ha^{-1})	Annual turnover (% total soil N)
Past	Present			
+200	+200	Drained	376	6.5
+200	+200	Undrained	317	5.7
+200	0	"	292	5.5
0	0	"	135	2.5
0	+200	"	270	5.3

Table 2. Effects of past and present sward mangement on soil microbial biomass contents and activities (from Lovell et al., 1995)

N treatment (kg N ha^{-1})		Drainage	Biomass contents (mg kg^{-1} dry soil)		Dehydrogenase activity (n mol INTF g^{-1} 2 h^{-1})	ATP content (μg g^{-1})	Basal respiration (μg CO$_2$-C (g^{-1} h^{-1}))
Past	Present		C	N			
+200	+200	Drained	1056 ($\pm$ 34.9)	159 ($\pm$ 3.0)	889 a	3.1 a	0.9 a
+200	+200	Undrained	1055 ($\pm$ 28.9)	169 ($\pm$ 4.0)	1343 b	4.7 b	1.8 b
+200	0	"	1009 ($\pm$ 21.6)	161 ($\pm$ 4.7)	1204 b	3.4 a	1.8 bc
0	0	"	1616 ($\pm$ 53.3)	255 ($\pm$ 7.7)	1362 b	5.7 b	2.6 d
0	+200	"	1716 ($\pm$ 49.9)	275 ($\pm$ 4.7)	1124 ab	7.3 d	2.1 be

Significant differences between activities ($p<0.05$) shown by values without same letter.

cess in many soils. In all tillage systems it is usually assumed that nitrification is not a limiting process and that nitrification rates exceed the rate of net mineralization and do not impose limitation to further transformation. For many grassland (Jarvis and Barraclough, 1991) and natural systems, however, there may be significant quantities of NH_4^+ accumulated in the soil profile (Table 3). Nitrification interacts strongly with local ambient soil conditions and there is a high degree of spatial compartmentalization of NH_4^+ production and consumption sites. These, and the diffusional constraints between microsites in the soil, exert controls over nitrification rates. Latent potentials can apparently develop which are dependent upon background management and which can be displayed over the relatively short (Jarvis and Barraclough, 1991) or longer term (Willison and Anderson, 1991). There are also important competition effects with plant uptake, for example (Barraclough et al., 1983). Interactions with denitrification (see later) may also be of some importance with respect to removal of NO_3^- and the production of trace gases.

Nitrification is central to the flows, transfers, losses or utilization of N. It is also a key interacting process which is linked to others by flows of substrates and spatial distribution in soil systems and a greater understanding will be important in maximizing, especially in grassland situations, N efficiency and reducing losses.

Nitrogen loss processes

Concerns over emissions will continue, not only because of direct concerns over environmental issues, but also because there is a growing desire to make nutrient cycles within farming systems more efficient from economic and self-sufficiency stand points as well. Much information has been gathered over recent years on many aspects of the loss processes but there are still major problems in being able to predict losses in many situations and over a range of scales. Research will be required to continue to improve our understanding of the mechanisms involved, to have the basis to extrapolate information to other circumstances, to take account of various interactive forces and factors and to be able to produce predictive estimates of effects over a range of scales through from field to farm to catchment and region and to accurate national budgets where

Table 3. Effects of fertilizer inputs to grazed grassland soils on soil mineral N in autumn (to 40 cm). The ratio $NH_4^+ : NO_2^-$ is taken as an indicator of nitrification potential (data taken from Jarvis and Barraclough, 1991)

Soil	Mineral N ($\mu g\ g^{-1}$ soil)	Fertilizer rate (kg ha^{-1})			
		100	250	450	750
Hurley	$NH_4^+ + NO_3^-$	24.1	22.7	64.3	207.3
(loam)	$NH_4^+ : NO_3^-$	2.2	2.0	0.7	0.4
Jealotts Hill	$NH_4^+ + NO_3^-$	30.3	27.5	48.7	310.3
(fine loam)	$NH_4^+ : NO_3^-$	1.7	1.5	0.4	0.1
Ravenscroft	$NH_4^+ + NO_3^-$	27.4	248.5	186.7	292.3
(clay)	$NH_4^+ : NO_3^-$	0.4	0.1	0.3	0.1
North Wyke	$NH_4^+ + NO_3^-$	89.6	55.1	76.5	239.8
(clay)	$NH_4^+ : NO_3^-$	7.0	6.7	2.1	0.1
Drayton	$NH_4^+ + NO_3^-$	2.6	1.8	3.1	60.7
(clay/loam)	$NH_4^+ : NO_3^-$	0.4	1.6	0.3	0.2

these are required. In order to do this with confidence some important gaps need to be filled. Some examples follow for each of the three major loss processes.

Leaching

Primarily because of legislative demands, much data on NO_3^- leaching has been accrued over the last 5 years (see for example, Archer et al., 1992). Even accepting the problems associated with sampling to provide accurate assessment of solute transport in many soils, data are usually limited to quantification of particular agronomic/environmental combinations and do not necessarily provide the means of describing, on a mechanistic basis, leaching losses for many important soil types. Optimization of land management to meet the demands of legislation and to be able to continue viable production systems will require the use of simulation models which describe N losses over larger scales. Such models will be dependent on accurate sub-models which define the accumulation of potentially leachable soil NO_3^- linked with others to describe NO_3^- transport. Many excellent models of leaching exist (e.g. Addiscott and Whitmore, 1991; Barraclough, 1989), but the use of mechanistic models to simulate N transport in a number of soil types, notably those that have a defined structure, have met with only limited success. A new approach has been developed recently to generate 'real-time' data with which to test and validate and

thereby modify and improve existing models. By the use of a range of tracers (Cl, ^{15}N, deuterium), transport of H_2O and solutes is being quantified at very fine spatial and temporal resolution in a large (3.4 × 5.4 × 1.2 m) undisturbed soil block (Scholefield and Holden, 1994). Initial results are already illustrating the degree of spatial variability in solute concentration fronts at depth in a well drained, structured soil which makes prediction very difficult. More such information will be required before effects at the larger scale can be predicted.

Previous studies have allowed a number of empirical relationships to be established and because these are of some considerable value, many of the existing data sets should be re-examined to develop further relationships. Thus long-term studies on grazed pasture (1 ha drained lysimeters) have provided clear relationships between recorded soil moisture deficients during summer and NO_3^- leaching losses during the following winter (Scholefield et al., 1993b) with more leaching with increasing soil moisture deficit. Other relationships between the soil load of potentially leachable NO_3^- and the peak concentration at which it leached were also established. These relationships were independent of drainage volume and only slightly influenced by drainage intensity. Other data from long term experiments have also been examined in the same way (Table 4) and demonstrate the possibility of employing such empirical approaches to account for preferential

Table 4. Relationships between peak NO_3^- leachate concentration and soil load of NO_3^- on different soils and under different managements (data from Scholefield and Holden, 1994)

peak $(\text{mg dm}^{-2}) = a \times NO_3^-$ leached $(\text{kg N ha}^{-1}) + b$

Soil type	Agricultural system	a	b	r^2
Loamy sand	Grassland	0.99	1.4	0.88
"	Arable	0.92	2.4	0.93
Loam	Grassland	0.62	3.5	0.99
Clay loam (undrained)	Grassland	0.58	3.9	0.96
Clay (cracking) (mole drained)	Arable	0.37	2.8	0.33
Clay loam (mole drained)	Grassland	0.28	5.6	0.97

Table 5. Potential denitrification rates to depth below a long-term grazed grass-clover sward. Data are calculated from laboratory measurements of samples incubated under anaerobic conditions in the presence of added NO_3^- and with (NC) or without (N) added C (sucrose) and are calculated from Jarvis and Hatch (1994)

Sample depth (m)	Potential denitrification $(\text{kg N ha}^{-1}\,\text{d}^{-1})$	
	N	NC
0 - 0.5	14.3	21.9
0.5 - 1.0	6.9	8.8
1.0 - 2.0	15.3	30.7
2.0 - 4.0	4.7	67.2
4.0 - 6.0	3.9	35.3
Total	45.1	163.9

Table 6. Total denitrification and nitrous oxide emission rates $(\text{g ha}^{-1}\,\text{d}^{-1})$ (determined over 4 hours using an enclosure method with or without acetylene, respectively) on a peat soil with added cattle slurry (from Jarvis et al., 1994)

Days after spreading	Total denitrification	N_2O
Week 1		
1	20 (7.5)	7 (4.5)
2	80 (34.1)	30 (14.5)
3	260 (63.1)	54 (29.9)
4	46 (15.9)	24 (6.0)
5	86 (51.0)	47 (19.9)
Week 2		
1	66 (31.7)	67 (5.4)
2	15 (3.7)	29 (5.7)
3	31 (10.5)	34 (14.9)
4	167 (16.9)	49 (2.6)
5	86 (9.5)	34 (11.5)

flow in modelling NO_3^- leaching which may be especially appropriate at the catchment and regional scales.

It will be essential to build into all such predictions an appreciation of the interactions that may occur. Again at an empirical level, it is clear that increased soil aerobicity (through, for example, drainage) can have a profound effect on NO_3^- leaching not only though reduction in denitrification but also, it is suggested, an enhanced rate of net mineralization (Scholefield et al., 1993b). Quantification of these effects will also require good mechanistic understanding so that overall impact at the larger scale can be determined.

The fate of NO_3^- moving away from agricultural systems also needs to be assessed. It is often assumed that NO_3^- leaching from grassland systems is non-reactive and is transported to the saturated zone in confined aquifers without change. There is, however, evidence from a number of studies which suggests that there may well be denitrification effects which influence the final concentrations of NO_3^- in ground water (Gillham, 1991). This denitrification may also be an important removal mechanism as NO_3^- moves through the unsaturated zone as well (Lind and Eiland, 1989). Recent studies have shown that considerable potential for denitrification existed at least to 6–8 m below long term grassland and was increased substantially when labile carbon was added to the experimental system (Jarvis and Hatch, 1994). Calculations indicate (Table 5) that even if only a small proportion of this potential for removal was achieved in practice, this would have implications for the movement of NO_3^- into aquifers. However, the consequences of this with regard to emissions of environmentally active gases (i.e. N_2O) must also be considered and a greater knowledge is again required to enable better management and policy decisions relating to potential pollution from agriculture.

Denitrification

The problems associated with determining denitrification in soil are considerable and well known. The spatial and temporal variability impose considerable restrictions on our current ability to measure and determine its impact (Smith and Arrah, 1990). Despite the very large research effort that has, and is going on, much is still uncertain about the impact of denitrification with regard both to its influence on overall N use efficiency and to its contribution to N_2O budgets.

Recent attempts to synthesize information to provide N budgets and flows at a farming system level indicated that at least 16.5% of an annual N input of 24.5 tonnes N to a 76 ha dairy farm would be lost through denitrification (Jarvis, 1993). It was suggested that, because of a current inability to estimate losses via this route a major proportion of the fraction of N still unaccounted for in this farming system could also be ascribed to denitrification losses. Research must continue to address this problem as a matter of urgency in order to increase efficiency and to reduce N_2O emissions. It is perhaps worth noting, however, that the ultimate fate of N over time is to re-enter the global N cycle through denitrification and gaseous transmission to the atmosphere. If increasing agricultural efficiency results in greater capture of N in products, then denitrification will become more important at a later stage of the food chain after consumption by humans. In the meantime, there is much to be learnt about the process and its interaction with environmental conditions.

Studies with a poorly drained clay soil under controlled environment conditions and using a flow-through system have enabled definition of effects of H_2O, O_2 and NO_3^- contents on the amounts and proportions of N_2 and N_2O produced (Scholefield et al., 1993a). The next stage will be to use this information to develop models for field use. Other recent laboratory and field studies (Bisson et al., 1994) have further underlined the complexity of the controls and demonstrated the need to provide mechanistic explanation. In this latter case, the release of NO was used as a fingerprint to define nitrification. NO is short-lived gas, but its emission from the soil should provide an indicator of process activities. Although nitrification was the dominant source of NO, release was obvious on addition of either NH_4^+ or NO_3^-. The preliminary conclusions from this study were that nitrification was the primary source of NO but that there was some evidence that denitrification was also an important source under certain moisture conditions. This raises questions of (i) the extent of coupling between the two processes, (ii) the possibility of sharing joint metabolites if the processes occur concurrently, (iii) the relationships and interactions between the amounts of substrates (NH_4^+ and NO_3^-), and (iv) the reactions of the relevant microbiological populations and their enzyme systems and the release of the trace gases N_2O and NO at various stages of either the oxidative or reductive transition. There may also be opportunities for losses of another short-lived intermediate, NO_2^-, directly into waters in some circumstances.

With this degree of complexity it is not surprising that results of field measurements can often be difficult to interpret. Studies of N_2O release and denitrification in a peat soil with added cattle slurry showed, over a range of time scales, a wide range in denitrification rates and $N : N_2O$ ratios as well as an expected high degree of spatial variability (Table 6). It is of interest to note that, over a range of treatments, the major proportion (0.635) of the daily variability in total denitrification could be accounted for by changes in soil moisture and NH_4^+ contents. Inclusion of soil temperature and NO_3^- contents in the regression model did not improve the R^2 value. This further demonstrates, at a practical level, the importance of considering the coupling of nitrification and denitrification processes in making predictions of losses. This is in addition to providing a better means of determining denitrification under field conditions. Whilst there are new approaches and techniques available to determine net N_2O fluxes and budgets (e.g. micro-meteorology methods, photoacoustic, infrared and laser instrumentation) these will not necessarily provide sufficient basic information to extrapolate information to other circumstances. The requirement will be for integration of controlled environment and field measurements to be coupled with appropriate model simulations.

NH₃ volatilization

Of the three major loss processes, the chemical and physical controls over the transfer of NH_3 *in simple systems* are probably the best understood. However, despite this and although the sources are generally well known, emission rates and transfers of NH_3 within many agricultural and natural systems are not well defined. NH_3 volatilization is an important loss process especially from animal production systems and in a typical dairy farm approximately 13.6% of a total annual input of 25 tonnes of N are estimated to be lost (Jarvis, 1993). Measurement of NH_3 in the field has only been possible relatively recently so that data from many aspects of agricultural managements are restricted. There are a considerable number of data sets from some components of agricultural production (e.g. spread farm wastes, see Nielsen et al., 1991) a restricted number from others, (e.g. losses from grazed swards) and even less from others (e.g. from stored wastes and animal houses). The current need to provide national budgets for NH_3 because of its implications for atmospheric chemistry and critical loads for terrestrial and aquatic systems will require (i) much greater

Table 7. Interaction of effects of various factors and influences on the impact of N cycling processes over a range of scales

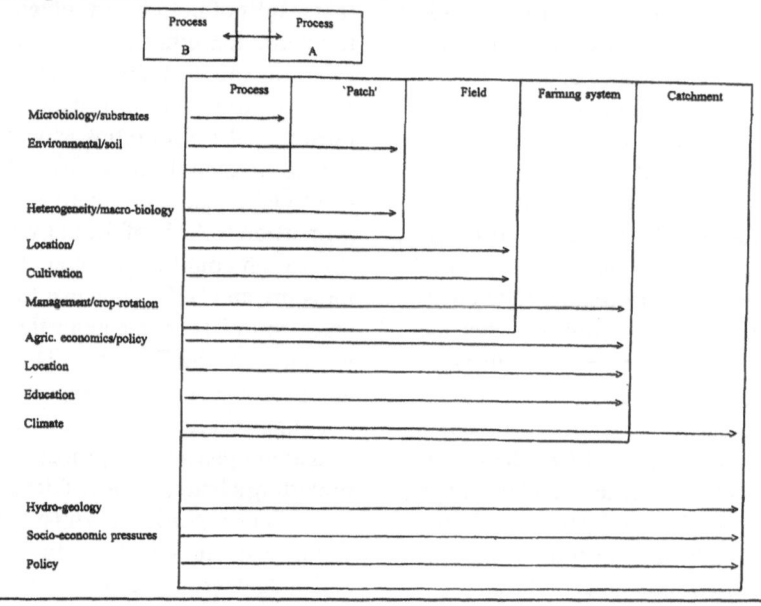

precision in estimates of emission factors from the major animal derived sources, especially excreta either deposited in fields and houses or stored, each under a wide range of conditions; (ii) a knowledge of the transfer coefficients for NH_3 from various sources over a range of scales, i.e. from urine patch to neighbouring sward, from fertilized, grazed or slurry-amended field to neighbouring areas, and from farming systems to adjacent natural systems; and (iii) some further definition of the role that crops have as sink or sources of NH_3. As remedial measures are introduced to reduce NH_3 emissions to the atmosphere (as they have been in the Netherlands), it is important that interactions with other processes and the consequences of these for the overall efficiency of the farming system are fully considered and quantified.

Interactions, integration and scaling of effects

Research approaches to N cycling have changed considerably over recent years and are opening up new insights to the transfer of N within various components and at various spatial levels within the whole complexity of the system. The demands of legislation and national policies have and, at least over the short term, will continue to raise issues which will require continued input into developing understanding and approaches to be able to do that. It is important on

the one hand that basic processes and mechanisms are understood and, on the other, that these are translated into simulation models appropriate to the question being raised. It is important that models of N flows continue to be developed at various levels of complexity. There is a need to define models operating for each of the various sub-systems involved so that simplification can be made further up the hierarchy but without the final model being a superficial treatment of the system. Changes in our perception about the various interactions and controls over individual processes have been made apparent through detailed experimental work and will continue with the use of new tools. For example, stable isotope definition of changes using enriched and natural abundance techniques; definition of controls over microbial activities in relation to community structure and location as determined by soil architecture using enzyme measurements and genetically manipulated organisms; long path detection systems for trace gas measurement; and near real time measurement of solute and water fluxes etc. are now, or becoming available. These will ultimately lead to increased accuracy in the models used for the management of agricultural systems and in the prediction of the impact of these systems on the environment.

There is a need therefore to integrate information which must be based on sound principles over a range of scales i.e. from microsite to patch, from patch to

Table 8. Estimated effects of management changes on nitrogen losses from dairy farming: initial desk studies, A: conventional 'model' management (see Jarvis, 1993); B: using 'tactical' fertilizer approach based on regular analysis of soil mineral N and injected slurry; C: grass/white clover swards; D: substituting 50% of grass silage by maize silage to (i) utilize slurry N and (ii) improve conversion of dietary N into product; E: B + D.

	A	B	C	D	E
N fertilizer use (kg ha^{-1})	250	200	0	185	148
Stocking rate livestock (units ha^{-1})	2.17	2.17	1.74	2.17	2.17
N losses (kg ha^{-1})	160	86	89	99	60

Table 9. Experimental management systems on 1 ha farmlets to determine the flows and losses of N from complete grassland production systems (from Pain et al., 1994)

Conventional management

N fertilizer:	Set times and amounts (280 kg ha^{-1} yr^{-1})
Slurry:	Surface applied in spring and post silage
Grazing:	May - October : beef steers
Silage:	3 times per year

Tactical nitrogen

N fertilizer:	'Tactical' adjustment based on fortnightly analysis of soil mineral N - to reduce excess soil mineral N and leachate concentrations (to below 11.3 mg L^{-1})
Slurry:	Injected - to reduce NH_3 volatilization - in spring and post-silage
Grazing:	May - August : beef steers
Silage:	4 times per year - 1 post grazing to remove excess soil mineral N

Grass/white clover

N fertilizer:	Nil
Slurry:	Injected in spring and autumn: + nitrification inhibitor to reduce denitrification/leaching
Grazing:	May - October: beef steers
Silage:	3 times per year

field, from field to farming system and from farming system to catchment (Table 7). Care must be taken to ensure that the models used to do this have the appropriate simplification and not superficiality. Throughout this discussion it has been clear that whilst a full understanding of the controls over a particular process is a prerequisite to be able to do this, full recognition must be made of interactive effects with other processes at the spatial/temporal level required. At the systems level the effects can be extremely important and require quantification.

Sufficient information and model development is available to be able to make estimates and calculations for the flows and transmission of N for existing farming systems (see Jarvis, 1993) and then to be able to manipulate managements to look at the impact on nitrogen losses. Information shown in Table 8 is a preliminary estimate of the impact that changing fertilizer and waste management on a dairy farm could have on leaching losses, taking into account the coincident effects that these changes have on mineralization, ammonia volatilization and denitrification for example. Exercises of this nature have value in that they provide an immediate perception of the overall impact and implications for a farming system, they identify and define gaps in knowledge and demonstrate the means of increasing N efficiency and reducing losses and environmental impact. Hand-in-hand with such desk studies, holistic system measurements should contribute to our understanding of N flows and transmission. As an example, a new series of experiments involving a series of grassland managements

56

(Table 9) has recently been initiated which is based on 1 ha farmlets and attempts to address the complexity of grassland production farm cycles within experimentally manageable areas. Each area is designed to contain elements of grazing, grass conservation and farm waste returns and measurements are made of herbage and animal production as well as intensive programmes to determine N losses and budgets. Whilst these farmlets and a range of managements (Table 9) provide a unique research opportunity for direct comparisons to be made in both environmental and production terms, it is critical that the controls over the processes that are deriving any changes in the flows of nutrients are fully appreciated. This will ultimately provide the means providing the model basis of extrapolating information to other systems or making prediction over longer time or large spatial scales.

References

Addiscott T M and Whitmore A P 1991 Simulation of solute leaching in soils of different permeabilites. Soil Use Manage. 7, 94–107.

Archer J R, Goulding K W T, Jarvis S C., Knott C M, Lord E, Ogilivy S E, Orson J, Smith K A and Wilson B (eds) 1992 Nitrate and Farming Systems. Asp. Appl. Biol. 30, 1–450.

Barraclough D 1989 A usable mechanistic model of nitrate leaching. I. The model. J. Soil Sci. 40, 543–554.

Barraclough D, Geens E L, Davies G P and Maggs J M 1983 Fate of fertilizer nitrogen III. The use of single and double-labelled ^{15}N ammonium nitrate to study nitrogen uptake by ryegrass. J. Soil Sci. 36, 593–603.

Bisson G, Jarvis S C and Barraclough D 1994 Sources of NO and N_2O in grassland soils. 8th Nitrogen Workshop, Ghent, Belgium.

Bristow A W and Jarvis S C 1991 Effects of grazing and nitrogen fertilizer on the soil microbial biomass under permanent pasture. J. Sci. Agric. 54, 9–21.

Dexter A R 1988 Advances in the characterisation of soil structure. Soil Tillage Res. 11, 199–238.

Gillham R W 1991 Denitrification in ground water. In: International Conference on N, P and Organic Matter. Ed. V Holter. pp 29–39. Danish Ministry of the Environment, Copenhagen, Denmark.

Gill K, Jarvis S C and Hatch D J 1994 Mineralization of nitrogen in long-term pasture soils: effects of management. Plant and Soil 172, 153–162.

Hasssink J, Bouwman L A, Zwart K B and Brusaard L 1993 Relationships between habitable pore space, soil biota and mineralization rates in grassland soils. Soil Biol. Biochem. 25, 47–55.

Jarvis S C 1993 Nitrogen cycling and losses from dairy farms. Soil Use Manage. 9, 99–105.

Jarvis S C and Barraclough D 1991 Variation in mineral nitrogen content under grazed grassland swards. Plant and Soil 138, 177–188.

Jarvis S C and Hatch D J 1994 The potential for denitrification at depth below long-term grass swards. Soil Biol. and Biochem. 26, 1629–1636.

Jarvis S C, Hatch D J, Pain B F and Klarenbeek J V 1994 Denitrification and the evolution of nitrous oxide after the application of cattle slurry to a peat soil. Plant and Soil 166, 231-224.

Jenkinson D S 1990 The turnover of organic carbon and nitrogen in soil. Philos. Trans. R. Soc., London. B 329, 361–368.

Lind A M and Eiland F 1989 Microbiological characterization and nitrate reduction in sub-surface soils. Soil Biol. Fertil. Soils 8, 197–203.

Lovell R D, Jarvis S C and Bardgett R D 1995 Soil microbial biomass and activity in long-term grassland: effects of management changes. Soil Biol. Biochem. 27, 969–975.

Nielsen V C, Voorburg J H and L'Hermite P (eds). 1991 Odours and Ammonia Emissions from Livestock Farming. Elsevier Applied Science London, UK. 222 p.

Pain B F, Jarvis S C and Laws J 1994 Reducing nitrogen losses from grassland production systems. IGER 1993 Annual Report. 83 p.

Powlson D S, Stockdale E A, Jarvis S C V and Shepherd M A 1994 A review of nitrogen mineralization and immobilization in UK agricultural soils. A review prepared for MAFF London, UK. 134 p.

Scholefield D, and Holden N 1994 Predicting nitrate leaching in structured soils. Institute of Grassland and Environmental Research 1993 Annual Report, pp 70–71.

Scholefield D, Hawkins J and Jackson S M 1993a The effects of nitrate, water content and temperature on denitrification in a grassland soil measure using a "flow-over" technique without acetylene blocking. 5th AFRC Meeting on Plant and Soil Nitrogen Metabolism 1993, Silsoe, UK.

Scholefield D, Tyson K C, Garwood E A, Armstrong A C, Hawkins J and Stone A C 1993b Nitrate leaching from grazed grassland lysimeters: effects of fertilizer input, field drainage, age of sward and patterns of weather. J. Soil Sci. 44, 601–613.

Smith K A and Arah J R M 1990 Losses of nitrogen by denitrification and emissions of nitrogen oxides from soils. Proc. Fertil. Soc. 299, 1–34.

Willison T. and Anderson J M 1991 Spatial patterns and controls of denitrification in a Norway spruce plantation. For. Ecol. Manage. 44, 69–76.

Section editor: R Merckx

1. ASPECTS RELATED TO ORGANIC MATTER TURNOVER

O. Van Cleemput et al. (eds.), Progress in Nitrogen Cycling Studies, 13–16, 1996.

Immobilization of labelled inorganic N and mineralization of applied organic N as reflected in electro-ultrafiltration and CaCl$_2$ soil extracts

Thomas Appel, Fuli Xu and Rolf Russow
Institute for Plant Nutrition, Justus-Liebig-University Giessen, Südanlage 6, D 35390 Giessen, Germany

Key words: CaCl$_2$ extraction, electro-ultrafiltration, extractable organic N, labelled nitrogen, nitrogen immobilization, nitrogen mineralization

Abstract

A soil incubation experiment was carried out over 80 days (aerobic, 20 °C). Mineralization of ^{15}N-labelled plant material (rape) and a microbial biomass suspension (*E. coli*) was monitored at various times during the incubation period by means of EUF and CaCl$_2$ soil extraction. An additional soil treatment (application of cellulose + labelled mineral N) was carried out in order to induce immobilization. Control treatments (soil without N application and soil with ^{15}N labelled NH$_4$NO$_3$ but without cellulose) were also incubated.

Both methods extracted decomposing plant residues with high selectivity. Not at any time was the mineralization potential of the green manure completely represented as extractable organic N. The applied microbial biomass mineralized mainly during the first week. It was not extractable as organic N by either EUF or by the CaCl$_2$ extraction. Transitory immobilization of inorganic N was then followed by a remineralization but was not reflected as extractable organic N.

It is concluded from these results that N mineralization indexes based on EUF and CaCl$_2$ extractable organic N do not completely account for all mineralizable N.

Introduction

The N supply to crops derives mainly from 3 sources: the inorganic N already present in the soil profile in spring, the N being mineralized during the vegetation period and the applied N. For N fertilizer recommendations to be given in advance, in spring, the soil inorganic N can be analysed. However, future N mineralization is unknown. Since most of the organic N present in the soil is highly resistant against N mineralization it appears to be more reasonable to estimate the amount of easily mineralizable N compounds in the soil. In Germany the electro-ultrafiltration (EUF) method has already been established for this purpose (Mengel, 1991). This method extracts not only inorganic N but also a low molecular weight organic N fraction which is assumed to represent easily mineralizable soil organic N. Recent research from field experiments has shown that similar results as those from the EUF can be obtained by a less expensive CaCl$_2$ extraction (Appel and Mengel, 1992). Although EUF is extensively used in Germany in practice to find N fertiliz-

er recommendations there is still no direct evidence to what extent EUF extractable organic N is actually mineralized. An incubation experiment was therefore carried out focusing how mineralization of ^{15}N labelled organic matter and immobilization of labelled inorganic N was reflected in EUF and CaCl$_2$ extractable N fractions.

Material and methods

The principle of the experiment was to mix soil and ^{15}N labelled organic and inorganic material and incubate the thus treated soil over 80 days in order to monitor release and immobilization of inorganic N and the extractability of organic N. The used soil derived from mottled sand stone sampled in early summer from the top layer of a farmers field in Hessia, Germany (pH 5.4 (CaCl$_2$), total N 1.58 g kg^{-1}, C:N ratio 11.3, clay 11%). The soil was sieved (1 mm) immediately after sampling and homogenized. For each treatment 39 topless poly-ethylene pots were each filled with 100 g field moist soil which had been treated as follows: treatment A = control without fertilization; B = green manure (4 weeks old rape, chopped to 1–5 mm pieces, 87% ^{15}N, 43.6 mg applied N kg^{-1} soil); C = *E. coli* suspension (21.8 mg applied N kg^{-1} soil); D = NH$_4$NO$_3$ solution (21.4 mg applied N kg^{-1} soil); E = NH$_4$NO$_3$ solution (21.4 mg applied N kg^{-1} soil) + cellulose (871 mg applied C kg^{-1} soil). Soil moisture was adjusted to 53% water capacity. From each treatment 3 replicate pots were reserved for immediate soil analysis (sampling date 0). The remaining pots were randomly put into an incubation cabinet at 20 °C. The soil was remoistened twice a week if necessary. Pots were sampled on days 0, 1, 2, 4, 6, 9, 12, 16, 20, 30, 40, 60, 80 of incubation. At each sampling date triplicate pots were taken from each treatment. Soil samples were air dried at 40 °C and ground (≤ 1 mm). Electro-ultrafiltration analysis of the soil samples was carried out according to Németh (1985) and fractionated CaCl$_2$ extraction as described by Appel and Mengel (1990). NO$_3$-N, NH$_4$-N and organic N in the extracts was determined by a colorimetric fashion by means of a continuous flow analyser. The N isotopic composition of the extracted N fractions was determined in EUF extracts of treatments B, D and E according to Appel and Xu (1995). Treatment effects were estimated either by isotopic dilution (only EUF method for the treatments B, D and E) or by a 'difference method'.

Results

During the first 40 days of incubation 38% of the applied organic N from the green manure (treatment B) had been mineralized, but no major mineralization took place after this date (Fig. 1). The course of labelled EUF-Norg peaked on day 2 followed by a decrease until day 40. After day 40 labelled EUF extractable organic N remained stable. This course clearly indicates that the EUF method extracted organic N from the green manure with a high selectivity. However, the major decrease of labelled EUF extractable organic N took place from day 2 to day 9, but about 80% of the increase of labelled inorganic N was observed after day 9. Similar results as for EUF (Fig. 1a) were found for the CaCl$_2$ extraction (Fig. 1b).

The organic N applied as microbial biomass suspension was mainly mineralized during the first week of incubation which was indicated by the rapid inorganic N accumulation (Fig. 2). Application of microbial biomass suspension did not affect CaCl$_2$ extractable

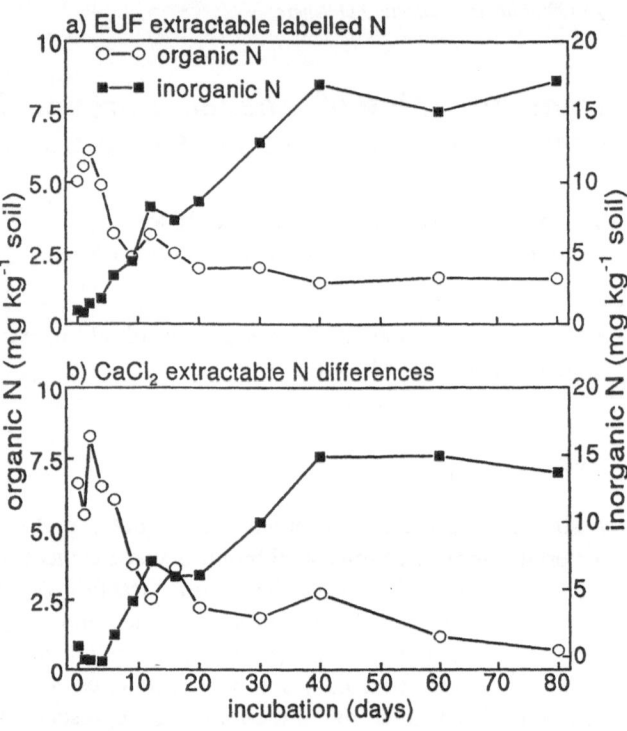

Figure 1. (a) EUF extractable labelled N from green manure (treatment B); (b) Differences of CaCl$_2$ extractable N between treatment B (green manure) minus treatment A (control).

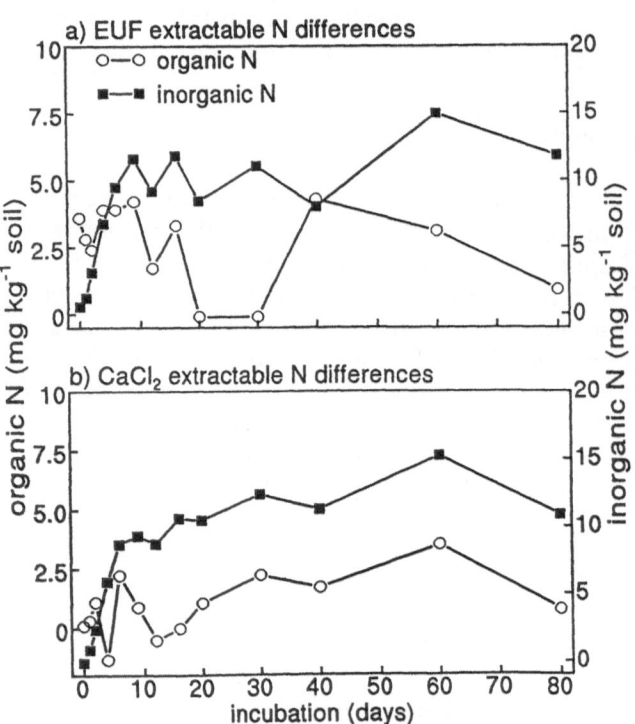

Figure 2. Differences of EUF (a) and CaCl$_2$ (b) extractable N between treatment C (*E. coli* suspension) minus treatment A (control).

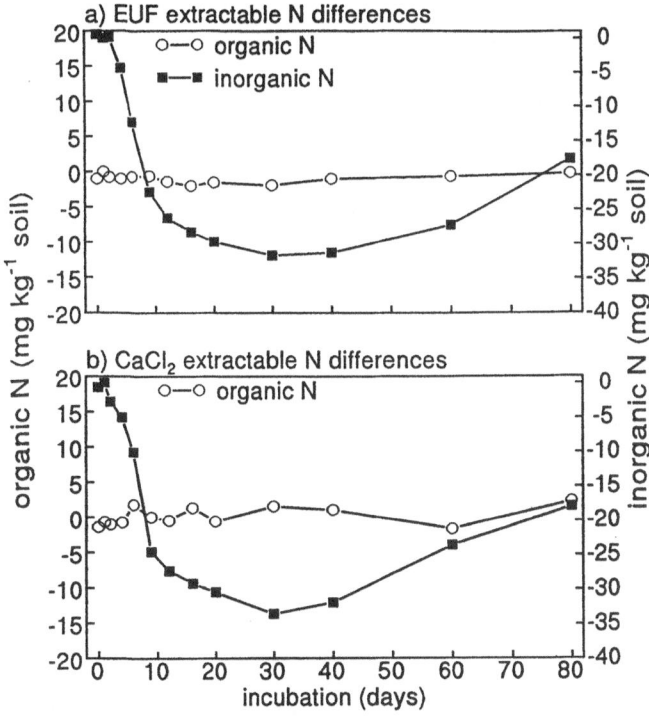

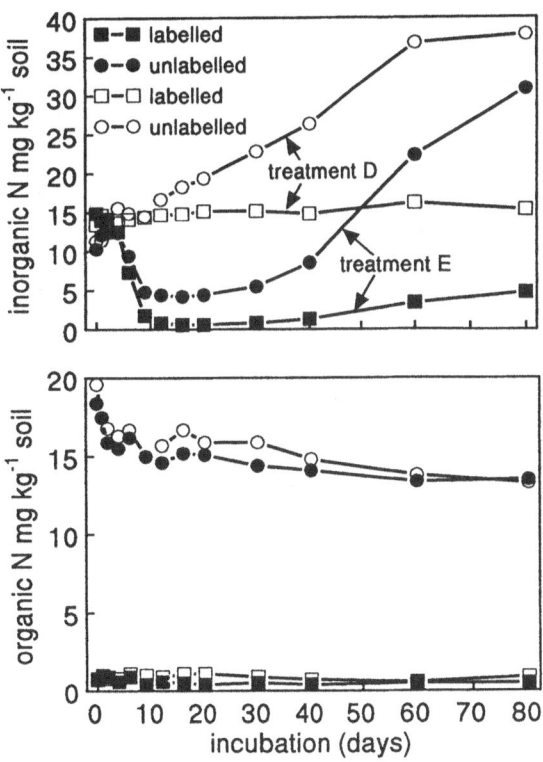

Figure 3. Differences of EUF (**a**) and CaCl$_2$ (**b**) extractable N between treatment E (mineral N + cellulose) minus treatment D (mineral N).

Figure 4. Labelled and unlabelled EUF extractable N fractions of treatment D (mineral N) and treatment E (mineral N + cellulose).

organic N on day 0. During a further incubation the extractable organic N did not decrease corresponding to the inorganic N increase. This was found for both methods, EUF and CaCl$_2$ extraction. Hence, the change of extractable organic N by both methods (EUF and CaCl$_2$) did not reflect the mineralization of added microbial biomass N.

The cellulose amendment caused a rapid immobilization of labelled and unlabelled inorganic N (Fig. 3). Re-mineralization of the transitory immobilized N occurred from day 30 on. The immobilization and the potential of re-mineralizable N was, in contrast to our expectation, not reflected as an increase of extractable organic N. This result was found for both extraction methods. Almost no labelled inorganic N was incorporated into the extractable organic N fractions (Fig. 4).

Discussion

Both extraction methods extracted with high selectivity organic N from green manure which was actually mineralized (Fig. 1). Not at any time was the N miner-

alization potential completely reflected by extractable organic N. The selectivity of the methods to differentiate between easily mineralizable and more resistant compounds of the green manure was poor. This can be concluded from the finding that considerable amounts of green manure N were extracted on day 40 and later, although no further mineralization of the green manure N took place after this date. It is therefore concluded that soil samples for N mineralization indexes based on EUF or CaCl$_2$ extractable organic N should be drawn before plant residues are incorporated into the soil.

The mineralization potential which was represented by the applied microbial biomass was not clearly reflected as extractable organic N, neither by the EUF method nor by CaCl$_2$ extraction (Fig. 2). We had hypothesized that the microbial biomass would be to some extent extractable due to the soil drying and grinding before extraction. Németh et al. (1988) suggested that a considerable amount of the EUF extractable organic N was derived from the soil microbial biomass. Organic N from microbe species other than *E. coli* present in soils under field conditions may be extractable by EUF. However, a more plausible explanation is that only low molecular decomposition products of the biomass are extractable by EUF and by $CaCl_2$. The soil drying and grinding procedure may not destroy the cell structures in so far that the organic compounds become extractable by mild extraction methods. This explanation is in line with the finding that immobilization and the potential of remineralizable N was not reflected as extractable organic N (Figs. 3 and 4). The inorganic N immobilization in the cellulose amended soil was presumably due to an increasing microbial biomass being unextractable by EUF and $CaCl_2$.

On balance the result of the investigation was that EUF and $CaCl_2$ extracted decomposing organic N from plant residues with a high selectivity but the N which was temporarily immobilized as microbial biomass was clearly not reflected as extractable organic N.

Acknowledgements

The investigation was funded by the 'Deutsche Forschunsgemeinschaft'. The study was dedicated to Professor Dr Dr h c Konrad Mengel for his 65th birthday.

References

Appel T and Mengel K 1990 Importance of organic nitrogen fractions in sandy soils, obtained by electro-ultrafiltration or $CaCl_2$ extraction, for nitrogen mineralization and nitrogen uptake of rape. Biol. Fert. Soils 10, 97–101.

Appel T and Mengel K 1992 Nitrogen uptake of cereals grown on sandy soils as related to nitrogen fertilizer application and soil nitrogen fractions obtained by electro-ultrafiltration (EUF) and $CaCl_2$ extraction. Eur. J. Agron. 1, 1–9.

Appel T and Xu F 1995 Extractability of ^{15}N labelled plant residues in soil by electro-ultrafiltration. Soil Biol. Biochem. 27, 1393–1399.

Mengel K 1991 Available nitrogen in soils and its determination by the 'Nmin-method' and by electro-ultrafiltration (EUF). Fert. Res. 28, 251–262.

Németh K 1985 Recent advances in EUF research (1980–1983). Plant and Soil 83, 1–19,

Németh K, Bartels H, Vogel M and Mengel K 1988 Organic nitrogen compounds extracted from arable and forest soils by electro-ultrafiltration and recovery rates of amino acids. Biol. Fertil. Soils 5, 271–275.

O. Van Cleemput et al. (eds.), Progress in Nitrogen Cycling Studies, 17–22, 1996.
© 1996 *Kluwer Academic Publishers.*

Nitrogen mineralization kinetics in sewage water irrigated and heavy metal treated sandy soils

D.K. Benbi[1] and Jörg Richter
Institut für Geographie und Geoökologie, TU Braunschweig, Langer Kamp 19c, 38106 Braunschweig, Germany.
[1]*Present address: Department of Soils, Punjab Agricultural University, Ludhiana 141004, India*

Key words: cadmium, copper, microbial processes, nitrification, organic matter, zinc

Abstract

Laboratory incubation experiments were conducted to study N mineralization kinetics of soils from north of Braunschweig, where sewage water has been displaced since the fifties. Three bulk top soil samples (0–30 cm depth) were collected from a field plot which had spatially variable concentrations of heavy metals. The soils were sandy (97% sand) with a pH of 5.3 and ranged in total Cu from 4.2 to 8.3, Cd from 0.73 to 1.3, Zn from 11.6 to 26.0 and Cr from 5.4 to 8.3 mg kg^{-1} soil. The N mineralization was not affected by the heavy metal concentration of the soil. The amount of N mineralized was related to total C and N concentration in soils. Addition of heavy metal salts, at rates of 50, 100 and 200 mg Cu kg^{-1}, 0.5, 1.0 and 5 mg Cd kg^{-1} and 200 and 400 mg Zn kg^{-1} soil, depressed N mineralization. Their effect was small until 8 weeks after incubation but the differences between heavy metal treated and untreated soils generally widened with the time of incubation. The heavy metals mainly inhibited nitrification and caused accumulation of NH_4-N in the soil. Addition of organic matter as finely ground radish leaves (0.55 g 100 g^{-1} soil) improved mineralization but did not alleviate the inhibitory effect of added heavy metals on nitrification.

Introduction

Heavy metals are known to exert toxic effects on microbial processes in soils and different aspects of their toxicity towards microorganisms and microbially mediated processes in soil have been studied. Reviews of these have been presented by Bääth (1989), Doelman (1986) and Smith (1991). The N transformation processes of organic N mineralization and nitrification are affected by high concentration of heavy metals in soils added through municipal or industrial wastes (Chang and Broadbent, 1982; Liang and Tabatabai, 1978; Tyler et al., 1974). The two processes were reported to be inhibited by soil metal concentrations of around 100 mg kg^{-1} for Cu, Ni and Zn; 100–500 mg kg^{-1} for Pb and Cr; 10–100 mg kg^{-1} for Cd and 1–10 mg kg^{-1} for Hg (Doelman, 1986). These generalizations were made based on the results from medium to fine textured soils. Since the heavy metal toxicity depends on soil physico-chemical characteristics (Doelman and Haanstra, 1979), it is likely that in coarse textured sandy soils the metals may become

toxic at relatively low concentrations. This may call for reappraisal of the critical concentrations for heavy metals in sandy soils.

Further, the deleterious effect of heavy metals added through municipal and industrial wastes is reported to be decreased because of their organic matter content (Brookes et al., 1984). However, the effect of addition of crop residues of narrow C: N ratio on heavy metal toxicity in soils is not known. To address the above questions, we selected a site near Braunschweig where sewage water has been added to sandy soils since the fifties. We collected soil samples from the site to conduct laboratory incubation experiments with the following objectives : (i) to study soil organic N mineralization and nitrification in the soils, (ii) to determine the effects of Cu, Cd and Zn when added at rates lower than or equal to the current maximum permissible concentrations in soils on the N transformation processes and (iii) to study the influence of organic matter addition (of narrow C:N ratio) on heavy metal toxicity to N mineralization and nitrification in soil.

Table 1. Selected soils properties

Soil property	Soil I	Soil II	Soil III
C (%)	0.65	0.94	0.56
N (%)	0.062	0.081	0.043
C/N	10.5	11.7	13.2
Heavy metals (mg kg^{-1} dry soil)			
Cd	0.7	1.3	0.9
Zn	14.9	25.9	11.6
Cu	4.2	8.3	4.5
Ni	3.7	5.9	1.9
Cr	5.4	8.3	5.8
Pb	6.2	17.5	11.8

Materials and methods

Bulk top (0–30 cm) soil samples were collected from the sewage water irrigated field at three spots (henceforth referred to as soil I, soil II and soil III) based on the relative heavy metal contamination. The soils were sandy (97% sand, 1% clay) with a pH of 5.3 (in 0.01 M CaCl$_2$) and had varying concentration of C and N (Table 1). For more details about the site and heavy metal distribution in the soil see Streck (1993).

Experiment I

Nitrogen mineralization kinetics of the three soils were studied in a laboratory incubation experiment. The incubation was carried out according to the method of Stanford and Smith (1972) with the modification that field-moist instead of air-dried soils were used. Triplicate samples were incubated at 35 °C for 19 weeks. Before the start of incubation, the mineral N initially present in the soil was removed by leaching the samples with 100 mL of 0.01 M CaCl$_2$ (added in five 20 mL portions) followed by 20 mL of an N-free nutrient solution. The leaching procedure was repeated after 1, 2, 4, 6, 8, 10, 13, 15, 17 and 19 weeks of incubation. At each leaching surplus water from the samples was removed by applying a suction of 60 kPa for 1 hour.

Experiment II

Independent effects of Cu, Cd and Zn application on soil N mineralization were studied in soil III. Each of the metals was applied at three different rates. The rates of application were 50, 100 and 200 mg Cu kg^{-1} soil; 0.5, 1.0 and 5 mg Cd kg^{-1} soil and 100, 200 and 400 mg Zn kg^{-1} soil. Aliquots of soil III were treated with the solution of metal salts (as chlorides) and then sub-samples were drawn for the incubation study. In another set of treatments an aliquot of soil III was amended with dried and finely ground radish leaves (composition 35.82% C and 4.89% N) at the rate of 0.55 g 100 g^{-1} soil (=25 t ha^{-1}/30 cm). Requisite amount of the plant material was mixed with the bulk soil which was then divided into four sub-samples. One sample was treated with 200 mg Cu kg^{-1}, the second one with 5 mg Cd kg^{-1} and the third one with 400 mg Zn kg^{-1} soil. The fourth sub-sample was kept as control. Each treatment was replicated three times. The incubation was carried out as for Experiment I.

Chemical analysis

The leachate at each sampling from both the experiments was analysed for NH$_4$- and NO$_3$-N (NO$_2$+NO$_3$) with a continuous flow autoanalyser (Chemlab instruments). Carbon and N in the soil and plant samples were determined by CN analyser (Carlo Erba). Metal concentration in the soils were determined in aqua regia digest using inductively coupled plasma.

Statistical analysis

Statistical significance of the differences in the cumulative N mineralized between soils (Exp. I) and the treatments (Exp. II) at each incubation interval was determined by analysis of variance using a completely randomised design. All the LSD values were tested at $p=0.05$.

Results and discussion

Soil effects

There were no differences in the time course of cumulative N mineralised between soil I and soil III. During the first eight weeks there were very little differences between soil II and the other two soils, however, the differences widened thereafter. No adverse effect of heavy metal concentration in soils on the amount and kinetics of N mineralization and nitrification was observed (Fig. 1). Soil II had relatively high concentration of heavy metals, still the cumulative amount of N mineralized from the soil was highest compared to the other two soils. The reasons for no apparent adverse effect of heavy metals on N mineralization in these soils may be the relatively low level of contamination.

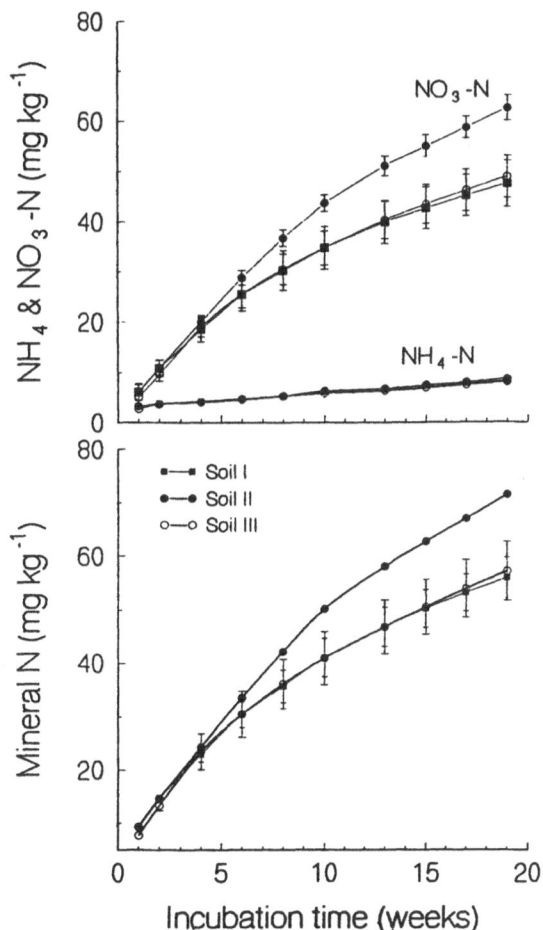

Figure 1. Time course of N mineralization and nitrification in sewage water irrigated sandy soils. The bars indicate the standard deviation of the measurements. Since the standard deviation bars for soil II (mineral N) are smaller than the symbols, these are not visible.

Treatment effects

Copper

Addition of Cu inhibited nitrification. Its inhibitory effect was observed even at 50 mg Cu kg^{-1} soil (Fig. 2). Addition of 50 mg Cu kg^{-1} soil caused 21% reduction in cumulative amount of NO$_3$-N produced in 19 weeks compared to the control soil. The inhibitory effect of Cu on nitrification increased with increasing rate of its application. The 100 and 200 mg Cu kg^{-1} treatments yielded 72 and 78% less NO$_3$-N in 19 weeks period as compared to the control. In earlier studies it has been reported that Cu may be toxic to microbial activity at concentration equal to or greater than 100 mg kg^{-1} soil (Doelman, 1986). But it appears that in the sandy soils the critical concentration for Cu toxicity may be lower than 50 mg kg^{-1} soil.

Due to inhibition of nitrification there was accumulation of NH$_4$-N in the soil. The NH$_4$-N accumulation was not in proportion to the decrease in NO$_3$-N and there was a net decrease in mineral N in the Cu treated soils as compared to the control (Fig. 2). Addition of 50 and 200 mg Cu kg^{-1} soil, decreased the cumulative amount of mineral N at 19 weeks by 19.7 and 21.8%, respectively. The decreased mineral N production in the Cu treated soils may be due to the inhibition of ammonification of soil organic N.

Cadmium

Addition of Cd affected both N mineralization and nitrification (Fig. 2). At 19 weeks, nitrification of the mineralized NH$_4$-N was inhibited by 18.2% in 0.5 and 1 mg Cd kg^{-1} and by 14.6% in 5 mg Cd kg^{-1} soil treatments. Inhibition of nitrification caused NH$_4$-N accumulation in 0.5 and 1 mg Cd kg^{-1} soil. However, in the 5 mg Cd kg^{-1} treatment there was no NH$_4$-N accumulation rather the cumulative amount of NH$_4$-N was significantly lower than in the untreated control soil. It seems that at this level of Cd application activity and population of both the diverse collection of N mineralizing microorganisms and the specific nitrifiers were affected.

Zinc

Addition of Zn, irrespective of its rate of application, inhibited nitrification of the mineralized NH$_4$-N (Fig. 2). As compared to the untreated soil, there was 15.4, 21.2 and 14.8% reduction in cumulative amount of NO$_3$-N produced in 19 weeks with 100, 200 and 400 mg Zn kg^{-1} soil, respectively. There was accumulation of NH$_4$-N, particularly in 100 mg Zn kg^{-1} treatment. With 200 mg Zn kg^{-1} soil addition the NH$_4$-N accumulation was small and the 400 mg Zn kg^{-1} soil treatment did not exhibit accumulation rather its amount was significantly less than in control. It implies that at 200 and 400 mg Zn kg^{-1} soil, ammonification was also affected.

Organic matter

Addition of Cu (200 mg kg^{-1} soil) along with the radish leaves inhibited N-mineralization by 85.7% during the first week of incubation (Fig. 3). As the incubation progressed, the inhibitory effect of Cu on N mineralization diminished and at 19 weeks the cumulative mineral N production in the Cu (plus plant material) treated soil was lower by only 14.6% as compared to the plant material amended soil alone. Addition of Cu

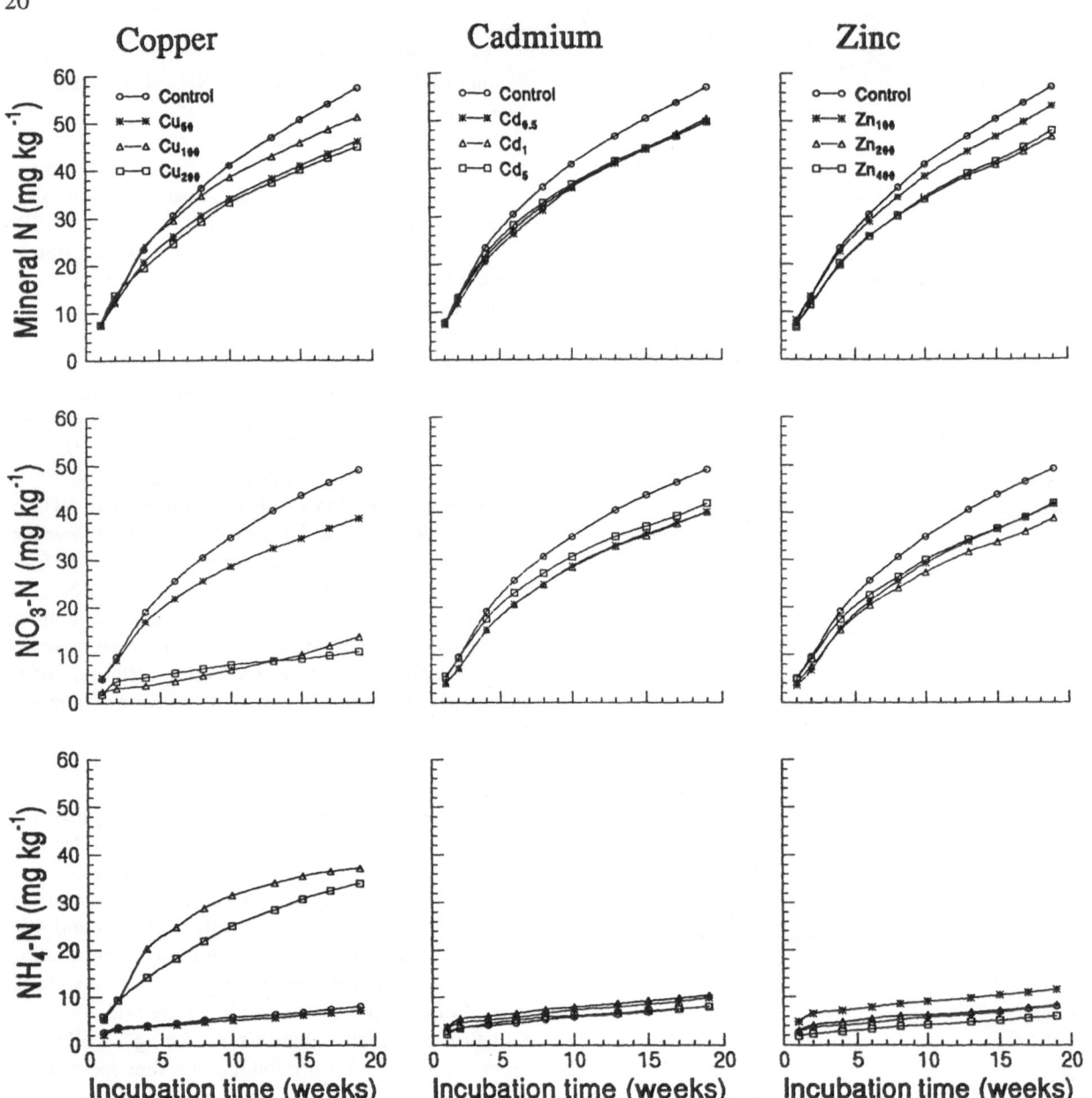

Figure 2. Effect of heavy metal (Cu, Cd and Zn) salts addition on N mineralization and nitrification of soil organic N. The numbers as subscripts to the metals in the legend indicate the rate of application (mg kg^{-1} soil).

inhibited nitrification of the mineralized NH$_4$ by about 85% during the first 4 weeks of incubation, however, the effect diminished to 63% by the end of incubation.

Addition of Cd (5 mg kg^{-1} soil) inhibited N-mineralization in the plant material amended soil by 18% and nitrification by 48% during the first two weeks of incubation. Similar to Cu, its effect diminished (to 8% and 29%, respectively) as the incubation proceeded.

Addition of Zn to the plant material amended soil did not affect N mineralization but the nitrification of mineralized NH$_4$-N was inhibited by 72% during the first week and it declined to 22% by the end of incubation (at 19 weeks). There was a significantly higher accumulation of NH$_4$-N in the Zn treated soil as compared to the plant material amended soil alone.

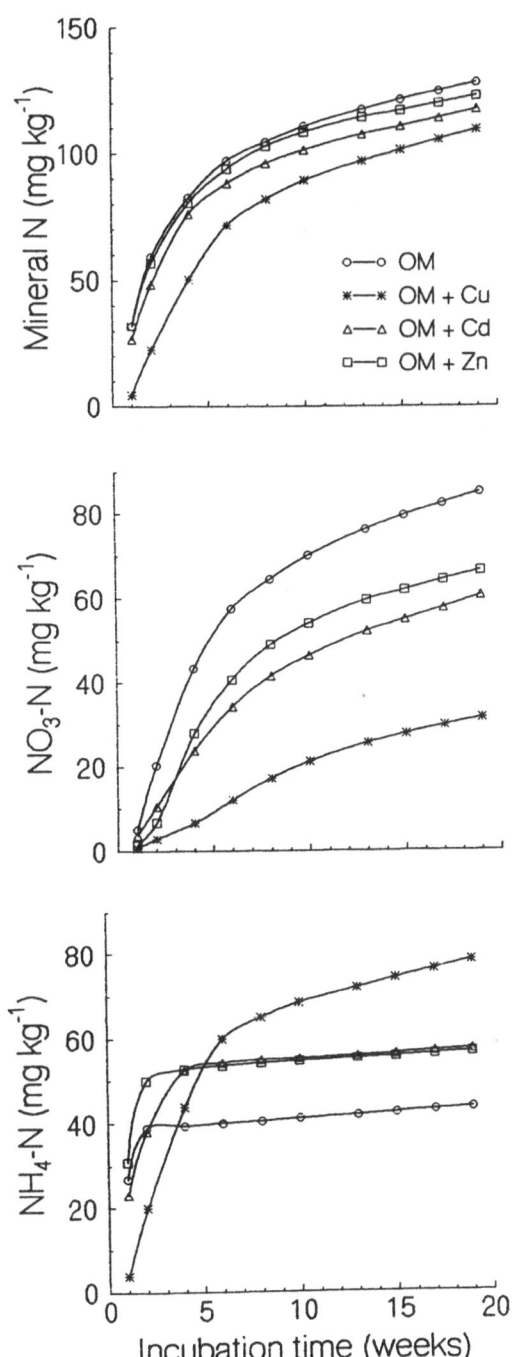

Figure 3. Effect of organic matter (OM) and heavy metal salts addition on N mineralization and nitrification of soil organic N.

of heavy metal accumulation. The elevated levels of heavy metals (soil II compared to the other two soils) did not detrimentally affect the N transformation processes of ammonification and nitrification of the native organic N. This may be explained either due to the relatively low bioavailability of the contaminants or to the adaptation of mineralizing and nitrifying organisms to elevated concentration of heavy metals (Rother et al., 1982). The differences in N mineralization and nitrification amongst soils could be explained as a function of soil C and N.

Apparently, in these sandy soils, heavy metals were toxic at concentrations much lower than the maximum permitted for soils in Germany. For example, against permitted concentrations of 60 mg Cu kg^{-1}, 1.5 mg Cd kg^{-1} and 200 mg Zn kg^{-1} soil (BGB, 1992), N transformation processes were inhibited at added metal concentrations of 50 mg Cu kg^{-1}, 0.5 mg Cd kg^{-1} and 100 mg Zn kg^{-1} soil. These were the lowest concentrations used in the study and it is likely that the heavy metals may be toxic at even lower concentrations. However, the elevated heavy metal concentrations in historically sludged soils may not be as detrimental to N transformation processes in soil as the added soluble metal salts. For example, soil II had an elevated Cd concentration of 1.3 mg kg^{-1} compared to 0.7 mg kg^{-1} in soil I, but no apparent detrimental effect on N transformation processes was observed. Whereas an addition of 0.5 mg Cd kg^{-1} to soil III inhibited mineralization by 12.6% and nitrification by 18.2% as compared to the control soil. This implies that two forms of metal concentration (total in the soil and water soluble metal salts) have different bioavailabilities. The bioavailability of heavy metals in sludged soils decreases because of a number of factors such as chemisorption, precipitation and complexation with organic matter. As is apparent from the results of this study, application of organic matter as plant material minimised the inhibitory effect of heavy metals on N-mineralization perhaps through the formation of organo-metal complexes. In general, mineralization process was less sensitive to the effects of heavy metals than was nitrification probably because mineralization can be carried out by a large diversity of microorganisms. Nitrification instead is performed by rather few specialised organisms only.

General discussion

After over three decades of irrigation with sewage water the experimental soils showed moderate levels

Acknowledgements

DKB is thankful to the Alexander von Humboldt-Stiftung, Jean-Paul-Straße 12, Bonn for a research fellowship for the studies.

References

Bääth E 1989 Effects of heavy metals in soil on microbial processes and populations (a review). Water Air Soil Pollut. 47, 335–379.

BGB (Bundesgesetzblatt) 1992 Klärschlammverordnung- AbfKlär V vom 15.04.1992. Bundesgesetzblatt (1992) I, 912–916.

Brookes P C, McGrath S P, Klein D A and Elliot E T 1984 Effects of heavy metals on microbial activity and biomass in field soils treated with sewage sludge. *In* Environmental Contamination (International Conference, London, July 1984). pp 574–583. CEP Ltd, Edinburgh, UK.

Chang F H and Broadbent F E 1982 Influence of trace metals on some soil nitrogen transformations. J. Environ. Qual.11, 115–119

Doelman P 1986 Resistance of soil microbial communities to heavy metals. *In* FEMS Symposium No. 33 Microbial communities in soil. Copenhagen, August 4–8, 1985. pp 369–398. Elsevier Applied Science Publishers Ltd, London, UK.

Doelman P and Haanstra L 1979 Effect of lead on soil respiration and dehydrogenase activity. Soil Biol. Biochem.11, 475–479.

Liang C N and Tabatabai M A 1978 Effects of trace elements on nitrification in soils. J. Environ. Qual. 7, 291–293.

Rother J A, Millbank J W and Thornton I 1982 Effects of heavy metal additions on ammonification and nitrification in soils contaminated with cadmium, lead and zinc. Plant and Soil 69, 239–258.

Smith S R 1991 Effect of sewage sludge application on soil microbial processes and soil fertility. Adv. Soil Sci. 16, 191–212.

Stanford G and Smith S J 1972 Nitrogen mineralization potentials of soils. Soil Sci. Soc. Am. Proc. 36, 465–472.

Streck T 1993 Schwermetallverlagerung in einem Sandboden im Feldmaßstab- Messung und Modellierung. Doctoral Thesis, TU Carolo-Wilhelmina, Braunschweig.

Tyler G, Mörnsjö B and Nilsson B 1974 Effects of cadmium, lead and sodium salts on nitrification in a mull soil. Plant and Soil 40, 237–242.

O. Van Cleemput et al. (eds.), Progress in Nitrogen Cycling Studies, 23–26, 1996.
© *1996 Kluwer Academic Publishers.*

Relationship between the fixed ammonium and the mineralization of the organic nitrogen in soil

A. Benedetti[1], L. Vittori Antisari[2], S. Canali[1,3], P. Gioacchini[2] and P. Sequi[1]

[1]*Istituto Sperimentale per la Nutrizione delle Piante Via della Navicella, 2–00184 Roma, Italia and* [2]*Istituto di Chimica Agraria - Università di Bologna Via di S. Giacomo, 7–46126 Bologna, Italia.* [3]*Corresponding author*

Key words: interlayer ammonium, organic nitrogen mineralization, potentially mineralizable nitrogen

Abstract

Potentially mineralizable nitrogen and interlayer ammonium-nitrogen pools of the soil were determined. The mineralization process has an influence on fixation and release of interlayer ammonium. During long-term incubation for determining of the potentially mineralizable nitrogen of soil, the mineralized nitrogen could be affected by the realease of inorganic nitrogen from the other pools, particularly from the interlayer ammonium-nitrogen pool.

Inorganic soil nitrogen pools (exchangeable and interlayer ammonium, nitrite, nitrate), biomass-N and total N before and after long-term incubation were investigated.

No significant difference between the values obtained before and after the incubation were observed for all the pools measured, indicating that they do not interfere with the measurement of potentially mineralizable nitrogen.

Abbreviations: N_O - potentially mineralizable nitrogen, SISS - Italian Society of Soil Science, NH_4^+-N_{int} - interlayer ammonium-nitrogen.

Introduction

Organic matter has been considered so far as the main nitrogen source for crops. In fact organic nitrogen mineralization plays a key role in regulating the amount of available N in soil (Stevenson, 1985).

The amount of mineral nitrogen available for crops during the whole cropping period can be estimated by the measure of the potentially mineralizable nitrogen (N_O).

Stanford and Smith (1972) and subsequently other authors (Addiscott, 1983; Beauchamp et al., 1986; Deans et al., 1986) have given valuable contributions to the study of this parameter. For Italian soils (total N 0.15% on average) Benedetti and Sebastiani (1996) found that N_O represents 8–12% of total nitrogen. Anyway, many uncertainties still remain on the evaluation of N_O in soil where agricultural practices are commonly carried out (Boone, 1990; Cabrera and Kissel, 1988; Hadas et al., 1989).

Another nitrogen soil reserve is represented by the ammonium fixed to inorganic soil components, with special regard to layersilicates. The ability of 2:1 clay minerals such as illite, vermiculite and montmorillonite to entrap interlayer ammonium ions between the silicate layers had been shown by many authors (Bremner, 1965a; Nommik and Vahtras, 1982; Stevenson and Dhariwal, 1959).

In the upper soil layers of most agricultural soils 3 to 14% of the total nitrogen occurs as interlayer ammonium. However, the percentage of this fractions in total N increases with depth (Uriyo and Singh, 1980) reaching as much as 80% (Dalal, 1977). Vittori Antisari and Sequi (1988) found that 200 to 400 mg interlayer ammonium kg^{-1} soil were common in Italian soils; in northern part of Germany soils contain between 150 to 850 mg interlayer ammonium kg^{-1} soil (Scherer and Weimar, 1993); this pool amounted to as much as 3000 kg NH_4^+ ha^{-1} in the rooting zone of representative soils of Germany (Keerthisinge et al., 1984). Considering the amounts of nitrogen present in interlayer ammonium form, it is important to study the availability of this fraction for crops.

In order to rationalize nitrogen fertilization it is important to understand correctly the relationship among the different nitrogen pools and their contribution to plant nutrition.

Aims of the present work have been i) to determine the mineralizable nitrogen pool and the interlayer ammonium pool of the soil and ii) evaluate if the measure of the potentially mineralizable nitrogen is affected by the realease of inorganic nitrogen from the other pools of the soil, particularly from the interlayer ammonium pool.

Materials and methods

The study has been carried out with a soil sample (typic Xerofluvent) collected from a peach orchard, located in Ravenna (Italy). This soil was selected because it is representative for a large proportion of the agricultural land in the Po Valley. The soil was fertilized with 200 kg N ha^{-1}, half of which was applied at the beginning of bud bursting (end of March), and the remainder applied at the time of pit hardening of the fruit (middle of May). Soil samples were carried out after the first fertilizer rate: at this time the soil shows the maximum fixation capacity (Marzadori, 1994).

The chemical and physical properties of the soil (Table 1) were determined by the SISS methods (1985), with the exception of the analysis for the layer silicates which were determined by X-ray diffraction analysis (Whitting et al., 1965).

The potentially mineralizable nitrogen was measured following the method proposed by Stanford and Smith (1972) partially modified by Benedetti (1983). In particular, 50 g of soil air-dried and sieved to a 2mm screen was mixed with quartz sand in a 1:1 ratio (particle size of the sand, 0.2–0.8 mm) and incubated in a Buchner funnel (outer diameter 13 cm), at 60% WHC and at 30 °C for 30 weeks. The mineral nitrogen in the soil was leached periodically (at 2, 4, 8, 16, 22 and 30 weeks), by adding 900 mL of 0.01 M CaSO$_4$ solution and 100 mL of nutrient solution "N minus" (0.002 M CaSO$_4$. 2H$_2$O, 0.005 M Ca(H$_2$PO$_4$), 0.0025 M K$_2$SO$_4$, 0.002 M MgSO$_4$). In the leachate, the NO$_3^-$-N, the NO$_2^-$-N, and the NH$_4^+$-N concentration were measured. N$_O$ was calculated using the non-linear regression method of the Statgraphics software (Version 6.1 - Statistical Graphics Corporation) by the Stanford and Smith equation:

$$N_t = N_O(1 - e^{-kt}) \qquad (1)$$

where N$_t$ is the nitrogen mineralized in the period t considered and k the velocity constant (0.054 week^{-1}).

Table 1. Chemical -physical characteristics of the soil (values refer to dry matter 105°C)

Silt(%)	58
Sand(%)	25
Clay(%)	17
pH (H$_2$O, 1:2.5)	8
Total Organic C(%)	0.7
Total CaCO$_3$(%)	20
Active CaCO$_3$(%)	6
CEC (cmol kg^{-1})	10
Total N (mg kg^{-1})	980
Available P$_2$O$_5$ (mg kg^{-1})	22
Exchangeable K$_2$O (mg kg^{-1})	128
Clay minerals	
Illite(%)	30
Smectite(%)	30
Intergrade(illite/smectite)(%)	30
Chlorite(%)	5
Kaolinite(%)	5

Before and after the 30 weeks incubation the interlayer ammonium-nitrogen pool, the exchangeable ammonium-nitrogen pool, the nitrate-nitrogen pool, the biomass nitrogen pool and the total nitrogen of soil was measured.

Determination of interlayer ammonium was carried out according to Silva and Bremner (1966). The soil sample was treated with a KOH-KOBr solution and the interlayer NH$_4^+$ was extracted from the resulting residue with an acid mixture (5M HF and 1M HCl) to dissolve the silicates. The NH$_4^+$ extracted was released with steam distillation, collected in 2% H$_3$BO$_3$, and titrated with 0.005 N H$_2$SO$_4$.

Some steps of the original method were modified slightly. The heat treatment of the KOH-KOBr-soil mixture, held in 150 mL Corex glass bottle (Beckmann), was carried out in a microwave system (CEM corp. MDS-81, Indian Trail, NC) equipped with a 600 W magnetron operating an 1-S duty cycle. The treatment involved heating the sample in two cycles: 5 min at 90% full power followed by 2 min at 80% full power.

The exchangeable ammonium-nitrogen (NH$_4^+$-N), nitrate-nitrogen (NO$_3^-$-N) and nitrite-nitrogen (NO$_2^-$-N) have been extracted from the soil according to Bremner (1965b).

Biomass N was determined using the CHCl$_3$ fumigation-extraction method (Brookes et al., 1985). Ten g of soil sample was saturated with purified liquid

CHCl$_3$ for 24 h. After fumigation the soil was extracted with $0.5 M$ K$_2$SO$_4$ (1:4 soil:extractant) for 30 min. The soil extracts were analyzed for total N using the Kjeldhal procedure. The flush of total N (K$_2$SO$_4$-extractable N in the fumigated soil minus that from the unfumigated soil) was derived by a k_N (fraction of biomass N extracted after CHCl$_3$ fumigation) value of 0.54.

For total N determination, the soil were digested by the Kjeldhal procedures.

The inorganic nitrogen forms in the Kjeldhal digestion, in soil extractant and leachate were measured by a continuous flow analyser (Autoanalyzer Technicon II) in accordance with Wall et al. (1975) for the NH$_4^+$-N, with Kampshake et al. (1967) for the NO$_3^-$-N and by a modification of the Griess-Ilosvay procedure (Griess, 1879; Ilosvay, 1889) for the NO$_2^-$-N (Keeney and Nelson, 1982).

Results

The potentially mineralizable nitrogen (N$_O$) was 37.5 mg kg^{-1} of soil and referred to the ploughing layer (5 10^6 kg of soil ha^{-1}) N$_O$ amounted to 188 kg N ha^{-1}.

In Table 2 the data of organic and inorganic nitrogen pools before (t$_0$) and after (t$_{30w}$) the 30 weeks incubation were reported.

The interlayer ammonium-nitrogen (NH$_4^+$-N$_{int}$) was 209 and 213 mg kg^{-1} of soil before and after the incubation respectively, and this difference was not significant.

No significant differences between the values obtained before and after the incubation were observed also for the biomass-N, exchangeable ammonium-nitrogen, nitrate-nitrogen and nitrite-nitrogen.

The total nitrogen of the soil lowered from 980 mg kg^{-1} before the incubation, to 933 mg kg^{-1} after the incubation.

Discussion

The potentially mineralizable nitrogen (N$_O$) represents about 4% of total nitrogen of the soil and this value can be considered of a low-medium level, commonly observed in many agricultural Italian soils.

The interlayer ammonium nitrogen (NH$_4^+$-N$_{int}$) represents 21% of total nitrogen of the soil. This can be considered a quite high value and represents, for this soil, the maximum fixation capacity (Marzadori, 1994).

Table 2. Interlayer ammonium, biomass, inorganic and total nitrogen before (t$_0$) and after (t$_{30w}$) the 30 weeks incubation (values in mg kg^{-1} soil refer to dry matter 105°C)

N pools	t$_0$	t$_{30w}$
Interlayer NH$_4^+$-N	209 ± 1.4	213 ± 2.3
Biomass N	78.7 ± 1.1	85.2 ± 0.9
Exchangeable. NH$_4^+$-N	7.5 ± 0.7	13.6 ± 0.7
NO$_3^-$-N	5.9 ± 0.3	2.8 ± 0.3
NO$_2^-$-N	0.7 ± 0.1	0.8 ± 0.1
Total N	980 ± 42	933 ± 36

The lack of significative differences between the values of the interlayer ammonium-nitrogen (NH$_4^+$-N$_{int}$) before and after the incubation allows us to affirm that this pool do not interfere with the evaluation of the potentially mineralizable nitrogen pool (N$_o$), at least when the fixation capacity of the soil is in saturated condition.

Analogous considerations are possible regarding the biomass-N pool, exchangeable ammonium-nitrogen pool, nitrate-nitrogen pool and nitrite-nitrogen pool.

The difference in total N content of the soil before and after the incubation (statistically significant) is due to the mineralization process.

In conclusion, our results show that the measure of the potentially mineralizable nitrogen is not affected by the release of inorganic nitrogen from the interlayer ammonium pool and from other nitrogen pools. Nevertheless, further researches are needed to understand the relationship between the potentially mineralizable nitrogen pool and interlayer ammonium pool when soils do not present the maximum fixation capacity.

Acknowledgement

This research was supported with funds provided by finalised project PANDA, paper no. 16, subproject 3.

References

Addiscott T P 1983 Kinetics and temperature relationships of mineralization in Rothamsted soils with different histories. J. Soil Sci. Soc. Am. J. 50, 1478–1483.

Beauchamp E G, Reynold W D, Brasche-Villeneuve D and Kirby K 1986 Nitrogen mineralization kinetics with different soil pretreatments and cropping histories. Soil Sci. Soc. Am. J. 50, 1478–1483.

Benedetti A 1983 Soil biological fertility and slow-release fertilizers. Annals XII Instituto Sperimentale por la Nutritione delle Plante, Roma, Italy.

Benedetti A and Sebastiani G 1996 Determination of potentially mineralizable nitrogen in agricultural soil. Biol. Fert. Soil 21, 114–120.

Boone R 1990 Soil organic matter as a potential net nitrogen sink in a fertilized cornfield South Deerfield, Massachussetts USA. Plant and Soil 128, 191–198.

Bremner J M 1965a Total nitrogen. In Methods of Soil Analysis, Part 2. Eds. C A Black, D D Evans, J L White, L E Ensminger and F E Clark. pp 1149–1176. Agronomy 9, Am. Soc. Agron., Madison WI, USA.

Bremner J M 1965b Nitrogen availability index. In Methods of Soil Analysis, Part 2. Eds. C A Black, D D Evans, J L White, L E Ensminger and F E Clark. pp 1324–1241. Agronomy 9, Am. Soc. Agron., Madison WI, USA.

Brookes P C, Landman A, Pruden G and Jenkinson D S 1985 Chloroform fumigation and release of soil N: a rapid direct extraction method to measure microbial biomass N in soil. Soil Biol. Biochem. 17, 837–842.

Cabrera M L and Kissel D E 1988 Evaluation of a method to predict nitrogen mineralized from soil organic matter under field conditions. Soil Sci. Soc. Am. J. 52, 1027–1031.

Dalal R C 1977 Fixed ammonium and sequential determination of exchangeable ammonium, organic nitrogen and fixed ammonium in soil. Soil Sci. Soc. Am. J. 52, 1020–1023.

Deans J R, Molina J A E and Clapp C E 1986 Models for predicting potentially mineralizable nitrogen and decomposition rate constants. Soil Sci. Soc. Am. J. 50, 323–326.

Griess P 1879 Bemerkungen zu der Abhandlung der HH. Weselski und Benedikt "Ueber einige Azoverbindungen". Chem. Ber. 12, 426–428.

Hadas A, Poigenbaum S, Feigin A and Portnoy R 1989 Nitrogen mineralization in field at various soil depths. J. Soil Sci. 40, 131–137.

Ilosvay M L 1889 L'acide azoteux dans la salive et dans l'aire exhale. Bull. Soc. Chim. 2, 388–391.

Kampshake L J, Hannah S A and Comen J M 1967 Automated analysis for nitrate by hydrazine reduction. Water Resour. Res. 1, 205–216.

Keeney D R and Nelson D W 1982 Nitrogen - Inorganic forms. In Methods of Soil Analysis, Part 2. pp 602–687. Agronomy 9, Am. Soc. Agron., Madison, WI, USA.

Keerthisnghe G, Mengel K and De Datta S K 1984 The release of nonexchangeable ammonium (^{15}N labelled) in wetland rice soils. Soil Sci. Soc. Am. J. 48, 291–294.

Marzadori C 1994 Variazioni stagionali della riserva di ammonio fissato nei fillosilicati in un terreno sottoposto alla coltivazione del pesco. Ph D Thesis, Univ. of Bologna, Italy.

Nommik H and Vahtras K 1982 Retention and fixation of ammonium and ammonia in soils. In Nitrogen in Agricultural Soils. Ed. F J Stevenson. Agronomy 22, 120-169.

Scherer H W and Weimar S 1993 Release of nonexchangeable NH_4-N after planting of ryegrass in relation to soil content and as affected by nitrate supply. Z. Pflanzenernahr. Bodenkd. 156, 142–148.

Silva J A and Bremner J M 1966 Determination and isotope-ratio analysis of different forms of nitrogen in soil. 5: Fixed ammonium. Soil Sci. Soc. Am. Proc. 30, 587–594.

SISS 1985 Metodi normalizzati di analisi del suolo. Edagricole (ed) Bologna, Italy.

Skjemstad J O, Vallis I and Myers R K J 1988 Decomposition of soil organic nitrogen. In Advances in Nitrogen Cycling in Agricultural Ecosystems. Ed. J R Wilson pp 134–144.

Stanford G and Smith S J 1972 Nitrogen mineralization potentials of soils. Soil Sci. Soc. Am. Proc. 36, 465–472.

Stevenson F J 1985 The internal cycle of nitrogen in soil. In Cycles of soil C,N,P,S, Micronutrients. Wiley J and Sons Inc. (eds), New York. pp 155–215.

Stevenson F J and Dhariwall A P S 1959 Distribution of fixed ammonium in soils. Soil Sci. Soc. Am. Proc. 23, 121–124.

Uriyo A P and Singh B R 1980 Distribution of native fixed ammonium in some tropical soil profile in Tanzania. J. Sci. Food Agric. 31, 526–531.

Vittori Antisari L and Sequi P 1988 Comparison of total nitrogen by four procedures and sequential determination of exchangeable ammonium, organic nitrogen and fixed ammonium in soil. Soil Sci. Soc. Am. J. 52, 1020–1023.

Wall L L, Gehrke C W, Neuner J E, Lathey R D and Rexnord P R 1975 Cereal protein nitrogen: evaluation and comparison of four different methods. J. Assoc. Off. Anal. Chem. 58, 811–817.

Whitting L D 1965 X-ray diffraction technique for mineral identification and mineralogical composition. In Methods of Soil Analysis, Part 1. Eds. C A Black, D D Evans, J L White, L E Ensminger and F E Clark. pp 837–842. Agronomy 9. Am. Soc. Agron., Madison, WI, USA.

O. Van Cleemput et al. (eds.), Progress in Nitrogen Cycling Studies, 27–30, 1996.
© 1996 *Kluwer Academic Publishers.*

Nitrogen turnover in alternatives to peat

Ian G. Burns and Mary K. Turner
Horticulture Research International, Wellesbourne, Warwick CV35 9EF, UK

Key words: immobilisation, incubation, nitrification, nitrogen, peat alternatives, volatilisation

Abstract

Samples of a sphagnum peat and 14 different alternatives were incubated aerobically at pH 6.0 to 6.5 for 12 weeks at 15 °C to study the transformations of added ammonium ions and losses by volatilisation. Turnover of N was generally greater in coir-based materials and in composted animal and domestic wastes than in most bark and paper products. There was little evidence of significant gaseous losses of NH_3; changes in ammonium concentration resulted largely from nitrification, but simultaneous immobilisation of ammonium and nitrate (and sometimes denitrification) restricted the accumulation of the latter. The decline in ammonium content was generally in direct proportion to the initial amount of ammonium present and was not clearly related to the organic N content or the C:N ratio of the material.

Introduction

Recent estimates put current use of peat in UK horticulture at around 2.7 million cubic metres per annum, of which about 60% is extracted from within the country (Bragg, 1990). Peat is not an easily renewable resource and, at this rate of extraction, existing reserves will be exhausted within 40 to 50 years. This has fuelled pressure to develop alternatives and recently attention has turned to recycled wastes and other products for soil amelioration, for mulching and for use as media for propagating and growing plants in containers (Scott and Burbridge, 1991). However, many of these materials are unstable (largely because of incomplete composting) and, although they could provide potentially useful constituents for use in growing media, they have often proved unreliable because of difficulties in maintaining a balanced nutrient supply to roots.

The objective of this project was to examine the factors affecting the turnover of N in a range of different alternatives to peat as a component of a broader programme to provide information on which to base advice for optimising their use in commercial growing media. The materials selected for study were provided by HRI Efford and were taken from bulk samples used in various combinations for the preparation of peat-free mixes as part of a study to evaluate their performance for the production of hardy nursery stock (Davies and Scott, 1994).

Experimental methods

Materials

Details of the materials used in the study together with their key properties are summarised in Table 1. These were derived from various wood and bark products and from a range of composted animal, crop and domestic wastes. The materials were taken from widely different sources and ranged from those suitable as constituents of composts (e.g. pig and cow manures, paper waste) including bulking agents (e.g. woodfibre, barks) to products which are sold commercially as complete composts. A sample of sphagnum peat was also included for comparative purposes as a control. The natural pH of the different materials varied from 4.5 to 9.0.

Prior to the start of the experiment each material was adjusted to between pH 6.0 and 6.5 using a small (predetermined) volume of dilute sulphuric acid or sodium hydroxide solution (as appropriate), loosely covered to minimise evaporation and equilibrated for about 7 days. The samples were then allowed to dry off partially in air before the application of nitrogen.

Table 1. Key properties of the materials

Type	Details of materials		Total N ($\mu g\ g^{-1}$)	C:N ratio	Incubation water content (% w/w)
Crop wastes	Coir	Goldengrow	3000	175.6	437
	Cocompost	Wessex	4000	121.7	515
	Novagrow	Hensby Biotech	2200	230.7	460
	Arable Waste	Fipro rape straw	12100	45.0	298
	Flaxin		14800	17.5	147
Animal and other wastes	Composted cow manure		28900	14.0	315
	Composted pig manure		22900	15.4	231
	Refuse derived humus (RDH)	Secondary Resources	23600	14.2	185
Bark, wood and paper products	Composted conifer bark	Melcourt	5000	91.3	224
	Granulated pine bark	Cambark 100	100	4883.9	157
	Woodfibre mix	Camlands	50	11439.8	298
	Woodfibre	Hortifibre medium	50	11439.8	298
	Bark/paper mix (SHL)	Sinclair Horticulture	3900	87.2	132
	Paper waste crumb	New Era Composts	700	378.3	210
Peat (control)	Medium Sphagnum peat	Shamrock	9000	57.7	298

Incubation experiment

N was applied in a spray as ammonium sulphate at a rate of either 0 or 250 mg N per lit of material, together with a cocktail of other nutrients to produce solution concentrations similar to that recommended by Hewitt (1966) in the final mix. The water content of each sample was then adjusted (by wetting) to 0.5 m suction (or to a slightly lower content where samples tended to slump), see Table 1. After mixing, weighed sub-samples of each material were incubated aerobically at 15°C in 'controlled atmosphere' chambers through which a steady stream of water-saturated air was passed. The outflow from the chambers was bubbled through 4% H_2SO_4 solution (v/v) to trap any volatilised ammonia gas. Three replicate sub-samples of each material were removed from the chambers for analysis after 1, 2, 4, 6, 8 and 12 weeks. The H_2SO_4 traps were renewed at the same times. Samples of each material were also analysed at the start of the incubation.

Analysis

Sub-samples of each material were extracted by shaking with K_2SO_4 solution (1:2 w/v) for 2 hours. The filtered extracts were analysed for ammonium (Crooke

and Simpson, 1971) and nitrate (Hunt and Seymour, 1985). Volatilised ammonia adsorbed in H_2SO_4 was measured in the same way as extracted ammonium.

Results and discussion

Measurements of ammonium concentration in the H_2SO_4 traps showed there was little or no loss of ammonium by volatilisation from the materials during the course of the experiment, irrespective of N treatment, so any changes in ammonium or nitrate concentrations were caused by transformations of N as a result of microbial activity. These transformations varied considerably between materials, with paper and bark products generally showing smaller net changes than coir-based materials or composted animal and domestic wastes. The pattern of the transformations was similar at each of the sampling dates, so only data from the final sampling are discussed here.

In materials with high microbial activity there was always a net decline in ammonium content, which was consistently greater in the high N treatment. Losses of ammonium in the latter were normally (but not always) associated with an accumulation of nitrate, as a result of nitrification, although the conversion was seldom quantitative, suggesting that significant immobilisa-

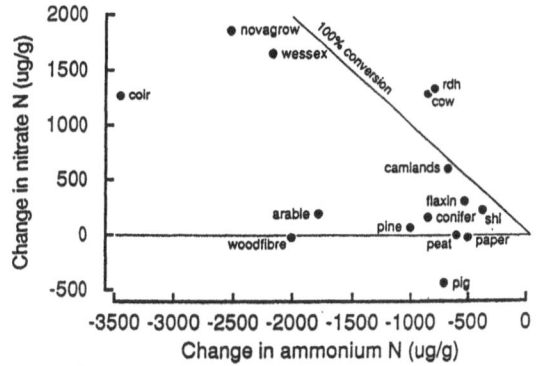

Figure 1. Graph of the change in nitrate content against the corresponding change in ammonium content in the N treated materials after incubation for 12 weeks. The diagonal line represents complete conversion of ammonium into nitrate.

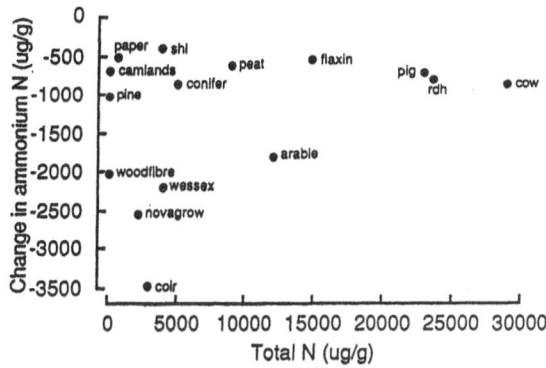

Figure 2. Graph of the change in ammonium content after incubation for 12 weeks against the total (organic + inorganic) N content of the materials before N application.

tion occurred in many materials. In the case of composted pig manure, it resulted in a net loss of nitrate, probably due to denitrification (Fig. 1). With woodfibre, on the other hand, the incomplete conversion was more likely to have been caused by immobilisation of either ammonium or nitrate (rather than by denitrification), because of its large C:N ratio, and highly open structure which would have helped to maintain aerobic conditions during the incubation. Some immobilisation of mineral N (either ammonium or nitrate) may also have occurred in the other materials.

However, comparison of the changes in ammonium contents with the chemical properties of the materials showed there was no clear relationship with either total N content or the C:N ratio of the material, see for example Figure 2. This suggests that a low N content (or high C:N ratio) did not inherently encourage greater immobilisation (at least during the relatively short period of this incubation), unlike the behaviour of crop residues with high C:N ratios (e.g. cereal straw) in the field. Part of the reason for this may have been due to the recalcitrant nature of a large proportion of the organic C in these heterogeneous materials, with only a small fraction involved in microbial activity.

Figure 3 shows there was a strong relationship between the change in ammonium content and the initial amount of ammonium in most materials. With the exception of woodfibre, composted arable waste and peat, virtually all of the ammonium present in each of the materials was utilised during the 12 week experiment. This suggests that transformation of ammonium (whether by nitrification or immobilisation) was generally limited by its availability during the incubation rather than by the underlying chemical composition of the organic component of the material.

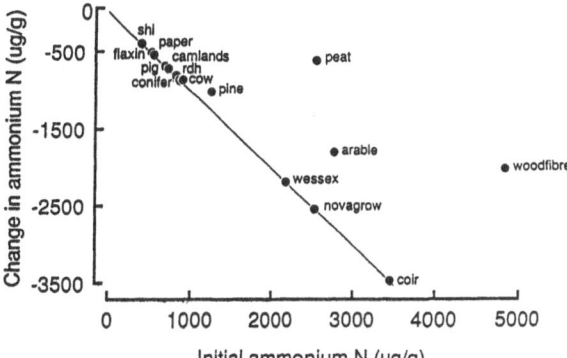

Figure 3. Graph of the change in ammonium content after incubation for 12 weeks against the initial ammonium content of the materials following application of N. The diagonal line represents complete loss of ammonium from the materials.

Conclusions

Turnover of N is generally greater in coir-based materials and in composted animal and domestic wastes than in most bark and paper products. Transformations result largely from nitrification, but simultaneous immobilisation of ammonium and nitrate, and occasionally denitrification of the latter, prevent quantitative conversion of ammonium to nitrate. The resulting decline in ammonium content is controlled more by its initial concentration than by the organic C and N contents of the materials. The high microbial activity of many of these materials means that they will require more careful handling than more stable products (such as peat) if they are to be used in growing media.

Acknowledgements

The authors wish to thank Miss M A Scott (HRI Efford) for supplying the various samples of peat and alternatives, Mr J P Spiller for technical assistance in the construction of the incubation chambers, and Mr J Hunt and Mr S Elliot for the N analyses. The financial support of the UK Ministry of Agriculture, Fisheries and Food is gratefully acknowledged.

References

Bragg N 1990 Peat and its alternatives. Horticultural Development Council, Petersfield, Hampshire, UK. 111 p.

Crooke W M and Simpson W E 1971 Determination of ammonium in Kjeldahl digests of crops by an automated procedure. J. Sci. Food Agric. 22, 9–10.

Davies E M and Scott M A 1994 Alternatives to peat for container HONS production 1991–1992 HDC HNS 28b, HDC Project Report (*In press*).

Hewitt E J 1966 Sand and Water Culture Methods used in the Study of Plant Nutrition. Technical Communication No 22. UK Commonwealth Bureau of Horticulture and Plantation Crops, Farnham Royal, UK. pp 431–432.

Hunt J and Seymour D J 1985 Method for measuring nitrate-nitrogen in vegetables using anion-exchange high-performance liquid chromatography. Analyst 110, 131–133.

Scott M A and Burbridge B 1991 Working to beat the peat problem. Grower 116, 15–18.

O. Van Cleemput et al. (eds.), Progress in Nitrogen Cycling Studies, 31–35, 1996.
© *1996 Kluwer Academic Publishers.*

Nitrogen dynamics on a soil with different tillage systems

M.C. Cameira, M.C. Magalhães and R.L. Pato
Escola Superior Agrária de Coimbra, Bencanta 3000, Coimbra, Portugal

Key words: mineralization, nitrogen, nitrate, organic matter, soil tillage

Abstract

The main objective of this work was to study the effects of conventional tillage and conservation tillage on the evolution and profile distribution of N. The study was carried out on a silty loam Fluvisol, under irrigated maize (FAO 700), on experimental fields of Escola Superior Agrária de Coimbra (Baixo Mondego).

Soil samples were collected before sowing (March 1991 and 1994) and after harvesting (October 1991 and 1993).

Soil analysis for total and mineral nitrogen and organic matter, were made. During the growing season, the influence of tillage on soil nitrogen seems to be important in the data of October 1991: the nitrate levels were higher in the conventional tillage system and the organic matter content was higher in the conservation tillage.

The nitrate evolution and distribution in the soil profile were similar in both tillage systems.

During the fallow period (October 1993 to March 1994) the nitrate levels do not seem to be different in both tillage systems. Some discussion is made about the precipitation and temperature data and their influence on the rate of mineralization.

Introduction

New seedbed preparation, planting techniques, and new tillage procedures will induce changes in soil physical, chemical and biological properties. These changes, outlined by several researchers (Dick, 1983; Doran, 1980), can influence the seasonal variations of the N cycle, by the effects on N mineralization, immobilization, nitrification and denitrification processes (Blevins et al., 1984). Soils under reduced or no tillage practices are generaly cooler, wetter and more compact than those under conventional tillage (Doran, 1980); plowing and cultivation will induce a more oxidized soil environment which will accelerate the organic matter mineralization (Blevins et al., 1977).

A field study established in March 1990 at an Eutric Fluvisol is conducted with three different tillage procedures: the conventional tillage (CT) which consists of a moldboard plow followed by secondary tillage with a disc; two conservation tillage treatments - minimum tillage (MT) , where the crop residues are maintained on the soil surface and the soil is disturbed only as needed to physically place the seed in the soil, being differences in the distance between ridges: under MT-S that distance is 75 cm and under MT-A is 150 cm.

The objective of this work is to study the effect of CT and MT on the evolution and profile distribution of N, under irrigated maize (cv. Lorena, FAO 700), at two reference periods: the fallow and the growing season.

Analyses of organic carbon, kjeldahl N and mineral N ($NO_3^- $-N and NH_4^+-N) were made in three soil layers: 0–15, 15–30 and 30–45 cm, at the two referred periods.

Materials and methods

The experiment was conducted at Coimbra, Portugal, in the fields of Escola Superior Agrária, under Mediterranean conditions, which are characterized by a quickly developing water deficit after spring, due to both a rapidly increasing evaporative demand and reduced rainfall. The monthly average values of air temperature and cumulative rainfall of the two reference periods are shown in Figures 1 and 2.

The soils of the experimental area are Eutric Fluvisols and their characteristics are shown in Table 1.

The experimental design was a randomized block with four replicates. Each plot was 6 by 80 meters and the corn hybrid was planted in 0.75 m rows on a 15

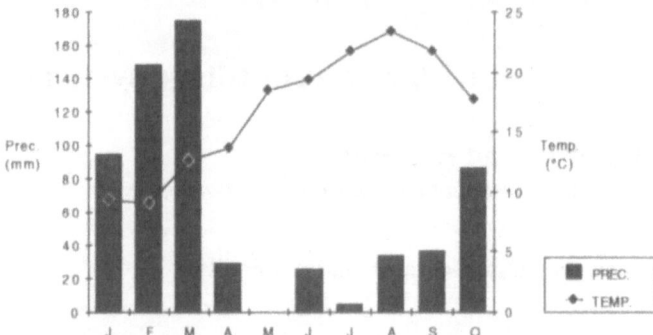

Figure 1. Cumulative precipitation and average temperature in the experimental site during the growing season (Mar 91–Oct 91).

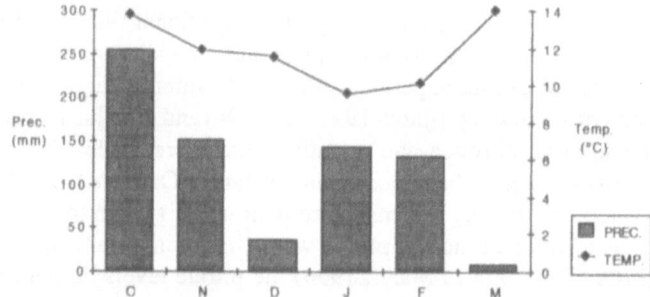

Figure 2. Cumulative precipitation and average temperature in the experimental site during the fallow period (Oct 93–Mar 94).

cm spacing, which provided a population density of 92,000 plants ha^{-1}.

Fertilizer was surface broadcasted in all plots at the rate of 214, 168 and 168 kg ha^{-1} of N, P and K at 1991. The following years, the amounts of fertilizers were reduced, and at 1993 the rate was 87, 114 and 114 kg ha^{-1} of N, P and K.

To control weeds, atrazine was sprayed at the rate of 5 liters ha^{-1} in all plots .

The yields of the three treatments were no significantly different.

The soil samples were collected at depths of 0–15, 15–30 and 30–45 cm at each plot, before the seedling and after the fallow period (March) and after harvesting and before the fallow period (October). The composite sample of each plot consisted of 10 single samples collected in rows.

Total organic carbon was analysed using the Tinsley wet combustion procedure (LQARS, 1977). Total kjeldahl N was determinated using the Kjeldahl method as described by Bremner (1979).

The NH$_4^+$-N and NO$_3^-$-N were determined by the steam destilation method, as described by Bremner (1979).

Table 1. Some soil characteristics before the establishement of experimental treatments (1990)

Depth	pH		Organic carbon	Sand	Silt	Clay
(cm)	H$_2$O	KCl	(g kg^{-1})	(%)		
0–15	6.1	5.4	14.4	51.2	36.0	12.8
15–30	6.1	5.4	14.3	50.3	35.0	14.7
30–45	6.2	5.6	12.3	49.4	36.7	13.9
45–60	6.5	5.8	10.8	54.3	32.7	13.0

The ANOVA was done by GENSTAT V and the significant differences, between tillage systems for various parameters were determined by LSD test.

Results

Results of analysis of variance showed a significant effect of depth and date of sampling on all parameters measured (Table 2).

The total organic carbon on the soil (0–45 cm) has increased since March 1991 to March 1994 (Table 3) and the major values were found at soil surface, specially under conservation tillage treatments (Fig. 3).

Table 2. ANOVA results for treatments effect on soil parameters

Source of variation	Organic C	Kjeldahl N	NO_3^--N	NH_4^+-N
Treatment	NS	NS	4.67*	NS
Date	3.34*	5.51**	113.42***	14.70***
Depth	6.74**	15.64**	11.60***	3.77*
Date × treatment	NS	NS	4.83*	NS
Date × depth	NS	NS	11.60**	NS

NS, *, ** and *** mean non-significant, or significant at $p<0.05$, $p<0.01$ and $p<0.001$ level, respectively.

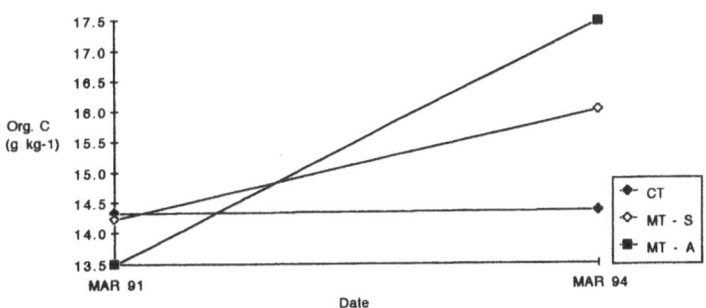

Figure 3. Evolution of Organic carbon values (g kg^{-1}) in the conventional and minimum tillage

Table 3. Mean values of organic carbon (g kg^{-1})

Date	Mar 91	Oct 91	Oct 93	Mar 94
Org. C (g kg^{-1})	13.90 a	14.28 a	15.41 b	15.94 b

Values followed by the same letter are not significantly different by LSD test, $p<0.05$.

Table 4. Mean values of kjeldahl N (g kg^{-1})

Date	Mar 91	Oct 91	Oct 93	Mar 94
Kjeldahl N (g kg^{-1})	1.61 a	1.60 a	1.73 b	1.73 b

Values followed by the same letter are not significantly different by LSD test, $p<0.05$.

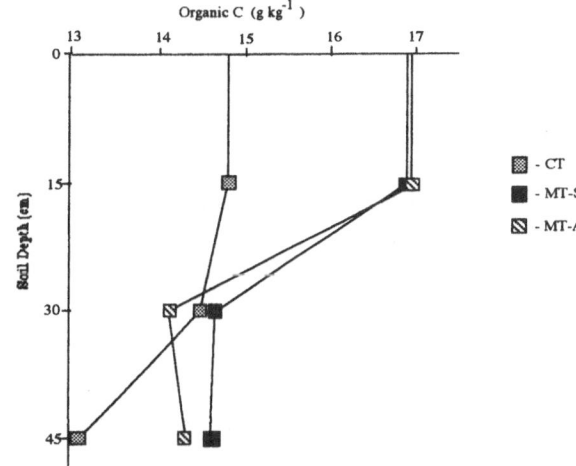

Figure 4. Distribution of organic carbon values (g kg^{-1}) in soil profile in the conventional and minimum tillage.

The organic carbon values under conservation tillage procedures decreased faster with the increase of soil depth, than under conventional tillage (Fig. 4).

Table 5. Mean values of kjeldahl N (g kg^{-1}) in soil profile

Depth (cm)	0–15	15–30	30–45
Kjeldahl N (g kg^{-1})	1.77a	1.68b	1.56c

Values followed by the same letter are not significantly different by LSD test, $p<0.05$.

The kjeldahl-N concentrations closely followed the pattern observed for organic carbon (Tables 4 and 5). Concentrations of soil surface increased, specially under conservation treatments, but the differences are not significant (Fig. 5).

The values of mineral N showed significant effects of the tillage procedures on the NO_3^--N amounts. In

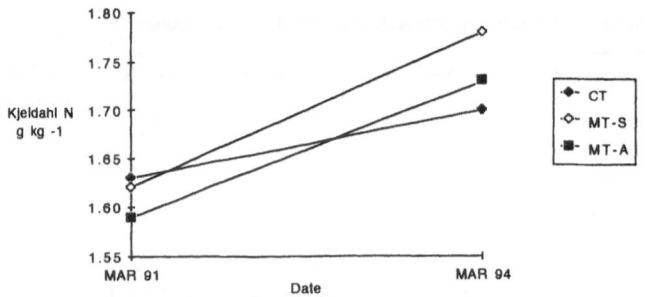

Figure 5. Evolution of Kjeldahl N values (g kg^{-1}) in the conventional and minimum tillage.

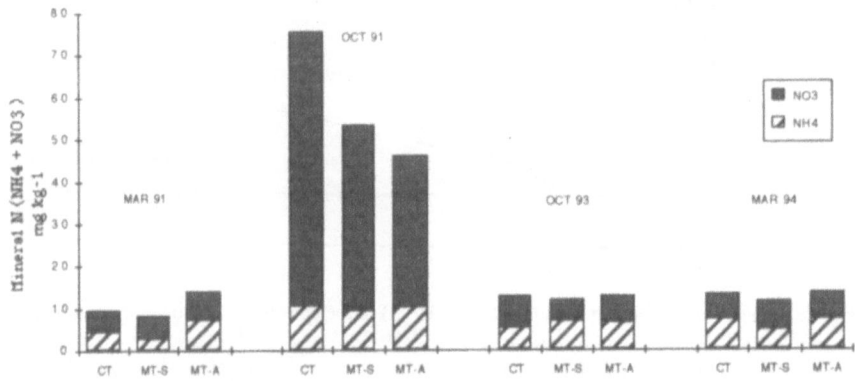

Figure 6. Mineral N (NH$_4^+$ + NO$_3^-$) evolution in the conventional and minimum tillage.

contrast, the reverse ocurred with the NH$_4^+$ -N values (Fig.6).

The NO$_3^-$ -N values were higher at the beginning of the growing season (before fertilization) under the conservation tillage practices; but they were smaller at the end of that period than those from conventional tillage procedures (Table 6).

All treatments showed an increase of NO$_3^-$ -N values during the growing season.

During the fallow period, the NO$_3^-$ -N values decreased 26 and 33% in the surface layers 0–15 and 15–30 cm, and increased 20% in the third layer sampled under conventional tillage procedures.

Different behaviour occurred under conservation tillage practices at the same period; NO$_3^-$ -N values increased 27 and 9% in 0–15 and 15–30 cm depth layers and decreased 23% in the third layer sampled under the MT-A. The NO$_3^-$ -N values from the MT-S treatments decreased 18% in the surface layer but increased 21 and 53% at 15–30 and 30–45 cm depth layers.

The NH$_4^+$ -N values, under the conventional tillage procedures, increased during the fallow period, specially in the surface layer (52%). Under the conservation tillage these values decreased 32 and 47% at MT-S

in the two surface layers and 19% at the 0–15 cm layer of the MT-A treatment.

Discussion

The increase of organic carbon and kjeldahl N amounts in the surface layer, under reduced tillage procedures, can be due to a lower rate of biological oxidation and less soil-residues interaction, as pointed by Dick (1983).

Under long term conventional tillage procedures, a major mineralization rate is expected as well as a decrease of organic carbon in soil.

Below the surface layer, the organic carbon and kjeldahl N contents of the reduced till plots were similar or lower than the conventional tilled soils. The increase in organic carbon with depth, in conventional tilled soils, results partially from burying plant residues with plowing. These results are in agreement with those found by Doran (1980). The accumulation of organic C under minimum tillage treatments represents a potential improvement of soil chemical and physical properties and consequently of the soil fertility at long term.

Table 6. Effect of the treatments on mean values of NO_3^--N (mg kg^{-1}) in growing season

Parameter	Tillage treatments					
	Conventional		Minimum tillage			
	CT		MT-S		MT-A	
	Mar 91	Oct 91	Mar 91	Oct 91	Mar 91	Oct 91
NO_3^--N(mg kg^{-1})	4.94 a	64.89 b	5.06 a	43.98 c	6.76 a	35.85 c

Values followed by the same letter are not significantly different by LSD Test, $p<0.05$.

Differences in mineral N levels between tillage systems were significant, partially due to the highs NO_3^--N values observed at the end of the growing season. During the fallow period the denitrification and higher NO_3^--N lixiviation will be due to the wet conditions of the soil as pointed by Fox and Bandel (1986).

The NH_4^+-N is not significantly affected by tillage treatment in both reference periods.

In 1991, the application of higher nitrogen rate to the soil and the usual conditions of soil water content and temperature accelerated the nitrogen cycle pathways, namely, organic N mineralization and nitrification. This was more evident in conventional tillage treatments.

Assuming that N uptake by crop under different tillage system did not differ substancially (Carter and Rennie, 1984), the lower nitrate values observed at the end of the growing period, under conservation tillage, could be due to the lower nitrification and mineralization, and to a great loss of nitrate by denitrification due to less aerobic conditions (Doran, 1980).

Although it is possible to observe the soil tillage treatments effects on the N dynamics, in this study they will only be conclusive after long term experiences.

Acknowledgement

The authors are grateful to C Gonçalves for her helpful collaboration concerning the field and laboratory work, to A Dias, J Bandeira and F Guerra for the soil chemical determinations and to H Abreu for his help in carrying out tables and graphics.

References

Blevins R L, Thomas G W and Cornelius P L 1977 Influence of no tillage and nitrogen fertilization on certain soil properties after 5 years of continuous corn. Agron. J. 69, 383–386.

Blevins R L, Smith M S and Thomas G W 1984 Changes in soil properties under no tillage. *In* No Tillage Agriculture - Principles and Practices. Eds. R E Phillips and S H Phillips. pp 190–225. Van Nostrand Reinhold, New York, USA.

Bremner J M 1979a Total nitrogen. *In* Methods of Soil Analysis - Chemical and Microbiological Properties, Agronomy 9, Part 2. Eds. C A Black, D D Evans, J L White, L E Ensminger and F E Clark. pp 1149–1176. American Society of Agronomy Publisher, Madison, WI, USA.

Bremner J M 1979b Inorganic forms of nitrogen. *In* Methods of Soil Analysis - Chemical and Microbiological Properties, Agronomy 9, Part 2. Eds. C A Black, D D Evans, J L White, L E Ensminger and F E Clark. pp 1179–1237. American Society of Agronomy Publisher, Madison, WI, USA.

Carter M R and Rennie D A 1984 Nitrogen transformations under zero and shallow tillage. Soil Sci. Soc. Am. J. 48, 1077–1081.

Dick W A 1983 Organic carbon, nitrogen, and phosphorus concentrations and pH in soil profiles as affected by tillage intensity. Soil Sci. Soc. Am. J. 47, 102–107.

Doran J W 1980 Soil microbial and biochemical changes associated with reduced tillage. Soil Sci. Soc. Am. J. 44, 765–771.

Fox R L and Bandel V A 1986 Nitrogen utilization with no tillage. *In* No-Tillage and Surface-Tillage Agriculture - The Tillage Revolution. Eds. M A Sprague and G B Triplett. pp 117–148. Wiley Interscience, New York, USA.

Laboratório Químico Agrícola Rebelo da Silva 1977 Sector de Fertilidade do Solo. DGSA, Ministério da Agricultura, Lisboa, Portugal. 39 p.

O. Van Cleemput et al. (eds.), Progress in Nitrogen Cycling Studies, 37–40, 1996.

N mineralization in undisturbed soil

K.K. Debosz[1], P. Schjønning and S.E. Simmelsgaard
Department of Soil Science, Danish Institute of Plant and Soil Science, Research Center Foulum, PO Box 23, DK-8830 Tjele, Denmark

Key words: aerobic incubation, intact soil cores, nonleached and leached soil, soil texture, winter wheat

Abstract

Two incubation methods were used to determine N mineralization in intact soil cores of a sandy soil. One included periodic leaching while the other was a nonleaching incubation technique at high relative humidity followed by analysis of soil mineral-N. Both methods were evaluated by applying them to soils of varying texture and depth during one month of incubation at 20 °C and at field capacity. Rates of N mineralization found in undisturbed soils during nonleaching incubation ranged from 0.66 to 1.20 kg N ha^{-1} day^{-1} in the plough layer and from 0.10 to 0.64 kg N ha^{-1} day^{-1} in the horizon just beneath the plough layer and were closely related to soil texture. The estimates of N-mineralization by the leaching incubation were comparable to these values. It is concluded that the nonleaching incubation is attractive as a non-laborious method for prediction of N mineralization potential.

Introduction

The productivity of cultivated soils in terms of the potential for delivering nitrogen to the crop during the growing season depends on topography, geology, pedology and cropping history. This means, that the optimum dose of fertilizers varies from place to place within the same field. It is now possible by means of satellites and advanced computer techniques (Differential Global Positioning system) to register continuously the position of the fertilizer distributor in the field and combine the position with a programmed differentiated fertilization plan for the field. A prerequisite for differentiated management is that the correct fertilization doses can be estimated from the soil characteristics. Based on these ideas a project entitled "Site specific crop management" was started in 1992 in Denmark. Two fields, with the study area about 10 ha each, were chosen for intensive studies of variations in soil and plant characteristics. The idea was to examine the relationships between soil characteristics and plant growth in order to produce a basis for a fertilization decision system for practical use (Olsen and Simmelsgaard, 1995).

Accurate N fertilizer recommendations depend upon knowing the amount of soil organic N that is mineralized during the growing season. In order to quantify

N mineralization under field conditions methods which minimize alteration of soil physical structure and moisture are required. A number of techniques have been successfully used to measure rates of N mineralization in situ (Debosz, 1994; Raison et al., 1987).

The objectives of the present study were to measure N mineralized in undisturbed soil, to compare two methods for this objective, to estimate the actual field mineralization, and to determine whether the measurements from undisturbed soil samples could be used to predict N mineralized under field conditions.

Materials and methods

Experimental site

The experimental field (400×300 m) was located at Vindum Overgaard Farm near Viborg, Denmark (56°22' N, 9°34' E). Geological origin of the soil is generally sandy morainic deposits from the Weichselian Ice Period. Elevation of the study site is 60 m, and the total annual precipitation averages 700 mm. In 1993 the crop was winter wheat (c. variety Haven) fertilized with 125 kg total N as calcium ammonium nitrate, in the spring. The field had a peas crop in the preceding season, with an application of 25 t ha^{-1}

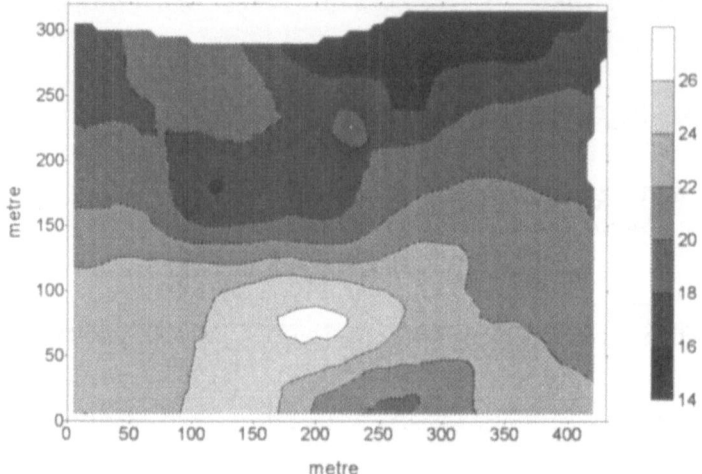

Figure 1. Contour map of the textural fraction < 0.063 mm (clay and fine silt content) for the field in the plough layer (0–25 cm), %.

pig slurry (150 kg total N). A grid of 20 × 20 m was established within the 10 ha area of the field for measurements of soil chemical and physical characteristics. Seventeen grid points were selected from the total number of about 285 grid points to represent soil variation within the field. The soil texture varied from coarse sand to loam soil (Fig. 1), organic C content 0.92–1.56% and total N content 0.067–0.141%.

Soil sampling

Undisturbed soil cores were sampled from the selected 17 points situated within the grid in March 1993, before fertilization. The method utilizes metal cylinders (6.1 cm id, 8.55 cm height) to sample and secure undisturbed soil cores at a water content of field capacity. The samples were taken from the 0.05 to 0.135 m and 0.30 to 0.385 m depths. Twelve cores of undisturbed soil samples were taken from each grid point and depth for the incubation without leaching. Twelve soil cores were taken right next to each undisturbed sample in order to describe physical and chemical properties of soil, and initial mineral-N content. The detailed sampling plan for each point in the grid is illustrated in Figure 2. From two points in the field an additional amount of 12 undisturbed samples was taken, at each of the two depths, for the leaching incubation. The total number of samples collected amounted to 912. In order to maintain field moisture, all soil cores were covered with a plastic cover until analysis could take place. The undisturbed samples were placed in a cold storage room (2 °C) until use.

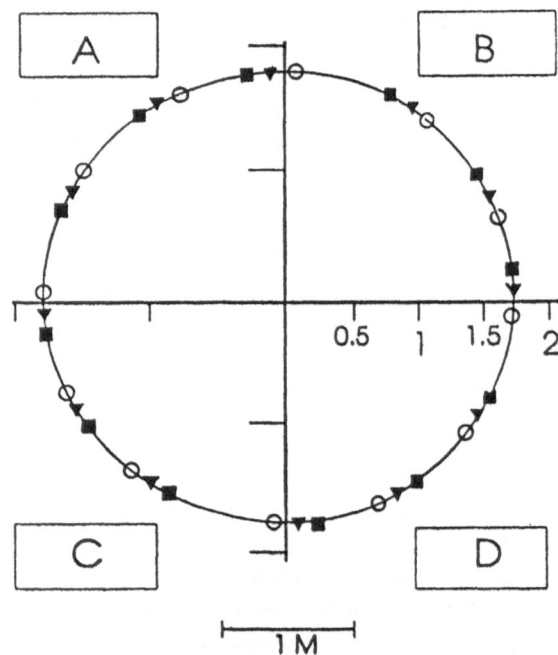

Figure 2. Plan of the sampling strategy at each grid point. The squares A - D represent plant sampling areas. ○: sample location, chemical characteristics, ■: sample location, nonleached method, ▼: sample location, leaching method (only in two grid points).

Laboratory incubations

Two incubation systems were used to determine N mineralization in intact soil cores. One included periodic leaching while the other was a nonleaching incubation technique.

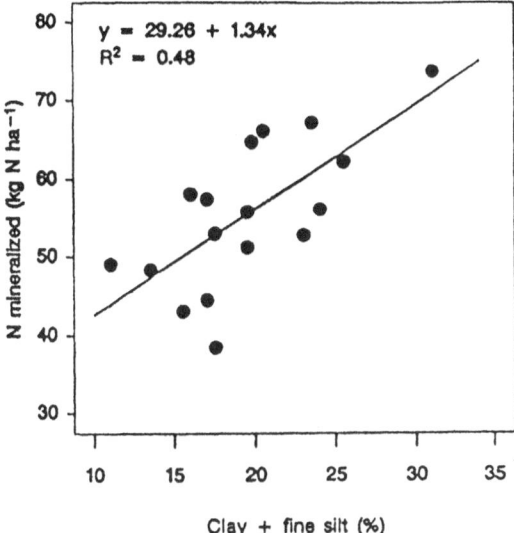

$$y = 29.26 + 1.34x$$
$$R^2 = 0.48$$

Figure 3. Amount of N mineralized (0–40 cm) related to the textural fraction < 0.063 mm (clay and fine silt content) in soils. Values significant at the 0.01 level of probability. n=17. The N mineralized for each grid point was estimated as the total amount of mineral-N present in soil after one month incubation of nonleached soil cores at 20 °C and a water content of field capacity.

Table 1. Net N mineralization rates (kg N ha^{-1} day^{-1}) during 28 days of nonleaching and leaching incubation of soil cores at 20°C

Soil type (Clay%)	Undisturbed soil	
	Nonleached	Leached
0–20 cm		
14.0	0.83	0.73
7.3	0.61	0.49
20–40 cm		
14.0	0.50	0.50
7.3	0.06	0.06

The nonleaching incubation

Twelve undisturbed soil cores were taken directly from the cold storage room and incubated in high humidity boxes at 20 °C. After an incubation period of 28 days the cores were taken apart and the mineral-N content was extracted with 2 *M* KCl from the bulk soil of 12 soil cores. The net rate of N mineralization was calculated as the difference between the initial mineral-N content and that at the end of the incubation period. Soil core weights were taken before and after the incubation period to evaluate moisture loss.

The multiple leaching incubation

The undisturbed soil cores were leached before incubation. 12 cylinders from a specific grid point and soil depth were placed on a ceramic plate connected to a vacuum line, and then allowing at least 200 mL of 0.01 *M* $CaCl_2$ to percolate through each soil core (250 cm^3), followed by 100 mL of N-free nutrient solution (Cabrera and Kissel, 1988). The leachate from all 12 cores was led to the same vessel as the microvariability within the grid point area was not a matter of this study. Preliminary work had showed that for the studied soil the leachate leaving the samples following percolation with 200 mL of solution (approx. 1.5 times the pore volume) had small and constant concentration of mineral-N. Furthermore, extraction of some

soil cores with 2 *M* KCl at the end of a leaching period also indicated that all leachable nitrate-N was removed from the soil. After the leaching procedure, the cores were allowed to drain to equilibrium at a potential of 100 cm water column. Cores were incubated in high humidity boxes at 20 °C. The undisturbed cores were retrieved from the incubator and leached as previously described at 7, 14 and 28 days of incubation. The net N mineralization rate was calculated from cumulative inorganic N found during the multiple leaching incubation.

Plant sampling

Samples of above ground parts of the crop were gathered four times during the growing season. The first samples were taken in March, before nitrogen fertilizers were added. The plant samples were cut at ground level from two 0.5 m^2 areas at each sampling time and taken from the same grid point in the field where soil samples were collected (Fig. 2).

Soil and plant analysis

Concentration of NO_3^--N + NH_4^+-N in 2 *M* KCl extracts (soil/solution ratio 1:2) of the soil samples and in $CaCl_2$ leachate were determined with a Technicon Autoanalyser (Henriksen and Selmer-Olsen, 1970). The plant material was dried at 80 °C and analyzed for total N by the Kjeldahl method, determined on a Tecator total-N analyzer. Particle size analysis was based on sieving and sedimentation.

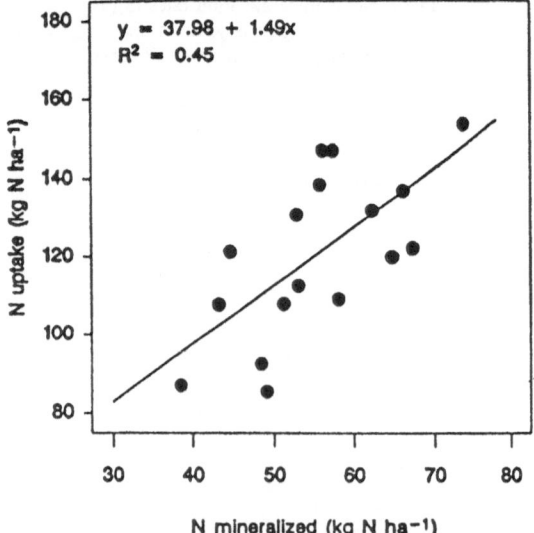

Figure 4. Actual N uptake in winter wheat at the beginning of flowering related to amount of N mineralized (0–40 cm). Values significant at the 0.01 level of probability. n=17. The actual N uptake at the beginning of flowering was calculated as the difference between the N uptake at the beginning of flowering (June 26 1993) and the N uptake in the early spring (tillering March 29 1993).

Results and discussion

The grid points varied widely in texture (ranging from sand to loam) (Fig. 1). All samples were taken in the spring (March), and therefore had a water content at field capacity. The undisturbed soil cores were incubated in high humidity boxes. The use of such boxes was important to maintain the soil samples at optimum water content. The incubator and humidity chamber eliminated any significant moisture loss from the soil cores during the incubation period. The weight loss from the samples during the 28 days period of incubation of nonleached cores corresponded typically to 0.1–0.4% v/v of water.

Rates of N mineralization in undisturbed soil analyzed by the nonleaching incubation ranged from 0.66 to 1.20 kg N ha^{-1} day^{-1} in the plough layer and from 0.10 to 0.64 kg N ha^{-1} day^{-1} in the horizon just beneath the plough layer. The range of values obtained depicts the variability in N mineralization that exists within an area of a field. Soil texture properties are very important factors affecting nitrogen mineralization. Significant correlation existed between N mineralization determined under nonleaching incubation and clay (< 0.002 mm) and fine silt (0.002–0.063 mm) content of soil (Fig. 3).

The rates of N mineralization determined by the leaching method were comparable to the values obtained by nonleaching incubation, especially for the samples from the horizon just beneath the plough layer (Table 1). Care should be taken not to extrapolate this conclusion to heavier soil types. Preferential flow during leaching of structured clay soils might give a wrong estimate of soil N content.

The top 40 cm soil will be the prime source of nitrogen for the crop in the beginning of the growing season. As can be seen from Figure 4, the proposed method based on a simple nonleaching incubation of undisturbed soil cores gives a good estimation of N pool potentially available for N uptake by plants.

The simple incubation method which involves incubation of undisturbed soil cores under controlled temperature and moisture followed by extraction of mineral-N seems to be a realistic approach to estimate the mineralization potential of soil. Soil disturbance and compaction are minimal, and mineralization proceeds under conditions similar to those existing in undisturbed soils. The variability in N mineralization that exists even within a small area of a field demands intensive sampling. Thus, the nonleaching incubation is attractive as a non-laborious method for prediction of N-mineralization potential of soils. Furthermore, the results presented here point out the possibility of using static parameters as the textural composition in pedotransfer functions to obtain agronomically important decision support in fertilization management (Debosz and Kristensen, 1995).

References

Cabrera M L and Kissel D E 1988 Potentially mineralizable nitrogen in disturbed and undisturbed soil samples. Soil Sci. Soc. Am. J. 52, 1010–1015.

Debosz K K 1994 Evaluation of soil nitrogen mineralization in two spring barley fields. Acta Agric. Scand. Sect B. 44, 142–148 .

Debosz K K and Kristensen K 1995 Spatial covariability of N mineralization and textural fractions in two agricultural fields. *In* Proceedings of the Seminar on Site Specific Farming, Danish Inst. of Plant and Soil Science. Ed. S E Olesen. SP Report no. 26, pp 174–180.

Henriksen A and Selmer-Olsen A R 1970 Automatic methods for determining nitrate and nitrite in water and soil extracts. Analyst 95, 514–518.

Olesen S E and Simmelsgaard S E 1995 Danish research on site specific farming. *In* Proceedings of the Seminar on Site Specific Farming, Danish Inst. of Plant and Soil Science. Ed. S E Olesen. SP Report no. 26, pp 39–55.

Raison R J, Connell M J and Khanna P K 1987 Methodology for studying fluxes of soil mineral-N in situ. Soil Biol. Biochem. 19, 521–530.

Plant and Soil **181**: 25–30, 1996.

Temperature effects on C- and N-mineralization from vegetable crop residues

S. De Neve, J. Pannier and G. Hofman
*University of Gent, Department of Soil Management and Soil Care, Division of Soil Fertility and Soil Data Processing, Coupure Links 653, 9000 Gent, Belgium**

Key words: C-mineralization, crop residues, incubation, N-mineralization, temperature

Abstract

Net N-mineralization and nitrification from soil organic matter and from vegetable crop residues (leaf-blades of cauliflower and stems of red cabbage) were measured at 4 temperatures during aerobic incubation in the laboratory. C-mineralization from leaf-blades of cauliflower was monitored at 3 different temperatures. N-mineralization from soil organic matter was best described by zero order kinetics $N(t)=kt$ whereas N- and C-mineralization from the crop residues were described by single first order kinetics. Stems of red cabbage mineralized much more slowly than leaf-blades of cauliflower. S-shaped functions were fitted to the relationship between the rate constants of both C and N-mineralization and temperature. The rate parameter κ of the S-shaped function reflects the temperature dependence of the mineralization rate k. The parameter κ for N-mineralization of the stem material ($\kappa=5.36$) was significantly higher than for the leaf-blades ($\kappa=3.38$), indicating that there is a strong interaction between temperature and resistance to degradation in the soil. N-mineralization from soil organic matter was least sensitive to temperature ($\kappa=2.63$). Temperature dependence of nitrification was not significantly different from mineralization over the temperature range considered. Rate constants for C-mineralization of cauliflower leaf-blades were higher than for N-mineralization, but the temperature dependence of the rate constants was not significantly different for both processes.

Introduction

Efficient use of N fertilizers is only possible if mineralization of organic N from different sources is fully understood. Most research on the dependence of N-mineralization on temperature has focused on mineralization from soil organic matter (Addiscott, 1983; Beck, 1983; Cassman and Munns, 1980; Kowalenko and Cameron, 1976; Lochmann et al., 1989; Stanford et al., 1973) or mineralization from residues of agricultural crops and green manures (Honeycutt et al., 1993; Pal et al., 1975). Mineralization from residues of vegetable crops and its relation to temperature, especially the quantitative aspects, has been studied in less detail. However, in regions with intensive vegetable production these residues contribute considerably to the mineral N content of the soil, which can be either beneficial (supply of N to subsequent crops) or harmful (N leaching to ground water in winter). To predict the release

of N from these vegetable residues it is most important to quantify the relationship with the most influential environmental factor, i.e. temperature. With respect to the prediction of leaching it is necessary to differentiate between mineralization and nitrification of these crop residues.

In this paper quantitative relationships are established between N- and C-mineralization parameters of vegetable crop residues and temperature.

Materials and methods

Pretreatment of soil and plant material

The soil used in the incubation trials is a loamy sand soil from Pittem (West-Flanders) which has been used several years for the growing of chicory. The soil was sampled at a moisture content well above FC and was allowed to dry to a moisture content of 80% of FC.

* FAX No.: +3292646247
Plant and Soil is the original source of publication of this article. It is recommended that this article is cited as: *Plant and Soil* **181**: 25–30, 1996.

The soil was not air dried in order to minimize disturbance of microbial activity. Visible plant material and stones were removed by hand and large soil aggregates were crumbled. Two kinds of crop residues were added: leaf-blades of cauliflower (considered to mineralize very fast) and the upper part of stems of red cabbage (considered more resistant to mineralization). The fresh crop residues were chopped into small pieces of $\pm$ 0.5 cm^2. The dry matter content was 14.1% and 19.3% and the total N content 3.46% and 1.90% for leaf-blades of cauliflower and stems of red cabbage respectively.

N-mineralization

For the N-mineralization trials plastic tubes with an inner diameter of 46.3 mm were used. Each tube was filled with 317 g moist soil mixed thoroughly with 6 g of fresh chopped crop residues. When calculated for a surface of 1 ha (ratio of 1 ha to surface of the tube) this corresponds to approximately 36 tons of fresh crop residues. For leaf-blades of cauliflower this is a realistic amount. For stems of red cabbage this is a high amount, but when incorporating lower amounts the mineralization process can not be followed accurately (especially at the lower temperatures when mineralization is very limited). The soil-crop residue mixture was slightly compacted to obtain a bulk density of 1.4 g cm^{-3}. The tubes were covered with a single layer of parafilm in order to keep moisture content at 80% of FC throughout the trial. Tubes were stored at 4 temperatures: 5.5, 10, 16 and 25 °C. For each temperature a series of blanks (no crop residues) was included. Sampling took place by removing intact tubes. Samples were removed in 3 replicates after 9, 23, 35, 54, 76, 93 and 112 days, extracted with KCl and analyzed for NO$_3^-$-N and NH$_4^+$-N with a continuous flow autoanalyzer. The sum of NH$_4^+$-N and NO$_3^-$-N measured at a certain time is further referred to as N-mineralization.

C-mineralization

C-mineralization was determined using glass jars with an inner diameter of 105 mm. Each jar was filled with a mixture of 300 g moist soil (80% of FC) + 10 g of fresh chopped crop residues (leaf-blades of cauliflower). For the C-mineralization trials more crop residue had to be incorporated per 100 g of soil than for the N-mineralization trials to allow accurate measurements towards the end of the incubation (when only very small amounts of C are being respired). Small vials

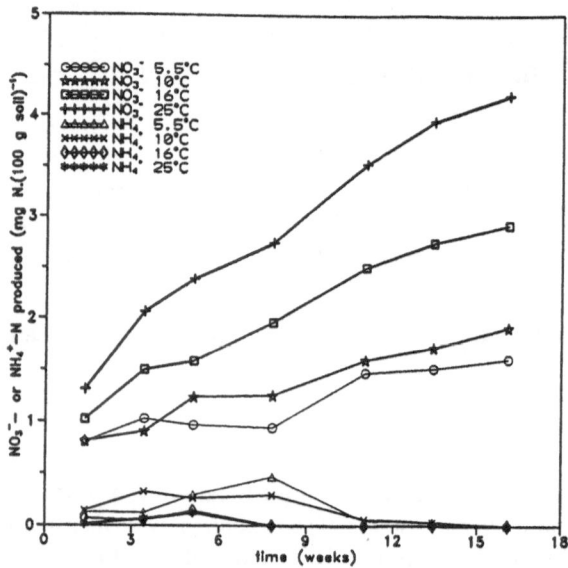

Figure 1. NO$_3^-$ - and NH$_4^+$-N produced in the blank soil.

filled with 15 mL of 1 N NaOH were placed into the jars to trap evolved CO$_2$. The jars were closed airtight and incubated at 3 different temperatures: 5.5, 10 and 16 °C. For each temperature a series of blanks was included. Analysis (in duplicate) of evolved CO$_2$ took place by removing the vials and titration with HCl after addition of BaCl$_2$.

Results

N-mineralization and nitrification

The N-mineralization from the blank soil (no crop residues) is given in Figure 1. Because the disturbance of the microbial biomass was kept minimal, there is hardly any flush of mineral nitrogen (Richter et al., 1982) at the start of the incubation. There is a relatively important rise in NH$_4^+$-N amounts at 5.5 and 10 °C between 2 and 8 weeks after the start of the incubation, but after 10 weeks the amounts of NH$_4^+$-N are negligible. The zero order kinetics model N(t) = k.t (with N(t) the amount of N mineralized after time t) was used to describe N-mineralization and nitrification from the blanks. The values of the rate constants k are given in Table 1. When calculated on field scale for a soil layer of 30 cm, N-mineralization for this soil ranges from 1.8 (at 5.5°C) to 7.9 kg N ha^{-1} week^{-1} (at 25 °C).

The N-mineralization from the crop residues is given in Figures 2 and 3. For leaves of cauliflower

Table 1. Values of the rate constants k and R^2 (R_a^2) values (zero or first order model) for the different temperatures (CaLb = leaf-blades of cauliflower, RcSu = upper part of stems of red cabbage).

| | Temperature (°C) | | | | | | | |
| | 5.5 | | 10 | | 16 | | 25 | |
	k	R^{2a}	k	R^{2a}	k	R^{2a}	k	R^{2a}
N-mineralization								
Blank soil	0.043	0.917	0.057	0.902	0.123	0.980	0.188	0.970
CaLb	0.123	0.958	0.141	0.963	0.290	0.922	0.635	0.993
RcSu	0.031	0.837	0.040	0.834	0.102	0.887	0.353	0.954
Nitrification								
Blank soil	0.057	0.871	0.075	0.972	0.130	0.981	0.193	0.987
CaLb	0.080	0.823	0.079	0.798	0.204	0.873	0.383	0.941
RcSu	0.031	0.845	0.033	0.725	0.098	0.886	0.304	0.910
C-mineralization								
CaLb	0.282	0.972	0.354	0.983	0.565	0.982	–	–

[a]For non-linear curve fitting R_a^2 values (adjusted R^2 values) are used rather than R^2 values as a measure of goodness of fit.

N-mineralization is fast at all temperatures and temperature changes above 16 °C do not further increase mineralization. At 5.5 and 10 °C important amounts of NH_4^+-N are present up to 2 months after the start of the incubation (more than 40 % of total N at 10 °C after 5 weeks, Fig. 4). The upper part of stems of red cabbage are mineralized much more slowly, especially at 5.5 and 10 °C. The lower rate of N-mineralization is also reflected in the smaller amounts of NH_4^+-N that are formed (less than 10% of total N). N-mineralization and nitrification from the crop residues was described using the single first order kinetics model:

$$N(t) = N_A \cdot (1 - e^{-kt}) \qquad (1)$$

with N_A the amount of mineralizable N expressed in % of total residue N (Table 1). Because immobilization (net negative values of NH_4^+-N or NO_3^--N) occurred only in very few cases for the stem material and was not significant, negative values of net NH_4^+-N or NO_3^--N were set equal to zero in the curve fitting procedure. First N_A was calculated for 25 °C and this value of N_A was then used for the other temperatures, i.e. it is supposed that only the rate constant k, and not N_A, is affected by temperature. The values of N_A are 82.2% and 66.7% of N_{tot} for the leaf-blades and the stems respectively.

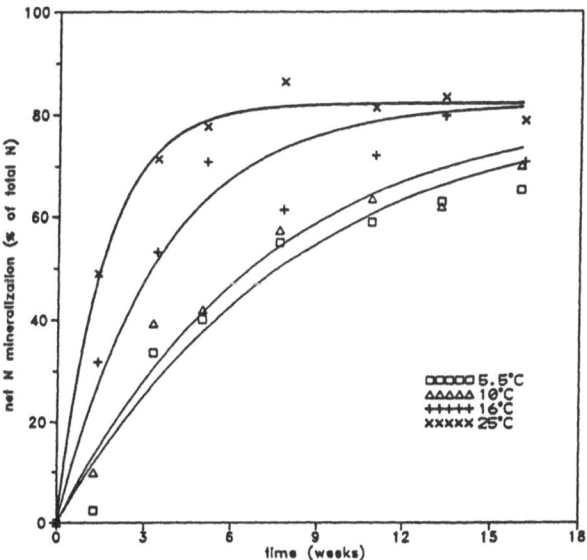

Figure 2. Net N-mineralization (NO_3^--N + NH_4^+-N) from leaf-blades of cauliflower with a first order model fitted to the data.

C-mineralization

C-mineralization of leaf-blades of cauliflower (Fig. 5) was also described by single first order kinetics, giving a very close fit (Table 1). Now the image is quite different. The rate constants for C-mineralization are much higher than for N-mineralization, but C_A (the amount

28

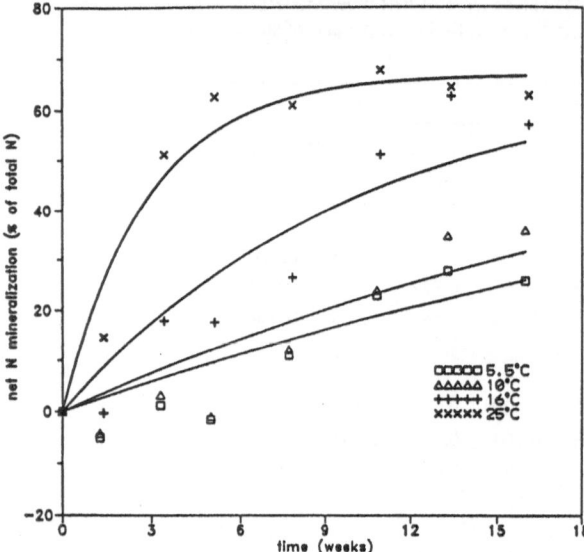

Figure 3. Net N-mineralization ($NO_3^--N + NH_4^+-N$) from the upper part of stems of red cabbage with a first order model fitted to the data.

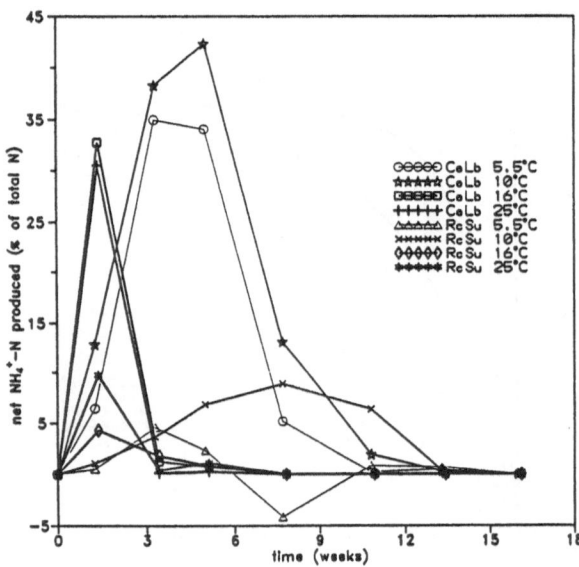

Figure 4. Net NH_4^+-N produced during the incubation of the crop residues (abbreviations: see Table 1).

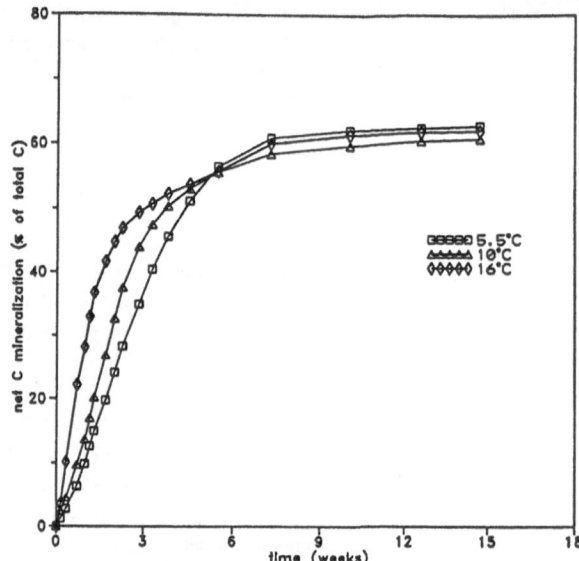

Figure 5. Net C-mineralization (expressed in% of total added C) from leaf-blades of cauliflower.

lost from straw incubated with soil at 22 °C than at 37 °C.

Quantification of the influence of temperature

For modelling the influence of temperature on both C- and N-mineralization an S-shaped function is used. We did not use an Arrhenius type relationship $k = A.e^{-B/T}$ because of several reasons. It has no biological significance as related to soil biomass (the optimum temperature for the Arrhenius equation is ∞), the fitting is poor if points close to 0 °C are included (Addiscott, 1983; Stanford et al., 1973) because the equation uses the absolute temperature, and its parameters, especially A, are not easy to interpret (Addiscott, 1983). The equation used here is:

$$k(T) = k_{opt}.e^{\left(-\kappa\left(1-\frac{T}{T_{opt}}\right)^2\right)} \qquad (2)$$

k(T) is the rate constant as a function of temperature, k_{opt} the maximum rate constant, T_{opt} the optimum temperature (in °C) and κ a rate parameter reflecting the temperature sensitivity of k. The number of data points here was too small to determine the optimum temperature in the equation from the curve fitting. According to Beck (1983) the optimum temperature for ammonification of organic N is $\pm$ 50 °C. T_{opt} in the equation here is set at 37 °C because above that temperature nitrification is severely restricted and eventually ceases at 45 °C (Beck, 1983; Stanford et al., 1973). For

of mineralizable C expressed in% of total residue C) amounts to 63% of total C, which is considerably smaller than N_A. It is remarkable that the final level of C mineralized is slightly higher at 5.5 °C as compared to the higher temperatures. This is comparable to results given by Pal et al. (1975), who found that more C was

Table 2. Parameters of the curve fitting of the rate constants k as a function of temperature and temperature quotients (Q_{10} values) (k_{opt} = value of the rate constant at optimum temperature; CaLb, RcSu: see Table 1). Bracketed values represent standard deviations

	k_{opt}	κ	R_a^2	Q_{10}	
				5.5–16 °C	16–25 °C
N-mineralization					
Blank soil	0.256 (0.017)	2.63 (0.14)	0.971	2.72	1.60
CaLb	0.900 (0.179)	3.38 (0.57)	0.986	2.26	2.39
RcSu	0.619 (0.193)	5.36 (0.93)	0.995	3.10	4.00
Nitrification					
CaLb	0.538 (0.116)	3.17 (0.46)	0.981	2.43	2.01
RcSu	0.512 (0.142)	5.00 (0.89)	0.995	3.01	3.52
C-mineralization					
CaLb	1.323 (0.008)	2.55 (0.01)	0.899	1.94	–

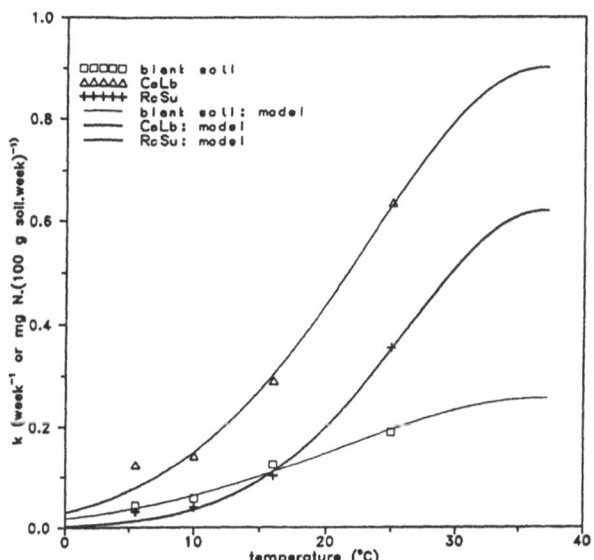

Figure 6. The S-shaped temperature model fitted to the N-mineralization rate constants of the blank soil and the crop residues (abbreviations: see Table 1).

C-mineralization the optimum temperature is also set at 37 °C (according to Roper (1985) C-mineralization is maximal between 25 and 45 °C). Although several authors found N-mineralization to occur at 0°C (Lochmann et al., 1989; Stadelmann et al., 1983), we considered it here to be negligible and the point k(0)=0 was included in the curve fitting procedure. The fitted curves for N-mineralization are given in Figure 6. Table 2 summarizes the results for N-mineralization, nitrification and C-mineralization.

Discussion

There is an obvious difference between the two kinds of crop residues. Indeed, the value of κ for the stems is significantly ($p=0.019$) higher than for the leaf-blades, which is equivalent to a significant interactive effect in a crop × temperature ANOVA of the rate constants k. This indicates that there is a strong interaction between temperature and kind of crop residue: N-mineralization from resistant crop residues will be more enhanced by a rise in temperature than N-mineralization from more easily degradable residues. This is comparable to results from Nuske and Richter (1981), in which temperature favoured N-mineralization from old (resistant) soil organic matter more than from fresh soil organic matter.

As pointed out by Addiscott (1983) for soil organic matter, there is no difference here in temperature sensitivity between N-mineralization and nitrification of crop residues. There is a lag of several weeks between N-mineralization and nitrification at temperatures of 10°C and below (Fig. 4), which means that leaching of NO_3^--N from these crop residues will also be delayed. For nitrification from the blank soil no values are given in Table 2 because the (small) differences in k values between nitrification and mineralization are merely a consequence of the pretreatment of the soil (some excess NH_4^+ is formed in the beginning of the trial). C- and N-mineralization from leaf-blades of cauliflower were not significantly different in their response to temperature. In the first weeks of the mineralization process the C/N ratio of the added cauliflower residues drops at the lower temperatures (5.5 and 10 °C). But eventually, because N-mineralization is more complete than C-mineralization, the material remaining after 15 weeks has a higher C/N ratio (the C/N ratio for leaf-blades of cauliflower at 16 °C rises from 10.1 at the start of the incubation to 19.5 after 15 weeks).

The Q_{10} values were also calculated. As can be seen in Table 2 these values heavily depend on the kind of organic matter (soil organic matter, crop residues). Especially the Q_{10} values for the stem material deviate much from the value $Q_{10} = 2$ accepted by Stanford et al. (1973) for soil organic matter.

In conclusion we can say that for modelling purposes not one but several relationships describing the influence of temperature on N-mineralization will have to be used, depending on the kind of crop residue or the kind of organic matter considered. Further research including more crop residues will have to clarify whether these crop residues can be classified into groups with a similar temperature dependence for N-mineralization.

Acknowledgements

Financial support by the EC (EC project No. 8001-CT91–0115) and the IWONL (Institute for Encouraging Scientific Research in Industry and Agriculture) is gratefully acknowledged.

References

Addiscot T M 1983 Kinetics and temperature relationships of mineralization and nitrification in Rothamsted soils with differing histories. J. Soil. Sci. 34, 343–353.

Beck T 1983 Die N-mineralisierung von Böden im Laborbrutversuch. Z. Pflanzenernähr. Bodenkd. 146, 243–252.

Cassman K G and Munns D N 1980 Nitrogen mineralization as affected by soil moisture, temperature and depth. Soil Sci. Soc. Am. J. 44, 1233–1237.

Honeycutt C W, Potaro L J, Avila K L and Halteman W A 1993 Residue quality, loading rate and soil temperature relations with hairy vetch (*Vicia Villosa* Roth) residue carbon nitrogen and phosphorus mineralization. Biol. Agric. Hortic. 9, 181–199.

Kowalenko C G and Cameron D R 1976 Nitrogen transformations in an incubated soil as affected by combinations of moisture content and temperature and adsorption-fixation of ammonium. Can. J. Soil Sci. 56, 63–70.

Lochmann R, van der Ploeg R R and Huwe B 1989 Zur Parametrisierung der Stickstoff-Mineralisierung in einem Ackerboden unter Feldbedingungen. Z. Pflanzenernähr. Bodenkd. 152, 319–324.

Nuske A and Richter J 1981 N-mineralization in löss-parabrownearthes: incubation experiments. Plant and Soil 59, 237–247.

Pal D, Broadbent F E and Mikkelsen D S 1975 Influence of temperature on the kinetics of rice straw decomposition in soils. Soil Sci. 120, 442–449.

Richter J, Nuske A, Habenicht W and Bauer J 1982 Optimized N-mineralization parameters of loess soils from incubation experiments. Plant and Soil 68, 379–388.

Roper M M 1985 Straw decomposition and nitrogenase activity (C_2H_2 reduction): effects of soil moisture and temperature. Soil Biol. Biochem. 17, 65–71.

Stadelmann F X, Furrer O J, Gupta S K and Lischer P 1983 Einfluß von Bodeneigenschaften, Bodennutzung und Bodentemperatur auf die N-Mobilisierung von Kulturböden. Z. Pflanzenernähr. Bodenk. 146, 228–242.

Stanford G, Frere M H and Schwaninger D H 1973 Temperature coefficient of soil nitrogen mineralization. Soil Sci. 115, 321–323.

Section editor: R Merckx

O. Van Cleemput et al. (eds.), Progress in Nitrogen Cycling Studies, 47–51, 1996.

The dissemination of *Pseudomonas fluorescens* ANP15, a plant-growth-stimulating microorganism, by *Lumbricus terrestris* in soil

W. Devliegher and W. Verstraete[1]
Laboratory Microbial Ecology University of Gent, Belgium. [1] *Corresponding author*

Key words: Lumbricus terrestris, Pseudomonas fluorescens ANP15, plant-growth-promotion, soil inoculation

Abstract

The earthworm *Lumbricus terrestris* makes deep vertical burrows in the soil and feeds overnight on organic material at the soil surface. Previous experiments showed that about 2 ton of organic matter dry weight, representing 100 kg of N, could be incorporated in the soil per year by earthworm populations of 1000 kg ha^{-1}. Earthworm burrowing furthermore provides channels for root growth, water infiltration and aeration.

Interestingly, the feeding behaviour of *L. terrestris* also provides the possibility to distribute beneficial microorganisms throughout the soil profile. Results demonstrated that *Pseudomonas fluorescens* ANP15 was efficiently distributed in the galleries of *Lumbricus terrestris* when the bacteria were mixed with the lettuce which served as a food for the worms. On average 5.10^5 CFU g^{-1} dry soil were found in the earthworm excreta taken from the galleries for a soil depth between 5 and 20 cm. In the upper soil layer (0–5 cm), the number of propagules in the casts was significantly higher and averaged 4.10^7 CFU g^{-1} dry soil. Root colonization of maize (7.2 ± 0.1 log CFU g^{-1} dry root), grown afterwards in the soil, was similar to a control treatment where the bacteria were manually mixed in the soil (7.5 ± 0.2 log CFU g^{-1} dry root).

It is proposed that a similar technique can be used for the distribution of other plant beneficial microorganisms or even microbiota that can support the bioremediation of a soil. It is discussed how this MIME-technique (Microbial Inoculation by Means of Earthworms) possibly could be applied and investigated in the field.

Introduction

General characteristics

Lumbricus terrestris is a deep-burrowing earthworm species, common in agricultural soils, meadows and orchards. Adult earthworms have an average length of 15–20 cm, with a diameter of about 0.5 cm. *L. terrestris* makes vertical galleries down to 2–3 m, and feeds overnight at the soil surface on organic material. During feeding, this organic material is pulled downwards into the galleries. Feeding and earthworm activity mainly occur during spring and autumn, when the ambient temperature and the soil moisture content are favorable.

The fresh earthworm biomass in a soil varies among different ecosytems (Table 1). On average, the total earthworm biomass ranges from 500 to 1000 kg fresh

Table 1. Fresh earthworm biomass in different ecosystems (After Lee, 1985 and Edwards, 1983)

Ecosystem	Fresh earthworm biomass (kg ha^{-1})
Old and sown pastures	500–3000
Orchards (with grass)	630–2300
Deciduous forests	20–2800
Natural grassland	30–1000
Wheat/Barley	30–700
Root crops	70–460
Average value	500–1000

weight ha^{-1}. Earthworm biomass is generally lower if the soil is often disturbed or tilled (e.g. when root crops are grown) or in ecosystems with a low organic matter input (e.g. tropical soils). *Lumbricus terrestris* is one of the most important representatives of the *Lumbricidae*,

48

which is the major group of earthworms found in temperate soils.

Agricultural value

Based on previous experiments, it was calculated that a *L. terrestris* population of 1000 kg fresh biomass ha^{-1} can incorporate about 2 ton organic matter dry weight and 100 kg N ha^{-1} y^{-1} (Devliegher and Verstraete, 1996). At the same time, during earthworm activity, nitrogenous wastes and mucus are excreted, yielding a nitrogen equivalent of 10–30 kg N ha^{-1} y^{-1} (Lee, 1985). A similar amount of nitrogen, i.e. 20–30 kg N ha^{-1} y^{-1}, is released from dead earthworm tissue.

Because of the peculiar feeding behaviour of *L. terrestris*, fertilizers, lime or pesticides that are supplied to the soil surface are also incorporated in the soil profile, especially if they are mixed or supplied with nutritious organic material.

In the presence of *Lumbricus terrestris*, the soil physical conditions are improved due to the formation of channels and the excretion of casts. Deep and wide burrows (=channels) are formed during earthworm movements. These channels allow a proper aeration, root penetration and water drainage in the soil. Similarly, casts are formed and deposited mainly on or near the soil surface during feeding. These casts (= excreta) improve the stability and the water holding capacity of the soil. It has been estimated that up to 5–10% of the top soil can be ingested and excreted each year (Lee, 1985).

New perspective - general concept

Several microorganisms have a beneficial effect on plant growth if they are brought into the soil (e.g. plant-growth-promoting bacteria (Kloepper et al., 1980) or microorganisms antagonistic against plant pests (Sivan and Chet, 1992)). It is however often very difficult to properly disseminate these desired microorganisms in the soil profile.

Due to the general behaviour of *Lumbricus terrestris* in soil, it was anticipated that this earthworm could be used for the distribution of microorganisms in soil. Since *L. terrestris* pulls the organic material into its galleries during feeding, microorganisms that are mixed with the organic material at the soil surface would be taken down and distributed concomitantly.

This paper reports about experiments set up with *Pseudomonas fluorescens* ANP15, a plant-growth-promoting, fluorescent *Pseudomonas* sp. (Iswandi et

al., 1987; Seong et al., 1991). The distribution of *P. fluorescens* ANP15 in the soil in the presence of *L. terrestris* was compared with a control where the bacteria were manually mixed in the soil. Afterwards, a pot experiment with maize was carried out in order to compare the root colonization in both treatments.

Materials and methods

Pseudomonas fluorescens ANP15 was grown in 250 mL Modified KingB (MKB)-medium. This MKB-medium contains (per L): proteose pepton n°3 (Difco): 5.0 g; MgSO$_4$.7H$_2$O: 1.5 g; K$_2$HPO$_4$.3H$_2$O: 1.57 g and glycerol: 4.0 mL. A freshly grown bacterial culture was inoculated in this liquid medium and grown for 3 days at 28°C on a rotary shaker. Thereafter, the cells were centrifuged (10′, 7200 g), washed with 0.1 M MgSO$_4$.7H$_2$O and centrifuged again. The pellet was finally resuspended in 20 mL 0.1 M MgSO$_4$.7H$_2$O. This suspension was then mixed with lettuce.

Prior to use, the lettuce was washed and the white veines were removed. The lettuce was then roughly cut into pieces of about 1 cm^2. The suspension of *P. fluorescens* ANP15 was mixed with these pieces of lettuce at a dose of about 2.10^7 CFU g^{-1} fresh weight. In the earthworm treatment, 10 or 15 g of this lettuce, mixed with the bacteria, was supplied (as a food source) on the surface of a soil core (1.5 kg of soil) containing two *L. terrestris* earthworms. In the control treatment without earthworms, the lettuce with *P. fluorescens* ANP15 was homogeneously mixed with the soil. This procedure (growth of *P. fluorescens* ANP15, mixing the bacteria with the lettuce and supply of lettuce on or in the soil) was repeated 5 times during incubation. The soil cores (20 cm high, 8 cm diameter) were incubated during 5 weeks at 15–20 °C in the dark at a relative humidity of almost 100%. In total, 55 g of fresh lettuce was supplied per kg dry soil during this 5 weeks incubation period. The experiment was carried out in 4 replicates.

After this feeding period, the number of *P. fluorescens* ANP15 was determined in the soil at 4 different depths (0–5, 5–10, 10–15 and 15–20 cm). For the earthworm treatment, the number of propagules was determined in the excreta of *L. terrestris*, lining the galleries (called 'Casts' in Fig. 1A). For comparison, the number of ANP15 was also counted in the soil that was not ingested by *L. terrestris* (called 'Non-cast soil' in Fig. 1A). In the control treatment where lettuce was manually mixed in the soil, the complete soil fractions were sampled (called 'Soil' in Fig. 1B).

A pot experiment with maize was carried out. Maize seeds were sown in the 4 soil columns of the earthworm treatment and in the 4 soil columns of the control treatment. Maize plants were grown for 4 weeks and incubated in a growth chamber at constant temperature and relative humidity (R.H.): during the day (16 h): 24 °C, 80% R.H. and during the night (8h): 16°C and 70% R.H. During that period, earthworms were kept in the soil. Afterwards, the number of *P. fluorescens* ANP15 colonizing the lower 10 cm of the roots was counted.

To allow detection of the inoculated strain in the soil, the genetically marked strain *P. fluorescens* JPB3 was used in this experiment instead of the wild-type ANP15. *P. fluorescens* JPB3 was derived from ANP15 by marking the strain with a Mud(lac) element (Höfte et al., 1990). *P. fluorescens* JPB3 was counted on a modified M9 medium supplied with X-gal (5-bromo-4-chloro-3-indolyl-ß-D-pyranoside). This MM9-X-gal medium contains (per liter): Na_2HPO_4: 6.0 g; KH_2PO_4: 3.0 g; NaCl: 0.5 g; $MgSO_4.7H_2O$: 49.3 mg; $CaCl_2.2H_2O$: 14.7 mg; NH_4Cl: 1.0 g; Na-succinate: 2.0 g and Km: 200 mg (Höfte et al., 1990). Colonies of *P. fluorescens* JPB3 turn blue on the MM9-X-gal medium.

For the enumeration of JPB3 in soil and casts, 4 times 1 g of fresh material was taken and added to 9 mL of sterile physiological solution. Serial dilution platings were prepared from these suspensions. For the enumeration of *P. fluorescens* JPB3 on the roots, the procedure described by Iswandi (1986) was followed. Plant roots were removed from the soil by gently pouring tap water over the root-soil system. At last, roots were rinsed 5 times with sterile demineralized water. Rinsed roots were blotted to dry and afterwards cut into small pieces. Two g of fresh roots was taken and thoroughly macerated in 9 mL of physiological solution. This suspension was further used to prepare a 10-fold serial dilution for counting *P. fluorescens* JPB3.

Results

The number of *Pseudomonas fluorescens* ANP15 in the earthworm casts was similar to the numbers found in the control soil for a soil depth between 5 and 20 cm (Fig. 1A, B). The average number of bacteria found in both treatments for these soil depths was about 5.10^5 CFU g^{-1} dry soil. The number of bacteria in the non-

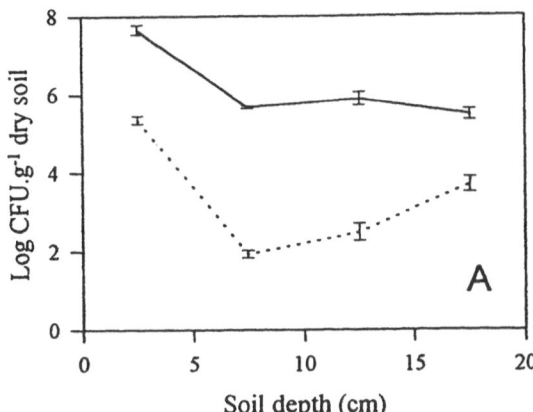

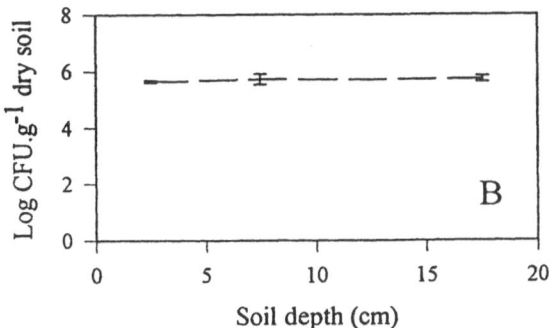

Figure 1. Distribution of *Pseudomonas fluorescens* ANP15 after soil inoculation. **A**. The earthworm treatment. **B**. The control treatment. '*Casts*' (———) represents the number of bacteria found in the excreta of *Lumbricus terrestris* collected in the earthworm galleries. '*Non-cast soil*' (·····) represents the number of bacteria found in the soil that was not ingested by *L. terrestris*. '*Soil*' (——— ———) represents the number of bacteria found in the soil after manually mixing the lettuce with *P. fluorescens* ANP15 in the soil.

ingested soil ('Non-cast soil') on the other hand was always 100 to 1000 times lower than the numbers found in the casts (Fig. 1A).

Unlike for the deeper soil layers, in the upper layer (0–5 cm) the number of ANP15 propagules in the earthworm casts was much higher than the number of bacteria in the upper soil layer from the control treatment (Fig. 1A, B). On average, in this layer, about 4.10^7 CFU g^{-1} dry soil were found in the earthworm casts. This is about 100 times higher than the number found in the soil of the control treatment (Fig. 1B).

After the pot experiment, the number of *P. fluorescens* ANP15 colonizing the roots of maize was the same, whether the bacteria were distributed by the earthworms (7.24 ± 0.10 log CFU g^{-1} dry root) or manually mixed in the soil (7.53 ± 0.21 log CFU g^{-1} dry root). Maize yield in both the earthworm treatment

$(4.19 \pm 0.22$ g DW kg^{-1} dry soil) and the control treatment $(4.42 \pm 0.39$ g DW kg^{-1} dry soil) was similar. Plant growth in both these treatments that had received a *P. fluorescens ANP15* inoculum, was also compared with a treatment that had lettuce manually mixed in the soil, together with an autoclaved *P. fluorescens* ANP15 inoculum. Non of the inoculated treatments had a higher maize dry weight yield in comparison with this autoclaved control (data not shown).

Discussion

The number of *P. fluorescens* ANP15 bacteria found along the earthworm burrowings averaged 5.10^5 CFU g^{-1} dry soil for a soil depth between 5 and 20 cm. This value was similar to the amount of bacteria found in the soil after manual mixing of the bacteria (and lettuce) in the soil (Fig. 1). These data demonstrated that *Pseudomonas fluorescens* ANP15 was efficiently distributed into the deeper soil layers by *Lumbricus terrestris*, at least along the galleries of *L. terrestris*. On the other hand, the number of bacteria found in the 'non-cast' soil in the earthworm treatment was always 100 to 1000 times lower than the numbers found in the casts (Fig. 1A). This showed that soil inoculation was limited to earthworm galleries.

The number of *P. fluorescens* ANP15 in the earthworm casts collected from the upper soil layer (0–5 cm) was much higher than in the equivalent soil sample (Fig. 1). This higher amount of bacteria was probably due to the high amounts of casts produced in this soil fraction (especially on the surface). During incubation, lettuce supplied on the soil surface was gradually taken and incorporated by *L. terrestris* and no residues accumulated on the surface. It is therefore assumed that leaching of bacteria into the casts in the upper soil layer was only of minor importance. According to the feeding behaviour of *L. terrestris*, the main reason for these high numbers is probably that the major part of the lettuce, with the desired bacterial strain, was taken at the soil surface, ingested and processed with soil and afterwards excreted near the soil surface again. Bacteria lining the galleries were then probably mainly distributed during movements of the earthworms up and down or as a result of residual excretion of casts.

The intense earthworm activity and concomitant processing of organic matter near the soil surface resulting in high amounts of target microorganisms in the upper soil layer would be especially advantageous for germinating seeds. Germinating seeds pref-

erentially cover this upper soil layer during initial growth. Therefore, potentially beneficial microorganisms should preferably be present and protect the seed in this zone. On the other hand, in the deeper soil layers, the average number of microorganisms found in the casts along the earthworm burrows (5.10^5 CFU g^{-1} dry soil) equalled the amount of bacteria found in the control soil. This could be important in a special way. Since the earthworm burrows can provide preferential paths for root growth, inoculation into these burrows would bring the bacteria into direct contact with the plant roots. The fact that root colonization was similar in both the earthworm and the control treatment, illustrates that the distribution of ANP15 in the earthworm galleries was as efficient as the manual distribution of ANP15 over the complete soil. In this experiment, the latter treatment was assumed to represent the best way of inoculation to yield a maximal root colonization. It was therefore very interesting to find that root colonization was equally well successfull in the earthworm treatment.

Even though root colonization was successfull, no increases in maize dry weight yield were reported due to inoculation. This finding corroborates earlier results showing that ANP15 was only successfull in stimulating plant growth in soils treated with sugars and amino acids prior to the plant growth tests. Under such conditions, the colonization of roots at a density of 10^7 to 10^8 CFU g^{-1} dry root gave a growth stimulation of 10 to 20% for different crops (Iswandi et al., 1987; Seong et al., 1991). Since the major purpose of this experiment was to observe root colonization in the presence of *L. terrestris*, normal, untreated Ardoyen soil was used. Therefore, no plant growth stimulation could be observed under these conditions.

The dissemination of microorganisms by earthworms has also been described for *Rhizobium meliloti* (Stephens et al., 1994) and for mycorrhizal fungi (Reddell and Spain, 1991). It is anticipated that this mechanism can be used for the distribution of any potentially biologically active microorganism in the soil. Other possibilies could be the distribution of *Azospirillum*, fungi or bacteria capable to control plant diseases (e.g. *Trichoderma harzianum*, Sivan and Chet, 1992) or even microorganisms aimed at bioremediation of the soil.

MIME: Microbial Inoculation by Means of Earthworms - Concept for further study

One of the main questions that should be answered is whether *Lumbricus terrestris* can be used as a vector to spread microorganisms in soil under field conditions. Further research should therefore also concentrate on the effect of the earthworms on the growth and the survival of the desired microorganisms.

On the other hand, since the population of earthworms should be large enough to efficiently distribute the microorganisms, it would be necessary to investigate how biomass growth of *Lumbricus terrestris* can be stimulated in the field, especially in combination with actual agricultural practises. Some possibilities could be to minimize soil tillage (e.g. direct drilling) or to supply attractive organic material on the surface in autumn or spring (e.g. straw, leaves, farmyard manure). A very interesting possibility is to investigate what kind of organic wastes can be used for the growth of earthworm biomass together with the distribution of microorganisms. Other possibilities to increase earthworm populations are the use of integrated pest control and to maximize soil coverage during the year to avoid bird preying. It is however still an open question whether *Lumbricus terrestris* can be manipulated economically to support an integrated, sustainable agriculture or a manageable soil bioremediation or restoration treatment.

Acknowledgements

The authors want to thank the NFWO (National Funds for Scientific Research) for financially supporting this research by providing a PhD scholarship to the first author.

References

Devliegher W and Verstraete W 1996 *Lumbricus terrestris* in a soil core experiment: effects of nutrient-enrichment processes (NEP) and Gut-associated processes (GAP) on the availability of plant nutrients and heavy metals. Soil Biol. Biochem. (*In press*).

Edwards C A 1983 Earthworm ecology in cultivated soils. *In* Earthworm Ecology: From Darwin to Vermiculture. Ed. J E Satchell pp 123–138. Chapman and Hall, London, UK.

Höfte M, Mergeay M and Verstraete W 1990 Marking the rhizopseudomonal strain 7NSK2 with a Mud(lac) element for ecological studies. Appl. Environ. Microbiol. 56, 1046–1052.

Iswandi A 1986 Seed inoculation with *Pseudomonas* spp. Ph.D. Thesis, State University of Gent, Faculty of Agricultural Sciences, Gent, Belgium. 139 p.

Iswandi A, Bossier P, Vandenabeele J and Verstraete W 1987 Deterioration and reactivation of beneficial rhizopseudomonads of barley (*Hordeum vulgare*). Biol. Fertil. Soils 4, 125-128.

Kloepper J W, Leong J, Teintze M And Schroth M N 1980 Enhanced plant growth by siderophores produced by plant growth-promoting rhizobacteria. Nature 286, 885–886.

Lee K E 1985 Earthworms: Their Ecology and Relationships with Soils and Land Use. Academic Press, Sydney, Australia. 411 p.

Reddell P and Spain A V 1991 Earthworms as vectors of viable propagules of mycorrhizal fungi. Soil Biol. Biochem. 23, 767–774.

Seong K Y, Höfte M, Boelens J and Verstraete W 1991 Growth, survival and root colonization of plant growth beneficial *Pseudomonas fluorescens* ANP15 and *Pseudomonas aeruginosa* 7NSK2 at different temperatures. Soil Biol. Biochem. 23, 423–428.

Sivan A and Chet I 1992 Microbial control of plant diseases. *In*: Environmental Microbiology. Ed. R Mitchell. pp 335–354. Wiley-Liss, John Wiley and Sons, Inc., New York, USA.

Stephens P M, Davoren C W, Ryder M H and Doube B M 1994 Influence of the earthworm *Aporrectodea trapezoides* (*Lumbricidae*) on the colonization of alfalfa (*Medicago sativa* L.) roots by *Rhizobium meliloti* L5–30R and the survival of *R. meliloti* L5–30R in soil. Biol. Fert. Soils 18, 63–70.

O. Van Cleemput et al. (eds.), Progress in Nitrogen Cycling Studies, 53–56, 1996.
© 1996 Kluwer Academic Publishers.

Estimating gross mineralisation of *Alnus glutinosa* residues, using ^{15}N mirror image experimentation *

Rebecca Clare Hood and Martin Wood
Department of Soil Science, SAC, West Mains Road, Edinburgh, EH9 3JG, Scotland UK and Department of Soil Science, University of Reading, P.O. Box 233, Whiteknights, Reading, RG6 2DW, UK

Key words: Alnus glutinosa, gross mineralisation, nitrogen, residues

Abstract

Alnus glutinosa is a nitrogen fixing tree frequently used for land reclamation. Studies of mineralisation were undertaken to understand the soil processes involved in restoration. Gross mineralisation rates were determined using ^{15}N tracer techniques and mirror image experimentation. The study involved adding *Alnus glutinosa* ^{15}N labelled and unlabelled, leaf and root material to the spoil and sequentially measuring gross mineralisation using ^{15}N and ^{14}N injection techniques. Using mineralisation equations it was possible to calculate gross mineralisation, from newly incorporated residues, as distinct soil organic matter. Results showed that the size of the ammonium pool in the labelled and unlabelled pots was identical. This suggested that the mineralisation rates were comparable. In the pots amended with leaf material the contribution to gross mineralisation from the leaves was 22, 41 and 14% after 20, 66 and 84 days, respectively. The relative contribution from the root material was less than 3% for all three sampling dates, this was the expected result and compared well with results from net mineralisation studies. In conclusion the technique proved a useful tool in determining the relative contributions from incorporated residues. It has great potential and may facilitate the unravelling of the mysteries of nitrogen cycling.

Introduction

The gross fluxes of nutrients during the transformations of plant material to soil organic matter are not well qualified (Paul and Juma, 1981). A better understanding of the processes controlling the specific transformations is required to manage nutrients released during decomposition.

Minerlisation is the release of ammonium during the decomposition of organic matter, immobilisation is the assimilation of the ammonium by the soil microbial biomass. Together they are termed mineralisation immobilisation turnover (MIT). Although MIT is of great importance the mechanisms remain elusive. This is due to the difficulty in measuring such complex processes. In these experiments analytical techniques were formulated to determin gross mineralisation.

Barraclough et al. (1985) and Nishio et al. (1985) independently published equations which enabled experimental determinations of gross mineralisation to

be made. The technique involves labelleling the ammonium pool with ^{15}N label and measuring the decline in ^{15}N abundance over period of time (less than 14 days). The N derived from the breakdown of organic matter is unlabelled and has a dilution effect. It is assumed that the N removed from the ammonium pool is in the same proportions of ^{15}N:^{14}N. By calculation and computation an average gross mineralisation rate may be determined. This is known as the ^{15}N dilution technique.

The mirror image technique is a paired experiment whereby identically prepared ^{15}N labelled and unlabelled residues are incorporated into the soil. By ^{15}N injection dilution techiques it is possible to estimate gross mineralisation from the soil and residues. By ^{14}N injection it is possible to substitute in the mineralisation rate and estimate the percentage mineralisation from the residues. This is achieved using procedures 1 and 2.

The aims of the work were to quantify the percentage of gross mineralisation derived from the root and leaf residues asdestinct from the soil organic matter.

* This work was funded by NERC (Natural Environmental Research Council) and the Forestry Authority.

Method

Alnus glutionosa seedlings were grown in sand culture supplemented with ^{15}N or ^{14}N Long Ashton nutrient solution, thus the plant material was of similar composition in all but the label. Treatments:A: labelled root B: Unlabelled root C: Labelled leaves D: Unlabelled leaves.

Residues containing 6.56 mg of N were incorporated in to 111g (100g dry weight equivalent) of fresh opencast coal spoil (Described as Medium Dunraven 2, Bending (1993)). Pots were incubated at 20/22 °C day/night and watered to weight daily. Four replicate samples were taken at t=6 t=42 and t=58 days, simultaneously a further 8 replicates were injected with 2 cm^3 of 18 mM unlabelled $(NH_4)_2SO_4$ (equivalent to 10 μg N g^{-1} soil). Labelled pots (A and C) received unlabelled solution and unlabelled pots (B and D) received 5 atom% ^{15}N solution (equivelent to 10 μg N g^{-1} soil), hence the mirror image. At 2 and 14 days subsequent to injection 4 replicates were sampled and analyzed for ammonium, nitrate and ^{15}N abundance. Nitrate and ammonium were measured from KCl extracts using flow injection analysis. For ^{15}N analysis the extract was prepared using the diffusion technique descibed by Brookes et al. (1989) and analysed using a Roboprep CN biological sample converter linked to a VG Micromass 622 mass spectrometer. Using the procedures outlined it was possible to determine the% of the gross N mineralised coming from the residues.

Procedure 1

Mineralisation is determined in pots receiving unlabelled residues by conventional ^{15}N dilution techniques following application of $(^{15}NH_4)_2SO_4$. From the decline in the ^{15}N abundance of the ammonium pool over the 12 day interval it is possible to calculate the gross mineralisation rate of all organic fractions of the soil. This is determined using Equation 1.

$$A_t^* = \frac{A_o^*}{\left(1 + \frac{\theta t}{N_o}\right)^{m/\theta}} \qquad (1)$$

where:
A_t^* = Atom% excess t=t
A_0^* = Atom% excess t=0
t = time
m = mineralisation rate

N_0 = Ammonium pool size at t=0

$$\theta = \frac{NH_{4t=t}^+ - NH_{4t=0}^+}{t}$$

Procedure 2

In the pots receiving labelled residues the size and enrichment of the ammonium pool is determined following application of unlabelled $(NH_4)_2SO_4$. If all other factors are assumed to be identical then gross rates of mineralisation determined using Equation 1 may be substituted into Equation 2 and the proportion of N derived from mineralisation of the residues α may be calculated.

$$\alpha = \frac{A_t^* \left(1 + \frac{\theta t}{A_0^*}\right)^{\frac{m}{\theta}}}{\left(\left(Res^* \left(1 + \frac{\theta t}{A_0}\right)^{\frac{m}{\theta}}\right) - Res^*\right)} \qquad (2)$$

Where:
α = proportion of mineralised nitrogen coming from the mineralisation of the residue. At* = Atom% excess t=t
A_0^* = Atom% excess t=0
Res*= Atom% excess of the residue incorporated
t = time
m = mineralisation rate

$$\theta = \frac{NH_{4t=t}^+ - NH_{4t=0}^+}{t} \qquad (3)$$

Results

It can be seen from the ammonium concentrations (Fig. 1 and 2) that the unlabelled and labelled systems were equivalent in all but the orientation of the ^{15}N label. These data suggest that the ammonium was rapidly lost from the ammonium pool subsequent to application of fertilizer, this could be due to immobilisation or nitrification.

Leaf residues

From Table 2 it can be shown that the gross rates of mineralisation differed for the 3 injections. It can also

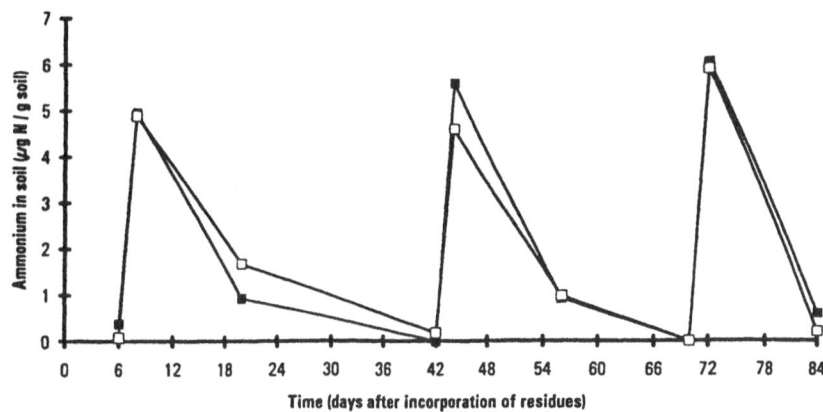

Figure 1. Average (n=4) net ammonium concentration in μg g^{-1} soil for opencast coal spoil amended with 6.54 mg N in the form of *Alnus glutinosa* leaf residues at T=0 and injected with the equivalent of 10 μg N g soil^{-1} of $(NH_4)_2SO_4$ at T=6 T=42 and T=70. □ = unlabelled residue ■ = labelled residue.

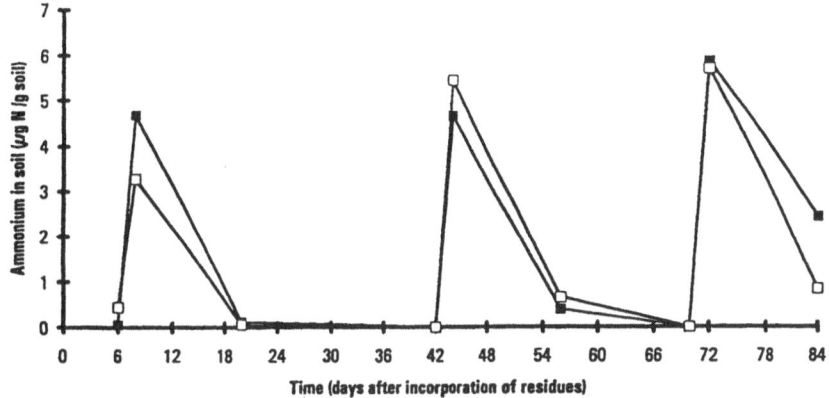

Figure 2. Average (n=4) net ammonium concentration in μg g^{-1} soil for opencast coal spoil amended with 6.54 mg N in the form of *Alnus glutinosa* root residues at T=0 and injected with the equivalent of 10 μg N g soil^{-1} of $(NH_4)_2SO_4$ at T=6 T=42 and T=70. □ = unlabelled residue ■ = labelled residue

Table 1. Composition of roots and leaves of labelled and unlabelled plant material

	%N	^{15}N Abundance	%C	C:N
Labelled				
Roots	1.25	17.1899	41.80	25.1
Leaves	4.75	20.4488	48.07	12.1
Unlabelled				
Root	1.26	0.3708	36.60	29.1
Leaves	4.82	0.3724	46.71	9.1

Table 2. The gross mineralisation and α values for opencast coal spoil amended with 6.54 mg N in the form of *Alnus glutinosa* leaf or root residues. Calculated using procedures 1 and 2. Each value shown is the mean of 4 replicates and the figures in parentheses are the standard deviations

	Mineralisation rate	$\alpha(\%)$
Leaves		
Injection 1	0.3859 (0.0630)	21.60 (4.22)
Injection 2	0.1276 (0.0226)	41.36 (8.72)
Injection 3	0.3096 (0.1017)	13.61 (7.73)
Roots		
Injection 1	0.1247 (0.0869)	0.31 (0.23)
Injection 2	0.1506 (0.0655)	1.61 (0.98)
Injection 3	0.0872 (0.4063)	2.18 (0.81)

be seen that the contribution of N to gross mineralisation reached a peak at the second injection, when approximately 41% of the the gross mineralised N products were derived from the mineralisation of leaf residues. After 70 days (as estimated by the 3rd injection) N contribution from the residues decreased to 13.6%. There are a number of possible explanations for this:

1. The readily utilisable low C:N products have been mineralised and the more recalcitrant N products are immobilised by the microbial biomass.

2. The increase in available NH_4 and NO_3 enhances mineralisation of the soil N.

Root residues

Mineralisation of soil amended with root residues was low. Contribution of root residues to mineralisation was very low about 3–4%. The% N contribution from root residues was significantly lower than that calculated for the residues of leaves. This is a well documented fact, C:N ratio is probably the main controlling factor in the breakdown of organic matter (Berg and Staaf, 1981; Harmsen and Van Schreven, 1955).

The below-ground nutrient interactions have been stressed as a possible source of N transfer in recent work. This work suggests that contribution of root N to mineralisation is minimal over a 84 day period.

Conclusions

The use of ^{15}N dilution equations and mirror image techniques gave a unique insight into the processes of the soil. However it may be that under low soil N conditions the addition of fertiliser has a substantial impact on the mineralisation potential of the soil. In an agricultural soil using these methods would be acceptable as the amount of fertiliser added is minimal compared to soil mineral N. In low N soils the methods for measuring gross rates of mineralisation per se are not completely satisfactory. However such methods allow us to compare effects of different regimes on gross rates of mineralisation, although they do not fully explain the complex processes operating in the natu-

ral system. The difficulty is in estimating gross rates of mineralisation without causing a substantial priming effect; the addition of even small amounts of fertilizer required for ^{15}N dilution experimentation may have a substantial effect, if the amount of fertilizer added is reduced further errors are introduced in calculating the mineralisation rates. The techniques described proved useful and effective tools in estimation of mineralisation from various soil N pools. ^{15}N methodologies may hold the key to unravelling the mysteries of N cycling the soil

Acknowledgements

We thank Declan Barraclough for his helpful disscussion and equations, we would also like to thank Andy Moffat and Bob Rees for their contributions.

References

Barraclough D, Geens E L, Davies G P and Maggs J M 1985 Fate of fertilizer nitrogen. III. The use of single and double labelled ^{15}N ammonium nitrate to study nitrogen uptake by ryegrass. J. Soil Sci. 36, 593–603.

Bending N A D 1993 Site factors affecting tree response on restored open ground in South Wales coal field. Unpublished Ph.D. thesis. University of East London.

Berg B and Staaf H 1981 Leaching accumulation and release of nitrogen in decomposing forest litter. *In* Terestrial Nittrogen Cycles. Eds. F E Clark and T Rosswall. Ecol. Bull. (Stockholm) 33, 163–173.

Brookes P D, Stark J M, Mc Inteer B B and Preston T 1989 Diffusion method to prepare soil extracts for automated ^{15}N analysis Soil. Sci. Soc. Am. J. 53, 1707–1711.

Harmsen G W and Van Scheven D A 1955 Mineralisation of organic matter in soil. Adv. Agron 7, 299–398.

Nishio T, Kanamori T and Fjuimoto T 1985 Nitrogen transformations in aerobic soils as determined by a $^{15}NH_4^+$ dilution technique. Soil. Biol. Biochem. 17, 149–154.

Paul E A and Juma N G 1981 Mineralization and immobilization of soil nitrogen by soil microorganisms. *In* Terestrial Nittrogen Cycles. Eds. F E Clark and T Rosswall. Ecol. Bull. (Stockholm) 33, 174–195.

O. Van Cleemput et al. (eds.), Progress in Nitrogen Cycling Studies, 57–61, 1996.

D- and L-amino acid metabolism in soil

D.W. Hopkins

Department of Biological Sciences, University of Dundee, Dundee, DD1 4HN, UK

Key words: amino acid isomers, ammonification, microbial biomass, mineralization, respiration

Abstract

The rates of D- and L-amino acid induced respiration in 11 different soils have been determined. D-alanine induced respiration was either less than or not significantly different from L-alanine induced respiration, depending on the soil. For all 11 soils, D-glutamine and D-glutamic acid induced respiration rates were significantly less than those of the corresponding L-amino acids. These observations are consistent with the slower rates of D-alanyl-D-alanine compared with L-alanyl-L-alanine induced respiration and of D-glutamine induced ammonification compared with L-glutamine induced ammonification. The rates of L-amino acid induced respiration for all 11 soils were significantly correlated with the size of the soil microbial biomass ($R^2 = 0.90$, 0.88 and 0.93 for L-alanine, L-glutamine and L-glutamic acid, respectively) and therefore provide an alternative basis for determining soil microbial biomass. Although the rates of D-amino acid induced respiration were also correlated with soil microbial biomass ($R^2 = 0.90$, 0.79 and 0.61, for D-alanine, D-glutamine and D-glutamic acid, respectively), the D-glutamine and D-glutamic acid induced respiration rates accounted for a smaller fraction of the variation in soil microbial biomass than the rates of respiration induced by the other amino acids.

Introduction

Amino acids are one of the largest identifiable groups of organic N compounds in soils (Stevenson, 1982). The fact that the relative abundance of the rarer D-isomers of amino acids in soils is greater than their occurrence in organisms would suggest (Wagner and Mutatkar, 1968) indicates that they are degraded less rapidly than the corresponding L-isomers (Stevenson, 1982). The principal natural occurrence of D-amino acids is in the cross-linking of bacterial peptidoglycan, one of the structural components of bacterial cell walls (Schleifer and Kandler, 1972), and their greater occurrence in bacteria compared with other organisms (Tschiersch and Mothes, 1963) has led to their use as chemical markers of bacteria (Gunnarson and Tunlid, 1986; Tunlid and White, 1992). The observation that many of the enzymes involved in D-amino acid metabolism are bacterial (Adams, 1972; Ulbricht, 1962) is consistent with the tentative suggestion that some aspects of D-amino acid metabolism in soil are bacterially-mediated (Hopkins and Ferguson, 1994; Hopkins et al., 1994).

The short-term rates of D-alanine, D-glutamine and D-glutamic acid induced respiration were significantly less than those of the corresponding L-amino acids when excess (saturating) amounts of the amino acids were added to soils (Hopkins and Ferguson, 1994; Hopkins et al., 1994). Furthermore, the rate of L-amino acid induced respiration in a restricted range of soils was well-correlated with microbial biomass (as determined by glucose induced respiration), but the rate of D-amino acid induced respiration varied in a manner which could not be related to total microbial biomass alone (Hopkins et al., 1994). Hopkins and Ferguson (1994) reported that the rate of D-alanine and D-glutamine induced respiration was increased by preincubation of the soil with D-alanine and D-glutamine, respectively, to a greater extent than when the soil was preincubated with L-alanine and L-glutamine, respectively, which indicated that the microbial community responded differently to D-amino acids than to L-amino acids. Furthermore, the enhancement effect of D-amino acid metabolism by preincubation with D-amino acids was highly sensitive to streptomycin but virtually insensitive to cycloheximide (Hopkins and Ferguson, 1994) which implicated the bacterial com-

58

ponent of the microbial community in D-amino acid metabolism.

In this contribution we report on the rates of metabolism of D- and L-amino acids in a wider range of soils than previously examined and summarise these and previous observations to show the relationships between soil microbial biomass and L- and D-amino acid induced respiration.

Materials and methods

Soils

Eleven soils with contrasting properties and/or management were used in this work. Details of the soil properties are summarised in Table 1. Soils 1–5 were sampled from the Palace Leas meadow hay plots and have received the following annual treatments since 1897: soil 2, farm yard manure (FYM) plus inorganic N, P and K; soil 3, FYM; soil 4, control; soil 5, $(NH_4)_2SO_4$; soil 6, inorganic N, P and K. Full details of the management of these plots and information about the biological properties of the soils have been published elsewhere (Hopkins and Shiel, 1996; Pawson, 1960). Soils 6–10 were the agricultural soils reported by Hopkins et al. (1994). Soil 11 has been used for most of the investigations described here.

Amino acid induced metabolism

The maximum rates of CO_2 evolution from soils following addition of D-alanine, L-alanine, D-alanyl-D-alanine, L-alanyl-L-alanine, D-glutamine, L-glutamine, L-glutamic acid and D-glutamic acid in the range 0–2.0 mg amino acid (or dipeptide) g^{-1} (d.w.) soil (previous experiments had shown that 2.0 mg amino acid g^{-1} soil was sufficient to give the maximum substrate induced respiratory response) over the period 0 to 5 (or in some cases 6) hours incubation at 22 °C (Hopkins and Ferguson, 1994). Miniaturized incubation devices similar to those described by Heilmann and Beese (1992) were used and CO_2 was determined by gas chromatography. The rate of NH_4^+ production over the period 0–6 hours was determined in shaken slurries of soil 11 to which between 0 and 4 mg of either D-glutamine or L-glutamine per gram of soil (d.w.) had been added after the background NH_4^+ had been reduced as described by Widmer et al. (1987). The soil slurries were centrifuged and filtered

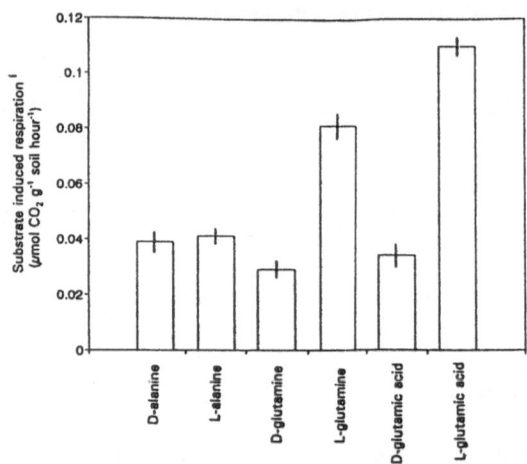

Figure 1. Comparison of substrate induced respiration rates in soil 11 amended with D- and L-alanine, glutamine and glutamic acid. The vertical bars represent ± standard errors.

and the NH_4^+ determined colorimetrically (Anderson and Ingram, 1993).

Results and discussion

For all three pairs of amino acid isomers, the substrate induced respiration was significantly less in soil 11 amended with D-isomers compared with L-isomers of amino acids. Figure 1 shows the substrate induced respiration data for soil 11 only; amino acid induced respiration was significantly lower in D-amino acid amended samples of soil 1–5 than in the corresponding L-amino acid amended soils (data not shown) and the data in Hopkins et al. (1994) for soils 6–10 showed the same trends. Furthermore, it is clear that in soil 11, the D-amino acid induced respiration was a different percentage of the L-amino acid induced respiration for each of the isomeric pairs (95% for alanine, 36% for glutamine and 31% for glutamic acid). The differences in C mineralization between D- and L-amino acids were matched both by differences in the rates of dipeptide induced respiration, which was lower in D-alanyl-D-alanine amended soil than in L-alanyl-L-alanine amended soil (Fig. 2) and by the rate of amino acid induced ammonification, which was lower in D-glutamine amended soil than in L-glutamine amended soil (Fig. 3).

The difference in the rate of metabolism of D- and L-amino acids (and dipeptides) indicates either that a restricted component of the soil microbial community was capable of utilizing the D-isomers compared with

Table 1. Soil properties

Soil	Soil type	Texture	Organic matter (%)	pH	Biomass C[a] ($\mu g\ g^{-1}$ soil)
1	Stagnogley	clay loam	5.4	5.5	730[b]
2	Stagnogley	Clay loam	5.7	5.4	980[b]
3	Stagnogley	Clay loam	5.0	5.0	590[b]
4	Stagnogley	Clay loam	8.1	3.8	220[b]
5	Stagnogley	Clay loam	3.7	4.7	530[b]
6	Iron podzol	Sandy loam	13	6.3	320[c]
7	brown forest soil	Silt loam	5.5	6.8	210[c]
8	Non-calcareous gley	Clay loam	4.9	4.9	140[c]
9	Non-calcareous gley	Sandy loam	4.0	4.8	60[c]
10	Iron podzol	Sandy loam	7.0	7.0	130[c]
11	podzol	loamy sand	5.8	7.0	80

[a]Biomass C determined by the glucose induced respiration method of Anderson and Domsch (1978).
[b]Biomass C data from Hopkins and Shiel (1996).
[c]Biomass C data from Hopkins et al. (1994).

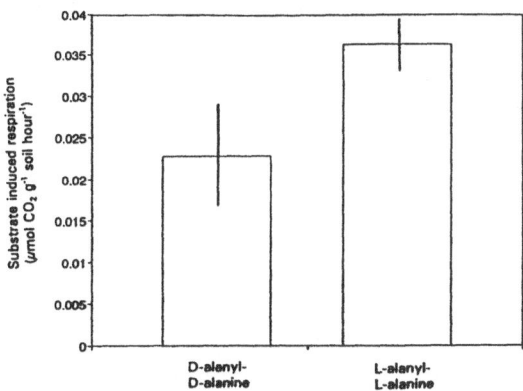

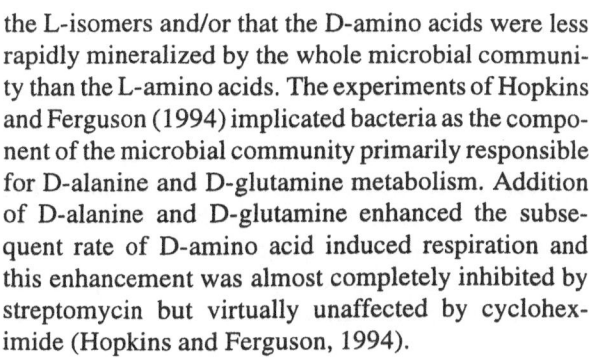

Figure 2. Comparison of substrate induced respiration rates in soil 11 amended with D-alanyl-D-alanine and L-alanyl-L-alanine. The vertical bars represent ± standard errors.

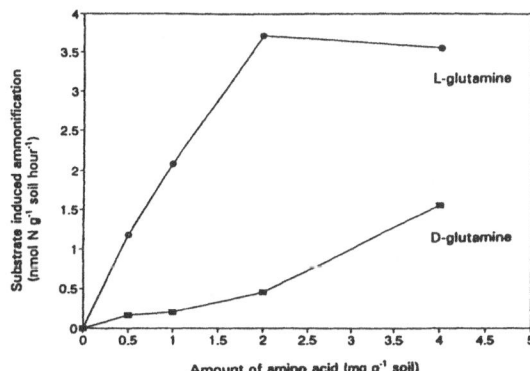

Figure 3. Comparison of substrate induced ammonification in soil 11 amended with D- and L-glutamine.

the L-isomers and/or that the D-amino acids were less rapidly mineralized by the whole microbial community than the L-amino acids. The experiments of Hopkins and Ferguson (1994) implicated bacteria as the component of the microbial community primarily responsible for D-alanine and D-glutamine metabolism. Addition of D-alanine and D-glutamine enhanced the subsequent rate of D-amino acid induced respiration and this enhancement was almost completely inhibited by streptomycin but virtually unaffected by cycloheximide (Hopkins and Ferguson, 1994).

When the data from soils 1–5 and 11 were combined with that in Hopkins et al. (1994) for soils 6–10, the relationships between soil microbial biomass (Table 1) and L-alanine, L-glutamine and L-glutamic acid could be derived (Fig. 4). These relationships provide alternative bases for estimating the size of the soil microbial biomass and, as such, were consistent with (L-)arginine induced ammonification (Alef and Kleiner, 1986; Alef et al., 1988), casamino acid induced respiration (Anderson and Domsch, 1978) and (L-glutamic acid (Norgren et al., 1988) measurements. The D-glutamine and D-glutamic acid induced respi-

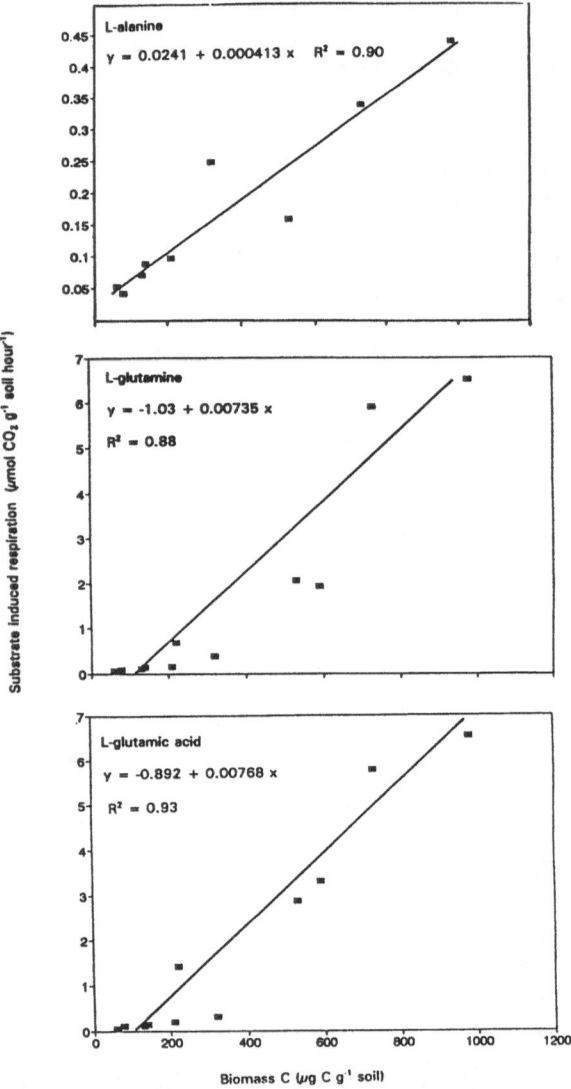

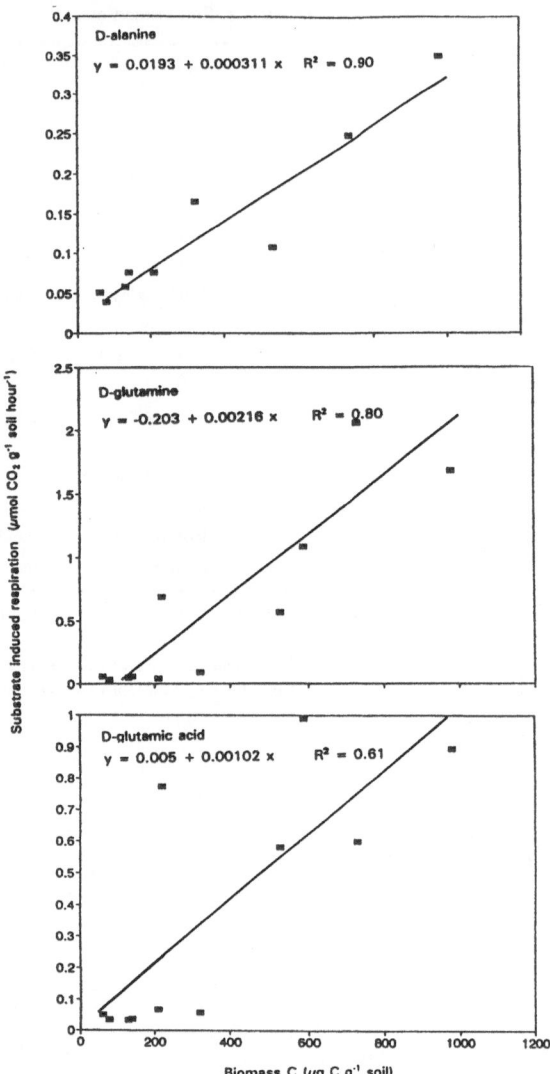

Figure 4. Relationships between L-alanine, L-glutamine and L-glutamic acid induced respiration rates and the soil microbial biomass for soils 1–11. L-alanine induced respiration was not determined for soils 3 and 4.

Figure 5. Relationships between D-alanine, D-glutamine and D-glutamic acid induced respiration rates and the soil microbial biomass for soils 1–11. D-alanine induced respiration was not determined for soils 3 and 4.

ration rates were not as well correlated with the size of the soil microbial biomass as L-amino acid induced respiration (Fig. 4 cf. Fig. 5). These observations represents an extension, using a larger number of soils, of those made by Hopkins et al. (1994). Although D-glutamine and D-glutamic acid induced metabolism was related to soil biomass, the size of the microbial biomass accounted for a smaller proportion of the variation in amino acid induced respiration for D-glutamine and D-glutamic acid than that for the other amino acids (Fig. 5).

The data presented here confirm the greater rates of L-amino acids than D-amino acids in a wider range of soils than previously reported and show that L-amino acid induced respiration was well correlated with microbial biomass, whilst D-amino acid induced respiration was less well correlated with biomass. The slower rates of D-amino acid metabolism in soil may, therefore, contribute to the relative accumulation of D-amino acids in soil. Two hypotheses can be proposed to account for the different in D- and L-amino acid metabolism:

1. D-amino acids are metabolised by a component (possibly the bacteria) of the whole (glucose- or L-amino acid-metabolising) microbial community in soils.

2. D-amino acids are metabolised by the whole (glucose- or L-amino acid-metabolising) microbial community, but the D-amino acids are less readily metabolised than the L-amino acids.

Acknowledgements

This work was supported in part by the UK AFRC/BBSRC and I am grateful to Dr RS Shiel for site access to collect soils 1–5, to Nicola Wilson for data from the dipeptide experiment and to Bruce Isabella, Susan Scott and Karin Ferguson for technical assistance.

References

Adams E 1972 Amino acid racemases and epimerases. *In* The Enzymes. Ed. P D Boyer. Volume 6, pp 479–507. Academic Press, New York, USA.

Alef K and Kleiner D 1986 Arginine ammonification, a simple method to estimate microbial activity potentials in soils. Soil Biol. Biochem. 18, 233–235.

Alef K, Beck T, Zelles L and Kleiner D 1988 A comparison of methods to estimate microbial biomass and N-mineralization in agricultural and grassland soils. Soil Biol. Biochem. 20, 561–565.

Anderson J M and Ingram J S I 1993 Tropical Soil Biology and Fertility: A Handbook of Methods, 2nd Edition. CAB International, Wallingford, UK.

Anderson J P E and Domsch K H 1978 A physiological method for the quantitative measurement of microbial biomass in soils. Soil Biol. Biochem. 10, 215–221.

Gunnarson T and Tunlid A 1986 Recycling of fecal pellets in isopods: microorganisms and nitrogen compounds as potential food for *Oniscus asellus* L. Soil Biol. Biochem. 18, 595–600.

Heilmann B and Beese F 1992 Miniaturized method to measure carbon dioxide production and biomass of soil microorganisms. Soil Sci. Soc. Am. J. 56, 596–598.

Hopkins D W and Ferguson K E 1994 Substrate induced respiration in soil amended with different amino acid isomers. Appl. Soil Ecol. 1, 75–81.

Hopkins D W, Isabella B L and Scott S E 1994 Relationship between microbial biomass and substrate induced respiration in soils amended with D- and L-isomers of amino acids. Soil Biol. Biochem. 26, 1623–1627.

Hopkins D W and Shiel R S 1996 Size and activity of soil microbial communities in long-term experimental grassland plots treated with manure and inorganic fertilizers. Biol. Fertil. Soils (*In press*).

Norgren A, Bååth E and Söderström B 1988 Evaluation of soil respiration characteristics to assess heavy metal effects on soil microorganisms using glutamic acid as a substrate. Soil Biol. Biochem. 20, 949–954.

Pawson C E 1960 Cockle Park Farm. Oxford University Press, London, UK.

Schleifer K H and Kandler O 1972 Peptidoglycan types of bacterial cells and their taxonomic implications. Bacteriol. Rev. 36, 407–477.

Stevenson F J 1982 Organic forms of soil N. *In* Nitrogen in Agricultural Soils, Agronomy Volume 22. Ed. F J Stevenson. pp 67–122. American Society of Agronomy, Soil Science Society of America, Crop Science Society of America, Madison, WI, USA.

Tschierich B and Mothes K 1963 Amino acids: Structure and distribution. *In* Comparative Biochemistry: A Comprehensive Treatise. Eds. M Florkin and H S Mason. pp 1–90. Academic Press, New York, USA.

Tunlid A and White D C 1992 Biochemical analysis of biomass, community structure, nutritional status and metabolic activity of microbial communities in soil. *In* Soil Biochemistry, Vol. 7. Eds. G Stotzky and J-M Bollag. pp 229–262. Marcel Dekker, New York, USA.

Ulbricht T L V 1962 The optical asymmetry of metabolites. *In* Comparative Biochemistry: A Comprehensive Treatise, Vol. 4. Eds. M Florkin and H S Mason. pp 125–152. Academic Press, New York, USA.

Wagner G H and Mutatkar V K 1968 Amino acid components of soil organic matter formed during humification of ^{14}C glucose. Soil Sci. Soc. Am. J. 32, 683–684.

Widmer P, Brookes P C and Parry L C 1989 Microbial biomass nitrogen measurements in soils containing large amounts of inorganic nitrogen. Soil Biol. Biochem. 21, 865-867.

O. Van Cleemput et al. (eds.), Progress in Nitrogen Cycling Studies, 63–67, 1996.
© *1996 Kluwer Academic Publishers.*

'Light' fraction mineralization potentials of humid tropical soils

Hans Imhof[1], Costas Ehaliotis[1], Ken Giller[1], Cesar Miranda[1,3], José M. Pereira[2], Segundo Urquiaga[4], Bob Boddey[4] and Georg Cadisch[1,5]

[1] *Department of Biological Sciences, Wye College, University of London, Wye, Ashford TN25 5AH, UK,* [2] *Animal Husbandry Station, ESSUL/CEPEC/CEP AC, Itabela, Bahia, Brazil,* [3] *CNPGC-EMBRAPA, CP 154, 79080 Campo Grande, Brazil and* [4] *CNPAB-EMBRAPA, Seropédica, 23851 Rio de Janeiro, Brazil.* [5] *Corresponding author*

Key words: Brachiaria decumbens, Desmodium ovalifolium, legumes, Ludox, sodium polytungstate, rainforest, tropical pastures

Abstract

Soil samples from a rainforest, a papaya plantation, a pure *Brachiaria humidicola* pasture and a *B. humidicola/Desmodium ovalifolium* mixture were analyzed for fertility parameters. Soils were sieved to >100 μm and the organic matter light fraction was obtained by 1.8 g mL^{-1} density separation in a sodium polytungstate solution which was subsequently incubated under anaerobic conditions. The effect of different density agents (sodium polytungstate, Ludox and sodium iodide) on mineralization was also tested on a temperate and a tropical soil and on maize tissue.

Anaerobic incubation of light fractions of tropical topsoils (0–2 cm) demonstrated a good differentiation between the compared management treatments whereas mineralization potentials of whole soil samples were less contrasting. Samples from the rainforest showed the highest mineralization values and those from the plantation the lowest values. With an increasing proportion of legumes in the pasture the mineralization potential increased and the C:N ratio of the light fraction decreased. Thus, although *D. ovalifolium* is considered to have poor quality attributes (relatively low %N and high tannin content), its litter contributed to increased soil fertility. Light fraction mineralization potentials of samples from 5–15 cm depth were much lower than the topsoil and not clearly affected by soil management. This confirms that rainforest soil fertility is very much restricted to the upper soil layer and the maintenance of soil fertility depends largely on the quantity and quality of above-ground litter inputs. None of the density agents tested were without interference on mineralization potential. Nevertheless, light fraction parameters appeared to be more sensitive to alterations in soil management than parameters of whole unfractionated soil samples and could be used as early warning indicators of changes in soil fertility.

Introduction

Evaluation of sustainability of newly developed systems requires sensitive methods which allow an early detection of changes in soil fertility before degradation becomes apparent. This holds similarly if the monitoring of changes in soil fertility is required during the transition from a natural ecosystem, i.e. rainforest, to more intensified production systems. Current methods of chemical analysis of whole soil samples are often unable to fulfil this requirement and allow detection of degradation only after substantial changes have occurred.

The objectives of the current investigations thus were: i) to test the effect of density fractionation agents on mineralization potential of light fractions; ii) to evaluate changes in soil fertility indexes after rainforest clearance and implementation of different soil management systems and iii) to identify early warning indicators of changes in soil fertility.

Table 1. Properties of the Bahian soils (0–15 cm)

	pH	Sand	Silt	Clay
		(%)		
Rainforest	5.0	82	2	16
Papaya plantation	6.1	78	2	20
B. humidicola pasture	5.9	82	1	17

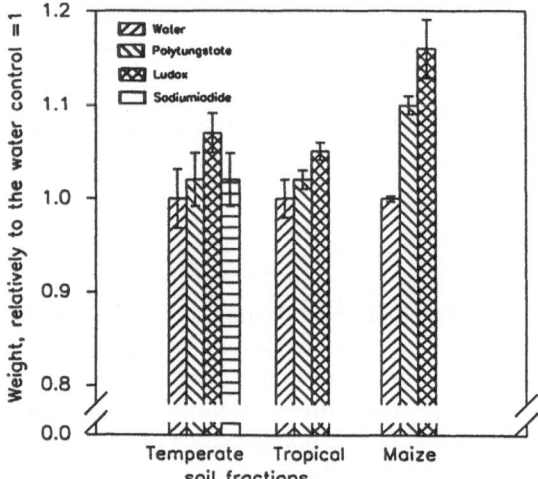

Figure 1. Influence of density liquids on the dry weight of soil fractions >100 μm and of maize tissue in relation to the water control = 1. (Bars = SE, n=3).

Materials and methods

Treatments

Soils samples were taken in October 1993 from a pasture experiment, from a rainforest (100–200 m away) and a nearby (5–50 m) papaya plantation at the Animal Husbandry Station, ESSUL/CEPEC/CEPLAC, Itabela, Bahia, Brazil. The soils were classified as Haplorthox (Oxisol) with characteristics as shown in Table 1.

Average rainfall is 1300 mm year^{-1} and average minimum/maximum temperatures are 19.3 °C and 28.6 °C, respectively. Samples in the pasture were taken from a pure grass (*B. humidicola*) and from a mixed grass/legume (*B. humidicola/D. ovalifolium*) treatment at the low stocking rate (2 animals ha^{-1}) established 1987/88 (Pereira, 1991). In the mixed pasture two samples were taken: i) from an area with an average legume content (20–40%) and ii) from a legume dominated patch (ca. 90% legumes). Twelve subsamples from a 20 × 20 m representative area within the replicated

(n=3) pasture plots or from three different areas (each 50 m apart) in the papaya plantation or rainforest were mixed.

Soil fractionation

Soils (40 g) were dispersed in a 5 g L^{-1} sodium hexametaphosphate solution for two hours (orbital shaker) and thereafter washed through a 100 μm sieve. The soil fraction >100 μm was further separated by a density separation (1.8 g mL^{-1}) in sodium-polytungstate. The yielded 'light fraction' was washed with distilled water and dried at 60 °C.

Comparison of density agents on mineralization potential

Soil samples of a temperate grassland (Wye College farm, sandy brown earth) and a tropical Brazilian grassland (CPAC, Brasilia, clayey Oxisol) were dispersed as above and thereafter sieved through a 100 μm sieve. To simulate a density separation treatment the fraction >100 μm was shaken for 30 min in either sodium-polytungstate, Ludox (an alkaline colloidal silica; Meijboom et al., 1995), sodium iodide density solutions (1.4 g mL^{-1}) or a water control. Thereafter, the >100 μm samples were sieved again, washed and dried followed by an anaerobic incubation test. Additionally, the 0.1–1 mm fraction of ground maize samples was tested in the same way.

Anaerobic incubation test

Light fraction (100 mg) covered by a sand layer and whole soil samples (2 g) were incubated at 40 °C for 7 days under anaerobic conditions followed by KCl extraction and ammonium analysis (Keeney, 1982).

Results and discussion

Comparison of density agents on mineralization potential

Measurements of dry weights of washed >100 μm samples after shaking in the density solutions revealed that all density agents contaminated the samples (Figure 1). Contamination with density agents was particularly evident with plant samples probably due to the penetration of reagent solution into cell wall compartments where they are protected from the subsequent

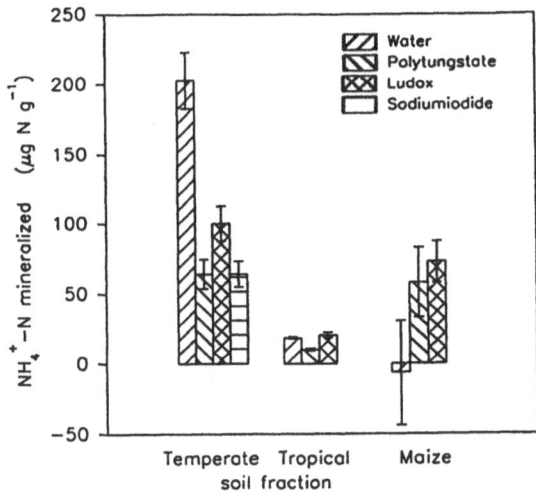

Figure 2. Influence of density liquids on the mineralization potential of soil fractions >100 μm and of maize tissue. (Bars = SE, n=3).

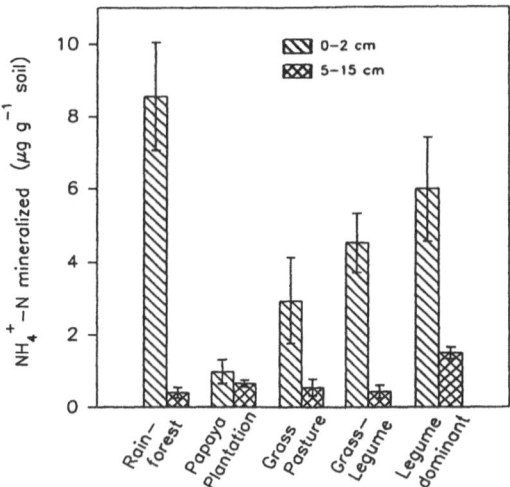

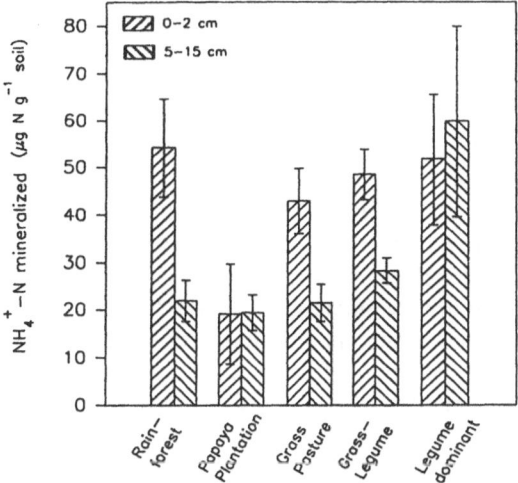

Figure 3. Anaerobic mineralization potential of **a.** light fraction (>100 μm, <1.8 g mL^{-1}) and **b.** whole soil samples of differently managed tropical soils. (Bars = SE, n=3).

washings with distilled water. This emphasises that light fraction values should be expressed on a carbon or nitrogen basis rather than on a dry weight basis unless an ash correction is applied. Contamination with density agents has also been observed by Christensen (1992). Light fraction may not only be contaminated by remains of the density solution but frequently also by organo-mineral complexes (Oades, 1988) or where dispersion had not been complete leaving intact micro-aggregates.

All density agents influenced the mineralization potential of the tested samples but the effect depended on the sample origin (Fig. 2). A particularly strong reduction in mineralization was found in the >100 μm fraction from the temperate soil with all density agents. The observed reduction of 68% in the mineralization potential caused by sodium iodide confirms similar results (70%) by Sollins et al. (1984). By contrast to the temperate soils, on the tropical soil only polytungstate had a negative effect on mineralization whereas Ludox treated soil fractions were similar to the control (sodium iodide was not included in this test). The reason for this apparent better behaviour of the Ludox treated samples could be due to the increased pH of the water solution above the anaerobic incubated samples (Table 2). This is likely be the result of the strongly alkaline nature of Ludox. The increase in pH can be regarded as a disadvantage in the handling of tropical soils which may alter the behaviour of samples. With plant materials both density agents appeared to stimulate mineralization of N from the plant residue com-

pared with the immobilization tendency of the control samples but the differences were not statistically significant. None of the density agents tested were thus without problems if further mineralization studies are envisaged on the separated fractions of organic matter.

Comparison of soil management on soil fertility

Anaerobic incubation data of light fractions (LF) of topsoil (0–2 cm) from the different tropical soils of Bahia demonstrated a good differentiation between soil management options (Fig. 3a) whereas mineralization potentials of whole soil samples were less contrasting

Table 2. pH values of the incubation water from the anaerobic incubation test of >100 μm soil fractions or maize plant samples and pH values of density solutions

	pH			Density solution
	Incubation water above incubated			
	Temperate soil fraction	Tropical soil fraction	Maize	
Water	6.6 (0.1)	6.6 (0.1)	6.4 (0.2)	
Polytungstate	7.0 (0.1)	6.7 (0.1)	6.6 (0.1)	6.5
Ludox	7.3 (0.1)	7.4 (0.1)	7.2 (0.1)	9.0
NaI	7.2 (0.1)			6.4

Values in brackets are standard errors of means.

Table 3. Whole soil and light fraction (LF) properties of 0–2 cm soil depth

	Whole soil		Light fraction		Specific mineralization (μg N mg^{-1} LF-C)
	%N	C:N	%N	C:N	
Rainforest	0.27	15	1.96	18	0.60
Papaya plantation	0.08	14	1.59	23	0.68
Grass alone	0.18	15	1.38	24	0.95
Grass-legume	0.19	13	1.96	18	1.08
Legume dominant	0.25	14	1.64	20	1.37

(Fig. 3b). Topsoil light fraction mineralization values from the rainforest had the highest ammonium-N values on a soil weight basis and those from the papaya plantation were smallest. This was largely due to the much greater amount of light fraction (13.9 mg C g^{-1} soil) present in rainforest topsoil samples compared to both plantation (1.4 mg C g^{-1}) and pasture plots (average 4.3 mg C g^{-1}) since the specific mineralization potential (μg N mg^{-1} LF-C) was the lowest in the rainforest (Table 3). The relatively poor specific mineralization (μg N g^{-1} LF-C) of rainforest light fraction was not due to a high C:N ratio but rather due to the high lignin content (45%) of the plant litter.

With an increasing proportion of legumes in the pasture the mineralization potential (μg N g^{-1} soil) derived from the top soil light fraction increased (Fig. 3a). This was due both to the increased amounts of light fraction present and improved relative mineralization of the light fraction (Table 3). The C:N ratios of whole soil, and particularly of the light fraction samples from the grass-legume swards were lower than from the pure grass treatment (Table 3) reflecting the quality of incoming plant material. Thus, although *D. ovalifolium* is considered to have poor quality attributes (relative low %N and high tannin/lignin content), its litter contributed to increased soil fertility in mixed pastures as has been shown with other legumes (CIAT, 1989).

Light fraction mineralization potentials of samples from 5–15 cm depth were much lower than of topsoil samples and less significantly affected by soil management. The decline in this parameter with soil depth was very marked in the rainforest samples. This confirms that rainforest soil fertility is very much restricted to the upper soil layer and that maintenance of soil fertility largely depends on the quantity and quality of above-ground litter inputs (Toledo and Navas, 1986).

Light fraction mineralization constituted less than 10% of the whole soil mineralization potential in the pasture soils. Considering that sodium polytungstate perhaps reduces the mineralization potential of light fraction by 50% (Fig. 1) it still would represent only 20% of the whole soil mineralization potential. However, this estimate has to be viewed with caution since the effect of sodium polytungstate or other density agents may affect mineralization of different materials differently making comparisons on mineralization potentials difficult. Nevertheless, the present investigation was able to show the development of marked differences in soil fertility potentials with changes in management. Light fraction parameters appeared to be more sensitive to early changes in soil management than parameters of whole unfractionated soil samples.

Thus separation of light fraction and measurements of its behaviour could be used as early warning indicators of changes in soil fertility.

Acknowledgements

We are greatful to the EC for their financial support through the International Scientific Cooperation Programme.

References

CIAT 1989 Annual Report Tropical Pastures Program. Working Document No. 10. CIAT, Cali, Columbia.

Christensen B T 1992 Physical fractionation of soil and organic matter in primary particle sizes and density separates. Adv. Soil Sci. 20, 1–90.

Pereira J M 1991 Avaliaçåo de pastagens formadas por *B. humidicola* (Rendle) Schweickt, em monocultivo ou consorciado com leguminosas e submetidas a diferentes taxas de lotaçåo, na regioa sul da Bahia. PhD thesis, Federal University of Vicosa, Brazil. 206 p.

Keeney D R 1982 Nitrogen availability indices. *In* Methods of Soil Analysis, Part 2. Ed. A L Page. pp 711–733. Am. Soc. Agron., Madison, USA.

Meijboom F W, Hassink J and van Noordwijk M 1995 Ludox density fractionation method. Soil Biol. Biochem. 27, 1109–1111.

Oades J M 1988 An introduction to organic matter in mineral soil. *In* Minerals in Soil Environment. Eds. B Weed and J B Dixon. pp 187–259. Soil Sci. Soc. Am., Madison, USA.

Sollins P, Spycher G and Glassmann C A 1984 Net nitrogen mineralization from light- and heavy-fraction forest soil organic matter. Soil Biol. Biochem. 16, 31–38.

Toledo J M and Navas J 1986 Land clearing for pastures in the Amazon. *In* Land Clearing and Development in the Tropics. Eds. P A Sanchez and R W Cummings. pp 97–116. AA Balkema, Rotterdam, the Netherlands.

Plant and Soil **181**: 39–45, 1996.

Nitrogen mineralization in relation to C:N ratio and decomposability of organic materials

B.H. Janssen
Department of Soil Science and Plant Nutrition, Wageningen Agricultural University, P.O. Box 8005, 6700 EC Wageningen, The Netherlands *

Key words: 'apparent initial age', C/N of microbes, humification coefficient, lignin, polyphenol, resistance index

Abstract

A desk study was made on N mineralization of various organic materials. Data were obtained from the literature containing information on mineralized organic N, C/N of the substrate and some index for decomposability. Per class of substrate decomposability, linear regression lines between N mineralization and substrate C/N were established, from which 'apparent initial age' and humification coefficient were calculated to indicate decomposability. For quantitative grading, a 'resistance index' was formulated comprising the concentrations of lignin and polyphenols. The results confirmed that the fraction of mineralized organic N is linearly related to substrate C/N for equally decomposable materials. The nature of the organic materials and the processes to which they had been subjected were reflected in the ranking by decomposability.

Introduction

Mineralization of organic nitrogen requires microbial conversion of the organic matter. Part of the converted matter is used for assimilation in microbial tissue and part for oxidation to gain energy (dissimilation). The dissimilation:assimilation ratio (D/A) differs with the type of micro-organisms. The often used term 'organic-matter decomposition', refers, strictly speaking, to dissimilation. The quantity of decomposed or dissimilated organic matter is thus the difference between the total amount of organic matter that is converted and the amount that is assimilated by the micro-organisms.

Similarly, part of the nitrogen that is present in the converted organic material is used in microbial tissue and part is released (mineralized) as inorganic nitrogen. If the converted organic matter is low in nitrogen, the amount of nitrogen that can be converted may be too low to satisfy the assimilation needs of the microbes. In such cases, microbes take inorganic nitrogen from their environment, i.e. the soil solution or the moisture in the organic material. This process, referred to as immobilization, results in an increase of organic N in the remaining organic material. After some time, the quantity of organic N that is converted

is sufficient for the assimilation requirements of the microbes, and then the result of mineralization turns from negative into positive. When the initial C:N ratio of a substrate ($C/N_{s,i}$) is higher (lower) than that of the microbes, the fraction of organic N that is mineralized (FMON) is less (higher) than the fraction of organic C that is dissimilated (FDOC), and the C:N ratio of the remaining substrate is decreasing (increasing) during decomposition until it has the same value as that of the microbes (CN_m). In case C/N_{si}, equals C/N_m, it retains this value, and FMON is equal to FDOC.

According to a simple model recently developed by Janssen and Noij (1996), a linear relationship exists between FMON and $C/N_{s,i}$, on condition that D/A, CN_m and decomposability (so, FDOC) are similar for all substrates under consideration. The linear equation is:

$$FMON = q - rC/N_{s,i} \qquad (1)$$

where q and r are regression constants. Their values depend on D/A and C/N_m, and on the decomposability of the substrates, and can be calculated directly from the input data. This is one possible use of Equation (1). Another possibility is to use the equation in the opposite direction, for the analysis of experimental results which is done in this paper. To this purpose, various papers on nitrogen mineralization were re-interpreted.

* FAX No.: +31 317 483766
Plant and Soil is the original source of publication of this article. It is recommended that this article is cited as: *Plant and Soil* **181**: 39–45, 1996.

Table 1. Calculation of FDOC, apparent initial age, and humification coefficient from the regression lines in Figure 1. C/N_m is set at 8, mineralization time (t) at 1 year, and temperature at 9 °C

Code	Equation	Lines	
		Upper	Lower
q	r.a.[a]	0.7385	0.4387
r	r.a.	0.0184	0.0149
FDOC	(2)	0.5913	0.3195
C_t/C_i	(4)	0.4087	0.6805
f_T	(5)	1.0	1.0
Age (years)	(4)	1.6	3.0
h.c.	(6)	0.41	0.68

[a] ra. = from regression analysis of experimental data.

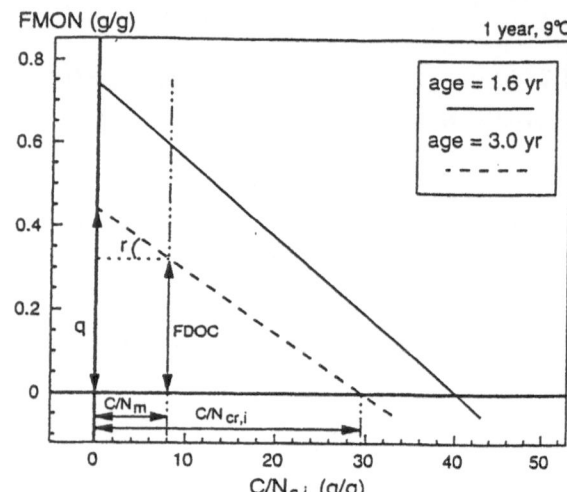

Figure 1. Theoretical relationship between FMON and $C/N_{s,i}$ for two organic materials with apparent initial ages of 1.6 and 3.0 years. Details as in Table 1.

The objectives of these exercises were to test whether the relation between FMON and $C/N_{s,i}$ is linear indeed, and to arrive at practical decomposability indices.

Methods

Papers were looked for containing data on FMON, $C/N_{s,i}$ and some index for decomposability of different substrates. Substrates were classified according to their decomposability, and for each decomposability class the regression line between FMON and $C/N_{s,i}$ was calculated. If no suitable decomposability index was available, it was attempted to group the substrates in such a way that standard deviations from the regression lines were minimized. Next, for each group, the three unknowns, C/N_m, D/A and 'apparent initial age' (age), a measure for decomposability (Janssen, 1984), were derived from q and r of the regression equations. Here only the procedure to find apparent initial age is illustrated on the basis of the lines in Figure 1, and the calculations in Table 1. A full explanation is given by Janssen and Noij (1996).

Because not more than two unknown variables can be solved from a linear relationship, it is necessary to estimate the value of one of the three unknowns. Since values of C/N_m, as given in literature, vary within a comparatively narrow range, it is most practical to estimate the value of C/N_m. It was assumed to have a value of 8, because, on average, this gave the best fit with experimental data. For the calculation of age, first FDOC has to be found. As mentioned above, FDOC

equals when $C/N_{s,i}$ equals C/N_m, and so:

$$FDOC = q - rC/N_m \qquad (2)$$

The complement to FDOC is the fraction of C_i that is remaining, so:

$$C_t/C_i = 1 - FDOC \qquad (3)$$

where C_t is the quantity of remaining C at time t (years), and C_i is the initial quantity of C.

The apparent initial age (expressed in years) is used in a equation, derived by Janssen (1984, 1986) to calculate C_t:

$$C_t = C_i \exp[4.7\{(age + f_T t)^{-0.6} - age^{-0.6}\}] \qquad (4)$$

where f_T is a correction factor for temperature:

$$f_T = 2^{(T-9)/9} \qquad (5)$$

and T is temperature in °C. Equation (5), based on the study by Jenkinson and Ayanabe (1977) predicts that the rate of dissimilation doubles for every 9 °C increase in temperature. The value of f_T is 1, when T is 9 °C. Equations (4) and (5) are also used to arrive at an expression for the humification coefficient (h.c.). This is the fraction of organic C remaining after $1/f_T$ years (which is after one year in temperate climates where T is 9 °C).

$$h.c = \exp[4.7\{(age + 1)^{-0.6} - age^{-0.6}\}] \qquad (6)$$

From Equation (1) it follows that FMON is zero, if q $= r \times C/N_{s,i}$, so if $C/N_{s,i} = q/r$. This value of $C/N_{s,i}$ is denoted as the critical C/N, indicated by $C/N_{cr,i}$, where cr stands for critical, and i for initial (Fig. 1). So:

$$C/N_{cr,i} = q/r \qquad (7)$$

It is shown by Janssen and Noij (1996) that q, r and q/r increase with increasing time of mineralization and with increasing decomposability. Accordingly, the line for age = 1.6 years in Figure 1 lies above and is steeper than the line for age = 3.0 years, and intersects the X-axis at a higher $C/N_{s,i}$ value.

Results

The results are presented in an order of increasing information on substrate decomposability. Only in the recent literature analyses of decomposability indices can be found.

Different soils and procedures, no decomposability analyses (Schulz, 1988)

Various plant materials were incubated in three soils of which two differed only a little in %C and the third was much lower in %C. Two methods were used to determine N mineralization: the ^{15}N-tracer technique, and the 'difference method' comparing treatments with and without addition of organic substrate. The tracer method gave consistently higher values for FMON than the difference method (Table 2), supposedly because the addition of organic material had reduced mineralization of soil organic N (Schulz, 1988). With the difference method a better fit for the negative relation between $CN_{s,i}$ and FMON was found than with the tracer method ($R^2 = 0.81$ and 0.01, respectively for the four substrates in the 1.73% C-soil). FMON was negatively related to %C too. This experiment illustrates that relationships between FMON and $C/N_{s,i}$ can show only under uniform incubation conditions. It also points to uncertainties in the methods used to measure mineralization. The difference method is used in almost all following examples.

Uniform conditions, only qualitative information on decomposability (Jensen, 1929)

The data of Figure 2 are from a half-year incubation in an alkaline loamy soil. A straight line was found

Table 2. Values of $C/N_{s,i}$ and of FMON measured with two techniques, for plant materials incubated in soils differing in %C. 25 °C, 54 days. Data from Schulz (1988)

Soil	%C	Substrate	$C/N_{s,i}$	FMON	
				Tracer	Difference
No. 1	1.73	Grass sprout	10.4	0.572	0.488
		Sugarbeet leaves	10.9	0.470	0.452
		Sunflower sprout	14.4	0.611	0.427
		Maize sprout	16.9	0.392	0.214
		Maize roots	22.4	n.d.	neg.
No. 2	1.90	Sunflower sprout	14.4	0.473	0.269
		Sugarbeet root	23.9	n.d.	neg.
No. 3	0.57	Sunflower sprout	14.4	0.807	0.545
		Sugarbeet root	23.9	n.d.	neg.

n.d.= not determined.

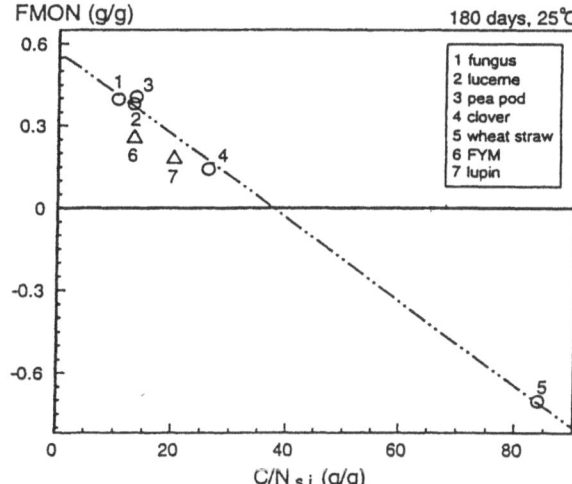

Figure 2. Relations between FMON and $C/N_{s,i}$ for FYM and six plant materials. Data from Jensen (1929).

for five plant materials of probably comparable chemical composition. The points below the line represent more resistant materials. One of them is blue lupin that may be less decomposable because it contains alkaloids. The other one is farmyard manure (FYM). It is more resistant than plant materials because the easily decomposable compounds in feed stuffs are digested when passing the animal digestive tract.

Uniform conditions, only qualitative information on processing of sewage sludges (Parker and Sommers, 1983)

The sludges of Figure 3 refer to sludges obtained from several regions of the USA. The FMON val-

42

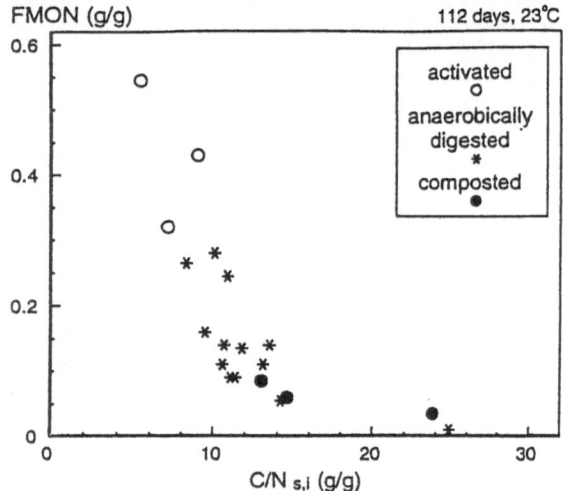

FMON (g/g)   112 days, 23°C

Figure 3. Relations between FMON and $C/N_{s,i}$ for differently processed sewage sludges. Data from Parker and Sommers (1983).

Table 3. Lignin concentrations (%), values of $C/N_{s,i}$ and FMON calculated from data by Melillo et al. (1982), and results of regression analysis. Litter layer, one year

Substrate	Lignin	$C/N_{s,i}$	FMON	Regr.anal.	
Pin cherry	19.3	41.7	−0.541	R^2	n.a.
Beech	24.1	55.6	−0.527	q	neg.!
				r	pos.!
Paperbirch	14.5	55.6	−0.323	R^2	0.998
Ash	12.2	55.6	−0.336	q	0.1234
Red mapple	10.1	71.4	−0.459	r	0.00815
Sugar mapple	10.1	83.3	−0.555		

ues are means of results of two incubation methods, one including leaching, the other not. The differences between the points in the graph can partly be attributed to the different processes to which the wastes were subjected. FMON decreased in the order: activated > digested > composted (Fig. 3), which agrees with the results of Raijmakers and Janssen (1993). Within each group, however, still a wide variation remained unexplained, maybe because of differences in decomposability. Values of R^2 from regression analysis were 0.234, 0.455 and 0.858 for activated, digested and composted wastes, respectively.

Field conditions, lignin concentrations for quantitative information on decomposability (Melillo et al., 1982)

An experiment was carried out with air-dried leaves in nylon bags that were placed in the litter layer of the forest floor for one year. For the present paper, FMON was calculated from a graph showing remaining organic N, and substrates were sorted in a group with 19–24 and one with 10–15% lignin (Table 3). The second group showed an excellent relationship between FMON and $C/N_{s,i}$. The first group had strongly negative values for FMON, and the results were unusual in view of $C/N_{s,i}$ and lignin concentrations, indicating that some other differences between beech and pin cherry leaves may have interfered. All materials in this study had a very high $C/N_{s,i}$, and it is possible that the rate of decomposition was reduced by both high lignin concentrations

in the substrates as well as by lack of immobilizable N in the inorganic-N poor environment of the litter bags.

Lignin and polyphenol concentrations for quantitative information on decomposability

Three examples are discussed, in which lignin (L) and polyphenols (PP) concentrations have been used to quantify decomposability.

Tian et al. (1992a, b)
PP nor L alone could satisfactorily predict decomposition rate, measured in a 98-days study with surface-placed litterbags in a maize field in Nigeria. In multiple regression analysis, the coefficient of PP was eight times that of L. Therefore, we propose to multiply PP by 8 and to use the following expression (L and PP in %) as resistance index (RI):

$$RI = L + 8 \times PP \qquad (8)$$

Equation (8) was applied to data of a 7-weeks N incubation experiment. Figure 4 illustrates that the relation between FMON and $C/N_{s,i}$ is affected by RI.

Palm and Sanchez (1991)
In this study the authors arrived at the conclusion that N-mineralization during an eight-weeks incubation experiment was better correlated to polyphenol concentration than to lignin or N concentration. For the present paper RI was calculated. The substrates could be subdivided into three classes with mean PP values of 1.06, 2.17 and 3.46%, and mean RI values of 16, 29 and 40, respectively (Table 4). In the first class, FMON and q were high, and in the third class low, in accordance with PP concentration and RI, but in the second class, where *Desmodium gyroides* behaved as

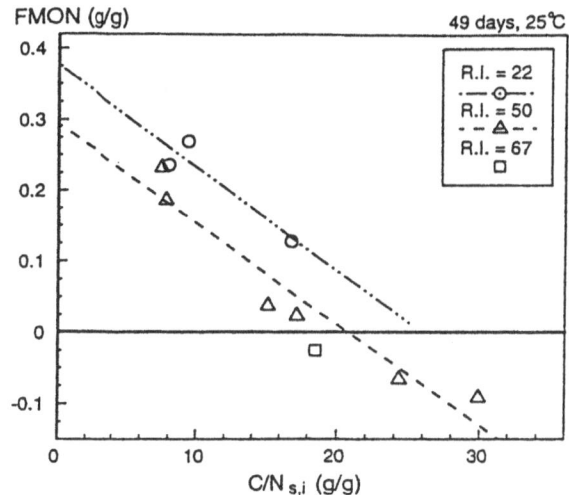

Figure 4. Relations between FMON and $C/N_{s,i}$ for plant materials of three RI classes. Data from Tian et al. (1992a, b).

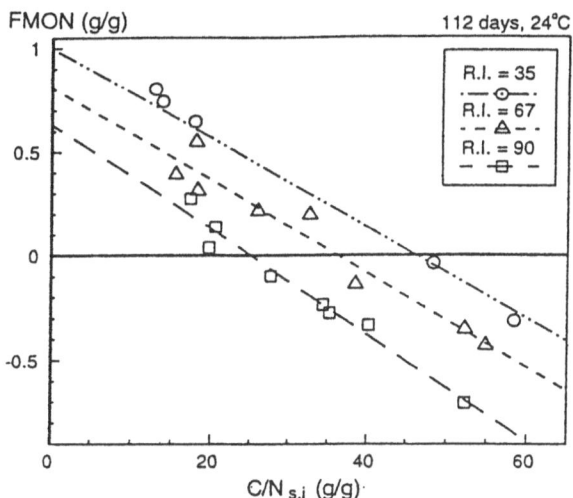

Figure 5. Relations between FMON and $C/N_{s,i}$ for three groups of plant materials with average RI of 35 67 and 90. Data from Constantinides and Fownes (1994).

an outlier, no relation between FMON and $C/N_{s,i}$ was found.

Constantinides and Fownes (1994)

In this study, leaf and litter samples of legumes and non-legumes were used. The authors report that FMON had highly significant correlation coefficients with substrate N, L, (L+PP):N and L:N ratios, but not with PP. The coefficients in multiple regression equations were highly, weakly and not significant for N, PP and L, respectively.

In Figure 5, the materials were subdivided in a way as to minimize standard deviations from the regression lines between FMON and $C/N_{s,i}$. Again RI was calculated. The R^2-values were 0.998, 0.939 and 0.970 for the groups with mean RI values of 35, 67 and 90, respectively. There was, however, some overlap in RI values of the lines; the ranges were 25–40, 27–94 and 46–180, respectively. No clear explanation is available for the relatively low FMON of some materials with a rather low RI. Nevertheless, also in this study the general trend is that, at given values of $C/N_{s,i}$, FMON decreases with increasing RI.

Discussion

The results obtained from this desk study confirm that N mineralization depends on both decomposability and N concentration of the substrate. Linear relations between FMON and $C/N_{s,i}$ were found only

when experimental conditions and decomposability were uniform for all substrates. From these linear relationships, and taking into account time and average temperature of incubation, FDOC, apparent initial age and humification coefficients of the substrates were calculated with Equations (2–6).

In Table 5, the relation between FDOC and age (or h.c.) is not very obvious. This is at least partly due to differences in experimental conditions, as can be illustrated with the results obtained for *Gliricidia* and *Leucaena* leaves, which were present in three studies. *Leucaena* always had higher values of PP, RI and h.c. (Table 6). The differences among the experiments, however, were quite large, although practically the same analytical procedures had been used. Many factors may have interfered like pH, moisture regime, accessibility for soil fauna, state of organic materials (intact or ground, dried or fresh) and position of litter bags (in or on top of the soil). This variabilty hampered the ranking of substrates by decomposability, and as a result the boundaries between the classes in Table 7 could not sharply be defined. The upper line in Figure 1 may be considered as the boundary between Classes 1 and 2, and the lower line as the boundary between Classes 2 and 3.

Combinations of lignin and polyphenols concentrations offer perspectives for quantitative evaluation of decomposability. Analytical procedures, especially those for PP, however, need further attention, to improve the reproducibility of the results. Another

Table 4. Polyphenol concentrations (PP,%), and values of FMON and $C/N_{s,i}$ as presented by Palm and Sanchez (1991), calculated RI, and results of regression analysis for three sets of materials with similar PP and RI values. 26 ºC, 56 days

Substrate	PP	RI	$C/N_{s,i}$	FMON	Regr.anal.	
Gliricidia sepium	1.02	16	12.0	0.456	R^2	0.943
Erythrina sp.	1.04	18	12.8	0.391	q	0.587
Albizia saman	1.48	18	14.1	0.296	r	0.0158
Oryza sativa	0.69	12	39.8	−0.035		
Leucaena leucocephala	2.94	29	11.4	0.017	R^2	0.064
Desmodium gyroides	1.91	27	15.1	−0.123	q	n.r[a]
Inga sp.	1.61	31	15.2	−0.034	r	n.r
Cassia reticulata	2.20	28	18.0	−0.011		
Cajanus cajan	3.34	37	12.9	−0.086	R^2	0.784
Inga edulis	3.43	44	14.2	−0.043	q	0.222
Desmodium ovalifolium	3.61	38	19.0	−0.194	r	0.0214

[a]n.r. = no realistic value

Table 5. Calculated fractions of dissimilated organic carbon (FDOC), equivalent mineralization time (= $f_T \times$ t) (years), apparent initial age (years), and humification coefficient (h.c.) of the organic materials with $R^2 > 0.45$ for the relation between FMON and $C/N_{s,i}$. Ranking according to apparent initial age and humification coefficient

Fig/Table	Code	FDOC	$f_T \times$ t	Age	h.c.
Fig. 5	RI 35	0.817	0.97	0.94	0.18
Table 2	1.73%C, difference	0.582	0.51	1.13	0.25
Table 4	RI 12–18	0.461	0.57	1.52	0.39
Fig. 5	RI 67	0.812	0.97	1.7	0.44
Fig. 5	RI 90	0.635	0.97	2.3	0.57
Fig. 4	RI 22	0.260	0.46	2.3	0.57
Fig. 2	Points 1 to 5	0.447	1.69	2.92	0.67
Fig. 4	RI 50	0.181	0.46	3.0	0.68
Fig. 3	Digested	0.198	0.90	4.2	0.79
Fig. 3	Composted	0.096	0.90	7.1	0.90
Table 4	RI 37–44	0.050	0.57	8.3	0.92
Table 3	Lignin 10–15	0.058	0.75	9.0	0.93

Table 6. Lignin and polyphenol concentrations of *Gliricidia* and *Leucaena* according to three references, and calculated RI and h.c.

Reference	*Gliricidia*				*Leucaena*			
	L	PP	RI	h.c.	L	PP	RI	h.c.
Constantinides et al. (1994)	7.2	2.27	25	0.18	5.9	9.03	78	0.44
Palm and Sanchez (1991)	7.8	1.02	16	0.39	5.2	2.94	29	n.r[a].
Tian et al. (1992b)	11.6	1.62	22	0.57	13.4	5.02	50	0.68

[a]n.r. = no realistic value.

Table 7. Preliminary decomposibility classification of organic materials with approximate values of RI, humification coefficient and apparent initial age

Class	RI	h.c.	Age	Description
1	< 35	< 0.4	< 1.6	Digestive leaves, young roots, rice straw
2	> 25	0.5–0.7	2.0–3.2	Other non-woody plant materials, animal manure, activated sludge
3	> 40	> 0.75	> 3.7	Wood, peat dust, compost, digested sludge

source of variation in the concentrations of PP may be the different physiological state of the plant material; in such cases decomposability should vary as well. Standardization is also needed for the technical conditions and procedures of incubation experiments as they are commonly used for the calibration of the results of chemical analysis.

Acknowledgement

I thank Mr W M F Raijmakers who kindly assisted in making regression analyses and drawing the figures.

References

Constantinides M and Fownes J H 1994 Nitrogen mineralization from leaves and litter of tropical plants: Relationship to nitrogen, lignin and soluble polyphenol concentrations. Soil Biol. Biochem. 26, 49–55.

Janssen B II 1984 A simple model for calculating decomposition and accumulation of 'young' soil organic matter. Plant and Soil 76, 297–304.

Janssen B H 1986 Een één-parameter model voor de berekening van de decompositie van organisch material. Vakbl. Biol. 66 (20), 433–436.

Janssen B H and Noij I G A M 1996 A simple model and a linear equation for the calculation of nitrogen and phosphorus mineralization. Plant and Soil.

Jenkinson D S and Ayanaba A 1977 Decomposition of carbon-14 labelled plant material under tropical conditions. Soil Sci. Soc. Am. J. 41, 912–915.

Jensen H L 1929 On the influence of the carbon:nitrogen ratios of organic material on the mineralization of nitrogen. J. Agric. Sci. 19, 71–82.

Melillo J M, Aber J D and Muratore J F 1982 Nitrogen and lignin control of hardwood leaf litter decomposition dynamics. Ecology 63, 621–626.

Palm C A and Sanchez P A 1991 Nitrogen release from the leaves of some tropical legumes as affected by their lignin and polyphenolic contents. Soil Biol. Biochem. 23, 83–88.

Parker C F and Sommers L E 1983 Mineralization of nitrogen in sewage sludges. J. Environ. Qual. 12, 150–156.

Raijmakers W M F and Janssen B H 1993 Assessment of plant-available nitrogen in processed organic wastes. *In* Optimization of Plant Nutrition. Eds. M A C Fragoso and M L van Beusichem. pp 107–115. Kluwer Academic Publishers, Dordrecht, The Netherlands.

Schulz E 1988 N-Tranformationsprozesse beim Abbau von organischer Primärsubstanz im Boden in Abhängigkeit von ihrer Stabilität und dem C/N-Verhältnis. Arch. Acker-Pflanzenbau Bodenkd., Berlin 32, 577–582.

Tian G, Kang B T and Brussaard L 1992a Biological effects of plant residues with contrasting chemical compositions under humid tropical conditions - decomposition and nutrient release. Soil Biol. Biochem. 24, 1051–1060.

Tian G, Kang B T and Brussaard L 1992b Effects of chemical composition on N, Ca and Mg release during incubation of leaves from selected agroforestry and fallow plant species. BiogeoChem. 16, 103–119.

Section editor: R Merckx

O. Van Cleemput et al. (eds.), Progress in Nitrogen Cycling Studies, 77–83, 1996.

Effect of wetting and drying cycles on N mineralization/immobilization in soil amended with organic materials

N. Jedidi[1], O. Van Cleemput[2] and A. M'Hiri[3]

[1]National Institute for Scientific and Technical Research. B.P. 95, 2050 Hammam Lif, Tunisia, [2]Faculty Agricultural and Applied Biological Sciences, Coupure Links 653, B-9000 Ghent, Belgium and [3]National Agricultural Institute of Tunisia, Cité Mahragène, 1082 Tunis, Tunisia

Key words: drying, immobilization, mineralization, organic amendment, wetting

Abstract

A study was undertaken to quantify N mineralization and immobilization in a loamy-clayey soil upon application of $K^{15}NO_3$-N and 4 types of organic amendments (2 composts of domestic waste of different age (C1 and C3), residual waste from a treatment plant (B) and farmyard manure (F)). The study was carried out during a one month incubation at 25 °C. In the first experiment, the soil was incubated at a moisture content of 2/3 field capacity. In the second experiment, we had a wetting and drying cycle: wetting at 2/3 field capacity for 10 days, followed by drying for 10 days and rewetting to 2/3 field capacity for 10 days.

The mineralization and immobilization processes depended on the nature of the organic amendments, their degree of stability and their N content. The N-mineralization was correlated with the C/N ratio, %N, Hemicellulose (NDF-ADF), (cellulose + hemicellulose + lignin)/N ratio (NDF/N) and (lignin + cellulose)/N ratio (ADF/N). The N-immobilization was only correlated with the %ADF (% lignin and cellulose). The cycles of wetting and drying considerably increased the N mineralization. Application of organic amendments favoured the process of N immobilization. It decreased with the degree of stability of the organic materials. In contrast, wetting and drying cycles markedly decreased the amount of N-immobilization. Loss of nitrogen (2–20%) was attributed especially to the denitrification.

Introduction

The effect of wetting and drying on the biological activity of the soil has been shown by several authors (Bottener, 1985; Rosacker and Kieft, 1990; Van Gestel, 1991). Drying provokes destruction of part of the soil flora. After wetting the microbial biomass is destroyed and degraded by other micro-organisms (Shields et al., 1974). The decomposition of plant residues by soil micro-organisms is affected by the wetting and drying cycle (Amato et al., 1984). Van Veen et al. (1985) observed a reduction of the biomass and a flash of mineralization. However, they noted that during the period of incubation the mineralization of organic matter was more important in the soil which had undergone a wetting-drying cycle than in the soil continuously humidified. The effect of wetting and drying cycles on the biological activity of soil and some amendments must be taken into consideration. In this context we quantified the mineralization and immobilization of four organic amendments in a loamy-clayey soil after wetting at 2/3 field capacity and a wetting and drying cycles. The objective of this survey consisted of: (1) quantification of the N-mineralization and gross immobilization upon addition of organic amendments, (2) comparison of different organic amendments by the indirect isotopic dilution method.

Materials and methods

The used soil had a loamy-clay texture with the following physico-chemical characteristics : clay, 27%; silt, 62%; sand, 11%; pH, 8.52; C, 0.87%; N, 0.095% and C/N, 9.15. Four organic amendments were used : (i) two composts of domestic waste of different age (C1, 2 months and C3, 8 months), (ii) residual waste from a treatment plant (B) and farmyard manure (F).

Table 1. Chemical analysis of the organic amendments

	C1	C3	F	B
pH	8.10	8.16	7.87	7.07
C(%)	17.83	14.04	29.18	27.24
N(%)	1.28	1.30	2.56	3.87
C/N	13.93	10.82	11.39	7.03
N.D.F(%)	22.0	16.9	25.0	19.5
A.D.F(%)	20.0	14.0	18.5	9.8
(N.D.F)-(A.D.F)(%)	2.0	2.90	6.5	9.7

N.D.F = Neutral Detergent Fiber (Hemicellulose, cellulose, lignin) ; A.D.F = Acid Detergent Fiber (Lignin, cellulose); (N.D.F)-(A.D.F) = Hemicellulose; C1 = Compost (2 months); C3 = Compost (8 months); F = Farmyard manure; B = Residual waste

The chemical characteristic of these amendments are given in Table 1. The study of the N-mineralization and immobilization was carried out during a one month incubation at 25 ° C. A quantity of 25 g of soil was air dried, passed the mesh to 2 mm, and than mixed with the organic amendments to reach an amount of 40 t ha^{-1} and a solution of $K^{15}NO_3$ with an isotopic excess of 60.54%, in doses of 100 kg N ha^{-1} (0.833 mg of N). In the first experiment, the soil was incubated at a moisture content of 2/3 field capacity. In the second experiment, we had wetting and drying cycles: wetting at 2/3 field capacity for 10 days, followed by drying for 10 days and rewetting to 2/3 field capacity for 10 days. In this study six treatments were compared and each one included three repetitions: soil without fertilizer and organic amendments (S), soil with fertilizer (S+N) and soil with fertilizer and organic amendments (S+N+C1, S+N+C3, S+N+F and S+N+B).

Analysis

At the end of the incubation the inorganic N (NH_4 and $NO_3 + NO_2$) was determined by the method of Bremner (1965). The organic nitrogen is determined by the Kjeldahl method (Bremner and Mulvaney, 1982). The labelled ^{15}N was determined by the method of Dumas (Hauck, 1982). The methodologies from characterization of soil and organic amendments were the following: the Kjeldahl method for the determination of total N; the method of Walkey and Black for determination of C; and the dry method for the determination of C in the organic amendments. The organic amendments were fractionated by the method of Van Soest (1967).

Result expression and methods of calculation

The calculation for net mineralization and gross immobilization method were done according to Guiraud (1984) and Jacquin and Vong (1990). The net mineralization (MIN) is the difference between the total quantity of mineral nitrogen calculated after and before incubation. The gross immobilization (OB) corresponds to the difference between the total amount of N derived from the mineral compartment (Ot) and the quantity lost by denitrification and volatilization. The gross immobilization is quantified by the formula suggested by Guiraud (1984). The calculation of (Ot) was done by measuring the reduction of the quantity of ^{15}N in the mineral nitrogen compartment for the period of incubation and by taking into account the mean of initial and final isotopic excess:

$$Ot = \frac{X_0 E_0 - XE}{(E_0 + E)/2}$$

E_0 and E: initial atom excess and final mineral nitrogen X_0 and X: initial and final quantities of the mineral nitrogen.

The calculation of the gross immobilization (OB) necessitates the acquaintance of the quantity of ^{15}N immobilized in the soil at the end of the incubation after elimination of the mineral ^{15}N by KCl:

$$OB + \frac{Q^{15}N \text{ immobilized}}{(E_0 + E)/2}$$

The losses (P) represent the difference between Ot and OB while the gross mineralization (Mt) is the sum of Ot and MIN.

Results and discussion

Characteristics of organic amendments

The result in Table 1 show that the farmyard manure (F) and the residual waste (B) have a low carbon content (29.2 and 27.2%, respectively). The nitrogen content is 2.56 and 3.87%, respectively. The carbon content of the composts decreases with the age, being 17.8% for C1 and 14.0% for C3. The C/N ratio of organic amendments varies from 7.03 for the residual waste to 13.93 for the C1 compost. The C1 compost and farmyard manure are richest in N.D.F (cellulose, hemicellulose and lignin) and A.D.F (lignin and cellulose) which are 25 and 18.5% for the farmyard manure respectively and 22 and 20% for the C1 compost respectively.

The hemicellulose (N.D.F-A.D.F) was more important for the farmyard manure (6.5%) and residual waste (9.7%) than for the composts (2 to 3%). The characteristics of the organic amendments applied to the soil influence their decomposition rate (Fox et al., 1990; Kachaka, 1993) and the process of mineralisation and immobilization of the soil nitrogen in the soil (Fox et al., 1990; Kachaka, 1993).

Net mineralization (MIN)

The net mineralization in the presence of the nitrogen fertilizer and organic amendments varied from -20.2 mg kg^{-1} for the treatment with the C1 compost to 94.3 mg kg^{-1} for the treatment with residual waste (Table 2). This net mineralization showed the following decreasing order: S+N+B > S+N+F > S+N+C3 > S+N > S+N+C1. The net mineralization was negative for the C1 compost treatment, which was not a stable amendment. Furthermore, it was established the nitrogen immobilization.

The wetting and drying cycle importantly increased the net mineralization. This mineralization varied from 13.54 mg kg^{-1} for the C1 compost treatment to 106.84 mg kg^{-1} for the waste residual treatment (Table 3). The increase of the mineralization was explained by the mineralization of the microbial biomass which died during drying. The sequence of mineralisation was the same as in the first experiment.

The mineral nitrogen derived from the fertilizer, organic amendments and soil (Table 4) was determed by the indirect isotopic dilution (Zapata, 1990). For the first experiment, the quantities of nitrogen derived from the organic amendments was showed the following sequence: C1 (8.5 mg kg^{-1}) < C3 (12.8 mg kg^{-1}) < F (21.0 mg kg^{-1}) < B (103 mg kg^{-1}). During the second experiment, the wetting and drying cycle increased (Table 5) the intensity of mineralization of the stable amendments (C3 and F) and maintained the order of importance : C1 (6.4 mg kg^{-1}) < C3 (28.9 mg kg^{-1}) < F (27.7 mg kg^{-1}) < B (76.7 mg kg^{-1}). The mineralization depended on the nature of the organic amendments, the degree of stability and their nitrogen content. However, the net mineralization was correlated with the C/N ratio, %N, NDF /N, NDF-ADF and ADF/N. The correlation coefficients at 2/3 field capacity were: 0.94; 0.89; 0.87; 0.89 and 0.87, respectively. The correlation coefficients for the wetting and drying cycles were: 0.94; 0.71; 0.79; 0.94 and 0.95, respectively. The net mineralization balance indicated that after one month of incubation, the mineral nitrogen derived from the organic amendments (NdfAm.) with regard to the total nitrogen presented the following order: (i) the first experiment: S+N+C3 (22%)< S+N+C1 (25%)< S+N+F (30%)< S+N+B (70%); (ii) the second experiment: S+N+C1 (10%) < S+N+C3 (25%)< S+N+F (31%)< S+N+B (48%).

Gross immobilization (OB)

The gross immobilization is the difference between the total amount of N derived from the mineral compartment (Ot) and the gaseous losses (P). This immobilization corresponds to the biological immobilization and physico-chemical adsorption on the soil (Chotte, 1986; Guiraud,1984). The gross immobilization (Table 2) was more important for the treatments with the organic amendments. The treatments with residual waste and C1 compost presented a more pronounced immobilization than the treatments with the farmyard manure and the C3 compost being stable amendments. Comparison of the values of the gross immobilization shows that the organic amendments stimulated immobilization. The effect decreased with the degree of the stability of the organic amendments. The gross immobilization also depended on the nature and the biochemical composition of the organic amendments, the correlation with the ADF (% lignin and cellulose) was (r= 0.607).

The extend of the gross immobilization decreased according to the following order: S+N+B (53.3 mg kg^{-1}), S+N+C1 (49.1 mg kg^{-1}), S+N+F (36.7 mg kg^{-1}), S+N+C3 (35.6 mg kg^{-1}) and S+N (29.1 mg kg^{-1}).

For the wetting and drying cycle, the quantities were less important than in the first experiment. However, a small disruption of order was noted especially for the treatments with the C3 compost and residual waste (B) (Table 3). The values of the gross immobilization were in the following order: S+N+C1 (38.4 mg kg^{-1}), S+N+F (24.1 mg kg^{-1}), S+N+B (22.3 mg kg^{-1}), S+N (17.6 mg kg^{-1}) and S+N+C3 (16.7 mg kg^{-1}).

Gross mineralization (Mt)

The gross mineralisation shows the quantities of nitrogen which are really mineralised (Tables 2 and 3). After one month of incubation, the gross mineralization was more pronounced in the treatments with the organic amendments. The values of the gross mineralization were in the following order: S+N+B (149 mg kg^{-1}), S+N+F (58.3 mg kg^{-1}), S+N+C3 (48.9 mg kg^{-1}),

Table 2. Evolution of net mineralization (MIN), gross immobilization (OB) and losses of nitrogen after addition of fertilizer and organic amendments (2/3 field capacity moisture)

	Init. val.	S+N	S+N+C1	S+N+C3	S+N+F	S+N+B
QN (mg kg^{-1})	53.61	53.30	33.44	58.53	67.94	147.91
E(%)	37.638	18.905	14.085	14.769	13.226	5.683
Q^{15}N	20.17	10.07	4.71	8.64	8.98	8.40
∂ Q^{15}N		10.10	15.46	11.53	11.19	11.77
E% mean		28.271	25.861	26.203	25.432	21.660
Ot		35.72	59.78	44.00	43.99	54.33
Q^{15}N im.		8.23	12.70	9.34	9.34	11.54
OB		29.11	49.10	35.64	36.72	53.27
MIN		-0.31	-20.17	4.92	14.33	94.30
Mt		35.41	39.61	48.92	58.32	148.63
Losses		6.61	10.68	8.35	7.27	1,06

QN = Amount of N mineral (mg kg^{-1}); E% = Isotope excess initial and final; E% moyen = Mean isotope excess

Q^{15}N = Amount of labelled mineral N (mg kg^{-1}); ∂ Q^{15}N = Difference between amount ^{15}N initial and final (mg kg^{-1})

Ot = Total amount N derived from mineral compartiment (mg kg^{-1}); OB = Gross immmobilization (mg kg^{-1})

MIN = Net mineralization (mg kg^{-1}); Mt = Gross mineralization (mg kg^{-1})

Table 3. Evolution of net mineralization (MIN), gross immobilization (OB) and losses of nitrogen after addition of fertilizer and organic amendments (wetting and drying cycles)

	Init. Val.	S+N	S+N+C1	S+N+C3	S+N+F	S+N+B
QN (mg kg^{-1})	53.61	94.08	67.15	117.85	122.30	160.45
E(%)	37.638	15.990	14.451	12.067	11.068	8.351
Q^{15}N	20.17	15.04	9.70	14.22	13.53	13.39
∂ Q^{15}N		5.13	10.47	5.95	6.64	6.78
E% mean		26.814	26.044	24.852	24.352	22.994
Ot		19.11	40.20	23.94	27.26	29.48
Q^{15}N im.		4.73	10.01	4.15	5.86	5.13
OB		17.64	38.43	16.69	24.06	22.31
MIN		40.47	13.54	64.24	68.69	106.84
Mt		59.58	53.74	88.18	95.95	51.79
Losses		1.47	1.77	7.25	3.2	7.17

QN = Amount of N mineral (mg kg^{-1}); E% = Isotope excess initial and final; E% moyen = Mean isotope excess

Q^{15}N = Amount of labelled mineral N (mg kg^{-1}); ∂ Q^{15}N = Difference between amount ^{15}N initial and final (mg kg^{-1})

Ot = Total amount N derived from mineral compartiment (mg kg^{-1}); OB = Gross immmobilization (mg kg^{-1})

MIN = Net mineralization (mg kg^{-1}); Mt = Gross mineralization (mg kg^{-1}).

S+N+C1 (39.6 mg kg^{-1}) and S+N (35.4 mg kg^{-1}). In the case of wetting and drying, the gross mineralization showed the following order : S+N+F (95.9 mg kg^{-1}), S+N+C3 (88.2 mg kg^{-1}), S+N (59.6 mg kg^{-1}), S+N+C1 (53.7 mg kg^{-1}) and S+N+B (51.8 mg kg^{-1}). The stable amendments had the most pronounced gross mineralization (F and C3) than the compost (C1) and the residuel waste (B).

The losses (P)

The losses were calculated by difference between (Ot) and the gross immobilization (OB) (Tables 2 and 3).

Table 4. Net mineralization (mg kg^{-1}) derived from the soil, fertilizer and organic amendments (2/3 field capacity moisture)

	Init. val.	S	S+N	S+N+C1	S+N+C3	S+N+F	S+N+B
N. min.	53.61	41.28	53.30	33.44	58.53	67.94	147.91
N. E	33.33		16.64	7.78	14.27	14.83	13.87
N. S	20.28		36.66	17.17	31.46	32.11	30.58
N. Am				8.49	12.80	21.00	103.46

N. E: N derived from fertilizer; N. S: N derived from soil ; N. Am : N derived from amendment.

Table 5. Net mineralization (mg kg^{-1}) derived from the soil, fertilizer and organic amendments (wetting and drying cycles)

	Init. val.	S	S+N	S+N+C1	S+N+C3	S+N+F	S+N+B
N. min.	53.61	62.19	94.08	67.15	117.85	122.30	160.45
N. E	33.33		24.84	16.02	23.48	22.35	22.12
N. S	20.28		69.24	44.68	65.47	72.62	61.67
N. Am				6.45	28.90	27.33	76.66

N. E: N derived from fertilizer; N. S: N derived from soil; N. Am: N derived from amendment.

For the two experiments, the losses varied between 1.06 and 10.7 mg kg^{-1}, which representing from 2 to 20% of the nitrogen which initial was mineralized. The organic amendment provoked a slight increase of loss, thus decreasing the gross immobilization. In this study the losses were attributed to denitrification and volatilization. On an average, the losses varied around 20% but they could atain 50% of the quantity of applied fertilizers (Kostov and Kaloianova, 1994; Vermoesen et al., 1992). The losses depend on the environmental conditions such as, pH, anaerobios is and high temperatures (Nögele and Conrad, 1990). They increase in the presence of easily degradable organic matter (Burford and Bremner, 1975).

Nitrogen balance of the added fertilizer

The nitrogen balance of the added fertilizer (Table 6) indicates that in the treatment without organic amendment, 50% of the fertilizer nitrogen remained in the soil with an average of 43% for the treatments with the residual waste, farmyard manure and the C3 compost, and 23% for the C1 compost. In contrast to the presence of organic amendments, the nitrogen from the fertilizer in the organic compartment was up to 63% for the C1 compost, 46% for the C3 compost and the farmyard manure, and up to 57% for the residual waste. For the treatment without organic amendments 40% of the fertilizer was immobilized. The wetting and drying

cycle (Table 7) increased the fertilizer nitrogen (48–75%) remained in the soil and decreased the N derived from fertilizer (20–50%) included in the organic compartment. Consequently the losses of nitrogen from the fertilizer were not high, from 1.3 to 13.7% . These results were in the range of 0 to 60%, with an average of 25 to 30%, as also reported by Guiraud (1984).

Conclusions

The mineralization and immobilization processes were influenced by the nature of the organic amendments, their degree of stability and their nitrogen content.

The N-mineralization correlated with the C/N ratio, %N, NDF-ADF, NDF/N and ADF/N. The N-immobilization only correlated with the %ADF (% lignin and cellulose, $r^2 = 0.61$). The amount of mineralized N after one month of incubation was in the following order: Soil+N+C1 < Soil+N < Soil+N+C3 < Soil+N+F < Soil+N+B. The cycles of wetting and drying considerably increased the N mineralization. Wetting and drying cycles had a stimulating effect on the mineralization of the soil organic nitrogen. Nitrogen mineralized from organic amendments was in the following order: C1< C3 < F <B.

Application of organic amendments favoured the process of N-immobilization. It decreased with the degree of stability of the organic materials. The amount

Table 6. Nitrogen balance of added fertilizer (2/3 field capacity moisture)

	N mineral				N organic				
	QN (mg kg^{-1})	Ndff (%)	QN E (mg kg^{-1})	%N fert.	QN (mg kg^{-1})	Ndff (%)	QN E (mg kg^{-1})	%N fert.	Losses (%)
S	41.28				891.96				
S+N	53.30	31.22	16.64	49.92	933.51	1.45	13.53	40.59	9.49
S+N+C1	33.44	23.26	7.78	23.34	1003.94	2.09	20.98	62.94	13.72
S+N+C3	58.53	24.39	14.27	42.81	1162.34	1.32	15.34	46.02	11.17
S+N+F	67.94	21.84	14.84	44.49	1128.83	1.36	15.35	46.05	9.46
S+N+B	147.91	9.38	13.88	41.64	1339.17	1.42	19.01	57.03	1.33

Table 7. Nitrogen balance of added fertilizer (wetting and drying cycles)

	N mineral				N organic				
	QN (mg kg^{-1})	Ndff (%)	QN E (mg kg^{-1})	%N fert.	QN (mg kg^{-1})	Ndff (%)	QNE (mg kg^{-1})	%N fert.	Losses (%)
S	62.19				891.75				
S+N	94.08	26.41	24.84	74.52	870.65	0.89	7.74	23.22	9.70
S+N+C1	67.15	23.87	16.02	48.06	1129.99	1.46	16.49	49.47	9.70
S+N+C3	117.85	19.13	23.48	70.44	931.40	0.73	6.79	20.23	9.08
S+N+F	122.30	18.28	22.35	67.05	1231.28	0.78	9.60	28.8	9.58
S+N+B	160.45	13.79	22.12	66.36	1223.18	0.69	8.43	25.29	9.16

QN = Amount of mineral N (mg kg^{-1}) ; Ndff% =% Nitrogen derived from fertilizer;
QN E = Amount N derived from fertilizer (mg kg^{-1}); %N =% Nitrogen of fertilizer.

of N-immobilization decreased as follows: Soil+N+B > Soil+N+C1 > Soil+N+F > Soil+N+C3 > Soil+N. Wetting and drying cycles markedly decreased the amount of N-immobilization but stimulate the mineralization. The losses of nitrogen (2–20%) were attributed mainly to denitrification.

References

Amato M, Jackson R B, Butler J H A and Ladd J N 1984 Decomposition of plant material in Austria soils. II. Residual organic ^{14}C and ^{15}N from legume plant parts decomposing under field and laboratory conditions. Aust. J. Soil Res. 22, 331–341.

Bottener P 1985 Response of microbial biomass to alternate moist and dry conditions in a soil incubated with ^{14}C and ^{15}N labelled plant material. Soil Biol. Biochem. 17, 329–33

Bremner J M 1965 Inorganic forms of nitrogen. *In* Methods of Soil Analysis, part 2, Agron. 9. Ed. C A Black. pp 1179–1239. Am. Soc. Agron, Madison, WI, USA.

Bremner J M and Mulvaney C S 1982 Nitrogen total. *In* Methods of Soil Analysis, part 2, Agron. 9. Chemical and Microbiological Properties. Ed. C A Black. pp 595–622. Am. Soc. Agron., Madison, WI, USA.

Burford J R and Bremner J M 1975 Relationship between the denitrification capacities of soils and total water soluble and readly decomposable soil organic matter. Soil Biol. Biochem. 7, 389–394.

Chotte J L 1986 Evolution d'une biomasse racinaire doublement marquée (^{14}C,^{15}N) dans un système sol-plante: étude sur un cycle annuel d'une culture de maïs. Thèse Doct., Univ. de Nancy I, France. 116 p.

Fox R H, Myers R J K and Vallis I 1990 The nitrogen mineralization rate of legume residues as influenced by their polyphenol, lignin and nitrogen contents. Plant and Soil 129, 251–259.

Guiraud G 1984 Contribution du marquage isotopique à l'évaluation des transferts d'azote entre les compartiments organiques et minéraux dans le système sol-plante. Thèse Doct. d'Etat, Univ. Pierre et Marie Curie, Paris VI, France. 335 p.

Hauck R D 1982 Nitrogen-isotope-ratio analysis. *In* Methods of Soil Analysis, Part 2, Agron. 9. Chemical and Microbiological Properties. Ed. C A Black. pp 735–779. Am. Soc. Agron., Madison, WI, USA.

Jacquin F and Wong P C 1990 Quantification des mécanismes d'organisation et de minéralisation de l'azote en sol cultivés. Sci. Sol 28, 349–362.

Kachaka S K 1993 Decomposition and N-mineralization of prunings of various quality and age. Ph D Thesis, KU Leuven, Belgium. 117 p.

Kostov O and Kaloianova N 1994 Nitrogen and carbon transformation during the decomposition of sunflowers stalks using ^{15}N. Trans. 15th World Congr. of Soil Sci. 56, 213–215.

Nägele W and Conrad R 1990 Influence of pH on the release of NO and N_2O from fertilized and unfertilized soil. Biol. Fert. Soils, 10, 139–144.

Rosacker L L and Kieft T L 1990 Biomass and adenylate energy charge of glass land soil during drying. Soil Biol. Biochem. 22, 1121–1127.

Shields J A, Paul E A and Lowe W E 1974 Factors influencing the stability of labelled microbial materials in soils. Soil Biol. Biochem. 6, 31–37.

Van Gestel M 1991 Microbial biomass responses to soil drying and rewetting. Ph. D., K. University, Leuven, Belgium. 144 p.

Van Soest P J and Wine R H 1967 Use of detergents in the analysis of fibrous feeds. IV Determination of plant cell-wall constituents. J. AOAC 46, 828–835.

Van Veen J A, Ladd J N and Amato M 1985 Turnover of carbon and nitrogen through the microbial biomass in a sandy loam and a clay soil incubated with $[^{14}C(U)]$ glucose and $[^{15}N](NH_4)_2$ SO_4 under different moisture regimes. Soil Biol. Biochem. 17, 747–756.

Vermoesen A, Demeyer P, Hofman G and van Cleemput O 1992 Field measurement of ammonia volatilisation upon application of differents NH_4 fertilizer and urea. Pédologie 42, 119–128.

Zapata F 1990 Isotope technique in soil fertility and plant nutrition studies. In Use of Nuclear Techniques in Studies of Soil-Plant Relationships. pp 61–127. IAEA, Training Course Serie n° 2, Vienna, Austria.

Plant and Soil **181**: 71–82, 1996.
© 1996 *Kluwer Academic Publishers.*

Interactions between decomposition of plant residues and nitrogen cycling in soil

B. Mary[1], S. Recous[1,4], D. Darwis[2] and D. Robin[3]
[1] *I.N.R.A., Unité d'Agronomie, rue F. Christ, 02007 Laon Cedex, France,* [2] *Fak. Pertanian Unhalu, Jl. Layjen parman, Kendari-Sul. Tenggara, Indonesia and* [3] *G.P.S.A., 22 Place des Vosges 92080 Paris La Défense 5, France.* [4] *Corresponding author* *

Key words: carbon, decomposition, immobilization, mineralization, residues

Abstract

The processes of N mineralization and immobilization which can occur in agricultural soils during decomposition of plant residues are briefly reviewed in this paper. Results from different incubation studies have indicated that the amounts of N immobilized can be very important and that the intensity and kinetics of N immobilization and subsequent remineralization depend on the nature of plant residues and the type of decomposers associated. However, most of the available literature on these processes refer to incubations where large amounts of mineral N were present in soil.

Incubations carried out at low mineral N concentrations have shown that the decomposition rate of plant residues is decreased but not stopped. The immobilization intensity, expressed per unit of mineralized C, is reduced and N remineralization is delayed. Nitrogen availability in soil can therefore strongly modify the MIT kinetics (mineralization-immobilization turnover) by a feed-back effect.

The mineralization and immobilization kinetics have been determined in a two-years field experiment in bare soil with or without wheat straw. Mineralization in plots without straw seemed to be realistically predicted by accounting for variations in soil temperature and moisture. Immobilization associated with straw decomposition was clearly shown. It was increased markedly by the addition of mineral N throughout decomposition. It is concluded that mineral N availability is an important factor controlling plant residues decomposition under field conditions. A better prediction of the evolution of mineral N in soil may therefore require description and modelling of the respective localization of both organic matter and mineral N in soil aggregates.

Introduction

The use of ^{15}N tracers in incubation studies has clearly demonstrated that mineralization and immobilization processes of nitrogen, can take place simultaneously in soil (Bjarnason, 1988; Jansson and Persson, 1982; Kirkham and Bartholomew, 1954). Microbial immobilization of mineral N frequently occurs whereas (positive) net mineralization is found in soil (Hart et al., 1994; Recous et al., 1992). Conversely, the isotopic dilution of mineral ^{15}N shows that ammonification continues even when N immobilization is the dominant process (Bjarnason, 1988; Mary and Recous, 1995). The existence of the two processes is particularly obvious during the initial decomposition of all

plant residues in spite of their diversity going from rhizodeposits produced during plant growth to unharvested plant parts (coming either from mature crops or young crops like cover-crops). However, most of these results have been established under laboratory conditions and it is not precisely known to what extent they can be extrapolated to field conditions. Answering this question is one of the present challenges : are the 'mechanistic' C and N models realistic enough to predict the evolution of mineral N in agricultural fields ? For example, these models often predict that gross rates of mineralization and immobilization are much higher than net rates (Houot and Chaussod, 1991; Paul and Juma, 1981). This paper first reviews the effect of decomposition processes on N transformations. It presents results from a field experiment

---

* FAX No.: +3323793615
Plant and Soil is the original source of publication of this article. It is recommended that this article is cited as: *Plant and Soil* **181**: 71–82, 1996.

where the kinetics of N mineralization and immobilization were determined. The second part of the paper deals with the reverse effect, i.e. the possible control of decomposition by inorganic N concentration.

Effect of plant residue decomposition on soil mineral N

Identification of C and N fluxes

During plant residue decomposition, C and N cycles in soil are strongly linked mainly because of the simultaneous assimilation of C and N by the decomposing microflora. The C assimilation rate depends on the rate of decomposition of plant material and the assimilation yield of the decomposed C by the microflora. The N assimilation requirements are then determined by this carbon flow and the C:N ratio of the decomposers (Fig. 1). The sources of nitrogen for the microbial biomass can be the plant residue itself, the mineral N already present in soil or recently mineralized, and the recycling soil biomass. It is often assumed that N coming from the residue and from the recycling biomass is mineralized before being assimilated by the newly-formed biomass. However, it has been shown that the soil microflora can directly assimilate significant amounts of organic N compounds coming from plant residues or from dead biomass (Barak et al., 1990; Hadas et al., 1992; Payne, 1980). An example of microbial recycling consists of re-use of N from less active by more active mycelia (Cowling and Merrill, 1966; Paustian and Schnürer, 1987). These later fluxes of nitrogen, which are actual components of the microbial assimilation of nitrogen during carbon decomposition, complicate the quantification of total N assimilation and the assessment of C-N relationships. Net microbial assimilation can be defined as the sum of gross immobilization of inorganic N and direct assimilation of organic N compounds (flux $i+j$ in Fig. 1). Total microbial assimilation also includes re-assimilation of N during biomass recycling (flux $i+j+a$).

Quantification under laboratory conditions

Measuring the variations in inorganic N with time allows to quantify net immobilization and net mineralization. The quality of plant residue added to soil determines both the rate of decomposition and the dynamics of mineral N. In the absence of recent additions of fresh organic matter, the soils generally present

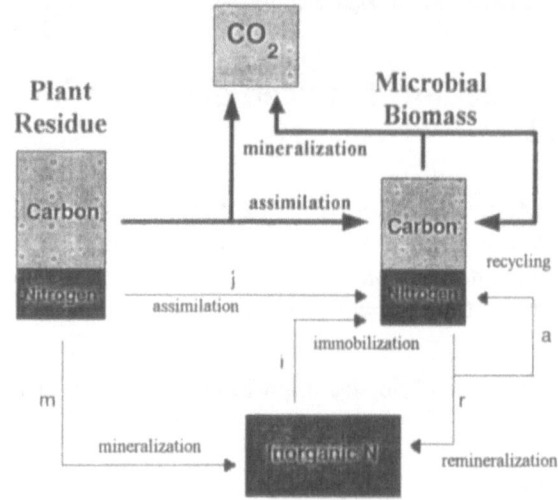

Figure 1. Flow chart of C and N transformations during initial decomposition of plant residues in soil. Thick lines : C fluxes ; thin lines : N fluxes

regular net mineralization kinetics during incubations. The addition of plant residues most often results in a net N immobilization phase followed by a net remineralization phase (these processes being evaluated by calculating the difference between an amended soil and a control soil) as illustrated in Figure 2A. The dynamics as well as the net amounts of N immobilized varied greatly in these experiments according to the nature of plant residues : the decomposition of root mucilage induced a very rapid and large immobilization, greater than that caused by glucose addition ; it was followed by a rapid release of nitrogen when the substrate was exhausted. At the opposite, decomposition of wheat leaves resulted in net mineralization only ; no net immobilization was observed. The amount of N immobilized appeared to be dependent on the type of substrate, particularly its C:N ratio (Table 1), in accordance with earlier work (e.g. Fog, 1988; Swift et al., 1979). But it is also clear from Table 1 that the C:N ratio of the residue explains only part of the differences found in maximum net immobilization. The kinetics of decomposition are even almost independent of the N:C ratio of the plant material (Fig. 2A). It is much more related to the biochemical composition of the residue, such as soluble C, cellulose and lignin contents (Jawson and Elliott, 1986; Kirchmann and Bergqvist, 1989; Knapp et al. 1983; Reinertsen et al., 1984). In order to better assess and predict the MIT kinetics during decomposition, it is necessary to deter-

Table 1. Maximum net immobilization of soil mineral N and microbial assimilation of nitrogen measured during decomposition of various plant residues in incubation studies at 25 °C

Plant material	C/N	N/C	Net immobilization	Microbial assimilation
		(mg N g^{-1} C)	(mg N g^{-1} C)	(mg N g^{-1} C)
Wheat leaves [d]	13	76	0	48
Maize roots [a]	14	72	19	66
Rye roots [b]	28	36	23	-
Wheat roots [d]	34	29	27	50
Rye grass roots [b]	51	19	24	-
Root mucilage [a]	64	16	72	88
Wheat straw [c]	100	10	28	32
Maize straw [d]	130	8	27	-
Glucose [a]	∞	0	61	61

[a] Data from Mary et al. (1993).
[b] Data from Mary (1988).
[c] Data from Robin (1994).
[d] Unpublished data.

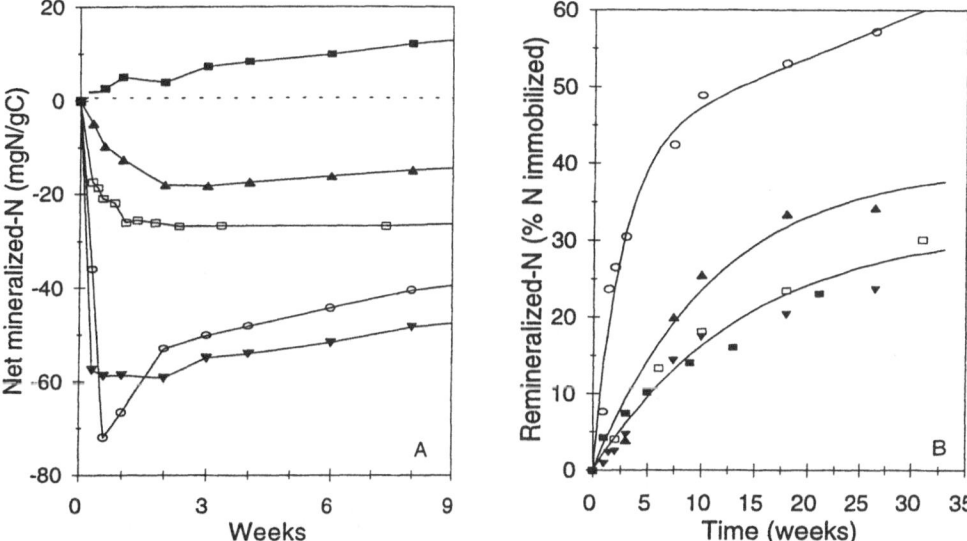

Figure 2. **A.** Kinetics of net mineralization resulting from the decomposition of different plant residues in laboratory experiments at 25 °C. Net mineralization is calculated by difference with an unamended soil. **B.** Kinetics of N remineralization during the decomposition of different plant residues in laboratory experiments at 25 °C. Net remineralization is calculated by difference with the unamended soil and expressed as % of maximum N immobilized. Day 0 is the time of maximum immobilization. ■ wheat leaves, ▲ maize roots, ○ root mucilage, ▼ glucose, □ maize straw. Data from [a] Mary et al. (1993), [b] Recous (1994), [c] Mary, unpublished results.

mine the exact fate of added residue-N and quantify microbial N assimilation.

Tracing soil mineral N or/and residue-N pools with ^{15}N can be done by measuring the variations in the N and ^{15}N mineral pools or N and ^{15}N organic pools (soil biomass) and by using calculations based on the isotopic dilution technique (Barraclough, 1991; Davidson et al., 1991; Kirkham and Bartholomew, 1954; Nishio

et al., 1985). This allows to describe the dynamics of gross N mineralization and immobilization after residue incorporation (Jensen, 1994a, c; Ocio et al., 1991; Recous et al., 1995; Sorensen, 1981). The use of 'paired' ^{15}N treatments such as those suggested by Bjarnason (1988) and Barraclough (1991), combining ^{15}N-NH$_4^+$ and ^{15}N-NO$_3^-$ applications, has been shown useful to assess simultaneously several N transforma-

[87]

tion rates and to improve the accuracy of N immobilization rate determinations. A similar approach can be adopted with plant residues, using combined treatments identical in total amount of added N and C but differing only in the N pool being labelled (^{15}N residues + [14]N mineral and ^{14}N residues + [15]N mineral). It could be fruitfully associated with improved methods of calculation like numerical techniques and optimization routines (Mary and Recous, 1994; Myrold and Tiedje, 1986; Wessel and Tietema, 1992). These tracing techniques are promising to better evaluate the various N fluxes (Fig. 1) and the confidence intervals for each of them, although they require very accurate data.

The N fluxes which do not enter the mineral N pool, i.e. direct assimilation of residue-N (*j*) and re-assimilation during biomass-N recycling (*a*), in fact escape quantification by the isotopic dilution technique. Evaluation of the direct assimilation of residue-N can be done by measuring the mean isotopic ^{15}N excess of the mineral N immobilized and that of the newly-formed biomass N (Mary et al., 1993). Using this method, we found that the net microbial assimilation (*i+j*) was much less variable between residues than net immobilization (Table 1). This suggests that direct assimilation of N may represent a significant part of microbial N assimilation and that this flux can explain a large part of the differences found in net N immobilization, particularly for residues rich in N. However, some differences in microbial assimilation still exist, probably arising from the nature of the decomposing microflora. These differences can be explained if we consider that fungi are mainly responsible for the decomposition of cellulose and lignin while bacteria are the main decomposers of soluble compounds and that the C/N ratio of the former is larger than that of the latter.

In the case of cereal straw which has a low N content, net immobilization is not very different from microbial assimilation. It is then easier to compare results reported in the literature. The calculations obtained from different experiments performed under laboratory conditions indicate a narrow range of N immobilization values : between 26 and 31 mg N g^{-1}C decomposed (Table 2). The lowest values observed by Reinertsen et al. (1984) and Bakken (1986) are likely to result from a limited availability of soil mineral N in these experiments. This suggests that the relationship between C decomposed and N immobilized can be markedly altered by the availability of N. This will be discussed in the second part of this paper.

Table 2. Maximum immobilization of soil mineral N measured during decomposition of wheat straw under laboratory conditions (values in mgN per g of added carbon)

Authors	Immobilized-N (mg N g^{-1} C)
Simon (1960)	28.5
Guiraud (1984)	30.8
Reinertsen et al. (1984)	18.0 [a]
	27.0 [b]
Bakken (1986)	15.1
Nieder and Richter (1986)	25.8
Robin (1994)	27.4

[a] Without N.
[b] With mineral N.

The remineralization kinetics, i.e. the rate of release of previously assimilated N, seems to be depending also on the type of substrate (Fig. 2B). Again one can hypothesize that the differences in remineralization rates between substrates arise from the nature of the decomposing microflora involved in the primary decomposition. For example, the rapid release of immobilized N observed during decomposition of root mucilage could be associated with a rather specific bacterial population, decaying rapidly (Mary et al., 1993). The remineralization rate decreased rapidly with time for all types of residues. The cumulative remineralization was always partial, confirming the results of many experiments done under laboratory or field conditions (Azam et al., 1985; Denys et al., 1990; Jensen, 1994b; Müller and Sundmann, 1988; Seligman et al., 1986).

Although the mechanisms of joint C and N evolution are well understood, there are still difficulties in generalizing existing data to other types of residues and in establishing stable relationships between C and N fluxes. The difficulties arise from :

— the rapid occurrence of biomass recycling process which, after a few days, can lead to the calculation of net rather than gross fluxes,

— the interaction between added N (mineral and organic) and native soil N which yields substitution effect during immobilization-mineralization turnover (Jenkinson et al., 1985), as shown by Bjarnason (1988), Fox et al. (1990), Ocio et al. (1991), Azam et al. (1993).

— the very different time-steps of investigation (day to year) in the literature and the relatively few data combining C and N evolution.

Table 3. Net immobilization of soil mineral N calculated during decomposition of wheat straw under field conditions

Straw added	N/C	Period of calculation	Net immobilization		References
(kg ha^{-1})	(mg N g^{-1} C)		(mg N g^{-1} C)	(kg N ha^{-1})	
3000	12.5	1 year	33.0	40	Powlson et al. (1985)
10000	20.8	14 days	11.5[a]	50[a]	Ocio et al. (1991)
			20.7 [b]	90[b]	
8000	8.0	1 year	11.5–13.0[a]	39–44[a]	This work
			24.3–32.0[b]	82–108[b]	

[a] Without N.
[b] With mineral N.

Quantification under field conditions

The intensity and kinetics of N mineralization and immobilization processes therefore seem to be satisfactorily characterized and understood in experiments performed under laboratory conditions. In contrast, the importance of these processes under field conditions is still poorly known. The scarcity of the available data results from two main difficulties : i) the heterogeneity of mineral N distribution in soil and ii) the interference of several other processes (nitrate movement in soil, root absorption, volatilization and denitrification, ...). In order to test the possibility of extrapolating laboratory results to field conditions, we have attempted to determine the kinetics of net N mineralization during two field experiments.

Methods

The two field experiments were carried out in Northern France in 1990–91 and 1991–92, comparing C and N transformations in a bare soil with or without wheat straw incorporated. Full details of the experiments are given by Darwis (1993). The experiments began after wheat harvest every year and were done on a loamy soil (18% clay, 70% loam; 8% sand; 4% stone, 0.11% organic N; 1.0% organic C; pH$_{H_2O}$ 8). Mature straw (C:N=125) was added at a rate of 8000 kg DM ha^{-1} and incorporated by rotavator or disk ploughing. Water and mineral N content in soil were measured over 10-12 months at 3–4 weeks intervals up to 150 cm depth. Straw decomposition was assessed by measuring residual coarse straw in the top soil (fraction > 1 mm).

Net N mineralization was calculated from the measurements of mineral N in the soil profiles, as indicated previously. It was assumed that the nitrogen losses by volatilization and denitrification were small and compensated by atmospheric deposition (which represents about 12 kg N ha^{-1} yr $^{-1}$). The rate of net mineralization was taken equal to the rate of accumulation of mineral N in the soil profile (0–150 cm), except during water drainage periods where the N leached out was added to the accumulated N. Leaching was calculated by using a model derived from Burns' model (1976) which has been found to be valid in loamy soils (Khanif et al., 1985).

The effect of possible deficiencies in soil mineral N on straw decomposition was investigated during the second experiment by adding two treatments : one receiving 180 kg NO$_3^-$-N ha^{-1} splitted in 6 times, and one receiving 330 kg NO$_3^-$-N ha^{-1} splitted in 11 times (after each soil sampling). In the latter treatment, each application of 30 kg N ha^{-1} was ^{15}N-labelled (atom% excess = 20.1%). Mineral and organic ^{15}N were measured on the 0–5, 5–10, 10–15 and 15–20 cm layers. Gross N immobilization was calculated using the enrichment isotope technique (Davidson et al., 1991; Mary and Recous, 1995).

The effect of soil temperature and water content on N mineralization was taken into account using a method derived from that of Andrén and Paustian (1987). A 'normalized' time (equivalent to Q$_{sum}$ in their work) was calculated as follows :

$$t_{normalized} = t_{real} \cdot f(T) \cdot g(\psi)$$

where $f(T)$ is a correction factor due to soil temperature, $g(\psi)$ is a reduction factor due to soil water potential. The factor $f(T)$ is a multi-exponential function of temperature. It is set at 1 at a conventional temperature of 25 °C. The rate constants (or coefficients Q$_{10}$) were those found by Recous (1995) in soils amended with maize residues and control soils. They vary according to the temperature range. The effect of soil moisture on N mineralization is described as an exponential function of soil water potential as proposed by Andrén et

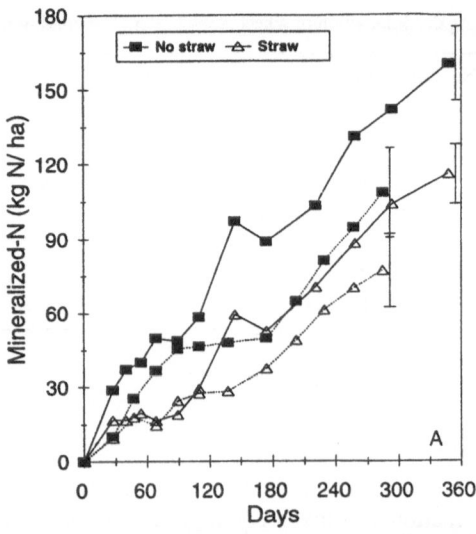 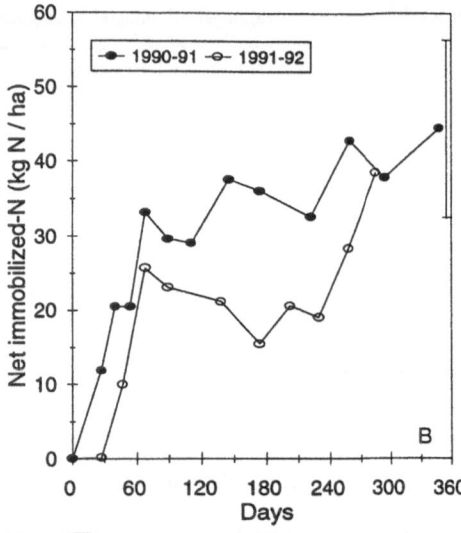

Figure 3. **A** Kinetics of net mineralization in a bare soil with ($\triangle$) and without (■) straw incorporated (8 t DM ha^{-1}), during two field experiments. First year : 07.26.1990 to 07.09.1991 (solid lines). Second year : 08.29.1991 to 06.09.1992 (dashed lines). Day 0 is the time of straw incorporation. Vertical bars indicate the mean standard deviation. **B**. Kinetics of net immobilization of mineral N due to straw decomposition during two field experiments (calculated from Figure 3A).

al. (1992). The combined effect was calculated as the product of the two terms, assuming that there were no interactions between temperature and moisture. The cumulative normalized time was calculated by summing each normalized day. This approach enables to compare field experiments differing in climatic conditions to laboratory experiments conducted under constant conditions of temperature and moisture.

Results

The calculated net mineralization for the two years and the two treatments (with or without straw) is shown in Figure 3A. Each point is the mean of six measurements (3 replicates × 2 soil tillage treatments) since the two modes of tillage (rotavator and disk ploughing) gave very similar results (the differences were never significant and randomly distributed with time). The accuracy of calculation of net mineralization mainly depends on the variability of mineral N distribution in soil and the validity of N leaching estimates. The average standard deviation on mineral N measurements was 12 and 16 kg N ha^{-1} for the years 1990–91 and 1991–92 respectively, corresponding to a similar coefficient of variation of 16%. Concerning leaching, the water balance indicated that the period of water drainage (below 150 cm) essentially took place in winter period (approximatively between days 140 and 240). The model indicated that the amount of nitrate leached out varied from 9 to 16 kg N ha^{-1} during the first year and 3 to 14 kg N ha^{-1} during the second year. Leaching was therefore

relatively small compared to the amount of N accumulated in soil, indicating that inaccuracies in the leaching model outputs would not affect the calculations of net mineralization.

In the bare soil without straw, net mineralization reached 160 ($\pm$ 21) kg N ha^{-1} after almost one year (348 days) in 1990–91 and 116 ($\pm$ 22) kg N ha^{-1} after 285 days in 1991-92. The mineralization was slower during the second year. Cumulative mineralization for both years was slightly higher than the reported values for similar pedoclimatic conditions : between 80 and 124 kg N ha^{-1} yr^{-1} (Hofman, 1988). The higher net mineralization in our experiments could be attributed to a lower immobilization due to the fact that a larger part of the readily decomposable carbon (wheat straw, stubble and chaff) had been removed.

Net mineralization was lower in the plots where wheat straw (8 t ha^{-1}) had been incorporated (Fig. 3A). The net N immobilization associated with straw decomposition was calculated by difference (Fig. 3B). The fluctuations in the kinetics were probably due to the relatively large inaccuracy inherent to the method of calculation. However, it appears that the kinetics consisted of a rapid immobilization phase after straw incorporation (in summer and autumn) followed by a slower immobilization phase in winter and spring. No apparent remineralization occurred, even one year after straw incorporation in soil. These results are consistent with our laboratory experiments. Net immobilization reached 44 ($\pm$ 16) and 39 ($\pm$ 9) kg N ha^{-1} at the end of

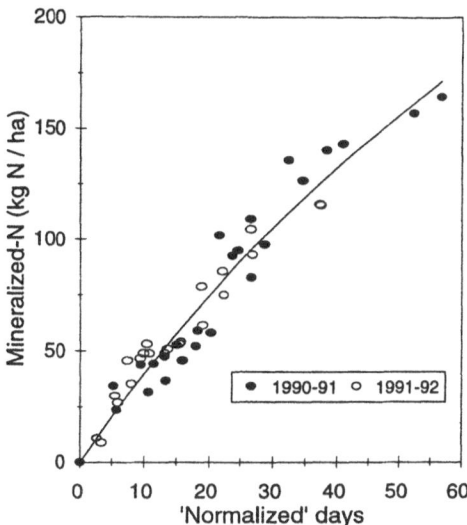

Figure 4. Evolution of net mineralization in a bare soil without straw versus 'normalized' time (equivalent days at 25 °C and optimum water potential)

the 1990–91 and 1991–92 experiments, respectively. This corresponds to 13.0 and 11.5 mg N g^{-1} added C. These values are in the lower range of previously reported estimates of net immobilization under field conditions, although those were variable and made at very different times (Table 3).

The two one-year experiments performed under our field conditions have represented 59–74 days of 'normalized' time (equivalent time at 25 °C temperature and optimal water conditions), corresponding to 16–20% of the real time. When the normalized days were substituted to real days, the differences in the kinetics of net mineralization observed between the 2 years in the plots without straw disappeared (Fig. 4). Furthermore, net mineralization could be fitted to a first order kinetics, similar to what can be found during incubation tests like those from Stanford and Smith (1972):

$$N_m = N_o \cdot \left(1 - e^{-k \cdot t}\right)$$

The best fit (calculated by non-linear regression procedure) was obtained for the following parameters : N_o=340 kg N ha^{-1} and k=0.087 week^{-1}. The fit was satisfactory : r^2=0.949 (n=50) and unbiased. However, each of these parameters could not be precisely determined since both were strongly correlated, in agreement with previous observations concerning mineralization kinetics in incubation studies (Dendooven, 1990). Indeed, a slightly less good fit was found when assuming that all organic nitrogen was 'potentially mineralizable N' : N_o=3950 kg N ha^{-1} and k=0.0062 week^{-1} (r^2=0.928). The product $k.N_o$

which can be much more precisely assessed was equal to 7.0 ($\pm$ 0.4) mg N kg^{-1}soil week^{-1} for the arable layer in these normalized conditions. Previous laboratory experiments using the incubation method of Stanford and Smith (1972) at 35°C had indicated that the $k.N_o$ values were around 9.5 mg N kg^{-1}soil week^{-1} in this type of soil (Mary and Rémy, 1979; Mary, 1988). If we consider a temperature coefficient Q_{10}=1.8 in the range 25–35°C (Campbell et al., 1981; Stanford et al., 1973) the $k.N_o$ found in the field would be 7.0 × 1.8=12.6 mg N kg^{-1} soil week^{-1} at 35 °C. Apparently, laboratory results would underestimate field mineralization by 25%. However, the unaccounted mineralization may have originated in the soil below the plough layer. The amount of organic N in the 30–60 cm layer still represents about 30% of the organic N in the 0–30 cm layer and may easily have been the source of the extra N mineralized. Therefore it seems safe to conclude that a reasonable agreement was found between field and laboratory results in plots without straw.

The kinetics of net N immobilization in the soils where straw was incorporated also were not significantly different between the two years, when the time was expressed in 'normalized' days (Fig. 5A). The kinetics also presented an exponential shape : the maximum net immobilization was calculated at 51 kg N ha^{-1}, the rate constant at 0.031 day^{-1}. The immobilization of soil mineral nitrogen was clearly correlated with C decomposition, since the residual straw-C (fraction > 1 mm) decayed exponentially with a rate constant (k = 0.027 day^{-1}) almost equal to the rate constant of N immobilization (Fig. 5B). This would indicate a ratio of immobilized N : decomposed C close to 6.5 kg N t^{-1} straw decomposed, corresponding to about 15 mg N g^{-1} C decomposed. This value is much smaller (around 50% less) than the N immobilization potential which has been found in laboratory incubations when mineral N is not a limiting factor of decomposition (Table 2). This suggests that soil mineral N (or more generally N available to soil microorganisms) could have limited the rate and the intensity of N immobilization in the field. This hypothesis is examined below.

Control of C decomposition by inorganic N availability

Characterization

The fact that nitrogen availability can be a limiting factor for microorganisms in soil responsible for decomposition process has been established for a long time

78

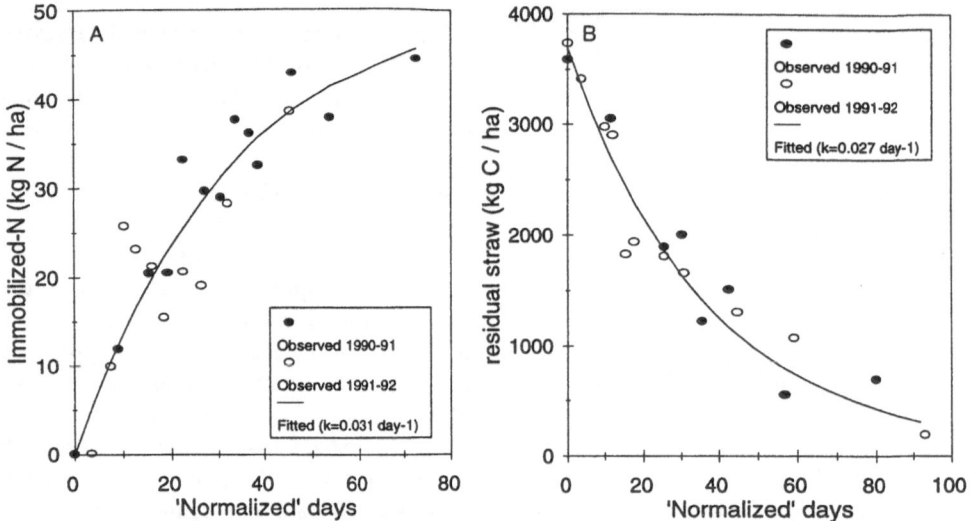

Figure 5. **A** Evolution of net immobilization of soil mineral N due to straw decomposition versus 'normalized' time. **B**. Evolution of residual straw C (fraction > 1 mm) measured in the field experiments versus 'normalized' time.

for various types of plant residues. This result is not surprising since the N:C ratio of soil microorganisms are greater than those of most plant residues (Swift et al., 1979) and therefore the external source of N can be rapidly exhausted.

The effect of N on the decomposition of carbon compounds was assessed in numerous investigations under laboratory conditions mainly by varying the initial C:N ratio of the soil+residue system by one of the following procedures: i) incubating one type of residue containing different amounts of nitrogen, ii) adding different amounts of mineral N in soil, and iii) using different types of residues with different initial N contents. The effect of N availability on C decomposition was assessed mainly by monitoring the rates of CO_2-C evolution over short or long time intervals. The results obtained have been quite variable as mentioned in Fog's review (1988), the variability depending mainly on the nature of the residue and the time scale used to evaluate the effect of added N. It has been shown also that the effect of N addition can be non-uniform during the decomposition process (Knapp et al., 1983). Cochran et al. (1988) concluded that the different C pools which co-exist in a residue create different responses to N addition : the addition of nitrogen clearly enhanced the initial rate of decomposition of the soluble fraction even though the N content of this fraction is high (Reinertsen et al., 1984; Bremer et al., 1991), but its effect can be less important, sometimes nil or even negative, in the later stages of decomposition (when the more recalcitrant com-

pounds are decomposing). In the long term, the humification process would be modified by N availability in terms of amount and quality of the humified products formed (Lueken et al.,1962; Fog, 1988). The complex interactions between N and type of residues, type of decomposers and chemical reactions during humification have been described in Fog's review (1988).

C and N relationships at low and high N levels

If the effect of N addition on straw decomposition has been clearly demonstrated, at least in the short term, little information exists on the C and N relationships during decomposition at different levels of N availability. The results obtained by Reinertsen et al. (1984), Bakken (1986) and Ocio et al. (1991) indicate that the C and N relationships can be modified at low N concentrations. Parker (1962) had already suggested that this could occur under field conditions.

The effect of N availability on the C and N relationships during decomposition of maize residue was investigated at five different levels of initial N content in soil after addition of $^{15}NH_4$ $^{15}NO_3$ (Recous et al., 1995). For the two initial lowest levels of mineral N (6 and 18 mg N g^{-1} added C), the disappearance of inorganic N in soil due to microbial immobilization caused a marked decrease of the C mineralization rate. For the three other initial levels of mineral N (35, 47 and 59 mg N g^{-1} added C), the immobilization process depleted only a part of the soil mineral N content. The kinetics of C mineralization were identical in these three treatments as were the kinetics of net N immobilization. The ratio between net immobilized N and decomposed C was calculated in each of the five treatments at times corresponding to the same degree of decomposition. For instance, when the mineralized C reached 30% of the added C, the ratio was identical for the three unlimiting N treatments, and was equal to 28.6 ± 1.3 mg N g^{-1} added C whereas it was only 19.5 and 8.8 mg N g^{-1} added C for the two lowest N levels. From these results it was concluded that there was no 'luxury' consumption of nitrogen by microorganisms at high N levels and that microbial N assimilation (flux $i+j$) per unit of decomposed C was reduced at low N availability. Several processes may explain such results: i) the modification of microbial succession (Bremer et al.,1991; Fog, 1988; Park, 1976), ii) the adaptation of the internal N content of fungi (Bremer et al., 1991; Levi and Cowling, 1966) and iii) the modification of C and N metabolism (energy allocated to growth or to maintenance). Other hypotheses concern the re-use of internal N (Bremer et al., 1991; Cowling and Merrill, 1969) and the acceleration of biomass recycling (Robin, 1994).

Control at field scale

Most of the reported experiments about the effect of N limitation on organic matter decomposition were conducted under laboratory conditions with ground residues well mixed to sieved soils. It was assumed that N and C were homogeneously distributed and equally available in each soil microsite throughout the whole period of incubation. This allowed to establish potential relationships between C and N under 'optimal' conditions, i.e. in which soluble C and N movements would not influence microbial activity. The conditions under which the availability of N would control the decomposition of C could be very different in intact soil cores and a fortiori under field conditions.

We have attempted to evaluate the effect of N availability on N immobilization in the previously described field experiment after straw incorporation (1991–92). The results are presented in Figure 6. Net immobilization was calculated similarly in the treatment N_0 without addition of mineral N (presented earlier, Fig. 3B) and in the treatment N_{180} which received 6×30 kg N ha^{-1}. Net immobilization was clearly stimulated during each 3 to 4 weeks period following each application of mineral N. There was a trend for a decrease in net immobilization in both treatments during winter (between normalized days 12 and 24). The decrease

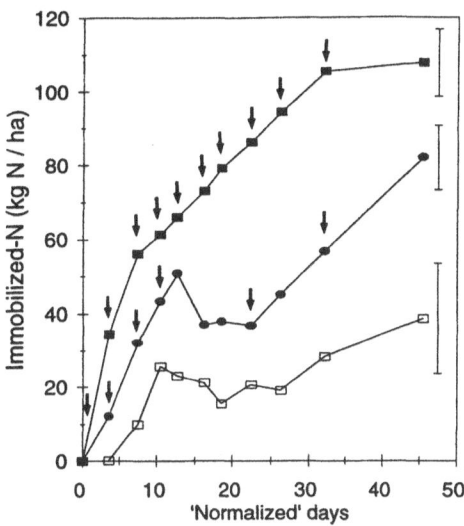

Figure 6. Evolution of N immobilization associated to straw decomposition versus 'normalized' time, with and without addition of mineral N. The arrows indicate the applications of NO_3^- (30 kg N ha^{-1}). The curves represent net immobilization in treatments N_0 and N_{180} and gross immobilization in treatment N_{330}.

appears to be in apparent contradiction with laboratory results which showed that net remineralization did not happen before most of the straw had decomposed (this condition was not achieved yet in the field). However, it can be observed that the decrease (or the stabilization) in net immobilization occurred when the amounts of soil mineral N were smallest : the average amounts of mineral N present in the straw layer during this period were 4 and 7 kg N ha^{-1} in treatments N_0 and N_{180} respectively, compared to 6 and 22 kg N ha^{-1} during the 0–12 days period and 14 and 40 kg N ha^{-1} during the 24–39 days period.

Gross immobilization was calculated in the treatment N_{330} which received the highest rate of mineral N: 11×30 kg N ha^{-1}. The kinetics of gross immobilization was close to what was found in laboratory incubations. The immobilization rate was high at the beginning of decomposition and declined regularly thereafter without reaching zero. The average amount of mineral N in the straw layer of this treatment remained always higher than 17 kg N ha^{-1}, even during the 12–24 days period. At the end of the experiment (day 285 or normalized day 46), net immobilization reached 44 ($\pm$ 21) and 82 ($\pm$ 7) kg N ha^{-1} in the N_0 and N_{180} treatments ; gross immobilization in the N_{330} treatment reached 108 ($\pm$ 11) kg N ha^{-1}. These figures represent 13.0, 24.3 and 32.0 mg N g^{-1} added C respectively, the latter value being equal to the immobilization potential found with straw in incubation studies. These results demonstrate that the availability of N in soil can control (slow down) the decomposition process and strongly reduce the N immobilization potential of plant residues (Table 3).

This effect is not properly accounted for by the present 'mechanistic'simulations models describing C and N transformations. In these models, mineral N is generally supposed to control decomposition by a switch function acting only on the decay rate of the C pool involved, when the amount of mineral N goes down a value close to 1 mg N kg^{-1} soil (Paul and Juma, 1981; Van Veen et al., 1984). An improved parameterization is necessary which will lead to reduce the importance of simulated immobilization rates.

Several authors have clearly demonstrated the interactions between N availability and parameters like residue size and residue placement. For example, the decomposition of residues left at the soil surface as a mulch can be enhanced by N addition while the rate of decomposition of the same residues well mixed in the upper soil layer may not be modified by the N treatment (Bremer et al., 1991; Parker, 1962; Schomberg et al., 1994).

A second important aspect concerns the localization of C in the soil. It has been shown that the distribution as well as the size and composition of the residues left at the soil surface and in the soil after harvesting and tillage operations are very heterogeneous. For example, the amount of straw left at the soil surface after combine harvesting may vary between 3 and 16 t DM ha^{-1} (Machet, pers. comm.). This implies that under normal agricultural conditions, the soil will present sites with large amounts of residues to be decomposed and others containing small amounts of decomposable C. In the former case the N availability will very probably limit the decomposition process while the decomposition can be optimal in the latter. In the previous example, the N immobilization potential in zones containing 16 t DM ha^{-1} would represent about 210 kg N ha^{-1}. Such an important amount of N cannot be found in the vicinity of straw environment during a one-year decomposition period. Our conclusions are contradictory with Bremer et al. (1991) who stated that microbial activity is unlikely to be limited by nitrogen under field conditions.

At the local scale, an interaction may occur with physical soil parameters (soil texture, water content). The local demand for nitrogen during decomposition could be supplied instantaneously or not, as a function of the rate of diffusion of nitrogen from the surrounding soil, even though the amount of N is high in the soil.

In the rhizosphere, this could be the more important factor, if we consider that mucilages can be shed at a much faster rate than mineral nitrogen can diffuse to the root surface.

Furthermore, the rapid transport of N from the arable to the deeper layers occurs during rainfall events, and may limit the availability of N in the upper layer where residues are incorporated, even though N mineralization remains active. This situation was clearly shown in our experiments during the autumn (Darwis, 1993).

Conclusion

Although basal mechanisms involved during organic residue decomposition have been clearly indentified, there are still difficulties to predict C and N evolution during the decomposition of plant residues, especially under field conditions. It is necessary to improve the combined description of C and N transformations and the quantification of total assimilation of N (gross immobilization, direct assimilation of organic N compounds and N recycling).

The effect of straw addition on N transformations was studied under field conditions. The intensity of N immobilization was generally smaller than under laboratory conditions, probably because N availability was limiting decomposition. The N immobilization potential associated with the decomposition of most plant residues must be compared to N availability in the soil layers where residues are incorporated. The immobilization potential is very high and probably often higher than available mineral N. Therefore plant residue decomposition could be frequently controlled by N. Under limiting N conditions, immobilization last longer but with lower rate. This could explain the relatively high immobilization potential measured with ^{15}N (Recous et al., 1992; Schimel et al., 1989).

According to this hypothesis, the distribution of plant residues and nutrients (N) would be important to account for and would explain a large part of the variability of mineral N distribution in soil. Mono-dimensional models appears to be insufficient to predict the mean amount of mineral N in soil. Studies of the effects of soil structure, plant residue localization and nitrate movement between aggregates on C and N transformations are necessary.

Acknowledgements

The investigation under field conditions was supported by the Hydro Agri France Company. We wish to thank J M Machet for his helpful collaboration and D. Angers for his useful comments, L Thouant, E Venet, D Varoteaux, O Delfosse and G Alavoine for their technical assistance.

References

Andrén O and Paustian K 1987 Barley straw decomposition in the field: a comparison of models. Ecology 68, 1190–1200.

Andrén O, Steen E and Raghai K 1992 Modelling the effects of moisture on barley straw and root decomposing in the field. Soil Biol. Biochem. 24, 727–736.

Azam F, Haider K and Malik K A 1985 Transformation of ^{14}C labelled plant components in soil in relation to immobilization and remineralization of ^{15}N fertilizer. Plant and Soil 86, 15–25.

Azam F, Simmons F W and Mulvaney R L 1993 Mineralization of N from plant residues and its interaction with native soil N. Soil Biol. Biochem. 25, 1787–1792.

Bakken L R 1986 Microbial growth, assimilation and mineralization of carbon and nitrogen during decomposition of barley straw. Sci. Rep. Agric. Univ. Norway 14, 1–14.

Barak P, Molina J A E, Hadas A and Clapp C E 1990 Mineralization of amino acids and evidence of direct assimilation of organic nitrogen. Soil Sci. Soc. Am. J. 54, 769–774.

Barraclough D 1991 The use of mean pool abundances to interpret ^{15}N tracer experiments. I- Theory. Plant and Soil 131, 89–96.

Bjarnason S 1988 Calculation of gross nitrogen immobilization and mineralization in soil. J. Soil Sci. 39, 393–406.

Bremer E, van Houtum W and van Kessel C 1991 Carbon dioxide evolution from wheat and lentil residues as affected by grinding, added nitrogen, and the absence of soil. Biol. Fertil. Soils 11, 221–227.

Burns I G 1976 Equations to predict the leaching of nitrate uniformly incorporated to a known depth or uniformly distributed throughout a soil profile. J. Agric. Sci., Camb. 86, 305-313.

Campbell C A, Myers R J K and Weier K L 1981 Potentially mineralizable nitrogen, decomposition rates and their relationship to temperature for five Queensland soils. Aust. J. Soil Res. 19, 323–332.

Cochran V L, Horton K A and Cole C V 1988 An estimation of microbial death rate and limitations of N or C during wheat straw decomposition. Soil Biol. Biochem. 20, 293-298.

Cowling E B and Merrill W 1969 Nitrogen in wood and its role in wood deterioration. Can. J. Bot. 44, 1539–1554.

Darwis S 1993 Effet des modalités de gestion de la paille de blé sur l'évolution du carbone et de l'azote au cours de sa décomposition dans le sol. Thèse de Doctorat, INA-PG, Paris, France. 195 p.

Davidson E A, Hart S C, Shanks C A and Firestone M K 1991 Measuring gross nitrogen mineralization, immobilization and nitrification by ^{15}N isotopic pool dilution in intact soil cores. J. Soil Sci. 42, 335–349.

Dendooven L 1990 Nitrogen mineralization and nitrogen cycling. Ph.D. thesis n° 191, KU Leuven, Belgium. 180 p.

Denys D, Muller J C and Mariotti A 1991 Conséquences de l'organisation de l'azote minéral d'un engrais sur la disponibilité pour la plante et sur la lixiviation. In Nitrates, Agriculture, Eau. Ed. R Calvet. pp 189–194. INRA éditions.

Fog K 1988 The effect of added nitrogen on the rate of decomposition of organic matter. Biol. Rev. 63, 433–462.

Fox R H, Myers R J K and Vallis I 1990 The nitrogen mineralization rate of legume residues in soil as influenced by their polyphenol, lignin, and nitrogen contents. Plant and Soil 129, 251–259.

Guiraud G 1984 Contribution du marquage isotopique à l'évaluation des transferts d'azote entre les compartiments organiques et minéraux dans les systèmes sol-plante. Thèse de Doctorat d'Etat, Université Pierre et Marie Curie, Paris VI, France. 335 p.

Hadas A, Sofer M, Molina J A E, Barak P and Clapp C E 1992 Assimilation of nitrogen by soil microbial population: NH_4 versus organic N. Soil Biol. Biochem. 24, 137–143.

Hart S C , Nason G E, Myrold D D and Perry D A 1994 Dynamics of gross nitrogen transformations in an old-growth forest: the carbon connection. Ecology 75, 880–891.

Hofman G 1988 Nitrogen supply from mineralization of organic matter. Biol. Wastes 26, 315-324.

Houot S and Chaussod R 1991 Rôle de la biomasse microbienne sur la minéralisation-réorganisation de l'azote dans les sols. Compte-rendu Ministère de l'Environnement, 82 p.

Jansson S L and Persson J 1982 Mineralization and immobilization of soil nitrogen. In Nitrogen in Agricultural Soils. Ed. F J Stevenson. pp 229–252. ASA Madison, WI, USA.

Jawson M D and Elliott L F 1986 Carbon and nitrogen transformations during wheat straw and root decomposition. Soil Biol. Biochem. 18, 15–22.

Jenkinson D S, Fox R L and Rayner J H 1985 Interactions between fertilizer nitrogen and soil nitrogen - the so-called priming effect. J. Soil Sci. 36, 425–444.

Jensen E S 1994a Dynamics of mature pea residue nitrogen turnover in unplanted soil under field conditions. Soil Biol. Biochem. 26, 455–464.

Jensen E S 1994b Availability of nitrogen in ^{15}N-labelled mature pea residues to subsequent crops in the field. Soil Biol. Biochem. 26, 465–472.

Jensen E S 1994c Mineralization-immobilization of nitrogen in soil amended with low C:N ratio plant residues with different particle sizes. Soil Biol. Biochem. 26, 519–521.

Kirchmann H and Bergqvist R 1989 Carbon and nitrogen mineralization of white clover plants (Trifolium repens) of different age during aerobic incubation with soil. Z. Pflänzenernähr. Bodenkd. 152, 283–288.

Kirkham D and Bartholomew W V 1954 Equations for following nutrient transformations in soil, utilizing tracer data. Soil Sci. Soc. Am. Proc. 18, 33–34.

Khanif Y M, Van Cleemput O and Baert L 1984 Evaluation of the Burns'model for nitrate movement in wet sandy soils. J. Soil Sci. 35, 511–518.

Knapp E B, Elliott L F and Campbell G S 1983 Carbon, nitrogen and biomass interrelationships during the decomposition of wheat straw: a mechanistic simulation model. Soil Biol. Biochem. 15, 455–461.

Levi M P and Cowling E B 1969 Role of nitrogen in wood deterioration. VII physiological adaptation of wood-destroying and other fungi to substrate deficient in nitrogen. Phytopathology 59, 460–468.

Lueken H L, Hutcheon W L and Paul E A 1962 The influence of the nitrogen on the decomposition of crop residues in the soil. Can. J. Soil Sci. 42, 276–287.

Mary B 1988 Rôle de la biomasse microbienne du sol dans la disponibilité d'azote minéral en conditions de plein champ. Compte-rendu Ministère de l'Environnement. 95 p.

82

Mary B, Fresneau C, Morel J L and Mariotti A 1993 C and N cycling during decomposition of root mucilage, roots and glucose in soil. Soil Biol. Biochem. 25, 1005–1014.

Mary B and Recous S 1995 Modélisation des flux d'azote dans les sols. Mesure par traçage isotopique ^{15}N. *In* Utilisation des isotopes stables pour l'étude du fonctionnement des plantes. Colloques l'INRA 70, 277–297.

Mary B and Rémy J C 1979 Essai d'appréciation de la capacité de minéralisation de l'azote des sols de grande culture. Ann. Agron. 30, 513–527.

Müller M M and Sundmann V 1988 The fate of nitrogen ^{15}N released from different plant materials during decomposition under field conditions. Plant and Soil 105, 133–139.

Myrold D D and Tiedje J M 1986 Simultaneous estimation of several nitrogen cycle rates using ^{15}N: theory and application. Soil Biol. Biochem. 18, 559–568.

Nieder R and Richter J 1986 Einfluß der Strohdüngung auf den Verlauf der N-Mineralization eines Löß Parabraunde-Ap Horizontes in Saülen Brut versuch. Z. Planzenernähr. Bodenkd. 149, 202–210.

Nishio T, Kanamori T and Fujimoto T 1985 Nitrogen transformations in an aerobic soil as determined bu a ^{15}NH$_4^+$ dilution technique. Soil Biol. Biochem. 17, 149–154.

Ocio J A, Martinez J and Brookes P C 1991 Contribution of straw-derived N to total microbial biomass N following incorporation of cereal straw to soil. Soil Biol. Biochem. 23, 655-659.

Park D 1976 Carbon and nitrogen levels as factors influencing fungal decomposers. *In* The Role of Terrestrial and Aquatic Organisms in Decomposition Processes. Eds. Anderson et al. MacFayden. pp 41–59. Blackwell Science Pub., Oxford, UK.

Parker D T 1962 Decomposition in the field of buried and surface-applied cornstalk residue. Soil Sci. Soc. Am. Proc. 26, 559–562.

Paul E A and Juma N G 1981 Mineralization and immobilization of soil nitrogen by microorganisms. Ecol. Bull. 33, 179–194.

Paustian K and Schnürer J 1987 Fungal growth response to carbon and nitrogen limitation : application of a model to laboratory and field data. Soil Biol. Biochem. 19, 621–629.

Payne J W 1980 Microorganisms and Nitrogen Sources. J Wiley and Sons, New York, USA.

Powlson D S, Jenkinson D S, Pruden G and Johnston A E 1985 The effect of straw incorporation on the uptake of nitrogen by winter wheat. J. Sci. Food Agric. 36, 26–30.

Recous S 1995 Effet de la température sur la minéralisation d'un résidu végétal (maïs) et de la matière organique d'un sol. *In* Ecosystèmes naturels et cultivés et changements globaux. Les Dossiers de l'Environnement. INRA, editions 8, 81–85.

Recous S, Machet J M and Mary B 1992 Partitioning of fertilizer-N between soil microflora and crop : comparison of ammonium and nitrate applications. Plant and Soil 144, 101–111.

Recous S, Robin D, Darwis D and Mary B 1995 Soil inorganic N availability : effect on maize residue decomposition. Soil Biol. Biochem. 27, 1529–1538.

Reinertsen S A, Elliott L F, Cochran V L and Campbell G S 1984 The role of available C and N in determining the rate of wheat straw decomposition. Soil Biol. Biochem. 16, 459-464.

Robin D 1994 Effet de la disponibilité de l'azote sur les flux bruts de C and N au cours de la décomposition des résidus végétaux dans le sol. Thèse INA-PG, Paris, France. 201 p.

Schimel J P, Jackson L E and Firestone M K 1989 Spatial and temporal effects on plant microbial competition for inorganic nitrogen in a California annual grassland. Soil Biol. Biochem. 21, 1059–1066.

Schomberg H H, Steiner J L and Unger P W 1994 Decomposition and nitrogen dynamics of crop residues : residue quality and water effects. Soil Sci. Soc. Am. J. 58, 372–381.

Simon G 1960 L'enfouissement des pailles dans le sol. - Etude générale et répercussions sur la microflore du sol. Ann. Agron. 11, 5–54.

Seligman N G, Feigenbaum S, Feinerman D and Benjamin R W 1986 Uptake of nitrogen from high C-to-N ratio, ^{15}N-labeled organic residues by spring wheat grown under semi-arid conditions. Soil Biol. Biochem. 18, 303–307.

Sorensen L H 1981 Carbon-nitrogen relationships during the humification of cellulose in soils containing different amounts of clay. Soil Biol. Biochem. 13, 313–321.

Stanford G, Frere M H and Schwaninger D H 1973 Temperature coefficient of soil nitrogen mineralization. Soil Sci. 115, 321–323.

Stanford G and Smith S J 1972 Nitrogen mineralization potentials of soils. Soil Sci. Soc. Am. Proc. 36, 465–472.

Swift M J, Heal O W and Anderson J M 1979 Decomposition in Terrestrial Ecosystems. Blackwell Scientific Publications, Oxford, UK.

Van Veen J A, Ladd J N and Amato M 1984 Turnover of carbon and nitrogen through the microbial biomass in a sandy loam and a clay soil incubated with ^{14}C-glucose and ^{15}N-(NH$_4$)$_2$SO$_4$ under different moisture regimes. Soil Biol. Biochem. 17, 747–756.

Wessel W W and Tietema A 1992 Calculating gross N transformation rates of ^{15}N pool dilution experiments with acid forest litter : analytical and numerical approaches. Soil Biol. Biochem. 24, 331–342.

Section editor: R Merckx

Plant and Soil **181**: 83–93, 1996.
© 1996 *Kluwer Academic Publishers.*

Turnover of organic nitrogen in soils and its availability to crops

K. Mengel
Institute of Plant Nutrition of the Justus Liebig University, Südenlage, 35390, Giessen, Germany *

Key words: amino sugars, fixed NH_4^+, immobilization, mineralization, nitrogen, proteins

Abstract

Major known fractions of soil nitrogen are amino nitrogen (proteins, peptides), polymers of amino sugars, and NH_4^+ fixed in interlayers of 2:1 minerals. Only a small percentage of the total soil organic N is easily mineralizable and contributes to the pool of mineral soil N. Predominant sources of mineralization are amino-N and polymers of amino sugars present in the soil microbial biomass. Influx into this pool occurs with the application of organic matter (green manure, straw), organic carbon released by plant roots, N_2 assimilation by leguminous species and inorganic nitrogen. Microbial metabolization of green manure proteins results in a partial mineralization of the applied organic N, microbial metabolization of straw in the assimilation (immobilization) of inorganic nitrogen.

Microbial biomass is characterized by a narrow C/N ratio (proteins, peptidoglycans, polymers of amino sugars). Its metabolization therefore is associated with a partial mineralization of the attacked organic nitrogen compounds. Nitrogen mineralization consists of a sequence of enzymatic processes for which the living microbial biomass provides the enzymes and the dead microbial biomass the substrate.

Introduction

Today the industrial fixation of N_2 amounts to more than 60% of the biological N_2 fixation (Isermann, 1987). Nearly all of this technically fixed nitrogen is finally used as fertilizer nitrogen and arrives to a large extent in the environment (Mengel, 1992). In order to reduce the amount of nitrogen which may be a hazard to the environment it is pertinent to adjust the nitrogen fertilizer rates to the level of available nitrogen in the soil. Nitrate and ammonium nitrogen are directly available to crops but substantial amounts of soil organic nitrogen may be mineralized during the growth period and thus contribute to crop nutrition (Thicke et al., 1993). The assessment of this mineralizable soil nitrogen in advance, e.g. at the beginning of the growth phase meets with difficulties. The objective of this paper is to present a concept on the turnover of organic nitrogen in soils with the aim of obtaining a better understanding of the net mineralization of organic nitrogen. This should be helpful for the elaboration of practical and reliable soil tests for available soil nitrogen.

Organic nitrogen fractions of soils

Catroux and Schnitzer (1987) analyzed a hydromorphic, humic soil for its nitrogen fractions. Their most important results are shown in Table 1. About 30% of the total soil N present in the upper soil layer was amino-N, mainly peptides, including proteins. The amount of amino sugars was relatively low, whereas the ammonium N represented a substantial proportion. The order of magnitude of the three nitrogen fractions listed in Table 1 agrees well with that found in arable soils (Schnitzer and Spiteller, 1986). About 50% of the total soil nitrogen could not be identified. A remarkable proportion of this non identified nitrogen is heterocyclic N present in purine and pyrimidine derivates, in indols, chinolin, isochinolin, aminobenzofuran, piperidine and pyrrolidine compounds. Presumably, most of this heterocyclic N is integrated humic and fulvic acids (Schnitzer and Spiteller, 1986). Heterocyclic compounds are introduced into soils predominantly by plant matter in the form of pyrrole rings from chlorophylls and cytochromes and the purine and pyrimidine bases of the nucleic acids. In the soil, these rings may undergo various transformations, but the ring seems to be very resistant and consequently, the same is true

---
* FAX No.: +4964172890
Plant and Soil is the original source of publication of this article. It is recommended that this article is cited as: *Plant and Soil* **181**: 83–93, 1996.

Table 1. Proportions of total nitrogen amounts in the upper layer of a humic loamy clay soil (AquoII). Data from Catroux and Schnitzer (1987)

	% of total soil N	kg N ha^{-1} 0–10 cm layer
Total N	100.0	5100
Amino N	30.0	1530
Amino sugars	4.0	204
NH_4^+-N	13.6	694
Non-identifiable	52.4	

Table 2. Effect of long term fertilizer application - organic and inorganic - on nitrogen fractions in the soil - Mollisol derived from loess (Garz and Chaanin, 1990)

Treatment	Total N (g kg^{-1})	Mineralizable N[+] (mg kg^{-1})	Stable N (g kg^{-1})
N$_0$ fertil.	1.05	58.8	0.95
NPK	1.07	59.5	0.93
NPK + FYM	1.17	70.3	1.06
NPK + Straw	1.13	72.3	1.00

[+] According to the method of Stanford and Smith (1972).

for its nitrogen atoms. Hence, heterocyclic nitrogen accumulates in soils.

As was recently shown by Schneider (1995) the most important sources for mineralizable nitrogen are peptides and amino sugars. This is in line with the observation that the concentration of amino N and amino sugars in soils decreases with their age (Schnitzer and Spiteller, 1986). A relatively high amount of amino N in soils is incorporated into humic acids of which some 50% may consist of peptides. The N concentration of the dry matter of humic acids is in the range of 3 to 6%. A substantial proportion of this N is present as heterocyclic N (Schulten and Schnitzer, 1992). Amino sugars on the other hand are hardly found in humic acids (Catroux and Schnitzer, 1987). It is assumed that the potentially available nitrogen in soils decreases, the more nitrogen is being extracted by plants provided that no nitrogen is recycled (Frederick and Klein, 1994). It is well known that cropping virgin prairie soils leads to a large decomposition of organic nitrogen, particularly in the first years of cultivation. The remaining organic compounds are much more resistant to decomposition (Schulten et al., 1994). Regular incorporation of N containing organic matter into soils as well as the application of mineral fertilizer increase the concentration of soil nitrogen as has been shown in field trials (Amberger and Schweiger, 1971; Jenkinson, 1991; Odell et al., 1984). Garz and Chaanin (1990) found in long term experiments in Germany, of which some important data are shown in Table 2, that about 90% of the organic nitrogen was very resistant to mineralization and only about 5 to 6% of the total soil N was accessible to mineralization during a long lasting incubation according to the technique of Stanford and Smith (1972). There is clear evidence that straw in combination with mineral fertilizer significantly increased the fraction of mineralizable soil nitrogen.

The quantity of ammonium found in the soil by Catroux and Schnitzer (1987) amounted to about 700 kg N ha^{-1} for a soil depth of 10 cm. This form of N is mainly NH_4^+ selectively bound by 2:1 clay minerals. In the given case of Catroux and Schnitzer (1987) NH_4^+ is fixed mainly by vermiculite since this was the dominant clay mineral. The amount of selectively bound (= fixed) ammonium differs considerably from soil to soil. In highly weathered soils with virtually no 2:1 clay minerals its concentration is practically nil. Nette and Resch (1992) found concentrations from 100 to 200 mg kg^{-1} NH_4^+-N in typical arable soils (alluvial soils, luvisols derived from loess). The concentration of selectively bound ammonium found in eight soils, analysed by these authors, ranged from 10 to 700 mg N kg^{-1} soil (Nette and Resch, 1992). The concentration depends mainly on soil type and type of clay minerals (Scherer, 1993). In contrast to organic soil nitrogen which mainly is present in the surface layer, selectively bound ammonium is present also in the deeper soil layers and may amount to about 100 kg N ha^{-1} in Luvisols derived from loess in the rooting zone of soils (Mengel and Scherer, 1981; Van Praag et al., 1980). According to Dressler and Mengel (1985) selectively bound ammonium of deeper soil layers may in particular contribute to crop nutrition.

Nitrogen input into soils and its turnover

Input of organic nitrogen

Arable soils generally receive nitrogen within short periods in heavy doses either in organic or inorganic forms. The fate of organic nitrogen incorporated into soils very much depends on the type of organic matter, particularly on its N concentration. This problem will be considered in the following with two examples, young leaves and straw, as shown in Fig. 1. Young leaves, e.g. sugar beet tops, have a N concentration in the range of 20 to 30 mg N g^{-1} dry matter. Around 80% of this N is protein N, the rest is mainly nucleic acid and inorganic N. Cell walls of young leaf tissue are poorly lignified and therefore can easily be decomposed by microbes. Generally such plant matter is at first attacked by fungi which themselves are later attacked by bacteria. Decomposing microbes feed from the organic matter from which they draw two principal components: the nitrogen required for the synthesis of their own proteins and further N containing organic molecules as well as organic carbon required as C skeletons and as energy source. In protein-rich vegetative plant matter the available organic carbon is generally a limiting factor and for this reason also amino acids will be used as an energy source. Proteins in the young leaves will be easily hydrolyzed and the resulting amino acids will directly serve the decomposer microbes for the synthesis of their own organic N but partially will also function as energy source. The latter proportion of amino acids will be deaminated while the resulting oxo acids are used as energy source and the resulting ammonium is the substrate for the nitrification process. Deamination (= ammonification) is carried out by numerous heterotrophic microbes (Beck, 1983) and is the crucial process for net mineralization. It depends on the available nitrogen and metabolizable organic carbon at micro sites where the microbial attack occurs. From this consideration it is obvious that the C/N ratio of total organic soil matter is only a very rough indicator for the net mineralization conditions in soils.

Ammonium produced by deamination is associated with the production of living biomass as can be easily derived from the concept considered above. It is for this reason that the nitrogen contained even in nitrogen-rich organic matter after incorporation into the soil can not be transformed 100% into inorganic nitrogen. Frequently less than half of the organic N applied is transformed into inorganic forms (Azam et al., 1993; Ke et al., 1990; Paul, 1994; Rees et al., 1993). From Table 3 it can be seen that in the case of incorporated pea residues about 35 to 45% of total N applied was mineralized in the following growth period under winter wheat. In the case of sugar beet tops, incorporated in the soil, only about 20% of the total organic N applied was mineralized during the total growth period of winter wheat (Paul, 1994). Only late in the summer after the harvest of the wheat the plots with incorporated sugar beet tops showed a remarkable nitrification.

Table 3. N quantities applied with organic residues and related mineralized N (Paul, 1994)

Residue and year	Applied	Mineralized	% of
	(kg N ha^{-1})		appl. N
Peas 1991/92	122	43.2	35
Peas 1992/3	126	56.9	45
Beet tops 1991/92	182	46.7	26
Beet tops 1992/93	232	31.8	14

Since in this field trial no substantial nitrate leaching was observed (Paul, 1994) it is assumed that the major part of the organic nitrogen present in the beet tops was assimilated by microbes and that this assimilation process was favoured by sugars still present in the beet tops. This interpretation is in line with research data of Ziegler et al. (1992) who found after sugar beet incorporation a high concentration of organic nitrogen extractable by electroultrafiltration (EUF).

Nitrogen is taken up by decomposer microbes in the form of amino acids (Barak et al., 1990) and also as NH$_3$. The nitrogen so assimilated (= immobilized) flows into the pool of living biomass (Fig. 1). Microbes attacking straw incorporated into the soil find a surplus of organic carbon and only low quantities of organic nitrogen which in straw of cereals amounts to only about 2% of the dry matter. It is protected by thick lignified cell walls. In analogy to animal nutrition one may assume that the 'digestibility' of this organic nitrogen is the poorer the higher the proportion of lignified cell wall material in the organic matter attacked. Under favourable conditions 80 to 90% of the straw is decomposed within one year (Thomsen, 1993). Obviously cellulose and hemicelluloses of straw can be easily used by microbes as energy source and as carbon skeletons for the synthesis of their cells. Numerous fungi and bacteria possess inducible enzymes (cellulases) which are synthesized as soon as cellulose (e.g. straw) is incorporated into soils (Quastel, 1965). In this situation the limiting factor for microbial growth is nitrogen which may be used in organic and inorganic form, including nitrate since many microorganisms can metabolize nitrate (Guerrero et al., 1981). Nitrate is very mobile in soils it can be easily transported by diffusion and mass flow to the microsites of straw attacking microbes. The frequent observation that straw incorporation into soils reduces soil nitrate substantially confirms this concept (Christensen, 1986; Freytag and Rausch, 1981; Göck and Ottow, 1988; Savant and

86

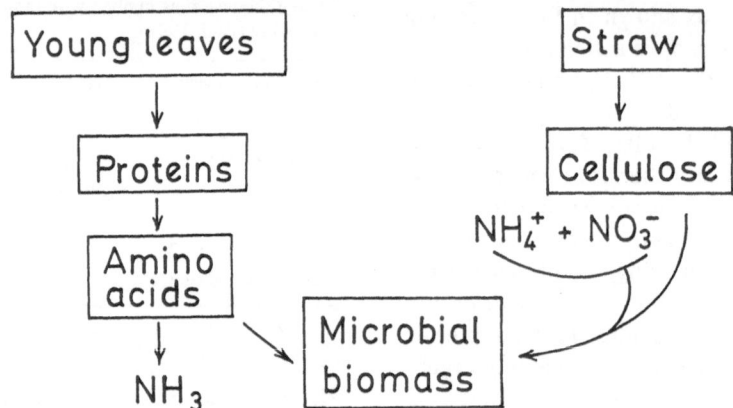

Figure 1. Mineralization of organic matter oncorporated into soils. Nitrogen rich matter (e.g. young leaves) produces inorganic nitrogen, matter poor in nitrogen (e.g. straw) consumes inorganic nitrogen. Both processes feed the microbial pool.

DeDatta, 1982; Schmeer and Mengel, 1984). If the availability of mineral N in the soil is very high it even may promote the mineralization of the organic nitrogen in the straw as found in incubation experiments carried out by Wojcik-Woytkowiak (1979) with ^{15}N labelled straw and by Schnier (1994) in field trials with flooded rice. In this process, known as the priming effect, the N source of microbial growth is mainly inorganic N and the organic N present is not needed and hence may be deaminated. The priming effect occurs also under field conditions (Gutser and Teicher, 1976). Assimilation of nitrogen by straw attacking microbes leads to the formation of biomass (Beck, 1983). Hence, also the application of straw finally feeds the pool of the living biomass (see Fig. 1). In an analogous way cellulose additions to soils may foster considerably the formation of biomass associated with the assimilation of inorganic nitrogen (Göck and Ottow, 1988; Olfs and Werner, 1989). Mineral fertilizer application together with straw may thus significantly increase the concentration of organic nitrogen in soils as shown by Garz and Chaanin (1990). Also, regular farmyard manure applications resulted in remarkable increases of soil N as was found in long term field experiments (Amberger and Schweiger, 1971; Garz and Chaanin, 1990; Jenkinson and Rayner, 1977). In the Rothamsted long-term field trials regular farmyard manure applications enriched the soil in available nitrogen so that crops did not respond to mineral fertilizers (Jenkinson, 1991).

Input of inorganic nitrogen

Nitrogen fertilizer additions imply a tremendous interference in the soil nitrogen turnover. Application rates

of 50 to 200 kg N ha^{-1} are in the order of the microbial biomass N content in soils. The fate of the nitrogen fertilizer depends to a large extent on the intensity of N uptake of crops and also on the N addition rate. Basically, higher plants and microbes may compete for fertilizer nitrogen which competition may even lead to a reduced plant growth (Azam et al., 1990). In vigorously growing crop stands and when the N fertilizer is well placed, chances for the N uptake by higher plants are high while assimilation of fertilizer N by microbes is limited and vice versa (Schnier et al., 1988). There are numerous examples from studies with labelled fertilizers that 10 to 20% of the added amounts are assimilated by soil microbes and thus flow into the organic soil nitrogen pool (Mahli et al., 1994; Riga et al., 1980; Schnier et al., 1988). The following crops benefit from the nitrogen of this pool. Labelled N taken up from this pool by the following crops, however, represents only a small percentage of the^{15}N labelled fertilizer applied as shown by several authors (Hart et al., 1993; Riga et al., 1980; Teske and Matzel, 1976).

Microbial assimilation depends on microbial activity and particularly on the availability of organic carbon. There is evidence that organic carbon released by plant roots feeds this assimilation process and thus promotes the propagation of microorganisms in the rhizosphere (Helal and Sauerbeck, 1989; Jensen and Sorensen, 1994). Joergensen et al. (1994) found an increase of biomass N from May until autumn which finding may be attributed to the high nitrogen fertilizer rate applied (250 kg N ha^{-1}) in combination with the organic carbon released by roots (Fig. 2). Presumably the release of photosynthates by roots has a direct impact on microbial assimilation of nitrate, ammoni-

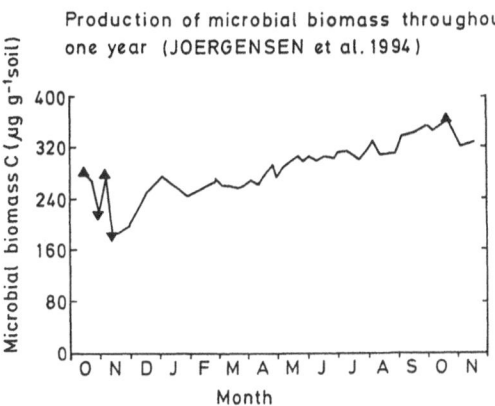

Production of microbial biomass throughout one year (JOERGENSEN et al. 1994)

Figure 2. Increase of microbial biomass under winter wheat well supplied with fertilized nitrogen (modified after Joergensen et al., 1994).

Table 4. Correlation coefficients for soil characteristics vs N uptake of rape and net N mineralization, n = 12, Sandy soils, pot experiments (Appel and Mengel, 1990)

Soil charact eristics	N uptake	N mineralization
Total C	−0.35*	−0.29*
Total N	−0.31*	−0.37*
C:N ratio	−0.72* *	−0.68* *
Sand proportion	−0.74* *	−0.75* *
Silt proportion	0.69* *	0.70* *
pH	0.73* *	0.73* *
CaCl$_2$ extr. -N	0.76* *	0.79* *
EUF extr. -N	0.76* *	0.75* *

um and N$_2$ (Neyra and Döbereiner, 1977). Appel and Mengel (1991) found in field trials on sandy soils an apparent recovery of fertilizer N of only about 44% although the N fertilizer rates were low and no major nitrate leaching was observed. These authors suggest that particularly on sandy soils relatively high proportions of fertilizer N may be assimilated by soil microbes supposedly due to the high mass flow rates of nitrate to plant roots. This assumption is supported by research data of Thies et al. (1977) and Peschke et al. (1984). The fertilizer N thus immobilized in early summer may be mineralized in autumn and may later be leached by winter rains and thus get lost from the system. This may be one reason why sandy soils are generally poor in organic nitrogen.

Mineralization and mineralization conditions

Mineralization of organic N in soils is basically a sequence of enzymatic processes, most important enzyme types involved are proteases and deaminases for the substrate peptide, and O-glycosidases, deaminases and acetyl hydrolases for the polymers of various amino sugars. In the case of peptides - including polypeptides and proteins - the interesting situation is that substrate and enzyme may be of the same biochemical nature, i.e. proteins. Presumably the enzymes originate from the living microbial biomass while the substrate is to a large extent dead microbial biomass (Jenkinson and Ladd, 1981). Enzyme activity is much dependent on pH and this is also true for enzymatic processes in soil. Until now not much is known about

the pH in microsites of soils whereas the bulk soil pH is only a very rough measure. Nevertheless, it is known that in acid soils the mineralization of organic nitrogen is retarded or even blocked (Kuntze and Bartels, 1979). In Table 4 some soil characteristics are shown having a significant impact on net mineralization studied in pot experiments with sandy soils (Appel and Mengel, 1990). One highly significant factor was soil pH. Also soil temperature and soil moisture influence N mineralization (Honeycutt et al., 1991; Tabatabai and Al-Khafaji, 1980). Kladiviko and Keeney (1987) found a six to seven fold higher N mineralization rate at 35 °C than at 10 °C. These results found in laboratory experiments overemphasize the processes occurring under field conditions at least if no extreme weather conditions prevail. According to field trials of Lochmann et al. (1989) nitrogen mineralization is not much influenced by normal oscillations of soil temperature and soil moisture during the growth period. Throughout the year highest N mineralization rates were found in summer and lowest in winter (Weller, 1983). Weller (1983) also found that dry periods in summer cause a drastic reduction in N mineralization. Lochmann et al. (1989) reported that the net N immobilization (microbial assimilation) was highest in spring while in May/June net N mineralization dominated.

There is evidence that soil texture has a strong impact on N mineralization. Proteins as a major substrate for N mineralization and proteases as starting enzymes of the mineralization process are both prone to binding to soil particles such as humus and clay minerals (Loll and Bollag, 1983). Binding to a surface affects the enzymatic process because the probability of enzyme-protein interactions is reduced and the adsorbed protein or peptide may be protected against enzymatic attack. Humic acids and fulvic acids may promote proteolysis but may also have an adverse effect while tannins always hamper or even block protein hydrolysis (Loll and Bollag, 1983). From this follows that N mineralization in sandy soils is more rapid than in soils with a higher clay content. This statement is in line with recent experimental data of Ke et al. (1990) who studied the N turnover of rape leaves in a sandy and in a loamy soil. Some relevant results of this pot experiment are shown in Fig. 3. It is evident that the ammonium peak in the sandy soil was higher and occurred earlier than the ammonium peak in the loam soil. (Fig. 3a). This difference is reflected by the subsequent nitrification; the amount of nitrate produced being considerably higher in the sandy than in the loamy soil (Fig. 3b). From this finding one may conclude that soils with a higher clay content are able to store organic nitrogen in the form of adsorbed polypeptides and thus may add to the potential of mineralizable soil N. Analogous results were reported by Chichester and Smith (1983) who found that from the labelled N applied the lowest amount remained in the sandy soil. The adsorption of polypeptides to clay minerals is predominantly due to electrostatic forces but also hydrogen bonds may link the polypeptide with the clay mineral (Loll and Bollag, 1983). Both binding forces are not very stable and are affected by changing soil conditions, e.g. soil moisture, temperature and pH which means that the protected polypeptide may later become susceptible to enzymatic break down. This behaviour is of agronomic relevance. Proteins adsorbed during late summer may be protected against mineralization and hence its nitrogen may not be leached by winter rainfall. In the following spring and summer these proteins may become available due to changing soil and weather conditions. In sandy soils in which the proteins of decomposing biomass are hardly adsorbed, mineralization will occur in late summer followed by nitrate leaching in winter.

The adsorption of proteins and/or polypeptides obviously leads to an enrichment of mineralizable nitrogen in soils as is shown in Figure 4 from the work of Hütsch and Mengel (1993). The figure shows the repeated extraction of organic N by electroultrafiltration (cumulated N) in form of curves. It is obvious that in the sandy soil the curves are much flatter than in the loam soil and it can also be seen that in the no tillage treatments (NT) the amounts of 'N$_{org}$' are higher than in the ploughed treatments (P). The assumption that in loamy soils the potential of mineralizable organic nitrogen is higher than in sandy soils is supported by

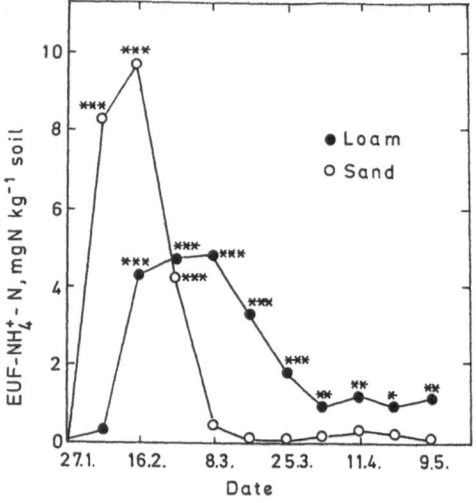

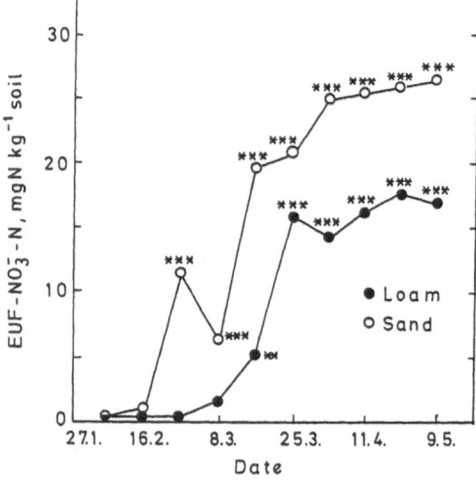

Figure 3. EUF (electroultrafiltration) extractable (upper) ammonium and (lower) nitrate after the incorporation of rape leaves in a sand and in a loam soil (Ke et al., 1990) *, **, *** significant from the treatment without rape leaves at the 5, 1 and 0.1 level, respectively.

experimental work of Matzel and Lippold (1990) and is also emphasized by Wehrmann and Scharpf (1986).

The microbial biomass contains an appreciable amount of organic N present in cell walls as polymers of amino sugars and acetylated amino sugars. Also, these N compounds are prone to mineralization and there is evidence that the N of amino sugars in soils in particular is quickly mineralized (Schnier et al., 1987). Fungi may contain cellulose or amino sugar polymers in their cell walls. Fungal cell walls with amino sugars as structural unit are at first hydrolyzed and then deaminated, the latter process releasing NH_3 than forming the starting substrate for nitrification. In Fig. 5 the putative

89

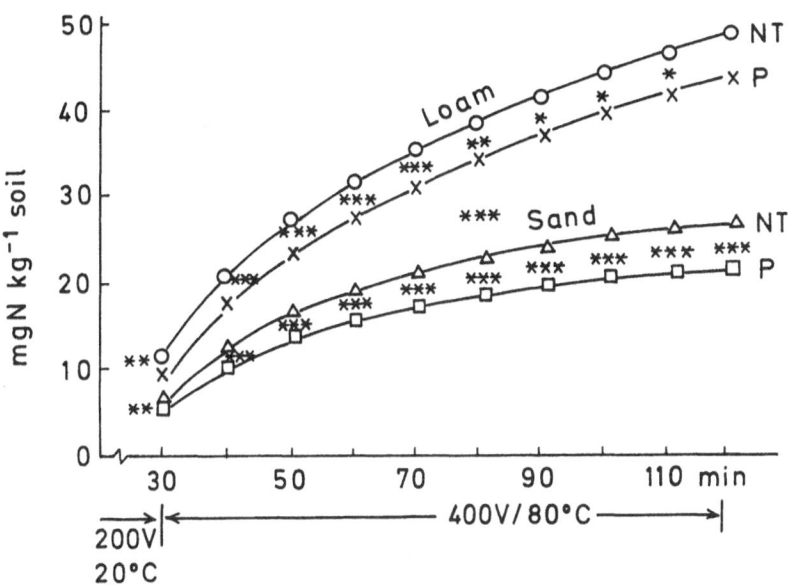

Cumulative EUF-Norg release of a sand and
loam soil (HÜTSCH and MENGEL 1993)

Figure 4. EUF-extractable organic nitrogen from a sand and a loam soil, shown as cumulative curves. NT = non-tillage treatment, P = plough treatment. Asterix indicate the significance level between the NT and P treatment (Hütsch and Mengel, 1993).

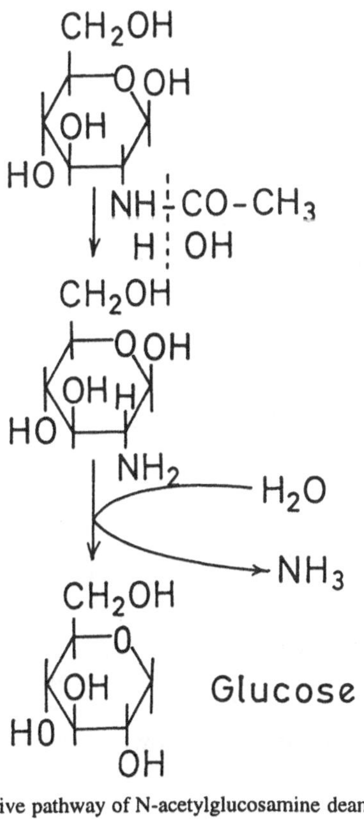

Glucose

Figure 5. Putative pathway of N-acetylglucosamine deamination.

pathway for the deamination of N-acetyl glucosamine is shown. The first step is a hydrolysis, giving acetic acid which can be well used by numerous species of bacteria and in the next step NH_3 is produced. The end product of the sequence is glucose which is an excellent energy source.

The C/N ratio of the N-acetyl glucosamine is about 7/1 which is narrow enough to give a net production of NH_3. Hence, such cell walls must be considered as potential sources for NH_3. This is not true for those fungi with cellulose as the major component in their cell walls. Cellulose will only serve as an energy source and when it is attacked by other organisms additional inorganic N may be required for the growth of the attacking organism. According to Jenkinson and Ladd (1981) fungal hyphae possess a C/N ratio of 10 to 12 and therefore may not represent a major potential for net N mineralization in all cases.

In bacterial cell walls muropeptide is a major building block of which the structure is shown below.

Muropeptide

The frame work of the muropeptide structure (peptidoglycans) consists of polysaccharide- like strands with the N-acetyl glucosamine as monomer (Fig. 6). These

[103]

strands are connected to numerous peptide strands with glycine as the dominating amino acid which has a C/N ratio of 1.7. Interestingly these peptides also contain D-alanine and D-glutamate. The C/N ratio of muropeptides is very narrow and hence they are likely potential sources for the production of NH_3. From this follows that if microbial biomass is decomposed there is a great chance that NH_3 is released and the nitrification process started. This conclusion is in line with the statement of Jenkinson and Ladd (1981) emphasizing that the microbial biomasss in soils is very labile. Besides the cell wall N also the protein N of fungi and bacteria is a potential source for NH_3. Presumably after the decomposition of the cell walls the proteins of cell membranes and cytosolic proteins are exposed to attacking enzymes. According to Schnürer and Rosswal (1987) fungi have a higher N mineralization rate than bacteria.

An important question related to nitrogen mineralization is whether it is mainly the dead biomass which is prone to mineralization. It is important to know which soil conditions may increase death of soil microbes and then promote the process of decomposition and thus also of nitrification. From Jenkinson's fumigation experiments one may conclude that predominantly the dead biomass is prone to biological decomposition (Jenkinson and Ladd, 1981). This assumption is in line with results of Azam et al. (1986) who found that mainly the dead microbial biomass is a source for NH_3 production. In a rough scheme one may assume that the dead biomass provides the substrate and the living biomass the enzymes for the N mineralization. According to Böhm (1993) substrate and enzymatic activity for the N mineralization are positively correlated. The provision of substrate depends on events which may lead to a major dying of soil microbes. Weller (1977) reported that particularly after dry periods, nitrification in soils was stimulated. It is feasible that soil dryness leads to the death of numerous microbes which are prone to mineralization after rewetting. Also the provision of oxygen can stimulate mineralization (Kaiser, 1994). In Figure 7 the course of mineral N production - mainly nitrate - is shown after the incorporation of green pea residues into soil by ploughing (Paul, 1994). Immediately after ploughing, N mineralization started and attained a peak in autumn. The following decline was due to N uptake by winter wheat and to some nitrate leaching. The maximum increase in mineral N was about $120 \, kg \, ha^{-1}$ from which only 1/3 originated from the pea residues. Hence it is evident that simply the process of ploughing considerably stimulated nitrification. Microbes may also

die due to starving. Even if they do not grow they need energy for maintenance. This energy may run out. Since in general a major microbial dying cannot be predicted, also the prediction of nitrification meets with difficulty. For this reason soils enriched in organic nitrogen, e.g. by regular farmyard application, may mineralize nitrogen at periods when it is not used by crops which may lead to nitrogen losses (Johnston and Powlson, 1994) presumably in form of nitrate leaching (Powlson et al., 1989).

Conclusion

Considering the nitrogen turnover processes described above a simplified scheme can be drawn which is shown in Figure 8. The scheme is based on the assumptions (1) that only organic matter with a narrow C/N ratio such as peptides (proteins, polypeptides, amino acids), polymers of amino sugars including acetylated amino sugars and peptidoglycans yield NH_3 when metabolized by soil microbes, (2) that the metabolization of energy rich organic matter low in nitrogen content (straw, mucilage of roots, lipids) requires inorganic N (NH_4^+, NO_3^-) and/or organic N, mainly in form of amino acids, (3) that only the dead biomass is prone to biological decomposition. Both the metabolization of peptides (proteins) and of organic carbon contribute to the build up of living microbial biomass the first process producing, the latter consuming inorganic nitrogen. The input of organic and inorganic nitrogen as well as organic matter in arable soils induces a quick turnover of nitrogen. The input of proteins (green manure) induces a flush of N mineralization within a few weeks, the input of organic carbon (straw) an asssimilation (immobilization) of inorganic nitrogen. If these inputs are well timed they may contribute to an efficient use of soil and fertilizer nitrogen.

The nitrogen mineralization potential is represented by the total microbial biomass (living + dead). The actual decomposable biomass is represented by the dead biomass. Hence soil and weather conditions inducing a major microbial mortality increase the potential of mineralizable N. Such soil and weather conditions could be dryness, lack of oxygen, or a lack of chemical energy for microbial life maintainance. These conditions may occur at various periods throughout the year. Growing crop stands are supposed to provide a substantial amount of organic carbon to rhizosphere microbes. As soon as this source

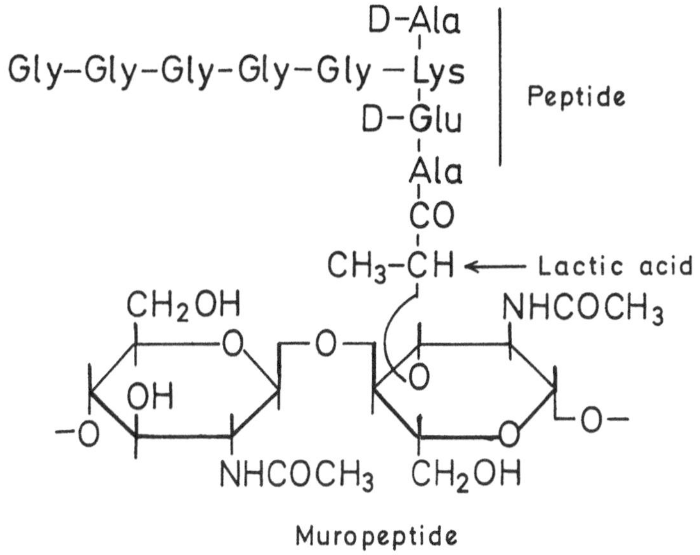

D–Ala
|
Gly–Gly–Gly–Gly–Gly — Lys   Peptide
|
D–Glu
|
Ala
|
CO
|
CH₃–CH ← — Lactic acid

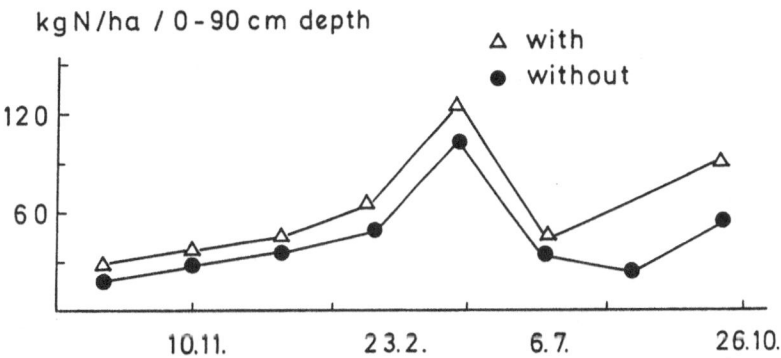

Muropeptide

Figure 6. Muropeptide structure.

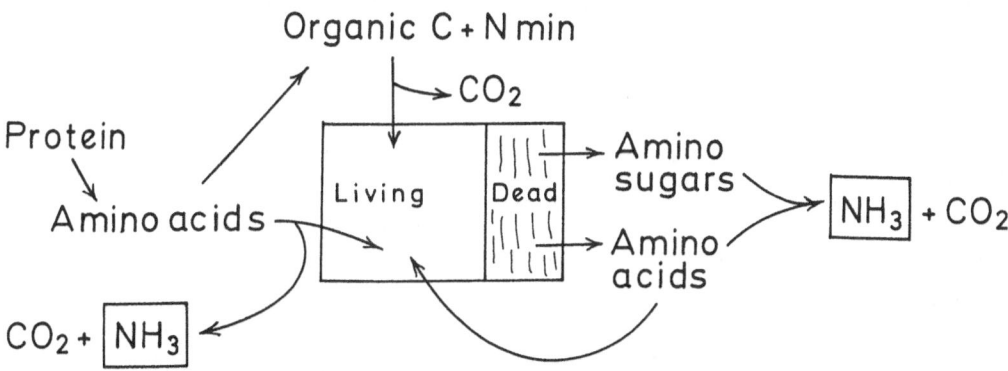

Figure 7. Formation of mineral nitrogen in soils after ploughing during one year. One treatment with and one without incorporated sugar beet tops (Paul, 1994).

Organic C + N min

Figure 8. Scheme of carbon and nitrogen turnover in soils. Major input of nitrogen comes from proteins of green manure, microbial N_2 assimilation and ftetilizer N: major input of organic C from plant residues and root exudates. Nitrogen mineralization starts with the deamination of organic compounds with a narrow C/N ratio.

92

is depleted a lack of energy will induce the death of microbes followed by a release of inorganic nitrogen.

References

Amberger A and Schweiger P 1971 Zur Wirkung einer kombinierten "Strohdüngung" mit Kalkstickstoff in langjährigen Feldversuchen. Z. Acker-Pflanzenb. 134, 323–334.

Appel T and Mengel K 1990 Importance of organic nitrogen fractions in sandy soils, obtained by electro-ultrafiltration or CaCl$_2$ extraction, for nitrogen mineralization and nitrogen uptake of rape. Biol. Fertil. Soils 10, 97–101.

Appel T and Mengel K 1992 Nitrogen uptake of cereals grown on sandy soils as related to nitrogen fertilizer application and soil nitrogen fractions obtained by electro-ultrafiltration (EUF) and CaCl$_2$ extraction. Eur. J. Agron. 1, 1–9.

Azam F, Malik K A and Hussain F 1986 Microbial biomass and mineralization-immobilization of nitrogen in some agricultural soils. Biol. Fertil. Soils 2, 157–163

Azam F, Ashraf M Asma Lohdi and Sajjad M I 1990 Availability of soil and fertilizer nitrogen to wheat (Triticum aestivum) following rice straw amendment. Biol. Fert. Soils 10, 134–138.

Azam F, Simmons F W and Mulvaney R L 1993 Mineralization of N from plant residues and its interaction with native soil N. Soil Biol. Biochem. 25, 1787–1792.

Barak P, Molina J A E, Hadas A and Clapp C E 1990 Mineralization of amino acids and evidence for direct assimilation of organic nitrogen. Soil Sci. Soc. Am. J. 54, 769–774.

Barekzai A, Becker A and Braschkat J 1993 Auswirkungen gesteigerter Güllegaben auf N-Entzug und N-Bilanz von Mais sowie auf die N Fraktionen des Bodens (NO$_3$, NH$_4$ und N$_{org}$-N): Arch. Acker-Pflanzenb. Bodenkd. 37, 341–350.

Beck T H 1983 Die Mineralisierung von Böden im Laborbrutversuch. Z. Pflanzenernähr. Bodenkd. 146, 243–252.

Böhm H 1993 Der Einfluß unterschiedlicher Bodenbearbeitungsverfahren auf mikrobielle Aktivitäten unter besonderer Berücksichtigung der N-Umsetzungen. Ph.D. Thesis, Agric. Fac., Justus-Liebig-Universität Giessen, Germany.

Catroux G and Schnitzer M 1987 Chemical, spectroscopic and biological characteristics of the organic matter in particle size fractions separated from an Aquoll. Soil Sci. Soc. Am. J. 51, 1200–1207.

Chichester F W and Smith S J 1983 Biological cycling of ^{15}N-labeled fertilizer nitrogen in lignite minesoil materials. Soil Sci. Soc. Am. J. 47, 676–682.

Christensen B 1986 Barley straw decomposition under field conditions: Effect of placement and initial nitrogen content on weight loss and nitrogen dynamics. Soil Biol. Biochem. 18, 523–529.

Dressler A and Mengel K 1985 Bedeutung des peripheren spezifisch gebundenen NH$_4^+$ von Löß- und Alluvialböden für die N-Düngerbedarfsermittlung. VDLUFA-Schriftenreihe, 16. Kongreßband, pp 137–146.

Frederick B A and Klein D A 1994 Nitrogen effects on rhizosphere processes of range grasses from different successional series. Plant and Soil 161, 241–250.

Freitag H E and Rausch H 1981 Zeitliche Veränderungen des löslichen Stickstoffs im Boden nach Zusatz verschiedener organischer Dünger im Feldmodellversuch ohne Bewuchs. Arch. Acker-Pflanzenb. Bodenkd. 25, 445–450.

Garz J and Chaanin A 1990 Zu den Wechselbeziehungen der Stickstoffdüngung und dem Umsatz der organischen Bodensubstanz. Tag. Ber. Akad. Landwirtsch. Berlin 289, 193–200.

Göck M and Ottow J C G 1988 Effect of cellulose and straw incorporation in soil on total denitrification and nitrogen immobilization at initially aerobic and permanent anaerobic conditions. Biol. Fertil. Soils 5, 317–322.

Guerrero M G, Vega J M and Losada M 1981 The assimilatory nitrate reducing system and its regulation. Ann. Rev. Plant Physiol. 32, 169–204.

Gutser R and Teicher K 1976 Veränderungen des löslichen Stickstoffes einer Ackerbraunerde unter Winterweizen im Jahresverlauf. Bayr. Landw. Jahrb. 53, 215–226.

Hart P B S, Powlson D S, Poulton P R, Johnston A E and Jenkinson D S 1993 The availability of the nitrogen in the crop residue of winter wheat to subsequent crops. J. Agric. Sci. (Cambridge) 121, 355–362.

Helal H M and Sauerbeck D 1989 Carbon turnover in the rhizosphere. Z. Pflanzenernähr. Bodenkd. 152, 211–216.

Honeycutt C W, Potaro L J and Haltemann W A 1991 Predicting nitrate formation from soil, fertilizer, crop residue and sludge with thermal units. J. Environ. Qual. 20, 850–856.

Hütsch B and Mengel K 1993 Effect of different soil cultivation systems, including no-tillage, on electro-ultrafiltration extractable organic nitrogen. Biol. Fertil. Soils 16, 233–237.

Isermann K 1987 Environmental aspects of fertilizer application. In Ullmann's Encyclopedia of industrial Chemistry, Vol. A 10. pp 400–409. VCH Verlagsgesellschaft, Weinheim, Germany.

Jensen L S and Sorensen J 1994 Microscale fumigation-extraction and substrate induced respiration methods for measuring microbial biomass in barley rhizosphere. Plant and Soil 162, 151–161.

Jenkinson D S 1991 The Rothamsted Long-Term Experiments: Are they still of use? Agron. J. 83, 2–10.

Jenkinson D S and Ladd J N 1981 Microbial biomass in soil: measurement and turnover. In Soil Biochem., Vol. 5. Eds. E A Paul and J N Ladd. pp 415-471. Marcel Dekker, New York, USA.

Jenkinson D S and Rayner J H 1977 The turnover of soil organic matter in some of the Rothamsted classical experiments. Soil Sci. 123, 298–305.

Joergensen R G, Meyer B and Mueller T 1994 Time-course of the soil microbial biomass under wheat: a one year field study. Soil Biol. Biochem. 26, 987–994.

Johnston A E and Powlson D S 1994 The setting-up, conduct and applicability of long-term, continuing field experiments in agricultural research. In Soil Resilence and substainable Land Use. Eds. D J Greenland and I Szabolcs. pp 395–421. CAB International, Budapest, Hungary.

Kaiser E A 1994 Significance of microbial biomass for carbon and nitrogen mineralization in soil. Z. Pflanzenernähr.. Bodenkd. 157, 271–278.

Ke F, Dou H and Mengel K 1990 Turnover of plant matter as assessed by electro-ultrafiltration and CaCl$_2$ extracts. Agrobiol. Res. 43, 337–347.

Kladiviko E J and Keeney D R 1987 Soil nitrogen mineralization as affected by water and temperature interactions. Biol. Fertil. Soils 5, 248–252.

Kuntze H and Bartels R 1979 Nährstoffversorgung und Leistung von Hochmoorgrünland. Landw. Forsch. 31, 1. Sonderh., 208–219.

Lochmann R, van der Ploeg R R and Huwe B 1989 Zur Parametrisierung der Stickstoffmineralisierung in einem Ackerboden unter Feldbedingungen. Z. Pflanzenern. Bodenkd. 152, 319–331.

Loll M J and Bollag J-M 1983 Protein transformation in soil. Adv. Agron. 36, 351–382.

Matzel W and Lippold H 1990 N application to winter wheat at tillering and shooting: N balance at different growth stages. Fert. Res. 26, 139–144.

Mengel K 1991 Nitrogen: agricultural productivity and environmental problems. In Nitrogen Metabolism of Plants. Eds. K Mengel and D J Pilbeam. pp 1–15. Oxford University Press, Oxford, UK.

Mengel K and Scherer H W 1981 Release of non-exchangeable (fixed) soil ammonium under field conditions during the growing season. Soil Sci. 131, 226–232.

Nette T and Resch H N 1992 Spezifisch gebundenes NH_4^+ in landwirtschaftlich genutzten Böden des Trierer Raumes. Agrobiol. Res. 45, 266–275.

Neyra C A and Döbereiner J 1977 Nitrogen fixation in grasses. Adv. Agron. 29, 1–38.

Odell R T, Melsted S W and Walker W M 1984 Changes in organic carbon and nitrogen of Morrow plot soils under different treatments, 1904–1973. Soil Sci. 137, 160–171.

Olfs H W and Werner W 1989 Veränderungen extrahierbarer "N_{org}"-Mengen unter dem Einfluß variierter C/N-Verhältnisse und Biomasse. VDLUFA-Schriftenreihe Kongressband 28, 15–26.

Paul R 1994 Feldversuche zur genaueren Erfassung der Stickstofflieferung aus Ernterückständen mittels Elektro-Ultrafiltration (EUF). Ph.D. Thesis, Agric. Fac., Justus-Liebig-Universität, Giessen, Germany.

Peschke H, Markgraf G, Oberdoerster U, Schmitt O and Görlitz W 1984 Zur Wirkung von Stickstoffdüngung und Beregnung bei Hafer (Avena sativa) L., Arch. Acker-Pflanzenbau Bodenkd. 28, 403–409.

Powlson D S, Poulton P R, Addiscott T M and Mc Cann D S 1989 Leaching of nitrate from soils receiving organic or inorganic fertilizers continuously for 135 years. In Nitrogen in Organic Wastes Applied to Soils. Eds. J A Hansen and K Hendriksen. pp 334–345. Academic Press, London, UK.

Quastel J H 1965 Soil metabolism. Annu. Rev. Physiol. 16, 217–240.

Rees R M, Yan L and Ferguson M 1993 The release and plant uptake of nitrogen from some plant and animal manures. Biol. Fertil. Soils 15, 285–293.

Riga A, Fischer V and van Praag H J 1980 Fate of fertilizer nitrogen applied to winter wheat as $Na^{15}NO_3$ and $(^{15}NH_4)_2SO_4$ studied in microplots through a four-course rotation: 1. Influence of fertilizer splitting on soil and fertilizer nitrogen. Soil Sci. 130, 88–99.

Savant N K and DeDatta S K 1982 Nitrogen transformation in wetland rice soils. Adv. Agron. 35, 241–302.

Scherer H W 1993 Dynamics and availability of the non-exchangeable NH_4-N- a review. Eur. J. Agron. 2, 149–160.

Schmeer H and Mengel K 1984 Der Einfluß der Strohdüngung auf die Nitratgehalte im Boden im Verlaufe der Wintermonate. Landw. Forsch. Band 1984, 214–229.

Schneider B 1995 Stoffliche Zusammensetzung extrahierbarer organischer Stickstoffverbin-dungen in Böden in ihrer Bedeutung für die Stickstoffmineralisation. Ph.D. Thesis, Agric. Fac., Justus-Liebig-Universität, Gießen, Germany.

Schnier H F 1994 Nitrogen-15 recovery fraction in flooded tropical rice as affected by added nitrogen interaction. Eur. J. Agron. 3, 161–167.

Schnier H F, DeDatta S K and Mengel K 1987 Dynamics of ^{15}N-labelled ammonium sulfate in various inorganic and organic soil fractions of wetland rice soils. Biol. Fertil. Soils 4, 171–177.

Schnier H F, DeDatta S K, Mengel K, Marqueses E P and Faronilo J E 1988 Nitrogen use efficiency, floodwater properties and nitrogen-15 balance in transplanted lowland rice as affected by liquid urea band placement. Fert. Res. 16, 241–255.

Schnitzer M and Spiteller M 1986 The chemistry of the "unknown" soil nitrogen. Intern. Soil Sci. Congr. Hamburg, Germany.

Schnürer J and Rosswal Th 1987 Mineralization of nitrogen from ^{15}N labelled fungi, soil microbial biomass and roots and its uptake by barley plants. Plant and Soil 102, 71–78.

Schulten H R and Schnitzer M 1992 Structural studies on soil humic acids by Curie-point pyrolysis-gas chromatography/mass spectrometry. Soil Sci. 153, 205–224.

Schulten H R, Monreal C M and Schnitzer M 1994 Effect of long term cultivation on the chemical structure of soil organic matter. Naturwiss.

Stanford G and Smith S J 1972 Nitrogen mineralization potential of soils. Soil Sci. Soc. Am. Proc. 36, 465–472.

Tabatabai M A and Al-Khafaji A A 1980 Comparison of nitrogen and sulfur mineralization in soils. Soil Sci. Soc. Am. J. 44, 1000–1006.

Teske W and Matzel W 1976 Stickstoffauswaschung und Stickstoffausnutzung durch die Pflanzen in Feldlysimetern bei Anwendung von ^{15}N-markiertem Harnstoff. Arch. Acker- Pflanzenbau Bodenkd. Berlin 20, 7, 489–502.

Thicke F E, Russelle M P, Heterman O P and Sheaffer C C 1993 Soil nitrogen mineralization indexes and corn response in crop rotations. Soil Sci. 156, 322–335.

Thies W, Becker K-W and Meyer B 1977 Bilanz von markiertem Dünger-N ($^{15}NH_4^+$ und $^{15}NO_3^-$) in natürlich gelagerten Sandlysimetern sowie zeitlicher Verlauf des Dünger- und bodenbürtigen N-Austrags im Vergleich Bewuchs-Brache. Landw. Forsch. Sonderh. 34/II, 55–63.

Thomsen I K 1993 Turnover of ^{15}N-straw and NH_4NO_3 in a sandy loam soil: Effects of straw disposal and N fertilization. Soil Biol. Biochem. 25, 1561–1566.

van Praag H J, Fischer V and Riga A 1980 Fate of fertilizer nitrogen applied to winter wheat as $Na^{15}NO_3$ and $(^{15}NH_4)_2SO_4$ studied in microplots through a four-course rotation: 2. Fixed ammonium turnover and nitrogen reversion. Soil Sci. 130, 100–105.

Wehrmann J and Scharpf H C 1986 The Nmin method - an aid to integrating various objectives of nitrogen fertilization. Z. Pflanzenernähr. Bodenkd. 149, 428–440.

Weller F 1977 Stickstoffnachlieferung und Stickstoffbilanz obstbaulich genutzter Böden. Erwerbsbau 19, 130–135.

Weller F 1983 Stickstoffumsatz in einigen obstbaulich genutzten Böden Südwestdeutschlands. Z. Pflanzenernähr. Bodenkd. 146, 261–270.

Wojcik-Wotkowiak D 1979 Nitrogen transformation in soil during humification of straw labelled with ^{15}N. Plant and Soil 49, 49–55.

Ziegler K, Nemeth K and Mengel K 1992 Relationship between electroultrafiltration (EUF) extractable nitrogen, grain yield and optimum nitrogen fertilizer rates for winter wheat. Fertil. Res. 32, 37–43.

Section editor: R Merckx

O. Van Cleemput et al. (eds.), Progress in Nitrogen Cycling Studies, 109–114, 1996.

Effect of plant residues on ammonium and nitrate content of soils during incubation

T. Németh, A. Abd El-Galil, L. Radimszky and Gy. Baczó
Research Institute for Soil Science and Agricultural Chemistry (RISSAC) of the Hungarian, Academy of Sciences, H-1022 Budapest, Herman Ottó út 15, Hungary

Key words: C/N ratio, environmental factors, incubation experiment, nitrogen transformation processes, plant residues

Abstract

Nitrogen transformation in the soil is a multifactorial process with several alternative paths and wide ranges of rates. These processes are the most uncertain part of the simulation models on the fate of nitrogen in soils. Practically nitrate and part of the ammonium and organic nitrogen content of the soil horizons can be considered to be the mobile fractions of nitrogen.

In order to study involvement of nitrate and ammonium in nitrogen transformation processes of soils under a wide range of various factors, a 168-day-long incubation experiment was carried out, using 27 combinations of treatments. The main factors as soil temperature, soil moisture content, nitrogen fertilizer, amount and C/N ratio of plant residues were applied at five levels each. Daily CO_2 production, NH_4^+ and NO_3^- concentrations were measured on selected days. The results of two calcareous sandy soils (Örbottyán - Hungary and El-Marashda - Egypt) are discussed in this paper. It was observed that the same processes took place in both soils, but the amounts of the NH_4-N and NO_3-N and the duration of the transformation were slightly different.

Introduction

The investigation of the transformation processes of different elements in the root-zone is very important not only for the purpose to follow the fate of a certain element, but also to formulate equations describing these processes. Such equations are essential for modeling the element cycles in the root-zone, in the case of nitrogen to develop a nitrogen transformation submodel. Satisfying the nitrogen demand of the cultivated crops during their growth and development requires easily uptakeable nitrogen in the root-zone at the time when this N-demand appears. On the other hand the surplus of the nitrogen – especially nitrate – in the root-zone after harvesting the crops means a significant hazard of the environmental contamination via leaching, erosion and denitrification. All of these processes strongly require the better understanding of the behaviour of nitrogen under different environmental conditions.

The N-supplying capacity of a certain soil layer could be characterized by N-mineralization of soil organic matter and by the accumulated mineral-N (Németh and Szebeni, 1987; Szebeni and Németh, 1987). For these studies the incubation techniques are one of the best tools. There are several biological (incubation) and chemical methods to forecast the rate of mineralization in the growing season. Several researchers have investigated the biological transformation processes during incubation of re-moistened soils using various lengths of time and temperatures under aerobic and anaerobic conditions (Addiscott, 1983; Dendooven et al., 1987; Kowalenko and Cameron, 1976).

To study the effect of soil moisture content, temperature, added nitrogen and two different plant residues (maize and alfalfa) on the mineral nitrogen (NH_4-N and NO_3-N) content and dynamics of different soils a series of half-year-long incubation experiment were carried out. Some of the results on the two sandy soils included in the experiments have already been published (Németh et al., 1993), while in this paper the effects of soil moisture content, temperature and the

Table 1. Multifactorial orthogonal experimental plan

Treatment code	Factors				
	Temperature (°C)	Soil moisture c. (% of MWC)	Added nitrogen (mg kg^{-1})	Alfalfa[a]	Maize[a]
				(g 100 g^{-1})	
1.	15	· 80	225	3	3
2.	35	80	225	3	1
3.	15	40	225	3	1
4.	35	40	225	3	3
5.	15	80	75	3	1
6.	35	80	75	3	3
7.	15	40	75	3	3
8.	35	40	75	3	1
9.	15	80	225	1	1
10	35	80	225	1	3
11.	15	40	225	1	3
12.	35	40	225	1	1
13.	15	80	75	1	3
14.	35	80	75	1	1
15.	15	40	75	1	1
16.	35	40	75	1	3
17.	45	60	150	2	2
18.	5	60	150	2	2
19.	25	100	150	2	2
20.	25	20	150	2	2
21.	25	60	300	2	2
22.	25	60	0	2	2
23.	25	60	150	4	2
24.	25	60	150	0	2
25.	25	60	150	2	4
26.	25	60	150	2	0
27.	25	60	150	2	2
28.	25	60	150	2	2
29.	25	60	150	2	2
30.	25	60	150	2	2
31.	25	60	150	2	2
32.	25	60	150	2	2
33.	25	60	150	2	2
34.	25	60	150	2	2
35.	25	60	150	2	2
36.	25	60	150	2	2

[a]Alfalfa and maize plant residues g 100 g^{-1} soil.

applied plant residues on the NH_4-N and NO_3-N contents of the two sandy soils are presented.

Materials and methods

For the 168 days long incubation experiments soil samples were collected from a non-fertilized field in Hungary at the Experimental Station of RISSAC (calcareous sandy soil – Örbottyán), and from an Egyptian site (calcareous sandy soil – El-Marashda). By applying the SITOBI program developed by Biczók and his coworkers (Szili-Kovács et al., 1993) it was easy to make an interactive design of multifactorial orthogonal experimental plans (Table 1) and execute the statistical analysis of data. In the factorial experiment 5 factors were applied each at 5 levels. Soil moisture content ranged between 20, 40, 60, 80 and 100% of the maximum water capacity (MWC), soil temperature was kept at 5, 15, 25, 35, 45 °C, nitrogen was added as 0, 75, 150, 225, 300 mg N kg^{-1} soil, and plant residues (alfalfa and maize) applied at rates of 0, 1, 2, 3, 4 g 100 g^{-1} soil.

To follow the transformation processes the mineral nitrogen content (exchangeable ammonium and nitrate) of the soil was measured 11 times during the incubation period (according to the method developed by Bremner and Keeney, 1966). Sampling was carried out more frequently (five times) in the first two weeks.

Results and discussion

Soil moisture content

Figure 1 shows the effect of soil moisture content (in the average of the other four factors) on the exchangeable ammonium-N and nitrate-N content during the investigated period on the calcareous sandy soil from Hungary. In the dry soil (20% MWC) the ammonium-N content was low during the whole incubation period, while in the more moist soils it increased in the first 6 weeks (extra mineralization). After this time nitrification occurred also in wet soil, and around the 84th day the nitrogen disappeared from the ammonium-form. In the dry soil the nitrate-N accumulated in the whole period, while at higher soil moisture contents no nitrate-N accumulation was observed. 300–350 mg NO$_3$-N kg^{-1} soil was measured in the first three moisture levels treatments (up to 60% of MWC). At higher soil moisture range the amount of nitrate-N decreased significantly with time. All nitrate disappeared from the soil in the 100% MWC treatments.

In the Egyptian calcareous sandy soil the trends of the processes were similar to those observed in the Hungarian soil (Fig. 1). The main difference was that in the Egyptian soil a slight ammonium-N accumulation occurred in the first three months in the whole soil moisture range with a local maximum of NH$_4$-N content (60–100 mg kg^{-1}) in the middle of this period. After 3 months the ammonium-N concentration decreased because of the nitrification. At the end of the experiment practically no nitrogen was found in this form. Nitrate-N increased significantly with time in the soil moisture range of 20–60% MWC, while no increase occurred in the 100% MWC treatments, where all the nitrogen disappeared also from this form. This may have been caused by denitrification. The maximum in the nitrate-N concentration was found in the 60% MWC treatments at the end of the incubation.

Temperature

Evaluating the results of the different temperature levels it can be seen, that in the Hungarian sandy soil the initial ammonium-N amount was nitrificated in the whole temperature range and in consequence practically no nitrogen was left in this form after 7 weeks (Fig. 1). The nitrate-N content of the soil samples showed an opposite pattern. A significant increase in nitrate-N was observed in every treatment. This was most pronounced in the high temperature zone, which refers to extra-mineralization in this range. The maximum nitrate-N concentration (over 350 mg NO$_3$-N kg^{-1} soil) was found at the highest temperature.

On the sandy Egyptian soil the nitrogen remained longer in the form of ammonium, than on the Hungarian soil, especially in the cooler treatments (Figure 1). At 5 °C nitrogen was detectable in this form till the end of the incubation. On the opposite side of the temperature range (45 °C) the ammonium-N concentration rapidly decreased, similarly as on the Hungarian sandy soil. In the other three treatments (15, 25 and 35 °C) at time while the nitrification could be observed, decreased with increasing temperature. In those treatments, where the nitrogen remained in the form of ammonium, no nitrate-N accumulation was observed. In the other temperature treatments the nitrate–N content increased due to nitrification (between 15–45 °C) and also due to extra-mineralization (between 25–45 °C). The maximum in nitrate-N concentration was detected at the end of the incubation at 25 °C.

Comparing the two sandy soils we conclude that there are only minor differences in the trend of the changes caused by soil moisture content and soil temperature between the Hungarian and Egyptian soils in the N-concentration, in the rate of the extra-mineralization and in the speed of the N-transformation. The high soil moisture content caused a significant N-loss. The optimal soil moisture range for the N-transformation processes (mineralization, nitrification, nitrate-N accumulation) was found to be between 20–60% MWC. The 5 °C temperature seems to be too low to start the N-transformation activities.

Őrbottyán   El-Marashda

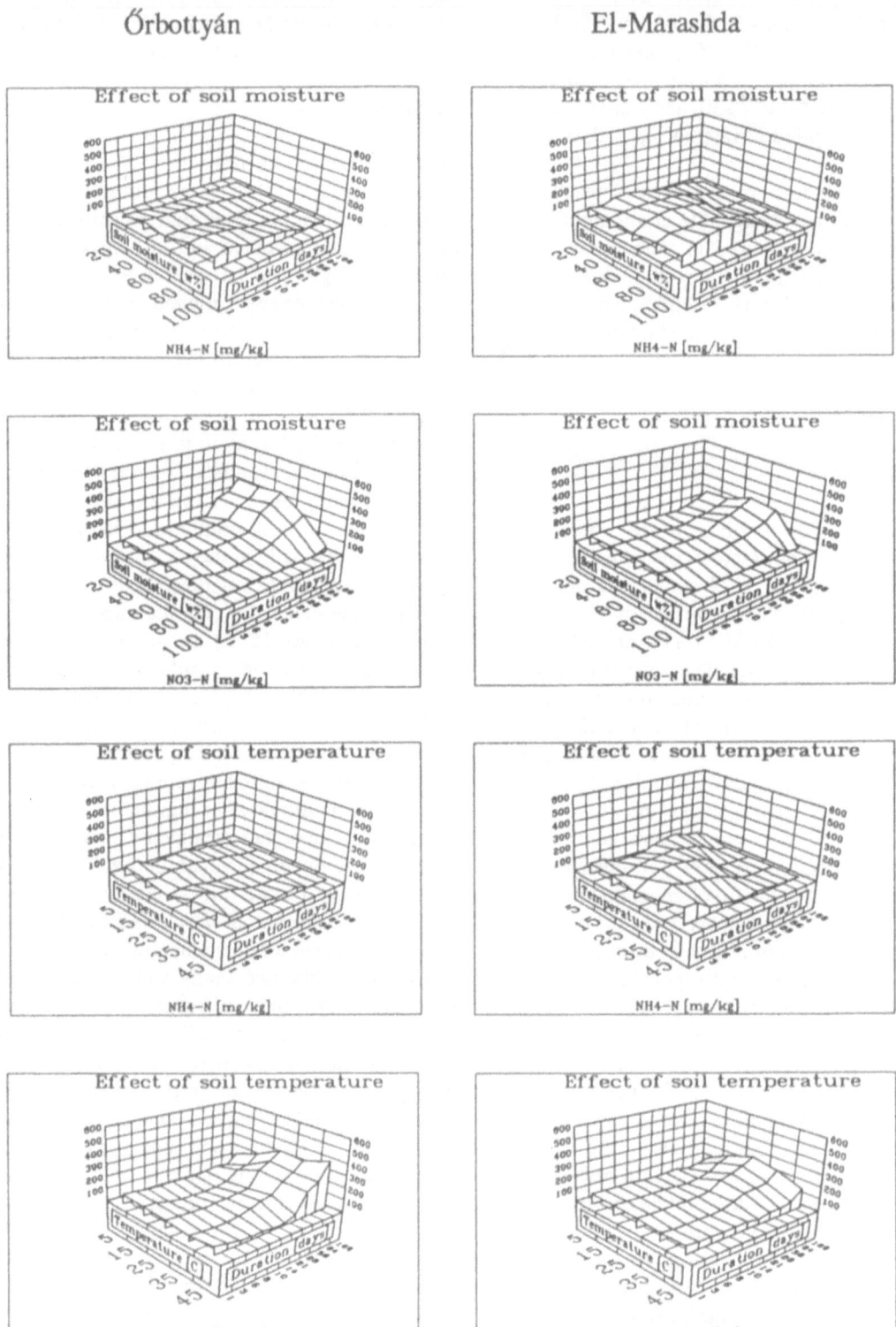

Figure 1. Effect of the soil moisture content and soil temperature on ammonium and nitrate content of two sandy soils during incubation.

Őrbottyán

El-Marashda

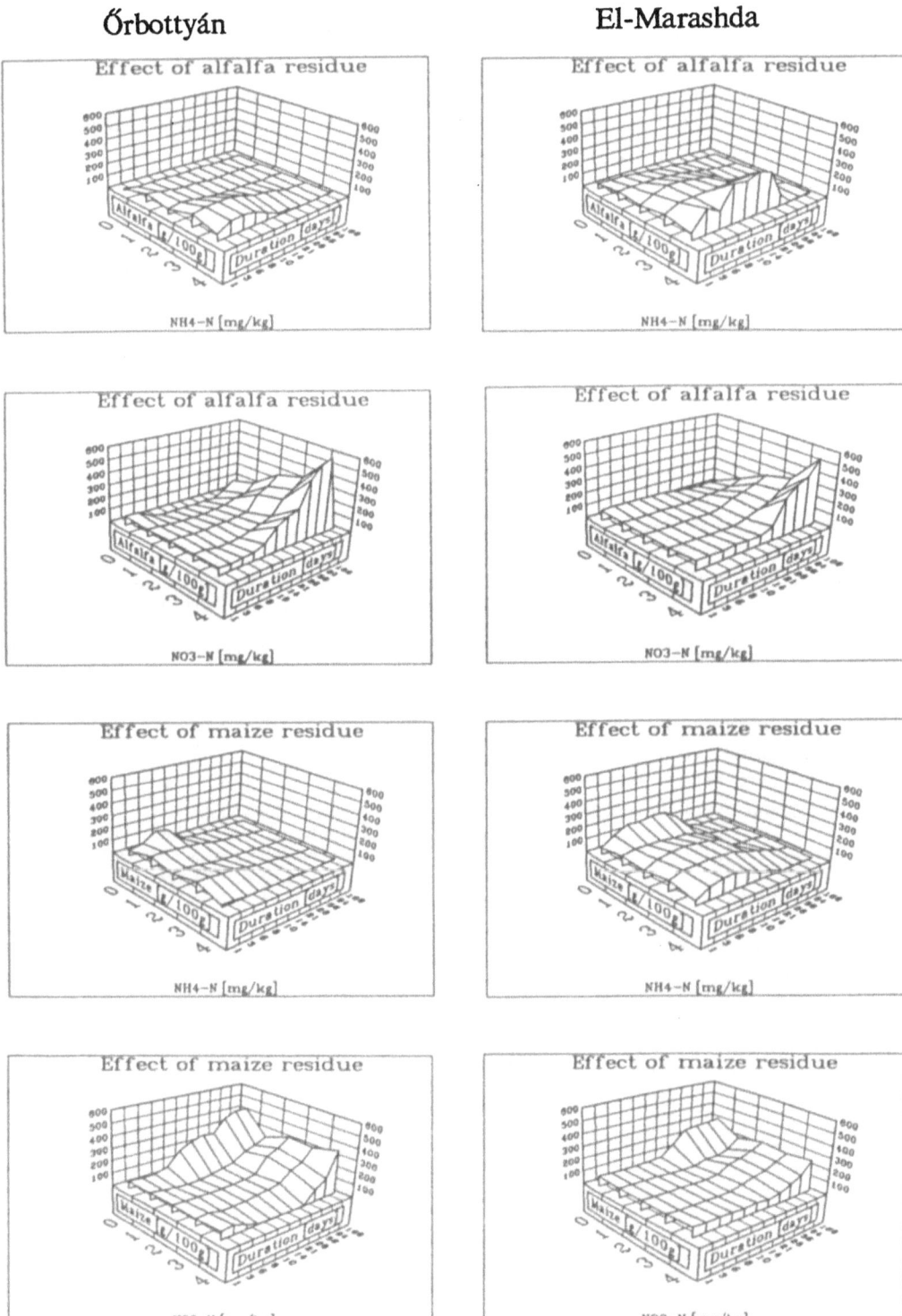

Figure 2. Effect of the mixed plant residues on ammonium and nitrate content of two sandy soils during incubation.

Plant residues

At the two highest application levels of the alfalfa plant residues the ammonium-N content of the Hungarian sandy soil increased in the first two weeks of the incubation (Fig. 2). After this period the nitrogen disappeared from this form in all treatments. On the Egyptian sandy soil the increase of the ammonium-N content at the higher alfalfa application rates was more pronounced (till the 56 day), compared to the Hungarian sandy soil. The accumulation of the nitrate-N was similar in both soils, increasing with time and with the dose of the applied alfalfa residues. The maximum in the nitrate-N concentration (around 500 mg kg^{-1} in both soils) was detected at the end of the incubation in the highest alfalfa treatments (4 g residue 100 g^{-1} soil), in the average of the other four factors. These high NO_3-N concentrations indicate high extra-mineralization from the alfalfa residues.

After mixing maize residues into the soils (the C/N ratio in maize was higher that in the young alfalfa residues) the ammonium-N content of the Hungarian sandy soil was low. In the Egyptian soil – which originally had lower organic matter content than the Hungarian soil – the ammonium-N content of the samples increased in all treatments till the end of the 3rd week. After the 3rd week the ammonium-N content of this soil also decreased. The nitrate-N content of both soils increased with time. There were no significant differences among the treatments and between the soils.

After mixing the alfalfa residues (characterized by a narrow C/N ratio) to the soil, the nitrate-N concentration increased on both soils with increasing applied rates. The nitrogen from the alfalfa mineralized quickly. After addition of maize residues (wider C/N ratio) the highest N-accumulation was detected in the control soils with no maize residue application, and the amount of the NO_3-N decreased with increasing applied rates

of maize (instead of the extramineralization nitrogen immobilization occurred after the maize application).

This half year incubation experiment, using different soils, and applying the investigated factors at five levels each, gave a good basis for modeling studies. The preliminary results of these were published by Kovàcs et al. (1993).

References

Addiscott T M 1983 Kinetics and temperature relationships of mineralization and nitrification in Rothamsted soils with differing histories. J. Soil Sci. 34, 343–353.

Bremner J M and Keeney D R 1966 Determination and isotope-ratio analysis of different forms of nitrogen in soils. 3. Exchangeable ammonium, nitrate and nitrite by extraction distillation methods. Soil Sci. Am. Proc. 30, 577–582.

Dendooven L, Verstraeten L M J and Vlassak K 1987 Temperature and N-mineralization. *In* Proc. 3rd Meeting for Assessment of N-Fertilizer Requirement. Ed. N E Nielsen. pp 3–13. Dep. Soil, Water and Plant Nutr., The Royal Veter. and Agric. Univ., Copenhagen, Denmark.

Kovàcs G, Németh T and Radimszky L 1993 Nitrogen transformation in the soil: computer simulation of the processes. *In* XXXV. Georgikon Days, Keszthely. pp 275–280. Pannon ATE, Keszthely, Hungary.

Kowalenko G G and Cameron D R 1976 Nitrogen transformation in an incubated soil as affected by combinations of moisture content and temperature and adsorption-fixation of ammonium. Can. J. Soil Sci. 56, 63–70.

Németh T and Szebeni I 1987 Mineralized soil nitrogen from a longterm, four-step-nitrogen trial. *In* Proc. of the 9th Int. Symp. on Soil Biol. and Conserv. of the Biosphere. Ed. J Szegi. pp 53–59. Akadémiai Kiadó, Budapest, Hungary.

Németh T, Abd El Galil A, Baczó Gy and Radimszky L 1993 Study of the ammonium-N and nitrate-N content of different soils during incubation. Agrokém. Talajtan 42, 173–178.

Szebeni I and Németh T 1987 Mineralization and immobilization changes on the effect of nitrogen fertilization in a pot experiment. *In* Proc. of the 9th Int. Symp. on Soil Biol. and Conserv. of the Biosphere. Ed. J Szegi. pp 43–52. Akadémiai Kiadó, Budapest, Hungary.

Szili-Kovács T, Radimszky L, Andó J and Biczók Cry 1993 CO_2 evaluation from soils formed on various parent materials in the Eastern Cserhát mountains (Hungary) during laboratory incubation. Agrokém. Talajtari 42, 140–146.

O. Van Cleemput et al. (eds.), Progress in Nitrogen Cycling Studies, 115–120, 1996.
© 1996 *Kluwer Academic Publishers.*

N turnover in soils with different mineralization potential after application of N fertilizer

H-W Olfs and Birgit Weyand

Institute of Agricultural Chemistry Friedrich-Wilhelm University Bonn, Meckenheimer Allee, 176 D-53115 Bonn, Germany

Key words: microbial biomass N, mineral N, N immobilization, N mineralization potential, N turnover

Abstract

A 40-day pot experiment was conducted to study the effect of N fertilizer application (0, 10 and 20 mg N kg^{-1} soil) on soil N turnover. Soil from a previous experiment was used, in which 4 different mineralization potentials (soil variants I - IV) had been established by stimulating microbial growth (simultaneous supply of C and N).

Mineral N and biomass N showed contrasting courses during the experiment. Towards the end of the experiment N mineralized from the soil N pool (as well as applied fertilizer N) was immobilized by microorganisms in fallow soils. Depending on the mineralization potential significant differences in N_{min} and biomass N were found for soil variants I - IV. Competition between microorganisms and plants for mineral N caused lower biomass N contents in cultivated soils. N uptake of grass varied according to the mineralization potentials of the soil variants. Mineralization from the soil N pool was slightly enhanced after fertilizer application ("added nitrogen interactions").

Introduction

Adjusting N fertilization to the nitrogen requirements of agricultural crops is difficult because N supply from the soil during the vegetation period varies enormously. Mineralization and immobilization always occur simultaneously during N turnover (Jansson and Persson, 1982). The distribution between these two contrasting processes is mainly influenced by the activity of the microbial biomass. Furthermore it has to be taken into consideration that soil N dynamic is modified by supplying N fertilizer to the soil (Engels and Kuhlmann, 1993).

Our investigations were focused on the influence of fertilizer N application on N turnover (changes in mineral N and biomass N) in soils with different N mineralization potentials and N uptake by plants in a 40-day pot experiment.

Materials and methods

Air dried soil (topsoil of a Luvisol derived from loess; C_t 1.4%, N_t 0.14%, pH(CaCl$_2$) 5.9, sand 7.4%, silt 75.7% and clay 16.9%) from a previous 3-year experi-

ment was used for a 40-day pot experiment. Microbial growth had been stimulated in this soil by simultaneous application of different amounts of carbon and nitrogen (Olfs and Werner, 1989). Due to this pre-treatment nitrogen was incorporated into various soil N pools and therefore different N mineralization was expected. As indices to characterize the modified mineralization potentials of these soil variants I - IV, the amount of N_{org} (CaCl$_2$ extractable organic N according to Houba et al., 1986) and the contents of biomass N (fumigation extraction method [FE]; Brookes et al., 1985) were detected (Table 1).

Air dried soils were rewetted and water holding capacity (WHC) was adjusted to 50% once per day. On day 16 grass (*Lolium multiflorum italicum* cv Turilo) was sown to 12 pots per soil variant. 8 days later N fertilizer was applied (0 ["N0"], 10 ["N10"], and 20 ["N20"] mg N kg^{-1} soil; NaNO$_3$ labelled with 11% ^{15}N) on cultivated and fallow pots. After a 24-day vegetation period grass was harvested. N content as well as ^{15}N ratio were measured via mass spectrometry (ANCA-MS, Roboprep-CN + Tracermass, Europa Scientific Ltd., Crewe, UK). Soil samples (4 replicates each time) were taken 2, 4, 16, 24 and 32 days after the

Table 1. Application rates of N (as calcium nitrate) and C (as cellulose) to induce different mineralization potentials (I - IV; N_{org} and biomass N as indices) in the soils used for the pot experiment

Variant	N- addition (g pot^{-1})	C-	N_{org} (mg N kg^{-1} soil)	Biomass N
I	0	0	4.9	9.2
II	2.4	0	5.4	14.0
III	0.6	30	5.3	14.0
IV	2.4	120	7.1	22.6

start of the experiment from fallow pots and on day 40 from fallow and planted pots.

Biomass N was determined by the FE method according to Brookes et al. (1985). Moist soil samples equivalent to 25 g oven dry soil (4 replicates) were fumigated with chloroform (stabilized with 20 μL 2-methyl-2-butene mL^{-1}; Merck, Darmstadt, Germany) in a desiccator for 24 h at laboratory temperature (20–22 °C). The chloroform was removed by repeated evacuation. The soils were then extracted for 30 min with 0.5 M K$_2$SO$_4$ (4 : 1 solution : soil ratio) and filtered through MN 261 G 1/4 filter papers (Macherey-Nagel, Düren, Germany). A similar set of non-fumigated soils were extracted at the time fumigation commenced. For fertilized soils and planted soils this procedure was modified using a pre-extraction and sieving step to eliminate high amounts of mineral N and plant roots from the soil (Müller et al., 1992). Briefly, soil samples (8 replicates) were extracted with 100 mL of 50 mM K$_2$SO$_4$ for 15 min and passed through a sieve (3 mm mesh). Roots on the sieve were washed carefully with 75 mL of 50 mM K$_2$SO$_4$. The soil suspensions were then filtered. Afterwards filter plus soil was treated as described above for FE, but additionally 250 μL of chloroform were supplied to the water saturated soil to ensure an efficient fumigation (Widmer et al., 1989). The filtered soil extracts were stored at - 18 °C prior to analysis. N_{min} (= NO$_3$-N + NH$_4$-N) and N_t (total oxidizable nitrogen) in the extracts were determined using a continuous-flow analyzer (Technicon AAII, Bran and Lübbe, Hamburg, Germany). Biomass N was calculated from N_t extracted from fumigated soils minus N_t extracted from unfumigated soils *without* taking into account a k_{EN} value because of the contradictory results reported up to now (Jenkinson, 1988). All results are expressed on an oven-dry basis.

Statistical evaluations were done using the software package "SPSS PC+" (SPSS Inc., Chicago, USA). Dif-

ferences in N_{min} and biomass N contents were checked with the "*Scheffé Test*" at the 5% level. The effect of plant growth compared to fallow soil on biomass N was tested with the "*t-Test*" procedure. To evaluate the influence of fertilizer rate on N uptake by plants from fertilizer N and native soil N the "*Tukey Test*" was used at the 5% level.

Results and discussion

Mineral nitrogen

Rewetting of the dried soil at the start of the experiment caused an increase in the contents of N_{min} for all soil variants (Fig. 1), an observation in line with many other studies (e.g. Scherer et al., 1992). This N flush can be explained by the fact that organic N compounds (e.g. microbial cell material), which had been accumulated during the previous drying process (Stevenson, 1956), were available for microbial decomposition (Marumoto et al., 1977). In addition, higher N_{min} levels may be partly explained by a better availability of organic substrate through desorption from soil surfaces (Seneviratne and Wild, 1985) and through an increase in organic surfaces exposed (Birch, 1959).

Results from Beauchamp et al. (1986) and Cabrera (1993) indicate that most of this N flush occurs in the first week after rewetting and no differences are found in the mineralization rates in rewetted and field moist soils afterwards. In our experiment N_{min} contents in the soil reached significant different plateaus after 8 days, depending more or less on the mineralization potential of the soil variant. Although no differences existed in N_{org} and biomass N levels (used to characterize the mineralization potential) in soil variant II compared to III, significantly higher N_{min} contents were found in the latter one. Obviously these indices were not sensitive enough to characterize N net mineralization after rewetting.

From day 28 onwards N_{min} levels declined for unfertilized (Fig. 1) as well as for fertilized (data not shown) soils under fallow until the end of the experiment. A possible explanation for this decrease is microbial driven immobilization. Gaseous N losses via N$_2$/N$_2$O are implausible, since conditions in the soil were not favorable for denitrification (e.g. soil moisture had been adjusted to 50% WHC throughout the experiment). Due to N uptake by plants, N_{min} levels in cultivated pots on day 40 were negligible for all treatments (data not shown).

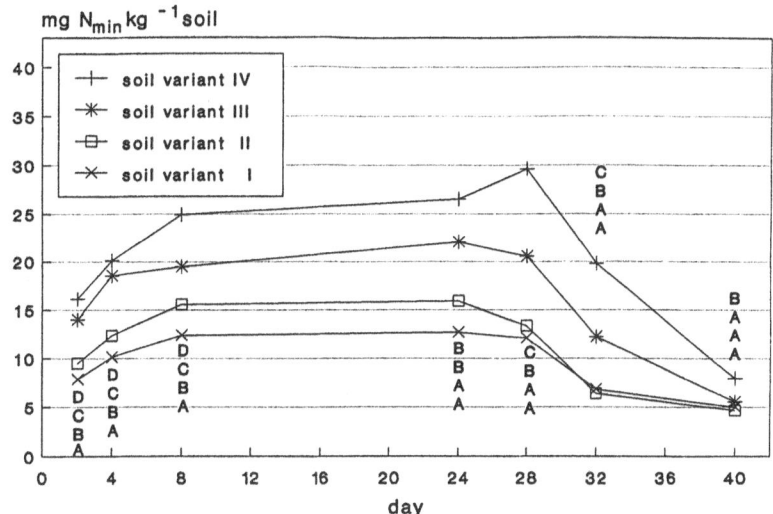

Figure 1. Changes in N_{min} contents (mg N_{min} kg^{-1} soil) for soil variants I - IV (unfertilized soils under fallow). Different letters in the same column indicate significant differences at the 5% level (*Scheffé Test*).

Biomass N for unfertilized soils under fallow

Biomass N contents in unfertilized soils under fallow showed a parallel, wave-like development for all soil variants (Fig. 2). Immediately after rewetting of the soils, C compounds which had become available during the drying process (e.g. C from microbial cells killed by drying) were rapidly metabolized, leading to significantly different amounts of biomass in the 4 soil variants. The first minimum after 4 days can be explained by the fact that readily decomposable substrate were becoming depleted and "zymogenous" microorganisms (Paul and Clark, 1989) decline in numbers. "Autochthonous" populations with slower and more constant growth rates can develop, because this microorganisms are able to make use of more recalcitrant substrate. Almost similar trends were observed in drying-rewetting studies by Van Gestel et al. (1993). Shortage of available C after a period of microbial growth also may be an explanation for the second minimum after 28–32 days.

Effect of fertilizer application on biomass N

After application of fertilizer N, biomass N contents did not change according to an uniform pattern in the 4 soil variants (Fig. 3). In soil variant I significantly higher biomass N levels occurred for fertilizer treatments (N10, N20) compared to unfertilized soils (N0) only on day 28. Significant effects of fertilizer application on biomass N values in soil variant II were

obtained on day 28 and 32, respectively. In soil variant III no differences in biomass contents were observed after fertilization. Significantly increased biomass levels after N application were found for soil variant IV on day 28 and 32, respectively. A reverse result occurred on day 40 for this soil variant when an unexpected high value for treatment N0 was determined. Overall, for each fertilizer level biomass N contents in the soil increased towards the end of the experiment.

Higher biomass N values for treatments with N fertilizer application in long term field experiments have been reported in several studies (e.g. Bonde et al., 1988; Deubel and Leithold, 1993). These effects are explainable due to stimulation of plant growth by N fertilization. As a result higher amounts of root exudates and more plant residues cause a better C supply for the microbial biomass. Also in pot experiments a positive influence of fertilizer application on microbial biomass was found in planted soils, but not in unplanted soil (Breland and Bakken, 1991). Since soils are typically C-limited environments (Dommergues et al., 1978), it has to be assumed that the short-term increase in biomass N after N supply in our experiments is due, at least in part, to accumulation of intracellular solutes (Kieft et al., 1987) and not connected with microbial growth. After cell lysis due to chloroform these compounds are extractable and contribute to the N flush after fumigation.

The different course of biomass levels in the soil variants I - IV indicates, that the development of the microorganism populations was not synchronous

118

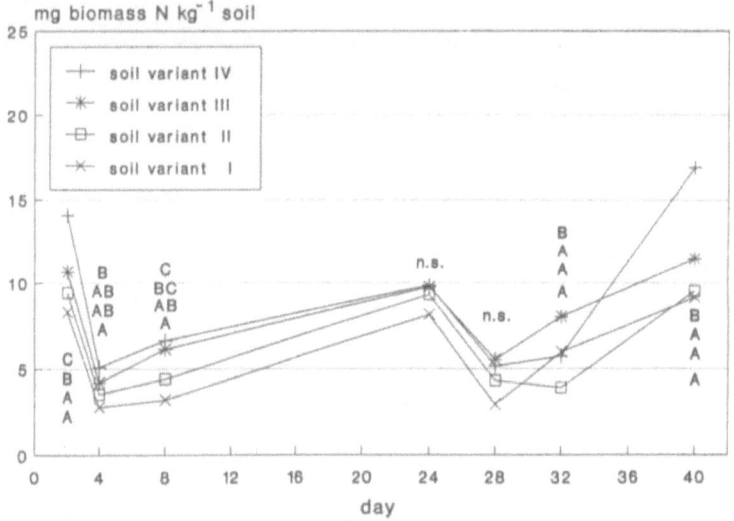

Figure 2. Changes in biomass N contents (mg biomass N kg^{-1} soil) for soil variants I - IV (unfertilized soils under fallow). Different letters in the same column indicate significant differences at the 5% level (*Scheffé Test*).

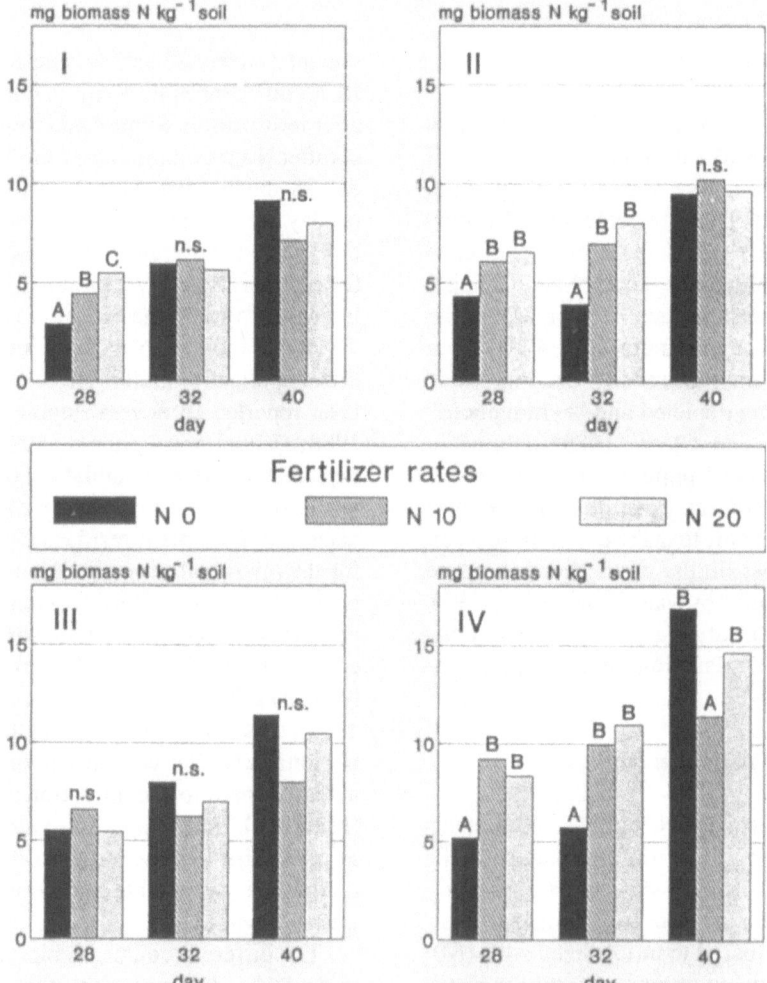

Figure 3. Biomass N contents (mg biomass N kg^{-1} soil) on day 28, 32 and 40 for soil variants I - IV (application of different fertilizer rates on day 24; soils under fallow). Different letters for one day indicate significant differences [n.s. = not significant] at the 5% level (*Scheffé Test*).

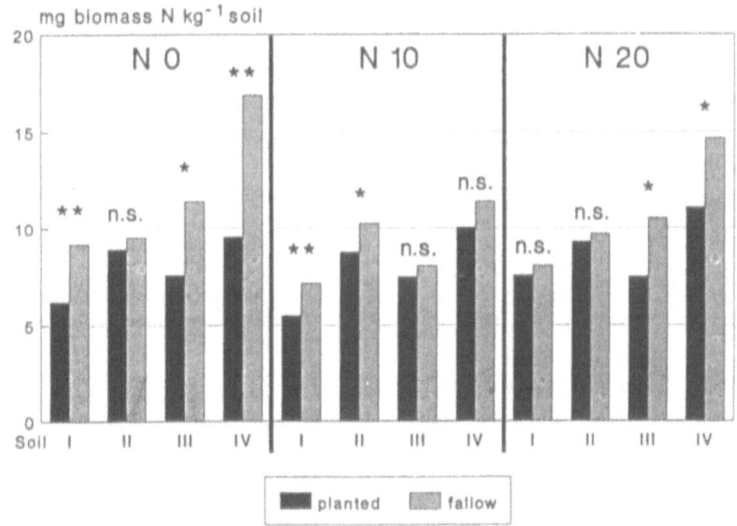

Figure 4. Biomass N contents (mg biomass N kg^{-1} soil) on day 40 for planted and fallow soils after application of different fertilizer rates on day 24. Differences between planted and fallow soils are indicated as follows: n.s. = not significant, * = significant at the 5% level and ** = significant at the 1% level, respectively (*t-Test*).

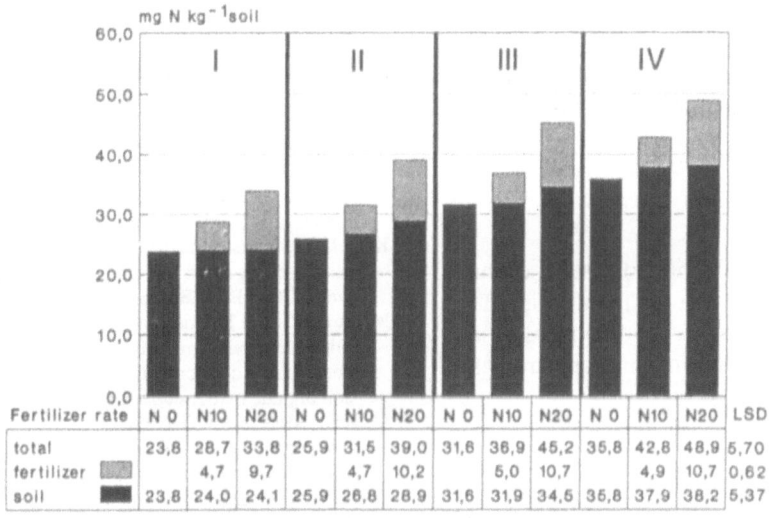

Fertilizer rate	N 0	N10	N20	N 0	N10	N20	N 0	N10	N20	N 0	N10	N20	LSD
total	23,8	28,7	33,8	25,9	31,5	39,0	31,6	36,9	45,2	35,8	42,8	48,9	5,70
fertilizer		4,7	9,7		4,7	10,2		5,0	10,7		4,9	10,7	0,62
soil	23,8	24,0	24,1	25,9	26,8	28,9	31,6	31,9	34,5	35,8	37,9	38,2	5,37

Figure 5. N uptake by grass (mg N kg^{-1} soil) from added ^{15}N labelled fertilizer and native soil N pool for soil variants I - IV on day 40. LSD = least significant difference between treatments at the 5% level (*Tukey Test*).

until the end of the experiment. It may be supposed that different growth stages and/or physiological states had been sampled. Data from Breland and Bakken (1991) verify that the size distribution of bacterial cells changed in response to fertilizer application.

Comparison of biomass N in fallow and cultivated pots

Biomass N contents were always lower (for some treatments not significantly) in cultivated soils in compari-

son to soils under fallow (Fig. 4). In contrast to results from Helal and Sauerbeck (1985), C input via root exudates in cultivated soils had no favorable effect on microbial growth. Obviously plants and microorganisms competed for mineral N. Due to N uptake by plants less N was available for microbial utilization and as a result biomass N contents were lower. Application of N may reduce this competition more or less (Breland and Bakken, 1991).

N uptake by plants

N uptake by plants on unfertilized soils (N0) was significantly higher for soil variants IV and III compared to II and I, respectively (Fig. 5). As noted earlier, N_{org} and biomass N levels were not sensitive enough for the prediction of N net mineralization. Addition of fertilizer N caused a slightly, but not significantly higher mineralization from the soil N pool (added nitrogen interaction = ANI). These differences could have been due, at least in part, to pool substitution processes (apparent ANI; Jenkinson et al., 1985). The development of a more elaborate root system, that absorbs nutrients from a larger volume of soil (real ANI; Azam et al., 1994) seems to be no suitable explanation, because the employed pots were rather small (900 mL) and completely rooted at the end of the experiment.

To summarize, the application of fertilizer N effected the amount of biomass N in soils with different N mineralization potential only for a short time. The rapid changes in biomass N during our 40-day experiment may be regarded as an indication of the "transitory" character of the microbial biomass.

Acknowledgements

We are grateful to Prof W Werner for useful discussions and valuable comments on the manuscript. The authors thankfully acknowledge the technical assistance of Carmen Berg in performing mass spectrometer analyses.

References

Azam F, Simmons F W and Mulvaney R L 1994 The effect of inorganic nitrogen on the added nitrogen interaction of soils in incubation experiments. Biol. Fertil. Soils 18, 103–108.

Beauchamp E G, Reynolds W D, Brasche-Villeneuve D and Kirby K 1986 Nitrogen mineralization kinetics with different soil pretreatments and cropping histories. Soil Sci. Soc. Am. J. 50, 1478–1483.

Birch H F 1959 Further observations on humus decomposition and nitrification. Plant and Soil 11, 262–286.

Bonde T A, Schnürer J and Rosswall T 1988 Microbial biomass as a fraction of potentially mineralizable nitrogen in soils from long-term field experiments. Soil Biol. Biochem. 20, 447–452.

Breland T A and Bakken L R 1991 Microbial growth and nitrogen immobilization in the root zone of barley (*Hordeum vulgare* L.), italian ryegrass (*Lolium multiflorum* Lam.), and white clover (*Trifolium repens* L.). Biol. Fertil. Soils 12, 154–160.

Brookes P C, Landman A, Pruden G and Jenkinson D S 1985 Chloroform fumigation and the release of soil nitrogen: a rapid direct extraction method to measure microbial biomass nitrogen in soil. Soil Biol. Biochem. 17, 837–842.

Cabrera M L 1993 Modeling the flush of nitrogen mineralization caused by drying and rewetting soils. Soil Sci. Soc. Am. J. 57, 63–66.

Deubel W D and Leithold G 1993 Ergebnisse zum Einfluß von langjährigen Fruchtfolge- und Düngemaßnahmen auf die mikrobielle Biomasse in alten Dauerfeldversuchen. VDLUFA-Schriftenr. 37, 193–196.

Dommergues Y R, Belser L W and Schmidt E L 1978 Limiting factors for microbial growth and activity in soil. Adv. Microbiol. Ecol. 2, 49–104.

Engels T and Kuhlmann H 1993 Effect of the rate of N fertilization on apparent net mineralization of N during and after cultivation of cereal and sugar beet crops. Z. Pflanzenernähr. Bodenkd. 156, 149–154.

Helal H M and Sauerbeck D 1985 Umsatz von ^{14}C-markierten Pflanzenresten und Veränderung der mikrobiellen Biomasse im Boden unter dem Einfluß von Maiswurzeln. Landwirtsch. Forsch. 38, 104–109.

Houba V J G, Novozamsky I, Huybregts A W M and van der Lee J J 1986 Comparison of soil extractions by 0.01 M CaCl$_2$, by EUF and by some conventional extraction procedures. Plant and Soil 96, 433–437.

Jansson S L and Persson J 1982 Mineralization and immobilization of soil nitrogen. *In* Nitrogen in Agricultural Soils. Ed. F J Stevenson. pp 229–252. Agronomy Monograph No 22, American Society of Agronomy, Madison, WI, USA.

Jenkinson D S 1988 Determination of microbial biomass carbon and nitrogen in soil. *In* Advances in Nitrogen Cycling in Agricultural Ecosystems. Ed. J R Wilson. pp 368–386. CAB International, Wallingford, UK.

Jenkinson D S, Fox R H and Rayner J H 1985 Interaction between fertilizer nitrogen and soil nitrogen - the so-called "priming" effect. J. Soil Sci. 36, 425–444.

Kieft L T, Soroker E and Firestone M K 1987 Microbial biomass response to a rapid increase in water potential when dry soil is wetted. Soil Biol. Biochem. 19, 119–126.

Marumoto T, Kai H, Yoshida T and Harada T 1977 Drying effect on mineralization of microbial cells and their cell walls in soil and contribution of microbial cell walls as a source of decomposable soil organic matter due to drying. Soil Sci. Plant Nutr. 23, 9–19.

Müller T, Jörgensen R G and Meyer B 1992 Estimation of soil microbial biomass C in the presence of living roots by fumigation-extraction. Soil Biol. Biochem. 24, 179–181.

Olfs H W and Werner W 1989 Veränderungen extrahierbarer "N_{org}"-Mengen unter dem Einfluß variierter C/N-Verhältnisse und Biomasse. VDLUFA-Schriftenr. 28, 15-26.

Paul E A and Clark F E 1989 Soil Microbiology and Biochemistry. Academic Press, San Diego, CA, USA. 275 p.

Scherer H W, Werner W and Rossbach J 1992 Effects of pretreatment of soil samples on N mineralization in incubation experiments. Biol. Fertil. Soils 14, 135–139.

Seneviratne R and Wild A 1985 Effect of mild drying on the mineralization of soil nitrogen. Plant and Soil 84, 175-179.

Stevenson I L 1956 Some observations on the microbial activity in remoistened air-dried soils. Plant and Soil 8, 170–182.

Van Gestel M, Merckx R and Vlassak K 1993 Microbial biomass responses to soil drying and rewetting: the fate of fast- and slow-growing microorganisms in soils from different climates. Soil Biol. Biochem. 25, 109–123.

Widmer P, Brookes P C and Parry L C 1989 Microbial biomass nitrogen measurements in soils containing large amounts of inorganic nitrogen. Soil Biol. Biochem. 21, 865–867.

O. Van Cleemput et al. (eds.), Progress in Nitrogen Cycling Studies, 121–125, 1996.

Nitrogen tranformation during the composting of different organic wastes

C. Paredes, M.P. Bernal[1], J. Cegarra, A. Roig and A.F. Navarro
Department of Soil and Water Conservation and Organic Waste Management, Centro de Edafología y Biología Aplicada del Segura, CSIC. P.O. Box 4195, 30080-MURCIA, Spain. [1]*Corresponding author*

Key words: ammonia volatilization, composting, mineralization, nitrification, organic wastes

Abstract

The transformation of nitrogen in different organic waste mixtures was studied during composting by the Rutgers static pile system. Three piles were prepared using cotton waste and sewage sludge as common factors in two of them. The greatest degradation of organic matter occurred in the pile of cotton waste and sewage sludge, which also showed the greatest degree of organic-N mineralization (26% organic-N), while the lowest OM degradation occurred in the pile of cotton waste and poultry manure with a total organic-N mineralization of 17.6% of the initial organic-N, indicating that the use of sewage sludge results in greater microbial activity in the composting mixture than when poultry manure is used. Substituting the cotton waste by maize straw decreased OM degradation, while no organic-N mineralization occurred. The organic-N mineralization augmented NH_4-N concentrations and pH values of the piles. Losses of N by NH_3-volatilization were observed when both high NH_4-N concentration and high pH value occurred simultaneously.

Introduction

The organic matter traditionally added to soils was animal manure composted with straw. However, the increased production of urban, agroindustrial and animal wastes has led to the substitution of manure and straw by other wastes in the production of composts. The transformation and loss of nitrogen has been widely studied in manures and slurries (Bernal et al., 1993; Bernal and Roig, 1993; Hansen et al., 1989; Kirchmann, 1985; Van Faasen and Van Dijk, 1979), and city refuse (Bhoyar et al., 1979), although there are not many studies of other wastes, especially when the Rutgers composting system is used. During the first phase of the process, the combination of high ammonium concentration, high temperature and elevated pH leads to substantial ammonia losses (Witter, 1986). Nitrogen losses during composting can be reduced by increasing the effectiveness of the system, which implies an appropriate selection of the wastes to be used and an adequate management of the composting piles. Also it might be possible to regulate organic-N mineralization and NH_3-volatilization from the composting material by controlling the pile's ceiling temperature during composting. The aim of this work is to evaluate nitro-gen transformation during the composting of different organic wastes by the Rutgers static pile system and to indicate the factors which might help reduce nitrogen loss.

Materials and methods

Three different mixtures were prepared using cotton waste and sewage sludge as common factors in two of them. Olive-mill wastewater (OMW) was also added to adjust the moisture of the piles up to 70%, which also served as a way of recycling this liquid waste. The proportions were as follows (on a fresh weight basis):

Pile 1: 65.4% cotton waste + 34.6% poultry manure + 1.60 L kg^{-1} OMW

Pile 2: 67.9% cotton waste + 32.1% sewage sludge + 0.94 L kg^{-1} OMW

Pile 3: 53.0% maize straw + 47.0% sewage sludge + 1.80 L kg^{-1} OMW

About 1500 kg of the mixtures were placed in trapezoidal piles of 1–1.5 m high with a 2 × 3 m base. The Rutgers static pile composting system was used, involving on-demand ventilation through temperature feedback control (Finstein et al., 1985). The air was blown from the base of the pile through the holes of three PVC tubes of 3 m length and 12 cm diameter. The timer was set for 30 sec. ventilation every 15 minutes, and the ceiling temperature for continuous air blowing was 55 °C. Piles 1, 2 and 3 were turned after 14, 35 and 49 days respectively, in order to improve the homogeneity of the material and the fermentation process. The biooxidative phase of composting was considered finished when the temperature of the pile was stable and near to that of the atmosphere, this stage being reached after 49, 84 and 63 days of composting for piles 1, 2 and 3, respectively. The air-blowing was then stopped to allow the compost to mature over a period of two months. The piles were sampled weekly during the biooxidative phase and after the maturation period. Each sample was subdivided into two subsamples, one of which was immediately frozen and kept for NH_4-N and NO_3-N analysis, while the other was air-dried and ground to 0.02 mm for analysis.

Moisture content of the samples was ascertained by drying at 105 °C, electrical conductivity and pH in water soluble extract 1:10 (w/v), organic matter (OM) by loss-on ignition at 430 °C during 24 hours (Navarro et al., 1993). Inorganic-N was extracted with 2 M KCl from the frozen subsamples and NH_4-N was determined by a colorimetric method based on Berthelot's reaction (Sommer et al., 1992), NO_3^--N was determined by the ultraviolet technique (APHA et al., 1987). Total nitrogen and organic carbon were determined by automatic microanalysis (Navarro et al., 1991). Losses of organic matter and nitrogen from the piles during composting were calculated from the initial (X_1) and the final (X_2) ash contents (Viel et al., 1987) according to the Equation (1). The same equation was used to determine N-losses by substituting the OM concentrations (values in brackets) for the corresponding N values.

$$OM - loss(\%) = 100 - 100[X_1(100 - X_2)]/[X_2(100 - X_1)] \tag{1}$$

Results and discussion

Organic matter concentration decreased at a similar rate in both piles containing cotton waste until the seventh week (Tables 1 and 2), except for the greater reduction of OM in pile 2 during the first week, which may have been due to the degradation of the labile organic compounds of the sewage sludge. However, when sewage sludge was used the fermentation phase of composting lasted almost double that when poultry manure was used (< 7 weeks in pile 1, 12 weeks in pile 2). Hence, the OM concentration of the former decreased by 20 %, while that of pile 2 decreased by 30 % over the whole period, indicating higher microbial activity in the latter pile brought about by the greater microbial load of the sewage sludge. A comparison of pile 2 and 3, which used the same sewage sludge shows that the degree of OM degradation in the mixture with cotton waste was greater than that in the pile with maize straw. Organic matter concentration decreased by only 11% in pile 3, the lowest of the three piles (Table 3), emphasising the lower biodegradability of the maize straw compared with the cotton waste.

In all piles total-N increased during the first phase of the process due to the concentration efect of the degrading organic-C compounds. The initial poultry manure mixture (pile 1) had a high concentration of total-N (Table 1), a considerable part of which was NH_4-N (12.5% of total-N), which was lost at a high rate during the first week as shown by the decrease of NH_4-N concentration and the total-N losses during the first 7 days of composting (Figs. 1a, 2a). The increase in pH during the same period strongly favoured these NH_4-N losses through NH_3-volatilization. From this period onwards, the NH_4-N value was almost constant until 28 days, when it started to decrease again. However, in pile 2 NH_4-N increased until day 49, when there was a sharp reduction, its concentration then remaining constant until the end of the process (Fig. 1b). The pH values also increased up to day 35 and decreased after 56 days of composting (Fig. 1b). Therefore, the initial production of NH_4-N through the mineralization of organic-N lasted up to day 49, and was followed by nitrification of the NH_4-N, as reflected by the increased NO_3-N level and the pH reduction. Since the pH values were similar in both piles (1 and 2) and the mineralization rate of pile 2 was only slightly higher than that of pile 1 during these first 49 days of composting (Fig. 2a and b), and since both used the same bulking agent, the only possible explanation for the difference in the NH_4-N evolution pattern is that a greater proportion of N was susceptible to volatilization in pile 1 than in pile 2. This was a consequence of the high initial NH_4-N concentration, the high pH and the extra ventilation due to the high temperature reached when fermentation started in pile 1, all these condition favoured the loss of NH_3 (Bhoyar et al., 1979). Pile 3, which had the lowest initial total-N concentration, the greatest

Table 1. Evolution of some parameters of cotton waste and poultry manure (pile 1) during composting by the Rutgers system

Days of composting	Ash (%)	OM (%)	C (g kg^{-1})	C/N	Total-N (g kg^{-1})	Organic-N (g kg^{-1})
0	21.58	78.42	407.2	15.03	27.1	21.6
7	24.79	75.21	392.2	13.29	29.5	25.1
14	29.26	70.74	371.2	11.75	31.6	27.1
21	30.37	69.63	359.5	11.52	31.2	27.1
28	33.22	66.78	353.3	10.48	33.7	29.9
35	35.63	64.37	340.7	10.23	33.3	29.4
42	35.99	64.01	345.6	10.13	34.1	30.2
49	36.87	63.13	334.3	9.66	34.6	30.4
Mature	37.10	62.90	337.3	9.72	34.7	30.2

Table 2. Evolution of some parameters of cotton waste and sewage sludge (pile 2) during composting by the Rutgers system

Days of composting	Ash (%)	OM (%)	C (g kg^{-1})	C/N	Total-N (g kg^{-1})	Organic-N (g kg^{-1})
0	19.26	80.74	405.4	21.11	19.2	15.9
7	26.22	73.78	388.5	16.05	24.2	21.1
14	26.47	73.53	351.0	13.98	25.1	21.8
21	31.67	68.33	332.5	12.41	26.8	23.2
28	33.86	66.14	329.7	11.37	29.0	24.9
35	37.48	62.52	336.8	10.79	31.2	25.5
42	38.53	61.47	333.1	10.96	30.4	26.3
49	40.58	59.42	316.7	10.74	29.5	24.6
56	41.53	58.47	320.8	10.35	31.0	28.4
63	39.33	60.68	337.2	10.57	31.9	28.9
70	41.46	58.54	314.3	10.24	30.7	27.4
77	45.20	54.80	301.9	9.83	30.7	27.3
84	43.67	56.33	300.8	10.06	29.9	26.7
Mature	43.57	56.43	393.7	9.44	31.1	27.2

Table 3. Evolution of some parameters of maize straw and sewage sludge (pile 3) during composting by the Rutgers system

Days of composting	Ash (%)	OM (%)	C (g kg^{-1})	C/N	Total-N (g kg^{-1})	Organic-N (g kg^{-1})
0	10.45	89.55	472.0	31.05	15.2	11.7
7	11.51	88.49	468.2	27.38	17.1	14.5
14	11.56	88.44	439.2	27.11	16.2	14.0
21	12.84	87.16	435.2	23.78	18.3	16.1
28	17.05	82.95	408.8	18.41	22.2	20.0
35	16.05	83.95	432.5	16.76	25.8	23.6
42	16.74	83.26	430.2	14.99	28.7	26.2
49	19.50	80.50	409.9	13.89	29.5	26.9
56	21.20	78.80	414.1	13.53	30.6	27.9
63	20.32	79.68	415.4	13.85	30.0	27.0
Manure	25.25	74.75	394.3	11.84	33.3	29.5

124

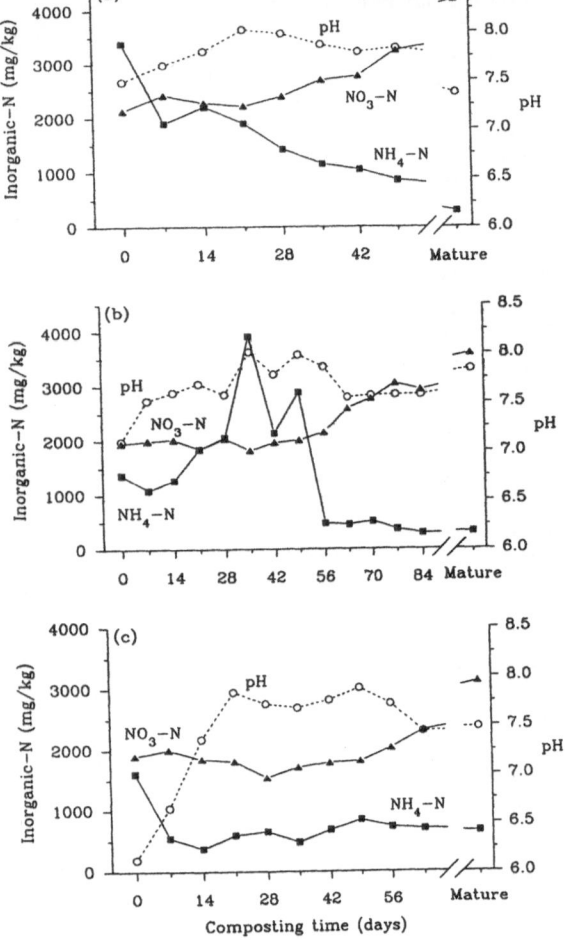

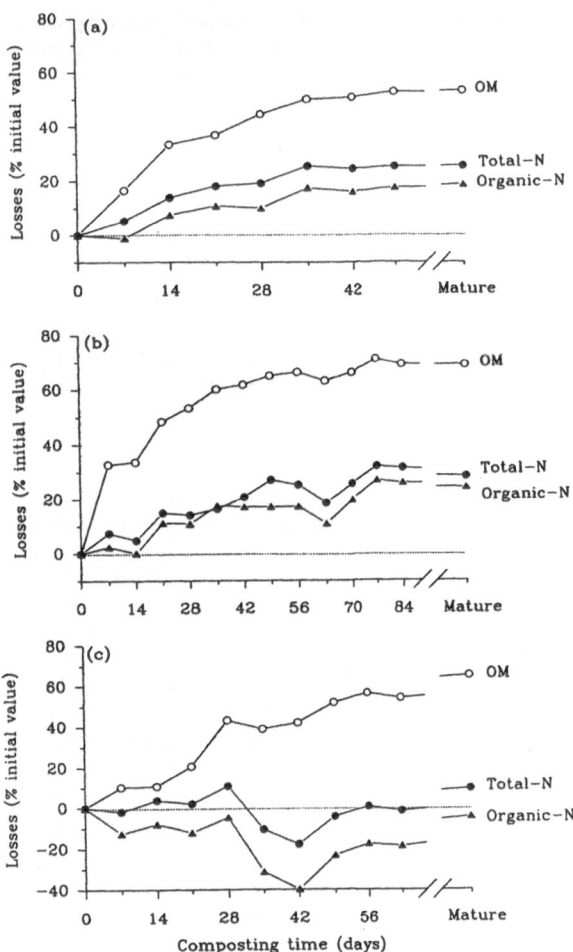

Figure 1. Changes in NH₄-N and NO₃-N concentrations and pH values during the composting of cotton waste with poultry manure (pile 1) (**a**), cotton waste with sewage sludge (pile 2) (**b**) and maize straw with sewage sludge (pile 3) (**c**).

Figure 2. Losses of organic matter, total-N and organic-N during the composting of cotton waste with poultry manure (pile 1) (**a**), cotton waste with sewage sludge (pile 2) (**b**), and maize straw with sewage sludge (pile 3) (**c**).

C/N ratio and initial pH < 7.0 (Table 3), showed a substantial reduction of NH₄-N during the first week (Fig. 1c). This was due to a microbial immobilization, which may have been associated with the high initial C/N ratio. The level of NH₄-N was almost constant after the second week, which indicates that the rate of NH₄-N production through mineralization was very low or similar to the rate of organic matter degradation.

The nitrate curve shapes for the three piles (Fig. 1) were similar with very high initial concentrations of NO₃-N in the three piles. As the increases of NO₃-N indicate, the nitrification process started after 28 days in pile 1 and after 49 days of composting in both piles 2 and 3. In all cases, nitrification was not observed until OM degradation had started to slow down (Fig. 2) and the temperature decreased to values close to 40

°C, which favours the presence of nitrifying bacteria (De Bertoldi, 1983). There was also an increase of NO₃-N in the three piles during the maturation period, which reflected the nitrification of the NH₄-N during this phase.

The OM and organic-N mineralization are shown by their precentage lost (Fig. 2). Pile 2 had an OM mineralization rate double that of pile 1 during the first week, and the total proportion mineralized over the whole the composting process was also higher (reaching 69.1% during its longer composting time). This supports the idea that the sewage sludge had a greater microbial activity than the poultry manure. Pile 3 showed lower OM mineralization than pile 2 especially during the first three weeks, revealing that the maize straw was more resistant to microbial degradation than the cotton waste. Organic-N mineralization followed the same pattern in piles 1 and 2 with similar values being observed as can be seen from the loss of organic-N (Fig. 2). However, pile 2 showed greater N-mineralization because of its longer composting time. The degradation rate of the OM and the organic-N mineralization rate depended more on the kind of bulking agent than on the nitrogen-rich organic waste. However, the duration of the composting process depended mainly on the nitrogen-rich waste which supplied most of the microbial population to the mixture.

Losses of total-N during composting are mainly due to NH_3-volatilization (Bishop and Godfrey, 1983). N-losses by denitrification were probably very low, since this process occurs under anaerobiosis and oxygenated conditions prevailed in the composting mixture. Also, very low lixiviation was observed from the piles, and so NO_3-lixiviation was presumably negligible. Nitrogen losses were always higher in the pile using poultry manure than in that using sewage sludge due to the higher NH_4-N concentration at the beginning of the process. In pile 3, total-N loss was very low, indicating little or no NH_3-volatilization. This was related to the low concentration of NH_4-N prevailing during the whole process, except during the first day which was mainly immobilized into organic fractions by the microorganisms, since a negative organic-N loss (mineralization) occurred (Fig. 2c). NH_4-N immobilization accounted for 43.7% of initial NH_4-N during 14 days of composting (5.5% of total-N). Bernal et al. (1993) found NH_4-N immobilization ranging from 58.8 to 86.9% of initial NH_4-N during 14 days of composting pig slurry with wheat straw. Mahimairaja et al. (1994) found a total-N loss of 11.19% after 12 weeks of aerobic incubation of poultry manure with maize straw, whereas in the present experiment the total-N loss during composting of poultry manure with cotton waste was higher (25.5%). Cereal straw was the most effective bulking agent in preserving NH_4-N (Mahimairaja et al., 1994). During the whole composting period of pile 3, both organic-N mineralization and losses of total-N were extremely low or zero (Fig. 2). N_2-fixation presumably occurred between days 35 and 49, when the temperature of the pile fell below 40 °C (De Bertoldi et al., 1985). The mature compost prepared with sewage sludge and maize straw had a very high total-N concentration, although the initial mixture had the lowest value. The nitrogen was preserved in this material, and so enriched the final product. The use of maize straw with a high C/N ratio as a bulking agent was more effective in reducing N-losses through NH_3-volatilization than the cotton waste, which had a lower C/N ratio.

References

APHA, AWWA, WPCF 1987 Standard Methods for the Examination of Water and Wastewater. 16th edition. American Public Health Association, Washington, USA.

Bernal M P, Lopez-Real J M and Scott K M 1993 Application of natural zeolites for the reduction of ammonia emission during the composting of organic wastes in a laboratory composting simulator. Biores. Technol. 43, 35–39.

Bernal M P and Roig A 1993 Nitrogen transformations in calcareous soils amended with pig slurry under aerobic incubation. J. Agric. Sci. 120, 89–97.

Bhoyar R V, Olaniya M S and Bhide A D 1979 Effect of temperature on mineralization of nitrogen during aerobic composting. Indian J. Environ. Health 21, 23–34.

Bishop P L and Godfrey C 1983 Nitrogen transformations during sludge composting. BioCycle, August, 34–39.

De Bertoldi M, Vallino G and Pera A 1983 The biology of composting: a review. Waste Manage. Res. 1, 157–176.

De Bertoldi M, Vallino G and Pera A 1985 Technological aspects of composting including modelling and microbiology. In Composting of Agricultural and other Wastes. Ed. J K R Gasser. pp 27–41. Elsevier Applied Science Publishers, Barking, Essex, England.

Finstein M S, Miller F C, MacGregor S T and Psaranos K M 1985 The Rutgers strategy for composting. Process design and control. EPA Project Summary (EPA/600/52–851059) U.S. EPA, Washington, USA.

Hansen R C, Keener H M and Hoitink H A J 1989 Poultry manure composting: an exploratory study. Trans. ASAE 36, 2151–2157.

Kirchmann H 1985 Losses, plant uptake and volatilization of manure nitrogen during a production cycle. Acta Agric. Scand. Suppl. 24, 1–77.

Mahimairaja S, Bolan N S, Hedley M J and MacGregor A N 1994 Losses and transformation of nitrogen during composting of poultry manure with different amendments: an incubation experiment. Biores. Technol. 47, 265–273.

Navarro A F, Cegarra J, Roig A and Bernal M P 1991 An automatic microanalysis method for the determination of organic carbon in wastes. Commun. Soil Sci. Plant Anal. 22, 2137-2144.

Navarro A F, Cegarra J, Roig A and García D 1993 Relationships between organic matter and carbon contents of organic wastes. Biores. Technol. 44, 203–207.

Sommer S G, Kjellerup V and Kristjansen O 1992 Determination of total ammonium nitrogen in pig and cattle slurry: sample preparation and analysis. Acta Agric. Scand., Sect. B, Soil Plant Sci. 42, 146–151.

Van Faasen H G and Van Dijk H 1979 Nitrogen conversion during the composting of manure/straw mixtures. In Straw Decay and its Effect of Disposal and Utilization. Ed. E Grossbard. pp 113–119. John Wiley and Sons, New York, USA.

Viel M, Sayag D, Peyre A and André L 1987 Optimization of in-vessel co-composting through heat recovery. Biol. Wastes 20, 167–185.

Witter E 1986 The fate of nitrogen during high temperature composting of sewage sludge-straw mixtures. PhD thesis, Wye College, University of London, UK.

O. Van Cleemput et al. (eds.), Progress in Nitrogen Cycling Studies, 127–132, 1996.

Decomposition of oil palm empty fruit bunches in the field and mineralization of nitrogen

A.B. Rosenani and S.F. Hoe

Department of Soil Science, Faculty of Agriculture, Universiti Pertanian Malaysia, 43400 Serdang, Selangor, Malaysia

Key words: decomposition, organic fertilizer, mineralization, N availability

Abstract

Currently, most of the EFBs (empty fruit bunches) that come out of the mill as waste are used as an organic fertilizer for oil palms. An experiment was conducted to study the decomposition of EFBs applied single-layered or double-layered under oil palm field conditions, and to investigate the availability of N as a result of mineralization in 15 weeks period. The experiment was set up using lysimeters filled with topsoil of the area (an Oxisol). The EFB biomass had reduced to 50% of initial DM (dry matter) weight in 7.5 weeks in the double-layered and 8.5 weeks in the single-layered EFB. After 15 weeks only 32% and 29% of the EFB initial dry matter weights were left in the single-layered and double- layered, respectively. The C/N ratios had reduced from 57 to 31 in the single-layered and from 55 to 24 in the double-layered EFB. About 50% of the initial EFB N was left after 15 weeks. The total nitrogen in the soil had increased from 0.23% to 0.27% in the single-layered and 0.28% in the double-layered EFB. The soil pH increased by 2–3 units. However, very small amounts of mineral N were found in the leachate indicating that in the 15 week period little mineralized N was available for plant uptake. This was probably due to immobilization by microbes, retention by the EFBs together with moisture, and also lost through NH_3 volatilization and denitrification.

Introduction

One of the main waste products of palm oil industry is the empty fruit bunch (EFB). For every tonne of oil palm fresh fruit bunches (FFBs) that goes to the mill, 20–25% of it are EFBs which are residues after FFBs are steamed and the fruits removed for oil extraction. With increasing production of palm oil, it is projected that in the year 2000, 9.5 million tonnes of EFBs would be produced (Vikineswary and Ravoof, 1990). The EFBs are rich in nutrients, particularly K, i.e. 0.65–0.94% N, 0.18–0.27% P_2O_5, 2.0–3.9% K_2O, 0.25–0.40% MgO and 0.15–0.48% CaO (Gurmit Singh et al., 1985). Traditionally, the EFBs were incinerated and the ash (high in K_2O) was applied in oil palm plantations as a K source for the palms. In recent years, however, with increasing awareness of the envinroment and the need to practise sustainable crop production, the EFBs are returned to the fields as an organic fertilizer and a green mulch to the palms.

The EFBs are placed in the oil palm fields in a heap in the middle of 4 mature palms at rates of 150–250 kg/palm/year or in a circle around young palms. Field trials had shown that EFB application could reduce the immature period of palms by several months and increase the yield when EFBs were applied in combination with inorganic fertilizers (Chan et al., 1991). EFB mulching, has also been found to improve the soil structure and water holding capacity, reduce soil temperature (Hoong and Nadarajah, 1988) and reduce soil erosion (Gurmit Singh et al., 1981).

Although the agronomic and economic benefits of EFB mulching in oil palm plantations have been proven in field trials, understanding and knowledge in the decomposition process of EFBs in the field and mineralization of nutrients is greatly lacking. This paper presents and discusses the results of a study which was conducted in an oil palm field aimed at investigating the decomposition of EFBs, applied single-layered and double-layered, and mineralization of nitrogen in a period of 15 weeks (September to December 1993).

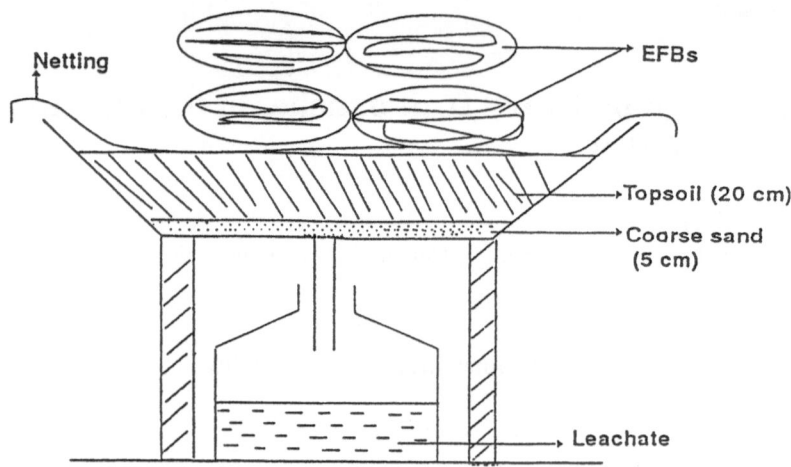

Figure 1. EFBs placed on soil surface in a lysimeter.

Materials and methods

This study was conducted using small lysimeters (made from round plastic containers of 55 cm diameter) which were filled with 20-cm layer of topsoil of the experimental area (as shown in Fig. 1). The soil used was classified as Gajah Mati series (Typic Hapludox), with 74.5% clay, 6.7% silt and 18.8% sand, pH 4.8 and CEC of 7.2 cmol(+) kg^{-1}. The soil surface was covered with a piece of fine mesh nylon netting before placing 2 EFBs for single-layered or 4 EFBs for double-layered treatments. A control, soil without EFBs, was also included. The treatments were replicated 4 times and laid out in a complete randomized design. Similar sized EFBs were used for the study; an average DM weight of 1.73 kg. The C/N ratio of the EFBs was 54–57, and the nutrient contents were 44–45% C, 0.80–0.90% N, 0.09–0.11% P and 2.40–2.50% K. The EFBs were allowed to decompose under natural field conditions under the oil palm canopy. As the EFBs decomposed, the mineralized nutrients were carried with the rain water into the soil and collected with the leachate in the bottle at the bottom of the lysimeter. N released from the decomposing EFBs into the soil was thus collected and determined.

Composite samples of EFB tissue and soil were taken every 3 weeks to determine change in DM weight, C/N ratio and N content of the EFB and soil pH. The total fresh weight of the EFBs was recorded and the moisture content determined. The DM weight of the tissue samples removed every 3 weeks was corrected for in the calculation of the total EFB dry matter weight in the following sampling time by taking into account the rate of weight loss in the 3 weeks period.

After 15 weeks of decomposition, larger amount of soil samples were taken to analyse for total N, mineral N, organic carbon, pH, and CEC (cation exchange capacity). Leachates that drained out of the lysimeters were collected and analysed weekly for mineral N.

Total N in the EFB tissue and soil were determined by the semimicro-Kjeldahl procedure using salicylic acid to include nitrate and nitrite (Bremner and Mulvaney, 1982). The organic carbon content of EFB tissue was determined according to Mebius method, and Walkley-Black method for organic carbon content of soil (Nelson and Sommers, 1982). The NO_3-N and NH_4-N concentrations in the leachates were determined by steam distillation and titration with hydrochloric acid (Bremner, 1965). Soil pH was measured in a soil to water ratio of 1:2.5, and the CEC was determined using the leaching method with ammonium acetate at pH 7. The data obtained was statistically analyzed using analysis of variance in the SAS-PC software.

Results

Changes in EFB DM weight

The loss in DM weight (Fig. 2a) indicates that the double-layered EFBs decomposed slightly faster than the single-layered. For both treatments the DM weight dropped rapidly in the first 9 weeks. After the 9th week decomposition was slower. The EFBs achieved 50% weight loss in 7.5 weeks for the double-layered and 8.5 weeks for the single-layered. After 15 weeks only 29% and 32% of initial DM weight was left in the double-

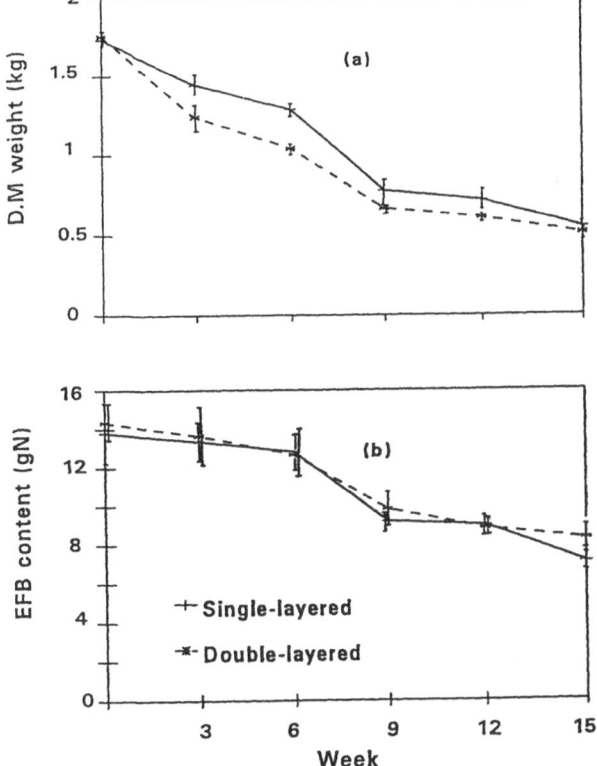

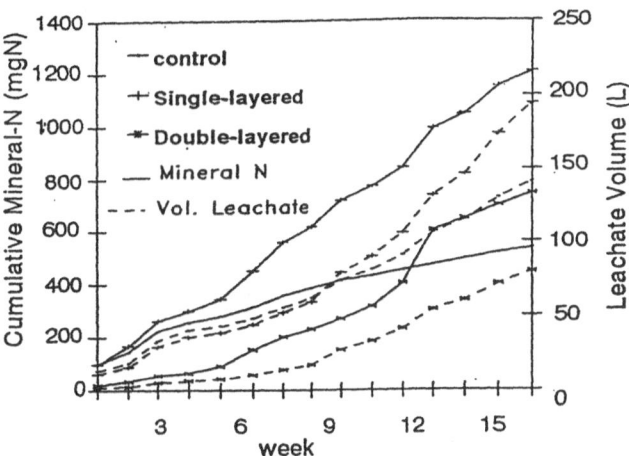

Figure 3. Cumulative mineral N in leachate and leachate volume during EFB decomposition.

Figure 2. Changes in DM weight of EFB (**a**) and total EFB N content (**b**), per bunch basis, with time of decomposition.

layered and single-layered, respectively. Although the weight loss was obvious the main structure of the EFB bunch had not disintegrated. During this 15 week period, the average weekly rainfall was 74 mm and temperature of 24–32 °C.

Changes in EFB nitrogen content and C/N ratio

The C/N ratio of the EFBs had reduced from 57.2 ±9.1 to 31.2 ±6.2 after 15 weeks of decomposition in the single-layered and 54.6 ±5.1 to 23.9 ±2.9 in the double-layered. This shows that the double-layered EFBs had decomposed to a slightly greater extent than the single-layered EFBs. As the decomposition process proceeded the percent N content of the EFBs increased steadily from about 0.83% to 1.62% in the doubled-layered and 1.30% in the singled layered. The amount of N in the EFB (Fig. 2b) had decreased more rapidly during the 6th to 9th week at the time when the weather was more consistently wet (average weekly rainfall of 82 mm). After 15 weeks there was a little more N left in the double-layered than the single-layered EFBs

(57.1% of intial N content in the former and 52.0% in the latter).

Mineral N in leachate

Mineral N in the leachate was considered as N in soil that was vulnerable to plant uptake as well as to leaching loss. Comparatively, the amount of mineral N leached from the double-layered EFBs throughout the 15 weeks period of decomposition was much lower than the single-layered EFBs, even lower than the control (soil without EFBs) up to the 12th week (see Fig. 3). Figure 4 shows the amounts of NH_4-N and NO_3-N found in the leachate each week. The higher amounts of NO_3-N after the 4th week indicate the occurrence of nitrification. Although there was continuous release of mineral N from the EFBs as found in the leachates, the total amounts of mineral N collected in the leachates after 15 weeks decomposition in both the EFB treatments were not more than 1.2 g N. The amounts of mineral N found in the leachate were only small percentages of the total amounts of N released from the EFBs during decomposition (Fig. 2b).

Soil pH, total N, mineral N and organic carbon

The soil pH in both the EFB treatments had increased rapidly in the first 3 weeks of decompostion (Fig. 5). The soil pH had increased by 2.3 units in the single-layered and 3.3 units in the double-layered EFB treatment. The CEC, total N and organic C of the soils in the EFB treatments had also increased slightly in this

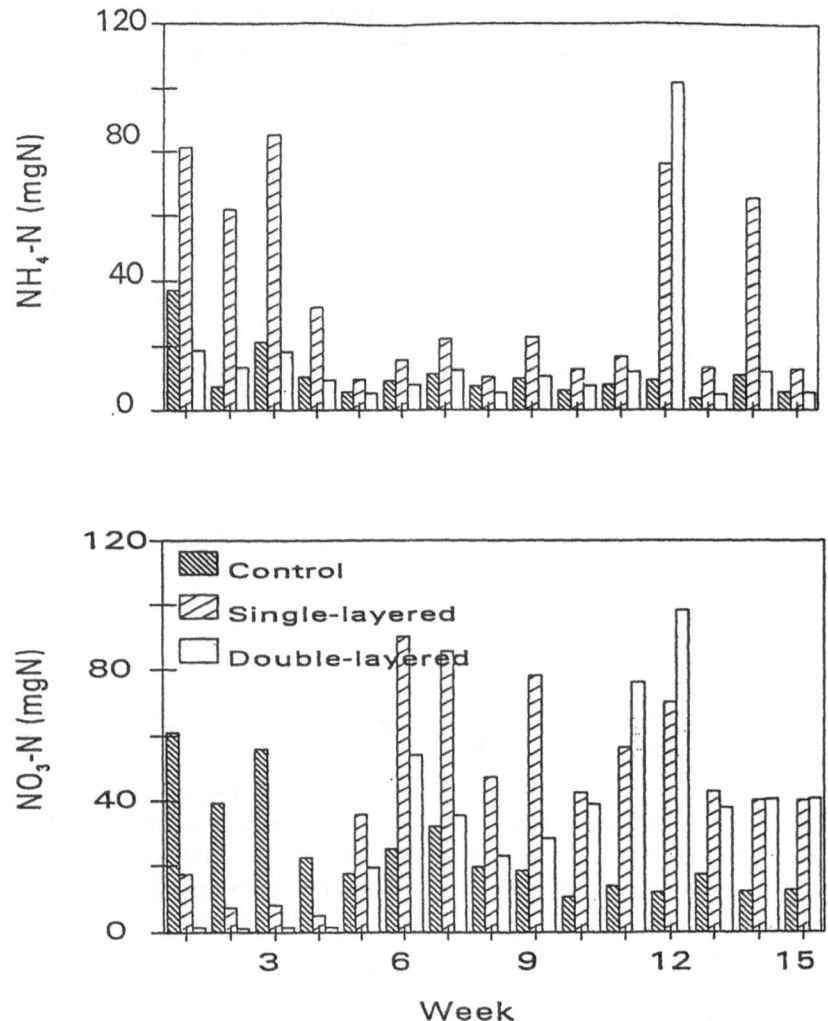

Figure 4. Weekly accumulated amounts of NH_4-N and NO_3-N in the leachates during EFB decomposition.

short period (Table 1). The C/N ratio of the soil had decreased slightly.

Discussion

Decomposition of an EFB was shown to be quite rapid with a half life of about 8 weeks . Decomposition rate of EFBs was slightly higher in the double-layered than the single-layered probably due to higher moisture retention and microenvironment temperature suitable for microbial activities. Although there was DM weight loss of 68–71% after 15 weeks, the physical structure of the bunch was still obvious. According to Yadvinder Singh et al. (1992), the easiliy decomposable parts of plant residues containing sugars, cellulose

and protein decomposed first, leaving the more resistant lignified structures. Decomposition rate of green manures or crop residues has been significantly correlated to the organic composition of lignin:N (Taylor et al.,1989; Thomas and Asakawa, 1993). Lim (1989) reported similar half life of EFB applied as a mulch on the ground, with 65% DM weight loss in 15 weeks.

In this study there was an average EFB N release of about 53% of initial N content in the 15 weeks period; slightly higher in the single-layered than in the double-layered. However, the total amounts of mineral N collected in the leachate were small, particularly in the double-layered EFB application. It could be deduced that in the short period of 15 weeks, very little of the released EFB N was available for plant uptake and or leaching. There could be a few reasons for this.

Table 1. Effects of EFB decomposition on soil chemical properties (mean ± std. deviation)

Soil chem. properties	Initial			> 15 weeks		
	Control	Single-layered	Double-layered	Control	Single-layered	Double-layered
CEC/cmol(+)kg^{-1}	7.2 ± 0.13	7.2 ± 0.21	7.2 ± 0.17	6.2 ± 0.14	8.3 ± 0.15	9.3 ± 0.19
pH	4.8 ± 0.10	4.8 ± 0.08	4.7 ± 0.05	4.7 ± 0.10	7.1 ± 0.08	8.0 ± 0.27
N(%)	0.17 ± 0.01	0.27 ± 0.01	0.28 ± 0.01	0.18 ± 0.01	0.27 ± 0.07	0.28 ± 0.02
C(%)	1.59 ± 0.04	1.58 ± 0.05	1.58 ± 0.03	1.38 ± 0.03	1.66 ± 0.03	1.76 ± 0.03
Soil moisture (% w/w)	19.9 ± 3.25	17.4 ± 1.86	16.1 ± 2.03	45.6 ± 2.80	58.2 ± 4.67	52.1 ± 1.83
NH_4-N (μg g^{-1})	–	–	–	26.0 ± 6.47	7.8 ± 2.89	7.6 ± 4.44
NO_3-N (μg g^{-1})	–	–	–	26.8 ± 3.82	16.2 ± 4.19	21.9 ± 0.71

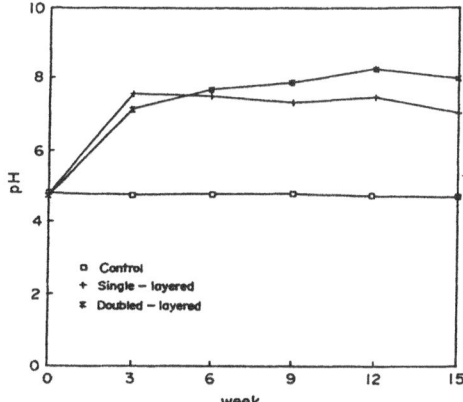

Figure 5. Changes in soil pH with time as affected by EFB applications.

Firstly, immobilization of N into organic forms by the microbial population. At the end of 15 weeks the C/N ratio of the EFBs was still in the range of 24 to 31, i.e. C/N ratios of net immobilization. Actually, several studies have shown that plant residues cannot have a constant decomposition rate, because they have 2 distinct N components, one decomposing rapidly and the other slowly (Yadvinder-Singh et al. 1992). The easily decomposable materials are partly synthesized into microbial products. Bolten et al. (1985) reported low leaching loss of N in soil receiving green manure due to immobilization. Also, part of the EFBs that disintegrated by microorganisms into <2mm sized particles was included in the soil organic matter fraction.

Secondly, an amount of the released NH_4-N could still be retained by the EFBs together with the moisture and therefore not released into the soil until a later period. The reason there was less mineral N found in the leachate from the double-layered lysimeters was probably because of higher moisture retention by the higher number of EFBs, compared to the single-layered application.

Thirdly, losses of mineralized N had possibly occurred through ammonia volatilization and denitrification. Moisture plays an important role in both the processes. During slightly drier days, the moisture content of the EFBs could be conducive for NH_3 volatilization. Nitrification may have occurred during this time too, evident by the NO_3-N in the leachates (Fig. 4). During wetter days, the EFBs may have retained more moisture and created anaerobiosis for denitrification of the NO_3-N. Also, readily decomposable organic matter stimulates microbial respiration which causes rapid oxygen consumption and accelerates onset of anaerobiosis. Bolten et al. (1985) reported signifigant increase in denitrifiers with green manure applications. The increase in soil pH (Fig. 5) could also increase the rate of volatilisation, nitrification and denitrification.

This study shows that EFBs applied in layers act as a slow release fertilizer with low possibility for N loss through leaching. However, N loss could probably occur through NH_3 volatilization and denitrification; this needs further investigation.

Acknowledgements

This study is part of a research project on oil palm funded by the Malaysian Government. We thank Golden Hope Plantation Banting for providing the EFBs and Miss Azizah Saad for preparation of the figures in this paper.

References

Bolten H, Jr. Elliot L F, Papandick R T and Bezdicek D F 1985 Soil microbial biomass and selected enzyme activities: Effect of fertilizers and cropping practices. Soil Biol. Biochem. 17, 297–302.

Bremner J M 1965 Inorganic nitrogen. *In* Methods of Soil Analysis, Part 2. Chemical and Microbiological Properties. Agron. Monogr. 9. Ed. C A Black. pp 1179–1237. ASA and SSSA, Madison, USA.

Bremner J M and Mulvaney C S 1982 Nitrogen - Total. *In* Soil Analysis Part 2: Chemical and Microbiological Properties, 2nd ed. Agron. Monogr. 9 Eds. A L Page et al. pp 595–614. ASA and SSSA, Madison, WI, USA.

Chan K W, Lim K C and Ahmad Alwi 1991 Fertilizer efficiency studies in oil palm. *In* Proc. 1991 PORIM international Palm Oil Development Conference, Kuala Lumpur, pp 302–311. Palm Oil Research Institute of Malaysia, Bangi, Malaysia.

Gurmit Singh, Manoharan S and Kanapathy K 1981 Commercial scale bunch mulching of oil palms. *In* Oil Palm in Agriculture in the Eighties, PORIM, Kuala Lumpur, pp 367–377.

Gurmit Singh, Manoharan, Toh Tai San and Thorairaj 1985 Optimal utilisation of empty fruit bunches and oil palm mill effluent. *In* Proc. Soil Science Dept., University, Pertanian Malaysia workhop, 27–29 June 1985, Serdang. pp 88–102.

Hoong H W and Nadarajah M N 1988 Mulching of empty fruit bunches(EFB) of oil palm. *In* SLDB/PORIM Workshop on Oil Palm Milling Technology, 1988, Kuala Lumpur, Malaysia. pp 38–50. Palm Oil Research Institute of Malaysia, Bangi, Malaysia.

Lim K H 1989 Trial on composting EFB of oil palm without prior shredding and liquid extraction. Proc. PORIM International Palm Oil Development Conference, Module II: Agriculture, 5–9 Sept. 1989, Kuala Lumpur. pp 212–224.

Nelson D W and Sommers L E 1982 Total carbon, organic carbon and organic matter. *In* Soil Analysis Part 2: Chemical and Microbiological Properties, 2nd Edition. Agron. Monogr. 9. Eds. A L Page et al. pp 539–579. ASA and SSSA, Madison, WI, USA.

Taylor B R, Parkinson D and Parsons W F J 1989 Nitrogen and lignin content as predictors of litter decay rates: a microcosm test. Ecology 70, 97–104

Thomas R J and Asakawa N M 1993 Decomposition of leaf litter from tropical forage grasses and legumes. Soil Biol. Biochem. 25, 1351–1361.

Vikineswary S and Ravoof A A 1990 Decomposition of oil palm empty fruit bunch in a sub-surface mulch. *In* Proc. The Regional Sem. on Management and Utilization of Agricultural and Industrial Wastes, 21–23 Mar.1990, Kuala Lumpur, Malaysia. Univ. of Malaya, Malaysia.

Yadvinder-Singh, Bijay-Singh and Khind C S 1992 Nutrient transformation in soils amended with green manures. *In* Advances in Soil Science, Vol. 20. Ed. B A Steward. pp 237–309. Springer-Verlag, New York Inc., USA.

O. Van Cleemput et al. (eds.), Progress in Nitrogen Cycling Studies, 133–139, 1996.

The effectiveness of the Rutgers system and the addition of bulking agent in reducing N-losses during composting

M.A. Sánchez-Monedero, M.P. Bernal[1], A. Roig, J. Cegarra and D. García
Department of Soil and Water Conservation and Organic Waste Management, Centro de Edafología y Biología Aplicada del Segura, CSIC. P.O. Box 4195, 30080-MURCIA Spain. [1]*Corresponding author*

Key words: composting, household biowaste, mineralization, NH_3-volatilization, sweet sorghum bagasse

Abstract

Two composting piles were prepared using a household biowaste and sweet sorghum bagasse mixed at a volume ratio of 50:50 (95: 5% fresh weight) with a C/N ratio of 17.7. One of them was composted by the Rutgers static pile system and the other by the turned system. A third pile was prepared with the same household biowaste alone and composted by turning (C/N = 13.9). The evolution of different nitrogen forms and the degradation of the organic matter were studied during the composting and maturation phases.

A very high ammonium concentration was produced in the biowaste when the turned pile technique was used due to mineralizatian of the organic fraction (85% of initial OM), although this was quickly lost through NH_3-volatilization during turning (63% of total-N). Adding the bulking agent to the turned pile reduced NH_4-N production, N-loss (57% total-N) and the OM degradation (76% initial OM). The addition of sweet sorghum bagasse as a bulking agent in the composting of biowaste is an effective way of reducing the nitrogen losses during the composting process.

However, use of the Rutgers system resulted in the lowest N-loss (41% total-N) and NH_4-production, due to the slower degradation of the organic fraction. The Rutgers system was more effective in retaining nitrogen during the composting process, thus preventing N-losses, than the traditional turned pile. This was due to the controlled temperature ceiling, which reduced the organic matter degradation ratio, and to the lack of turnings during the first stage of the process, when OM decomposition was at its highest. As a result, the final total-N concentration of the matured compost from the biowaste + bagasse mixture was 2.0% using the Rutgers system, and 1.5% using the turned pile technique.

Introduction

The composting of organic wastes is a biooxidative process involving the mineralization and partial humification of the organic matter, leading to a stabilized final product. During the first phase of the process the simple organic carbon compounds are easily mineralized and metabolized by the microorganisms, producing CO_2, NH_3, H_2O, organic acids and heat. The accumulation of this heat raises the temperature of the pile.

The composting of organic wastes rich in easily biodegradable nitrogen compounds leads to the formation, accumulation and subsequent loss of nitrogen mostly through ammonia volatilization. During the first phase of composting, the combination of high temperatures, high ammonium concentration and pH levels may lead to high losses of ammonia (Witter, 1986). Thus, losses of NH_4-N during composting depend on the rate of its production. The addition of carbon sources to wastes rich in NH_4-N results in its partial incorporation into the organic fractions or its immobilization to form such fractions (Van Faasen and Van Dijk, 1979). It may thus be possible to reduce ammonia losses during the composting of materials rich in nitrogen compounds. The loss of nitrogen from the compost piles also depends on the diffusion of NH_3 through the pile into the atmosphere, and frequent turning of the pile may facilitate this NH_3-volatilization (De Bertoldi et al., 1982).

The aim of this paper is to study the influence of the bulking agent and the composting system on the

Table 1. Chemical analysis of the organic waste materials

	Household biowaste	Sweet sorghum bagasse	Mixture
Ash (%)	34.71	3.78	37.48
Organic matter (%)	65.28	96.22	62.52
Organic-C (g kg^{-1})	325.0	520.6	321.4
Total-N (g kg^{-1})	23.3	14.2	18.2
C/N	13.95	36.66	17.66
Organic-N (g kg^{-1})	20.5	14.2	15.7
NH_4-N (mg kg^{-1})	1918	152	1611
NO_3-N (mg kg^{-1})	909	483	835
pH	6.64	7.0	6.73
Electrical conductivity (S m^{-1})	0.550	0.027	0.460
Lignin (%)	12.18	25.08	13.24
Cellulose (%)	22.7	34.4	20.4
Hemicellulose (%)	23.9	32.2	21.5

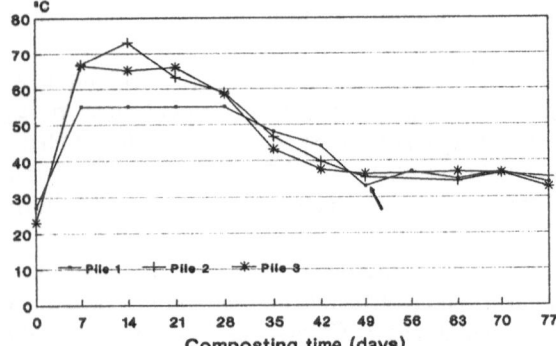

Figure 1. Temperature evolution of the piles during composting. The arrow indicates the turning of pile 1.

evolution of the nitrogen forms and on the reduction of N-losses during the composting of household biowaste.

Materials and methods

Two waste materials, a household biowaste and a sweet sorghum bagasse (Table 1), were used to prepare three composting piles:

Pile 1: Mixture of 50:50 v/v biowaste + bagasse (95% and 5% fresh weight, respectively) in a static pile (Rutgers system).

Pile 2: The same mixture as in pile 1 but in a turned pile.

Pile 3: 100% biowaste (no bagasse added) in a turned pile.

The piles (1–1.5 m high with a 2 × 3 m base) held about 1.500 kg of the waste materials. The Rutgers

static pile composting system used in pile 1, involves on-demand ventilation through temperature feedback control (Finstein et al., 1985). The air was blown from the base of the pile through the holes of three PVC tubes of 3 m length and 12 cm diameter. The timer was set for 30 sec ventilation every 15 minutes, and the ceiling temperature for continuous air blowing was 55 °C. Piles 2 and 3 were turned every two days during the first week of composting, twice a week during the second week, and once a week during the rest of the biooxidative phase. The biooxidative phase of composting was considered finished when the temperature of the pile was stable and near to that of the atmosphere. This stage was reached after 77 days of composting in all three piles (Fig. 1). The air-blowing was then stopped to allow the compost to mature over a period of two months. The piles were sampled weekly during the active phase and after the maturation period. Six subsamples were taken from the complete depth of the composting material and homogenously distributed in the pile at each sampling time. The total amount of material sampled was approximately 80 L. A representative composite sample of approximately 7 kg was obtained from the six subsamples after mixing, homogenizing them together and partitioning by 'quartering'. The excess of material was returned to the composting pile. Each sample was subdivided into two parts, one of which was immediately frozen and kept for NH_4-N and NO_3-N analysis, while the other subsample was air-dried and ground to 0.02 mm for analysis. All analyses were carried out in duplicate.

The following analytical methods were used: moisture content was ascertained by drying at 105 °C, electrical conductivity and pH in water soluble extract 1:10 (w/v), organic matter (OM) by loss-on-ignition at 430 °C during 24 hours (Navarro et al., 1993). Inorganic-N was extracted with 2 M KCl from the frozen subsample and NH_4-N was determined by a colorimetric method based on the Berthelot's reaction (Sommer et al., 1992), adding sodium nitrate to complex divalent cations (Kempers and Zweers, 1986), NO_3-N was determined by the ultraviolet technique (APHA et al., 1987) after treatment with activated charcoal to avoid any organic matter interference. Total nitrogen and organic carbon were determined by automatic microanalysis (Navarro et al., 1991). Lignin and cellulose were determined by the American National Standard Methods (ANSI/ASTM, 1977a and b) and holocellulose according to the method of Browning (1967). Losses of organic matter and nitrogen from the piles during composting were determined from the decomposition of the organic fraction. The OM losses were calculated from the initial (X_1) and the final (X_2) ash contents (Viel et al., 1987) according to the Equation 1.

$$OM - loss(\%) = 100 - 100\,[X_1\,(100 - X_2)]\,/\,[X_2\,(100 - X_1)] \quad (1)$$

Similar equation was used to determine N-losses by substituting the factor of OM concentrations (between brackets) for the final and initial N concentrations.

Results

Increases in the ash concentration and decreases in both organic matter (OM) and organic-C concentrations pointed to the decomposition of the organic wastes during composting in the three piles (Tables 2, 3 and 4). A greater reduction in OM concentration occurred in pile 2 using the turned system (from 62.52 to 28.20%) than in pile 1 which was composted with the static system (from 62.52 to 30.13%). The lowest OM values were obtained in pile 3 consisting of biowaste alone (from 65.28 to 22.49%), because the OM of the biowaste was more easily degradable by the microorganisms than that of the sweet sorghum bagasse, as shown by the concentration of lignin in the latter (Table 1). A similar pattern was observed with respect to the organic-C concentration.

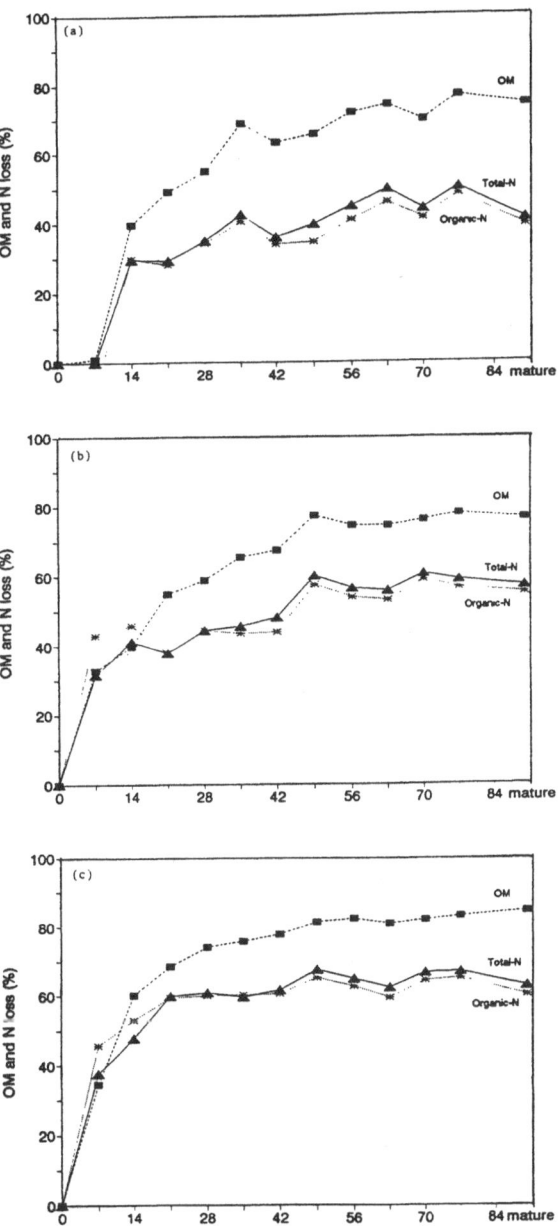

Figure 2. Losses of organic matter, total-N and organic-N during the composting of household biowaste with sweet sorghum bagasse by the Rutgers system (pile 1) (**a**), by turned pile (pile 2) (**b**) and of household biowaste by turned pile (pile 3) (**c**).

The total OM-loss followed the order pile 1 < pile 2 < pile 3 (Fig. 2), indicating the mineralization of the organic fraction of the piles, and thus the weight loss during composting. More than 84% of the OM of the biowaste was degraded during composting (pile 3). However, adding sweet sorghum bagasse to the household biowaste (pile 2) reduced the OM degra-

Table 2. Evolution of the organic matter and nitrogen of a household biowaste and sweet sorghum bagasse mixture during composting by the Rutgers system of static pile (pile 1)

Composting time (days)	OM (%)	C (g kg^{-1})	Total-N (g kg^{-1})	C/N	Organic-N (g kg^{-1})	pH	NH$_4$-N (mg kg^{-1})	NO$_3$-N (mg kg^{-1})
0	62.52	321.4	18.2	17.66	15.7	6.7	1611	817
7	62.30	313.9	18.1	17.34	17.0	6.8	2041	841
14	50.08	280.4	17.0	16.49	14.7	7.2	1529	792
21	45.93	232.8	18.6	12.52	16.3	7.5	1463	770
28	42.91	208.3	18.0	11.58	15.7	7.6	1614	646
35	38.23	216.6	18.3	12.97	16.3	7.9	1319	653
42	38.10	213.7	19.0	11.25	17.1	7.9	1366	664
49	36.44	207.0	18.5	11.19	17.4	8.0	245	866
56	32.14	195.1	18.1	10.78	16.8	8.3	220	1066
63	30.46	170.9	17.0	10.05	15.8	8.0	203	1012
70	30.83	181.6	18.0	10.09	16.2	7.8	266	1457
77	28.05	161.7	17.4	9.29	15.6	7.9	182	1559
Mature	30.13	156.7	20.0	7.84	17.8	8.0	184	2000

Table 3. Evolution of the organic matter and nitrogen of a household biowaste and sweet sorghum bagasse mixture during composting by turned pile (pile 2)

Composting time (days)	OM (%)	C (g kg^{-1})	Total-N (g kg^{-1})	C/N	Organic-N (g kg^{-1})	pH	NH$_4$-N (mg kg^{-1})	NO$_3$-N (mg kg^{-1})
0	62.52	321.4	18.2	17.66	15.7	6.7	1611	817
7	52.92	274.7	15.6	17.61	11.3	6.9	3522	824
14	50.09	256.8	14.2	18.08	11.4	7.6	2074	781
21	42.93	235.8	17.2	13.71	14.9	7.9	1558	702
28	40.73	211.4	15.9	13.30	13.8	8.4	1441	686
35	36.42	195.6	16.7	11.71	15.0	8.2	1010	656
42	35.11	206.4	16.3	12.66	15.2	8.0	368	693
49	27.45	151.8	14.0	10.84	12.9	8.2	296	812
56	29.92	169.4	14.7	11.52	13.6	8.1	306	850
63	29.85	154.7	15.0	10.31	13.8	8.1	280	889
70	28.63	156.7	13.6	11.61	12.3	8.1	258	1020
77	26.98	154.5	14.5	10.66	13.3	8.1	261	944
Mature	28.20	136.0	14.9	9.13	13.5	8.4	192	1194

dation to 76.45%, showing the greater resistance of the bagasse OM to microbial degradation than the OM of the household biowaste. When the same mixture was composted by static pile (pile 1) the OM loss was reduced to 74.15%.

The total-N concentration of pile 1 was almost constant during composting (Table 2), whereas in pile 2 total-N decreased during the first 14 days with small changes later on (Table 3), which can be due to the weight loss of the pile as a consequence of the OM degradation. The initial concentration of total-N in pile 3 was the highest, since it did not have any bulking agent reduced the total-N and NH$_4$-N concentrations and raised the C/N ratio of the initial composting material because of the low N concentration of the bagasse. A general decrease in total-N in pile 3 was observed during composting. Losses of total-N during the biooxidative phase occurred through NH$_3$-volatilization and accounted for 50.08% of the initial value in pile 1, 59.08% in pile 2 and 66.93% in pile 3 (Fig. 2). The losses of organic-N indicated its mineralization and occurred in the same order as that of total-N, the lowest being in pile 1 (40% of initial organic-N), the highest in pile 3 (60%).

Table 4. Evolution of the organic matter and nitrogen of a household biowaste during composting by turned pile (pile 3)

Composting time (days)	OM (%)	C (g kg^{-1})	Total-N (g kg^{-1})	C/N	Organic-N (g kg^{-1})	pH	NH$_4$-N (mg kg^{-1})	NO$_3$-N (mg kg^{-1})
0	65.28	325.0	23.3	13.95	20.5	6.4	1918	909
7	55.16	281.8	18.8	14.99	14.4	6.8	3540	878
14	42.81	222.5	20.1	11.07	15.9	7.3	3371	828
21	37.27	201.0	16.9	11.89	15.0	7.8	1164	755
28	32.73	176.1	17.7	9.95	15.9	7.9	1117	678
35	31.29	154.0	18.6	8.28	16.1	8.1	1846	691
42	29.57	155.9	18.2	8.57	16.4	8.1	1107	688
49	25.93	136.1	16.2	8.40	15.2	8.4	298	663
56	24.78	133.9	17.7	7.56	16.6	8.4	501	653
63	26.28	155.3	18.7	8.30	17.6	8.5	397	703
70	25.13	135.8	16.8	8.08	15.8	8.5	322	735
77	24.20	134.1	16.9	7.93	15.6	8.2	348	889
Mature	22.49	125.8	19.5	6.45	1.83	8.8	150	1019

In piles 1 and 2, a substantial amount of NH$_4$-N was produced during the first week of composting as showed by the increases in NH$_4$-N concentration (Tables 2 and 3). This was greater in pile 2 as a consequence of the faster organic-N mineralization (42.9% of initial organic-N, Fig. 2) since this process release NH$_3$. The total-N also decreased in the same period, indicating the losses of N through NH$_3$-volatilization caused by turning (Fig. 2). However, the NH$_4$-N concentration decreased slowly in pile 1 and greatly in pile 2 up to day 42. As no nitrification was observed, NH$_3$-volatilization may have occurred, since this is usually high during turning (De Bertoldi, 1982) and is also favoured by a high pH (> 7.5). After this time, NH$_4$-N fell greatly in pile 1 since nitrification started, aided by the drop of temperature below 35 °C. Losses of N through NH$_3$-volatilization may have happened in pile 1 on day 49, when the pile was turned in order to homogenize the materials in the internal and external parts of the pile. In addition, losses of total-N increased after 42 days of composting (Fig. 2). The concentration of NO$_3$-N increased and NH$_4$-N reached very low values in both piles from day 49 until the end of the process. The maximum NO$_3$-N concentration was found in pile 1 after the maturation phase, with 2000 mg kg^{-1} NO$_3$-N in the compost.

In pile 3 maximum NH$_4$-N production occurred during the first 14 days of composting (Table 4) due to organic-N degradation, which accounted for 52.96% of the initial organic-N (Fig. 2c). From day 14 onwards the NH$_4$-N concentration decreased due to NH$_3$-volatilization, since the pH values were higher than 7.5 and no NO$_3$-N was produced. Again, from day 49 onwards, the NH$_4$-N concentration was very low, although the NO$_3$-N concentration did not increase before day 63. The final NO$_3$-N concentration was the lowest of the three composts. Nevertheless, the concentrations of total and organic-N of the final compost were very close to those of the compost from pile 1. Since the starting value of total-N and organic-N in pile 3 were the greatest, higher N-losses occurred in pile 3 using the turning system than in pile 1 using the static system with bulking agent.

Discussion

The use of a lignocellulosic material such as sweet sorghum bagasse as a bulking agent in the composting of household biowaste led to a final product containing a high level of organic matter. The bulking agent was more resistant to microbial degradation due to its high lignin content. Maximization of the decomposition rate is a goal of a controlled process based on temperature feedback (Rutgers system). According to Finstein et al. (1985), a high rate of organic matter decomposition occurs through avoiding high pile temperatures by removing the excess heat produced by forced ventilation. However, the OM concentration of the compost produced by the Rutgers system (pile 1) was higher than that of the turned pile (pile 2,). Since analysis of the OM and organic-C cannot discern between fresh and stabilized OM (Finstein et al., 1986), humification and mineralization of the OM may have happened with the Rutgers system, leading to a high content of stabilized organic matter in the final compost. The compost produced under these conditions will be highly valuable due to its low ash concentration.

In the pile prepared with only household biowaste and using the turning system (pile 3), both total-N and NH_4-N concentrations decreased strongly during the first 21 days of composting and more slowly from then on. The losses of organic-N were continuous during the process, which led to increased pH values, the greatest changes occurring during the first 21 days of composting. All these factors indicated a continuous organic-N mineralization and therefore NH_4-N formation. A high pH and turning of the pile favoured N-loss through NH_3-volatilization, as revealed by the losses of total-N (Fig. 2).

However, when the sweet sorghum bagasse was added to the biowaste as a bulking agent, the initial total-N and NH_4-N concentrations decreased, giving a higher C/N ratio. The initial decrease in total-N was due to NH_3-volatilization of the NH_4-N formed by organic-N mineralization, since both total and organic-N diminished, the pH values increased and NH_4-N was produced. The slight increases in organic-N concentration after 14 days could be due to the weight loss of the pile, since organic-N losses did not increase during this period (Fig. 2), which indicated low or no organic-N mineralization. However, pH values rose, which may indicate mineralization. Therefore mineralization-immobilization of N might have occurred simultaneously. Bernal et al. (1993) found that NH_4-N immobilization can account for up to 87% of the initial NH_4-N during the composting of pig slurry with chopped wheat straw in a laboratory composting simulator. In the present experiment, the subsequent reduction of available organic-C led to a new and slow mineralization of the organic-N (Fig. 2b) with little changes in pH values (Table 3). Adding bulking agent to the household biowaste reduced the N-losses through NH_3-volatilization by decreasing the organic-N mineralization rate and therefore NH_4-N formation. A similar situation was described by Morisaki et al. (1989) during the composting of sewage sludge. These authors also found some immobilization of the NH_4-N formed. Nitrogen conversion was decreased by the addition of rice husks to sewage sludge, and so they concluded that NH_3-volatilization during sewage sludge composting could be reduced by increasing the amount of rice husks as a bulking agent. The compost prepared of household biowaste in a turned pile was more valuable as a N-fertilizer than that prepared with sweet sorghum bagasse under the same system, since the former had a greater concentration of nitrogen. However, its lower organic matter concentration meant it had a lower value as an organic substrate or as soil organic amendment.

When the household biowaste and bagasse mixture was composted by the Rutgers system the concentration of total-N changed very little. During the process, the mineralization of the organic-N was delayed. This started 7 days after the beginning of composting and was constant throughout the composting time, as shown by the loss of organic-N (Fig. 2a) and the increases in pH values (Table 2). In this way, peaks in NH_4-N production were avoided and its concentration in the pile was lower than in the other piles. At the same time, losses of total-N were lower than under the turned system, and so NH_3-volatilization was reduced by using the Rutgers system. Ashbolt and Line (1982) found that mixing the composting materials increases the rate of volatilization of the ammonia formed, and this is what happened in the turned pile, Also, De Bertoldi et al. (1982) found greater losses of N during the composting of urban waste with sewage sludge in a turned pile than in a static pile with forced aeration. Conditions that would maximize the build up of NH_3 in solution would also increase the probability of NH_3-volatilization. Thus, the higher N-losses from the turned pile through NH_3-volatilization were due to greater pH values, higher NH_4-N production through organic-N mineralization and the frequent turning. In the static pile the air was blown from the bottom to the external part of the pile. The NH_3 and H_2O produced during OM decomposition would be diffused with the air through the pile. Because the surface of the pile was cooler than the interior, the H_2O may condense and NH_3-N be absorbed by the composting material, helping to prevent N-losses. Thus, NH_4-N could be immobilized or nitrified by the microorganisms. It can be concluded that the Rutgers system was more effective in retaining the nitrogen and therefore in avoiding N-losses during the composting process than the traditional system of turned pile. The final compost obtained by static pile was of higher agronomical value as an organic fertilizer, soil organic amendment or organic substrate.

References

ANSI/ASTM 1977a Standard test method for lignin wood. D 1106. American National Standard.

ANSI/ASTM 1977b Standard test method for alpha-cellulose in wood. D 1103. American National Standard.

APHA, AWWA, WPCF 1987 Standard Methods for the Examination of Water and Wastewater. 16th edition. American Public Health Association, Washington, USA.

Ashbolt N J and Line M A 1982 A bench-scale system to study the composting of organic wastes. J. Environ. Qual. 11, 405–408.

Bernal M P, Lopez-Real J M and Scott K M 1993. Application of natural zeolites for the reduction of ammonia emission during the composting of organic wastes in a laboratory composting simulator. Biores. Technol. 43, 35–39.

Browning B L 1967 Methods of Wood Chemistry. Interscience Publ, New York, USA. 395 p.

De Bertoldi M, Vallini G, Pera A and Zucconi F 1982 Comparison of three windrow compost system. BioCycle 23, 45–50.

Finstein M S, Miller F C, MacGregor S T and Psaranos K M 1985 The Rutgers strategy for composting. Process design and control. EPA Project Summary (EPA/600/52-851059) US. EPA, Washington, USA.

Finstein M S, Miller F C and Strom P F 1986 Monitoring and evaluating composting process performance. J. Water Pollut. Control Fed. 58, 272–278.

Kempers A J and Zweers A 1986 Ammonum determination in soil extracts by the salicylate method. Commun. Soil Sci. Plant Anal. 17, 715–723.

Morisaki N, Phae C G, Nakasaki K, Shoda M and Kubota H 1989 Nitrogen transformation during thermophilic composting. J. Ferment. Biogen. 67, 57–61.

Navarro A F, Cegarra J, Roig A and Bernal M P 1991 An automatic microanalysis method for the determination of organic carbon in wastes. Commun. Soil Sci. Plant Anal. 22, 2137–2144.

Navarro A F, Cegarra J, Roig A and García D 1993 Relationships between organic matter and carbon contents of organic wastes. Biores. Technol. 44, 203–207.

Sommer S G, Kjellerup V and Kristjansen O 1992 Determination of total ammonium nltrogen in pig and cattle slurry: sample preparation and analysis. Acta Agric. Scand. Sect. B, Soil Plant Sci. 42, 146–151.

Van Faasen H G and Van Dijk H 1979 Nitrogen conversion during the composting of manurestraw mixtures. In Straw Decay and its Effect of Disposal and Utilization. Ed. E Grossbard. pp 113–119. John Wiley and Sons, New York, USA.

Viel M, Sayag D, Peyre A and André L 1987 Optimization of in-vessel co-composting through heat recovery. Biol. Wastes 20, 167–185.

Witter E 1986 The fate of nitrogen during high temperature composting of sewage sludge-straw mixtures. PhD thesis. Wye College, University of London, UK.

O. Van Cleemput et al. (eds.), Progress in Nitrogen Cycling Studies, 141–145, 1996.
© 1996 *Kluwer Academic Publishers*.

Short-term anaerobic storage of ^{15}N-labelled sheep urine does not influence the mineralization of nitrogen in soil

Peter Sørensen
Plant Nutrition, Environmental Science and Technology Department, Risø National Laboratory, DK-4000 Roskilde, Denmark

Key words: biomass N, immobilization, manure, soil texture, turnover

Abstract

The turnover of ^{15}N-labelled fresh and anaerobically stored urine was studied in two soils of different texture. ^{15}N-labelled urine was obtained by feeding a sheep on ^{15}N-labelled hay. A sandy (4.5% clay) and a sandy loam soil (17% clay) were incubated at 23 °C with 100 μg N g^{-1} soil in ^{15}N-labelled sheep urine or in (^{15}NH$_4$)$_2$SO$_4$. After 2 days of incubation, there was no significant effect of urine storage on the net mineralization of N. After 55 days of incubation, the total recovery of labelled urine-N was 96% in the sandy soil and 92% in the sandy loam soil. In the same soils 85% and 67% of the labelled urine-N was recovered as inorganic N, respectively. The net mineralization of urine-N was lower in the sandy loam than in the sandy soil because there was a larger increase of the biomass N concentration in the sandy loam soil after urine application.

Introduction

Urine nitrogen (N) is an important component of the N-cycle in agricultural systems since more than 60% of N excreted from cattle and sheep is in urine (Haynes and Williams, 1993). Most urine-N is found as urea or other easily decomposable compounds (Bristow et al., 1992). However, the utilization of urine-N in pastures is often low (Cuttle and Bourne, 1993; Thomas et al., 1988; Whitehead and Bristow, 1990), which may be due to losses of N by NH$_3$ volatilization (Whitehead and Bristow, 1990) and leaching (Cuttle and Bourne, 1993). Results reported by Williams and Haynes (1994) indicated that a significant mineralization-immobilization turnover (MIT) is taking place in urine patch affected soil. Thus MIT may also influence the utilization of urine-N.

In Denmark most livestock urine is collected in animal-houses and stored anaerobically in tanks, either alone or mixed with dung as a slurry. Stored urine can be injected or incorporated in soil which may reduce losses of N by NH$_3$ volatilization (Sommer and Christensen, 1990; Thompson et al. 1987).

The MIT in soil amended with animal manure is significantly influenced by the soil texture (Castellanos and Pratt, 1981; Sørensen et al. 1994b; Van Faassen and van Dijk, 1987) and it may also have an effect on the turnover of urine-N. ^{15}N-labelled urea has been used either alone (Keeney and Macgregor, 1978) or after mixing a small amount of ^{15}N-labelled urea with cattle urine (Whitehead and Bristow, 1990; Williams and Haynes, 1994) to study the cycling of urine-N in pasture. However, only 60–90% of urine-N from sheep and cattle urine is urea-N (Bristow et al., 1992), and the other organic N compounds in urine may be important for the total turnover of urine-N.

In this study ^{15}N-labelled sheep urine, obtained by feeding a sheep on ^{15}N-labelled hay, was used. The objectives were to determine 1) the mineralization of ^{15}N-labelled urine-N in a sandy and a sandy loam soil, and 2) the effect of anaerobic short-term urine storage on the mineralization of urine-N in the soil.

Materials and methods

Urine and soil

Urine was collected from an adult castrated sheep fed on ^{15}N-labelled hay of Italian ryegrass (4.52 atom% ^{15}N excess) as described by Sørensen et al. (1994a). Nine days after starting ^{15}N feeding, urine was collected during 24 h without being in contact with faeces and frozen. A subsample (500 mL) was anaerobically stored for 19 days at 22 ± 2 °C. Data on the fresh (frozen) and the stored urine are shown in Table 1. Soil was sampled from the plough layer of two arable fields, air-dried and sieved (2 mm). The sandy soil from Lundgård contained 0.11% total N, 1.4% total C, 4.5% clay, 4.2% silt and 89% sand (in dry matter). The water holding capacity (WHC) was 0.36 g g^{-1} dry soil, the CEC was 9.9 cmol(+) kg^{-1} and pH (H$_2$O) was 6.0. The sandy loam soil from Risø contained 0.14% total N, 1.3% total C, 17% clay, 16% silt and 65% sand. The WHC was 0.42 g g^{-1} dry soil, the CEC was 15 cmol(+) kg^{-1} and pH (H$_2$O) was 6.8.

Incubation procedure

Fifty gram soil samples (oven-dry weight basis) were preincubated in 250 mL polyethylene bottles at 40% WHC and 23 ± 1 °C. To prevent loss of soil moisture, the bottles were covered with aluminium foil with holes for aeration. After 7 days the soils were given one of the following treatments: 1) 100 μg N g^{-1} soil in ^{15}N-labelled ammonium sulfate (1.981 atom % ^{15}N enrichment) 2) 100 μg N g^{-1} soil in 1.01 mL ^{15}N-labelled fresh urine 3) 100 μg N g^{-1} soil in 1.01 mL ^{15}N-labelled stored urine. The treated soils were carefully mixed, water was added to 60% WHC and the incubation was continued. After 0, 2, 7, 14, 28 and 55 days three replicates of each treatment were analysed for NO$_3^-$-N and NH$_4^+$-N and ^{15}N. After 55 days, seperate soil samples were assayed for biomass and total N.

Chemical analysis

Total N and C in air-dried soil were determined by elemental analysis (Carlo Erba NA1500 N/C analyzer), and the ^{15}N enrichment was determined on a mass spectrometer (Delta, Finnigan MAT) coupled on-line to the elemental analyzer (Jensen, 1991). Total N in urine was determined using a semi-micro Kjeldahl method as described by Jensen (1991). Inorganic N was extracted by shaking 1 mL of urine or 10 g of soil with 100 mL 2M KCl for 1 h. Ammonium-N and NO$_2^-$-N + NO$_3^-$-N were measured on a Technicon Auto-analyzer II.

Inorganic N in extracts was concentrated for ^{15}N analysis using a diffusion procedure; after adding MgO, N in KCl extracts was diffused as NH$_3$ to an acidified glass filter enclosed in teflon tape (Sørensen and Jensen, 1991). If NO$_3^-$-N was included, Devarda's

alloy was also applied to the solution. Ammonium-N in Kjeldahl extracts was concentrated by diffusion as described by Jensen (1991).

Biomass N in soil was determined by fumigation-extraction (Brookes et al., 1985). Total N in extracts was analysed after persulfate oxidation (Cabrera and Beare, 1993), and biomass N (B$_N$) was calculated from $B_N = 2.22 \cdot E_N$, where E_N is the fraction of total N extracted after fumigation with CHCl$_3$. All results are expressed on an oven-dry basis (soil 105 °C, 24 h; urine 80 °C, 24 h).

Results and discussion

The urine was assumed to be homogeniously labelled with ^{15}N, since there was no significant difference in the ^{15}N enrichment of total N and NH$_4^+$-N before and after storage (Table 1). After 19 days of storage, 77% of total urine-N was NH$_4^+$-N (calculated from Table 1), which is comparable to observations by Whitehead and Raistick (1993) on short-term storage of cattle urine. After storage, all the urea and probably also some other easily decomposable N compounds were hydrolyzed to NH$_4^+$-N (Whitehead and Raistick, 1993).

After two days of incubation, no effect of urine storage was observed on the concentration of inorganic N in soil (Fig. 1a, b). Consequently, the urea-N was quickly hydrolysed in the soils receiving fresh urine, and the hydrolysis of N compounds during storage did not affect the percentage of urine-N that was present as inorganic N after 2 days. Similarly, Sørensen and Jensen (1995) found no effect of anaerobic storage of sheep faeces on the final concentration of inorganic N in soil, when similar amounts of fresh and stored faecal N had been applied.

The recovery of ^{15}N at the end of the experiment was 96% in the sandy soil and 91–93% in the sandy loam soil (Table 2). Losses of N from (NH$_4$)$_2$SO$_4$ and urine by NH$_3$ volatilization were probably negligible since the soil pH was low. The hydrolysis of urea in fresh urine usually causes a temporary rise in soil pH (Vallis et al., 1982), which may cause losses of N by NH$_3$ volatilization, especially when the concentration of urine is high in soil. However, the recovery of ^{15}N from fresh and stored urine was not significantly different. Soil pH decreased due to nitrification in all treatments, but the reduction was highest in the soil receiving (NH$_4$)$_2$SO$_4$ (Table 2). At the end of the experiment there was no significant effect of urine storage on soil pH. Minor losses of ^{15}N by denitrification

Table 1. Composition of fresh (frozen) and anaerobically stored urine (standard deviation in parentheses)

	Total N	NH$_4^+$-N	% d.m.	pH	Atom % ^{15}N excess	
	(mg mL^{-1})				Total N	NH$_4^+$-N
Fresh urine	4.95 (0.14)	0.217(0.002)	4.81	8.9	2.49 (0.06)	2.56 (0.02)
Stored urine	4.95 (0.08)	3.79 (0.01)	3.86	9.1	2.56 (0.04)	2.58 (0.02)

could not be excluded as not all the applied ^{15}N was accounted for, but losses were probably similar in all treatments since the total ^{15}N recoveries were similar.

After 2 days of incubation, the sandy soil amended with urine contained nearly as much inorganic nitrogen as the (NH$_4$)$_2$SO$_4$ treated soil (Fig. 1a), indicating that nearly all the urine-N was net mineralized. At day 55 the urine treated sandy soil contained 6 μg N g^{-1} less than the control. The difference in inorganic N between the sandy loam soil receiving urine and (NH$_4$)$_2$SO$_4$ was larger, and after 55 days the urine treated soil contained 20–23 μg N g^{-1} less than the control (Fig. 1b). Thus the net mineralization of urine-N was lower in the sandy loam soil. After 7 days, 71% of the applied ^{15}N in fresh urine was on inorganic form in the sandy soil, and 57% was on inorganic form in the sandy loam soil (Fig. 2). The concentration of inorganic ^{15}N continued to be lowest in the sandy loam soil, and at day 55, 85% of the ^{15}N applied in urine was recovered on inorganic form in the sandy soil while 67% was on inorganic form in the sandy loam soil (Table 2). At the end of the experiment, the concentration of soil biomass N in the urine treated sandy loam soil was 22–25 μg N g^{-1} soil higher than in the (NH$_4$)$_2$SO$_4$ treated soil, while the biomass N concentration of the urine treated sandy soil was only 12 μg N g^{-1} soil higher than in the (NH$_4$)$_2$SO$_4$ treated sandy soil (Table 2). Thus the lower net mineralization of urine-N in the sandy loam soil was mainly due to a larger increase of biomass N in the sandy loam soil with the highest clay content. The soil microbial biomass was probably more protected in the soil with the highest clay content (Amato and Ladd, 1992; Van Veen et al., 1985). Urine storage had no significant effect on the soil biomass N after 55 days, which indicate that most of the urine carbon utilized by the soil microbial biomass was not mineralized during storage. The microbial biomass utilizing carbonaceous compounds in urine would preferably immobilize NH$_4^+$-N in soil (Jansson, 1958). Since nearly all NH$_4^+$-N in soil was enriched with ^{15}N during the initial 7 days after urine application when MIT was largest, most immobilized

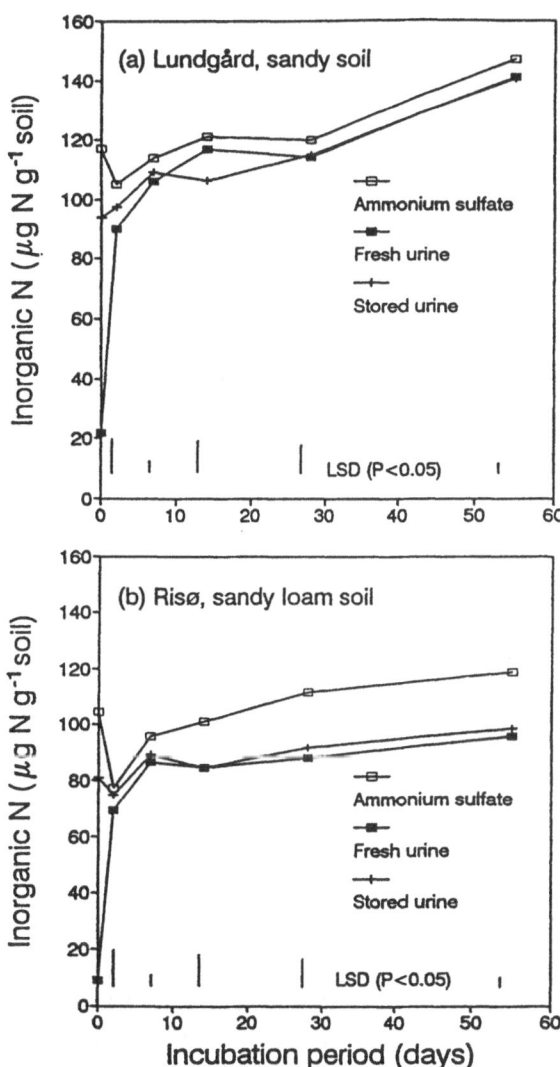

Figure 1. Concentration of inorganic N in two soils during 55 days of incubation with 100 μg N g^{-1} soil in fresh or stored urine or (NH$_4$)$_2$SO$_4$.

N would be ^{15}N-labelled. Thus the lower recovery of inorganic ^{15}N from urine in the sandy loam soil was mainly due to a higher amount of ^{15}N in the microbial biomass in this soil.

Table 2. Distribution of ^{15}N, soil biomass N and soil pH after 55 days of incubation of ^{15}N-labelled $(NH_4)_2SO_4$, fresh or stored urine in two soils

Soil	Treatment	% recovery of ^{15}N			Biomass N	pH(H$_2$O)
		Total N	Inorganic N	Non-exchangeable N[a]	(μg N g^{-1} soil)	
Sandy (Lundgård)	$(NH_4)_2SO_4$	95.8	94.1	1.7	6.6	4.5
	Fresh urine	95.5	84.4	11.1	19	5.4
	Stored urine	95.5	85.6	9.9	19	5.4
Sandy loam (Riso)	$(NH_4)_2SO_4$	93.0	83.9	9.1	19	5.7
	Fresh urine	91.1	66.4	24.7	41	6.7
	Stored urine	92.8	67.4	25.4	44	6.6
LSD ($p<0.05$)		1.8	1.9	ND[b]	4.2	ND

[a]Non-exchangeable ^{15}N = Total ^{15}N - inorganic ^{15}N. [b]ND = not determined.

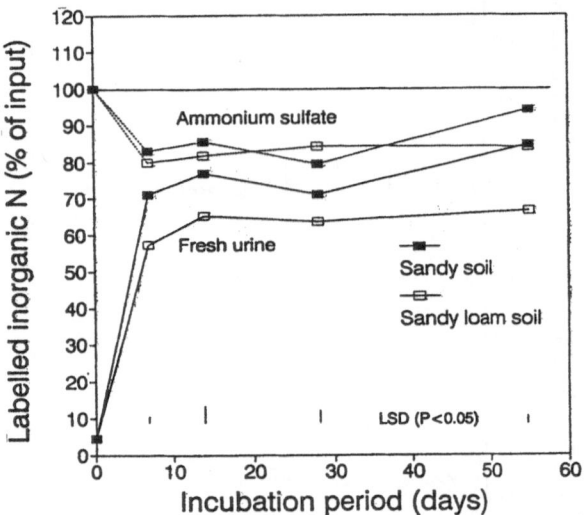

Figure 2. Recovery of labelled N as inorganic N in two soils after incubation with ^{15}N-labelled fresh sheep urine (solid line) or $(NH_4)_2SO_4$ (dashed line).

In another experiment the same soils were incubated one week with $(^{15}NH_4)_2SO_4$ and 0.6% and 18% of the applied ^{15}N was recovered as non-exchangeable clay-fixed NH_4^+-N in the sandy and the sandy loam soil, respectively (data not shown). Most of the clay-fixed $^{15}NH_4^+$ apparently was released again since only 9.1% of ^{15}N from $(NH_4)_2SO_4$ was non-exchangeable in the sandy loam soil at day 55 (Table 2). Clay fixation of ^{15}N from urine would probably be sligtly lower, since the initial concentration of $^{15}NH_4^+$ was lower in the treatments with urine, but clay fixation was considered to be of little importance for the observed differences between treatments.

It is concluded that the net mineralization of urine-N was influenced by soil, because soil biomass N increased more after urine application in the soil with the highest clay content, resulting in a lower net mineralization of N. The net mineralization of urine-N was close to 100% in the sandy soil, while less than 80% was net mineralized in the sandy loam soil. Thus the plant utilization of urine-N may be influenced by mineralization/immobilization after application. Short-term anaerobic storage of urine had no significant effect on the formation of inorganic N in soil, indicating that most of the urine carbon utilized by the soil microbial biomass is not mineralized during short-term anaerobic storage.

Acknowledgements

I thank M Brink Jensen and U Weinreich for skilled technical assistance, Dr E S Jensen for useful comments on the manuscript and the staff at The Danish Institute of Plant and Soil Science, Department of Roughage for performing the feeding and urine sampling. The study was financially supported by the research programme: Sustainable Farming, The Ministry of Agriculture, Denmark.

References

Amato M and Ladd J N 1992 Decomposition of ^{14}C-labelled glucose and legume material in soils: Properties influencing the accumulation of organic residue C and microbial C. Soil Biol. Biochem. 24, 455–464.

Bristow A W, Whitehead D C and Cockburn J E 1992 Nitrogenous constituents in the urine of cattle, sheep and goats. J. Sci. Food Agric. 59, 387–394.

Brookes P C, Kragt J F, Powlson D S and Jenkinson D S 1985 Chloroform fumigation and the release of soil nitrogen: The effects of fumigation time and temperature. Soil Biol. Biochem. 17, 831–835.

Cabrera M L and Beare M H 1993 Alkaline persulfate oxidation for determining total nitrogen in microbial biomass extracts. Soil Sci. Soc. Am. J. 57, 1007–1012.

Castellanos J Z and Pratt P F 1981 Mineralization of manure nitrogen - correlation with laboratory indexes. Soil Sci. Soc. Am. J. 45, 354–357.

Cuttle S P and Bourne P C 1993 Uptake and leaching of nitrogen from artificial urine applied to grassland on different dates during the growing season. Plant and Soil 150, 77–86.

Haynes R J and Williams P H 1993 Nutrient cycling and soil fertility in the grazed pasture ecosystem. Adv. Agron. 49, 119–199.

Jansson S L 1958 Tracer studies on nitrogen transformation in soil with special attention to mineralization-immobilization relationships. K. Lantbrukhögsk. Annaler 24, 101–361.

Jensen E S 1991 Evaluation of automated analysis of ^{15}N and total N in plant material and soil. Plant and Soil 133, 83–92.

Keeney D R and Macgregor A N 1978 Short-term cycling of ^{15}N-urea in a ryegrass-white clover pasture. N.Z.J. Agric. Res. 21, 443–448.

Sommer S G and Christensen B T 1990 Ammonia volatilization from solid manure and raw fermented and separated slurry after surface application, injection, incorporation into the soil and irrigation. Tidskr. Planteavl. 94, 407–417.

Sørensen P and Jensen E S 1991 Sequential diffusion of ammonium and nitrate from soil extracts to a polytetrafluoroethylene trap for ^{15}N determination. Anal. Chim. Acta 252, 201–203.

Sørensen P and Jensen E S 1995 Mineralization of carbon and nitrogen from fresh and anaerobically stored sheep manure in soils of different texture. Biol. Fertil. Soils 19, 29–35.

Sørensen P, Jensen E S and Nielsen N E 1994a Labelling of animal manure nitrogen with ^{15}N. Plant and Soil 162, 31–37.

Sørensen P, Jensen E S and Nielsen N E 1994b The fate of ^{15}N-labelled organic nitrogen in sheep manure applied to soils of different texture under field conditions. Plant and Soil 162, 39–47.

Thomas R J, Logan K A B, Ironside A D and Bolton G R 1988 Transformations and fate of sheep urine-N applied to an upland U.K. pasture at different times during the growing season. Plant and Soil 107, 173–181.

Thompson R B, Ryden J C and Lockyer D R 1987 Fate of nitrogen in cattle slurry following surface application or injection to grassland. J. Soil Sci. 38, 689–700.

Vallis I, Harper L A, Catchpoole V R and Weier K L 1982 Volatilization of ammonia from urine patches in a subtropical pasture. Austr. J. Agric. Res. 33, 97–107.

Van Faassen H G and Van Dijk H 1987 Manure as a source of nitrogen and phosphorus in soils. In Animal Manure on Grassland and Fodder Crops. Eds. H G van der Meer, R J Unwin, T A van Dijk and G C Ennik. pp 27–45. Martinus Nijhoff Publ, Dordrecht, the Netherlands.

Van Veen J A, Ladd J N and Amato M 1985 Turnover of carbon and nitrogen through the microbial biomass in a sandy loam and a clay soil incubated with $[^{14}C(U)]$glucose and $[^{15}N](NH_4)_2SO_4$ under different moisture regimes. Soil Biol. Biochem. 17, 747–756.

Whitehead D C and Bristow A W 1990 Transformations of nitrogen following the application of ^{15}N-labelled cattle urine to an established grass sward. J. Appl. Ecol. 27, 667–678.

Whitehead D C and Raistick N 1993 Nitrogen in the excreta of dairy cattle: changes during short-term storage. J. Agric. Sci., Camb. 121, 73–81.

Williams P H and Haynes R J 1994 Comparison of initial wetting pattern, nutrient concentrations in soil solution and the fate of ^{15}N-labelled urine in sheep and cattle urine patch areas of pasture soil. Plant and Soil 162, 49–59.

O. Van Cleemput et al. (eds.), Progress in Nitrogen Cycling Studies, 147–151, 1996.

Monitoring of N-uptake by green manures and of the influence of N-release on N-availability, production and quality of sugar beet

H. Vandendriessche[1], L. Vanongeval[2], E. Smeets[2] and M. Geypens[2]
[1]*Comité voor Toegepaste Bodemkunde, W. de Croylaan 48, B-3001 Leuven, Belgium and* [2]*Soil Service of Belgium, W. de Croylaan 48, B-3001 Leuven, Belgium*

Key words: green manures, N-leaching, N-monitoring, N-release, sugar beets

Abstract

To study N-cycling in a crop system with green manures, on two field trials, a monitoring programme was set up to evaluate the N-uptake and the N-release capacity of several types of green manure. The studied green manures vary not only with the efficiency with which they take up nitrate but also with which they release it again when ploughed-in. The N-release depends on the time of dying of the green manure (frost, chemical damage or mechanical cut), the time of ploughing-in, and, the type of green manure (easily decomposable or not). Important is whether this N is made available quickly enough to the following commercial sugar beet crop and if there is no reduction in yield and quality of the sugar beet. Monitoring of nitrate content in soil during the experiments, allows to classify green manures into several classes. One class are the winter-hardy green manures, like ryegrass and reply of winter barley. Notwithstanding their N-uptake capacity, there is seldom N available for the following sugar beet crop. Yield and sugar beet quality are not significantly influenced. A second class are green manures with a good developed aerial plant part like phacelia and mustard. A clear net mineralization of these green manures is determined but crop technical and climatic factors determine whether this mineralization results in a higher N-availability for the following crop. A third class are the legumes like vetch. In the two field trials the N-uptake by vetch is the highest and N is made available during a longer period in comparison with phacelia and mustard. The root and sugar yield were higher, but not significantly. No significant difference in technological quality of the beets is measured.

Introduction

Because of their N-uptake capacity, green manures are used as a catch-crop for nitrate-N (Addiscott et al., 1991). In that way, the use of green manures can be considered as an important technical measure to prevent nitrate from leaching during the winter period. Besides, the N released as a result of the mineralization of the green manure is available for the following sugar beet crop. This N, released before or during the growing period of the sugar beet crop, has to be taken into account by calculating the N fertilization. Soil samples of sugar beet parcels from practice, which are taken in February-May, show that the use of green manures influences the distribution of mineral N in the soil profile : on average, the parcels with green manures in the crop rotation have a higher mineral N content in the top layers (0–30 cm and 30–60 cm), which has

an impact on the N recommendations for sugar beets (Geypens et al., 1994).

The N-uptake capacity of several types of green manures and the influence of N-release on N-availability, production and quality of the following crop (sugar beet) is studied in two field trials. Determining the N-uptake of the green manure and monitoring the nitrate content of the soil profile (0–90 cm) allows to study N-cycling in a crop system with green manures.

Materials and methods

Regarding this research, two field trials on a loamy soil were set up in the period 1991–1993. On each trial the green manures were sown after winter barley and ploughed-in in winter. Thereafter, sugar beets are sown in spring.

Table 1. General information about both field trials

	1991–1992	1992–1993
Sowing of green manures	09.08.91	03.08.93
Nitrate content of soil profile at sowing (kg N ha^{-1})	100.9	70.3
N-fertilization green manures (kg N ha^{-1})	–	40
Cutting of mustard	23.10.91	06.11.92
Ploughing-in of green manures	18.12.91	30.01.93
Sowing of sugar beets	10.04.92	28.03.93
N-fertilization sugar beets	09.04.92	22.03.93
Harvest of sugar beets	21.09.92	20.10.93

On both trials, ryegrass, mustard, phacelia and vetch were sown. In addition, an object with reply of winter barley and a fallow object were provided. Every treatment (12 × 15 m plots, randomly dispersed) was set up in four replications. Two times during the growing period of the green manures, the yield and the N-uptake of their aerial parts were measured.

After the ploughing-in of green manures, the sugar beets were sown on every object, one part without any N-fertilization, another part with N-fertilization following the N-recommendation according to the N-index of the parcel (Vandendriessche et al., 1992). During the growing period, soil samples are taken regularly until 90 cm of depth by layers of 30 cm to analyse for mineral N. At harvest, yield and quality parameters (such as sugar content, content of α-amino-N, potassium and sodium) were determined. General information about the field trials is given in Table 1.

Results

In 1991 the mineral N residues in the soil profile (0–90 cm) at sowing time of green manures are clearly higher than in 1992 (101 kg N ha^{-1} vs. 70 kg N ha^{-1}). Therefore, only in the second trial a limited N-fertilization (40 kg N ha^{-1}) is applied to the green manures. In that way, the growth of green manures in both trials was moderate, but sufficient. The N-uptake of the aerial parts at the end of the growing season varies from 28 to 98 kg N ha^{-1}.

Figure 1 illustrates the N-uptake capacity of the different green manures : in October the mineral N reserve in the soil is clearly lower on the parcels with green manure (especially in the layers 0–30 cm and 30–60 cm). The same conclusions can be drawn from the 1992-trial, despite the fertilizer application of 40

kg N ha^{-1}. In both trials, vetch and mustard are characterized by a high N-uptake.

Comparatively to the fallow object, the parcels with ryegrass show no net N-mineralization during the growing season of the sugar beet (Fig. 2). Only at the end of the season some N-mineralization can be remarked. The green manures mustard and phacelia, on the other hand, are characterized by a net mineralization in spring. The mineralization of vetch is totally different from the other green manures. In both trials a clear net mineralization of vetch can be distinguished during the growing season of the following crop. This net mineralization starts early and persists for a longer period. The mineralization on the winter barley parcel is comparable with that on the ryegrass parcel. The net mineralization of the winter barley during the growing season of the sugar beets is negligible.

The sugar beet yield reflects the N-availability during the growing season. Considering the unfertilized sugar beets, only the vetch parcels got the highest beet and sugar yields on both trials. Of all green manures, vetch had the most important net mineralization during several months. The total N-uptake of the unfertilized sugar beets (roots and crowns) was 192 kg N ha^{-1} on the vetch parcel, clearly higher than on the other objects : 148, 149, 160 and 161 kg N ha^{-1} for the ryegrass, mustard, phacelia and winter barley object respectively. The fertilized sugar beets got on average higher yields than the unfertilized ones. The use of a green manure in the crop rotation had a small positive effect on the fertilized sugar beet yield, except for ryegrass and for reply of winter barley (Table 2). The technological quality of the sugar beets is reflected by the extractability index of Van Geijn (Van Geijn et al., 1983), taking into account the α-amino-N, potassium and sodium content of the beets. No significant difference in technological quality is measured (Table 2).

1991-1992 1992-1993

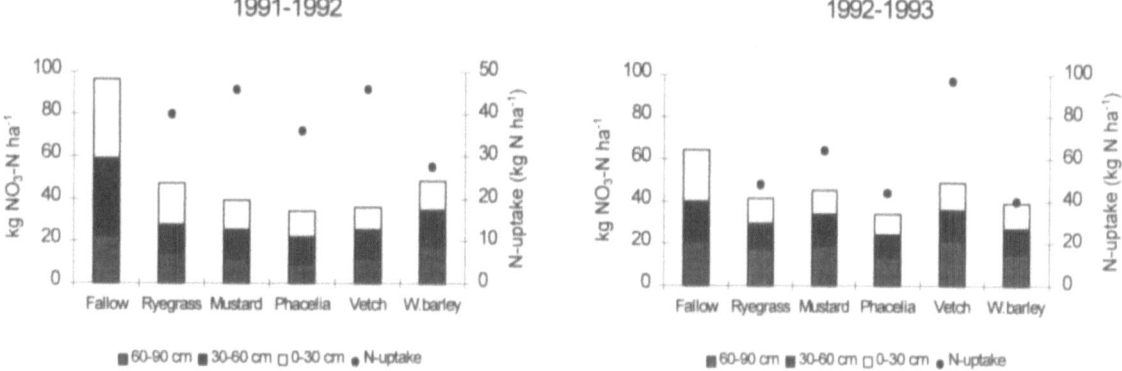

Figure 1. Nitrate reserve (kg NO_3-N ha^{-1}) of the soil profile on 23.10.91 and on 04.11.92 (before ploughing-in) and N-uptake (kg N ha^{-1}) of the green manures.

Table 2. Root yields, sugar content and quality of sugar beets after different green manures

	1992			1993		
	Root yield (ton ha^{-1})	Sugar (%)	EI[a] (%)	Root yield (ton ha^{-1})	Sugar (%)	EI[a] (%)
Unfertilized sugar beets						
Fallow	67.3	15.4	83.1	70.5	17.8	91.6
Ryegrass	57.2	16.5	85.4	64.6	17.9	91.2
Mustard	56.8	16.0	84.5	64.8	17.7	91.3
Phacelia	60.8	16.2	84.8	66.3	18.0	91.3
Vetch	69.8	16.3	84.5	74.9	17.6	91.2
W. barley	62.4	16.2	85.1	68.0	17.9	91.1
Fertilized sugar beets						
Fallow	60.0	15.2	83.8	76.9	17.7	90.9
Ryegrass	60.4	16.1	85.0	75.8	17.7	91.1
Mustard	70.5	16.1	85.2	79.2	17.7	90.9
Phacelia	67.2	16.1	84.9	77.0	17.8	91.1
Vetch	64.4	15.9	83.8	77.0	17.4	90.5
W. barley	63.1	16.1	84.7	76.7	17.9	90.5

[a] extractability index of Van Geijn (Van Geijn et al., 1983)

Discussion

The situation in the two field trials was certainly not identical (rainfall, temperature, N- content of the soil at sowing, date of ploughing-in, N-fertilization...). Nevertheless, both trials confirm the N-uptake capacity of the green manures and show their capacity to function as a catch-crop for nitrate-N, however the development of the green manure was moderate. A higher N-uptake by the green manures would have been possible by applying more mineral or organic N-fertilizer, but that was not the aim of the experiment (to avoid higher mineral N residues in the soil profile during winter period).

The amount of N-release and the moment of net N-mineralization are influenced by several factors : time of ploughing-in, time of cutting or dying of the green manure, climatic factors and type of green manure. This means that too early mineralization can be prevented by the use of crop technical measures.

The results of the soil profile analysis on the green manure parcels, compared with the fallow object, suggest that the green manures can be classified into three groups, considering their rate and time of mineralization.

150

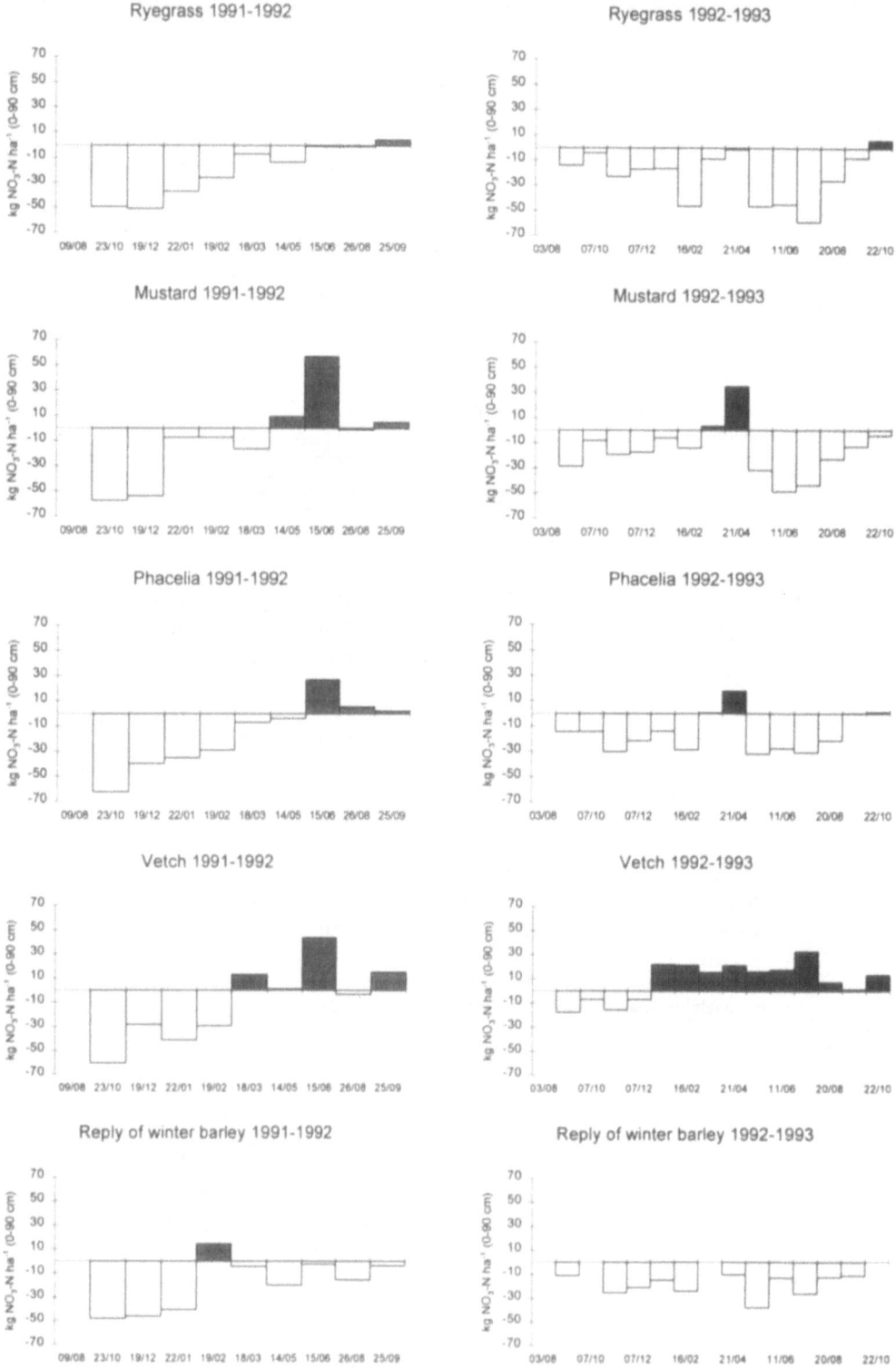

Figure 2. Evolution of the mineral N reserve of the soil profile (0–90 cm) comparatively to the fallow object.

Ryegrass and reply of winter barley

The mineral N reserve of the soil profile after ploughing-in of these green manures is continuously lower than on the fallow object. Ryegrass and winter barley are winter-hardy green manures with moderate N-uptake in their aerial parts, but with a relatively good developed root system. Possibly, because of the relatively high C/N-ratio of the root system, a part of the mineralised N from the soil organic matter is used for further decomposition of the green manure. As a result, less N is available for the following crop (sugar beets). However, the sugar beet yield and quality was hardly influenced.

Mustard and phacelia

These green manures have (compared with the first group) a good developed aerial plant part. After ploughing-in, a weak net mineralization was determined. The rate and time of mineralization may be influenced by climatic factors and crop technical measures (such as time of sowing and ploughing-in). The N-availability for the following sugar beet crop depends on the time of net mineralization. For these green manures, it will be important to prevent too early mineralization by the use of crop technical measures.

Vetch

Vetch is the only legume in the experiment. This green manure provides a net mineralization during a relatively long period. So, the released N can be used by the following crop. The results of the sugar beet trials show a higher N-uptake by the sugar beet crop and also a higher root and sugar yield.

As a result, the use of green manures as a catch-crop for N is only significant when too early mineralization of the green manure is prevented by the use of crop technical measures (choice of the green manure, ploughing date,...). Because N releases before or during the growing period of the following crop, this N has to be taken into account by the calculation of the N-fertilization of the following crop. If not, the use of green manure is only a delay of N-leaching in time.

Acknowledgements

This research is financially supported by the IWONL (Institute for Encouraging Scientific Research in Industry and Agriculture).

References

Addiscott T M, Whitmore A P and Powlson D S 1991 Farming, fertilizers and the nitrate problem. C A B International, Wallingford, Oxon, UK. 170 p.

Geypens M, Vandendriessche H, Bries J and Hendrickx G 1994 Experience with the N-index expert system : a powerful tool in N recommendation. Commun. Soil Sci. Plant Anal. 25, 1223–1238.

Vandendriessche H, Geypens M and Bries J 1992 N-index : an expert system for N fertilization of arable crops. *In* Nitrogen Cycling and Leaching in Cool and Wet Regions of Europe. Proceedings of the COST Workshop, October 22-23, 1992, Gembloux, Belgium. Eds. E François, K Pihan and N Bartiaux-Thill. pp 55–57.

Van Geijn N J, Giljam L C and De Nie L H 1983 Alfa-amino-N in sugar processing. Proceedings of the I.I.R.B. Symposium 'Nitrogen and Sugar-beet', February 16-17, 1983, Brussels, Belgium. pp 13–24.

O. Van Cleemput et al. (eds.), Progress in Nitrogen Cycling Studies, 153–157, 1996.
© *1996 Kluwer Academic Publishers.*

Soil litter dynamics and N cycling in alley cropping systems

B. Vanlauwe[1], S. Van den Bosch, M. Van Gestel and R. Merckx
Laboratory of Soil Fertility and Soil Biology Faculty of Agricultural and Applied Biological Sciences K.U. Leuven, Kardinaal Mercierlaan 92, 3001 Heverlee, Belgium. [1] Present address: International Institute of Tropical Agriculture Soil Microbiology, Ibadan (Nigeria) c/o Lambourn and Co, Carolyn House 26 Dingwall Road, Croydon CR9 3EE, UK

Key words: alley cropping, nitrogen cycling, soil litter, soil organic matter fractionation

Abstract

Many tropical alley cropping systems rely on an efficient transfer of nitrogen from organic residues to food crops. Several reports indicate that N released from decomposing residues is poorly recovered by food crops. This has been ascribed to a bad synchronization between supply and demand of nutrients, and to a temporarily immobilization of N not taken up by the crop. To improve the N use efficiency by the accompanying crop in alley cropping systems, a better understanding of the processes involved is required. Here, we report on two studies carried out to trace the fate of N in such systems.

In a first (pot) experiment, the accumulation of N in soil litter fractions (according to TSBF methodology) following a single application of ^{15}N-labelled shoots or roots of *Leucaena leucocephala*, *Flemingia congesta* and *Dactyladenia barteri* was determined. The results show that major shifts in N-distribution are restricted to the first two weeks following application. Furthermore, it is shown that the final distribution of residual N is controlled to a large extent by the biochemical quality of the residues.

In a second (field) experiment, microplots were amended with ^{15}N-labelled residues of *Leucaena leucocephala*, and the fate of N in soil and soil litter fractions was followed. The results again indicate a rapid release of N from applied residues during the first two weeks. After 236 days of decomposition, 8% of *Leucaena*-N remained in the soil. This residual ^{15}N was largely associated with the more stable, non-litter soil organic matter fraction.

Introduction

Alley cropping systems are cropping systems where food crops are grown between hedges of preferably nitrogen-fixing trees. The philosophy behind such practice is that upon regular pruning the residues may provide nutrients to the growing food crops. The benefit then is to be obtained either from the nitrogen fixed by the trees or from nutrients recovered by the tree roots from soil layers beyond the reach of the roots of the crop (e.g. Kang et al., 1989).

Over the last years a consensus has been growing that such systems may be characterized by a so called bad synchronization between supply and demand of the nutrients. Although many reports indicate that substantial amounts of nitrogen can be fixed, the recovery of it by a food crop like maize for instance is very low, sometimes below 10% (Van der Meersch et al., 1993).

Since in many situations no valid alternative can be forwarded, an urgent need to increase the N use efficiency of these and all residue-based farming systems is obvious. Improving the efficiency can however only be achieved through a proper understanding of the processes involved. In this paper therefore a number of unanswered problems are investigated. We believe that especially the dynamics of the various soil organic matter fractions need a thorough study as they are responsible for the long-term release of nitrogen, and constitute a buffer which has to be replenished every season. In this paper we report on the dynamics of surface and soil litter fractions following the amendment of different ^{15}N labelled prunings to the soil in pot and microplot experiments. This enables to trace the fate of the nitrogen in these farming systems in an unequivocal way.

154

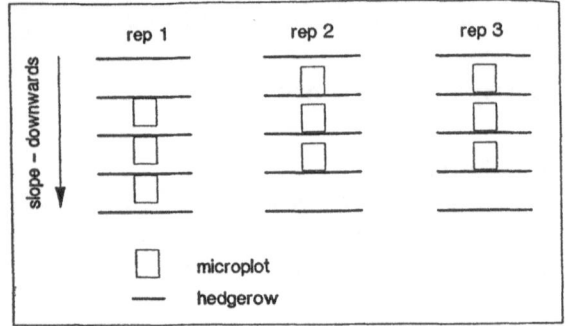

Figure 1. Location of the *Leucaena* main plots and the microplots in the field.

Materials and methods

Pot experiment

The soil used was sampled from the 0–10 cm layer of an Alfisol (oxic Paleustalf) at the International Institute of Tropical Agriculture (IITA) in Ibadan (Nigeria). The soil was of $pH(H_2O)$ 5.7, and contained 0.85%C, 0.08%N and 9% clay (Vanlauwe et al., 1994). Replicate samples of 1200g of air-dried, sieved (2 mm) soil were weighed into plastic pots (cylindric, 12.85 cm diam.). Samples were remoistened to obtain a moisture content of 40% WHC, and preincubated for 30 days in a glasshouse with open sides. Every 4 or 5 days, water was added to compensate for evaporation losses.

Soil samples were amended with 1.162 g (on ash-free basis) of dried, chopped (2–6 mm), ^{15}N-labelled leaves or roots of 3 alley cropping species (*Leucaena leucocephala*, *Dactyladenia barteri*, *Flemingia congesta*) of different quality. Characteristics of plant material are given in Table 1. A treatment consisting of the addition of a solution of ^{15}N-labelled $(NH_4)_2SO_4$ fertilizer was also included. In this case, similar amounts of total N and ^{15}N were added as to the soils amended with *Leucaena* leaves. Pots were placed again in a glasshouse, and moisture content was adjusted daily. After 14, 30 and 92 days, replicate pots (3 per treatment) were emptied and soils were analyzed.

Field experiment

A field experiment was conducted at the site from which the soil used in the pot experiment was sampled. In brief, 3 microplots (1 × 2.4 m) were installed in each of 3 replicate alley cropping plots (5 × 18 m) with *Leucaena* as hedgerow (4.5 m inter-row space)

(Fig. 1). Freshly pruned ^{15}N-labelled *Leucaena* leaves (3.93%N, 1.159 atom% ^{15}N, C/N ratio: 8.22) were distributed equally over the microplot surface at a rate of 1,980 kg (dry matter) per hectare. Two days later, maize (*Zea mays*) was planted on the microplots (3 rows of 5 plants). Three microplots were harvested 45 days and 129 days after application of *Leucaena* prunings. Forty five days after residue addition, the litter remaining on the surface of the microplots was removed by hand ('surface litter'). At day 45, 129 and 236 after residue application, soil samples were taken from the 0–5 cm layer of one microplot in each main plot. Samples were air-dried and sieved (4 mm).

Soil litter fractionation

The light soil organic matter fraction was separated from the bulk soil by flotation (using a brass root washing apparatus), followed by wet sieving (TSBF, 1993). Two kg of air-dried soil was placed in the fractionation bucket, and a water flow was passed through during 2 h. The soil litter was collected on a 2 mm- and a 0.25 mm-sieve. The soil litter fractions larger than 2 and 0.25 mm were called the 'coarse' and 'fine soil litter' respectively. These two organic fractions were dried and weighed before analysis.

Soil samples, litter fractions and surface litter were ground and analyzed for total N and ^{15}N enrichment with a combined dry combustion apparatus-mass spectrometer system (Europa Scientific - Roboprop CN analyzer, VG Micromass 602).

Results and discussion

Pot experiment

Figure 2a and b represent the percentages of residue-derived N remaining in the litter fraction >0.25 mm at different sampling times. For all types of plant material, ^{15}N is rapidly released from the >0.25 mm fraction during the first two weeks of the incubation period. Thereafter, the changes with time of the amounts of ^{15}N remaining in the soil litter are small or non-existent.

Differences between the various plant species in the percentage of ^{15}N remaining in the soil litter were larger for leaf than for root material. The highest percentages were observed for *Dactyladenia* leaves, followed by *Flemingia*; the *Leucaena* amended soils retained least of residue-derived N in their litter fraction.

Table 1. Selected characteristics of plant material

Plant species	%N	[15]N atom%	C/N	Lignin[a] (%)	(Hemi)cell[a] (%)
Leaves					
Leucaena	4.81	2.749	7.57	3.2	12.2
Dactyladenia	2.00	3.836	16.69	21.9	10.4
Flemingia	3.40	3.228	11.33	18.4	20.1
Roots					
Leucaena	2.30	2.749	13.83	29.7	19.4
Dactyladenia	2.17	1.644	17.05	40.3	11.1
Flemingia	3.05	2.278	11.22	27.5	19.8

[a] Determined by a method described by Van Soest (1963)

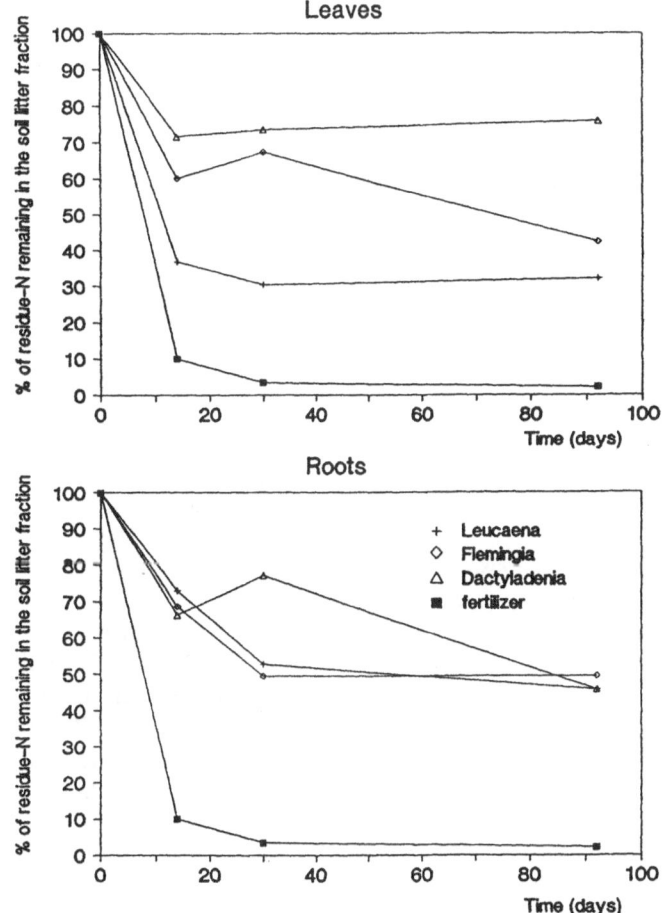

Figure 2. Percentages of residue-derived N remaining in the soil litter fraction (>0.25 mm) during incubation of soils amended with residues of different quality.

The quality of residues influenced the retention of ^{15}N in the soil organic fractions. Correlations with different residue quality parameters (%N, C/N, lignin/N, (lignin + (hemi)cellulose)/N) were calculated. For the lignin/N and (lignin + (hemi)cellulose)/N ratio, the relationships with percentage ^{15}N remaining in the soil litter were only significant for leaf material, but not for the roots. As an example, the relations between per-

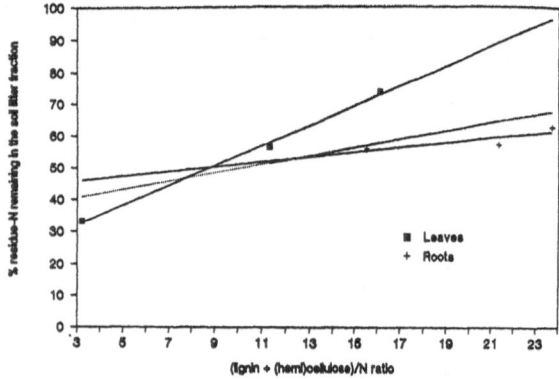

Figure 3. Relationships between the (lignin + (hemi)cellulose)/N ratio and the percentage of N remaining in the soil litter fraction (> 0.25 mm). Regression is done between mean of 3 sampling dates and residue quality. (Dotted line represents regression for all types of plant materials (leaves and roots)).

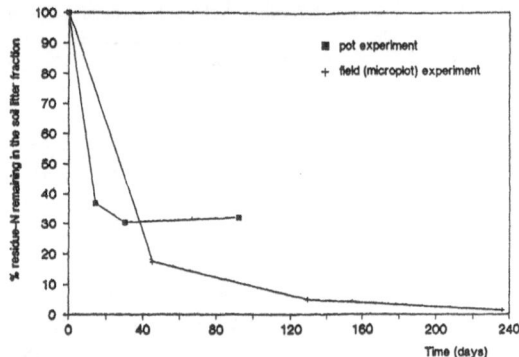

Figure 4. Percentage of *Leucaena*-N remaining in the soil litter fraction (> 0.25 mm) after addition of residues to soils in a pot experiment or in the field.

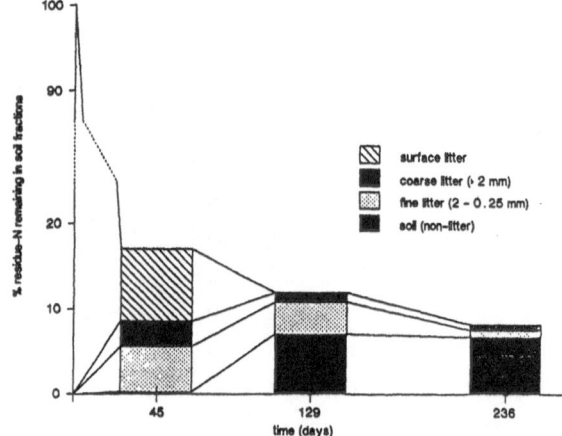

Figure 5. Distribution of residue-derived N amongst soil fractions after different periods of decomposition in the field.

centage ^{15}N remaining in the soil litter and the (lignin + (hemi)cellulose)/N ratio are shown in Figure 3.

The discrepancy between leaf and root quality versus incorporation of plant N into soil litter may be explained by the difference in way of application of root and leaf material (on the surface - mixed with the soil). It is also possible that the difference in morphology of roots and leaves influences fragmentation and incorporation in the soil litter.

Microplot experiment

Also for *Leucaena* prunings applied to soils in the field, residue-derived N is moving very fast through the soil litter (Fig. 4). Initially, ^{15}N disappears from the litter fraction at a similar rate as observed for the pot experiment. But in a second phase, residue-derived N remaining in the soil litter decreases much faster. This is most probably due to leaching and washing down of residues by rainfall on the sloping surface, or by faunal activity.

Figure 5 shows the accumulation and disappearance of residue-derived N in the different soil fractions with time. The surface litter is decomposed very fast. Labelled N, derived from added plant material, enters into the soil litter fraction and also in the non-litter soil fraction. Forty five days after addition of residues, the portion of ^{15}N associated with the latter fraction is marginal. But at day 129, more than half of the ^{15}N remaining in the soil is associated with the non-litter fraction. The amounts of residue-derived N present in the two soil litter fractions decrease with time from day

45 onwards, and at a faster rate for the >2 mm fraction than for the 2–0.25 mm litter fraction.

It is important to note that the residue-derived N remaining in the soil can be mineralized and taken up by a crop or by the hedgerow during following seasons. In view of the often reported small short-term benefits of alley cropping systems it can be advocated that rather these long-term benefits should be emphasized when evaluating the efficiency of alley cropping systems.

Acknowledgements

Results presented here are obtained through a collaborative project with the International Institute of Tropical Agriculture financed by the Belgian Administration for Development Cooperation. We acknowledge the skilful and timely ^{15}N analyses by Mrs Christine Coorevits and Ms Hilde Van den Broeck.

References

Kang B T, van der Kruijs A C B M and Couper D C 1989 Alley cropping for food crop production in het humid and subhumid tropics. *In* Alley Farming in the Humid and Subhumid Tropics: Proceedings of an International Workshop held at Ibadan, Nigeria, 10–14 March 1986. Eds. B T Kang and L Reynolds. pp 16–26. IDRC, Ottawa, Canada.

Tropical Soil Biology and Fertility 1993 Handbook of Methods. Eds. J M Anderson and J S I Ingram. CAB International, Oxon, UK.

Van der Meersch M K, Merckx R and Mulongoy K 1993 Evolution of plant biomass and nutrient content in relation to soil fertility changes in two alley cropping systems. *In* Soil Organic Matter Dynamics and Sustainability of Tropical Agriculture. Eds. K Mulongoy and R Merckx. pp 143–153. John Wiley and Sons, Chichester, UK.

Vanlauwe B, Vanlangenhove G, Merckx R and Vlassak K 1994 Impact of rainfall regime on the decomposition of leaf litter with contrasting quality under subhumid tropical conditions. Biol. Fertil. Soils 20, 8–16.

Van Soest P J 1963 Use of detergents in the analysis of fibrous feeds. II. A rapid method for determination of fibre and lignin. J. AOAC 46, 829–835.

Plant and Soil **181**: 163–167, 1996.
© 1996 *Kluwer Academic Publishers.*

The production of volatile organic compounds during nitrogen transformations in soils

R.E. Wheatley[1], S.E. Millar[1] and D.W. Griffiths[2]
[1]*Unit of Integrative Bioscience, Cellular and Environmental Physiology Department** and* [2]*Chemistry Department, Scottish Crop Research Institute, Invergowrie, Dundee DD2 5DA, UK*

Key words: nitrogen transformations, soil, volatile organic compound

Abstract

The volatile organic compounds produced during a sequence of soil incubations under controlled conditions, with either added NH_4^+-N or NO_3^--N, were collected and identified. The nature and relative amounts of the volatile organic compounds produced by the microorganisms in the soils were remarkably reproducible and consistent.

Introduction

There is a large variety of volatile organic compounds (VOCs), in a wide range of concentrations, in soil atmospheres. These VOCs may come from either internal sources such as plants or microorganisms, or external sources as soil is a major sink for many atmospheric constituents.

Most VOCs in atmospheres are probably microbial in origin (Stotzky and Schenck, 1976). So, production of them will probably be influenced by factors that influence either population dynamics or microbial activities.

Total soil microbial biomass is relatively stable throughout the growth of crops (Ritz et al., 1992), while specific activities such as nitrification and denitrification show very dynamic temporal changes in rates (Wheatley and Williams, 1989; Wheatley et al., 1991). These changes in activity rates can be both widespread and rapid throughout the bulk soil. Variations in the pattern of VOC production may reflect the type and amount of activity occurring in soils, and also changes in cultural conditions. These compounds may also have a role in microbial communication in the rhizosphere.

Materials and methods

Soil

Samples were taken from a field of Carbrook association soil cropped to potatoes. The soil was an estuarine silty-clay loam; pH 6.1, total C 2.23% (w/w), total N 0.22% (w/w) and water-holding capacity (WHC) 0.47 mL H_2O g^{-1} dry soil. P at a rate equivalent to 85 kg ha^{-1} and K at 220 kg ha^{-1}, had been applied to the soil. No N was applied to these plots at any time.

Soil incubation and volatile product collection

After sieving, <4 mm, 2.5 kg fresh soil was placed in each of triplicate 5 L quickfit vessels, then amended with water and appropriate solutions. Soya broth (Oxoid, Ltd.; 250 mL) was added to all the treatments, and 25 mL of a 25 g L^{-1} solution of NH_4NO_3 and 25 mL of glucose, 30% (w/v) to one set and 25 mL $(NH_4)_2SO_4$, 0.3% (w/v) to another. The third received no further amendments. Water to 50% of the WHC was added to all. Air and O_2-free N_2 were used alternatively, for 24 h periods, to purge the headspace products from the soils through a tube of adsorbant, 0.4 g Haysep Q, 800–100 mesh,(Analytical Polymers Inc., Bandera, Texas). Then the tube was removed and dried by reverse flushing with O_2-free N_2 at 20 mL min^{-1} for 10 min. All gases used in the incubations were passed through filters of molecular sieve and activated

---
* FAX No.: +44382562426
Plant and Soil is the original source of publication of this article. It is recommended that this article is cited as: *Plant and Soil* 181: 163–167, 1996.

charcoal that had been previously conditioned in a He stream, 25 mL min^{-1}, for 24 h at 180 °C, then a sterile sintered-steel filter, before use. Similarly the collection tubes were conditioned in a He stream, 10 mL min^{-1}, for 24 h at 180°C.

Analyses of the volatile organic compounds produced

The VOCs adsorbed by the porous polymer were analysed on an integrated system consisting of a Perkin Elmer ATD50 automated thermal desorber connected to a Hewlett Packard 5890 gas chromatograph interfaced to a VG Trio 1000 quadrupole mass spectrometer. The sample tubes were desorbed at 130 °C for 15 min with an outlet split ratio of 7.5:1. The mixture was then separated on a DB1701 chromatographic column, 60 m long by 0.25 mm i.d., 1 μm film, utilising He at 1 mL min^{-1} as the carrier gas. The oven was temperature programmed from 40 °C to 240 °C at 5 °C min^{-1}, final isothermal period 20 mins. Compounds were identified by retention time and comparison of the mass spectra with standards or published mass spectral data bases.

Compounds in the C_3 to C_{10} range could be identified in this study. The relative proportions of such VOCs produced by soil micro-organisms, under both aerobic and anaerobic conditions, were then determined.

Results

A total of 35 VOCs were collected and identified from the soil samples. The range of compounds detected (Table 1) included aliphatic alcohols, aldehydes, ketones and esters, polysulphides and variously substituted simple aromatic compounds. The production of these VOCs in each of the soils studied was reproducible under defined conditions, both with respect to the compounds identified and their relative proportions.

Aerobic incubations

The predominant compounds produced under aerobic conditions were sulphides, accounting for more than 65% of the total VOCs in all incubations (Table 2). The most abundant S-compound was dimethyl disulphide, representing over 90% of the total sulphides identified. Smaller quantities of dimethyl sulphide and dimethyl trisulphide were also present in the aerobic incubations

Table 1. Volatile organic compounds detected in the headspace of aerobically and anaerobically incubated soil

Alcohols	Ketones
Ethanol	Propan-2-one
Propan-1-ol	Butan-2-one
Propan-2-ol	Pentan-2-one
Butan-1-ol	Pentan-3-one
Butan-2-ol	4-Methyl pentan-2-one
2-Methyl propan-1-ol	5-Methyl heptan-2-one
2-Methyl butan-1-ol	3-Hydroxy butan-2-one
3-Methyl butan-1-ol	
Aldehydes	**Aromatics**
2-Methyl-butan-1-al	Benzene
3-Methyl-butan-1-al	Ethyl benzene
Sulphides	Dimethyl benzene
Dimethyl sulphide	Methylethyl benzene
Dimethyl disulphide	Trimethyl benzene
Dimethyl trisulphide	Benzaldehyde
2-Methyl propylsulphide	
Methyl Esters	**Ethyl Esters**
2-Methyl butanoic acid	Acetic acid
3-Methyl-butanoic acid	Butanoic acid
Butyl Esters	2-Methyl propanoic acid
Acetic acid	2-Methyl-butanoic acid
	3-Methyl-butanoic acid

and the relative proportions of each did not vary with amendment.

The relative proportions of volatile ketones were significantly increased by the addition of either KNO_3 or $(NH_4)_2SO_4$. In all but the $(NH_4)_2SO_4$-enriched incubations only 2 ketones were detected, propan-2-one and butan-2-one, the latter predominated in all cases. In the $(NH_4)_2SO_4$-enriched incubations trace amounts of both pentan-2-one and 4-methyl pentan-2-one were also identified.

Adding $(NH_4)_2SO_4$ also significantly increased the proportion of aldehydes, and uniquely under aerobic conditions produced 2-methyl butan-1-al. 3-methyl butan-1-al was the predominant aldehyde in the $(NH_4)_2SO_4$-enriched incubation and the only aldehyde detected in the 3 other aerobic treatments.

An increase in the amount of ethanol, the predominant alcohol in all the aerobic treatments, caused a significant increase in the proportion of alcohols produced (Table 2) when $(NH_4)_2SO_4$ was added. However, other alcohols including propan-2-ol, 2-methy propan-1-ol, butan-1-ol and 3-methyl butan-1-ol, were also detected in these incubations. In the aerobic KNO_3 plus glucose

Table 2. The effect of nutrient additives on the distribution by chemical functional groups of the volatile organic compounds (% total scan area) identified in the head space of aerobically incubated soil

Chemical group	Additive 0	KNO_3	KNO_3 + Glucose	$(NH_4)_2SO_4$	SED*
Alcohols	0.6	1.2	1.1	4.3	0.48
Ketones	3.9	18.6	6.2	17.8	3.44
Aldehydes	0.3	0.5	0.4	3.2	0.54
Esters	nd	nd	nd	0.3	NS
S-Compounds	75.7	73.1	79.1	66.1	NS
Aromatics	15.4	3.9	9.5	4.8	NS
Unidentified	4.1	2.8	3.6	3.6	NS

* Standard error of difference; nd = not detected, NS = not statistically different.

tions without N additions only ethanol and butan-2-ol were detected; these together with propan-2-ol were also produced when KNO_3 was added to the soil.

The proportions of aromatic compounds were not significantly different between incubations. Ethylbenzene, dimethyl benzene and benzene were detected in all the aerobic incubations, and the relative concentration of benzene was constant, at 1–2% of the total VOCs. The observed fluctuations in the relative proportions of total aromatic compounds were largely due to differences in the relative amounts of substituted benzenes, although trace quantities (<0.2% of total VOCs) of trimethyl benzenes, ethyl methyl benzenes and benzaldehyde were occasionally detected.

Esters were not found in the aerobic incubations, except for one replicate of the $(NH_4)_2SO_4$-enriched incubations, when butyl acetate was present as <1% of the total VOCs,

Anaerobic incubations

A greater proportion of the VOCs detected in the anaerobic incubations were alcohols. When soils were incubated with either added KNO_3 plus glucose or $(NH_4)_2SO_4$ (Table 3) alcohols accounted for almost 70% of the total VOCs. All 8 alcohols in Table 1 were detected in all the treatments, with the sole exception of soil incubated with KNO_3 plus glucose when butan-2-ol was not found. The predominant alcohol changed according to the N-supplementation (Fig.1). When no N was added 3-methyl butan-1-ol was the major alcohol produced, and the addition of KNO_3, KNO_3 plus glucose or $(NH_4)_2SO_4$ altered the predominant alcohol to ethanol, butan-1-ol or 2-methyl butan-1-ol respectively.

With the exception of pentan-3-one, all the monofunctional ketones listed in Table 1 were detected in all treatments. As in the aerobic incubations the major ketones were propan-2-one or butan-2-one with pentan-3-one only present in incubations with either no added N or added KNO_3. Also the bifunctional ketone, acetoin (3-hydroxybutan-2-one), previously reported as a volatile product of *Erwinia* infected potatoes was also detected in the KNO_3-enriched anaerobic soil incubations.

The amounts of aldehydes detected were not affected by any N-supplementation, and the major aldehyde produced was 3-methyl butan-1-al.

Dimethyl disulphide was again the main S-compound and, as in the aerobic samples, both the mono and trisulphides were found in all the anaerobic soil incubations. The proportion of sulphides was significantly affected by N-supplementation, with the levels found being in inverse proportion to the values found for the alcohols.

The same aromatic compounds were detected as in the aerobic incubations and the increases in total aromatic content in both the KNO_3 only and unsupplemented incubations were similarly due to an increase in the relative amounts of dimethyl benzenes.

Although there were no significant differences in total esters detected in the anaerobic incubations (Table 3), there were considerable differences in diversity. In particular, the $(NH_4)_2SO_4$-supplemented incubations yielded all the methyl and ethyl esters listed in Table 1, whilst in the KNO_3 plus glucose incubation all the ethyl and butyl esters were detected. In contrast the KNO_3-supplemented incubations yielded only ethyl acetate and in the unsupplemented samples only ethyl,

[161]

166

Table 3. The effect of nutrient additives on the distribution by chemical functional groups of the volatile organic compounds (% total scan area) identified in the head space of anaerobically fermented soil

Chemical group	Additive 0	KNO₃	KNO₃ + Glucose	NH₄)₂SO₄	SED*
Alcohols	37.7	16.3	71.9	69.2	8.11
Ketones	13.3	21.1	8.5	6.2	2.37
Aldehydes	1.3	2.4	1.3	2.5	NS
Esters	0.2	0.4	6.2	3.9	NS
S-Compounds	24.2	23.2	7.3	11.7	2.95
Aromatics	23.3	28.7	2.9	6.2	6.54
Unidentified	nd	7.8	1.9	0.4	1.44

* Standard error of difference; nd = not detected, NS = not statistically different.

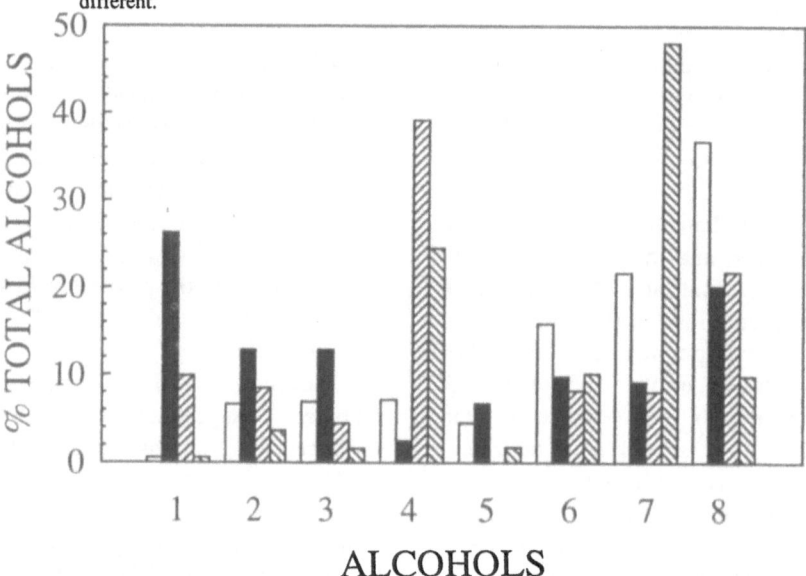

Figure 1. The alcohols produced during the anaerobic incubation of amended field soils. Amendments:- □, None; ■, KNO₃; ▨ KNO₃ plus Glucose; ▨, (NH₄)₂SO₄. Alcohols:- 1, ethanol; 2, propan-1-ol; 3, propan-2-ol; 4, butan-1-ol; 5, butan-2-ol; 6, methyl propan-1-ol; 7, 2-methyl butan-1-ol; 8, 3-methyl butan-1-ol.

2-methyl butanoate and methyl, 3-methyl butanoate were present.

Discussion

A greater diversity of VOCs was produced under anaerobic compared to aerobic conditions, 27 individual compounds were produced in the anaerobic incubation of soil compared with 13 in a similar aerobic incubation. Despite this wide diversity the VOCs produced in any particular incubation were consistent for those conditions. This consistent response to environmental conditions could also be particularly subtle, for example down to the single compound level in the alcohols. The responses to changes in aeration may well have been expected, but the responses to minor changes in N status, particularly of N-species probably would not.

These responses could be useful in monitoring specific microbial activities in situ, but it may well be more interesting to explore the possible role of the diverse range of VOCs identified here in communication between microorganisms in the rhizosphere, and the effects that this phenomenon may have on soil processes.

Consistency of production, both in terms of groups and individual compounds together with responsiveness to the environment, would be required if these products are to act as signals between microbial groups, as this would allow for some constancy in action.

Dimethyl disulphide has been previously reported (Bremner and Bundy 1974; McCarty and Bremner 1991) to be an effective inhibitor of nitrification. Other compounds can also act as effective inhibitors of other processes. Fiddaman and Rosswall (1993) reported that a strain of *Bacillus subtilis* produced volatile compounds that severely impaired the growth of *Rhizoctonia solani* and *Pythium ultimum* . However, these effects can be variable. Ko and Hora (1972) reported that ammonia significantly reduced the germination rate of the conidia of *P. chrysogenum*, but when similar concentrations were combined with other volatiles the growth of some fungi such as *Fusarium solani* could be stimulated. Or indeed the effects can be selective. Schisler and Linderman (1989) demonstrated selectivity in action when they transferred volatiles between systems and showed that the VOCs produced by ectomycorrhizal Douglas Fir seedlings caused a significant increase in bacterial numbers, in receiver soils, but did not significantly affect populations of either actinomycetes, *Fusarium* sp., extracellular chitinase producers, facultative anaerobes or phosphate-solubilising bacteria. The same direct, but selective effect could also occur in cultivated soils. So it is possible that the rapid and widespread changes in the rates of specific microbial activities in the bulk soil may be a result of the VOCs produced by one part of the biomass influencing another, and this requires further investigation.

Acknowledgement

This work was funded by the Scottish Office Agriculture and Fisheries Department.

References

Bremner J M and Bundy L G 1974 Inhibition of nitrification in soils by volatile sulfur compounds. Soil Biol. Biochem. 6, 161–165.
Fiddaman P J and Rossall S 1993 The production of antifungal volatiles by *Bacillus subtilis*. J Appl. Bact. 74, 119–126.
Ko W H and Hora F K 1972 The nature of a volatile inhibitor from certain soils. Phytopathol. 62, 573–575.
McCarty G W and Bremner J M 1991 Inhibition of nitrification in soil by gaseous hydrocarbons. Biol. Fertil. Soils 11, 231–233.
Ritz K, Griffiths B and Wheatley R E 1992 Soil microbial biomass and activity under a potato crop fertilised with N with and without C. Biol. Fertil. Soils 12, 265–271.
Schisler D A and Linderman R G 1989 Response of nursery soil microbial populations to volatiles purged from around Douglas fir ectomycorrhizae. Soil Biol. Biochem. 21,397-401.
Stotzky G and Schenck S 1976 Volatile organic compounds and microorganisms. CRC Crit Rev. 4, 333–381.
Wheatley R E, Griffiths B S and Ritz K 1991 Variations in the rates of nitrification and denitrification during the growth of potatoes (*Solanum tuberosum* L.) in soil with different carbon inputs and the effect of these inputs on soil nitrogen and plant yield. Biol. Fertil. Soils 11, 157–162.
Wheatley R E and Williams B L 1989 Seasonal changes in rates of potential denitrification in poorly-drained reseeded blanket peat. Soil Biol. Biochem. 21, 355–360.

Section editor: R Merckx

O. Van Cleemput et al. (eds.), Progress in Nitrogen Cycling Studies, 165–169, 1996.

Effect of zinc on nitrogen transformation during composting process and in soil

P. Zaccheo, L. Crippa and P.L. Genevini

Dipartimento di Fisiologia delle Piante Coltivate e Chimica Agraria - Università degli Studi di Milano, Via Celoria 2, I- 20133 Milano, Italia

Key words: compost, N mineralization, organic N, soil N availability, zinc

Abstract

The influence of Zn on nitrogen partitioning in different organic fractions during plant composting and on nitrogen transformation in compost amended soil added with ammonium sulphate, was investigated. ^{15}N labelled maize plants (stems and leaves) were composted with and without high level of $ZnSO_4$ so obtaining a Zn enriched compost, (4000 mg kg^{-1} of Zn), and a low Zn compost (control compost). The carbon and nitrogen recovery in the alkali insoluble, alkali soluble-acid insoluble and alkali and acid soluble fractions showed that two different end-products were obtained. In fact the presence of Zn induced a decrease on the rate of carbon compounds humification and the lowering of alkali insoluble nitrogen level.

In order to study the mineralization rate of the two composts and to compare the effect of the Zn-enriched compost with the direct effect of inorganic zinc on nitrogen transformation in soil, two incubation experiments were conducted over 3 months under a controlled greenhouse, one without plants, the second with a test plant (wheat). At different incubation time KCl soluble NO_3-N and NH_4-N, and biomass-N were determined in soil and yield and nitrogen concentrations were determined in plants.

Introduction

Organic wastes usually applied to soil, such as animal slurry or sewage sludge, contain substantial amount of zinc that, if present in high concentration, may be toxic to soil organisms. Biological transformation of nitrogen in soil is known to be affected by high level of heavy metals, even if experimental data are not always in agreement (Bååth,1989)

Moreover, zinc may interfere with the biological processes that carry over during composting of organic waste, a stabilization process that is improving as a way to convert organic wastes to relatively stable humus (Chen et al., 1994). As the compost microbiota determine the rate of composting and the quality of the product, it is of particular interest to evaluate the possible influence of heavy metal on microbial activity in composting organic residues.

The aim of this study is to investigate nitrogen and carbon partitioning in humic and non-humic fractions of a ^{15}N enriched laboratory compost, and to evaluate the influence of Zn on redistribution of carbon and nitrogen compounds after composting. Moreover, the influence of Zn-enriched compost on mineral nitrogen immobilization and on compost derived-N plant uptake was investigated in a greenhouse study.

Experimental design

Composting material

Aerial parts of 90 day old maize (*Zea mays L.*), grown in a nutrient solution including tagged $Ca(NO_3)_2$ (11.5 atom% ^{15}N) as the only N source, were dried and chopped. Two laboratory composters were filled with approximately 300 g (dry weight) of plant material, moistened at 50% of water holding capacity. Detailed methods for plant growth and compost production were reported in a previous paper (Zaccheo et al., 1993); the experiment was carried out with two treatments, a control treatment and a treatment with inorganic zinc. Zinc was added as a solution of $ZnSO_4$, to reach an expected final concentration of approximately 4000 mg

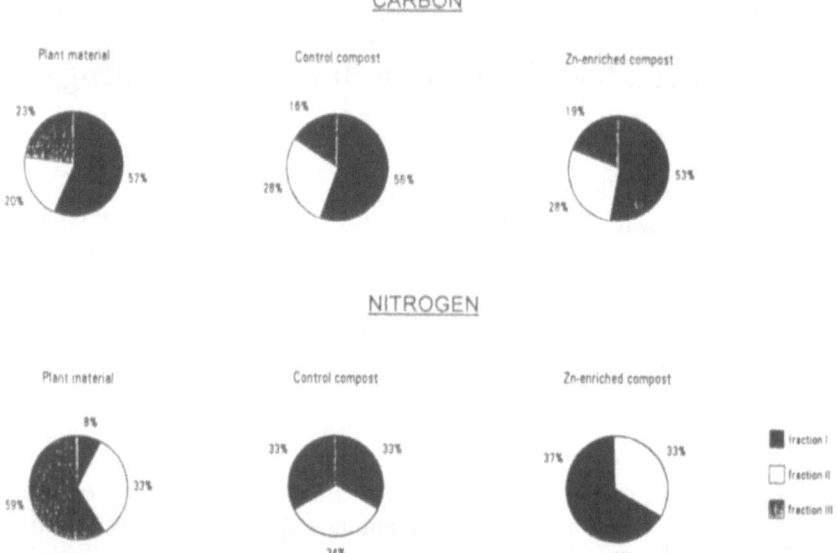

Figure 1. Distribution of organic C and total N among different fractions (Fraction I = alkali insoluble; Fraction II = alkali soluble-acid insoluble; Fraction III = alkali soluble-acid soluble).

kg $^{-1}$ of dried compost. Composting was performed at 40 °C in aerobic condition for 110 days.

The composts were then oven dried at 40 °C, weighed, finely ground and analysed, together with the original material, for their pH, ash, CEC, organic carbon, total nitrogen and %^{15}N enrichment. Carbon and nitrogen compound distributions were determined in alkali insoluble (fraction I), alkali soluble-acid insoluble (fraction II) and alkali soluble-acid soluble (fraction III) fractions, and C/N ratios for all the fractions were computed.

Greenhouse study

Two experiments were conducted under a controlled greenhouse environment, without and with test plants, in 80 plastic pots containing 350 g of 1 mm-sieved dry soil. The soil was a sandy soil with pH (H_2O) 7.0, organic C 9 g kg^{-1}, kjeldahl N 0.91 g kg^{-1}, cation exchange capacity 6.1 cmol$^{(+)}$ kg^{-1}, total zinc 68 mg kg^{-1}.

Three compost treatments equivalent to 500 kg ha^{-1} of total N, on the basis of the N content of the compost, were established, as well as a control (0 treatment). The compost treatments were: C-treatment = soil amended with control compost; Z-treatment = soil amended with Zn-enriched compost; M-compost = soil amended with control compost and $ZnSO_4$ to reach the same soil zinc concentration of Z-treatment.

A NPK nutrient solution was applied to the surface of all the pots (100 kg ha^{-1} N as unlabelled ammonium sulphate, 44 kg ha^{-1} P, 80 kg ha^{-1} K). At different times, two pots for each treatment were removed, and two subsamples for each pot were analyzed. In the incubation experiment (without plant), at 0, 3, 13, 27, 58, 90 days, KCl soluble NH_4 and NO_3, and biomass N were determined. The experiment with plant was performed in 32 pots; in each pot 6 seedlings of wheat (*Triticum aestivum L.*) were transplanted and, after 14, 28, 58 and 90 days, harvested cutting the plant tops 1 cm above the soil surface. Plant N and fresh and dry yields were determined, together with soil KCl soluble NH_4 and NO_3.

Each N form was transformed to NH_4 and steam-distilled: two distillations were carried out, the second one was used for isotopic analysis, performed by emission spectrometry (Jasco ^{15}N Analyzer). The recovery of compost-derived N (labelled N) was calculated according to He et al. (1988).

Results

During composting, for both treatments intense losses of organic matter occurred (Table 1), and the pH increased, as the result of the microbial decomposition of organic acids and the release of ammonia from proteins.

Table 1. Chemical characterization of the composts compared with the starting material (maize stems and leaves)

		Plant material	Control compost	Zn-enriched compost
pH		5.6	6.6	6.4
Ashes	(mg g^{-1})	97.2	156.4	162.4
Organic C	(mg g^{-1})	453	434	427
Total N	(mg g^{-1})	21.4	20.6	24.1
C/N ratio		21.19	21.09	17.73
Total Zn	(mg kg^{-1})	24.3	45.6	4177

Table 2. Microbial biomass N (mg kg^{-1})

Days	Z-treatment	M-treatment	C-treatment	Control
3	15.1	7.9	6.7	0.4
13	17.9	13.7	14.5	9.1

Zinc was shown to affect total nitrogen content, with a decrease in C/N ratio of the Zn-enriched compost.

The original carbon and nitrogen distributions in alkali insoluble (fraction I), alkali soluble-acid insoluble (fraction II) and alkali soluble-acid soluble (fraction III) fractions (Fig.1) were modified by the composting process and affected by zinc.

The percentage of total carbon was only little affected in fraction I, enhanced in fraction II (+ 37% in both the composts) and lowered in fraction III (-30% in control compost and -16% in Zn-enriched compost) by composting.

A different trend was seen for nitrogen distribution: composting affected nitrogen content of fraction I that was significantly increased leading to much lower C/N ratios (36.6 and 31.6 respectively in control compost and Zn-enriched compost) than in plant material (C/N 150). Nitrogen in fraction II was unaffected, and lowered in a large extent (-44% in control compost and -37% in Zn-enriched compost) in fraction III.

Table 3. Nitrogen uptake by wheat (mg pot)

Days after transplanting	C-treatment	Z-treatment	M-treatment	Control
14	4.16	4.68	4.07	5.45
28	8.85	12.52	10.93	12.00
56	9.15	19.46	11.92	11.87
90	14.23	18.35	13.19	16.84

Table 4. Labelled nitrogen uptake by wheat (mg pot))

Days after transplanting	C-treatment	Z-treatment	M-treatment
14	0.74	0.97	0.61
28	2.43	3.51	2.93
56	2.42	3.83	2.65
90	2.75	3.87	2.83

By the presence of high level of zinc, two different end products were obtained. Zinc seemed to induce a decrease on the carbon compounds humification rate, as well shown by the higher level of carbon and nitrogen in fraction III of the Z-compost, and a lowering of alkali insoluble nitrogen. Probably, zinc didn't induce a widespread toxic effect on microbial activity, rather it involved differences in decomposing populations. So, it could be hypothesized that the two composts would have different behaviour when added to soil, related to their different chemical complexity.

The effects of the two composts on soil and fertilizer nitrogen availability, and their mineralization in soil are reported in Figure 2. M-treatment allows to know the direct effect of inorganic zinc on these processes.

Both composts induced a higher net immobilization of fertilizer N, related to the control. In fact, while in the control about 93% of N-NH$_4$ was nitrified after 13 d, in the Z-treatment and in the C-treatment the percentage was lower (about 70%).

A higher mineralization of the nitrogen compounds of the Zn-enriched compost was observed, and, moreover, nitrification occured more rapidly than in the C-treatment. In fact, values of biomass-N were higher in the Z-treatment during the first 13 d of incubation (Table 2).

Inorganic zinc induced a significative effect on compost degradation and on nitrification of native and fertilizer N only after 60 d of incubation in soil. It seems that it induced a similar effect of the Zn-enriched compost on soil nitrogen turnover, but delayed.

Probably, in the Z-treatment we can see both the previous effect of zinc on composting and the direct effect of zinc on soil nitrogen dynamic.

In the experiment with wheat, inorganic N rapidly decreased in all treatments, due to the rizosphere effect (Fig. 3). Nevertheless, during the first period, labelled inorganic N reached a higher level than in the experiment without plants. Moreover the two zinc-treatments (Z- and M-treatments) showed that there is a direct effect of zinc on this enhancing.

168

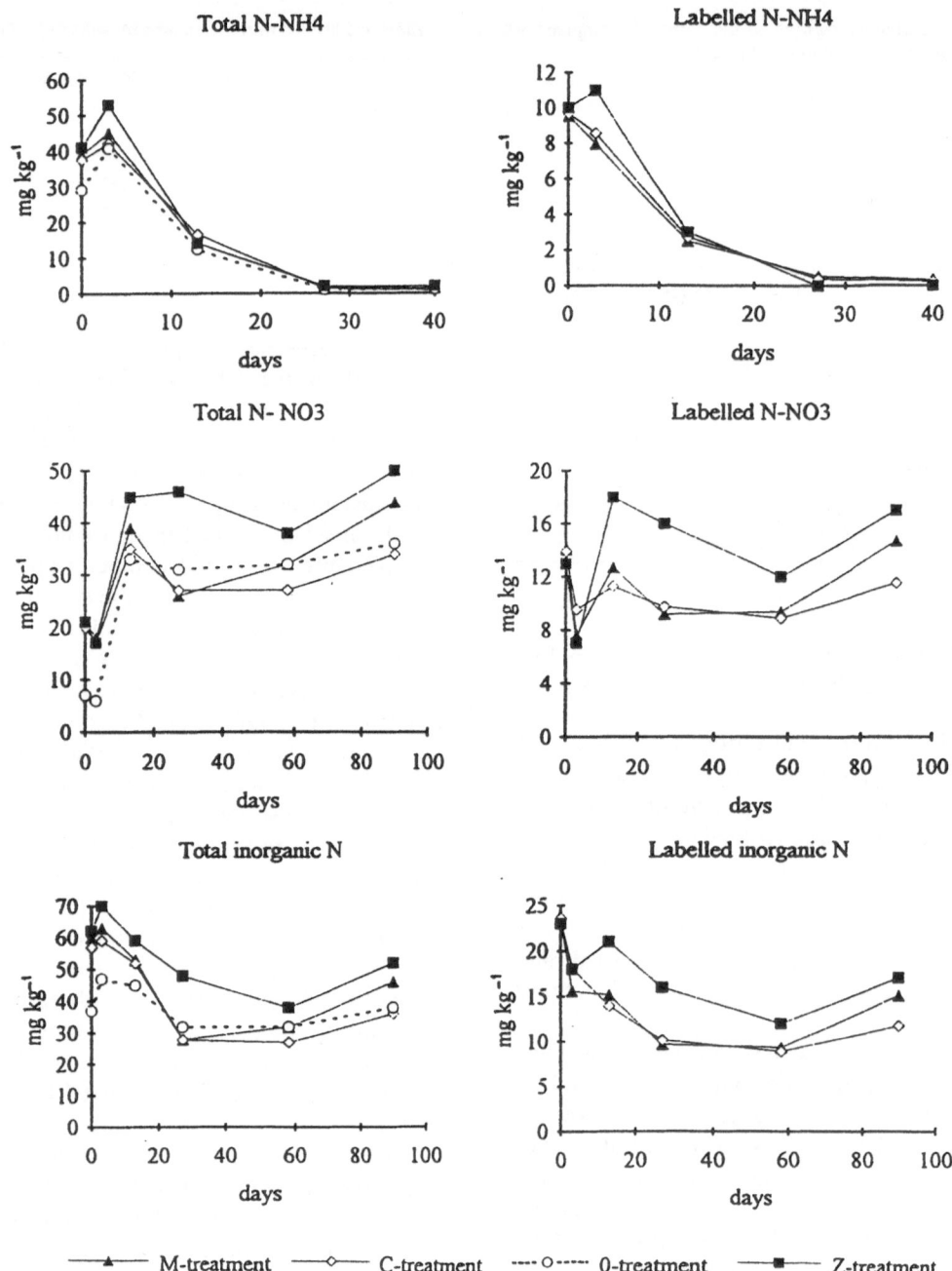

Figure 2. Changes in total and labelled soil inorganic N during incubation experiment without plants.

In the first period plant growth was depressed by composts, probably because of some critical physical characteristics of the mixtures (excess of softness) or for a decrease in O_2 concentration in the rizosphere (Fig. 4). Nevertheless, after 90 days, the Zn-enriched compost induced a yield significantly higher than those obtained in the other treatments.

Plant N uptake (Tables 3 and 4) confirmed that the Zn-enriched compost induced an increase in soil available N, as showed in the incubation experiment, and labelled N uptake strongly indicated that Zn-enriched compost provided significantly larger amounts of available N than control compost.

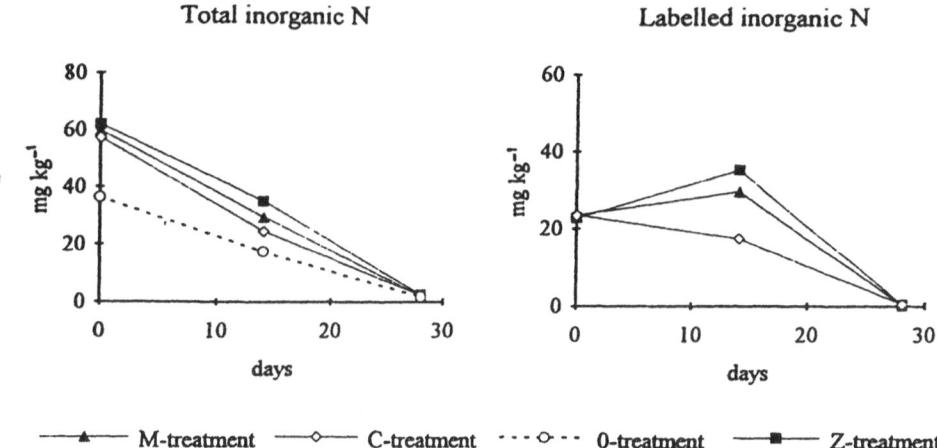

Total inorganic N   Labelled inorganic N

───▲─── M-treatment   ───◇─── C-treatment   ···○··· 0-treatment   ───■─── Z-treatment

Figure 3. Changes in total and labelled soil inorganic N during incubation experiment with plants.

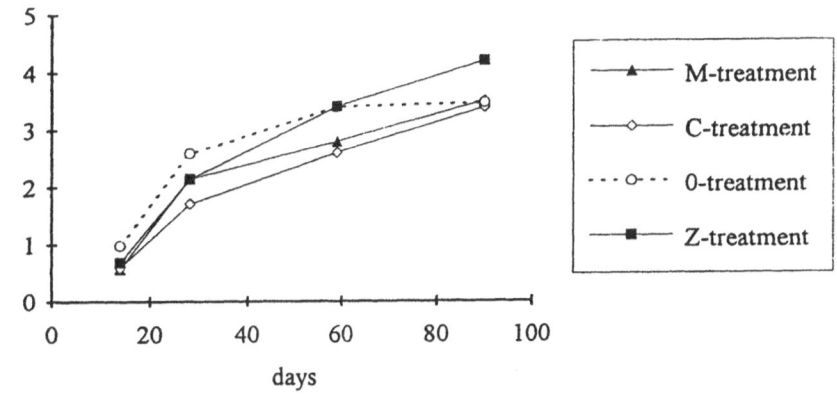

Figure 4. Fresh yields of wheat (g/pot).

References

Bååth E 1989 Effects of heavy metals in soil on microbial processes and populations (a review). Water Air Soil Pollut. 47, 335–379.

Chen Y, Magen H and Riov J 1994 Humic substances originating from rapidly decomposing organic matter: properties and effects on plant growth. *In* Humic substances in the global environment and implications on human health. Ed. N Senesi and T M Miano. pp 427–444. Elsevier Publishers, Amsterdam, The Netherlands.

He X-T, Stevenson F J, Mulvaney R L and Kelley K R 1988 Incorporation of newly immobilized ^{15}N into stable organic forms in soil. Soil Biol. Biochem. 20, 75–81.

Zaccheo P, Crippa L and Genevini P L 1993 Nitrogen transformation in soil treated with ^{15}N labelled dried or composted ryegrass. Plant and Soil 148, 193–201.

2. ASPECTS RELATED TO FERTILIZATION:
   UPTAKE, AVAILABILITY, EFFICIENCY, TECHNIQUES

O. Van Cleemput et al. (eds.), Progress in Nitrogen Cycling Studies, 173–177, 1996.

Effects of environmental factors and Mn, Zn, Cu trace elements on the available N content of two soils

A. Anton[1], L. Radimszky[1], T. Szili-Kovács[1], G. Füleky[2] and F. Gulyás[1]

[1]*Institute for Soil Science and Agricultural Chemistry of Hungarian Academy of Sciences, Herman Otto street 15, H-1022 Budapest, Hungary and* [2]*Department of Soil Science and Agrochemistry, Gödöllö University of Agricultural Sciences, Páter K. street 1, Gödöllö, Hungary*

Key words: available nitrogen, modelling, multifactorial experiment, N-immobilization, N-transformation, soil incubation

Abstract

A soil incubation experiment was carried out in a complex experimental system to simulate the effect of environmental factors (temperature, soil moisture, lucerne shoot powder as plant residue with mineral N, and mineral P addition), and 3 trace elements (Mn, Zn, Cu) on the available N content (NH_4^+-N and NO_3^--N) of two soils. The applied DISITOBI model assures information about the experimental object characterized by a multidimensional response function.

After a 33 days laboratory incubation period the available N content in a calcareous chernozem soil was directly influenced by the plant residue + mineral N addition, the incubation temperature and the soil moisture content. The lucerne powder had a positive linear effect, while the temperature - and moisture content - available N content relationships showed maximum-curves. The temperature - soil moisture, Zn - plant residue with mineral N, Mn-P and Cu-P pair-interactions also had significant effects on the available soil N.

According to the results the main relationships are similar comparing the two model soil. In contrast to Chernozem, the moisture content had a significantly negatively linear effect on the available N content in the acid rusty Brown Forest soil.

Results demonstrate that the DISITOBI model is a good initial basis for developing new experimental methodology approaching soil biological and biochemical processes.

Introduction

The fate of nitrogen fertilizers added to soils has been examined in numerous investigations. Scientific recognition of biological, biochemical transformations of different C- and N-sources in soil is an important objective (Beck, 1970; Cervelli et al., 1978; Frankenberger and Dick, 1983).

It is important to include in models that these transformation processes are influenced by different environmental conditions (Bjarnason, 1987; Jansen and Kucey, 1988; Novak, 1972).

Our objective was to investigate the effects of environmental factors and 3 trace elements on the available N content of 2 soils by means of short-term laboratory incubation test in a complex experimental system.

Materials and methods

Soils and analysis

Soil samples investigated in the experiment were collected from the ploughed layer of a calcareous Chernozem from Nagyhörcsök (pH/KCl/ = 7,0 contains 3,3% $CaCO_3$, 3,6% organic matter, 36 mg kg^{-1} available N and 123 mg kg^{-1} Al- soluble P_2O_5) and a rusty Brown Forest soil from Gödöllö (pH/KCl/ = 3,1 contains 0,0% $CaCO_3$, 1,5% organic matter, 22 mg kg^{-1} available N and 80 mg kg^{-1} Al-soluble P_2O_5).

On the 33rd day of incubation, the available N content (NH_4^+-N, NO_3^--N) of the soil samples was determined in the 2 *M* KCl extract, according to Bremner (1965).

Experimental design and evaluation

The DISITOBI model assures information about the experimental object characterized by a multidimensional response-function. The response of the object is described by a hypersurface (Szili-Kovács et al., 1993).

By means of the SITOBI program, based on the DISITOBI model, it is easy to make an interactive design of multifactorial orthogonal experimental plans and statistical analysis of data.

Form of the DISITOBI model in the present experiment:

$$
\begin{aligned}
Y = {} & B_O + B_1T + B_2W + B_3L + B_4P \\
& + B_5Mn + B_6Zn + B_7Cu + B_8TxW \\
& + B_9L \times T + B_{10}L \times W + B_{11}P \times T \\
& + B_{12}P \times W + B_{13}PxL + \times B_{14}MnxT + \\
& + B_{15}Mn \times W + B_{16}Mn \times L + B_{17}Mn \times P \\
& + B_{18}Zn \times T + B_{19}Zn \times W + \\
& + B_{20}Zn \times L + B_{21}Zn \times P + B_{22}Zn \times Mn \\
& + B_{23}Cu \times T + B_{24}Cu \times W + \\
& + B_{25}Cu \times L + B_{26}Cu \times P + B_{27}CuxMn \\
& + B_{28}CuxZn + B_{29}T^2 + \\
& + B_{30}W^2 + B_{31}L^2 + B_{32}P^2 + B_{33}Mn^2 \\
& + B_{34}Zn^2 + B_{35}Cu^2
\end{aligned}
$$

where Y = /available N, NH_4^+-N and NO_3^-N content of soil/

T = temperature (°C, range: 5–45)

W = soil moisture (% of max. water holding capacity, range: 20–100)

L = lucerne shoot powder (%, range: 0–1, and NH_4NO_3, according to 10:1 C:N ratio)

P = KH_2PO_4-P (mg kg^{-1}, range: 0–240, and $KHCO_3$ to compensate K-effect)

Mn, Zn, Cu = in sulphate form (mg kg^{-1}, range: 0–500)

B0 - B35 = parameters of the model

5–5 levels of the seven studied factors were combined in an orthogonal way by means of the program with respect to the normalized matrix-plan. The minimum-maximum values of factors were chosen on the basis of our previous studies.

Measured values of available N as experimental data were processed by means of the SITOBI program. Parameters of the models (equations), regression, correlation and other statistical characteristics were determined.

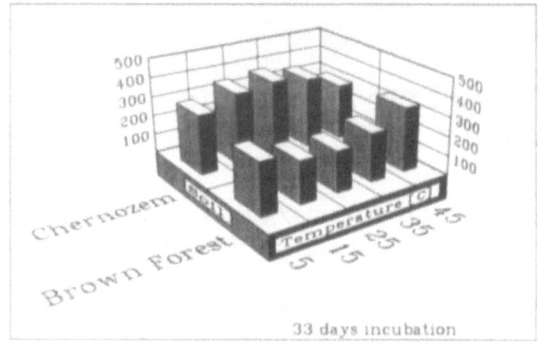

Figure 1. [NH_4^+ + NO_3^-]-N content of soil (mg kg^{-1}).

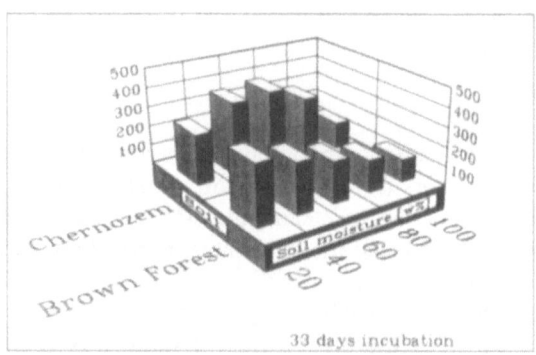

Figure 2. [NH_4^+ + NO_3^-]-N content of soil (mg kg^{-1}).

Results

Parameters (linear, quadratic components and pair-interactions) of computed models based on the measured available N content data taken as a function of investigated factors are demonstrated in the Tables 1 and 2.

Note: values of other experimental factors on the Figures 1–8 illustrating some main relationships are on the central level of their investigated range.

Discussion

Considering the limited size of this paper we can make only some short comments on the results.

The direct effects of the temperature and the soil moisture on the available N content showed maximum curves in the Chernozem in contrast to the Brown Forest soil, in which the moisture content had a significantly negatively linear effect (Fig. 1 and 2). It can be stated that the advantageous levels of the moisture content for microbial N immobilization are different

Table 1. Parameters of the model and the determination coefficients (R^2). Chernozem soil (Nagyhörcsök), 33 days incubation. The treatments and explanation of abbreviations are described in Materials and methods

	Available N	NO_3^--N	NH_4^+-N
B_0	-46.3699[xxx]	-191.2377[xxx]	144.3037[xx]
T	1.7692	4.8190	-3.0432[xxx]
W	1.5043[xx]	2.6466	-1.1325[x]
L	116.2541[xxx]	149.7719[xxx]	-33.4046[xxx]
P	0.3353	0.6440[x]	-0.3064
Mn	-0.1964	-0.0302	-0.1639
Zn	0.1050[x]	0.1789[xx]	-0.0744[xx]
Cu	-0.1809	-0.0199	-0.1605
T×W	-0.0311[xxx]	-0.0408[xxx]	0.0098[xxx]
L×T	-0.0934	-0.4581	0.3524[x]
L×W	0.2055	0.2527	-0.0408
P×T	0.0035	0.0012	0.0022[xx]
P×W	0.0018	0.0005	0.0013[xxx]
P×L	-0.1634[x]	-0.2168[xx]	0.0530
Mn×T	0.0009	-0.0005	0.0013[xxx]
Mn×W	0.0003	-0.0003	0.0006[xx]
Mn×L	0.0151	-0.0157	0.0297[x]
Mn×P	-0.0005[xxx]	-0.0008[xxx]	0.0003[xxx]
Zn×T	0.0008	0.0010	-0.0002
Zn×W	-0.0001	-0.0005	0.0004[x]
Zn×L	-0.1988[xxx]	-0.2610[xxx]	0.0629[xxx]
Zn×P	-0.0001	-0.0002	0.0001[xx]
Zn×Mn	-0.0001	-0.0001	0.0001
Cu×T	0.0017	0.0004	0.0013[xxx]
Cu×W	0.0008	0.0003	0.0005[xx]
Cu×L	-0.0180	-0.0518	0.0342[xx]
Cu×P	-0.0008[xxx]	-0.0011[xxx]	0.0003[xxx]
Cu×Mn	0.0008	0.0006	0.0002
Cu×Zn	-0.0002[x]	-0.0002[xx]	0.0001
T^2	-0.0233[x]	-0.0543[xxx]	0.0309[xxx]
W^2	-0.0121[xxx]	-0.0150[xxx]	0.0028[xx]
L^2	-5.4249	-2.9679	-2.7540
P^2	-0.0005	-0.0005	-0.0001
Mn^2	0.0001	0.0001	-0.0000
Zn^2	0.0001	0.0001	-0.0000
Cu^2	0.0001	0.0001	-0.0000
R^2	82.4%	78.7%	79.8%

[xxx] = $p < 0.01$ [xx] = $p < 0.05$ [x] = $p < 0.10$

Table 2. Parameters of the model and the determination coefficients (R^2). Brown Forest soil (Gödöllö), 33 days incubation. The treatments and explanation of abbreviations are described in Materials and methods

	Available N	NO_3^--N	NH_4^+-N
B_0	123.8156[xxx]	99.4295[xxx]	26.4397[xxx]
T	-1.6886	-1.7038[x]	-0.0850
W	-0.7952[xxx]	-0.6403[xxx]	-0,1776
L	27.8206[xxx]	-0.5170[xxx]	27.3466[xxx]
P	-0.1698	-0.1785	0.0156[x]
Mn	-0.2023[x]	-0.0932[xx]	-0.1150
Zn	-0.0335	-0.0462	0.0138[x]
Cu	-0.2456	-0.1469	-0,0973
T×W	0.0130[xx]	0.0103[xxx]	0.0028
L×T	0.6799	0.8784[xxx]	-0.2031
L×W	-0.2779	-0.1400	-0.1127
P×T	0.0029	0.0016	0.0015
P×W	-0.0007	0.0007	-0.0013[xx]
P×L	0.0307	-0.0126	0.0404
Mn×T	-0.0010	0.0001	-0.0012[xx]
Mn×W	0.0004	0.0000	0.0004
Mn×L	-0.0506	-0.0145	-0.0378
Mn×P	0.0003[x]	0.0002[xx]	0.0001
Zn×T	-0.0011	-0.0006	-0.0005
Zn×W	0.000	0.0009[xxx]	-0.0002
Zn×L	0.0831[xx]	0.0662[xxx]	0.0181
Zn×P	-0.0003[x]	-0.0002	-0.0001
Zn×Mn	-0.0000	-0.0000	0.0000
Cu×T	-0.0007	0.0006	-0.0013[xx]
Cu×W	0.0007	0.0004	0.0004
Cu×L	-0.0650[x]	-0.0362	-0.0330
Cu×P	0.0003[xx]	0.0003[xxx]	0.0001
Cu×Mn	0.0007	0.0003	0.0004
Cu×Zn	0.0000	-0.0000	0.0000
T^2	0.0217[x]	0.0108	0.0120
W^2	0.0001	-0.0011	0.0011
L^2	19.6300	1.0417	19.9575[x]
P^2	0.0002	0.0001	0.0001
Mn^2	0.0000	0.0000	0.0000
Zn^2	0.0000	0.0000	0.0000
Cu^2	0.0001	0.0001	-0.0000
R^2	83.3%	69.9%	85.2%

[xxx] = $p < 0.01$ [xx] = $p < 0.05$ [x] = $p < 0.10$

in the Chernozem and the Brown Forest soil (60% and 100% of max. water holding capacity, respectively).

The shape of curves illustrating the changes in the NH_4^+ and NO_3^- contents taken as functions of the temperature and the soil moisture are different from one another in the Chernozem soil (Fig. 3 and 4). Based on these data the optimal temperature and moisture content for nitrification process can be estimated. In contrast to the above the intensity of nitrification in the Brown Forest soil was unchanged in the investigated range of above mentioned factors (Table 2).

The available N content of both soils were directly influenced (linear increase) by the mineral N added to

176

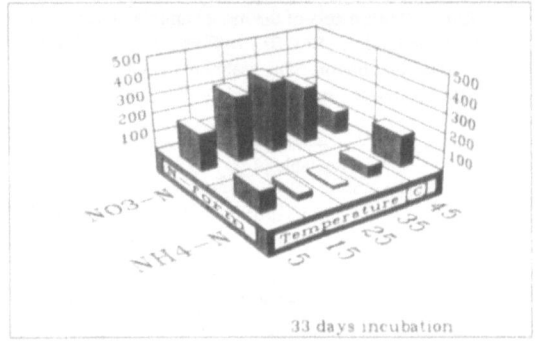

Figure 3. NH$_4^+$ -N and NO$_3^-$ -N content of calcerous Chernozem soil (mg kg^{-1}).

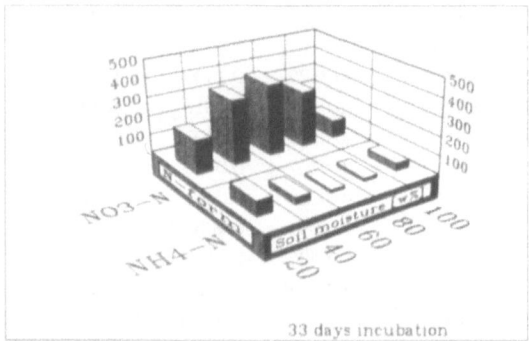

Figure 4. NH$_4^+$ -N and NO$_3^-$ -N content of calcerous Chernozem soil (mg kg^{-1}).

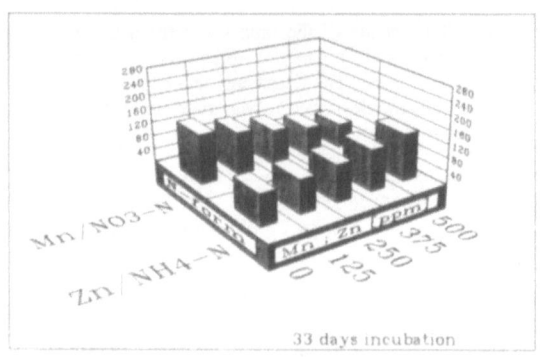

Figure 6. NH$_4^+$ -N and NO$_3^-$ -N content of rusty Bown Forest soil (mg kg^{-1}).

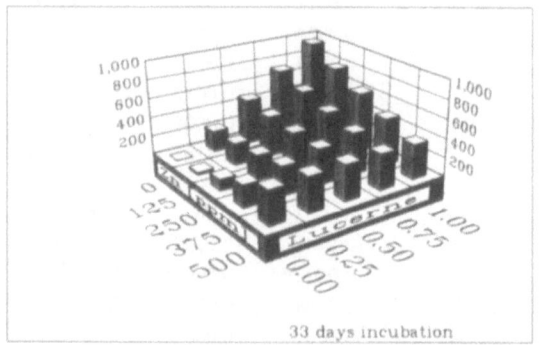

Figure 7. [NH$_4^+$ + NO$_3^-$]-N content of calcareous Chernozem soil (mg kg^{-1}). Zn-plant residue interaction.

the soil together with the lucerne residue. It can be concluded, according to the results, that the microbial N immobilization was more intensive in the Brown Forest than in the Chernozem soil (Tables 1 and 2).

The applied trace elements had little significant direct effects on the soil available N (Fig. 5 and 6). Their influences operated through interaction with other factors. One of the mutual effects is the Zn - Lucerne residue+mineral N interaction which operated differently in two soils characterized by different physico-chemical properties (Fig. 7 and 8).

Our results demonstrate that the DISITOBI model is a good initial basis for developing new experimental methodology approaching soil biological and biochemical processes, collecting complex information about interactive effects of various factors on the soil metabolism.

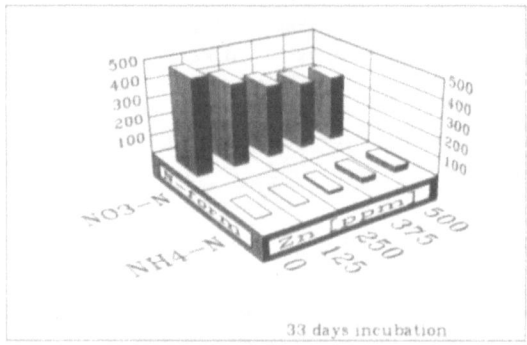

Figure 5. NH$_4^+$ -N and NO$_3^-$ -N content of calcareous Chernozem soil (mg kg^{-1}).

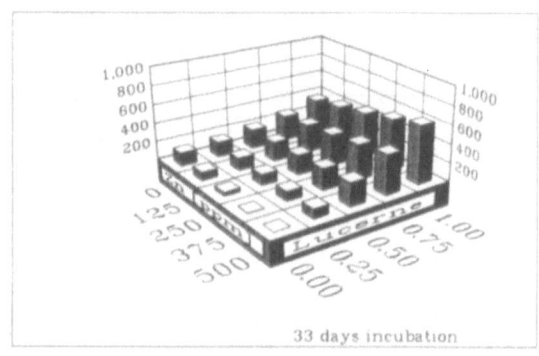

Figure 8. [NH$_4^+$ + NO$_3^-$]-N content of rusty Brown Forest soil (mg kg^{-1}). Zn-plant residue interaction.

Acknowledgements

This research was supported by Hungarian National Scientific Research Fund (OTKA). The excellent technical assistance of Ms Kathie Wohlmuth is gratefully acknowledged.

References

Beck T 1970 Der mikrobielle Abbau von Herbiziden und ihr Einfluss auf die Mikroflora des Bodens. Z. Bakteriol. II. 124, 304–313.

Bjarnason S 1987 Immobilization and remineralization of ammonium and nitrate after addition of different energy sources to soil. Plant and Soil 97, 381–389.

Bremner J M 1965 Inorganic forms of nitrogen. *In* Methods of Soil Analysis. Agronomy 9, Part 2. Ed. C A Black. pp 1179–1237. American Society of Agronomy, Madison, USA.

Cervelli S, Nannipieri P and Sequi P 1978 Interaction between agrochemicals and soil enzymes. *In* Soil Enzymes. Ed. R G Burns. pp 251–293. Academic Press, London, UK.

Frankenberger W T and Dick W A 1983 Relationship between enzyme activities and microbial growth and activity indices in soil. Soil Sci. Soc. Am. J. 7, 945–951.

Jansen H H and Kucey R M N 1988 C, N, S mineralization of crop residues as influenced by crop species and nutrient regime. Plant and Soil 106, 35–41.

Novák B 1972 Effect of increasing amounts of nitrogen on the microbial transformation of straw in soil. *In* Proceedings Symp. on Soil Microbiology. Ed. J Szegi. pp 49–53. Publishing House of Hungarian Academy of Sciences, Budapest, Hungary.

Szili-Kovács T, Radimszky L, Andó J and Biczók Gy 1993 CO_2 evolution from soil formed on various parent materials in the eastern Cserhát mountains (Hungary) during laboratory incubation. Agrokémia Talajtan 42, 140- 146.

O. Van Cleemput et al. (eds.), Progress in Nitrogen Cycling Studies, 179–182, 1996.

Maize supplementation for grazing cows on the N content of milk, faeces and grass

L. Carlier and I. Verbruggen
Rijksstation voor Plantenveredeling (R.v.P.) (CLO-Gent), Burg. Van Gansberghelaan 109, 9820 Merelbeke, Belgium and Nationaal Centrum voor Grasland- en Groenvoederonderzoek, 1e Sectie (R.v.P.) - (I.W.O.N.L.), Burg. Van Gansberghelaan 109, 9820 Merelbeke, Belgium

Key words: herbage, maize silage, nitrogen excretion, supplementation

Abstract

During the grazing season of 1993 one group of cows (n=34) grazed day and night while an other group (n=40) grazed only during the day ; these were housed at night receiving maize silage (5.3 kg per cow per day). Although N-input and N-output, in kg N per ha, was different for both grazing systems, the N-surplus was the same. The area of grassland covered by dung pats and urine scorch, the N-content of the cow faeces and the grass and the urea content of the milk (Azotest) were significantly higher on the day + night grazing system. From these results, it can be concluded that the choice of the grazing and management system has a real influence on the nitrogen cycle on grassland.

Introduction

Nowadays the N and P fertilization of crops is widely discussed (Van Der Meer, 1994).

In Belgium and especially in Flanders the surplus of minerals on the NPK balances of dairy farms is rather large (Verbruggen et al., 1993). The Flemish Government will decrease the N- and P-emission in agriculture by the limitation of the N-P fertilization on the crops. Otherwise practical measures are suggested to improve the N efficiency by e.g. combining protein rich herbage with energy rich maize silage (Van Vuuren and Meys, 1987).

In this paper the effects of housing cattle at night, supplemented with maize silage on the N content of the herbage, faeces and milk of the cows are discussed.

Materials and methods

On a practical farm in Bornem the dairy cows were split up into two groups during 1993.

During the grazing season one group of 34 cows could graze day and night, while the second group of 40 cows was housed during the night, receiving a supplement of 5.3 kg dry matter from maize silage per cow. Above the production of 15 kg of milk each cow received a supplement of 1 kg concentrates per 1.5 kg milk.

On a year basis, the group of cows grazing day and night, got a bigger area of grassland for both grazing and mowing while the other group, supplemented at night, got a larger area to grow maize.

The mown grassland received more fertilizer than the grazed one and therefore this group of cows got a higher NPK input on their mineral balance. On the other hand the supplemented group got more protein rich concentrates during winter time and this is reflected in their total N input.

During the grazing season, every two weeks, samples of the herbage and fresh faeces were taken for the determination of their NPK content. The areas covered by dung pats and urine scorch were measured monthly. The Azotest (determination of the urea content with test strips) on milk samples was carried out weekly in order to evaluate the imbalance in the ration of the cows.

Table 1. N, P and K content of herbage and faeces of different grazing systems

Management	Grazing day + night system I			Grazing during the day housing at night + 5.3 kg d.m. from maize silage Systems II		
	N	P	K	N	P	K
N, P and K of faeces (g kg^{-1} d.m.)	3.82	2.46	1.02	2.49	2.31	0.70
N, P and K of grass (g 100^{-1} g d.m.)	2.96	0.43	3.29	2.73	0.37	2.95

Results and discussion

In Table 1 the results of the chemical analyses of the herbage and faeces samples are given.

The corresponding figures in Table 1 are all significantly different at the $p = 0.01$ level. These results show that, during the grazing season, housing the cows at night with a supplement of energy rich, protein poor roughage may decrease significantly the N, P and K content of the cow faeces and the grazed herbage. From results of Van Vuuren and Meys (1987) it can be concluded that the N excretion of cows fed a combination of herbage and maize silage (51/49 on d.m. basis)

reached only 72% of that of cows fed fresh herbage alone. This means a much smaller return of N and so total N input on the herbage of this management system.

Figure 1 shows the evaluation of the urea content in the milk of both systems. The Azotest gives an indication of the urea content in milk. Date above the optimal urea content of 0.2 g L^{-1} shows an imbalance of the N/energy ratio in the ration of the cows.

Lantinga et al. (1987) concluded from their data that urine scorch is a more serious problem on sandy soils than on peat and clay soils and that the extent is strongly correlated with the rate of fertilizer N.

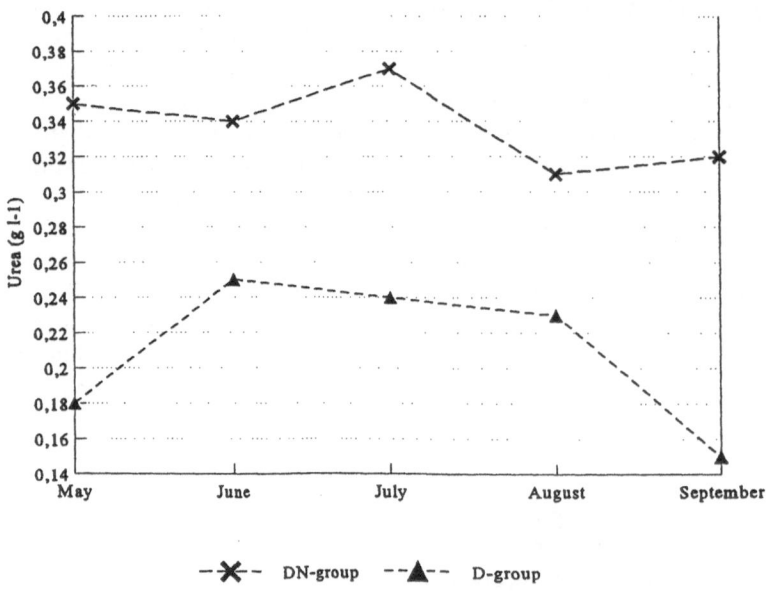

Figure 1. Urea content in milk (g L^{-1}).

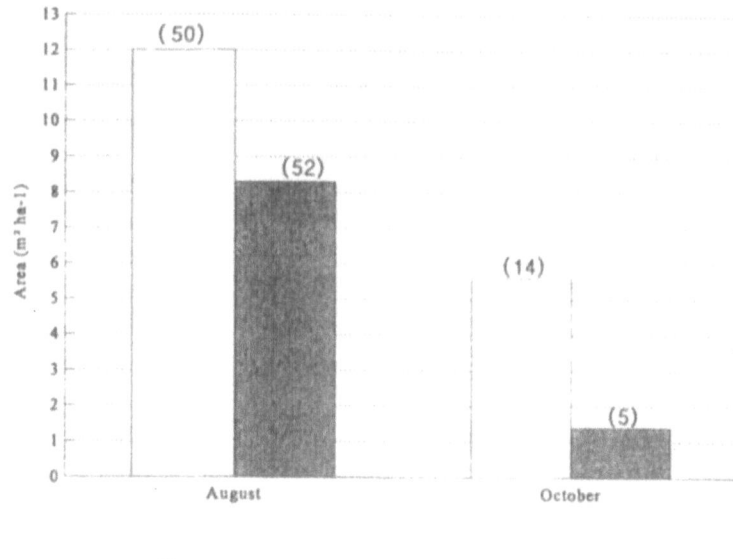

Figure 2. Area of urine scorch (m^2 ha^{-1}) (number of scorch between brackets).

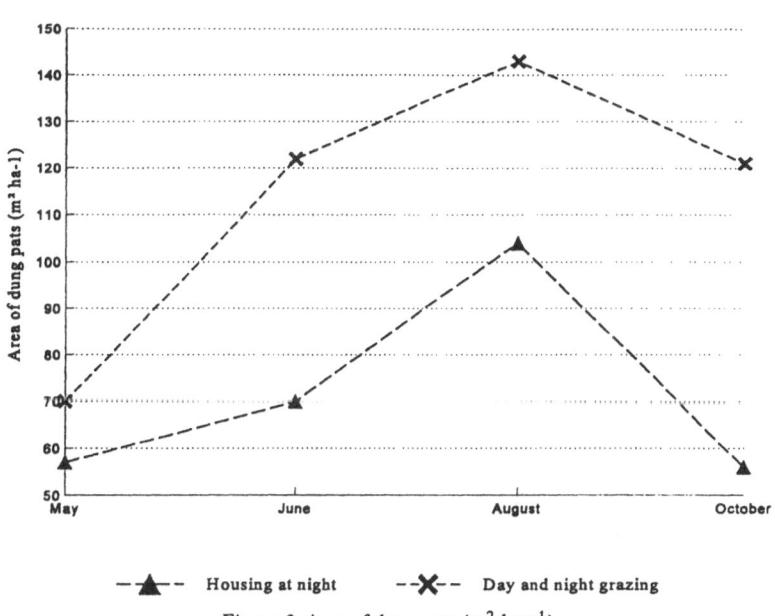

Figure 3. Area of dung pats (m^2 ha^{-1}).

In Figure 2, the total area covered by urine scorch on both management systems is shown and the number of scorch areas in August and September '93 are reported too.

Figure 3 gives the evolution of the area covered by dung pats on both grazing systems. Both figures show clearly the positive influence of housing cattle at night on the damaged area, due to a lower total N input from animal excrements and mineral fertilizers.

The option to grow more maize and less grass resulted also in a lower input of grassland products (pre wilted silage) during winter time. Therefore concentrates with a higher protein content were necessary to fulfil the feed requirements of the cows on this management system (system II).

To measure the influence of both systems on the NPK surplus per hectare the mineral year balances had to be calculated. The results are summarised in Table 2.

182

Table 2. Mineral year-balances for two grazing systems

Management	System I			System II		
	N	P	K	N	P	K
Input (kg ha⁻¹)						
Concentrates	105.7	22.4	53.5	134	24.1	63.3
Mineral fertilizers	174.9	26.0	49.4	155.5	20.2	38.3
Deposition	40	0.9	4.1	40	0.9	4.1
Output (kg ha⁻¹)						
Milk	79.1	14.3	24.6	84.2	14.8	26.2
Meat	21.3	6.2	1.4	23.2	6.8	1.6
Total input (kg ha⁻¹)	320.6	49.3	107	329.5	45.2	105.7
Total output (kg ha⁻¹)	100.4	20.5	26	107.4	21.6	17.8
Surplus (kg ha⁻¹)	220.2	28.8	81	222.1	23.6	77.9

On a year basis there was no significant difference in N surplus per hectare on both grazing systems. Only the P and K surplus of the day and night grazing system is somewhat higher.

The advantage of the lesser N input of "the housing at night system" during the grazing season on the mineral balance is completely lost during the winter time because more protein rich concentrates are necessary.

Conclusion

Housing milking cows during the night, with the supplementation of maize silage had a significant influence on the chemical composition of the herbage and the produced milk.

The areas covered by dung pats and urine scorch became smaller all over the grazing season.

However the total mineral NPK balance over the year shows no significant difference in N surplus per hectare, mainly because of the fact that 'the housing at night' system needs more protein rich concentrates during winter time because much less prewilted silage was available.

References

Lantinga E A, Keuning J A, Groenwold J and Deenen P J A G 1987 Distribution of excreted nitrogen by grazing cattle and its effects on sward quality, herbage production and utilization. *In* Animal Manure on Grassland and Fodder Crops Fertilizer or Waste ? Eds. H Van Der Meer et al. pp 103–117. Martinus Nijhoff Publishers, Dordrecht, the Netherlands.

Van Der Meer H 1994 Grassland and society. *In* Grassland and Society. Eds. L 't Mannetje and J Frame. pp 19–32. Wageningen Pers, Wageningen, The Netherlands.

Van Vuuren A M and Meys J A C 1987 Effects of herbage composition and supplement feeding on the excretion of nitrogen in dung and urine by grazing dairy cows. *In* Animal Manure on Grassland and Fodder Crops Fertilizer or Waste ? Eds. H Van Der Meer et al. pp 103–117. Martinus Nijhoff Publishers, Dordrecht, the Netherlands.

Verbruggen I, Carlier L en Van Bockstaele E 1993. Beperking van de N-P-K overschotten op het melkveebedrijf. Mededeling RvP 674. 26 p.

O. Van Cleemput et al. (eds.), Progress in Nitrogen Cycling Studies, 183–190, 1996.

Residual effects of poultry manure and fertiliser nitrogen applications

B.J. Chambers[1], J.R. Williams[1] and K.A. Smith[2]
ADAS Arable Research Centre, [1]Anstey Hall, Maris Lane, Trumpington, Cambridge, CB2 2LF, UK and [2]Wergs Road, Woodthorne, Wolverhampton, WV6 8TQ, UK

Key words: potatoes, poultry manure, residual nitrogen, sugar beet

Abstract

The effect of fertiliser nitrogen (N) and poultry manure N applications to sugar beet and potatoes on residual soil mineral nitrogen (SMN) supply and yield of following cereal crops was studied in four field experiments in 1991 and 1992. In both seasons, fertiliser N applications to sugar beet (range 0–180 kg N ha^{-1}) had no effect on residual SMN supply or yield of the following cereal crops. However, poultry manure applications supplying > 600 kg total N ha^{-1} increased residual SMN supply and yield of the following cereal crops. Fertiliser N applications to potatoes in 1991 (range 0–300 kg N ha^{-1}) increased residual SMN and yield of the following cereal crop where 300 kg N ha^{-1} was applied. However, in 1992, no increase in residual SMN or yield of the following cereal crop was recorded. Poultry manure N applications > 800 kg total N ha^{-1} increased residual SMN supply and > 400 kg total N ha^{-1} increased following cereal crop yields. These results demonstrate the ability of sugar beet to scavenge soil N and low residual N value of poultry manure applications at agronomically sensible rates.

Introduction

In the UK, an estimated 4.6 million tonnes of poultry manure containing around 113,000 tonnes of nitrogen is produced annually (Smith, 1991). Poultry manure is applied before 9% of potato and sugar beet crops grown in the UK, with little apparent allowance being made by farmers in their inorganic fertiliser policies for the manures N value (Anon., 1994). ADAS cereal experiments (Chambers et al., 1994) have shown that approximately 35% of the total N content of spring applied poultry manure is plant available (principally ammonium and uric acid N) in the season following application.

The residual effect of poultry manure N applications from the plant available N supplied and mineralisation of organic N over subsequent years/months will affect the potential for nitrate leaching losses and nitrogen supply for subsequent crops. Work at Rothamsted showed that the residual effects of large repeated dressings of solid manures persisted for many years (Johnson, 1970). Lande Cremer (1985) suggested that the efficiency of N supplied from poultry manure was 65% in the first year after application and that efficiency increased to 76% in subsequent years because of

the residual effects. To achieve full economic benefit from manure applications and to minimise nitrate losses to surface and ground waters, a greater understanding of the residual N effects of manure applications is required. This paper reports results from experiments to investigate the effects of poultry manure and fertiliser N applications to potatoes and sugar beet on soil nitrogen supply and yield of the following cereal crop.

Materials and methods

Two field experiments were established in Suffolk during harvest years 1991 and 1992 on sandy loam textured soils of the Methwold Association (Soil Survey of England and Wales, 1983), and two experiments in Yorkshire on a loamy sand soil (Wick Association) in 1991 and sandy loam textured soil (Escrick Assoc.) in 1992. Poultry manure was applied in winter and spring before crops of sugar beet in Suffolk, and potatoes in Yorkshire. At each site the winter and spring applications were taken from the same manure source, which had been stored overwinter under a plastic sheet to minimise N losses by leaching and volatilisation. However, at the Yorkshire site during winter 1990/91,

Table 1. Nitrogen analysis of applied poultry manures

Site Name / Manure timing	Poultry manure	Dry matter (%)	Nitrogen (kg tonne^{-1} fresh weight)		
			Total	Ammonium	Uric acid
1. Suffolk 1991					
December 1990	Turkey litter	54.3	31.2	9.0	4.8
March 1991	Turkey litter	55.5	30.7	6.9	11.0
2. Suffolk 1992					
January 1992	Turkey litter	57.3	34.4	9.4	4.2
March 1992	Turkey litter	57.7	36.2	8.7	3.8
3. Yorkshire 1991					
October 1990	Broiler litter	58.3	23.5	7.1	N.D.
April 1991	Broiler litter	33.8	10.4	1.5	0.11
4. Yorkshire 1992					
November 1991	Broiler litter	53.7	23.1	6.4	4.2
March 1992	Broiler litter	60.3	32.1	8.7	5.1

Table 2. Cereal crop nitrogen uptake at harvest in 1992 and 1993 following poultry manure applications to preceding sugar beet crops, in 1991 and 1992

N application to previous sugar beet crop (kg ha^{-1})		Crop N uptake (kg ha^{-1})	
Fertiliser		Harvest 1992	Harvest 1993
0		84	78
35		99	74
70		77	74
105		99	80
140		89	86
175		99	82
210		109	83
Winter poultry manure			
1990: 123	1992: 150	94	87
246	301	91	115
615	753	133	147
Spring poultry manure			
1991: 100	1992: 158	89	88
201	317	97	87
502	793	101	139
p		<0.001	<0.001
SED		10.6	14.1
CV (%)		12.5	16.7

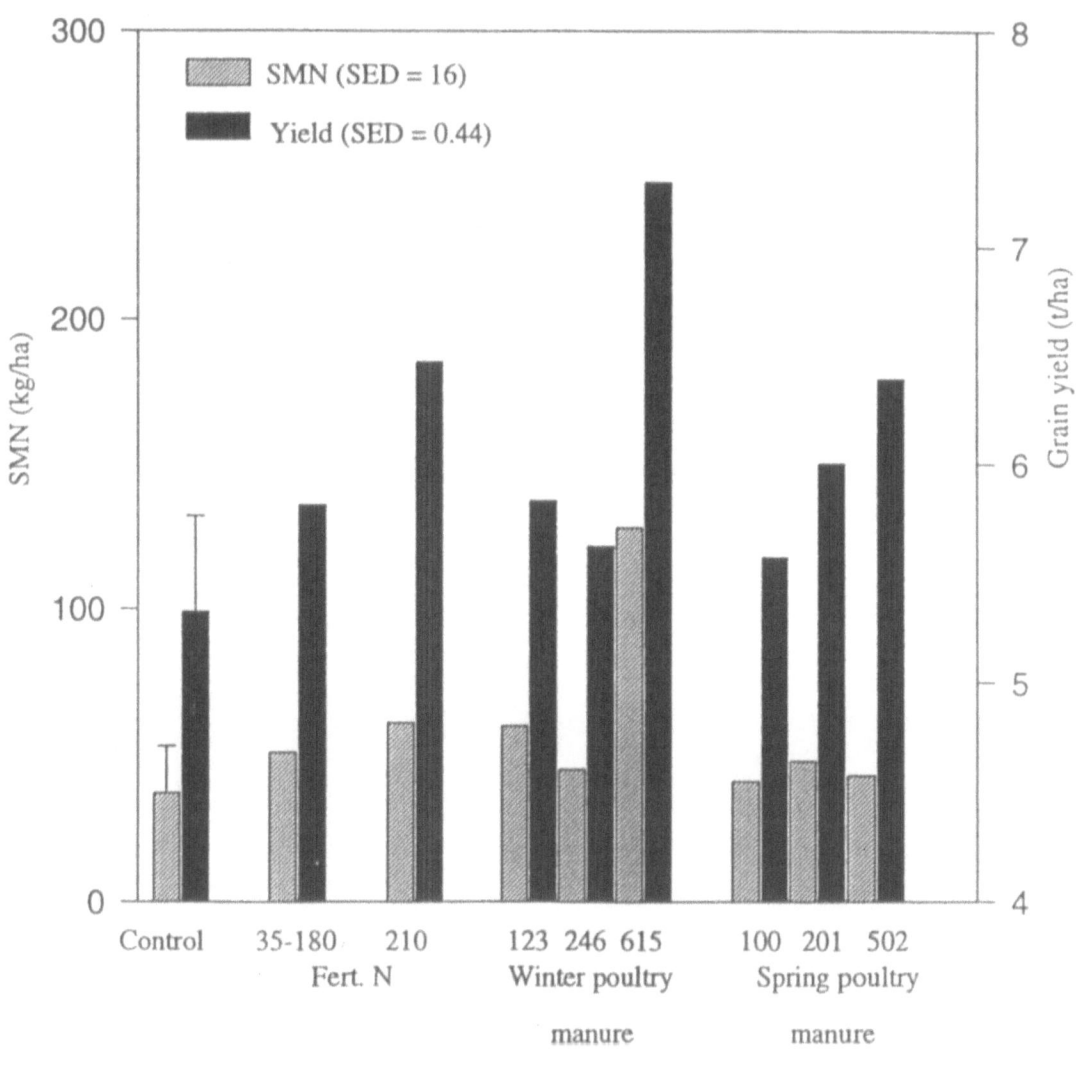

Figure 1. Residual soil mineral nitrogen supply and grain yield of 1992 cereal crop following fertiliser and poultry manure N applications to sugar beet in 1990/91.

rainfall penetrated the sheet reducing the manure dry matter content. For both timings target manure total N application rates were 100, 200 and 600 kg N ha^{-1} to sugar beet and 150, 300 and 750 kg N ha^{-1} to potatoes. The poultry manures were applied by hand and incorporated into the soil within 6 hours of application to minimise ammonia volatilisation losses. At each application date poultry manure samples were analysed to determine their total N, ammonium N (i.e. inorganic N), uric- acid N (i.e. readily mineralisable organic N) and dry matter contents, using standard ADAS techniques (MAFF,1986), Table 1.

Each treatment was replicated three times in a randomised block design. Plot sizes were 6 m × 18 m. All sites had satisfactory soil pH and nutrient status, and where necessary maintenance phosphate and potash fertiliser applications were made to ensure that neither nutrient was limiting. Spring fertiliser N applications in the range 0 to 210 kg N ha^{-1} were applied to the sugar beet crops (35 kg N ha^{-1} to the seedbed, with the balance at the 2–4 leaf stage), and 0 to 300 kg N ha^{-1} (half to the seedbed and the remainder at tuber initiation) to the potato crops.

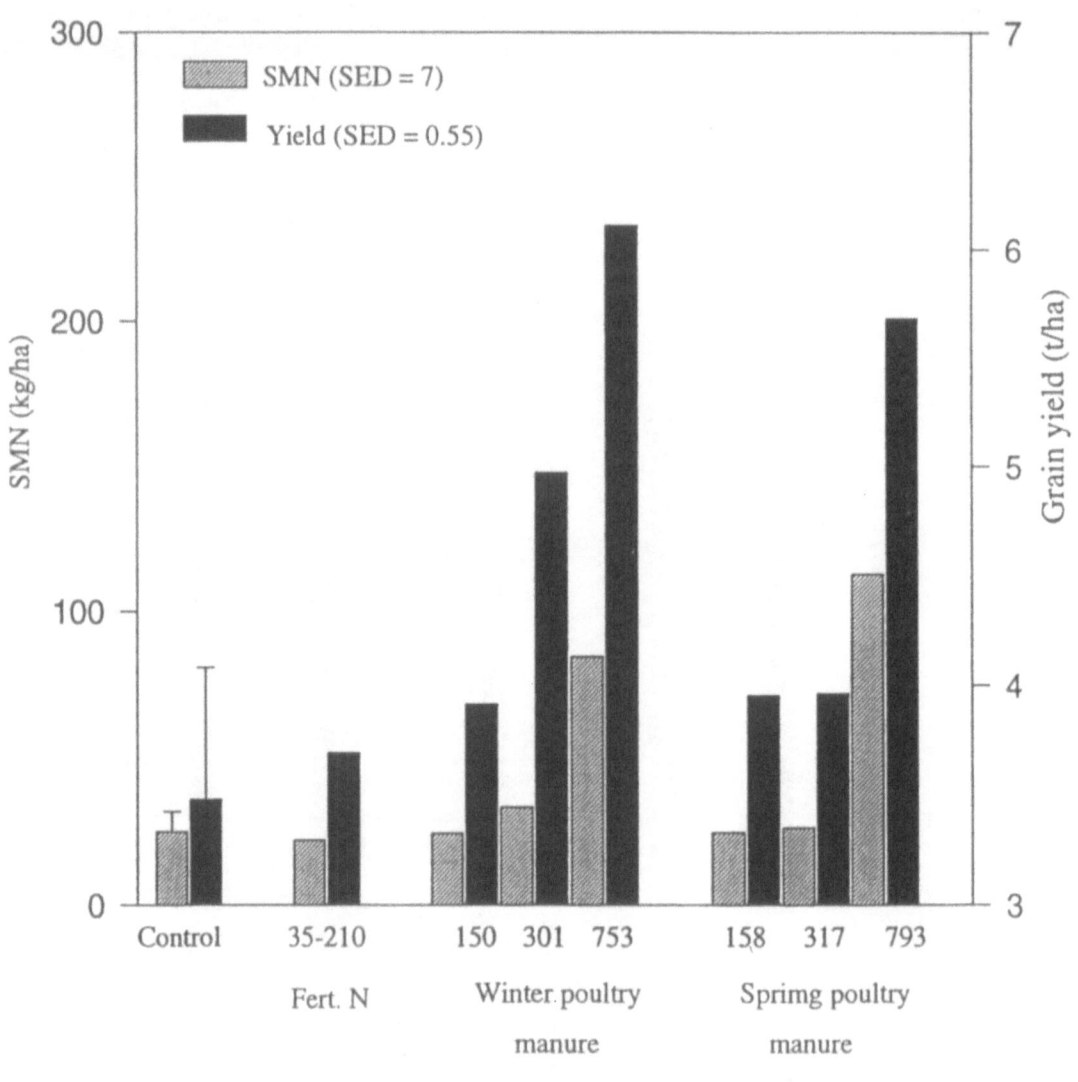

Figure 2. Residual soil mineral nitrogen supply and grain yield of 1993 cereal crop following fertiliser and poultry manure N applications to sugar beet in 1992.

Sugar beet yields (roots and tops), potato tuber yields and nitrogen uptakes were measured at harvest. Grain yields (85% dry matter) and N uptakes (grain and straw) were determined for the following cereal crops. Plant and grain N concentrations were determined using Kjeldahl digestion (MAFF, 1986). Soil mineral nitrogen samples were taken in four incremental depths; 0–15 cm, 15–30 cm, 30–60 cm and 60–90 cm (6 cores per plot) at harvest of the sugar beet and potato crops. Soil mineral nitrogen concentrations were determined using standard ADAS methods (MAFF, 1986) and converted to kg ha^{-1} (0 cm to 90 cm depth) assuming a soil bulk density of 1.3. Hydrologi-

cally effective rainfall (i.e. drainage) was calculated for each site using the Meteorological Office Rainfall and Evaporation Calculation System (MORECS, 1981).

Results

Sugar beet

In both years, fertiliser N applications in the range 0–180 kg N ha^{-1} had no effect ($p > 0.05$) on residual autumn SMN supply, nor on yield of the following cereal crops (Figs. 1 and 2). Fertiliser N applications

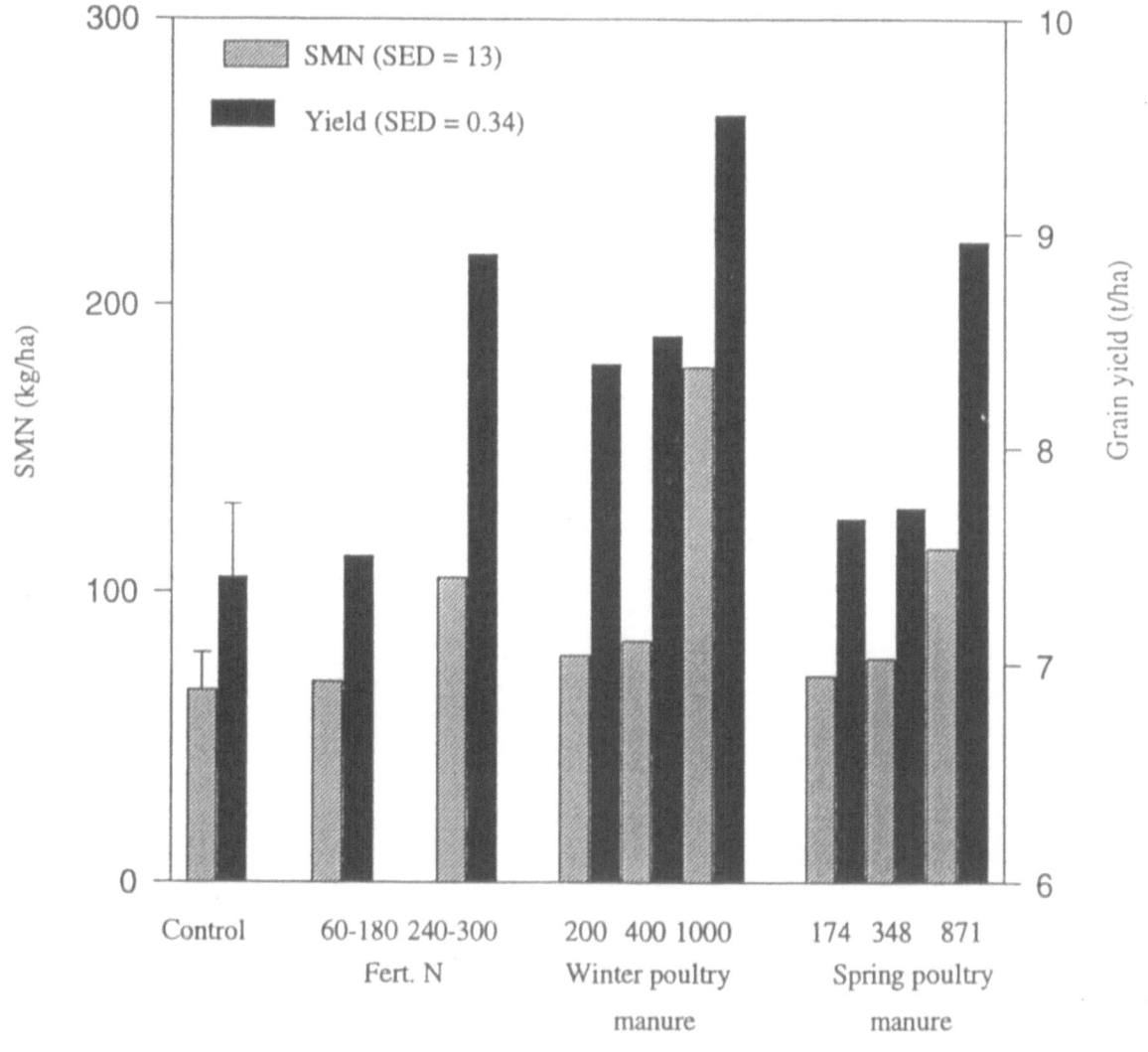

Figure 3. Residual soil mineral nitrogen supply and grain yield of 1992 cereal crop following fertiliser and poultry manure N applications to potatoes in 1990/91.

of 210 kg N ha^{-1} in 1991 increased following cereal crop yield by over 1.1 t ha^{-1}, compared to the untreated control. However, fertiliser N applications had no effect on SMN, or either of the measured crop parameters in 1992. Optimum economic fertiliser N rates (calculated assuming that 10 kg of adjusted clean beet yield (16% sugar) is required to cover the cost of 1 kg of fertiliser N) were 65 kg N ha^{-1} in 1991 and 79 kg N ha^{-1} in 1992.

Poultry manure applications supplying > 600 kg total N ha^{-1} increased residual SMN ($p < 0.01$) and cereal N uptake at harvest ($p < 0.01$) (Table 2), and >

500 kg total N ha^{-1} increased following cereal yields ($p < 0.01$) when compared to untreated control plots. Similar residual effects for the winter and spring poultry manure applications is a reflection of the low rainfall following the winter timings in 1991 (112 mm) and 1992 (73 mm) which was insufficient to cause significant N losses by leaching prior to the establishment of the sugar beet crops. Hydrologically effective rainfall (i.e. drainage) was also low over winters 1990/91 (42 mm) and 1991/92 (36 mm).

188

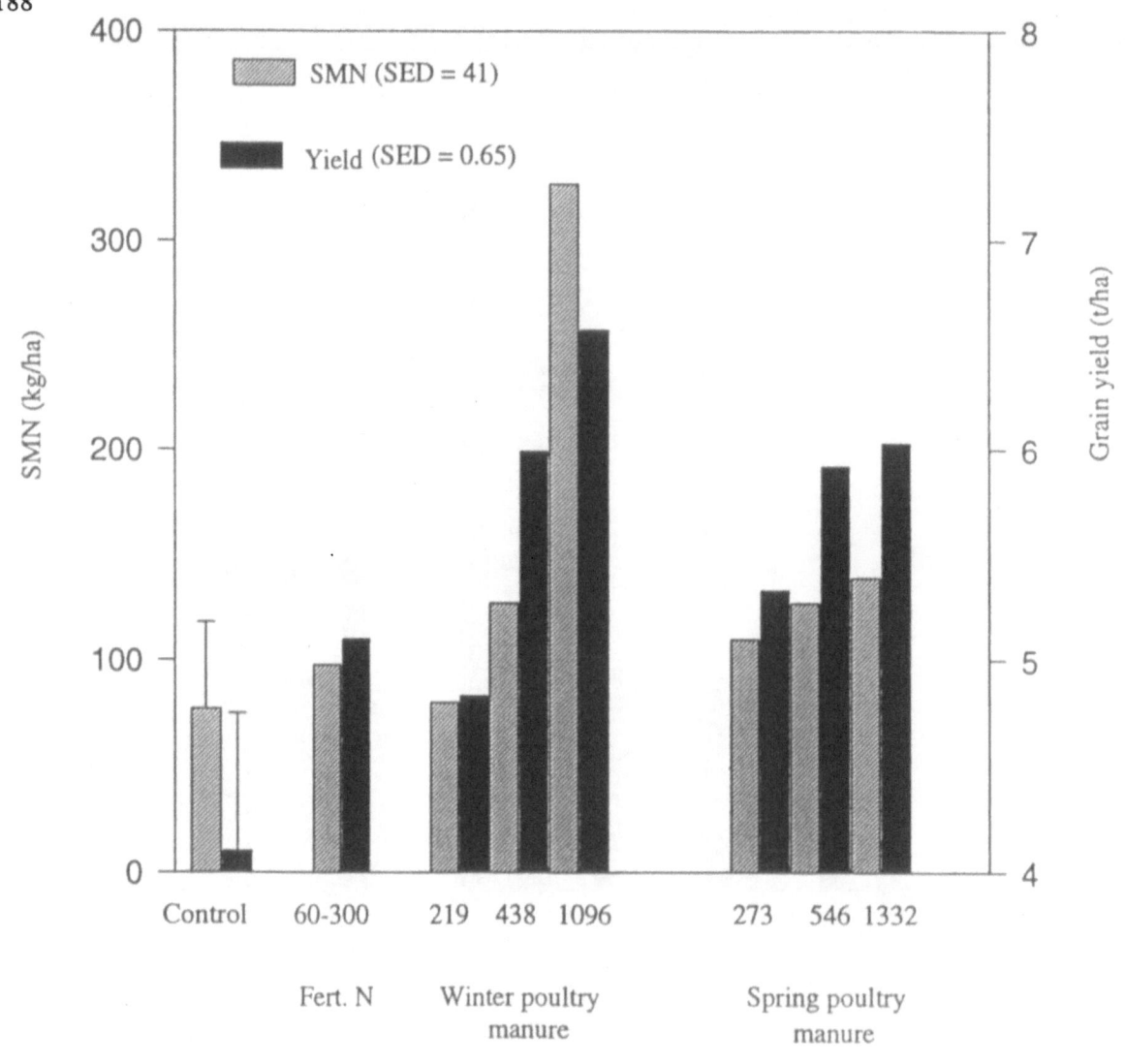

Figure 4. Residual soil mineral nitrogen supply and grain yield of 1993 cereal crop following fertiliser and poultry manureN applications to potatoes in 1991/1992.

(ii) Potatoes

Fertiliser N applications to potatoes in 1991 increased residual SMN ($p < 0.001$) where 300 kg N ha^{-1} was applied. Yield (Fig. 3) and N uptake (Table 3) was increased ($p < 0.05$) where > 240 kg N ha^{-1} was applied. However, fertiliser N applications in 1992 had no effect on residual SMN, or yield (Fig. 4) and N uptake (Table 3) by the following cereal crop in 1993. The absence of residual effects from the 1992 fertiliser applications was probably caused by greater overwinter nitrate leaching losses, with 92 mm of effective

rainfall in 1992/93 compared to 57 mm in 1991/92. Also in 1992, soil mineral N samples after the potato crop were not taken until mid December and soils returned to field capacity in mid November. Optimum economic fertiliser N rates for the potato crops (calculated assuming that 4 kg of ware yield is required to cover the cost of each 1 kg of fertiliser) were >300 kg N ha^{-1} in 1991 and 253 kg N ha^{1} in 1992.

Poultry manure applications supplying >800 kg total N ha^{-1} increased ($p < 0.001$) residual SMN in 1991 and winter applications supplying > 1000 kg N ha^{-1} increased residual SMN in 1992. In both seasons,

Table 3. Cereal crop nitrogen uptake at harvest in 1992 and 1993 following poultry manure applications to preceding potato crops

N application to previous potato crop (kg ha^{-1}) Fertiliser	Crop N uptake (kg ha^{-1})	
	1992	1993
0	120	62
60	124	98
120	115	71
180	138	57
240	149	89
300	183	80
Winter poultry manure		
1990: 200 1991: 219	140	75
400 438	151	102
1000 1096	197	117
Spring poultry manure		
1991: 174 1992: 273	124	103
348 546	121	86
871 1332	159	100
p	<0.001	<0.001
SED	11.4	5.1
CV(%)	9.3	20.3

yields of the following cereal crop and crop N uptake was increased ($p < 0.05$) following applications >400 kg N ha^{-1} (Table 3).

Discussion

The results demonstrate that the effect of fertiliser N applications to sugar beet on residual SMN and yield of the following cereal crop is low even when rates upto 180 kg N ha^{-1} are applied. These application rates were considerably in excess of the optimum economic fertiliser N rates determined in the two experiments of 65 kg N ha^{-1} (1991) and 79 kg N ha^{-1} (1992). In contrast, for the 1991 potato crop N, applications of 300 kg N ha^{-1} increased residual SMN and yield of the following cereal crop, where the economic optimum fertiliser rate was above 300 kg N ha^{-1}.

Chaney (1990) reported that there was no significant increase in autumn SMN contents following the application of ammonium nitrate to cereals up to the economic optimum for yield. However, applications above the optimum significantly increased SMN contents. The data reported here shows that fertiliser N

applications to sugar beet up to 100 kg N ha^{-1} in excess of optimum economic rates had little effect on residual SMN supply or yield of the following cereal crop. However, fertiliser N applications around the economic optimum rate for potatoes can significantly increase SMN supply. Sylvester-Bradley and Chambers (1992) showed that on average, fertiliser N applied to cereals and oilseed rape resulted in only small increases in SMN supply upto the economic optimum rate. Mean effects for 21 cereal experiments were 10 kg ha^{-1} SMN increase per 100 kg N ha^{-1} applied, and for 4 oilseed rape experiments, 13 kg ha^{-1} SMN increase per 100 kg N ha^{-1} applied. In the case of 4 potato experiments studied, the effect was much greater with 26 kg ha^{-1} SMN increase per 100 kg N ha^{-1} applied. Similarly, Wadman et al. (1990) reported mean elevations in SMN at harvest of 20 kg N ha^{-1} per 100 kg N ha^{-1} applied for 18 potato experiments receiving upto 180 kg N ha^{-1}.

Poultry manure N applications at agronomically sensible rates (<400 kg ha^{-1} total N) had no measurable effect on SMN supply or yield of following cereal crops. These results are in agreement with those reported by Pain and Saunders (1980) who showed that cattle slurry applications to maize at a rate of 276 kg N ha^{-1} had no effect on residual soil N supply. Unwin et al. (1986) also found that cattle slurry applications in the range 200–420 kg N ha^{-1} had only a small effect on grass growth in the second year following application. Marked residual effects from poultry manure N applications were only evident where application rates exceeded 400 kg total N ha^{-1}. On the 6 treatments from the 4 experiments where poultry manure N applications increased ($p < 0.05$) residual SMN supply, the mean elevation in SMN content was 11 kg ha^{-1} (range 6–18 kg ha^{-1}) per 100 kg total N ha^{-1} supplied by the poultry manure. Wadman et al. (1990) measured the residual effect of winter and spring poultry slurry applications to potatoes. They reported mean elevations in SMN after harvest of 14 kg N ha^{-1} per 100 kg ha^{-1} applied for the winter timings and 17 kg N ha^{-1} per 100 kg ha^{-1} N applied for the spring timings.

Overall, the results reported here are in agreement with those of Smith et al. (1985) who reviewed the data from slurry experiments in the UK between 1960 and 1980, and concluded that the residual effects of manure applications were generally small, other than those following relatively high applications. The results do however emphasise the need to match N supply from fertilisers and organic manures to crop requirements, to maximise N use efficiency and minimise nitrate leaching losses. The ability of sugar beet to scavenge soil N will limit the potential for nitrate leaching losses even when applications are moderately above economic optimum rates. However, N applications to potatoes even at the economic optimum rate are likely to increase leaching losses. These losses will increase markedly where optimum rates are exceeded following high rates of poultry manure dressings or fertiliser N applications.

Acknowledgement

Financial support for this work by the UK Ministry of Agriculture Fisheries and Food (MAFF) is gratefully acknowledged.

References

Anon 1994 The British Survey of Fertiliser Practice. Fertiliser Use on Farm Crops, 1993. HMSO, London, UK.

Chambers B J, Smith K A and Cross R B 1994 Effects of poultry manure application timing on nitrogen utilisation by cereals. *In* Animal Waste Management. Proc. of the Seventh Technical Consultation on the ESIORENA network on Animal Waste Utilization. REUR Tech. Ser. 34. Ed. J E Hall. pp 199–205. FAO, Rome, Italy.

Chaney K 1990 Effect of nitrogen fertiliser rate on soil nitrate nitrogen content after harvesting winter wheat. J. Agric. Sci. 114, 171–176.

Johnson A E 1970 The value of residues from long-period manuring at Rothamsted and Woburn. II. A summary of the results of experiments started by Lawes and Gilbert. Rothamsted Report for 1969, part 2, pp 7–21. Rothamsted Experimental Station, Harpenden, Hertfordshire, UK.

Lande Cremer L C N 1985 Long-term effects of farm slurry applications in the Netherlands. *In* Long Term Effects of Sewage Sludge and Farm Slurry Applications Eds J H Williams, G Guidi and P L' Hermite pp 84–90 , Elsevier, London, UK.

MAFF 1986 The Analysis of Agricultural Materials. Ministry of Agriculture Fisheries and Food. Reference Book 427, third edition. HMSO, London, UK.

MORECS 1981 Meteorological Office Rainfall and Evaporation Calculation System. Hydrological Memorandum No 45. Meteorological Office, UK.

Pain B F and Saunders L T 1980 Effluents from intensive livestock units fertiliser equivalent of cattle slurry for grass and forage maize. *In*: Effluents from livestock. Ed. J K R Gasser. pp 300–311. Applied Science Publications Limited, London, UK.

Smith K A 1991 Waste production from UK livestock. Internal report to the UK Ministry of Agriculture Fisheries and Food, London, UK.

Smith K A, Unwin R J and Williams J H 1985 Experiments on the fertiliser value of animal waste slurries. *In* Long Term Effects of Sewage Sludge and Farm Slurry Applications. Eds. J H Williams, G Guidi and P L' Hermite. pp 124–135, Elsevier, London, UK.

Soil Survey of England and Wales 1983 Soil Map of England and Wales, Scale 1:250000. Soil Survey of England and Wales Harpenden, England.

Sylvester-Bradley R and Chambers B J 1992 The implications of restricting use of fertiliser nitrogen for the productivity of arable crops, their profitability and potential pollution by nitrate. Aspects Appl. Biol. 30, 85–95.

Unwin R J, Pain B F and Whinham W N 1986 The effect of rate and time of application of nitrogen in cow slurry on grass cut for silage. Agric. Wastes 15, 253–268.

Wadman W P Neeteson J J and Winen G 1990 Effects of slurry with and without the nitrification inhibitor dicyandiamide on soil mineral nitrogen and nitrogen response of potatoes. *In* Nitrogen in Organic Wastes Applied to Soils. Eds. J A Hansen and K Henriksen. pp 304–314. Academic Press, London, UK.

O. Van Cleemput et al. (eds.), Progress in Nitrogen Cycling Studies, 191–194, 1996.

Effect of herbicides on the urea transformation in soil

Gábor Csitári, Katalin Debreczeni and István Sisák
Pannon University of Agricultural Sciences, Keszthely, Hungary

Key words: herbicides, nitrification, soil-microorganisms

Abstract

Nitrogen availability is one of the most important factors for plant performance in agricultural systems. Any compound that alters the number or activity of microorganisms could affect the soil biochemical processes mainly the nitrogen cycle and ultimately the plant growth.

A soil incubation experiment was carried out to study the influence of soil herbicides on the nitrification activity and mineralization-immobilization processes in the soil. In this experiment we have investigated the changes of soluble and easily hydrolizable N content in the soil. The conditions were to be optimised for the aerobe nitrification. Four N levels were used in the experiment adding 0–50–100–200 mg urea-N kg^{-1} soil. The soil used in this experiment was a Ramann's brown forest soil. Nitrogen forms were determined after 1, 2, 4 and 8 weeks of incubation.

It was established that the three herbicides used in the experiment (atrazine, EPTC, alachlor) had affected the soil microorganisms. EPTC and atrazine inhibited the nitrification.

Introduction

Herbicides may affect the nitrogen uptake by plants in two ways. They can affect the biochemical processes of plants and they can alter the number or activity of soil microorganisms. Studies have been available on the impact of herbicides in agricultural soils since a long time (Brown, 1978, Grossbard, 1976): Enzyme activities, CO_2 evolving, O_2 consumption and a diversity of species of microorganisms were studied thoroughly. Results are controversial, they strongly depend on soil type and the dose of herbicide.

The present paper examines the influence of three herbicides on the soil N fractions. The three herbicides were atrazine (Hungazin 440FW), EPTC (Alirox 80EC) and alachlor (Satoklór 480EC), recommended for maize. The soluble (inorganic) nitrate and ammonium content and ratio play a great role in the growth of maize (Arnon, 1975). The soluble ammonium, nitrate and a certain part of the organic N (referred as easily hydrolizable N) content were determined. We have assumed that the easily mineralizable fraction is related to the biologically active soil nitrogen (Duxbury et al., 1991). This experiment was carried out for 8 weeks. It was long enough for studying the effect of changes of

these N forms on plant growth in our planned experiments.

These measurements were carried out at four nutrient levels. Urea was chosen, based on its low cost, relatively low salt index and compatibility in tank-mixed solutions with numerous pesticides.

Materials and methods

In a soil incubation experiment 100 g of air dried soil was put in each pot. The soil characteristics are: Ramann's brown forest soil (FAO taxonomy: Eutric cambisol), soil plasticity index (K_A) 37, humus 1.7%, pH $_{(KCl)}$ 6.30, soluble N 12.4 mg kg^{-1}, easily hydrolizable N 66.5 mg kg^{-1}.

The moisture content of the soil was kept at 60% of maximum water holding capacity. For ensuring aeration, a layer of river-gravel was under the soil and a glass pipe was placed in it. Four soil-N levels were used in the experiment: 0–50–100–200 mg urea-N kg^{-1} soil, in 4 replications. Thereafter, they are referred as (nutrient) levels 1, 2, 3 and 4, respectively. Incubation was carried out at 28 °C for 8 weeks. Soil samples were analysed after 1, 2, 4 and 8 weeks of incubation.

Soluble (inorganic) N (NH_4^+, $NH_4^+ + NO_3^-$) was extracted from 40 g soil sample, shaking for 1 hour with 100 mL 1% KCl as extractant. Soil extracts were filtered, steam distilled with or without reduction and the distillates were titrated with 0.01 N H_2SO_4 (Ballenegger, 1953).

For the determination of acid soluble N, 20 g soil was mixed with 100 mL 0.5 N H_2SO_4 overnight. Then the soil extracts were filtered. Dewarda alloy (0.5 g) was added to 50 mL filtrate and boiled for 5 minutes. After cooling, 5 mL cc. H_2SO_4 was added and boiled until white SO_2 gases appeared (about 10 min). After cooling 2.5 mL 10% $K_2Cr_2O_7$ were added to the solution and boiled for about 10 min until it had turned green (Tyurin, 1954). The quantity of produced ammonium was determined by steam distillation. The soluble N content plus a certain part of organic N was measured in this way. Acid soluble N minus soluble N have been considered as easily hydrolizable N in this paper.

The herbicides used were provided by the manufacturers (Hungazin from Budapesti Vegyimûvek Co., Alirox and Satoklór from Sagrochem Ltd., Sajóbábony, Hungary). Hungazin 440 FW contained 38.2% (v/w) atrazine, Alirox 80 EC contained 73.4% (v/w) EPTC and Satoklór contained 45.5% (v/w) alachlor. The recommended doses of commercial products were 6.6–10 kg ha^{-1}, 5.0–8.0 L ha^{-1} and 3.5–5.5 L ha^{-1}, respectively. Manufacturer's recommended doses in kg ha^{-1} or L ha^{-1} were converted to mg cm^{-2}. Pot surface was 20 cm^2. Maximum doses were applied to be equivalent to 10 kg ha^{-1} Hungazin, 8 L ha^{-1} Alirox and 5.5 L ha^{-1} Satoklór, respectively. Experimental data were analysed using the LSMLMW and MIXMDL PC-2 version program (Harvey, 1990).

Results

Regression curves were fitted on the measured values. The linear regression model (y=a+bx) was used in case of nitrate and easily hydrolizable N (Table 1) and the quadratic regression model (y=a+bx+cx^2) in case of ammonium (Table 2) because it proved to be the best in the former analysis. Parameters of the regression curves were compared to reveal the significant impact of herbicides.

The time curves of nitrate (Fig. 1a) were similar at each level except level 4. Slopes of regression lines did not differ from each other by treatment (Table 1a). Quadratic curves were fitted on the values of level 4,

Table 1. Slopes (b value) of the regression lines (y=a+bx) of the soluble nitrate (**a**) and the easily mineralizable N (**b**). Least significant differences (LSD) are given at $p=0.05$ probability level

Treatments	Nutrient levels			
	1	2	3	4
a. Soluble nitrate				
Control	3.64	3.44	4.40	9.84
Atrazine	4.66	4.93	4.61	16.08
EPTC	4.63	5.23	7.00	11.97
Alachlor	3.31	2.99	2.45	7.59
LSD	1.68	2.91	4.45	7.36
b. Easily mineralizable N				
Control	1.44	3.40	9.14	-1.52
Atrazine	0.15	-3.05	-6.10	-4.95
EPTC	3.05	2.21	0.40	-4.67
Alachlor	-0.22	-0.25	-3.13	-3.92
LSD	3.89	3.59	6.22	10.05

too. They showed that the linear parameter of atrazine differed significantly from the control (data are not shown).

The time curves of easily hydrolizable N differed by nutrient levels. There was a decrease until the end of the 2nd week (Fig. 1b). The slopes of the regression lines are shown in Table 1b. The changes of the slopes show a similar tendency but they differ significantly at levels 2 and 3 . Atrazine and alachlor decreased the slopes, but atrazine did it to a greater extent.

The time curves of the soluble ammonium are given in Figure 1c. They were similar at each level, except level 4. The parameters of the quadratic regression are given in Table 2. The quadratic parameters (Table 2b) did not differ significantly from the control. The linear parameters of herbicides were usually smaller than those of the control at each level. In case of atrazine and EPTC this difference was significant. The parameters of alachlor were similar to the control.

Quantities of ammonium and nitrate after the first week were analysed too (Table 3). Amount of ammonium were higher in case of atrazine and EPTC. Differences were significant at levels 3 and 4. EPTC decreased the amount of nitrate at each level. Atrazine decreased it at levels 3 and 4 and increased it significantly at level 1. Alachlor increased it at levels 2, 3 and 4, it proved to be significant at level 3.

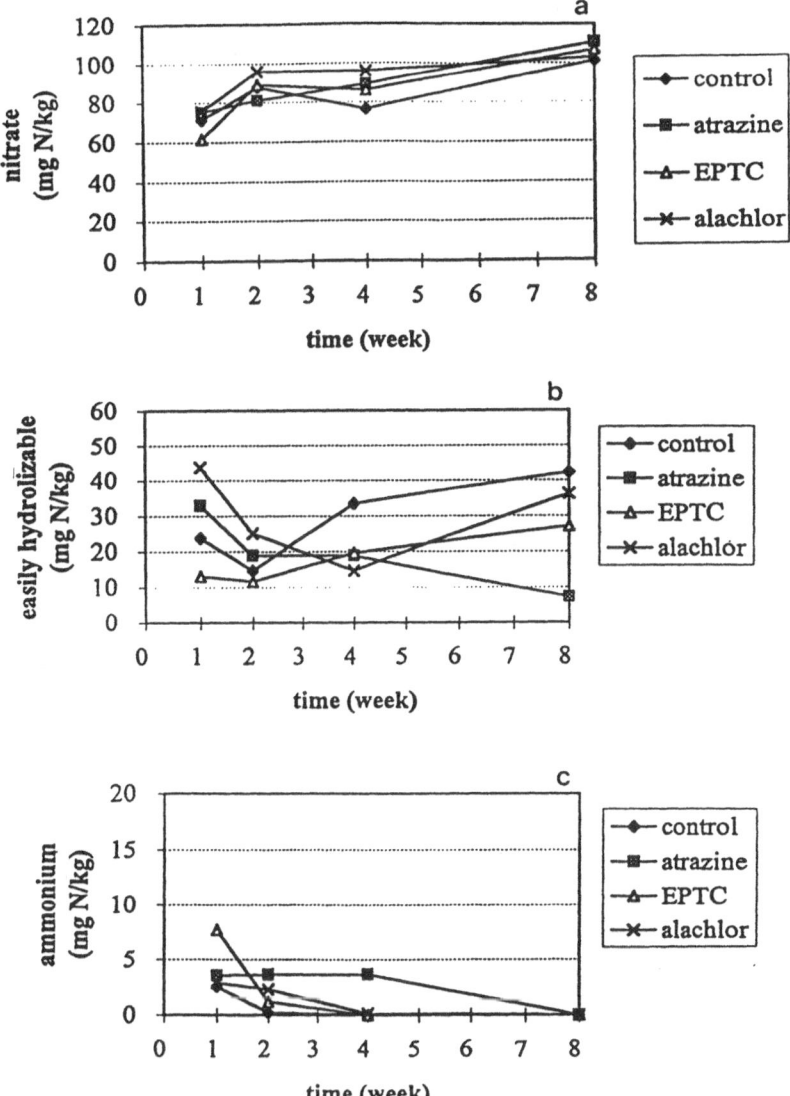

Figure 1. Time curves of nitrate (**a**), easily hydrolizable N (**b**) and ammonium (**c**) at nutrient level 2.

Discussion

Urea is transformed to ammonium in the soil within a short period of time (Antal, 1989). The produced ammonium can be transformed to nitrate or immobilized by soil microorganisms. Leaching, volatilization and fixation by clay minerals is not considerable in experimental conditions. The immobilized ammonium and a part of the organic N can be re-mineralized. The ammonium and nitrate levels were determined in during nitrification, mineralization and immobilization.

A decrease of the easily mineralizable fraction was a general tendency until the end of the 2nd week. It seemed that the extent of decrease depended on the nutrient level: the higher the nutrient level the greater and longer was the decrease. In addition, all three herbicides decreased the easily hydrolizable fraction, especially atrazine. We have assumed that herbicides inhibited soil microorganisms. Therefore, this fraction could not be recovered. This inhibition was probably transient. Figure 1b shows that curves of alachlor and EPTC were rising at the 8th week while the slope of alachlor was -0.25.

Table 2. Linear (b value) parameters (**a**) and quadratic (c value) parameters (**b**) of quadratic regression curves ($y=a+bx+cx^2$) of the soluble ammonium. Least significant differences (LSD) are given at $p=0.05$ probability level

Treatments	Nutrient levels			
	1	2	3	4
a. Linear (b value) parameters				
Control	-0.20	-0.50	-1.12	-7.64
Atrazine	-0.29	-0.24	-3.77	-15.43
EPTC	-0.06	-1.52	-2.59	-12.33
Alachlor	-0.13	-0.63	-1.73	-9.15
LSD	0.47	0.90	1.28	5.02
b. Quadratic (c value) parameters				
Control	0.06	0.14	0.24	2.06
Atrazine	0.08	-0.12	0.72	3.96
EPTC	-0.02	0.40	0.67	3.41
Alachlor	-0.03	0.12	0.47	2.65
LSD	0.19	0.35	0.50	1.97

Table 3. Quantities of ammonium (**a**) and nitrate (**b**) after 1 week (mg kg^{-1}). Least significant differences (LSD) are given at $p=0.05$ probability level

Treatments	*Nutrient levels*			
	1	2	3	4
a. Ammonium				
Control	1.06	2.59	6.89	41.26
Atrazine	1.47	3.55	21.81	82.84
EPTC	0.00	2.91	13.01	63.96
Alachlor	0.06	7.68	10.44	50.61
LSD	2.10	6.50	5.90	14.46
b. Nitrate				
Control	38.59	71.62	100.63	137.65
Atrazine	42.61	75.12	89.33	108.01
EPTC	31.46	61.46	85.09	121.03
Alachlor	36.90	75.80	118.33	146.12
LSD	3.75	8.04	10.04	18.98

The increase of soluble ammonium could be observed as a general tendency, compared to the control at the end of the first week. It was significant in the case of atrazine and EPTC at levels 3 and 4. EPTC decreased the amount of soluble nitrate at the end of the first week at each level. For this reason we can state that EPTC inhibits the nitrification. Atrazine inhibited the nitrification to a lesser degree because the inhibition appeared at higher nutrient levels only.

Summing up it can be stated that all three herbicides affected the soil microorganisms. Atrazine and EPTC partially inhibited the nitrification.

Acknowledgement

Appreciation is expressed to Katalin Sárdi for reviewing the manuscript.

References

Antal M, Anton A, Németh T and Biczók G 1989 Effect of C-sources and urea on the available N content and urease activity of different soils. Agrokém. Talajtan 3–4, 399–403.

Arnon I 1975 Nutritional requirements of maize. *In* Mineral Nutrition of Maize. pp 100–112. International Potash Institute, Bern, Switzerland.

Ballenegger R 1953 Talajvizsgálati módszerköny. Mezõgazdasági kiadó, Budapest, Hungary.

Brown A W A 1978 Herbicides and the soil microflora. *In* Ecology of Pesticides. Ed. A W A Brown. pp 362–401. John Wiley and Sons, New York, USA.

Duxbury J M, Lauren J G and Fruci J R 1991 Measurement of the biologically active soil nitrogen fraction by a ^{15}N technique. Agric. Ecosyst. Environ. 34, 121–129.

Grossbard E 1976 Effects on the soil microflora. *In* Herbicides (Physiology, Biochemistry, Ecology) Vol. 2. Ed. L J Audus. pp 99–147. Academic Press, London, UK.

Harvey W R 1990 Mixed model least-squares and maximum likelihood computer program (LSMLMW and MIXMDL PC-2 version). Purdue University, West Lafayette, Indiana, USA.

Tyurin F W 1954 Metodü opredelenija szojedinenij azota v pocsve. *In* Agrohimicseszkije metodü isszledovanija pocsv. Eds. A V Sokolov, D L Askinazi and I P Serdobolski. pp 47–62. Izdatelstvo Akademii Nauk, Moszkva, USSR.

O. Van Cleemput et al. (eds.), Progress in Nitrogen Cycling Studies, 195–197, 1996.

Fate of nitrogen applied to grasses and its influence on forage quality

J.P. Destain[1], P Lecomte[1], B. Toussaint[2] and R. Lambert[2]
[1]Centre de Recherches Agronomiques, Ministère Fédéral de l'Agriculture, B-5030 Gembloux, Belgium and [2]Unité d'écologie des prairies, U.C.L. - I.R.S.I.A., B-6600 Michamps, Belgium

Key words: forage quality, N budget, ^{15}N, rye-grass, timothy

Abstract

In the cool and wet belgian Ardennes (altitude 500 m) grassland is predominant. As growth resumption generally occurs very late in spring (mid April-May), fine tuning nitrogen fertilization to plant needs is critical.

The aim of this paper was to draw up a critical appraisal of the established fertilization practice by means of N budgets (^{15}N). If for the 1st of the 3 annual cuts of rye-grass and timothy, DM and N taken up increased with N rate (0–120 kg N ha^{-1}), these parameters were very dependent of water availability for the 2nd and 3rd cuts. True recovery (^{15}N) of applied N generally increased with N rate of the 1st cut (extreme values were 31 and 54%), apparent recovery decreased but sometimes took values exceeding 100%. Added nitrogen interactions is suspected. Recoveries of N applied for the 2nd and 3rd cuts averaged between 18 and 47% for rye-grass; after effects were very negligible. Over-all recovery of fertilizer N (3 applications of 40 or 80 kg N ha^{-1} for the 3 cuts) by rye-grass did not exceed 52% while 45 to 50% of applied N remained in soil in organic form; losses were less than 15%. If timothy seemed to be more productive in these climate conditions, VEM and DVE values (feeding value of the forage calculated according to NIR prediction) were lower than for rye-grass.

Introduction

In the cool and wet Ardennes (altitude 500 m) in Belgium, grassland is predominant. Growth resumption generally occurs very late in spring (mid April-May) owing to unfavorable conditions for mineralization (soil temperature below 5 °C). It is thus of prime importance to assess the needs of the grass to ensure timely applications of nitrogen. Too early applications would cause N losses.

The aim of this work was to draw up a critical appraisal of the usual fertilization practice of grassland (regarding forage productivity and quality) by means of nitrogen budgets (using ^{15}N).

Material and methods

Soil and climate

The experiments were carried out on a dystric Cambisol with a pH 5.5 and a carbon content between 2 and 2.5%. The climate of the Ardennes is characterized by an annual mean temperature of 7 °C (with possible long periods of frozen soil in winter) and a rainfall of 1,000 mm or more rather well distributed all over the year.

Experimental design

The influence of fertilizer N (0 to 120 kg ha^{-1} for each of the 3 cuts) on yield and quality of forages (rye-grass (*Lolium perenne*) and timothy (*Phleum pratense*) was studied in 10 m^2 plots in a complete randomized block design with 4 replications). ^{15}NH$_4$ ^{15}NO$_3$ at 2.5 atom percent ^{15}N was split applied in 3 equal dressings for each cut in microplots (steel cylinders - 35 cm in length and 30 cm in diameter - pressed into the soil in early spring).

Sampling procedure and sample preparation

Dry matter production and N uptake were estimated in the plots. Grass samples were very finely ground for NIR determination of protein content, energy (VEM) and DVE (feed and ruminally synthesised digestible

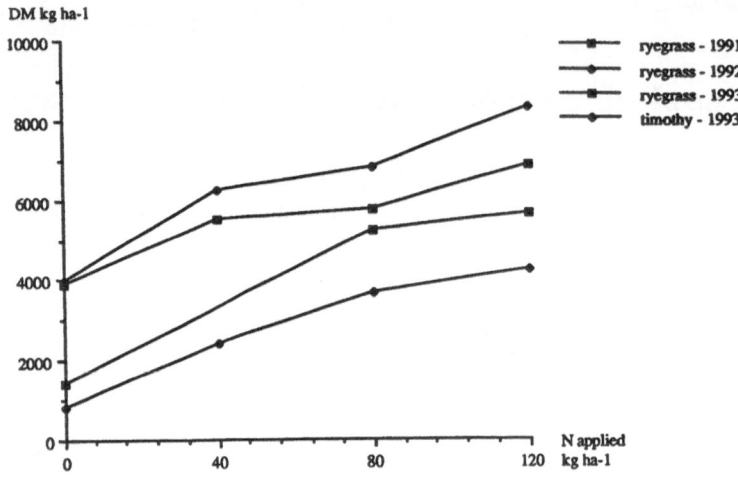

Figure 1. Dry matter production by rye-grass and timothy (1st cut).

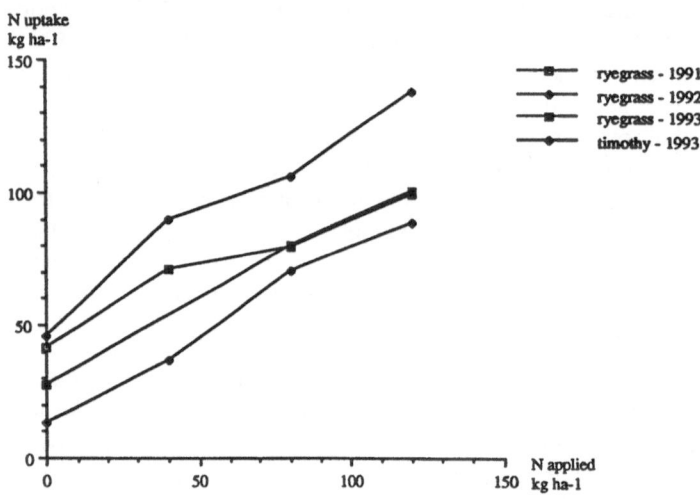

Figure 2. N uptake by rye-grass and timothy (1st cut).

Table 1. N removed by grass (3 cuts) from an unfertilized plot

Plant	Year	N removed (kg ha^{-1})
Rye-grass	1991	54
	1992	29
	1993	44
Timothy	1993	51

protein). In the microplots, forage was collected at each cut and soil removed completely (by 10 cm layers) after the growing season and sampled after thorough mixing. Plants and soil samples were freeze dried and finely ground with a hammer mill. Analysis for total N was carried out using Kjeldahl method modified to include nitrate (Bremner, 1965) or by Dumas technique (Roboprep, Europa Scientific, UK). The isotopic composition of nitrogen was determined by mass spectrometry (VG SIRA 12, UK).

Results and discussion

N removed by grass from a 0 N plot has varied from 29 to 54 kg N ha^{-1} indicating a very low potential of soil mineralization (Table 1). The highest values were observed for the 1st year after ploughing and establishing the grass.

For the 1st cut of rye-grass and timothy, dry matter increased with N rate (0–120 kg N ha^{-1}), while N

Table 2. Recovery of fertilizer N by grasses (example of the 1st cut) (% N applied)

N applied (kg ha[-1])	Site	Type of grass			
		Rye-grass		Timothy	
		1[a]	2	1	2
40	Michamp	66.4	34.0	109.5	43.6 ± 5
80		59.9	40.1	75.2	53.0 ± 6
120		55.5	46.8	76.6	
40	Libramont	35.0 ± 2			
60		35.4 ±			
100		41.2 ± 4			

[a]1 - Difference method, 2 - True recovery (^{15}N).

Table 3. Influence of harvest stage on recovery of fertilizer N by grasses

N applied (kg ha[-1])	Type of grass	Recovery d(% N applied) at:		
		Stem elongation (15 cm)	Heading	Flowering
80	Rye-grass	38 ± 1	48 ± 3 d	43 ± 5
80	Timothy	44 ± 6	58 ± 7	57 ± 3

taken up increased linearly with a steeper slope for timothy (Fig. 1 and 2); both parameters were site - and year - dependent.

While apparent recovery by plant (sometimes exceeding 100% e.g. 109%) generally decreased with N rate, true recovery determined with ^{15}N seemed to increased and was lower (max. 53%) indicating ANI effects (added nitrogen interaction) as already mentioned by different authors (Jenkinson et al., 1985; Westerman and Kurtz, 1974) (Table 2).

Recovery (using ^{15}N) was maximum at harvest heading stage (Table 3) when the most favorable VEM and DVE values were observed, with a superiority of rye-grass to timothy (for a similar protein content of 14% DM, VEM and DVE were respectively 1,014 and 94.5 for rye-grass and 906 and 82.2 for timothy). By using ^{15}N, it was also possible to estimate recoveries of following N dressings and to dissociate direct effect from after effects (only one of 3 microplots received a labeled dressing at each cut, the 2 others received

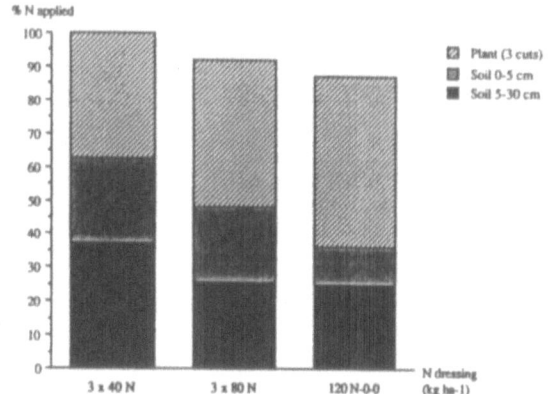

Figure 3. Balance sheet of fertilizer N (example of Michamps 91–3 cuts of rye-grass).

a non labeled dressing). Direct effects of 2nd and 3rd cuts ranged from 18 to 47% depending mainly on water availability. After-effects (effects of N applied for the 1st cut on the 2nd and 3rd cuts and effect of N applied for the 2nd cut on the 3rd cut) were very low (2 to 17%). Total recovery of the 240 kg N ha^{-1} dressing (80 kg N ha^{-1} at each cut) was 44% in 1991 and 52% in 1992 for rye-grass, while for the 120 kg N ha^{-1} dressing (40 kg N ha^{-1} at each cut) it was only 38% in 1991 and 40% in 1992. Not accounted for was generally less than 15% but were higher for the 240 kg N dressing than for the 120 kg N dressing (Fig. 3). Residual N in soil, exclusively organic, has varied between 40 and 60% of N applied and between 1/3 and 1/2 of the total amount was located in the 5 cm upper layer of soil indicating a very high immobilization of N in the root system.

It may be concluded that 60 kg N ha^{-1} per cut might be recommended in this cool and wet region as the optimal N rate.

References

Jenkinson D S, Fox R H and Rayner J H 1985 Interactions between fertilizer nitrogen and soil nitrogen - the so-called "priming effect". J. Soil Sci. 36, 424–444.

Westerman R L and Kurtz L T 1974 Isotopic and non isotopic estimations of fertilizer nitrogen uptake by Sudangrass in field experiments. Soil Sci. Soc. Am. Proc. 38, 107–109.

O. Van Cleemput et al. (eds.), Progress in Nitrogen Cycling Studies, 199–202, 1996.
© 1996 *Kluwer Academic Publishers.*

^{15}N Characterisation of immobilisation of fertiliser N in blanket peat under improved grass pasture

J.M. Hall[1], B.L. Williams[2] and K. Killham[1]

[1]*Department of Plant and Soil Science, University of Aberdeen, Aberdeen, AB9 2UE, Scotland and* [2]*Macaulay Land Use Research Institute, Craigiebuckler, Aberdeen, AB9 2QJ, Scotland*

Key words: biocides, fertiliser, immobilisation, microbial biomass, peat

Abstract

In North Scotland, reseeded grass pasture on blanket peat responds poorly to Spring applied fertiliser N. Microbial immobilisation may represent a possible explanation for this phenomenon. The application of fertiliser N and C to homogenised peat samples resulted in a peak of microbial immobilisation 2 days after addition. The amount of fertiliser N found in the microbial biomass amounted to 30% of that applied, a substantial proportion which is unavailable to plants for uptake. By applying fertiliser N with the bacterial inhibitor streptomycin, it was possible to reduce the quantity of N immobilised by the microbial biomass. In acidic, blanket peat with a fluctuating water table, bacterial populations therefore appear to be the main immobilisers of applied N, and factors affecting their sink strength will greatly contribute to the efficiency of fertiliser N recovery in herbage in these systems.

Introduction

Productivity of upland grass pastures can be improved by reseeding, drainage, liming and the application of fertilisers, particularly N and P (Frame et al., 1985). Early application of N is recommended to boost grass production during the Spring and early Summer, although a number of workers (e.g. Rangeley, 1988; Williams and Wheatley, 1992) have found that fertiliser N use efficiencies are low (usually between 10–25%) in upland reseeded grass pastures, particularly for Spring applied N.

One possible explanation for the poor recovery of fertiliser N in above ground herbage in Spring may be due to immobilisation of applied N by the soil microbial biomass. The microbial biomass has been shown to be an effective sink for applied N, particularly under conditions which are prevalent in Spring (Hall et al., 1994) and therefore may outcompete plant roots for available N and limit plant growth. The short-term dynamics of the microbial biomass has been found to be important in controlling N flow in mineral soils (Schimel et al., 1989), although very little work has been carried out in highly organic soils.

The application of selective biocides to soil has been used as an approach to repress specific con-stituents of the microbial population (Ingham et al., 1986) and therefore study the role of fungi and bacteria in biological transformations occurring in soil. Jamieson and Killham (1994) successfully repressed the fungal population in the LFH layer of a coniferous forest and diverted a greater proportion of applied N into tree seedlings (10 fold increase), compared to where no biocide had been applied. The competition for N between plant roots and the microbial population can be investigated using selective biocides and may lead to a greater understanding of N dynamics in reseeded grass pastures.

A laboratory based microcosm experiment was set up to determine i) the timecourse of microbial N immobilisation following the addition of N and C, and ii) whether fungi or bacteria are the dominant immobilisers of N in a blanket peat.

Materials and methods

The peat was sampled from a reseeded upland grass pasture overlying blanket peat at Slethill, Forsinard, Highland Region, Scotland (NC 924464). More details about the site can be found in Williams and Wheatley (1992). The peat in both experiments, was incubated without amendment to water content (i.e. field state) thus differences in the water holding capacity percentage in Experiment 1 and 2, are due to changes in field moisture content at the time of sampling.

Experiment 1

Homogenised peat (pH 4) was transferred into 250 mL flasks which were immediately sealed with parafilm to maintain a constant moisture content (75% water holding capacity (WHC)) and were then incubated at 22 °C in the light. Fertiliser N (as $^{15}NH_4^{15}NO_3$ at 5 atom %) was added to the peat at a rate of 200 mg N 100 g^{-1} O.D. peat, as a solution (volume 5 mL). The solution was spotted onto the peat surface (100 μL aliquots) and the peat well mixed to ensure even distribution of N. Carbon (as glucose) was applied at a rate of 500 mg 100 g^{-1}O.D. peat, as a solution (5 mL), in the manner described above for N. Untreated samples received 10 mL of deionised H_2O. Sampling took place 1, 2, 3, 4, 6 and 15 days after N and C addition.

Biomass N concentrations were determined by the chloroform fumigation direct extraction method (Brookes et al., 1985), where K_2SO_4 extracts containing ^{15}N were freeze dried and then analysed using a pyrolysis mass spectrometer. Available N concentrations were determined using 2 M KCl as an extractant.

Experiment 2

Peat was mixed with either benlate (2 g kg^{-1} O.D. soil) or streptomycin (2 g kg^{-1}O.D. soil), using talc as a carrier (10:1 talc to biocide). Control samples were simply mixed with talc. The peat was then transferred to 250 mL flasks for incubation as above (maintained at 65% WHC). Two days after biocide additions, fertiliser N (as $^{15}NH_4$ $^{15}NO_3$) was added at a rate of 150 mg N 100 g^{-1} O.D. peat (total volume 1 mL, added in 50 μL aliquots) and C was added as glucose at a rate of 3.9 mg C 100 g^{-1} O.D. peat (total volume 2 mL, added in 50 μL aliquots). Sampling for biomass N and available N concentrations occurred 1, 2, 3, 4 and 8 days after N and C addition. Determinations were carried out as detailed above.

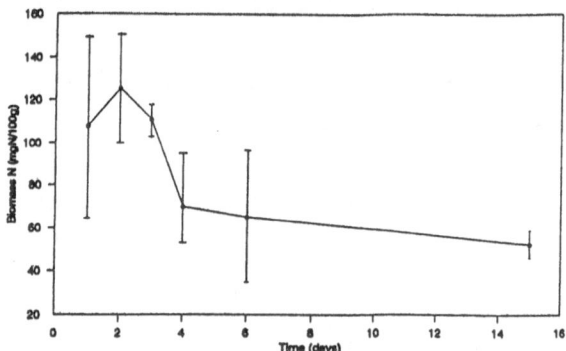

Figure 1. Timecourse of microbial biomass N after the application of N and C. Values represent means of 3 and the bars show standard errors of the mean.

Table 1. Percentage of applied fertiliser N found within the microbial biomass in the presence and absence of C. Means of 3, ± standard error of the mean

Day	+ C	- C
1	20.7 ± 9.36	18.2 ± 5.72
2	31.0 ± 6.36	42.4 ± 7.63
3	27.0 ± 1.76	21.1 ± 7.78
4	18.3 ± 6.43	6.05 ± 6.05
6	18.0 ± 8.09	12.7 ± 3.92
15	14.7 ± 2.04	12.7 ± 2.72

Results

Experiment 1

In fertilised and C-amended peat, microbial biomass N concentrations peaked at day 2 and then declined slowly until the end of the incubation (Fig. 1). By day 2, approximately 30% of applied fertiliser N could be found in the microbial biomass. The addition of C appeared to prolong the peak of immobilisation of N compared to unamended peat (Table 1). In the presence of available C, available N concentrations declined over the course of the incubation from 193 mg N 100 g^{-1} at day 1 to 165 mg N 100 g^{-1} at day 15.

Experiment 2

Microbial biomass N concentrations were very variable over time in benlate, streptomycin and control samples and hence the results will not be discussed here.

Total available N concentrations declined over the period of the incubation to different levels depend-

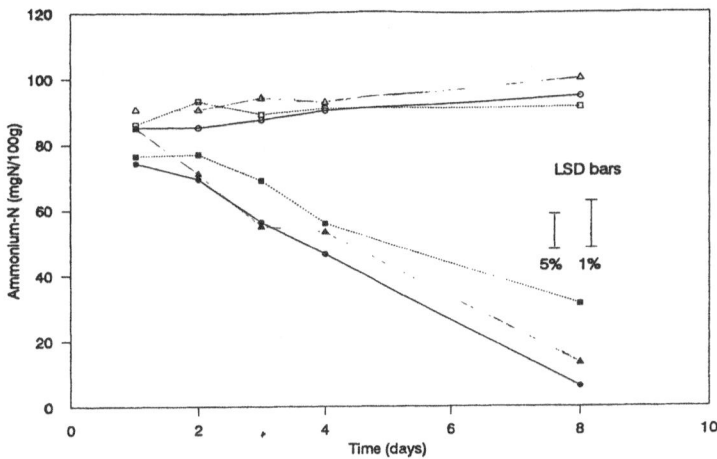

Figure 2. Ammonium-N concentrations over time after the addition of N and C. Values represent means of 3. Least significance difference bars shown at $p<0.05$ and $p<0.001$ significance levels. Key: Benlate(+C) —●—; Benlate (-C) —○—; streptomycin (+C) —■—; streptomycin (-C) —□—; control (+C) —▲—; control (-C) —△—

ing upon the biocide treatment. As expected, available N concentrations were lower in the presence of C (a readily available energy source). In C-amended peat, biocide treatment effects could be seen (Fig. 2). Ammonium-N concentrations differed significantly ($p<0.05$) in streptomycin amended samples compared to control samples. Nitrate-N concentrations followed the same trend but differences were not significant.

Discussion

Immobilisation of N occurs rapidly in peat and can remove a substantial proportion of fertiliser N out of the available N pool, restricting the quantity of N available for plant uptake. By day 2, up to 30% of applied fertiliser N can be found within the microbial biomass. For an established grass sward of perennial ryegrass, Bristow et al. (1987) found approximately 37% of labelled fertiliser N was recovered in the microbial biomass two days after fertiliser addition.

The rapid immobilisation of fertiliser N by the microbial biomass emphasises the need for further short-term studies investigating the controls on the partitioning of N. Schimel et al. (1989) found that within 12 hours, the microbial biomass had assimilated more NH_4^+ than plant roots in a Californian annual grassland. These authors found that competition for available N between plant roots and the microbial biomass not only controlled the dynamics of N flowing within the soil but also the fate of N.

The use of biocides to selectively suppress a proportion of the microbial population has been exploited by a number of workers investigating the role of fungal and bacterial populations in biological transformations. In this study, the application of streptomycin (a bactericide) suppressed the amount of N immobilised, resulting in higher available N concentrations in the soil. This suggests that the bacterial population may be the main immobiliser of applied N in blanket peat. Ingham et al. (1986) also found that by reducing the bacterial population in a forest soil, N immobilisation decreased and available N concentrations increased. The application of biocides may also affect non target populations (e.g. fungi and soil fauna), thereby leading to difficulties in the interpretation of experiments (Ingham and Coleman, 1984). However, in a blanket peat ecosystem subject to intermittent flooding, soil animal populations will probably be low and not contribute greatly to nutrient transformations. Therefore, in this type of ecosystem, the addition of biocides in combination with glucose, may be very effective in the selective inhibition of target populations with few non target populations affected.

Once applied N has been immobilised by the microbial biomass, incorporation into more stable organic forms will restrict availability for plant uptake. Hulm and Killham (1990) found that after the addition of fertiliser N (as urea) to a Sitka spruce stand, approximately 90% of applied N was organically bound in the LFH layers of the soil profile. Rapid immobilisation of N in organic soils with its subsequent slow remineralisation may be important processes controlling the quantity of N available for plant uptake (e.g. see Strickland et al., 1992).

To conclude, bacterial populations appear to be largely responsible for the immobilisation of applied N in a reseeded grass pasture on blanket peat. Although the acidity of the site (pH 3.8) may suggest that fungal populations would be predominant, the fluctuating water-table resulting in periodic anaerobism, limits the activity of fungi and favours the bacterial population. As previous work (Hall et al., 1994) has shown, the competitive sink strength of the microbial biomass is strongly affected by a combination of abiotic and climatic factors. Therefore, the N dynamics of a reseeded grass pasture on blanket peat may be controlled by the extent of microbial immobilisation of N and the climatic conditions prevalent at the time of fertiliser application. Plant N uptake at this site will largely be determined by the competitive sink strength of the microbial and perhaps more importantly, bacterial biomass.

Acknowledgements

The Ministry of Agriculture, Fisheries and Food provided J Hall with a studentship to conduct this research.

References

Bristow A W, Ryden J C and Whitehead D C 1987 The fate at several time intervals of ^{15}N-labelled ammonium nitrate applied to an established grass sward. J. Soil Sci. 38, 245–254.

Brookes P C, Landman A, Pruden G and Jenkinson D S 1985 Chloroform fumigation and the release of soil nitrogen: a rapid direct extraction method to measure microbial biomass nitrogen in soil. Soil Biol. Biochem. 17, 837–842.

Frame J, Newbould P and Munro J M M 1985 Herbage production from the hills and uplands. *In* Hill and upland livestock production. Occ. Pub. No. 10 Eds. T J Maxwell and R G Gunn. pp 9–37. Br. Soc. Anim. Prod., Edinburgh, UK.

Hall J M, Williams B L and Killham K 1994 Investigation of N mineralisation immobilisation dynamics in blanket peat to optimise the N-economy of improved grass pasture. Eur. J. Agron. (*In press*).

Hulm S C and Killham K 1990 Response over two growing seasons of a Sitka spruce stand to ^{15}N-urea fertiliser. Plant and Soil 124, 65–72.

Ingham E R and Coleman D C 1984 Effects of streptomycin, cycloheximide, fungizone, captan, carbofuran, cygon and PCNB on soil microorganisms. Microbial. Ecol. 10, 345–358.

Ingham E R, Cambardella C and Coleman D C 1986 Manipulation of bacterial, fungi and protozoa by biocides in lodgepole pine forest soil microcosms: effects on organism interactions and nitrogen mineralisation. Can. J. Soil Sci. 66, 261–272.

Jamieson N and Killham K 1994 Biocide manipulation of N flow to investigate root/microbe competition in forest soil. Plant and Soil 159, 283–290.

Rangeley A 1988 The utilisation of N fertiliser applied to perennial ryegrass/white clover pasture growing on humus iron podzol in N.E. Scotland. Grass Forage Sci. 43, 363–369.

Schimel J P, Jackson L E and Firestone M K 1989 Spatial and temporal effects on plant-microbial competition for inorganic N in a California annual grassland. Soil Biol. Biochem. 21, 1059–1066.

Strickland T C, Sollins P, Rudd N and Schimel D S 1992 Rapid stabilisation and mobilisation of ^{15}N in forest and range soils. Soil Biol. Biochem. 24, 849–855.

Williams B L and Wheatley R E 1992 Mineral nitrogen dynamics in poorly drained blanket peat. Biol. Fertil. Soils 13, 96–101.

O. Van Cleemput et al. (eds.), Progress in Nitrogen Cycling Studies, 203–205, 1996.
© 1996 *Kluwer Academic Publishers.*

Nitrogen uptake by cover crops at 7 sites in the UK: 1990–1993

R. Harrison[1], D.B. Davies[1] and S. Peel[2]
[1]*ADAS Arable Research Centre, Anstey Hall, Maris Lane Trumpington, Cambridge, CB2 2LF, UK and* [2]*ADAS Dairy Research Centre, Martyr Worthy, Winchester, Hants., SO21 1AP, UK*

Key words: cover crop, nitrate, N uptake, sowing date, winter rye

Abstract

An experiment to compare the performance of a range of cover crop species at 7 sites in the UK was established in the autumn of 1990. At each site treatments included 3 sowing dates (late August, mid-September, mid-October). Nitrogen (N) uptake by each cover crop was determined at the beginning of December and the beginning of February.

Cover crop growth in the first two years of the experiment was rather small; this was ascribed to the dry autumns in those years. However, the third year (1992/93) autumn was moist and, in general, much larger N uptake was recorded. Of the species tested, winter rye proved the most consistent cereal with N uptake (averaged over 3 sowing dates and 7 sites) ranging from 8 to 27 kg ha^{-1} for the three years.

Linear regression analysis indicated that there was a significant ($p<0.05$) relationship between N uptake by winter rye and day.degrees (T>0 °C) from sowing. For a particular year, between 20 and 50% of the variation in N uptake could be explained in terms of this simple regression model. However, when nominal variables were introduced to represent sites, ca. 90% of the variation was accounted for. It appeared that sites could be grouped on the basis of factors such as N status, available moisture and ease of establishment of cover crops.

Introduction

The 1980 EC Directive on the Quality of Water Intended for Human Consumption set a maximum allowable concentration of 50 mg L^{-1} nitrate in drinking water. Subsequently the 1991 Nitrate Directive required member states to protect all water sources at risk of exceeding this limit. Most nitrate leaching in the UK occurs during autumn and winter, and sowing an autumn cover crop represents an important management strategy to immobilise nitrate which would otherwise be liable to leaching (Powlson and Davies, 1993).

In autumn 1990, an experiment was established to provide practical information on the management of cover crops in UK situations. The objectives were to: (a) evaluate a range of species on land vunerable to over-winter nitrate leaching preceding a spring sown crop; (b) evaluate appropriate husbandry techniques; and (c) measure the effect of cover crops on following crops.

Materials and methods

The trial was established at 7 ADAS research centes (Table 1) representing a wide range of soil and climatic conditions. The crop preceding the cover crop was always winter wheat, so that the trial was located in a different part of each research centre each year. Cultivation and sowing was carried out using methods appropriate to the soil type, whilst attempting to conserve moisture by rolling. In the second (1991/92) and subsequent years of the experiment, ploughing was used as a first cultivation to reduce competition from volunteers and to remove straw. The target dates for drilling were late August, mid-September and mid-October. In addition to drilled crops, wheat was also established as soon as possible after harvest by broadcasting and disc-harrowing. The experiment layout was randomised block with two replicates. Total above ground matter and %N were determined on 1 December and 1 February, prior to cover crop destruction, using four 0.25 m^2 quadrats per plot, and standard methods of analysis (Anon, 1986). The 1 February

Table 1. Sites, soils and climate data

| | ADAS research centre | Soil type | Average for 1 Sep - 31 Nov | | |
			Rainfall (mm)	Max. T (°C)	Min. T (°C)
AR	Arthur Rickwood	Peaty	136	14.5	5.8
BR	Bridgets	Calcareous loam over chalk	220	14.4	6.3
BX	Boxworth	Clay	152	14.2	6.7
GT	Gleadthorpe	Loamy sand	158	13.7	5.6
HM	High Mowthorpe	Calcareous clay loam over chalk	198	12.1	6.1
RM	Rosemaund	Silty clay loam	174	13.8	5.8
TT	Terrington	Alluvial silty clay loam	153	14.1	6.6

destruction date plots were also sampled on 1 December except where a large area had to be sacrificed. Soil samples (6 cores per plot) were also taken for mineral N analysis in depth increments of 30 cm (Anon, 1986).

Results and discussion

The N uptake by cover crops to 1 December averaged across all 7 sites and 3 sowing dates (for the drilled species) are given in Table 2. N uptake in the first two years of the experiment (1990/91 and 1991/92) was considerably less than that in the third year (1992/93). Both 1990 and 1991 were dry autumns, and it is likely that cover crop growth was limited by available moisture. In comparison, autumn 1992 was much wetter and all the cover crop species performed well. Some caution should be exercised in intepreting the data in Table 2, however, as the data for drilled species are the average of 3 sowing dates whereas that for broadcast wheat is for a single (early) date of establishment. The N uptake associated with the mid-October sowing date was markedly lower than that associated with the two earlier sowing dates in all years, and in the third year of the experiment N uptake from the late August sowing was greater than that from the mid-September sowing. In fact for 1992/93, N uptake by the drilled species varied between 38.3 and 54.1 kg ha^{-1} N from the late August sowing compared to 28.6 kg ha^{-1} N for broadcast wheat.

The data in Table 3 show that there were also marked differences between sites, particularly in 1992/93. Greatest N uptake occurred at Arthur Rickwood on a peaty soil, and at Rosemaund and Terrington on silty clay loam soils, respectively. Lower N uptake occurred at Bridgets, High Mowthorpe and Gleadthorpe, on calcareous loam over chalk, calcareous clay

Table 2. N uptake by cover crop species to 1 December

| | Average of 7 sites and 3 sowing dates[a] | | |
| | N uptake (kg ha^{-1}) | | |
Cover crop	1990/1	1991/2	1992/3
Winter rye	8.1	16.4	27.2
Winter barley	9.6	11.5	24.8
White mustard	13.0	15.9	29.5
Phacelia	9.2	12.8	19.4
Forage rape		9.6	23.6
Stubble turnips		15.0	25.1
Broadcast wheat	5.5	12.7	28.6

[a] Wheat broadcast as soon as possible after harvest

Table 3. N uptake to 1 December at each site

| | Average of all species at all sowing dates | | |
| | N uptake (kg ha^{-1}) | | |
Site	1990/1	1991/2	1992/3
AR	13.9	22.5	58.9
BR	4.6	12.6	15.1
BX	4.2	11.7	9.9
GT	7.4	14.9	10.3
HM	6.2	5.0	11.4
RM	19.2	1.7	33.9
TT	12.9	10.0	37.2

loam over chalk and loamy sand, respectively. At Boxworth on clay, a site where it proved difficult to obtain good cover crop establishment, N uptake was also very low.

In addition to available N and moisture, differences in temperature are likely to contribute to the variation in cover crop N uptake between sites. In order to investigate further the reasons for variations in N uptake

Table 4. Coefficients for the regression equation N uptake = k * day.degrees (T>0 °C) for winter rye

	AR	BR	BX	GT	HM	RM	TT	R^2 (%)
1990/1	0.012	n.d.	0.012	0.012	0.012	0.028	0.012	91
1991/2	0.033	0.018	0.018	0.033	0.018	n.d.	0.018	93
1992/3	0.118	0.020	0.020	0.020	0.020	0.049	0.049	89

between sites, the relationship between N uptake and day.degrees (T>0 °C) was examined by linear regression analysis. Using N uptake by winter rye (the cereal which performed most consistently at each site and in each year) at both 1 December and 1 February for each of the 3 sowing dates, for each year a single regression equation was constructed involving all sites. The variation accounted for was 35%, 50% and 27% for 1990/91, 1991/92 and 1992/93, respectively. However, when nominal variables were introduced into the regression equation to represent individual sites, the variation accounted for increased significantly. Using an F-test, all terms in the regression equation were then tested ($p < 0.05$) for inclusion in a modified regression equation. The F-test had the form:

$$F = \{\Delta(SS\ error)/\Delta(df)\}/\{SS\ error/df\} \qquad (1)$$

where the numerator term represents the difference between the modified and full regression models, and the denominator term refers to the full regression model. This analysis indicated that the constant term for each year was not significantly different from zero, and that the sites could be grouped on the basis of their regression equation coefficients (Table 4).

Regression coefficients for 1990/91 and 1991/92 were generally similar and small compared to those obtained for 1992/93. This reflects the poor growing conditions of these two years due to moisture limitation. The regression coefficients for 1992/93 indicate three site groups: namely, Arthur Rickwood; Rosemaund and Terrington; and, Bridgets, Boxworth, Gleadthorpe and High Mowthorpe. This grouping probably reflects a combination of N status, available soil moisture and establishment characteristics. There is a strong correlation between the individual site regression coefficients and soil mineral N in the 0–90 cm depth increment of bare ground plots sampled on 1 December, but this is in large part due to the high N uptake and high N status of the Arthur Rickwood site. With the exception of Boxworth where considerable establishment difficulties were encountered, this grouping also represents a trend in available soil moisture capacity.

Conclusions

Overall, this limited analysis indicates that with good establishment, adequate moisture and day.degrees (T>0°C) in excess of 600, a range of cover crop species are capable of immobilising between 12 and 45% of mineral N in the 0–90 cm bare ground. For winter rye there appears to be a relatively simple linear relationship between N uptake and day.degrees (T>0 °C), with a slope depending on N status and available moisture.

Acknowledgements

Financial support for this work from the Ministry of Agriculture, Fisheries and Food is gratefully acknowledged.

References

Anon 1986 The Analysis of Agricultural Materials, MAFF Reference Book No. 427. MAFF Publications, London, UK.

Powlson D and Davies D B 1993 Autumn and winter land management. In Solving the Nitrate Problem. pp 15–19. MAFF Publications, London, UK.

O. Van Cleemput et al. (eds.), Progress in Nitrogen Cycling Studies, 207–212, 1996.

Influence of cadmium on the NO_3^- assimilation of two cultivars of *Zea mays*

L.E. Hernández, R. Carpena-Ruiz and A. Gárate[1]
Departemento Química Agrícola, Geología y Geoquímica, Facultad de Ciencias, Universidad Autónoma de Madrid, Canto Blanco E-28049 Madrid, Spain. [1]*Corresponding author*

Key words: cadmium, maize, nitrate assimilation, nitrate reductase, *Zea mays*

Abstract

The presence of Cd (control, 0.15, 1.50 and 15.0 mg Cd L^{-1}) in the growing medium of two maize cultivars (Dekalb XL 72 AA and Paolo) lead to a decrease in the NO_3^- content in plant tissues. As a consequence, there was a loss of nitrate reductase activity. The results of a shock treatment with 15 mg Cd L^{-1} (after 24 h) and the assesment of the effect of Cd on the nitrate reductase de novo synthesis and stability suggest that Cd influences on the NO_3^- root uptake and translocation to the shoot, thereafter nitrate reductase activity loss.

Abbreviations: $LSD_{0.05}$ – less significant difference with 0.05 of limit of confidence, calculated by using the test of Duncan; NR – nitrate reductase; g_{DW} – dry weight; g_{FW} – fresh weight.

Introduction

Nitrogen is the most important nutrient of crops, limiting quantitative and qualitatively their production. Among different nitrogen fertilizers, nitrate is the main source of nitrogen for higher plants, which is reduced to NH_4^+ and subsequently incorporated into carbon skeletons as amino acids. The keystep of the NO_3^- assimilation is its reduction to NO_2^-, catalyzed by the enzyme nitrate reductase (NR, EC 1.6.6.1.)(Guerrero et al., 1981). Nitrate is absorbed in the root epidermis and transported to the shoot via xylem, mediated by an active transport (Rufty et al., 1986). NR activity depends on de novo synthesis of the enzyme, induced by the NO_3^- concentration in the cytosol (Hoff et al., 1992; Tischner et al., 1993).

On the other hand, Cd is one of the most toxic heavy metal pollutants, because it is not present in biotic systems in normal conditions and its toxic effects appear even at low concentrations (van Assche and Clijters, 1990). Mainly, its presence in polluted environments is in most cases due to human activities (e.g. application of fertilizers and sewage sludges).

Among the wide range of symptoms of toxicity described in higher plants, one of the most interesting might be the inhibition of the assimilation of nitrate observed by different authors (Khater et al.,

1991; Mathys, 1975; Singh et al., 1988). A decrease of assimilated nitrogen might limit the synthesis of N-containing organic compounds (e.g. proteins or nucleotides) leading to a more severe toxic effect. In the present work, maize plants were treated with Cd in order to evaluate its effect on the nitrate assimilation.

Materials and methods

Plant material

Maize (*Zea mays* cv. Dekalb XL 72 AA and Dekalb Paolo) were germinated in moistened paper for 3 days at 28 °C. The seedlings were cultivated in pure hydroponic solution (macronutrients (mmol L^{-1}): 2 KH_2PO_4, 1.5 $MgSO_4$, 0.1 NaCl, 1 $Ca(NO_3)_2$, 1.5 KNO_3, and micronutrients (mg element L^{-1}): 2.5 Fe(EDDHA), 1 $MnSO_4$, 0.25 $CuSO_4$, 0.5 $ZnSO_4$, 0.5 H_3BO_3, 0.2 $Mo_7O_{24}(NH_4)_6$. $4H_2O$, pH was adjusted to 5.00) in a controlled environment chamber at 28 °C day/20 °C night, with 16 h of light (10 Sylvania Cool White VHO lamps of 120 W m^{-2} each) and 70–75% of relative humidity.

Plants of the Experiment 1 were supplied with 4 concentrations of Cd: control (0.0), 0.15, 1.5 and 15.0

mg L^{-1}, and samples were collected at 10, 15 and 20 days of growing.

In the Experiment 2 (shock treatment of Cd) maize seedlings were placed in plastic pots containing control nutrient solution. After 7 days, half of the plants were transfered to nutrient solution containing 15 mg Cd L^{-1}. Samples were harvested after 0, 24 and 48 h.

In order to test the effect of Cd on de novo synthesis of NR, the Experiment 3 was carried out with maize scutella. Maize seeds were germinated in perlite with deionized water at 28 °C, and allowed to ethiollate for 5 days. Scutella were separeted from the softened endosperms, rinsed with deionized water and 0.5 g per sample were placed in NR aereated induction medium (30 mM KNO$_3$, 50 mM KH$_2$PO$_4$, pH 4.0 and containing control (0.0), 0.5 or 5.0 mg Cd L^{-1}) for 5 h at 28°C. The in vitro application of Cd was undertaken by assaying NR extracts of control induced scutella in NR reaction buffers containing control (0.0), 0.5 and 5.0 mg Cd L^{-1}.

In vitro NR assay

Was carried out according to the method described by Ramón et al. (1989). Aproximately 0.5 g of fresh tissue was frozen in liquid N$_2$ and homogenized in extraction buffer (25 mM KH$_2$PO$_4$/K$_2$HPO$_4$, 5 mM EDTA, 1% PVP and 1 mM cystein, pH 7.7). The extract was centrifuged at $10000 \times g$ for 15 min at 4°C. The supernatant contained the solubilized NR and was kept in ice bath till enzyme activity determination. NR was assayed by adding 100 μL of 0.1 M KNO$_3$, 500 μL of reaction buffer (100 mM KH$_2$PO$_4$/K$_2$HPO$_4$, 1 mM EDTA, pH 7.5) and 100 μL of 1 mg NADH mL^{-1} as reductant substrate. In the Experiment 3, EDTA was absent in the NR reaction buffer of the in vitro application of Cd. The NR in shoots were assayed using the latest appearing leaf, which in previous assays had the highest activity (inedit). NO$_2^-$ was analysed with fresh prepared colorimetric reagent (sulfanilamide 1% in 3 M HCl and 0.02% N-(1-naphtyl)ethylenediamide dihydrochloride, 1:1). Absorbance was read at 540 nm.

NO$_3^-$ determination in plant tissues

Was determined essentialy as described by Gojon et al. (1991). 100 mg of dried sample were digested with 5 mL of 0.1 M HCl for 30 min at 70 °C. The extract was filtered and volume brougth to 25 mL. NO$_3^-$ was analysed by using a colorimetric autoanalizer Technicon Acta II with a Cd/Cu reductant col-

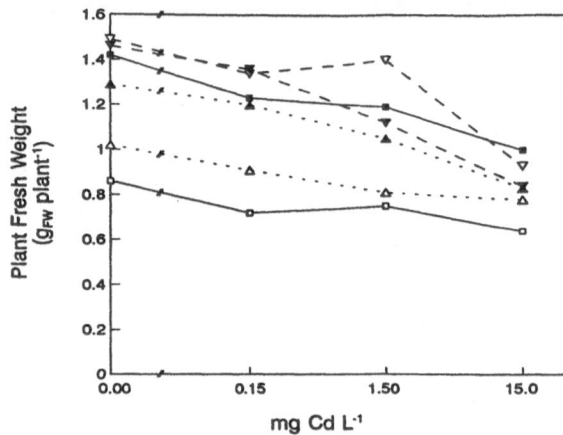

Figure 1. Fresh weight (g$_{FW}$ plant^{-1}) of maize plants treated with several levels of Cd in Experiment 1. Cv Dekalb *XL 72 AA*: 10 days (■ LSD$_{0.05}$ 0.13), 15 days (▲ LSD$_{0.05}$ 0.21) and 20 days (▼ LSD$_{0.05}$ 0.17). Cv Dekalb *Paolo*: 10 days (□ LSD$_{0.05}$ 0.23), 15 days (△ LSD$_{0.05}$ 0.11) and 20 days (▽ LSD$_{0.05}$ 0.25). Means of 4 replicates are shown.

umn. The sample was diluted in reduction buffer (10 g L^{-1} NH$_4$Cl, pH 8.5) and reacted with the colorimetric reagent (1% sulfanylamide, 14 M H$_3$PO$_4$, 0.05% N-(1-naphtyl)ethylendiamide dihydrochloride and 0.5 mg mL^{-1} Brij 35). Absorbance was read at 520 nm.

Cd and K analysis

Plant tissues were dried at 70 °C until constant weight was achieved. After homogenization, samples were digested with a mixture of acids HNO$_3$:H$_2$SO$_4$:HClO (5:1:2), at <200°C. Cd and K were analysed by spectrophotometric atomic absorption and emission (Perkin-Elmer 4000), repectively, as described Gárate et al. (1992).

Results

Plant fresh weight decreased when the Cd level increased in the growing medium in both maize cultivars and at every sampling period (Fig. 1). Plants of cv. Dekalb XL 72 AA showed a more intense decrease than cv. Dekalb Paolo, being almost half the weight of control plant in the treatment with 15 mg Cd L^{-1}, although after 20 days both cultivars suffered a similar decrease ratio. Plants of this treatment appeared very damaged, with short darkened roots and few thick secondary root hairs.

Cd content increased in shoot and root according to Cd supply in every sampling day (Table 1). Cd

Table 1. Cd (μg g_{DW}^{-1}) and K (mg g_{DW}^{-1}) content of shoot and root maize plants (Dekalb XL 72 AA and Dekalb Paolo) treated in the Experiment 1. (* different to control sample with $p < 0.05$, n = 3)

| (mg Cd L^{-1}) | XL 72 AA | | | | Paolo | | | |
| | Shoot | | Root | | Shoot | | Root | |
	Cd	K	Cd	K	Cd	K	Cd	K
10 days								
Control	0.0	415.7	4.5	408.2	0.0	291.5	20.4	264.8
0.15	34.6*	382.2*	189.4*	345.8*	17.0*	343.2	236.8*	311.9
1.50	89.1*	476.7	1546.2*	369.4*	69.1*	290.2	577.6*	280.8
15.0	315.4*	258.9*	2923.1*	138.7*	156.2*	209.7*	2197.5*	142.6*
15 days								
Control	5.1	428.3	52.3	319.3	0.0	384.7	21.3	283.7
0.15	85.6*	427.9	414.4*	369.2	39.0*	395.7	435.7*	254.4
1.50	119.0*	428.0	1077.9*	344.4	90.5*	321.3	1075.9*	227.3*
15.0	607.2*	331.4*	3899.6*	235.6*	284.5*	215.7*	2926.1*	126.8*
20 days								
Control	0.0	479.2	6.9	378.7	3.0	361.3	33.5	280.5
0.15	85.0*	408.6	445.8*	340.0*	45.4*	280.3*	820.5*	263.2
1.50	125.0*	399.9*	679.9*	310.6*	80.9*	289.4*	1191.9*	186.2*
15.0	767.3*	391.8*	5434.8*	189.6*	457.5*	187.8*	3257.0*	138.8*

was accumulated mainly in the root (approx. 10 times more than in shoot) in both studied cultivars. As Cd concentration increased in the treatments the content of K in plant tissue decreased, which was severly reduced in root of both maize cultivars (Table 1).

The content of NO_3^- diminished in shoot as the concentration of Cd supplied to the medium was higher in maize plants of cv. Dekalb XL 72 AA (Fig. 2a). Also, a consistent decrease in the NO_3^- content of root was observed for Cd levels up to 1.5 mg Cd L^{-1}. In the presence of 15 μg Cd mL^{-1}, NO_3^- analysis revealed values similar to control plants (Fig. 2a). The same trendency was observed in plant tissues of cv. Dekalb Paolo (data not shown).

In vitro NR activity of Dekalb XL 72 AA root decreased for every Cd treatment and sampling period (Fig. 2b). The addition of 15 μg Cd mL^{-1} inhibited severely NR activity, by a 90% of the control at 20 days of growth. No consistent effect was observed in shoot (Fig. 2b). NR activity of Dekalb Paolo plants showed a similar response to the presence of Cd (data not shown).

After 24 h of shock treatment with 15 mg Cd L^{-1} (Exp. 2) nitrate content diminished in shoot and root of maize Dekalb XL 72 AA (Fig. 3a). This behaviour was maintained also after 48 h of treatment (data not

shown). Also, in Dekalb Paolo plants the content of NO_3^- diminished in shoot and root after 24 h (Fig. 3a), but after 48 h of treatment there was no significant difference of the NO_3^- accumulation in root of control plants (data not shown). Besides, the Cd shock treatment led to a loss of root NR activity of both maize cultivars after 24 h of Cd treament (Fig. 3b), but after 48 h in Dekalb Paolo root NR recovered the activity of control plants (data not shown).

In the Experiment 3 we observed that de novo NR synthesis was not affected by the presence of Cd in the NR induction medium of maize scutella of cultivars Dekalb XL 72 AA and Paolo. Moreover, addition of Cd to the reaction buffer of the in vitro NR assay did not influence on its activity (data not shown).

Discussion

The content of Cd in the shoot and the root reflected the Cd levels supplied to the growing medium. Similar Cd accumulation was reported in other maize cultivars (Florijn and van Beusichem, 1993a; Siedlcka and Baszynski, 1993). A group of maize cultivars named 'shoot Cd excluders' accumulate between 10 to 20 times more Cd in root that in shoot (Florijn and

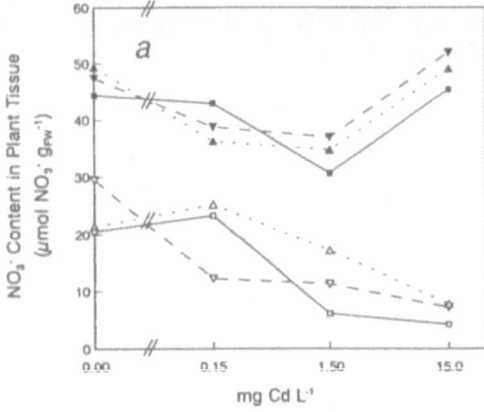

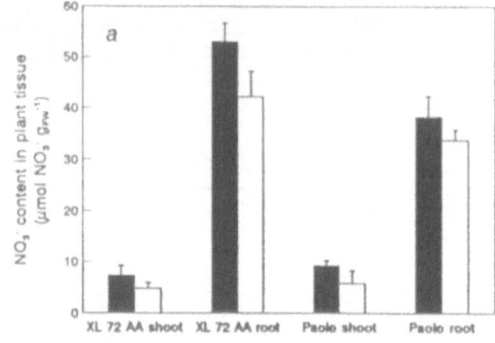

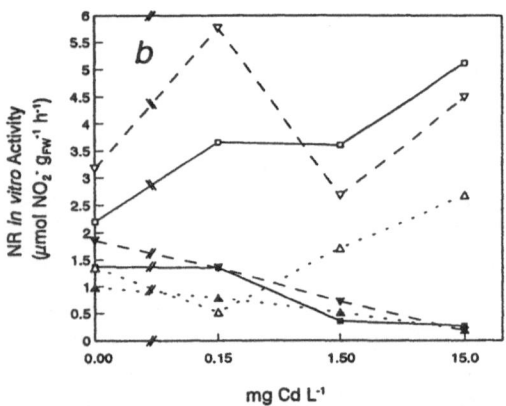

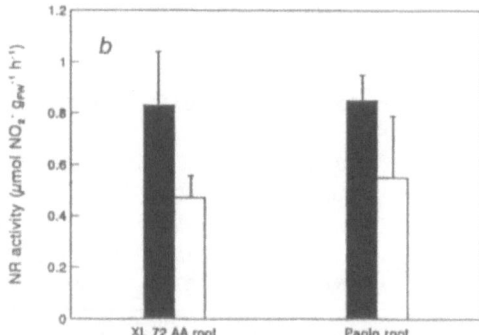

Figure 3. Shock application of Cd in Experiment 2 sampled after 24 h of treatment (■ control and □ 15 mg Cd L^{-1}). (a) NO$_3^-$ content in plant tissue (μmol NO$_3^-$ g$_{FW}^{-1}$). (b) In vitro NR activity (μmol NO$_2^-$ g$_{FW}^{-1}$ h^{-1}). Means and standard deviations ($p < 0.05$) of 4 replicates are shown.

Figure 2. Maize cv. Dekalb XL 72 AA in Experiment 1 sampled after 10 days (□), 15 days (△) and 20 days (▽). (Shoot empty symbols and root filled symbols). (a) NO$_3^-$ content in plant tissue (μmol NO$_3^-$ g$_{FW}^{-1}$), □ LSD$_{0.05}$ 1.90, △ LSD$_{0.05}$ 1.78, ▽ LSD$_{0.05}$ 3.61, ■ LSD$_{0.05}$ 2.65, ▲ LSD$_{0.05}$ 2.67 and ▼ LSD$_{0.05}$ 2.49. (b) In vitro NR activity (μmol NO$_2^-$ g$_{FW}^{-1}$ h^{-1}), □ LSD$_{0.05}$ 0.70, △ LSD$_{0.05}$ 0.53, ▽ LSD$_{0.05}$ 0.50, ■ LSD$_{0.05}$ 0.22, ▲ LSD$_{0.05}$ 0.20, ▼ LSD$_{0.05}$ 0.23. Means of 4 replicates are shown.

van Beusichem, 1993b). According to our results the studied maize Dekalb XL 72 AA and Paolo might be included in that group.

As a consequence of the presence of Cd a whole negative effect on the nitrogen nutrition parameters determined was observed. Similar results were obtained in other plant species: cucumber (Burzynski, 1988), sugar beet (Petrovic et al., 1990) and broad bean (Khater et al., 1991) plants accumulated less NO$_3^-$ after treatment with various levels of Cd. The increment of the NO$_3^-$ content in root observed in the treatment with 15 mg Cd L^{-1} (Fig. 2a) might be the result of less translocation of NO$_3^-$ to the shoot, as we have observed in pea plants (inedit). NR activity decreased

in root of both maize cultivars due to Cd treatments (Fig. 2b), in agreement with results described by other authors (Burzynski, 1988; Chugh et al., 1992; Mathys, 1975; Petrovic et al., 1990; Singh et al., 1988). On the contrary, NR activity data from shoot was no consistent with the level of Cd supplied (Fig. 2b). This phenomenon might be a consequence of the growing status of the sampled leaf, as evidenced that the emerging leaf of control plants weighted almost twice that of 15 mg Cd L^{-1} treated plants. As was observed in our previous experiments and by other authors (King et al., 1992) Teyker et al., 1991), young tissues had higher NR activity. Nussbaum et al. (1988) reported an increase of shoot NR activity in *Zea mays* plants treated with relative low Cd levels (up to 5 μM), probably related to differences of tissue growing. Nevertheless, at higher Cd concentrations (over 50 μM) NR activity decreased after 3 days of treatment.

De novo NR synthesis was not affected by the presence of Cd in the induction medium. Also, the treatments of Cd in the reaction buffer of the in vitro

NR assay did not alter NR activity, in agreement with results obtained in pea (Chugh et al., 1992) and cucumber (Burzynski, 1988) using similar Cd levels. Those results suggest that the negative effects of Cd on the NR activity was not due to NR inactivation or decrease in NR synthesis. Therefore, NR might be related with the decrease of NO_3^- content, substrate that mediates in the NR genes expression (Hoff et al., 1992; Tischner et al., 1993). This hypothesis is in concordance with that postulated by Burzynski (1988), where a reduction of NO_3^- uptake, produced by the presence of Cd in the growing medium, would lead to a decrease in the cytosolic NO_3^- and, therefore, in the NR activity. This is supported by the reduction in the content of NO_3^- and in the NR activity found after 24 h of the shock treatment with 15 mg Cd L^{-1} (Fig. 3a and b). Furthermore, Cakmak and Horst (1991) and Durieux et al. (1993) have reported a reduction of the permeability of root membranes to NO_3^- in the presence of Al after few hours of treatment. In addition, K^+ is considered the NO_3^- counterion in the processes of root absorption and transport to the shoot (Rufty et al., 1981; Tremblay et al., 1988). Thus, the less K plant content due to Cd supply (Table 1) might be also a response to the alteration of the permeability of root membranes to nutrients as has been reported for NO_3^-.

Plants of maize Dekalb Paolo accumulated less Cd than Dekalb XL 72 AA (Table 1). In general, Dekalb Paolo showed less effect of Cd on the several kinds of parameters determined, such as plant weight (Fig. 1) or NO_3^- content in plants of the Experiment 2 (Fig. 3a). This might indicate that cv. Dekalb Paolo is slightly less sensitive to Cd than cv. Dekalb XL 72 AA.

In conclusion, our results suggest that Cd affects the NO_3^- uptake mechanism, decreasing the amount of available NO_3^- to the NR. Subsequently, the NR activity loss might be due to a decrease of de novo synthesis of the enzyme, mediated by the relative low level of NO_3^- in the Cd treated plant cell cytosol.

Acknowledgements

We thank the financial support of the Spanish D.G.I.C.Y.T. through the project PB–1097. Maize seeds were a gift from Dekalb Ibérica, SA.

References

Burzynski M 1988 The uptake and accumulation of phosphorous and nitrate and the activity of nitrate reductase in cucumber seedlings treated with Pb and Cd. Acta Soc. Bot. Poloniae. 57, 349–359.

Cakmak I and Horst W J 1991 Effect of aluminium on the net efflux of nitrate and potassium from root tips of soybean (Glycine max, L.). J. Plant. Physiol. 138, 400–403.

Chugh L K, Gupta V K and Sawhney S K 1992 Effect of cadmium on enzymes of nitrogen metabolism in pea seedlings. Phytochem. 31, 395–400.

Durieux R P, Jackson W A, Kamprath E J and Moll R H 1993 Inhibition of nitrate uptake by aluminium inmaize. Plant and Soil. 151, 97–104.

Florijn P J and van Beusichem M L 1993a Cadmium distribution in maize inbred lines: effects of pH and level of Cd supply. Plant and Soil 153, 79–84.

Florijn P J and van Beusichem M L 1993b Uptake and distribution of cadmium in maize inbred lines. Plant and Soil 150, 25–32.

Gárate A, Ramos I and Lucena J J 1992 Efecto del cadmio sobre la absorción y distribución de manganeso de distintas variedades de Lactuca. Suelo Planta 2, 581–591.

Gojon A, Wakrim R, Passama L and Robin P 1991 Regulation of NO_3^- assimilation by anion availability in excised soybean leaves. Plant Physiol. 96, 398–405.

Guerrero M G, Vega J M and Losada M 1981 The assimilatory nitrate-reducing system and its regulation. Ann. Rev. Plant Physiol. 32, 169–204.

Hoff T, Stummann B M and Henningsen K W 1992 Structure, function and regulation of nitrate reductase in higher plants. Physiol. Plant. 84, 616–624.

Khater A, Abou-Seeda M, Soliman S and Salem N 1991 Growth, nodule formation, chlorophill content and the uptake of some nutrients by broad bean plants as affected by cadmium application. Agrochim. 35, 434–440.

King B J, Siddiqui M Y and Glass D M 1992 Studies of the uptake of nitrate in barley. V. Estimation of root cytoplasmic nitrate concentration using nitrate reductase activity -Implications for nitrate flux. Plant Physiol. 99, 1582–1589.

Mathys W 1975 Enzymes of heavy-metals-resistant and non-resistant populations of Silene cucubalus and their interactions with some heavy metals in vitro and in vivo. Physiol. Plant. 33, 161–165.

Nussbaum S, Schmutz D and Brunold C 1988 Regulation of assimilatory sulfate reduction by cadmium in Zea mays L. Plant Physiol. 88, 1407–1410.

Petrovic N, Kastori R and Rajcan I 1990 The effect of cadmium on nitrate reductase activity in sugar beet (Beta vulgaris). In Plant Nutrition-Physiology and Applications. Ed. M L van Beusichem. pp 107–109. Kluwer Academic Publishers, Dordrecht, the Netherlands.

Ramón A M, Carpena-Ruiz R O and Gárate A 1989 In vitro stabilization and distribution of nitrate reductase in tomato plants. Incidence of boron deficiency. J. Plant. Physiol. 135, 126–128.

Rufty T W, Thomas J F, Remmler J L, Campbell W H and Volk R J 1986 Intercellular localization of nitrate reductase in roots. Plant Physiol. 82, 675–680.

Siedlecka A and Baszynski M L 1993 Inhibition of electron flow around photosystem I in chloroplasts of Cd-treated maize plants due to Cd-induced iron deficiency. Physiol. Plant. 87, 199–202.

Singh JP, Singh B and Karwasra SPS 1988 Yield and uptake response of lettuce to cadmium as influenced by nitrogen application. Fert. Res. 18, 49–56.

Teyker R H, Dallmeir K A, Aubin G R S T and Lambert R J 1991 Seedling nitrate reductase activity and nitrate partitioning in maize (*Zea mays* L.) strains divergently selected for post-anthesis leaf-lamina nitrate reductase activity. J. Exp. Bot. 42, 97–102.

Tischner R, Waldeck B, Goyal S P and Rains W D 1993 Effects of nitrate pulses on the nitrate-uptake rate, synthesis of mRNA coding for nitrate reductase, and nitrate reductase activity in roots of barley seedlings. Planta 189, 533–537.

Tremblay N, Gasia M C H, Ferowge M T H, Gosselin A and Trudel M J 1988 Influence of photosynthetic irradiance on nitrate reductase activity, nutrient uptake and partitioning in tomato plants. J. Plant. Nutr. 11, 17–36.

Van Assche F and Clijters H 1990 Effects of metals on enzyme activity in plants. Plant Cell Environ. 13, 195–206.

O. Van Cleemput et al. (eds.), Progress in Nitrogen Cycling Studies, 213–215, 1996.
© *1996 Kluwer Academic Publishers.*

[15]N manipulations in a forest ecosystem subjected to an experimental decrease in nitrogen deposition

C.J. Koopmans and A. Tietema
Department of Physical Geography and Soil Science, University of Amsterdam, Nieuwe Prinsengracht 130, 1018 VZ Amsterdam, The Netherlands

Key words: atmospheric deposition, forest, ^{15}N, tracer

Abstract

Nitrogen deposition in a nitrogen saturated ecosystem has been reduced to natural background levels by means of a roof construction. The same amount of ^{15}N was applied to the throughfall water of a high and low nitrogen deposition plot leading to enrichments of 3400‰ and 43000‰ [15]N, respectively. After nine months of manipulation with ^{15}N major ecosystem compartments where ^{15}N enriched. In the plot under the reduced nitrogen deposition more ^{15}N was retained in the soil and vegetation compared to the plot receiving the high natural deposition.

Introduction

The overall objective of the study is to quantify nitrogen transformation processes and fluxes in a coniferous forest in the Netherlands subject to a manipulated decrease in nitrogen inputs. The stable isotope ^{15}N is used as a tracer. The study proposes to support the discussion about the implications of a reduction in N deposition for nitrogen cycling, soil acidification and forest decline and recovery.

Experiments are carried out within the European research project NITREX (NITRogen saturation EXperiments). Within NITREX comparable nitrogen manipulations are carried out at various nitrogen deposition levels (Dise and Wright, 1992). Nitrogen and Carbon cycles in a defined system have been described (Van Dam and Van Breemen, 1995) by a process orientated model NICCCE (Nitrogen Isotopes and Carbon Cycling in Coniferous Ecosystems).

The overall objectives of the ^{15}N experiment are:

– to quantify the fate of the incoming nitrogen in terms of the whole N budget on a plot scale.

– to calibrate and validate the dynamic nitrogen model developed on conventional nitrogen data (NICCCE).

– to achieve a cross site comparison (Kjønaas *et al.*, 1993) to test whether excessive additions of nitrogen may lead to the retention of the heavier ^{15}N isotope due to fractionation processes in the microflo-

ra using the pollution gradient within the NITREX sites. Preliminary results on natural abundance and enrichment levels of soil and vegetation after nine months of ^{15}N manipulation are presented in this paper.

Materials and methods

The effects of high nitrogen deposition in a nitrogen saturated Scots pine (*Pinus sylvestris* L) forest are being studied (Boxman et al., 1995) for five years. Nitrogen deposition (58 kg N ha^{-1} yr^{-1}) has been reduced to natural background levels (4–6 kg N ha^{-1} yr^{-1}) by means of a roof construction. The water is given to the plots under the roof by means of an automated sprinkling system in an almost real time watering regime. Three plots can be distinguished. The first plot, the low-deposition plot, receives artificial throughfall water where nitrogen is added at natural background levels. The second plot under the roof, the high-deposition plot receives the throughfall water intercepted by the roof. The third plot, the ambient-plot outside the roof, receives unaltered throughfall water.

To enable the fate of one year's input to be followed, the ^{15}N was applied over one full year to the plots under the roof. The ^{15}N isotope was applied as 99% enriched $(NH_4)_2SO_4$. During one year the throughfall water to the high deposition plot was enriched, in natural abun-

Soil

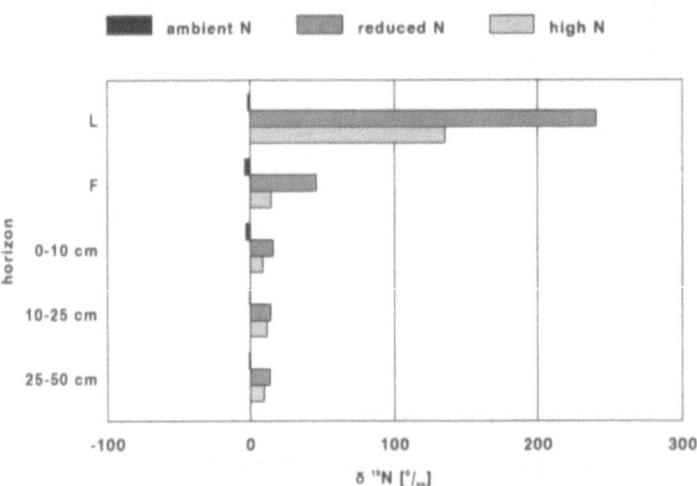

■ ambient N   ■ reduced N   ☐ high N

Figure 1. δ^{15}N values of the soil after nine months of ^{15}N addition to the high and low nitrogen deposition plot. The ambient plot served as a control.

Table 1. Estimated retention of the ^{15}N in the ecosystem compartments. ^{15}N was applied at a rate of about 43000‰ in the low deposition plot and at 3400‰ ^{15}N at the high deposition plot. See text for details.

Ecosystem pool	natural δ^{15}N (‰)	low deposition plot N pool (kg ha^{-1})	δ^{15}N (‰)	%N retained	high deposition plot N pool (kg ha^{-1})	δ^{15}N (‰)	%N retained
L horizon	−2.62	78	240.3	18.4	77	135.5	10.3
F horizon	−4.10	626	46.2	30.5	898	15.0	17.8
Soil 0–10 cm	1.59	1494	14.3	18.5	1523	10.0	12.4
Soil 10–25 cm	5.27	1276	12.1	8.4	1429	11.5	7.2
Soil 25–50 cm	6.47	2032	8.8	4.5	2032	9.6	6.3
Current year	−4.92	77	19.1	1.8	90	33.9	3.3
1st year needles	−5.19	22	12.0	0.4	26	19.9	0.6
Twigs	−3.75	2	29.9	0.1	3	42.4	0.1
Branches	−3.99	18	22.1	0.5	22	36.1	0.8
Total ^{15}N retained				83			59

dance units to 3400 ‰ δ^{15}N[1]. The same amount of ^{15}N, 6.8 g ha^{-1} was added to the reduced nitrogen plot resulting in a δ value of about 43000 ‰ δ^{15}N. Both treated and ambient plots were analyzed to determine also any temporal variations in ^{15}N natural abundances.

The fate of the nitrogen was followed into trees, ground vegetation, soil and forest floor, roots, precipitation, soil waters and outflow, litter and microbes.

[1] The isotopic composition is expressed in terms of δ values: δ^{15}N = $((^{15}N/^{14}N$ sample : $^{15}N/^{14}N$ standard)-1)1000‰. The standard is usually the atmospheric N$_2$ standard of 0.3676 atm% ^{15}N.

Results and Discussion

After nine months of ^{15}N manipulation major compartments of the ecosystem were enriched and therefore served as a sink for the added ^{15}N. The soil compartments, for instance, became enriched (Fig 1). δ Values of the forest floor (F) horizon at the low-deposition plot increased from about −4.1‰ up to 46‰ ^{15}N. Vegetation too served as a sink for the added ^{15}N (Fig 2). δ Values for the current year needles increased from −4.9‰ to about 19‰ in the nine months period. Old-

Needles

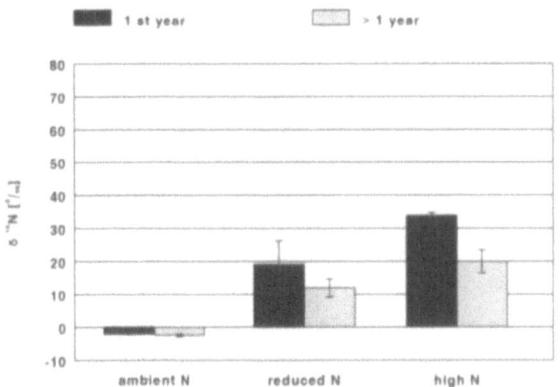

■ 1 st year   ▢ > 1 year

Figure 2. Mean δ^{15}N values of the needles (average of 4 trees, n=3) after nine months of ^{15}N addition to the high and low nitrogen deposition plot. The ambient plot served as a control.

er needles were more depleted in N, indicating the fact that younger needles are more active than the older needles, and thus represent a greater sink for nitrogen.

Mass balancing techniques (Nadelhoffer et al., 1993) and the simulation model NICCCE are required for estimating fluxes into the different ecosystem compartments. Preliminary results on the estimated retention of the ^{15}N-labelled ammonium in the various compartments of the ecosystem show that the forest floor and upper mineral soil (0–10 cm) served as major sinks for the added ^{15}N. The vegetation retained only a limited amount of nitrogen, 3–6% of the added ^{15}N was estimated in the needles, current year twigs and branches after nine months of ^{15}N manipulation. Although the total ^{15}N assimilated by the vegetation may increase if wood, bark and roots are included in the mass balance.

A higher percentage of the added ^{15}N was retained in the plots where nitrogen in the throughfall was reduced compared to the high-deposition plot. About 83% of the added ^{15}N could be accounted for in the soils and vegetation of the Scots pine plot were nitrogen in throughfall was highly reduced (Table 1). In the high-deposition plot about 59% of the ^{15}N could be

accounted for in these compartments, indicating that at this plot less N was retained in the soil and vegetation. Most of the N may be leached out of the soil. This hypotheses will be tested after analysis of soil water will be completed.

Acknowledgements

This study was funded by the EC Commission of European Communities (STEP-CV-90-0056). The authors are also indebted to A. Boxman and H. van Dijk of the department of Ecology of the University of Nijmegen, for providing the research site and their help with the experimental work. We thank T. Veltkamp and B. Beemsterboer of the Netherlands Energy Research Foundation for the ^{15}N analysis work.

References

Boxman A W, Van Dam D, Van Dijk H F G, Hogervorst R F and Koopmans C J 1994 Ecosystem responses to reduced nitrogen and sulphur inputs into two coniferous forest stands in the Netherlands. For. Eco. Manage. 71: 7–29 (*In press*).

Dise N B and Write R F 1992 The NITREX project (Nitrogen Saturation Experiments). Ecosystems Research Report 2. Commission of the European Communities, Brussels, Belgium

Kjønaas O J, Emmett B A, Gundersen P, Koopmans C J and Tietema A 1993 ^{15}N approach within NITREX: 2. Enrichment studies. *In*: Experimental Manipulations of Biota and Biogeochemical Cycling in Ecosystems. Approach, Methodologies, Findings. Eds. L. Rasmussen et al. pp 235–237. Ecosystems Research Report 4. Commission of the European Communities, Brussels, Belgium.

Nadelhoffer K J, Aber J D, Downs M R, Fry B and Melillo J M 1993 Biological sinks for nitrogen additions to a forested catchment. *In*: Experimental Manipulations of Biota and Biogeochemical Cycling in Ecosystems. Approach, Methodologies, Findings. Eds. L. Rasmussen et al. pp 64–70. Ecosystems Research Report 4. Commission of the European Communities, Brussels, Belgium.

Van Dam D and Van Breemen N 1995 NICCCE: A model for Nitrogen Isotopes and Carbon Cycling in Coniferous Ecosystems. Ecological Modelling. 79:255–275.

O. Van Cleemput et al. (eds.), Progress in Nitrogen Cycling Studies, 217–222, 1996.

Release of NH_4^+ from synthetic zeolite P_c added to the soil

František Kovanda and Pavel Ružek

Department of Solid State Chemistry, Institute of Chemical Technology, 16628 Praha 6, Technicka' 5, Czech Republic and Research Institute of Crop Production, 161 06 Praha 6 - Ruzynšek, Drnovská 507, Czech Republic

Key words: ammonium, ion exchange, zeolite

Abstract

The ion exchange reactions that are running in the soil after adding the NH_4-enriched zeolite were observed in model binary solution systems. The synthetic zeolite P_c produced by the hydrothermal alteration from the ashes of the coal power stations has been used. The synthetic zeolite P_c has been enriched in 54 mg NH_4^+-N (in 1 g dry matter) by NH_4Cl solution. In laboratory experiments, the NH_4-enriched zeolite P_c was shaken with K, Na, Ca and Mg solutions and the balance isotherms of the ion-exchange reactions were observed. The experimental temperatures (0, 10 and 22 °C) and the cation concentrations were chosen according to the usual conditions of the soil solutions in the field. The used zeolite P_c was high selective for Ca^{2+} and K^+ ions, less for Na^+ ions, no selectivity was observed for Mg^{2+} ions. The release of NH_4^+ entrapped in NH_4-enriched zeolite applied to the soil was observed also in pot experiments. The soil solutions from pots were sampled at several day intervals and NH_4^+, NO_3^-, K^+, Na^+, Ca^{2+} and Mg^{2+} concentrations were determined. The relatively rapid release of entrapped NH_4^+ from zeolite was observed in respect of the cation exchange ability of the zeolite-soil system and the soil solution.

Introduction

Zeolites represent the group of minerals with unique physical-chemical properties due to their ability to entrap or release different substances as a consequence of cation-exchange reactions, adsorption etc. Application of zeolites into soils can improve some physical-chemical properties of the soil system as pH, cation exchange capacity and water holding capacity (Králová et al., 1994). Natural zeolites as clinoptilolite or phillipsite entrap NH_4^+ ions very well, which can be used in the waste water purification processes (Ciambelli et al., 1988; Collela and Aiello, 1988; Hagiwara and Uchida, 1978). The synthetic zeolite of phillipsite structure can be obtained by hydrothermal alteration from the ashes of the coal power stations (Koloušek et al., 1990). The synthetic zeolite prepared in this way was 2.5 times more effective in the ammonium entrapping from water solutions than the natural clinoptilolite (Kovanda et al., 1994).

Zeolites enriched by NH_4^+ ions after using in the waste water purification processes could be utilized as a cheap nitrogen fertilizer. Regarding to the ion-exchange properties of zeolites and their ability to well

entrap NH_4^+ ions, the utilization of zeolites as slow released fertilizers was considered (Lewis et al., 1984; Pirela et al., 1984; Weber et al., 1984). The preliminary microplot experiments studying the effect of the NH_4-enriched zeolite addition into the soil realized in the Research Institute of Crop Production in Prague did not demonstrate the presumed effect of the slow release of ammonium nitrogen. It is evident that the release of NH_4^+ ions from the zeolite into the soil is influenced by kinetics of cation-exchange reactions between the zeolite and the soil solution as well as by the conditions in which the dynamic equilibrium in the soil system is reached. Therefore, the experiments studying the ion-exchange reactions between NH_4-enriched zeolite and solutions of cations present in significant concentrations in soil solution (K^+, Na^+, Ca^{2+} and Mg^{2+}), were carried out. The followed up pot experiments should answer the question, which processes take place in the soil after addition of NH_4-enriched zeolite and how rapid they are.

Materials and methods

Preparation of zeolite

The synthetic zeolite P_c was prepared by the hydrothermal alteration of power-station ashes produced by combustion of brown coal. This synthesis was realized in NaOH solution and in spite of washing the product was very alkaline (pH of the water extract was equal to 12). Therefore the zeolite was washed with 3% HCl and distilled water. The value of pH was then reduced to pH of 6. Further the zeolite was mixed with 5 M NH_4Cl. The suspension was stirred for 24 h and then filtrated and washed with distilled water. This procedure was repeated 6 times to reach the maximum Na-NH_4 exchange. Distillation of the NH_4-enriched zeolite sample with NaOH solution was used to determine the ammonium content in the zeolite. In this way the content of 69.3 mg NH_4 g^{-1} of dry zeolite was found. A part of ammonium was found in a water extract (2.57 mg NH_4^+ g^{-1} dry zeolite).

NH_4 metal cation exchange measurements in model solutions

Ion-exchange reactions between NH_4-enriched synthetic zeolite P_c and the solutions of KNO_3, $NaNO_3$, $Ca(NO_3)_2$ and $Mg(NO_3)_2$ were studied. The measurements were carried out in the batch arrangement at 0 °C, 10 °C and 22 °C. Different weights of NH_4-enriched zeolite were added to 150 mL of the K, Na, Ca or Mg solution and the suspension was shaken in a closed polyethylene bottle in the thermostatic box for the time required to reach the equilibrium. The final equilibrium concentrations of metal cations in solutions should correspond to levels usually found in the real soil solution. Therefore, the initial concentrations of cations were chosen as follows: 50 mg K L^{-1}, 60 mg Na L^{-1}, 190 mg Ca L^{-1} and 50 mg Mg L^{-1}. Time required for the equilibrium was determined experimentally. Three days were needed for NH_4-K and NH_4-Na exchange and 6 days for NH_4-Ca and NH_4-Mg exchange. After this time the suspension was filtrated. The concentration of metal cations in the solution was determined by means of AAS and the concentration of ammonium was determined by means of FIA. The contents of cations in the zeolite were calculated from the balance equations.

Pot experiments

Pot experiments were carried out under laboratory conditions to study the changes of ion concentrations in soil solution, which occurred after the addition of ammonium nitrogen into the soil. The soil from the locality Prague-Ruzyne (orthic Luvisol) was used in these experiments. There were placed 5 kg of soil mixed with a fertilizer into the pots (calculation on the dry basis). The dose of nitrogen was 1.5 g NH_4^+-N for one pot. Four sets of the experiment were tested: (i) soil mixed with crystallic ammonium sulphate, (ii) soil mixed with the NH_4-enriched synthetic zeolite P_c, (iii) soil mixed with NH_4-enriched natural clinoptilolite, and (iiii) the control set without fertilizer. To avoid the water evaporation from the pots, the soil was covered by a 5 cm layer of washed coarse-grained sand. All pots were weighed and then 800 mL of distilled water to each pot were added. The water content in the soil was kept at 25%. The first sampling of soil solution were carried out 24 h after water addition and then samplings continued in intervals of 3–4 days. The pots were weighed before each sampling and distilled water was added to replace the water losses due to evaporation. The soil solution was sampled directly from pots using the Rhizon Soil Moisture Samplers. The concentrations of K, Na, Ca and Mg were determined using AAS and the concentrations of NH_4^+ and NO_3^- were determined by means of FIA.

Results and discussion

Ion exchange reactions between NH_4-enriched synthetic zeolite P_c and solutions containing K^+, Na^+, Ca^{2+} and Mg^{2+} ions

The observed ion-exchange reactions were very rapid especially in the case of the monovalent cations (K^+ and Na^+). Already after the first two hours the reaction was completed more than 50%. Obvious changes of concentrations of the exchanged ions were observed during the first 48 h, later the ion concentrations practically did not show any further change. The course of the reaction was slower for Ca^{2+} and Mg^{2+} ions, 50% of the reaction was completed after 12–16 h and concentration changes could be observed also after 96 h. On the basis of these results the time required for reaching of the equilibrium was chosen as follows: 72 h for NH_4-K and NH_4-Na exchange and 144 h for NH_4-Ca and NH_4-Mg exchange, respectively. During observed

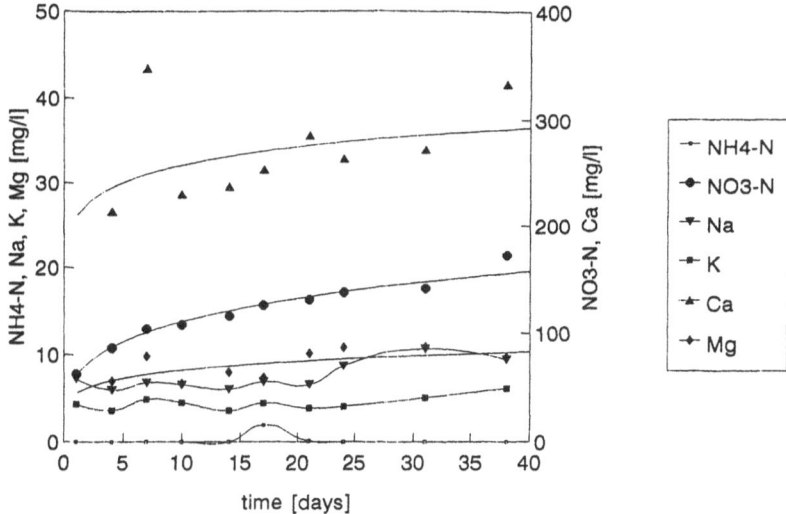

Figure 1. The changes of ion concentrations in soil solution in time observed in pots without NH_4^+-N addition.

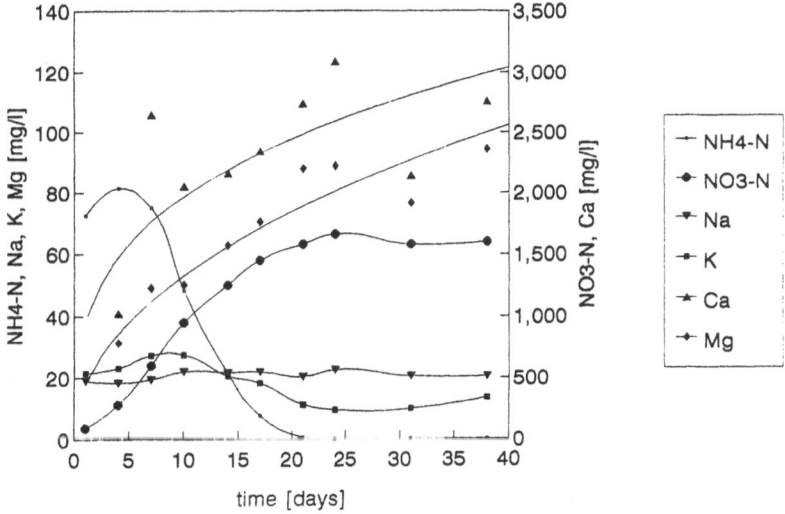

Figure 2. The time relations of ion concentrations in soil solution in pots with addition of ammonium sulphate. (Dose 1.5 g NH_4-N per pot).

ion-exchange reactions not only NH_4^+ ions, but also Na^+ ions were released into the solution. A part of Na^+ ions (about 10%) remained in ion-exchange positions in synthetic zeolite P_c after Na-NH_4 exchange in 5 M NH_4Cl, i.e. the full exchange was not obtained. These remained Na^+ ions participated in ion-exchange reactions together with NH_4^+ ions.

The equilibrium isotherms of the ion-exchange reactions between the NH_4-enriched synthetic zeolite P_c and the solutions of K, Na, Ca and Mg nitrates were measured in the concentration ranges usually found for these ions in the real soil solution. Experimental temperatures were chosen with respect to the real soil temperature during the major part of the year (0, 10

and 22 °C). In the case of the ion exchange in the solution containing K^+ ions at low temperature (0 °C) the synthetic zeolite P_c had approximately the same selectivity to exchanged ions as well as to entering ions. With the increasing temperature the selectivity of the zeolite towards entering ions increased, i.e. the exchange NH_4-K was preferred and NH_4^+ ions passed into the solution. In case of solution containing Na^+ ions the zeolite showed a negative selectivity. Sodium ions occupied about 10% of ion-exchange positions already at the start of reaction but this fact cannot influence the character of the NH_4-Na exchange reaction. It means that in defined experimental conditions a full exchange between NH_4^+ and Na^+ ions could not be

220

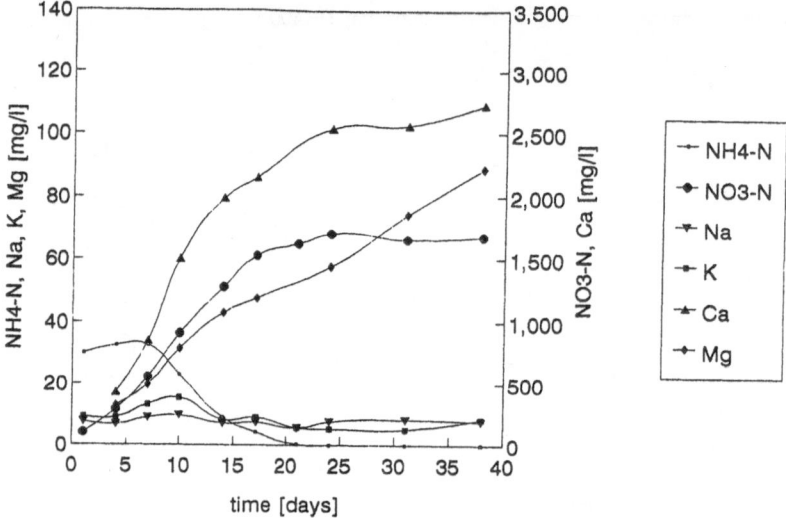

Figure 3. The time relations of ion concentrations in soil solution in pots with addition of NH₄-enriched natural clinoptilolite. (Dose 1.5 g NH₄-N per pot).

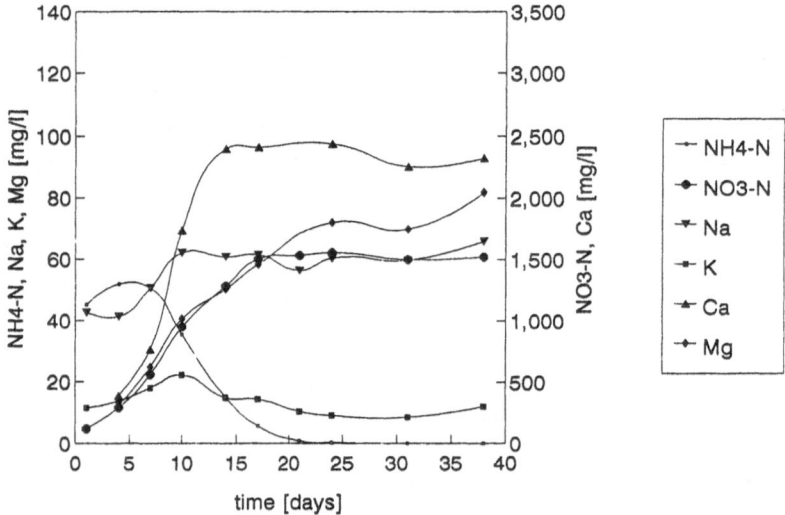

Figure 4. The time relations of ion concentrations in soil solution in pots with addition of NH₄-enriched synthetic zeolite P_c. (Dose 1.5 g NH₄-N per pot).

completed in studied zeolite. In opposition to Na^+ ions, the Ca^{2+} ions were exchanged selectively and practically all NH_4^+ ions present in the zeolite passed into the solution. The exchange between NH₄-enriched zeolite and the solution containing Mg^{2+} ions was absolutely non-selective and in all cases not more than 50% of present NH_4^+ ions were released from zeolite. The measured data indicate, that under experimental conditions defined above the zeolite is not selective towards Mg^{2+} ions.

The results presented above can be summarized as follows: in solutions containing Ca^{2+} and K^+ ions,

the rapid transfer of NH_4^+ ions from the NH₄-enriched synthetic zeolite P_c to solution was preferred under conditions defined above.

Release of NH_4^+ ions from the NH_4 - enriched zeolite in the soil

The way of the release of ammonium-nitrogen from NH₄-enriched zeolite was observed in pot experiments. The pots were placed in the room, where the temperature ranged during the followed time from 15 °C to 30 °C. The set with the NH₄-enriched syntethic zeolite P_c

(54 mg NH_4^+-N g^{-1} of dry zeolite), the set with natural clinoptilolite (Eastern Slovakia) enriched with ammonium (12 mg NH_4-N g^{-1} of dry zeolite) and the set with ammonium sulphate (i.e. without the influence of zeolite) were observed. The dose of nitrogen applied into pots was identical for all sets (1.5 g NH_4-N per pot with 5 kg of dry soil). At the same time the control set without the addition of nitrogen was also observed. Each set consisted of three replication.

The concentrations of NH_4^+-N, NO_3^--N, K^+, Na^+, Ca^{2+} and Mg^{2+} were measured in samples of the soil solution. Figures 1–4 demonstrate the changes of the concentrations of above-mentioned ions in time. No or only minimum amount of NH_4^+-N was found in the soil solution of the control set (Fig. 1). The content of NO_3^--N gradually increased in the soil solution, probably due to mineralization of the soil organic matter and following nitrification. The content of Na^+ and K^+ ions in soil solution varied only a little and remained practically constant. The increase of the concentration of Mg^{2+} and Ca^{2+} ions in the soil solution was remarkable and followed the increase of the NO_3^--N concentration. A quite high concentration of NH_4^+-N in the soil solution (70–80 mg NH_4^+-N L^{-1}) in consequence of the dissolution of crystal ammonium sulphate added into the soil was found in the beginning of the experiment in the set with ammonium sulphate (Fig. 2). But the concentration of NH_4^+-N began to decrease rapidly after 5–7 days and after 20 days was equal to zero. The decrease of the NH_4^+-N concentration was joined by the rapid increase of the NO_3^--N concentration in soil solution, predominantly due to the mineralization of soil organic matter and followed up nitrification. The increase of the NO_3^--N concentration was again accompanied by the increase of Ca and Mg concentrations in the soil solution. These concentrations reached several times higher values than ones in the control set. The concentration of Na^+ remained practically constant. At first the content of K^+ ions in the soil solution gently increased, then followed a small decrease. The concentration changes in soil solution in experimental sets with NH_4-enriched natural clinoptilolite (Fig. 3) and with NH_4-enriched synthetic zeolite P_c (Fig. 4) had practically the same tendency like the set with ammonium sulphate. Relatively high concentration of NH_4^+-N in the soil solution in the initial period of the experiment (first 5–7 days) testifies to a rapid release of NH_4^+ ions from the zeolite due to ion-exchange reactions with the soil solution. Then probably the occured nitrification of the released NH_4^+-N and a rapid decrease of its concentration in the soil

solution followed. In opposite to the other experimental sets, in case of the set with the synthetic zeolite P_c, a significant increase of the Na concentration in the soil solution could be observed. As already presented above, also Na^+ ions remaining in synthetic zeolite P_c after ammonium enrichment, participated in ion-exchange reactions. The full exchange of all Na^+ ions present in the zeolite was accomplished already after approximately 7 days as latter the concentration of Na^+ ions in the soil solution remained stable.

It can be concluded that in the soil, a rapid ion-exchange reaction between NH_4^+ ions present in the NH_4-enriched zeolite and cations present in soil solution, especially of Ca^{2+} and K^+ ions, takes place. For this reason, the influence of zeolites as a factor retarding the release of ammonium-nitrogen into the soil did not manifest and therefore the utilization of ammonium enriched zeolites as slow-released nitrogen fertilizer is questionable.

References

Ciambelli P, Corbo P, Liberti L and Lopez A 1988 Ammonium recovery from urban sewage by natural zeolites. In Occurence, Properties and Utilization of Natural Zeolites. Eds. D Kalló and H S Sherry. pp 501–510. Akadémiai Kiadó, Budapest, Hungary.

Collela C and Aiello R 1988 Ammonium removal from tannery sewages by selective ion exchange using natural phillipsite. In Occurence, Properties and Utilization of Natural Zeolites. Eds. D Kalló and H S Sherry. pp 491–500. Akadémiai Kiadó, Budapest, Hungary.

Hagiwara Z and Uchida M 1978 Ion-exchange reactions of processed zeolite and its application to the removal of ammonia-nitrogen in wastes. In Natural Zeolites: Occurence, Properties, Use. Eds. L B Sand and F A Mumpton. pp 463–470. Pergamon Press, New York, USA.

Koloušek D, Procházková E, Seidl V and Šmejkalová M 1990 Zpusob přípravy zeolitu. Czech Patent Application No 04530–90R, C–01–B–33/34.

Kovanda F, Koloušek D, Seidl V, Procházková E and Obšasníková J 1994 Utilization of synthetic phillipsite as a means for decreasing the concentration of ammonium ions in waste water. Ceramics-Silikáty 38, 75–79.

Králová M, Hrozinková A, Ružek P, Kovanda F and Koloušek D 1994 Synthetic and natural zeolites affecting the physical-chemical soil properties. Rostl. Výroba 40, 131–141.

Lewis M D, Moore F D and Goldsbery K L 1984 Ammonium-exchanged clinoptilolite and granulated clinoptilolite with urea as nitrogen fertilizers. In Zeo-Agriculture: Use of Natural Zeolites in Agriculture and Aquaculture. Eds. W G Pond and F A Mumpton. pp 105–111. Westview Press, Boulder, Colorado, USA.

Pirela H J, Westfall D G and Barbarick K A 1984 Use of clinoptilolite in combination with nitrogen fertilization to increase plant growth. In Zeo-Agriculture: Use of Natural Zeolites in Agriculture and Aquaculture. Eds. W G Pond and F A Mumpton. p. 113. Westview Press, Boulder, Colorado, USA.

Weber M A, Barbarick K A and Westfall D G 1984 Application of clinoptilolite to soil amended with municipal sewage sludge. *In* Zeo-Agriculture: Use of Natural Zeolites in Agriculture and Aquaculture. Eds. W G Pond and F A Mumpton. pp 263–271. Westview Press, Boulder, Colorado, USA.

O. Van Cleemput et al. (eds.), Progress in Nitrogen Cycling Studies, 223–229, 1996.
© 1996 Kluwer Academic Publishers.

Maize nitrogen fertilization in fluvisols of the Lower Mondego Valley - field trials

Maria Isabel Ferreira Magalhaes-Martins
Direccao Regional de Agricultura da Beira Litoral, Avenida Fernao de Magalhaes, 465, 3000 Coimbra, Portugal

Key words: curve-plotting, field trials, leaf analysis, maize, nitrates, nitrogen

Abstract

With the main purpose of increasing the knowledge about the nitrogen fertilization of a very high productive variety of maize in the Lower Mondego Valley, trials were carried out under field conditions, with the application of five nitrogen levels (0, 100, 200, 300 and 400 kg N ha^{-1}), during two years (1992 and 1993).

The field trials included the analysis of the production, the nutritive conditions of the maize plants, the NO_3^--N content in the soil, and the precipitation. It was found that (1) the contribution of the mineralisable organic nitrogen can be very important, depending on the climatic conditions, (2) the total nitrogen in the leaves, at two sampling dates, showing a significant correlation with yield, and (3) the most economic amount was between 250 and 320 kg N ha^{-1}, for productions of 14.4 and 12.4 t ha^{-1}, respectively, for low and high precipitation in winter and spring. It also influenced the nitrate-N content in soil.

Introduction

In the last decade, the Portuguese regions of greatest aptitude for maize cultivation have witnessed a notable increase in the productivity, based mainly on the utilization of higly productive strains. The region of Lower Mondego (of about 14,000 ha) is located at the coastal-centre zone (with strong influence of the Atlantic climate), at the agrarian region of Beira Litoral, comprising the Valley of the Mondego River itself (from Coimbra to Figueira da Foz) and the valleys of its tributaries. The yield of this crop at this region, after several tests in alluvial soils of high fertility, reached levels higher than 14,000 kg ha^{-1}, being 10,000 kg ha^{-1} the mean productivity of the crop in this region.

To get these high yields, it is necessary to establish with great strictness the answer of these highly productive strains of maize to the nitrogeneous fertilization in order to avoid the use of excessive amounts which could result in adverse effects to the environment and human health, and which could also reduce the yield of the crops.

One experiment with maize (grain) carried out for two years (1992 and 1993), in an alluvial soil, with a silt loam texture, representing about 45% of the soils

of the Lower Mondego Valley, allowed the increase of the N fertilization knowledge of the maize crop in this type of soils (Magalhães Martins, 1994).

Indeed, in previous tests carried out with similar soils, very high yields were obtained, during several years, without any use of N fertilizers, which shows the need for a better knowledge of some aspects of the nutrition and fertilization of the maize plant. This study allowed to establish the curve-plotting from a strain of hybrid maize (FAO 600), highly productive during two years under very different climatic conditions. In 1992, the weather conditions were quite normal and very favourable for a good growing of the crop. In 1993, there were very intensive rains during the months of April, May, June, September and October, which affected the availability of the nitrogen in soil, the growth of the crop and the yields obtained.

The main purposes of this study were: (1) determination of the best economical level of the N fertilization of maize in a representative soil of Lower Mondego Valley, (2) evaluation of the nutritive state of the crop using the leaf diagnostic as a means of checking the availability of the nutrients in the soil, (3) knowledge of the evolution of the soil N content, as another means of foreseeing the crops needs and improve fertilization recommendations.

Table 1. Mean temperature and rainfall occurred in 1941–79, 1992 and 1993, at Coimbra

Months	Mean temperature (° C)			Rainfall (mm)		
	1941–1970	1992	1993	1941–1970	1992	1993
January	9.5	5.7	5.7	124.0	65.8	34.6
February	10.2	8.2	9.0	89.0	27.4	51.9
March	12.8	10.4	11.3	129.0	37.2	22.9
April	14.5	13.6	11.8	72.0	89.2	126.1
May	16.3	18.0	14.8	70.0	40.2	153.9
June	19.3	17.7	18.5	36.0	40.7	51.0
July	21.0	19.7	20.2	10.0	6.2	0.5
August	21.2	18.5	18.6	17.0	50.6	0.3
September	20.0	16.8	16.6	43.0	37.1	115.8
October	16.4	13.4	14.7	83.0	114.8	254.8
November	12.8	11.9	10.5	102.0	43.7	151.7
December	10.0	9.8	9.7	140.0	112.2	36.2
Mean and total annual	15.3	13.6	13.4	915.0	665.1	999.7

Table 2. Mean yields per treatment obtained in 1992 and 1993. Summary of analysis of variance between the grain yields. F estimated and its significance, corresponding to the sources of variation

Treatments	Grain yield (kg ha^{-1})		Source of variation	1992		1993	
	1992	1993		f.d.	F estimated	f.d.	F estimated
N0	11,923	6,253	Total	19		14	
N1	13,381	9,472	Nitrogen	4	6.29 **	4	13.80 **
N2	14,404	11,378	Replications	3	1.45 NS	2^a	2.87 NS
N3	14,072	12,216	Error	12		8	
N4	14,385	12,462	C.V.(±)(%)		6.1		11.6
			L.S.D. (±) (kg)	1354		2262	

C.V.-coefficient of variation; LSD -Least Significant Difference, between N level; N.S. - $p > 0.05$; ** - $p \leq 0.01$.
[a] In 1993, one replication was abandoned owing to the weather conditions.

Materials and methods

The experiments were done on a representative Eutric Fluvisol of alluvial type, in a field with the following geographical coordinates - latitude 40° 13' 35"N, longitude 8° 29' 32"W and altitude 10 m, which for many years was cultivated with rice. The soil has a silt loam texture with water table located at depths between 0.90 and 1.20 m.

Before starting the experiments, a composite soil sample was taken at several depths. The layer 0 to 0.30 m shows a pH (H$_2$O) value of 5.5, a medium content of organic matter (2.5–3.0%), a medium content of available phosphorus (45–60 mg kg^{-1} P$_2$O$_5$, Egner-Rhiem method) and a high content of available potassium (90–130 mg kg^{-1} K$_2$O, Egner-Rhiem method). The

typical temperature and rainfall of the region (mean of 1941 to 70) and those occured in 1992 and 1993, are shown in Table 1.

The fertilizations consisted of five N levels (0, 100, 200, 300 and 400 kg of N ha^{-1}) with four replications. The experimental layout was a randomized complete block design. The experiments were done for two consecutive years, running on the same plot. Each plot, 6.75 m wide and 15 m long, was sown with nine rows of maize, of the cultivar BLACK (a simple hybrid, FAO 600), in a spacing of 0.75 × 0.165 m (80, 800 seeds ha^{-1}). The yields were evaluated with the three central rows, on each plot.

The rate of fertilization, before sowing, was 135 kg ha^{-1} of P$_2$O$_5$, 50 kg ha^{-1} K$_2$O and one third of nitrogen. The remaining two thirds of nitrogen, of each

Table 3. Mean contents per treatment of total nitrogen (N) and nitrates ($N-NO_3$) in the dry matter on two seasons of plants sampling, during the two years of the trial: the entire plant, at the fourth-leaf stage and the leaf opposite and just below the primary ear shoot, while the silks were green

Treat-ments	Total nitrogen		- N (%)		Nitrates $N-NO_3$		$(mg.kg^{-1})$	
	Entire plant		Leaf near ear shoot		Entire plant		Leaf near ear shoot	
	1992	1993	1992	1993	1992	1993 [a]	1992	1993 [a]
N0	2.42	3.03	2.41	1.50	1 092	1 800	681	·267
N1	3.08	3.22	2.70	2.17	947	1 397	739	199
N2	4.27	3.48	2.88	2.70	1 020	2 850	796	166
N3	3.88	3.39	2.74	2.79	1 305	2 376	739	165
N4	3.32	3.54	2.70	3.01	1 293	2 928	739	214
Mean	3.40	3.33	2.68	2.43	1 131	2 270	739	202

[a] Mean from some determinations in some treatments and replications, therefore it wasn't made the analysis of variance.

Table 4. Summary of analysis of variance between the mean contents of total nitrogen and nitrates, in the dry matter of the plants sampling, in 1992 and 1993. F estimated and its significance, corresponding to the sources of variation

Source of variation	g.l.	1992				g.l.	1993 [b]		
		Total Nitrogen -N		Nitrate - NO_3-N			Total Nitrogen - N		
		Entire plant	Leaf near ear shoot	Entire plant	Leaf near ear shoot		Entire plant	Upper leaf	Leaf near ear shoot
Total	19					14			
Nitrogen	4	0.84 NS	2.07 NS	0.42 NS	1.62 NS	4	1.56 **	9.76 **	33.9***
Replications	3	5.01 *	2.41 *	12.9***	4.57 *	2	48.5***	0.39 NS	1.18 NS
Error	12					8			
C.V.($\pm$)%		46.3	8.9	43.9	8.7		8.7	11.4	7.4

N.S. $p>0.05$; ** $p \leq 0.01$; *** $p \leq 0.001$.
[b] In 1993, it wasn't made the nitrates analysis of variance.

treatment, were applied as top-dressing, before raising, at the 8th leaf stage. The necessary phytosanitary treatments were done to control the main pests existing in this field.

Irrigation was done by independent furrows in each treatment. Each year three irrigations were done with variable amounts of water from 540 m^3 ha^{-1} to 930 m^3 ha^{-1}. The harvest, threshing and evaluation of the yields were done during the last ten days of October and the first fortnight of November. The grain yields took to account an humidity of 14%.

Plant sampling was done at the following growth stages: (1) at the fourth-leaf stage, cutting the entire plant of 25 plants per treatment, (2) one week before tasseling, the upper leaf (the youngest leaf) was sampled, of 20 plants per treatment, (3) the leaf opposite and just below the primary ear shoot, of 20 plants per treatment, while the silks of the ear shoots were green.

The soil sampling was done before top-dressing with N, several times during the growth season and after the maize harvest, at several depths.

In plant and soil samples, mean values of total-N and nitrate-N were determined, per treatment. Total-N was evaluated by Kjeldahl standard procedure (AOAC, 1975) and nitrate-N were determined by the potentiometric method of the ion-specific electrode (Byrne, 1979; Carranca, 1988) and by the hydrazine reduction method with an autoanalyzer.

Data were checked using ANOVA and correlation analysis. The mean value of treatments were evaluated using the LSD multiple range test and regression analysis to estimate the curve-plotting to the nitrogen applications. The homogeneity of regression coefficients was tested using F-test.

Results and discussion

Analysis of yields

Table 2 shows the mean yields per treatment obtained in 1992 and 1993, and the summary of the analysis of variance. During the two years of the experiment the effect of the N applications was highly significant ($p \leq 0.01$).

In 1992, without any N application, the mean yield in the treatment with N0 reached about 12,000 kg ha^{-1}, being smaller than the yields obtained with the four levels of N application, not importables varying among each other. In 1993, in the treatments N0, the mean yield was about 6,250 kg ha^{-1}, which was about half that obtained in 1992. However, the difference between the maximum yields in these two years was smaller (about 2,000 kg ha^{-1}).

It was shown that the climatic conditions had a great influence on the yields. In 1992, the rainfall in winter and spring was quite lower than normal, which could lead to the accumulation of organic nitrogen mineralized in the soil, justifying the yields of about 12, 000 kg ha^{-1} without N fertilization. In 1993, the high rainfall in April, May and June may have caused a high N leaching, specially that which resulted from the mineralization of organic nitrogen.

In this soil the contribution of the mineralizable organic nitrogen seems to have been much greater in the year with a reduced rainfall in spring. But in the year with a rainy spring it seems that there was a greater loss of mineralizable nitrogen and the importance of fertilization was much greater. In 1992, the efficiency of the N fertilization was 8 kg maize per 1 kg of N applied. In 1993, the N efficiency was double of the previous year, reaching about 17.6 kg maize per 1 kg of N applied. According to Mengel and Kirkby (1987), a high efficiency of N fertilization is obtained when the increment of yield per unit of fertilizer applied is high. This usually happens when the content of available nitrogen in the soil is low and the rate of nitrogen application is not too high.

These results are in agreement with several authors, namely Recous et al. (1988a, b), Macdonald et al. (1989), Powlson et al. (1992) who state that the availability of nitrogen in the soil is ruled by the mineralization of the organic nitrogen, more than by the quantity of N fertilizer used. They also stated that the N fertilizer applied is absorbed by the crops during a short period, while the absortion of the nitrogen in soil is effective during the growing period. Powlson et al.

(1992) stressed that the total losses of nitrogen were influenced more by the climates than by the type of soil, or the cultivation history. Specially the great rainfall during the three weeks after the application of the nitrogen fertilizers were important.

The equations of regression which best fit to mean yields obtained in 1992 and 1993 are as follows:

First year – 1992
$$Y = 11992 + 16.04N - 0.02608N^2$$
$$R^2 = 94.5\% \text{ N.S.} (p > 0.05)$$

Second year – 1993
$$Y = 6321 + 35.20N - 0.050098N^2$$
$$R^2 = 99,8\% \ **(p \leq 0.01)$$

Figures 1 and 2 show the curves correlating maize yield with the amount of N applied. For each year, the maximum production of the cultivation (agronomic answer) and the economic optimum (economic answer) were calculated:

1992
$$N\mathit{max}. = 307 \text{ kg ha}^{-1}; Y\mathit{max}. = 14\,458 \text{ kg ha}^{-1};$$
$$N\mathit{optim}. = 253 \text{ kg ha}^{-1};$$
$$Y\mathit{econ}. = 14\,380 \text{ kg ha}^{-1}$$

1993
$$N\mathit{max}. = 351 \text{ kg ha}^{-1}; Y\mathit{max}. = 12\,505 \text{ kg ha}^{-1};$$
$$N\mathit{optim}. = 323 \text{ kg ha}^{-1};$$
$$Y\mathit{econ}. = 12\,465 \text{ kg ha}^{-1}$$

Analysis of the mean content of the nutrients in the vegetable material

The results of the contents in total nitrogen and nitrates in the samples of the vegetable material collected in 1992 and 1993 are shown in Tables 3–5, as well as a bridged analysis of variance.

In 1993, the several levels of N applied had a highly significant effect on the contents of nitrogen, at the second sampling of vegetable material (last leaf) and on the nitrogen content at the third sampling, in relation to the leaf opposite to the main ear. Regarding

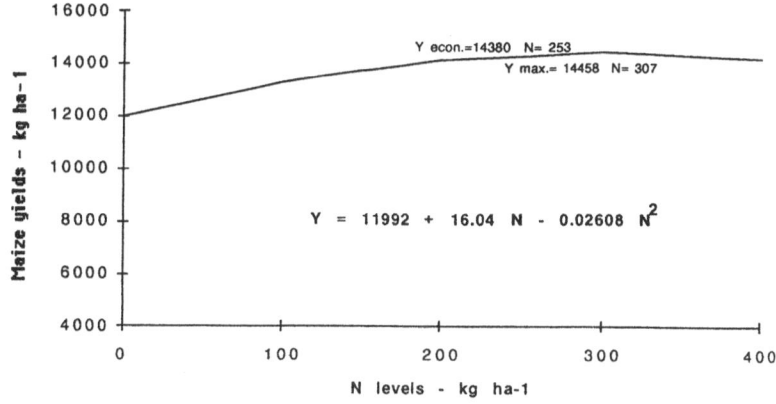

Figure 1. 1992 - Curve-plotting to the nitrogen applications

$$Y = 11992 + 16.04 \ N - 0.02608 \ N^2$$

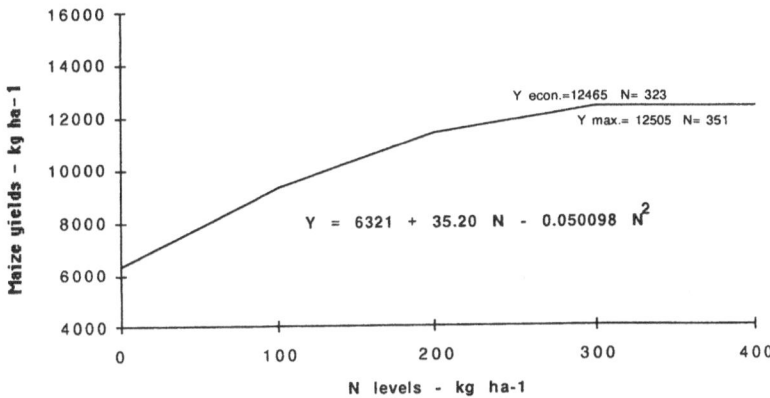

Figure 2. 1993 - Curve-plotting to the nitrogen applications

$$Y = 6321 + 35.20 \ N - 0.050098 \ N^2$$

Table 5. Mean contents per treatment of total nitrogen (N), nitrates (N-NO₃), in the dry matter of the plants sampling in 1993: the upper leaf (the youngest leaf), one week before tasseling

Treatments	N0	N1	N2	N3	N4	Mean
Nitrogen - N (%)	1.78	2.11	2.57	2.87	2.97	2.46
Nitrates [a] - N-NO₃ (mg kg^{-1})	194	172	181	187	200	187

[a] Mean from some determinations in some treatments and replications, therefore it wasn't made the analysis of variance.

the nitrogen content of the ear's leaf, the majority of authors suggest, as the critical level, a value of about 3.0% (between 2.7 and 3.1%).

It was noticed that in 1992, in the treatments without a significant answer, the N contents showing in the leaf varied between 2.7 and 2.9%, and in the level N0 a mean content of 2.4%. In 1993, the leaves at N0

had a mean content of 1.5%, the plants showing typical symptoms of N deficiency. At level N1 the mean N contents were around 2.17%, showing a significant answer to nitrogen applications. From level N2, where the nitrogen contents in the leaf reached 2.7% there was no significant yield answer to the N application.

Correlation matrices were set up for the two years between yields and nutrient contents in the plants of each of the samples of vegetable materials (Tables 6 and 7).

Both in 1992 and 1993, the total nitrogen concentrations in the leaf just below and opposite to the upper ear shoot, sampled during flowering, were significantly correlated ($p \leq 0.01$) with the yields. The same occurred with the second sampling in 1993 (the youngest leaf), one week before tasseling. In the two years, the higher correlation coefficients, noticed in all plants sampled, occurred between potassium and magnesium contents, being always very negatively significant ($p \leq 0.01$). This confirms the ionic antagonism

Table 6. Correlation matrices in 1992, between yields and nutrients contents in the plants in two samples of vegetable material: 1st sampling - entire plant, with four leaves; 2nd sampling - leaf opposite and just below the primary ear shoot, during tasseling

	Yield	Total N	Phosphorus	Potassium	Calcium	Magnesium	Nitrate
1st sampling							
Total N	0.2553						
Phosphorus	0.2981	0.3811					
Potassium	0.0845	0.2549	0.4145				
Calcium	0.5880**	0.2709	0.4665*	0.1807			
Magnesium	0.0430	-0.2510	-0.2921	-0.6936**	-0.1173		
Nitrate	0.3346	0.6854**	0.4942*	0.3546	0.5706**	-0.2281	1.0000
2nd sampling							
Total N	0.6244**						
Phosphorus	0.3984	0.4703*					
Potassium	0.4340	0.2193	0.6252**				
Calcium	0.1479	0.4533*	0.5822**	0.4299			
Magnesium	-0.3866	-0.1732	-0.4467*	-0.9311**	-0.2422		
Nitrate	0.4652*	0.4157	0.2831	0.4992*	0.3854	-0.5208*	1.0000

*$p \leq 0.05$; ** $p \leq 0.01$.

Table 7. Correlation matrices in 1993, between yields and nutrients contents in the plants in two samples of vegetable material: 1st sampling - entire plant, with four leaves; 2nd sampling - the upper leaf (the youngest leaf), one week before tasseling; 3rd sampling - leaf opposite and just below the primary ear shoot, during tasseling

	Yield	Total N	Phosphorus	Potassium	Calcium	Magnesium
1st sampling						
Total N	0.4847					
Phosphorus	-0.6109*	-0.1557				
Potassium	-0.0432	0.6391*	0.3332			
Calcium	-0.0789	-0.0310	-0.1096	-0.0403		
Magnesium	0.0217	-0.8165**	-0.2348	-0.8576**	0.1207	1.0000
2nd sampling						
Total N	0.8051**					
Phosphorus	0.6618**	0.8051**				
Potassium	-0.1198	-0.1691	-0.4877			
Calcium	0.0640	0.2389	0.3914	-0.3787		
Magnesium	0.2537	0.1958	0.4419	-0.8371**	0.3571	1.0000
3rd sampling						
Total N	0.9204**					
Phosphorus	0.0377	0.1875				
Potassium	0.1951	0.0118	-0.7163**			
Calcium	0.5807*	0.6965**	0.2683	-0.2764		
Magnesium	0.4852	0.6500**	0.5688*	-0.6383**	0.7425**	1.0000

* $p \leq 0.05$; ** $p \leq 0.01$.

between these two cations, which was also found by several authors.

Nitrate content in the soil

In 1992, before starting the experiment, the nitrate-N content in the layer 0–0.30 m was about 20 mg NO_3^--N kg^{-1}. On 9 July (before top-dressing fertilization) it ranged from 3.8 mg NO_3^--N kg^{-1} at N0 to 6.6mg NO_3^--N kg^{-1}at N4. On 27 August, it ranged from 8.4 to 9.3 mg NO_3^--N kg^{-1}, at those two treatments, and on 3 December, from 1.0 to 2.5 mg NO_3^--N kg^{-1}. In 1993, the nitrate-N content on 18 May (before fertilization and sowing) ranged from 3.7 mg NO_3^--N kg^{-1}at N0 to 6.3 mg NO_3^--N kg^{-1} at N3 and N4 (applied the previous year). On 27 August it ranged from 11.0 mg NO_3^--N kg^{-1} at N0, 13.5 mg NO_3^--N kg^{-1} at N1, 29.2 mg NO_3^--N kg^{-1} at N2, 34.1 mg NO_3^--N kg^{-1} at tN3 and 48.5 mg NO_3^--N kg^{-1} at N4. On 21 December, the range was from 2.6 to 4.9 mg NO_3^--N kg^{-1}.

The contents of NO_3^--N, in 1992, at sowing time, were available between 15 to 20 mg NO_3^--N kg^{-1}, which means about 50 to 90 kg NO_3^--N ha^{-1} in the layer 0 to 0.30 m. In 1993, this values were much lower.

The samples collected in August also reflected the different climatic conditions in the two years. In 1992, the rainfall occurred in July and August, reaching about 57 mm, while in 1993 it was about 0.8 mm.

In 1993, the nitrogen contents ranged from about 10 to 49 mg NO_3^--N kg^{-1}, at N0 and N4. It means, by the end of August there was in soil between 38 and 188 kg NO_3^--N ha^{-1}, in spite of two irrigations.

In December, in both years, practically all nitrate in the soil had been leached, in spite of the rainfall during autumn being very different (195 mm in 1992 and 531 mm in 1993).

Conclusion

The economic optimum, for this type of soil, at extremely different conditions, is situated between 250 and 320 kg ha^{-1}, respectively for the low and high rainfall occurring in winter and spring.

The contribution of the mineralizable organic nitrogen in the soil, in spring, can be rather high, depending on the climatic conditions, mainly the rainfall occurring in winter and spring. This may have influenced the nitrate concentrations in the soil and the yields obtained, specially those where no nitrogen fertilizer was applied.

The results of the analysis of leaves showed that the total nitrogen contents, in the youngest leaf, harvested before the appearance of the panicula and in the leaf immediately below and opposite to the high ear, harvested when the silks were still green, had a very good correlation with the yields, which may be considered as a good indicator of the maize nutritive state. In these soils, with a high level of fertility, and with a water table rather near the surface, the curve-plotting, based on the criteria of economical optimum, may lead to the recommendation of high fertilization. It must bearing in mind, besides fertilizing profitability, also the impact on the environment, specially the contents of nitrates in soil and in the water table.

References

Association of Official Analytical Chemists (AOAC) 1975 Official Methods of Analyses, 12th ed. AOAC, Washington DC,USA.

Byrne E 1979 Chemical Analysis of Agricultural Materials. An Foras Talunntais, Jonhstown Castle Research Centre, Wexford, Ireland. 320 p.

Carranca C F 1988 Determinação do teor em nitratos pelo método do eléctrodo selectivo, em solos, plantas e águas. Pedologia 23, 135–153.

Macdonald A J, Powlson D S, Poulton P R and Jenkinson D S 1989 Unused fertiliser nitrogen in arable soils - its contribution to nitrate leaching. J. Sci. Food Agric. 46, 407–419.

Magalhães Martins M I F 1994 A fertilizaçào azotada do milho em solos do Baixo Mondego. Thesis for M.Sc. *In* Plant Nutrition, Soil Fertility and Fertilization. Instituto Superior de Agrononia, Universidade Tècnica de Lisboa, Lisboa, Portugal. 203 p.

Mengel K and Kirkby E A 1987 Principles of plant nutrition. 4th ed. International Potash Institute, Bern, Swiss. 687 p.

Powlson D S, Hart P B S, Poulton P R, Johnston A E and Jenkinson D S 1992 Influence of soil type, crop management and weather on the recovery of ^{15}N-labelled applied to winter wheat in spring. J. Agric. Sci. Cambridge 118, 83–100.

Recous S, Fresneau C, Faurie G and Mary B 1988a The fate of labelled ^{15}N urea and ammonium nitrate applied to winter wheat crops. I. Nitrogen transformations in soil. Plant and Soil 112, 205–214.

Recous S, Machet J M and Mary B 1988b The fate of labelled ^{15}N urea and ammonium nitrate applied to winter wheat crops. II. Plant uptake and N efficiency. Plant and Soil 112, 215–224.

O. Van Cleemput et al. (eds.), Progress in Nitrogen Cycling Studies, 231–235, 1996.
© 1996 *Kluwer Academic Publishers.*

Nitrogen uptake by *Brachiaria* spp. and its effects on soil mineral N transformations

Cesar H.B. Miranda[1], Segunda Urquiaga[2], Georg Cadisch[3] and Ken E. Giller[3]
[1]*EMBRAPA-CNPGC, CP154, 79. 002, 970, Campo Grande, MS, Brazil,* [2]*EMBRAPA-CNPAB, Seropedica, RJ, Brazil and* [3]*Wye College, University of London, Wye, Ashford, Kent TN25 5AH, UK*

Key words: Brazilian Cerrados, mineralization, nitrification, tropical grasses

Abstract

An experiment with six *Brachiaria* spp. (1 variety and 1 ecotype of each *B. decumbens, B. humidicola* and *B. brizantha*) was conducted on a Dark Red Latosol from the Brazilian Cerrados, to determine plant N uptake potential and the possible influence of the presence of plants on the soil mineral N transformations (immobilization and mineralization-nitrification). After 60 days of growth under controlled conditions, 3 replications from the planted and bare soil were harvested and sampled for moisture, pH, ammonium, nitrate, and total N determinations. Soil cores were also carefully collected for a further 7 day incubation. The plant shoots and roots were collected, and their dry weight and total N content determined. To another 6 replications of each treatment an ammonium solution was added, and further harvests were made 3 and 12 days after the N fertilizer addition. There were significant differences between species on the rates of N uptake and N use efficiency, with both varieties of *B. decumbens* showing the best performance. This fast N uptake may be an advantageous factor for growth and competition in low N-providing soils. After the fertilizer N addition, immobilization was the main pattern of the soil mineral N transformations in the presence of the plants (as determined at each harvest) and in their absence (as determined in the soil cores incubated for further 7 days). There was no indication that the species tested inhibit nitrification in this soil. Soil N transformations were dynamic, varying over a short period of time. Therefore, conclusions about effects of plants on mineralization-nitrification based on a single harvest, or harvests largely spaced in time, could be misleading.

Introduction

Brachiaria is a genus of tropical grasses well adapted to soil and climatic conditions of the Brazilian Cerrados (the savannah-like area of Central Brazil). However, few species of this genera are commonly used, and there is an urgent necessity for new varieties. To address that demand, a program of selection and breeding is underway at the National Centre for Beef Cattle Research (CNPGC), EMBRAPA, Campo Crande, MS, Brasil (Valle et al., 1993a). The mineral nutrition of these grasses is one aspect being examined in detail in this program, especially nitrogen (N). N is considered to be the main limiting factor to production in the area in the long-term (Seiffert, 1981), although phosphorus is more necessary for establishment. As nitrogen fertilizers are uneconomical for use by most of the farmers in these extensive rangelands, and the use of forage legumes is uncommon, the plants have to rely on an efficient cycling of N to obtain their nutritional requirements. As these grasses evolved in areas of low-N providing soils, they may have developed the ability to regulate mineralization or immobilization processes. Thus, they would have an advantage in competition with soil biomass for the soil mineral N. Inhibition of nitrification is one aspect sometimes associated with tropical grasses. Sylvester-Bradley et al. (1988), for example, reported a lower concentration of nitrate in soils under *B. humidicola* than the same soil under *B. decumbens* or *B. dyctioneura*. If the plant is able to reduce nitrification, it may be maintaining in the system a proportion of N that otherwise could be lost by leaching, which is potentially severe in this area during the rainy season.

In this paper an experiment is described in which 3 different ecotypes and 3 commercial varieties of

Brachiaria were tested for their response to a sudden change in soil mineral N availability, simulating a flush of N mineralization. In parallel, evaluations of the transformations occurring in the soil mineral N pool were followed, either by direct measurements in the presence of the plants, or after incubation of extracted soil cores.

Materials and methods

Sixty-three pots with 300 g of a Dark-Red Latossol soil from the Cerrados of Brazil were prepared for this experiment. This soil has a content of 51% sand, 13% silt, and 36% clay; a pH in water of 5.3; 2.45% organic matter, 0.10% total N, and 1.2% total C; 0.7, 0.7, 0.6, 012, and 0.01 cmol$^{(+)}$L^{-1} of Ca, Mg, K, and P, respectively. During the preparation, a basic fertilization with macro (except N) and micro-nutrients was provided, as described by Miranda (1994). After 2 weeks of soil incubation they were separated into sets of 9 pots each. One set was left bare and the remaining sets each received 3 pre-germinated seedlings of one of the following species: the Brazilian commercial varieties of *B. decumbens*, *B. brizantha* and *B. humidicola*. And the ecotypes *B. decumbens* D1; *B. brizantha* B44; and *B. humidicola* H6.

The experiment was conducted in a growth-chamber, with 12 h of light (150 μE) per day, and temperature varying from 30 °C by day to 25 °C by night. The soil moisture was adjusted daily to 25% (w/w), which is equivalent to the field moisture capacity of this soil.

Sixty days after the transplanting 3 pots from each treatment were harvested. To the remaining ones a solution containing 45 mg of N as $(NH_4)_2SO_4$ enriched with 4.524 atom% ^{15}N excess was added, and further harvests (3 replications each time) were done 3 and 12 days after the N addition. At each harvest the following parameters were evaluated: shoot and root dry weight, total N and ^{15}N enrichment (ground subsamples analysed directly in a Roboprep CN coupled to a mass spectrometer Micromass 622); the soil pH, moisture, total N and ^{15}N enrichment; initial content of mineral N and after 7 days of incubation of soil cores. For the mineral N, soil samples varying from 30 to 50 g each were shaken for 2 h with a 2 N KCl solution (ratio 3 to 5:1), filtered and analysed for ammonium and nitrate using colourimetric methods. Further details may be found in Miranda (1994). The total mineral N was considered as the sum of both fractions.

An index of N uptake efficiency of the roots was determined in parallel, using the concept of specific absorption rate (SAR) described by Hunt (1990), expressed by the following equation:

$$SAR = \frac{(M_2 - M_1)}{H_2 - H_1)} \times \frac{(\log_e R^2 - \log_e R_1)}{(R_2 - R_1)}$$

Where M is the whole plant N content, H is the harvest (day), R is the root dry weight, and the subscripts 1 and 2 indicates sequential harvests. As the harvests were destructive, the value of an individual sample on a given harvest was compared against the mean of the replications at the previous harvest, to estimate the variation between samples. Using the ^{15}N and total N data, the plant fertilizer use efficiency (% FUE) was determined, using the following equation:

$$\%FUE = \frac{(\text{sample atom \% } ^{15}\text{N excess} \times \text{plant total N})}{(\text{fert atom \% } ^{15}\text{N excess} \times \text{N application rate})} \times 100$$

Results and discussion

Plant production and N uptake

The species tested showed considerable differences in their dry weights 60 days after planting, indicating different rates of establishment in this sandy soil. *B. brizantha* showed the largest initial weight (Table 1), while *B. humidicola* had the lowest initial dry weight. All species increased their total dry weight in relation to the initial weight immediately after the N addition, indicating a fast assimilation of the added N.

All species showed a similar N specific absorption rate 3 days after the N addition (Table 2), with a high variability in the data. There was a tendency, however, for both *B. decumbens* and *B. decumbens* D1 to have a high rate at this harvest. At day 12, *B. humidicola* H6 and *B. decumbens* D1 showed a significantly higher rate than the other species, which did not differ among themselves and had rates similar to those at the first harvest. *B. humidicola* H6 had the highest rate on average, followed by *B. decumbens* D1 and *B. decumbens*. *B. brizantha* showed the slowest rates of all species, but was in the same range as *B. brizantha* B44, *B. humidicola* and *B. decumbens*.

B. brizantha showed the highest shoot fertilizer use efficiency of all species tested at the harvest at day 3, but it was lower than that of *B. decumbens* D1 and

Table 1. Total dry weight (g pot^{-1}) and total N (mg pot^{-1}) of *Brachiaria* spp.

Treatment	Days after N addition			Days after N addition		
	0	3	12	0	3	12
	Dry weight			Total N		
B. decumbens	1.0	1.1	2.2	18.5	25.5	52.0
B decumbens D1	1.2	1.5	2.5	18.7	24.1	61.7
B. brizantha	2.4	2.5	3.9	20.2	37.7	55.7
B. brizantha B44	1.4	1.5	2.4	17.5	27.5	55.4
B. humidicola	0.7	0.8	1.6	18.8	21.9	44.1
B. humidicola H6	0.9	1.1	2.0	16.0	22.5	44.0
	CV (%) = 16.1 SED = 0.2			CV (%) = 22.1 SED = 4.1		

Table 2. N specific absorption rate (μg of N mg dry root^{-1} day^{-1}) and fertilizer use efficiency (% of the added N) of *Brachiaria* spp. (n= 3 $\pm$ SEM)

Treatment	Growth interval		Days after N addition	
	0–3	3–12	3	12
	N specific absorption rate		Fertilizer use efficiency	
B. decumbens	13.5 $\pm$ 1.6	15.1 $\pm$ 0.2	18.4 $\pm$ 1.3	69.2 $\pm$ 1.7
B decumbens D1	19.9 $\pm$ 1.6	99.4 $\pm$ 7.7	24.1 $\pm$ 2.0	72.6 $\pm$ 5.0
B. brizantha	7.5 $\pm$ 0.6	3.2 $\pm$ 0.5	35.9 $\pm$ 1.8	57.3 $\pm$ 3.6
B. brizantha B44	9.5 $\pm$ 1.9	10.9 $\pm$ 0.2	21.4 $\pm$ 3.7	66.6 $\pm$ 1.7
B. humidicola	8.3 $\pm$ 5.0	11.2 $\pm$ 21.6	13.7 $\pm$ 3.6	53.5 $\pm$ 6.8
B. humidicola H6	3.8 $\pm$ 7.1	131.9 $\pm$ 35.7	13.2 $\pm$ 1.6	51.0 $\pm$ 5.9

B. decumbens at the last harvest (Table 2). Despite its high N absorption rate *B. humidicola* H6 still had the lowest fertilizer use efficiency of all species, followed by *B. humidicola*.

Overall, these results indicated large variation between the species tested in response to the variation in the soil mineral N availability imposed. Both *B. decumbens* genotypes seemed to be more responsive to this variation, increasing their N absorption faster than the other species tested. This faster response may be an advantageous factor for growth and competition in low N-providing soils. In such a condition, the plants need to exploit effectively the nitrogen which is mineralized when conditions of soil moisture and temperature, for example, are favourable. These conditions typify the rainy season. However, they are also favourable for the growth of soil micro-organisms, thus enhancing the competition for such a valuable resource. A fast N uptake could result in the absorption of a large proportion of the mineralized N, a better growth in that season and more reserves for regrowth in the following rainy season. Conversely, a plant which is slow to absorb N may lose a significant proportion of what is mineralized, produce less and accumulate fewer reserves. If

such an effect is cumulative, in the long term this plant should be less able to compete in the system than the first case considered.

Soil mineral N transformations

A large concentration of total mineral N was measured in the bare soil at 60 days after planting (Table 3). Most of it was nitrate, indicating active nitrification in this soil (Table 4). The soil under all species tested showed a total mineral concentration at least 5 times smaller than of the bare soil (Table 3), and ammonium was the predominant fraction found (Table 4). It seemed that the plants were very effective in using up the available ammonium or any nitrate that was formed, and as a result, no residual nitrate could be detected.

There was not a standard pattern of transformation in the mineral N of the soil under the different species tested. For instance, under *B. decumbens* and *B. brizantha* an increase in the ammonium concentration after the incubation of cores taken at the harvest at day 3 was detected (Table 4), which indicates that ammonification was active in the soil under these species. There was also a positive variation in the soil

Table 3. Total mineral N concentrations (mg g dry soil^{-1}) of the sandy soil under *Brachiaria* spp. or the bare soil. (n = 3 ± SEM)

Treatment		Days after N addition		
		0	3	12
Bare soil	Harv[a]	11.0 ± 1.0	45.5 ± 0.6	33.1 ± 6.2
	Inc	11.5 ± 1.3	44.6 ± 1.4	31.8 ± 6.0
	Var	0.5	−0.9	−1.3
B. decumbens	Harv	2.1 ± 0.6	25.3 ± 1.7	10.4 ± 1.1
	Inc	2.3 ± 0.1	35.0 ± 5.6	5.6 ± 0.8
	Var	0.2	9.7	−4.8
B decumbens DI	Harv	2.8 ± 0.3	36.8 ± 5.5	5.6 ± 0.1
	Inc	2.1 ± 0.3	33.5 ± 1.5	7.5 ± 0.9
	Var	−0.7	−3.3	1.9
B. brizantha	Harv	1.4 ± 0.3	15.0 ± 1.6	3.7 ± 1.7
	Inc	1.9 ± 0.2	31.3 ± 0.7	4.7 ± 1.0
	Var	0.5	16.3	1.0
B. brizantha B44	Harv	2.6 ± 0.6	36.5 ± 4.1	6.1 ± 0.1
	Inc	2.0 ± 0.1	30.7 ± 4.6	3.7 ± 0.5
	Var	−0.6	−5.8	−2.4
B. humidicola	Harv	3.3 ± 0.3	34.2 ± 4.1	10.3 ± 3.8
	Inc	2.4 ± 0.2	31.5 ± 1.7	15.7 ± 1.0
	Var	−0.9	−2.7	5.4
B. humidicola H6	Harv	1.9 ± 0.2	35.7 ± 2.2	11.6 ± 3.4
	Inc	2.3 ± 0.3	39.5 ± 6.4	5.3 ± 0.4
	Var	0.3	3.8	−6.3

[a] Var – Variation, Inc – Concentration after incubation, Harv – Concentration at harvest.

nitrate concentration under these species during the incubation related to the harvest at day 3. This result indicated that nitrification was occurring as well during this period, and that the nitrifiers responded positively and immediately to the increase in ammonium availability. This was confirmed by the increase in the nitrate concentration at the next harvest. However, in the incubation related to the last harvest there was a decrease in the nitrate concentration, suggesting that either nitrate was being immobilized preferentially by other soil micro-organisms, that there was insufficient substrate to induce nitrification, or that nitrate was lost. An examination of the total recovery of the added N indicated that between 10 and 20% was lost from the plant/soil system. There was nearly a total recovery in the bare soil, with the losses occurring in the planted soil, especially under both *B. humidicola* (Miranda, 1994).

In the soil under *B. decumbens* D1, on the other hand, a decrease in the ammonium concentration during the incubation period related to the harvest at day 3 was observed. However, there was an increase dur-

ing the incubation period related to the harvest at day 12. Contrary to what was observed in the soil under *B. decumbens*, there was very little variation in the nitrate concentration after incubation of cores taken at both harvests. On the other hand, there was a significant increase from the harvest at day 3 to the harvest at day 12, indicating that in the presence of the plant some nitrification occurred. A pattern similar to that, but with even stronger immobilization or loss of nitrate during the incubation period was observed for the soil under *B. brizantha* B44.

Another pattern of transformation was observed for the soil under *B. humidicola*. There was immobilization of ammonium during the incubation of cores taken at day 3, and some ammonification during the incubation of cores taken at day 12. The nitrate concentration increased from the harvest at day 3 to the harvest at day 12, and even more throughout the incubation period related to both harvests. This result suggests that the nitrifiers were very active in the soil under this species throughout the experimental period. Perhaps this effect was due simply to the higher concentration of ammonium available at the second harvest, compared with that available in the soil under the other species.

With these contrasting patterns of N transformation in the soil, it is difficult to conclude that there was any significant effect of the species tested. These transformations were dynamic, and seemed to vary with fluctuations in the activity of the microbial biomass. There was no indication of an inhibition of nitrification in the soil under any of the species tested. The concentrations of nitrate decreased in some cases at the first harvest in relation to concentration measured in the bare soil. However, by the following harvest, or even during the incubation of cores taken at that harvest, higher concentrations of nitrate were measured, indicating simply a delay in one or another situation. Therefore, conclusion about potential inhibition of these grasses based on a single harvest, or harvests largely spaced on time, could be misleading. Depending on the peak of soil micro-organism activity at the time of harvests, different interpretations could be reached. This was further demonstrated in studies with different soils of the Brazilian Cerrados, in which more frequent harvests were made (Miranda, 1994).

Table 4. Ammonium and nitrate concentrations (mg g dry soil^{-1}) of the sandy soil under *Brachiaria* spp. or the bare soil. (n = 3 $\pm$ SEM)

Treatment		Days after N addition					
		0	3	12	0	3	12
		Ammonium			Nitrate		
Bare soil	Harv[a]	1.4 ± 0.3	41.4 ± 0.9	26.3 ± 6.8	9.6 ± 1.2	4.1 ± 1.1	6.8 ± 1.1
	Inc	0.3 ± 0.6	39.4 ± 1.2	26.1 ± 5.8	11.2 ± 1.0	5.2 ± 0.5	5.7 ± 1.3
	Var	−1.3	−2	−0.2	1.6	1.1	−1.1
B. decumbens	Harv	1.9 ± 0.7	25.2 ± 1.7	4.2 ± 0.8	0.2 ± 0.1	0.1 ± 0.0	6.2 ± 0.4
	Inc	0.5 ± 0.1	30.1 ± 5.1	1.9 ± 0.5	1.8 ± 0.1	4.9 ± 0.5	3.7 ± 0.5
	Var	−1.4	4.9	−2.3	1.6	4.8	−2.5
B decumbens DI	Harv	2.7 ± 0.3	36.7 ± 5.5	2.6 ± 0.2	0.1 ± 0.0	0.1 ± 0.0	3.0 ± 1.0
	Inc	0.7 ± 0.3	33.1 ± 1.2	4.6 ± 1.1	1.4 ± 0.2	0.4 ± 0.1	2.9 ± 0.4
	Var	−2.0	−3.6	2.0	1.3	0.3	−0.1
B. brizantha	Harv	1.2 ± 0.2	14.8 ± 1.4	2.3 ± 0.4	0.2 ± 0.1	0.2 ± 0.1	1.4 ± 0.5
	Inc	0.7 ± 0.1	24.6 ± 0.7	2.8 ± 0.4	1.2 ± 0.2	6.7 ± 1.4	1.9 ± 0.6
	Var	−0.5	9.8	0.5	1.0	6.5	0.5
B. brizantha B44	Harv	2.0 ± 0.6	35.6 ± 4.2	2.2 ± 0.1	0.6 ± 0.0	0.9 ± 0.3	3.9 ± 0.6
	Inc	0.6 ± 0.1	30.5 ± 3.1	2.7 ± 0.5	1.4 ± 0.1	0.2 ± 0.0	1.0 ± 0.6
	Var	−1.9	−5.1	0.5	−0.2	−0.7	−2.9
B. humidicola	Harv	1.5 ± 0.3	34.1 ± 4.2	8.5 ± 3.7	1.8 ± 0.2	0.1 ± 0.0	1.8 ± 0.2
	Inc	0.6 ± 0.1	30.0 ± 1.8	10.0 ± 1.8	1.8 ± 0.2	1.5 ± 0.1	5.7 ± 0.3
	Var	−0.9	−4.1	1.5	0.0	1.4	3.9
B. humidicola H6	Harv	1.8 ± 0.2	34.8 ± 2.1	7.2 ± 3.1	0.1 ± 0.1	0.9 ± 0.0	4.4 ± 0.5
	Inc	0.7 ± 0.1	38.6 ± 5.9	1.4 ± 0.5	1.6 ± 0.3	0.9 ± 0.5	3.9 ± 0.5
	Var	−1.1	3.8	−5.8	1.5	0	−0.5

[a] Var – Variation, Inc – Concentration after incubation, Harv – Concentration at harvest.

Acknowledgements

C H B Miranda gratefully acknowledges a scholarship assistance from the Brazilian National Council of Research (CNPq).

References

Hunt R 1990 Basic Growth Analysis. Unwin Hyman Ltd., London, UK. 112 p.

Miranda C H B 1994 Nitrogen uptake by *Brachiaria* spp. growing on Brazilian Cerrados soils, and their effect upon soil mineral N transformations. Ph D Thesis. Wye College, University of London, Wye, Ashford, Kent, UK. 308 p.

Seiffert N F 1981 Nitrogen availability in *Brachiaria decumbens* under continuous grazing. *In* Biological Nitrogen Fixation Technology for Tropical Agriculture. Eds. P H Grahan and S C Harris. pp 387–393. CIAT, Cali, Colombia.

Sylvester-Bradley R, Mosquera D and Mendez J E 1988 Inhibition of nitrate accumulation in tropical grassland soil: effect of nitrogen fertilization and soil disturbance. J. Soil Sci. 39, 407–416.

Valle C B, Maass B L, Almeida C B and Costa J C G 1993a Morphological characterisation of *Brachiaria* germoplasm. *In* Proceedings of the XVII Inter. Grassland Congress, 18–21 February, 1993. pp 208–209. New Zealand Grassland Association, Rockhampton, Australia.

Valle C B, Calixto S and Amezquita M C 1993b Agronomic evaluation of *Brachiaria* germoplasm in Brazil. *In* Proceedings of the XVII Inter. Grassland Congress, 18–21 February, 1993. pp 511–512. New Zealand Grassland Association, Rockhampton, Australia.

O. Van Cleemput et al. (eds.), Progress in Nitrogen Cycling Studies, 237–241, 1996.

Effect of weather conditions and time of N application on the uptake of soil and applied N by winter wheat

H. Mouchová[1], J. Klír[1] and H. Lippold[2]

[1]Research Institute of Crop Production, Praha, Czech Republic and [2]Institute of Soil Science and Plant Production, Leipzig, Germany

Key words: labelled nitrogen, N leaching, N uptake, time of N application, winter wheat

Abstract

Leaching of mineral nitrogen under winter wheat and the N uptake by winter wheat were studied by means of ^{15}N in field lysimeter and microplot experiments carried out on a orthic Luvisol soil. Analysing the mineral N in lysimeter water, it was found, that as much as 95% of the total mineral nitrogen leached till harvest including 8–70% of the dose of 50 kg N applied in autumn was leached out of the 30 cm layer of topsoil till following spring. When the doses of 100–135 kg N had been applied, one third, on the average, of the total amount of nitrogen in winter wheat grain was derived from fertilizer. The recovery of the applied N by winter wheat grain varied from 16 to 42% of the total nitrogen dose.

Introduction

The efficiency of applied nitrogen fertilizers as well as their losses are influenced by the interaction of many factors, such as type of soil, crop, agricultural practice, including the dose, the form, the time of fertilization and the weather of the year. Field microplot (1975 – 1992) and lysimeter (1982 – 1991) experiments were established to study the effect of N added on yield, N uptake by winter wheat and recovery as well as on the leaching of mineral nitrogen from the top soil layer. In these experiments, the different doses of labelled fertilizer nitrogen were applied in various ways and at different times.

Material and methods

Model microplot experiments with winter wheat were carried out on a orthic Luvisol, on the field of the Research Institute of Crop Production in Prague. Microplots, sized 50×50 cm, were situated inside a winter wheat field, where only P and K fertilizers had been applied. The rate of 100 kg N or 135 kg N ha^{-1} were used as single dose either in autumn or in spring. Later a N rate of 100 kg was divided for application at both times. In the last series, a N rate of 120

kg was applied in three splits in spring. The results refer to the time of the winter wheat harvested at full ripeness.

The leaching of mineral nitrogen from soil under winter wheat was investigated in relation to precipitation. Lysimeters with an area of 530 cm^2 were filled every year in autumn with 30 cm of topsoil. The soil was the same as that in the microplot experiments. The lysimeters were placed into the soil so that their soil surface was at the same level as the surrounding field with winter wheat. The nitrogen fertilizer was applied to the 5 cm top soil layer before sowing or to the soil surface in spring. The lysimeter water samples were analyzed for mineral nitrogen, usually three times till the wheat harvest. In the years 1975–1980 the winter wheat cv. Mironovska was sown, cv. Juna in 1981 and, later, in 1984–1990, cv. Regina.

Results and discussion

The microplot experiments

The microplot experiment scheme and the yield of winter wheat grain and straw are summarized in Table 1. The sum of rainfall from the sowing of winter wheat or from the beginning of April to the harvest at full

Table 1. The yields of winter wheat grain and straw in the microplot experiments

Year of harvest	Scheme of experiments		Yield	
	N dose application		Grain	Straw
	(kg N ha^{-1})		(g m^2)	(g m^2)
1976	100	At sowing	464	756
1977	100	At sowing	276	470
1978	135	At sowing	470	1252
	135	In spring	782	1364
1979	135	At sowing	468	548
	135	In spring	392	424
1981	100	At sowing	488	652
	100	In spring	601	744
	100	Div.50 + 50	511	684
1982	100	In spring	756	784
	100	Div.50 + 50	648	662
1985	100	Div.50 + 50	505	602
1986	100	Div.50 +50	656	678
1990	120	3 splits	883	873
1991	120	3 splits	959	859
1992	120	3 splits	772	604

Table 2. The sum of precipitation during the microplot experiments

Year of harvest	Sum of precipitation till wheat harvest		Precipitation in	
	From sowing	From April	April	May
	(mm)	(mm)	(mm)	(mm)
1976	245	145	25	46
1977	408	269	33	39
1978	311	224	30	106
1979	391	225	34	18
1980	518	207	85	63
1981	431	320	18	65
1982	397	152	11	49
1985	473	356	28	79
1986	431	261	26	141
1990	223	134	41	38
1991	483	238	47	55
1992	299	158	21	5

Table 3. Sum of precipitation, average temperature during the winter period and amount of mineral nitrogen in kg N ha^{-1} leached out from the 30 cm topsoil layer under winter wheat

Season	Period from October to March (6 months)		Mineral nitrogen leached applied in autumn		
	(mm)	(°C)	Control	50 kg N in fertilizer applied in autumn	
				Sum	From fertilizer
1982/83	152	3.3	79	129	35
1983/84	117	1.1	12	28	9
1984/85	138	0.5	35	55	8
1985/86	170	0.8	74	102	25
1986/87	202	0.3	62	88	23
1987/88	186	4.2	68	94	28
1988/89	122	2.6	8	28	4
1989/90	91	4.1	3	33	4
1990/91	145	2.9	39	83	19

ripeness are given in Table 2. The amount of rainfall in April and May is added for the reason of its influence upon plant growth and its relation to grain and straw yields.

The recovery of labelled fertilizer nitrogen by the winter wheat grain varied from 16 to 34% (23% in average) of the 100 or 135 kg N dose applied as single dose before sowing, or 28 to 34% (30% in average) when the same doses were applied in spring. When the dose of 100 kg N ha^{-1} has been applied in 50 kg N splits both at sowing and in spring, the recovery of the added fertilizer was in the range of 17–40% (30% in average) and it reached 38–42% when a dose of 120 kg N was added in spring in three splits. The recovery of N applied by the winter wheat straw was 5–7% of the N dose. These recovery data are demonstrated in Figure 1 in relation to the sum of rainfall from April to the winter wheat harvest. The fertilizer and soil nitrogen uptake by the winter wheat grain and straw are shown in Figures 2–5. In our experiments, when 100 or 135 kg N ha^{-1}had been applied as single dose, the amount of N derived from fertilizer (Ndff) in the total nitrogen uptake varied from 22 to 49% (35% in average). The values of Ndff were between 27 and 35% when 50 kg N had been applied both at sowing and in the spring, and it reached 32–39% for the 120 kg N dose added in three splits in spring.

The lysimeter experiments

The amount of mineral nitrogen leached under winter wheat from the 30 cm soil layer within six months (from October to the beginning of April) was approximately 95% of the total mineral nitrogen leached during the whole period (till harvest). The sum of precipitation and average air temperature in the period October–March as well as the data of mineral nitrogen leached out from the non-fertilized and fertilized soil and from fertilizers are presented in Table 3. In 5 of

239

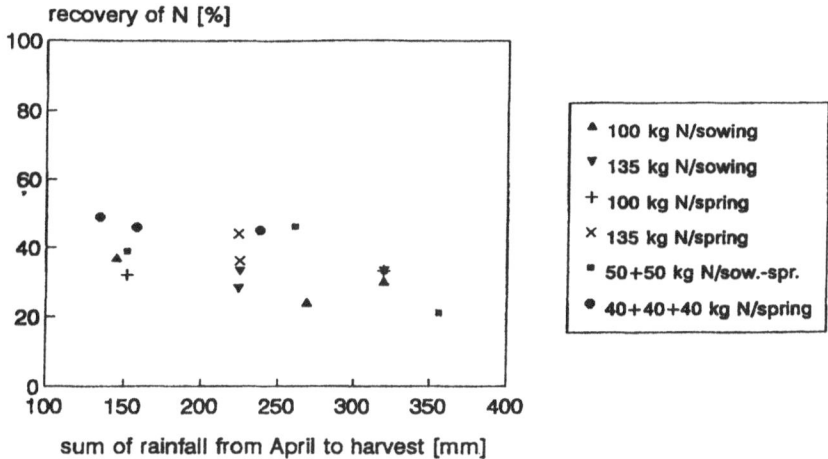

Figure 1. The recovery of N fertilizer added in winter wheat top at full ripeness.

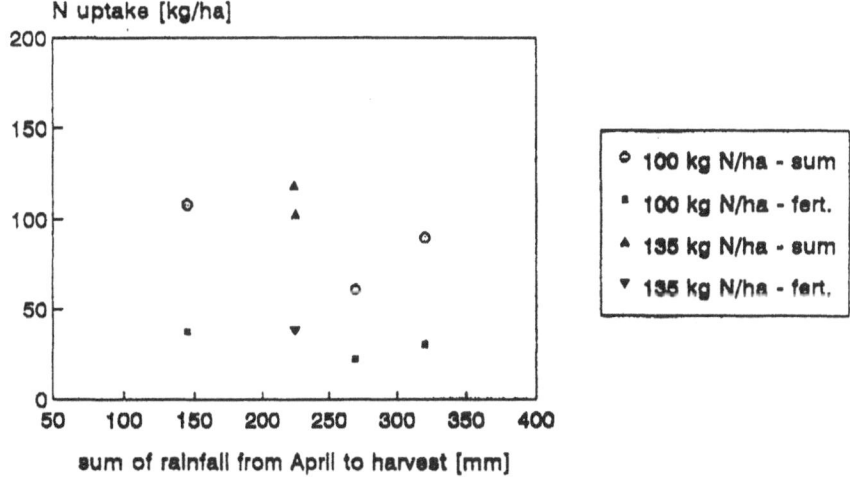

Figure 2. N uptake by winter wheat top at full ripeness (application of N at sowing).

the 9 experiments as much as 38 to 70% of the autumn applied 50 kg N was leached out. A lower amount of leached mineral nitrogen, 8–18% of this dose, was found when extremely dry or cold weather conditions occurred (Fig. 6). Leaching of mineral nitrogen under winter wheat within growth season was considerably lower and the leaching of spring applied low fertilizer doses was only very limited.

Conclusion

The rate of 120 kg N applied to winter wheat in three splits in spring resulted in a higher N uptake slightly influenced by spring rainfall as compared with other ways of fertilization. Most of the added N under winter wheat in autumn was leached out of the topsoil till following spring, on the contrary, the leaching of spring applied N was slight.

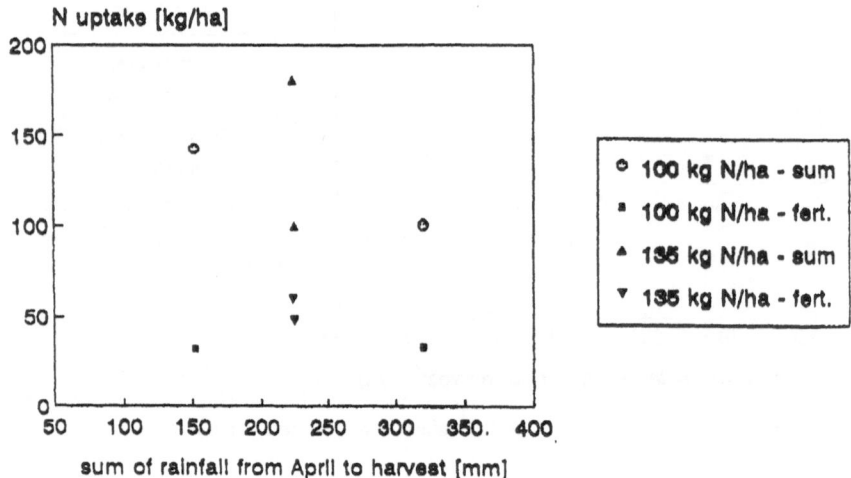

Figure 3. N uptake by winter wheat top at full ripeness (application of N at spring).

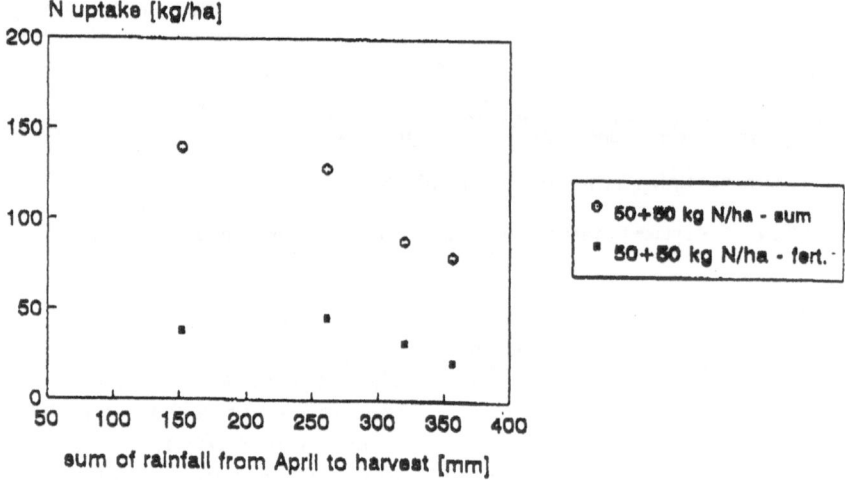

Figure 4. N uptake by winter wheat top at full ripeness (application of 50 + 50 kg N ha^{-1} in spring).

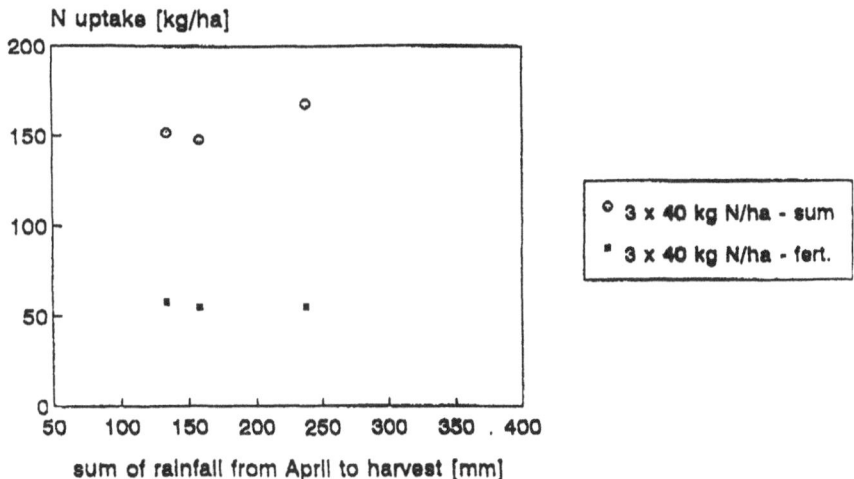

Figure 5. N uptake by winter wheat top at full ripesness (application of 3×40 kg N ha^{-1} in spring).

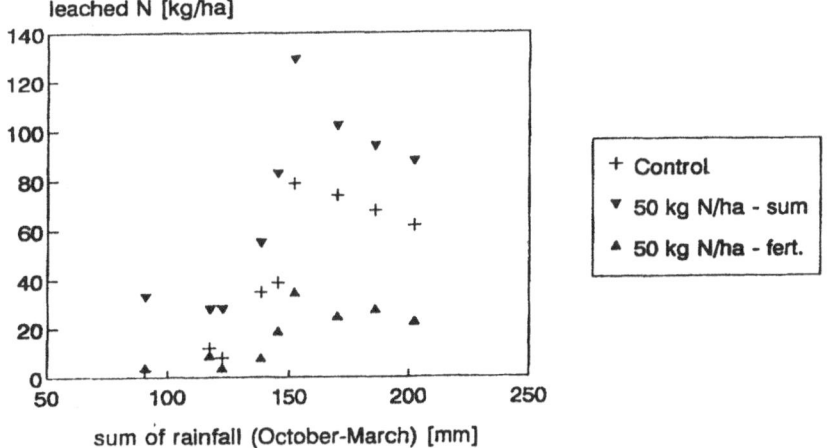

Figure 6. Leaching of mineral N under winter wheat during winter period.

References

Mouchová H, Klír J and Benešová J 1990 Nitrogen leaching from the soil lysimeters after autumn and spring nitrogen application to winter wheat. Rostl. Výroba 36, 785–790 (*In Czech*).

Mouchová H, Klír J and Benešová J 1988 Utilization of nitrogen by plants and nitrogen leaching from the soil and fertilizer in lysimeters. Rostl.Výroba 34, 951–959 (*In Czech*).

Mouchová H, Apltauer J, Lippold H and Matzel W 1987 The balance of supplied nitrogen in the soil and plant determined by means of ¹⁵N in winter wheat with the use of a nitrification inhibitor. Rostl. Výroba 33, 1173–1182 (*In Czech*).

Mouchová H and Lippold H 1994 Environmental balance of ¹⁵N-labelled fertilizer nitrogen applied to winter wheat in spring Arch. Acker- Pflanzenbau Bodenkd. 38, 83–88.

Mouchová H, Lippold H, Apltauer J and Matzel W 1990 Effect of nitrification inhibitor on the utilization of added nitrogen by winter wheat. Rostl. Výroba 36, 39–46 (*In Czech*).

O. Van Cleemput et al. (eds.), Progress in Nitrogen Cycling Studies, 243–246, 1996.
© 1996 *Kluwer Academic Publishers.*

Nitrogen nutrition of tomato derived from the use of sewage sludge and almonds residue as fertilizers

J. Navarro Pedreño[1], I. Gómez[2,3], R. Moral[2] and J. Mataix[2]
[1]*Dept. Química Agrícola, Geología y Geoquímica, Univ. Autónoma de Madrid, Cantoblanco, 28049-Madrid, Spain and* [2]*División de Agroquímica, Facultad de Ciencias, Universidad de Alicante, P.O. Box 99, 03080-Alicante, Spain.* [3]*Corresponding author*

Key words: almonds residue, growth, nitrogen, sewage sludge, tomato, yield

Abstract

Sewage sludge and epicarp-mesocarp residue of almonds were used as fertilisers for tomato cultivated under greenhouse conditions in pots containing 15 kg of a calcareous soil. Three treatments were established: without organic fertilisation (W), with the addition of 0.5 kg of dry sewage sludge (SS) and almonds residue (AR). The nitrogen content of soil, stem+branches and leaves and nitrate reductase activity were determined. Growth and yield were also determined. Results showed that sewage sludge produced a general increment of N both in soil and plant tissue and improved growth and yield when compared with the other treatments. However, almonds residue did not produce significant differences between AR and W treatments.

Introduction

Application of organic wastes such sewage sludge to soils could improve the growth of plants (Falahi-Ardakani et al., 1988). In Mediterranean areas, the cultivation of almond trees produces several residues, the most important being the epicarp-mesocarp of the almonds which contains considerable amounts of nutrients (Gómez et al., 1989).

The organic matter content and organic nitrogen in the soils of the Mediterranean region are usually low. Both wastes mentioned above could be used as organic amendments in order to find an adequate destination for these residues and to improve the conditions of crop production. But first, the effects of these additions on the nutrition of the plants must be studied. Deficiency of availability of nitrogen is most often the limiting factor of growth (Papadopoulos and Rendig, 1983). In this work we evaluate the role of these residues on nitrogen nutrition, growth and yield of tomato (*Lycopersicon esculentum* Mill), an important plant for many countries of southern Europe.

Materials and methods

Seeds of tomato (cv. Muchamiel) were sown in black peat. At the first true leaf stage the plants were transplanted to pots (one plant/pot) and cultivated under greenhouse conditions for 135 days. The pots were cylindrical (28.5 cm diameter and 30 cm height) and contained 15 kg of a calcareous soil (Table 2). Plants were drip irrigated (0.5 L six times a week) with tap water (Table 1).

The experimental design included three treatments: **W** (no addition of organic fertiliser), **AR** (application of 0.5 kg of dry almonds residue per pot) and **SS** (addition to the soil of 0.5 kg/pot of dry sewage sludge), and four replicates for each treatment and sample. No addition of inorganic fertilisers was done. The main properties of the sewage sludge and the almonds residue are given in Table 2.

Nitrogen status in soil and plant tissue was determined every 45 days, corresponding with the periods of flowering (**1**), harvesting (**2**) and beginning of senescence (**3**) (De Koning, 1989). Nitrate reductase activity (NR-activity) was also measured. Four plants were taken and analysed separately as well as four soil subsamples (soil column of 0 to 20 cm deep) composing the samples of the stages 1, 2 and 3.

Table 1. Characteristics of the tap water used for irrigation

Properties	Unit	Value	CONF. INT.[a]
Electrical conductivity (20 °C)	mS cm^{-1}	0.9	0.1
pH		7.7	0.2
Chloride	mg L^{-1}	115	25
Bicarbonate	mg L^{-1}	218	19
Sulphate	mg L^{-1}	143	20
Sodium	mg L^{-1}	91	12
Potassium	mg L^{-1}	3	1
Calcium	mg L^{-1}	38	4
Magnesium	mg L^{-1}	29	2

[a] CONF. INT. - Confidence interval.

Nitrogen in the soil was determined by the Kjeldahl method (using block digestor, distillation and titration with HCl); plant tissue (leaves, stem and branches) was dried at 60 °C, then the nitrogen was determined in leaves and in stem+branches using the same soil procedure.

The activity of the enzyme was determined in vivo with a method proposed by several authors (Heuer and Plaut, 1978; Mauriño et al., 1986). Five disks (5 mm diameter) from mature leaves situated at the second level of the plant from the top to the bottom were incubated with 2 mL of KNO_3 solution (0.25 M) in darkness at 30 °C (pH=7.8 with tris-HCl 0.05 M). After 70 min, the nitrite formed was measured by extracting 1 mL of the previous solution and adding 1 mL of sulfanilamide (1% w/v in HCl 2 N) and 1 mL of N-(1-naphtyl)-ethylenediamine dihydrochloride (0.02% w/v in aqueous solution). Absorbance at 540 nm was read after 15 min and the nitrite formed was determined by comparing with standards measured under the same conditions.

Plant growth was evaluated measuring length and fresh weight of stems and all the branches formed, and fresh weight of total leaves and fruits obtained during the experiment.

The confidence interval (p=0.05) was calculated for each mean obtained from four replications (**CONF. INT.**).

Table 2. Properties of the soil, almonds residue and sewage sludge

Properties	Units	Soil	Almond residue	Sewage sludge
Clay (<0.002 mm)	%	15	-	-
Silt (0.02–0.002 mm)	%	28	-	-
Sand (> 0.02 mm)	%	27	-	-
Total Carbonate	%	55	-	-
Active lime	%	2.1	-	-
E.C. (1:5 soil-H_2O)	mS cm^{-1}	0.39	7.13	6.35
pH (1:5 soil-H_2O)		7.6	8.9	5.5
Organic matter	g kg^{-1}	13.6	830	566
N (Kjeldahl)	g kg^{-1}	1.31	10.3	29.9
P	g kg^{-1}	0.022	2.3	17.9
K	g kg^{-1}	0.42	42.5	2.6
Na	g kg^{-1}	0.19	3.7	0.66
Ca	g kg^{-1}	4.40	42.6	49.4
Mg	g kg^{-1}	0.54	4.4	5.6
Fe	mg kg^{-1}	1.9	995	9700
Mn	mg kg^{-1}	1.2	86	115
Cu	mg kg^{-1}	0.7	390	272
Zn	mg kg^{-1}	1.6	72	905
B	mg kg^{-1}	-	85	79
Cd	mg kg^{-1}	-	-	4
Cr	mg kg^{-1}	-	-	12
Hg	mg kg^{-1}	-	-	1
Ni	mg kg^{-1}	-	-	18
Pb	mg kg^{-1}	-	-	2

Nitrogen in soil

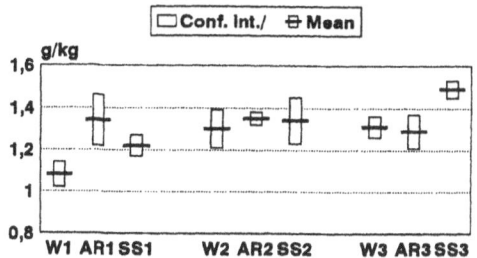

Nitrogen in stem+branches

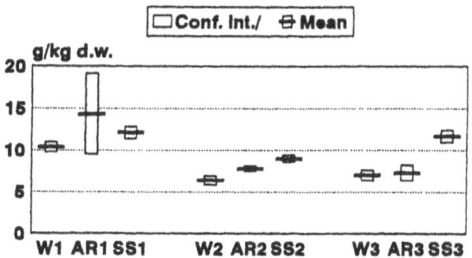

Nitrogen in leaves

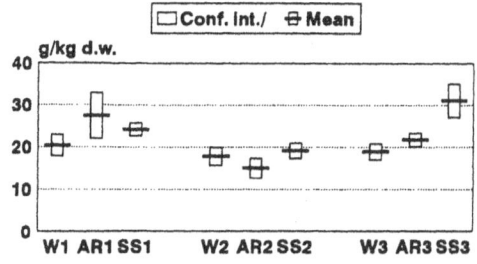

NR activity in leaves

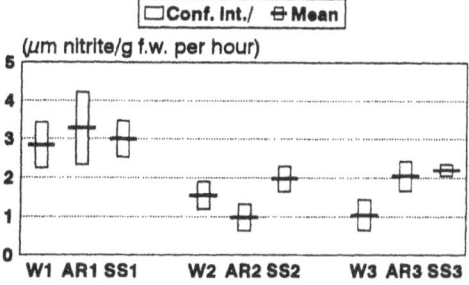

Figure 1. Nitrogen concentration in soil, stem+branches and leaves in dry weight basis. Nitrate reductase activity measured in vivo in leaves.

Results and discussion

The results are showed in Figures 1 and 2. Sewage sludge produced an small increment of nitrogen concentration in soil and plant tissue. This residue is commonly used as N-fertiliser because of this fact (Menelik et al., 1991).

A reduction of N-Kjeldahl in soil was expected during the course of the experiment but in fact, it was not observed in these conditions probably because there was an input of organic matter from the development of the roots and exudates from the plant in the wet zone where it was measured. The SS treatment maintained until the end of the experiment the highest level of N in soil which corresponded with the highest initial presence of this element in sewage sludge.

The concentration of N in the plant tissue was clearly enhanced by sewage sludge as can be seen in Figure 1. At the beginning of the experiment there were no statistical differences among the treatments (stage 1), then probably due to the high mineralization rate of

sewage sludge producing inorganic nitrogen (Devitt et al., 1990), it was significative.

Both organic treatments modified the activity of the enzyme, nitrate reductase. As happened with nitrogen concentration in the plant tissue, NR-activity was similar in the first sample and differences among treatments were increased along with the time of the experiment. Sewage sludge promoted the highest mean values in stages 2 and 3. The presence of nitrates, due to the mineralization of this residue, could favour the maintenance of the NR activity in these levels.

The input of nitrogen due to sewage sludge in the SS treatment was the responsible for the greatest plant development that can be observed in Figure 2 (length and fresh weight of stem+branches as well as fresh weight of leaves). Yield was also enhanced and total production was higher where sewage sludge was applied.

Almonds residue had a very poor capacity for improving plant development and increasing N concentration. There was very little difference between W and AR treatments as Figures 1 and 2 demonstrate.

246

Fresh weight of stems+branches

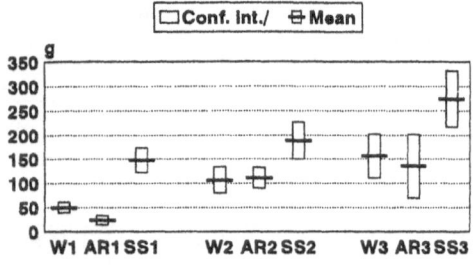

Fresh weight of leaves

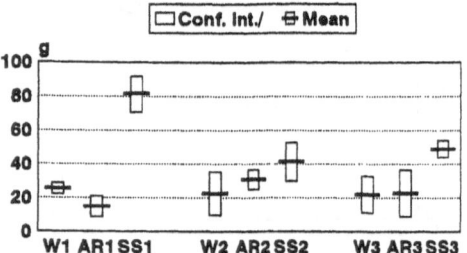

Yield/plant

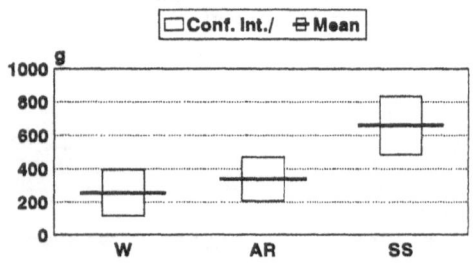

Length of stems+branches

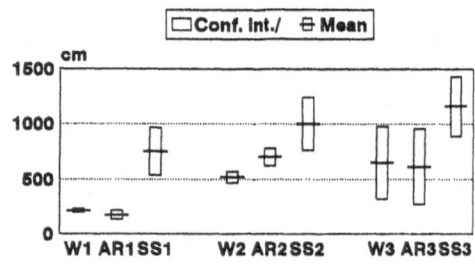

Figure 2. Growth parameters (length and fresh weight) and yield of tomato plant.

Sewage sludge stimulated plant growth and increased the presence of N in the soil and plant tissue in this experiment. The effects of this nitrogen fertilisation were to increase the growth, yield and N concentration. These results suggest that the residue from wastewater treatment, when containing low amounts of pollutants could be used as nitrogen fertiliser for horticultural purposes but not the almonds residue.

References

De Koning A N M 1989 Development and growth of a commercially grown tomato crop. Acta Hortic. 260, 267–273.
Devitt D A, Morris R L and Bowman D C 1990 Response of tall fescue to composted sewage sludge used as a soil amendment. J. Plant Nutr. 13, 1115–1139.

Falahi Ardakani A, Bownkamp J C, Gonin F R and Chaney R L 1988 Growth response and mineral uptake of lettuce and tomato transplants grown in media amended with composted sewage sludge. J. Environ. Hortic. 6, 130–132.
Gómez I, Gómez B and Mataix J 1989 Evaluación mediante un sistema E.U.F. de la fertilidad potásica de un suelo compostado con piel de almendra. Agrochimica 33, 458–467.
Heuer B and Plaut Z 1978 Reassessment of the in vivo nitrate reductase assay in leaves. Physiol. Plant. 43, 306–312.
Mauriño S G, Echevarría C, Mejías J A, Vargas M A and Maldonado J M 1986 Properties of the in vivo nitrate reductase assay in maize, soybean and spinach leaves. J. Plant Physiol. 124, 123–130.
Menelik G, Reneau R B, Martens D C and Simpson T W 1991 Yield and elemental composition of wheat grain as influenced by source and rate of nitrogen. J. Plant Nutr. 14, 205–217.
Papadopoulos I and Rendig V V 1983 Interactive effects of salinity and nitrogen on growth and yield of tomato plants. Plant and Soil 73, 47–57.

O. Van Cleemput et al. (eds.), Progress in Nitrogen Cycling Studies, 247–250, 1996.

Recovery of green manure N by succeeding sugar beet

V. Ninane[1], J.P. Goffart[2], J.P. Destain[2] and E. François[2]
[1]Institut pour l'Encouragement de la Recherche Scientifique dans les Industries Agricoles (I.R.S.I.A), 115 Chaussée de Wavre, B-5030 Gembloux, Belgium and [2]Centre de Recherches Agronomiques, Ministère Fédéral de l'Agriculture, B-5030 Gembloux, Belgium

Key words: green manure, mustard, ^{15}N, phacelia, rye-grass, sugar-beet

Abstract

In the loam region of Belgium, a three-course rotation (sugar-beet, winter wheat and winter barley) prevails and the green manures are sown in late July immediately after the harvest of winter barley.

N dressings of 80 or 120 kg ha^{-1} were applied to green manures of mustard, rye-grass and phacelia grown in 65 m^2 plots; 12 m^2 plots received comparable dressings labelled with ^{15}N (6 at.%). At the end of November they were turned under. They had taken up 110 to 180 kg N ha^{-1}, depending on the year (three years of experimentation). In the following spring, sugar-beets were sown. During the growing season, from 25 to 100 kg NO$_3$-N were produced from green manure; mustard being characterized by the highest decomposition rate. Only 20 to 60 kg N ha^{-1} had been taken up by the beets; from 43 to 68% of green manure N was still present in the soil, in organic form; not-accounted for ranged from 10 to 32%, highest values were obtained for mustard. Only once in 3 years did green manures affect sugar yield (less sugar was produced after mustard than after bare soil in 1992). It is supposed that mustard has made unavailable part of mineral N to the crop.

Introduction

In the loam region of Belgium, a three-course rotation (sugar-beet, winter wheat and winter barley) prevails and the green manures are sown in late July immediately after the harvest of winter barley. If vetch (*Vicia sativa*) is always largely sown, mustard (*Sinapis alba*), rye-grass (*Lolium perenne* and *L. multiflorum*) and phacelia (*Phacelia tanacetifolia*) are gaining ground. For these last three crops, it is of prime importance to apply a nitrogen dressing. The aim of this work is to study the influence of the green manures on sugar-beet growth and nitrogen balance sheet.

Material and methods

Soil and climate

The experiments were carried out from 1990 to 1993 in the loamy region of Belgium where the soils (Hapludalf) are characterised by a clay content of less than 15% and a silt fraction of about 82%. The average N concentration was 0.11% in the upper 0–30 cm layer and the C concentration was ±1%.

The climate is temperate (average temperatures are 5 °C in winter and 17 °C in summer). The rainfall of 800 mm is regularly distributed all over the year.

Experimental design

After the harvest of winter barley (late July or August), green manures of mustard, rye-grass, phacelia and vetch were sown in 1990, 1991 and 1992, in plots of 65 m^2 in a randomized block design with 4 replications. These plots were compared to bare soil and, except those sown with vetch, they received a dressing of 80 or 120 kg N ha^{-1}. In smaller plots of 12 m^2 as in microplots (steel cylinders of 0.07 m^2 and 30 cm in length pressed into the soil), an equivalent N dressing labelled with ^{15}N (6 at.%) was applied on the same green manures except in 1991 when the latter received 20 kg N ha^{-1} less than the large plots. In November, the crops were turned under in the 65 m^2 and 12 m^2 plots which were sown with sugar-beets in the following spring (March–April). The microplots were harvested

Table 1. Dry matter production and N uptake by green manure

Type	Year	N dressing (kg ha^{-1})	DM (t ha^{-1})	N uptake (kg ha^{-1})
Mustard	1990	120	6.6 ± 0.8^{az}	127 ± 22^a
Rye-grass	1990	120	6.2 ± 1.0^a	164 ± 11^a
Mustard	1991	80	5.0 ± 0.9^b	125 ± 16^a
Rye-grass	1991	80	6.8 ± 0.8^a	108 ± 13^a
Phacelia	1991	80	5.8 ± 0.9^b	122 ± 16^a
Vetch	1991	0	4.5 ± 0.8^b	125 ± 19^a
Mustard	1992	80	6.9 ± 0.9^a	178 ± 26^a
Rye-grass	1992	80	8.8 ± 0.6^b	169 ± 14^a
Phacelia	1992	80	8.3 ± 1.0^b	155 ± 28^a
Vetch	1992	0	4.6 ± 0.6^c	176 ± 29^a

[z] Values within the same column followed by the same letter are not significantly different at the 0.05 probability level.

Table 2. Origin of N taken up by green manure (% of N taken up)

Type	Year	^{15}N dressing (kg ha^{-1})	Soil	Fertilizers
Mustard	1990	120	57	43
Rye-grass	1990	120	59	41
Mustard	1991	60	75	25
Rye-grass	1991	60	78	22
Phacelia	1991	60	78	22
Mustard	1992	80	66	34
Phacelia	1992	80	69	31

Table 3. Sugar-beet yields (t sucrose ha^{-1}). N dressing applied to sugar-beet (kg ha^{-1})

Beets after	0	60	120	180
Mustard 90	12.7 ± 1.3	13.0 ± 1.4	12.2 ± 1.4	12.2 ± 1.3
Rye-grass	12.7 ± 1.2	12.8 ± 0.9	12.2 ± 1.2	12.7 ± 0.8
Bare soil	12.4 ± 1.3	12.4 ± 0.6	12.4 ± 1.1	12.3 ± 0.8
Mustard 91	11.8 ± 1.1	13.1 ± 1.1	12.7 ± 0.9	12.6 ± 1.0
Rye-grass	11.9 ± 0.6	12.6 ± 0.6	12.9 ± 1.1	12.6 ± 0.8
Phacelia	13.0 ± 0.9	13.1 ± 0.9	12.9 ± 1.6	13.7 ± 0.5
Vetch	12.5 ± 1.0	12.7 ± 0.5	12.3 ± 1.3	12.7 ± 0.6
Bare soil	11.5 ± 0.8	13.1 ± 0.9	13.3 ± 1.2	12.7 ± 0.7
Mustard 92	12.3 ± 0.9^{cz}	14.8 ± 0.3	15.5 ± 0.7	15.6 ± 1.5
Rye-grass	13.3 ± 0.8^{bc}	14.4 ± 0.5	13.8 ± 1.3	14.2 ± 0.5
Phacelia	13.0 ± 0.9^{bc}	14.4 ± 1.4	14.8 ± 0.7	14.7 ± 1.6
Vetch	14.8 ± 0.8^a	14.7 ± 1.7	15.5 ± 1.2	15.7 ± 1.5
Bare soil	13.7 ± 0.4^{ab}	15.3 ± 0.6	15.5 ± 1.1	15.6 ± 1.4

[z] Values within the same column followed by the same letter are not significantly different at the 0.05 probability level.

Table 4. N taken up by the beets in unfertilized plots

Beets after	N taken up (kg ha^{-1})
Mustard 92	130 ± 12
Rye-grass	156 ± 8
Phacelia	137 ± 6
Vetch	199 ± 22
Bare soil	156 ± 5

Table 5. Balance sheet of fertilizer N applied to green manure (% N applied)

Type and year	^{15}N applied (kg ha^{-1})	At ploughing time of the green manure			At harvest of sugar-beet		
		N in green manure	N in soil	N not-accounted for	N in beets	N in soil	N not-accounted for
Mustard 90	120	41 ± 2	19 ± 1	40 ± 3	29	44	27
Rye-grass	120	55 ± 1	27 ± 1	18 ± 1	32	48	20
Mustard 91	60	76 ± 2	22 ± 2	2 ± 2	35 ± 4	43 ± 3	22 ± 5
Rye-grass	60	55 ± 8	23 ± 5	22 ± 11	22 ± 1	68 ± 7	10 ± 5
Phacelia	60	65 ± 3	21 ± 5	14 ± 6	17 ± 3	62 ± 2	21 ± 12
Mustard 92	80	86 ± 4	19 ± 2	0 ± 5	25 ± 2	43 ± 6	32 ± 4
Rye-grass	80	74 ± 8	12 ± 4	15 ± 6	21 ± 4	66 ± 3	13 ± 7
Phacelia	80	74 ± 8	15 ± 3	12 ± 8	18 ± 5		

Table 6. Fate of green manure N at sugar-beet harvest. Results in kg ha^{-1} (rounded values) and in % of green manure N in brackets (estimate)

Type and year	N in the green manure	N in beets	N in soil	N losses
Mustard 90	127	50 (39)	75 (61)	0
Rye-grass	164	60 (38)	95 (57)	10 (5)
Mustard 91	125	45 (36)	55 (44)	25 (20)
Rye-grass	108	25 (24)	80 (76)	0
Phacelia	122	25 (20)	90 (72)	10 (8)
Mustard 92	178	45 (25)	75 (43)	55 (32)
Rye-grass	169	40 (24)	130 (76)	0
Phacelia	155	30 (20)	-	-

in November, at ploughing time of the green manures. In the 65 m^2 plots, the plants were quantitatively collected on 0.25 m^2 areas (4 replications) in order to measure DM production and N uptake while the recovery of fertilizer N was determined in the microplots (plants and the 30 cm upper soil layer). Sugar-beet yields at 4 N rates (0–60–120–180 kg N ha^{-1}) and, for the last experiment in unfertilized plots, N uptake were deter-

mined the following year in the 65 m^2 plots while a complete balance sheet of N was drawn up in the 12 m^2 plots. To this end, four central 0.90 m^2 areas corresponding to 2 rows of beets were harvested; the upper 30 cm soil was completely removed in 10 cm layers and thoroughly mixed before sampling; a mechanical auger was then used to 150 cm. Plant and soil samples were freeze dried and finely ground with a hammer mill.

Table 7. Soil N in the unfertilized beets at harvest (kg ha^{-1}) (estimate)

Soil N	
Mustard 92	85
Rye-grass	116
Phacelia	107
Bare soil	156

Analysis for total N was carried out using the Kjeldahl method modified to include nitrate (Bremner, 1965) or by the Dumas technique (Roboprep, Europa Scientific UK). The isotopic composition of N was determined by mass spectrometry (SIRA 12 VG Isogas UK).

Results and discussion

Depending on the year, DM production and N uptake by the green manures varied considerably: they ranged respectively from 4.5 to 8.8 t ha^{-1}, with the lowest values for vetch, and from 108 to 178 kg N ha^{-1} (Table 1). By using ^{15}N, it was possible to specify the origin of N uptake (fertilizer or soil). Although decreasing with rate, N originating from soil was always higher (Table 2).

In 1991 and 1992, an intermediate crop did not influence sugar-beet yields (Table 3). Indeed, the optimum N rate for beet was low (60 kg N ha^{-1}) owing to a large mineral nitrogen production by the soil. In 1993 (green manure sown in 1992), in the unfertilized plots, sugar yields were higher after vetch (14.8 t sucrose ha^{-1}) and bare soil (13.7 t sucrose ha^{-1}); the lowest yield was observed after mustard (12.3 t sucrose ha^{-1}). The N content of the beet was maximum after vetch (199 kg N ha^{-1}) and minimum after mustard (130 kg N ha^{-1}) (Table 4).

If green manures (except vetch) did not seem to have a positive effect on sugar-beet yield, we nevertheless deemed interesting to examine their impact on the balance sheet of nitrogen at ploughing time and at beet harvest. At ploughing time, N recovered by rye-grass and phacelia averaged between 55 and 74% of the N applied (Table 5) and was more variable for mustard (41 to 86%). N remaining in the soil did not change very much (20–25%) while not accounted for, except for mustard in 1990 (with a possible too high N dressing), did not exceed 22%. At harvest of sugar-

beet, a maximum of 35% of N applied to green manure could be recovered in the plant, the lowest value (17%) being observed after phacelia. The remaining N in the soil varied from 43 to 68%, not accounted for from 13 to 27%. In this respect, N applications to green manure seem to be inefficient. But, as most of green manure N, turned under in autumn, stems mainly from soil, this conclusion deserves further consideration. Table 6 describes the fate of green manure N at sugar-beet harvest assuming that it is no different from that of labelled N recovered (in plants and soil) at ploughing time. From 20 to 39% of green manure N was recovered by sugar-beets. Similar values were obtained in a spring barley grown in a soil amended with ^{15}N rye-grass (Jensen, 1922; Thomsen, 1993). Sugar-beet had taken up from 25 to 60 kg N originating from green manure, the lowest values being observed for phacelia (1990 and 1991) and for rye-grass in 1991, the highest for rye-grass in 1990 (after the highest N dressing of 120 kg N ha^{-1}). N left in soil, entirely under organic form, was higher after rye-grass and phacelia and lower after mustard which was thus characterized by the highest decomposition rate. In parallel experiments, mustard turned under released mineral N more quickly than rye-grass and phacelia (Goffart, 1993; Meeùs-Verdinne, 1994). Highest losses (25 and 55 kg ha^{-1}) occured twice with mustard. As shown in Table 7, soil N uptake by the unfertilized beets at harvest in 1993 may then be deduced from Table 4 and 6. These values and (sugar) yields presented in Table 3 strongly suggest that green manures may have made unavailable part of mineral N for the succeeding crop.

References

Bremner J M 1965 Total nitrogen. *In* Methods of Soils Analysis. Part 2. Eds. C-A Black, D D Evans, White, L E Ensmigner and F E Clark. Agronomy 9, pp 1149–1178. Am. Soc. of Agron., Madison, WI, USA.

Goffart J P, Ninane V and Guiot J 1993 Catch crops as a means of decreasing nitrate leaching in winter. *In* Proceedings of Environmental Platform, organized at Leuven, May 14. pp 199–202. Laboratory of Industrial Microbiology, Heverlee, Belgium.

Jensen E S 1992 The release and fate of nitrogen from catch-crop materials decomposing under field conditions. Soil Sci. 43, 335–345.

Meeùs-Verdinne K, Ninane V, Goffart J P and Francois E 1994 Gestion raisonnée de l'azote: sa restitution par la minéralisation de trois engrais verts marqués. Coll. Int. AIEA, Vienne, Austria, 17–21 Octobre.

Thomsen I K 1993 Nitrogen uptake in barley after spring incorporation of ^{15}N-labelled Italian ryegrass into sandy soils. Plant and Soil 150, 193–201.

O. Van Cleemput et al. (eds.), Progress in Nitrogen Cycling Studies, 251–254, 1996.

Efficiency of fertilizer use by a rain-fed wheat crop following split-application of fertilizer nitrogen

C.J. Pilbeam[1], A.M. McNeill[2], D. Court[1], H.C. Harris[3] and R.S. Swift[4]
[1]Department of Soil Science, The University of Reading, Reading, RG6 6DW, UK, [2]CLIMA, Nedlando, Perth, WA 6907, Australia, [3]ICARDA, P.O. Box 5466, Aleppo, Syria and [4]Division of Soils, CSIRO, Adelaide, SA 5064, Australia

Key words: fertilizer recovery, ^{15}N-labelled fertilizer, N fertilizer, rainfed cropping, split-application, wheat

Abstract

The efficiency of use of a split-application of nitrogen fertilizer was measured in wheat plots at ICARDA, northern Syria. Nitrogen (60 kg N ha^{-1}) was applied as a split-dressing; 30 kg N ha^{-1} at sowing (November/December) and 30 kg N ha^{-1} at tillering (March). Urea was used in 1991/92 and ammonium sulphate in 1992/93. Fertilizer applications were (i) ^{15}N-labelled at both sowing and tillering, (ii) ^{15}N-labelled fertilizer at sowing, unlabelled fertilizer at tillering, (iii) unlabelled fertilizer at sowing, ^{15}N-labelled fertilizer at tillering.

Dry matter production and total nitrogen content of the above-ground plant material were unaffected by the time of ^{15}N-label application. The mean amounts of nitrogen recovered in the grain were 61.5 kg N ha^{-1} in 1992 and 69.9 kg N ha^{-1} in 1993. Fertilizer nitrogen recovered by the wheat averaged 11.3% after ammonium sulphate and 17.2% after urea, with greater recoveries from an autumn application in both cases. Generally less than 10% of the nitrogen in the plant at harvest was derived from the ^{15}N-labelled fertilizer. Mineralization of soil nitrogen provided on average 82.9 kg N ha^{-1} (1992) and 93.3 kg N ha^{-1} (1993) to the wheat.

Under these conditions N fertilizer made only a small contribution to current production, most of the plant nitrogen was derived from soil. Fertilizer use efficiency was maximized by application at sowing.

Introduction

Much less nitrogen is used on dryland wheat than on wheat grown under more favourable moisture regimes and there are indications that even these modest applications are used less efficiently. Our aim in this paper is to examine the fate of fertilizer N applied to two successive dryland wheat crops grown in N.W. Syria, using ^{15}N-labelled fertilizer as a tracer. Field experiments with ^{15}N-labelled fertilizer can give information on the quantities of fertilizer N taken up by the crop and remaining in the soil, on N losses from the soil/crop system, and on the amount of unlabelled (i.e. soil derived) N taken up by the crop (e.g. Addiscott and Powlson, 1992; Macdonald et al, 1989; Powlson et al., 1986, 1992).

Materials and methods

The field site was at the ICARDA station, Tel Hadya, Syria (36°1'N, 36°56'E). The soil is classified as a Calcixerollic Xerochrept, with <1.0% organic matter, a pH of 8.2 and a high clay content (>60%). The climate is mediterranean with rain falling mainly in the months from November to March, inclusive.

As part of a larger Two Course Rotation experiment (Harris, 1990), wheat (*Triticum turgidum* var. *durum*) was sown in late November 1991 and in early January 1993 in 12 m × 37.5 m plots, which were replicated 3 times. Both wheat crops were preceded by a one-year fallow. Phosphate fertilizer was drilled with the wheat at a rate of 50 kg P$_2$O$_5$ ha^{-1}. Nitrogen (60 kg N ha^{-1}) was applied to each plot in both seasons as a split application of nitrogen fertilizer; 30 kg N ha^{-1} was broadcast prior to sowing and an additional 30 kg N ha^{-1} was broadcast at the tillering stage (early

March). Urea was used in the first season (1991/92) and ammonium sulphate in the second (1992/93).

Three different ^{15}N fertilizer treatments were established as three 2 m × 2 m microplots along the central axis of each main plot, prior to sowing and the broadcasting of nitrogen fertilizer. These microplots were covered when unlabelled fertilizer was broadcast. Applications were: (i) ^{15}N-labelled fertilizer at sowing (30 kg N ha^{-1}), ^{15}N-labelled fertilizer at tillering (30 kg N ha^{-1}); (ii) ^{15}N-labelled fertilizer at sowing (30 kg N ha^{-1}), unlabelled fertilizer at tillering (30 kg N ha^{-1}); (iii) unlabelled fertilizer at sowing (30 kg N ha^{-1}), ^{15}N-labelled fertilizer at tillering (30 kg N ha^{-1}). Fertilizer was always applied in solution to the microplot. In 1991/92, urea with an abundance of 10.424 atom% was used on both 1 December and 12 March. In 1992/93, the ammonium sulphate had an abundance of 10.289 atom% at the first application (5 January) and an abundance of 10.322 atom% at the second application (15 March).

At harvest, 3 June in both seasons, plants from a 70 cm × 50 cm area were cut at ground level from the middle of each microplot. They were separated into ears and stem plus leaf, dried at 80 °C for 48 h, and weighed. The ears were then threshed and the grain yield determined. Plant material was finely ground with a TEMA mill. Total nitrogen and ^{15}N/^{14}N ratios were determined after complete combustion on a linked GC-IRMS system (Europa Instruments).

Soil samples were taken to a depth of 40 cm from the sampled area within each microplot in both seasons. A 30 cm × 30 cm square of soil was dug to a depth of 20 cm. Two further 20 cm soil cores were taken from the bottom of the 30 cm × 30 cm hole with a 6.5 cm auger. All soil samples were passed through a 13 mm sieve and thoroughly mixed before being subsampled and finely ground. Total nitrogen and ^{15}N/^{14}N ratios were determined as for the plant material.

The amount of fertilizer recovered in the whole plant was calculated by multiplying the amount of nitrogen in the plant by the atom% excess of the plant. The product was then divided by the atom% excess of the fertilizer. This was then expressed as a percentage of the amount of N applied as fertilizer. Similar calculations were made for component tissues of the plant, and for the soil in the 0–20 cm and 20–40 cm layers.

Results

Grain yields averaged 2.69 t ha^{-1} and 3.23 t ha^{-1} in 1991/92 and 1992/93, respectively (Table 1). The nitrogen content of the whole crop was greater in 1992/93 (98 kg N ha^{-1}, on average) than in 1991/92 (89.7 kg N ha^{-1}, on average) reflecting greater dry matter in 1992/93. Most (>68%) of the nitrogen in the crop at harvest was in the grain (Table 1).

None of the measurements in Table 1 (grain yield, whole crop yield, grain N uptake or whole crop N uptake) should be affected by *when* the labelled fertilizer was applied. It should not matter whether the crop receives labelled or unlabelled N fertilizer, since both are given at the same time and at the same rate. For the most part this is true, but Table 1 does contain a few exceptions. For example, both the whole crop dry matter and the total N uptake by grain in 1991/92 was significantly greater for ^{15}N given at tillering than if it were given at sowing. No explanation can be offered for these discrepancies.

The amount of nitrogen in the crop derived from labelled fertilizer was inevitably greater after a double rather than a single labelled fertilizer application. With the exception of the double applications of ^{15}N-labelled urea, less than 10% of the nitrogen in the crop at harvest was derived from labelled fertilizer. The remainder of the nitrogen in the crop was derived from soil and (for the two treatments receiving 30 kg N ha^{-1} unlabelled fertilizer) unlabelled fertilizer. In both seasons more of the crop nitrogen was derived from soil and unlabelled fertilizer following an application of labelled fertilizer at tillering (93.8 kg N ha^{-1} on average) than at sowing (81.8 kg N ha^{-1} on average).

In both seasons, percentage fertilizer recovery in the crop was small (Table 2), although greater in 1991/92 (17.2% on average) than in 1992/93 (11.3% on average). In 1991/92 percentage fertilizer recoveries in the crop were greater if the labelled fertilizer was applied at sowing (Table 2). There was no significant difference in 1992/93. Percentage fertilizer recovery was greater in the soil than in the crop (Table 2). Most of the fertilizer in the soil was found in the 0–20 cm soil layer (Table 2), with on average, <5% of the fertilizer in the 20–40 cm layer. Percentage fertilizer recovery in the soil was greater after an application of labelled fertilizer at tillering. They were greater in 1991/92 (42.9% on average) than in 1992/93 (31.5% on average). Overall, percentage fertilizer recoveries were poor, < 50% in 1992/93. They were greater following fertilizer applications at tillering than at sowing.

Table 1. Dry matter and nitrogen content of grain and above-ground biomass of wheat grown in 1991/92 and 1992/93. Nitrogen fertilizer (60 kg N ha^{-1}) was applied as a split-dressing at sowing and tillering, 30 kg N ha^{-1} on each occasion. This fertilizer was labelled either at sowing, at tillering or at both sowing and tillering. Standard errors of differences of means are presented

	Time of ^{15}N-labelled fertilizer application							
	1991/92				1992/93			
	Sowing	Tillering	Both	SED	Sowing	Tillering	Both	SED
Dry matter								
Grain (kg ha^{-1})	2586	2947	2542	134	2980	3040	3660	467
Whole crop (kg ha^{-1})	8214	9283	7923	155.9	8530	9110	9910	1159
Nitrogen								
Grain (kg ha^{-1})	58.0	67.5	59.0	1.87	64.4	66.6	78.8	11.2
Whole crop (kg ha^{-1})	83.9	98.6	86.6	3.3	90.3	95.6	108.1	13.0

Table 2. Percentage fertilizer recoveries in the grain and whole crop, or 0–20 cm and 20–40 cm soil layers from wheat crops grown in 1991/92 and 1992/93. Nitrogen fertilizer (60 kg N ha^{-1}) was applied as a split dressing at sowing and tillering, 30 kg N ha^{-1} on each occasion. This fertilizer was labelled either at sowing, at tillering or at both sowing and tillering

	Time of ^{15}N-labelled fertilizer application		
	Sowing	Tillering	Both
1991/92			
Grain	15.6	7.5	10.2
Whole crop	23.6	12.0	16.0
0–20 cm	20.4	55.1	37.0
20–40 cm	5.3	4.8	6.1
Total recovery	49.3	71.9	59.1
Unaccounted for	50.7	28.1	40.9
1992/93			
Grain	8.0	6.8	9.2
Whole crop	11.5	9.8	12.7
0–20 cm	22.7	33.1	25.8
20–40 cm	3.9	4.2	4.9
Total recovery	38.1	47.1	43.4
Unaccounted for	61.9	52.9	56.6

Discussion

The objective of sustainable cereal production is to balance nutrient inputs, often as inorganic fertilizer, with those removed in the harvested grain. Grain yields at Tel Hadya in these two seasons were comparable to those recorded for previous seasons with similar rainfall at this site (Anderson, 1985). By harvest most of the nitrogen in the crop had been translocated to the grain. Consequently between 58 and 79 kg N ha^{-1} were removed from the system in the grain, which corresponds approximately to the 60 kg N ha^{-1} applied as inorganic fertilizer.

Using ^{15}N-labelled fertilizer it is possible to quantify the amount of nitrogen taken up by the crop, retained in the soil or lost from it. In these experiments uptake of labelled fertilizer by the crop accounted for < 10 kg N ha^{-1} of the applied fertilizer. The remainder of the crop nitrogen (at least 77 kg N ha^{-1}) was derived from non-fertilizer sources, mainly mineralized soil nitrogen, although there will also have been contributions from rain, dry fixation of NH_3, and so on. A possible explanation of the relatively low uptake of labelled N could be that an Added Nitrogen Interaction (Jenkinson et al., 1985) occurred, i.e. the addition of ^{15}N-labelled fertilizer caused a real or apparent increase in plant uptake of unlabelled N. This usually occurs by pool substitution, with inorganic N from labelled fertilizer standing proxy for ^{14}N that would have been removed from the soil inorganic N pool by processes such as immobilization, or denitrification. However, pool substitution requires adequate mixing of the labelled and unlabelled N. Even though the labelled fertilizer was applied in solution to a moist soil which subsequently received rain, mixing may have been limited. Pool substitution by immobilization will result in increased retention of labelled N in the soil. The salient feature of Table 2 is not the retention of large quantities of labelled N in the soil, but rather its loss from the whole plant/soil system. This suggests that pool substitution, if it occurred, was driven by denitrification, not immobilization.

254

A common management strategy for increasing fertilizer uptake is to apply the fertilizer around the time of maximum crop growth (e.g. Penny et al., 1986). However, in 1991/92 fertilizer recoveries were greater following a fertilizer application at sowing rather than at tillering. There was no difference between timings in 1992/93, which may reflect the later sowing date in 1992/93, caused by continuous rainfall in November/December, which made the soil unworkable. Clearly, in dryland agriculture where crop growth is dependent upon rainfall, uptake of fertilizer by the crop is maximized by applying the fertilizer at sowing.

Fertilizer that is not recovered by the crop in the season of application is either retained in the soil or lost. In this experiment between 25–60% (37% on average) of the labelled fertilizer was retained in the top 40 cm of soil, and most (>79%) of this was in the top 20 cm. The proportion of this residual fertilizer available to subsequent crops is the subject of current investigation. However it may be small: a total of only 16% of the labelled fertilizer nitrogen remaining in the soil at harvest of the year of application was recovered by winter wheat crops over 4 successive seasons on a silty clay loam in England (Hart et al., 1993).

Less labelled fertilizer (26% on average) was retained in the soil when the fertilizer was applied at sowing only, compared to when it was applied either twice or at tillering only. The data presented here show that losses of labelled fertilizer from the crop/soil system were greater when the fertilizer was applied at sowing (50% and 61% in 1991/92 and 1992/93, respectively) than when it was applied at tillering (28% and 52% in 1991/92 and 1992/93, respectively). Such large losses of labelled fertilizer represent an economic and environmental cost, and require explanation. Leaching is unlikely: at Tel Hadya there is insufficient rainfall to cause drainage from the soil profile. Furthermore, only small amounts of ^{15}N were found below 20 cm. Denitrification following nitrification is a strong possibility, as is volatilization of NH_3. Volatilization could well occur from the calcareous soil at Tel Hadya from both urea and (to a lesser extent) ammonium sulphate in the period immediately following application. These fertilizers may be particularly vulnerable not only because they were applied as solution to the soil surface, but also because, generally they were applied to a warm wet soil. The significance of these loss processes to the nitrogen balance of wheat crops in these locations requires further investigation.

Acknowledgements

We thank the Overseas Development Administration for financial support under grant number R4779(H). We thank Prof D S Jenkinson for his contribution to this work, and his comments on an earlier draft of this paper.

References

Addiscott T M and Powlson D S 1992 Partitioning losses of nitrogen fertilizer between leaching and denitrification. J. Agric. Sci., Camb. 118, 101–107.

Anderson W K 1985 Differences in response of winter cereal varieties to applied nitrogen in the field. 1. Some factors affecting the variability of responses between sites and seasons. Field Crops Res. 11, 353–367.

Harris H C 1990 Productivity of crop rotations. In Farm Resource Management Program Annual Report 1989. pp 137–166. ICARDA, Aleppo, Syria.

Hart P B S, Powlson D S, Poulton P R, Johnston A E and Jenkinson D S 1993 The availability of the nitrogen in the crop residues of winter wheat to subsequent crops. J. Agric. Sci., Camb. 121, 355–362.

Jenkinson D S, Fox R A and Rayner J H 1985 Interaction between fertilizer nitrogen and soil nitrogen - the so-called 'priming effect'. J. Soil Sci. 36, 425–444.

Macdonald A J, Powlson D S, Poulton P R and Jenkinson D S 1989 Unused fertiliser nitrogen in arable soils - its contribution to nitrate leaching. J Sci. Food Agric. 46, 407–419.

Penny A, Widdowson F V and Jenkyn J F 1986 Results from experiments on winter barley measuring the effects of amount and timing of nitrogen and some other factors on the yield and nitrogen content of the grain. J. Agric. Sci., Camb. 106, 537–549.

Powlson D S, Hart P B S, Poulton P R, Johnston A E and Jenkinson D S 1992 Influence of soil type, crop management and weather on the recovery of ^{15}N-labelled fertilizer applied to winter wheat in spring. J. Agric. Sci., Camb. 118, 83–100.

Powlson D S, Hart P B S, Pruden G and Jenkinson D S 1986 Recovery of ^{15}N-labelled fertilizer applied in autumn to winter wheat at four sites in eastern England. J. Agric. Sci., Camb. 107, 611–620.

O. Van Cleemput et al. (eds.), Progress in Nitrogen Cycling Studies, 255–258, 1996.

Prediction of nitrogen fertilizer requirement with the HRI WELL-N computer model

C.R. Rahn, D.J. Greenwood and A. Draycott
Horticultural Research International, Wellesbourne, Warwick CV 35 9EF, UK.

Key words: computer model, fertiliser, nitrogen, prediction, recommendation

Abstract

HRI WELL-N is an easy to use computer model, which can predict crop nitrogen requirements for a wide range of agricultural and horticultural crops.

Growers and farmers have relied on their own experience, or the advice of specialist consultants, to estimate the amount of fertiliser required by their crops. Estimating the amount of nitrogen fertiliser required by a growing crop with any degree of accuracy is extremely difficult. The amount of nitrogen available to the crop depends on many factors including soil type and structure, temperature, local weather conditions, leaching, etc, while the crop's actual requirement is dependent on its growth and the efficiency with which the individual plant root system is able to take up the nitrogen from the soil. A dynamic simulation model was developed to gain a greater understanding of the processes affecting crop nitrogen requirement and soil supply. More recently a user friendly version was developed, WELL-N, capable of providing fertiliser recommendations with input data readily available to growers.

The reliability of the model and its subroutines have been tested by comparing predicted values with actual data obtained from a wide range of field crops grown in the UK and Northern Europe.

WELL-N calculates the nitrogen fertiliser requirements for most field crops grown in the UK, together with the likely final nitrogen content of the crop and the amount of nitrate which could potentially be leached into ground water from specific fertiliser applications.

Introduction

With around 750,000 t of nitrogen applied to tilled crops in England and Wales it is important to optimise fertiliser nitrogen applied. Vegetables particularly Brassicas, receive large quantities of fertiliser nitrogen, for example 36% of the area of Brussels sprout crops in England and Wales received more than 250 kg ha^{-1} N (British Survey of Fertiliser Practice 1993, pers. comm.). The economic penalties for failing to meet market criteria are so high relative to the cost of fertiliser nitrogen that growers can be tempted to apply insurance dressings of nitrogen above the optimum. On the other hand excessive amounts of fertiliser can cause cereal crops to lodge, bitterness and lodging in brussels sprouts, reduce storability of produce as well as increase the risk of nitrate leaching. In order to allow environmental sustainability of both arable and horticultural crops there is a great need to maximise the efficiency of nitrogen use and match it to nitrogen demand.

The methods for predicting fertiliser requirement include:

1. Experience
2. Applying the same amount of N to all fields
3. A simple table system
4. Using measurements of soil mineral nitrogen
5. Computer models

Experience is subjective and may not always lead to sound decision making. If growth is poor there is always the temptation of adding additional bags of nitrogen, but the limiting factors may be poor soil structure, pests, soil moisture, or disease problems rather than nitrogen.

Systems based on one rate of fertiliser for all crops may be provide satisfactory yields for crops (Neeteson et al., 1987), but may give rise to nitrate leaching and could increase the risk of variable quality produce. An improvement is to use simple tables such as provided by ADAS (MAFF, 1988) for UK use, with previous cropping history being taken account of as a nitrogen index. However, in high residue situations where large applications of manure have been applied (Shepherd, 1993), or in intensive Brassica rotations (Rahn et al., 1993) timely measurements of soil mineral nitrogen allow more balanced fertiliser predictions to be made. Fertiliser recommendation systems such as the 'KNS' system (Lorenz et al., 1989) provide a more comprehensive system of fertiliser advice but does relies on the ability to make more than one measurement of soil mineral N to take account the release of nitrogen from crop residues, and the loss of nitrogen due to leaching. The 'KNS'system also assumes that irrigation will support the availability of late applications of nitrogen fertiliser. Another system is provided by computer 'expert' systems such as 'N Expert' which makes more allowance for the release of nitrogen from crop residues and soil organic matter (Fink and Scharpf, 1993).

Getting the fertiliser application right involves understanding the contribution of many interacting factors such as previous crop residues, release of nitrogen from soil organic matter, rainfall, temperature, soil type, crop demand, rooting depth, planting and harvest times. This paper presents a possibility of using a mechanistically based computer simulation model to predict the nitrogen requirements of a range of crops.

Structure of the model

In order to achieve a greater understanding of the interacting soil and plant processes a dynamic simulation model was developed for a wide range of crops, such as potatoes (Greenwood et al., 1985a, b) and wheat (Greenwood et al., 1987a, b). The model includes generalised relationships for growth and its dependence on plant size, nitrogen content and temperature, the development of roots, nitrogen uptake, release of nitrogen from soil organic matter, evapotranspiration, soil water content and leaching. Since 1987 new subroutines have been added to allow for the release of nitrogen from incorporated crop debris. The structure of the model is shown in Figure 1, which operates on a daily timestep. This basic model was modified to run on IBM Compatible PCs and extended to include user-friendly input and output to allow farmers, growers and their advisors to use it. The model provides fertiliser recommendations for 22 different crops including wheat, potatoes and sugar beet.

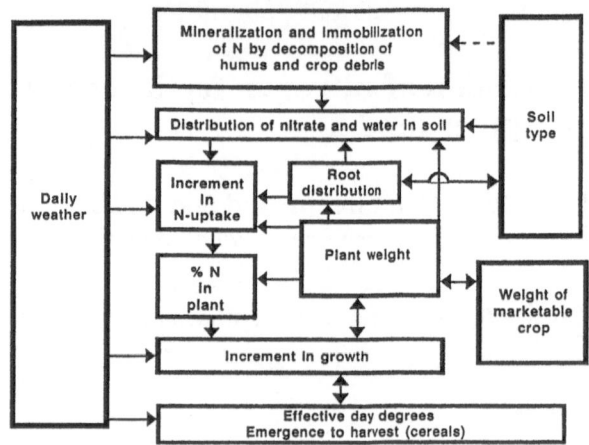

Figure 1. Flow diagram of WELL-N computer model.

Model - evaluation

The derivation and testing of the important functions in the model have been outlined in a series of papers. These include; growth rate and nitrogen concentration (Greenwood et al., 1986, 1990, 1991) root development (Greenwood et al., 1982); apparent fertiliser recovery (Greenwood et al., 1989) and leaching (Burns, 1974). The whole model has been tested for potatoes (Neeteson et al., 1987), wheat (Greenwood et al., 1987a), onion (Greenwood et al., 1992), cabbage (Riley and Guttormsen 1993, 1994) and for a wide range of different arable and vegetable crops (Greenwood and Draycott, 1989).

In addition to testing the validity of the model in conventional field experiments it was also "commercially" tested by a number of farm consultants and growers. The feedback revealed a number of unexpected shortcomings in the model which were subsequently corrected and provided suggestions for further improvement to enable the model to more easily meet their requirements. For example, improvements were made to the calculations for the distribution of mineral N down the soil profile during winter and to the growth of overwinter cauliflower. In addition improvements were made to the presentation of input and output data.

Model inputs

The inputs to the model are stored on the computer with a file for each field. The input file holds information on weather, soil type, previous crop, future crop, soil moisture deficit and available measurements of soil

Table 1. Crops for which WELL-N is designed to produce fertiliser recommendations

Broad beans	Chinese cabbage	Potato early
Brussels sprouts	French beans	Potato late
Dutch white cabbage	Leek	Radish
Cabbage summer	Lettuce butterhead	Red beet
Cabbage winter/spring	Lettuce crisp	Spinach
Calabrese	Onion	Sugar beet
Carrot	Parsnip	Swede
Cauliflower	Peas	Turnip
		Wheat

Table 2. Examples of fertiliser recommendations for cauliflower crops grown on silt soils using WELL-N and ADAS (MAFF, 1988)

Previous crop	Soil mineral N (kg ha^{-1})			WELL-N	ADAS
	0–30 cm	30–60 cm	60–90 cm		
Brussels sprouts	25	100	150	125	200
Brussels sprouts	150	100	25	25	200
None	25	100	150	200	250
Wheat	25	100	150	200	250

mineral nitrogen. Data entry is assisted by the provision of help screens which provide guide values for data not readily available to the grower. Data entry is numeric or by choice of an option.

Future weather data, such as mean air temperature, rainfall and open water evaporation are estimated from historic records by the model for 8 different regions of England and Wales. Estimations can be improved if the annual rainfall is known for the field, and if fertiliser predictions are required for spring grown crops, it is possible to include field, or farm measurements of monthly, or daily measurements of overwinter rainfall if available. Soil texture is entered, to enable calculations of available water within the soil profile. Limits to rooting depth such as rock or the presence of a soil pan can be entered. The release of nitrogen from soil organic matter is calculated, with a default mineralisation rate of 0.7 kg ha^{-1} N per day at 15.9 °C. The yield of the previous crop is entered together with the method of disposal of the crop residues in order to calculate the contribution of nitrogen from them to the next crop.

Recommendations can be provided for 22 crops, Table 1. The intended crop is selected from a menu as is the intended fertiliser application method (top or base dressing). The date of drilling, or transplanting, date of intended fertilisation, duration of growth, and expected yield are also required. The assessment of prospective yield should be based on previous experience of marketable yields in the area. Soil moisture deficit values have to be entered, the simplest is the date when drains begin to run when the soil moisture deficit can be assumed to be 0 mm. The model can accomodate measurements of soil mineral nitrogen in a very flexible way, as the model stores mineral nitrogen in up to 20 × 5cm layers. Pre planting/drilling amounts of irrigation water can also be entered.

The model can accept additional information collected during crop growth such as rainfall temperature, evaporation, irrigation, soil mineral N, soil moisture deficit, crop size and nitrogen content to further improve crop management by checking the need for any further topdressings of nitrogen.

Model running and outputs

The model runs with different levels of N fertiliser and an iterative procedure identifies an 'optimum' level defined as that level where there is a 2.5% increase in yield for every 25 kg N ha^{-1} increase in fertiliser application, this single figure recommendation is chosen to satisfy growers, though in practice such single 'optima' are difficult to establish (Sutherland et al., 1986). The model simulates the values of marketable fresh weight, total dry weight, the N content of the crop residues and the soil mineral N remaining at harvest for a range of levels of fertiliser N. The latter providing estimates of nitrogen potentally available to leaching before the next crop is planted.

Table 2 shows a number of recommendations derived using WELL-N. The model is very sensitive to the presence of previous crop even where it has been partly accounted for by measurements of soil mineral N. The model is also sensitive to the distribution of soil mineral N through the soil profile. For example where most of the mineral N is below 30 cm, the immediate rooting depth of the young crop there is a need to apply 125 kg ha^{-1} N, but where most of the mineral N is in the surface layer the recommendation is only for 25 kg ha^{-1}.

Conclusions

Computer modelling has lead to a greater understanding of processes and their interactions within the nitrogen cycle. This has led to the development of a scientific research model which has now been converted into a user friendly form for farmers and growers to use.

Acknowledgements

Financial support from the UK Ministry of Agriculture, Fisheries and Food for strategic research and Horticultural Development Council for the development work is greatly appreciated.

References

Burns I G 1974 A model for predicting the redistribution of salts applied to fallow soils after excess rainfall evaporation . J. Soil Sci. 25, 165–168.

Fink M and Scharpf H C 1993 N Expert - A decision support system for vegetable fertilization in the field. Acta Hortic. 339, 67–74

Greenwood D J and Draycott A 1989 Experimental validation of an N-response model for widely different crops. Fert. Res. 18, 153–174

Greenwood D J, Gastal F, Lemaire G, Draycott A, Millard P and Neeteson J J 1991 Growth rate and %N of field grown crops: Theory and experiments. Ann. Bot. 67, 181–190.

Greenwood D J, Gerwitz A, Stone D A and Barnes A 1982 Root development of vegetable crops. Plant and Soil 68, 75–96.

Greenwood D J, Kubo Ken-Ichi, Burns Ian G and Draycott A 1989 Apparent recovery of fertilizer N by vegetable crops. Soil Sci. Plant Nutr. 35, 367–381.

Greenwood D J, Neeteson J J and Draycott A 1985a Response of potatoes to N fertilizer: Quantitative relations for components of growth. Plant and Soil 85, 163–183.

Greenwood D J, Neeteson J J and Draycott A 1985b Response of potatoes to N fertilizer: Dynamic model. Plant and Soil 85, 185–203.

Greenwood D J, Neeteson J J and Draycott A 1986 Quantitative relationships for the dependence of growth rate of arable crops on their nitrogen content, dry weight and aerial environment. Plant and Soil 91, 281–301.

Greenwood D J, Neeteson J J, Draycott A, Wijnen G and Stone D A 1992 Measurement and simulation of the effects of N-fertilizer on growth, plant composition and distribution of soil mineral-N in nationwide onion experiments. Fert. Res. 31, 305–318.

Greenwood D J, Stone D A and Draycott A 1990 Weather, nitrogen-supply and growth rate of field vegetables. Plant and Soil 124, 297–301.

Greenwood D J, Verstraeten L M J and Draycott A 1987a Response of winter wheat to N-fertilizer: Quantitative relations for components of growth. Fert. Res. 12, 119-137.

Greenwood D J, Verstraeten L M J, Draycott A and Sutherland R A 1987b Response of winter wheat to N-fertilizer: Dynamic model. Fert. Res. 12, 139–156.

MAFF 1988 Fertiliser recommendation. Reference Book 209. HMSO, London, UK.

Lorenz H P, Schlaghecken J, Engl G, Maync A, Ziegler J and Strohmeyer K 1989 Kulturbegleitendes Nmin sollwert (KNS) - System, Ministerium fur Landwirtschaft, Weinbau und Forsten, Mainz, Germany.

Neeteson J J, Greenwood D J and Draycott A 1987 Proceedings No. 262, The Fertiliser Society. pp 1–29. A dynamic model to predict yield and optimum nitrogen fertiliser application rate for potatoes. Paper read before the Fertiliser Society of London on 10 December 1987.

Rahn C R, Paterson C D and Vaidyanathan L V V 1993 Improving the use of nitrogen in brassica rotations. Acta Hortic. 339, 207–218.

Riley H and Guttormsen G 1993 N requirements of cabbage crops grown in contrasting soils. I: Field trials. Norw. J. Agric. Sci. 7, 275–291.

Riley H and Guttormsen G 1994 N requirements of cabbage crops grown on contrasting soils in southern Norway. II. Model predictions. Norw. J. Agric Sci. 8, 99–113.

Shepherd M A 1993 Measurement of soil mineral nitrogen to predict the response of winter wheat to fertiliser nitrogen after application of organic manures or after ploughed - out grass. J. Agric. Sci., Cambridge 121, 223–231.

Sutherland R A, Wright C C, Verstraeten L M J and Greenwood D J 1986 The deficiency of the 'economic optimum' application for evaluating models which predict crop yield response to nitrogen fertiliser. Fert. Res. 10, 251–262.

O. Van Cleempu: et al. (eds.), Progress in Nitrogen Cycling Studies, 259–264, 1996.
© 1996 *Kluwer Academic Publishers.*

Formation of nitrate pool in spinach grown on different soils (with ^{15}N)

L.L. Raikova and P.V. Petkov
N Poushkarov Institute of Soil Science and Agroecology, 7 Chaussee Bankya, 1080 Sofia, Bulgaria

Key words: fertilization, ^{15}N isotope, nitrate pool, soils, spinach

Abstract

The influence of soil type and fertilization on the formation and origin (participation of soil and fertilizer nitrogen) of nitrate pool was studied in spinach grown on six typical soils of the country under greenhouse conditions.The results show that fertilization with NPK affects the formation of nitrate pool in spinach most strongly, while the influence of the soil type is more slightly pronounced. With the treatments which gave highest yields, the share of soil nitrogen in the composition of nitrate pool was 10.6–18.1% under ammonium nitrate application and 14.5–23.0% under urea, while that of the fertilizer nitrogen was 81.9–89.4% and 77.0–85.5%, respectively.

Introduction

Under intensive agriculture a number of ecological problems arose, including the pollution of crop production with nitrates (the vegetable and fodder crops in particular). The accumulation of nitrate, toxical to people and animals, is determined by a variety of factors: plant species and crop varieties, soil properties, climatic features, agrochemical and agrotechnical measures, conditions under which the produce is processed and preserved (Andrushcenko, 1983; Breimer, 1982; Paschold and Hunt, 1986; Pokrovskaya, 1988; Raikova and Rankov, 1984; Vulstake and Briston, 1978). The influence of soil properties which to a great extent determine the processes of nitrogen transformation in soil, its assimilation and metabolism in plants, as well as the formation of nitrate pool in crop production, have not yet been studied sufficiently (Breimer, 1982; Geyer, 1978; Guillard and Allinson, 1988; Paschold and Hunt, 1986; Pokrovskaya, 1988; Semenov and Sokolov, 1984). The participation of fertilizer N and soil N in the nitrate pool depends on the properties of the soils. Realistic and exact assessment of the quantitative participation of soil N and fertilizer N in nitrate pool of crop produce can only be obtained by using ^{15}N isotope method (Agaev et al., 1986; Semenov and Sokolov, 1984; Smirnov et al., 1984).

The aim of this study was to investigate the influence of soil type and fertilization on the formation and origin (participation of soil N and fertilizer N) of the nitrate pool in spinach production. It was sought for the relationship fertilizing-yield-nitrate accumulation-origin of nitrate pool and the factors likely to affect it in spinach grown on different soils.

Materials and methods

The study was conducted in the spring of 1992 with spinach (Matador variety) as a test crop with regard to nitrate accumulation. The experiment was carried out under greenhouse conditions in 3 replications in pots holding 3 kg of soil each, and with 6 plants in each pot. The investigation involved 6 typical virgin soils of the country, presenting a rich variety of soil properties (Table 1). The basic properties of these soils are not affected by cultivation. The fertilization was effectuated with NPK (N = 300, P = 240 and K = 150 mg pot^{-1}) in increasing rates while keeping the same correlation (N:P:K = 1.0: 0.8:0.5). The following fertilizing treatments were tested: 1) 0NPK, 2) 1NPK, 3) 2NPK, 4) 4NPK and 5) 8NPK. Two nitrogen fertilizer forms were used (ammonium nitrate and urea), labelled with ^{15}N at 10 atom % enrichment. Plant samples were taken from the above-ground mass at harvesting and were then dried and prepared to be analysed. The content of nitrates was determined in water extract through distillation by the method of Bremner and Keeny. The isotope composition of the nitrogen in the nitrates was determined with the NO1-3 apparatus.

Table 1. Agrochemical characteristics of soils

Soils	Particles (%)		CEC (meq 100 g^{-1})	Base saturation (%)	pH		Organic C (%)	Total N (%)	Mineral N (mg kg^{-1})			P$_2$O$_5$ (mg 100 g^{-1})	K$_2$O (mg 100 g^{-1})
	<0.01 mm	<0.001 mm			H$_2$O	KCl			NH$_4$-N	NO$_3$-N	Σ		
1. Leached chernozem, G. Dabnik, Lovech area	52.1	37.7	34.1	88.1	6.3	5.4	2.42	0.260	6.3	1.0	7.3	1.4	33.4
2. Calcareous chernozem D. Mitropolia, Lovech area	38.9	23.9	30.2	100.0	7.9	7.2	1.04	0.130	3.7	4.0	7.7	4.6	22.9
3. Leached smolnitza Bojuritche, Sofia area	71.4	56.2	52.2	92.1	6.6	5.6	2.19	0.171	5.0	1.0	6.0	0.5	33.3
4. Leached cinnamonic forest D. Izvor, Plovdiv area	52.9	36.0	32.3	86.6	6.1	5.2	2.04	0.185	6.0	1.0	7.0	1.0	45.3
5. Dark-grey forest Radichevo, Lovech area	45.3	26.8	23.9	77.4	5.6	4.8	1.99	0.195	4.7	2.0	6.7	1.6	15.7
6. Alluvial-meadow N. Iskar, Sofia area	57.5	39.1	20.2	84.9	6.4	5.7	1.49	0.160	4.5	2.0	6.5	0.9	12.4

The experimental and analytical data were processed by the methods of mathematical statistics (analysis of variance and analysis of regression).

Results and discussion

Yields

The results obtained for spinach yield (Fig.1) show that the natural fertility of the virgin soils was rather different. In the non-fertilized treatments, yields rank in a descending order as follows: leached chernozem, calcareous chernozem and dark-grey forest soil (2.77, 2.51 and 2.39 g pot^{-1} air-dry matter); leached cinnamonic forest soil (1.75 g pot^{-1}); alluvial-meadow and leached smolnitza (0.30 and 0.24 g pot^{-1} respectively). Yield variations are conditioned by the soils, different agrochemical properties which determine their natural fertility. In all soils increasing NPK rates result in increased spinach yields. Highest increase was observed in the leached smolnitza (17.4–64.5 times) and the alluvial- meadow soil (1.3–43.3 times) followed by the leached cinnamonic forest soil (1.7–9.3 times), the leached chernozem (1.4–5.3 times), the calcareous chernozem (3.8–5.3 times) and the dark-grey forest soil (3.1 – 5.3 times). It should be noted that, with the change of plant's nutrient conditions resulting from fertilization, yields in the soils with low

natural fertility increase considerably. In the leached chernozem, the calcareous chernozem and the leached smolnitza highest yields are obtained when fertilized with the 8NPK rate (14.8; 13.9 and 13.2 g pot^{-1} for the ammonium nitrate and 13.6; 136 and 15.5 g pot $^{-1}$ for the urea), while in the leached cinnamonic forest soil, the alluvial-meadow and the dark-grey forest soil - when the 4NPK rate is applied (15.1; 12.3 and 10.4 g pot^{-1} for ammonium nitrate and 16.2; 13.0 and 12.8 g pot^{-1} for the urea, respectively). In the non-chernozem soils fertilizaton with the 8NPK rate reduces the yield because of their unfavourable sorption and agrochemical properties. The high concentration of the fertilizers in the soil has had a depressive effect on the sprouting seeds, root system development and yield formation. Yield differences (between the control and the fertilized treatments) in LSD at $p = 0.1\%$ have been statistically proven in all soils. The effectiveness of the tested nitrogen fertilizer forms (ammonium nitrate and urea) does not differ significantly in the different soils and fertilizing rates, and in the majority of cases the yield differences have not been statistically proven. The differences are significant only in the calcareous chernozem and the alluvial-meadow soil when LSD at $p = 0.1\%$.

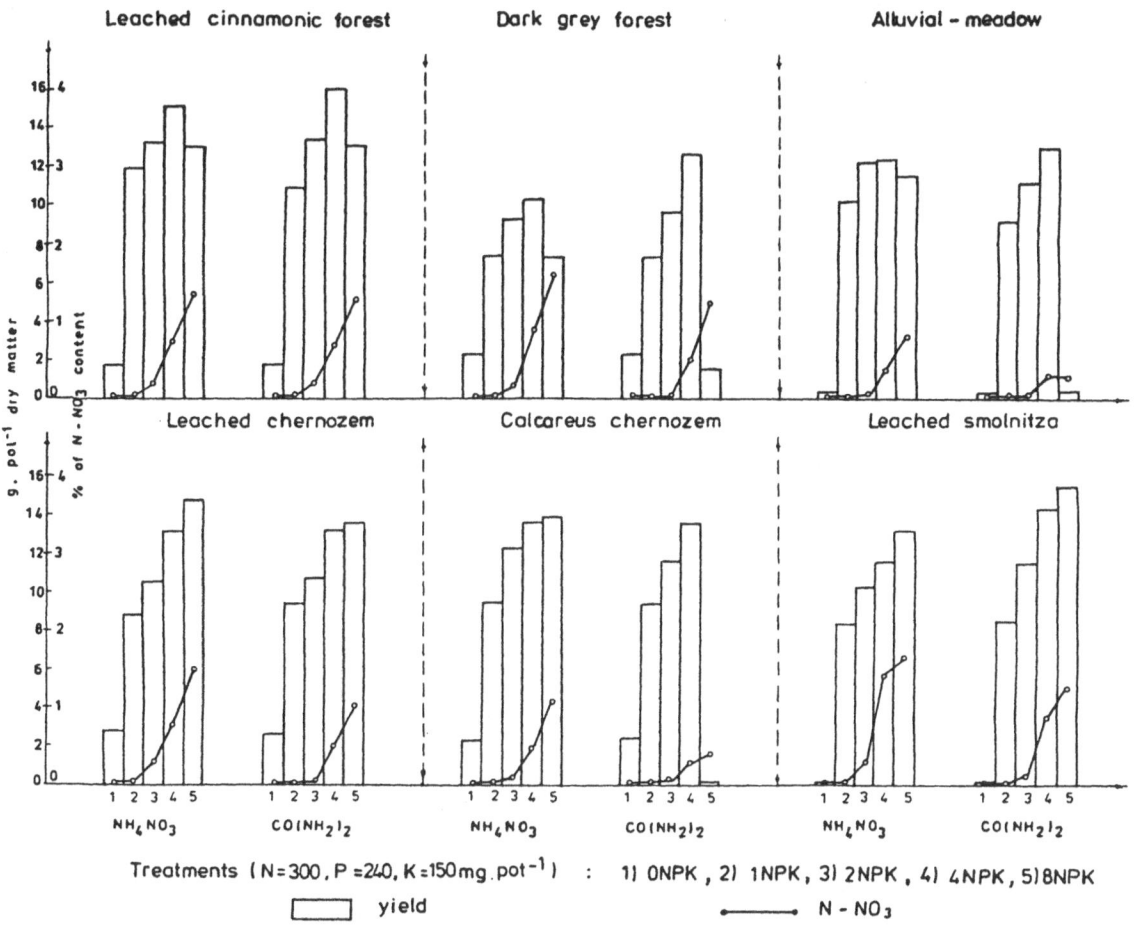

Figure 1. Yield and nitrate content in spinach grown on different soils.

Content of nitrates

The content of nitrates (N-NO₃) in spinach at the non-fertilized treatments is low and varias in the different soils from 0.03 to 0.08% air-dry matter (Fig.1). Fertilizing with increasing rates of NPK enhances the nitrate content in all soils, the highest value being established at the 8NPK rate. Highest nitrate content increase is found in the dark-grey forest soil (to 1.66% for the ammonium nitrate and 1.01% for the urea), the leached chernozem (to 1.47 and 1.01%), the leached smolnitza (to 1.71 and 1.33%) and the leached cinnamonic forest soil (to 1.34 and 1.26%), followed by the lower values in the calcareous chernozem (to 1.08 and 0.37%) and the alluvial-meadow soils (to 0.84 and 0.23%). The differences in the nitrate content established in the individual soils are determined by their agrochemical properties. Heavy textured soils, having lower pH and higher humus and nitrogen content, accumulate greater nitrate amounts. Almost in all soils nitrate concentration in spinach after urea application is lower compared to the one after ammonium nitrate application. The results comply with other investigations (Breimer, 1982; Poschold and Hunt, 1986; Raikova and Rankov, 1986; Raikova and Shevkenov, 1988; Titz, 1989) and confirm the inference that, when ammonium and amide fertilizer forms are used, the nitrate content in yields decreases.

There is a positive correlation between the fertilization with increasing NPK rates and the content of nitrates in spinach. This can be seen from the regression equations calculated (for the various soils r = 0.927 - 0.991 under ammonium nitrate and r = 0.726 - 0.920 when urea is applied).

Formation of the nitrate pool

The ^{15}N labelled nitrogen fertilizers involved in the investigation and the analysis of the isotope composition of nitrares allowed to determine the actual par-

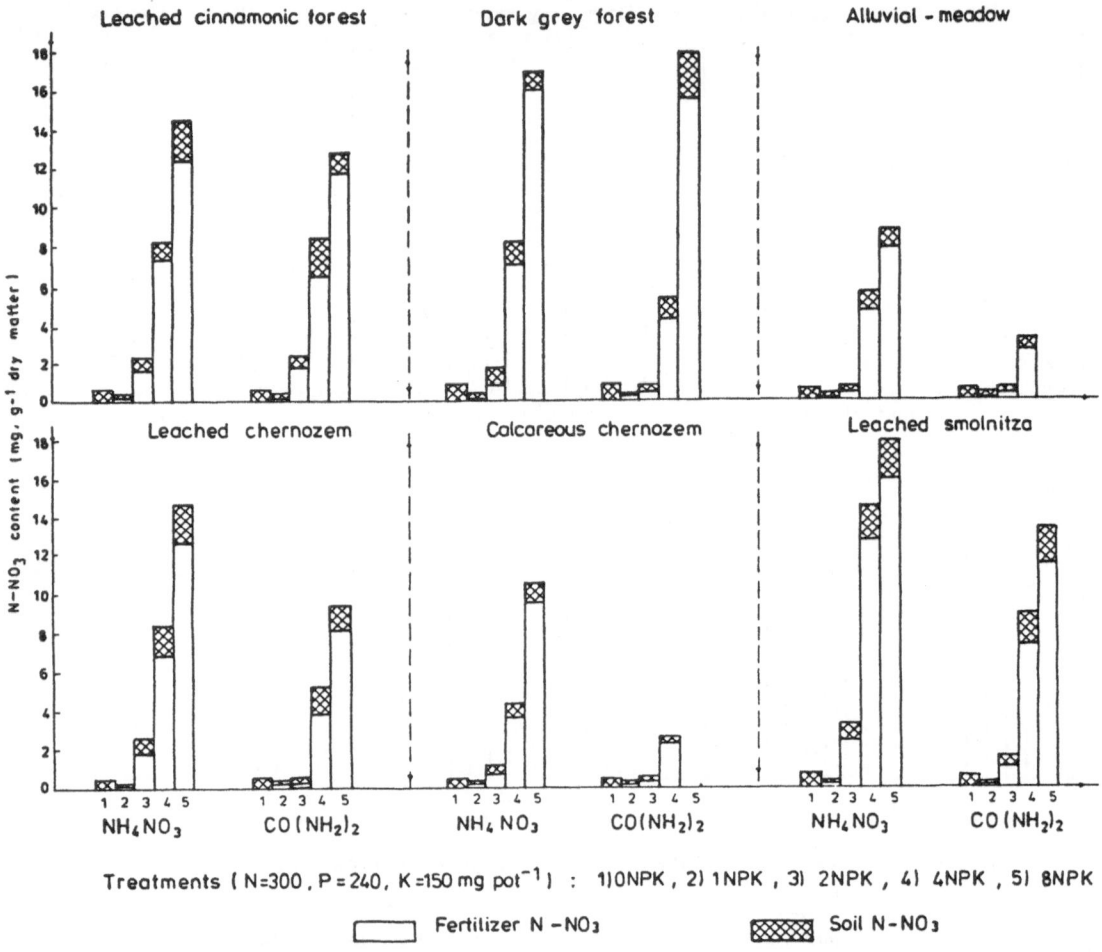

Figure 2. The fertilizer and soil nitrogen participation in the formation of nitrate pool in spinach grown on different soils.

ticipation of fertilizer N and soil N in the formation of the nitrate pool in spinach. The results (Fig. 2) show that the nitrate pool was formed at the expense of both the fertilizer nitrogen and the nitrogen of the soil stock. In all soils the amount of fertilizer N and soil N, making the nitrate pool, increase under the influence of fertilization with increasing NPK rates (from 0.04 to 16.15 and from 0.11 to 2.52 mg g^{-1} dry matter). The increase is less strongly pronounced in the calcareous chernozem and the alluvial-meadow soil compared to the remaining soils. The increase is also lower when introducing urea in comparison with the ammonium nitrate application. Besides the effect of the fertilizers additional amounts of nitrates accumulate in spinach coming from the stock of soil nitrogen. As the fertilizer rate increases the amount of "extra-nitrates" also increases (from 0.05 to 1.89 mg g^{-1} dry matter). These amounts are greater in the chernozem soils, when ammonium nitrate was applied. The results

from the conducted investigation show that the determinative effect on the formation of the nitrogen pool in spinach belongs to fertilization, while the role of the soil type is less significant. These results comply with the investigations of other authors (Agaev et al., 1986, 1988; Semenov and Sokolov, 1984; Semenov et al., 1985, 1986), who also confirm the same inferences, namely that the nitrate pool in plant produce is formed with the participation of both the fertilizer and soil nitrogen.

Origin of the nitrate pool

The use of the ^{15}N isotope method in the study allowed to determine the origin of nitrate pool in spinach (soil N and fertilizer N participation) grown on different soils. The results show (Fig. 3) that the share of soil nitrogen varies from 3.5 to 75.0% under the various soils and fertilizer treatments, while that of the fer-

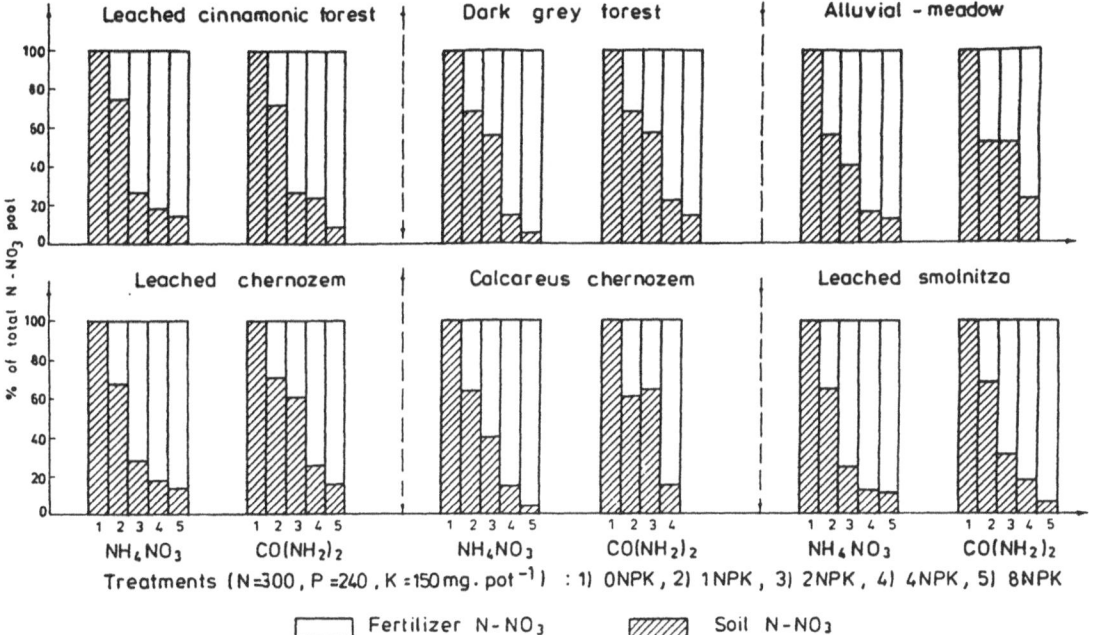

Figure 3. The fertilizer and soil nitrogen ratio in the nitrate pool of spinach grown on different soils.

tilizer nitrogen -from 25.0 to 96.5%. The increase of fertilizer rate strongly reduces the participation of soil nitrogen and enhances that of the fertilizer nitrogen for all soils. The form of the nitrogen fertilizer does not affect substantially the composition of the nitrate pool in spinach. The influence of the soil type on the correlation soil N - fertilizer N in the nitrate pool is lower in comparison with that of the fertilization. As to the treatments where highest yields were obtained (8NPK for the chernozems and 4NPK for the remaining soils) the share of the soil nitrogen is 10.6–18.1% under ammonium nitrate application and 14.5-23.0% - when urea is introduced. For the fertilizer nitrogen the share is 81.9–89.4 and 77.0–85.5%, respectively.

The results from the investigation, conducted as a greenhouse experiment, show the composition of the nitrate pool in spinach to be determined mainly by fertilization, and to a lesser extent - by the soil type. This is confirmed by the investigations of other authors (Smirnov et al., 1984), who have found that 60–80% of the nitrates accumulating in cabbage originate from the fertilizers, and 20–40%-from the soil. On the other hand, there are references (Agaev et al., 1986, 1988; Agaev, 1989, 1990; Semenov et al., 1985, 1986; Semenov and Sokolov, 1984; Sokolov et al., 1989) indicating that the share of soil N in the formation of nitrate pool in spinach, lettuce and radishes is

70–85%, while the remaining 15-30% are represented by fertilizer N.

Conclusion

The formation and origin of nitrate pool in spinach is determined mainly by fertilization, while the effect of the soil type is pronounced less strongly. The obtained quantitative parameters for the correlation soil N - fertilizer N in the nitrate pool of spinach grown on different soils can serve as a basis to develop models for predicting and regulating the nitrate pollution of crops. This correlation can be used as a diagnostic criterion to predict the participation of fertilizer and soil nitrogen in the formation of nitrate pool in plant production.

References

Agaev V A 1989 Formation of nitrate pool in plants as a result of the nutrious conditions. Proc. Conf. Ecological problems of nitrate accumulation in the environment, Poushcino. pp 46–47.

Agaev V A 1990 Significance of nitrogen fertilizers in the formation of nitrate pool in plants. Proc. Conf. Nitrogen-fertilizer-soil-plant, Prague. pp 37–141.

Agaev V A, Semenov V M, Movsumov Z P and Sokolov O A 1988 Incorporation of fertilizer nitrogen in plant nitrates. Proc. Soviet Union Acad. Sci., Biol. Ser. 5, 733–739.

Agaev V A, Semenov V M and Sokolov O A 1986 Nitrate accumulation in spinach. Agrochem. 12, 7–14.

Andrushcenko V K 1983 Nitrate content in vegetables and ways for its decreasing. Review Inform., Rio Niinti, Kishinev. 54 p.

Breimer T 1982 Environmental factors and cultural measures affecting the nitrate content in spinach. Fert. Res. 3, 191–292.

Geyer B 1978 Untersuchungen zur wirkung hoher stickstoffgaben auf den nitratgehalt von freilandgemuse. Arch. Gartenbau 26, 1–13.

Guillard K and Allinson D W 1988 Effect of nitrogen fertilization on a Chinese cabbage hybrid. Agron. J. 8, 21–26.

Paschold P J and Hunt I 1986 Production von spinat and mohren mit reduziertem nitrate gehalt. Obzor MS Agroinform, Berlin. 53 p.

Pokrovskaya S F 1988 Ways for decreased nitrate levels in vegetables. Obzor MS Agroinform, Moscow. 61 p.

Raikova L and Rankov V 1984 Conditions contributing to the accumulation of nitrates in the vegetable and feed crops at intensive inorganic fertilization. Review Inform., Sofia. 80 p.

Raikova L and Rankov V 1986 Relationship of nitrogen fertilizer forms and N-serve application to nitrate content in radish plants. Summ. XIII Congress of International Society Soil Science, Hamburg. pp 919–920.

Raikova L and Shevkenov M 1988 Effect of nitrogen and potassium fertilizer forms on nitrate accumulation in spinach. Proc. Conf. Mineral fertilizer-production-application, Varna. pp 68–69.

Semenov V M, Knop K, Agaev V A, Matoula I, Balik I, Matoush O and Sokolov O A 1985 Nitrogen fertilizer application and nitrate content regulation in plants. Agrochem. 9, 6–16.

Semenov V M, Prugar V, Knop K, Pekhova A V, Agaev V A and Sokolov O A 1986 Nitrates accumulation by plants under conditions of intensive application of nitrogen fertilizers. Proc. Soviet Union. Acad. Sci., Biol. Ser. 2, 201–210.

Semenov V M and Sokolov O A 1984 Transformation of nitrogen fertilizer in grey forest soils at different ways of application. Agrochem. 9, 3–10.

Smirnov P M, Bazilevich S D, Obukhovskaya L V and Kudriashova L A 1894 Pe-tsai cabbage yielding capacity and nitrate content under different nitrogen levels and nitrification inhibitors application. Proc. Timiriazev Acad. Agric. Sci. 3, 75–79.

Sokolov O A, Timchenko A B and Bubanova T L 1989 Fertilizer and soil nitrogen in formation of nitrate pool of plants. Proc. Conf. Ecological problems of nitrate accumulation in the environment, Poushcino. pp 111–112.

Titz R 1989 Stickstoffdungung bei frilandgemuse in der ammonium zur einschrankung von nitratanreicherungen in pflanzen und boden. Diss. Universitat Bonn. 173 p.

Vulstake C and Briston K 1978 Factors affecting the nitrate content in field grown vegetables. Qual. Plant. 28, 71–87.

O. Van Cleemput et al. (eds.), Progress in Nitrogen Cycling Studies, 265–270, 1996.

Testing of different ways of N-fertilizer application for decreasing the nitrate content in some vegetable crops

L.L. Raikova, D.A. Stoichev, G.K. Angelov and D.I. Stoicheva
N. Poushkarov Institute of Soil Science and Agroecology, 7 Chaussee Bankya, 1080 Sofia, Bulgaria

Key words: fertilization, inhibitor, mineral nitrogen, nitrates, rotation, vegetable crops

Abstract

Under the conditions of a field experiment in vegetable crop rotation (potatoes, cabbage, spinach), we studied the influence of different ways of nitrogen fertilizer application on the crop productivity, the accumulation of nitrates in the crops, and the mineral nitrogen contents in the soil profile. Results show that yields depend mainly on the nitrogen rate, and are not significantly influenced by the way of fertilizer application, as well as by the form of the potassium fertilizer. The nitrate content in the crop decreases when the nitrogen fertilizer is split applied, by an inhibitor of the nitrification, and under the form of potassium chloride. The amount of the non-utilised mineral nitrogen in the soil profile after the rotation is insignificant, and is not the precondition for an ground water nitrate pollution.

Introduction

The nitrate concentration in crops has been intensively studied during the last years, both internationally and in our country. The main sources of nitrate uptake by humans are vegetables (70%), drinking water (21%) and meet and related products (6%) (Breimer, 1982; Danek - Jesik, 1990; Loggers, 1979; Pokrovskaya, 1988). The already introduced nitrates in the human body are reduced to nitrites, and cause methemoglubinemia and the formation of cancerogeneous nitrocompounds, lower the immunity and the resistance towards cancerogeneous and mutageneous agents (Bryson, 1984; Loggers, 1979; Opopol and Dobrynskaya, 1986; Pokrovskaya, 1988).

A large part of the studies is connected with the various factors influencing the nitrate content in the vegetable crop production, and the development of technological solutions for regulation of their concentrations under the maximum permissible levels (Andruschenko, 1983; Breimer, 1982; Kelly, 1990; Poschold and Hunt, 1986; Pokrovskaya, 1988; Prugar and Prugarova, 1985; Raikova and Rankov, 1984; Sindelarova, 1986; Titz, 1989; Vulstake and Byston, 1978). The importance of the conditions of nitrogen plant nutrition (rates, forms, terms and types of application of the nitrogen fertilizer, balance of the nitrogen with the rest

of the macro and micro elements) for the process of nitrogen accumulation has also been noted (Breimer, 1982; Pokrovskaya, 1988; Titz, 1989).

The general conclusion from those experiments is that the nitrate pollution of vegetable crops has a complex character, and it depends on the biological peculiarities of the species and the diversity of the plants, the soil-climatic conditions, the level of application of the agrochemical and agrotechnical practices.

The purpose of the present study is to test different ways of application of the nitrogen fertilizer in order to obtain lower nitrate contents in vegetable crops and to minimise the accumulation of non-utilised nitrogen in the soil profile.

Materials and methods

The study was carried out under the conditions of a field fertilizer experiment, on alluvial-meadow soil in South Bulgaria. The soil profile has a light texture, and low water-holding capacity. The ploughing horizon is characterised by a low cation exchange capacity (7.9 cmol (+) kg^{-1}, acid soil reaction (pH in water = 6), organic C- (0.41%), and total nitrogen (0.052%). Mineral nitrogen in the 0-30, 30- 60, and 60-90 cm layers was 6.0, 4.1, and 2.1 mg kg^{-1} soil respectively at the start of the experiment i.e. spring 1992. The trial was set up in three replications, the size of the experimental plots was 28 square meters, and the crop rotation was as follows: early potato of the Sante variety, late cabbage of the Kyosse variety, and winter-spring spinach of the Matador variety.

In order to find out the parameters of nitrate accumulation in the crop on a background of $P_{100}K_{100}$ kg ha^{-1}, we applied a three-factor experimental scheme: nitrogen rates (N_0, N_{60}, N_{120} and N_{180} kg ha^{-1}); way of application of the fertilizer (1 – one-time spread before sowing; 2 – one-time spread before sowing with an inhibitor of nitrification CMP (1-carbamoyl-3(5)-methylpyrazol); 3 – split -1/2 of the rate spread before sowing, and 1/2 in rows as a dressing); form of the potassium fertilizer (potassium chloride and potassium sulphate). The mineral fertilizers for potatoes were applied as urea, potassium chloride, potassium sulphate and super-phosphate, while the cabbage and the spinach were grown as an after-effect with no fertilization.

Plant samples were taken during the process of vegetable production collecting, and soil samples were taken before the experment setting and after the cabbage and spinach harvest. The nitrate levels in the fresh plant samples after extraction with 1% [KAI $(SO_4)_2$]. $12H_2O$ (1:5) were determined by an ion-selective electrode and MIN – 100 measuring microprocessor. Soil N_{min} was extracted by 1 N KCl (1:2.5) and measured by ADM-300 analyser. The resulting experimental and analytical data were processed by the method of mathematical statistics, and their credibility was evaluated by LSD at p=0.05.

Results and discussion

Yields

Yields data are represented in Figure 1a, b, c. For early potatoes (Fig. 1a) maximum yields are obtained for the N_{120} rate, as the increase compared to the non-fertilized control is from 10.4 to 20.6%. There are no significant differences between the yields in the control plots and the experimental nitrogen rates, the ways of application and forms of the potassium fertilizers. The recommended rate of N_{180}, according to the model for fertilization recommendations in our country, provides a lower yield and results in the accumulation of residual fertilizer nitrogen in the soil. Cultivating late cabbage as a second crop after potatoes, comes up as an

effective technological method for maximum uptake of the applied fertilizer nitrogen into the biological cycle. Data in Figure 1b show that the highest cabbage yields are obtained from the after-effect of potato fertilization with the N_{180} rate (an increase in relation to the control of 22.3 – 54.5%), but significant differences are observed only of split application of this rate on a potassium sulphate form. For all other variants, yields are also higher as compared to the control, but are not significantly different. The inclusion of spinach as a third crop gives a final proof for the residual mineral nitrogen. Data for the yields (Fig. 1c) show highest values for the fertilization with N_{120} and N_{180} rates, as the increase in relation to the control varies from 16.4 to 43.1%. The mentioned differences are significant for the one-time spread with an inhibitor of nitrification for both forms of potassium fertilizers, and in parts, for the potassium sulphate. For all other variants differences have not been proved on a significant basis.

Results from the study show that the productivity of the crops grown under crop rotation (potatoes, cabbage, spinach) is determined mainly by the nitrogen rate, and is not significantly influenced by the way of nitrogen application and the form of the potassium fertilizer. These results are in accordance with other studies we have made (Raikova and Petkova, 1982; Raikova and Rankov, 1986) and they prove the leading role of nitrogen in the process of yield formation.

Nitrates content in the vegetables

The nitrate content in the crops grown under crop rotation (Fig. 2a, b, c) gives grounds to the assumption that the tested nitrogen rates and ways of fertilization do not lead to a nitrate accumulation. It is worth mentioning that in all variants there is no vegetable production received with nitrate contents above the accepted permissible level for Bulgaria (potatoes - 250, cabbage - 500, and spinach - 1200 mg kg^{-1} fresh weight). The cultivation of potatoes, cabbage, and spinach in a crop rotation provides maximum inclusion of the applied nitrogen into the process of yield formation. For all of these crops, the nitrate contents in the production increases with higher nitrogen rates, without reaching maximum allowed levels. Reduction of the model-determined rate also decreases the nitrate accumulation. The nitrate content in the crop is decreased by pre-sowing fertilization with a nitrification inhibitor, and split nitrogen application for both tested potassium forms. The CMP inhibitor suppresses the nitrification and maintains nitrogen from the urea in ammo-

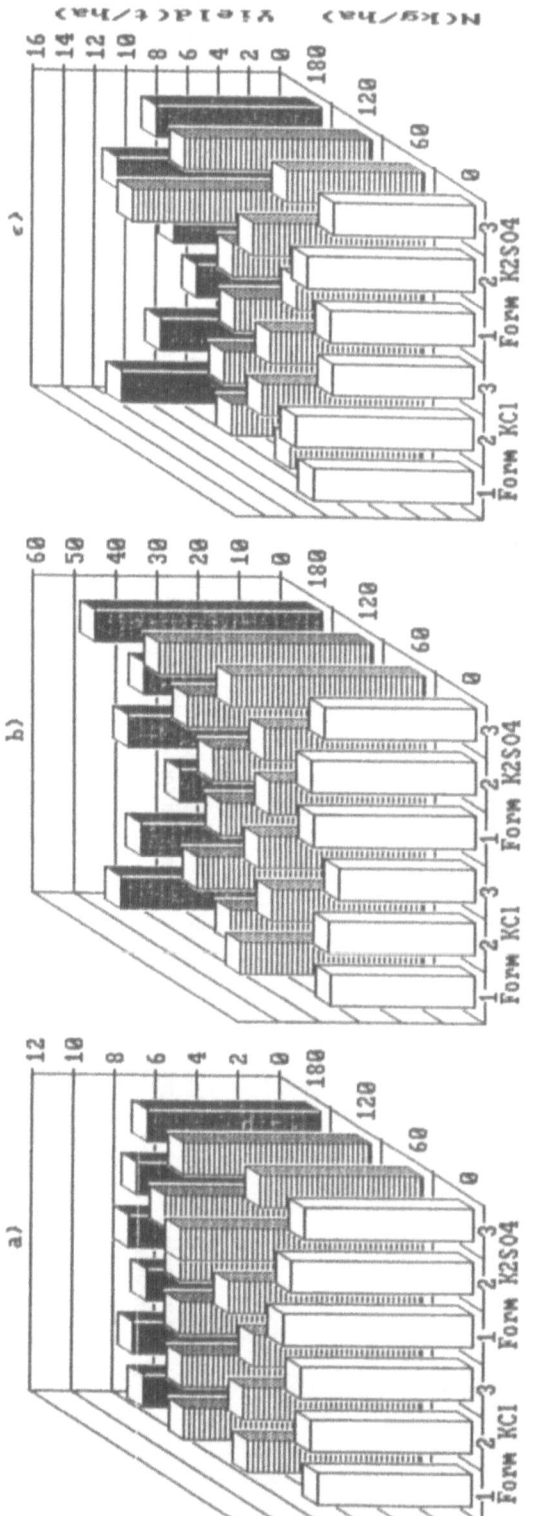

Figure 1. Yields of potatoes (**a**), cabbage (**b**), and spinach (**c**). Ways of application: 1 - one-time spread before sowing; 2 - one-time spread before sowing with an inhibitor of nitrification (CMP); 3 - split -1/2 of the rate spread before sowing, and 1/2 in rows as a dressing.

268

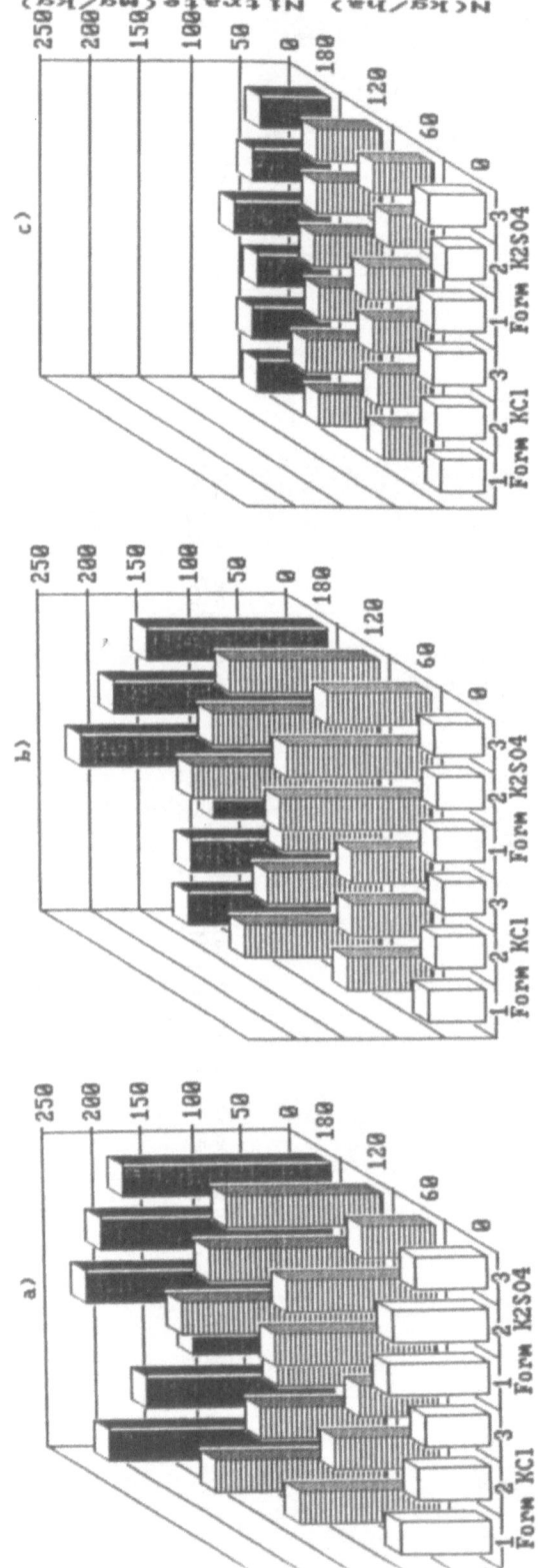

Figure 2. Nitrate content in the vegetable production of the potatoes (a) cabbage (b), and spinach (c). Ways of application:1 - one-time spread before sowing; 2 - one-time spread before sowing with an inhibitor of nitrification (CMP); 3 - split -1/2 of the rate spread before sowing, and 1/2 in rows as a dressing.

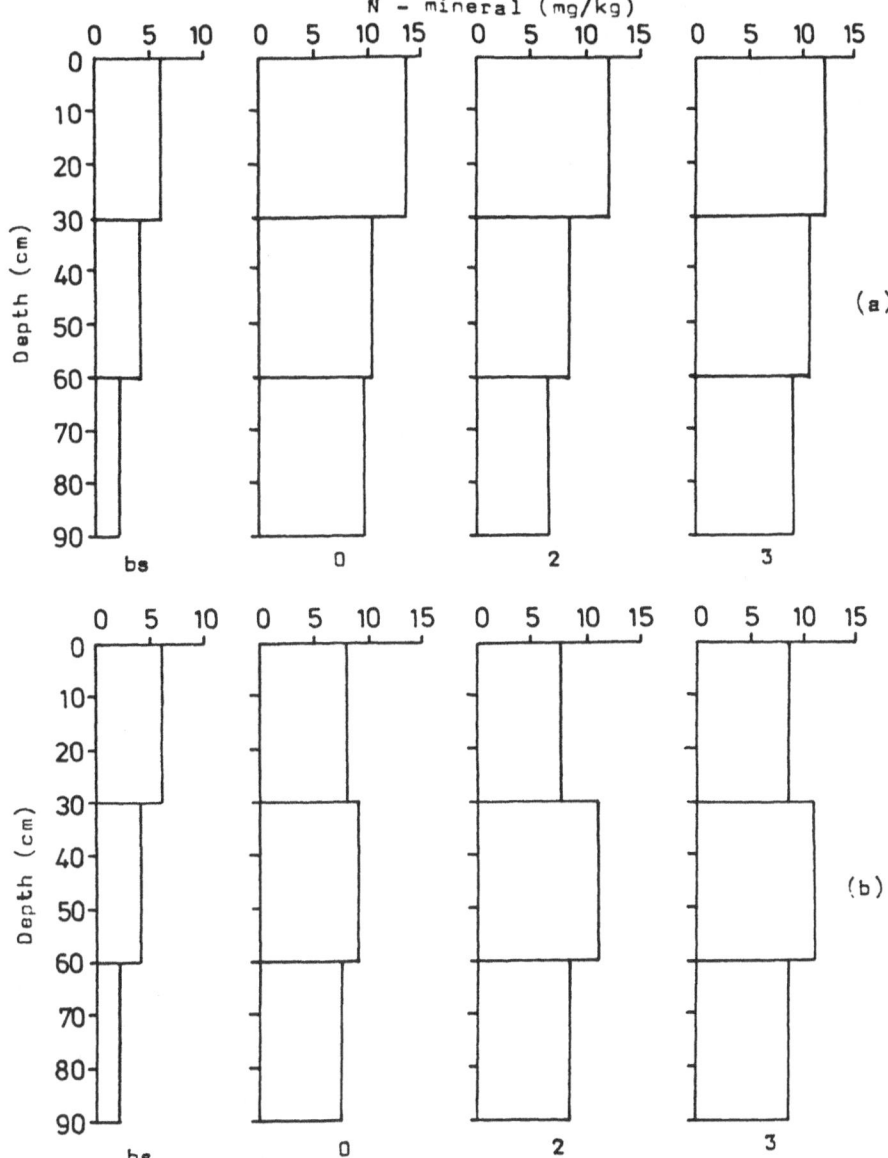

Figure 3. Mineral nitrogen content in the soil profile after cabbage (**a**), and spinach (**b**). Ways of application: bs - before sowing; 0-background ($P_{100}K_{100}$); 2-N_{180}-one-time spread before sowing with an inhibitor of nitrification (CMP); 3-N_{180} split-1/2 of the rate spread before sowing and 1/2 in rows as a dressing.

nium form for a certain period of time, thus decreasing the uptake of additional nitrate over-abundance. The split application of the urea provides a smoother nitrate availability for the plants, and decreases the nitrate contents in the production. The nitrate amounts of all tested rates and ways of application, when introducing potassium chloride, are lower than in the case of potassium sulphate application. The chlorine repels the nitrates in their osmoses processes, it activates the nitrate reduction, suppresses the soil nitrification, and

decreases the nitrate uptake from plants and accumulation in the crop.

Results from the study show that the nitrate content in the various vegetable crops is regulated under the maximum permissible concentrations. Differences in the nitrate contents for the majority of the variants in the experiment are mathematically proved. The tested technology of cultivation (crop-rotation, reduced nitrogen rate, urea application - in parts, and with a nitrification inhibitor, application of potassium chloride)

270

ensures high yields with a low nitrate content. The various parts of the experimental technology have been estimated by a number of scientists (Breimer, 1982; Hageman, 1983; Mouravin, 1989; Nurzinski, 1976; Raikova and Rankov, 1986; Raikova and Shevkenov, 1988; Rankov and Raikova, 1989; Semenov, 1990; Seiz, 1986; Titz, 1989).

Mineral nitrogen content in the soil profile

Data for the changes in the mineral nitrogen content in the soil profile are presented in Figure 3. For a fertilization with a N_{180} rate after harvest of the cabbage, there is an increase in comparison with that of the initial soil: in the 0-30 cm layer of 6.7-7.4, in the 30-60 cm layer of 4.3-6.4, and in the 60-90 cm layer of 4.8-7.7 mg kg^{-1} of soil. This implies that there are still some quantities of non-utilised mineral nitrogen in the soil profile, with a predominance of the ammonium form. After the harvest of spinach, the increase is lower: in the 0-30 cm layer of 2.1-2.6, in the 30-60 cm layer of 2.8-7.0, and in the 60-90 layer of 5.6-6.5 mg kg^{-1} of soil. When the rotation ends, the amount of the non-utilised mineral nitrogen in the soil is insignificant. For both crops, there is a minor influence of the way of nitrogen application (with a nitrification inhibitor, split, with potassium chloride, and with potassium sulphate) on the amounts of mineral nitrogen in the soil profile.

Results from the study show that the applied technology of vegetable cultivation prevents the accumulation of non-utilised mineral nitrogen in the soil, as a potential source of nitrate pollution of the ground water.

Conclusion

The tested technology for the cultivation of vegetable crops (crop rotation, reduced N-rate, one-time urea application with a nitrification inhibitor, and split application, use of potassium chloride) ensure high yields with low nitrate content in the crop, and insignificant amounts of mineral nitrogen in the soil profile of the end of the rotation.

References

Adruschenko V K 1983 Nitrate content in vegeables and ways for its decreasing. RIO NIINTI, Kishinev. 54 p.

Breimer T 1982 Environmental factors and cultural measures affecting the nitrate content in spinach. Fert. Res. 3, 191–292.

Bryson D D 1984 Nitrates and health. Fert. Soc. 228, 1–16.

Danek-Jesic K 1990 Nitraten Baden und Gemuse. Gartenbauwirtschaft 45, 1–12.

Hageman R H 1983 Effect of form of nitrogen on plant growth. *In* Nitrification Inhibitors - Potentials and Limitations. Eds. J J Meisinger, G W Randall and M L Vitosh. pp 47–62. ASA and SSSA, Madison, WI, USA.

Kelly W C 1990 Mineral use of synthetic fertilizers in vegetable production. Hort. Sci. 25, 168–169.

Loggers G 1989 Nitraat, overdaad schaadt. Voeding (the Hague) 40, 431–433.

Mouravin E A 1989 Nitrification inhibitors. VO AGROPROMIZDAT, Moscow. 248 p.

Nurzynski J 1976 Effect of the chloride and sufphate of the yields of some vegetable crops and garden peat. Bul. Warzywniczy 19, 105–118.

Opopol N I and Dobrynskaya E V 1986 Nitrates. Shtintsa, Kishinev. 113 p.

Poschold P J and Hunt I 1986 Production von Spinat und Mohren mit reduziertem Nitratgehalt. AL and ILID, Berlin. 53 p.

Prugar J and Prugarova A 1985 Dusicany v zelenine. Priroda, Bratisiava. 150 p.

Raikova L and Rankov V 1982 Effect of mineral fertilizing on nitrate accumulation in late head cabbage. Soil Sci. Agrochem. Plant Prot. 22, 40–48.

Raikova L and Petkova V 1984 Condition contributing to the accumulation of nitrates in the vegetable and feed crops at intensive inorganic fertilization. CNTII AGROHRANINFORM, Sofia. 80 p.

Raikova L and Rankov V 1986 Nitrate accumulation in spinach as a result of mineral nutrition. Soil Sci. Agrochem. Plant Prot. 22, 34–42.

Rankov V and Raikova L 1989 Effect of nitrification inhibitor N-serve on nitrate content in late head cabbage. Proc. Conf. Soil Science Problems Influence by Intensive Agriculture. PB BEC, Sofia. pp 262–268.

Raikova L and Shevkenov M 1988 Effect of nitrogen and potassium fertilizer forms on nitrate accumulation in spinach. Proc. Int. Conf. Mineral Fertilizer - Production and Application, Varna. pp 68–69.

Semenov V M 1990 Agrochemistry placement ways application of nitrogen fertilizers. Proc. Int. Conf. Nitrogen - Fertilizer - Soil - Plant, Prague. pp 141–146.

Siez P 1986 La problemetatica dei nitrate in orticultura. Cult. Prot. 15, 15–21.

Sindelarova J 1986 The content of nitrates and nitrates in vegetables. UVIZ, Prague. 92 p.

Titz R 1989 Stickstoffdundung bei Freilandgemuse in der Ammoniumform zur Einschrankung von Nitratanreicherung in Pflanzen und Boden. Dissertation, Universitat Bonn. 173 p.

Vulstake C and Briston K 1978 Factors affecting nitrate content in field grown vegetables. Qual. Plant. 28, 71–87.

O. Van Cleemput et al. (eds.), Progress in Nitrogen Cycling Studies, 271–276, 1996.

Impact of soil pH on nitrogen mineralization in grassland soils

Barbara Sapek

Institute for Land Reclamation and Grassland Farming at Falenty, 05-090 Raszyn, Poland

Key words: fertilizer form, grassland soil, in situ method, nitrogen mineralization, soil pH

Abstract

The aim of the investigations was to evaluate the impact of soil pH on the efficiency of nitrogen mineralization for two forms of nitrogen fertilizer - ammonium nitrate (AN) and calcium nitrate HYDRO (CN). The investigations were performed in long-term grassland experiments on mineral soils. Unlimed objects (Ca_0) and those limed in 1981 with calcium carbonate (Ca_2) were chosen. Nitrogen content was determined by the in situ method. $N-NO_3$ and $N-NH_4$ content was determined in soil samples and the total amount of mineral nitrogen (N-min) was calculated. A higher pH increased the efficiency of mineralization. A lower pH favoured the accumulation of ammonium ions. This phenomenon was not observed when calcium nitrate was used as the fertilizer. Differences in soil pH affected the total mineralization efficiency only to a small extent. Calcium carbonate promoted nitrogen mineralization at a higher soil pH.

Abbreviations: AN – ammonium nitrate, CN – calcium nitrate HYDRO, J – experiment in Janki, L – experiment in Laszczki, Ca_0 – unlimed object, Ca_2 – limed object, Hh – hydrolytic acidity.

Introduction

The process of mineralization of nitrogen from soil organic matter is affected by many factors due to both the soil and the climate, but the most important ones are the humidity and temperature (Schepers and Mosier, 1991; Raison et al., 1987). The investigations of Adams et al. (1989) and Raison et al. (1987) indicate that the soil structure and its state have a fundamental effect on the yield of the mineralization process. Another factor which affects the course of this process is soil pH. The quoted authors in addition to laboratory incubation experiments determined the yield of in situ mineralization. Using these papers as an example, similar work was initiated in 1991 in permanent grassland in the aspect of the effect of pH changes due to nitrogen ammonium nitrate fertilization on N-min content in the top layer of grassland soil (Sapek and Sapek, 1993). The results presented here are a continuation of that research and encompass the years 1991–1993. Their aim, besides the evaluation of pH effects on net soil mineralization is the comparison of the course of the investigated process on the basis of the action of two nitrogen fertilizers - ammonium nitrate and calcium

nitrate HYDRO - during three successive regrowths and in the autumn-winter period.

Materials and methods

The investigations are performed in two long-term grassland experiments on mineral soil conducted since 1981, in the localities Janki (J, C_{org} = 1.9%, particles < 0.02 mm - 18.4%, N_{tot} = 0.15%) and Laszczki (L, C_{org} = 3.8%, particles < 0.02 mm = 22.4%, N_{tot} = 0.32%). For the investigations objects which had not been limed (C_0) and those limed in 1981 (Ca_2) with calcium, carbonate were chosen, with doses according to 2Hh (4.6 - J and 7.2 - L t ha^{-1} CaO). Nitrogen fertilization in the form of ammonium nitrate (AN) and calcium nitrate (CN) in the amount 240 kg ha^{-1} $year^{-1}$ was used in doses of 100, 80 and 60 kg ha^{-1} for three successive regrowths.

In each plot (each in four repeats) with an area of 25 m^2 plastic tubes covered by a cap were installed in the soil before each regrowth and after fertilizing. In plots where CN was used the tubes were installed from 1993. The tubes were left in the soil during each of the three regrowths and in the autumn- winter period. PVC tubes were used with an internal diameter of 27 mm and a length of 125 mm, with holes bored at a height of 11 mm to allow equilibration of the temperature and humidity with the environment. In research so far the effects of these two factors on mineralization efficiency had not been considered. After each removal of the tubes from the grassland, before spring fertilization and the successive swaths, soil was pushed out from the tubes and samples with a volume of 20 cm^3 were formed from them with the current field humidity. In soil samples nitrate nitrogen (N-NO$_3$) and ammonium nitrogen (N-NH$_4$) were determined in a 1% K_2SO_4 extract by the colorimetric method using a TECHNICON autoanalyzer. The results of the determinations are given in mg dm^{-3} of soil in relation to the 0 - 10 cm layer. The thus calculated contents are given in kg ha^{-1} N-NO$_3$ and N-NH$_4$. From the sum of N-NO$_3$ and N-NH$_4$ the total mineral nitrogen content (N-min kg ha^{-1}) was calculated. The measurement of soil pH was performed in a 1 M potassium chloride solution.

Results

The change in the pH of the top soil layer from strongly acidic to neutral caused an increase in N-NO$_3$ content during the vegetative season, for each of the three regrowths and in the autumn- winter period. On the average yearly in the soils with acid pH about 131 to 181 kg ha^{-1} N-NO$_3$ were released and in soils with a pH of 7.1 241 to 289 kg ha^{-1} N-NO$_3$ (Table 1). In the J experiment the pH increase caused a rise of N-NO$_3$ content by 13.5% per pH unit. In experiment L this increase was 12.8% per pH unit (Table 1). It may be assumed that the increase of N_{tot} and C_{org} in the soil did not significantly modify the way pH change affected nitrification yields. During the autumn-winter period the effect of pH increase on nitrification efficiency was much smaller - 11.9% (J) and 7.9% (L) per pH unit. During the whole year regardless of soil pH a constant decrease of N-NO$_3$ in the soil was observed (Table 1). It was particularly clear and almost linear in Experiment L. The lowest yield of nitrification always occurred in the autumn-winter period.

The use of CN in comparison with AN increased nitrification yield in the soil in Experiment J especially at acid soil pH (increase of N-NO$_3$ by 27% at pH 3.7 and by 19% at pH 7.1) (Table 1). The effect of the change of soil pH from acidic to neutral regardless of

the form of fertilizer used was always the increase of the N-NO$_3$ content in the soil.

A low pH favoured an increase in the yield of ammonification (Table 2). Under these conditions in the soil on the average 107 kg ha^{-1} N-NH$_4$ accumulated at pH 4.2 and up to 153 kg ha^{-1} N-NH$_4$ at pH 3.7 (Table 2). The increase in soil pH in both experiments to 7.1 caused a considerable decrease in N-NH$_4$ content in the soil (from 18.3% in experiment J to 24.6% per pH unit in Experiment L).

Similarly, though not as clearly as in the case of N-NO$_3$ a decrease in the accumulation of N-NH$_4$ during the year was observed (Table 2). The efficiency of ammonification was not always the lowest in the autumn-winter period and in comparison to the vegetative season did not differ as significantly as the yield of nitrification. The use of CN in comparison with AN decreased the accumulation of N-NH$_4$ to a larger extent at an acid pH, particularly in soil from the J experiment - by about 61% (Table 2). No such difference in the action of CN and AN on this soil was observed at pH 7.1.

The change in soil pH from acidic to neutral affected the efficiency of nitrogen mineralization to a small extent especially that poorer in C_{org} from Experiment J (Table 3). At acid soil pH the N-min content was on the average 284 kg ha^{-1} to 288 kg ha^{-1} per year. After changing the pH to 7.1 the N-min content did not change significantly (Table 3).

The mineralization efficiency decreased during the year, especially clearly in soil from Experiment L (Table 3). The lowest amount of organic nitrogen underwent mineralization in the autumn-winter period, on the average about 34 kg ha^{-1}.

With the exception of this single case the form of nitrogen fertilizer did not have a greater effect on the total amount of nitrogen mineralized in the soil both during the vegetation season and in the autumn-winter period (Table 3). Only in the soil less abundant in C_{org} and N_{tot} from experiment J and under conditions of neutral pH the use of CN increased the amount of N-min by about 18% (Table 3).

Discussion

These investigations only partly confirm the results indicated among others by the work of Adams et al. (1989) and Raison et al. (1987) on the effect of pH on the course of the mineralization process. We have shown that the change in grassland soil pH only to a

Table 1. Nitrification efficiency (N-NO$_3$ content) in the function of soil pH and nitrogen fertilizer forms (AN-ammonium nitrate, CN-calcium nitrate) in the years 1991–1993 on grassland experiments Janki and Laszczki

Date	N-NO$_3$ (kg ha^{-1})							
	Janki				Laszczki			
	Ca - pH 3.7		Ca - pH 7.1		Ca - pH 4.2		Ca - pH 7.1	
	$\bar{x}$	SD	$\bar{x}$	SD	$\bar{x}$	SD	$\bar{x}$	SD
AN								
91.05.21	46.0	18.2	93.4	20.6	71.2	23.6	116.9	42.1
91.07.16	34.9	9.3	60.3	21.1	41.4	14.8	68.3	49.1
91.09.24	47.0	13.6	77.7	18.2	35.7	9.2	51.4	32.4
92.03.30	9.9	3.8	23.6	5.7	18.2	6.2	21.9	6.4
Total	Σ 137.8		Σ 255.5		Σ 166.5		Σ 258.4	
92.05.26	32.1	8.8	68.6	18.4	59.5	13.6	102.3	20.1
92.07.23	35.9	12.8	63.4	21.6	74.8	18.4	101.2	28.7
92.09.28	31.1	6.7	61.5	29.1	43.4	14.7	84.4	35.2
93.03.23	16.9	6.7	29.2	7.2	28.8	10.5	37.5	6.7
Total	Σ 116.0		Σ 222.4		Σ 206.5		Σ 325.4	
93.05.28	45.5	7.2	78.9	21.1	62.3	31.8	102.5	30.7
93.07.26	39.0	17.6	81.7	51.6	51.9	25.3	98.3	49.1
93.09.27	35.1	10.6	53.9	17.0	33.8	9.3	50.3	26.3
94.04.06	19.6	10.2	30.1	9.6	23.6	11.1	32.1	14.1
Total	Σ 139.2		Σ 244.6		Σ 171.6		Σ 283.2	
1991–1993	131.0	13.0	240.8	16.8	181.5	31.4	289.0	33.9
CN								
93.05.28	71.4	34.4	109.8	45.6	68.9	25.3	108.8	67.0
93.07.26	60.4	17.4	95.7	29.6	58.4	40.1	93.1	40.4
93.09.27	38.0	11.4	62.2	41.5	28.1	12.1	49.5	23.8
94.04.06	21.5	11.4	35.3	14.3	14.4	8.4	31.0	13.7
Total	Σ 192.3		Σ 303.0		Σ 169.8		Σ 282.4	

small extent differentiates the total amount of mineralized nitrogen in soil. It, however, affects to a considerable extent the efficiency of the components of the mineralization process - nitrification and ammonification. As can be inferred from the work of the quoted authors the amount of nitrogen taken up by the plants from the pool of nitrogen which undergoes mineralization under field conditions is difficult to assess. The reason is in the accuracy of the measurement methods of net mineralization including in situ measurements which is limited by the variation of soil and climatic conditions. This is confirmed by the values of standard deviations (SD) given in Tables 1–3.

In this work a particularly large impact of pH changes on nitrification efficiency has been shown. At acid soil pH the participation of N-NO$_3$ in the total amount of N-min is on the average 56% to 63% whereas at neutral pH it is 77% to 90%. This increase in N-NO$_3$ content caused by an increase in pH is particularly large in soils rich in humus, such as the soil of grasslands.

Table 2. Ammonification efficiency (N-NH$_4$ content) in the function of soil pH and nitrogen fertilizer forms (AN-ammonium nitrate, CN-calcium nitrate) in the years 1991–1993 on grassland experiments Janki and Laszczki

Date	N-NH$_4$ (kg ha^{-1})							
	Janki				Laszczki			
	Ca - pH 3.7		Ca - pH 7.1		Ca - pH 4.2		Ca - pH 7.1	
	$\bar{x}$	SD	$\bar{x}$	SD	$\bar{x}$	SD	$\bar{x}$	SD
AN								
91.05.21	48.3	39.7	15.5	12.6	38.4	37.8	6.5	6.0
91.07.16	26.5	7.2	13.5	11.7	22.9	13.9	7.8	12.6
91.09.24	48.2	30.7	9.8	6.2	15.0	11.0	7.0	10.1
92.03.30	27.4	8.7	3.8	4.7	7.0	8.1	3.1	1.3
Total	Σ 150.4		Σ 42.7		Σ 83.3		Σ 24.5	
92.05.26	110.0	32.7	47.0	28.6	61.4	35.2	11.9	9.0
92.07.23	34.7	15.3	18.3	8.0	62.5	42.2	20.4	21.2
92.09.28	55.2	29.3	27.6	9.2	45.7	27.9	17.0	10.5
93.03.23	25.1	13.1	8.6	9.3	10.1	10.7	4.1	6.4
Total	Σ 225.0		Σ 101.1		Σ 179.7		Σ 53.4	
93.05.28	34.7	8.7	10.9	10.2	29.9	13.2	2.9	3.6
93.07.26	21,9	17.9	3.0	5.2	14.4	14.5	2.8	6.4
93.09.27	13.8	15.0	7.7	8.0	8.6	11.0	3.4	3.8
94.04.06	12.7	10.2	7.1	7.7	4.6	4.5	4.8	7.3
Total	Σ 83.1		Σ 28.7		Σ 57.5		Σ 13.9	
1991–1993	152.8	71.0	57.5	38.4	107.0	64.3	30.6	20.4
CN								
93.05.28	15.6	13.7	4.1	7.5	10.1	7.1	1.8	0.8
93.07.26	3.2	5.2	21.2	4.8	10.4	15.6	2.0	3.7
93.09.27	6.7	5.3	4.8	4.4	2.9	3.1	1.8	1.5
94.04.06	7.1	6.8	2.9	5.8	3.3	3.0	0.8	0.9
Total	Σ 32.6		Σ 33.0		Σ 26.7		Σ 6.4	

The used method permitted the evaluation of the total amount of nitrogen which may be used by the vegetation as it excludes the taking up of nitrogen by the plant and its leaching. The data for the autumn-winter period correspond to the efficiency of nitrogen mineralization without the participation of nitrogen from fertilizer. It is much lower, especially the amount of released N-NO$_3$. As is shown by the investigations of Raison et al. (1989) this may also be caused by the accumulation of large amounts of N-NO$_3$ during the vegetative season which will increase the yield of den-

itrification. This process is also favoured by the higher humidity and lower temperature at that time.

The decreasing amounts of N-NO$_3$ and N-NH$_4$ and N-min during the vegetative season probably are also a result of the decreasing doses of nitrogen fertilizer for the subsequent regrowths. In preliminary investigations (Sapek and Sapek, 1993) in which the efficiency of mineralization before and after treatment with ammonium nitrate was analyzed it was shown that a dose of 240 kg ha^{-1} N increased the amount of N-min on the average by 42%, including N-NO$_3$ by 43% and

Table 3. Mineralization efficiency (N-min content) in the function of soil pH and nitrogen fertilizer forms (AN-ammonium nitrate, CN-calcium nitrate) in the years 1991–1993 on grassland experiments Janki and Laszczki

	N-min (kg ha^{-1})								
	Janki				Laszczki				
Date	Ca -	pH 3.7	Ca -	pH 7.1	Ca -	pH 4.2		Ca -	pH 7.1
	$\bar{x}$	SD	$\bar{x}$	SD	$\bar{x}$	SD		$\bar{x}$	SD
AN									
91.05.21	94.3	52.4	109.8	26.1	109.6	49.4		123.4	41.7
91.07.16	61.4	13.7	73.9	28.0	64.4	24.6		76.0	59.8
91.09.24	95.2	37.5	87.5	21.6	50.7	15.7		58.4	38.0
92.03.30	37.3	9.9	27.4	5.8	25.2	11.7		25.0	6.5
Total	Σ 288.2		Σ 297.7		Σ 249.8			Σ 282.9	
92.05.26	142.1	30.8	115.6	32.3	120.8	42.0		114.2	21.0
92.07.23	70.6	22.6	81.5	24.0	137.3	44.3		121.6	39.4
92.09.28	86.4	29.9	88.8	34.5	89.2	35.0		101.5	39.7
93.03.23	42.0	16.3	37.9	12.6	39.0	16.2		41.6	9.5
Total	Σ 341.1		Σ 323.8		Σ 386.3			Σ 378.9	
93.05.28	80.1	8.9	89.8	25.4	92.2	36.1		105.4	31.8
93.07.26	60.9	30.8	84.7	55.6	66.3	34.3		101.1	49.1
93.09.27	48.9	23.8	61.6	19.0	42.4	17.1		53.7	28.8
94.04.06	32.3	17.3	37.2	15.8	28.2	10.8		36.9	19.5
Total	Σ 222.2		Σ 273.3		Σ 229.1			Σ 297.1	
1991–1993	283.8	59.6	298.3	25.2	288.4	85.4		319.6	51.8
CN									
93.05.28	87.0	45.8	113.9	49.9	79.0	22.2		110.6	67.0
93.07.26	63.6	20.4	116.6	29.8	68.8	49.4		95.0	40.2
93.09.27	44.7	12.2	67.0	42.7	31.0	13.0		51.3	23.9
94.04.06	28.6	13.7	38.2	14.5	17.6	9.5		31.7	14.1
Total	Σ 223.9		Σ 335.7		Σ 196.4			Σ 288.6	

N-NH$_4$ by 47%. Including fertilization in the measurement of the nitrogen mineralization process is justified in the practical aspect of the evaluation of the potential productivity of a grassland. The decrease in the efficiency of mineralization during the vegetative season may also be the result of the progress in time of nitrogen immobilization, especially in the conditions of higher pH, which favours the increase in the activity of soil microorganisms.

Acidity and a lower pH favour a higher yield of ammonification and N-NH$_4$ accumulation in grassland soils, especially those poorer in humus. The decrease in N-NH$_4$ content at a pH close to neutral may also be due to the loss by ammonium volatilization.

In comparison with AN, CN fertilization increase the amount of mineralized nitrogen, including the nitrate which is easily available to plants but also easily undergoes leaching, in soils poorer in C$_{org}$ and N$_{tot}$ - especially at a neutral soil pH. The use of CN counteracts the increase in N-NH$_4$ accumulation in soils with an acid pH. The efficiency of ammonification is also lower when this form of fertilizer is used.

Summing up we may state that the used method of in situ investigations may be an approximate indicator for the evaluation of the amount of mineral nitrogen available for grassland vegetation as well as the evaluation of the effects of pH change and form of nitrogen fertilizer on the course of mineralization processes in grassland soil.

Acknowledgements

It is my pleasant duty to thank Mrs Jadwiga Bojdo for help in the analytical work, the group of the Experimental Department of the Institute for Land Reclamation and Grassland Farming headed by Jerzy Barszczewski, M Eng for performing the field work and for technical assistance and my boss Professor Andrzej Sapek for joint initiation of the researches and valuable and critical comments.

References

Adams M A et al. 1989 In situ studies of nitrogen mineralization and uptake in forest soils; some comments on methodology. Soil Biol. Biochem. 21, 423–429.

Raison R J et al. 1987 Methodology for studying fluxes of soil mineral-N in situ. Soil Biol. Biochem. 19, 521–530.

Sapek B and Sapek A 1993 Impact of liming on mineral nitrogen content in grassland soil. *In* Problemy Wapnowania Uzytkow Zielonych. Materialy Seminaryjne 32, pp 74–79. IMUZ, Falenty, Poland (*In Polish*).

Schepers J S and Mosier A R 1991 Accounting for nitrogen in nonequilibrium soil-crop systems. *In* Managing Nitrogen for Groundwater Quality and Farm Profitability. Eds. R F Follet, D R Keeney and R M Cruse. pp 125–138. Soil Science Society of America, Inc., Madison, WI, USA.

Plant and Soil **181**: 131–137, 1996.
© 1996 *Kluwer Academic Publishers.*

131

Regulation of nitrogen uptake on the whole plant level

M.K. Schenk

Institute of Plant Nutrition, University of Hannover, Herrenhäuser Str. 2, D-30419 Hannover, Germany *

Key words: cucumber, growth rate, lettuce, maximum inflow (I_{max}), model, nitrate, regulation, uptake

Abstract

The largest part of nitrogen requirements of crops is mostly covered by nitrate. The uptake of this ion is thermodynamically uphill and thus dependent on metabolism. This article considers regulation of N uptake in higher plants putting emphasis on NO_3^- and the whole plant level.

In field conditions the transport rate depends on the concentration at the root surface in Michaelis-Menten-Kinetics. Maximum net influx of NO_3^- (I_{max}) was often reported at concentrations of 100 μM NO_3^- and even lower. There are indications that for unrestricted growth the NO_3^- concentration at root surface has to be in the order of magnitude allowing I_{max} if plants are not able to compensate for lower NO_3^- concentrations by increasing root surface per unit of shoot.

I_{max} is not a constant but depends for a given variety on N status of plants, the availability of NO_3^- and plant age. The decrease of I_{max} with increasing plant age is closely related to relative growth rate as long as the relationship between N demand and new growth is linear and the root:shoot ratio keeps constant. It seems that I_{max} is a meaningful physiological characteristic of NO_3^- uptake reflecting absolute N demand. There is evidence that shoot demand is linked to NO_3^- uptake of the root through an amino acid transport pool cycling in the plant via phloem and xylem.

The N demand of a crop depends on increase of dry mass and might not be linear if the "critical level" of nitrogen in plant dry matter changes during crop development or if retranslocation of nitrogen from older leaves to meristematic tissue occurs. Radiation and temperature drive plant growth and thus N demand of crops. These relationships can be described by mathematical models.

Introduction

Plants cover their nitrogen requirements mostly by taking up NO_3^- and NH_4^+. This uptake is metabolically dependent and seems to be driven by transmembrane electric potential differences as is discussed by Clarkson (1986) and Glass (1988) in review articles. The dominant form of nitrogen available to plants is NO_3^- in all but very acidic or anaerobic soils, because of soil microorganisms oxidizing NH_4^+ to NO_3^-.

The first step of NO_3^- assimilation is the reduction to NH_4^+ which in most species occurs in the leaves, because of the ready availablility of carbon skeletons, reductants and energy (Oaks, 1986). In contrast to NO_3^-, NH_4^+, as such, is not normally transported to the shoot via the xylem but incorporated into the amide

nitrogen of glutamine and/or asparagine in the root (Lewis, 1986).

Nitrogen demand of plants likely depends on the growth because N is the major constituent of numerous products of plant metabolism but it is also, in form of NO_3^-, an important osmoticum in some species. The emphasis of this contribution will be on the regulation of NO_3^- uptake because of its major contribution to nitrogen nutrition of plants while focusing on the whole plant level.

Kinetics of NO_3^- uptake

In field conditions net nitrate uptake depends on its concentration at the root surface as described by Michaelis-Menten Kinetics. It has to be stressed that most reports of NO_3^- uptake are based on depletion studies in nutri-

---

* FAX No.: +495117623611
Plant and Soil is the original source of publication of this article. It is recommended that this article is cited as: *Plant and Soil* **181**: 131–137, 1996.

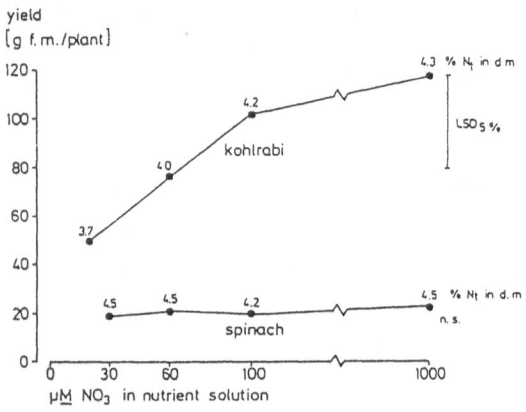

Figure 1. Influence of nitrate concentration at the root surface on fresh matter yield and total nitrogen (N_t) content of kohlrabi and spinach (from Steingrobe and Schenk, 1991).

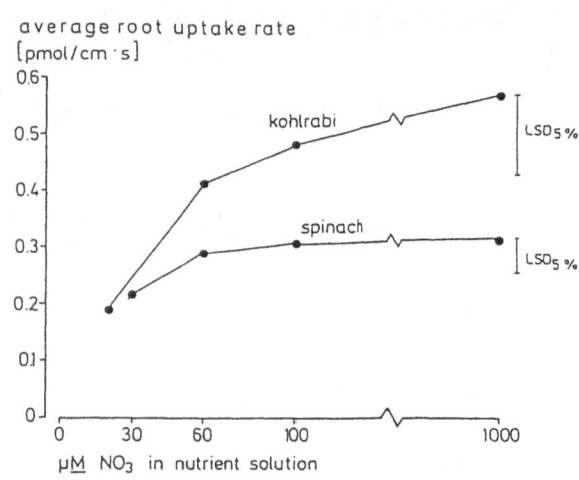

Figure 2. Influence of nitrate concentration at the root surface on the average root uptake rate of kohlrabi and spinach (from Steingrobe and Schenk, 1991).

ent solution or on use of ^{15}N. These methods supply information on net NO_3^- uptake which equals total NO_3^- influx minus efflux. One parameter of Michaelis-Menten Kinetics is V_{max}, the maximum uptake rate. The term I_{max} was suggested for situations where net nitrate uptake (inflow) is considered (Barber, 1984). The maximum NO_3^- uptake rate per unit of roots, normally root length or root surface area, is obtained at a certain critical concentration at the root surface. Maximum nitrate inflow for maize was observed at concentrations below 20 μM. Edwards and Barber (1976a) and Olsen (1950) found no influence on nitrate absorption of ryegrass even at 3 μM. For spinach and kohlrabi a concentration of about 100 μM was necessary for I_{max}. These investigations also showed that the average inflow of nitrate in soil culture was of the same order as the maximum inflow found in nutrient-solution cultures, thus suggesting that roots had to absorb at their maximum rate (Heins and Schenk, 1987).

To prove this hypothesis, the nitrate concentration in flowing nutrient solution was varied from K_m (20–30 μM) to ten times the concentration needed for maximum inflow (1000 μM). This technique was used to keep the NO_3^- concentration at root surface constant. Figure 1 shows that fresh matter yield and N_{total} content of kohlrabi increased with nitrate concentrations increasing from 20 μM up to 100 μM. A ten-times higher nitrate concentration did not affect yield significantly. However, yield and N_t content of spinach was not influenced by variations in nitrate concentration.

In accordance with growth reaction, the average inflow of kohlrabi per cm root length was decreased by low nitrate concentration (Fig. 2); the highest aver-

age inflow was reached at about 100 μM. The average inflow of spinach was approximately half that of kohlrabi at high NO_3^- concentration. In contrast to yield, the average inflow of spinach was decreased at low concentration, but less than that of kohlrabi. The highest average inflow for spinach occured also at 60–100 μM. Both species increased the root: shoot ratio by about 30% but this was only sufficient for spinach to compensate for lower inflow.

Microscopical investigations showed that the higher uptake efficiency of spinach was due to increased root hair surface at low N supply but not to improved absorption efficiency per unit of root (Steingrobe and Schenk, 1991). These results indicate that kohlrabi roots had to take up nitrate at their maximum rate to achieve highest yield, whereas spinach could compensate and achieve highest yield at less than maximum inflow.

The ecological significance of the induction of root hair growth might be that plant roots in nitrate-rich soil zones might compensate for insufficient uptake by other parts of the root system. This may be supported by a parallel increase of root length and physiological capacity (induction of NO_3^- uptake).

Maximum inflow rate during plant growth

The maximum inflow rate of nitrate is not a constant but depends for a given variety on N status of plants and the availability of NO_3^-. Furthermore, it is known that I_{max} declines with increasing plant age (Edwards and Barber, 1976a; Pitman and Cram, 1976). However, it is unlikely that this is caused by a reduced capacity of older roots, though it could be shown in split-root experiments that the capacity of older roots can be enhanced (Kuhlmann and Barraclough, 1987). It is more reasonable that I_{max} depends on the N demand, which has to be covered by a unit of root (Edwards and Barber, 1976b; Nye and Tinker; 1977; White, 1973). The N demand of a plant results from the N amount in new growth and changes in N_t content of the whole plant. The new growth of a plant can be described in terms of growth rates.

Figure 3 shows the dependance of I_{max} on relative growth rates (RGR) of lettuce. Different RGR resulted with varied plant age as well as climatic conditions (temperature, radiation). I_{max} was determined in depletion studies. Statistical analysis showed that a linear function described the data set most accurately. A close correlation between relative growth rate and I_{max} is also reported for P uptake by Wild and Breeze (1981) and Rodgers and Barneix (1988) for NO_3^- uptake of wheat. The close correlation is probably caused by the fact that both RGR as well as I_{max} are related to the existing plant: for calculation of RGR new growth is related to the existing fresh weight whereas for I_{max} the N demand caused by the new growth is related to the existing root length. Therefore, the correlation must be close if the relationship between N demand and new growth is linear and the root:shoot ratio keeps constant. However, statistical analysis showed that the root:shoot ratio (cm root g^{-1} fresh matter shoot) varied according to growth conditions. Thus, the equation to calculate I_{max} from the relative growth rate had to be expanded by a term to consider the root:shoot ratio. The good correlation between measured and calculated I_{max} for plants grown at varied air temperature and radiation can be seen in Figure 4a and b, respectively. Integration of N uptake of plants based on this model for calculation of I_{max} resulted in total N uptake/plant. The relationship between measured and calculated uptake was good (Fig. 5a and b). This also indicates that I_{max} equals the actual NO_3^- uptake rate of plant roots. It seems that I_{max} is a meaningful physiological characteristic of NO_3^- uptake reflecting absolute N demand, which is in accordance with the conclusions of the previous chapter.

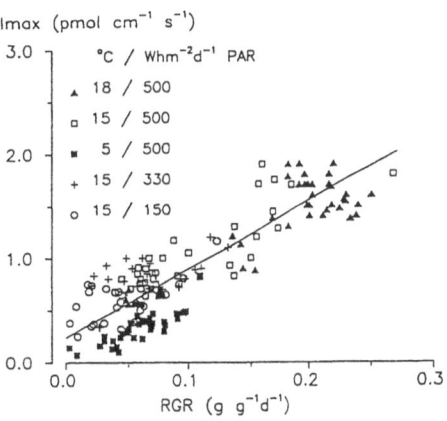

Figure 3. Relationship between relative growth rate of the whole plant (RGR) and maximum nitrate inflow (I_{max}) of lettuce at various temperatures and radiation levels. The equation of the line is: $I_{max} = 0.24 + 6.57$ RGR ($R^2 = 0.94$) (from Steingrobe and Schenk, 1994).

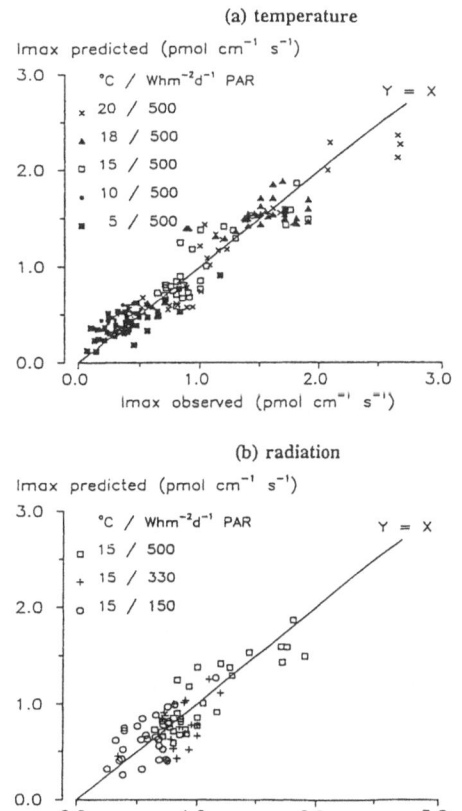

Figure 4. Comparison of observed and predicted I_{max} of lettuce grown at various temperatures (**a**) and radiation levels (**b**). I_{max} was predicted by means of this equation (1): $I_{max} = (0.31 + 9.5$ RGR$)$ $e^{(-0.0014 \text{ RSR})}$ (RGR = relative growth rate [g g^{-1} d^{-1}], RSR = root:shoot ratio [cm g^{-1} FM]) (from Steingrobe and Schenk, 1994).

(a) temperature

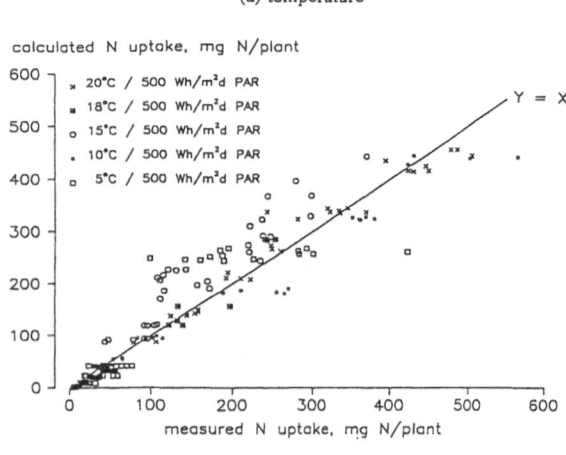

(b) radiation

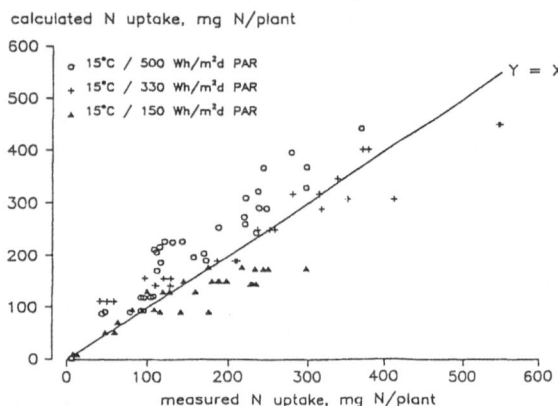

Figure 5. Comparison of observed and predicted N uptake of lettuce grown at various temperatures (**a**) and radiation levels (**b**). N uptake was obtained by integrating I_{max} calculated according to Equation (1) in Figure 4 by means of models for relative growth rate (RGR) and root:shoot ratio.

Interaction between shoot and root

NO_3^- uptake capacity per unit of root depends on growth rate of the shoot. Three sites of active transport processes are obviously involved in the regulation of net inflow: plasmalemma, tonoplast and the interface symplast/xylem. Clarkson (1986) pointed out that the amino acid pool might play a significant part in regulating the transport processes. Cooper and Clarkson (1989) suggested that an amino acid transport pool cycling in the plant via phloem and xylem might provide a mechanism by which shoot demand can affect N uptake. A schematic representation of these interdependencies is given in Figure 6.

Circulation of nitrogen from root to shoot and vice versa is a generally accepted model to relate

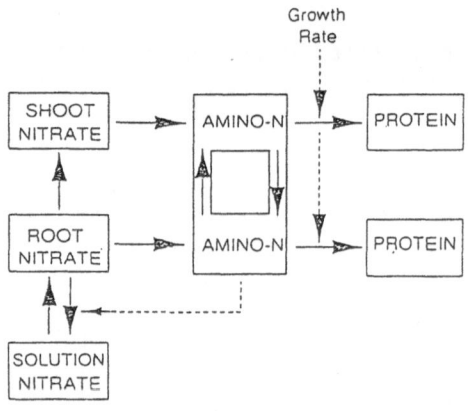

Figure 6. A scheme for the regulation of N uptake by a transport pool of amino-N, common to roots and shoots. The amino-N that is cycling between shoots and roots is conceived as a single pool. Amino-N is removed from this pool for protein synthesis in growth of shoots and roots and enters this pool following uptake and reduction of nitrate. It is postulated that the rate of net uptake of N is regulated by the level of amino-N in the pool. Thus, uptake rate is determined by the demand for N in growth (from Cooper and Clarkson, 1989).

shoot demand and uptake for nutrients being actively absorbed such as P, K and N (Glass and Siddiqi, 1984). Large portions of these nutrients in the xylem are reported to recirculate. In cereal seedlings up to 50% of the N flux in the xylem can be cycling nitrogen (Cooper and Clarkson, 1989; Simpson et al., 1982). Cooper and Clarkson (1989) discuss the efficiency of this high cycling rate to regulate N uptake. They calculate that N cycling would enhance the photon and carbon costs of NO_3^- uptake and assimilation by about 20% to 30%, which is quite considerable in these terms. However, it is of minor importance in relation to total growth costs, which are increased by only 1%–2%.

N demand of a crop and growth rate

The cycling amino acid pool might be the link between N demand of the shoot and NO_3^- uptake of roots. Plant N demand depends on increase of dry mass, as is confirmed by the result of an experiment given in Figure 7, where several butterhead lettuce crops were grown throughout the vegetation period and N uptake per head was determined frequently during the growth of a crop. The relationship between dry matter production and N uptake might not be linear as in this case, if the "critical level" of nitrogen in plant dry matter changes during crop development or if retranslocation of nitrogen from older leaves to meristematic tissue occurs. However, for practical purposes like determining nitrogen ferti-

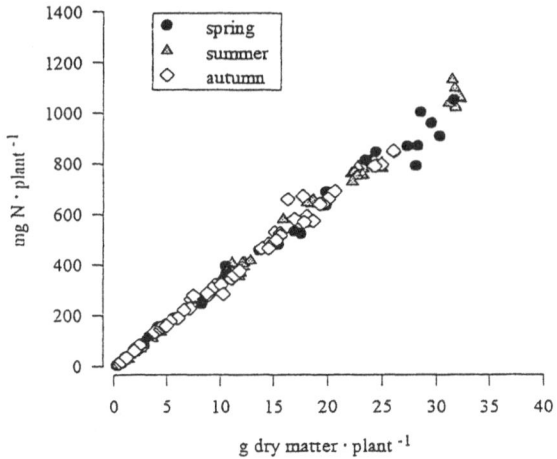

Figure 7. N uptake in relation to dry matter yield of butterhead lettuce at optimal N supply grown in the field throughout the growing season.

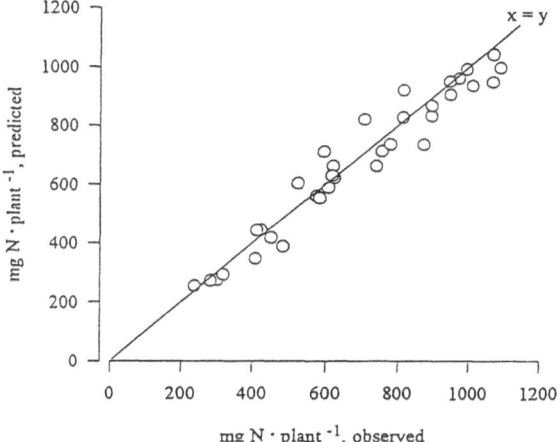

Figure 8. Comparison of observed and predicted N uptake of butterhead lettuce grown in the field throughout the growing season. N uptake was calculated by this equation : mg N plant^{-1} = A·$(1-e^{-P_1 \times FM}).(1-e^{-P_2 \times T})$ (FM = fresh matter in g plant^{-1}, T=24 h average of air temperature in °C, A = 1505, P_1 = 0.00247, P_2 = 0.12856).

lization in both amount and time plant growth might be described more conveniently in terms of fresh matter, which applies for a large number of horticultural crops. Most vegetable crops like butterhead lettuce are traded on a fresh matter basis.

The relationship between fresh matter increase and N uptake is close for one specific crop, but the slope varies between crops grown under different climatic conditions. This is due to variations of dry matter content in fresh matter. Butterhead lettuce crops grown in spring and in autumn have about 25% less dry matter than during summer. For other plants variations with time of the year are also reported (Jungk, 1970).

Generally, it can be said that dry matter content of plants is lowered during low radiation conditions. This implies for farming practice that N uptake of lettuce crops giving normal yields of 60 t fresh matter ha^{-1} varies from 110 to 130 kg N ha^{-1}, which means that N supply necessary for optimum growth varies to the same extent. These relationships have to be considered for ecologically-friendly plant production. The application of growth models might contribute to this aim.

As long as growth models are based on fresh matter production, then changes of dry matter content during the growth period as well as during crop development will have to be included. Dry matter content generally decreases when plants become older and can increase again during senescence. Figure 8 shows the result of the use of such a growth model for the estimation of N uptake of butterhead lettuce. The growth model was based on air temperature and radiation. N uptake could be described by an exponential relationship to fresh matter multiplied by a term reflecting average of air temperature to include the influence of the season. The N uptake measured during the growth of 10 different crops throughout the vegetation period was quite precisely estimated by the N uptake model in these experiments which were not used for model development. The deviation was not more than 10 kg N ha^{-1}. The use of this N uptake model to control N fertilization in split application resulted in optimum yield and less NO_3^- leaching during periods of heavy precipitation compared to conventional fertilization only at planting.

The existence of a valid growth model is not necessarily a prerequisite for relating N demand to growth conditions. In a more controlled environment of a greenhouse it was possible to relate daily N uptake of cucumber plants grown in a soilless culture system directly to radiation (see Fig. 9). N absorption per plant during the stem fruit period was at a higher level than in the following sucker fruit period. Lower N uptake in the latter period of crop development was mainly due to a retranslocation of nitrogen from senescing leaves to growing plant parts. Another reason might be that growth efficiency declines because of plant senescence (Liebig, 1989). In contrast to the field grown lettuce, temperature did not determine N uptake of the greenhouse crop cucumber because air temperature of the greenhouse was only varied within a suitable range. Control of daily N fertilization by means of the simulation model based on the relationships of Figure 9 gave optimum yield. Taking into account the regulation of N uptake by growth rate and thus finally by growth

136

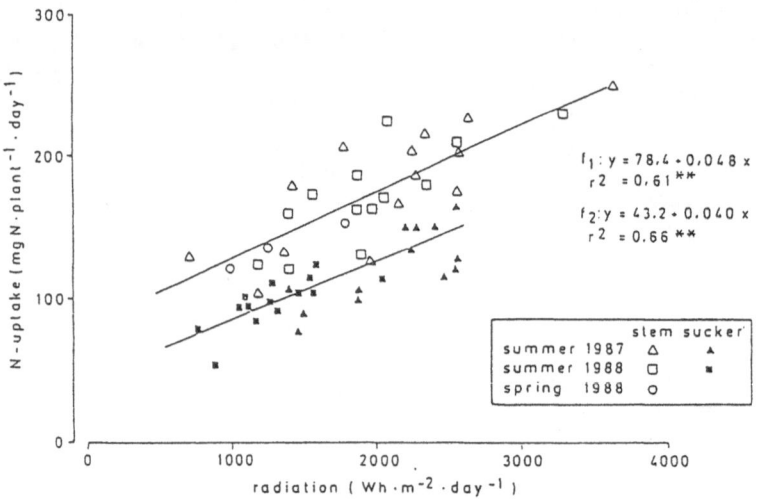

Figure 9. Relation between radiation *in* the greenhouse and daily N uptake of cucumber during period of stem fruit and sucker fruit harvest (from Schacht and Schenk, 1990).

factors could lead to an improvement of fertilization practice.

Conclusion

The presented picture of regulation of nitrate uptake is quite well accepted, but there are still uncertainties. It is reasonable that NO_3^- inflow rate of roots is linked to plant demand in the shoot via the amino acid pool cycling in phloem and xylem. However, the understanding of symplastic radial transport and its regulation by the cycling amino acid pool needs to be improved. Furthermore, it seems worthwhile to include into concepts for fertilizer management nitrogen demand of crops regulated via growth rate which depends on actual growth conditions. This could contribute to increase efficiency of N fertilizer and safe ecological farming.

References

Barber S A 1984 Soil Nutrient Bioavailability - A Mechanistic Approach. Academic Press, New York.

Clarkson D T 1986 Regulation of the absorption and release of nitrate by plant cells: A review of current ideas and methodology. *In* Fundamental, Ecological and Agricultural Aspects of Nitrogen Metabolism in Higher Plants. Eds. H Lambers, J J Neeteson and I Stulen. pp 3–12. Martinus Nijhoff Publ., Boston, USA.

Cooper H D and Clarkson D T 1989 Cycling of amino-nitrogen and other nutrients between shoots and roots in cereals - A possible mechanism integrating shoot and root in the regulation of nutrient uptake. J. Exp. Bot. 40, 753–762.

Edwards J H and Barber S A 1976a Nitrogen uptake characteristics of corn roots at low N concentration as influenced by plant age. Agron. J. 68, 17–19.

Edwards J H and Barber S A 1976b Nitrogen flux into corn roots as influenced by shoot requirement. Agron. J. 68, 471–473.

Glass A D M and Siddiqi M Y 1984 The control of nutrient uptake rates in relation to the inorganic composition of plants. *In* Advances in Plant Nutrition 1. Eds. P B Tinker and A Läuchli, pp 103–147. Praeger, New York, USA.

Glass A D M 1988 Nitrogen uptake by plant roots. ISI Atlas of Plant Science: Plants and Animals 1, 151–156.

Heins B and Schenk M K 1987 Root growth and nitrate uptake of vegetable crops. J. Plant Nutr. 10, 1743–1751.

Jungk A 1970 Mineralstoff- und Wassergehalt in Abhängigkeit von der Entwicklung von Pflanzen. Z. Pflanzenernähr. Bodenkd. 125, 119–129.

Kuhlmann H and Barraclough P B 1987 Comparison between the seminal and nodal root systems of winter wheat in their activity for N and K uptake. Z. Pflanzenernähr. Bodenkd. 150, 24–30.

Lewis O A M 1986 Plants and Nitrogen. Edward Arnold, London.

Liebig H P 1989 Models to predict crop growth. Acta Hortic. 248, 55–68.

Nye P H and Tinker P B 1977 Solute Movements in the Soil-root System. Studies in Ecology Vol. 4. Blackwell Scientific Publications, Oxford, UK.

Oaks A 1986 Biochemical aspects of nitrogen metabolism in a whole plant context. *In* Fundamental, Ecological and Agricultural Aspects of Nitrogen Metabolism in Higher Plants. Eds. H Lambers, J J Neeteson and I Stulen. pp 3–21. Martinus Nijhoff Publ., Boston, USA.

Olsen C 1950 The significance of concentration for the rate of ion absorption by higher plants in water culture. Physiol. Plant. 3, 152–164.

Pitman M G and Cram W J 1976 Regulation of ion content in whole plants. Symp. Soc. Exp. Bot. 31, 391–424.

Rodgers C O and Barneix A J 1988 Cultivar differences in the rate of nitrate uptake by intact wheat plants as related to growth rate. Physiol. Plant. 72, 121–126.

Schacht H and Schenk M K 1990 Control of nitrogen supply of cucumber (*Cucumis sativus*) grown in soilless culture. *In*

Plant Nutrition - Physiology and Application. Ed. M L v Beusichem. pp 753–758. Kluwer Academic Publishers, Dordrecht, the Netherlands.

Simpson R J, Lambers H and Dalling M J 1982 Translocation of nitrogen in a vegetative wheat plant. Physiol. Plant. 56, 11–17.

Steingrobe B and Schenk M K 1991 Influence of nitrate concentration at the root surface on yield and nitrate uptake of kohlrabi and spinach. Plant and Soil 35, 205–211.

Steingrobe B and Schenk M K 1994 A model relating the maximum nitrate inflow of lettuce (*Lactuca sativa* L.) to the growth of roots and shoots. Plant and Soil 162, 249–257.

White R E 1973 Studies on mineral ion absorption by plants. II. The interaction between metabolic activity and the rate of phosphorus uptake. Plant and Soil 38, 509–523.

Wild A and Breeze V G 1981 Nutrient uptake in relation to growth. *In* Physiological Processes limiting Plant Productivity. Ed. D E Johnson. pp 331–344. Butterworths, London, UK.

Section editor: R Merckx

O. Van Cleemput et al. (eds.), Progress in Nitrogen Cycling Studies, 285–291, 1996.

Plant response and environmental aspects as affected by rate and pattern of nitrogen release from controlled release N fertilizers

A. Shaviv

Faculty of Agricultural Engineering, Technion-IIT, Haifa, Israel

Key words: controlled-release, environmental pollution, N-recovery, use-efficiency

Abstract

The effect of N release rate and pattern on plant growth, N uptake and losses was assessed by comparing five urea-based controlled release nitrogen (CRN) fertilizers significantly differing in release mechanisms and characteristics and split applied conventional urea. Profiles of N release in water and in an incubation test in a sandy loam were compared to results of a pot experiment in which N losses (4 leachings) and N uptake (4 cuttings) by rye grass were determined.

CRN exerting high initial release in water ("burst effect") caused losses of up to 45–50% of the applied N, mainly in the first leaching. This inflicted in some cases damage to young seedlings.

The release rate of N from some of the CRN was significantly reduced after 40 to 60% of the N was released ("tailing effect"). Consequently, growth rate and N uptake by plants was reduced in comparison to other treatments. Fertilizers having a linear or sigmoidal release pattern and with almost no "tailing" at the final release stage produced highest yields and N uptake with minimal losses of N by leaching.

Using CRNs which well match the pattern of plant demand resulted in total uptake of 70 to 80% of the applied N, while losses by leaching ranged between 1 to 15%. On the other hand, the use of fertilizers which poorly fitted the pattern of plant demand resulted in uptake of only 20 to 30% of the applied N and the losses due to leaching exceeded the amount taken up by the plant! Split application of urea in a manner which minimized leaching losses resulted in uptake of 35 to 55% of the applied N.

Results clearly indicate that increasing N use efficiency and lowering environmental damage by using CRN can be critically affected by the ability to match release characteristics of a CRN to the pattern of N demand of the crop.

Introduction

Controlled release nitrogen (CRN) fertilizers have the potential of optimal supply of N during the growth period of crops by a single application. This results in enhanced economic use (reduced losses and application cost); improved growth conditions (enhanced germination and growth conditions) and effective protection of the environment (minimizing nitrate pollution, nitrous oxide and ammonia emission) (Allen, 1984; Hauck, 1985; Shaviv and Mikkelsen, 1993).

The production and use of coated CRN specially those based on urea is steadily increasing due to the fact that the release of N from coated fertilizers is less sensitive to microbial activity than urea-formaldehyde and they offer higher flexibility in release control (Gertz, 1991, 1994).

Coated CRN can be divided into two main groups according to their release mechanisms (Raban, 1994; Shaviv et al., 1994)): a. CRN in which catastrophic (failure) release occurs, like sulfur coated urea (SCU) and to some extent even the polymer sulfur coated urea (PSCU); and b. CRN with diffusion controlled release as is the case with polymer or resin coated fertilizers.

The release pattern in case of the failure mechanisms is characterized by an initial "burst" of the fertilizer from the poorly coated granules, followed by a period of linear release and the last 20 to 50% of the content are released over a very long period (denoted "tailing" or "lock-off effects"). The tailing is typical to the sulfur coating and occurs with too thickly coated granules.

The release from polymer coated urea takes usually a linear or a sigmoidal form. The (initial) lag in sigmoidal curves is due to the time needed for water to penetrate into the coated core, dissolve the urea and induce its transport out. This is followed by a period of linear release (as long as the internal solution is saturated) and finally by a period of first order (declining) release, which may extend over relatively long periods. Little efforts in assessing the effects of release rate and pattern of CRNs on their agronomic efficiency and their effectiveness in preserving a safer environment were done so far.

The objective of this work was to test effect of several CRNs differing in pattern and release rate on plant response (dry matter yield and N uptake of rye grass) and losses (leaching) in a pot experiment with rye grass grown on sandy loam.

Materials and methods

Fertilizers

Five different urea based CRN fertilizers were tested and compared with split application of urea during the growth period.

Three of the CRN (denoted CRU-932, CRU-948, and CRU-1049) are coated with a new type of organically based coating having no burst effect and releasing their content without any significant tailing. The other two were PSCU (e.g. Goertz, 1994) and PULC - polyurethane like coating of a urea core (e.g. RLC coating in Goertz, 1991). Nitrogen content in the CRNs ranges between 36 to 40%.

Characterization of CRN fertilizers was performed by determining their release in de-ionized water and by incubation in a sandy loam.

Water Release was determined by placing 5 g fertilizer samples in 50 mL water at 30 °C. The solution was decanted from the granules at pre-determined time intervals and replaced by fresh water. Urea concentration was determined by using urease to transform it to ammonium which concentration was determined by an auto-analyzer.

Incubations were performed with the same sandy-loam used in the pot experiment. Soil moisture was maintained constant at a potential of 0.2 atmospheres and at a temperature of 30 °C. Soils were sampled in duplicates at pre-determined time intervals and ammonium, nitrate and urea concentrations determined with an auto-analyzer.

Pot experiment

Rye grass (*Lolium perenne*) was grown on a sandy loam in 3-L pots, in a greenhouse with controlled heating. The crop was grown for 175 days starting on 20th October 1993. Plants were cut four times: 54, 90, 129, and 175 days after sowing. Three days before each cutting the pots were leached with 500 mL (about 1 pore volume of the water held in soil at field capacity). CRU-932 was harveted only 3 times. Average temperature in the green house was 20 °C in the first 4 months and then it steadily increased to 28 °C during the last 2 months.

Nitrogen was applied at three levels of 0.8, 1.2, and 1.6 g N pot^{-1}. The 5 CRNs were all mixed in the soil bulk prior to seeding. The split applied urea was provided in three equal portions-in the first granular urea was mixed in soil bulk and the other two were applied in solutions a week after the first and the second cutting. The harvested plants were dried, weighted and digested after millng. Concentration of the reduced N was determined in the digest (Thomas et al., 1967) and its accumulation in plants calculated.

At the end of the experiment, several pots were opened and the roots separated from the soil, washed, dried and used for determination of N content and for estimating the ratio between N in roots and shoots. Nitrogen recovery was estimated on the basis of N accumulation in shoots, roots and leachates.

Results and discussion

Release characteristics

The five CRN fertilizers differ significantly in their release pattern and rates (Fig. 1). The PSCU has an initial burst of about 22% whereas all the others show linear to sigmoidal release with low initial release. The PSCU and the PULC show a very prominent tailing in water unlike the three CRNs with the organically based coating which end their release without tailing. Release in the incubation experiment is slower than in water and specially with the organically based coatings (Fig.1).

A criteria (used by manufacturers of membrane coated CRF) for defining release rate is to determine the time needed for 80% water release of the initial content, at a given temperature (usually around 20 °C).

Accordingly, the release time of 80% of the urea from CRU-932, CRU-948, CRU-1049, PSCU and

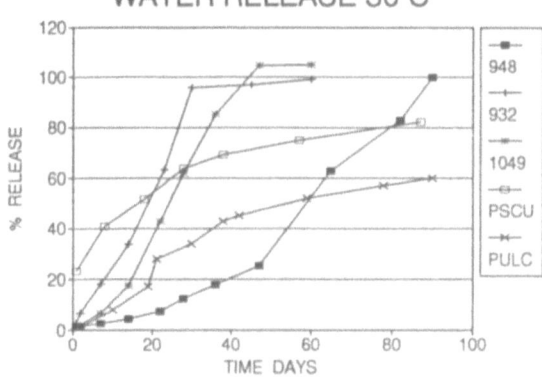

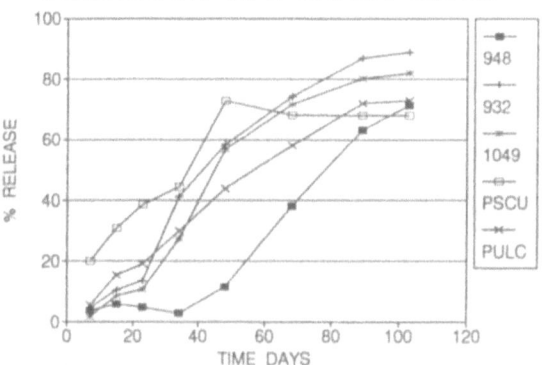

Figure 1. Urea release from five different CRN fertilizers in water (upper graph) and in an incubation exper. with sandy loam (lower graph), at 30 °C.

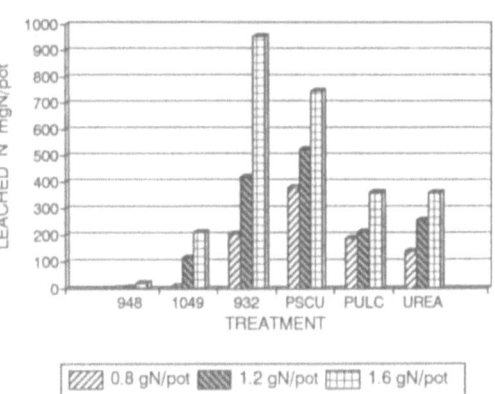

Figure 2. Cumulative leaching of mineral N, obtained for 6 N fertilization treatments and three levels of applied N after 3 leachings.

PULC was estimated as: 4, 1.1, 5, 11 and 14 weeks respectively, at 30 °C. The release time at 20 °C is about twice as long.

Noteworthy is the fact that reporting only the period of 80% release may not be informative enough, since it does not indicate anything regarding the undesired burst and tailing effects or about the release pattern (Fickian, linear or sigmoidal).

Leaching losses

Figure 2 shows cumulative losses, during the three first leachings (in the fourth one, losses were negligible). Highest losses were with CRU-932 and with PSCU ranging between 25 to 60% of the applied N, despite the fact that according to the release data (Fig. 1) the two CRNs significantly differ in their 80% release time (4 vs. 11 weeks, respectively). This obviously demon-strates the contribution of the initial burst in case of the PSCU which considerably increased the initial losses. Moreover, with CRU-1049 which apparently released its 80% in 5 weeks the losses were only 12%. Leaching losses with CRU-948 were 1–2% of the applied N!

In all cases the highest losses were in the first leaching 54 days after seedling. According to Figure 1, a large portion of the urea (about 40%) of the PSCU was released within the first two weeks, namely before and during the germination period!

Dry matter yield

Figure 4 compares total DM in 4 cuttings and in the first one. In first harvest, a sharp decline of the DM is observed when increasing N levels with PSCU, indicating a poisoning effect in this case. This conforms with burnings on the seedlings which were observed only with this CRN source. Noteworthy is the fact, that despite the high losses observed with CRU-932, the yields were much higher than with the PSCU! This can be explained by the fact that PSCU releases immediately after its application too high amounts of urea whereas the initial release from CRU-932 is low and therefore the damage to seedlings negligible. Consequently, the highest yields and N losses in the first cutting were with CRU-932. However, highest total yields and lowest losses during 4 cuttings were obtained with CRU-948 and CRU-1049.

Figure 5 shows changes in accumulated DM yield with time. Noteworthy is the relative increase of yield with the split applied urea in the two last cuttings. This is mainly due to the fact that the side dressings of urea

TOTAL N ACCUMULATION
4 HARVESTS

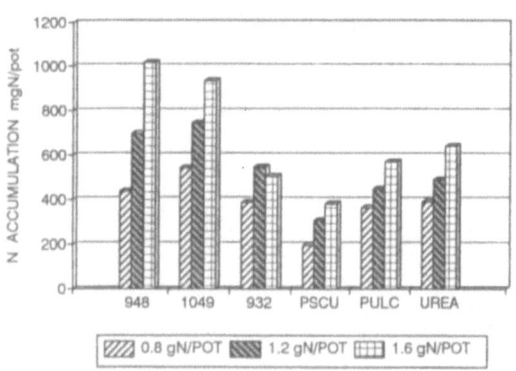

Figure 3. Total N accumulation in rye grass shoots after 4 cuttings, obtained for 6 N fertilization treatments and three N levels.

DRY MATTER YIELD
FIRST HARVEST

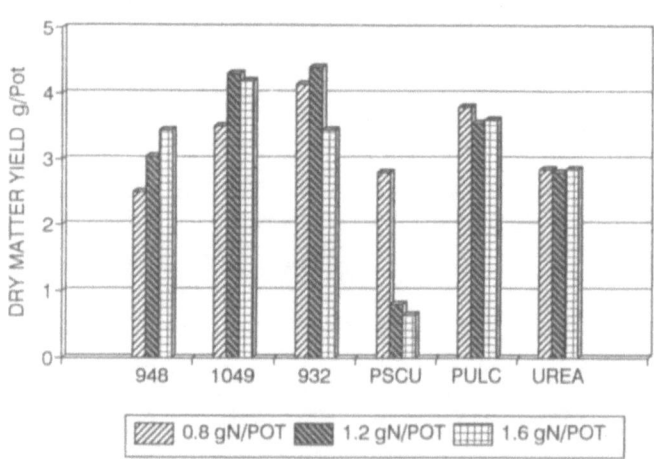

solutions was a week after each of the first and second leachings, thus minimizing the losses from this specific treatment!

N accumulation in plants

Figure 3 shows results of total N accumulated in shoots during 4 cuttings. It conforms well with N leaching losses (Fig. 2), except in case of the PSCU. In that case an additional adverse effect due to damage to seedlings reduced N accumulation more than expected from leaching losses.

Changes with time in N accumulation are shown in Figure 6. It is noteworthy that at the two lower levels of N application CRU-1049 induced the highest yields and only at the highest level of N application CRU-948 produced similar results. All other treatments were significantly inferior.

N recovery

Recovery of the applied N was evaluated by summing over the N accumulated in shoots, roots and leachates. Figure 7 presents changes with time of the recovered N. Those can be used as an (under) estimate of the released N from the CRNs. The actual release rate is assumably somewhat faster since we did not account for possible losses by de-nitrification or by biological fixation.

Highest recoveries were obtained with CRU-932 which according to Figures 1 and 7 released its content within two months. Lowest values were obtained with CRU-948 and regular urea. According to the release

TOTAL DRY MATTER YIELD
4 HARVESTS

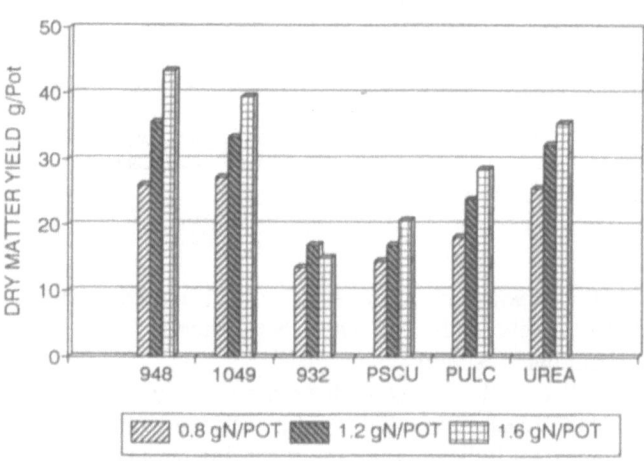

Figure 4. Dry matter yield of rye grass at first harvest (upper graph); and total dry matter yield after 4 harvests (lower graph), both obtained for 6 N fertilization treatments and three N levels.

profile of CRU-948, it assumably did not release its original content during the experiment. The poor recovery with urea can be attributed to un-accounted for losses or transformations.

Figure 8 demonstrates the basic difference between CRU-1049 and PSCU which differ significantly in their release profiles. The losses with CRU-1049 are small,

289

ACCUMULATED DRY MATTER YIELD

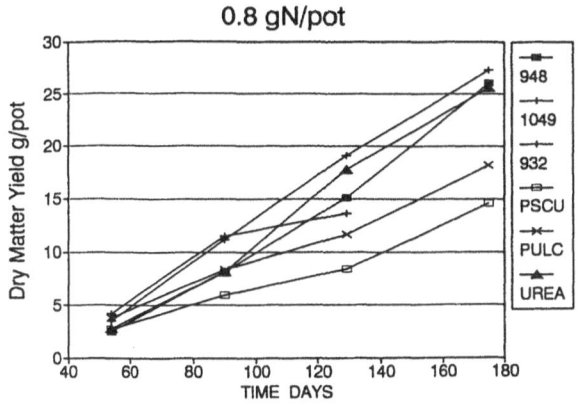

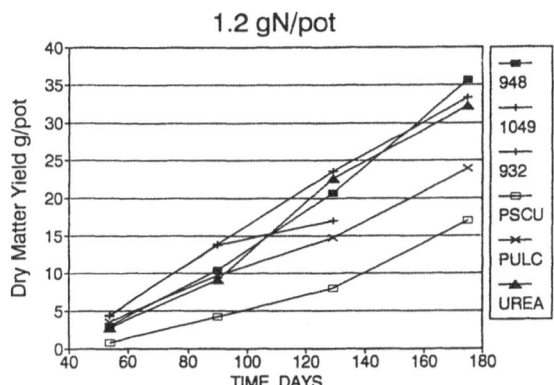

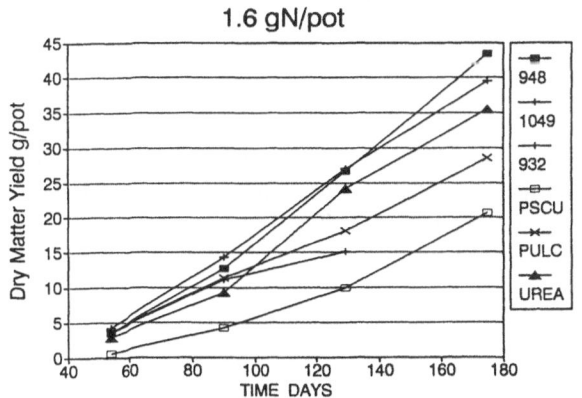

Figure 5. Accumulation of dry matter yield with time, obtained for 6 fertilization treatments with N application levels of: 0.8, 1.2 and 1.6 g N pot^{-1}.

N ACCUM. in SHOOTS

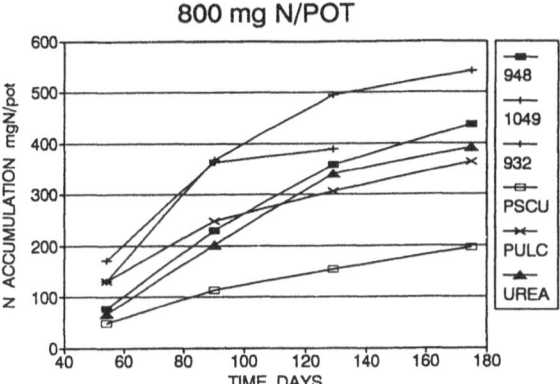

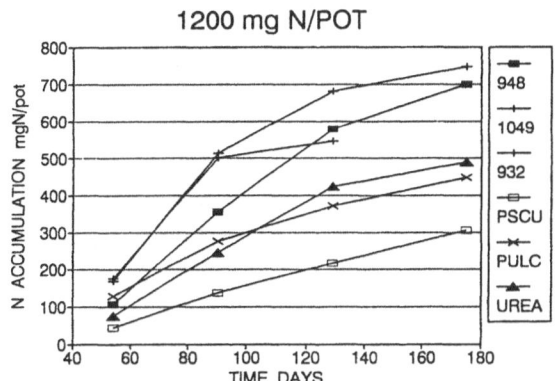

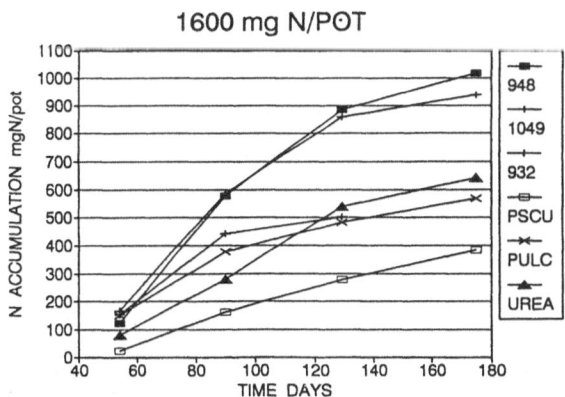

Figure 6. Accumulation with time of N in shoots, obtained for 6 N fertilization treatments and three application levels of: 0.8, 1.2 and 1.6 g N pot^{-1}.

mainly due to its sigmoidal release profile which well matches the N demand of rye grass. On the other hand, losses with PSCU are very large due to the exceeding release over demand in the initial growth stages. These

results are contradictory to the release periods of 80% obtained in water (Fig. 1), 5 vs. 11 weeks for CRU-1049 and PSCU, respectively.

N - RECOVERY

800 mg N/pot

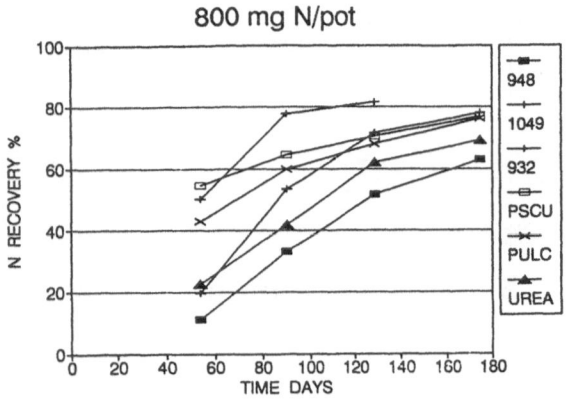

1200 mg N/pot

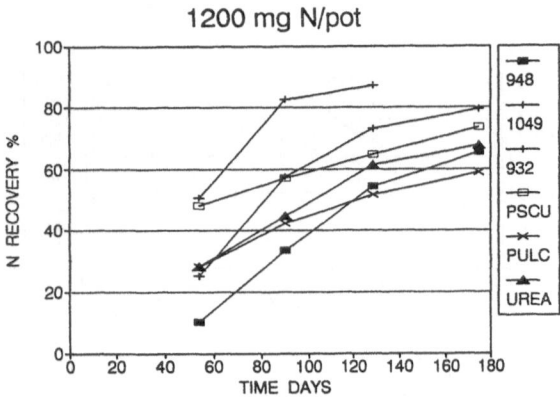

1600 mg N/pot

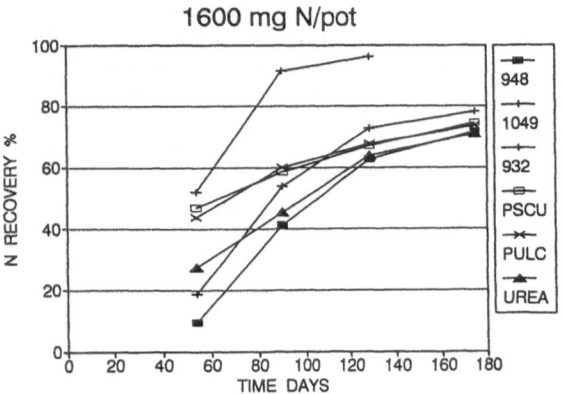

Figure 7. N - Recovery (shoots, roots, leachates) with time, obtained for 6 fertilization treatments and three application levels: 0.8, 1.2 and 1.6 g N pot^{-1}.

PSCU N - BALANCE
1200 mgN/pot

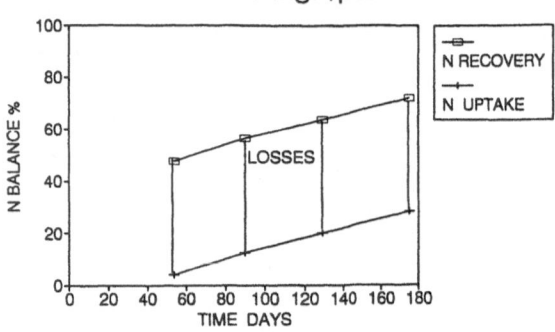

CRU-1049 N - BALANCE
1200 mgN/pot

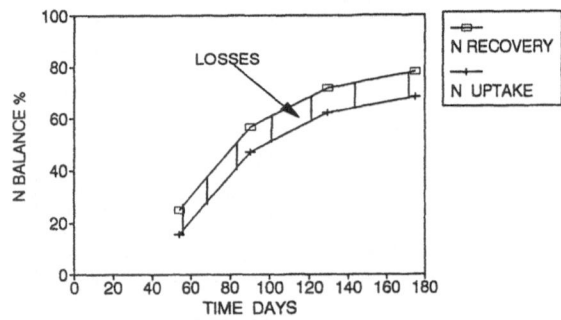

Figure 8. N - balance - Time changes of recovery vs. uptake as obtained for PSCU and for CRU-1049 at a N level of 1.2 g N pot^{-1}.

The rate of N accumulation between first and last cutting is about 5 times larger with CRU-1049 than with PSCU. This is in accordance with the release profiles. Between first and fourth cuttings the CRU-1049 is mainly in the linear phase of the sigmoidal pattern and the PSCU in predominantly in the "tailing" phase (compare Figs. 1, 7 and 8).

Conclusion

The effects of release rate and pattern of different CRNs on rye grass yields, uptake of N and N losses by leaching have been demonstrated. Two main factors seem to have the most striking effects on N use efficiency and losses: the existence of a high initial burst: and the tailing effect. The first induces both high losses and in extreme cases inflicts damage to seedlings and young plants; The second causes a situation in which N supply rate is lower than plant demand. CRN with a linear or sigmoidal release pattern and with average rates of water release between 1.2 to 2.8 %/ day induced best results from both agronomic and environmental aspects. It has been shown, at least with CRU-1049, that using a suitable CRN source should result in both high yields and minimal adverse effects on the environment. Characterization of CRN just by the time needed for release in water of 80% of their content does not provide good enough information for optimal management of plant nutrition and reliable control over N losses to the environment. Detailed information regarding the release pattern and relations between release in water and in soil are extremely important for good management of N fertilization.

Acknowledgement

Thanks are due to Ronit Faran, Abraham Moses and Smadar Nedan for their devoted assistance.

References

Allen S E 1984 Slow-release nitrogen fertilizers. *In* Nitrogen in Crop Production. Ed. R D Hauck. pp 195–206. ASA, Madison, WI, USA.

Goertz H M 1991 Commercial granular controlled release fertilizers for specialty markets. *In* Controlled Release Fertilizer Workshop - 1991, Proceedings. Ed.R M Scheib. pp 51–67. TVA, NFERC, AL, USA.

Goertz H M 1995 Technology developments in coated fertilizers. *In* Dahlia Gredinger Memorial International Workshop On Controlled Release Fertilizers, Proceedings. Eds. J Hagin and J Mortvedt. Haifa, March 1993. Technion.

Hauck R D 1985 Slow release and bio-inhibitor-amended nitrogen fertilizers. *In* Fertilizer Technology and Use Ed. O P Engelsted. pp 293–322. Third ed. SSSA, Madison, WI, USA.

Raban S 1994 Mechanisms of controlled release from coated fertilizers. M. Sc. Thesis. Faculty Agric. Engn. Technion-IIT, Haifa, Israel.

Shaviv A and Mikkelsen R L 1993 Controlled-release fertilizers to increase efficiency of nutrient use and minimize environmental degradation - A review. Fert. Res. 35, 1–12.

Shaviv A, Zlotnikov E and Zaidel E 1995 Mechanisms of nutrient release from controlled release fertilizers. *In* Dahlia Gredinger Memorial International Workshop on Controlled Release Fertilizers, Proceedings. Eds. J Hagin and J Mortvedt Haifa, March 1993. Technion.

Thomas R L, Sheard R W and Moyer J R 1967 Comparison of conventional and automated procedures for nitrogen, phosphorus and potassium analysis of plant material using a single digestion. Agron. J. 59, 240–243.

O. Van Cleemput et al. (eds.), Progress in Nitrogen Cycling Studies, 293–297, 1996.
© 1996 *Kluwer Academic Publishers.*

The fate of inorganic N additions to an ombrotrophic bog

D.J. Silcock and B.L. Williams
Macaulay Land Use Research Institute, Craigiebuckler, Aberdeen, AB9 2QJ, Scotland, UK

Key words: microbial biomass, N deposition, ombrotrophic bogs, organic N, peat, *Sphagnum*

Abstract

Undisturbed peat cores (50 cm depth) were maintained in situ in a *Sphagnum* dominated ombrotrophic bog for four months. Solutions of NH_4NO_3 at four application rates were added at two week intervals, the total amount of N added over the experimental period was 0, 2.7, 8.1 and 26.8 kg N ha^{-1} (equivalent to 0, 10, 30 and 100 kg N ha^{-1} yr^{-1}). On harvesting, cores were cut into 5 cm slices for analysis. Mineral-N was present in all N treatments predominantly as NH_4-N with only trace amounts of NO_3-N and there was no significant ($p < 0.05$) effect of treatment on either of the mineral N forms. Soluble organic N extracted with 0.5 M K_2SO_4 was present in greater amounts (a mean value across the treatments of 47.2 kg N ha^{-1}) to 50 cm depth than inorganic-N (13.3 kg N ha^{-1}) throughout the peat profile, and the microbial biomass N pool, determined by fumigation extraction, (113.7 kg N ha^{-1}) was the largest of the fractions measured. All N pools varied significantly ($p < 0.005$) with depth and a greater value for the microbial N flush in the 20–25 cm layer than at other depths suggested that this was the zone where decomposition and immobilization were on average most active in this peat profile.

Introduction

Peatland ecosystems are perceived as sinks for inputs of inorganic N in atmospheric deposition either because the N deficient vegetation absorbs N from the rain and dry deposition (Edwards et al., 1985) or because microorganisms immobilise inorganic N reaching the underlying peat (Damman, 1988). Bog vegetation is adapted to utilise NH_4- and NO_3-N in rainwater (Woodin et al., 1985), but at high input levels mosses become damaged and respond physiologically by the production of amino acids (Baxter et al., 1992). Inorganic N penetrating the canopy may be transformed in the peat by several different processes which will have an important bearing on how peatland ecosystems respond to increasing additions of N in atmospheric deposition (Hemond, 1983) especially with an increasing trend of total N deposition across Europe (Pearson and Stewart, 1993) derived from industrial and agricultural practices.

The major pathway for N to enter peat is in the form of plant litter and Ingram (1978) has defined the acrotelm as the horizon in the peat profile where aerobic decomposition occurs. Clymo (1992) defined four zones in the surface 50 cm of ombrotrophic bogs

which changed in terms of structure and microbiological activity with increasing depth in the profile. In this scheme, the second layer beneath the living vegetation was regarded as the horizon where aerobic decomposition was active and below this there was a transition zone, where both aerobic and anaerobic processes occurred, to the permanently anoxic peat. For inorganic N passing through the vegetation canopy, immobilisation could be expected to occur in the zone where aerobic decomposition was active.

There is a sparsity of information concerning the N dynamics in peat and its vegetation and the aim of the work presented is to determine the effects of N deposition on vegetation and peat under natural conditions. In this paper, we describe the results of an experiment where NH_4NO_3 was applied at different rates to undisturbed peat cores in situ inside pvc tubes. The different N fractions, extractable inorganic, organic and microbial N pools were examined for changes after four months and the results and fate of the applied N are discussed.

Materials and methods

Experimental site

The site (Moidach More, 760 ha, National Grid Ref. NJ030420) is a raised bog situated 160 km N.W. of Aberdeen, Scotland. The experimental site is located on an area of undrained deep peat (2.5–6.0 m depth), altitude 275 m, where disturbance from cutting and drainage is minimal. The vegetation is dominated by a variety of *Sphagnum* species, *Erica tetralix* L., *Trichophorum cespitosum* (L.), *Eriophorum vaginatum* (L.) and *Calluna vulgaris* (L.).

Twelve pvc cores (15 cm diameter × 50 cm length) were inserted into the peat in a randomised block design (3 blocks × 4 N treatments). Areas of *Sphagnum magellanicum* were identified as dull crimson hummocks and these areas of nearly 100% cover were selected for location of the cores. Associated species (*Sphagnum papillosum, S. capillifolium, Aulacomnium palustre, Drosera rotundifolia* and vascular plants) were removed from the cores. A rainfall collector, and bore hole to measure depth of the water table were installed near each block of tubes.

Nitrogen was applied as NH_4NO_3 at 14 day intervals for four months (June–October 1993). Total amounts applied over the four month period were 0, (ambient N inputs from rainfall), 2.7, 8.1 and 26.8 kg N ha^{-1} which are equivalent to rates of 0, 10, 30 and 100 kg N ha^{-1} yr^{-1}, respectively. Solutions made up in high purity water (Elgastat UHP) were applied dropwise to simulate rainfall using a 50 cm^3 polythene syringe modified by incorporation of five needles (1.2 mm diameter × 40 mm length). To avoid edge effects of application, N solutions were applied over an area (20 × 20 cm) containing each core. Lower quantities of N were applied than expected, (81% of 10, 30 and 100 kg N ha^{-1} yr^{-1}) as on two of the nine treatment dates the site was severely waterlogged. Cores were harvested four weeks after the final N application. Rainfall was collected for analysis, (mean values determined from three samples) and the water table measured at two week intervals.

Chemical analysis

Cores were removed from the field site after four months and peat extruded from the pvc tubes and cut into 5 cm thick slices to a depth of 50 cm. Samples were stored at 4°C until required for analysis. Each slice was weighed fresh and moisture content determined from subsamples dried to a constant weight at 105 °C. Fresh subsamples (10 g) were extracted by shaking in 0.5 M K_2SO_4, filtering the suspension under suction through Whatman GF/A glass fibre filters and 0.45 μm membrane filters.

Ammonium-N concentrations were determined colorimetrically (Crooke and Simpson, 1971), following addition of MgO and diffusion of NH_3 into 0.01 M H_2SO_4 (Bremner, 1965). NO_3-N was analysed colorimetrically as NO_2^- after reduction with copperized cadmium (Henriksen and Selmer-Olsen, 1970). Total N in the extract was determined by oxidation of N to NO_3 using alkaline potassium persulphate and subsequent colorimetric analyses of NO_3 as described above, (Williams et al., 1995). Soluble organic-N in the extracts was calculated as the difference between the total extractable N and inorganic N (NH_4-N and NO_3-N).

Subsamples of fresh peat were fumigated for 18 h with chloroform vapour (Brookes et al., 1985). Vapour was removed by repeated evacuations and samples were extracted immediately by shaking with 0.5 M K_2SO_4. The suspensions were filtered and prepared as described for mineral-N extraction above. Total concentrations of soluble N in the extract were determined after oxidation with alkaline potassium persulphate and analysis of NO_3-N as described previously. The microbial N flush was determined as the difference between total extractable N of fumigated and unfumigated peat samples.

Statistical analysis

The results of the chemical analyses were expressed as weight per unit area (kg ha^{-1}) calculated from the area of cross section and fresh weight of each core. The significance of differences between the treatments and depth were tested using an analysis of variance (Genstat, 1987). Total amounts of each N fraction in the peat cores (0–50 cm) were also analysed by ANOVA using Mintab (Ryan et al., 1985).

Results and discussion

Nitrogen in rain

The total amount of NH_4 and NO_3 input from rain was 5.5 and 1.7 kg N ha^{-1}, respectively, for the duration of the experiment (127 days) with lower levels of organic N, 1.2 kg N ha^{-1}. These values can be compared with

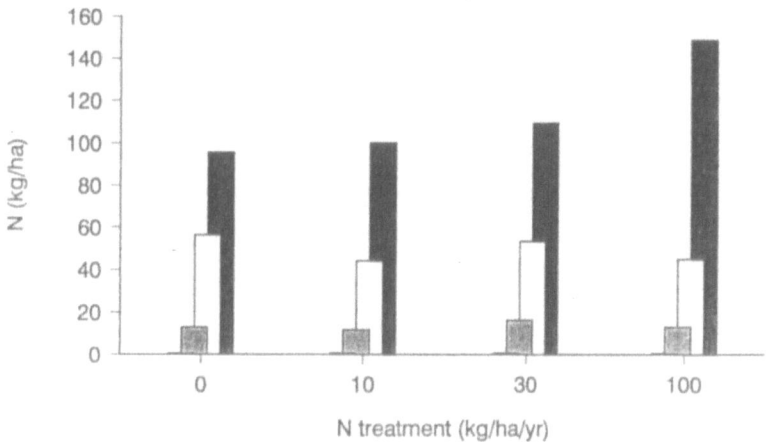

Figure 1. Total content (kg ha^{-1}) to 50 cm depth for NO$_3$-N (black shading, SED=0.36), NH$_4$-N (light shading, SED=1.52) extractable organic N (no shading, SED=10.9) and the flush of microbial N (heavy shading, SED=13.8). SED = Standard Error of Difference.

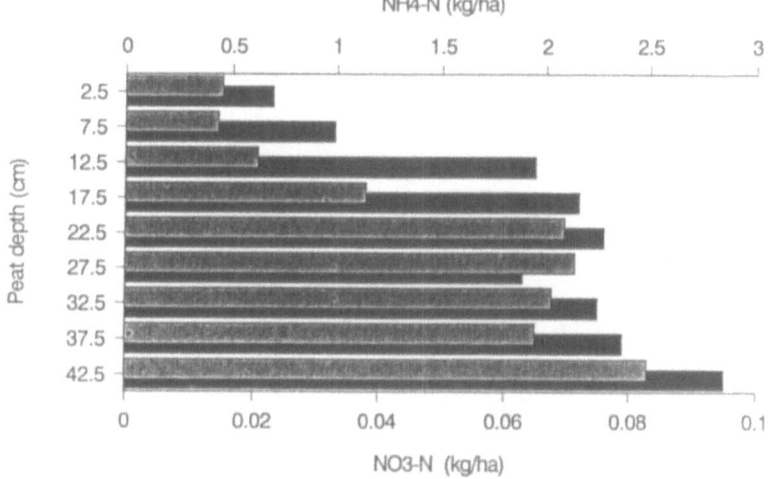

Figure 2. Mean NH$_4$-N (light shading) and NO$_3$-N (black shading), (kg ha^{-1}) for all N treatments at 5 cm intervals within the peat profile.

the actual amounts of either NO$_3$-N or NH$_4$-N that were applied during the four month experimental period, 1.3, 4.0 and 13.4 kg N ha^{-1} in the 10, 30 and 100 kg N ha^{-1} yr^{-1} treatments, respectively.

Peat mineral-N

The total amount of mineral-N in the profile to 50 cm depth ranged from 7.4 to 6.0 kg N ha^{-1} in the 30 and 100 kg N ha^{-1} yr^{-1} treatment, respectively and there was no significant effect of treatment (Fig. 1). Ammonium was the predominant form of mineral-N with NO$_3$ present in only trace amounts.

Averaged over all depths for each treatment there was no significant effect of N treatment on the NH$_4$-N content, (Fig. 1) but values averaged for all treat-

ments increased significantly ($p < 0.001$) with depth in the profile (Fig. 2). Even though the quantities of NO$_3$-N were small, they were significantly ($p < 0.001$) greater below 10 cm depth than in the surface (Fig. 2). The lower contents of NH$_4$ and NO$_3$-N in the surface 15 cm compared with lower depths probably reflects the greater utilization of mineral-N by the living moss and by microorganisms in the surface. Denitrification activity in acid peats is not considered to be great unless the pH has been increased by liming (Klemmedtsson et al., 1977), but *Sphagnum* mosses have adapted to utilize inorganic N in rainwater (Woodin et al., 1985).

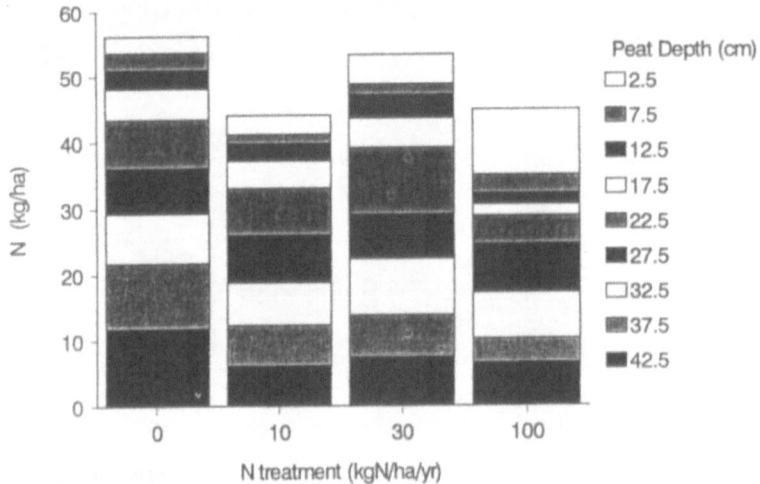

Figure 3. Mean soluble organic N (kg ha^{-1}) for each treatment (kg N ha^{-1} yr^{-1}) at 5 cm intervals within the peat profile.

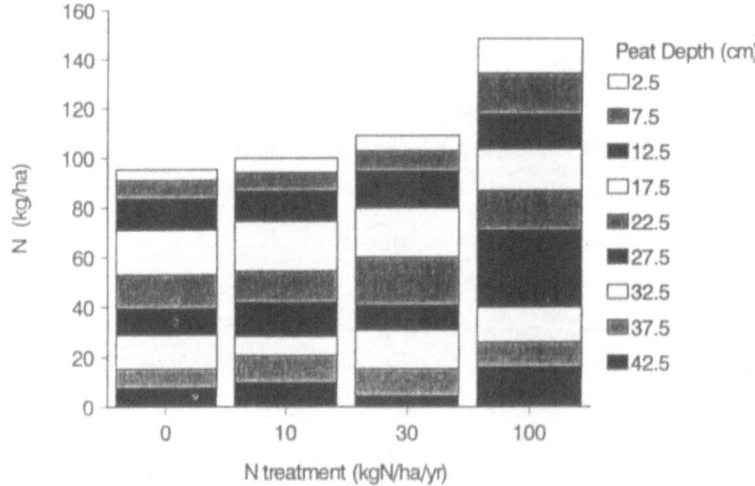

Figure 4. Mean flush of microbial N (kg ha^{-1}) for each N treatment (kg N ha^{-1} yr^{-1}) at 5 cm intervals within the peat profile.

Extractable organic-N

The total quantities of organic N in the profile extracted by 0.5 M K$_2$SO$_4$ ranged from 53.6 to 40.5 kg N ha^{-1} for control and 100 kg N ha^{-1} yr^{-1} treatments and was much greater than the mineral-N fraction (Fig. 1). The amounts fluctuated significantly ($p < 0.001$) with depth, increasing below 20 cms (Fig. 3). Averaged over all depths the extractable organic N was not affected by N treatments, but in the surface 5 cm there were indications that extractable organic N increased in cores receiving the two highest additions of N. This suggests that the moss had released organic N in response to additions of inorganic N which would be consistent with the response of *Sphagnum* spp. to high rates of N addition (Baxter et al., 1992).

Microbial N

The flush of microbial N was the largest of the pools measured in the peat cores and accounted for 100.0 to 144.3 kg N ha^{-1} in the surface 50 cm of the cores of 10 and 100 kg N ha^{-1} yr^{-1} treatments, respectively (Fig. 1). Assuming a value of 0.45 for the recovery factor, k (Williams and Sparling, 1984) yields a microbial biomass N value in this peat of approx. 252.7 kg N ha^{-1} in the surface 50 cm. The N flush fluctuated significantly ($p < 0.05$) with depth and was greater at 15–20 cm layer than in samples from the surface 10 cm and lower down in the profile (Fig. 4). The concept of a horizon in the peat profile where aerobic decomposition is active and called the acrotelm (Ingram, 1978) has been developed by Clymo (1992) and this horizon can be found beneath the living moss. In the profile from Moidach More, the higher microbial N flush values suggest that this zone may spread from 10 to 30 cm averaged over all N treatments.

The microbial N flush averaged over all depths was not significantly altered by additions of N which were small compared with the size of the flush even assuming no uptake by the vegetation. However, there were indications that biomass N in the surface 10 cm and at 25–30 cm may have been increased by the highest level of N addition (Fig. 4).

Conclusions

There was little evidence that NH_4NO_3 additions to peat cores could be detected as inorganic N pools after four month treatment indicating rapid transformation either by the vegetation or the microbial biomass. The extractable organic N pool was much greater than the mineral fraction, but the microbial pool contained the most N. Variations in the sizes of these pools with depth in the profile indicated greatest activity in the acrotelm, the zone beneath the living plants where aerobic decomposition occurs. Increases in the microbial and soluble N pools at the highest rates of N addition warrant further investigation and experiments using ^{15}N labelled NH_4NO_3 are ongoing to enable the fate of the various ^{15}N-labelled pools to be determined within the peat profile and consequently elucidate the uptake by vegetation versus microbial biomass in response to increased N inputs.

Acknowledgements

This work was funded by the Scottish Office Agriculture and Fisheries Department and by the European Commission, Contract No EV5V-CT92–0099. The authors thank Braemoray Estates and Scottish National Heritage for allowing access and use of the site on Moidach More, Betty Duff for statistical advice and Miriam Young for chemical analyses.

References

Baxter R, Emes M J and Lee J A 1992 Effects of an experimentally applied increase in ammonium on growth and amino-acid metabolism of *Sphagnum cuspidatum* Ehrh. ex. Hoffm. from differently polluted areas. New Phytol. 120, 265–274.

Bremner J M 1965 Inorganic forms of nitrogen. *In* Methods of Soil Analysis, Part 2. Ed. C A Black. pp 1179–1237. American Society of Agronomy, Madison, WI, USA.

Brookes P C, Landman A, Pruden G and Jenkinson D S 1985 Chloroform fumigation and the release of soil nitrogen: a rapid direct extraction method to measure microbial biomass nitrogen in soil. Soil Biol. Biochem. 17, 837–842.

Clymo R S 1992 Productivity and decomposition of peatland ecosystems. *In* Peatland Ecosystems and Man: an Impact Assessment. Eds. O M Bragg, P D Hulme, H A P Ingram and R A Robertson. pp 3–16. Department of Biological Sciences, University of Dundee, Dundee, UK.

Crooke W M and Simpson W E 1971 Determination of ammonium in Kjeldahl digests of crops by an automated procedure J. Sci. Food Agric. 22, 9–10.

Damman A W H 1988 Regulation of nitrogen removal and retention in *Sphagnum* bogs and other peatlands. Oikos 51, 291–305.

Edwards A C, Creasy J and Cresser M S 1985 Factors influencing nitrogen inputs and outputs in two Scottish upland catchments. Soil Use Manage. 1, 83–88.

Genstat 5 1987 Reference Manual, Clarendon Press, Oxford, UK.

Hemond H A 1983 The nitrogen budget of Thoreau's bog. Ecology 64, 99–109.

Henriksen A and Selmer-Olsen A R 1970 Automatic methods for determining nitrate and nitrite in water and soil extracts. Analyst 95, 514–518.

Ingram H A P 1978 Soil layers in mires: function and terminology. J. Soil Sci. 29, 224–227.

Klemmedtsson L, Svensson B H, Lindberg T and Rosswall T 1977 The use of acetylene in quantifying denitrification in soils. Swedish J. Agric. Res. 1, 179–185.

Pearson J and Stewart G R 1993 Tansley Review No. 56 The deposition of atmospheric ammonia and its effects on plants. New Phytol. 125, 283–305.

Ryan B F, Joiner B L and Ryan T A 1985 Minitab Handbook, 2nd ed. Duxbury Press, Boston, USA.

Williams B L, Shand C A, Hill M, O'Hara C, Smith S and Young M E 1995 A procedure for the simultaneous oxidation of total soluble N and P in extracts of fresh and fumigated soils and litters. Commun. Soil Sci. Plant Anal. 26, 91–106.

Williams B L and Sparling G P 1984 Extractable N and P in relation to microbial biomass in UK acid organic soils. Plant and Soil 76, 139–148.

Woodin S, Press M C and Lee J A 1985 Nitrate reductase activity in *Sphagnum fuscum* in relation to wet deposition of nitrate from the atmosphere. New Phytol. 99, 381–388.

O. Van Cleemput et al. (eds.), Progress in Nitrogen Cycling Studies, 299–302, 1996.

Field measured net N mineralization pattern in an alluvial Chernozem soil in the Szigetkoz Region of Hungary

M. Szucs[1], O. Van Cleemput[2] and G. Hofman[2]

[1]*Pannon University of Agricultural Sciences, 9200 Mosonmagyarovar, Hungary and* [2]*University of Ghent, Coupure 653, 9000 Ghent, Belgium*

Key words: Chernozem, mineralization, nitrogen, nitrate

Abstract

Long-term observation on the changes of the mineral N content in the 0–150 cm profiles for an unfertilized Chernozem soil was made in field conditions. Because of the high nitrification rate of the soil only the nitrate content is reported. Under the influence of the downward solute movement the mean nitrate content of the soil profiles decreased 5–7 years after stopping the N fertilization. In the unfertilized soil the relative amplitude of the nitrate content changes became large, giving low values during the cropping season and an important rise when fallowing. The year round average net N mineralization was 0.45 kg ha^{-1} day^{-1}. Low values were obtained for the winter and the dry period of the summer. The net N mineralization decreased sharply with soil depth. One third of it occurred in the top 10 cm layer of the soil.

Introduction

Mineralization of soil organic nitrogen and of plant residues is the major natural source of available nitrogen to plants and has been thoroughly investigated in many experiments. Numerous chemical methods have been developed to estimate the potentially mineralizable N of the soil. The main problem of the purely chemical approaches is that they cannot estimate the influence of the microbiological activity changing greatly under field conditions. Therefore, the generalization of the results and the forecast of the mobilization for different seasons of the year are rather difficult.

The laboratory incubation technique has a long history. However, a more successful period emerged with the mathematical description and interpretation of the net mineralization as a first order kinetic process (Deans et al., 1986; Standford and Smith, 1972). Under field conditions, an estimate of the net N mineralization can be obtained by measuring the cumulative uptake of nitrogen by a crop grown on unfertilized soil, on condition of minimum loss. Using this method, Ten Holte and Van Keulen (1989) mention the value of net N mineralization to be 0.37–0.5 kg ha^{-1}day^{-1}. Generalizing their data obtained with maize, Oberle and Keeney (1990) stated that for a wide range of soils the

Figure 1. The location of the experiment in Hungary

net N mineralization for the growing season is about 3.5% of the soil organic N. The main restrictions of this method are that the results depend on the plant and the measured values are valid only for the period of the growing season when the previously accumulated N has been depleted. Furthermore, these data can be underestimated because of the not accounted or uncertain N-uptake by the roots.

In our experiment, the direct measurement of the mineral N accumulation was used. This method

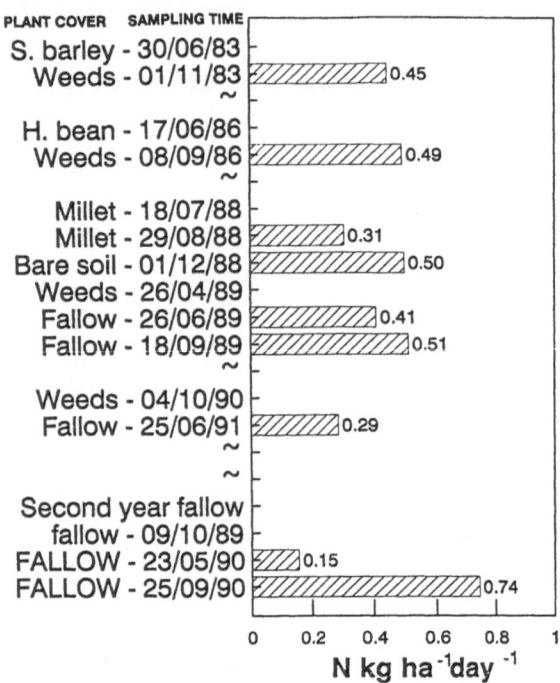

Year round average net N mineralization:
- for cropped field = 0.45 kg ha⁻¹ day⁻¹
- for second year fallow field = 0.36 kg ha⁻¹ day⁻¹

Figure 2. Daily net N mineralization of the unfertilized soil.

Table 1. Nitrate content changes in the unfertilized soil until a depth of 150 cm

Plant[a] - Sampling time	NO_3^--N kg ha^{-1}
S. barley - 04/05/83	281
S. barley - 30/06/83	157
Weeds - 01/11/83	212
Pea - 04/05/84	215
Pea - 06/06/84	281
Weeds - 20/09/84	296
Bare soil - 31/10/84	245
W. wheat - 24/04/85	141
Weeds - 01/10/85	90
H. bean - 23/04/86	152
H. bean - 17/06/86	29
Weeds - 08/09/86	70
S. barley - 15/04/87	83
S. barley - 08/06/87	73
S. barley - 24/06/87	31
Weeds - 17/08/87	34
Weeds - 27/10/87	164
Weeds - 14/04/88	165
Weeds - 25/04/88	151
Millet - 07/07/88	133
Millet - 18/07/88	51
Millet - 29/08/88	64
Bare soil - 01/12/88	111
Weeds - 26/04/89	85
Fallow - 26/06/89	110
Fallow - 18/09/89	153
Fallow - 09/10/89	204
W. barley - 20/03/90	113
W. barley - 03/05/90	23
W. barley - 23/05/90	79
Bare soil - 17/07/90	29
Bare soil - 01/08/90	126
Weeds - 25/09/90	22
Weeds - 04/10/90	39
Fallow - 25/06/91	115

[a] Letters before plant names: S. - Spring; W. - Winter; H. - Horse.

requires the selection of periods, when no crop is present and no considerable leaching occurs. Because of these conditions, sufficient results can be obtained only after several years of observation, but they are closely linked to the farming practice. A year round mineralization pattern and mineralization versus soil depth curves can be produced and used as an important input for simulation models.

Materials and methods

The field experiment was carried out from 1983 until 1991 on an alluvial Chernozem soil, typical for the area shown in Figure 1. The soil has 70% particles less then 0.02 mm and its texture is classified as a clayey loam. The organic matter content is 3.4% in the top 30 cm layer and the mollic epipedon has a thickness of 55 cm. The calcium carbonate content is 21%, giving a soil pH of about 8.0. During the investigation period, the field was regularly cropped, dominantly with cereals. All conditions were similar to usual farmland except the fertilizer use. No fertilizer

was applied to the field starting from 1978, five years before the observation started, and until the end of the experiment. The soil was regularly sampled until a depth of 150 cm with 10 cm layers. The nitrate and ammonium N content were determined immediately after sampling. As the ammonium ion content was low and constant, the nitrate N content has been used. The net N mineralization has been calculated as the difference in nitrate N content of every 10 cm soil layer for the periods when no uptake by crop and no

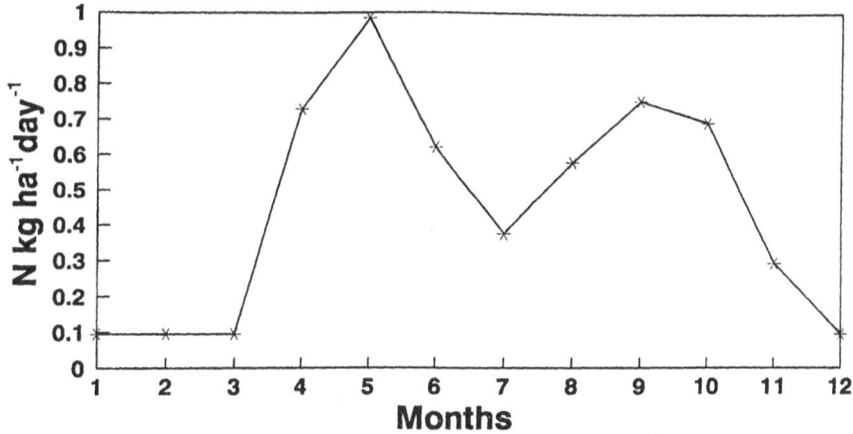

Figure 3. The net N mineralization pattern of the soil

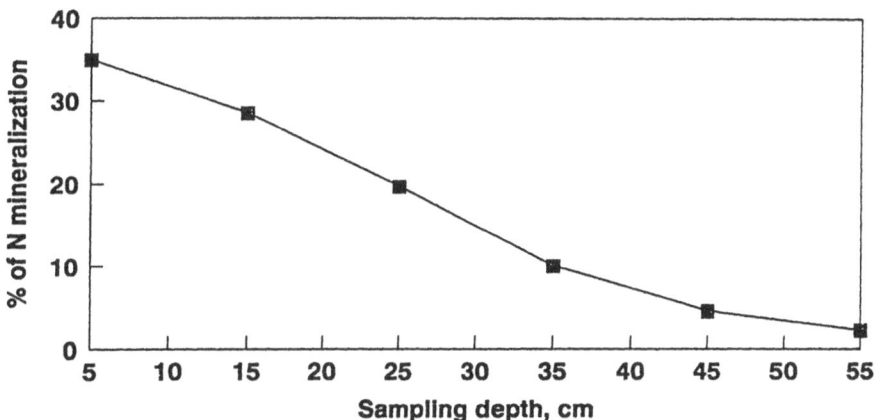

Figure 4. Different soil depths' contribution to the net N mineralization

Results

During the first two years of the experiment the nitrate content of the soil profile was high (Table 1). This is the result of the fact that the last portion of the original fertilizer N left the lower level of the profile only during the wet spring of 1985. This has no effect on our results even during the first years, because this excess nitrogen, at the time of the start of the observation, was below 1 m depth and the mineralization is not important below 60 cm depth. Table 1 shows that without fertilizer the relative changes in nitrate content became large, giving low values under crops and a significant increase during the fallow period.

Selected sampling times were used to calculate the net N mineralization. Sampling was done, when crop was present or the N uptake decreased near zero because of the ripening of the plant (spring barley in 1983) or when the crop was damaged by a long period of drought (millet in 1988). Short periods were omitted, especially when the samples were taken after ploughing. The aeration, caused by human activity, sharply increased the mineralization, but these data were not characteristic for longer periods. The mean value calculated from Figure 2 gives an average net N mineralization equal to 0.45 kg ha^{-1}day^{-1}. The value for the second year fallow calculated by the method of weighed averages was smaller (0.36 kg ha^{-1} day^{-1}). The large value for the summer of 1990 was the result of the preceding cultivation.

considerable leaching could occur. The mean annual temperature is $10\,°C$ and the mean annual precipitation is about 600 mm. Due to the high evapotranspiration the drainage is minimal and occurs only in early spring.

The change of the net N mineralization during the year (Fig. 3) followed the pattern observed in many other experiments, having low values in winter time, because of the low temperature, increasing in early summer, decreasing in midsummer presumably because of the dryness of the soil and having a second increase in early autumn.

Figure 4 illustrates the change of the net N mineralization with the soil depth. The decrease at the depths of 15 and 25 cm can be attributed to less intense aeration, as the organic matter content of the upper 30 cm layer is uniformly distributed because of regular ploughing.

Discussion

The obtained results for the average net N mineralization and for its change during the year were in good agreement with the literature data. These values and the distribution of the mineralization rate at different soil depths can be used for model inputs. The con-

siderable changes in nitrate content of the soil during the year and especially the rather low values during the growth of unfertilized crops indicated that the soil itself cannot provide enough nitrogen to get high yield. Without fertilizer use the depletion of the labile organic fraction of the soil leads to further reduction of the mineralization.

References

Deans J R, Molina J A E and Clapp C E 1986 Models for predicting potentially mineralizable nitrogen and decomposition rate constants. Soil Sci. Soc. Am. J. 50, 323–326.

Oberle S L and Keeney D R 1990 Factors influencing corn fertilizer N requirements in the Northern U.S. corn belt. J. Production Agric. 3, 527–534.

Standford G and Smith S J 1972 Nitrogen mineralization potentials of soils. Soil Sci. Soc. Am. Proc. 36, 465–472.

Ten Holte L and Van Keulen H 1989 Effect of white and red clover as a green manure crop on growth, yield and nitrogen response of sugar beet and potatoes. In Legumines in farming systems. Eds. P Plancquaert and R Hagar. pp 16–24. Kluwer Academic Publishers, Dordrecht, The Netherlands.

O. Van Cleemput et al. (eds.), Progress in Nitrogen Cycling Studies, 303–306, 1996.

Sustainability of nitrogen use in two dryland farming systems

M. Wood, A.M. McNeill, C.J. Pilbeam[4], R.S. Swift[1], H.C. Harris[2] and P.G. Mugane[3]
Department of Soil Science, The University of Reading, Reading, RG6 2DW, UK.[1] *Division of Soils, CSIRO, Adelaide, SA 5064, Australia,*[2] *ICARDA, P.O. Box 5466, Aleppo, Syria and* [3]*KARI, P.O. Box 30148, Nairobi, Kenya.* [4] *Corresponding author*

Key words: biological nitrogen fixation, crop rotation, dryland agriculture, fertilizer recovery, intercropping, N fertilizer

Abstract

In the drier environments of rainfed agricultural systems both water and nitrogen are major limitations to crop productivity. In the Mediterranean region the traditional cereal-fallow system is being replaced by a cereal-legume rotation. An experiment to measure crop N uptake and biological nitrogen fixation in wheat/lentil and wheat/chickpea rotations was conducted in 1991/92. Recovery of ^{15}N-labelled urea fertilizer by wheat was poor. Significant amounts of crop nitrogen came from biological nitrogen fixation in lentil (>50%) and chickpea (ca. 30%).

In Kenya farmers are attempting to grow intercrops of maize and bean in semi-arid regions. Experiments in four seasons measured crop N uptake and biological nitrogen fixation by maize/bean and maize/cowpea intercrops. Recovery of ^{15}N-labelled ammonium sulphate fertilizer was poor in all crops. There was no measurable N fixation by bean. In contrast, cowpea fixed 50% of their N, and so provided a net input of N into the system.

These two dryland soils are characterized by rapid rates of mineralization-immobilization turnover and poor recoveries of N fertilizer. In the long-term the N status of these cropping systems can only be maintained or increased either by the use of N fertilizer, or by the inclusion of a legume. Lentil and cowpea appear well suited to replace the need for N fertilizer in the 2 areas described here.

Introduction

In the drier environments of rainfed agricultural systems both water and nitrogen are major limitations to crop productivity. Research efforts are being directed to increasing short-term productivity whilst also attempting to reduce the risk of jeopardising the long-term capacity of the land to provide food. In addition, the development strategies employed in the studies described here have been based as closely as possible on existing farming practises with minimal additional inputs. The use of supplementary irrigation has therefore been avoided and attention has been focused on inputs of nitrogen as the major variable in terms of quantities and sources. The experiments have involved the use of ^{15}N-labelled fertilizer to estimate fertilizer use by the cereal and nitrogen fixation by the legumes. Measurements have also been made of water use, but these data are not presented here.

Data are included from experiments in two contrasting traditional dryland farming systems. First, a cereal-legume mixed cropping system used in East Africa and second, a cereal-fallow rotation system used in the West Asia/North Africa region.

In Kenya, pressure of population expansion is forcing farmers to attempt to grow maize (*Zea mays*) intercropped with bean (*Phaseolus vulgaris*), which are the preferred crops, in semi-arid regions to which these crops are not ideally suited. A variety of other crops such as cowpea (*Vigna unguiculata*), which is better suited though less preferable as food, is also grown. These experiments were carried out in collaboration with the Kenya Agricultural Research Institute (KARI) at Kiboko Research Station, which has a bimodal rainfall pattern with approximately 700 mm per year divided between the Long Rains (March to July) and the Short Rains (October to February). The soil has been described as an acri-orthic Ferralsol and contains 1%

organic matter. Experiments were carried out, during a four-year period, on nitrogen and water use by mixtures and monocultures of these two crops.

In Syria, pressure is increasing to replace the fallow phase of the traditional system with a legume such as lentil (*Lens culinaris*) or chickpea (*Cicer arietinum*) or in extreme cases to attempt to grow wheat continuously. These experiments were carried out at Tel Hadya Research Station near to Aleppo. The climate is characterized by wet, cold winters and hot, dry summers with an average annual rainfall of approximately 330 mm. The soil has been described as a Calcixerollic Xerochrept and contains 1% organic matter. The experiments described here, which will extend over four years, form part of a long-term two-course rotation study started in 1983 by ICARDA. Preliminary results are reported for a single season.

Material and methods

Kiboko, Kenya

The response of maize (cultivar Makueni Composite) to the equivalent of 10 or 80 kg N ha^{-1} as calcium ammonium nitrate (CAN) was measured during the Long Rains 1990, Short Rains 1990 and Long Rains 1991. Bean (cultivar B9) treated with the equivalent of 10 kg N ha^{-1} was also included as separate plots, and in the Short Rains 1990 an additional treatment was included in which bean seed was inoculated with Rhizobium leguminosarum biovar phaseoli strain 9C (Karanja and Wood, 1988). In the Short Rains 1992 cowpea (a local seed mixture) was included and treated with the equivalent of 10 kg N ha^{-1}. Phosphate fertilizer was applied to all plots at a rate equivalent to 40 kg P ha^{-1}. There were four replicate plots (6 × 4 m) per treatment.

In all main plots 1 m^2 microplots were treated not with CAN but with ^{15}N-labelled ammonium sulphate (5.3 atom% in all cases, except 6.0 atom% in the short rains 1992) at a rate equivalent to 10 kg N ha^{-1}.

Tel Hadya, Syria

In the 1983/84 season, as part of an on-going long-term study, a range of different cropping sequences were established. Four of the sequences will be discussed here, namely wheat following fallow (w/f), wheat following lentil (w/l), wheat following chickpea (w/c) and continuous wheat (w/w). Both phases of the rotation

were included. There were three replicate plots (0.54 ha) in each phase.

In the first two years no inputs were used, but from the 1985/86 season fertilizer, and weed, pest and disease control measures have been applied, and improved cultivars used. Four levels of nitrogen (0, 30, 60 or 90 kg N ha^{-1}) were applied as a split-plot design to sub-plots in the wheat phase in order to assess the long-term responses to N and the N benefits derived from the presence of the legumes.

In the 1991/92 season in all wheat plots 4 m^2 microplots were treated with ^{15}N-labelled urea (10 atom%) at the same rate of N application as the main plot, in order to measure fertilizer use efficiency. In addition, in the legume plots 1 m^2 microplots were treated with ^{15}N-labelled ammonium sulphate (10 atom%) at a rate equivalent to 30 kg N ha^{-1} in order to estimate the rate of nitrogen fixation.

Sample analysis

At final harvest dry matter, total nitrogen and ^{15}N/^{14}N ratio of plant samples were determined. The proportion of N in the plant derived from fertilizer was calculated from the ratio of the enrichment of the plant to the enrichment of the fertilizer, and the proportion of N in the legume derived from fixation was estimated from the ratio of the enrichment of the legume to the enrichment of the cereal (Fried and Middleboe, 1977).

Results and discussion

Kiboko, Kenya

Maize showed no response to N fertilizer in terms of grain yield (Table 1) and total dry matter (data not shown) in three seasons with a wide range of rainfall (Table 1). Furthermore, the recovery of the applied fertilizer in the crop was very low (Table 1), with a trend towards higher recovery in wetter seasons. This indicated that maize was not limited by N under these conditions. Attention therefore focused on the legume component.

One of the major benefits associated with the presence of a legume in a cropping system is the input of atmospheric nitrogen derived from nitrogen fixation. Although there are difficulties in obtaining accurate estimates of rates of fixation under field conditions, a comparison of the total N yield of a cereal and legume provides a rough estimate, and the use of ^{15}N isotope

Table 1. Grain yield (kg ha^{-1}) for maize in response to two levels of nitrogen fertilizer at Kiboko during 3 seasons. Data are included for fertilizer recovery (proportion of N applied recovered in the crop) based upon ^{15}N data, and for total rainfall during the season

Season	N Rate (kg N ha^{-1})		Fertilizer recovery (%)	Rainfall (mm)
	10	80		
Long Rains 1990	2305	2237	6	179
Short Rains 1990	1679	1588	18	258
Long Rains 1991	144	199	1	135

Table 2. Total shoot nitrogen yield and ^{15}N enrichment of bean (uninoculated) and maize at Kiboko during the Long Rains 1990

Crop	Shoot N (kg N ha^{-1})	^{15}N atom% excess
Bean	34	0.107
Maize	33	0.092

Table 4. Total shoot nitrogen yield, ^{15}N enrichment and fertilizer recovery (proportion of N applied recovered in the crop) for bean (uninoculated), maize and cowpea at Kiboko during the Short Rains 1992

Crop	Shoot N (kg N ha^{-1})	^{15}N atom % excess	Fertilizer recovery (%)
Bean	53	0.152	14
Maize	87	0.104	16
Cowpea	394	0.052	37

Table 5. Sources of N (kg N ha^{-1}) for bean (uninoculated), maize and cowpea at Kiboko during the Short Rains 1992

Crop	Fertilizer	Soil	Fixation
Bean	2	51	0
Maize	2	85	0
Cowpea	4	193	197

dilution increases the accuracy of estimates. Data for maize and bean from a single season (Table 2) show no difference in either total N yield or ^{15}N enrichment between these crops indicating the absence of nitrogen fixation by bean plants. This was confirmed in other seasons, and supported by the observation that the plants were very poorly nodulated.

A possible solution to the poor nodulation by the bean plants was to inoculate the seed with a suitable strain of root nodule bacteria. Earlier screening work had identified strains which might be useful in Kenya (Karanja and Wood, 1988), therefore an inoculation treatment was included. The results were once again disappointing with no increase in N yield or reduction in ^{15}N enrichment of bean plants as a result of inoculation (Table 3). It was therefore concluded that beans were of no benefit to maize cropping systems in this region in terms of N cycling.

Table 3. Total shoot nitrogen yield and ^{15}N enrichment of bean (either uninoculated, or inoculated with *Rhizobium* strain 9C) and maize at Kiboko during the Short Rains 1990

Crop	Shoot N (kg N ha^{-1})	^{15}N atom % excess
Bean – 9C	45	0.153
Bean + 9C	50	0.142
Maize	71	0.127

In the final season, cowpea was included as a legume which is better adapted to semi-arid environments. Data in Table 4 indicate that cowpea was superior to bean in terms of (a) total N yield, (b) nitrogen fixation as indicated by ^{15}N enrichment and (c) fertilizer recovery. Calculations indicated that the cowpea crop derived 50% of its nitrogen from fixation, equivalent to 200 kg N ha^{-1}, and derived considerably more N from the soil mineral pool than did maize or bean (Table 5).

If the seed alone of maize is removed from the land, then the seasonal input of N required to sustain this level of productivity is approximately 50 kg N ha^{-1}. It is clear that bean plants cannot supply this, and under these conditions bean can be considered to be similar to maize in terms of sustainability of N use. Cowpea can supply this N however, and is able to provide a net input per season of approximately 150 kg N ha^{-1} assuming that only the seed is removed from the land.

Cowpea also assimilated approximately 200 kg N ha^{-1} from the soil mineral N pool during the season. This indicates a rapid rate of mineralization in this soil. Laboratory estimates of gross rates of N mineralization in this soil at different matric potential values extrapolated to field soil moisture data for the short rains 1990 and long rains 1991 indicate annual rates of mineralization of 150 kg N ha^{-1} in the upper 20 cm of soil at Kiboko (Pilbeam et al., 1993). These rates of mineralization of soil N would more than satisfy the N demand

Table 6. Total shoot dry matter yield (kg ha^{-1}) for wheat in response to three levels of nitrogen fertilizer, and total rainfall for each season, at Tel Hadya during three seasons (ICARDA, 1990)

Season	N rate (kg N ha^{-1})			Rainfall (mm)
	0	30	60	
1986/87	4600	5100	5700	333
1987/88	7800	10000	12300	485
1988/89	3700	5800	5900	238

Table 7. Total shoot nitrogen yield and ^{15}N enrichment for wheat, chickpea and lentil treated with 30 kg N ha^{-1} at Tel Hadya during the 1991/92 season

Crop	Shoot N (kg N ha^{-1})	^{15}N atom % excess
Wheat	37	0.607
Chickpea	82	0.513
Lentil	81	0.311

Table 8. Sources of N (kg N ha^{-1}) for wheat, chickpea and lentil treated with 30 kg N ha^{-1} at Tel Hadya during the 1991/92 season

Crop	Fertilizer	Soil	Fixation
Wheat	2	35	0
Chickpea	4	65	13
Lentil	3	39	39

by maize, and if they were associated with rapid rates of immobilization of fertilizer N in particular, they could explain the lack of crop response to N fertilizer and the poor fertilizer recovery. The inclusion of cowpea in a maize (or maize and bean) cropping system would appear to offer great prospects for sustainable, low input food production in this region.

Tel Hadya, Syria

At Tel Hadya, wheat responded to N fertilizer in three season when inputs began (Table 6). The rainfall during the first season was equal to the long-term average, and in the second and third seasons approached the upper and lower extremes respectively (ICARDA, 1990). These data indicate the need for an input of N in this system. Detailed studies of the use of N by wheat and two legumes (Table 7) indicated that the recovery of N fertilizer in the wheat crop was low (8%). Also, based on N difference and isotope dilution estimates, both legumes fix significant amounts of N. Lentil appeared to be superior to chickpea at this site, deriving almost 50% of its N from fixation, equivalent to about 40 kg N ha^{-1} (Table 8).

If the seed alone of wheat is removed from the land, then the seasonal input of N required to sustain this level of productivity is approximately 50 kg N ha^{-1}. These preliminary data for a single season indicate that lentil can provide part of this N requirement, but

that additional inputs in the form of fertilizer will be necessary to maintain productivity in the long term.

Conclusions

1. Different responses to N fertilizer were observed in these two dryland farming systems.
2. There was low recovery of fertilizer nitrogen in the season of application in both systems, and evidence (in the absence of any significant loss mechanisms) of rapid immobilization of fertilizer N.
3. The presence of a legume in a cropping system (e.g. bean at Kiboko) is not necessarily associated with a net input of N into the system.
4. The use of the appropriate legume in each of these systems could lead to an input of N, and reduce the requirement for N fertilizer.

Acknowledgements

The work in Kenya was funded by the European Community (Project Number TSA-A-0193-UK(TT)), and the work in Syria was funded by the Overseas Development Administration (Project Number R4997(H)). We are grateful for the technical assistance provided by KARI and ICARDA.

References

Fried M and Middelboe V 1977 Measurement of the amount of nitrogen fixed by a legume crop. Plant and Soil 47, 713–715.

ICARDA 1990 Farm Resource Management Programme annual report 1989. ICARDA, Aleppo, Syria.

Karanja N K and Wood M 1988 Selecting *Rhizobium phaseoli* strains for use with beans (*Phaseolus vulgaris* L.) in Kenya: tolerance of high temperatures and antibiotic resistance. Plant and Soil 112, 15–22.

Pilbeam C J, Mahapatra B S and Wood M 1993 Soil matric potential effects on gross rates of nitrogen mineralization in an orthic ferralsol from Kenya. Soil Biol. Biochem. 25, 1409–1413

3. ASPECTS RELATED TO BIOLOGICAL FIXATION

O. Van Cleemput et al. (eds.), Progress in Nitrogen Cycling Studies, 309–313, 1996.
© 1996 Kluwer Academic Publishers.

Field symbiotic fixation using ^{15}N : II. *Vicia faba* L. and *Pisum sativum* L.

C. Carranca[1], D. Eskew[3], A.S. da Silva[1], E. Ferreira[1], M.T. de Sousa[2], M.R. Gusmão[1], M.L. Fernandes[1] and E.M. Sequeira[1]
[1]*Estação Agronómica Nacional, - Quinta do Marquês, 2780 Oeiras, Portugal,* [2]*Estação Nacional de Melhoramento de Plantas, 7350 Elvas, Portugal and* [3]*FAO/IAEA, Vienna, Austria*

Key words: barley, fababean, ^{15}N dilution technique, pea, symbiotic fixation

Abstract

In Portugal, results on the amount of N_2-fixed and efficiency of grain legumes-rhizobia symbiosis under field conditions are scarce. A two-years experiment in two Orthic Luvisols of Elvas (ENMP and CV), with uninoculated and inoculated fababeans and peas, was set in randomized blocks, replicated five times. Barley was used as control crop. At sowing, 20 kg N ha^{-1} as ^{15}(NH$_4$)$_2$SO$_4$ 4.8 atom% ^{15}N excess were applied in each microplot. Plants were harvested at physiological maturity. In the 1st year experiment, dry-matter, N yield and fixation in both crops were higher at CV soil, probably due to its higher fertility, namely the higher Mo content. In fababean,% Ndfa varied from 65% at ENMP soil to 90% at CV, corresponding to N_2-fixed of 80 and 125 kg N ha^{-1}; as to pea,% Ndfa ranged from 35% at ENMP to >85% at CV, corresponding 22 and 107 kg N_2-fixed ha^{-1}. Sites and years statistically affected crops, mainly by the deficient Mo level at ENMP, and drought stress in the 2nd year. Inoculation did not significantly affect fixation of fababean, but significantly affected pea. Values of% Ndfa were >60%, for both crops, at both sites, but due to the hydric stress in the 2nd year, N_2-fixed was lower than in the 1st year: an average of 70 kg N ha^{-1} by fababean, and 37 to 55 kg N ha^{-1}, respectively by uninoculated and inoculated peas at CV. At ENMP, pea poorly fixed N_2, and negative values for% Ndfa in the pea straw were found.

Introduction

Crop management, nutritional factors, climate, soil characteristics (mainly soil moisture, pH, P, K, B, and Mo) and higly effective rhizobia under field conditions are important factors for N_2-fixation (Holding, 1982; Meisinger and Randall, 1991).

Difficulties of introducing efficient strains of *Rhizobium* spp. that are compatible with agronomicaly good cultivars and have the ability to compete successfully with native bacteria to form nodules and fix N_2 may justify the failure of inoculation efficiency (Greenwood, 1982).

The ^{15}N dilution technique has been widely accepted to measure N_2-fixed in the symbiosis because of its capability to discriminate between nitrogen fixed from the atmosphere and that assimilated from the soil, making sure that the reference crop was the appropriate (Barrie, 1991; Danso, 1986).

Portuguese environmental conditions seem to be good for most legume symbioses but data on the amount of N_2-fixed in the symbiosis are scarce (Carranca et al., 1993).

In this paper, results of symbiotic fixation over two-years with field grain legumes (fababean and pea), in two Portuguese soils are evaluated through measurement of the% N derived from the atmosphere (% Ndfa) and the N_2-fixed by the ^{15}N isotope dilution technique.

Material and methods

A two-years experiment in two Portuguese Orthic Luvisols (FAO-UNESCO) (Cardoso, 1974) of Elvas, with uninoculated and inoculated fababeans and peas, was set in randomized blocks replicated five times. Sites were located at Estação Nacional de Melhoramento de Plantas (ENMP) and Casas Velhas (CV).

Every year, and before sowing, 60 kg of either P$_2$O$_5$ and K$_2$O ha^{-1} were applied as basal dressing. After this fertilization, soil samples were taken at 0–20 cm depth for soil characterization. Organic matter

Table 1. Some physical and chemical characteristics of the soils at both sites (ENMP and CV)

Site	Date	Soil type	Texture	pH	Org. matter (g kg^{-1})	Total N (g kg^{-1})	NO$_3^-$-N (mg kg^{-1})	NH$_4^+$-N (mg kg^{-1})	CEC (cmol(+)kg^{-1})	Ca	Na	K	Mg	Avail. P (mg kg^{-1})	Avail. K	Ext. B	Ext. Mo
ENMP	1990	Orthic Luvisol	Loamy clay	7.43± 0.22	7.26± 1.38	0.67± 0.12	4.73± 2.47	2.51 1.71	20.45± 0.85	13.34± 1.38	0.09± 0.02	0.34± 0.04	2.73± 0.12	141± 12	144± 12	0.25± 0.07	0.029± 0.009
"	1991	"	Sandy clay loam	7.92± 0.07	9.77± 1.46	0.83± 0.08	8.79± 1.53	6.13± 1.94	18.30± 1.17	14.30± 0.18	0.20± 0.05	0.50± 0.07	3.90± 0.57	350± 24	81± 12	0.56± 0.00	0.040± 0.000
CV	1990	"	Loamy clay	7.01± 0.04	22.7± 2.87	1.57± 0.08	9.63± 4.16	nd	21.40± 1.33	13.06± 0.55	0.08± 0.01	0.67± 0.06	3.59± 0.07	160± 15	206± 30	0.62 0.04	0.150± 0.030
"	1991	"	Loam	6.96± 0.09	17.08± 1.91	1.45± 0.07	12.62 1.89	6.65± 2.23	21.25± 1.21	12.58± 0.78	0.10± 0.02	0.65± 0.09	3.79± 0.08	139± 26	100± 20	0.92± 0.05	0.130 0.014

Org. - organic, Avail. - available, Ext. - extractable, n.d. - not detectable.

was determined by wet combustion, total and inorganic N according to Bremner (1965), CEC and exchangeable cations according to Metson (1961), available P and K according to Riehm (1958), extractable B by boiling water (Berger and Truog, 1939), extractable Mo according to Sequeira and Lucas (1968), and pH in a 1:2.5 soil:water ratio.

At sowing time, 20 kg N ha^{-1} as (^{15}NH$_4$)$_2$SO$_4$ 4.8 atom% ^{15}N excess were applied in each microplot (1.2 m^2) in a liquid form (fertilizer dissolved in 633 mL).

Crops were sown in rows. Fababean (*Vicia faba* L. cv. Favel) was sown at a rate of 40 seeds m^{-2}; pea (*Pisum sativum* L.cv.Ballet) was sown at a seeding rate of 70 seeds m^{-2}, and barley (*Hordeum vulgare* L. cv. Sereia) was sown as 300 seeds m^{-2}. Barley was used as non-fixing control crop.

In 1990/91, only uninoculated fababean and pea were tested. In 1991/92, inoculation of both legumes was also used.

Inoculation was prepared using a mixture of two strains of *Rhizobium leguminosarum* bv. *viceae*, from University of Granada (Spain). Before sowing, the inoculum was applied to the soil as a *Rhizobium* suspension containing an equivalent to 10^5 bacteria seed^{-1}.

Plants were harvested at physiological maturity in 1 m^2 of each microplot. They were separated into straw and pods or spikes. Dry-weights were determined by oven-drying at $\approx$75 °C for 48 hours, and dry-matter was ground to pass through a 0.5 mm sieve.

Total N was determined by Kjeldahl (Bremner, 1965) and atom% ^{15}N excess was determined by mass spectrometry, either at Europa Scientific Inc. (USA) or Seibersdorf Lab. (Austria). N yield per hectare was estimated as the product of total N by the dry-matter yield, and N$_2$-fixed was determined by the product of% Ndfa and N yield.

Data for each crop was statistically analyzed, after transforming the% results through "arc sin", and using the analysis of variance and the means comparison by LSD.

Results

Both Orthic Luvisols showed neutral reaction, but CV soil presented higher levels of organic matter, total N (though in low amounts), CEC, B and Mo (Table 1). The soil at ENMP was Mo deficient. P, K and B levels were adequated in both soils.

Table 2. Comparison of means for each crop and plant characteristics at both sites (1990/91)

Site	Crop	Treatment	Dry-matter yield (g m^{-2})		Total N (%)		N yield kg N ha^{-1}		Ndfa (%)		N$_2$-fixed kg N ha^{-1}	
			Straw	Pod	Straw	Pod	Straw	Pod	Straw	Pod	Straw	Pod
ENMP	Fababean	0	229.86 b	334.26 a	0.73 a	3.01 a	16.81b	99.64 a	63.50 b	69.43 b	11.05 b	69.24 b
CV	"	0	330.57 a	372.90 a	0.89 a	2.91 a	29.97 a	108.60 a	87.80 a	90.52a	26.42 a	98.47 a
ENMP	Pea	0	146.41 a	116.56 b	0.99 a	2.52 a	14.47a	29.91b	35.20 b	56.36 b	5.05 a	16.67 b
CV	"	0	202.39 a	328.27a	0.97a	2.90 a	21.72 a	97.02 a	84.96 a	89.39a	19.03 a	87.58 a

0 = uninoculated seeds; for each crop and characteristics, columns with the same letter are not significantly different at $p<5\%$.

Table 3. Results of analysis of variance for each crop, in 1991/92, as affected by sites and treatment, and means comparison for each crop, plant characteristic and site (ENMP and CV), in 1991/92, as affected by treatment

Source of variation	Crop	Dry-matter yield		Total N		N Yield		Ndfa		N$_2$-fixed	
		Straw	Pod	Straw	Pod	Straw	Pod	Straw	Pod	Straw	Pod
Site	Fababean	1.45 ns	0.47 ns	0.50 ns	0.70 ns	0.74 ns	0.03 ns	2.90 ns	7.70*	0.72 ns	0.93 ns
Treatment		5.95 ns	6.07*	4.20 ns	1.30 ns	1.25 ns	2.49 ns	0.20 ns	0.10 ns	0.42 ns	2.71 ns
Site	Pea	11.23**	9.02*	39.10***	1.50ns	0.05 ns	6.26*	-^{z}	175.03***	-^{z}	20.41**
Treatment		0.46 ns	3.70 ns	1.20 ns	2.90 ns	17.03**	10.94**	-^{z}	1.00 ns	-^{z}	8.15*

Site	Crop	Treatment[b]	Dry-matter yield (g m^{-2})		Total N (%)		N yield (kg N ha^{-1})		Ndfa (%)		N$_2$-fixed (kg N ha^{-1})	
			Straw	Pod	Straw	Pod	Straw	Pod	Straw	Pod	Straw	Pod
ENMP	Fababean	0[y]	166.48 a	204.14 a	1.57 a	3.32 a	26.84 a	69.39 a	43.61 a	59.39 a	13.17 a	41.90 a
		I	227.20 a	225.66 a	1.77 a	3.28 a	40.05 a	74.28 a	60.10 a	73.71 a	24.39 a	54.99 a
	Pea	0	134.29 a	256.25 a	0.77 b	2.19 a	10.58 b	55.87 a	-^{z}	17.48 a	-^{z}	11.35 a
		I	188.48 a	336.51 a	1.03 a	2.73 a	19.20 a	88.31 a	-^{z}	29.39 a	-^{z}	27.28 a
CV	Fababean	0	150.35 a	189.46 b	2.09 a	3.31 a	30.86 a	62.90 a	70.76 a	81.49 a	21.27 a	50.61 a
		I	192.49 a	268.35 a	1.40 b	3.09 a	28.05 a	83.47 a	49.47 b	69.07 b	14.46 a	58.62 a
	Pea	0	94.56 a	164.72 a	1.40 a	2.67 a	13.04 a	43.05 a	51.32 a	71.74 a	6.49 b	30.58 a
		I	140.47 a	219.18 a	1.25 a	2.81 a	17.67 a	62.24 a	54.85 a	74.66 a	9.36 a	45.84 a

ns,*,** and *** = F-values non-significant and significant at 5%, 1% and 0.1% probabilities.
^{z} Values not shown because negative values of fixation were not statistically analyzed.
^{y} O= uninoculated, I= inoculated.
For each crop and characteristics, columns with the same letter are not significantly different at $p<5\%$.

The mean values of rainfall that occurred at Elvas during both growth stages were around 500 mm and 270 mm, respectively in the 1st and 2nd year experiment. In the 1st year there was a normal rainfall distribution but in the 2nd one such distribution was quite irregular and the amount of precipitation was lower than the average ($\approx$ 600 mm) for the 1951–1980 period (Martins and Gonçalves, 1986).

To evaluate the effects of sites on fixation, results of the 1st year experiment were compared for each plant characteristic (Table 2). Dry-matter, N yield and N$_2$-fixation in both legumes were significantly higher at CV soil. The %Ndfa in fababean varied from about 65% to 90%, respectively at ENMP and Cv, corresponding to about 80 and 125 kg N$_2$-fixed ha^{-1}; as to pea,% Ndfa varied from 35% in the straw at ENMP to 89% in the pods at CV, corresponding to 22 and 107 kg N$_2$-fixed ha^{-1}.

Results of analysis of variance to evaluate the effects of site and treatment on both crops in 1991/92 are shown in Table 3.

Data for %Ndfa in the straw of peas at ENMP are not shown due to the negative values found.

In this 2nd year, sites did not significantly affect most characteristics of fababean, but affected highly significantly most pea characteristics.

Table 4. Results of analysis of variance for each uninoculated crop, in both sites (ENMP and CV) and years (1990/91 and (1991/92)

Crop	Source of variation	Dry-matter yield		Total N		N Yield		Ndfa		N_2-fixed	
		Straw	Pod	Straw	Pod	Straw	Pod	Straw	Pod	Straw	Pod
Fababean	Site	3.14 ns	0.18 ns	10.51**	0.13 ns	4.42 ns	0.02 ns	28.17***	50.79***	13.10**	6.54*
	Year	26.01***	31.98***	91.72***	6.24*	1.79 ns	13.99**	14.52**	9.90**	0.22 ns	25.68***
	Site × Year	5.98*	0.93 ns	2.77 ns	0.10 ns	1.25 ns	0.58 ns	0.08 ns	0.03 ns	1.26 ns	1.91 ns
Pea	Site	0.18 ns	5.63**	8.80*	8.46*	1.76 ns	9.22*	4.66 ns	15.00*	-[z]	25.27***
	Year	9.78**	0.01 ns	0.69 ns	3.36 ns	3.76 ns	1.38 ns	0.12 ns	29.32**	-[z]	11.68**
	Site × Year	7.27**	27.19**	11.34*	0.16 ns	0.75 ns	17.13**	8.45 ns	34.21**	-[z]	5.30*

ns,*,** and*** = F-values non-significant and significant at 5%, 1% and 0.1% probabilities. [z] Values not shown because negative values of fixation were not statistically analyzed.

As to treatment, inoculation did not significantly affect fababean, but in case of peas, treatment significantly affected dry-matter and N yield, %Ndfa and N_2-fixed.

Comparing the mean values of plant characteristics in 1991/92 (Table 3) we can see that fababeans had a %Ndfa around 60% at ENMP and >70% at CV, which corresponded to 67 and 75 kg N_2-fixed ha^{-1}; peas presented a very variable fixation at ENMP with some negative values of %Ndfa, and an average of 60% Ndfa at CV, which corresponded to 37 and 55 kg N_2-fixed ha^{-1}, respectively by uninoculated and inoculated peas.

Analysing the effects of both years and sites on uninoculated fababean and pea (Table 4), analysis of variance showed that both factors significantly affected both crops, but their interaction was especially significant for peas.

In both crops, N_2-fixed was accumulated in the pods.

at ENMP, Mo deficiency in the soils probably affected symbiosis.

Peas were especially affected by both adverse factors. Under favourable conditions, peas fixation reached more than 100 kg N_2-fixed ha^{-1}, but under unfavourable situations, low values as 22 kg N_2-fixed ha^{-1} were found. Rennie and Dubetz (1986) and Bremer et al. (1988) found averages %Ndfa by inoculated field peas varying from 50% to 80%, which corresponded to 50 and 185 kg N ha^{-1}.

Bremer et al. (1988) also verified that fababean fixed the most N_2 under wetter conditions, but for peas they found higher symbiotic fixation under drought stress conditions. They still found that N_2-fixation by inoculated grain legumes under dryland conditions was mainly dependent on soil moisture.

In both sites, inoculation of fababean did not significantly affect its fixation capacity. It seems that native *Rhizobium* spp. was efficient for N_2-fixation. Unlike, at CV, inoculation significantly affected fixation by peas.

Discussion

As previously reported by several authors (Carranca et al., 1993; Duc et al., 1988; Rennie and Dubetz, 1986; Zapata and Van Cleemput, 1986; Zapata et al., 1987, 1993), under the present field conditions, also fababeans showed the highest fixing capacity, even with the native *Rhizobium* spp. The referred authors found values of around 80% Ndfa for fababeans, corresponding to annual rates of N_2-fixed as high as 60–310 kg N ha^{-1}, in temperate regions.

The lowest values found in the present experiments (2nd year) may be explained by the drought stressed conditions under which crops were developed. Also,

Acknowledgements

This study was part of a co-ordinated research program supported by Food Agricultural Organization (FAO) and International Atomic Energy Agency (IAEA) on the use of ^{15}N isotope techniques in improving nitrogen efficiency by different crops.

The authors acknowledge Europa Scientific (USA) and Dr H Axmann from Seibersdorf Lab (Austria) for the ^{15}N results, and O Romero, A V Oliveira, M F Guimarães, M L Cravo, M C Campos and M A Pinto for other analyses.

They also acknowledge J Sardinha from ENMP for the field work.

References

Barrie A 1991 New methodologies in stable isotope analysis. *In* Stable Isotopes in Plant Nutrition, Soil Fertility and Environmental Studies. pp 3–25. IAEA, Vienna, Austria.

Berger K C and Truog E 1939 Boron determination in soils and plants. Ind. Eng. Chem. Anal. Ed. 11, 540–544.

Bremer E, Rennie R J and Rennie D A 1988 Dinitrogen fixation of lentil, field pea and fababean under dryland conditions. Can. J. Soil Sci. 68, 553–562.

Bremner J M 1965 Total nitrogen. *In* Methods of Soil Analysis 2. Chemical and Microbiological Properties. Eds. C A Black et al., pp 1149–1178. American Society of Agronomy Inc., Madison, WI, USA.

Cardoso J C 1974 A classificação dos solos de Portugal. Bol. Solos 17, 14–46.

Carranca C F, Eskew D, Sousa M T, Ferreira E, Gusmão M R, Silva A S da, Sequeira E M and Sardinha J 1993 Utilização da metodologia ^{15}N para avaliação da fixação biológica do azoto em leguminosas para grão: I - *Vicia faba* L. e *Pisum sativum* L. *In* Actas de Horticultura do II Congreso Ibérico de Ciencias Hortícolas. Identificación. Cultivos in vitro. Horticultura, 10. pp 1603–1608. Soc. Española de Ciencias Hortícolas, Zaragoza, Spain.

Danso S K A 1986 Review: estimation of N$_2$-fixation by isotope dilution: an appraisal of techniques involving ^{15}N enrichment and their application-comments. Soil Biol. Biochem. 18, 234–244.

Duc G, Mariotti A and Amarger N 1988 Measurements of genetic variability for symbiotic dinitrogen fixation in field grown fababean (*Vicia faba* L.) using a low level N-tracer technique. Plant and Soil 106, 269–276.

Greenwood D J 1982 Nitrogen supply and crop yield: the global scene. *In* Nitrogen Cycling in Ecosystems of Latin America and Caribbean. Eds. G P Robertson, R Herrera and T Rosswall. Plant and Soil 67, 45–59.

Holding A J 1982 Some priority research areas in nitrogen studies. *In* Nitrogen Cycling in Ecosystems of Latin America and the Caribbean. Eds. G P Robertson, R Herrera and T Rosswall. Plant and Soil 67, 81–90.

Martins J C and Gonçalves M C 1986 Avaliação dos caudais de drenagem do solo e das necessidades hídricas das culturas em algumas regiões de Portugal (Cova da Beira, Baixo Mondego e Alentejo). Pedologia, Oeiras 21, 1–77.

Meisinger J J and Randall G W 1991 Estimating nitrogen budgets for soil-crop systems. *In* Managing Nitrogen for Groundwater Quality and Farm Profitability. Eds. R F Follett, D R Keeney and R M Cruse. pp 85–124. Soil Sci. Soc. Am., Inc, Madison, WI, USA.

Metson A J 1961 Methods of Chemical Analysis for Survey Samples. Soil Bureau (12). New Zeland. 208 p.

Rennie R J and Dubetz S 1986 Nitrogen-15-determined nitrogen fixation in field-grown chickpea, lentil, fababean, and field P. Agron. J. 78, 654–660.

Riehm H 1958 Die ammoniumlaktatessigsaüre - methode zur bestimmuing der leichtloslichenphosphorsaür in karbonathaltigen Böden. Agrochimica 3, 49–65.

Sequeira E and Lucas M D 1968 Determinação do molibdénio extraível dos solos. Pedologia, Oeiras 3, 371–373.

Zapata F and Van Cleemput O 1986 Fertilizer nitrogen recovery and biological nitrogen fixation in fababean - sugarbeet cropping sequences. Fert. Res. 8, 269–278.

Zapata F, Danso S K A, Hardarson G and Fried M 1987 Nitrogen fixation and translocation in field-grown fababean. Agron. J. 79, 505–509.

O. Van Cleemput et al. (eds.), Progress in Nitrogen Cycling Studies, 315–319, 1996.
© 1996 Kluwer Academic Publishers.

Nitrogen fixation, nitrification and denitrification in ombrotrophic bog

Irina K. Kravchenko

Institute of Microbiology, Russian Academy of Sciences, Prospect 60-let Octyabrya 7/2, Moscow 117811, Russia

Key words: acetylene reduction, denitrification, N_2O flux, nitrification, nitrogen fixation, peat microcosm

Abstract

The nitrogen fixing activity estimated in peat cores was low and didn't exceed 0.014–0.022 mg N kg^{-1} h^{-1} in the most active 10–20 cm depth layer, but all examined samples possesed a high nitrogen fixing potential determined in glucose amended samples in optimum conditions. Numerically the leading groups of diazotrophs are methylotrophic bacteria (aerobic) and enterobacteria (anaerobic).

The rate of nitrification was 0.05 mg $N-NO_3^-$ kg^{-1} day^{-1} only in samples from upper the 0–10 cm layer amended by ammonium. It was negligible in all other samples. The *Nitrosomonas* and *Nitrobacter* were present in aerobic ammonium oxidizing enrichment cultures from the top layer.

Denitrification rates were estimated in anaerobic acetylene inhibited intact flooded peat cores upon fertilization with nitrate. Addition of 10 mg glucose g^{-1} to the peat samples led to an increased N_2O production from 5,3 to 14,7 mg N kg^{-1} day^{-1} in 10–20 cm layer and from 2,7 to 4,7 in the 20–30 cm layer.

Introduction

Nitrogen fixation, nitrification and denitrification are the most important microbial processes involved in nitrogen metabolism in soils. The nitrogen pool of ombrotrophic bog is negligible and the only nitrogen fixation is the natural source of combined nitrogen. The coupling of nitrification with denitrification near the air-water interface in the bog may be a significant factor for the decrease of nutrient loading.

Both nitrification and denitrification are responsible for nitrous oxide production. N_2O is important in atmospheric chemistry and the radiation budget of the Earth. It contributes to the destruction of stratospheric O_3, affects the oxidation capacity of the troposphere and water budget of the stratosphere. In addition, N_2O is a greenhouse gas whose atmospheric concentrations is increasing at an annual rate of 0,25% (Crutzen, 1983). We have a relatively good understanding of the emission from antropogenic sources, but biotic sources and sinks are less well known.

The objectives of this study were:

1 . to measure the intensity of microbiological processes of the nitrogen cycle in peat soils in laboratory experiments with microcosms (peat cores, composed samples, enriched cultures);

2 . to examine some environmental factors controlling these processes;

3 . to determine the quantity and diversity of the main microorganisms participating in the nitrogen cycle.

Materials and methods

Sampling site

We measured trace gas fluxes, soil nitrogen dynamics and microbiological characteristics in a stand of the Tver region in Northeastern Russia. This stand was located in the central part of the ombrotrophic bog. The intact peat cores were taken with soil corer of 75 mm diameter at a depth from 0 to 100 cm. Cores were sampled in Sept 1993. Before the start of the experiments the cores were stored at 4 °C in the dark. Actual values during sampling were 16 °C in air and 13 °C in the soil.

Experiments with intact cores

Cores were transferred to 500 mL glass vials of 78 mm inner diameter. The vials were capped with latex stoppers and incubated statically in the dark at 20 °C.

The head space was replaced every two days (aerobic conditions) or flushed with ultra-pure nitrogen or argon (anaerobic conditions). The main experiment was repeated twice using 3 cores for each run.

The inhibition of nitrous oxide reductase by acetylene was used to determine the potential activity of denitrification The acetylene reduction technique was used to determine the rates of nitrogen fixing activity. Acetylene (C_2H_2) was added to a final concentration of 10 kPa to flasks used for denitrification and N_2 fixation determination (Knowles, 1979).

Gases were analyzed by gas chromatography. N_2O was analyzed on a GC (3700 model, Russia) equipped with an electron capture detector and a Porapak Q column. C_2H_4 concentrations were determined on the GC (3700 model, Russia) with a flame ionization detector and Poraplot U column.

In a pilot study we found that the nitrate and ammonium pool in the untreated peat was negligible (less than 2 mg N kg^{-1}) in all layers examined. In experiments with N-amended samples NO_3^-, NH_4^+, NO_2^- were determined by ion chromatograph in 1 mL membrane-filtered samples (pore size 0.2 mm), that had been frozen until analysis.

Experiments with peat samples

For the following experiments, the layers of peat were removed from peat cores and transferred into 50 mL vials. The vials were capped with a latex stopper, flushed with nitrogen or argon and incubated at 20–22 °C. Nitrogen fixing activity was determined by both the acetylene reduction and ^{15}N incorporation method (Chistyakova and Kalininskaya, 1984). To estimate the total size of the viable nitrogen-fixing community the most probable number (MNP) procedure was applied as previously described (Chistyakova, 1985).

Results

Nitrogen fixation

The nitrogen fixing activity estimated in untreated peat cores and samples was low and didn't exceed 0.014–0.022 mg N kg^{-1} h^{-1} in the most active 10–20 cm depth layer. We used glucose addition to stimulate nitrogen fixation and to determine the dynamics of the process in peat microcosms. The results of such experiments with peat samples are shown at Figure 1. A lag period of about 18–26 hours was observed after

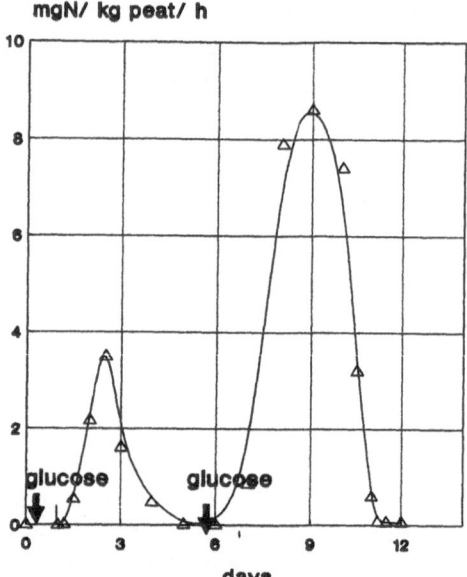

Figure 1. Dynamics of nitrogen fixation in composite samples from the 10–20 cm layer of ombrotrophic peat soil amended with 10 mg C-glucose g^{-1}. The arrows indicate the glucose addition.

glucose addition. Then the rate of nitrogen fixation increased and reached a maximum at 60–70 hours.

The rate of N_2-fixation was significantly higher in peat samples then in the cores. It reached 3.0–3.3 mg N kg^{-1} h^{-1} in samples and never exceeded 0.12–0.18 mg in peat cores of the same depth.

The process of nitrogen fixation began when about one half of glucose was consumed or transformed to fermentation products. The maximum rate of N_2-fixation was practically synchronized with the respiration maximum. When the glucose was added a second time there was practically no lag period. The rate was higher and reached 8.5 mg N kg^{-1} h^{-1}.

The summarized quantity of fixed nitrogen was determined by the ^{15}N incorporation method (Table 1). It was negligible in the untreated samples and reached 8.13 mg kg^{-1} in samples amended by glucose after 15 days of incubation.

The MNP procedure was used to determine the number of diazotrophs of different trophic and systematic groups (Table 2). The aerobic ones predominated numerically in all samples examined including the waterlogged ones. The metylotrophic bacteria and enterobacteria were the leading groups and their numbers were 2.0–2.5 10^6 cells g^{-1}. Usually most probable numbers of diazotrophs did not correlate with the potential N_2-fixation.

Table 1. Nitrogen fixing activity (^{15}N incorporation) in peat and forest soils

Soil type, depth (cm)	N content (mg g^{-1})	Untreated soil		+ 10 mg glucose g^{-1}	
		σ^{15}N at.%	(mg N kg^{-1} per 15 d)	σ^{15}N at.%	(mg N kg^{-1} per 15 d)
Ombrotrophic bog					
0–10	8.92	0.001	0.17	0.048	8.13
10–20	9.52	0.005	0.91	0.015	2.71
Gray forest					
0–10	1.94	0.000	0.00	0.100	3.68
10–20	1.63	0.001	0.03	0.003	0.07

Table 2. Numbers of diazotrophic bacteria from ombrotrophic peat soil

Depth, (cm)	Cells per g of wet peat				
	Aerobic			Anaerobic spore formed	
	Glucose utilised	Malate utilised	Methylo-trophic	Strike	Facultative
0–10	6.0 10^4	2.5 10^4	2.5 10^6	1.2 10^5	2.0 10^6
10–20	3.0 10^4	2.5 10^4	6.0 10^5	1.2 10^5	2.0 10^6
20–30	2.5 10^4	6.0 10^4	4.5 10^5	2.5 10^5	4.5 10^5

Nitrification

The dynamics of formation of the intermediate, NO_2^-, and terminal, NO_3^-, products of nitrification was estimated in aerobic peat samples amended by ammonium. The process was very slow and only after 5 days, the quantity of nitrate stabilized at the 0.06–0.08 mg N-NO_3 kg^{-1}. The NO_2^- was found only during a very short period and its quantity didn't exceed 0.01 mg N-NO_2^- kg^{-1} at 72 hours after ammonium addition. The overall recovery rate of N-NH_4^+ as N-NO_3^- was about 0.007%.

Denitrification

The production of N_2O in the presence of acetylene was measured in closed bottles. The highest rates of potential denitrification were determined in the 40–50 cm depth layer and reached 7.8 mg N kg^{-1} day^{-1} (Fig. 3). Glucose addition led to a significant increase of denitrification rates, for example, from 5.3 to 14.7 mg N kg^{-1} day^{-1} in 10–20 layer.

In experiments with peat samples we, obtained practically 100% recovery of N-NO_3^- as N-N_2O as long as nitrate was still present.

Discussion

Wetlands are of great interest due to their microbial role in trace gases formation. There are plenty of papers connected with methane and carbon dioxide fluxes from peat soils and environment factors controlling these processes (Andrea and Schimel, 1989; Cicerone and Oremland, 1988). Microbiological processes of the N-cycle in peat soils are significantly less well known. The nitrogen pool of peat soils is negligible and direct measurements of in situ rates of nitrification and denitrification are very rare, because analytical tools for such low concentrations are lacking.

Nitrogen fixation is of great importance for peat soils because it is a single natural source of combined nitrogen. Peat soils possess a great supplies of undestroyed organic matter, but the quantity of water soluble carbon compounds is negligible. As a result, addition of glucose to peat samples led to an increase of 3–40 times of the nitrogen fixing activity. The effectiveness of microbial utilization of glucose for nitrogen fixation was about 0.2–1.4 mg N 1 g^{-1} of glucose.

The lag period in nitrogen fixation and nitrification (Fig.1) may be the result of an induction effect of substrates on non-constitutive enzymes. When the enzymes are in active form, the substrates are used practically immediately and more effectively.

NaN318

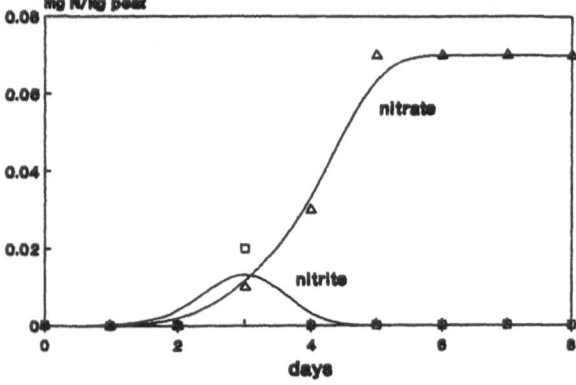

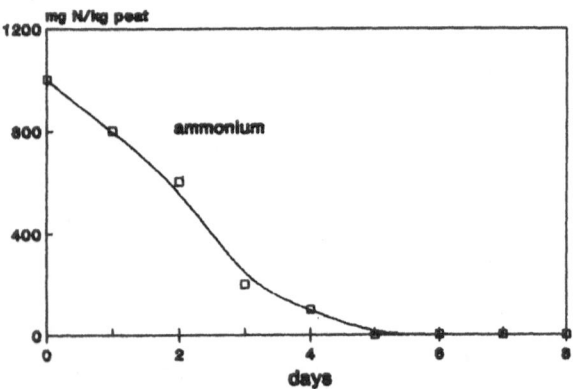

Figure 2. Dynamics of nitrification (NO_2^- and NO_3^- formation) in composite samples from the upper 0–10 cm layer amended with 1 mg N-NH_4^+ g^{-1} during a laboratory incubation at optimal conditions.

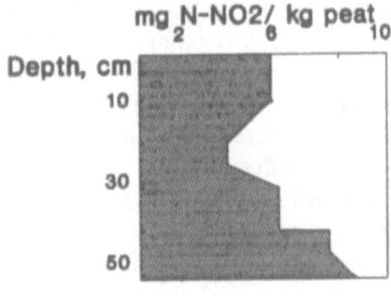

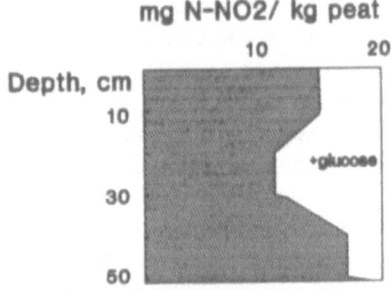

Figure 3. Profile distribution of denitrification activity (flux of N_2O) under anaerobic conditions after fertilization by 1 mg N-NO_3^- g^{-1} in peat samples without glucose and with 10 mg C-glucose g^{-1}.

Low net nitrification rates may have been cased by many factors including pH, redox potential, inhibition, high rates of microbial assimilation of nitrate. Our data of potential nitrification rates are lower than was found in grassland soil (Both et al., 1992). The main reason for this is the very low pH (3.5–3.8) in peat soil. The optimum for nitrification is about 7.0, but the autotrophic nitrifiers *Nitrosomonas* and *Nitrobacter* were identified in enrichment cultures from the upper 0–10 cm layer. We assume the existence of a close relationship between nitrifiers and, for example, soil algae which are able to increase the pH during the growth.

Empirical evidence suggested that exchange of N_2O between the atmosphere and soils is affected by the soil nitrogen cycling. A strong relationship was found between N_2O produced by denitrification and NO_3^- in the soil (Melillo et al., 1983). In many ecosystems nitrogen fertilization has been shown to increase the soil nitrogen cycling and the emissions of N_2O to the atmosphere. The peat soils are the ecosystems with low nitrogen deposition but now they are affected by long-term chronic addition of nitrogen from the atmosphere and underground waters.

Conclusion

We conclude that the microbial agents of the main N-cycle processes, including nitrification, denitrification and N_2-fixation, are present in the acid peat soil from forested ombtrotrophic bog.

Denitrification and nitrification are considered to be the most important ones in N_2O formation in the soils. We found that both nitrification and denitrification activity in the peat samples were very low due to the lack of accessible substrates and significantly increased after fertilization. This result suggests that under nitrogen-stressed conditions acid peat soils may have high potential for N_2O formation.

References

Andrea M O and Schimel D S 1989 Exchange of Trace Gases between Terrestrial Ecosystems and Atmosphere. John Wiley and Sons, New York, USA. 347 p.

Both G J, Gerards S and Laanbrock H J 1992 Temporal and spatial variation in the nitrite oxidizing bacterial community of a grassland soil. FEMS Microbiol. Ecol. 101, 99–112.

Cicerone R and Oremland R 1988 Biogeochemical aspects of atmospheric methane. Global Biogeochem. Cycles 2, 299–327.

Chistyakova I K 1985 Nitrogen-fixing bacteria assimilating mono-carbon compounds in soil under rice. Mikrobiologiya 54, 384–388.

Chistyakova I K and Kalininskaya T A 1984 Nitrogen fixation in takyr-like soils under rice. Mikrobiologiya 53, 123–128.

Crutzen P J 1983 Atmospheric interactions - homogeneous gas reactions or C, N and S containing compounds. *In* The major Biogeochemical Cycles and their Interactions. Eds. B Bolin and R B Cook. pp 66–112. John Wiley and Sons, New York, USA.

Knowles R 1979 Denitrification, acetylene reduction and methane metabolism in lake sediment exposed to acetylene. Appl. Environ. Microbiol. 38, 486–493.

Melillo J M, Aber J D, Steudler P A and Shimell J P 1983 Denitrification potential in successional sequence of northern hardwood stands. Environ. Biogeochem. 35, 217–228.

O. Van Cleemput et al. (eds.), Progress in Nitrogen Cycling Studies, 321–326, 1996.
© 1996 *Kluwer Academic Publishers.*

Field measurement of N_2 fixation by *Gliricidia sepium*

A.R. Zaharah, H.A.H. Sharifuddin and R. Anuar
Soil Science Department, Universiti Pertanian Malaysia, 43400 UPM, Serdang, Selangor, Malaysia

Key words: cuttings, *Gliricidia sepium*, ^{15}N, nitrogen fixation

Abstract

A field experiment was conducted to assess the N_2 fixation capacity of *Gliricidia sepium* subjected to three cutting regimes (4, 2 and 1 cut per year) using the ^{15}N isotope dilution technique. The reference plant used was *Cassia siamea*. Total dry matter produced by *G. sepium* cut 4 times yr^{-1} was 23,054 kg ha^{-1}, while those trees cut 2 times yr^{-1} was 21,128 kg ha^{-1}. The trees cut only once in a year produced 9,338 kg dry matter ha^{-1} $year^{-1}$. Total N was highest in trees cut 4 times yr^{-1}(439 kg N ha^{-1}), with most of the N contributed by the leaves. Trees cut twice yr^{-1} produced only 315 kg N ha^{-1}, since most of the dry matter fraction was the stem, which had a lower N concentration. Trees cut once yr^{-1} had the lowest N yield (140 kg N ha^{-1}). Trees with the highest frequency of cuts yr^{-1} (4 times yr^{-1}) fixed about 56 kg N ha^{-1} yr^{-1} as compared to 25 kg N ha^{-1} yr^{-1} for trees cut twice yr^{-1} and no N_2 fixation in trees without any cuts made yr^{-1}. *G. sepium* is a poor N_2 fixer when grown under field conditions in Malaysia.

Introduction

Nitrogen fixing trees have been recognised to play an important in maintaining and improving soil fertility (Charreau and Vidal, 1965; Virginia, 1986), help in soil conservation (Lal, 1989) and in providing fuel wood. Since nitrogen is one of the major limitation for intensive crop production in the humid tropics, studies have shown that the use of prunings from nitrogen fixing trees as a substitute to inorganic N fertilizers increased yield of food crops (Ladha et al., 1989; Sanginga et al., 1986). Thus, nitrogen fixed by nitrogen fixing trees forms a good and free contribution of N to agroforestry systems and enables farmers to grow their own nitrogen. *Gliricidia* spp. have been regarded as having a great potential in agroforestry. It is a fast growing leguminous tree, which is currently being integrated into farming systems in the tropics (Attah-Krah et al., 1986; Onkuwa, 1984). Its nitrogen fixing ability as measured by acetelyne reduction assay suggested that it is a poor N_2 fixer. However, Awonaike et al. (1992) reported that N_2 fixed by *Gliricidia sepium* was comparable to *L. leucocephala* which is regarded as high N_2 fixer, where most of the genotypes studied derived more than 50% of their total N from the atmosphere. Danso et al. (1991) reported 72% Ndfa in *G. sepium*, while

Liya et al. (1991) reported 85% Ndfa when *G. sepium* was grown in concrete cylinders sunk 1 meter into the ground. This study uses the ^{15}N isotope dilution method in estimating the amounts of N_2 fixed by *Gliricidia sepium* over a period of one year grown under field conditions and the effects of successive cutting on N_2 fixed by this legume.

Materials and methods

A field experiment was conducted at Universiti Pertanian Malaysia, Puchong farm. The soil is classified as Kandic Tropudult having the following chemical characteristics: $pH_{(water)}$ is 5.24; total N of 0.16%; Bray-2 extractable P of 29.18 mg kg^{-1}; ammonium acetate exchangeable K, Ca and Mg of 120, 376 and 90.5 mg kg^{-1}, respectively. Plots of 5 m × 5 m were marked with 0.5 meters distance between plots. Basal fertilizer of 100 kg P ha^{-1} and 100 kg K ha^{-1} was applied to each plot. Three months-old *Gliricidia sepium* seedlings which were established in the glasshouse were transplanted into the plot at 1 m × 1 m distance. *Cassia siamea* plants, which was used as the reference crop, were also established and planted in the field at the same time and same planting distance as the *Gliricidia*. The centre 3 m × 3 m area in each plot was trenched up to 1 m depth, lined with plastic sheet and covered with the original soil. This area was evenly applied with 3g ^{15}N as ammonium sulphate enriched with 10% ^{15}N. The same amount of N as ordinary ammonium sulphate was applied in the area outside this trenched area. Trees in the trenched area were devided into one row of three trees to be cut at one meter above the ground every three months (4 times yr^{-1}), one row of three trees to be cut every six months (2 times yr^{-1}), and the remaining row was left uncut for the whole year. The same procedure was made on the reference trees of *Cassia siamea*. All the plots had five replications and they were randomly arranged. At the end of the year, one tree from each cutting regime from two replications were dug out to obtain the top as well as root samples. Plant samples were devided into leaves, stem and root and weighed. These were chopped using a garden shredder and subsampled. The subsamples were weighed and dried in a forced-air oven at 65 °C until constant weight was achieved. The samples were finely ground and mixed again. Total N and N isotope ratios were determined in the IAEA Seibersdorf laboratory by the Automatic Nitrogen Analyser 1500 Carlo Erba coupled to a SIRA mass spectrometer (Fiedler and Proksch, 1970). The isotope dilution equation (Fried and Middelboe, 1977) was used to calculate N$_2$ fixed, with *Cassia siamea* as the reference non-fixing tree. Analysis of variance and LSD at 5% was calculated to compare means using SAS-PC computer package.

Results

Plant growth and total N

The total dry matter produced by *Gliricidia* cut four times in the year amounted to 23,054 kg ha^{-1}, in which 7334 kg came from leaves, 9715 kg from stem and 6005 kg from roots. The stem contributed 42.14% of the total dry matter weight, 26.05% from roots and 31.81% from leaves. The same dry matter distribution was found by Liyanage et al. (1994) while working with four *Gliricidia* provenances in Sri Lanka. When two cuttings were made per year, the total dry matter produced was 21,128 kg ha^{-1}, in which 4681 kg came from leaves, 12,020 kg from stem and 4,427 kg from roots. It is seen that with two cuttings yr^{-1}, the trees have enough time to produce a higher stem growth compared to leaves and roots. Here the stem contribut-

ed 56.89% of the total dry matter weight. When the trees were not pruned during the year, the total dry matter produced was only 9338 kg ha^{-1}. Here again the stem contributed the highest fraction, i.e, 5758 kg ha^{-1} (61.66%), while the leaves and roots were 2192 kg ha^{-1} (23.47%) and 1388 kg ha^{-1} (14.86%), respectively (Table 1).

The dry matter yield of *Cassia siamea* which is used as the reference non-fixing tree is found to equivalent to the *Gliricidia* in the treatment with 4 cuttings made in the year. A total of 21,881 kg of dry matter yield was produced, in which 32.01% was contributed by the leaves (7022 kg ha^{-1}), 52.19% from stem (11,419 kg ha^{-1}) and 15.72% from roots (3440 kg ha^{-1}). From the dry matter distribution, it is seen that *Cassia siamea* may be a suitable reference plant in this study. But with only two cuttings made on the plant the dry matter distribution of *Cassia siamea* was 31.28% from leaves, 43.91% from stem and 24.81% from roots. With no cuts made on the trees during the year, most of the dry matter was contributed by the leaves (40.34%) while the stem and the roots contributed about 30% each for the total dry matter produced. This growth pattern is in reverse of the *Gliricidia*, where with less frequent cuttings made, the increase in plant growth was mainly in the stem, while the roots and leaves remain constant.

The N concentration in the leaves, stem and roots of *Gliricidia* at all regime of cuttings were significantly different ($p < 0.05$) than the leaves, stem and roots of *Cassia siamea*. The average N concentration in *Gliricidia* was 3.57% in the leaves, 0.95% in the stem and 1.40% in the roots, while in *Cassia siamea* the leaves had an average of 3.04% N, the stem had 0.72% N and the roots had 0.55% N respectively for the plants with four cuttings made per year. The total N contributed by *Gliricidia* was 439 kg N ha^{-1} yr^{-1} by the plants cut four times yr^{-1} compared to 315 kg N ha^{-1}yr^{-1} by *Cassia*. The amounts of N contributed by *Gliricidia* cut twice yr^{-1} also exceeded those contributed by *Cassia* (315 kg ha^{-1} yr^{-1} as compared to 194 kg N ha^{-1} yr^{-1}). But when the plants were not pruned for the whole year, *Cassia* produced a much higher total N (289 kg ha^{-1} yr^{-1}) as compared to *Gliricidia*, which produced only 140 kg N ha^{-1} yr^{-1}. The main contribution of N in *Cassia* was from the leaves (Table 1).

^{15}N enrichment and N$_2$ fixation

The percent ^{15}N enrichment of plant samples from the first two cuttings made from the treatments with four cuttings per year and the first cut from the treatment of

Page 323

Table 1. Dry matter yield (kg ha^{-1}), %N and total N yield of plants (kg ha^{-1})

Treatments	Dry matter yield (kg ha^{-1})					% N				Total N				
	Harvest No					Harvest No				Harvest No				
	1	2	3	4	Total	1	2	3	4	1	2	3	4	Total
4 cuts year^{-1}														
G.sepium														
Leaf	1134	2607	1018	2575	7334	4.46	3.55	2.98	3.28	50.58	92.55	34.01	84.46	261.60
Stem	1155	1439	3164	3958	9715	1.41	1.01	1.10	0.68	16.29	14.53	34.80	26.91	92.53
Root	-	-	-	6005	6005	-	-	-	1.40	-	-	-	84.07	84.07
														438.20
C.siamea														
Leaf	1405	2930	829	1858	7022	3.47	3.18	3.32	2.38	48.75	93.17	27.52	44.22	213.66
Stem	1255	3319	3052	3793	11419	1.07	0.81	0.89	0.40	13.43	26.88	27.16	15.17	82.64
Root	-	-	-	3440	3440	-	-	-	0.55	-	-	-	18.92	18.92
														315.22
LSD $\leq$0.05	n.s	n.s	1024	n.s						10.26	31.54	23.47	59.07	99.15
2 cuts year^{-1}														
G.sepium														
Leaf	-	3263	-	1418	4681	-	3.63	-	3.17	-	118.45	-	44.95	163.40
Stem	-	4182	-	7838	12020	-	0.80	-	0.81	-	33.46	-	63.49	96.95
Root	-	-	-	4427	4427	-	-	-	1.23	-	-	-	54.45	54.45
														314.80
C. siamea														
Leaf	-	2820	-	2301	5121	-	2.58	-	2.27	-	72.85	-	52.39	125.24
Stem	-	3320	-	3869	7189	-	0.69	-	0.47	-	22.87	-	18.37	41.24
Root	-	-	-	4061	4061	-	-	-	0.59	-	-	-	23.88	23.88
														190.36
LSD $\leq$0.05	-	2241	-	4410	-	-	-	-	-	-	44.34	-	101.8	133.53
1 cut year^{-1}														
G. sepium														
Leaf	–	–	–	2129	2192	–	–	–	3.07	–	–	–	67.57	67.57
Stem	–	–	–	5758	5758	–	–	–	0.94	–	–	–	55.17	55.17
Root	–	–	–	2129	1388	–	–	–	1.21	–	–	–	-16.79	16.79
														139.53
C.siamea														
Leaf	–	–	–	10701	10701	–	–	–	2.08	–	–	–	223.69	223.69
Stem	–	–	–	7834	7834	–	–	–	-0.42	–	–	–	32.90	32.98
Root	–	–	–	7995	7995	–	–	–	0.41	–	–	–	32.78	32.78
														289.37
LSD $\leq$ 0.05	–	–	–											

two cuttings per year showed higher values in *Gliricidia* compared to *Cassia* (Table 2). Only the third and fourth cuttings showed a lower ^{15}N enrichment in the leaves and stem of *Gliricidia* as compared to *Cassia*. The percentage of N_2 derived from fixation was calculated using the isotope dilution method of Fried and Middelboe (1977) and found to be about 15% from the third cutting from both the leaf and stem. The final cut showed 39.46% of the fixed N_2 to be in the leaves, 23.50% in the stem and 6.12% in the roots. The total amount of N_2 fixed was 55.48 kg N ha^{-1} in which 38.25 kg N came from the leaves, 12.08 kg N from stem and 5.15 kg from the roots. The trees which were cut twice per year showed a small amount of N_2 fixed in the leaves (1.44%), which amounted to only 1.71 kg N ha^{-1}. At the final harvest, 38.14% of the N_2 fixed was found in the leaves, and 10.64% in the stem which correspond to 17.14 kg N ha^{-1} and 6.79 kg N ha^{-1} respectively (Table 2). The total amount of N_2 derived from fixation was 25.64 kg N ha^{-1}.

Gliricidia trees which were not cut for the whole year showed higher ^{15}N atom excess in the leaves, stem and roots compared to *Cassia*. This may be due to *Cassia* trees being a more vigirous plant growing under the same conditions. The total nitrogen present in *Cassia* was also significantly higher than *Gliricidia* and this has diluted the ^{15}N in *Cassia* plant samples. Thus no N_2 fixation can be calculated.

Discussion

This study demonstrates that dry matter yield of *Gliricidia* is highest when four successive cuttings were made on the trees compared to two and one cut yr^{-1}. The proportion of the dry matter produced were found to be different too. When four cuts were made, most of the dry matter produced were in the leaves, while less frequent cuttings results in most of the dry matter fraction being in the stem. While *Cassia* showed that stem contributed more to the dry matter when frequent cuts were made, and leaves contributed the most in plants where no cutting regime was imposed. Total N distribution were also found to be high in the roots of *Gliricidia* trees that had undergone four successive cuttings in the year, with the leaves and roots contributing most of the total N. The total N in the roots decreased with fewer cutting regime. The root N is of significant importance for crop management under alley cropping as was explained by Sanginga et al. (1990b).

^{15}N declined significantly in all plant parts following regrowth for all cuttings made to the plant. This has been explained by Sanginga et al. (1990a) as being due to the decline in $^{15}N:^{14}N$ ratio in the soils between the cutting periods.

The percentage of N_2 fixed was not detected in the first and second cut of *Gliricidia*. About 15% of the total N in the third cut was found both in the stem and leaves. At the final harvest the leaves showed the highest percentage of N_2 fixed followed by the stem. Roots showed the lowest fraction of N derived from fixation. This is in contrast to the data by Sanginga et al. (1990b), where in *L. leucocephala*, the roots had a significant contribution of N_2 fixed. In the trees which were cut twice per year, there was no N_2 fixed in the roots.

The total amount of nitrogen that was contributed through N_2 fixation in the whole plant was 55.48 kg ha^{-1}, which amounted to only 12.65% of the total N in the trees pruned four times/year, while in trees pruned twice per year, the contribution was only 8.14%. Thus this study did not show that *Gliricidia* grown under field condition in Malaysia to be a high N_2 fixer as was reported by Awonaike (1992), even though these plants all showed to be vigirously nodulating at planting time, and *Gliricidia* has been reported to be easily nodulated with indigenous soil Rhizobium strains (Awonaike et al., 1992). Liya et al. (1991) who reported 85% Ndfa grew *Gliricidia sepium* without innoculations. All the reports made above were on *Gliricidia* grown in pots in the glasshouse (Awonaike et al., 1992; Danso et al., 1991) or in concrete cylinders buried into the ground (Liya et al., 1991), which might induce a higher N_2 fixation.

Acknowledgements

The authors are indebted to International Atomic Energy Agency for funding this research under the 'Coordinated research programme on the use of nuclear and related techniques in management of nitrogen fixation by trees for enhancing soil fertility and soil conservation in fragile tropical soils'. The technical assistance of Ms Helga Axmann and the staff of the Soil Science Unit, IAEA Seibersdorf Laboratory is gratefully acknowledged.

Table 2. ^{15}N enrichment and nitrogen derived from fixation

Treatment	^{15}N enrichment Harvest No				%NdfA Harvest No				Total NdfA (kg ha^{-1}) Harvest No			
	1	2	3	4	1	2	3	4	1	2	3	4
4 cuts year^{-1}												
G.sepium					(*C.siamea* as reference crop)							
Leaf	1.028	0.425	0.195	0.112	–	–	14.47	39.46	–	–	4.92	33.33
Stem	1.141	0.425	0.206	0.345	–	–	16.60	23.50	–	–	5.76	6.32
Root	–	–	–	0.491	–	–	–	6.12	–	–	–	5.15
										Total = 55.48 kg N ha^{-1} yr^{-1}		
C.siamea												
Leaf	0.685	0.347	0.228	0.185								
Stem	0.751	0.406	0.247	0.451								
Root	–	–	–	0.523								
LSD ≤0.05	0.496	0.120	0.061	0.123	–	–	–	–	–	–	–	–
2 cuts year^{-1}												
G.sepium					(*C.siamea* as reference crop)							
Leaf		0.519		0.133		1.44		38.14		1.71		17.14
Stem		0.674		0.351		–		10.69		–		6.79
Root		–		0.604		–		–		–		–
										Total = 25.64 kg N ha^{-1} yr^{-1}		
C.siamea												
Leaf		0.430		0.215								
Stem		0.503		0.393								
Root		–		0.364								
LSD ≤0.05		0.114		0.116		–		–		–		–
1 cut year^{-1}												
G.sepium												
Leaf				0.132			No fixation					
Stem				0.254								
Root				0.709								
C.siamea												
Leaf				0.101								
Stem				0.156								
Root				0.204								
LSD ≤0.05				0.104								

References

Atta-Krah A N, Sumberg J E and Reynolds L 1986 Leguminous fodder tree in farming systems. *In* Farming Systems in Sub-Sahara Africa. Eds. I Haque, S Jutsi and P J H Neafe. International livestock Centre for Africa (ILCA), Addis Ababa, Ethiopia.

Awonaike K O, Hardarson G and Kumarasinghe K S 1992 Biological nitrogen fixation of *Gliricidia sepium/Rhizobium* symbiosis as influenced by plant genotype, bacterial strain and their interactions. Trop. Agric. 69, 381–385.

Charreau C and Vidal P 1965 Influence de l'*Acacia albida* Del sur le sol, nutrition minéral et redements des mils *Pennisetum* au Sénégal. Agron. Trop. 20, 600–626.

Danso S K A, Zapata F, Bowen G D and Sanginga N 1991 Applications of ^{15}N methods for measuring nitrogen fixation in trees. *In* Proc. Symp. Stable Isotopes in Plant Nutrition, Soil Fertility and Enviromental Studies. pp 155–168. IAEA, Vienna, Austria.

Fiedler R and Proksch G 1975 The determination of nitrogen-15 by emission and mass spectrometry in biochemical analysis: A review. Anal. Chim. Acta 78, 1–62.

Fried M and Middelboe V 1977 Measurement of amount of nitrogen fixed by a legume crop. Plant and Soil 47, 713–715.

Ladha J K, Miyan S and Garcia M 1989 *Sesbania rostrata* as a green manure for lowland rice. Biol. Fertil. Soil 7, 191–197.

Lal R 1989 Agroforestry systems and soil surface management of a tropical alfisol. II. Water runoff, soil erosion and nutrient loss. Agroor. Systems 8, 97–111.

Liya S M, Odu C T I, Agboola A A and Mulongoy K 1991 Estimation of N_2 fixation by nitrogen fixing trees in the subhumid tropics using ^{15}N dilution and difference methods. *In* Proc. Symp. Stable Isotopes in Plant Nutrition, Soil Fertility and Enviromental Studies. pp 240–242. IAEA, Vienna, Austria.

Liyanage M de S, Danso S K A and Jayasundara H P S 1994 Biological nitrogen fixation in four *Gliricidia sepium* genotypes. Plant and Soil 161, 267–274.

Onwuka C F I 1984 *Gliricidia sepium* as dry season feed for goat production in Nigeria. *In* Proc. of the first AABNF Meeting, Nairobi. Eds. H H Ssali and S O Keya. pp 533–539. The Nairobi *Rhizobium* MIRCEN, Nairobi.

Sanginga N, Mulongoy K and Ayanaba A 1986 Inoculation of *Leucaena leucocephala* (Lam.) de Wit with *Rhizobium* and its nitrogen contribution to a subsequent maize crop. Biol. Agric. Hortic. 3, 341–347.

Sanginga N, Zapata F, Danso S K A and Bowen D G 1990a Effect of successive cutting on nodulation and nitrogen fixation of *Leucaena leucocephala* using ^{15}N dilution and the difference methods. *In* Plant Nutrition-Physiology and Applications. Ed. M L van Beusichem. pp 667–674. Kluwer Academic Publishers, Dordrecht, The Netherlands.

Sanginga N, Zapata F, Danso S K A and Bowen G D 1990b Effect of successive cuttings on uptake and partitioning of ^{15}N among plant parts of *Leucaena leucocephala*. Biol. Fertil. Soils 9, 37–42.

Virginia R A 1986 Soil development under legume tree canopies. For. Ecol. Manage. 16, 69–79.

4. ASPECTS RELATED TO NITROGEN ADVICE SYSTEMS AND N-BALANCE

Plant and Soil **181**: 31–38, 1996.

Advisory systems for nitrogen fertilizer recommendations

Maarten Geypens and Hilde Vandendriessche
Soil Service of Belgium, W. de Croylaan 48, B-3001 Leuven, Belgium*

Key words: models, N advisory systems, nitrogen dynamics, nitrogen fertilizer recommendation, plant analysis, sap test

Abstract

During the last decades several methods have been proposed to optimize N fertilization, some of them being implemented in advisory systems.

We focus on nitrogen fertilization of some important arable and vegetable crops. The recommendation systems can be divided into three groups: systems mainly based on soil analysis, systems based on plant analysis and systems based on simulation models. For these three systems we will discuss the possibilities, drawbacks and, if possible, ways to overcome the difficulties.

Most systems based on soil analysis lean on the determination of mineral nitrogen content at the end of the winter period, sometimes also taking into account nitrogen mineralization during the crop growth period. Different approaches are possible.

Plant analysis determines the actual nitrogen status of crops. As a single basis for nitrogen recommendation it may never satisfy, but it may be helpful in optimizing split applications of nitrogen fertilizer.

Simulation models are interesting tools to assess nitrogen dynamics in the soil. Mechanistic models try to incorporate the best possible description of the known processes. As a consequence these mechanistic models are in general rather complex. Functional models, on the other hand, aim to give a reasonably good general description of the nitrogen dynamics in the soils without going into great detail. Modelling in combination with soil analysis can possibly result in more accurate nitrogen recommendations on field scale and predictions of the risk of nitrogen residues in the soil after harvest.

Introduction

Nitrogen fertilization not only aims at a high economic return of the investment through optimized yield and quality, but also at minimizing environmental hazards through leaching of residual nitrogen towards the ground or surface waters. Some crops need a low nitrate content due to its alleged role in human and animal health, sometimes resulting in legislative restrictions.

Although nitrogen is, from a quantitative point of view, the most important nutrient in crop production, in comparison with phosphorus and potassium, nitrogen fertilizer recommendations were for a long time the most uncertain and the least scientifically justified.

Attempts to use soil parameters such as the total nitrogen content or the C/N ratio of soil organic matter as a base for nitrogen fertilizer recommendations failed. From 1950 onwards, large series of field experiments with different nitrogen fertilizer levels were carried out on a regional scale to establish optimum rates of nitrogen fertilizer for different crops based on mineral nitrogen reserve in the soil layers. These recommendations could be refined for cereals by taking into account site specific characteristics such as previous crop, humus content of the soil, etc. (Braun, 1980; Laloux et al., 1975; Sturm, 1974). In UK, ADAS developed an advisory system for cereals based on soil type (depth and structure) and on a so-called 'soil N index', which is based on previous cropping. Also, other fac-

* FAX No.: +3216224206
Plant and Soil is the original source of publication of this article. It is recommended that this article is cited as: *Plant and Soil* **181**: 31–38, 1996.

tors related to crop, climate and soil conditions are taken into account (MAFF, 1984).

According to Neeteson (1989), Russell (1914) was probably the first to recognize that soil mineral nitrogen affects the fertilizer nitrogen requirement of arable crops. Nevertheless, until the seventies no serious attention has been paid to this statement.

From the seventies onwards in several West-European countries a great effort has been done to develop N-recommendation systems for arable crops based on the measurement or the estimation of mineral nitrogen in the soil at the beginning of the vegetation period (Boon, 1981; Remy and Viaux, 1982; Ris, 1974; Wehrmann and Scharpf,1979). It is interesting to note that in most of these countries, initially only winter wheat was involved and that N-recommendation systems for other arable crops were elaborated later.

Advisory systems

Advisory systems based on soil mineral nitrogen content

Investigations in the Netherlands (Ris, 1974) indicated a negative relation between the amount of mineral nitrogen in the profile around the first of March and the optimum N-dose (N_{op}) needed for maximum yield of the relevant crop of that year. The negative relation could be described, with sufficient confidence, by a straight line. In this way recommendation equations of the general form:

$$N_{op} = C - m.N_{min} \qquad (1)$$

for winter wheat, sugar beet and potatoes were obtained (Kolenbrander et al., 1981). In the above equation C stands for the optimum N-dose (N_{op}) in kg N ha^{-1}, if the profile contained no mineral nitrogen in spring. The factor m, by which the stored amounts of N_{min} in the profile have to be multiplied, depends on the utilization of N_{min} in the profile by the crop. Uptake of nitrogen from unsampled soil layers, and/or capillary rise of nitrogen will cause the value of factor m to increase. For winter wheat for instance the above equation (1) reads: $N_{op} = 140-1.0 \ N_{min}$; with N_{min} the amount of mineral nitrogen down to 1 m in the profile. Depending on the amount of mineral nitrogen in the soil profile, a second nitrogen application of 60 or 30 kg N will be carried out at stem elongation or the second application will be omitted.

A similar method for winter wheat was developed in the southern Hannover region (Wehrmann and Scharpf, 1979). The method involves the determination of the amount of mineral N in the rooting zone until 90 cm depth in spring. Results of field trials in the region indicate that N fertilizer should be applied to raise the total available amount ($N_{min} + N_{fert}$) up to 120 kg N, (Sollwerte or target value) per ha. This is followed by normal rates of 20–30 kg N per ha during stem elongation and of 50–60 kg N per ha at ear emergence. According to Wollring and Wehrmann (1981, 1990), it is possible to use sap analysis to assist in deciding the level and time of the additional nitrogen applications. Experience with this method shows better results when the target value is fine-tuned to the actual site. Knittel and Sturm (1986) found a negative correlation between the optimal demand of winter wheat and the soil fertility index (Ackerzahl).

Although a high correlation was obtained between the N_{min} in the soil profile and the optimum N-dose, considerable deviations from the average optimum N-dose may occur. Ris et al. (1981) reported the following regression coefficients -0.53, -0.65, -0.68 for winter wheat, potatoes and sugar beet, respectively. Preliminary fertilization experiments in Belgium showed that the correlation between N_{min} and the optimal fertilization rate could be improved by taking into account a number of factors, determining the availability of nitrogen during the growing season (Boon, 1981). This improved N_{min} value has been called 'N-Index'. 'The N-Index' may be considered as the sum of different factors. These factors may be divided in three groups:

i. Factors representing the amount of mineral nitrogen in the soil available for the crop and the amount of nitrogen already taken up by the crop at the time of soil sampling. The depth of the sample is determined by the rooting depth and rooting intensity. Nitrogen already taken up by the crop at the sampling time is estimated from plant development.

ii. Factors estimating the supply of mineral nitrogen by the soil during crop growth. Several parameters from different origins contribute to the mineralization and so the mineral nitrogen supply. The total amount of mineralized nitrogen is estimated by the summation of the mineral nitrogen coming from mineralization of the humus in the arable layer, of crop residues from preceding crops and of mineralization of applied animal and green manures, organic wastes, etc. Soil texture, structure and regional climatic conditions are evaluated to estimate possible influences on the mineralization process.

iii. Losses resulting in a diminished availability of mineral nitrogen. In temperate regions, leaching of nitrate may mainly occur between sampling time and the start of the active growth period. It depends on soil texture and the amount and the distribution of mineral nitrogen in the soil profile. Denitrification and volatilisation losses, although existing, are not taken into account because of a lack of reliable information.

The 'N-Index' is a calculated measure of available nitrogen for a specific crop on a specific site (Geypens et al., 1994; Vandendriessche et al., 1992). Based on this "N-Index", a nitrogen fertilizer advice expressed in kg N ha^{-1}, can be generated which is generally formulated in the following equation:

$$N - advice = A - b \text{ N-Index} \qquad (2)$$

In Equation (2) the value b is specific for each arable crop in a well defined region. In winter wheat the value of A depends on the lodging resistance of the cultivar and of its tillering capacity. For potatoes, e.g. the value A depends on the properties of the respective variety and on the final destination of the harvested product: processing potatoes, starch production or fresh market.

As an example, the N-advice expressed in kg N ha^{-1}, for a lodging resistant winter wheat variety in the loamy region of Belgium reads like this:

$$N - advice = 305 - 0.77 \text{ N-Index} \qquad (3)$$

for n = 74 and r^2 = 0.897. Besides, an advice of split application is given for winter wheat on the basis of the total N-advice, the distribution of N$_{min}$ in the soil profile, the management of the crop like the use of growth regulators, and some other properties of the wheat cultivar such as tillering capacity and lodging resistance. The effect of splitting fertilizer nitrogen on yield components of winter wheat is well known. Although the effect of different methods of splitting the total amount of nitrogen on yield is usually small (Heyn and Witzel, 1986; Sylvester-Bradley et al., 1987), in some situations a significant effect of the method of splitting has been observed (Dilz et al., 1982; Geypens et al., 1991). Moreover, an appropriate splitting will increase the reliability of the yield. Under temperate growing conditions 18,000–20,000 grains m^{-2} are required for maximum grain yield of wheat. Due to its compensatory ability, wheat may obtain this number in different ways (tillering, spikelets, florets) but a prolonged grain filling period is needed for a high 1000-grain weight.

In France, the balance sheet method (bilan prévisionel) is widely used (Remy and Viaux, 1982). Before the growing season, a balance is established of inputs and outputs of nitrogen. Thus, nitrogen requirement is calculated from the following equation:

$$bY = (Nm + Ms + Mr + Mo + F)C \qquad (4)$$

where Y = expected yield;

b = total absorbed N per unit yield of grain;

Nm = mineral nitrogen in soil at the end of the winter;

Ms = N mineralised from soil organic matter;

Mr = N mineralised from residues of previous crop;

Mo = N mineralised from organic manure;

F = fertilizer N to be applied;

C = efficiency of nitrogen utilisation.

This method is based on the assumption that every 100 kg of grain to be produced will require the uptake of 3 kg of N. However, this assumption has been found approximative and dependent on climatic and soil conditions. There is also no method for accurate forecasting of yield and in this respect the information coming from the farmer is not always reliable or realistic. In the early stages, the method was used without the coefficient C and proved satisfactory in the Northeast of France but unsatisfactory in other regions (Viaux, 1980).

During the first years of application of the balance sheet method, Nm was estimated but due to the unreliability of this estimation, Nm now is measured in the soil profile before the start of a new vegetation period.

A similar balance method (Hofman, 1983) for more general use or an alternative balance method (Neeteson et al., 1988) has been proposed for the calculation of the 2nd or 3th split-dressing in winter wheat.

The sugar beet industry in Germany and Austria has introduced the Electro Ultra Filtration (EUF) technique for fertilization advisory purposes in sugar beet crops. Later on this technique has also been applied to other arable crops. Nitrogen fertilizer recommendation is based on soil analysis of the arable layer in the summer or early autumn preceding sugar beet crop. By this technique both mineral (Nmin) and soluble low molecular nitrogen compounds (N$_{org}$) are analysed. In this approach it is hypothesised that N$_{org}$ provides a

34

good estimate of labile organic nitrogen, most likely to be mineralized during the next growing season (Németh and Wiklicky, 1982). However, the reliabilty of this method is still widely debated (Heyn and Brüne, 1989). Claims by the sugar beet industry that following the introduction of the EUF-technique, sugar yields have increased are not substantiated. Reasons for the ongoing debate are due to the fact that mineral nitrogen is subject to large seasonal fluctuations, i.e. leaching during the winter period. As a consequence in the field trials of the Soil Service of Belgium, not any useful correlation was found between mineral nitrogen in the arable layer during the summer months and mineral nitrogen in 0–30 or 0–60 cm layers in February. Whereas a very good correlation exists between mineral nitrogen in the soil layers in spring and the optimal nitrogen rate for optimal sugar yield (Vanstallen and Boon, 1983). Also the thesis that fluctuations in mineral nitrogen are balanced by fluctuations in N_{org} (Németh and Fürstenfeld, 1985) is contradicted by other authors (Kohl and Werner, 1986). This is not surprising because N_{min} and N_{org} are partially subject to independent processes. Moreover, the usefulness of N_{org} as a reliable predictor of mineralisable N can be questioned (Gutser et al., 1989; Hege et al., 1990; Severin et al., 1985).

Due to the poor reproducibility and high cost of EUF (Houba et al., 1994) research on an alternative method, using $CaCl_2$-extraction, is continuing (Heyn et al., 1990; Houba et al., 1994) but soil sampling will also be done late in autumn or at the end of the winter.

Some of the actual nitrogen fertilization recommendation systems and research are directed towards dynamic optimization of nitrogen supply, e.g. the KNS-system (Kulturbegleitende N_{min}-Sollwerte-System) in Germany (Lorenz et al., 1985) and the NBS-system (N-bijmestsysteem) in the Netherlands (Breimer, 1989) and in Belgium (Demyttenaere, 1991). These systems have mainly been introduced for open-air vegetables, characterised by a limited rooting depth. The main reason for this approach is the difficulty to predict nitrogen mineralization during crop growth. Essential in this system is the knowledge of the uptake curve of the crop during the growing season. The purpose of the system is to ensure a sufficient mineral nitrogen buffer in the soil to match the nitrogen requirement of the crop during the growth period. Apart from an initial fertilization before the start of the crop growth, one or some additional mineral nitrogen analyses of the soil will be carried out. Depending on the outcome of each measurement, the mineral nitrogen

buffer will be replenished until a target value. This so called dynamic optimization method will however not result in miracles. In the advisory methods based on N_{min}, no attention is paid to nitrogen supply by the soil. In recent field trials, we did not find any advantage using the dynamic optimization method as compared to the N-Index expert system for N-fertilizer advices to leek or Brussels sprouts.

Recommendations based on plant or sap analysis

Plant analysis can be very useful as a basis for assessing the nitrogen status of the crop but can not be considered as a substitute for total N-recommendation. This is because plant and sap analysis very well reflect the actual plant uptake of nitrogen but not the N-availability of the soil because nitrate uptake will also be influenced by other factors. As a consequence, these factors have to be taken into account when using sap analysis as a tool in N recommendation. Nevertheless, plant analysis may be helpful in some situations.

For the assessment of the nitrogen status of the crop, either the total N-content or the nitrate-N concentration has been proposed as a sensitive indicator. Total nitrogen may be chosen because of its straightforward analysis. However, the interpretation of total nitrogen contents is complicated by the variation of N due to the stage of development, the genetic characteristics and the climatic conditions during plant growth (Greenwood, 1966). For corn, relationships between concentrations of N in young plants and fertilizer N applied were not consistent across 14 site years (Binford et al., 1992). Overall, the results showed that a tissue test based on the concentration of N in young plants would not be a reliable indicator of the N availability. Partly due to the availability of quick nitrate tests, nitrate analysis in plant sap has become an interesting alternative research subject. Generally, plant growth increases have been found with an increase of nitrate up to 1000 mg L^{-1} (Greenwood et al., 1980; Siman 1974; Vlassak 1984). So, this value can be considered as a critical threshold below which the crops should receive additional nitrogen. Thus, nitrate content in plants may be used as a diagnostic tool to detect nitrogen deficiencies in plants in an early stage.

The potential use of leaf nitrate concentrations to allow growers to apply additional nitrogen when the crop is deficient, is investigated by Doll et al. (1971). They concluded that the nitrate concentrations of petioles decreased rapidly during the season and that no valid interpretation of these levels could be made unless the age of plants was taken into account.

Haverkort and van de Waart (1994) investigated trials with starch potatoes on reclaimed peat soils from 1979 to 1989 with varying nitrogen fertilizer rates, either applied prior to planting or split in starter and additional applications. They concluded that increased starter nitrogen applications not always result in increased tuber yields, but increased always the leaf nitrate concentration early in the season, and this increase varied markedly with year and site. They concluded that all nitrogen to be applied should preferably be given early in the season: prior to planting or before hilling at the latest. Later applications were less effective in increasing yield. Their results do not support the view that leaf nitrate concentration offers potential of monitoring the crop nitrogen needs, nor do they indicate that additional nitrogen dressings will improve crop yields. The latter is also found in Belgian experiments (Geypens et al., 1991). These experiments revealed that both for yield and for quality parameters no general or overall advantage could be assigned to split applications of nitrogen. Splitting the total rate, recommended on the basis of N-index, had no influence on apparent recovery of nitrogen nor on the nitrogen residue after harvest. In our regions, potatoes reach their maximal nitrogen uptake rate early in the growing season. This is the reason why in general late additional dressings do not make sense.

Despite the above mentioned field trial results, farmers in the Netherlands are adviced to split the recommended N-dose for potatoes into a major starter (60–65% of the complete dose) and one or two minor additional applications based on petiole nitrate concentration or soil sampling for mineral nitrogen. For the recommendation of the additional dressing, a relation between the nitrate content in petioles and optimum yield of potatoes has been established. Splitting nitrogen application aims to improve apparent recovery, avoid salt damage, diminish leaching during early crop growth and decrease nitrogen residue after harvest. Gardner and Jones (1975) reported that the petiole nitrate status reflected well the amount of N-fertilizer applied to potato crops and found a correlation between the early season nitrate concentration and total yield.

Wollring and Wehrmann (1981, 1990) developed a recommendation system for the level and time of N-top dressings for winter wheat and winter barley based on the nitrate content of the basal internode of the stems. According to these authors, advices of top dressings based on soil N_{min} are unsatisfactory due to a lack of

information on the availability of mineral nitrogen during crop growth. On the other hand they claim that the plant nitrate test is unsuitable for N-recommendation at the start of the growing season but should be looked at as complementary to the N_{min} method.

Simulation models as a tool for N fertilizer recommendation

The use of any laboratory-based analytical method for the prediction of nitrogen requirement on an individual field basis may pose logistical problems of sampling, sample preparation and analysis. Moreover, at the suitable time of sampling soil may still be frozen. A possible alternative to predict mineral nitrogen is to use computer modelling. In more advanced research situations, simulation models for crop growth, crop nitrogen dynamics and soil nitrogen availability may be used.

Simulation is a technique to study the behaviour of real systems on the basis of models. These models imitate the operation of the system. In agriculture, mostly biological processes are involved. However, they are rather difficult to be described mechanistically, leading to a frequent use of empirical relationships.

Models built for N fertilizer recommendations on a field scale are usually deterministic, although some input parameters are subject to variability.

Although considerable progress has been made in the simulation of crop growth and of processes involved with nitrogen dynamics in soil and plant systems, until now, as far as the authors are aware, not a single integrated model for N fertilizer recommendation appears to be used in practice. Perhaps one of the reasons could be, as Jones (1989) described about modelling, that many researchers seem to be reluctant to try their products in the real world (the system always seem to need just a little more refinement). Another reason can be the lack of accuracy or sensitivity of the model, although each increase in model sensitivity generally results in increased complexity of the model and in an increased demand for data to run it. For practical purposes, models should therefore only need a minimum set of easily available input data. Apparently, such complete models which are easy to run do not exist up to now. Nevertheless, some rather simple submodels may be used to improve N fertilizer recommendations as is applied in the N-Index expert system for the leaching and mineralization part.

As we already reported, logistic and climatic problems may occur when using a recommendation method based on the measurement of mineral nitrogen in the soil in spring. Computer modelling may offer an interesting alternative approach allowing the prediction of the amount and the distribution of mineral nitrogen in the soil in spring starting from measurements made in autumn. Such models exist (Anonymous, 1989; Whitmore et al., 1987) and although encouraging results are obtained, they are not yet implemented, apparently because advisers fear that accuracy and reliability of N fertilizer recommendation on a field scale will decrease when using these models.

Neeteson et al. (1987) described a dynamic model to predict yield and optimum nitrogen fertilizer application rate for potatoes. The input data for the model were limited to readily available data, so the model could be used for advisory purposes. The model was used to predict tuber yields and the optimum level of nitrogen fertilizer for each of 61 different field experiments. Although predicted optima on occasion differed considerably, according to the authors these discrepancies could be largely attributed to the considerable error in the measured data. If the predicted application rate had been applied to the 61 experiments, the yield deficits from the yields obtained with the measured optima were less than 2% in 84% of the experiments. The fertilizer advice based on this model was about as good as that obtained with the N_{min}-method in the Netherlands. However, the amount of mineral nitrogen in the soil in early spring was also used as an input for the model. The merit of the model is to provide a better understanding of what is happening in the soil-plant system in the period between the start of the growth period and the final harvest of the crop. Similar models may also help in the dynamic approach for on line advices. Moreover, they might be able to predict the nitrate residue in the soil after harvest in a reasonable way and as a consequence, appropriate management measures can be taken.

In Belgium, the Institute for Encouraging Scientific Research in Industry and Agriculture (Dutch acronym I.W.O.N.L.) funded an extensive research to convert, in addition to the traditional soil fertility research, the nitrogen advice system for winter wheat and sugar beet into a more dynamic tool (Anonymous, 1989). This research led to the description and validation of several simulation models either related to one or more aspects of the climate-crop-soil system. However, one of the conclusions was again that the usefulness of the described models is limited in practice by the number and the nature of input data required. So, it should be made possible to generate more complex input data from easily measurable or available data.

Perspectives and conclusions

In general, most nitrogen recommendation systems purely based on mineral nitrogen analysis before the start of the growing season or before planting perform very well. This is especially the case on sites where mineral nitrogen before the growing season constitutes the major part of nitrogen available for the crop. This is partly due to fact that the basic equations in the recommendation are derived from field trials on comparable soils. In this case all other processes of nitrogen dynamics during crop growth are implicitly included in the target value. In situations, more or less different from average field conditions, refinement of N recommendation will result in a better performance. Nevertheless, some prediction uncertainty will remain. A major part of this uncertainty is due to the difficulty to calculate a reliable optimum application rate of nitrogen in experiments with various rates of nitrogen. Crop response to fertilizer N can be described by different models, e.g. linear response, quadratic response, polynomials, exponentials and split lines. For each experiment, the most appropriate response curve should be used from which the optimum rate should be derived. If no reliable optimum can be obtained, results should not be used for calculation of the relationship between soil characteristics and optimum application rate of nitrogen. Also the comparison of the performance of advisory methods for nitrogen fertilization should be done with field experiments fulfilling the statistical requirements and on sites with a wide range of characteristics. Unfortunately, this is not always the case and as a consequence wrong or unjustified conclusions may be drawn.

Plant and sap analysis is an interesting tool to know the actual nitrogen status of the plant. As the nitrogen content is influenced by several factors, the practical use of this measurement is limited. Taking into account these factors in interpreting the nitrogen content, plant analysis may be helpful in deciding the timing of splitting nitrogen dressings.

In general, mathematical models have proved to be very useful tools to formalize our knowledge of various processes and their interactions within a given system. So far simulation models are used for research purposes but are not applied in practice for fertilization recommendations. Submodels, i.e. for leaching or mineralization may be integrated in fertilizer recommendation systems like 'N-Index' or 'N balance sheet method'. In our opinion, only simplified models with a limited number of readily available input data will be suitable for N recommendation in practice. However, partly due to the error of the calculated optimum, no substantial progress will be expected in the reliability of N fertilizer recommendation performed by means of simulation models. Nevertheless, these models may be interesting tools for policy makers when applied on a larger scale.

References

Anonymous 1989 Simulatie als hulpmiddel bij het stikstofbemestingsadvies voor de teelten wintertarwe en suikerbieten. Comité voor Toegepaste Bodemkunde I W O N L, Brussel. 257 p.

Binford G D, Blackmer A M and Cerrato M E 1992 Nitrogen concentration of young corn plants as an indicator of nitrogen availability. Agron. J. 84, 219–223.

Boon R 1981 Stikstofadvies op basis van profielanalyse voor wintergraan en suikerbieten op diepe leem- en zandleemgronden. Pedologie 31, 347–363.

Braun H 1980 Die Stickstoffdüngung des Getreides. DLG-Verlag, Frankfurt/M, Germany.

Breimer T 1989 Stikstofbijmestsysteem (NBS) voor enige vollegrondsgroenten. Consulentschap voor Bodem- Water-, en Bemestingszaken in de Akkerbouw en Tuinbouw, Wageningen, the Netherlands. 58 p.

Demyttenaere P 1991 Stikstofdynamiek in de bodems van de Westvlaamse groentestreek. PhD. Thesis, Universiteit Gent, Belgium. 203 p, bijlage 40 p.

Dilz K, Darwinkel A, Boon R and Verstraeten L M J 1982 Intensive wheat production as related to nitrogen fertilisation, crop protection and soil nitrogen: experience in the Benelux. Proc. Fert. Soc. (London) 211, 93–124.

Doll E C, Christenson D R and Wolcott A R 1971 Potato yields as related to nitrate concentrations in petioles and soil. Am. Potato J. 48, 105–112.

Gardner B R and Jones J P 1975 Petiole analysis and the nitrogen fertilization of Russet Burbank Potatoes. Am. Potato J. 52, 195–200.

Geypens M, Vandendriessche H and Bries J 1991 Verwachtingen en resultaten met N-deelgiften in de akkerbouw. Studiedag: Optimale bemestingstechnieken voor landbouw en milieu. Meise 2, 1–14.

Geypens M, Vandendriessche H, Bries J and Hendrickx G 1994 The N-Index Expert System, a Tool for Integrated N-Management. 15th World Congress of Soil Science, Volume 5a, pp 165–173.

Greenwood E A N 1966 Nitrogen stress in wheat - Its measurement and relation to leaf nitrogen. Plant and Soil 24, 279–288.

Groot J J R and Van Keulen H 1990 Prospects for improvement of nitrogen fertilizer recommendations for cereals: a simulation study. In Plant Nutrition-Physiology and Applications. Ed. M L van Beusichem. pp 685–692.

Gutser R, Vilsmeier K, Teicher K and Beck Th 1989 Aussagekraft des N_{org}-Stickstoffs für die N-Nachlieferung von Böden. VDLUFA-Schriften. 30, 187–194.

Haverkort A J and van de Waart M 1994 Yield and leaf nitrate concentration of starch potato on sandy humic soils supplied with varying amounts of starter and supplemental nitrogen fertilizer. Eur. J. Agron. 3, 29–41.

Hege U, Süss A and Maier S 1990 Optimierung der N-Düngung ohne und mit Berücksichtiging des N_{org}-Stickstoffs. VDLUFA-Schriften. 32, 249–255.

Heyn J and Witzel D 1986 Ergebnisse einer fünfjährigen feldversuchsserie in Hessen zur Frage der Stickstoff-Düngung des Winterweizens. VDLUFA-Schriften. 20, 389–413.

Heyn J and Brüne H 1989 Ein Vergleich zwischen N-Düngempfelungen zu Zuckerrüben nach N_{min}- und EUF-Bodenuntersuchungen an hand hessischer Feldversuche. VDLUFA-Schriften. 30, 195–200.

Heyn J, Ellinghaus R, Schaaf H and Witzel D 1990 Der Einfluss von Beprobungszeit, Entnametiefe und Trocknung auf die Aussage von $CaCl_2$-N-Bodenuntersuchungen gemessen an Ergebnissen von Feldversuchen. VDLUFA-Schriften. 32, 321-329.

Hofman G 1983 Minerale stikstofevolutie in zandleemprofielen. Aggregaatsthesis. Universiteit Gent, Belgium. 183 p.

Houba V J G, Novozamsky I, Huijbregts A W M and Van Der Lee J J 1994 Sugar beet Res. Symposium Brussels, Belgium. IIRB, Brussels, Belgium.

Jones P 1989 Agricultural Applications of Expert Systems Concepts. Agric. Syst. 31, 3–18.

Kohl A and Werner W 1986 Untersuchungen zur saisonalen Veränderung der EUF-N-Fraktionen und zur Charakterisierung des leicht mobilisierbaren Bodenstickstoffs durch Elektro-Ultrafiltration (EUF). VDLUFA-Schriften. 20, 333–341.

Kolenbrander G J, Neeteson J J and Wijnen G 1981 Investigation in the Netherlands of optimum nitrogen fertilization on the basis of the amount of N_{min} in the soil profile. Pedologie 31, 365–377.

Knittel H and Sturm H 1986 Wechselwirkungen zwischen N-Düngung zu Winterweizen und Standort bezogener Bodenfruchtbarkeit. VDLUFA-Schriften. 20, 415–424.

Laloux R, Poelaert J and Falisse A 1975 Stikstofbemesting bij graangewassen. Landbouwtijdschr. 28, 1155–1184.

Lorenz H, Schlaghecken J and Engl G 1985 -Gezielte Stickstoffversorgung-das kulturbegleitende N_{min}-Sollwerte-System (KNS-System). Dtsch. Gartenbau 13, 646-648.

MAFF 1984 The nitrogen requirement of cereals. Reference book, 385. Min. Agric. Fish. Food, London, UK. 260 p.

Neeteson J J, Greenwood D J and Draycott A 1987 Dynamic model to predict yield and optimum nitrogen application rate for potatoes. Proc. Fert. Soc. 262. London, UK. 31 p.

Neeteson J J 1989 Assessment of fertilizer requirement of potatoes and sugar beets. Ph.D Thesis. Landbouwuniversiteit Wageningen, the Netherlands. 141 p.

Németh K and Wiklickly L 1982 Bestimmung pflanzenverfügbarer Stickstoff-fractionen im Boden und Beurteilung des Stickstoff-Düngerbedarfs für die Zuckerrübe mit EUF. Zuckerindustrie 107, 958–962.

Németh K and Fürstenfeld F 1985 Changes of EUF-N and EUF-K fractions in three deep grey-brown luvisols during 18 months under fallow. Plant and Soil 86, 248–256.

Remy J C and Viaux Ph 1982 The use of nitrogen fertilisers in intensive wheat growing in France. Symposium on fertilisers and intensive wheat production in the EEC. The Fertiliser Society, London, UK. pp 67–92.

Ris J 1974 Stikstofbemestingsadviezen voor bouwland. Stikstof 7, 169–173.

Ris J, Smilde K W and Wijnen G 1981 Nitrogen fertilizer recommendation for arable crops based on soil analysis. Fert. Res. 2, 21–32.

38

Russell E J 1914 The nature and amount of the fluctuation in nitrate contents of arable soils. J. Agric. Sci. (Cambridge) 6, 18–57.

Severin K, Kersebaum K C and Richter J 1985 Die simulation der Stickstoffdynamik im Winterhalbjahr zur Berechnung des anorganischen N-Vorrats zu Vegetationsbeginn im Vergleich mit unterschiedlichen Messverfahren. VDLUFA-Schriften. 16, 129–135.

Siman G 1974 Nitrogen status in growing cereals. The Royal Agricultural College of Sweden, 93 p.

Sturm H 1974 Hilfstabelle zur N-Dungüng. Was muss bei Getreide in Frühjahr beachtet werden. Mitt. DLG 97, 5–6.

Sylvester-Bradley R, Bloom T M, Vaidyanathan A and Murray A W A 1987 The quest for the optimum: a comparison of nitrogen recommendation systems for winter wheat. In Assessment of Nitrogen Fertilizer Requirement. Ed. N E Nielsen. pp 113–133. The Royal Veterinary and Agricultural University, Copenhagen, Denmark.

Vandendriessche H, Geypens M and Bries J 1992 N-index: an expert system for nitrogen fertilization of arable crops. Workshop Cost, Gembloux, pp 55–57.

Vanstallen R and Boon R 1983 Betteraves sucrières. La fumure pour betteraves sucrières. Fumure azotée et valeur industrielle de la betterve sucrière. Rev. Agric. 36, 725–730.

Viaux Ph 1980 Fumure azotée des céréales d'hiver. Perspect. Agric. 43, 10–26.

Vlassak K and Verstraeten L M J 1984 Saptesten als hulp voor bemestingsadvies. Studiedag: Bemesting van intensieve akkerbouwteelten, Heverlee, 2.1–2.15.

Wehrmann J und Scharpf H C 1979 Der Mineralstickstoffgehalt des Bodens als Massstab für den Stickstoffdüngerbedarf (Nmin-Methode). Plant and Soil 52, 109–126.

Whitmore A P, Milford G F J and Armstrong M J 1987 Evaluation of a nitrogen leaching/mineralization model for sugar beet. Plant and Soil 101, 61–65.

Wollring J and Wehrmann J 1981 Der Nitrat-Schnelltest-Entscheidungshilfe fur die N-Spätdüngung. Mitt. DLG 96, 449–450.

Wollring J and Wehrmann J 1990 Der Nitratgehalt in der Halmbasis als Massstab für den Stickstoffdüngerbedarf bei Wintergetreide. Z. Pflanzenernähr. Bodenk. 153, 47–53.

Section editor: R Merckx

Plant and Soil **181**: 65–69, 1996.

Nitrogen inputs and losses from New Zealand dairy farmlets, as affected by nitrogen fertilizer application: year one

S.F. Ledgard[1], M.S. Sprosen[1], G.J. Brier[1], E.K.K. Nemaia[1] and D.A. Clark[2]
[1]*AgResearch, Ruakura Research Centre, Private Bag 3123, Hamilton, New Zealand* * and [2]*Dairying Research Corporation, Private Bag 3123, Hamilton, New Zealand*

Key words: ammonia volatilization, denitrification, grazed pasture, leaching, nitrogen, N_2 fixation

Abstract

Inputs and losses of nitrogen (N) were determined in dairy cow farmlets receiving 0, 225 or 360 kg N ha^{-1} (in split applications as urea) in the first year of a large grazing experiment near Hamilton, New Zealand. Cows grazed perennial ryegrass/white clover pastures all year round on a free-draining soil. N_2 fixation was estimated (using ^{15}N dilution) to be 212, 165 and 74 kg N ha^{-1} yr^{-1} in the 0, 225 and 360 N treatments, respectively. The intermediate N rate had little effect on clover growth during spring but favoured more total pasture cover in summer and autumn, thereby reducing overgrazing and resulting in 140% more clover growth during the latter period.

Removal of N in milk was 76, 89 and 92 kg N ha^{-1} in the 0, 225 and 360 N treatments, respectively. Denitrification losses were low (7–14 kg N ha^{-1} yr^{-1}), increased with N application, and occurred predominantly during winter. Ammonia volatilization was estimated by micrometeorological mass balance at 15, 45 and 63 kg N ha^{-1} yr^{-1} in the 0, 225 and 360 N treatments, respectively. Most of the increase in ammonia loss was attributed to direct loss after application of the urea fertilizer.

Leaching of nitrate was estimated (using ceramic cup samplers at 1 m soil depth, in conjunction with lysimeters) to be 13, 18 and 31 kg N ha^{-1} yr^{-1} in a year of relatively low rainfall (990 mm yr^{-1}) and drainage (170–210 mm yr^{-1}). Drainage was lower in the N fertilized treatments and this was attributed to enhanced evapotranspiration associated with increased grass growth.

Nitrate-N concentrations in leachates increased gradually over time to 30 mg L^{-1} in the 360 N treatment whereas there was little temporal variation evident in the 0 (mean 6.4 mg L^{-1}) and 225 (mean 10.1 mg L^{-1}) N treatments. Thus, the 360 N treatment had a major effect by greatly reducing N_2 fixation and increasing N losses, whereas the 225 N treatment had little effect on N_2 fixation or on nitrate leaching. However, these results refer to the first year of the experiment and further measurements over time will determine the longer-term effects of these treatments on N inputs, transformations and losses.

Introduction

New Zealand dairy farmers depend on N_2 fixation by white clover as the main source of nitrogen (N) input to their pastures. The rate of N fixed has been estimated at between 100 and 300 kg N ha^{-1} yr^{-1} (Ledgard et al., 1990). A small proportion of farmers also apply moderate rates of fertilizer N (50–200 kg N ha^{-1} yr^{-1}), mainly in the form of urea. There have been no measurements of the effects of these various N inputs on N transformations and losses from New Zealand dairy farming systems. However, indirect evidence suggests that significant nitrate leaching losses may occur. Restricted ad hoc surveys within the two main dairying regions of New Zealand have indicated that up to 20% of shallow wells contain groundwater with nitrate-N concentrations exceeding 10 mg L^{-1} (Ledgard, 1993). In June 1993, a long-term dairy grazing trial with self-contained farmlets varying in the rate of N fertilizer application was started. This is being used to examine the effects of varying N inputs on N_2 fixation, N transformations and losses.

* FAX No.: +617 8385 160
Plant and Soil is the original source of publication of this article. It is recommended that this article is cited as: *Plant and Soil* **181**: 65–69, 1996.

Methods

Farmlets

Three farmlets (6.47 ha each) received either 0, 225 or 360 kg N ha^{-1} (as urea), and were located on a free-draining silt loam soil on D.R.C. Number 2 Dairy near Hamilton, New Zealand. Pastures (predominantly perennial ryegrass and white clover) were rotationally-grazed throughout the year by Friesian dairy cows at 3.24 cows ha^{-1}, with the rotation length varying between 24 days in spring and 128 days in winter.

The dairy cows calved in July or August and were milked twice-daily after calving for 247, 284 and 288 days in the 0, 225 and 360 N farmlets, respectively. The 360 kg N ha^{-1} was applied as 90 kg N ha^{-1} on 20 July 1993, 45 kg N ha^{-1} on 2 and 27 September, 22 kg N ha^{-1} on 3 November, 7 December, 21 December and 11 January, and 90 kg N ha^{-1} on 8 April 1994. The 225 kg N ha^{-1} farmlet had half the rates of N application on these same dates, except on 8 April when it also received 90 kg N ha^{-1}.

Measurements

Consumption of pasture by cows was estimated using visual assessment of pasture before and after grazings with calibration by cutting. Removal of N in milk was determined from the volume and total N concentration of milk collected after each milking. N$_2$ fixation was determined using a ^{15}N dilution method (Ledgard et al., 1990) with 6 replicate plots in each farmlet. In the N fertilizer treatments, the ^{15}N microplots received ^{15}N-labelled urea at the same rate as the rest of the paddock and the fate of the added N was examined by analysing pasture prior to each grazing and soil at the end of the first 12 months.

Denitrification was measured using an acetylene-inhibition technique (Ryden et al., 1987) involving 24 h incubation of soil cores (25 mm diameter, 0–75 mm depth). Six replicate jars with 12 cores per jar were analysed at approximately 2-weekly intervals. Volatilization of ammonia was measured using a micrometeorological mass balance method with samplers as described by Leuning et al. (1985). Masts containing 5 samplers were located upwind in ungrazed paddocks and 20 m within paddocks which had been grazed or had received N fertilizer.

Leaching losses were determined using ceramic cup samplers (30 per farmlet) located at 1 m soil depth. Samples of solution were collected at approximately 2-weekly intervals and analysed for nitrate and ammonium. Drainage was determined from the volume of water passing through lysimeters containing intact soil cores (0.4 m diameter, 1 m depth) which received 0 or 360 kg N ha^{-1} yr^{-1}.

Results

N$_2$ fixation by white clover decreased relative to the 0 N farmlets by 22 and 65% (SE = 11) in the 225 and 360 N farmlets respectively (Fig. 1). The mean proportion of total clover herbage N obtained from N$_2$fixation was 79, 55, and 46% in the 0, 225 and 360 N farmlets, respectively.

The total N content of above-ground mixed pasture was 521, 675 and 714 kg N ha^{-1} in the 0, 225 and 360 N treatments respectively. Likewise, there was an increase in intake of pasture N by cows in the N-fertilized farmlets but it was similar for both the 225 and 360 N farmlets (Fig. 1). However, the error in estimating N intake was relatively high with coefficients of variation of about 15–20%.

Milk production was 4120, 4860 and 5040 L cow^{-1} in the 0, 225 and 360 N farmlets, respectively, and this represented the main form of N removal/loss (Fig. 1). Removal of N by transfer of cow excreta to raceways and to dairy sheds during milkings was estimated as 9% of the difference between N intake and milk N, since this was the average proportion of time that the cows spent away from grazing.

Denitrification losses were increased (Fig. 1, SED = 1.9) in the N-fertilized farmlets but were small on an annual basis. There was a marked seasonal pattern of denitrification with most loss occurring during winter (Fig. 2). In addition, there was a short sharp increase in the rate of denitrification at about 5–10 days after most applications of N fertilizer (Fig. 2).

Ammonia volatilization increased by three- and four-fold in the 225 and 360 N farmlets, respectively (Fig. 1). Measurements using duplicate masts indicated a typical coefficient of variation of 6- 9%. There was no obvious seasonal pattern to the amount of ammonia loss after each grazing. Measurements of daily ammonia loss indicated that in most cases over 80% of the total loss occurred within 4 days of grazing or application of urea.

Leaching was determined by analysing samples of soil solution collected at 1m depth. The samples showed significant nitrate (Fig. 3) and negligible ammonium. The nitrate concentration was initially

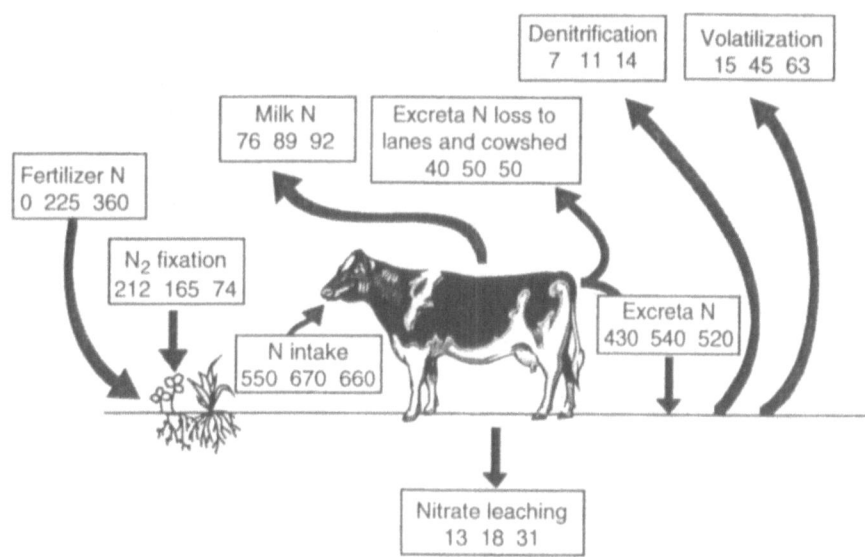

Figure 1. Nitrogen transformations in pastoral farmlets grazed by dairy cows. Data refers to the annual flux in kg N ha^{-1} and values from left to right are for treatments receiving N fertilizer at 0, 225 or 360 kg N ha^{-1} yr^{-1} respectively.

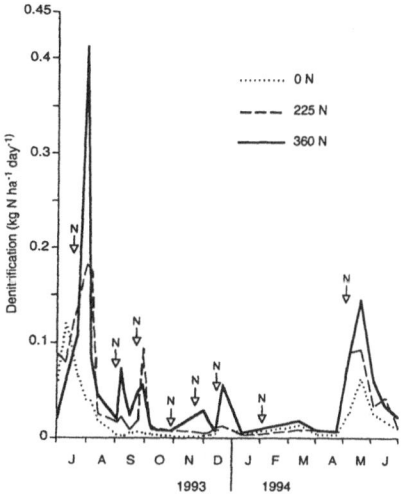

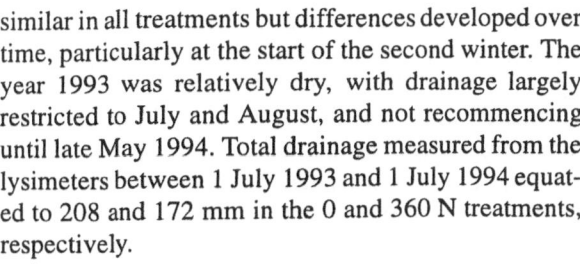

Figure 2. Temporal pattern of denitrification from grazed pastoral soils (0 -75 mm depth) as affected by application of N fertilizer at 0, 225 or 360 kg N ha^{-1} yr^{-1}. The arrows show times of N application.

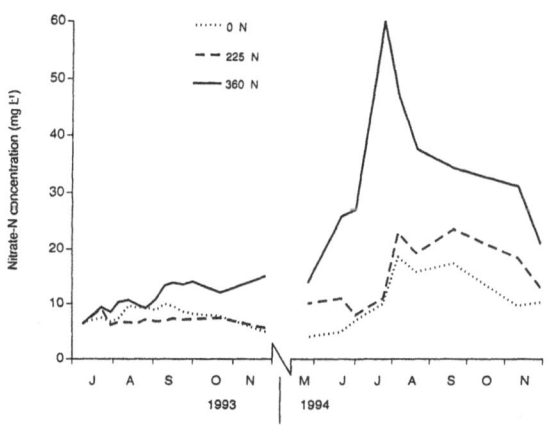

Figure 3. Nitrate-N concentration of leachates sampled at 1 m soil depth from dairy cow farmlets receiving 0, 225 or 360 kg N ha^{-1} yr^{-1}. The average SED was 2.8 (range 1.9–4.1) in 1993 and 5.0 (range 4.1–6.1) in 1994.

similar in all treatments but differences developed over time, particularly at the start of the second winter. The year 1993 was relatively dry, with drainage largely restricted to July and August, and not recommencing until late May 1994. Total drainage measured from the lysimeters between 1 July 1993 and 1 July 1994 equated to 208 and 172 mm in the 0 and 360 N treatments, respectively.

Within any one farmlet there was a wide range in nitrate-N concentrations (e.g. varying between 0 and 120 mg L^{-1}), presumably reflecting the distribution of urine patches. The average concentration of nitrate-N in leachate over the period 1 July 1993 to 1 July 1994 was 6.4, 10.1 and 17.9 mg L^{-1} in the 0, 225 and 360 N treatments, respectively. The corresponding values for the amount of nitrate-N leached were 13.3, 18.4 and 30.8 kg N ha^{-1} (SED = 5.2), respectively.

Discussion

Application of N fertilizer at 360 kg N ha^{-1} resulted in a 65% reduction in N$_2$ fixation by white clover. This was caused by a decrease in clover growth of 34% as competition from associated grasses increased, and by direct substitution of uptake of fertilizer N for N$_2$ fixation. In contrast, there was much less effect of 225 kg N ha^{-1} on N$_2$ fixation. This was due to greater amounts of pasture being carried through the summer period, thereby minimizing the effects of over-grazing on clover growth which occurred in the 0 N treatment. Consequently, clover growth in late-summer/autumn was 140% higher in the 225 N treatment than in the 0 N treatment.

Application of 225 kg N ha^{-1} resulted in a 30% increase in plant N accumulation and a 24% increase in intake of pasture N by cows. However, 360 kg N ha^{-1} only produced an extra 7% N uptake by plants and no apparent increase in N intake by cows, although the error associated with these intake estimates was relatively high. This could reflect adequate feeding in the 225 N farmlet for the stocking rate of 3.2 cows ha^{-1}.

Removal of N in milk in the unfertilized farmlet was equivalent to 36% of the N input from N$_2$ fixation. However, the increase in N removal in milk in the 225 and 360 N farmlets was equal to only 6 and 5%, respectively of that applied as fertilizer N. Similarly, only 10–14% of the increased N intake by cows was harvested in milk and the remainder was returned in excreta.

Loss of N by denitrification was low (7–14 kg N ha^{-1}) and largely confined to the winter period. This is likely to be an underestimate of the long-term denitrification loss, since it was a dry year (990 mm rainfall compared to the average of 1200 mm) and soils were at or near field capacity for about 6 weeks less than normal. Ruz-Jerez et al. (1994) identified high soil moisture status as the primary requirement for denitrification in grazed pastoral soils. A secondary factor limiting denitrification was soil nitrate, as rates increased in the N fertilized farmlets.

Ammonia volatilization was the main form of gaseous N loss, and in the unfertilized farmlet it represented 7% of the N input from N$_2$ fixation or 3% of the N returned in excreta, which is of a similar order to that measured by Bussink (1992) and Jarvis et al. (1989). However, ammonia loss increased markedly in the N fertilized farmlets (from 15 to 45–63 kg N ha^{-1}) and this was much greater than the 20% increase in excreta N returns in these farmlets. Associated measurements using an enclosure technique on ungrazed, N-fertilized plots showed direct losses of ammonia-N from added urea fertilizer of between 7 and 26% (Sprosen and Ledgard, unpublished). Thus, the increased ammonia loss from the N fertilized farmlets was probably mainly due to direct loss after fertilizer application and was equivalent to about 14% of the urea-N applied in both treatments.

Leaching of nitrate from the unfertilized farmlet was low (13 kg N ha^{-1}) and was comparable with past estimates from sheep-grazed pastures in New Zealand (typical values 10–30 kg N ha^{-1} yr^{-1}; Ledgard, 1993). However, the measured leaching losses were likely to be lower than normal, because of the smaller amount of drainage water percolating through the soil from only 75% of normal winter rainfall.

Nitrate leaching increased in the N fertilized farmlets (from 13 to 18 and 31 kg N ha^{-1}), with the largest increase at the high N rate. Similar effects have been measured in other studies with sheep and cattle (Field et al., 1985; Scholefield et al., 1993). This could have been due to a number of factors including increased excreta N loss or direct N leaching from fertilizer.

Increased consumption of pasture N by cows in N fertilized farmlets led to greater return of excreta N, which will have resulted in some increase in leaching. The major contribution from excreta to nitrate leaching was clearly evident from the lack of leaching in ungrazed lysimeters and from the skewed, wide range of nitrate concentrations measured in ceramic cup samplers under grazing. However, estimates of cow N intake and excreta return were similar for the 225 and 360 N farmlets, which would suggest similar leaching from excreta N. This needs confirmation since the estimates of N intake, and therefore excreta N return, had large errors associated with them.

[15]N data from the microplots revealed almost complete recovery (98% in plant, soil and gaseous losses) of fertilizer N in the 225 N treatment, but lower recovery (89%) in the 360 N treatment (Ledgard and Sprosen, unpublished data). Thus, the main cause of increased nitrate leaching at the high N rate appeared to be direct loss of fertilizer N.

Differences between farmlets in the average nitrate-N concentration of leachates were 20% greater than differences in the amount of N leached. This occurred because N fertilizer enhanced pasture growth and evapotranspiration, and resulted in 17% less drainage in the 360 N treatment. This reduction in drainage was evident at the start and end of the drainage period. Thus, it is important to obtain separate estimates of drainage for determining nitrate leaching in experiments with N fertilizer treatments, rather than the commonly used procedure of obtaining a single estimate from soil water balance calculations. In the current experiment, the amount of nitrate-N leached from the 360 N farmlet would have been *overestimated* by 20% if drainage had been calculated using average values for evapotranspiration.

Nitrate-N concentrations were measured in soil solution collected from 1 m soil depth and should reflect N transformations in the surface soil over the previous 12–24 months. The volume of total available soil water contained in the 0–1 m soil layer was estimated to be equivalent to 160 mm at field capacity, and if leaching were mainly by piston displacement this volume of soil water would have been fully displaced within the measurement year. Thus, the impact of N fertilizer applications on nitrate leaching below 1 m depth should increase over time, and this is apparent from greater treatment differences in the second winter (Fig. 3).

Conclusions

The 360 N rate appeared excessive for the stocking rate of 3.24 cows ha^{-1} in that it gave little extra milk production (relative to 225 kg N ha^{-1}), severely reduced clover growth and N$_2$ fixation, and enhanced N losses. Leaching of nitrate increased steadily over time and by the second winter the nitrate-N concentration of leachate peaked at 6 times the World Health Organization recommended limit for drinking water of 10 mg L^{-1}.

In contrast, 225 kg N ha^{-1} had only a minor impact on N$_2$ fixation and N losses, while producing an 18% increase in milk production. The low rates of N application in spring had a beneficial effect on clover growth in summer and autumn by reducing overgrazing, with the net result of little decrease in annual N$_2$ fixation relative to the 0 N pasture. There was little effect of 225 kg N ha^{-1} on the nitrate-N concentration of leachate during the first year of the experiment. The long-term effects of N applications on the extent of N$_2$ fixation and N losses are being determined.

Acknowledgements

We thank J W Penno, A M Bryant and K A Macdonald for managing the trial, Dairying Research Corporation for experimental support, and the Foundation for Research, Science and Technology for funding.

References

Bussink D W 1992 Ammonia volatilization from grassland receiving nitrogen fertilizer and rotationally grazed by dairy cattle. Fert. Res. 33, 257–265.

Field T R O, Ball P R and Theobald P W 1985 Leaching of nitrate from sheep-grazed pastures. Proc. N.Z. Grassl. Assoc. 46, 209–214.

Jarvis S C, Hatch D J and Roberts D H 1989 The effects of grassland management on nitrogen losses from grazed swards through ammonia volatilization; the relationship to excreta N returns from cattle. J Agric. Sci., Cambridge 112, 205–216.

Ledgard S F 1993 Nitrate contamination of ground and surface water in pastoral agriculture in New Zealand. Proc. Int. Workshop Nitric-acid Based Fertilizer Environment. pp 337–348. I.F.D.C., Alabama, USA.

Ledgard S F, Brier G J and Upsdell M P 1990 Effect of clover cultivar on production and nitrogen fixation in clover-ryegrass swards under dairy cow grazing. N.Z. J. Agric. Res. 33, 243–249.

Leuning R, Freney J R, Denmead O T and Simpson J R 1985 A sampler for measuring atmospheric ammonia flux. Atmos. Environ. 19, 1117–1124.

Ruz-Jerez B E, White R E and Ball P R 1994 Long-term measurement of denitrification in three contrasting pastures grazed by sheep. Soil Biol. Biochem. 26, 29–39.

Ryden J C, Skinner J H and Nixon D J 1987 Soil core incubation system for the field measurement of denitrification in grassland soils. Soil Biol. Biochem. 19, 753–757.

Scholefield D, Tyson K C, Garwood E A, Armstrong A C, Hawkins J and Stone A C 1993 Nitrate leaching from grazed grassland lysimeters: effects of fertilizer input, field drainage, age of sward and patterns of weather. J. Soil Sci. 44, 601–613.

Section editor: R Merckx

O. Van Cleemput et al. (eds.), Progress in Nitrogen Cycling Studies, 343–346, 1996.
© 1996 *Kluwer Academic Publishers*.

Nitrogen balance studies in a long-term crop rotation field experiment

T. Németh
Research Institute for Soil Science and Agricultural Chemistry (RISSAC) of the Hungarian Academy of Sciences, H-1022 Budapest, Herman Ottó út 15, Hungary

Key words: deep drilling, field experiments, nitrate-N accumulation, nitrogen balances, plant nitrogen uptake

Abstract

The influence of fertilization with different N doses on nitrate-N content of the soil profiles and on nitrogen balances of the different N-treatments were studied in long-term field experiments on a calcareous chernozem soil formed on loess. After 12 and 17 years of the experiments, soil samples were taken with deep-drilling to 600 cm depth on the unfertilized plots and on the plots which had received 100, 200 and 300 kg N ha^{-1} yr^{-1} nitrogen fertilization.

The amount of nitrogen taken up by the harvested crops was also measured. On the basis of N-fertilization data, N-uptake of the cultivated plants and nitrate-N content of the profiles a simplified nitrogen balance was calculated. The nitrate-N distribution and accumulation along the profiles showed that a great part of the surplus (residual) nitrogen could be detected in the form of nitrate on those plots which had more nitrogen by fertilization than nitrogen demand of the crops. The nitrogen balances of these treatments were also positive.

Introduction

Among the essential elements in plant nutrition nitrogen has a special feature, most of the nitrogen in the soil-plant system (around 96%) is bound in organic forms. The nitrogen can be released from this form via mineralization thus become available for plants. The mineral nitrogen forms: exchangeable ammonium-N and nitrate-N are the plant available forms. Above a certain production limit the amount of nitrogen originating from the soil organic matter and from other natural sources may not be enough to cover the nitrogen demands of the cultivated plants. In these cases nitrogen fertilizers must be applied to better supply the crops.

Hungary has a continental type of agriculture which focuses on crop production in the agricultural land use, and has a high rate of cereal production in mono- and biculture. The maintanence of soil fertility has to be based mainly on the application of mineral fertilizers (Németh, 1993, 1994; Németh and Kádár, 1991). During the last few decades farmers broadcasted more nitrogen than the crops needed. The budgetary approach of the Hungarian agricultural land use shows that the overall nitrogen balance became positive in the early seventies (with the average of 7 kg N ha^{-1})

and remained positive till the end of the eighties (the positive balances varied between 16–27 kg N ha^{-1} in this period) (Csathó, 1994; Kádár, 1987).

In the last decades, public attention turned to the environmental protection. This is one of the reasons why the fate and behaviour of the chemicals applied in agriculture are still in the interest of research. As was mentioned, the nitrogen plays a special role among the nutrient elements. While that part of the other nutrients which is not taken up by the crops can be accumulated in the upper soil layers (mainly P and K) increasing both the total and the available nutrient contents of the soil, a significant part of the surplus nitrogen may leave the rooting zone via erosion, surface run-off, leaching, denitrification and volatilization. Also, there are some internal nitrogen transformation processes in the soil, like mineralization - immobilization. For an agricultural system mineralization and immobilization are usually in equilibrium, these fluxes will be balanced during a longer period. For non-leguminous arable cropping systems a simplified nitrogen balance is often constructed (Goss et al., 1994; Hill, 1986; Lund, 1982; Tanji et al., 1977). In this case usually the N-fertilization (organic and mineral) and the N-removal with harvested yield are taken into account to calculate the simplified nitrogen balance.

The influence of fertilization with different N doses on nitrate-N content of the soil profiles and on nitrogen balances of the different N-treatments were studied in a long-term field experiment on a calcareous chernozem soil formed on loess.

Materials and methods

The experimental site is located 110 km SW from Budapest. The soil of the site is migration type (calcareous chernozem), with deep organic matter layer. The organic matter content is 3.0% in the ploughed layer and 1.1% at 100 cm depth. The average depth of the groundwater table is 13–15 m. The average mean temperature is 10.9 °C, the annual precipitation is 550–600 cm. The soil moisture regime is also migration (chernozem) like, the annual precipitation is not enough to wash through the soil profile down to the water table.

On selected plots (unfertilized plots and the plots which have received 100, 200 and 300 kg N ha^{-1} yr^{-1} nitrogen fertilization) deep-drillings were carried out twice, first in the 12th then in the 17th year of the experiments, in both years in July after the harvest of the cultivated crops. Soil samples were taken from the profiles at every 20 cm to a depth of 600 cm. Soil moisture content of the samples were determined immediately, the mineral-N and other chemical analyses were done after air-drying the soils.

From the beginning of the experiments till the 17th year, the following crops were cultivated: winter wheat, winter wheat, maize, maize, potato, winter barley, oat, sugarbeet, sunflower, poppy, winter oilseed rape, mustard, spring barley, oil-flax, soybean, hemp and peas (Kádár and Németh, 1993).

The balances were calculated from the amount of applied fertilizer, from the nitrogen taken up by the plants and from the nitrogen which was found in the profiles in the form of nitrate.

Results and discussion

In Figure 1 the nitrate-N distribution in the profiles of the different nitrogen treatments can be seen in the 12th year of the experiment, while Figure 2 shows similar results for the 17th experimental year. In 1985 (after 12 years) significant nitrate-N accumulation was found between 60 and 200 cm in the intensively fertilized treatments. These curves differed significantly from each other, while practically no difference was

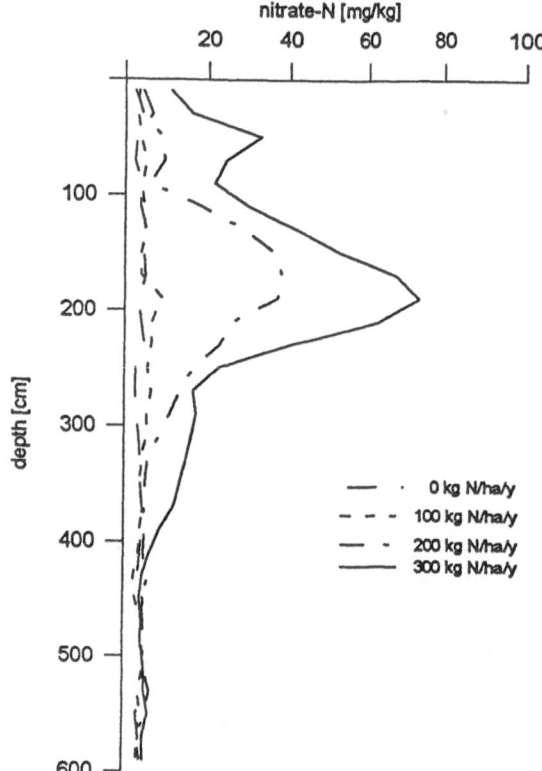

Figure 1. Nitrate-N accumulation in the soil profile (1985).

measured between the nitrate-N content of unfertilized plots and of the plots which had received the lower application rate (100 kg N ha^{-1} yr^{-1}). The amount of nitrogen accumulated in the profiles of the plots fertilized yearly with 300 kg N ha^{-1} was ten times higher than those measured in the unfertilized plots. The effect of the overfertilization in the residual nitrate-N form could be detected down to 350–400 cm (Németh et al., 1987–1988).

After 17 years significant changes can be observed in the nitrate-N accumulation. These were due partly to the crop rotation in the last five years (the nitrogen taken up by the cultivated plants during this period was less than earlier) and to the drought occurring during the growing seasons. The changes may be summarized as follows:

– more nitrogen was measured in the profiles down to 600 cm than at the time of the first drilling, i.e. with 2.9 mg kg^{-1} more (in the average of the 600 cm) after the yearly application of 100 kg N ha^{-1}, with 8.7 mg kg^{-1} more in the 200 kg ha^{-1} yr^{-1} treatment and with 12.5 mg kg^{-1} in the 300 kg N ha^{-1} yr^{-1} treatment. These values show - when we calculate with 1.4 kg dm^{-3} soil bulk density of

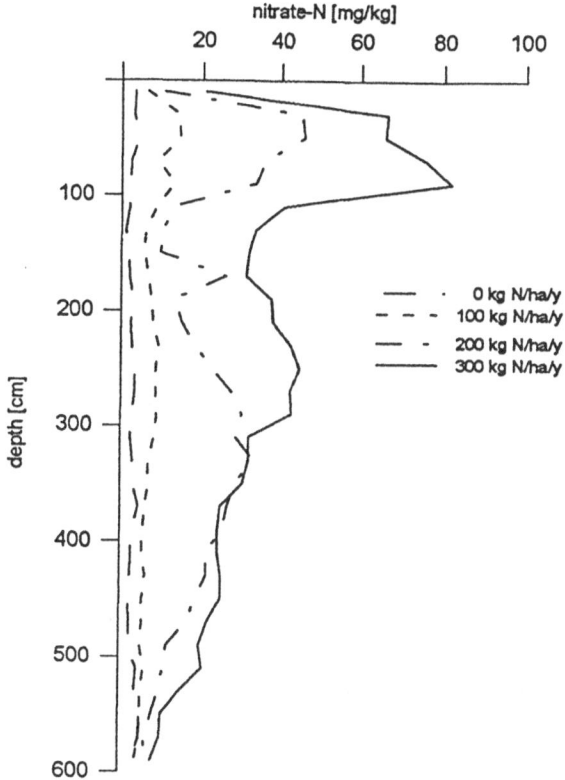

Figure 2. Nitrate-N accumulation in the soil profile (1990).

Table 1. Estimated N-balances (kg ha^{-1}) of the long-term fertilizer experiment

Items of the balance	Code of treatments			
	N_0	N_{100}	N_{200}	N_{300}
July, 1985 (after 12 experimental years)				
Added-N	–	1200	2400	3600
N taken up by crops	1318	1804	1941	2043
Balance	−1318	−604	459	1557
Difference to N_0		714	1777	2875
NO$_3$-N found in the soil				
(0–600 cm)[a]		41	664	1466
July, 1990 (after 17 experimental years)				
Added-N	–	1700	3400	5100
N taken up by crops	1557	2177	2301	2476
Balance	−1557	−477	1099	2623
Difference to N_0		1080	2656	4180
NO$_3$-N found in the soil				
(0–600 cm)[a]		382	1495	2613

[a] = Amount of nitrate-N in the 0–600 cm soil layer (difference to N_0).

the experimental site - that 244 kg more nitrogen accumulated during this 5-year-period in the profiles of the lowest nitrogen treatments than during the first 12 years, as well as 731 kg and 1050 kg more nitrogen in the other two treatments,

– a difference was found in the nitrate-N concentration between the unfertilized plots and the plots which received 100 kg N ha^{-1} yr^{-1} during the last five years (after 12 years these two concentration values were the same),

– the peak of the nitrate-N accumulation was not so well-defined than five years ago, the surplus nitrate-N reached the depth of 550 cm in the profiles of overfertilized plots.

In Table 1 the results of the nitrogen budgetary approach are summarized. On N-fertilized plots the average annual N uptake by the plants was about 150–160 kg N ha^{-1}. During the first 12 years the plants grown on the plots fertilized with 300 kg N ha^{-1} yr^{-1} took up only 240 kg more nitrogen (20 kg N yearly) then the plants which received only 100 kg N ha^{-1} yr^{-1}. The balance calculated only from the amounts of added fertilizer and nitrogen taken up by plants show a negative value at the lowest fertilization rate (100 kg N ha^{-1} yr^{-1}), after the yearly application of 200 kg N ha^{-1} the balance was positive with 459 kg N (38 kg N ha^{-1} yr^{-1}) and after fertilizing with 300 kg N ha^{-1} yr^{-1} with 1557 kg N ha^{-1} (130 kg N ha^{-1} yr^{-1}), respectively. If we take into account in all treatments, the same amount of nitrogen coming into the system from other sources (the amount of nitrogen taken up by plants on unfertilized plots indicates this value - 1318 kg N ha^{-1} 12 yr^{-1}) the balances become positive even in the lowest N-treatment. According to this calculation the yearly positive balances are 60 kg N ha^{-1} at the lowest N-fertilization treatment, 148 kg N ha^{-1} in the next N-treatment, and after yearly application of 300 kg N ha^{-1} the balance was 240 kg N ha^{-1}. As the amount of the nitrogen coming from other sources is decreasing with increasing fertilizer application, the amount of this nitrogen is probably between these two calculated values, for the lowest N-fertilization rate between 30 and 60 kg N ha^{-1} yr^{-1} (30 kg N ha^{-1} yr^{-1} because of the negative balance based only on the applied fertilizer dose and the nitrogen content of the cultivated plants), for the 200 kg N ha^{-1} yr^{-1} treatment between 0 and 148 kg N ha^{-1} yr^{-1}, while for the highest N-treatment between 0 and 240 kg N ha^{-1} yr^{-1}.

The balance calculation after 17 experimental years shows that during the last 5 years the cultivated plants took up yearly less nitrogen, then the previous crops (Kádár and Németh, 1993). This was due partly to the lower N-demand of the last five cultivated crops, and partly to the drought during this period. This feature can also be seen from the differences between the fertilizer added nitrogen versus nitrogen taken up by the plants. After 12 experimental years the yearly differences in the two positive balanced nitrogen treatments were 38 and 130 kg N ha^{-1} as mentioned above, while in average of the last five years these values went up to 128 and 213 kg N ha^{-1}, respectively.

The average yearly nitrogen uptake in the first 12 years in the lowest treatments was 150 kg N ha^{-1}, in the 200 kg N ha^{-1} yr^{-1} treatments 162 kg ha^{-1}, while in the 300 kg N ha^{-1} yr^{-1} treatments 170 kg ha^{-1}. The N-uptake in the average of the last five years was measured as 75, 72 and 87 kg N ha^{-1}, respectively. This lower nitrogen uptake caused that in the 100 kg N ha^{-1} yr^{-1} treatment the balance is less negative than 5 years ago, and the balances in the other two treatments became more positive. A great part of this 5-year surplus nitrogen can also be found in the soil profiles in the form of nitrate (Fig. 2).

Conclusions

The effects of N-fertilization on the amount of residual-N and on N-balances of different nitrogen treatments were investigated in a long-term fertilizer experiment. In those treatments where more nitrogen was applied than the N-demand of the plants a great part of the surplus-N had accumulated in the form of nitrate under the specific environmental conditions (migration type soil, deep groundwater table). In the profile of the highly overfertilized plots (300 kg N ha^{-1} yr^{-1}) as compared to the unfertilized plots 1466 kg more nitrogen

was found in the form of nitrate after 12 experimental years and 2613 kg more after 17 years. The results of balance calculations also show the great effects of N-uptake of crops and of the drought during the growing season on the amounts of the nitrogen remaining in the soil after harvest.

References

Csathó P 1994 NPK balances of the Hungarian soils in 1990 and in 1991. Növénytermelés 43, 551–561 (*In Hungarian*).

Goss M J, Beauchamp E G and Miller M H 1994 A farming system approach to minimizing nitrogen losses to the environment. *In* Transactions 15th World Congress of Soil Science, July 10–16, 1994, Acapulco, Mexico. Volume 5a, pp 123–137. International Society of Soil Science, Acapulco, Mexico.

Hill A R 1986 Nitrate and chloride distribution and balance under continuous potato cropping. Agric. Ecosys. Environ. 15, 267–280.

Kádár I 1987 Nutrient regime of arable land use. Növénytermelés 36, 517–526 (*In Hungarian with English Summary*).

Kádár I and Németh T 1993 Study on nitrate leaching in long-term fertilization trial. Növénytermelés 42, 331–338 (*In Hungarian with English summary*).

Lund L J 1982 Variations in nitrate and chloride concentrations below selected agricultural fields. Soil Sci. Soc. Am. J. 46, 1062–1066.

Németh T 1993 Fertilizer recommendations - Environmental aspects. Zesz. Probl. Postepow Nauk Roln. 400, 95–104.

Németh T 1994 Nitrate-N accumulations in the soil profiles of long-term fertilizer experiments. Agrokém. Talajtan 43, 231–238.

Németh T and Kádár I 1991 Macro- and micronutrients in Hungarian soils. *In* Cycling of Nutritive Element in Geo- and Biosphere. Ed. I Pais. pp 19–52. Univ. Horticult. Food Ind., Budapest, Hungary.

Németh T, Kovács G and Kádár I 1987–1988 Nitrate, sulphate and "water soluble salt" accumulation in the soil profiles of long-term fertilization experiment. Agrokém. Talajtan 36–37, 110–126 (*In Hungarian*).

Tanji K K, Fried M and Van De Pol R M 1977 A steady-state conceptual nitrogen model for estimating nitrogen emmissions from cropped lands. J. Environ. Qual. 6, 155–159.

O. Van Cleemput et al. (eds.), Progress in Nitrogen Cycling Studies, 347–351, 1996.

Nutrient management in integrated nursery stock production

C. Oele

Research Station for Nursery Stock, P.O. Box 118, 2770 AC Boskoop, the Netherlands

Key words: integrated production, nitrogen leaching, nursery stock

Abstract

Integrated production systems mean using low inputs of herbicides, pesticides and fertilizers to achieve emission levels that are environmentally acceptable, while ensuring good economic performance. Two prototype nurseries (one for forestry and hedging plants, the other for ornamental conifers and shrubs) were started in 1991 in the Netherlands, to see if these goals could be attained via crop rotation, mechanical weed control, integrated pest management, the use of nematode-suppressing green manures which can replace chemical soil disinfection, and integrated nutrient management. A strategy was developed to adjust fertilizer application to plant growth. The methods used were row fertilization and split applications, based on the actual nitrogen availability during the growing season as measured by soil sampling. One aspect of this management is fertilizing with manure. The organic matter can reduce nutrient leaching by increasing the buffer for nutrients and water in the soil. It was studied whether the chosen strategy reduced the loss of nutrients to acceptable levels. For three years the nitrogen balance for the two prototypes in the growing season was worked out and checked against the amount of available nitrogen. The research indicated that the nitrogen uptake of nursery stock is relatively low. Under laboratory conditions, high potential mineralization rates were measured. These two factors explain why nitrogen accumulated in the top soil. This available nitrogen subsequently leached out, either later - as measured in the dry summer of 1992 - or sooner - as measured in the wet summer of 1993. No leaching occurred in systems with high plant densities and in crops with high nitrogen demands. Concluded is that fertilizer use could be reduced by taking the amount of mineralized nitrogen into account. In nursery stock production, the loss of nutrients by leaching can be reduced to acceptable levels if organic matter low in available nutrients is applied.

Introduction

The need for knowledge on sustainable production methods that emerged in the mid 1980s in Dutch arable farming (Vereijken, 1990) has since been recognized in other branches of Dutch agriculture. Therefore, in 1991, two experimental integrated nurseries were started for the nursery stock industry; one for field-grown forestry and hedging plants, and one for ornamental conifers and shrubs (Dolmans, 1992). In these production systems, low inputs of fertilizers, herbicides and pesticides are used, to keep emissions to environmentally acceptable levels without adversely affecting profitability.

Integrated production involves crop rotation, mechanical weed control, integrated pest management, the use of nematode-suppressing green manures (*Tagetes* spp.) which can replace chemical soil disin-

fection, and integrated nutrient management. Nursery stock requires high inputs of organic matter, to maintain plant quality. If plants with root balls are to be produced, the supply of organic matter needs extra attention. Nowadays, nurseries use large amounts of manure to achieve the desired organic matter content. This practice has resulted in many nursery soils becoming phosphate-saturated. More than 70% of Dutch nurseries have been found to have over 60 mg P_2O_5 (per 100 g dry soil, ammonium-lactate acitric acid extraction) in the top layer of soil which suggests that phosphate leaching occurs. Nitrogen leaching may also be a problem. Nevertheless, the official Dutch recommendation for fertilizing nursery stock is to apply 100 kg N ha^{-1} yr^{-1} before the growing season.

In integrated nutrient management, high inputs of fertilizer must be reduced, so leaching can be avoided. The fertilizer practices in this management are to use manure and fertilizer on the basis of soil sampling. The study reported below examined whether the two prototype integrated nurseries could match the EC standards for groundwater quality and whether they showed changes in soil fertility parameters in the period 1991 to 1993. The EC standards for nitrogen and phosphorus are, respectively, less than 50 mg NO_3 L^{-1} and less than 0.16 mg P L^{-1} in the groundwater. The relationship between nutrient content in the top layer of soil and that in the underlying groundwater is not yet clear. Therefore in this research the limit for nitrogen adopted was that residual nitrogen in the uppermost 30 cm of soil must not exceed 45 kg N ha^{-1} at the end of the growing season.

Materials and methods

Prototype nurseries

On prototype nursery A, crops are cultivated for forestry and hedging, and for rose rootstocks. The nursery is situated in the north of the Netherlands and it has an area of 1.5 ha divided into 24 plots. The soil type is described as a black 'enk' eerd soil (de Bakker and Schelling, 1966) or reclaimed peat, with typical properties of high organic matter content (6–10%) and high sand fraction. The quality of the (old) organic matter seems poor, with a high C/N ratio of 18. The water table is at a depth of 1.5 m.

In the prototype A nursery, the crops are rotated every year. The plant species are divided into four main rotation groups. The fifth group, *Tagetes erecta* is used as a nematode-suppressing crop and as a green manure. The main groups are rose rootstocks, other Rosaceae, sown forestry and hedging plants, and transplanted forestry and hedging plants. The plant density depends on the kind of crop and germination in the field of the seed, but is more than 100,000 plants ha^{-1}.

On prototype B, ornamental shrubs and conifers are grown on 1.0 ha divided into 24 plots. The nursery is situated in the south of the Netherlands. The soil type is a brownish 'enk' eerd, a sandy soil low in organic matter content (up to 3%) but with a C/N ratio of 14. The water table is at a depth of 4 m. *Tagetes* spp. are also grown as a green manure. The crops are planted for two years, at densities of 40, 000 to 66, 700 plants per hectare. The main rotation groups are *Rosaceae*, evergreen shrubs, deciduous shrubs and three groups of conifers.

Fertilizer practice

The basis for integrated nutrient management is the use of manure. Because the phosphate nutrient status of both nurseries is high, the organic matter supplied must have a low nutrient content but a high percentage of organic matter. Therefore, compost was applied at a maximum of 6 t ha^{-1} yr^{-1} dry weight. This maximum amount is set by the Dutch government on the basis of heavy metal content. The compost originated from household waste (Vegetable, Fruit and Garden waste: VFG compost). After controlled processing, this material has an organic matter content of about 35%. The total input of nutrients from 6 t VFG was 70 kg N and 40 kg phosphate per hectare. The nutrient availability is low (Van Lune and Hassink, 1991). In addition to the nutrient management, artificial fertilizer was applied, based on soil sampling. A nitrogen content of 70 kg N ha^{-1} in the top 30 cm of soil was assumed to be sufficient for plant growth. If the content was above this value, no fertilizer was applied. The soil sampling date for nitrogen fertilizing was at the end of May. Sampling was repeated at the beginning of July, to see if the application had been sufficient. If it was found to be insufficient, fertilizer was applied again, using row application.

Soil fertility

During the three-year research period, parameters indicating changes in soil fertility were analysed in samples of the top 30 cm of soil. In both nurseries, samples were taken from all plots, twice a year. The soil parameters analysed were pH-KCl, percentage organic matter, elemental C, P-Al and P-water and total N. The samples were analysed according to standard methods of the Industrial Laboratory for Ground and Crop Analysis.

Nitrogen balance study

The recorded inputs of the N balance in this study were fertilizer gifts in the manure, artificial fertilizer, input of nitrogen with irrigation water, and the N mineralization of the soil. The outputs were plant uptake and leaching of N into the deeper soil layers. During the growing season, the available N was measured in the uppermost 30 cm of soil six times. Other inputs/outputs such as deposition, nitrification and denitrification were considered to be of minor importance in this study. The uptake was measured as the amount of nitrogen at end of the growing season, minus the amount of nitrogen at the outset, using whole-plant analysis for N-Kjeldal by destroying five plants per plot. The potential mineralization was measured by incubating soil samples at 21 °C. These values were corrected for soil temperature measured on the nursery, to estimate the amount of mineralized N in the field. N leaching was determined by sampling the horizons below the rooted area at the end of the growing season. In addition, the water balance was constructed with the SWATRE model (Belmans et al., 1983), to reveal information about the moment of leaching.

Results

Changes in soil fertility

The soil parameters were monitored for three years. The parameters P-water, P-Al, elemental C and total N did not change statistically significantly. The parameters which showed relevant changes are presented in Table 1. Both prototypes A and B showed a small, but significant ($p < 0.05$) decrease in organic matter content and a significant ($p < 0.05$) increase in soil pH from 1991 to 1993. The decrease in soil organic matter was greater on nursery A than on nursery B.

Available nitrogen

The three years monitored showed the same trend in development of available N in the uppermost 30 cm of soil. The values in 1993 are presented in Table 2a for prototype A and in Table 2b for prototype B. At the start of the growing season the available N was low: less than 10 kg N ha^{-1}. As a result of mineralization, in 1993 the amount of available N increased during the season until June. The measured potential mineralized nitrogen per 100 g dry soil during 90 days was 7 mg in prototype A and 5 mg in prototype B. Given a dry bulk density of 1.3 kg L^{-1}, this means that in the first 15 cm of soil the potential mineralization at 21 °C was 1.5 kg ha^{-1} day^{-1} in prototype A and 1.1 kg ha^{-1} in prototype B. The net mineralized nitrogen until July, after temperature correction, was estimated to be 135 kg N ha^{-1} in prototype A and 99 kg N ha^{-1} in prototype B. Fertilizer N was applied at the start of June at rates of 55 kg N ha^{-1} for rose seedlings and 75 kg N ha^{-1} for transplanted crops and *Tagetes erecta*. On prototype B the inputs of fertilizer were 0 kg N ha^{-1} for the first-year crops and 50 kg N ha^{-1} for the second-year crops and *Tagetes*. The estimated amount of mineralized N

on prototype B is less than the value for available N measured in June. The large amounts of available N and their subsequent decline can be attributed to plant uptake or leaching, as will be explained below.

Plant uptake

Data on plant uptake during the research period are given in Table 3. The values are means of different species in a group with more or less the same uptake and circumstances, but is rather arbitrary. On prototype A, the values for rose rootstocks are means of *Rosa multiflora*, *Rosa canina* 'Inermis' and *Rosa corymbifera* 'Laxa' within the range of 95 to 102 kg N ha^{-1}. Similar uptake was achieved by sown forestry and hedging plants of the species *Quercus robur*, *Fagus sylvatica*, *Carpinus betulus* and *Malus* within the range of 71–124 kg N ha^{-1}, the highest being for *Quercus robur*. In forestry species sown on beds or when seeds have a poor germination the uptake may also be low: for *Alnus glutinosa*, *Betula pubescens*, *Robinia pseudoacacia* and *Crataegus monogyna*, the mean uptake was 31 kg N in the range of 17 to 40 kg N ha^{-1}. The group of transplanted forestry and hedging plants also includes very demanding crops, such as *Prunus* 'St. Juliën', *Betula pubescens*, *Quercus robur* and *Fraxinus excelsior*, ranging between 85 and 163 kg N ha^{-1}. The uptake of second-year *Quercus robur* was 156 kg N ha^{-1}. Species with low uptake are *Pinus sylvestrus*, *Fagus sylvatica* and *Betula pubescens*, with a mean uptake of 23 kg N ha^{-1}. The uptake of the green manures *Tagetes* 'Nemanon' (350 kg N ha^{-1}) and *Tagetes erecta* (279 kg N ha^{-1}) was high.

In prototype B a group of conifers with a low nitrogen demand in their first year of growth can be distinguished: *Juniperus communis* 'Repanda', *Picea omorika*, *Abies koreana*, *Taxus media* 'Hicksii' and *Pinus mugo mughus*, with an uptake of 34 to 52 kg N ha^{-1}. But in the second year of growth their uptake doubled, with a mean of 88 kg N ha^{-1} in a range between 60 kg and 105 kg N ha^{-1}. The crops *Chamaecyparis lawsoniana* 'Columnaris', *Thuja occidentalis* 'Brabant' and *Thuja plicata* 'Excelsa' have a higher demand: between 53 and 64 kg N ha^{-1} in the first year and between 132 and 175 kg N ha^{-1} in the second year. The evergreen shrubs are grouped together with the crops *Mahonia aquifolium* 'Apollo' and *Viburnum davidii*. Their uptake was 32 and 44 kg N ha^{-1} in the first year and 78 to 95 kg N ha^{-1} in the second. *Prunus laurocerasus* 'Otto Luyken' demanded much more nitrogen: 61 kg N ha^{-1} in its first year and 163 kg

Table 1. Changes in soil parameters on integrated nursery
prototypes A and B in the period 1991–1993 (n=48)

	Year	% Organic matter	pH-KCl
Prototype A	1991	6.44a	4.77ab
	1992	6.02b	4.74a
	1993	5.87b	4.78b
Prototype B	1991	2.51a	4.79a
	1992	2.45b	4.81b
	1993	2.37c	4.93b

Table 2. Available nitrogen (kg N ha^{-1}, top 30 cm of soil) measured on Prototype A (**a**) and Prototype
B (**b**) in 1993

	Transplanted		Seedlings		Tagetes
	High uptake	Low uptake	High uptake	Low uptake	
a.					
April	7	7	7	7	7
May	37	36	60	53	41
June	52	51	35	57	40
July	59[a]	101[a]	70[a]	86[a]	122[a]
August	8	17	17	13	0
September	5	8	8	4	0
b.	First-year crops	Second-year crops	Tagetes		
April	5	5	5		
May	77	23	37		
June	148	200[a]	112[a]		
July	122	126	75		
August	15	20	5		
September	5	5	5		

[a]Including fertilizer gift.

N ha^{-1} in its second. Deciduous crops were also distinguishable into low-demanding crops (*Malus* 'Golden
Hornet') and crops with high uptake (*Prunus serrula-
ta* 'Kanzan' and *Rosa* 'Queen Elizabeth'). The green
manure used in prototype B, *Tagetes patula*, demanded
190 kg N ha^{-1}.

Nitrogen leaching

In Table 4 the amounts of available (dissolved) N in
different soil layers of prototype B in 1993 are presented. Leaching was expected only on prototype B,
because on nursery A there was no excess of available
N was measured except in the plots of the transplanted
species were uptake was low. The results in 1991 and
1992 were similar; only the time of leaching differed.
As noted in Table 2a and 2b, the level of available

nitrogen increased until July because mineralization
and fertilizer application exceeded plant uptake. In
July the rainfall was 150 mm, so leaching could be
expected. Indeed, as shown in Table 4, in August all
the available N in the first 30 cm of soil disappeared.
Assuming that the amounts of N in the 30–60 cm and
60–90 cm layers of soil, the are lost for plant uptake by
leaching, as is measured in case of the nursery crops.
Less nitrogen disappeared under the second-year crops
than under the first-year crops. No leaching occurred
under the *Tagetes*.

Discussion

It can be concluded that in general the plant uptake
of nursery stock is low, compared with other crops.

Table 3. Plant uptake (kg N ha^{-1}) in nursery stock production

	Prototype	First year	Second year
Rose rootstocks	A	101	
Sown forestry plants - high uptake	A	91	
Sown forestry plants - low uptake	A	31	
Transplanted forestry plants - high	A	116	156
Transplanted forestry plants - low	A	23	
Conifers - low uptake	B	43	88
Conifers - high uptake	B	59	159
Evergreen shrubs - low uptake	B	38	93
Evergreen shrubs - high uptake	B	61	163
Deciduous crops - low uptake	B	20	70
Deciduous crops - high uptake	B	101	140

Table 4. Amount of available nitrogen (kg N ha^{-1}) in August 1993, prototype B

Depth	First-year crops	Second-year crops	*Tagetes*
0–30	15	20	5
30–60	63	54	6
60–90	108	45	6

The nitrogen uptake differs widely between the cultivated species, as Alt has demonstrated (Alt, 1990), because of the difference in species, age and plant density. When plant requirements are low, fertilizer input can be minimal, as is the case in the integrated nurseries. But if the net mineralized nitrogen exceeds plant uptake, nitrogen may be lost (prototype B). The EC standards will then be exceeded. If the grower does not know the plant uptake, the method of fertilizer application based on soil sampling seems to be satisfactory. The amount of nitrogen mineralized will be taken in account, not too much nitrogen is available or is susceptible to leaching, and plant growth does not suffer from a nitrogen deficit.

The decrease in organic matter and therefore the higher mineralization cannot be explained easily. The use of VFG compost may be responsible for the higher decomposition rate, because of its higher pH. Other reasons for the decrease might be the change to another equilibrium, or the integrated farming practice itself. Practices such as mechanical weed control and crop rotation can stimulate soil microbiological processes, thereby increasing the rate of organic matter decomposition. A longer research period is needed to evaluate structural changes in organic matter (Van Faassen and Lebbink, 1990). From the findings presented above it can be inferred that it is premature to draw conclusions about changes in soil fertility under integrated management in nursery stock. However, it is clear that more inputs of organic matter are needed, especially in prototype B, where organic matter is removed with the root balls of lifted crops. The correct choice of organic fertilizers and careful decisions about fertilizer application techniques allow integrated nursery stock production to meet standards for environmental emissions, but more needs to be known about the behaviour of organic matter in the soil.

References

Alt D 1990 Düngen in der Baumschule. Taspo praxis. Verlag B, Thalacker.

De Bakker H and Schelling J 1966 System for soil classification in the Netherlands. Soil Survey Institute, Wageningen, The Netherlands.

Belmans C, Wesseling W G and Feddes R A 1983 Simulation model of the water balance of a cropped soil: SWATRE. J. Hydro. 63, 271–286.

Dolmans N G M 1992 Integrated nursery stock production. Neth. J. Agric. Sci. 40, 269–275.

Van Lune P and Hassink J 1991 Higher potato yield and less diseases with VAM-VFG compost. VAM-Mededelingen 2, 12–15.

Van Faassen H G and Lebbink G 1990 Nitrogen cycling in arable farming systems. Neth. J. Agric. Sci. 38, 265–282.

Vereijken P 1990 Integrated nutrient management for arable farming. Rech. Agron. Suisse 29, 359–365.

O. Van Cleemput et al. (eds.), Progress in Nitrogen Cycling Studies, 353–358, 1996.

Optimization of a nitrogen advice system: Target values as a function of N-mineralization rates

J. Pannier[1], G. Hofman[1] and L. Vanparys[2]

[1]*University of Ghent, Faculty of Agricultural and Applied Biological Sciences, Department of Soil Management and Soil Care, Coupure Links 653, B-9000 Gent, Belgium and* [2]*Centre for Agricultural and Horticultural Research, Ieperseweg 87, B-8800 Roeselare, Belgium*

Key words: latent N_{min}-residue, nitrogen advice system, nitrogen mineralization, nitrogen uptake pattern, vegetables

Abstract

The proposed N advice system for cauliflower, leek and Brussels sprouts is the Dutch NBS system, derived from the German KNS (Kulturbegleitendes N_{min}-Sollwerte)-system. In the KNS-system, the mineral nitrogen content in the soil profile is measured at fixed intervals during the growing season and compared with target values. These target values depend on the crop's nitrogen uptake pattern and the latent N_{min}-residue in the soil profile. The result is a fractionated nitrogen fertilization. However, in its present form this system is not quite adequate because it incorporates the nitrogen mineralization solely in retrospect. The system does not anticipate nitrogen mineralization occurring till the next measurement date or harvest. This may lead to a sub-optimal nitrogen use, especially for crops with a long growing season. Therefore, a correction of this N advice system where the target values also depend on the estimated nitrogen mineralization for the coming interval is suggested. In co-operation with the Centre for Agricultural and Horticultural Research (Roeselare), nitrogen fertilizer trials were carried out with cauliflower, leek and Brussels sprouts. For these crops, the target values are presented as a function of nitrogen mineralization rates.

Introduction

An adequate nitrogen advice system is an important tool for a cost-effective and environmentally sound farming system. N fertilizer recommendations should aim at bridging the gap between the nitrogen demand of the crop and what the soil is able to provide during any given period of the growing season. This requires knowledge on both the nitrogen uptake pattern of the crop and the nitrogen dynamics in the soil profile. An example of such a nitrogen advice system is the Dutch NBS (N-bijmestsysteem)-system (Gröninger and Soorsma, 1991) derived from the German KNS (Kulturbegleitendes N_{min}-Sollwerte)-system (Lorenz et al., 1985; Scharpf and Weier, 1988). It involves the measurement of the residual mineral nitrogen in the profile at fixed intervals during the growing season and the comparison with target values. These target values depend on the nitrogen uptake pattern of the crop and the latent N_{min}-residue (Hofman, 1983; Hofman et al., 1984) in the soil profile. If needful, soil nitrogen reserves are replenished up to the target val-

ue. A shortcoming of the system in its present form is that nitrogen mineralization occurring up to the next measurement date or harvest is not taken into account. Especially for crops with a long growing season this may lead to suboptimal nitrogen use and consequently to higher mineral N-residues at harvest.

The present study aimed at deriving a NBS-system for cauliflower, leek and Brussels sprouts based on nitrogen fertilization experiments. Target values will be corrected depending on the nitrogen mineralization rate as gained from earlier field trials.

Materials and methods

Between 1990 and 1993 nitrogen fertilization experiments with cauliflower, leek and Brussels sprouts were carried out in the sandy loam area of Belgium. A summary of the different field experiments is given in Table 1. The crops were grown according to common practice. Before planting, the experimental plots received 0, 50, 100, 200 or 300 kg N ha^{-1} (NH_4NO_3, 27% N) in

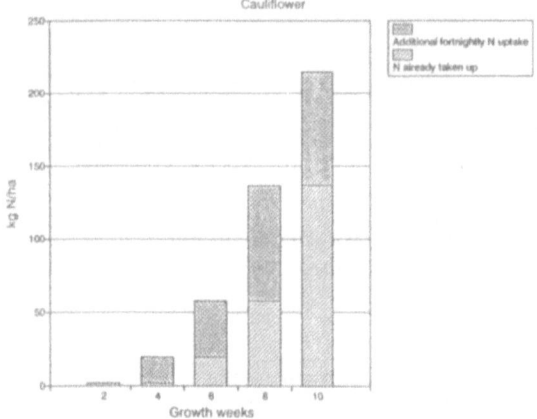

Figure 1. N uptake pattern for cauliflower.

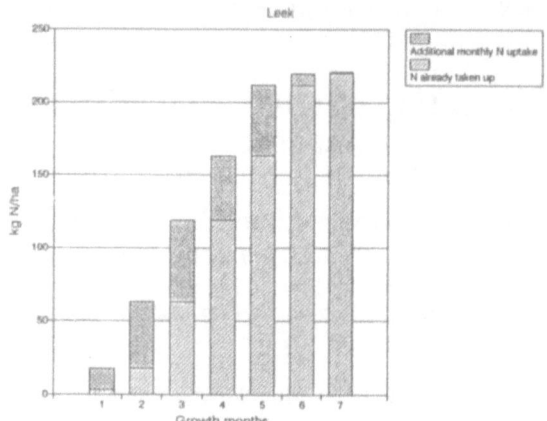

Figure 2. N uptake pattern for leek.

4 replicates. Plant growth and nitrogen uptake pattern were measured for the current application rate of 200 kg N ha^{-1}. The crops were harvested every two weeks (cauliflower) or monthly (leek, sprouts). At the same time, soil samples were taken every 30 cm down to 120 cm depth.

For the upper 30 cm, nitrogen ($NO_3^- + NH_4^+$) was extracted with 1 N KCl from the moist soil samples and measured colorimetrically by a continuous flow auto analyzing system. Concerning the other layers, nitrogen (NO_3^-) was extracted with 1% $KAl(SO_4)_2$ and measured potentiometrically with a nitrate specific electrode (Keeney and Nelson, 1982).

Fresh and dry weights were determined for the different plant parts (roots, stem and leaves). The plant parts were dried at 60 °C, weighed, ground and stored. Total nitrogen was measured with the Kjeldahl method (Bremner and Mulvaney, 1982).

Results

The presented results are mean values calculated from the results of two field experiments per crop.

Nitrogen uptake pattern

Cauliflower

Based upon the nitrogen demand of the crop two growth phases could be distinguished (Fig. 1). A first growth phase, covering the first six weeks, showed an increasing nitrogen demand. In this initial growth phase ca. 58 kg N ha^{-1} was needed. A second growth phase was linear with a constant nitrogen demand of ca. 5.5 kg ha^{-1} day^{-1}. The total nitrogen uptake at har-

vest was ca. 215 kg ha^{-1}. These results corresponded with a marketable produce of ca. 51 ton ha^{-1} (Table 2).

Leek

Three growth phases could be distinguished (Fig. 2). The first month after planting showed a modest demand for nitrogen (15 kg ha^{-1}). A second growth phase followed in which leek showed an almost constant nitrogen demand of ca. 1.5 kg ha^{-1} day^{-1}. During winter this demand decreased sharply and nitrogen uptake was near to insignificant (9 kg ha^{-1}). The total nitrogen uptake at harvest was ca. 220 kg ha^{-1}. These results corresponded with a marketable produce of ca. 45 ton ha^{-1} (Table 3).

Brussels sprouts

Three growth phases could be distinguished (Fig. 3). The first month after planting showed a small demand for nitrogen (39 kg ha^{-1}). The second month, a steep increase of accumulated nitrogen (120 kg ha^{-1}) was measured. A third phase consisted of an important but gradual decreasing nitrogen demand. The total nitrogen uptake at harvest was ca. 340 kg ha^{-1}. These results corresponded with a marketable produce of ca. 18 ton ha^{-1} (Table 4).

Latent N_{min}-residue

The latent N_{min}-residue is defined as the mineral nitrogen amount present in the soil profile at the moment when the crop reaches its maximum N-content. This residue is determined to the normal depth of the rooting system, under optimal growth circumstances and

Table 1. Summary of the different field experiments

Crop	Plant density	Planting date	Harvest date
Cauliflower	24700	21–04–1990	09–07–1990
		27–04–1991	17–07–1991
Leek	166700	11–07–1991	21–01–1992
		09–06–1992	28–01–1993
Brussels sprouts	39000	19–05–1992	10–11–1992
		12–05–1993	17–11–1993

Table 2. Mean marketable produce and N_{min}-residue at harvest of cauliflower as a function of the N application rate

Nitrogen fertilizer (kg ha^{-1})	Marketable produce (ton ha^{-1})	N_{min}-residue (0–60 cm) (kg ha^{-1})
0	46.0	107
50	48.3	174
100	48.8	210
200	51.3	270
300	50.9	335

with an adequate N fertilization (Hofman et al., 1984). It reflects in this way the efficiency of a crop to extract nitrogen from the soil, i.e. the needed N-amount in the rooting zone to ensure optimum N-transport to the roots (Wehrmann and Scharpf, 1986). An approximate value for the latent N_{min}-residue can be derived from the relation between crop yield and the amount of residual nitrogen present in the soil profile at harvest. As a reference depth 60 cm (cauliflower, leek) and 90 cm (sprouts) were chosen according to the mean rooting depth of the studied crops.

Except for the highest N demanding crop, e.g. sprouts, there were no significant differences in marketable crop production between the various N levels (Table 2–4). This can be explained by the regular use of large amounts of pig slurry, resulting in very high mineralization rates from this 'young' organic matter. Two consequences are linked with this observation:

1. Excessive nitrogen residues in the soil profile at harvest. For cauliflower, the N_{min}-residue increased gradually with the higher N levels. From literature (Böhmer et al., 1981) and own research the latent N_{min}-residue must be of the order of 75 kg ha^{-1}. The N_{min}-residues for winter leek were rather low already and the difference between the various N levels small. Drainage losses during winter are responsible for these results. A latent N_{min}-residue of about 50 kg ha^{-1} is accepted as a critical value for an optimal production (Gröninger and Soorsma, 1991). For sprouts, it is clear that the latent N_{min}-residue is low and reaches about 20 kg ha^{-1};

2. The need for the incorporation of the expected N mineralization into the N advice.

Nitrogen mineralization and target values

The observed N uptake pattern and latent N_{min}-residue were used to formulate a nitrogen fertilization advice based upon the NBS-system. Therefore the growing season was divided into two intervals with different nitrogen requirements. Just before each interval the mineral nitrogen reserve in the soil profile has to be

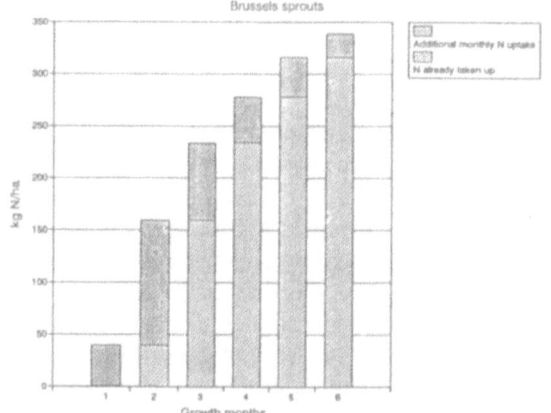

Figure 3. N uptake pattern for Brussels sprouts.

356

Table 3. Mean marketable produce and N_{min}-residue at harvest of leek as a function of the N application rate

Nitrogen fertilizer (kg ha^{-1})	Marketable produce (ton ha^{-1})	N_{min}-residue (0–60 cm) (kg ha^{-1})
0	45.2	55
50	45.3	55
100	45.4	66
200	44.4	85
300	45.3	86

Table 4. Mean marketable produce and N_{min}-residue at harvest of sprouts as a function of the N application rate

Nitrogen fertilizer (kg ha^{-1})	Marketable produce (ton ha^{-1})	N_{min}-residue (0–90 cm) (kg ha^{-1})
0	11.0	17
50	14.1	16
100	16.0	17
200	18.0	23
300	17.8	20

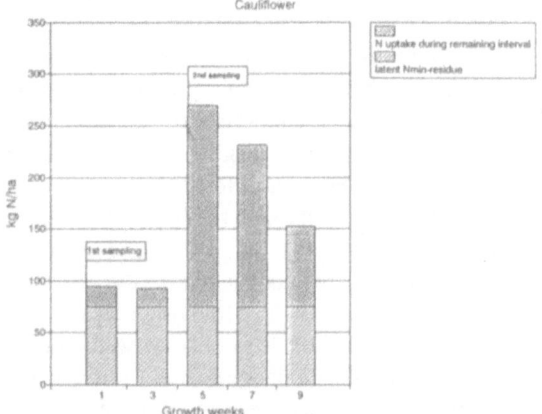

Figure 4. N target values for cauliflower.

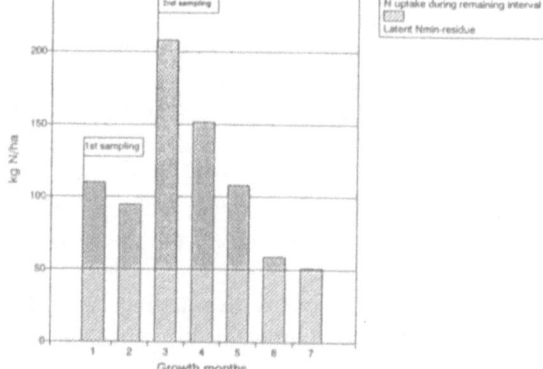

Figure 5. N target values for leek.

Table 5. Target values as a function of N mineralization rate for cauliflower

Nitrogen mineralization (kg ha^{-1} day^{-1})	Target value (kg N ha^{-1}) 1st sampling	2nd sampling
0.6	78	245
0.8	75	236
1.0	75	228
1.2	75	220
1.4	75	211
1.6	75	203
NBS-system	95	270

determined. This result has to be compared with the target values which can be read from the graphs (Figs. 4–6). The target value is the necessary mineral nitrogen reserve in the soil profile at that particular point of time in view to reach maximum marketable yield. It is the sum of the latent N_{min}-residue and the nitrogen uptake rate during the next interval. To reach the target value the mineral nitrogen reserve in the soil profile has to be replenished with the appropriate amount. Measuring dates should be fixed at the start of the growing season and associated target values calculated.

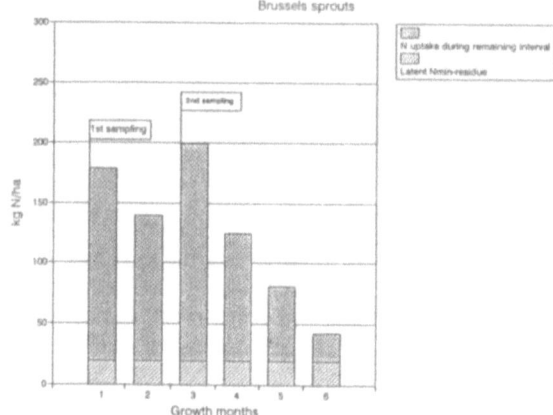

Figure 6. N target values for Brussels sprouts.

Table 6. Target values as a function of N mineralization rate for leek

Nitrogen mineralization (kg ha^{-1} day^{-1})	Target value (kg N ha^{-1})	
	1st sampling	2nd sampling[a]
0.6	74	138
0.8	62	114
1.0	50	90
1.2	50	66
1.4	50	50
1.6	50	50
NBS-system	110	210

[a] From the 6th growth month (November) onwards the N mineralization rate was halved.

Cauliflower

The growing season (10 weeks) is divided into two periods: soil samples have to be taken before planting and after the first month. Depending on the rooting depth the measuring depth should be 30 cm before planting and 60 cm for the second sampling. The latent N_{min}-residue is set at 75 kg N ha^{-1}. Target values are 95 kg N ha^{-1} before planting and 270 kg N ha^{-1} at the beginning of the second month (Fig. 4).

Leek

The growing season (7 months) is divided into two periods: soil samples have to be taken before planting and after the second month. The measuring depth should be 30 cm before planting and 60 cm for the second sampling. The latent N_{min}-residue is set at 50 kg N ha^{-1}. Target values are 110 kg N ha^{-1} before planting and 210 kg N ha^{-1} at the beginning of the third month (Fig. 5).

Table 7. Target values as a function of N mineralization rate for sprouts

Nitrogen mineralization (kg ha^{-1} day^{-1})	Target value (kg N ha^{-1})	
	1st sampling	2nd sampling
0.6	144	128
0.8	132	104
1.0	120	80
1.2	108	56
1.4	96	32
1.6	84	20
NBS-system	180	200

Brussels sprouts

The growing season (6 months) is divided into two periods: soil samples have to be taken before planting and after the second month. The measuring depth should be 30 cm before planting and 90 cm for the second sampling. The latent N_{min}-residue is set at 20 kg N ha^{-1}. Target values are 180 kg N ha^{-1} before planting and 200 kg N ha^{-1} at the beginning of the third month (Fig. 6).

Discussion

The target values as calculated according to the NBS-system, only take into account the nitrogen mineralization of the previous interval. As was shown for cauliflower this can lead to a less than optimal nitrogen use. The efficiency of the method can be improved if the expected nitrogen mineralization for the coming interval is incorporated in the calculation of the target values. To illustrate this, corrected target values are presented for the studied crops as a function of nitrogen mineralization rates for the coming interval (Tables 5–7). The range of N mineralization rates was assessed through past field trials (higher mineralization rates reflect frequent amendments of organic manure) (Demyttenaere, 1991). Comparison of corrected and uncorrected target values shows that an important reduction of nitrogen input can thus be achieved. A remark is made on the latent N_{min}-residue which is the minimum nitrogen reserve that has to be maintained during the growing season in order to maximize production. In some cases, e.g. leek, this may still cause some nitrogen surplus in the soil profile at harvest, depending on the N mineralization rate.

358

Conclusion

By way of the classical NBS-system a first attempt was made to optimize the N_{min}-method fertilization regime, which is normally used for vegetables. A new approach to this NBS-system was made by taking into account the N-mineralization during the period after the sampling date. This can lead to a more optimal use of fertilizers and a lesser loss of nitrogen to the environment. Though promising, further field trials should be laid out in view of evaluating the corrected NBS-system and gain more experience with this method.

Acknowledgements

Financial support by IWONL (Institute for encouraging Scientific Research in Industry and Agriculture, Brussels) is gratefully acknowledged.

References

Böhmer M, Wiebe H and Wehrmann J 1981 Zur Stickstoffdüngung bei Blumenkohl. Gemüse 17, 44–47.

Bremner J M and Mulvaney C S 1982 Nitrogen - total. *In* Methods of Soil Analysis, part 2, Agronomy 9. Eds. A L Page, R H Miller and D R Keeney. pp 595–624. American Society of Agronomy, Inc., Soil Science Society of America, Inc., Madison, WI, USA.

Demyttenaere P 1991 Stikstofdynamiek in de bodems van de Westvlaamse groentestreek. Doctoraal proefschrift, Faculteit van de Landbouwwetenschappen, Rijksuniversiteit Gent, Belgium. 203 p.

Gröninger H and Soorsma H E 1991 Stikstofbijmestsysteem (NBS) voor enige vollegrondsgroentegewassen. IKC-agv, Lelystad, The Netherlands. 86 p.

Hofman G 1983 Minerale stikstofevolutie in zandleemprofielen. Aggregaat voor het Hoger Onderwijs, Faculteit van de Landbouwwetenschappen, Rijksuniversiteit Gent, Belgium. 183 p.

Hofman G, Ossemerct C, Ide G and Van Ruymbeke M 1984 Significance of the latent mineral N-residue in the soil profile in nitrogen fertilization advices. *In* Fight against Hunger through improved Plant Nutrition, 9th World Fertilizer Congress Proceedings, Budapest, Hungary, Vol. 2. Eds. E Welte and I Szabolcs. pp 225–229. CIEC and Hungarian Academy of Sciences, Goeltingen.

Keeney D R and Nelson D W 1982 Nitrogen - inorganic forms. *In* Methods of Soil Analysis, part 2, Agronomy 9. Eds. A L Page, R H Miller and D R Keeney. pp 643–698. American Society of Agronomy, Inc., Soil Science Society of America, Inc., Madison, WI, USA.

Lorenz H, Schlaghecken J and Engl G 1985 Gezielten Stickstoffversorgung - das kulturbegleitende N_{min}-Sollwerte-System (KNS-System). Dtsch. Gartenbau 13, 646–648.

Scharpf H C and Weier U 1988 Abgestimmte Sollwerte für die Stickstoffdüngung. Gemüse, 4–5.

Wehrmann J and Scharpf H C 1986 The N_{min}-method - an aid to integrating various objectives of nitrogen fertilization. Z. Pflanzenernaehr. Bodenkd. 149, 428–440.

O. Van Cleemput et al. (eds.), Progress in Nitrogen Cycling Studies, 359–363, 1996.
© 1996 *Kluwer Academic Publishers.*

Potential nitrogen availability and fertiliser recommendations

Robert M. Rees, Iain P. McTaggart and Keith A. Smith
Soil Science Department, SAC, School of Agriculture, West Mains Road, Edinburgh EH9 3JG, UK

Key words: available-N, cereals, fertiliser-N, nitrogen uptake

Abstract

In a series of trials, uptake of N derived from soil by cereals was compared with potentially available N in the soil, as estimated by boiling with 2 M KCl for four hours and measuring the ammonium-N released. In previous work, between 1988 and 1990, nitrogen fertilisers were applied to spring barley at recommended rates, and microplots using ^{15}N were used to discriminate between soil-derived and fertiliser-derived nitrogen. In 1991 and 1992 the trials were extended to include winter cereals (barley and wheat) with different rates of fertiliser. Where fertiliser additions were reduced below the normal recommended additions, no reductions in yield or N content of the cereal grain were observed. The uptake of soil-derived nitrogen by the grain in spring barley varied between 34 and 108 kg ha^{-1}. It was correlated significantly with potentially available N over the four trial years (r=0.56; $p<0.01$). The uptake of soil-derived N by winter cereals was found not to correlate with potentially available N. When sites with a high soil organic matter content were excluded, the correlation between soil N uptake and soil organic-carbon was significant between 1988 and 1990 (r=0.96; $p<0.001$); however, this relationship was not apparent in 1991 and 1992 at sites with a wider range of organic-carbon contents and management histories.

Introduction

Studies using ^{15}N-labelled fertilisers applied to cereal crops in northern Europe have shown a wide variation in the supply of soil-derived N (Powlson et al., 1986; Recous et al., 1988; Smith et al., 1984). On the basis of present knowledge, it is not possible to identify in advance the amount of nitrogen that will be released by a particular soil. Such information would be useful in allowing fertiliser applications to be targeted more precisely, thereby reducing losses to the environment and improving the quality of specialist crops such as malting barley. The purpose of this study was to evaluate the use of a simple chemical index, previously developed by McTaggart and Smith (1993), to adjust standard fertiliser-N recommendations.

Materials and methods

Nitrogen uptake by spring barley was studied between 1988 and 1990 at 10 experimental sites, in work described by McTaggart and Smith (1993). In 1991 and 1992 a further 11 sites from across the arable regions

of Scotland were chosen, with the inclusion of winter wheat and winter barley at some sites. Details of the sites and the crops grown are given in Table 1.

In 1991, NH$_4$NO$_3$ was applied to all crops in three replicate plots at each site, at recommended rates (SAC, 1990). Labelled fertiliser N was applied to microplots to discriminate between soil- and fertiliser-derived N in the crop. Soil was sampled to a depth of 200 mm in December 1990 and March 1991 at all sites, and analysed for potentially available N according to the method of McTaggart and Smith (1993), based on earlier work by Gianello and Bremner (1986). Briefly, this involved measuring the amount of NH$_4^+$-N released after boiling 12 g of soil in 80 mL of 2 M KCl for four hours. At harvest, dry matter production and N uptake by the grain were measured. The isotope ratio of the N in plant samples was determined using a VG Isogas MM622 mass spectrometer, and the results used to calculate soil and fertiliser derived N, as described by McTaggart and Smith (1993).

In 1992, NH$_4$NO$_3$ was again applied at recommended rates, but in addition rates were adjusted to take account of the potentially mineralisable N at each site. This was achieved by first calculating the differ-

Table 1. Details of sites and treatments used for studies of N uptake in 1991 and 1992

Site/Crop	Year	Soil texture	Organic-C (g kg^{-1})	pH	Fertiliser-N application (kg N ha^{-1})
Whitehouse					
Spring barley	1991	Sandy clay loam	23	6.3	130
Winter barley	1991	Sandy clay loam	23	6.3	180
Winter wheat	1991	Sandy clay loam	23	6.3	200
Bush, Fulford					
Spring barley	1991	Loam	22	6.5	120
Winter barley	1991	Loam	22	5.7	180
Winter wheat	1991	Loam	22	5.7	230
Treaton					
Spring barley	1991	Sandy loam	33	6.3	130
Winter barley	1991	Sandy loam	33	6.3	180
Winter wheat	1991	Sandy loam	33	6.3	200
South Road					
Winter barley	1991	Loam	33		120
Crosshall					
Spring barley	1992	Sandy clay loam	23	5.9	120[a]
Winter barley	1992	Sandy clay loam	23	6.0	180[a]
Winter Wheat	1992	Sandy clay loam	23	6.0	160[a]
Bush, Cowloan					
Spring barley	1992	Loam	43	5.7	120[a]
Winter barley	1992	Loam	43	5.7	180[a]
Winter Wheat	1992	Loam	43	5.7	200[a]
Bush, Crofts					
Spring barley	1992	Clay loam	27	5.7	120[a]
Winter barley	1992	Clay loam	27	5.7	180[a]
Winter Wheat	1992	Clay loam	27	5.7	200[a]
Panlathy					
Spring barley	1992	Sandy clay loam	18	6.4	110[a]
Winter Wheat	1992	Sandy clay loam	18	6.4	200[a]
Aldrochty					
Spring barley	1992	Sandy loam	19	6.0	120[a]
Craibstone Bed C					
Spring barley	1992	Loam	12	6.2	90[a]
Craibstone Sunnybrae					
Spring barley	1992	Sandy loam	18	6.2	60[a]

[a]Nitrogen applications also adjusted to account for mineralisable-N (see Table 2).

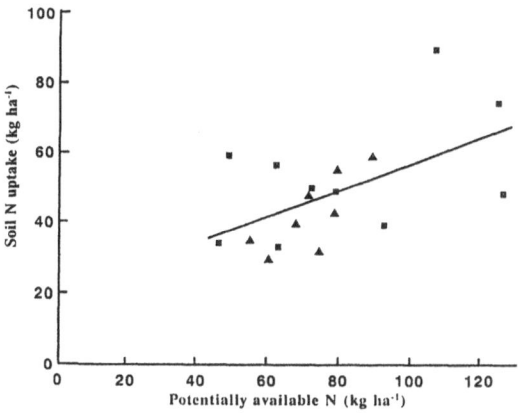

Figure 1. The relationship between potentially available nitrogen and soil nitrogen uptake by the grain of spring barley. The data from 1988–1990 are represented by triangles, and the data from 1990 and 1991 are represented by squares. Regression line y = 19.65 + 0.38x, r = 0.56, $p < 0.05$.

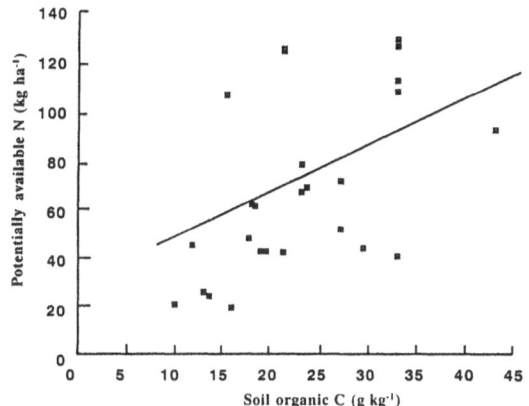

Figure 2. The relationship between soil organic-C and potentially available-N for all sites studies between 1988–1992. Regression line y = 46.4 + 1.59x, r = 0.46 $p < 0.01$.

ence between the potential N availability at a particular site and the mean potential N availability from a number of similar sites. This value was then added to 0.9 times the fertiliser recommendation (see Results and discussion) i.e.:

Adjusted fertiliser = (Fertiliser recommendation × 0.9) – (PANs – PANx)

Where:

PANs = Potentially available N at a particular site

PANx = Average potentially available N

This value of PAN_x (69.6 kg ha^{-1}) was taken as an average of 23 sites in arable regions of SE Scotland (unpublished).

Three replicate plots were used for both the recommended and adjusted rates at each site. Details of the amounts of N applied to each crop are given in Table 1. Soil samples were taken for determination of potentially available N in the first two months of 1992, and determinations of dry matter production and N uptake were made at harvest.

Results and discussion

The uptake of soil-derived N by spring barley was found by McTaggart and Smith (1993) to be closely related to potentially available N as determined by the method used in this study. When the results of the 1991 and 1992 seasons were plotted together with those of McTaggart and Smith (Fig. 1) there was a significant linear relationship over the five years (r = 0.56; $p < 0.05$). The relationship between the uptake of soil

N and potentially available N under the winter cereals was, however, not significant. This may be a consequence of variable amounts of mineral N remaining in the soil when the crop was sown, or of differences in sowing date, both of which would affect the supply of N to the plant independently of the amount of potentially mineralisable N present.

The amounts of potentially available N in December 1990 and March 1991 were more or less constant, with a difference of less than 6% between the two sets of values. This meant that the relationship between soil N uptake and potentially available N was unaffected by the date of sampling. Studies by Stockdale and Rees (1994) have shown that the method used to determine potentially available N measures a relatively stable fraction of the soil organic matter that is distinct from the microbial biomass. This fraction has been found to be unaffected by short term changes in management, such as the addition of manures (Rees, unpubl. data). In this respect it is very useful as a guide to adjusting fertiliser recommendations, since it can be used over a reasonably long period before crop establishment, providing adequate time for soils to be analysed and fertiliser requirements calculated.

Between 1988 and 1990 a strong linear correlation (r=0.96) was observed between soil N uptake and soil organic-C when soils with a high organic-C were excluded (McTaggart and Smith, 1993). In 1991 and 1992 this relationship was not apparent, despite the linear relationship between soil organic-C and potentially available-N (r=0.46; $p < 0.01$, Fig. 2). The sites used in the later experiments had a more varied history of previous cropping, and a wider range of organic-C con-

Table 2. Dry matter production and nitrogen uptake by cereals grown in 1992 at recommended and adjusted fertiliser application rates

Site/Crop	Fertiliser application (kg N ha^{-1})	Grain yield (t ha^{-1})	Total N uptake by grain (kg N ha^{-1})	Soil N uptake by grain (kg N ha^{-1})
Spring barley				
Crosshall	120 (R)	6.8	116	90
	108 (A)	7.9	137	108
Panlathy	110 (R)	5.9	83	59
	120 (A)	6.1	85	59
Crofts	120 (R)	4.9	80	50
	85 (A)	5.2	83	58
Cowloan	120 (R)	4.5	70	39
	123 (A)	4.5	70	43
Aldrochty	120 (R)	6.1	98	57
	115 (A)	6.2	103	65
Craibstone Bed C	90 (R)	4.6	56	34
	79 (A)	4.3	55	34
Craibstone	60 (R)	5.8	77	64
Sunnybrae	60 (A)	5.5	71	59
Winter wheat				
Crosshall	160 (R)	9.4	144	102
	144 (A)	9.3	145	100
Panlathy	200 (R)	8.3	107	59
	201 (A)	8.3	101	54
Crofts	200 (R)	7.5	125	69
	180 (A)	7.5	112	69
Winter barley				
Crosshall	180 (R)	8.1	123	59
	162 (A)	7.9	111	57
Crofts	180 (R)	6.6	111	74
	139 (A)	7.3	113	78

R = Recommended fertiliser rate.
A = Adjusted fertiliser rate.

centrations and at these sites there was no significant relationship between soil N uptake and soil organic-C. It is interesting to note that despite the correlation between soil organic-C and potentially available-N, only the latter is able to predict soil-N uptake by cereal crops. A possible explanation is that it measures a fraction of the organic matter that is readily mineralisable, and in that context it provides a more useful means of adjusting fertiliser recommendations than soil organic-C contents.

During the 1992 growing season, fertiliser recommendations were adjusted using measurements of potentially available N. The difference between potentially available N at a particular site and its average value was calculated. In order to reduce the average fertiliser application rate this was subtracted from 90% of the fertiliser recommendation for each site. If this were not done then the changes would not lead to an average reduction in fertiliser application; however, given the improved targeting, some reduction should be possible without loss in yields. This hypothesis is supported by the results, which show that there was little or no reduction in yield or N uptake as a result of these changes (Table 2). It might be argued that average reductions in fertiliser applications of the order of 10% would not be expected to result in any significant change in crop growth. However, the reduction of fertiliser applications at one site (Crofts, spring barley) was much larger, 29%, but even this reduction resulted in a 6% increase in grain weight and no change in the grain N concentration.

Acknowledgements

The authors wish to thank R Howard and F Wright for technical assistance, and to staff from the Crop Systems Department at Scottish Agricultural College for the management of field sites. Funding was provided by the Scottish Office Agriculture and Fisheries Department.

References

Gianello C and Bremner J M 1986 A simple chemical extraction method for assessing potentially available organic nitrogen in soil. Commun. Soil Sci. Plant Anal. 17, 195–214.

McTaggart I P and Smith K A 1993 Estimation of potentially mineralisble nitrogen in soil by KCl extraction. II. Comparison with soil N uptake in the field. Plant and Soil 157, 175–184.

Powlson D S, Pruden G, Johnson A E and Jenkinson D S 1986 The nitrogen cycle in the Broadbalk wheat experiment: Recovery and losses of ^{15}N-labelled fertilizer in spring and inputs from the atmosphere. J. Agric. Sci., Camb. 107, 591–609.

Recous S, Machet J M and Mary B 1988 The fate of labelled ^{15}N urea and ammonium nitrate applied to a winter wheat crop. II. Plant uptake and N efficiency. Plant and Soil 112, 215–224.

SAC 1990 Recommendations for cereals: Nitrogen. Technical Note, Fertiliser Series No. 5. The Scottish Agricultural College. Edinburgh, UK.

Smith K A, Elmes A E, Howard R S and Franklin M F 1984 The uptake of fertilizer nitrogen by barley growing under Scottish climatic conditions. Plant and Soil 76, 49–57.

Stockdale E A and Rees R M 1994 Relationships between biomass nitrogen and nitrogen extracted by other nitrogen availability methods. Soil Biol. Biochem. 26, 1213–1220.

O. Van Cleemput et al. (eds.), Progress in Nitrogen Cycling Studies, 365–369, 1996.

Nitrogen balances of ecosystems, landscapes, and watersheds

E.-W. Reiche, C.-G. Schimming, R. Mette and J. Schrautzer
Projektzentrum Ökosystemforschung, Univers; Lat, Kiel, Schauenburger Str. 112, D-24118 Kiel, Germany

Key words: ecosystems, modelling, nitrogen cycling, nitrogen transformations, watersheds

Abstract

The dynamics of different quantities of nitrogen cycling on the scale of a typical agricultural ecosystem in North Germany (Schleswig-Holstein) were investigated. With special reference to microbial transformations, a process-orientated model was applied to simulate the inorganic (Nmin), the dissolved nitrogen content of the soil and leaching. The use of deduced and statistical data permits scaling nitrogen cycling up from a specific point up to a complete catchment area, in this case the watershed of Lake Belau because of the spatial reference of the model. Therefore, nitrogen cycling of a forest ecosystem belonging to the area was also considered. While measurements generally represent single points at selected times, simulations permit a continuous and extensive description of nitrogen cycling on any scale. Thus, simulation makes it easier to forecast the effects of agricultural management and supports the development of basic principles in order to plan a more efficient use of nitrogen. This use of nitrogen was observed to be very poor when the exports in agricultural products were related to the total level of imports in the area.

Introduction

Nitrogen losses to surface waters, to ground water and to the atmosphere can be related to the structure of the cropping system, to the soil conditions as well as to the input level of fertilizer. Environmental influences of nitrogen emissions such as the eutrophication and acidification of sensitive terrestrial and aquatic systems as well as long-range climatic change are well known. To develop an economically and environmentally sustainable agricultural crop production system knowledge about the interrelationships between energy, water, and the nitrogen cycles on different spatial scales (ecosystem, catchment area, landscape, etc.) under various external condititons is needed.

One of the main objectives of the ecosystem research project in Kiel is to develop an environmental information system, which allows environmental impacts related to human land use and natural resources to be predicted. Long-term measurements have been taken in order to investigate the nitrogen fluxes in a beech forest and in a crop field. In comparison, a scientifically validated nitrogen model was used in order to be able to discuss the deviations which were mainly caused by different investigation scales. By integrating

methods of landscape regionalization, the simulation system enables simulations for small catchment areas $(1–100 \text{ km}^2)$ to be carried out. Nitrogen budgets have been calculated for the agricultural and forest ecosystem investigated in this study as well as for the catchment area of a small lake in which they are found. This paper will discuss the results of this research.

Material and methods

The area which was investigated is sited in the weichselian-moraine landscape of Schleswig-Holstein about 30 km south of the city of Kiel at the westernmost outskirts of the glaciation. The pronounced variety of ecosystems is typical for landscapes of North Germany (Hörmann et al., 1992). Extensive experimental data was gained from a typical agroecosystem with consecutive growing of maize oat and fodder beet. Additionally, data obtained from monitoring nitrogen cycling within a beech forest ecosystem was taken into account. The parent materials of the Cambic Arenosols (FAO) are highly waterpermeable glacial sands. Both sites are part of the watershed belonging to Lake Belau. The area (4.5 km^2) is characterized by a rural structure

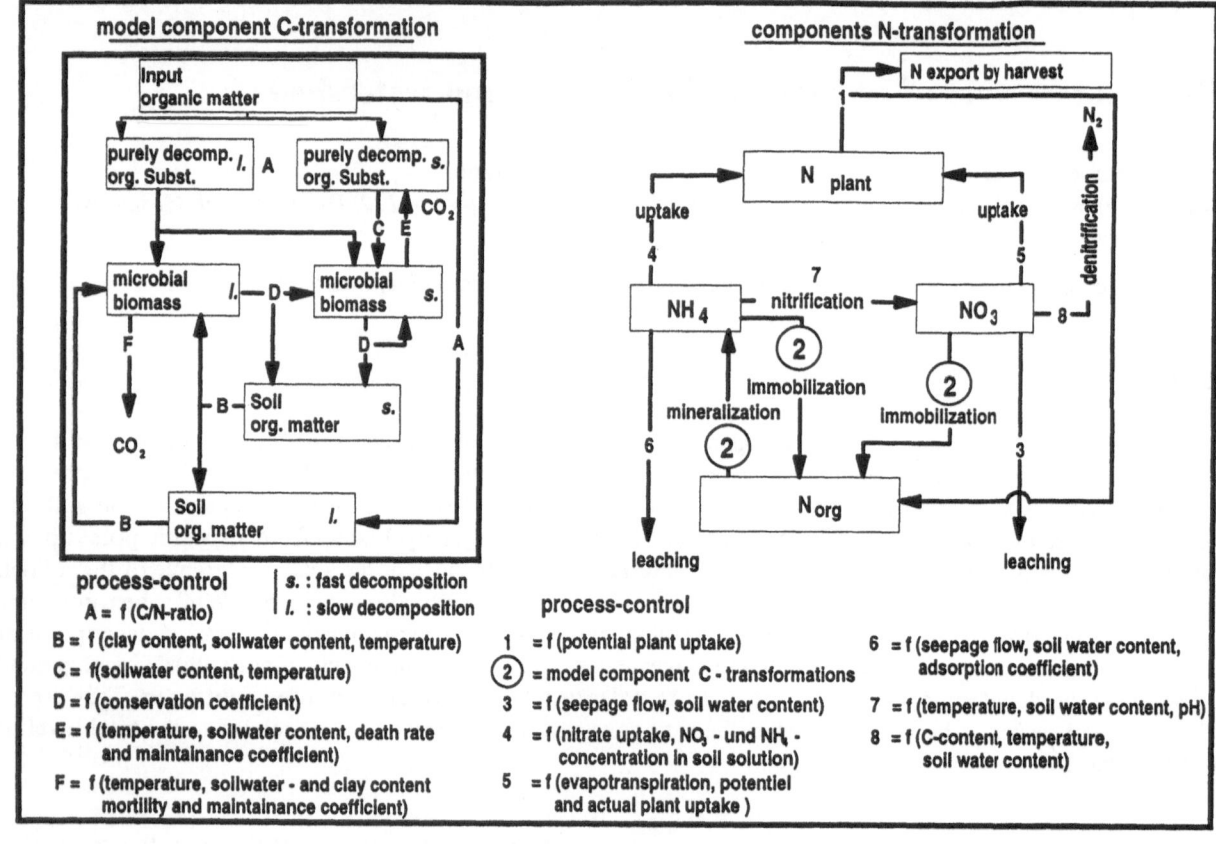

Figure 1. Partial structure of the model WASMOD and STOMOD (Reiche, 1991, 1994) designed for nitrogen cycling of ecosystems.

with agricultural land and small woods next to each other.

The nitrogen dynamics of the agricultural site are characterized by inorganic nitrogen concentrations of soil solution (ceramic cup samplers) and bulk deposition. Net-mineralization was determined by breeding soil samples in polythene bags. The content of the inorganic soil nitrogen was measured before and after breeding by extraction with 1 *N* KCl-solution (see Gerlach, 1973). Ammonia volatilization was measured once during the application of manure (in 1989) and on other occasions was calculated based on the method devised by Hoffmann and Ritchie (1993) using climatic parameters. The nitrogen content of the manure and the biomass at specific times were determined. Additionally, the development of root length density was examined (Tennant, 1975). Soil nitrogen fluxes were calculated from concentrations of nitrogen in the soil solution and the seepage water flows described in the model (Bornhöft, 1993).

The finite difference model WASMOD and STOMOD (Fig. 1) is based on different modules simulating

the interactions of soil processes. The spatial reference is established by measurements at specific points or by using deduced or statistical data for any area. The model was calibrated and tested at different sites. To arrive at a set of characteristic relationships between different Corg fractions and C/N ratios, data of real weather and land-use conditions over a preceeding nine-year period were fed into the model.

Results

The main method of fertilizing was the application of manure every year before tillage in the spring time (Fig. 2). The amount of inorganic nitrogen which was applied depended on the expected yield. Measured net-mineralization transformed 220 kg N ha^{-1} yearly. Depending on the soil temperature and the soil moisture, the transformation rates reach their highest point beginning in the springtime and decrease sharply with end of the vegetation period. In 1989, growing a crop of maize resulted in a fertilization of

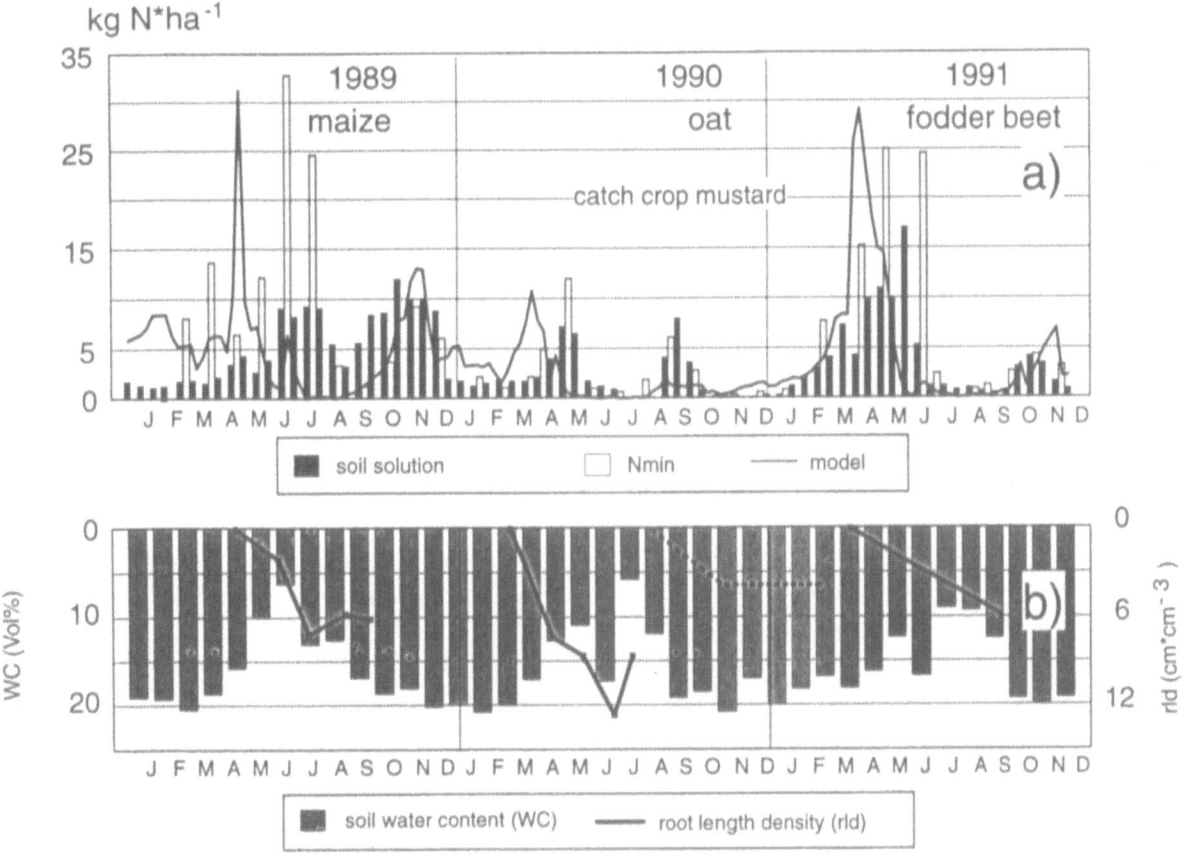

Figure 2. Dynamics of dissolved inorganic nitrogen (soil solution) and extractable inorganic nitrogen (Nmin) influenced by crop in the soil layer from 0 - 10 cm compared to the modelled values (**a**). Development of root length density and the dynamics of soil water content (**b**).

318 kg N ha^{-1} a^{-1} with resultant high levels of dissolved and extractable nitrogen (Fig. 2). Therefore the next year fertilization was reduced by reducing the inorganic nitrogen. Consequently, the mobile nitrogen content also decreased sharply. The soil nitrogen dynamics conformed satisfactorily with the interacting factors. Coincident reactions of fertilization, mineralization, development of root system, plant uptake, and soil water content became obvious. The dynamic of dissolved nitrogen appearing in the model deviated from the measurements because the manure applied around the ceramic cup samplers differ from application of the manure on the total crop field. In addition to net-mineralization, important internal transformation rates were calculated (1, 2, 3 in Fig. 3) based on the measured rates. Despite the difficulties with measuring net mineralization (1b) there is an acceptable accord between the different ways (A, B, C in Fig. 3) of calculating plant uptake (2) (according to Powlson et al., 1986) and the recycling of harvest residues (3). Diver-

gences can be mainly explained due to the spatial references. Measurements can represent only single points at selected times but the applied simulation performs a spatial extrapolation.

Actual deposition rates of the beech-forest ecosystem which was studied can be assumed to be 40 kg N ha^{-1} a^{-1} (Fig. 4) based on the measured fluxes of bulk deposition, throughfall and steam flow and assuming an increase of 20 kg N ha^{-1} a^{-1} in woodmass. Simulation of internal cycling features nitrogen saturation and instabilization of the ecosystem. Half of the external nitrogen input (18 kg ha^{-1} a^{-1}) in the ecosystem is leached.

Introducing statistical data from the area permits the scaling up of nitrogen cycling to the extent of the watershed (Fig. 4). Relating the inputs to the watershed of 123 kg N ha^{-1} a^{-1} (including the import of foreign forage) to the exports of 19 kg N ha^{-1} a^{-1} demonstrates the general inefficiency of nitrogen use. Recent losses from the watershed result from the high inter-

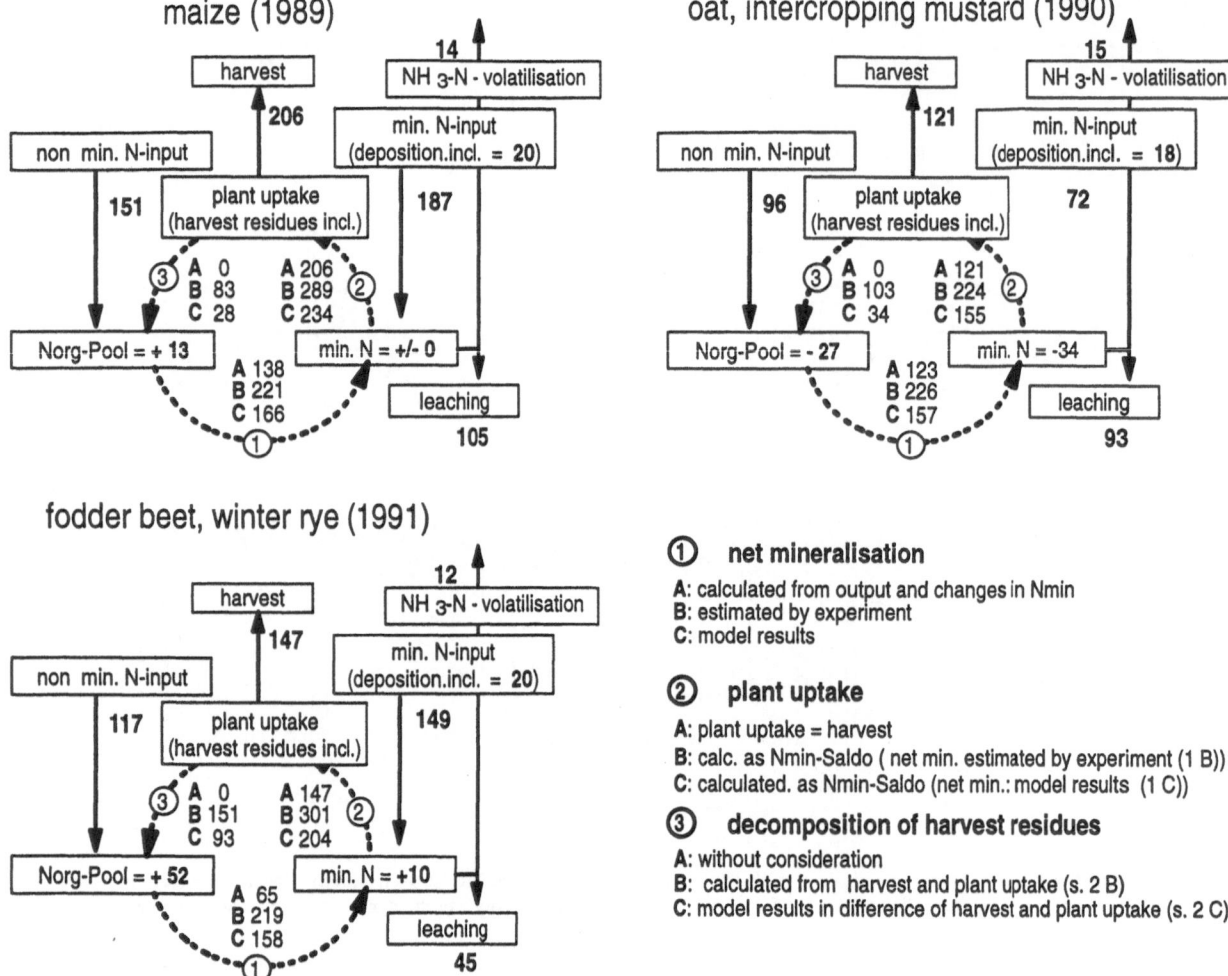

Figure 3. Subtly differentiated N-budgets (kg N ha^{-1} a^{-1}) with different crop growing in the years 1989, 1990 and 1991 (Explanations partly in the text).

nal cycling rates of nitrogen in mobile forms. Internal cycling of forest ecosystems must be taken into account, because 17% of 75 kg of residual nitrogen in animal production is in the form of ammonia. This contributes to the decrease in stability of the forest ecosystems due to atmospheric deposition. Nitrate leaching is a serious problem within the area whereas the concentrations of nitrates in the groundwater and in many wells are generally above 100 mg L^{-1}. The model calculates leaching of 50 to 60 kg N ha^{-1} a^{-1} from the total area of the specific soil properties in the area and the seepage rates are considered. The average concentrations in the groundwater will reach 70 mg NO$_3$ L^{-1} within the next 15 years. Recent values are already above the critical value of 50 mg NO$_3$, L^{-1}.

Discussion

Assessments and mapping on more extensive scales are often desired tools in planning and management in order to be able to deal with ecological problems. In the case of nitrogen, there is little sense in only assessing the input/output balances (Richter et al., 1992) because ecological relevance is highly dependant on the mobile or 'mobilizable' organic forms. It is necessary to collate knowledge about processes and the resulting fluxes up to the respective scale. This was satisfactorily done by WASMOD and STOMOD. The results presented demonstrated the importance of specific process dynamics. These aspects do not always receive enough attention. German regulations governing land use and thus the purity of the groundwater are generally based

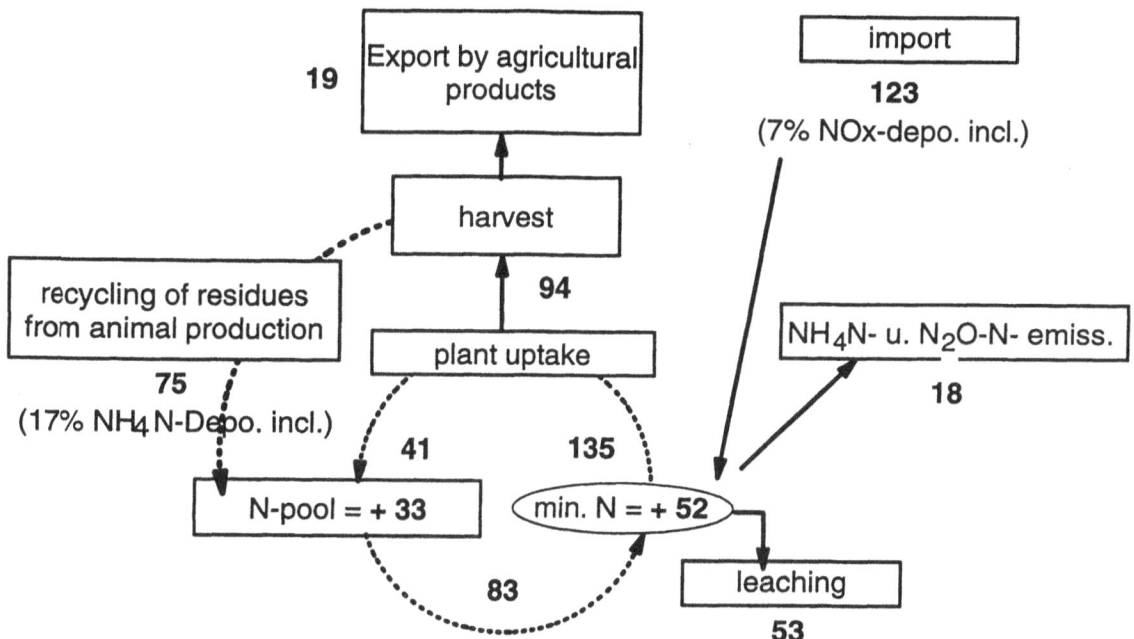

Figure 4. Subtly differentiated N-budgets (kg N ha^{-1} $^{-1}$) of the watershed belonging to Lake Belau (area 4.5 km^2) simulated based on statistical data under consideration of large scale soil properties.

on the extractable nitrogen content which is measured once a year (SchALVO, 1991). It became obvious that mobilization of nitrogen took place in the spring time while the crop cover had not yet developed . Consequent leaching from the soil environment occurred later. It is advisable to prevent the sandy soils in the agricultural site from lying fallow. To unterstand the subsequent nitrogen transformations, intensive studies based on measurements are necessary. Such studies are, due to the cost involved, generally carried out by scientific institutions. Modelling however can be an effective tool for planning fertilization and cropping. The results presented urge the consideration of microbial nitrogen transformations.

References

Bornhöft D 1993 Untersuchungen zur Beschreibung und Modellierung des Bodenwasserhaushaltes entlang einer Agrar- und einer Wald Catena im Bereich der Bornhöveder Seenkette (SchleswigHolstein). EcoSys Suppl. 6, 134.

Gerlach A 1973 Methodische Untersuchungen zur Bestimmung der Nettomineralisation. Scripta Geobot 5. 115 p.

Hoffmann F and Ritchie J T 1993 Model for Slurry and Manure in CERES and similar methods. J. Agron. Crop Sci. 170, 330–340.

Powlsen D S, Pruden G, Johnston A E and Jenkinson D S 1986 The nitrogen cycle in Broadbalk Wheat experiment: Recovery and losses of ^{15}N-labelled fertilizer applied in spring and inputs of nitrogen from the atmosphere. J. Agric. Sci., Cambridge 107, 591–609.

Hörmann G, Irmler U, Müller F, Piotrowski J, Pöpperl R, Reiche E W, Schernewski G, Schimming C-G, Schrautzer J and Windhorst W 1992 Ökosystemforschung im Bereich der Bornhöveder Seenkette. Arbeitsbericht 1988–1991. EcoSys 1, 329.

Reiche E-W 1991 Entwicklung, Validierung und Anwendung eines Modellsystems zur Beschreibung und flächenhaften Bilanzierung der Wasser- und Stickstoffdynamik in Böden. Kieler Geographische Schriften. Bd. 79, 150 S.

Reiche E -W 1994 Modelling water and nitrogen dynamics on catchment scale. Ecol. Model. Proc. 75/76, 371–384.

Richter G M, Beblik A J, Kersebaum K Chr and Richter J 1992 Modellierung des Nitrataustrages - Beratungsinstrument für den GW-Schutz in Niedersachsen. Mitt. Dtsch. Bodenkundl. Ges. 68, 115–118.

SchALVO 1991 Verordnung des Umweltministeriums über Schutzbestimmungen in Wasser- und Quellschutzgebieten und die Gewährung von Ausgleichszahlungen. Gesetzblatt Baden Wuerttemberg, Nr. 22.

Tennant D 1975 A test of modified line intersect method of estimating root length. J. Ecol. 63, 95–101.

O. Van Cleemput et al. (eds.), Progress in Nitrogen Cycling Studies, 371–375, 1996.
© 1996 *Kluwer Academic Publishers.*

Nitrogen balance on national, regional and farm level in Poland

Andrzej Sapek
Institute for Land Reclamation and Grassland Farming at Falenty, 05–090 Raszyn, Poland

Key words: farm, nitrogen balance, nitrogen surplus, Poland

Abstract

The impact of changes of the economical system on nitrogen surplus and losses in Polish agriculture was described. For this aim, an estimation of nitrogen balances on national, regional and farm basis was performed. The introduction of a market economy results in an average decrease of fertilizer use and in a limitation of agricultural production on most area in Poland. In spite of much smaller fertilizer use, the nitrogen surplus amounted to 69 kg N ha^{-1} in 1991, what was equal to 1.3 million tons of N losses from Polish agriculture. The nitrogen surplus varies a lot in Polish agriculture depending on the regions and farmer skill. Dairy farms grounded on permanent grassland can have rather small nitrogen surplus, but the mineralization of soil nitrogen should be considered. The nitrogen surplus in one of the best farms in Poland does not exceed 200 kg N ha^{-1}.

Introduction

The structural changes in Polish agriculture resulted also in an alteration of the fertilizer use on farms. The large-state owned farms, subsidized by the government, disappeared and the agricultural production is now based mostly on private, family farms. The relative prices of the previous subsidized commercial fertilizers increased manifold in the meantime. The new market economy canceled the wasteful application of fertilizers, and a drastic reduction of this application in most of family farms is observed nowadays.

The aim of this paper is to describe the impact of changes of the economical system on nitrogen surplus and losses in Polish agriculture. This is possible to achieve by calculating the nitrogen balance in agriculture on national, regional and farm basis.

The surplus of nitrogen can be easily calculated grounding upon the inputs and outputs base considering the inputs of nitrogen with commercial fertilizers, fodder imported from outside the farm, and other sources as nitrogen fixation by legume plants and the nitrogen content in precipitation. The only considered productive outputs of nitrogen from agriculture are connected with the animal and plant products sold from farms. The nitrogen content in soil is considered to be constant. The difference between the inputs and outputs is a surplus, which can also be supposed as nitrogen losses from the agricultural system. Such a balance is mostly accounted on an annual basis according to the cycle of agricultural production and is presented in kg ha^{-1} N to enable the comparison between farms. Various balances, on a national scope, for nitrogen and phosphorus have been made for some European Countries (Isermann, 1991) and Poland (Sapek and Sapek, 1993).

Materials and methods

The years 1985, 1990 and 1991 have been selected to estimate the balance of nitrogen on national and regional basis. Data available from the National Statistical Office were used. The nitrogen inputs by legumes were calculated relatively to the area of legume crops. The nitrogen entering with precipitation was taken as a mean value from the concentration in rain water measured at a point located in Falenty during the last eight years.

Data from demonstration farms established in the project 'Polish Agriculture and Protection of Water Quality' were used to estimate the nitrogen balance on farm level.

Results

National level

The use of commercial nitrogen fertilizers was rather high in the 1980s, but has sharply decreased in 1991 and it is maintained on this low level up till now (Table 1). The contemporary political changes in Poland, and the economical crisis in agriculture caused this evident decrease of fertilizer use to the lowest possible rate. However, an increase of this use is necessary and is expected in the forthcoming years.

The commercial fertilizers are the main sources of nitrogen inputs to the Polish agriculture, and the second one is the atmospheric precipitation, where the ammonia and nitrate concentrations are nearly of the same ratio. The inputs of nitrogen with imported fodder of high protein content was always low, and was the lowest in 1991. Legume crops are a permanent source of nitrogen, and an increase of legume crops is now advised to farmers for economical as well as environmental reasons.

The surplus of nitrogen amounted to about 95 kg ha^{-1} yr^{-1} N as estimated for 1985 and 1990, and was high enough to be harmful to the environment, particularly to the water quality. This surplus of nitrogen decreased since 1991, due greatly to the production crisis observed at the beginning of this year, but it was still as high as 68 kg ha^{-1} N (Table 2). Even this meaningfully lower nitrogen surplus can still be harmful to the environment since about 1.2 million ton of nitrogen is lost in Polish agriculture each year.

The surplus of nitrogen corresponded to about 86% of the inputs during all years under question. The depressed nitrogen inputs in 1991 was compensated by the belittled sale of farm products. The sale of farm products was steady during the 1980s, but dropped evidently from 1991 on. The decreased disposal of all main products was noticed and it resulted in lowering of the nitrogen outputs from Polish agriculture by about 25% in 1991.

Regional level

Poland is divided in 49 regional provinces - voivodeships and for these political units some statistical data are available. The same data as used for the estimation of the national balance were collected for regional balances, but the imported fodder do not consider fodder purchased from other voivodeship but only the really imported from abroad (Table 2).

There are evident differences in the regional nitrogen balance. Higher nitrogen surplus was noticed in the west part of Poland during 1980s, and this tendency is maintained up to 1991 (Table 2).

Farm level

Three kinds of farms have been selected for calculation. In the first two groups were farms localized on sandy soils (arable fields) and peat soils on sand (permanent grassland). For productivity and soil conservation reasons only dairy farming can be advised. The sandy soils need a constant input of organic matter, and the mineralization peat is slowest under permanent grassland. In one group of farms, relatively higher rates of fertilizers were used. This is double as compared to the mean national rate in 1991. In another group low rates of fertilizers, if any, were used. The main sold product was milk.

On this kind of soil rich in organic matter and nitrogen, the effectiveness of nitrogen inputs does not only depend on nitrogen fertilization, but also on the farming system and the mineralization rate of the soil nitrogen, as in the case of dairy farms in Ostroęka region (Tables 3 and 4). In each of these farms, the nitrogen surplus was greater than the input with commercial fertilizer. The efficiency on nitrogen inputs on farm with low input was rather high with the highest value about 67% on one farm, where also the nitrogen surplus amounted only to 12 kg ha^{-1} N (Table 4). The mineralization of peat soils in Rupin is presumed to be great and thus has a main impact on nitrogen balance.

The third group of farms consists of two farms on good soils and farmed by educated farmers. Both farms are in good economical condition. Distinctively, the swine farm uses also as nitrogen sources legume crops such as green peas and French bean. The swine feeding is based also on purchased fodder, which is also the main nitrogen source on this farm. The nitrogen surplus on this farm was not far from 200 kg ha^{-1} N in 1992. On this farm the nitrogen outputs in sold products was about the same as the input with commercial fertilizers, and the efficiency of total nitrogen inputs was about 30% (Table 5). This farm can be supposed as model in developing of Polish agriculture based on family farm system.

The second used much lower rates of nitrogen fertilizers and purchased fodder. On this farm too, the nitrogen outputs in sold products was about the same as the input with commercial fertilizers, but the efficiency of total nitrogen inputs was higher (Table 5).

Table 1. Nitrogen balance in Polish agriculture

	1985 (kg ha^{-1} N)	1990 (kg ha^{-1} N)	1991 (kg ha^{-1} N)	Ratio (1985+1990)/ (2×1991)
Inputs	112.40	109.13	79.77	0.72
Commercial fertilizers	65.40	67.40	39.40	0.59
With precipitation	17.00	17.00	17.00	1.0
Fixation by legumes	9.25	5.35	3.59	0.98
Fixation by free-living microorganisms	5.00	5.00	5.00	1.0
Outputs	15.70	15.76	11.36	0.72
Plant products	9.49	10.10	7.05	0.72
- Sold small grain	5.10	5.57	4.26	0.80
- Sold potatoes	0.77	0.57	0.25	0.37
- Sold vegetable	0.24	0.15	0.10	0.56
- Sold fruits	0.04	0.02	0.02	0.67
- Sold sugar beets	1.54	1.87	1.20	0.7
- Sold rapeseeds	1.80	1.92	1.21	0.65
Animal products	6.21	5.66	4.31	0.73
- Meat	2.84	2.84	2.12	0.74
- Milk	3.12	2.74	2.18	0.74
- Eggs	0.22	0.06		
- Wool	0.03	0.02	0.01	0.74
Nitrogen surplus	**96.70**	**93.37**	**68.41**	**0.72**
Nitrogen surplus in per cent of inputs	86.0%	85.6%	85.8%	1.00

Table 2. Nitrogen surplus in selected voivodeships with extreme values (kg ha^{-1} N)

Voivodeships	1985 (kg ha^{-1} N)	1990 (kg ha^{-1} N)	1991 (kg ha^{-1} N)	Ratio (1985+1990)/ (2×1991)
Higher surplus				
Bydgoszcz	97.5	106.5	102.5	1.00
Leszno	113.0	120.7	93.2	0.80
Lódź	115.9	94.1	101.6	0.97
Opole	109.6	109.3	94.9	0.87
Szczecin	118.8	115.0	93.8	0.80
Toruń	113.6	94.8	91.5	0.88
Lower surplus				
Bielsk Podlaski	72.5	75.0	51.8	0.70
Białystok	75.1	87.8	49.6	0.61
Chełm	69.7	63.9	46.9	0.70
Tarnobrzeg	84.9	72.5	45.5	0.58
Zamość	67.3	62.9	42.7	0.66

Table 3. Nitrogen balance on dairy farms with higher input Ostrolenka region. Data from 1993

Farm number	1	2	3
Village	Łady	Łyse	Łyse
Farm area (ha)	14.5	24.3	21.7
	$(kg\ ha^{-1}\ N)$	$(kg\ ha^{-1}\ N)$	$(kg\ ha^{-1}\ N)$
Inputs	127.5	91.3	104.2
Commercial fertilizers	88.0	69.9	78.2
Purchased fodder	17.3	0.7	2.1
Precipitation	17.0	17.0	17.0
Fixation by legume	5.2	3.7	6.9
Outputs	31.9	11.6	12.4
Plants products	9.3	0	0
Animal products	22.6	11.6	12.4
Surplus	95.6	79.7	91.8
Outputs/Inputs (%)	25.0	12.8	11.9

Table 4. Nitrogen balance on farms with high inputs Ostrolenka region. Data from 1993

Farm number	4	5	6
Village	Rupin	Rupin	Rupin
Farm area (ha)	16.3	23.5	24.0
	$(kg\ ha^{-1}\ N)$	$(kg\ ha^{-1}\ N)$	$(kg\ ha^{-1}\ N)$
Inputs	35.6	215	51.9
Commercial fertilizers	4.7	1.9	31.3
Purchased fodder	9.7	1.0	0.3
Precipitation	17.0	17.0	17.0
Fixation by legume	4.1	16	3.3
Outputs	23.9	12.6	19.1
Plants products	0	0	1.8
Animal products	23.9	12.6	17.3
Surplus	11.6	8.9	32.8
Outputs/Inputs (%)	67.3	58.7	36.7

This short review of farm nitrogen balances gave a picture of how much Polish agriculture differs at the present stage of its transformation, and how useful the method of nitrogen balance can be for predicting nitrogen losses and developing an agricultural system, which could stand the demands of high productivity and abating water and air pollution with nitrogen compounds.

Table 5. Nitrogen balance on farms with high inputs Elblag region

Farm number	7		8	
Farm kind	Dairy		Swine	
Village	Bogaczewo		Pastwa	
Farm area (ha)	26.3		21.3	
	1992	1993	1992	1992
Inputs	105.2	93.7	268.2	251.4
Commercial fertilizers	45.8	44.9	74.2	78.9
Purchased fodder	38.3	25.9	167.0	136.6
Precipitation	15.8	17.0	15.8	17.0
Fixation by legume	5.4	5.9	11.3	18.9
Outputs	44.4	52.3	72.0	77.0
Plant products	33.2	38.5	28.9	16.5
Animal products	11.2	13.8	43.1	60.4
Surplus	60.8	41.4	196.3	174.5
Outputs/Inputs (%)	42.2	55.8	26.8	30.6

Conclusion

1. The nitrogen balance in Polish agriculture displays considerable surplus, which could be identified as nitrogen losses. The introduced market economy results in an average decrease of commercial fertilizer use and in a limitation of agricultural production on most areas in Poland. In spite of abated fertilizer use the nitrogen surplus amounted to 69 kg ha^{-1} N in 1991, amounting to 1.29 million ton of N losses from Polish agriculture.

2. The nitrogen surplus varies much in Polish agriculture depending on the regions and farmers skill. That should be considered in the proposal of development of sustainable agriculture in Poland.

3. The nitrogen balance on farm level exemplifies the need of a development of fertilization advice service which can help in increasing the benefits of farmers, and protecting the water quality.

Acknowledgements

I am grateful to T Marcinkowski and S Pietrzak for completing the nitrogen balance on farm level and P Nawalany for framing a computerized data basis.

References

Isermann K 1991 Nitrogen and Phosphorus balances in Agriculture - A Comparison of several Western European Countries. Proceedings International Conference on Nitrogen, Phosphorus and Organic Matter, May, 13–15, Helsingor, Denmark. pp 1–20.

Sapek A and Sapek B 1993 Assumed non-point water pollution based on the nitrogen budget in Polish Agriculture. Water Sci. Technol. 28, 483–488.

O. Van Cleemput et al. (eds.), Progress in Nitrogen Cycling Studies, 377–380, 1996.

N-balances on Flemish dairy farms

I. Verbruggen and L. Carlier
Nationaal Centrum voor Grasland- en Groenvoederonderzoek 1e sektie (R.v.P.) - (I.W.O.N.L.), Burg. Van Gansberghelaan 109, 9820 Merelbeke, Belgium and Rijksstation voor Plantenveredeling (R.v.P.) (CLO-Gent), Burg. Van Gansberghelaan 109, 9820 Merelbeke, Belgium

Key words: dairy farm, fodder beets, nitrogen budget, organic manure

Abstract

From 40 Flemish dairy farms the nitrogen balances have been calculated, based on sampling and analysing of the inputs (concentrates, litter, ...) and outputs (milk, roughage, ...). The average N surplus was 329, 296 and 290 kg N ha^{-1}, respectively, for the years 1991, 1992 and 1993. The decline of the surplus was the result of using less mineral fertilizer, feeding less moist concentrates and producing more milk per ha. The most important measures (resulting from a multiple regression analysis) are minimizing the input of manure from pig or poultry farms and using less N fertilizer on grassland. Fodder beets on the farm seems to be a better alternative than increasing the maize area.

Introduction

In the north of Belgium, farms cause major environmental problems, also some dairy farms. In some regions there are problems with to high nitrate concentrations in the drinking-water. The Belgian government has taken measures to protect the environment and to diminish these problems. The polluting risks can be calculated from nitrogen balances. The N surplus is the sum of losses of a dairy farm which contains the N-accumulation, which is only a small part, ammonia volatilization, denitrification, leaching and run off-losses. Registration and calculation of nitrogen surpluses has important advantages over measurement and calculation of separate emissions. Measurement of all emissions on every farm is impossible.

Materials and methods

A nutrient budget shows the difference between total nutrient input by way of fertilizers, purchased concentrates and roughage, atmospheric deposition, etc. and total nutrient output from the farm by way of milk, meat, roughage and manure. N budgets for 40 Flemish dairy farms have been calculated for the years 1991,

1992 and 1993. Some characteristics of these farms are given in Table 1.

At regular time intervals, samples of various products were taken and analysed for N to make the N budget per farm more precise (Verbruggen et al., 1992). Different farm characteristics, such as N fertilizer use on grassland and the level of concentrate intake per cow were recorded in order to calculate their influence on the N surplus.

Results and discussion

Compared to 1991, the average surplus of N in 1993 was decreased by 39 kg (11.9%) per ha (Table 2). This was the result of using less mineral fertilizer on the grassland as well as on the maizeland, using less moist concentrates and of a greater milk production per ha. The most important factors affecting the N surplus in 1993 were deduced from multiple regression analysis, according to:

$$Y = -94.067 + 0.6445X_1 + 0.0528X_2 + 0.1945X_3 + 1.484X_4 + 0.8886X_5$$
$$r^2 = 0.84; \text{ standard error} = 48.05 \qquad (1)$$

where:
Y = N surplus (kg ha^{-1})

Table 1. Mean characteristics of 40 Flemish dairy farms during the years 1991, 1992 and 1993

Characteristics	1991	1992	1993
Area (ha)	30.28	30.60	30.45
Grassland (% area)	64.82	62.85	60.04
Fodderbeet (% area)	0.95	1.22	1.12
Milk production (l cow^{-1})[a]	6 126	6 192	6 346
Milk production (l ha^{-1})	10 860	10 958	11 550
SCU ha^{-1}[b]	2.90	2.87	2.95
Dry concentrates (kg SCU^{-1})	922	921	917
Moist concentrates (kg SCU^{-1})	177	194	158
Dry concentrates (kg cow^{-1})	1 291	1 244	1 295
kg N ha^{-1} grassland	–	278	266
kg N ha^{-1} maizeland	–	62.3	46.5
purchased organic manure (kg N ha^{-1})	21.4	25.9	27.9

[a] Milk with 4% fat.
[b] SCU (standard cow unit): one cow with a production of 4000 L (4% fat) per year; for each 1000 L milk extra, SCU increases by 0.1 unit.

Table 2. Mean input, output and surplus of nitrogen on 40 Flemish dairy farms for the years '91, '92 and '93 (in kg ha^{-1})

	1991	1992	1993
N-input			
Cattle	1.98	2.58	2.19
Dry concentrates	92.50	92.75	96.00
Moist concentrates	16.22	15.60	13.97
Roughage	10.58	6.90	14.32
Litter (straw)	5.21	4.04	2.51
Mineral fertilizer	217.31	195.43	176.93
Organic manure	29.81	34.21	36.07
Fixation by legumes	2.87	1.05	1.44
Atmospheric deposition	40	40	40
N-output			
Cattle	14.92	14.34	14.95
Milk	56.32	56.82	60.01
Roughage	8.05	16.92	10.63
Organic manure	8.44	8.27	8.22
Total input	416.48	392.56	383.43
Total output	87.73	96.35	93.81
Surplus	329	296	290

X_1 = application of mineral N fertilizer on grassland (kg ha^{-1})

X_2 = purchased dry concentrates (kg cow^{-1})

X_3 = purchased moist concentrates (kg DM cow^{-1})

X_4 = grassland (% of total farm area), and

X_5 = purchased organic manure (kg N ha^{-1})

A decrease of 1 kg of N on the nutrient balance may be obtained via:

– a decrease of the application of mineral N fertilizer on grassland with 1.55 kg N ha^{-1} ;

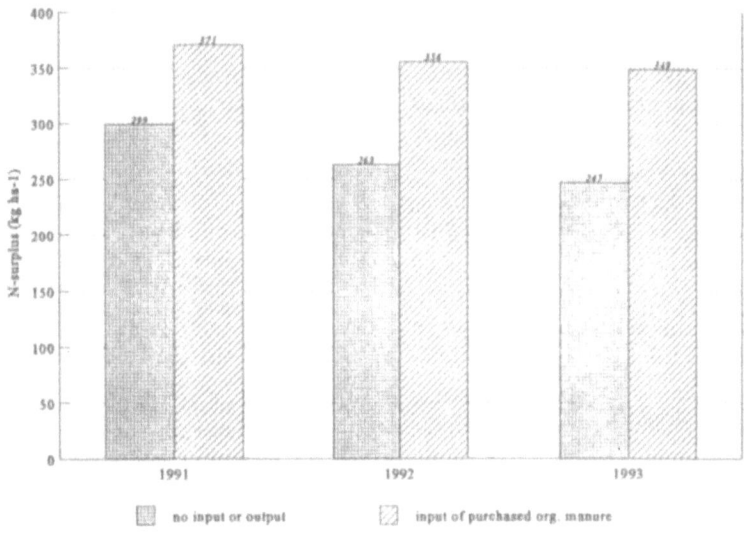

Figure 1. N-surplus (kg ha^{-1}) from farms with or without input of purchased organic manure in the years '91, '92 and '93.

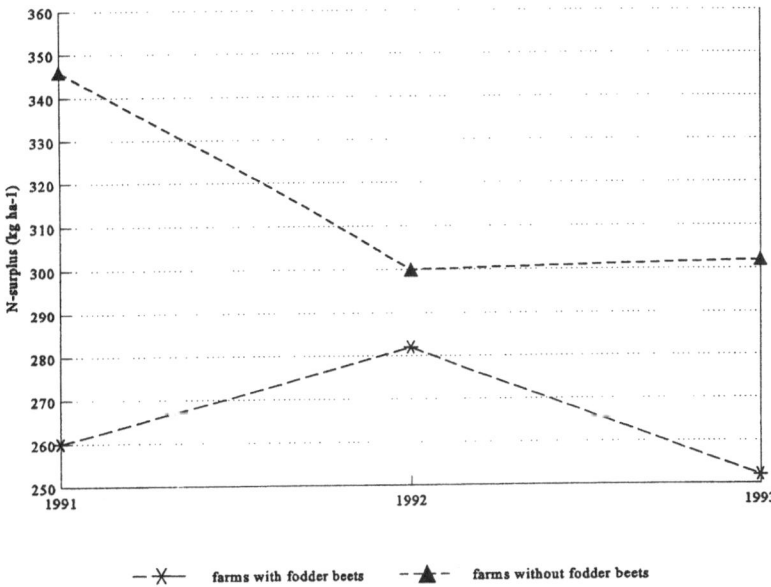

Figure 2. The effect of fodder beets on the N-surplus in the years '91, '92 and '93.

– a decrease of the dry concentrate intake per cow with 18.94 kg ;

– a decrease of the moist concentrate intake per cow with 5.14 kg DM ;

– a decrease of the grassland area on the farm with 0.67% and

– a decrease of purchased organic manure by 1.125 kg N ha^{-1}.

The multiple regression analysis for the year 1992 gives as result:

$$Y = -40.98 + 0.6674X_1 + 0.043641X_2$$
$$+0.1238X_3 + 0.76X_4 + 0.8790X_5$$
$$r^2 = 0.86; \text{ standard error} = 40.49. \quad (2)$$

In agreement with Simon et al. (1994) the fertilizer N inputs are highly correlated with the N surplus. Also the input of purchased organic manure has a negative effect on the N surplus as it is shown in Figure 1.

In 1993 the average N-input via purchased organic manure was 100 kg N ha^{-1} on the farms which make use of organic manure which is not of their own farm. The N surplus on these farms was also 100 kg ha^{-1} higher.

In agreement with Van der Ham (1992), having more maizeland at the expense of grassland has a small influence on the N surplus especially in the year 1992 (see regression analysis). A greater effect is to be expected from fodder beets. Farms with this crop had an average N surplus that was 51 kg N ha^{-1} lower than that of farms without fodder beets (see Fig. 2). These farms use less concentrates which has a positive effect on the nutrient balance.

Conclusions

The N surplus was diminishing during the years. The input of purchased organic manure has a negative effect on the nutrient balance. Introduction of fodder beets will diminish the N surplus. The effect is higher than having more maizeland at the expense of grassland.

References

Simon J-C, Le Corre L and Vertes F 1994 Nitrogen balances on a farm scale: results from dairy farms in north west France. *In* Grassland and Society. Eds. L 't Mannetje and J Frame. pp 429–433. Wageningen Press, Wageningen, The Netherlands.

Van der Ham A 1992 Effects of farm management on environment and income. Proceedings of the 14th General Meeting of the European Grassland Federation, Lahti, Finland. pp 358-362. EGF, Lahti, Finland.

Verbruggen I, Carlier L, van Lierde D and Van Bockstaele E 1992 Mineralenoverschotten in de melkveehouderij. KVIV-studiedag, Graslandgebruik en ruwvoederwinning met beperkingen, Gent, Belgium. 4.1–4.14. KVIV, Antwerpen, Belgium.

O. Van Cleemput et al. (eds.), Progress in Nitrogen Cycling Studies, 381–385, 1996.
© 1996 *Kluwer Academic Publishers*.

System of Adjusted Nitrogen Supply (SANS) for fertilization of grassland

Th.V. Vellinga, A.P. Wouters and R.G.M. Hofstede
Research Station for Cattle, Sheep and Horse Husbandry (PR), Runderweg 6, 8219 PK Lelystad, The Netherlands. [1]*Corresponding author*

Key words: fertilizer recommendations, N mineralization, N recovery, N uptake, soil mineral N

Abstract

To reduce N losses and fertilization costs on grassland, a System for Adjusted Nitrogen Supply (SANS) has been developed, taking into account N uptake, N mineralization, N recovery and soil mineral N content. On sand and clay soils, application of N according to SANS showed satisfying results per year and per cut, when compared to current N recommendations. The results on peat soils were variable. Despite the positive results, the estimation of the parameters used in the SANS equation for calculation of the N application still can be improved.

Introduction

N losses from agriculture have to be reduced to meet environmental goals. Both N losses and fertilization costs on grassland could be reduced if N is applied according to the actual requirements per cut, taking into account N uptake, N mineralization and available soil mineral nitrogen. Therefore the System of Adjusted Nitrogen Supply (SANS) is being developed and tested (Unwin and Vellinga, 1994; Wouters and Vellinga, 1994).

The present study aimed to evaluate if SANS can optimize N application per cut and per year to reduce N losses without a reduction in yield compared to the present N recommendations (Unwin and Vellinga, 1994). Additionally, the accuracy of the estimation of parameters in the SANS equation is tested.

Methods

N application per cut according to SANS (N-SANS) is based on predicted N requirement of the cut and on measured available soil mineral nitrogen. N-SANS is determined as follows:

$$\boxed{N-SANS} = \boxed{\frac{\text{N Yield} - \text{Apparent N Supply}}{\text{Apperent N Recovery}}}$$

(model calculations)

$$-\boxed{\text{Available Soil Nitrogen}}$$

(laboratory analysis)　　(1)

Model calculations are done using extensively tested empirical models of grass production and N-uptake, apparent N supply (ANS, mineralization) and apparent N recovery (ANR), which are based on average growing conditions. Available Soil Mineral Nitrogen (SMN) is estimated from laboratory analysis of soil samples (0–30 cm) from representative reference plots (CaCl$_2$ extraction of field-humid samples).

N-SANS was compared with the present N recommendations in cutting trials on sand, clay, poorly and well drained peat soils since 1992 (completely randomized block design; variance analysed by ANOVA and F-test). Dry matter (DM) and N yields were determined per cut. SMN (soil layer 0–30 cm) was measured before every cut. In November 1992 and 1993, SMN was also determined in the layer 0–100 cm to evaluate if SANS meets environmental goals (<70 kg N ha^{-1}; MANMF, 1990).

Besides SMN, the equation to calculate the adjusted N application is also dependent on a set of estimations, which are all based on average growing conditions:

the planned N uptake, the estimated ANS and ANR. Therefore, not only an evaluation of the results per year is done, but also a comparison per cut of the estimated values and the realized values is necessary to evaluate the usefulness of the equation. Here, we present the results per cut for the clay site only.

Results

DM yield and applied N per year

In 1992 the amount of N applied according to SANS was lower than N recommendations while DM yields were not significantly different ($p > 0.1$) and ANR was improved (Table 1). SMN at the end of the growing season (layer 0–100 cm) was lower at the SANS treatment. The growing season of 1992 was relatively dry when compared to the 30 year average.

In 1993, the growing season was relatively warm and dry until July 1st, then the season became rather cold and wet. The amount of N applied according to SANS was higher than N recommendations, which resulted in significantly ($p < 0.001$) and economically higher DM yields (> 7.5 kg DM yield per kg extra N applied) on sand and clay. ANR was similar or higher on the SANS treatment than with N recommendations. The SMN content at the end of the growing season of 1993 was somewhat higher at the SANS treatment. On peat soils the SANS treatment gave poorer results than N recommendations in terms of DM yield per N applied and residual SMN in November.

DM yield and applied N per cut

At the site on clay soil, the DM yield at the SANS treatment was in 1992 and 1993 similar or higher than at the N recommendations treatment (Fig. 1A). Only one cut had a significantly lower DM yield ($p < 0.001$), but the N application for that cut was 34 kg ha^{-1} less than the recommended amount. In three cuts the DM yield was significantly higher at the SANS treatment, in two of these the extra yield is unlikely to be caused by the small amount of extra N applied (2 and 3 kg ha^{-1}, respectively).

In five out of 13 recorded cuts the N application according to SANS was lower than the recommended amounts, mostly corresponding to a high SMN content in the soil (Fig. 1B). Most of the reductions in applied N occurred in 1992. In 1993 the SANS treat-

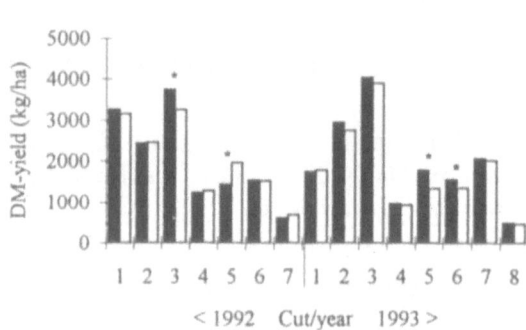

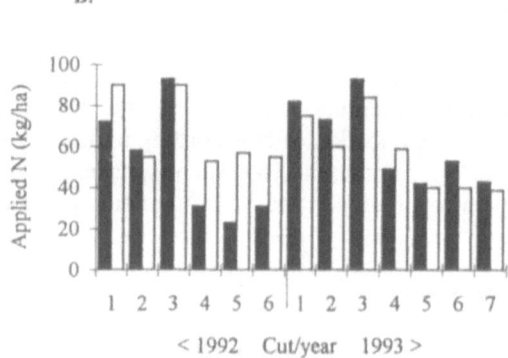

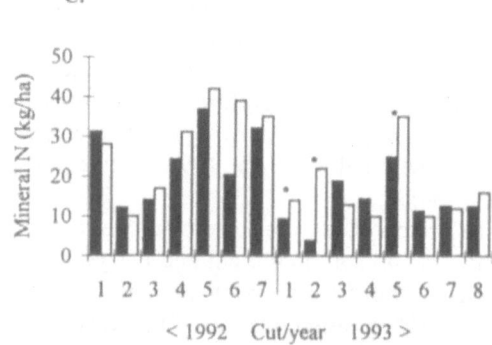

Figure 1. The DM yield (**A**), the amount of applied N (**B**), and the soil mineral N content (**C**) on plots treated according to SANS and to the current N recommendations. Results of 15 cuts in 1992 and 1993 on clay soil, asterisks indicate significant differences between SANS (■) and N recommendations (□) (F-test, $p < 0.05$).

ment received more N than the recommended amounts in almost every cut.

The SMN content at the SANS treatment (0–30 cm) was lower than at N recommendations in 9 out of 15 recorded cuts, but only in three of these the difference was significant ($p < 0.05$, Fig. 1C). In 1992, SANS generally resulted in lower SMN contents than the N recommendations. In 1993 however, SMN at the SANS treatment was higher before cuts 3, 4, 6 and 7.

Table 1. The total amount of N applied, the DM yields, the Apparent N Recovery (ANR; weighted average) and the soil mineral N (SMN) content (0–100 cm) in plots fertilized according to SANS and to the current recommendations (N-Rec.), in the experimental years 1992 (six cuts) and 1993 (seven cuts). All data in kg ha^{-1}, asterisks indicate significantly diffferent DM yield between SANS and N-Rec. (F-test, $p<0.05$)

Year/soil type	N applied		DM yield		ANR		SMN	
	SANS	N-Rec.	SANS	N-Rec.	SANS	N-Rec.	SANS	N-Rec.
1992								
Clay	308	400	14.2	14.3	0.81	0.65	47	61
Sand	356	423	13.0	13.5	0.70	0.68	67	81
Peat wet	294	365	12.5	13.5	0.67	0.66	52	56
Peat dry	190	229	13.4	13.0	1.09	0.83	59	51
1993								
Clay	448	398	15.6*	14.5	0.87	0.86	18	14
Sand	468	405	14.3*	13.5	0.67	0.64	25	25
Peat wet	465	380	16.1	15.5	0.73	0.76	113	99
Peat dry	349	240	14.0	14.1	0.92	0.91	101	99

On the other hand, SMN contents were lower in 1993 compared to 1992.

Evaluation of the SANS equation

Two levels of N uptake were planned: cuts for silage (DM yield 3000 kg ha^{-1}, N uptake ca. 100 kg ha^{-1}; third cut of both years) and cuts for grazing (DM yield 1700 kg ha^{-1}, N uptake 60–70 kg ha^{-1}; all other cuts). The N uptake of the first three cuts in both growing seasons was underestimated in the SANS equation (Fig. 2A). The later cuts had a N uptake between 20 and 50 kg N ha^{-1}, which were all below the planned uptake. In the first half of both growing seasons the growing conditions were good: relative high temperatures and no water limitations. The second half of the growing season was relatively dry in 1992 and wet and cold in 1993. Therefore, the planned yields were not realized within the planned growing period of three or four weeks. It might have been feasible to wait for the planned yields, but this would result in growing periods of about six weeks, which are not common practice. On the other hand in optimal growing periods, when there is a higher N-uptake than expected, the growing period can be easily adjusted.

Prediction of ANS and ANR per cut was not optimal. The ANS per cut was often overestimated, especially in 1993 (Fig. 2B). In that year, ANS in most cuts was lower than the ANS values in 1992. In 1992 the ANS in the first three cuts were highest and underestimated. The realized ANR per cut was very variable and not related to the estimated ANR values (Fig. 2C).

Only the estimate for the first cut was acceptable. The high ANR values in the fifth cut are in both years preceded by a relatively low ANR in the fourth cut.

Discussion

Results per year

On sand and clay the use of SANS meets the economical and environmental goals, if the total results per year are evaluated. The adjusted N application resulted in similar DM yields on three sites in 1992, while in 1993 the extra N resulted in significantly higher DM yields on three sites. On peat soil the results were variable, and offered opportunities for improvement.

Results per cut

The analysis of the data for 1992 showed that in four cuts less N was applied according to SANS than the recommended amount. Nevertheless, in three of these cuts the DM yield was not significantly lower. In 1993, higher N applications per cut resulted in a significantly higher DM yield in three cuts. However, in two cases the extra amount of N was very small and could hardly be the reason for the higher DM yield. These higher N applications were a result of lower SMN contents and of a larger correction for non-available SMN in 1993. Since the mineralization in 1993 lacked behind the estimated average, the higher N application according to SANS formed an (unplanned) advantage compared

384

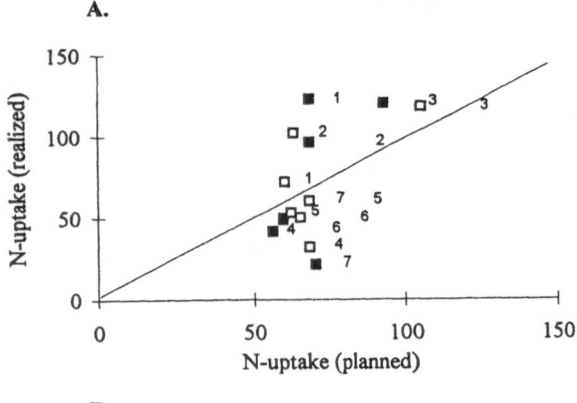

A.

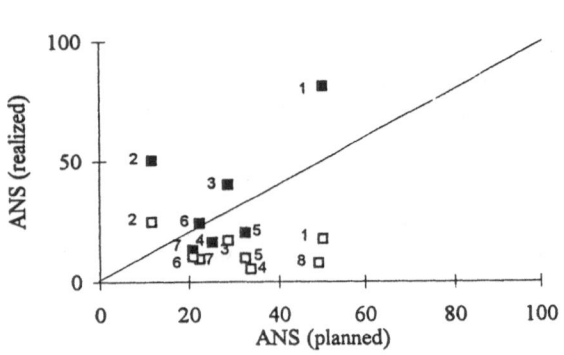

B.

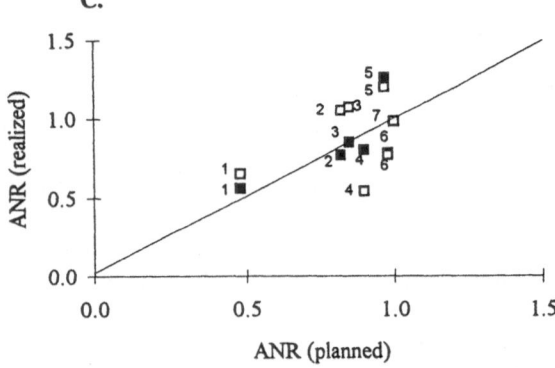

C.

Figure 2. Realized and planned N uptake (**A**), Apparent Nitrogen Supply (ANS; **B**) and Apparent Nitrogen Recovery (ANR; **C**) on plots treated according to SANS during two growing seasons (1992 ■; 1993 □) on clay soil. Number of cut and line y=x are indicated.

to the N recommendations in terms of optimization of N fertilizer.

It is likely that in our experiments, mineralization was influenced more by spatial and temporal variation than by fertilization. ANS was measured on unfertilized control plots, and according to Hassink (1991) the net N mineralization of a grassland soil was not influenced by fertilization or grassland use. Consequently, the higher DM yields in cuts where only 2 or 3 kg

ha^{-1} more N was applied might be caused by local variations in net mineralization.

The data showed that the possibilities for adjustment of the applied amount of N depend on the weather conditions. When growth is limited by drought (1992), the applied N is still available for the next cut, while this is not the case under wet and cold conditions (1993). Hence, especially under dry conditions SANS is a useful tool on sand and clay soil.

Evaluation of the SANS equation

Although we consider SANS, applied on clay, as a satisfactory and accurate fertilization tool, the parameters used in the equation were not estimated very accurately. In both recorded years, the N uptake was underestimated in the first half and overestimated in the second half of the growing season. This showed some relationship with the ANS per cut, which was also higher than expected in the first three cuts of 1992. The later cuts had lower ANS values, mainly caused by drought. In 1993 the ANS was overestimated during the entire season, except for the second cut. The ANS values for the cuts 1–3 in 1993 also were higher than those for the latter. The large difference in ANS between 1992 and 1993 shows that estimates based on average conditions, are of low predictive value for the purpose of SANS.

The favourable growing conditions in the cuts 1–3 are reflected in the combination of a high N uptake, a high ANS and a relatively high ANR, especially in 1992. The lower values for N uptake, ANS and ANR in the later cuts indicated poorer growing conditions, when other factors than N availability determine plant growth. The high ANR in the fifth cut was caused by residual N from the forth cut, which showed a rather low ANR. For the fate of the N which is not taken up by the plant (= not harvested) and is unknown, it is very difficult to make a right estimate of the expected ANR.

For the second half of the growing season we might have overestimated the potential for grass production, although experimental data showed a good grass production for that period (Prins et al., 1980).

The main reason for the inaccuracy in estimating N uptake, ANS and ANR is that these factors are very dependent on weather conditions. Improvement of the estimation of the ANS on managed grasslands, based on actual growing conditions, is investigated (Hassink, pers. comm.). Another improvement of the accuracy of the equation might be a closer achievement of the desired DM yield and N uptake. This can be realized by correcting the values for N uptake and ANR or by adjusting the growing period of the cuts. However, growing periods of about six weeks for a grazing cut are not desired for economic reasons.

Conclusions

SANS offers opportunities to optimize the N application per cut to actual requirements, especially on sand and clay soils. The amount of N applied with SANS depends on growing conditions. If poor, N applied with SANS appears lower and if favourable, higher than N recommendations. Use of SANS needs to be improved by a better estimation of the mineralization rate (ANS), N recovery (ANR) and N uptake based on actual growing conditions.

References

Hassink J 1991 Effect of grassland management on N mineralization potential, microbial biomass and N yield in the following year. Neth. J. Agric. Sci. 40, 173–185.

MANMF 1990 Advies van de commissie Stikstof. Eds. F R Goossensen and P C Meeuwissen. Ministry of Agriculture, Nature Management and Fisheries, The Hague, The Netherlands.

Prins W H, Van Burg P F J and Wieling H 1980 The seasonal response of grassland to nitrogen at different intensities of nitrogen fertilization, with special reference to methods of response measurements. *In* Proceedings of an International Symposium of the EGF on the Role of Nitrogen in Intensive Grassland Production. Eds. W H Prins and G H Arnold. pp 35–49. Pudoc, Wageningen, The Netherlands.

Unwin R and Vellinga Th V 1994 Fertilizer recommendations for intensively managed grassland. *In* Grassland and Society. Proceedings of the 15th General Meeting of the EGF. Eds. L 't Mannetje and J Frame. pp 590–602. Wageningen Press, Wageningen, The Netherlands.

Wouters A P and Vellinga Th V 1994 Development of System of Adjusted Nitrogen Supply (SANS) for grassland. *In* Workshop Proceedings of the 15th General Meeting of the EGF. pp 258–261. Neth. Soc. for Grassl. and Fodder Crops, Wageningen, The Netherlands.

5. ASPECTS RELATED TO MODELLING

Plant and Soil **181**: 13–17, 1996.

Nitrogen dynamics of crop and soil subjected to different water and nitrogen inputs, including daily irrigation and steady-state fertilization - measurements and modelling

Olof Andrén, Thomas Kätterer, Roger Pettersson, Malin Flink and Ann-Charlotte Hansson
Department of Ecology and Environmental Research, Swedish University of Agricultural Sciences, P.O. Box 7072, S - 750 07 Uppsala, Sweden *

Key words: cereals, clay soil, grass ley, roots, simulation modelling

Abstract

Major carbon and nitrogen fluxes through crop and soil were studied in a series of field experiments. Barley, winter wheat, a grass mixture cut for hay and the energy crop reed canary-grass *(Phalaris arundinacea)* were studied.

The treatments ranged from drought to daily irrigation/fertilization with high doses of water and nitrogen. Crop biomass and nitrogen dynamics above and below ground and incident light as well as soil temperature, moisture and mineral N content were monitored. Litter decomposition experiments were also performed in the field.

The results were used to parameterize, validate and improve a set of soil/plant simulation models. Selected experimental results and experiences gained from the water, C and N budgeting and modelling work are presented.

Introduction

Crop production must be efficient; i.e. the yield should be as high as possible and the cost of cultivation as low as possible. In addition, positive effects on the environment should be maximized and negative effects minimized. Although these two constraints may seem to be incompatible, this is not always the case. For example, increasing the precision of nitrogen fertilization can increase protein yields while simultaneously decreasing N losses to the environment.

Recently, models and methods for fertilization according to the daily needs of a plant have been developed (Ingestad, 1988; Ingestad and Ågren, 1988, 1992). These have been applied in laboratory growth chambers where the roots have been subjected to frequent sprayings of low-concentration nutrient solutions (Ingestad and Lund, 1986). Such procedures result in plants with a well-defined nutrient status, which are necessary for high-precision studies such as assessments of metal toxicity (Göransson and Eldhuset, 1991) or the effects of increased CO_2 concentrations (Pettersson et al., 1993a). In the field, the nutrient buffering capacity of the soil (e.g., N minera-

lization/immobilization) as well as weather influences make the modelling of water and nutrient dosage more complicated. However, field experiments with daily irrigation/fertilization according to plant growth rate and nutrient demand have been performed in forest (Aronsson and Elowson, 1980; Christersson, 1986; Linder, 1987) as well as in arable land (van Koninckx-loo, 1990).

Based on this background, we decided to perform a series of experiments with daily irrigation/fertilization of arable crops in Sweden. The difference between the assumed crop N uptake and net mineralization was supplied as fertilizer to meet the calculated daily N demand of the crop. We measured the dynamics of plant biomass and N contents above (Flink et al., 1995; Pettersson et al., 1993b) and below ground (Andrén et al., 1993a; Hansson et al., 1992; Kätterer et al., 1993) as well as soil water dynamics (Andrén et al., 1991). Further, we studied litter decomposition and N dynamics in relation to soil temperature and moisture dynamics (Andrén et al., 1992, 1993b) and mapped the field with regard to soil moisture, texture and nutrient contents (Andrén et al., 1990; Rajkai and Rydén, 1992). The present paper is an overview of the results obtained and the methods and models used in the series of experiments made during the period 1988–1994.

* FAX No.: +46 18 673430
Plant and Soil is the original source of publication of this article. It is recommended that this article is cited as: *Plant and Soil* **181**: 13–17, 1996.

14

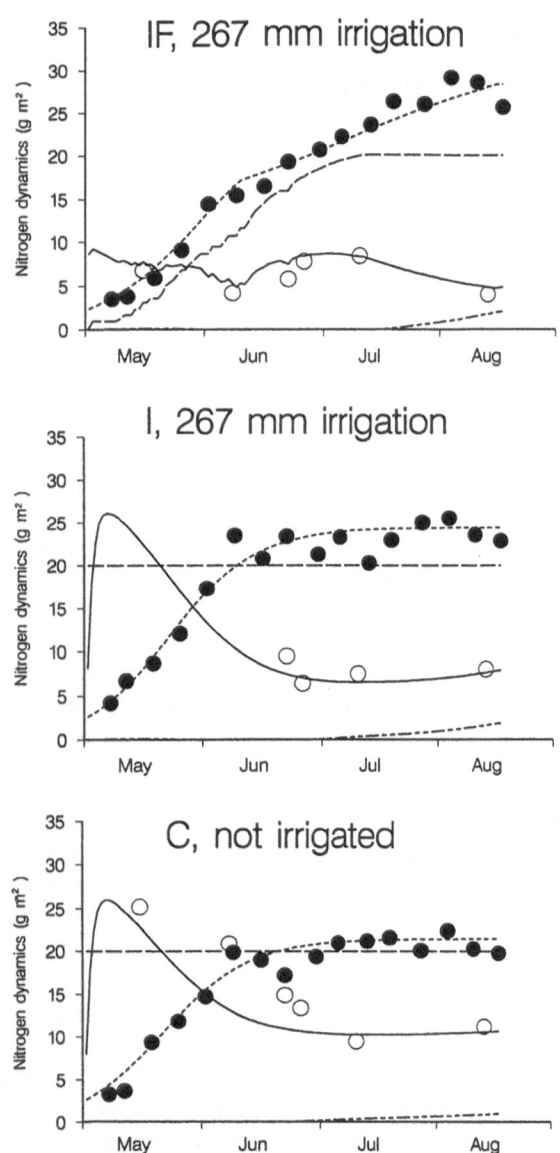

IF, 267 mm irrigation

I, 267 mm irrigation

C, not irrigated

Figure 1. Measured (symbols) and simulated (lines) dynamics of nitrogen in winter wheat and soil between 2 May and 18 August 1989 (Kätterer and Andrén, 1995). N in crop, including roots (●, - - -); Cumulative N fertilization (- - -) according to Ingestad (1988) in IF or as a single dose on 2 May in C (Control) and I (Irrigated daily); Soil mineral N (0–90 cm, ○, —); Net N mineralization (-- — --).

Materials and methods

Site, treatments and measurements

The experiments were carried out at Ultuna, about 5 km south of Uppsala, in central Sweden, 59°48' N and 17°38' E. Uppsala has a cold-temperate and semi-humid climate with a mean annual precipitation of 520 mm and an annual mean screen temperature of +5.4

°C. The coldest month is January (−5.3 °C) and the warmest is July (+16.7°C). The soil of the experimental field is a Fluventic Eutrochrept (Andrén et al., 1990; Soil Survey Staff, 1987) with 53% clay, 2.8% C, 0.28% N, and a pH of 7.4 in the topsoil (0–30 cm).

The experimental layout was a randomized block with four replicates, and the same basic layout was used all years. However, to avoid residual effects of earlier treatments the experiment was moved within the field each time a new crop was grown. Before the first experiment in 1988, the field had been subjected to three years of black fallow, i.e. cultivation without a crop. At suitable intervals, the entire field was fertilized with PK and micronutrients to minimize the risk of deficiencies occurring in nutrients other than N. A series of experiments were performed with barley, winter wheat, a grass ley mixture and reed canary-grass, *Phalaris arundinacea* (Table 1). Applications of water (I treatment) and fertilizer (in the IF treatments) were made through a drip-tube system, controlled by a personal computer (Biotronic, Uppsala, Sweden; Ingestad, 1988). Irrigation tubes with drip holes spaced 50 cm apart were placed in each row (12.5 cm spacing), resulting in 16 dripping points m^{-2}.

Above-ground biomass was sampled by making repeated clippings (Flink et al., 1995; Pettersson et al., 1993a), and roots were monitored using soil coring, ingrowth cores and rhizotrons (Andrén et al., 1993a; Hansson et al., 1992; Kätterer et al., 1993). Straw and root litter mass loss and N dynamics were studied using litter-bags (Andrén et al., 1992, 1993b). Soil moisture was monitored gravimetrically as well as by using capacitance probes (Andrén et al., 1991) and time-domain reflectometry (Rajkai and Rydén, 1992). Soil mineral N was monitored by making repeated soil corings to 90 cm depth. In each treatment, soil temperature was measured at 10 cm depth with thermocouples connected to a Campbell CR 10 logger (Campbell Inc., Logan, Utah, USA). Incident light and light at the soil surface were also monitored using the logger.

Models used

The nitrogen fertilizer dosage model used for the daily fertilization is based on the logistic growth model, which is commonly written as (Von Bertalanffy, 1957):

$$dM/dt = R_{G,max}M(1 - M/M_{max}) \qquad (1)$$

where M = plant biomass, $R_{G,max}$ = the maximum relative growth rate and M_{max} = maximum crop biomass. In the dosage model (Ingestad, 1988), the nitrogen

Table 1. Crops and treatments used within the project. Treatments: D = Drought, induced by a plastic screen over the crop; C = Control, fertilized with solid N-fertilizer in the spring (grass ley in the spring and after each cutting); I = Irrigated daily, fertilized as C; IF = Irrigated and Fertilized daily through a drip-tube system. Irrigation (mm), N fertilization (g m^{-2})

Growing season	Crop	Treatment$_{irrigation, \text{ total nitrogen dose}}$				
		1	2	3	4	5
1988	Barley	$D_{0,0}$	$C_{0,0}$	$I_{266,0}$	$IF_{266,26}$	–
1989	Winter wheat	$D_{0,20}$	$C_{0,20}$	$I_{267,20}$	$IF_{267,20}$	–
1990	Grass ley	$C1_{0,0}$	$C2_{0,20}$	$I_{205,20}$	$IF1_{205,20}$	$IF2_{205,20}$
1991	Grass ley	$C1_{0,0}$	$C2_{0,32}$	$I_{370,32}$	$IF1_{370,32}$	$IF2_{370,48}$
1992	Reed-grass	*Whole field sown in June, 6 g N m^{-2} fertilization*				
1993	Reed-grass	$C1_{0,0}$	$C2_{0,15}$	$I_{500,15}$	$IF1_{500,15}$	$IF2_{500,42}$
1994	Reed-grass	*Not yet analyzed; treatments similar to those in 1993*				

uptake rate and, consequently, the optimum N fertilizer dosage (if no contribution from the soil is assumed) is expressed as:

$$dN/dt = R_{G,max}N - bN^2 \qquad (2)$$

where b is an aging factor specific for the crop.

According to the nutrient productivity theory (Ingestad and Ågren, 1992), $R_{G,max}$ is equal to the maximum relative N uptake rate and b is related to $R_{G,max}$ and and the maximum amount of nitrogen in the crop (N_{max}):

$$b = R_{G,max}/N_{max} \qquad (3)$$

Equations (2) and (3) can be combined to give

$$dN/dt = R_{G,max}N - R_{G,max}N^2/N_{max} \qquad (4)$$

which is equivalent to Equation (1). This equation can be integrated and rearranged to:

$$N = N_0 N_{max}/(N_0 + (N_{max} - N_0)e^{-R_{G,max}t}) \qquad (5)$$

where N_0 = the initial N mass and t = time. This form of the logistic equation was used to calculate the N fertilization rate in the daily fertilized treatments. It was also fitted to crop biomass and N data to obtain $R_{G,max}$, M_{max} and N_{max} (Flink et al., 1995).

Straw and root litter decomposition dynamics were modelled by assuming first-order kinetics (a constant proportion lost per day) affected by soil temperature and water tension:

$$dM/dt = kMf(T, \Psi) \qquad (6)$$

where M = remaining litter mass, k = the rate constant (day^{-1}), T = measured daily soil temperature and Ψ = measured soil water tension. Both T and Ψ were converted to multiplication factors ranging from 0 to 1; T by assuming Q_{10} = 2 and Ψ by assuming a log-linear influence on mass loss rate (Andrén and Paustian, 1987; Andrén et al., 1992, 1993b).

Volumetric soil water content was converted to soil water tension using a three-parameter model fitted to soil water retention data (Andrén et al., 1991, 1993b). Converted values were corrected for drying-related shrinkage of the clay soil using the model devised by Jarvis et al. (1991). To obtain a comprehensive description of heat and water fluxes in the soil, including daily values of soil temperature and moisture in different depth strata, we used (Andrén et al., 1993b) the physically based SOIL simulation model (Jansson, 1991). Soil N fluxes were described using the SOILN model (Johnsson et al., 1987; Paustian et al., 1990) using heat and water flows from the SOIL model as input. We are currently trying to validate and refine the SOILN-CROP model (Eckersten and Jansson, 1991). The CROP submodel is based on the light interception concept and empirical allometric functions. Growth is the driving force for nitrogen uptake, but is itself limited by the availability of mineral N in the soil.

Results and discussion

Except for 1988, when only IF received N fertilizer, there were at least three treatments each year receiving the same amount of N fertilizer (Treatments 2–4; see Table 1). Perhaps the most interesting comparison of

Table 2. Maximum dry mass and N mass of the crops, including below-ground parts. Relative values: IF or IF1 set to 100. Sums over the growing season (1 May to 30 September) of rain (mm) and daily mean air temperature (°C). ND = not yet analyzed/compiled

Growing season	Crop	Treatment$_{dry\ mass,\ N\ mass}$					Sums 1 May–30 Sept	
		1	2	3	4	5	Rain	Temperature
1988	Barley	$D_{75,\ 49}$	$C_{81,\ 62}$	$I_{90,\ 59}$	$IF_{100,\ 100}$	-	310	2190
1989	Winter wheat	$D_{66,\ 60}$	$C_{79,\ 75}$	$I_{97,\ 84}$	$IF_{100,\ 100}$	-	152	2170
1990	Grass ley[a]	$C1_{43,\ 29}$	$C2_{111,\ 95}$	$I_{111,\ 102}$	$IF1_{100,\ 100}$	$IF2_{118,\ 114}$	315	2040
1991	Grass ley[b]	$C1_{74,\ 52}$	$C2_{93,\ 92}$	$I_{107,\ 100}$	$IF1_{100,\ 100}$	$IF2_{102,\ 117}$	391	1990
1992	Reed-grass	*Whole field sown in June, minor samplings only*					275	2200
1993	Reed-grass	$C1_{74,\ ND}$	$C2_{85,\ ND}$	$I_{97,\ ND}$	$IF1_{100,\ ND}$	$IF2_{96,\ ND}$	229	1890
1994	Reed-grass	*Samples not yet analyzed*					ND	ND

[a] Establishment year; one cut only, root biomass not sampled.

[b] Above-ground data from the 3rd of four cuttings (5 August), below-ground data from 11 September.

maximum dry mass and N mass can be made between treatments 3 and 4, which differ only in when and how the fertilizer was applied (Tables 1, 2). Clearly, the only major difference found was in the N mass of winter wheat during 1989 (1988 excluded, see above), where the application of N according to the steady-state concept increased the N uptake. This increase was attributed to the relatively more extensive root system in IF, induced by the relative shortage of N early in the growing season, and post-anthesis N uptake by the crop in IF (Flink et al., 1995; Kätterer et al., 1993; Kätterer, 1995). The small differences in grass ley biomass and N uptake were to be expected for two reasons: First, 1990 was an establishment year and I was fertilized four times in 1991, making I fairly similar to IF, which received daily fertilization. Second, only one harvest is shown here, but the sum of harvests would be a better measure. For example, the total nitrogen harvest in 1991 was in IF1 10% higher than in I.

In a strict sense, crop comparisons are not possible since the weather differed between growing seasons. For example, the precipitation in 1989 was less than 40% of that in 1991 (Table 2). However, one advantage of having simulation model descriptions of biomass and N dynamics is that when the model is parameterized for one year, weather data for another year can be used as driving variables. Then one can answer questions like: What would have happened if we had grown a grass ley in 1989?

An example of simulated N dynamics is given in Figure 1. The measured N in the crop agreed well with the output of a logistic model, although in IF it was necessary to decrease $R_{G,max}$ in early June, probably due to to N shortage (Kätterer and Andrén, 1995). Simulated soil mineral N amounts also agreed reasonably

well with measured mineral N amounts. Root turnover was included in the simulations (Kätterer and Andrén, 1995). However, since data on the cumulative net N mineralization were lacking the models could not be validated in this regard. Concerning treatment differences, it is clear that the daily, sigmoidal N dosage strategy (IF) combined with irrigation increased crop N uptake and reduced soil mineral N contents. The mineral N content in the irrigated/fertilized treatment remained lower through the following autumn, winter and early spring, thereby reducing the risk for N leaching (Kätterer, 1995; Kätterer and Andrén, 1995). The differences in net N mineralization due to irrigation are partly corroborated by the litter-bag experiment by Andrén et al. (1993b), wherein less straw litter N immobilization was found in IF than in C.

In conclusion, the computer-controlled dosage system functioned well, allowing high-precision daily additions of water and nutrients. The effects of the treatments, assessed both as measured data and as model parameters and variables, have been reasonably consistent. The daily addition of nitrogen seems to increase N uptake while reducing the soil mineral N content. One area where more information is needed concerns the contribution of nitrogen resulting from the mineralization of soil organic matter during the growing season. Although modelled in SOILN-CROP, validation data, e.g. from incubations of soil under controlled conditions, would be of great interest.

To take advantage of the increased production potential due to the increasing atmospheric CO_2 concentration more precision will be required in water and nutrient application regimes. The concepts described in the present paper should therefore increase in merit in the future.

Acknowledgements

This work was supported by grants from the Swedish Council for Forestry and Agricultural Research, the Oscar and Lili Lamm Foundation and the Swedish Foundation for Plant Nutrient Research. We are grateful to D Tilles for improving the language.

References

Andrén O, Hansson A-C and Végh K 1993a Barley nutrient uptake, root growth and depth distribution in two soil types in a rhizotron with vertical and horizontal minirhizotrons. Swedish J. Agric. Res. 23, 115–126.

Andrén O and Paustian K 1987 Barley straw decomposition in the field: a comparison of models. Ecology 68, 1190–1200.

Andrén O, Rajkai K and Rajkai Végh K 1990 Spatial variation of soil physical and chemical properties in an arable field with high clay content. Swedish University of Agricultural Sciences, Department of Ecology and Environmental Research, Uppsala. Report 40. 17 p.

Andrén O, Rajkai K and Kätterer T 1991 A nondestructive technique for studies of root distribution in relation to soil moisture. Agric. Ecosys. Environ. 34, 269–278.

Andrén O, Rajkai K and Kätterer T 1993b Water and temperature dynamics in a clay soil under winter wheat - influence on straw decomposition and N immobilization. Biol. Fert. Soils 15, 1–8.

Andrén O, Steen E and Rajkai K 1992 Modelling the effects of moisture on barley straw and root decomposition in the field. Soil Biol. Biochem. 24, 727–736.

Aronsson A and Elowson S 1980 Effects of irrigation and fertilization on mineral nutrients in Scots pine needles. In Structure and Function of Northern Coniferous Forests - an Ecosystem Study. Ed. T Persson. Ecol. Bull. 32, 219–228.

Christersson L 1986 High technology biomass production by Salix clones on a sandy soil in southern Sweden. Tree Physiol. 2, 261–272.

Eckersten H and Jansson P-E 1991 Modelling water flow, nitrogen uptake and production for wheat. Fert. Res. 27, 313–329.

Flink M, Pettersson R and Andrén O 1995 Growth dynamics of winter wheat in the field with daily fertilization and irrigation. J. Agron. Crop Sci. 174, 239–252.

Göransson A and Eldhuset T D 1991 Effects of aluminium on growth and nutrient uptake of small Picea abies and Pinus sylvestris plants. Trees 5, 136–142.

Hansson A-C, Steen E and Andrén O 1992 Root growth of daily irrigated and fertilized barley investigated with ingrowth cores, soil cores and minirhizotrons. Swedish J. Agric. Res. 22, 141–152.

Ingestad T 1988 A fertilization model based on the concepts of nutrient flux density and nutrient productivity. Scand. J. For. Res. 3, 157–173.

Ingestad T and Lund A-B 1986 Theory and techniques for steady state nutrition and growth of plants. Scand. J. For. Res. 1, 439–453.

Ingestad T and Ågren G I 1988 Nutrient uptake and allocation at steady-state nutrition. Physiol. Plant. 72, 450–459.

Ingestad T and Ågren G I 1992 Theories and methods on plant nutrition and growth. Physiol. Plant. 84, 177–184.

Jansson P-E 1991 Soil water and heat model. Technical description. Swedish University of Agricultural Sciences, Division of Hydrotechnics, Uppsala. Report 165. 72 p.

Johnsson H, Bergström L, Jansson P-E and Paustian K 1987 Simulated nitrogen dynamics and losses in a layered agricultural soil. Agric. Ecosys. Environ. 18, 333–356.

Jarvis N, Jansson P-E, Dik P E and McAffee M 1991 Modelling water and solute transport in macroporous soil. I. Model description and sensitivity analysis. J. Soil Sci. 42, 59–70.

Kätterer T 1995 Nitrogen dynamics in soil and winter wheat subjected to daily fertilization and irrigation - measurements and simulations. Agr.D. Thesis. Swedish University of Agricultural Sciences, Department of Ecology and Environmental Research, Uppsala. Report 81. 32 p.

Kätterer T, Hansson A-C and Andrén O 1993 Wheat root biomass and nitrogen dynamics - effects of daily irrigation and fertilization. Plant and Soil 151, 21–30.

Kätterer T and Andrén O 1995 Measured and simulated nitrogen dynamics in winter wheat and a clay soil subjected to drought stress or daily irrigation and fertilization. Fert. Res. (In press).

Linder S 1987 Responses to water and nutrients in coniferous ecosystems. In Potentials and Limitations of Ecosystem analysis. Eds. E-D Schulze and H Zwölfer. Ecol. Stud. 61, 180–202.

Paustian K, Bergström L, Jansson P-E and Johnsson H 1990 Ecosystem dynamics. In Ecology of Arable Land - Organisms, Carbon and Nitrogen Cycling. Eds. O Andrén, T Lindberg, K Paustian and T Rosswall. Ecol. Bull. 40, 127–180.

Pettersson R, Andrén O and Végh K 1993a Growth and nutrient uptake of spring barley under different water and nutrient regimes. Swedish J. Agric. Res. 23, 171–179.

Pettersson R, McDonald A J S and Stadenberg I 1993b Response of small birch plants (Betula pendula Roth.) to elevated CO_2 and nitrogen supply. Plant Cell Environ. 16, 1115–1121.

Rajkai K and Rydén B E 1992 Measuring areal soil moisture distribution with the TDR method. Geoderma 52, 73–85.

Soil Survey Staff 1987 Keys to Soil Taxonomy (third printing). SMSS technical monograph no. 6, Ithaca, New York, USA. 280 p.

Van Koninckxloo M 1990 Ecophysiological approach of the fertilization of crops on the basis of the nutrients productivity and nutrients flux density concepts. Agr.D. Thesis, Université libre de Bruxelles, section interfacultaire d'agronomie, Bruxelles, Belgium.

Von Bertalanffy L 1957 Quantitative laws in metabolism and growth. Quart. Rev. Biol. 32, 217–231.

Section editor: R Merckx

Plant and Soil **181**: 19–23, 1996.

Modelling the uptake of nitrate by a growing plant with an adjustable root nitrate uptake capacity

I. Model description

Jan Buysse, Erik Smolders and Roel Merckx
*Laboratory of Soil Fertility and Soil Biology, K.U. Leuven, Kardinaal Mercierlaan 92, 3001 Heverlee, Belgium**

Key words: diffusion, model, nitrate, plant growth, spinach, uptake

Abstract

A new model is presented to predict the plant uptake of nitrate supplied by diffusion and mass flow to its roots. Plant growth, root-shoot ratio and the plant's nitrate uptake capacity are all set dependent on the plant's N nutrition state. By thoroughly integrating processes occurring in both plant and soil, the model enables to control the relative importance of both under a wide range of different nutritional scenarios.

Abbreviations: Soil parameters: D_0 – diffusion coefficient in water (m^2 day^{-1}), D_e – diffusion coefficient in soil (m^2 day^{-1}), C – nitrate concentration in soil (mol m^{-3}), f – tortuosity (-), θ – volumetric moisture content (-), R – radial distance from root axis (m), *Plant parameters*: b_1, b_2 – parameters of biomass partitioning Equation (10), IR – interroot distance (m), K_mU – Michaelis-Menten constant of the uptake system (mol m^{-3}), K_mNRA – Michaelis-Menten constant of nitrogen reduction system (mol g^{-1}), k_1, k_2, k_3 – parameters of growth model Equation (9), L_v – Root length density (m m^{-3}), $NO_{3\,set}^-$ – Set point of the cytoplasmatic nitrate pool (mol g^{-1} dw), $NO_{3\,c}^-$ – cytoplasmatic nitrate concentration (mol g^{-1} dw), $NO_{3\,v}^-$ – vacuolar nitrate concentration (mol g^{-1} dw), NRA_{max} – maximum nitrate reductase activity (mol g^{-1} dw day^{-1}), N_{re} – reduced nitrogen content (mol), N_{remax} – maximum reduced N concentration in the plant (mol g^{-1} dw), P – partitioning coefficient of nitrate between cyplasm and vacuole, R(1)–root radius (m), RGR–relative growth rate (day^{-1}), U – uptake rate (mol day^{-1} m^{-2}), U_{max} – maximum uptake rate (Eq. 6) (day^{-1} m^{-2}), Vo – water flux at root surface (m day^{-1}), W_r – root dry weight (g), W_{sh} – shoot dry weight (g), X – model parameter: number of root compartments, Y–model parameter: number of nodes.

Introduction

Many modellers have concentrated on the description of nutrient uptake by roots from soils. Together with these models, preliminary assumptions are outlined giving an indication of the processes that the authors consider to have only negligible influence on the model output. However, when comparing various models, it can observed that these preliminary assumptions can differ considerably between the models. Often and depending on the author's background, either soil processes or the nutrient uptake as regulated by the plant, are described very sketchy. As a consequence, the over-

all relevance of these models for various nutritional situations can be questioned.

In line with these considerations, we advance in this paper a model for nitrate uptake by growing plant roots expliciting a link between the overall nitrogen nutrition state, the overall plant growth and the partitioning of biomass between root and shoot parts of the plant. Additionally, both the effect of increasing root density in a limited soil volume and the effect of the plant's nitrogen nutrition state on the kinetic nitrate uptake properties are taken into account. By integrating all these factors into one model, it becomes possible to assess their relative importance under a given set of environmental conditions.

---

* FAX No.: +3216321997
Plant and Soil is the original source of publication of this article. It is recommended that this article is cited as: *Plant and Soil* **181**: 19–23, 1996.

20

Model description

Soil compartment

All nitrate is supposed to be present in the soil solution at the start of the simulation. N mineralisation or NO_3^- leaching are not included in this model. All nitrogen in the soil solution is present as nitrate. When nutrient deficiency occurs, nitrogen is supposed to be the only limiting nutrient.

Nitrogen transport towards the roots occurs by both diffusion and mass flow. Diffusion and mass flow are described using the equation of continuity of De Willigen and Van Noordwijk (1987)

$$\theta \frac{\delta C}{\delta t} = \frac{D_e}{R} \frac{\delta}{\delta R} \left[R \frac{\delta C}{\delta R} \right] - V_r \frac{\delta C}{\delta R} \qquad (1)$$

The effective diffusion constant is defined as

$$D_e = D \times f \times \theta \qquad (2)$$

The equation of continuity was solved numerically according to Smolders (1993) using the finite differences method and integration with respect to time with explicit Euler's method with a time step of 0.00001 day. Finite differences were calculated for a limited number of nodes (initial number = 200) which were separated at equal radial distance dR from each other.

Boundary conditions at R=R(0) and R=R(200) were explicited by calculating the balance between root uptake, diffusion, mass flow and nitrate present at the corresponding soil volume.

Roots were supposed to grow maximally dispersed, which corresponds in a homogeneous soil to new roots growing in the zones with the highest remaining nitrate. It was anticipated that no root competition would occur before a certain root density L_v0 had been reached. Root length density was calculated as

$$L_v = L/V \qquad (3)$$

Average interroot distance at root length density L_v was calculated as

$$IR = \frac{2}{(\pi L_v)^{0.5}} \qquad (4)$$

The root length density L_v0 corresponds with an average interroot distance of two times the thickness of 200 root nodes. The assumption of no root competition at this point can be controlled after simulation, and in this way either dR or the number of nodes can be adjusted before a new simulation.

Before root length density L_v0 was attained, total root length was subdivided in parts corresponding to 1% of the root length L_0 at root density L_v0 (Fig. 1). Within these root length parts an identical nitrate concentration profile around the root was assumed. Total root length thus consisted of maximally 99 parts with equal length (1% of L_0) and a residual growing part with total root length of less than 1% of L_0. When root length of this residual part exceeded 1% of L_0, a new root length compartment was started, the soil surrounding it having the initial nitrate concentration.

Once the total root length L_0 was reached, another way of subdivisioning root length was chosen. As a first action, and for reasons of simplicity, the nitrate profile around the already existing root length was made equal. This was done by calculating the average of the nitrate concentration at all nodes at that moment.

Since at root length L_0, the volume of the soil cilinders around the roots equalled the total potted soil volume, each new extension of the root length had to be accompanied with a reduction of the soil cilinder surrounding the roots. The moment of reduction of the soil cilinder and of simultanous addition of a new root length compartment was chosen at the moment when

$$IR = R(1) + 2 \times (y + 1/2) \times dR \qquad (5)$$

At that time, analogous to the method of Hoffland et al. (1990), all posterior root growth was localised in a soil compartment, where the nitrate concentration was set at the concentration which was calculated by averaging over the whole root length the concentration of the soil shell to be eliminated.

Root uptake

The root uptake of nitrate was calculated using the nitrate uptake model for spinach of Smolders (1993). This nitrate uptake model relates actual nitrate uptake rate to the nitrate concentration at the root surface and to internal nitrate concentration in the cytoplasma using an error activated feedback control model.

Principle of the model is that the actual maximal uptake capacity of the root surface depends on the physiologically possible maximal uptake rate and on the deviation of the actual cytoplasm nitrate concentration from a set-point optimal nitrate concentration. A similar concept but for potassium was proposed by Siddiqi and Glass (1986). Formulating this principle

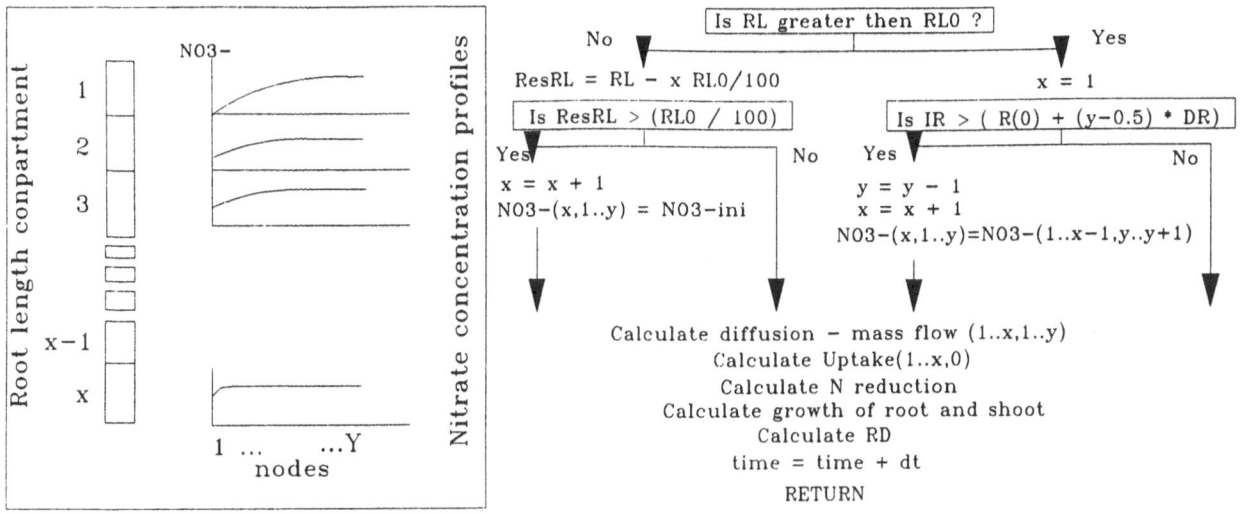

Figure 1. Flowchart and illustration of the algorithm used to subdivide the total soil volume into homogeneous soil shells surrounding the root length compartments.

mathematically results in the following equation

$$U = \frac{U_{max}[NO_{3\ set}^- - NO_{3\ c}^-]}{K_m U + C} \times C \qquad (6)$$

Plant growth

The nitrate pool in the plant is subdivisioned into nitrate in the cytoplasm and nitrate in the vacuole. A partitioning coefficient P determines the amounts actually present in both compartments.

$$NO_{3\ v}^- = P \times NO_{3\ c}^- \qquad (7)$$

Only nitrate present in the cytoplasm affects uptake and only this pool is available for reduction.

Nitrate reduction is described using Michaëlis-Menten kinetics.

$$\frac{dN_{re}}{(W_r + W_{sh})dt} = \frac{NRA_{max} \times NO_{3\ c}^-}{(K_m NRA + NO_{3\ c}^-)} \qquad (8)$$

Nitrate is reduced in the plant until a maximum level of reduced nitrogen compounds is reached. This maximum concentration is a parameter to be inputted in the model. The reduced nitrogen concentration determines the growth potential of the plant and the partitioning of the growth between root and shoot. It has been proven by several authors that in the exponential growth phase and until the maximal growth rate is reached, growth is linearly related with the internal reduced nitrogen concentration. However, when plant weight increases, growth rate per unit reduced nitrogen decreases due to senescence effects and internal shading. This can mathematically be formulated as (Smolders and Merckx, 1992):

$$RGR = \frac{k_1 \times (N_{re} - k_2)}{k_3 + (W_r + W_{sh})} \qquad (9)$$

The partitioning of biomass depends upon the reduced nitrogen concentration but is also influenced by the growth medium (Smolders et al., 1993). The biomass partitioning function is expressed as

$$\frac{dW_{sh}}{dW_r} = b_1 + b_2 \times N_{re} \qquad (10)$$

At each timestep, the biomass increase is thus divided over root and shoot following this equation. Root biomass is then converted in root length using typical root diameter values. The model as described above was explicitated in Fortran and run on a Vax workstation.

Discussion

This paper describes the transport towards and the uptake of nitrate by the root in a new, partially more complete and realistic model, hereby overcoming some of the inadequaties of other models.

A first important mechanism which is often overlooked in uptake models concerns the effects of an increasing root density. Barber (1984) does not consider increasing root density: in his models the distance between two roots is a constant, not varying in time. All root growth is thus supposed to explore new soil and to attain immediately the same density as elsewhere. Baldwin et al. (1973) on the other hand do consider increasing root density. An important difference between the presented model and the Baldwin model however is that our model assumes new root growth to occur in the zones with maximal nitrate concentration, whereas the Baldwin model assumes new root growth to occur at random.

The method which is used in our model to simulate the increasing root density by decreasing the number of soil shells around the root, is similar to the method used in the model of Hoffland et al. (1990). In the present model however, an additional mechanism is included to avoid abrupt discontinuities at the lower root densities.

A second frequent imperfection in uptake models concerns the zone of solute depletion around the roots. In steady state analytical models, such as the Baldwin-model, it is assumed implicitly, that depletion occurs over the whole soil volume. This assumption may however not hold at very low root densities and a low diffusability of the solute. Therefore, time dependent corrections of the zone of root exploitation are needed (Baldwin and Nye, 1974).

A third important mechanism which should not be neglected, is the variability of the kinetic parameters of root uptake, which is now generally accepted in plant physiology. The group of Siddiqi and Glass for example, has reported in numerous publications that the kinetic parameters of NO_3^- and K^+ are related to the internal N or K nutrition state respectively (Glass, 1989, and references therein). Also, Scaife (1989) took into account the internal nitrate concentration when calculating the net uptake rate of nitrate. This relationship, and with it the role of the plant's nutrition state, is neglected in many other uptake models. Approximations of the uptake function used instead are the zero sink assumption (Robinson et al., 1991), a linear relationship between the external concentration and uptake (Baldwin et al., 1973), a constant uptake independent of the external nutrient concentration (De Willigen and Van Noordwijk, 1987) or simple Michaelis-Menten kinetics with its parameters being constant in time (Barber, 1984).

General conclusions arising from these models may as a result be variant as well. Barber (1984) concludes that the maximum uptake rate per unit root surface has a major influence on nitrate uptake. A zero sink assumption as used by Robinson et al. (1991) excludes this influence from the beginning.

Analogous to the feedback signal of internal nitrate status on the nitrate uptake mechanism, most models do not take into account the effect of nitrate on root growth. In many cases, root growth is assumed to be constant, independent of either plant age or plant nutrient status.

Finally, also the differences between analytical and numerical models need to be considered. Advantages and disadvantages of numerical methods, such as the present one, to analytical methods have already been discussed several times (Nye and Tinker, 1977; Passioura and Cowan, 1968; Smethurst and Comerford, 1993). The most important disadvantage of numerical models is their complexity and the high computer requirements which are needed. An important advantage on the other hand is that numerical models enable to calculate the different solute concentrations at different zones of the total root length. Most analytical models assume an identical (steady-state) concentration of the solute over the whole root surface. The uptake by new roots growing in undepleted soils can therefore not be modelled in analytical models, unless, again, correction factors are added such as proposed by Yanai (1994).

Acknowledgements

J B and E S are indebted to the Nationaal Fonds voor Wetenschappelijk Onderzoek for their research position. This research was financed partially by a grant of the K U Leuven Onderzoeksfonds (OT/91/22).

References

Baldwin J P, Nye P H and Tinker P B 1973 Uptake of solutes by multiple root systems from soil. III. A model for calculating the solute uptake by a randomly dispersed root system developing in a finite volume of soil. Plant and Soil 38, 621–635.

Baldwin J P and Nye P H 1974 A model to calculate the uptake by a developing root system or root hair system of solutes with concentration variable diffusion coefficients. Plant and Soil 40, 703–706.

Barber S A 1984 Soil Nutrient Availability. A mechanistic approach. Wiley-Interscience, New York, USA. 398 p.

De Willigen P and Van Noordwijk M 1987 Roots, plant production and nutrient use efficiency. Thesis, Agricultural University, Wageningen, The Netherlands.

Glass A D M 1989 Plant Nutrition: An Introduction to Current Concepts. Jones and Bartlett, Boston, MA, USA. 212 p.

Hoffland E, Bloemhof H S, Leffelaar P A, Findenegg G R and Nelemans J A 1990 Simulation of nutrient uptake by a growing root system considering increasing root density and interroot competition. Plant and Soil 124, 149–155.

Nye P H and Tinker P B 1977 Solute Movement in the Soil-Root System. Blackwell Scientific Publications, Oxford, UK.

Passioura J B and Cowan I R 1968 On solving the non-linear diffusion equation for the radial flow of water to the roots. Agric. Meteorol. 5, 129–134.

Robinson D, Lineham D J and Caul S 1991 What limits nitrate uptake from soil? Plant Cell Environ. 14, 77–85.

Scaife A 1989 A pump/leak/buffer model for plant nitrate uptake. Plant and Soil 114, 139–141.

Siddiqi M Y and Glass A D M 1986 A model for the regulation of K^+ influx, and tissue potassium concentrations by negative feedback effects upon plasmalemma influx. Plant Physiol. 81, 1–7.

Smethurst P J and Comerford N B 1993 Simulating nutrient uptake by single or competing and contrasting root systems. Soil Sci. Soc. Am. J. 57, 1361–1367.

Smolders E 1993 Kinetic aspects of the soil-to-plant transfer of nitrate. Thesis, Faculty of Agricultural Sciences, KU Leuven, Belgium.

Smolders E and Merckx R 1992 Growth and shoot:root partitioning of spinach plant as affected by nitrogen supply. Plant Cell Environ. 15, 795–807.

Smolders E, Buysse J and Merckx R 1993 Growth analysis of soil-grown spinach plants at different N regimes. Plant and Soil 154, 73–80.

Yanai R D 1994 A steady state model of nutrient uptake accounting for newly grown roots. Soil Sci. Soc. Am. J. 58, 1562–1571.

Section editor: R Merckx

O. Van Cleemput et al. (eds.), Progress in Nitrogen Cycling Studies, 401–405, 1996.
© 1996 *Kluwer Academic Publishers.*

Measurement and modelling of the effects of crop residues on nitrate loss from clay loam and sandy loam soils

Katrina Castle, Jonathan Arah and Andrew Vinten
Soils Department, SAC, School of Agriculture, West Mains Road, Edinburgh, EH9 3JG, UK

Key words: denitrification, leaching, model, nitrate, residue, texture

Abstract

Drained plots of different textures were fertilised with ^{15}N labelled tracer, following incorporation of a range of organic matter residues, to compare measurements of denitrification and leaching with predictions from a mechanistic model. Leaching from a clay loam soil was well simulated, while the model overestimated leaching from sandy loam plots. The model failed to predict the extent or the time course of denitrification on both soils. Improving the model may require fuller treatments of gas transport through the soil profile, structural and organic matter heterogeneity, and the relationship between organic matter decomposition and soil temperature.

Introduction

Prediction and control of nitrate (NO_3^-) leaching to ground and surface waters depends upon understanding and modelling the transport and reduction of NO_3^- in soil. ^{15}N-labelled fertiliser was used in a tracer experiment on drained plots of different textures subjected to different organic matter incorporation treatments to compare measurements of denitrified gas emissions and NO_3^- leaching with predictions from a model that utilises a mechanistic approach to simulate N losses.

The model SLIM (Addiscott and Whitmore, 1991) simulates the transport of solute through structured soils in which part of the water can be considered mobile and part immobile. The soil is modelled as a series of layers: a fraction of the mobile water moves on to the next layer each day. Solute concentrations in the mobile and immobile water fractions are partially equilibrated at each time step, the extent of the equilibration being determined by the equivalent aggregate size. Bypass flow allows transport of water and solute from the top layer to the drainage depth on days when the rainfall exceeds the available pore space at the surface. This model was reasonably successful in simulating leaching behaviour (Vinten and Redman, 1990) at the clay loam site used in the current work.

We linked SLIM to a capillary bundle treatment of denitrification used in the model ANIMO (Rijtema

and Kroes, 1991). This treatment pictures the air-filled pore space at any particular moisture content as a collection of randomly-distributed cylindrical pores of a (uniform) radius determined by the water release characteristic of the soil. Around each equivalent air-filled pore the radius of oxic saturated soil is calculated by steady-state solution of Fick's Law. Denitrification occurs within the remaining (anoxic) soil at a rate equal to the potential rate of organic matter decomposition, provided sufficient NO_3^- is present.

Materials and methods

Field sites

Field work was carried out between July 1993 and March 1994 at two sites on the Bush Estate, some 10 km south of Edinburgh: Glencorse Mains (a clay loam soil derived from glacial till) and No.3 Field (a sandy loam soil derived from partially sorted glacial till). Figure 1 shows soil temperatures and rainfall data from a weather station situated near both field sites.

At Glencorse Mains, two 0.5 m² microplots were established on July 14th 1993 on plots with high organic residues from ploughed-in grass (approximately 3400 kg ha⁻¹ dry weight; 0.79% N) and two on low residue plots (bare fallow). Similar microplots at No.3 Field were set up on 29 October 1993 on

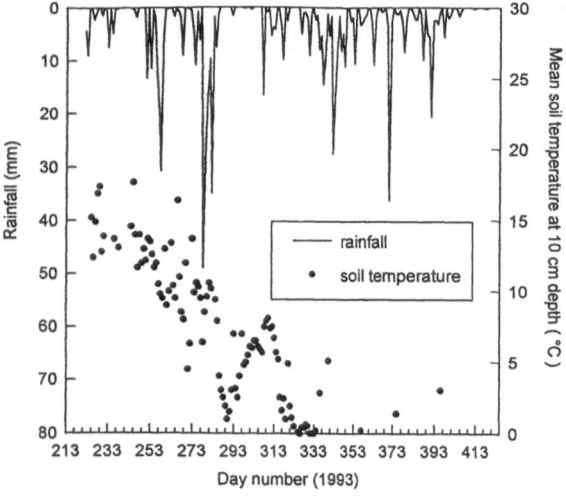

Figure 1. Rainfall and soil temperature at 10 cm depth.

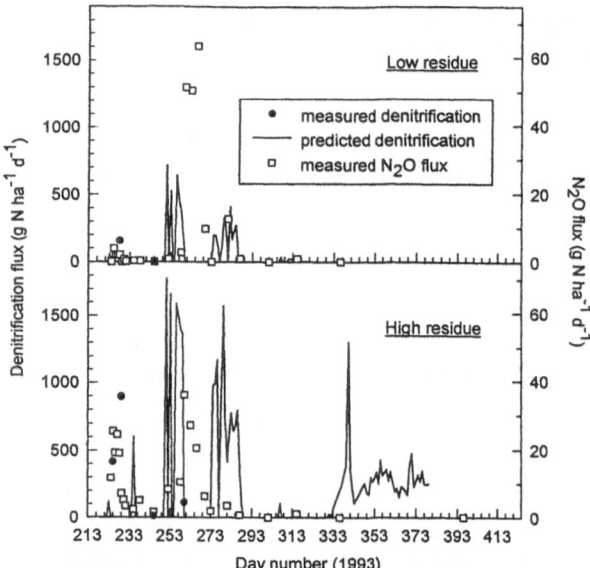

Figure 2. Gaseous N fluxes: Glencorse Mains (clay loam).

two high organic residue plots (approximately 3700 kg ha^{-1} straw fresh weight; 0.59% N), one plot with medium residues (ploughed-in barley stubble, approximately 1500 kg ha^{-1}; 1.5% N), and a low residue plot (bare fallow). Plots were ploughed and rotavated prior to fertilisation. Glencorse Mains was sown to grass, while No.3 Field remained unsown. Plots were fertilised with 60 kg N ha^{-1} as Ca(NO$_3$), and microplots with 60 kg N ha^{-1} 99 atom% ^{15}N-KNO$_3$. Glencorse Mains was fertilised on 12 August 1993 and No.3 Field on 5 November.

Gas sampling and analysis

Gas fluxes were measured by the closed chamber method. Nitrous oxide (N$_2$O) concentrations in syringe samples drawn from the chambers were analysed using electron capture gas chromatography. The ^{15}N content was analysed using a single-inlet VG Micromass triple detector mass spectrometer coupled to an elemental analyser (Carlo Erba 1400 Automatic Nitrogen Analyzer) which converts nitrogenous gases to N$_2$ by the Dumas oxidation-reduction procedure. Calculations of pool enrichments and denitrified gas fluxes were based on the ion current ratios of the 28, 29 and 30 N$_2$ isotopes, using the formulae of Arah (1992).

Drainage sampling and analysis

Plots at both sites were hydrologically isolated (Vinten et al., 1991), allowing monitoring of fertiliser N as it appeared in drainage. Concentrations of NO$_3^-$ and ammonium (NH$_4^+$) were determined by continu-

ous flow analysis using the methods of Henriksen and Selmer-Olsen (1970), copper and hydrazine replacing cadmium as a reducing agent, and Crooke and Simpson (1971), respectively. Concentrations of NH$_4^+$ were insignificant throughout the field trial.

Drainage samples for analysis of ^{15}NO$_3^-$ were acidified and concentrated on a sand bath. The ^{15}N enrichment was then determined by steam distillation followed by mass spectroscopy (Hauck, 1982). No ^{15}N data is available for the low residue plot at the sandy loam site due to flooding of the drainage pit through much of the winter. Samples for NO$_3^-$ analysis from this plot were taken directly from the drainage pipes.

Results

Denitrification

We had difficulty measuring total denitrified gas fluxes as the direct ^{15}N technique pushes the mass spectrometer to its limits (Arah et al., 1993); moreover, it is almost certain to give an underestimate of the true flux (Arah, 1992). N$_2$O emissions are not the same as predicted denitrification fluxes, but in the absence of reliable ^{15}N data the pattern may be relevant in comparing predicted and observed values; we therefore present observed N$_2$O fluxes in Figures 2 and 3 (note the different scales).

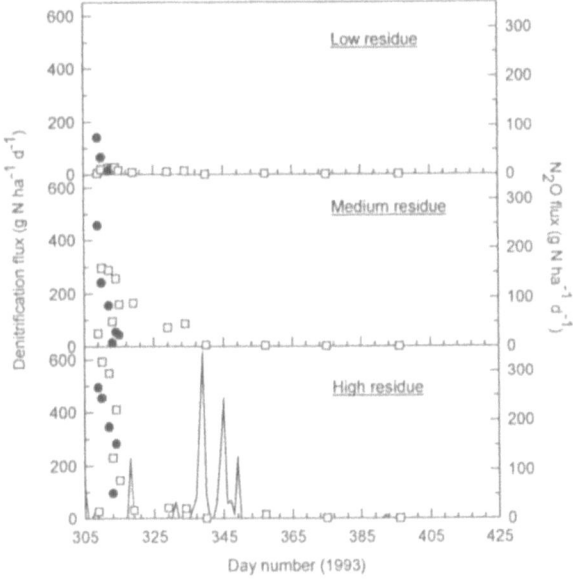

Figure 3. Gaseous N fluxes: No.3 Field (sandy clay loam). Symbols as for Figure 2.

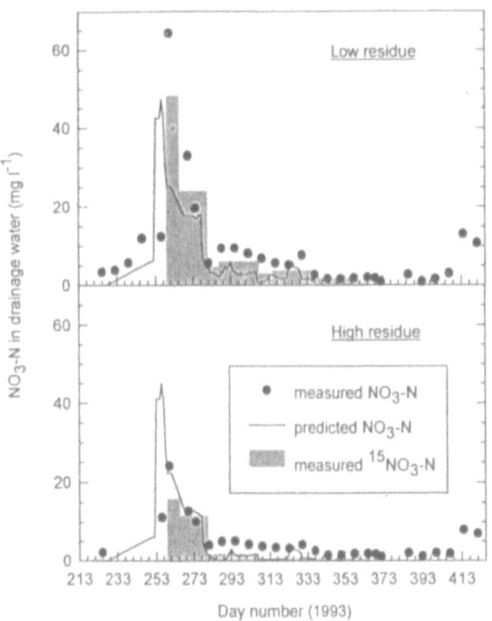

Figure 4. Nitrate leaching: Glencorse Mains (clay loam).

At Glencorse Mains the highest observed denitrification fluxes occurred immediately after fertiliser application on day 224 (Fig. 2). Predicted fluxes were lower than measured denitrification peaks, and shorter-lived than measured denitrified and N_2O emissions. Predictions of significant denitrification losses following heavy rain on day 253 slightly preceded and greatly exceeded observed N_2O emissions. The simulated peaks again continued for a shorter period of time.

At No.3 Field there were considerable losses of denitrified gases immediately after fertiliser application; these were not predicted, except as a small pulse on the high residue plots (Fig. 3).

Periods of significant denitrification loss were predicted later in the season from high residue plots at Glencorse Mains (day 273 to 285; day 333 to 373), and No.3 Field (day 338 to 350). Soil temperatures fell below 5 °C during these periods, and the simulated fluxes were not accompanied by any significant measured N_2O emissions (no ^{15}N measurements were made at this time).

The effect of high residues was largely as expected, with higher denitrified gas fluxes observed and predicted from the high residue plots at both sites.

Leaching

Leaching of NO_3^- from plots at Glencorse Mains occurred very rapidly, with concentrations in drainage down to near background by November (Fig. 4). The time course was well simulated. On the low residue plots measured and predicted leachate NO_3^- and $^{15}NO_3^-$ compared well, while predicted concentrations exceeded those measured from the high residue plots. Observed differences between high and low residue plots were greater than those predicted.

Simulations of NO_3^- in leachate from No.3 Field showed a sharper peak and a higher concentration than observed (Fig. 5). The model underestimated denitrification from this site, especially during the first few weeks after fertiliser addition (Fig. 3), and this may partly explain the overestimated NO_3^- concentrations.

At Glencorse Mains unlabelled NO_3^- concentrations were similar to those of $^{15}NO_3^-$, suggesting that most of the leached NO_3^- came from fertiliser. By contrast, differences between unlabelled and labelled NO_3^- concentrations from the high and medium residue plots at No.3 Field indicate high background NO_3^-, the effect of residual N from the summer's cropping.

404

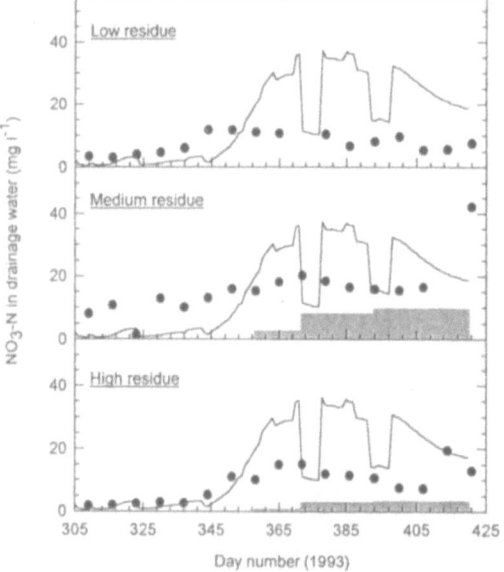

Figure 5. Nitrate leaching: No.3 Field (sandy clay loam). Symbols as for Figure 4.

Discussion

The time course of NO_3^- leaching from the low residue plots at Glencorse Mains was quite well simulated and concentrations were broadly comparable. Differences between simulated and observed NO_3^- leaching from the high residue plots might be explained by immobilisation of NO_3^- caused by addition of high C:N ratio organic substrate (immobilisation does not feature in the model). This is less likely to have been the case, however, in the low residue plot at No.3 Field, where leaching losses were also overestimated. Part of this mismatch may be ascribed to an anomalous delay in the labelled NO_3^- reaching the drains (Vinten et al., 1994). A more probable explanation is that the model simply underestimated denitrification losses in all except the low residue plots at Glencorse Mains. We are unable to demonstrate this conclusively because of the inevitable uncertainties in field measurements of denitrification flux (of which N_2O emissions are just a variable fraction).

How might the model systematically underestimate denitrification? It does not allow for structural aggregates or patchy organic matter distribution. These are both sources of anaerobiosis, with structural aggregates (Smith, 1980) and rapidly denitrifying microsites of high organic matter concentration (Parkin, 1987) leading to a localised demand for oxygen exceeding the supply. The creation of anaerobic microsites promotes

denitrification. The likely presence of such anaerobic microsites, even in the absence of large rainfall events, may explain the high denitrification fluxes observed but not predicted by our model. This would be especially important early after fertilisation, when the applied NO_3^- is close to the surface and therefore in macroscopically well-aerated volumes of soil. To improve the temporal match between measured and predicted denitrification fluxes the model may also need to account for the time taken for diffusion of gas through the soil profile, especially in heavier soils such as the clay loam at Glencorse Mains. The simulations also appear less sensitive to soil temperature than might be expected, so there may be a need to look more closely at the relationship between organic matter decomposition and soil temperature.

Acknowledgements

This work was supported by the Joint Initiative on Pollution Transport through Soils and Rocks of the Agricultural and Food Research Council and the Natural Environment Research Council (Award GST/02/590). We would like to thank Ian Crichton, Rab Howard, Frances Wright and Lesley Swan of SAC for their help in field and laboratory.

References

Addiscott T M and Whitmore A P 1991 Simulation of solute leaching in different soils of differing permeabilities. Soil Use Manage. 7, 94–102.

Arah J R M 1992 New formulae for mass spectrometric analysis of nitrous oxide and dinitrogen emissions. Soil Sci. Soc. Am. J. 56, 795–800.

Arah J R M, Crichton I J and Smith K A 1993 Denitrification measured directly using a single-inlet mass spectrometer and by acetylene inhibition. Soil Biol. Biochem. 25, 233–238.

Crooke W M and Simpson W E 1971 Determination of ammonium Kjeldhal digests of crops by an automated procedure. J. Sci. Food Agric. 22, 9–10.

Hauck 1982 Nitrogen-isotope-ratio analysis. *In* Methods of Soil Analysis - Part 2. Ed. A L Page. pp 735-779. Am. Soc. Agron., Madison, WI, USA.

Henriksen A and Selmer-Olsen A R 1970 Automatic methods for determining nitrate and nitrite in water and soil extracts. Analyst. 95, 514–518.

Parkin T B 1987 Soil microsites as a source of denitrification variability. Soil Sci. Soc. Am. J. 51, 1194–1199.

Rijtema P E and Kroes J G 1991 Some results of nitrogen simulations with the model ANIMO. Fert. Res. 27, 189–198.

Smith K A 1980 A model of the extent of anaerobic zones in aggregated soils, and its potential application to denitrification. J. Soil Sci. 31, 263–277.

Vinten A J A and Redman M H 1990 Calibration and validation of a model of non-interactive solute leaching in a clay-loam arable soil. J. Soil Sci. 41, 199–214.

Vinten A J A, Howard R S and Redman M H 1991 Measurement of nitrate leaching losses from arable plots under different nitrogen input regimes. Soil Use Manage. 7, 3–14.

Vinten A J A, Vivian B J, Wright F and Howard R S 1994 A comparative study of nitrate leaching from soils of differing textures under similar climatic and cropping conditions. J. Hydrol. 159, 197–213.

O. Van Cleemput et al. (eds.), Progress in Nitrogen Cycling Studies, 407–411, 1996.
© 1996 *Kluwer Academic Publishers.*

Combining GIS and a dynamic nitrogen simulation model to assess nitrogen leaching susceptibility on a regional scale: A case study

Karen Christiaens, Katja Vander Poorten, Marnik Vanclooster and Jan Feyen
Institute for Land and Water Management, K.U. Leuven, Vital Decosterstraat 102, B-3000 Leuven, Belgium

Key words: GIS, nitrate leaching, nitrogen modelling, regional scale

Abstract

In this study a GIS-modelling framework is developed for the assessment of the regional environmental impact of nitrogen fertiliser use in terms of nitrate leaching. To identify the maximum allowed fertiliser dressing as a function of the soil-crop system, a process-based model, simulating the fate of nitrogen in this continuum at the point scale (WAVE) and a geographic information system (GIS) integrating the point model data at the regional scale, were combined. The GIS is required to manage the vast amount of data which is involved. The GIS enables to extract the necessary model input data from the available digitised soil map and the soil profile databases. In addition, the GIS is used to present the modelling results in a cartographic way. The modelling framework is applied to asses the vulnerability for nitrate leaching of a pilot study region in Flanders. The resulting vulnerability map illustrates how policy makers can diversify the nitrogen leaching standards as a function of the soil type.

Abbreviations: WAVE – Water and Agrochemicals in the soil and Vadose Environment, GIS – Geographic Information System, a.o. ARC-INFO MAP – Mestaktieplan (Policy framework to control the use of slurry in agriculture), KMI – Koninklijk Meteorologisch Instituut (Royal meteorological institute)

Introduction

The use of agrochemicals endangers the quality of land and water resources in many countries and necessitates measures. In Flanders e.g., the nitrate drinking water standard of 50 mg L^{-1} is exceeded in many surface and groundwater systems. This is mainly a result of the nitrate-nitrogen leaching from excessive fertilised agricultural land (Geypens and Rutten, 1994). Legislation measures for controlling the fertiliser application and reducing the nitrate pollution are currently considered at regional level (VLM, 1993). These measures are based on rough estimates of the nitrogen leaching hazard, which do not reflect the interaction between climate, crop, soil and geo-hydrological situation. Even if those effects are considered it is still not clear what the long term effect of the regulative measures will be. Both the development of regulations and the assessment of their impact would be substantially simplified by the availability of a comprehensive and reliable simulation framework able to analyse the fate of nitrogen at the regional scale in terms of different unsteady state boundary conditions.

In this study, a distributed modelling approach is presented to assess the nitrogen leaching risk at the regional scale. The approach combines a GIS with a point-scale mechanistic model simulating the fate of nitrogen in the soil-crop system. The point model calculates, amongst other things, the daily nitrate leaching and crop growth rates for a given soil, crop, climate, geo-hydrological and farming scenario. The GIS preprocesses the regional databases to collect data for the model runs and summarizes the main simulation results in a cartographic way. The modelling framework was applied to establish the nitrogen leaching vulnerability map for the community of Lubbeek (Flanders, Belgium), assuming a constant simplified crop and fertiliser management scenario.

Materials and methods

The WAVE-model

The simulation model WAVE (Water and Agrochemicals in the soil and Vadose Environment; Vanclooster et al., 1994), is an integrated mechanistic deterministic model describing the transport of water, solutes and heat, the transformation of nitrogen and crop growth in the soil-crop environment. The model is a revised and extended version of the SWATNIT-model (Vereecken et al., 1990). The model encompasses a module for the description of water flow in the unsaturated zone of the soil profile based on the Darcian flow theory. This module originates from the SWATRER-model (Dierckx et al., 1986). The solute transport module is based on the convection-dispersion approach, and includes a physical non-equilibrium concept (Vanclooster et al., 1992). The model incorporates a heat transport module, which changes the organic and inorganic nitrogen transformation constants. The nitrogen fate module includes first order nitrification, denitrification, volatilisation and organic matter turnover from three organic matter pools. The crop growth module in the WAVE model is derived from the SUCROS87-model (Spitters et al., 1988). The WAVE-model is a point model and the model output is representative for a small field if effective parameters are used. The model solves the systems differential equations numerically and integrates the matter and energy fluxes on a daily basis. Components of the model have been thoroughly screened, calibrated and evaluated in different studies, both at the laboratory and at field scale.

Data requirements and processing

The soil map of Lubbeek was converted into simulation units. For each unit the climatic-, soil- and crop conditions and the nitrogen transformation rate data were determined and used as input for the WAVE model. Since the fate and transport of water and nitrogen are dependent on the climatic conditions, calculations were performed for thirty-one subsequent years (1960–1990). To this end, time series of the weather, monitored by the KMI and incorporated in the database BLIKSEM, were processed. The time series include daily readings of precipitation, minimum and maximum temperature, global radiation and the calculated potential evapotranspiration of a reference crop (grass - Etref, Raes et al., 1986). Soil information was derived from the digitised soil map using ARC-INFO. Based on the soil series observed in the region, statistical profiles were selected from the soil information system AARDEWERK-BIS (Van Orshoven and Vandenbroucke, 1993). For the statistical profiles the geometry of the soil profile (boundaries of the pedogenetic horizons), the soil hydraulic parameters and soil organic carbon content are available in the database. The position and fluctuation of the groundwater level was derived from the soil drainage class. Solute, heat and nitrogen transformation data were adopted from compiled literature data. The case study was limited to one crop (maize) and one fertiliser scenario. Crop growth parameters were a.o. taken from Boons-Prins et al. (1993). The fertiliser scenario was based on the proposed application standards (VLM, 1993) and consists of an organic (170 kg N ha^{-1} on 25th of April) and inorganic (105 kg N ha^{-1} on the 5th of May) fertiliser application.

The simulations (1364 simulation years) were performed in a UNIX environment. Shell script programmes were written to compile the different data sources in the required input file format. Similar script programmes were developed, to process the simulation output. The GIS package ARC-INFO was finally used to represent these results in a cartographic way.

Results and discussion

Figures 1 and 2 illustrate the temporal variability of the simulated leaching and crop return for a given soil type Lbp01 within the Lubbeek community. These figures reveal that the model needs some time to equilibrate the different system state variables. Hence, the simulation results for the first two years were not considered in the subsequent discussion. The simulated total dry matter production of the maize crop varies between 13.4 and 19.1 tonnes ha^{-1} yr^{-1}. The average yield amounts to 16.6 tonnes ha^{-1} yr^{-1}. The corresponding simulated nitrate-nitrogen leaching ranges between 0.0 (year 1976) and 134.8 kg ha^{-1} yr^{-1} (year 1977), with an average of 70.8 kg ha^{-1} yr^{-1}. From this analysis it can be stated that for the given soil and crop on average the proposed target (an allowed leaching of 70 kg ha^{-1} yr^{-1}, VLM, 1993) is reached.

Figures 3 and 4 illustrate the effect of the soil type on nitrate leaching and crop production. The mean dry matter production, calculated over the simulated period, varies between 12.2 (Zafe soil type, suitability index for maize = 4) and 19.8 (Lep01 soil type, suitability index for maize = 1) tonnes ha^{-1} yr^{-1}. The mean

total crop yield (maize) for Lbp01 soil

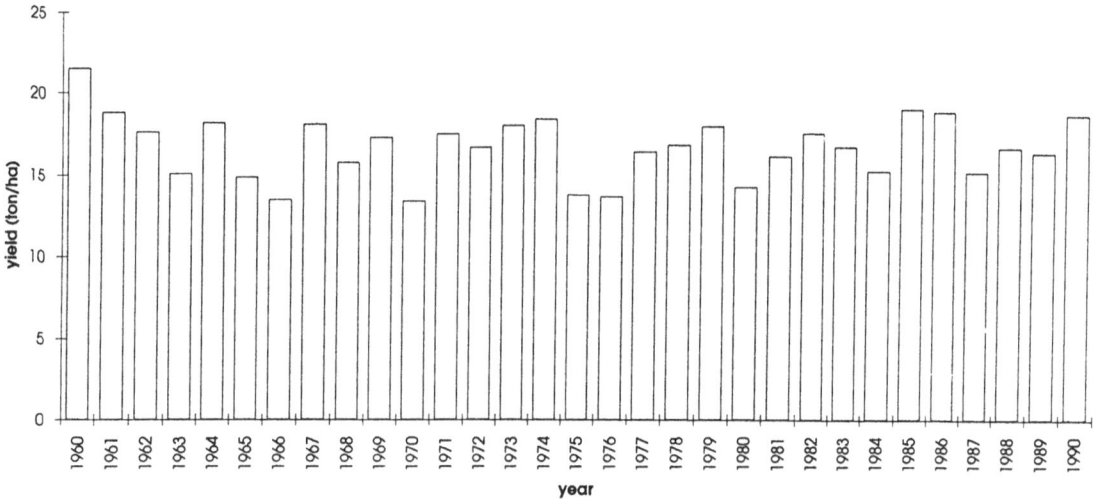

Figure 1. Temporal variability of the simulated crop yield for a Lbp01 soil unit.

nitrate leaching for Lbp01 soil

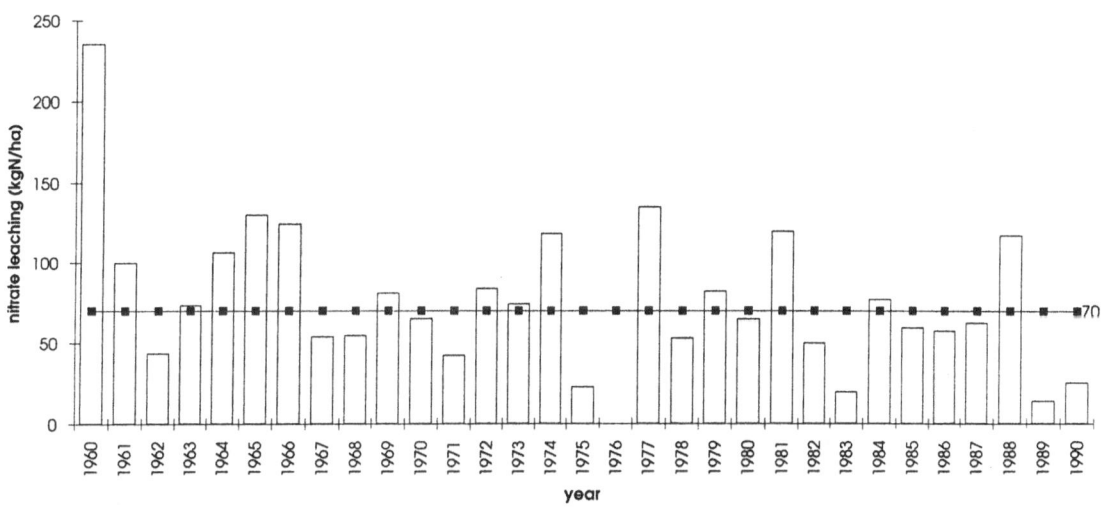

Figure 2. Temporal variability of the simulated nitrate leaching for a Lbp01 soil unit.

nitrate-nitrogen leaching varies between 21.0 (Udx soil type) and 158.2 kg ha^{-1} yr^{-1} (wZafe2 soil type). The different soil types were clustered in 5 different classes to yield the nitrogen leaching risk vulnerability map. Figure 5 illustrates the spatial distribution of the different clusters found within the pilot region.

Conclusion

From the study it can be concluded that both the simulated nitrate leaching and calculated crop production vary considerably in time and space. Hence, the soil type and the fluctuations in climate should be considered when formulating nitrogen fertiliser measures that meet the standards for nitrogen leaching imposed by environmentalists. The study further illustrates that a GIS in combination with a mechanistic point model

410

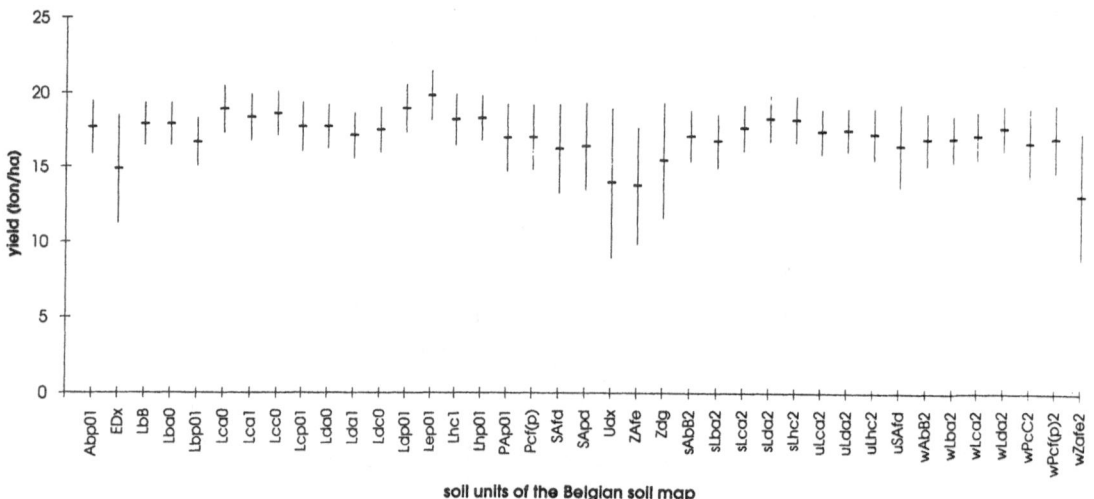

Figure 3. Variability of the simulated crop yield for different soil units within the pilot study region.

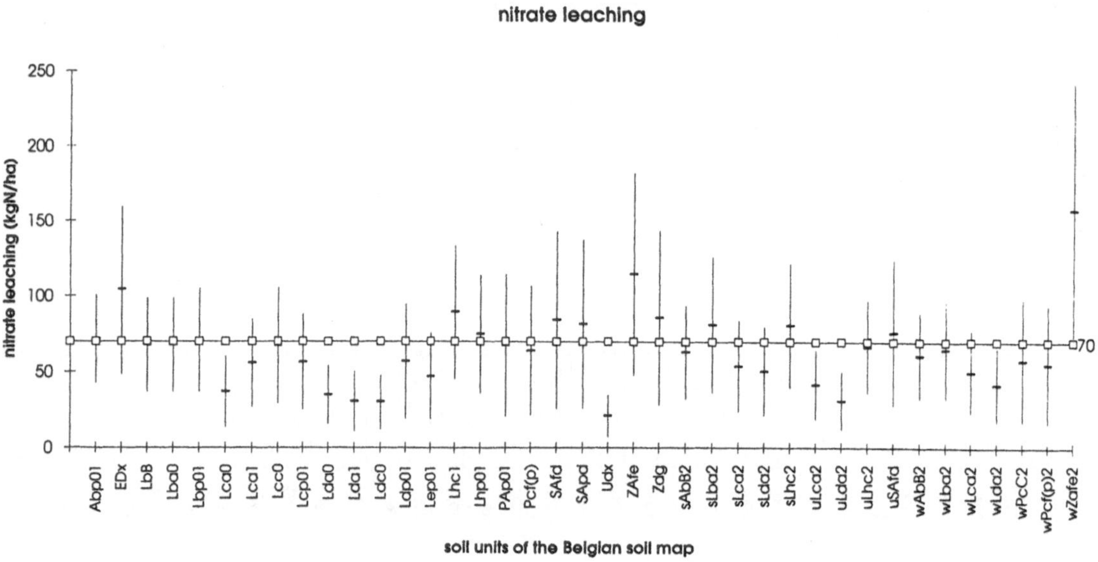

Figure 4. Variability of the simulated nitrate leaching for different soil units within the pilot study region.

is a powerful and suitable tool to analyse the nitrogen leaching risk at regional scale for different land use scenarios.

Acknowledgements

This research was financed by the Belgian Institute for the Encouragement of Scientific Research in Agriculture and Industry (IWONL).

References

Boons-Prins E R, de Koning G H J, van Diepen C A and Penning de Vries F W T 1993 Crop specific simulation parameters for yield forecasting across the European Community. Simulation Reports CABO-TT, nr.32. Wageningen Agricultural University, Wageningen, the Netherlands. 43 p.

Dierckx J, Belmans C and Pauwels P 1986 SWATRER, a computer package for modelling the field water balance, Reference manual. Soil and Water Engineering Laboratory, Katholieke Universiteit Leuven, Leuven, Belgium. 114 p.

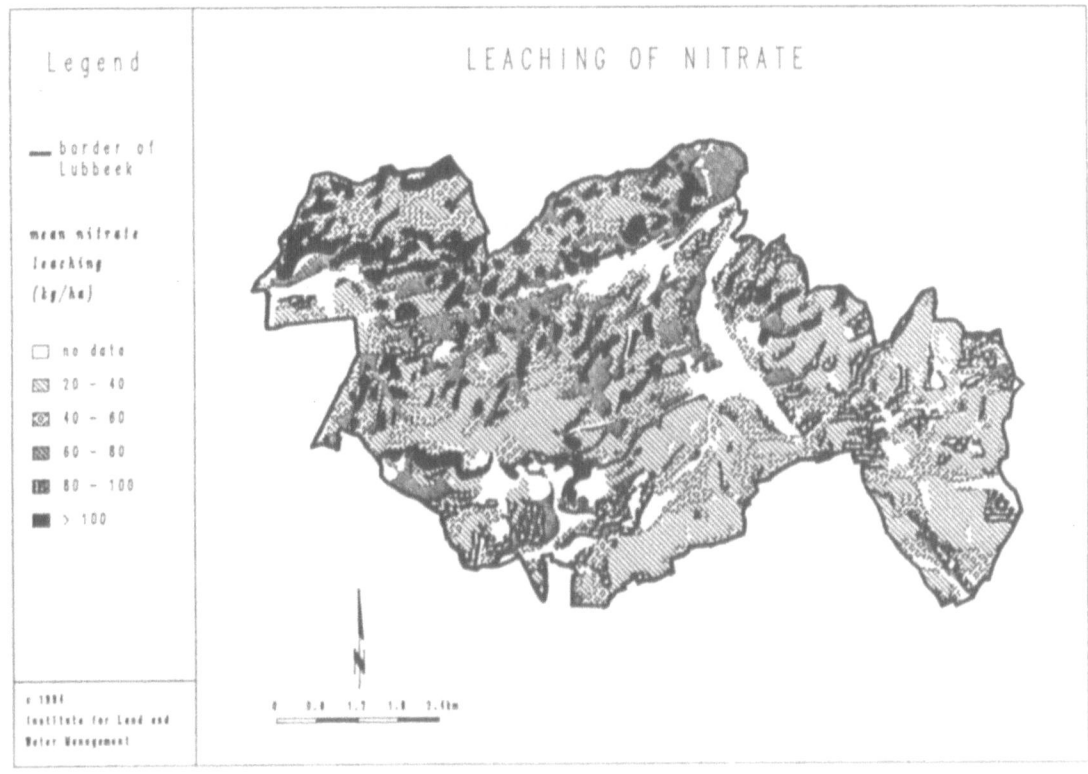

Figure 5. Nitrate-nitrogen leaching risk vulnerability map for the community of Lubbeek.

Geypens M and Rutten J 1994 Vermesting. *In* Leren om te keren: Milieu en natuurrapport Vlaanderen. Ed. A Verbruggen. pp. 245–268. VMM en Garant, Leuven, Belgium.

Raes D, Van Aelst P and Wyseure G 1986 ETREF - a computer simulation package for calculating crop water requirements. Reference Manual. Laboratorium voor bodem- en waterbeheersing, Reference Manual. KU Leuven, Leuven, Belgium. 104 p.

Spitters C J T, Van Keulen H and Van Kraailingen D W G 1988 A simple but universal crop growth simulation model, SUCROS87. *In* Simulation and Systems Management in Crop Protection. Eds. R Rabbinge, H Van Laar and S Ward. Simulation Monographs. Vol 32, pp 147–181. PUDOC, Wageningen, The Netherlands.

Vanclooster M, Vereecken H, Diels J, Huysmans F, Verstraete W and Feyen J 1992 Effect of mobile and immobile water in predicting nitrogen leaching from cropped soils. Model. Geo-Biosph. Proces. 1, 23–40.

Vanclooster M, Viaene P, Diels J and Christiaens K 1994 WAVE: a mathematical model for simulating water and agrochemicals in the vadose environment. Reference and user's manual, release 1.0, Institute for Land and Water Management, Katholieke Universiteit Leuven, Leuven, Belgium. 147 p.

Van Orshoven J, Maes J, Vereecken H and Feyen J 1991 A procedure for the statistical characterisation of the units of the Belgian soil map. Pedologie XLI-3, 193–212.

VLM Vlaamse Landmaatschappij 1993 De mestbank - Mestaktieplan 1992. Uitgave van de Vlaamse Landmaatschappij, Brussel, Belgium. 98 p.

Vereecken H, Vanclooster M and Swerts M 1990 A model for the estimation of nitrogen leaching with regional applicability. *In* Fertilizer and the Environment. Eds. R Merckx, H Vereecken and K Vlassak. pp 250–263. Leuven Academic Press, Leuven. Belgium.

O. Van Cleemput et al. (eds.), Progress in Nitrogen Cycling Studies, 413–418, 1996.
© 1996 *Kluwer Academic Publishers.*

Simulation by NCSWAP of the nitrogen dynamics under crop amended with sewage sludge in two soils

S. Houot, L. Cadot and J.A.E. Molina
I.N.R.A., Unité de Sciences du Sol, 78850 Thiverval Grignon, France

Key words: leaching, model, nitrogen, organic matter, sewage sludge, soil

Abstract

The model NCSWAP describes C and N dynamics in soil - plant - water systems. It includes different sub-models concerning the crop growth, the water movements (INFIL and REDIS), C and N evolutions in soil (NCSOIL). In NCSOIL, two organic pools are considered: the microbial biomass and a fraction of the humified organic matter still biodegradable (pool II). Other pools can be added to represent the organic inputs: crop residues or organic amendments such as sludge. A sewage sludge was applied to a loamy soil (Typic Eutrochrept) in Grignon and a drained hydromorphic clay soil (Aquic Eutrochrept) in Rambouillet. NCSOIL parameters describing the sludge evolution were calibrated using laboratory incubations. The sludge rate of degradation was high (0.5 day^{-1}). So was its efficiency factor (0.8), proportion of the degraded sludge incorporated into the microbial biomass. After sludge addition, maize was grown in Grignon and sorghum in Rambouillet. The soil water and NO_3^- contents were measured during 1.5 year after sludge spreading. NCSWAP simulation was rather accurate in Grignon, giving satisfactory simulations of the soil water and NO_3^- contents. In Rambouillet, discrepancies were observed between observed and simulated values of soil NO_3^- contents because the model did not simulate the fast water movements through soil cracks present in autumn.

Introduction

Land application of sewage sludges on agricultural soils have been frequently used to recycle nutrients such as nitrogen through crop production. The rates of application have to be adjusted to the crop needs to avoid production of excess NO_3^-. Laboratory incubations are often used to estimate the N available in organic amendments (Houot et al., 1994). Simulation models allow to better understand N fluxes in soils which are not available through measurements by considering simultaneously the numerous processes occurring in the N dynamics in soil and soil - water - plant systems (Tanji, 1982). Some of the models have been specially adapted for manure amendments (Hoffmann and Ritchie, 1993). In a precedent work (Houot et al., 1994), we used the model NCSOIL to characterize sludge evolution in 2 soils. The parameters describing sludge evolution in soils were calibrated. The sludge had a very high rate of degradation (0.47 d^{-1}) with incorporation of 80% of the decomposed fraction into the microbial biomass. Field experiments have been

conducted on the same soils and with the same sludge. Water and mineral N contents have been measured in soils during 1.5 year after sludge application and the N exported in crop has been determined. The NCSWAP model describes N and C dynamics in the soil - water - plant system. It has been elaborated under Minnesota climatic and soils conditions. The first validation by the authors gave satisfactory results (Clay et al., 1985a, 1985b, 1989). When used in different conditions, the validation required previous field calibrations of parameters (Lengnick and Fox, 1994). The objectives of the present work were to (1) validate NCSWAP in the 2 studied soils with sludge amendments using the results got with NCSOIL from the laboratory incubations but without supplementary field calibration and (2) use the model to better understand the N dynamics and determine when N leaching could be observed. The NCSWAP validation was realized by comparison of observed and simulated data for soil water and mineral N contents.

Materials and methods

Field experiments

The sewage sludge came from the Achères sewage plant, near Paris. It was made from domestic-industrial source and anaerobically digested. Its dry matter content was 28.5 g L^{-1}. The main characteristics were on a dry matter basis: organic C, 28.0%; organic N, 3.0%; total N, 7.1%; organic C/ organic N, 9.5; organic C/ total N, 4.0 .

Two field experiments were conducted in Grignon on a loamy soil (Typic Eutrochrept) under a wheat - corn rotation and Rambouillet on an hydromorphic clay soil (Aquic Eutrochrept) under continuous corn, both locations near Paris (France). The main soil characteristics are reported in Table 1. The sludge was applied (400 Mg ha^{-1}) at the beginning of November 1989 and incorporated after 3 days by plowing at a depth of 30 cm in Grignon, but only after one month in Rambouillet at a depth of 20 cm. The rate of application was doubled compared to the recommended rates as it was required by other works concerning trace elements dynamics in soils. Control plots without organic amendment were also studied in both locations. In 1990, corn (Zea mays L.) was sowed at the beginning of May in Grignon and Rambouillet and harvested in Grignon at the beginning of October. In Rambouillet, the corn germination was very low and it was replaced by sorghum harvested in October.

Three soil cores, 5 cm in diameter were taken to 120 cm in Grignon and 90 cm in Rambouillet peri-odically during 1.5 year after sewage application in the control and the amended plots. The soil water was measured gravimetrically and mineral N extracted with KCl 1 M and determined by automated analysis (Skalar SA 40). The fumigation-incubation method (Jenkinson and Powlson, 1976 modified by Chaussod et al., 1986) was used for soil microbial biomass estimation in the surface soil.

Models

The NCSWAP model (Molina and Richards, 1984) includes four major submodels describing the crop growth, the water infiltration (INFIL) and redistribution (REDIS) in soil, and the C and N dynamics in soil (NCSOIL). A reference crop growth (dry matter and N percentage), representative of the optimum crop growth without N or water limitations in the considered environment is defined as controllable input. The simulated crop growth is calculated comparing the N and water availabilities to the reference needs. Water flow through microporosity is simulated with a vertical, one-dimensional water infiltration and redistribution model driven by Green and Ampt infiltration. Detailed description of NCSOIL can be found elsewhere (Houot et al., 1989, 1994; Molina et al., 1990). It models C and N flows between active organic pools. Two soil organic pools are considered: the microbial biomass and a fraction of the humified organic matter still labile (pool II). Other organic pools can be added as organic inputs (litter, sludge amendments...). The N dynamics follows the C one function of the C/N

Table 1. Main soils characteristics: granulometry, C and N content, bulk density and water content at field capacity (F.C.), saturation (Sat.) and permanent wilt (P.W.)

Depth	Granulometry (%)			Organic matter		Bulk density	Soil water content (g g^{-1})		
(cm)	Clay	Silt	Sand	C (%)	N (%)		F.C.	Sat.	P.W.
Grignon Soil : *Typic Eutrochrept*									
0–30	22	73	5	1.11	0.12	1.33	0.22	0.34	0.10
30–60	23	73	4	0.59	0.08	1.53	0.21	0.32	0.10
60–90	27	71	2	0.35	0.05	1.46	0.21	0.33	0.13
90–120	21	77	2	0.32	0.04	1.50	0.20	0.30	0.13
Rambouillet soil: *Aquic Eutrochrept*									
0–30	21	35	44	0.89	0.09	1.33	0.22	0.34	0.10
30–50	26	33	41	0.92	0.08	1.40	0.28	0.40	0.15
50–90	55	18	27	0.24	0.03	1.50	0.40	0.50	0.20

Table 2. Initial sizes and C/N ratios of the active organic pools in the NCSWAP simulations. The microbial biomass values were experimentally measured. The initial levels of Pool II were calculated from the results of the laboratory incubations. The sludge initial level corresponded to the applicated rate

	Microbial biomass	Pool II	Sludge
Grignon			
Initial level (mg C kg^{-1})	118	850	800
C/N ratio	6.0	20.0	9.5
Rambouillet			
Initial level (mg C kg^{-1})	92	700	800
C/N ratio	6.0	11.0	9.5

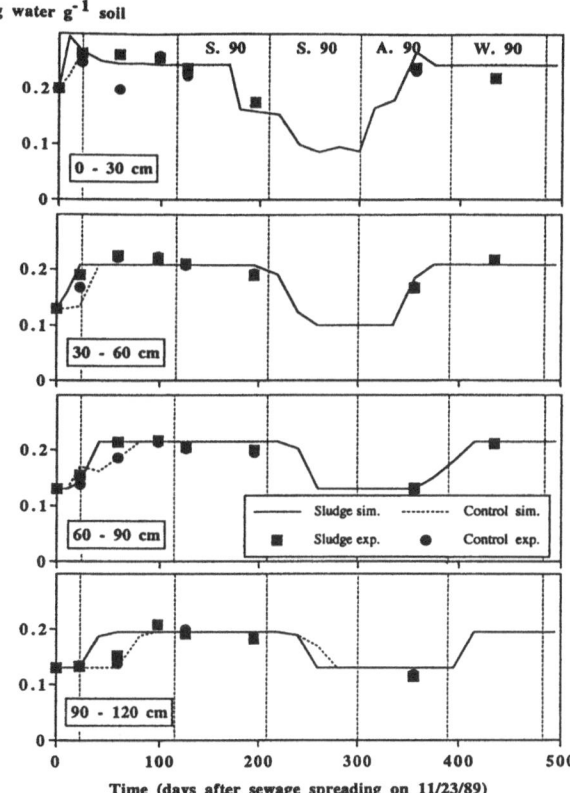

Figure 1. Soil water content (in g g^{-1}) evolution in the control and amended plots in Grignon: simulated (lines) and observed (symbols). The four soil depths were considered.

ratios of the pools. All the biological transformation rates interact with temperature and soil water content as described by Clay et al. (1985b).

The input file of NCSWAP has been summarized by Lengnick and Fox (1994). In Grignon, all the climatic data came from a weather station adjacent to the experimental field. In Rambouillet, only air temperature and rainfall came from the field experiment. For the potential evaporation and the soil temperature, data of Grignon was used as the two experiments were close from each other. In Grignon, the soil water contents at saturation, field capacity and permanent wilt, the soil bulk density were experimentally measured. Only the saturated hydraulic conductivity was estimated from the literature (Hillel, 1984). In Rambouillet, they were estimated from a study about soil water dynamics realized in the control plot (Coulomb, 1992). The reference crop growth was calibrated for corn on a previous experiment on the same soil in Grignon.

Active organic pools were only considered in the surface soil (Table 2). The flush of C mineralization obtained after fumigation in the November 89 samples was used as initial value for pool I. Initial values for pool II were optimized with NCSOIL from laboratory incubations (Houot et al., 1994), using the Marquardt algorithm modified to accept data from simulation models (Barak et al., 1990). So were the constant rates of degradation of sludge and the pool II and sludge efficiency factors. The constant rate of degradation of Pool II (0.003) was modified compared to the value usually used (0.006), following suggestions made by Menasseri et al. (1994).

Measurement of simulation accuracy

The Loague and Green's (1991) root mean square error (RMSE = [(Σ(simulated - observed)^2/number of observations)]$^{1/2}$ × [100/observed mean]) was used to measure the deviation of simulated values from observed values reported on an observed mean basis (Lengnick and Fox, 1994).

The mean difference (MD = [Σ(simulated - observed)/number of observations]) has been proposed by Addiscott and Whitmore (1987) to measure the average deviation of the simulated values from the observed ones which can be overestimated (MD>0) or underestimated (MD<0). A test-t is used to determine if MD is significantly different from 0 (Addiscott and Whitmore, 1987).

Table 3. Accuracy of the simulations of mineral N and water content evolution in Grignon soils throughout the simulated period after sewage spreading: mean deviation (MD in kg N ha^{-1} for mineral N and g g^{-1} for water) and root mean square error (RMSE in%)

	Control		Sludge	
	MD	RMSE (%)	MD	RMSE (%)
Total mineral N, 0–120 cm (kg N ha^{-1})	-11.1	56	174.2[a]	46
Mineral N, 0–30 cm (kg N ha^{-1})	-7.6	75	-61.0	61
Mineral N, 30–60 cm (kg N ha^{-1})	-1.1	61	89.7	80
Mineral N, 60–90 cm (kg N ha^{-1})	-2.7	87	72.9	129
Mineral N, 90–120 cm (kg N ha^{-1})	0.3	36	43.3	195
Soil Water Content, 0–30 cm (g g^{-1})	0.016	14	0.003	7
Soil Water Content, 30–60 cm (g g^{-1})	-0.005	9	0.002	7
Soil Water Content, 60–90 cm (g g^{-1})	0.005	6	0.002	5
Soil Water Content, 90–120 cm (g g^{-1})	-0.001	5	0.010	12

[a] MD different from 0 at the level of 5%

Results

Grignon

The simulated values of soil water content well coincided with the observed ones (Fig. 1) as shown by the low MD and RMSE (Table 3). The soil water content remained at field capacity when soil was bare in winter and spring then decreased to permanent wilt during the corn growing season in summer and autumn. In the surface horizon, water evaporation decreased the soil water content below permanent wilt. The evolution of soil humidity was identical in the control and the amended plots.

During winter and spring after sewage application, the total mineral N in the soil profile of the amended plot was rather well simulated as NCSWAP results were within the confidence interval of the observed values (Fig. 2). The measured variability was very important. Then, during the growing season, the observed soil mineral N content decreased more than the simulated one. On the contrary in the control plot, the simulated total mineral N was overestimated during winter and spring then underestimated during the crop growth. In the two plots, the simulated values of total mineral N coincided with the observed values similarly with equivalent RMSE (Table 3). In the control plot, the simulated values were globally underestimated but overestimated in the amended plot, with MD significantly different from 0. When looking at the mineral N content in the different horizons of the amended plot (Fig. 3), the simulated values were underestimated in the surface horizon but overestimated in the 3 others, mostly in the 2 deepest ones. A decrease of the simu-

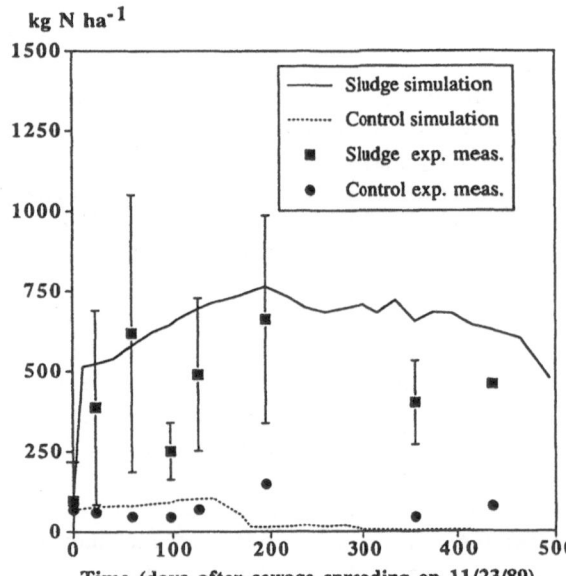

Figure 2. Total soil mineral N storage (in kg N ha^{-1}) evolution in the control and the amended plots in Grignon: simulated (lines) and observed (symbols). Standard errors of the experimental values are shown when larger than the symbol size.

lated mineral N content during the growing season was observed in the "30–60 cm" and "60–90 cm" horizons, because of the corn use of N, but not in the surface horizon. The simulation estimated a more important lixiviation of mineral N during the second winter after sludge application (235 kg N ha^{-1}) than during the first one (86 kg N ha^{-1}).

The simulated corn yield was 9172 kg ha^{-1} when the actual yield was 8600 kg ha^{-1} (total above ground dry matter). The simulated total N exported in the

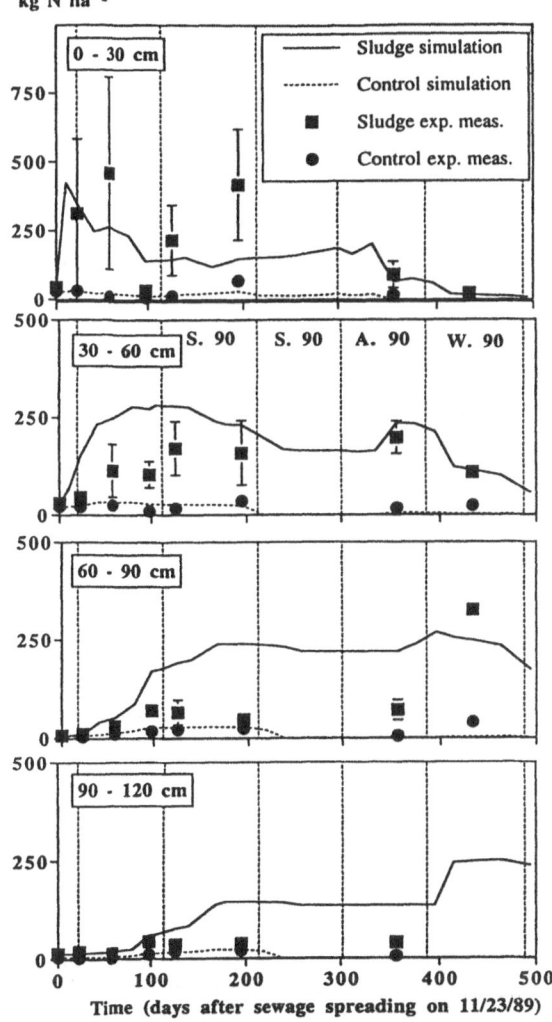

kg N ha^{-1}

Figure 3. Soil N-NO$_3^-$ content (in kg N ha^{-1}) evolution in the control and amended plots in Grignon: simulated (lines) and observed (symbols). The four soil depths were considered. Standard errors of the experimental values are shown when larger than the symbol size.

above ground plant part was 109 kg ha^{-1} when the observed one was 112 kg ha^{-1}.

Rambouillet

Only the first drainage period after sludge application and before corn sowing was simulated. The soil water content (data not shown) was well simulated by NCSWAP (MD varying between - 0.04 and 0.02 g g^{-1}, RMSE varying between 13 and 17% in the 3 considered horizons). On the other hand, total mineral N simulation did not coincide with the observed evolution (Fig.

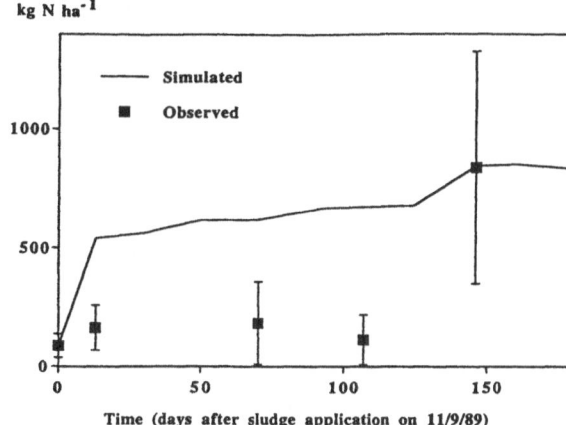

kg N ha^{-1}

Figure 4. Total soil mineral N storage (in kg N ha^{-1}) evolution in the amended plot in Rambouillet: simulated (line) and observed (symbols). Standard errors of the experimental values are shown when larger than the symbol size.

4). The mean mineral N total content in soil was 325 kg ha^{-1} over the period going from November 1989 to April 1990. NCSWAP overestimated this nitrogen amount to 670 kg ha^{-1} with a MD of 345 kg ha^{-1} and a RMSE of 124% when compared to the experimental values.

Discussion

NCSWAP validation gave better results in Grignon than in Rambouillet. A previous validation work realized by Clay et al. (1985a, 1985b, 1989) gave satisfactory results in a Waukegan silt loam (fine-silty over sandy or sandy-skeletal, mixed, mesic Typic Hapludoll). On the other hand, Lengnick and Fox (1994) found some limitations in the water flow simulation and NO$_3^-$ production in the model. They worked on a Hargestown silt loam (fine, mixed, mesic Typic Hapludalf). The overestimation of the NO$_3^-$ leaching was related to the water flow simulation, because of the non differentiation of the macropore and micropore flows. In Grignon, the soil water and NO$_3^-$ contents were satisfactory simulated. In the amended plot, the simulated total NO$_3^-$ storage in soil remained within the measured confidence interval because of the important spatial variability. Nevertheless, at the end of the simulation period, it was overestimated. The results were estimated satisfactory as obtained without field calibration. In Grignon, the NO$_3^-$ content was lightly underestimated in the surface horizon but overestimated in the three others during the first drainage period but not during the second one : NO_3^- production was probably overestimated at the beginning of the simulated period. Lengnick and Fox (1994) found another limitation in the overestimation of the NO_3^- production in the surface soil. In our study, the pool II calibrated from the laboratory incubations represented 3.5% of total N, similar proportion as calibrated from field experiments by Lengnick and Fox (1994). The pool II constant rate of degradation has been decreased as suggested by Menasseri et al. (1994) reducing the pool II important decrease observed during previous works.

An inaccurate water flow simulation could explain the bad simulation results obtained in Rambouillet. A detailed study of the water circulation in soil has been realized by Coulomb (1992). At the end of summer, large cracks are formed in soil. When raining, all the water goes down these cracks until they disappear. The water reaches very rapidly the drain and the day following the rainfall, the water flow is maximum, then decreases exponentially within one day. The sludge was applied during a dry autumn, when cracks were present in soil. As water circulation through soil cracks was not considered by NCSWAP, the model could not give accurate simulations of the N dynamics in Rambouillet.

Acknowledgement

This work was financially supported by the "Agence de l'Environnement et de la Maitrise de l'Energie" (ADEME).

References

Addiscott T M and Whitmore A P 1987 Computer simulation of changes in soil mineral nitrogen and crop nitrogen during autumn, winter and spring. J. Agric. Sci., (Cambridge) 109, 141–157.

Barak P, Molina J A E, Hadas A and Clapp C E 1990 Optimization of an ecological model with the Marquardt algorithm. Ecol. Model. 51, 251–263.

Chaussod R, Nicolardot B and Catroux G 1986 Mesure en routine de la biomasse microbienne des sols par la méthode de fumigation au chloroforme. Sci. Sol 2, 201–211.

Clay D E , Clapp C E, Linden D R and Molina J A E 1989 Nitrogen-tillage-residue management: III Observed and simulated interactions among soil depth, nitrogen mineralization and corn yield. Soil Sci. 147, 319–325.

Clay D E, Clapp C E, Molina J A E and Linden D R 1985a Nitrogen-tillage-residue management: I Simulating soil and plant behavior by the model NCSWAP. Plant and Soil 84, 67–77.

Clay D E, Clapp C E, Molina J A E and Linden D R 1985b Nitrogen-tillage-residue management: II Calibration of potential rate of nitrification by model simulation. Soil Sci. Soc. Am. J. 49, 322–325.

Coulomb C 1992 Etude de la circulation de l'eau dans un sol argileux drainé. Approche hydrodynamique, isotopique et géochimique. PhD diss. Université Paris XI Orsay, France.

Hillel D 1984 L'eau et le sol. Cabay, Louvain-La-Neuve, Belgique. 288 p.

Hoffmann F and Ritchie J T 1993 Model for slurry and manure in CERES and similar models. J. Agron. Crop Sci. 170, 330–340.

Houot S, Cadot L and Molina J A E 1994 Simulation with the model NCSOIL of the temperature and humidity impact on sewage sludge nitrogen evolution in soils. In Modelling the fate of Agro-chemicals and Fertilizers in the Environment. Eds. C Giupponi, A Marani and F Morari. pp 273–293. ESA and Unipress, Padova, Italy.

Houot S, Molina J A E, Chaussod R and Clapp C E 1989 Simulation by NCSOIL of net mineralization in soils from the Deherain and 36 parcelles fields at Grignon. Soil Sci. Soc. Am. J. 53, 451–455.

Jenkinson D S and Powlson D S 1976 The effect of biocidal treatments on metabolism in soil. V. A method for measuring soil biomass. Soil Biol. Biochem. 8, 209–213.

Lengnick L L and Fox R H 1994 Simulation by NCSWAP of seasonal nitrogen dynamics in corn: I soil nitrate. Agron. J. 86, 167–175.

Loague K and Green R E 1991 Statistical and graphical methods for evaluating solute transport models: overview and application. J. Contam. Hydrol. 7, 51–73.

Menasseri S, Houot S and Molina J A E 1994 Field calibration of the decomposition rate of organic pools in the NCSWAP model. In Modelling the fate of Agrochemicals and Fertilizers in the Environment. Eds. C Giupponi, A Marani and F Morari. pp 295–314. ESA and Unipress, Padova, Italy.

Molina J A E, Hadas A and Clapp C E 1990 Computer simulation of nitrogen turnover in soil and priming effect. Soil Biol. Biochem. 22, 349–353.

Molina J A E and Richards K 1984 Simulation model of the nitrogen and carbon cycle in the soil-water-plant system, NCSWAP. Soil Series n° 116. Department of Soil Science, University of Minnesota, Saint Paul, MN.

Tanji K K 1982 Modeling of the soil nitrogen cycle. In Nitrogen in Agricultural Soil. Ed F J Stevenson. pp 721–772. Agron. Monogr. 22 ASA, CSSA and SSSA, Madison, WI, USA.

O. Van Cleemput et al. (eds.), Progress in Nitrogen Cycling Studies, 419–424, 1996.

Modelling of nitrogen transformation in the soil

G.J. Kovács, T. Németh, L. Radimszky and T. Szili-Kovátcs
Research Institute for Soil Science and Agricultural Chemistry, 1022 Budapest Herman u. 15, Hungary

Key words: calibration, incubation, nitrogen, pot-experiment, simulation, validation

Abstract

Nitrate nitrogen leaching studies raised a specific demand to know the actual nitrate concentration in the soil layers any given time of the season. The biggest uncertainties of soil nitrate concentration were observed in cases of significant changes in the soil, like introduction of fertilisers or large amount of residue or fast changes in temperature and moisture status.

Experiments and simulations were carried out simultaneously to describe the dynamics of the soil-nitrate concentration applying different treatments. CO_2 production, NH_4, NO_3 concentration were measured and used for gaining parameter estimates.

Sequences and daily rates of processes of soil-nitrogen transformations were characterised in a model for the conditions of high fresh C and N combinations, 5 moisture and 5 temperature levels, wide ranges of C/N ratios, pH and humus conditions. Twenty seven treatments were introduced - both experimentally and in simulation - and kept unchanged for 168 days on 3 (sandy, loamy, clayey) soils.

The living microbial biomass is a central part of the simulation. The following rates of N-transformation are estimated by the model: microbial immobilisation, demineralisation, denitrification, nitrification, ammonification of fresh and humus-N, humification of residue-N and microbial-N.

Calibration was made to compare the simulated and the observed data for optimal temperature, moisture, C/N ratio, and N fertiliser dose and for the extreme conditions as well.

Abbreviations: DNRATE–rate of denitrification of nitrogen (mg $kg^{-1}day^{-1}$), FOM–fresh organic matter (residue)(mg kg^{-1}), FON–fresh organic nitrogen (N of residue)(mg kg^{-1}), MB–microbial biomass (mg kg^{-1}), MBN–microbial biomass nitrogen (mg kg^{-1}), RDC–rate of decomposition of carbon (mg $kg^{-1}day^{-1}$), RIN–rate of immobilisation of mineral N (mg $kg^{-1}day^{-1}$), RIN1–rate of immobilisation of ammonium N (mg $kg^{-1}day^{-1}$), RIN2–rate of immobilisation of nitrate N (mg kg^{-1} day^{-1}), RNTRF–rate of nitrification (mg kg^{-1} day^{-1}), RRN–rate of demineralisation of N (mg $kg^{-1}day^{-1}$).

Introduction

Nitrogen transformation studies are presently motivated by not only the key role of mineral nitrogen in plant nutrition but by the nitrate pollution in ground water. In order to predict the nitrate leaching from the soils, the crucial problem is to predict the nitrate concentration of the soil solution in the moment of a water flux. Nitrogen transformation in the soil is a multifactorial process with several alternative paths and wide ranges of rates. Multitudes of papers, reviews and books deal with modelling of N-transformation and transport in soils (Addiscott and Whitemore, 1987; Berghuijs, 1985;

Frere et al., 1975; Frissel et al., 1981; Godwin et al., 1985; Hadas et al., 1986; Hansen et al., 1991; Jones et al., 1974, 1986; Johnsson et al., 1987; Molina et al., 1983; Smith, 1979; Tanji, 1982; Vinten and Smith, 1992). Still, the variability of the concentration of nitrate is the most uncertain part of the simulation models of complex soil-plant-atmosphere continuum (Kovács, 1986).

The focus of the paper is the first 5–6 months of N-transformation after introducing a larger amount of residue to the soil. Under the circumstances of the Hungarian soils the flux of ammonium ions is not great even on the "flush-days" since the soils in this region have a considerable cation-exchange capacity and the ammonium is mostly absorbed on the colloid surfaces. In contrary, the changes of nitrate concentration in the soil profile can really be dramatic. Drops of 15–20 mg kg^{-1}week^{-1} have frequently been observed under field conditions. That corresponded to a 200–250 kg ha^{-1}week changes for a 1 meter soil horizon. In respect of the leaching estimation this magnitude of changes is significant. N-deficiency can also be measured in plants and significant effects on yields in sensitive growing periods (Kovács, 1982).

Material and methods

A theoretical nitrogen transformation model was created on the bases of literature of the nitrogen transformation (Stevenson, 1982) and its modelling (listed above). Special attention was given to the microbial biomass (MB). The model has interactively been developed using our experiments as follows. The form of the program was made in a way that it can be used without any change as an integral part of CERES models (Jones,1986) and the DISNIT model (Fehér et al.,1991)

As an aid of simulating the processes of nitrogen transformation a five factorial pot experiment was carried out with three different (sandy, sandy loam, and clay) soils of Hungary. They were sieved (2 mm) and homogenised. 400 grams per pot air-dried soil was treated with nitrogen fertiliser in the form of ammonium-nitrate, and fresh organic matter as powder of soybean leaf and maize stem. The pots then were incubated on different but constant temperature and moisture levels for 168 days with 27 treatment combinations. The levels of treatments were as follows:

Temperature	(°Celsius)	5	15	25	35	45
Moisture	(% of WC$_{SAT}$)	20	40	60	80	100
Fertiliser	(mg kg^{-1})	0	75	150	225	300
Soybean residue	(% of dry soil)	0	1	2	3	4
Maize residue	(% of dry soil)	0	1	2	3	4

On selected days CO_2 evolved was measured by gas chromatography (Szili Kovács et al.,1993) NH$_4$- and NO$_3$-N were analysed by distillation (Bremner and Shaw,1955). Denitrification was also measured by gas chromatograph (Szili Kovács et al., 1988).

Results and discussion

Model for N-transformations in soils

The simulation includes mineralisation, nitrification, biological immobilisation, demineralisation, denitrification, humification and humus decomposition of nitrogen (Fig. 1).

Exponential curves were fitted to the measured daily CO_2 production data of each of the 27 treatment combinations of the experiments on all 3 soils. A ratio of 1 to 1 was assumed to be between maintenance respiration and assimilation of carbon by microbes. The rate of humification was considered as proportional to the rate of oxidation of fresh organic carbon. The *rate* of *de*composition of *c*arbon is calculated as:

$$RDC = 2.2(aEXP(bt) + c) \qquad (1)$$

Where t is time in days. The a, b and c parameters have been defined by stepwise regression analyses. The factors were: *f*resh *o*rganic *m*atter (FOM), that is the residue of the previous crop, temperature, C/N ratio, relative soil water content (relative to saturation) and water content at wilting point.

The daily **rate of carbon assimilation of microbial biomass** then was estimated by these functions based on the CO_2-production measurements. Assuming a C:N ratio of 8:1 in the body of microbes, a daily **rate of microbial nitrogen assimilation** was estimated depending on the treatments and soils. Since microbes basically use inorganic forms of nitrogen for their growth, this later rate equals to the **rate of microbial** *i*mmobilisation of *n*itrogen (RIN). Since ammonium is the preferred source of nitrogen RIN is partitioned as 70% from ammonium source (RIN1) and 30% from nitrate source (RIN2), as long as both are available. Ammonium, though, has a limit of 5 mg kg^{-1} dry soil and nitrate can be utilised exhaustively.

The **rate of demineralisation** of microbial *n*itrogen (RRN) has been defined as follows. The RIN curve (with a little bit smoothened beginning) is restarted in the model but this time instead of being added to the *m*icrobial *b*iomass *n*itrogen (MBN) as growth component it is used to reduce the MBN, since the RRN is a consequence of the senescence of living microbes. In case of each experimental nitrogen concentration curves a local minimum was observed before the mineral nitrogen concentration started to grow rapidly. This time was used as a signal to start demineralisation in the simulation.

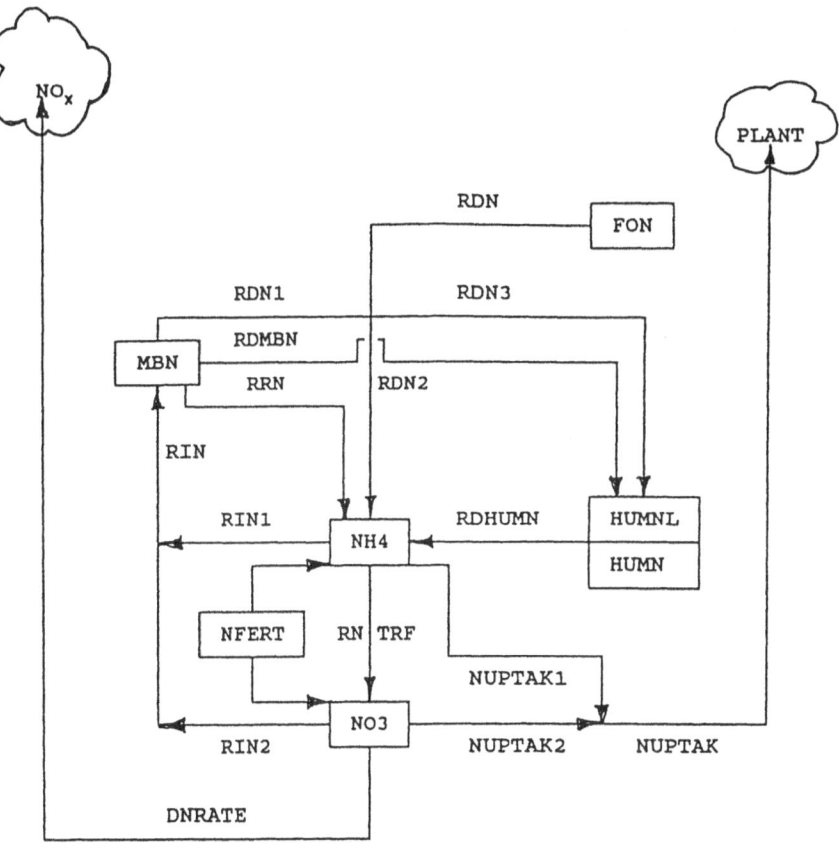

```
FON     =  fresh organic N (crop residue)
MBN     =  microbial N (living microbes)
HUMN    =  humus N
HUMNL   =  labile humus N
NH4     =  soil ammonium
NO3     =  soil nitrate
NFERT   =  fertilizer N
RDN     =  rate of decomposition of FON
RIN     =  rate of immobilization of N
RNTRF   =  rate of nitrification
NUPTAK  =  N uptake
RDMBN   =  rate of decomposition of microbial N
```

Figure 1. Model for N-transformation

Using the concept of the above paragraph the model results MBN curve with a pick after C and N additions to the soil (fertilisation, straw incorporation, etc) but MBN decreases close to the initial values as the estimated CO_2-production in the soil also goes back to normal.

Denitrification rate (DNRATE) is estimated as follows:

$$DNRATE = TFD\ MFD\ NO_3 \qquad (2)$$

where NO_3 is nitrate concentration, TFD is temperature factor and MFD is moisture factor. The *rate of nitrification* (RNTRF) is estimated by first order kinetics of ammonium concentration and the minimum of the corresponding moisture or the temperature factors:

$$RNTRF = NH_4 min(TFN, MFN) \qquad (3)$$

The functions describing temperature and moisture factors do not change within an optimum range (25–35 Celsius degree and 40–60% of saturation) but there are different functions above and under these ranges. There are different natural factors being effective, ie. the moisture factor expresses rather an oxygen deficiency above 60% of saturation water content than a direct moisture effect.

The *rate of decomposition of the fresh organic (residue) nitrogen* (RDN) has 3 components. RDN1 would be the direct uptake of some organic forms of N. In this version RDN1 is set to 0. **The rate of ammonification (mineralisation)** (RDN2) was estimated as follows:

$$RDN2 = aEXP(-b/NDAY) - 100 \quad (4)$$

Where a and b values gained from curve fitting to the CO_2 production in the experiment and then stepwise regression analyses based on the assumption that RDN is proportional to the microbial activity. Parameters a and b are functions of fresh organic matter (FOM), C/N ratio of the residue or fresh organic nitrogen (FON), moisture (SW in vol% or SAT in% of saturation), pH, humus (HUM) and temperature (T). NDAY is number of days after start of decomposition.

$$a := -0.00035FOM + 0.01C/N - 3.0SW + 0.87pH - 0.1HUM - 0.018T + 96.4;$$
$$b := 0.00017FOC - 0.005FON - 0.007T + 0.16pH + 0.036HUM - 0.022SAT - 1.2;$$

Humification of residue nitrogen (RDN3) is assumed to be 20% of the daily RDN2. Humification of microbial nitrogen (RDMBN) is assumed to be 10% of the daily demineralisation. The decomposition of humus-N (RDHUMN) is defined by a rate constant and a temperature and a moisture factor and can be modified by an (0–1 type) input basic mineralisation factor.

The boundary conditions are defined by the lower and upper limits of the treatments. There is no evidence for the validity of the model under 5 or above 45 °C, under 20% of the saturation moisture content, above 300 kg per ha of N fertiliser and less then 2% or more than 6% of residue, a C/N ratio lower than 13 or a ratio higher than 50.

Calibration

Calibration of the model was done by all 27 treatment combinations and on the 3 soils. The optimum and the extreme conditions are represented in the selections of the Figure 2.

The following treatment combinations were used:

Treatment No.	Temperature (°C)	Moisture (% of sat)	Fertiliser (N kg ha^{-1})	Soya (%)	Maize (%)	C/N ratio ratio
27	25	60	150	2	2	21
26	25	60	50	2	0	13
24	25	60	150	0	2	48
22	25	60	0	2	2	21
21	25	60	300	2	2	21
20	25	20	150	2	2	21
19	25	100	150	2	2	21
18	5	60	150	2	2	21
17	45	60	150	2	2	21

Treatment No. 27 represents the optimum conditions. It is obvious from the figures that the model works the best around the optimum conditions. The ammonium concentration was estimated near to the measured values in all cases. Fertiliser amounts can be extreme (high and low). It does not cause significant biases in the prediction of nitrate level. The NO_3-N estimation is better in nitrogen dominated soil than on a carbon dominated soil. In extreme temperature conditions the model overestimated the nitrate concentration.

Validation

Validation has been carried out using two independent incubation experiments on soils of Dömös and Örbottyán (Hungary). The first is a meadow alluvial loamy soil, the latter one is a calcareous sandy soil. The results are nearly identical in the two experiments. Similarly to the results of the calibration the treatment combination of all factor at optimum approached the observed values the best here, too. The prediction was fairly good in case of the extreme fertiliser doses. A better estimation was reached on the nitrogen dominated soil (low C/N ratio) in contrary to the carbon dominated soil (high C/N ratio) as it was experienced at the calibration. With oxygen deficiency, high ammonium concentration was predicted but a low concentration was observed. In drought and in case of very high and very low temperature the model overestimated the observed values.

Acknowledgements

This publication has been sponsored by US-Hungarian Joint Fund (495) and the Hungarian Scientific Research

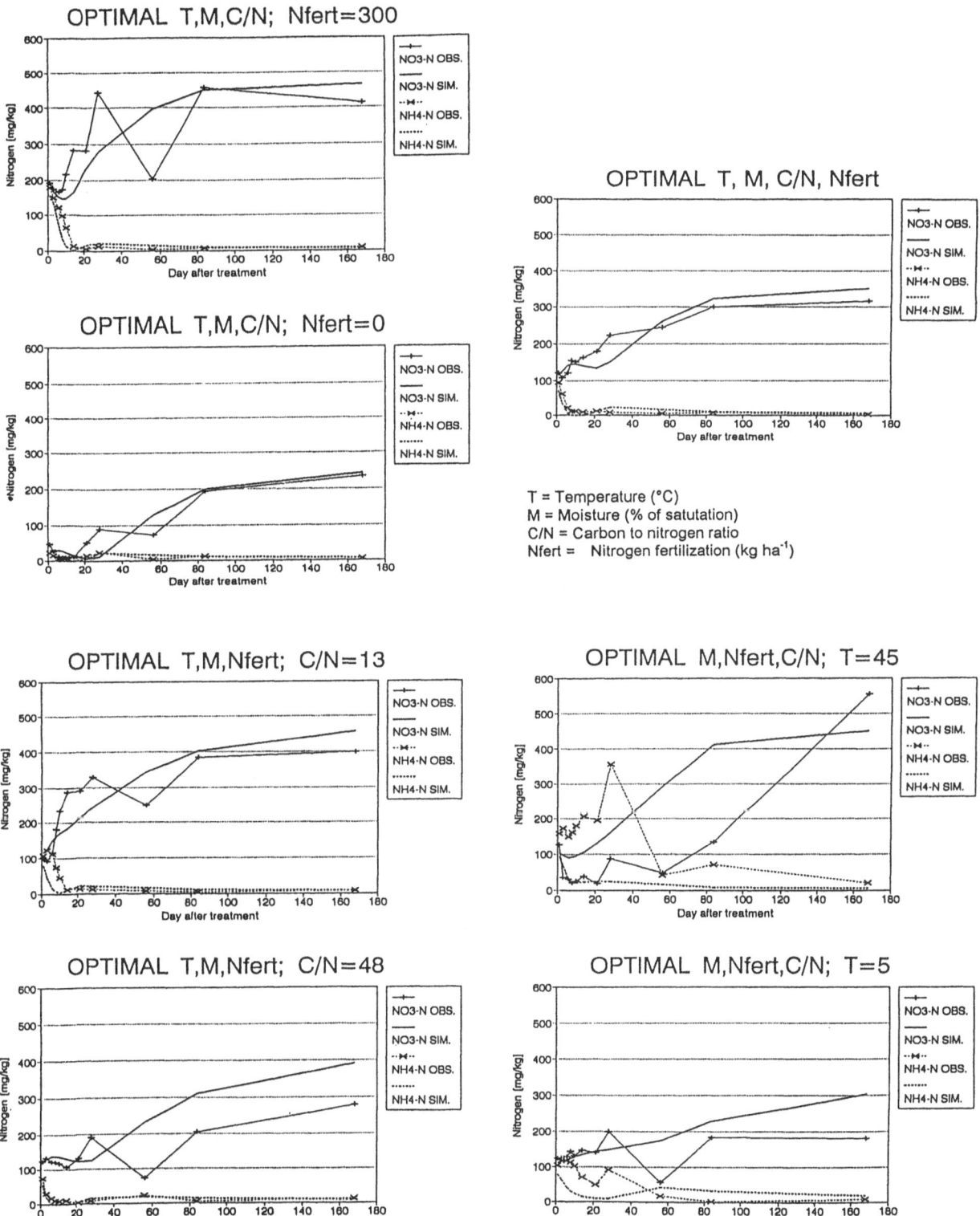

Figure 2. Observed and simulated NO_3^--N and NH_4^+-N concentrations in time of incubation on different optimal and extreme conditions.

Fund OTKA (TO17646). Special thanks to J Békéssy, and M Gyimesi for computer programming and assistance.

References

Addiscott T M and Whitmore A P 1987 Computer simulation of changes in soil mineral nitrogen and crop nitrogen during autumn, winter and spring. J.Agric. Sci., Cambridge 109, 141–157.

Berghuijs J T, Rijtema P E and Roest C W J 1985 ANIMO, agricultural nitrogen model. NOTA 1671, Inst. for Land and Water Management Research, Wageningen, the Netherlands.

Bremner J M and Shaw K 1955 Determination of ammonia and nitrate in soil. J. Agric. Sci. 46. 320.

Fehér J, van Genuchten M Th, Kienitz G, Németh T, Biczók Gy and Kovács G J 1991 DISNIT2, a root zone water and nitrogen management model. In Hydrological Interactions Between Atmosphere, Soil and Vegetation Eds. G Kienitz et al. IAHS Publ. 204, 197-205.

Frere M H, Onstad C A and Holtan H N 1975 ACTMO, an agricultural chemical transport model. USDA, ARS-H-3, US Gov't Printing Off.,Washington DC, USA. 54 p.

Frissel M J and van Veen J A 1981 Simulation of nitrogen behaviour of soil-plant systems. PUDOC, Wageningen, the Netherlands. 277p.

Godwin D C and Vlek G 1985 Simulation of nitrogen dynamics in wheat cropping systems. In Wheat Growth and Modelling. Eds. W Day and R K Atkin. Plenum Publ. Corp., New York, USA.

Hadas A, Feigenbaum S, Feigin A and Portnoy R 1986 Nitrification rates in profiles of differently managed soil types. Soil Sci. Soc. Am. J. 50, 633-639.

Hansen S, Jensen H E, Nielsen N E and Svendsen H 1991 Simulation of nitrogen dynamics and biomass production in winter wheat using the Danish simulation model DAISY. Fert. Res. 27, 245–259.

Jones C A and Kiniry J R (ed) 1986 CERES-Maize. A simulation model of maize growth and development. Texas A and M Univ. Press, College Station, Texas, USA.

Jones J W, Hesketh J D, Kamprath E J and Bowen H D 1974 Development of nitrogen balance for cotton growth models: a first approximation. Crop Sci. 14, 541–546.

Johnsson H, Bergström L, Jansson P E and Paustian K 1987 Simulated nitrogen dynamics and losses in a layered agricultural soil. Agric. Ecosys. Environ. 18, 333–356.

Kovács G J 1982 Ecophysiological relationship between water-and nutrition dynamism in maize. Növénytermelés 31.4, 355–365.

Kovács G J 1986 Seasonal relation of growth and nitrogen assimilation in maize to changes in the condition of soil. Acta Agron. Hung. 35, 133–145.

Molina J A E, Clapp C E, Shaffer M J, Chichester F W and Larson W E 1983 NCSOIL, a model of nitrogen and carbon transformations in soil: description, calibration, and behaviour. Soil Sci. Soc. Am. J. 47, 85–91.

Smith O L 1979 An analytical model of the decomposition of soil organic matter. Soil Biol. Biochem. 11, 585–606.

Stevenson F J (Ed) 1982 Nitrogen in Agricultural Soils. Agronomy series No.22 ASA-CSSA-SSSA Publisher, Wisconsin, USA.

Szili Kovács T, Szegi J and Kovács G J 1988 Potential denitrification in relation to irrigation. Zentralbl. Mikrobiol. 143, 447–451.

Szili Kovács T, Radimszky L, Andó J and Biczók Gy 1993 CO_2 evolution from soils formed on various parent materials. Agrokém. Talajtan 42, 140–146.

Tanji K K 1982 Modelling of the nitrogen cycle. In Nitrogen in Agricultural Soils. Ed. Stevenson. Agronomy 22, 721–772.

Vinten A J A and Smith K A 1992 Nitrogen cycling in agricultural soils. In Nitrate: Processes, Patterns and Management. Eds. T P Burt, A L Heathwaite and S T Trudgill. Wiley and Sons, New York, USA.

O. Van Cleemput et al. (eds.), Progress in Nitrogen Cycling Studies, 425–430, 1996.

Kinetic mechanisms of N transformation by soil microorganisms: Experimental studies and mathematical modelling

N.S. Panikov

Institute of Microbiology, Russian Academy of Sciences, Prospect 60-Let Octyabrya, 7/2, Moscow, 117811, Russian Federation

Key words: kinetics, microbial N, nitrification, nitrogenase, N uptake, turnover

Abstract

Suggested mechanistic mathematical models describe dynamics of microbial growth and activity in soil samples amended with available C-source or(and) mineral N. The models take into account the possibility of transition from C- to N-limitation, variation of microbial physiological state and intracellular N content, repression and derepression of nitrogenase activity, maintenance requirements and turnover of cell macromolecules. Kinetic analysis revealed that conventional models based on first-order rate equations are justified only for narrow range of experimental conditions characterized by starvation of soil microbial populations.

Introduction

Kinetics is a branch of natural science that deals with the rates and mechanisms of dynamic processes. The kinetic approach is based on a combination of dynamic experimental studies and field observations with mathematical modeling. The model describes the postulated mechanism of studied reactions, so the comparison of observation and prediction allows to discard wrong assumptions (Cornish-Bowden, 1976).

There are many attempts to use mathematical models to study N transformations in soil. Most of these models are aimed to assess turnover times of different fractions of soil N and prediction of the dynamics of N-transformations as being dependent on environmental conditions (Honeycutt and Potaro, 1990; Jenkinson and Parry, 1989; Reddy et al., 1990; Van Veen et al., 1984). The aim of the present study was to use a mathematical simulation for better understanding the real mechanisms involved in microbial transformations of N.

The Model

Concept and assumptions

Adequate mechanistic simulation of N-transformation may be accomplished only if we consider simultaneously the C- and N-cycles. Fluxes of C are needed to explain growth and activity of heterotrophic microorganisms which carry out energy-dependent reactions of N_2-fixation, denitrification, N-assimilation etc.

Let us ignore the trivial effects of environmental factors on the rates of microbial reactions. Suppose that moisture, temperature, redox potential etc. are physiologically suitable and kept constant. These restrictions are met in soil incubation experiments initiated by mixing of soil samples with some known amount of nutrient(s). We will consider 3 examples of such experiments.

Let us ignore also the spatial heterogeneity of the incubated soil by assuming that the added substrate is uniformly distributed throughout the soil mass. It is a quite realistic requirement in the case of soluble or volatile chemicals.

Aerobic incubation of soil with C-substrate

The governing equations are summarized in Table 1. It was assumed that added glucose is oxidized directly to

Table 1. Mathematical simulation of N and C transformation by soil microorganisms

Variable	Equation	
C-substrate (e.g. glucose)	$\begin{array}{c}\text{net}\\\text{change}\\\dot{s}\end{array} = \begin{array}{c}\text{oxidative}\\\text{consumption}\\q_s x\end{array} - \begin{array}{c}\text{fermentative}\\\text{consumption}\\q_s' x\end{array}$	(1)
Fermentation products	$\begin{array}{c}\text{net}\\\text{change}\\\dot{m}\end{array} = \begin{array}{c}\text{formation}\\\\q_s' x\end{array} - \begin{array}{c}\text{oxidative}\\\text{uptake}\\q_m x\end{array} + \begin{array}{c}\text{cell}\\\text{turnover}\\arx\end{array}$	(2)
Microbial biomass	$\dot{x} = x \cdot min\{\begin{array}{c}\mu_2 \quad \text{underC} - \text{limitation}\\\mu_2 \quad \text{underN} - \text{limitation}\end{array}$	(3)
Specific growth rate under C-limitation	$\mu_1 = \begin{array}{c}\text{assimi}\\\text{lation}\\\overline{Y_s(q_m + q_s)}\end{array} - \begin{array}{c}\text{maintanance}\\\text{of NGase}\\\overline{\Theta Y_s(q_s + q_m)}\end{array} - \begin{array}{c}\text{other maint}\\\text{functions}\\\overline{Y_s m}\end{array} - \begin{array}{c}\text{turnover}\\ar\end{array}$	(4)
Specific growth rate under N-limitation	$\mu_2 = \begin{array}{c}\text{from}\\\text{N-uptake}\\\overline{Y_N q_n}\end{array} + \begin{array}{c}\text{from}\\\text{N}_2\text{-fixation}\\\overline{Y_N Y_a \Theta(q_s + q_m)}\end{array} - \begin{array}{c}\text{turnover}\\ar\end{array}$	(5)
Soil mineral N (NO_3^-, NH_4^+)	$\begin{array}{c}\text{net change}\\\dot{N}\end{array} = \begin{array}{c}\text{cell turnover}\\ar\sigma x\end{array} - \begin{array}{c}\text{uptake}\\(\mu + ar)x/Y_N\end{array}$	(6)
Index of physiological state	$\dot{r} = (\mu + ar)(\bar{r} - r), \bar{r} = \min\left(\frac{s}{K_s + s}, \frac{N - N^*}{K_N + N}\right)$	(7)
CO_2 evolution (respiration)	$\dot{p} = \begin{array}{c}\text{C-uptake}\\\overline{(q_s + q_m)x}\end{array} - \begin{array}{c}\text{C-release}\\\overline{arx}\end{array} - \begin{array}{c}\text{C-assimi}\\\text{lation}\\\overline{\mu x}\end{array}$	(8)
N-content in microbial cells	$\dot{\sigma} = (\mu + ar)(1/Y_N - \sigma)$	(9)
Microbial yield per N consumed	$Y_N = \frac{K_N + N - N^*}{\sigma_0 K_N + \sigma_m(N - N^*)}$	(10)
NA (N_2-fixation)	$A = Y_a \Theta (q_s + q_m)x, \Theta = \frac{1 - r}{1 + K_a N}$	(11)
Specific metabolic rates	$q_s = rQ_s \frac{s}{K_s + s}; q_s' = rQ_s' \frac{s}{K_s' + s}; q_m = rQ_m \frac{m}{K_m + m}$	(12)
	$q_N = rQ_N(N - N^*)/(K_N + N)$	

CO_2 and via some intermediates m (Eqs. 1–2). Microbial cells consume C- and N-containing compounds (Eqs. 1, 2, 6). C-compounds are used for growth and maintenance (osmoregulation, defense of NA from O_2). Besides, there is a continuous turnover of short-living polymeric cell components (proteins, nucleic acids, cell wall polymers). Their relative amount is characterized by a single variable r called an index of the physiological state (Eq. 7). The decay products are released into extracellular liquid and may be reutilized (Eqs. 2, 4, 5, 6).

Microbial growth rate may be either C- or N-limited, depending on the concentration of respective substrates in the soil. Under N-limitation, two reactions of the microbial community are indicated: variation of the N-content of microbial biomass σ, and derepression of the nitrogenase activity (NA) (N_2-fixation). Variation of cell N-content, σ, may be substantial (Fig. 1) and can be adequately explained by Equations 9–10. They account for the hyperbolic relation-

ship between σ and the concentration of available N in the soil solution. In biological terms, this relationship implies an adaptive metabolic response of microbial cells to N-limitation: the lower the availability of N in the environment, the higher the intracellular content of the N-free cell components (starch, glycogen, poly-β-oxybutyrate), which partly replace N-containing polymers such as RNA and proteins. The lowest limit of cellular N is attained under N-starvation ($\sigma \rightarrow \sigma_0$, if N→0), while maximum values, σ_m, occur under N-sufficient conditions (N→ ∞).

Synthesis of nitrogenase was assumed to be directly related to the total true growth rate (Eq. 11 contains the variable Θ which is a fraction of the whole C-flux directed to synthesis of this enzyme) and inversely related to the N-availability (imitation of the repression phenomenon).

All specific metabolic rates (Eq. 12) are given as Michaelis-Menten equations extended by the variable r which simulates the dependence of the instant metabol-

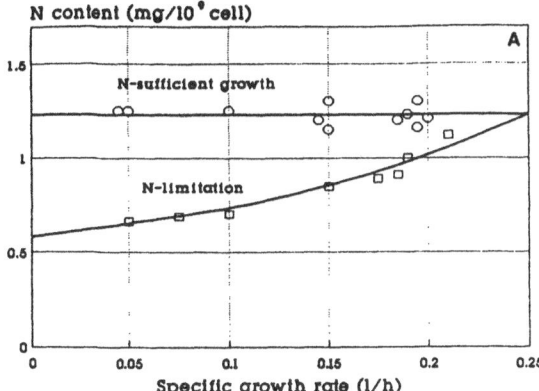

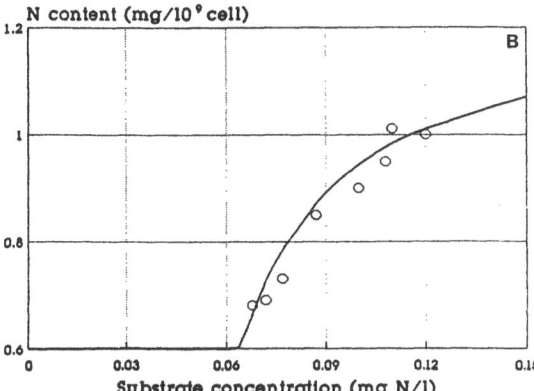

Figure 1. The effect of growth conditions on N-content of microbial cells. Algae *Chlorella vulgaris* were grown in chemostat culture at different dilution rates *D* under conditions of N- and P-limitation (Panikov and Pirt, 1978). **A**. Relationship between N-content of microbial cells σ and steady-state growth rate μ (=*D*). **B**. Dependence of σ on residual concentration of N (urea) in N-limited chemostat culture. The curves were calculated according to model (Eqs. 9–10).

ic activity on the physiological state of cells, in particular on the content of enzymes responsible for metabolic activity. N-Uptake was assumed to be stopped at ambient N-concentration below the threshold level N^* as a result of N leakage from cells.

Conservation conditions are formulated for both C and N as follows:

$$C - balance \quad \dot{s} + \dot{m} + \dot{x} + \dot{p} = 0 \qquad (1)$$

$$N - balance \quad \dot{N} + \dot{N}_{cell} = \dot{N} + \sigma\dot{x} + x\dot{\sigma} = A$$
$$(N_2 - fixation) \qquad (2)$$

The first equation implies that the sum of all accounted for C-compounds (residual glucose *s*, intermediates *m*, biomass *x*, and CO_2 *p*) should be at any time exactly equal to the initial amount of glucose-C added to the soil. Some slow reactions (decay and de novo formation of humus) are neglected. The second equation states that the total content of N in the studied soil system should be governed by the rate of N_2-fixation. If this rate is zero, then the sum of soil available N and microbial N remains to be constant. Again, the model neglects slow processes of humus mineralization and N incorporation into recalcitrant substances. Besides, we ignore denitrification losses as soon as the soil incubation is aerobic.

Anaerobic incubation of soil with C-substrate

The structure of the model which describes denitrification is basically the same as above (Table 1). The major modification which has to be made is an account for NO_3^- reduction during anaerobic oxidation of available C-compounds. The derivation could be made from the mass balance condition:

total	amount required to	amount required
amount of	= oxidize consumed	- to oxidize grown
reduced NO_3^-	C-substrate	biomass

To explain the formation of N_2O as intermediate product (denitrification sequence: $NO_3^- \rightarrow N_2O \rightarrow N_2$) let us assume that high NO_3^- concentrations inhibit the second reduction step.

Aerobic incubation of soil with mineral N-substrate

Here we will consider only autotrophic nitrification (Table 2). The derivation of the model is straightforward and we need to make only two comments.

1. Equation 14 contains the term *F* which is the input of ammonium. It is essential that normally most of it comes from microbial biomass ($F=ar\sigma x$, Equation 6), the mineralization process being driven by turnover of cell biopolymers.

2. Oxidation of ammonium to nitrite is associated with the accumulation of toxic products (mainly H^+). Their formation is directly related to the oxidation reaction (Eq. 19). Inhibition of microbial growth is supposed to be a non-competitive one.

Table 2. Mathematical simulation of nitrification in amended soil

Variable	Equation	
NH_4^+	$\dot{s}_1 = F - q_1 x_1$	(14)
NO_2^-	$\dot{s}_2 = q_1 x_1 - q_2 x_2$	(15)
NO_3^-	$\dot{s}_3 = q_2 x_2$	(16)
Biomass of NH_4^+-oxidizing bacteria	$\dot{x}_1 = Y_1 q_1 x_1 = \mu_1 x_1,\ \mu_1 = \mu_m \frac{s}{(K_s + S)(1 + H/K_H)}$	(17)
Biomass of NO_2^--oxidizing bacteria	$\dot{x}_2 = Y_2 q_2 x_2 = \mu_2 x_2,\ \mu_2 = \mu'_m \frac{s}{K'_s + s}$	(18)
Toxic products	$\dot{H} = \alpha q_1 x_1$	(19)

Materials and methods

Samples of arable grey forest soil under barley were taken from a depth of 0–10 cm (Poushchino, Moscow Area). After removal of plant roots and sieving through a 2 mm screen the soil samples were mixed with 4 mg glucose g^{-1} dry soil and incubated aerobically under controlled conditions (25 °C, water content 50% WHC). During incubation, the following parameters were measured: CO_2 evolution rate by GC with a catharometer, residual glucose with the glucoseoxidase, the sum of residual glucose and exometabolites by dichromate oxidation of the water extract of the soil, the content of NO_3^- and NH_4^+ in the KCl extract with ion-selective electrodes and indophenol, respectively. NA was assayed by acetylene-reduction technique.

In the second experiment, the soil was amended with $(NH_4)_2SO_4$ or KNO_2 (100 mg N kg^{-1} soil) and incubated as described above.

Results and discussion

The experimental data and simulated curves are presented on Figure 2 and 3. Addition of glucose induced a significant acceleration of soil respiration and microbial consumption of inorganic N. The resulted decline of *N* down to the threshold level *N** led to derepression of NA which sharply increased and then dropped to almost zero values. Thus, NA may be observed only transiently for a short period of time when simultaneously 2 conditions are met: i) low N concentration, and ii) elevated supply of available C-substrates. Deple-

tion of C-substrates (*s+m*) resulted in a decline of the starving microbial population and an accumulation of mineral N originated from the decayed biomass.

Thus, we may distinguish 3 successive stages which take place in the soil after glucose amendment: i) rapid microbial growth on glucose with concomitant N-immobilization, ii) retarded N-limited growth associated with N_2-fixation, and iii) decline in biomass with release of mineral N. The model adequately simulates the observed dynamics. It gives also a realistic imitation of what we are not able to measure directly and accurately, in particular the N-content of microbial cells σ. This variable varied in a predictive way attaining the highest level during the most intensive microbial growth (curve 6, Fig. 2).

Dynamics of NH_4^+ and NO_2^- conversion to NO_3^- (Fig. 3) were described by sigmoid and exponential curves respectively. The exponential pattern is a clear indication of the non-limited growth of nitrifying bacteria on nitrite. Deviation from this pattern in the case of growth on ammonium agrees with the postulate that the first nitrification step is subjected to autoinhibition.

In numerous soil-incubation studies mineralization and nitrification are described as first order reactions (see as examples Mary et al., 1996). It follows from our data that kinetics of N-transformation is actually much more complicated and could not be reduced to one particular 'kinetic order'. However, long-term laboratory incubations do provide such specific conditions that observed dynamics formally obey first order kinetics. The most essential feature of long-term incubation is C-starvation of heterotrophic microbial populations. In our model, we may put such a situation: $\mu \rightarrow 0$,

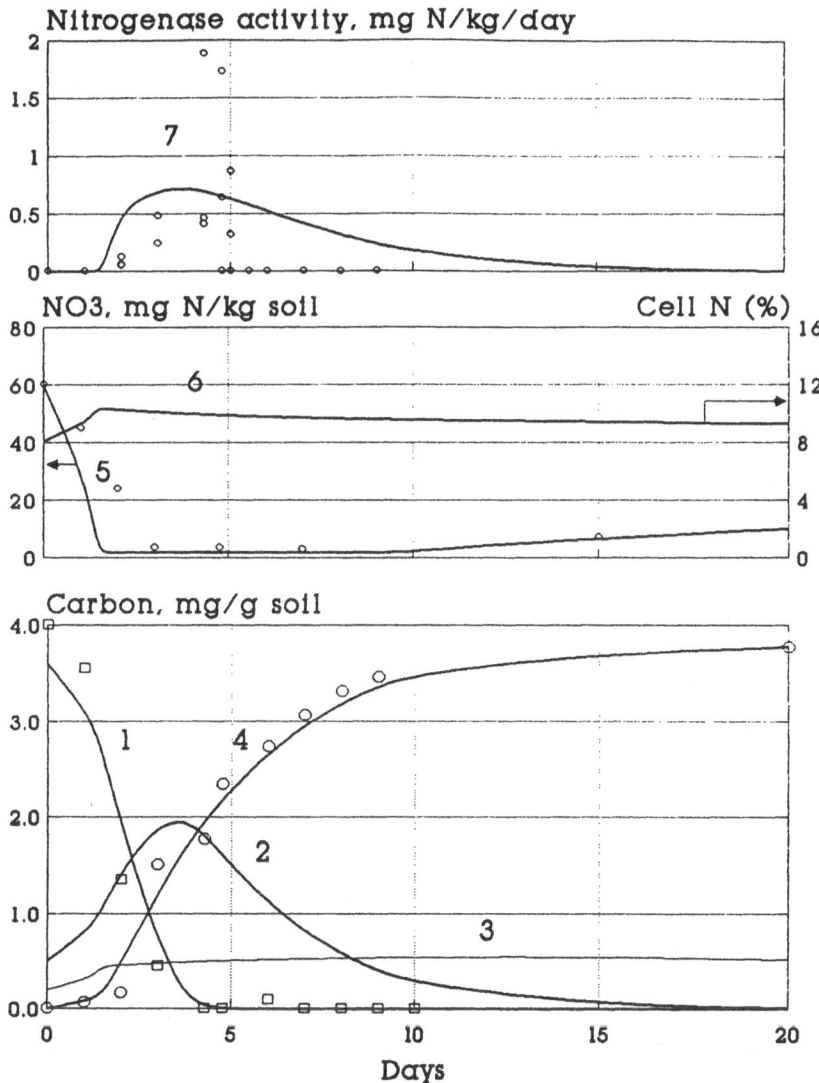

Figure 2. Microbial transformation of C and N in soil amended with glucose. Kinetic model (Table 1) was used to simulate dynamics of the following variables: residual glucose (1), fermentation products (2), microbial biomass (3), CO_2 (4), mineral N (5), N-content in microbial biomass (6), N_2-fixation rate (7). Identified parameters: Q_s=5, Q'_s=17, Q_m=15, m=0.03 (mg C mg^{-1} C biomass day^{-1}), Q_N=2 (mg N mg^{-1} C biomass day^{-1}), K_s=K'_s=0.5, K_m=2.5 (mg C g^{-1} soil), K_N=5 mg N g^{-1} soil, Y_s=0.6, Y_a=0.03, K_a=3400, a=0.02 day^{-1}, σ_0=0.06, σ_m=0.30.

$1/Y_N \approx \sigma_0$, and $F=arx(\sigma-\sigma_0)=arN_{cell}$, where N_{cell} is changeable fraction of microbial N. During starvation this fraction is susceptible to decay with a pseudo-first order rate constant ar. Then, the observed dynamics of NO_3^- formation from some point of incubation (e.g. from 10th day, Fig. 3) may be approximated by the following expression:

$$[NO_3^-] = [NO_3^-]_\infty (1 - e^{-kt}), \qquad (3)$$

where $k=ar$, and $[NO_3^-]_\infty=N_{cell}$. However, it is clear that the use of the derived Equation 20 is justified only for a restricted range of experimental conditions (late stages of soil incubations). Its application in in situ situations (open systems with more or less continuous supply of substrates) is questionable.

Acknowledgement

The research described in this publication was made possible in part by Grant No MIQ 000 from the International Science Foundation, and by Grant No 93–04–

430

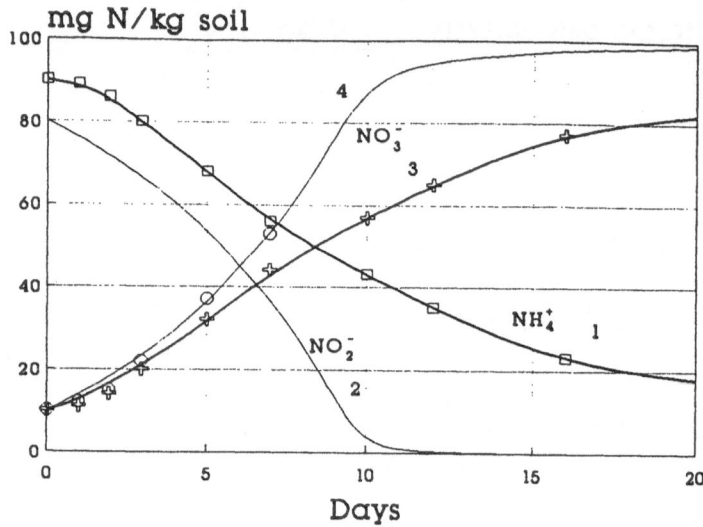

Figure 3. Simulation of nitrification dynamics. Kinetic model (Table 2) was used to simulate dynamics of nitrification in samples of grey forest soil amended with $(NH_4)_2SO_4$ (1,3) or KNO_2 (2,4): residual ammonium (1), residual nitrite (2) and nitrate formed from ammonium (3) and nitrite (4).

7202 from the Russian Fund of Fundamental Research. I thank I K Kravchenko and M V Sizova for analytical data.

References

Cornish-Bowden A 1976 Principles of enzyme kinetics. Butterworths, London, UK. 206 p.

Honeycutt C W and Potaro L J 1990 Field evaluation of heat units for predicting crop residue carbon and nitrogen mineralization. Plant and Soil 125, 213-220.

Jenkinson D S and Parry L C 1989 The nitrogen cycle in the Broadbalk wheat experiment: a model for the turnover of nitrogen through the soil microbial biomass. Soil Biol. Biochem. 21, 535-541.

Mary B, Recous S, Darwis S and Robin D 1996 Interactions between decomposition of plant residues and nitrogen cycling in soil. Plant and Soil 181.

Panikov N S and Pirt J S 1978 The effects of cooperativity and growth yield variation on the kinetics of nitrogen or phosphate limited growth of *Chlorella* in a chemostat culture. J. Gen. Microbiol. 108, 295-303.

Reddy K R, Rao P S C and Jessup R E 1990 Transformation and transport of ammonium nitrogen in a flooded organic soil. Ecol. Model. 51, 205-216.

Van Veen J A, Ladd J N and Frissel M J 1984 Modelling C and N turnover through the microbial biomass in soil. Plant and Soil 76, 257–274.

Plant and Soil **181**: 109–121, 1996.

Modeling of nitrogen transformations and translocations

Jörg Richter and Dinesh Kumar Benbi
Institute of Geography and Geoecology, Technische Universität Carolo-Wilhelmina, Braunschweig, Germany*

Key words: microbial immobilization, net N-mineralization, nitrate leaching, plant N-uptake, simulation

Abstract

Different submodels within complex model packages on N regimes - for plant N-uptake, net N-mineralization, nitrate leaching and microbial N immobilization - are critically reviewed mainly with regard to their prediction ability on the basis of three comparative papers. Only for some of the processes adequate statistical evaluation of the models was possible. Compared to the other statistically evaluable process, nitrate leaching, modeling of plant N-uptake yields the better results. Most models for mineralization use arbitrary approaches rather than empirical ones. Although only approximate estimates of N mineralisation were at hand, the models generally behave expectedly poor. Only one model - DAISY - out of 16 involved in the comparison uses an explicit microbial biomass sub-model including microbial growth, decline and maintenance terms. So DAISY is the only model coupling C and N cycles. But what is true for an individual model describing the C and N transformation of a lab incubation experiment seems to be valid for most of the complex simulation work on the C and N regimes: this model was said to be overparameterized with respect to the available data.

Introduction

Thirty years ago, in Europe we used about one fourth of the mineral nitrogen fertilizer amounts of today. Since yields only doubled in this period we have surplus nitrogen problems in intensive farming leading to high amounts of vagabonding nitrogen in the plants, the atmosphere and the water bodies. Now, science is asked to solve the problems caused by misinterpretation of the role of nitrogen by its practioners.

Process simulation is believed to be a comprehensive and integrative tool not only for understanding but also for managing problems with nitrogen in the environment. To use process simulation effectively for research as well as for advisory purposes of nitrogen regimes, we should have a basic and quantitative understanding of all the essential processes involved in the respective situation. For the temperate climate of central Europe, our deterministic N-regime models for arable crops now usually comprise nitrogen uptake by plants, the processes of ammonification and nitrification of organic nitrogen, nitrate leaching and denitrification. For ammonification-nitrification, there is now a great body of basic knowledge from incuba-

tion experiments, which should be used in simulations of field situations. For root uptake, even approaches using simple root distribution patterns usually perform satisfactorily, although they can not simulate translocations within the plants or disappearance of surplus ammonia by evaporative processes. Also leaching is a process which may be simulated simply and quite correctly even in rather heterogeneous fields. Denitrification is thought to be of minor importance in well drained arable field soils in temperate climate.

To reach a generally higher degree of agreement between simulation and measurement, future development in modeling seems to concern essentially processes like denitrification, microbial immobilization, ammonia volatilization and ammonium exchange and fixation-defixation. A comprehensive picture of the nitrogen cycle in the soil is given in Figure 1. With regard to denitrification, we may be in a phase where it may still be worthwhile to compare different simulation approaches on the basis of their relative ability to represent experimental losses. Only recently we ourselves incorporated the description of ammonium exchange and fixation-defixation into our approaches. The majority of NH_4^+ ions may be withdrawn from soil solution by ion exchange if nitrification is slow.

---

* FAX No.: +495313918170
Plant and Soil is the original source of publication of this article. It is recommended that this article is cited as: *Plant and Soil* **181**: 109–121, 1996.

110

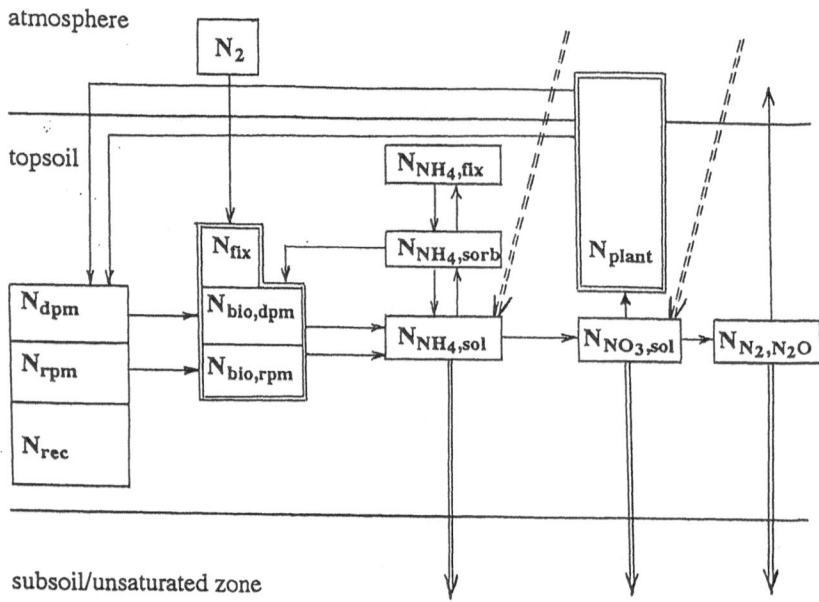

atmosphere

topsoil

subsoil/unsaturated zone

Figure 1. A presentation of the nitrogen cycle.

This ammonium may also be fixed within relatively short periods (days) and defixate within decades, when the ammonium concentration in the solution decreases again. Moreover, the microbial immobilization sub-model may fill a gap, but till now we have only limited quantitative knowledge of the real processes. This is especially so because we are unable to treat the process separately from other processes, as we apparently can do for net N mineralization. Applying our model approach to understand nitrogen regimes in arid cal-careous loess soils in China, ammonia loss has to be included in the model with special regard to the forms in which most nitrogen comes into the field, namely as urea and ammonium bicarbonate.

In this paper we will not demonstrate our newest approaches mentioned in the last paragraph since there is only limited experimental proof for them. Instead we would like to stick to the more easily described processes on which comprehensive work has already been done. Especially, we would like to confront some concepts with their use in related approaches of some well known model simulation packages. In two recent reports these different simulation packages have extensively been compared regarding their ability to describe N-regimes in part as well as in toto. - The reports are:

1. European Communities - Commission EUR 13501, 1991 - hereafter referred as report 1.

2. Groot J J R, de Willigen P and Verberne E L J, 1991 - hereafter referred as report 2.

Materials and methods

In report 1, Vereecken et al. (1991), selected 5 out of 9 complex models used to compare simulation results with measured or estimated output components of the water and nitrogen regimes: N-mineralization, den-itrification, nitrate leaching, and crop N-uptake at harvest. Measurements or estimates were based on data sets for two to five years of three differently cropped sandy soils, two from Denmark and one from the Netherlands. Five statistical criteria were used to evaluate simulations (Table 1). The coefficient of determination-equivalent was regarded as an expression of the scattering of the simulated values relative to the scattering of the measurements. Negative model-ing efficiencies indicate that the mean of the measured values is a better estimate than the mean of the simu-lated values. Correspondingly, positive coefficients of residual mass indicate a tendency of the simulations to underestimate the measured values, and vice versa. No comparison has been done based on measured time series of nitrate-N profiles. The models were originally created for different purposes but will not be character-ized here besides their abilities evaluated below. The 5 models in the review of Vereecken et al. (1991), are ANIMO, DAISY, EPIC, RENLEM and SWATNIT.

Table 1. (=Table 1 of Vereecken et al.) Statistical criteria for evaluation of the simulated results according to Loague et al. (1991)

Criterion	Symbol	Calculation formula[a]	Range	Optimum
Maximum error	ME	$\text{Max } |P_i\text{-}O_i|_{i=1}^{n}$	≥ 0	0
Root mean square error	RMSE	$(\sum_{i=1}^{n}(P_i - O_i)2/n)^{1/2} \cdot 100/\bar{O}$	≥ 0	0
Coefficient of determination	CD	$\sum_{i=1}^{n}(O_i - \bar{O})^2 / \sum_{i=1}^{n}(P_i - \bar{O})^2$	≥ 0	1
Modelling efficiency	EF	$(\sum_{i=1}^{n}(O_i - \bar{O})^2 - \sum_{i=1}^{n}(P_i - O_i)^2)/(\sum_{i=1}^{n}(O_i - \bar{O})^2)$	≤ 1	1
Coefficient of residual mass	CRM	$\sum_{i=1}^{n}O_i - \sum_{i=1}^{n}P_i)/(\sum_{i=1}^{n}O_i)$	≤ 1	0

[a] P_i = predicted value i, O_i = observed value i, $\bar{O}$ = mean of the observed values, n = number of data pairs,

In report 2, de Willigen (1991) compared 14 models with regard to the submodels involved and their descriptions of the different processes. The results of simulations based on data sets from the Netherlands given to all the modellers were also included. In addition to other models, ANIMO, DAISY and SWATNIT also participated in this comparison. In the same report, Otter-Nacke and Kuhlmann (1991) compared three out of the same 14 models with respect to their ability to predict N_{min}-contents in early spring under farmers field conditions. The models were run between harvest and spring of the following year for two different winter periods for at least 70 farmers field plots. The model performance was evaluated according to the number of cases in which prediction was within ± 20 kg N_{min} ha^{-1} of the measured profile contents. This is the observed average difference between two measurements repeated on the same field plot for the relatively homogeneous loess field soils if sampling is done with great care (Richter et al., 1984). Neither ANIMO, DAISY nor SWATNIT were included in the latter comparison.

Results and discussion

Beginning with the paper of Otter-Nacke and Kuhlmann (1991), results may simply be presented by showing i) the worst case of the first year, where no initial conditions were at hand, and ii) the results of the second year using measured data on N_{min} in soil after harvest of the preceding crop as initial condition (Fig. 2). There was an agreement of around 66% for the first and around 75% cases for the second year. The latter improvement is essentially due to the fact, that initial values were at hand for this year. Detailed information on the results are contained in Table 2. According to the authors, the Mean Bias Error (MBE) indicates if the model gives well balanced results. Only model B yielded well balanced simulations, while model C

overestimated measured values by 6 to 7 kg N ha^{-1} for both years. If such an increase in agreement of roughly 10% for the relatively simple winter models marks an essential progress - can we expect a similar progress by extending the simulation period to the whole season, with the much more complex situations and processes? It should be clearly stated that these authors proved the suitability of models for advisory purposes.

On the basis of the other review papers of the two reports, a more detailed picture on the modeling approaches and a rough evaluation of the abilities to predict the different processes such as **nitrogen uptake by plants, net-N mineralization, nitrate leaching** and **microbial immobilisation** may be given. Time series of N_{min} profiles and sums could also help to qualify an approach. However, as in the paper of Otter-Nacke and Kuhlmann (1991) where no measured N_{min} time series were at hand, both review papers (de Willigen, 1991; Vereecken et al., 1991) use time series only scarcely. The following evaluation will essentially follow these original reviews.

N plant uptake: the soil process modelers can do best

According to Vereecken et al. (1991), the modeling efficiency (Fig. 3 and Table 3) for plant N-uptake is high in comparison to nitrate leaching, and is small only for the Jyndevad site with the poor sand. Considering the substantial variation of measured N uptake the performance of the relevant submodels seems to be good. One would expect that even relatively small variation in simulated N-uptake may have relatively large effects on the simulation quality of the other processes, e.g. leaching. However, for the Jyndevad site the modeling efficiency for nitrate leaching is highest.

Also de Willigen (1991) concluded, that above ground variables had been simulated more accurately than below ground ones. Some of the compared models calculated dry matter production and N-uptake in an independent way, i.e. without parameter fitting.

Table 2. Statistical parameters for the comparison of the N_{min}-profile content in spring of three selected models in Otter-Nacke and Kuhlmann (1991)

Year	Model	n	Differences <10(kg(N)ha^{-1})		>20(kg(N)ha^{-1}		ME^a	MBE	MAE
			no.	(%)	no.	(%)			
1988	A	114	50	43.9	39	34.2	76	-11.03	18.0
	B	126	50	39.7	37	29.4	95	-1.60	17.0
	C	84	25	29.8	32	38.1	65	0.02^b	18.9
1989	A	68	19	27.9	21	30.9	71	-1.17	19.2
	B	69	32	47.1	12	17.6	68	1.37	14.3
	C	53	29	54.7	17	32.1	47	6.25	13.7

^{a} ME – Maximal Error, MBE – Mean Bias Error, MAE – Mean Absolute Error.
^{b} Probably misprinted in Otter–Naoke and Kuhlmann (1991); should read 6.02.

Table 3. (= Table 6 of Vereecken et al., 1991) Statistical criteria for the performance of the different models on simulation of crop N-uptake (kg N ha^{-1} a^{-1}) for the different sites

Site	Model	n^a	ME^b	RMSE^c	CD^d	EF^e	CRM^p
Askov	ANIMO	6	31.6	25.9	0.60	0.70	0.10
	DAISY	5	23.2	16.5	0.66	0.89	-0.10
	EPIC	6	38.0	21.3	0.64	0.65	0.08
	SWATNIT	6	24.7	21.1	0.54	0.77	-0.09
Jyndevad	ANIMO	12	43.5	27.5	0.54	0.30	0.03
	DAISY	12	47.3	27.8	0.48	0.30	-0.17
	EPIC	12	106.0	33.8	0.28	-0.45	-0.10
	SWATNIT	12	56.3	32.1	0.39	0.04	-0.10
Ruurlo	ANIMO	25	172.2	16.5	1.31	0.86	0.04
	DAISY	12	60.0	11.4	0.91	0.97	-0.06
	EPIC	25	191.0	19.7	0.82	0.80	0.01
	SWATNIT	20	120.9	14.1	1.36	0.90	-0.05
All^g	ANIMO	44	172.2	19.4	1.17	0.94	0.04
	DAISY	29	60.0	15.2	0.92	0.97	-0.08
	EPIC	44	191.0	22.9	0.92	0.90	0.00
	SWATNIT	39	120.9	17.0	1.06	0.95	-0.05

^{a} n: number of observations.
^{b} ME: maximum error.
^{c} RMSE: root mean squared error.
^{d} CD: coefficient of determination.
^{e} EF: modeling efficiency.
^{p} CRM: coefficient of residual mass.
^{g} Including Pittem, Belgium.

The problem with fertilizer recommendations of today is, however, much more related to below ground processes like leaching.

Net N mineralization: "arbitrary" vs. empirical modeling

Table 4 (adopted from Vereecken et al., 1991) contains a description of the different organic matter pools used

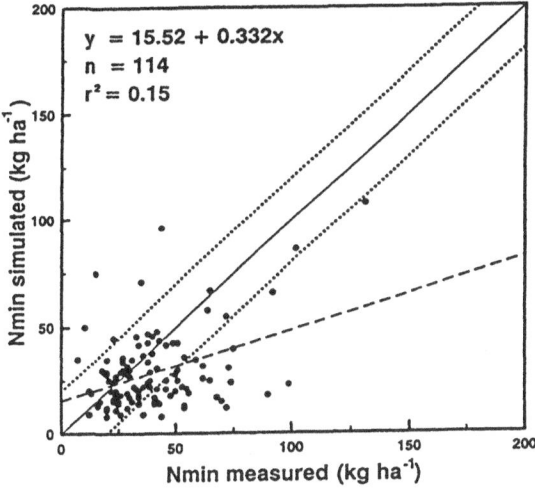

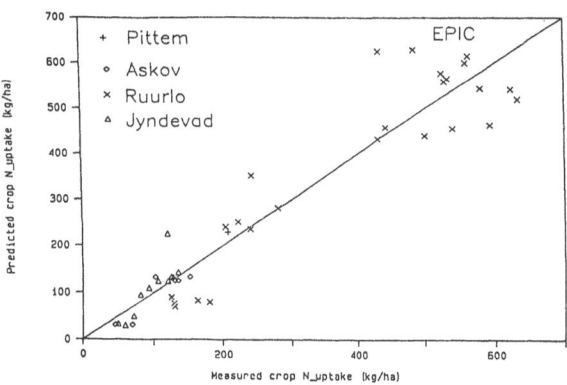

Figure 3. Comparison of measured with simulated N uptake by crops with EPIC.

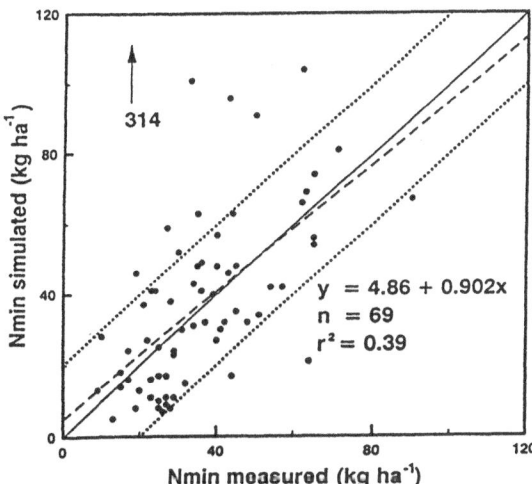

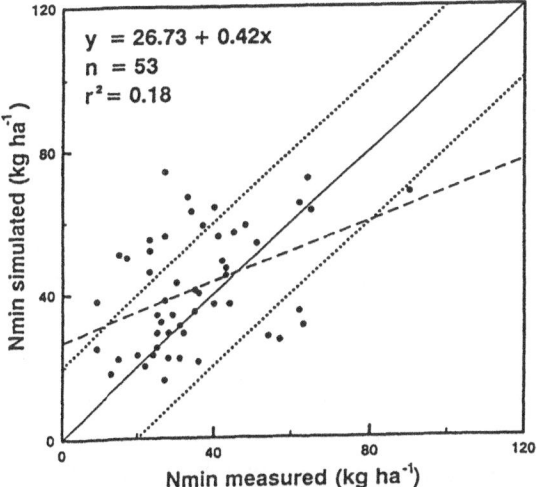

Figure 2. The worst case (model A) of the first year, where no initial conditions were at hand (top), compared with results of the second year using data after harvest of the preceding crop as initial condition (model B and C), after Otter-Nacke and Kuhlmann (1991).

in the different models of report 1 with respect to C/N ratio,% of organic C and half-life times of the usually first order mineralization of the organic material. RENLEM is not mentioned in Table 4, since REN-LEM cannot treat short term mineralization. In this model the mineralization of newly added organic matter is assumed to be complete within one year and no change in the organic N content of the soil occurs. The pools of the different models vary greatly by number, N- and C- content and half-life times. Therefore, it is not amazing that the simulated mineralization figures in Table 5 are so different. The question arises what may be the reasons for the chosen pool numbers, sizes and time behaviour and how the submodels have been parameterized by the authors or users.

Although perhaps decisive for the judgement of simulation quality it was not possible to obtain the answers from the report. This is quit clear with regard to the summarizing report of Vereecken et al. (1991). But also from the more detailed model descriptions in the individual reports only very little can be drawn. Given pool names and attributed half-life times are presented almost always without reference to the scientific basis. For ANIMO it is merely stated that "a historical run of the program is performed" prior to any actual simulation in order to get a realistic initial condition. EPIC uses a modification of the mineralization model of Seligman et al. (1981) originally developed for pastures. Although this paper appeared before all publications suggesting two N-pools and reviving the incubation technique of Stanford and Smith (1972), the model already uses a two pool approach. Even the kinetic parameters used are very close to the estimates we ourselves regard as reliable. Inter alia, ANIMO seems to be the four compartment model instead of

Table 4. (= Table 3 of Vereecken et al., 1991) Characteristics of the different organic matter pools distinguished in the four dynamic models for the Jyndevad site, Danmark

Model	Pool	C/N ratio	% of organic C[a]	Half life time
SWATNIT	Litter	8	8-1	693 d
	Manure	10	±1	693 d
	Humus	12	92-99	189 y
EPIC	Stable humus	-[b]	-	-
	Fresh organic N	-	-	23 d
	Active organic N	-	-	2500 d
DAISY	Biomass pool 1	6	0.28	693 d
	Biomass pool 2	10	0.04	49.5 d
	Soil organic pool 1	11	±80	515 y
	Soil organic pool 2	11	±20	10 y
AMINO	Humus	16	99.2	50 y
	Fraction 2	12	<0.5	77 d
	Fraction 3	58	<0.5	3 y
	Fraction 4	76	0.5	130 d
	Fraction 5	76	<0.5	37 d
	Fraction 6	24	<0.5	65 d
	Fraction 7	24	<0.5	590 d

[a] Figures are approximative.
[b] No information given.

Table 5. Comparison of the estimated with the simulated net N-mineralization (in kg N ha^{-1} a^{-1}) for four models and three sites in the comparison of Vereecken et al. (1991)

	Askov	Jyndevad	Ruurlo
Experimental estimates	75-49	82-97	85-126
EPIC	24-45	24-45	72-92
SWATNIT	77-78	102-103	110-230
ANIMO	40-60	60-90	200
DAISY	40-60	60-90	105

DAISY. For DAISY's subpool of soil organic matter it is stated qualitatively that the rate coefficients depend on the clay content of the soil. For SWATNIT the choice of the three compartments - a fast cycling soil litter pool, a slow cycling soil humus pool and a manure pool - is based on the referenced paper of Johnsson et al. (1987) which by the way may answer the above questions.

Although only approximative estimates for net N-mineralization are at hand (Table 5) the comparison with the simulated figures tells how well the different models reflect the figures of the different field experiments. Only SWATNIT seems to be able to predict the level of expected amounts, though only for the Danish experiments. For the Dutch experiments over 5 years, some of the SWATNIT simulations greatly overestimate net mineralization. This inability was judged by Vereecken et al. (1991) with the statement that "the organic N system in the SWATNIT simulation is clearly not yet in equilibrium." However, according to our own experience, the estimated figures for the Ruurlo plots after five years of continuous grazing and changing back to arable cropping with high mineral fertilizer N inputs, seem to be too small. EPIC strongly underestimated net mineralization in all experiments, while this error is less pronounced for ANIMO and DAISY.

Table 6. (= Table 2 of de Willigen, 1991) Soil biological processes considered in different models

Model	(a1)[a]	(a2)[b]	(a3)[c]	(b)[d]	(c)[e]
[A]	1	1	-	+	-
[B]	1	0	-	+	+
[C]	2	1	-	+	+
[D]	4	1	+	+	+
[E]	6	1	+	+	+
[F]	1	1	-	+	-
[G]	2	1	-	+	+
[H]	7	1	+	+	+
[I]	2	1	+	-	-
[J]	2	1	-	-	-
[K]	4	1	+	-	-
[L]	1	0	-	-	-
[M]	1	0	-	-	-
[N]	2	1	-	+	+

[a] (a1): Number of pools in mineralization/immobilization reacion.
[b] (a2): Order of the rate equations.
[c] (a3): +One or more pools of biomass; - no biomass.
[d] (b): +Nitrification considered.
[e] (c): +Dentrification considered.

For the Dutch experiment, only DAISY predicted the estimated mineralization by a relatively constant value. Which parameterization renders such a success of a 7 pool mineralization model?

Also the models compared in report 2 used quite different pool numbers for mineralization (Table 6): only 5 out of 14 use two compartments, 5 only one compartment and the residual 4 use between 4 and 7 pools: Model [D] is ANIMO, here with only four mineralizable N pools, model [H] is DAISY with 7 pools. Within the five models using two compartments there is SWATNIT (model [G]), which in the comparison of Vereecken et al. (1991) was said to use three compartments. Infact, the kinetic coefficients of two pools are identical (Table 4), so it is justified to speak of only two pools. Two of the other models use the submodel of Bergström the pools of which are characterized as litter and faeces. The remaining two models also use similar pool designs, namely crop residues and soil organic matter. However, for one of them the same reference as SWATNIT, Johnsson et al. (1987) is cited.

The obviously arbitrary choice of number, size and kinetic coefficients of organic pools - for nitrogen at least - is puzzling all the more because for the description of net mineralization safe progress has been made

during the last decade by an empirical approach. Since the often cited trial of van Veen (1977) to evaluate the extent and decomposability of the N-pools according to their chemical composition, the renewed interest in Stanford and Smith's (1972) incubation technique brought a real step forward in the evaluation of pool numbers, sizes and kinetics. Initially, there were two independent papers asserting that in incubation experiments with field moist soils mineralization is best described by superposing two first order decomposition curves whose coefficients differ roughly by one order of magnitude (Molina et al., 1980; Nuske and Richter, 1981). Now it seems to be well justified to describe N mineralization by superposing the decay of two fractions or pools often referred to as decomposable (dpm) and resistant plant material (rpm; for a review see e.g., Benbi and Richter, 1994). The kinetic description used is the first order degeneration for small substrate concentrations of the more general Michaelis-Menten approach. The kinetic coefficient is essentially dependent on temperature, moisture and pH. Typical half-life times at 10 °C average soil temperature are around 0.35 years ($\approx$ 130 days) for the fast decomposing dpm and about 4.5 years ($\approx$ 1650 days) for the slow rpm pool. Pool sizes of the dpm fraction are, depending on the type of crop, between 1/2 to 2/3 of the N of the plant residues left on the field after harvest. The size of the rpm N-pool depends primarily on organic matter and clay content of the soil. It ranges from 500 to 800 kg N ha^{-1} topsoil^{-1} for luvisols (Nordmeyer and Richter, 1985; Richter et al., 1994). The pool itself is replenished essentially by the other 1/3 to 1/2 of residual N in crop residues which is not dpm N and may vary also according to management conditions. Only very little - around 1% of the organic residues - is directed via this pool to more recalcitrant pools, the decomposition of which is extremely slow and which are, therefore, neglected in simulation. The two-pool-estimates have been derived independently from incubation experiments and show little resemblance to the pool numbers, sizes and kinetic coefficients in Table 3. A similar approach exists for organic carbon (Jenkinson and Rayner, 1977).

Nitrate leaching: "mechanistic" vs. "phenomenological" modeling of the water movement

Nitrate leaching essentially seems to "balance" the surplus N fertilization in humid climates. It depends primarily on drainage rates (or vice versa, on evapotrans-

116

piration), soil texture and fertilization rates and distribution. Correct simulation depends, therefore, first on the quality of estimates of evapotranspiration ET and to a much smaller extent on the quality and method of modeling the water movement. There are principally two ways of modeling water movement: the so-called "mechanistic" approach, (which is not well described by this term and binds us to the mechanistic Newton paradigma) and the capacity approach (for a review see Addiscott and Wagenet, 1985). The mechanistic approach uses the well-known Richards equation and needs a complete data set of parameters, $\psi(\theta,z)$ and $K(\theta,z)$. Since these are not at hand in normal field situations, these parameter functions have to be calculated from easy-to-obtain characteristics by co-called pedotransfer-functions. Instead, capacity models use a more rigorous approach: they avoid the use of conductivities and make only use of information on the possible limits of the water storage. - What is to be expected from these different approaches?

In the comparison of Vereecken et al. (1991), it is stated that actual ET varied greatly from one model to the other. But the reviewers felt unable to judge the quality of the ET parts of the models by the measured moisture profiles. Regarding the water movement and storage, however, the models showed, for the Danish experimental sites, systematic differences between the capacity type model (EPIC) and the "mechanistic" models (ANIMO, DAISY, SWATNIT). The mechanistic models, all using the same hydraulic model SWATRE, predicted substantial changes in water storages for the two years, whereas storage was not much changed in EPICs capacity model. Unfortunately, there were no measurements at the beginning and the end of the simulated periods to judge these results.

At the Dutch experimental site considerable capillary rise can be expected from the shallow and varying water table. EPICs capacity type model is expected (Vereecken et al., 1991) not to be able to reflect this situation, but instead it does. Perhaps the explanation for this is simply that the reviewers may have disregarded the "presence" of the groundwaters in the upper equilibrium limit - similar to field capacity - which, therefore, exerts a tendency to pull water upwards according to ET.

In report 2, 14 models are compared. In two of them a water transport model is not included but instead an external water transport calculation scheme is used. This is perhaps true for others apart from these two, because a similar explanation holds for ANIMO. Anyhow, of the remaining 12 models, 6 use mechanistic

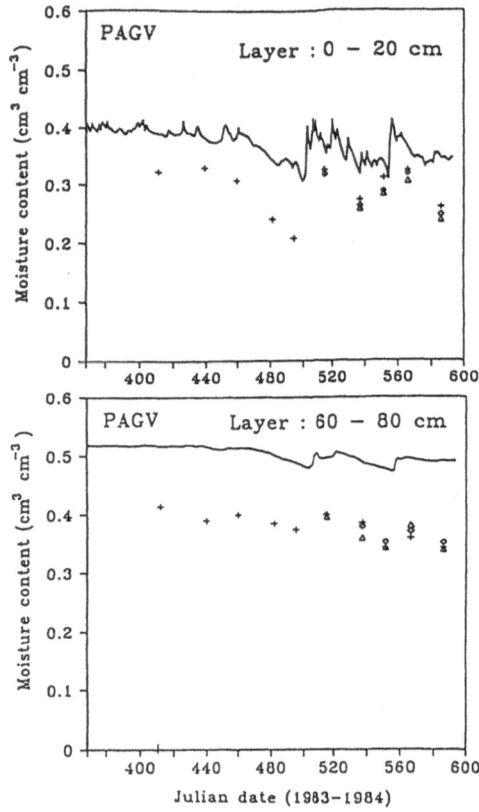

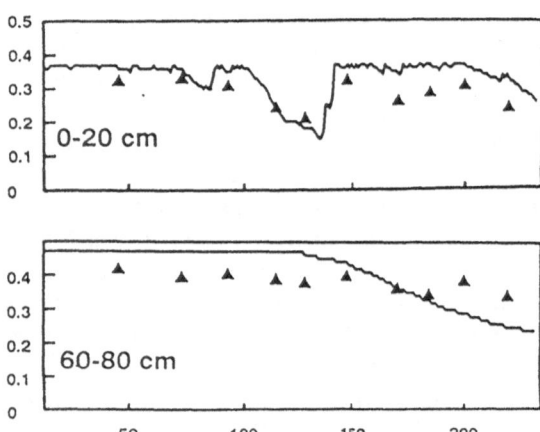

Figure 4. A comparison between measurements and simulations of soil moisture content in two layers with a "mechanistic" (model [G], above) and a capacity based submodel (model [I]) of water transport for the PAGV-site for 1984, given in Groot et al. (1990).

models for water transport, the other 6 rely on capacity models. According to the reviewers, the evaluation with regard to the water movement on this basis reveal a slight superiority of the capacity based models above the mechanistic approaches. As an example, the comparison of the two model performances done for the

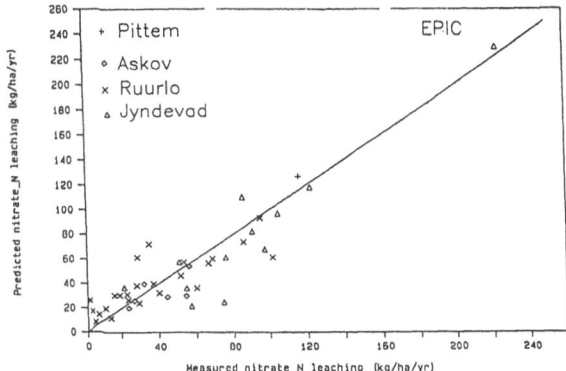

Figure 5. Annual nitrate leaching: comparison of the simulated (EPIC) with the measured field observations (Fig. 3a of Vereecken et al., 1991).

PAGV-site for two layers, 0 to 20 and 60 to 80 cm depth is shown in Figure 4.

Despite the different approaches for N-mineralization and water movement a comparison with measured nitrate leaching in report 1 reveals (Table 7), that ANIMO, DAISY and EPIC are similar in quality. DAISY seems to be a little superior in behaviour, because it explains 89% (EF) of the measured variability with a RMSE of only 27%, but it does so on a reduced sample of only 29 data sets out of 43. RENLEM showed a tendency to overestimate leaching. SWATNIT did the opposite. EPIC (Fig. 5) overestimated by about 40 kg ha^{-1}.

In report 2, evaluation of the models regarding leaching has been done on the basis of comparisons with the mineral N contents or depth profiles in a much less exhaustive way as in the first report. Reflecting the combined result of all N processes, no clear distinction can be made regarding the way of modeling N-mineralization or water movement. This may be seen as supporting the conclusions drawn by the first reviewers (Vereecken et al., 1991) that there is obviously no difference in simulation quality in the different approaches.

Following this conclusion the question may arise whether the quality of estimates by simulation of nitrate leaching has really improved in comparison e.g. to the empirical graphical estimates of Kolenbrander (1981) (Fig. 6). Or, in other words, is all the simulation work on nitrogen processes really justified, at least with regard to leaching? Although we can not give an adequate answer, the application of Kolenbrander's approach to data gathered more recently by Walther et al. (1985) shows that it is still a valuable approach.

Therefore, we may doubt great progress in the field of simulation of nitrate leaching.

Microbial N-immobilisation: the neglected process may be most essential

Based on a general comparison of the results from 14 models, de Willigen (1991) draw conclusions similar to the ones in an earlier report in 1985 based on only 6 participants and their models. He stated that the essential error may stem from the lack of simulating the soil microbial processes. None of the models could account for the apparent mineral nitrogen loss observed shortly after application of fertilizers in late spring or early summer.

Microbial immobilization in soils will not play a role in long term approaches except where changes of land use or management take place. But regarding short term and seasonal changes only one - DAISY - out of originally 9 models in the review of the report 1 (Vereecken et al., 1991) and, at a first look, four out of 14 models in report 2 (de Willigen, 1991) take explicitly into account one (or more) soil microbial biomass pools. However, also in report 2 only one - again DAISY - model is making efforts in using two separate biomass pools describing biomass turnover by growth and decline rates as well as by maintenance. The others do not really model growth and maintenance of microbial biomass explicitly, but instead regard microbes as consumers of substrates only. One other modeler seems to make use of Jenkinson's approach, but it was not possible to get a clear picture of how this has really been done. None of these four models, however, showed especially good agreement with measured soil nitrate profiles or contents despite their special efforts.

Our own comparisons between measured and simulated soil nitrate profiles show short term discrepancies after soil frost and after manuring and tillage procedures as well as long term deviations during the vegetation season. Based on such discrepancies we have realised since a long time ago the lack of a microbial pool and submodel (Kersebaum and Richter, 1991; Richter et al., 1987). Besides possible explanations of seasonal variation of soil mineral N on the basis of changing the relative extent of ammonification and nitrification, together with a change of NH_4^+-sorption or fixation, seasonal variation of microbial biomass and/or N seems to be of major importance. In order to get a sound basis for simulations, expensive measurements are necessary to follow microbial biomass N

118

Table 7. (=Table 4 of Vereecken et al., 1991) Statistical criteria of the performance of the five different models on simulation of nitrate leaching (kg N ha^{-1} a^{-1}) for the three different sites

Site	Model	n[a]	ME[b]	RMSE[c]	CD[d]	EF[e]	CRM[f]
Askov	ANIMO	6	13.4	20.2	1.29	0.60	-0.15
	DAISY	5	24.5	30.0	0.95	0.28	0.10
	EPIC	6	20.4	24.4	0.69	.42	-0.13
	RENLEM	6	55.4	64.9	0.20	-3.09	-0.18
	SWATNIT	6	16.0	32.9	6.38	-0.05	-0.07
Jyndevad	ANIMO	12	41.3	23.0	1.02	0.82	-0.10
	DAISY	12	26.5	19.4	1.25	0.87	0.01
	EPIC	12	49.6	28.0	0.72	0.74	0.17
	RENLEM	12	45.2	35.5	0.68	0.58	0.17
	SWATNIT	12	54.1	28.9	0.82	0.72	-0.07
Ruurlo	ANIMO	24	61.8	56.5	1.87	0.44	0.17
	DAISY	12	39.7	47.1	1.29	0.68	0.15
	EPIC	24	53.7	52.0	1.46	0.53	-0.13
	RENLEM	24	55.4	76.1	1.15	-0.01	0.52
	SWATNIT	19	85.7	86.6	0.33	-0.31	-0.45
ALL[g]	ANIMO	43	61.8[h]	36.2	0.96	0.78	0.00
	DAISY	29	39.7	27.2	1.09	0.89	0.05
	EPIC	43	53.7	37.1	1.17	0.77	0.01
	RENLEM	43	55.4	53.4	0.83	0.52	0.27
	SWATNIT	38	85.7	52.6	0.63	0.53	-0.27

[a] n : Number of observations.
[b] ME : Maximum error.
[c] RMSE : Root mean squared error.
[d] CD : Coefficient of determination.
[e] EF : Modelling efficiency.
[f] CRM : Coefficient of residual mass.
[g] Including Pittem, Belgium.
[h] This column repeats erroneously the previous one in Vereecken et al. (1991).

changes accurately at field scale (Lindloff et al., 1994; Widmer, 1993).

Modeling a system showing different activities with changing environmental conditions seems to be quite a difficult task. Following the original idea of Vinogradsky (1949), microbial biomass needs to be split into 2 pools, a fast responding zymogeneous and a rather indolent autochthoneous one. Further, also two metabolic states may be distinguished, an active and a dormant one. According to this line, a Pirt type model (Pirt, 1975) may be developed in different ways: one perhaps according to Kersebaum and Richter (1994), another one according to Yevdokimov et al. (1993) in order to see, which one gives better response to the few data over the season, directly measured or indirectly estimated as stated in Kersebaum and Richter

(1991). The Kersebaum and Richter (1994) approach uses two different biomass pools, but up to now only one state of activity. Furthermore it uses two organic pools differing in size and decomposability, according to our experience with mineralisation experiments. Differences to usual biomass growth approaches reside in the use of a nonlinear death or decline rate evolving essentially logistic growth on both biomass pools. Blagodatsky's approach (used in Yevdokimov et al., 1993) differs in using only one organic pool but two different states, an active and an inactive one. Furthermore, according to the experience of soil microbiologists, a purely soluble organic matter pool is used. Figure 7 shows the pool and flux diagram of that model.

Parameterization of such a model, even for laboratory experiments, is a laborious task. Figure 8 shows

[440]

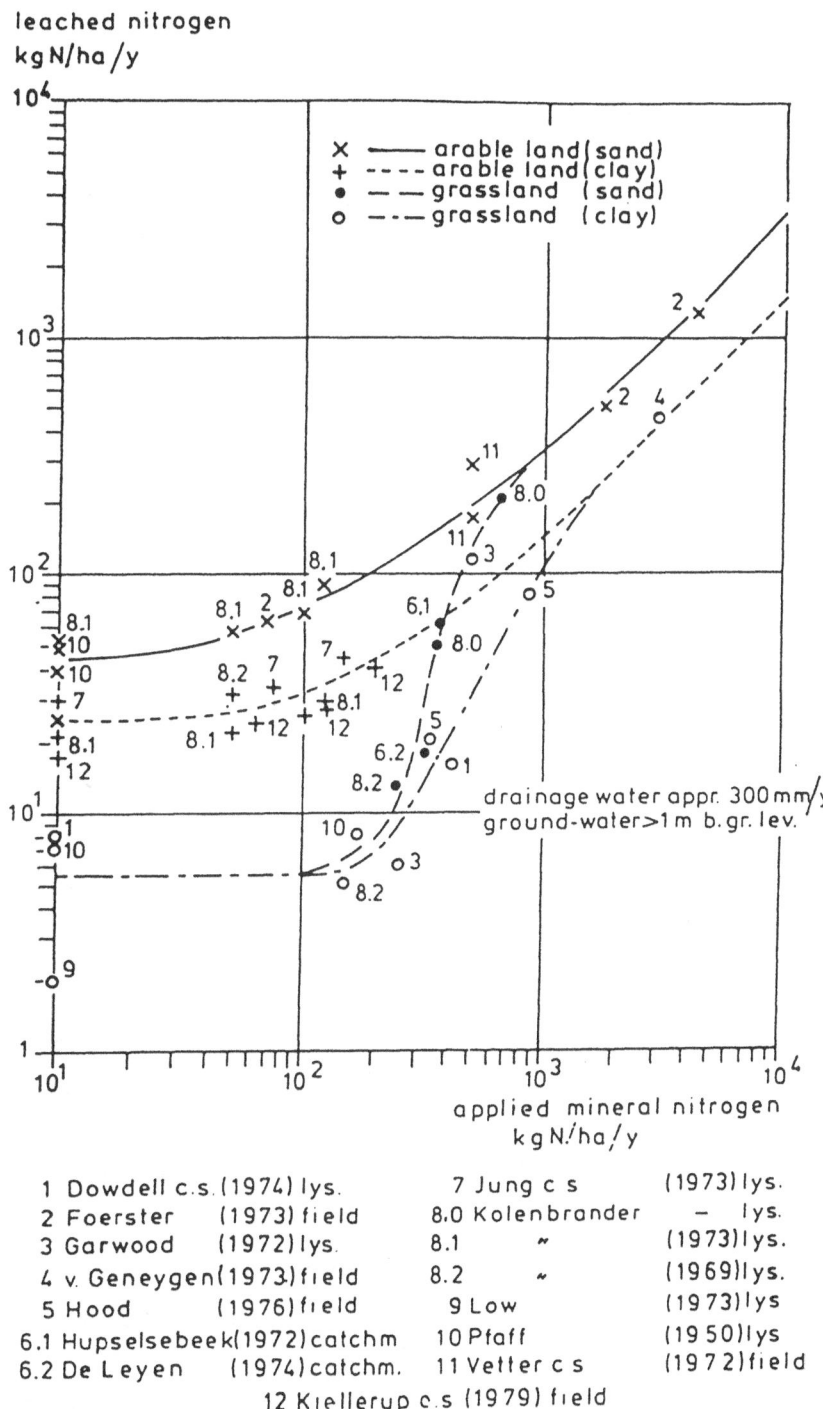

Figure 6. Annual nitrate leaching rates in arable and grassland, depending on applied fertilizer nitrogen and drainage, according to Kolenbrander (1981).

the simulation with fitted parameters using modern parameter estimation techniques (here embedded in the statistical program package BMDP) of a steady state

incubation experiment of Smith et al. (1986), according to Kersebaum and Richter (1994). The essential task of the experiment of Smith et al. (1986), was to

Microbial growth in soil

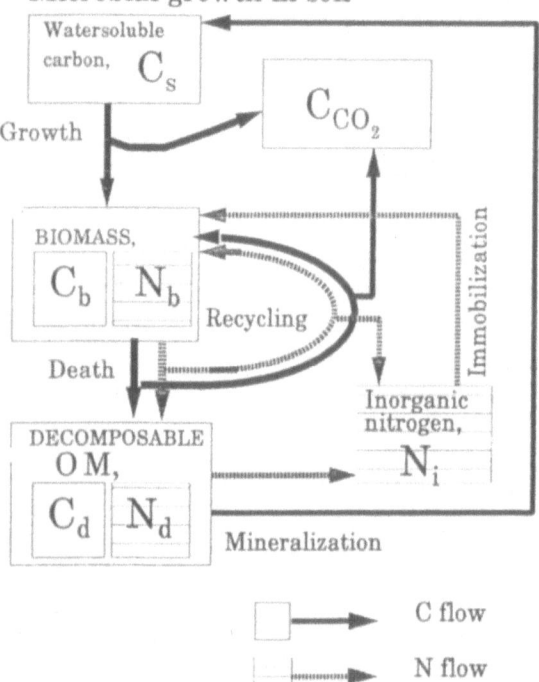

Figure 7. A sketch of the model of Blagodatsky according to Yev-dokimov et al. (1993).

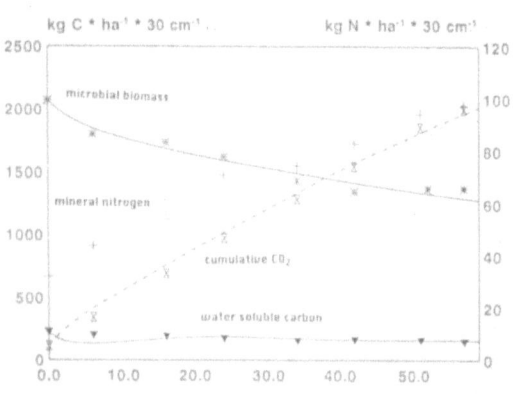

Figure 8. Simulation with fitted parameters using modern parameter estimation techniques, according to Kersebaum and Richter (1994).

estimate maintenance parameters. We would be happy to have similar results of field observations, where we have additional difficulties with the changing environmental conditions. But since this example also shows that the model was overparameterized with respect to available experimental data, the situation is usually worse for conventional field experiment designs. In the terminology of mathematicians, the design decides whether the involved mathematical problem is "well posed" or "ill conditioned" (Kersebaum and Richter, 1994).

Concluding remarks

All the above evaluations were done without special regard to the spatial variability of the investigated fields, either structured or non-structured ones. Heterogeneity of the soil properties as well as of the inputs like precipitation, nitrogen or management practices may have a strong influence on the quality of the results. This explains the large scattering of individual measurements. This should not discourage us from trying to get a more quantitative understanding using process simulation. We should, however, remain aware of the fact that modeling does not necessarily entail the accurate quantitative description of all involved processes - but rather requests to find the essential ones.

References

Addiscott T M and Wagenet R J 1985 Concepts of solute leaching in soils: a review of modeling approaches. J. Soil Sci. 36, 411–424.

Benbi D K and Richter J 1994 Concepts of nitrogen mineralization kinetics and potentially mineralizable nitrogen in soils: A review of the modeling approaches. Soil Biol. Biochem. (*Submitted*).

de Willigen P 1991 Nitrogen turnover in the soil-crop system: comparison of fourteen simulation models. Fert. Res. 27, 141–149.

European Communities 1991 Commission EUR13501. Soil and ground-water res. report II - Nitrate in soils, Luxembourg 1991, 544 p.

Groot J J R, de Willigen P and Verberne E L J 1991 Nitrogen turnover in the soil-crop system - Modelling of biological transformations, transport of nitrogen and nitrogen use efficiency. Proceedings of a workshop held at the Institute for Soil Research, Haren, The Netherlands, 5–6 June 1990. Fert. Res. 27, 141–383.

Jenkinson D S and Rayner J H 1977 The turnover of soil organic matter in some of the Rothamsted classical experiments. Soil Sci. 123, 298–305.

Johnsson H, Bergström L, Jansson P E and Paustian K 1987 Simulated nitrogen dynamics and losses in a layered agricultural soil. Agric. Ecosyst. Environ. 18, 333–356.

Kersebaum K C and Richter J 1991 Modelling nitrogen dynamics in a plant-soil system with a simple model for advisory purposes. Fert. Res. 27, 272–281.

Kersebaum K C and Richter O 1994 A model approach to simulate C and N transformations through microbial biomass. *In* Nitrogen Mineralization in Agricultural Soils. Eds. J J Neeteson and J Hassink. pp 221–229. AB-DLO Thema's, Wageningen, the Netherlands.

Kolenbrander G J 1981 Leaching of Nitrogen in Agriculture. *In* Nitrogen Losses and Surface Run-Off. Ed. J C Brogan. pp 199–216. Den Haag, the Netherlands.

Lindloff A, Nieder R and Richter J 1994 Temporary microbial immobilisation of nitrogen in an arable loess soil. *In* Nitrogen Mineralization in Agricultural Soils. Eds. J J Neeteson and J Hassink. pp 221–229. AB-DLO Thema's, Wageningen, the Netherlands.

Molina J A E, Clapp C E and Larson W E 1980 Potentially mineralizable nitrogen in soil: the simple exponential model does not apply for the first 12 weeks of incubation. Soil Sci. Soc. Am. J. 44, 442–444.

Nordmeyer H and Richter J 1985 Incubation experiments on nitrogen mineralization in loess and sandy soils. Plant and Soil 83, 433–455.

Nuske A and Richter J 1981 N mineralization in loess-parabrownearthes: incubation experiments. Plant and Soil 59, 237–247.

Otter-Nacke S and Kuhlmann H 1991 A comparison of the performance of N simulation models in the prediction of Nmin on farmers' fields in the spring. Fert. Res. 27, 341–347.

Pirt S J 1975 Principles of Microbe and Cell Cultivation. Blackwell, Oxford, UK.

Richter J, Nordmeyer H and Kersebaum K C 1984 Zur Aussagesicherheit der N_{min}-Methode. Z. Acker- Pflanzenbau 153, 285–296.

Richter J, Kersebaum K C and Utermann J 1988 Modelling of the Nitrogen Regime of Arable Field Soils for Consultation Purposes. In Nitrogen Efficiency in Agricultural Soils. Eds. D S Jenkinson and K S Smith. pp 371–383. Elsevier, Amsterdam, the Netherlands.

Richter G M, Widmer P and Richter J 1994 Biological relevance of easily extractable organic nitrogen (EUF-N_{org}) for the estimation of nitrogen mineralization. Eur. J. Agron. 3, 281–289.

Seligman N G and van Keulen H 1981 PAPRAN: A simulation model of annual pasture production limited by rainfall and nitrogen. In Simulation of Nitrogen Behaviour of Soil-Plant Systems. Eds. M J Frissel and J A van Veen. pp 192–221. Pudoc, Wageningen, the Netherlands.

Smith J L, McNeal B L, Cheng H H and Campbell G S 1986 Calculation of microbial maintenance rates and net nitrogen mineralisation in soil at steady state. Soil Sci. Soc. Am. J. 50, 332–338.

Stanford G and Smith S J 1972 Nitrogen mineralization potentials of soils. Soil Sci. Soc. Am. Proc. 36, 465–472.

Van Veen J A 1977 The behaviour of nitrogen in soil. Doctoral thesis, Amsterdam, the Netherlands.

Vereecken H, Jansen E J, Hack-ten Brooke M J D, Swerts M, Engelke R, Fabrewitz S and Hansen S 1991 Comparison of simulation results of five nitrogen models using different datasets. In European Communities - Commission EUR13501: Soil and groundwater res. Report II - Nitrate in soils, Luxembourg 1991. pp 321–338. See reference European Communities, 1991.

Vinogradsky S 1949 Microbiologie du Sol: Problemes et Methods. Masson et Cie, Paris, France.

Walther W, Scheffer B and Teichgräber B 1985 Ergebnisse langjähriger Lysimeter-, Drän- und Saugkerzenversuche zur Stickstoffauswaschung bei landbaulich genutzten Böden und ihre Bedeutung für die Belastung des Grundwassers. Veröff. Inst. Stadtbauwesen TU Braunschweig, 40. 340 p.

Widmer P 1993 Zeitliche und räumliche Variabilität der mikrobiellen Biomasse in niedersächsischen Böden. Doctoral thesis, Hannover, Germany.

Yevdokimov I V, Blagodatski S A and Kudeyarov V N 1993 Microbiological immobilization, remineralization and plants uptake of fertilizer nitrogen. Eurasian Soil Sci. 25, 64–72.

Section editor: R Merckx

O. Van Cleemput et al. (eds.), Progress in Nitrogen Cycling Studies, 445–450, 1996.
© 1996 *Kluwer Academic Publishers.*

Measurement and modelling of soil organic matter decomposition using biochemical indicators

L.J. Sanger, M.J. Whelan, P. Cox and J.M. Anderson
Department of Biological Sciences, University of Exeter, Devon, UK

Key words: biomass fractionation, decomposition, lignin signature, model

Abstract

One problem with many mechanistic models of soil organic matter decomposition and nitrogen MIT is that the conceptual pools used to represent fractions of material which decompose at different rates are not easy to measure directly. A range of biochemical techniques for fractionating soil organic matter and the microbial biomass are presented in this paper and their potential for inclusion in models of carbon and nitrogen turnover discussed. These techniques include analysis of lignin de-polymerisation, a quantification of total and labile carbohydrate and a fractionation of the microbial biomass between fungi and bacteria using phospholipid and ergosterol analysis. Preliminary comparisons between the biochemical signatures of different soil types under different land uses are discussed in relation to the physical and biological factors affecting decomposition rates.

Introduction

One problem with many mechanistic models of soil organic matter (SOM) decomposition and nitrogen Mineralisation Immobilisation Turnover (MIT) is that the conceptual pools used to represent fractions of material which decompose at different rates are not easy to measure directly. Consequently detailed model dynamics become difficult to validate. By constructing a model which uses measurable soil organic matter fractions instead of conceptual ones, a better understanding of the system may be gained.

Basic model structure

The model structure is similar to other models of SOM decomposition (e.g. Jenkinson and Rayner, 1977; Van Veen and Paul, 1981) except that the pools relate explicitly to components which can be measured using a range of techniques (primarily involving gas chromatography selected to allow quantitative distinction between salient soil organic fractions (Fig. 1). Experimental results should allow us to determine the nature of the inter-relations between the biochemical indicators measured under different environmental regimes.

Quantification and fractionation of the microbial biomass

Models of SOM decomposition commonly lump soil organisms into a "microbial biomass" black box. However, it is well recognised that fungi and bacteria have different substrate preferences (for example, most lignin decomposition is attributed to fungi) and respond differently to environmental stresses in soils (fungi, for example, by virtue of their hyphal systems, are better equipped to cope with high soil moisture tensions). Furthermore, different soils have different microbial community structures, some more fungi dominated than others. Phospholipid determination from soil microbial cell walls may provide a powerful way of examining the composition of the soil microbial biomass (Vestal and White, 1989). Phospholipids are good potential indicators of live biomass since they are relatively short lived outside of live cells. In addition, compounds such as ergosterol (a sterol peculiar to fungi) may also corroborate estimations of the proportion of the microbial biomass comprised of fungi (e.g. Davis and Lamar, 1992).

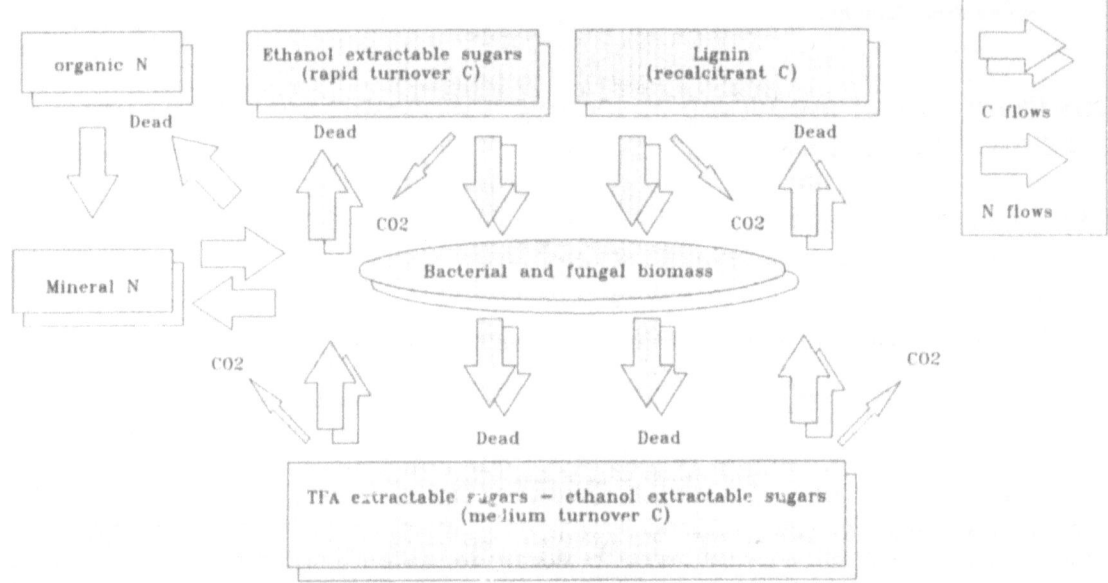

Figure 1. Provisional model structure

Table 1. Chemical charateristics of the soils supporting mature *Pinus sylvestris* stands at sampling sites in Devon, UK

Site	Grid ref.	Soil series	Soil type	pH (H_2O)	Ca^{2+} (meq $100g^{-1}$)	Mg^{2+} (meq $100g^{-1}$)	K^+ (meq $100g^{-1}$)	CEC (meq $100g^{-1}$)	% BS
Charles	SX 743 895	Bridford series	Brown earth	4.3	0.5	0.2	0.07	16.7	5
Haldon	SX 872 725	Dunsford series	Brown earth	4.6	0.7	0.1	0.13	16.2	7
Yarner	SX 783 792	Yarner series	Indurated podzol	4.8	0.7	0.1	0.04	3.9	23
Parke	SX 805 785	Teign complex	Brown earth	6.5	10.2	0.7	0.09	12.5	90
University	SX 918 943	nd*	Brown earth	nd*	nd*	nd*	nd*	nd*	nd*

Soil chemical data for each sampling site from the 10-20 cm depth B horizon (after Clayden, 1971).
*nd - no chemical data available for this site.

Lignin

A significant structural component of SOM is lignin derived. This fraction is resistant to rapid microbial decomposition. It has a complex phenylpropanoil structure which often encrusts the cellulose - hemicellulose matrix. It provides a good indicator for the size of the recalcitrant (slow turnover) SOM fraction. Using a copper-oxidation technique (Hedges and Ertel, 1982) we can determine the extent of de-polymerisation and decomposition of soil lignin. Individual degradation products can be identified and quantified. Such a lignin

"signature" can provide evidence of the provenance and age of a lignin derived fraction in the soil.

Preliminary results from analysis of the lignin signatures in soils from four different land uses (coniferous and deciduous woodland, arable and grassland; Fig. 2) suggest that the type of vegetation cover can exert a strong influence on the breakdown products found in the profile. Relative concentrations of ferulic acid, for example, are higher in grassland soils; syringyl and vanillyl are high in deciduous woodland soil and vanillyl is high in coniferous woodland soils.

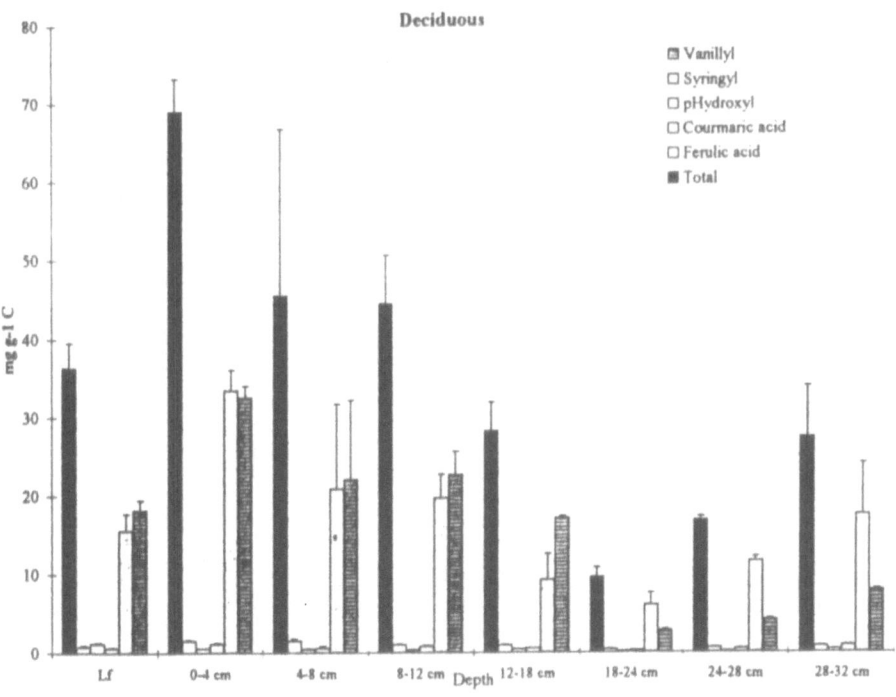

Figure 2. Concentrations of lignin derivatives (per g C) with depth in soils under different land uses

Figure 3 shows the total concentration of lignin and lignin derivatives (95% confidence limits) measured in fresh needles and litter from *Pinus sylvestris* at several sites over a range of soil types (Table 1). The results suggest that the availability of base cations especially Ca, may influence the lignin signatures observed at different sites under the same tree species. Calcium ions affect peroxidase concentrations in plant cells which is

448

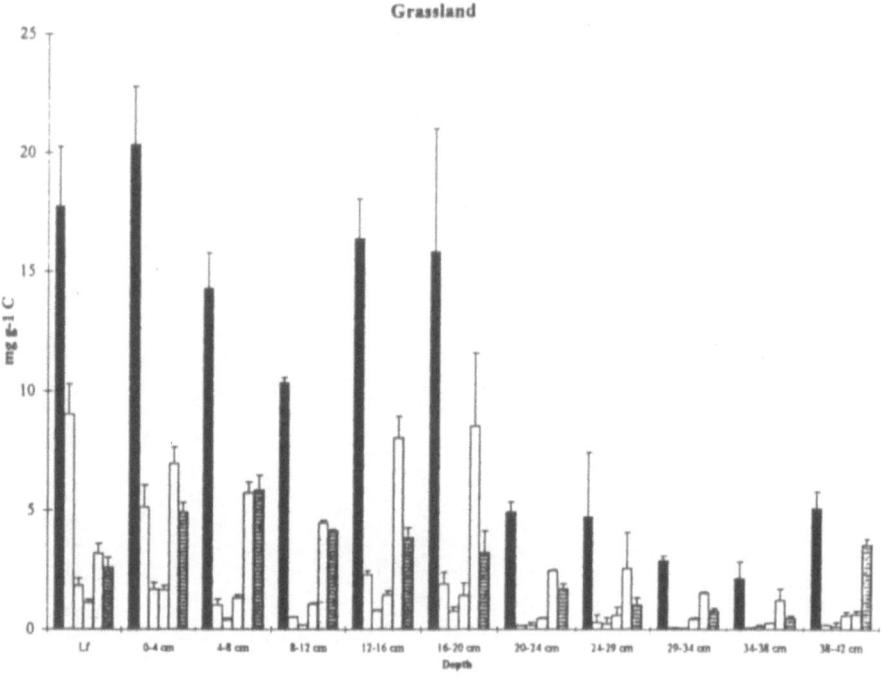

Grassland

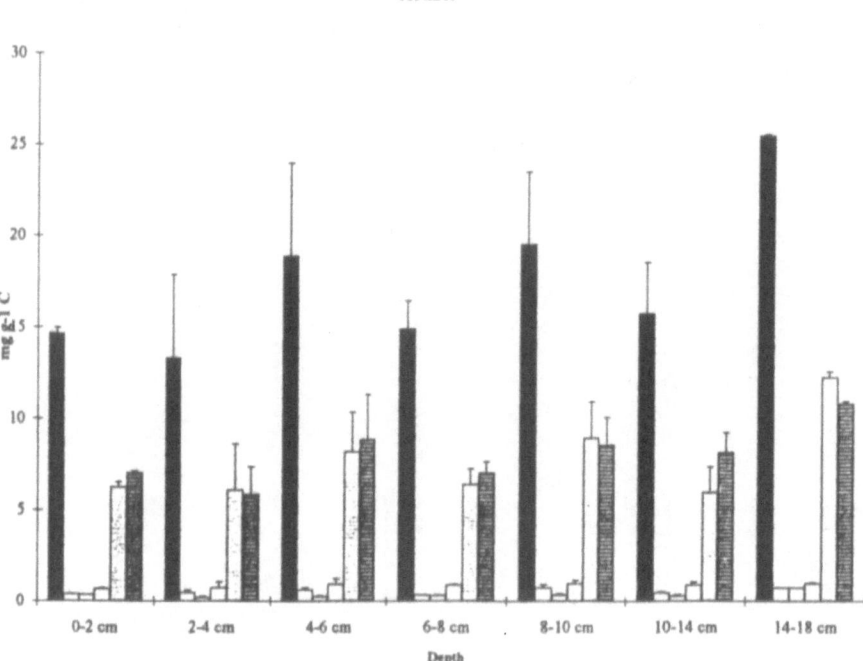

Arable

Figure 2. Continued.

thought to initiate the formation of free radical lignin precursors (Brett and Waldron, 1990). Earlier studies by Lipetz (1962) have shown that Ca ions can control the degree of total plant tissue lignification. In addi-

tion, marked seasonal differences in lignin signatures were observed in fresh needles sampled from trees on two sampling occasions less than three months apart. It is difficult to explain satisfactorily the increase in

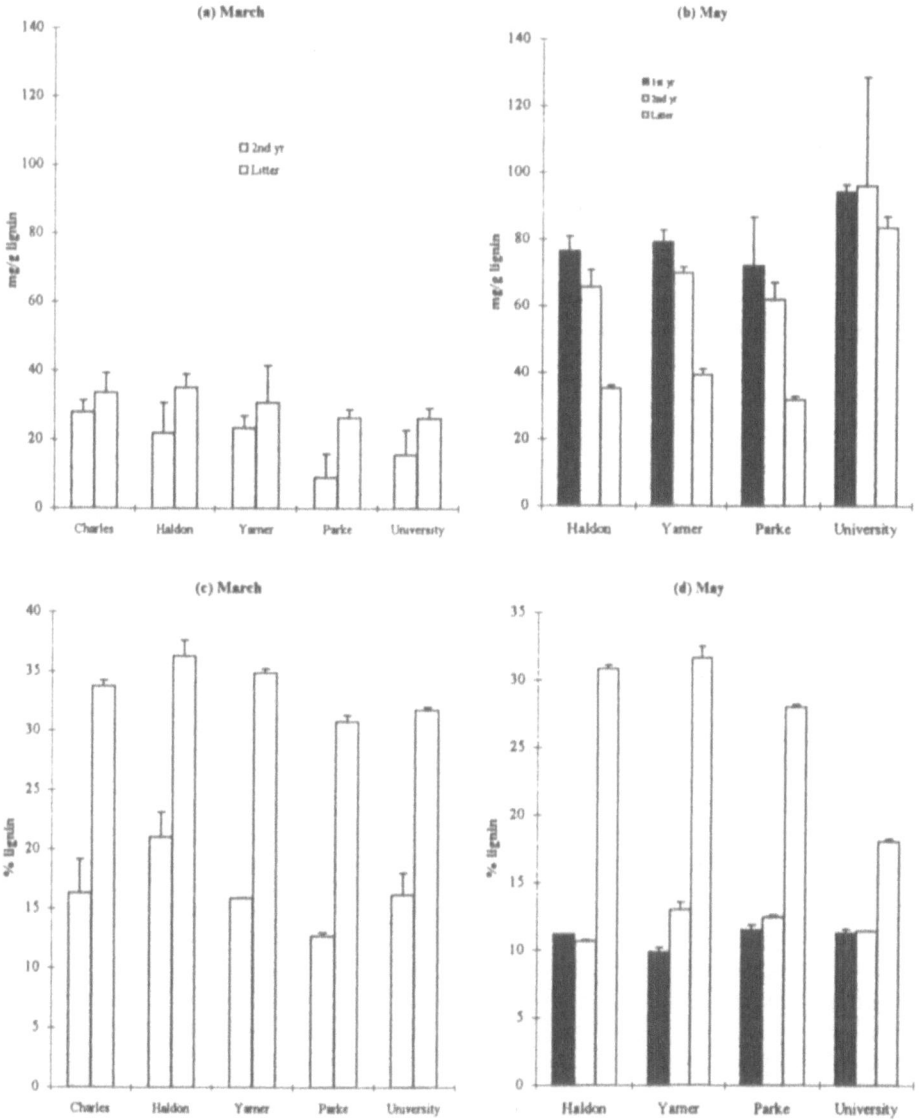

Figure 3. Concentration of lignin and lignin derivatives in needles and litter collected in early March and late May from stands of *Pinus sylvestris* grown on different soil types.

derivatives in needles sampled from the trees in May cf. March, at this stage. The patterns may reflect natural seasonal changes in the lignification process. Seasonal changes in the concentration of lignin derivatives in litter samples may be attributed to fluctuations in the activity of endophytic and soil fungi. Overall these results show that changes in litter quality in response to environmental factors can be measured in detail and may be included as inputs to a model of C dynamics in soils.

Carbohydrates

Carbohydrates represent the most available energy supply for soil micro-organisms. The quantity of total non crystalline carbohydrate fraction of the SOM may be determined using a trifluoracetic acid (TFA) hydrolysis technique, in which monosaccharide components from polysaccharide compounds (e.g. cellulose and hemicellulose) are released (Guggenberger and Zech, 1994). Using GC after hydrolysis enables a quantification of the ratio of pentose to hexose sugars from which we may be able to infer plant and microbial origin (e.g. Murayama, 1984; Oades, 1984). Hexose sugars are known to originate predominantly in plants whereas pentose sugars are produced primarily by microbes. Separate extraction of sugars with 80% ethanol provides a measure of the immediately available carbohydrates (very rapid turnover pool) and the difference between the two (TFA extractable - ethanol extractable sugars) gives the concentration of amorphous hollocellulose which we imagine will provide a measure of medium turnover C for the model.

Current experimental work

Laboratory experiments are currently underway to follow the fate of the above indicator compounds under different environmental regimes (temperature, moisture content and wetting and drying cycles) in reconstituted cores over approximately 18 months. Results from these experiments will be used to identify important trends and to calibrate and validate the model. Soil water and KCl extractable ammonium and nitrate concentrations and CO_2 efflux will also be determined periodically to give a measure of mineral N dynamics and total soil respiration respectively.

Acknowledgement

This work is part of a project funded by the Commission of the European Communities. Grant number EV5VCT920141.

References

Brett C and Waldron K 1990 Physiology and Biochemistry of Plant Cell Walls. Topics in Plant Physiology. Eds. M Black and J Chapman. Unwin Hyman, London.

Clayden B 1971 Soils of the Exeter District. Memoirs of the Soil Survey of Great Britain: England and Wales, Harpenden, Herts., UK.

Davis M W and Lamar R T 1992 Evaluation of methods to extract ergosterol for quantitation of soil fungal biomass. Soil Biol. Biochem. 24, 189–198.

Guggenberger G and Zech W 1994 Composition and dynamics of dissolved carbohydrates and lignin degradation products in two coniferous forests, N.E. Barvaria. Soil Biol. Biochem.

Hedges J I and Ertel J R 1982 Characterisation of lignin by gas capillary chromatography of cupric oxide oxidation products. Anal. Chem. 54, 174–178.

Jenkinson D S and Rayner J J 1977 The turnover of soil organic matter in some of the Rothamsted classical experiments. Soil Sci. 123, 298–305.

Lipetz J 1962 Calcium and the control of ligninfication in tissue cultures. Am. J. Bot. 49, 460–464.

Murayama S 1984 Changes in the monosaccharide composition during the decomposition of straws under field conditions. Soil Sci. Plant Nutr. 30, 367–381.

Oades J M 1984 Soil organic matter and structural stability: Mechanisms and implications for management. Plant and Soil 76, 319–337.

Van Veen J A and Paul E A 1981 Organic carbon dynamics in grassland soils 1. Background information and computer simulation. Can. J. Soil Sci. 61, 185–201.

Vestal J R and White D C 1989 Lipid analysis in microbial ecology: Quantitative appoaches to the study of microbial communities. BioScience 39, 535–541.

O. Van Cleemput et al. (eds.), Progress in Nitrogen Cycling Studies, 451–457, 1996.

Application of nitrogen cycle modelling to land use policy issues: nitrate leaching from winter and spring cereal cropping

Andrew J.A. Vinten
Soils Department, SAC, West Mains Road, Edinburgh, EH9 3JB, UK

Key words: modelling, nitrate leaching, spring cereals, winter cereals

Abstract

Under some circumstances nitrate leaching may be more serious under winter cereals than under spring cereals. We have evaluated the factors controlling the relative efficiency of N use by winter and spring cereals under cool, moist climatic conditions using simulation modelling, field trials data and literature survey. The most important factors which may cause winter cereals to be more leaky are larger fertiliser application rates for winter cereals, not fully compensated by larger N offtake; smaller potential for post-harvest denitrification of excess N from cultivated land sown to winter cereals than from uncultivated stubble land; and greater accumulation of unused mineralised N because of earlier harvesting of winter cereals. Simulation modelling over 3 years under local weather conditions showed mean annual Nitrate-N concentrations in drainage water were 5 mg L^{-1} higher on a sandy loam soil, and 3 mg L^{-1} higher on a clay loam soil with winter instead of spring cereals at recommended fertiliser rates.

Introduction

The UK Nitrate Sensitive Areas scheme is designed to reduce the amount of nitrate leaching occurring in areas where ground or surface waters may become polluted by high nitrate concentrations. The requirements for arable cropping under the 'Basic Scheme' include the establishment of a winter cereal by 15 October or a catch crop (MAFF, 1994). This recommendation is made on the basis that the uptake of mineralised N by newly established winter cereals in the autumn prevent that N from being leached. However, this paper investigates the hypothesis that, under some circumstances, it may be more effective in the long term to encourage a switch from winter to spring cereals as a means to reduce nitrate leaching. In the cooler, more northerly parts of the UK, later harvests and later winter cereal establishment give less opportunity for N uptake by a winter cereal or a cover crop. Fisher et al. (1991) suggested that the difference between the applied N and the harvested N (representing N susceptible to leaching loss) may be smaller under spring cereal cropping. This is illustrated in Figure 1. Data from trials (Rees et al., 1995) in southern and eastern Scotland with spring barley (10 sites in 2 years), winter wheat (7 sites in two years), winter barley (3 sites in two years) and from a long term N balance experiment at Bush Estate, near Edinburgh, over 5 years (Vinten et al., 1992) were used. The total mineral N available to the crop was taken as the sum of the fertiliser N input and the measured soil N contribution (derived from ^{15}N data for the trials sites and N uptake from the zero N plot in the N balance experiment). The 'unaccounted for N' was the difference between this amount and the observed total crop N uptake (including root N, estimated by assuming 10% of dry matter at harvest (Barraclough and Leigh, 1984), at 1.3% N). The 'unaccounted for' N increased with increasing N application for both winter and spring cereals, and the losses from winter cereals were higher than those for the spring barley at the trials sites. Average field rates of N to spring barley, winter barley and winter wheat in south-east Scotland in 1990–93 were 101, 173 and 200 kg ha^{-1} respectively (Russell and Wilson, 1994). Figure 1 therefore implies that excess N liable to leaching would normally be smaller in spring barley.

We have investigated a number of factors influencing the relative efficiency of N use by winter and spring cereals under climatic conditions of the northern UK by simulation modelling, using the Agricultural Nitrogen Model (ANIMO – Rijtema and Kroes, 1991), calibrated for local weather conditions and for two local soil

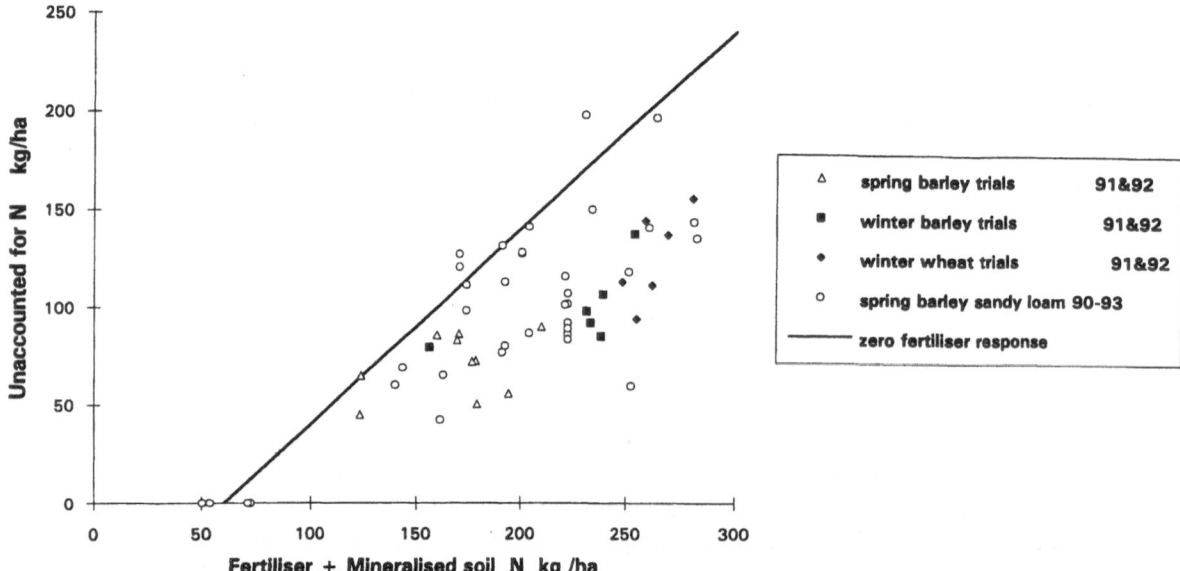

Figure 1. Unaccounted for N in spring and winter cereal N response experiments in eastern Scotland as a function of total crop available N (fertiliser + soil mineralised N). The unaccounted for N is the difference between N uptake and crop available N. Soil mineralised N is estimated from ^{15}N recovery data (trials sites) or crop uptake in zero N plots (spring barley, sandy loam 1990–93).

types: a sandy loam and a clay loam, where drainage water interception and long term N balance studies (Vinten et al., 1992, 1994) allow some validation of the model predictions to be made.

Simulation of N cycle using ANIMO

For a full description of the model ANIMO the reader is referred to other publications (Berghuis-van Dijk et al., 1985; Rijtema and Kroes, 1991). The model describes the transformation of various organic and inorganic forms of N in soils. Four kinds of organic matter are present in the system: fresh organic matter, soluble organic matter, root exudates and humus. Each type of fresh organic matter is defined by a number of fractions, each with its own decomposition rate and N content. During decomposition, fresh organic matter is decomposed partly into soluble organic matter and partly into humus, with release of CO_2 and mineral N. Soluble organic matter is further broken down to humus and CO_2. Root exudates decompose directly to humus and CO_2. Root exudate production rate depends on N uptake. Decomposition follows first order kinetics and is influenced by temperature according to an Arrhenius-type relationship. The supply of oxygen in the soil is simulated using Fick's Law and an aerobic radius around each randomly distributed, cylindrical air filled pore is calculated for each layer. The anaerobic fraction in each layer is calculated and, within this fraction, decomposition continues, using nitrate as an electron acceptor until it is consumed. Once the nitrate is consumed, decomposition in the anaerobic fraction is assumed to stop. Crop growth and development occurs according to preset maxima of N uptake during two growth periods, with uptake of nitrate occurring by mass flow to the roots. Soil hydrology is simulated using a simple water balance model, WATBAL (Berghuis-van Dijk, 1985). The soil is divided into a root zone and a sublayer below the root zone. The model calculates, per time step, the change in water content of each layer. Precipitation, evapotranspiration, capillary rise, surface runoff, deep percolation and drainage to a maximum of four drainage systems are included. Each drainage system is characterised by a drainage spacing, an effective soil depth below the drain contributing to flow to the drain, a saturated hydraulic conductivity. Calibration details for two local sites, a sandy loam and a clay loam, are given in Table 1.

Results

Comparison of ANIMO with observations

We have few data for winter barley, but we have data over 3–4 years for spring barley, receiving 0–210 kg N ha^{-1}. Measured drainage water (NO_3-N from the

Table 1. Soil parameters for ANIMO calibration

	Unit	Sandy Loam Eutric Cambisol		Clay Loam Gleysol	
		Root Zone	Sublayer	Root Zone	Sublayer
A. Soil parameters[a]					
Thickness	m	0.7	1.8	0.3	1.2
Dispersion length	m	0.05	0.2	0.075	0.2
Bulk density	$kg\,m^{-3}$	1350	1780	1190	1470
Water content(− 5kPa)	m^3m^{-3}	0.34	0.23	0.46	0.43
(−1500kPa)	m^3m^{-3}	0.16	0.12	0.19	0.19
Saturated conductivity	$m\,d^{-1}$	0.22	0.05	11.2	0.0036
Drainage pipe spacing	m		7		7.5/15
depth	m		1.5		0.5/1.0
Backfill	m		1.5–0.3		0.5/1.0–0.3
texture		Sandy loam/ Loamy sand	Sandy clay loam	Clay loam	Clay loam
Stone content	%	16	28	11	14
pH		6.5	6.5	6.2	6.1
N content of rainfall	$kg\,m^{-3}$	0.008		0.008	
humus content	$mg\,kg^{-1}$ soil	43	5	55	12
(4.8% N)	(fine earth)	(0–0.3 m)	(0.3–0.7 m)		
NH_4^+ adsorption coefficient	m^3kg^{-1}	0.0035	0.0002	0.0049	0.0028
degradation rate constants	humus	0.012	0.0024	0.0065	0.0013
at 8 °C[c]	straw	2.4		2.4	
(y^{-1})	roots	1.9		1.9	
annual air temperature[d]	Mean °C	8		8	
	Amplitude °C	6		6	
B. Crop parameters		Spring Barley		Winter Barley	
Sowing	julian day	95		275	
Fertiliser applied	julian day	95		80/110	
Fertiliser amount	$kg\,ha^{-1}$	100		80/80	
Harvest	julian day	245		215	
Ploughing	julian day	345		245	
period of full cover[e]	julian day	165–205		345–175	
Max ET/ETo[f]		0.7		0.7	
Potential N uptake[g]	$kg\,ha^{-1}$	174	at harvest	234	at harvest
		28	day 145	94	day 131
Residue: roots[h]	$kg\,ha^{-1}$	1660	1.3% N	2000	1.3% N
stubble	$kg\,ha^{-1}$	1125	0.6% N	1125	0.6% N

[a] See Vinten et al. (1992) and Vinten et al. (1994) for more details of physical measurements.

[b] Measured for clay loam top and subsoil. Estimated coefficients for clay and organic matter, by simultaneous equations, and applied to sandy loam.

[c] Fitted double first order kinetics to $CO_2 - C$ release from incubated soils. Corrected to 8 °C using Arrhenius equation.

[d] Based on monthly records for Bush Estate, Midlothian, Scotland. Recommended fertiliser rates from SAC (1992).

[e] Crop cover function based on Francis and Pidgeon (1982).

[f] Estimated from neutron probe data for No.3 site, summer 1991.

[g] Observed spring barley yield maximum at No.3 field; add 60 kg ha^{-1} for winter barley maximum. Intermediate data from Elmes (1985).

[h] 10% of total dry matter at harvest (based on Barraclough and Leigh, 1984).

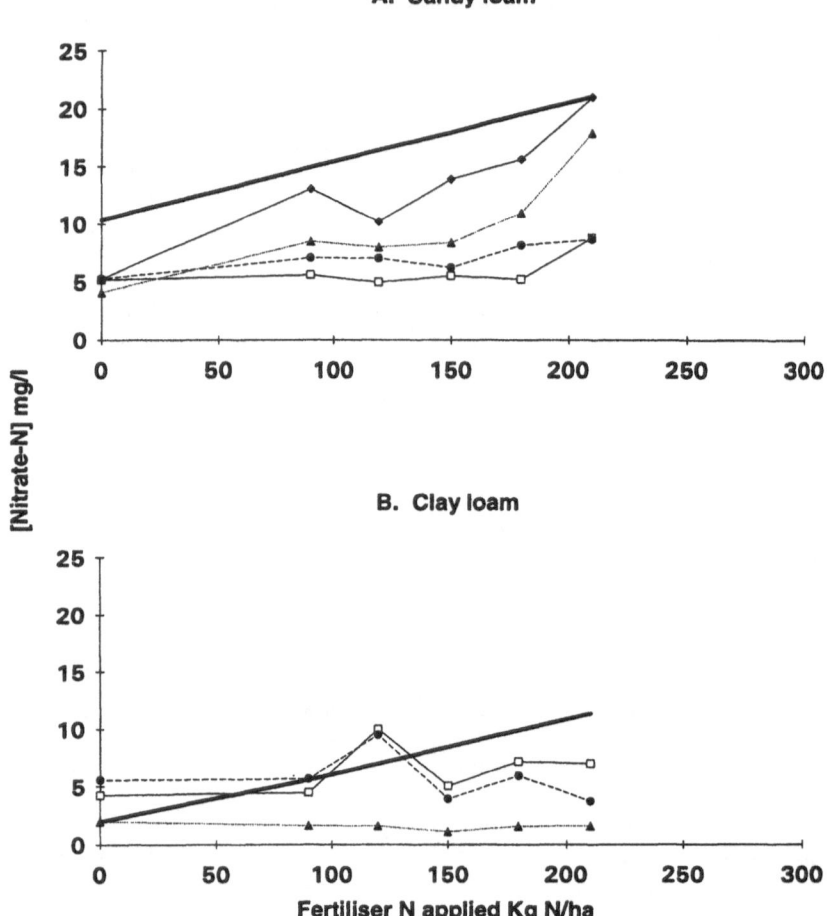

Figure 2. Comparision between measured annual mean nitrate-N concentrations in drainage water (-□ -1990/91; - - - ●- - -1991/91; - - -▲- - -1992/93;- - -◆- - -1993/94) and ANIMO simulations (———mean of 1990–93) for (**A**) Sandy loam (1990–94) and (**B**) Clay loam (1990–93) soils at Bush Estate, near Edinburgh, Scotland. Years run from julian day 90.

sandy loam increased with fertiliser rate, and in successive years from 1990–94 (Fig. 2a), but only in the final year did the (NO_3-N) approach that predicted by ANIMO. This suggests either a physical delay or accumulation of N in organic form in the soil. Measured drainage water nitrate concentrations from the clay loam (Fig. 2b) were lower and showed no increase with fertiliser rate. ANIMO simulations were also lower than for the sandy loam site, but showed a steady increase in (NO_3-N) in drainage water with increasing fertiliser N.

ANIMO predictions of [NO_3-N] *from winter and spring cereals*

Figure 3 shows predictions of annual (NO_3-N in drainage water for the two sites and three years. The (NO_3-N) was about 5 mg L^{-1} (sandy loam) or 3 mg

L^{-1} (clay loam) higher in the winter barley receiving 160 kg ha^{-1} than the spring barley receiving 100 kg ha^{-1} (both recommended rates). Even if both crops received the same rate of N application the (NO_3-N) for spring barley was still lower than for winter barley on the sandy loam site, though there was little difference on the clay loam site.

Discussion

Autumn cultivation

Nitrate leaching loss from the clay loam soil growing spring barley was 32 kg N ha^{-1} after autumn chisel ploughing and subsoiling of a clay loam soil, but only 18 kg N ha^{-1} if cultivations were delayed until

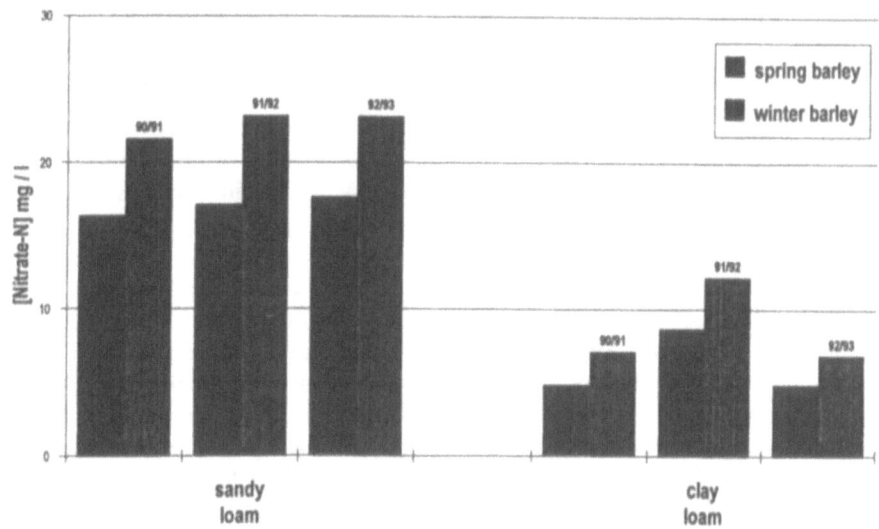

Figure 3. Simulated mean annual nitrate-N concentrations (1990–93) in drainage water from sandy loam and clay loam plots growing cereals at recommended rates (spring barley 100 kg ha^{-1}, winter barley 160 kg ha^{-1}). Years run from julian day 90.

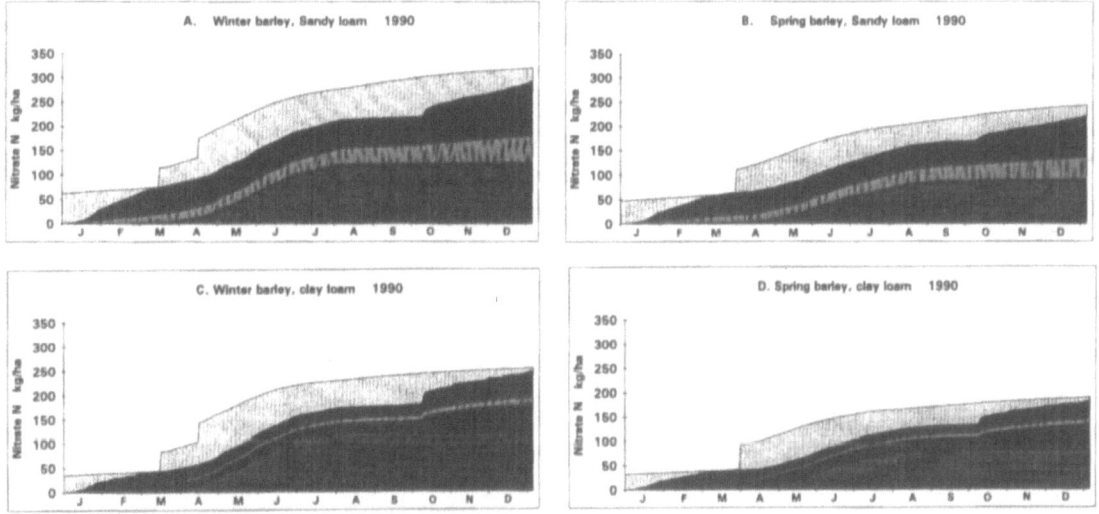

Figure 4. Annual nitrate budget for sandy loam and clay loam plots growing cereals at recommended rates (spring barley 100 kg ha^{-1}, winter barley 160 kg ha^{-1}). Years run from julian day 90. (▦ - nitrate-N in soil; ▦ - nitrate N leached to drains; ▥ - nitrate N leached beyond drainage depth; ▨ - N denitrified; ▨ - N uptake by crop.

January (Vinten et al., 1992). We have measured the release of CO_2 from this soil during a 9 month incubation of both intact cores and sieved soil at 23 °C and constant moisture. The samples were taken from plots which had either recently been ploughed and rotivated after a 2 year grass ley (high residue) or had been in an over-winter unsown fallow period (low residue). using a consecutive double first order kinetic model, we estimated the humus and fresh organic matter decay constants. The humus decay constant obtained was unaltered by sieving. The effective size of the labile pool was increased by sieving, but the decay

constant was only marginally higher. This suggests that the main effect of cultivation was not the exposure of stable humus to microbial degradation but improved aeration allowing more rapid, aerobic decomposition of fresh organic matter. Leaving this soil uncultivated in autumn promotes the anaerobic decay of roots etc, allowing denitrification of excess nitrate present in the soil. ANIMO simulations for spring and winter barley (Fig. 4) predict that the bulk of the denitrification loss occurs in autumn following harvest, when fresh C rich residue supply is high. Early autumn cultivation

456

reduces the effectiveness of this method of harmless disposal of excess N as N_2.

Nitrogen uptake pattern

Autumn/winter
Barraclough (1984) has evaluated the growth of roots and shoots of winter wheat as a function of thermal time (accumulated degree days from sowing). Assuming the shoot and root N contents are 5% and 1.3% respectively, the expected N in the crop at the end of February, using local daily air temperature data for ten years (1981–90), was estimated to be only $18\pm10\,kg\,ha^{-1}$, following sowing of winter wheat on 1 October. This agrees well with observations made by Redman (1991) (7–$14\,kg\,ha^{-1}$ in winter barley in March 1988), Elmes (1985) ($12\,kg\,ha^{-1}$ in winter barley in March 1982, Aberlady, East Lothian) and Vaidyanathan (1984) (13 and $36\,kg\,ha^{-1}$ in winter wheat at end of February), and shows that for cooler conditions, winter cereals are not always a very effective sink for residual mineral N. Weeds left in stubble after cereals may also accumulate some N, and Redman (1991) found 5–$20\,kg\,ha^{-1}$ in the green material in spring barley stubble in November.

Spring and summer
ANIMO simulations (Fig. 4) illustrate the difference in the uptake pattern of winter and spring cereals. Although the early spring uptake of N by the winter cereals was underestimated, the pattern of N mineralisation fitted the uptake pattern of spring barley better, so post-harvest mineral N accumulation was lower.

Early N application

Simulations of ANIMO showed little effect of splitting the N application on leaching or denitrification of N from winter cereals at either site in the years 1990–1991. Smith et al. (1987) found that recovery of spring fertiliser N in winter barley was low (16–30%) for both early and late split N on the clay loam soil (Winton series) of poor structure, but recovery of later applied N was better on the sandy loam (Macmerry series). The losses in spring were attributed mainly to denitrification, in contrast with the ANIMO simulations on the clay loam site, where most of the denitrification was post-harvest.

Carbon returns by winter and spring cereals

Barraclough and Leigh (1984) found that, for winter wheat, the ratio of dry matter in roots to that in the whole plant decreased from 0.4 to 0.1 from early spring to anthesis. Biscoe et al. (1975), found, for spring barley, this ratio decreased from 0.4 to 0.25 at anthesis and further to 0.15. The latter work was with a variety (Proctor) known to have a high root productivity (Atkinson, 1989), so we have assumed the ratio falls from 0.4 at the first N uptake date to 0.1 at anthesis in both crops, continuing unchanged to harvest. This means that extra C returned directly from root residues is only about $58\,kg\,ha^{-1}$ for every tonne difference in dry matter yield between spring and winter cereals. Assuming an assimilation factor of 0.25 and a C/N ratio of 12 for soil humus, this extra C would only humify about $1\,kg\,ha^{-1}$ of N. Root turnover may increase this factor but de Willigen and Noordwijk (1987) found that in winter wheat only 13% of the root length formed had disappeared by the end of the growing season. Exudation of carbon containing compounds from the roots may be important. ANIMO assumes that 40% of the N taken up by the roots is returned as labile organic N (2.5% N) to the soil. If net N uptake for winter cereals is $40\,kg\,ha^{-1}$ larger, this leads to the return of $600\,kg\,ha^{-1}$ of C as root exudate which could contribute to the conversion of about $12\,kg\,ha^{-1}$ of extra N into humus.

Conclusions

Simulation modelling supports the hypothesis that spring cereals may be less prone to leaching of N than winter cereals. Where the potential for substantial post-harvest denitrification is present, and potential for N uptake by a winter cereal or cover crop is small, delaying cultivation may be more desirable than growing a winter cereal or cover crop.

More experimental work, comparing losses from winter and spring cereals directly, is needed to confirm the hypothesis. The uptake and turnover of N by the plants needs to be better simulated and time dependence of soil physical properties may need to be included.

Acknowledgements

The support of Scottish Office Agriculture and Fisheries Department is acknowledged. The Royal Society provided a grant for a study visit to The Netherlands. Dr J Kroes of The Staring Center, Wageningen provided the executable version of ANIMO. N M Fisher made helpful comments.

References

Atkinson D 1989 Root growth and activity: Current performance and future potential. Asp. Appl. Biol. 22, 1–13.

Barraclough P B 1984 The growth and activity of winter wheat roots in the field: Root growth of high yielding crops and relation to shoot growth. J. Agric. Sci. Cambridge 103, 439–442.

Barraclough P B and Leigh R A 1984 The growth and activity of winter wheat roots in the field: The effect of sowing date and soil type on root growth of high yielding crops. J. Agric. Sci. Cambridge 103, 59–74.

Berghuis-van Dijk J T 1985 WATBAL: A simple water balance model for an unsaturated-saturated soil profile. Institute for Land and Water Management Research, Wageningen, Note No 1670. 23 p.

Berghuis-van Dijk J T, Rijtema P E and Roest C W J 1985 ANIMO: Agricultural Nitrogen Model. Institute for Land and Water Management Research Wageningen, the Netherlands. Note No 1671. 86p.

Biscoe P V, Scott R K and Monteith J L 1975 Barley and its environment. III. Carbon budget of the stand. J. Appl. Ecol. 12, 269–293.

Elmes A 1985 The influence of soil, tillage and other factors on the uptake of nitrogen by barley. PhD Thesis, University of Edinburgh, Edinburgh, UK.

Fisher N M, Davies D H K and Atkinson D 1991 Implications for wildlife, landscape and the environment of farming without pesticides. 1991 Brighton Crop Prot. Conf. pp 745–752.

Francis P E and Pidgeon J D 1982 A model for estimating soil moisture deficits under cereal crops in Britain. J. Agric. Sci. Cambridge 98, 651–661.

MAFF 1994 The Nitrate Sensitive Areas scheme. HMSO, London, UK. 43p.

Redman M H 1991 Nitrogen dynamics of an arable soil under different agronomic practices. PhD Thesis, University of Edinburgh, Edinburgh, UK.

Rees R M, McTaggart I and Smith K A 1996 Potential nitrogen availability and fertiliser recommendations. In Progress in Nitrogen Cycling Studies. Eds. O van Cleemput, G Hofman and A Vermoesen. Kluwer Academic Publishers, Dordrecht.

Rijtema P E and Kroes J G 1991 Some results of nitrogen simulations with the model ANIMO. Fert. Res. 27, 189–98.

Russell G and Wilson G W 1994 An agro-pedological knowledge base of wheat in Europe. Joint Research Centre/CEC, Luxembourg, Luxembourg (In press).

SAC 1992 Farm Management Handbook, Thirteenth edition. Scottish Agricultural College, 447p.

Smith K A, Howard R S and Crichton I J 1987 Efficiency of recovery of nitrogen fertiliser by winter barley. In Nitrogen Efficiency in Agricultural Soils. Eds. D S Jenkinson and K A Smith. pp 73–84. CEC Elsevier Applied Science Publishers, Barking, England.

Vaidyanathan L V 1984 Soil and fertiliser use by winter wheat. In The Nitrogen Requirement of Cereals. pp 69–77. HMSO, London, UK.

Vinten A J A, Vivian B J and Howard R S 1992 The effect of fertiliser on the nitrogen cycle of two upland arable soils of contrasting textures. Proc. Fert. Soc. 329.

Vinten A J A, Vivian B J, Wright F and Howard R S 1994 A comparative study of nitrate leaching from soils of differing textures under similar climatic and cropping conditions. J. Hydrol. 159, 197–213.

De Willigen P and van Noordwijk M 1987 Roots plant production and nutrient use efficiency. PhD thesis. Agricultural University, Wageningen the Netherlands. 282p.

Plant and Soil **181**: 169–173, 1996.

Alternative kinetic laws to describe the turnover of the microbial biomass

A.P. Whitmore

*DLO-Research Institute for Agrobiology and Soil Fertility (AB-DLO), Oosterweg 92, P.O. Box 129, NL-9750 AC Haren, The Netherlands**

Key words: kinetics, microbial biomass, model

Abstract

Many models that describe the turnover of the microbial biomass in soil use either first order kinetics where the rate of turnover is directly proportional to the microbial mass, or a variant of the Michaelis-Menten law that describes enzyme kinetics. To account for the different rates of microbial turnover observed at different times after the addition of substrate, some authors have suggested the existence of more than one pool of biomass. Each pool obeys the same kinetic law but with a different rate. In other experiments a disproportionately large increase in the turnover of native organisms has been observed relative to the amount of fresh substrate added. A change in the kinetic law describing the turnover of organisms can account for these observations and yet retain the simplicity of a single pool of micro-organisms. However where multiple pools of organisms are justified a mixed kinetic law with both first and second order terms may be more appropriate; in other words one pool of micro-organisms but two rate constants. The advantage of retaining a single pool of microbial biomass is that models may more readily be constructed in relation to the routine measurements of total microbial mass.

Introduction

In the last 20 years or so techniques for measuring the microbial biomass based on the effects of fumigating soil (Jenkinson and Powlson, 1976) have enjoyed a great deal of success and have advanced our knowledge of the behaviour of this important, living, part of the soil organic matter. At the same time, and partly as a result of our increased understanding of the role played by micro-organisms in soil organic matter dynamics, computer simulation models of organic matter in soil have also proliferated (e.g. Jenkinson and Rayner, 1977, Molina et al., 1983, Verberne et al., 1990). Conceptually modellers have visualized the existence of 'pools' of organic matter in soil. A pool for these purposes consists of a quantity of organic matter all of which can be said to behave in the same way and which is distinct or distinguishable from other pools. A pool decomposes at the same rate, has a more or less uniform chemical composition, is always derived from other pools in the same way and so on. It is therefore highly logical to build models with different pools of micro-organisms too, for example: fungi, bacteria

and actinomycetes (e.g. McGill et al., 1981), different trophic levels (e.g. de Ruiter et al., 1993) or organisms that are protected or not-protected (e.g. Hansen et al., 1991; Verberne et al., 1990). These models have all been quite successful but have the disadvantage that their predictions of changes in the different microbial pools cannot easily be verified. Fumigation techniques measure all of the soil organisms at once without regard to kingdom, class or phylum. De Ruiter et al. (1933) and Rutherford and Juma (1992) took the trouble to count micro-organisms and distinguish predator from prey in building their models, but this data is unfortunately not widely available or simple to obtain. Measurements based on fumigation techniques are widely available and this makes them a more attractive proposition to model builders.

Most of the models have assumed that organic matter (living or dead) follows a first order kinetic law as it decomposes; that is to say the amount decomposing is directly proportional to the current mass of that pool and to that alone. In this article it is shown that by returning to the derivation of the rate law describing the turnover of microbial biomass and assuming that a part of the microbial biomass interacts in some way with

* FAX No.: +31 505337291
Plant and Soil is the original source of publication of this article. It is recommended that this article is cited as: *Plant and Soil* **181**: 169–173, 1996.

170

the rest, the turnover of biomass measured by fumigation can be better described with second order kinetics. The description of microbial turnover can be fully validated because the single pool of microbe modelled is equivalent to the total fumigation mass.

The equations

In most of the biomass models introduced above the rate of decomposition or turnover of microbial biomass is assumed to be proportional to the mass of organisms. So where B is the microbial mass, k_1 the first order rate constant, and t time the rate can be written as follows:

$$-\frac{dB}{dt} = k_1 B \qquad (1)$$

This has solution:

$$B = B_0 e^{-k_1 t} \qquad (2)$$

where B_0 is the amount of biomass at time 0.

It is possible to envisage other descriptions of the rate of turnover of micro-organisms; for example where an interaction determines turnover. This interaction might be through predation, through competition with other organisms for space or a protected niche, for food, through succession at the surface of recalcitrant substrate or even reproduction. Let the one part of the biomass be αB, and the other $(1-\alpha)B$; where α is a fraction $(0 < \alpha < 1)$, assumed constant. Then the rate of turnover can be expressed as:

$$-\frac{dB}{dt} = k_2 \alpha(1-\alpha)B^2 \qquad (3)$$

This is a second order rate law with rate constant k_2; it has solution:

$$B = \frac{B_0}{1 + k_2\alpha(1-\alpha)B_0 t} \; for \; 0 < \alpha < 1 \qquad (4)$$

If α is one or zero there is no interaction and Equations (1) and (2) apply. If α is a constant then $k_2\alpha(1-\alpha)$ is the resultant second order rate constant k_2'.

Equation (3) assumes that all the turnover of biomass taking place in time t does so because of the competition or predation suffered by micro-organisms. In fact this may not be the case. Many organisms are resting in soil; others are out of reach of predators from a different trophic level. Conceptually, organisms which exist in plenty of space, have more than sufficient substrate or exist in some part of the soil

where toxic excretions are removed directly they are produced, enjoy a similar kind of protection to each other: protection against themselves or against a part of what is measured with fumigation techniques. Equation (5) describes a mixed kinetic law where one part of the biomass experiences competition or predation $(1-\beta)B$; this part decomposes according to second order kinetics as above so that it consists itself of two parts: $\alpha(1-\beta)B$ and $(1-\alpha)(1-\beta)B$. The turnover of the unrestricted or unpredated part (βB) obeys a first order rate law.

$$-\frac{dB}{dt} = k_2\alpha(1-\alpha)(1-\beta)^2 B^2 + k_1\beta B \qquad (5)$$

Equation (5) can be integrated with the help of partial fractions and has solution:

$$B = \frac{k_1\beta B_0 e^{-k_1\beta t}}{k_1\beta + k_2\alpha(1-\alpha)(1-\beta)^2 B_0(1 - e^{-k_1\beta t})} \qquad (6)$$

If β is constant it is only necessary to derive the two composite rate constants, $k_1\beta$ and $k_2\alpha(1-\alpha)(1-\beta)^2$

Testing the equations

Predictions from these equations were compared with some data of the formation of microbial biomass derived from sugar beet residues. To each of 24 pots containing 300 g sand soil (14.8% moisture by weight) were added both 1.8 g of dry sugar beet leaves (35.1% C, 2.9% N, 15 g fresh matter) and 3.6 g sugar beet crowns (35.5% C, 0.85% N, 15 g fresh matter). The microbial biomass N (Brookes et al., 1985) and C (Vance et al., 1987) were measured in the soils after 1, 3, 6, 12, 24 and 48 weeks incubation during which time the pots were kept at 20 °C and constant moisture content. Four replicates were thus analyzed at each time. The two rate laws described above and a first order kinetic law were each used in turn in a model to describe the turnover of microbial biomass derived from crop residues (Whitmore and Groot, 1994) assuming that the decomposition of the residues follows first order kinetics. The model predicts mineralization of C and N as well as growth (as a result of the decomposition of the residues) and decline of the microbial biomass. Optimum values of the parameters k_1 and or k_2' were obtained using a computer program for reducing the uncertainty in parameter values (Stol et al., 1992) in which the goodness of fit of simulations is assessed against measured data (C and

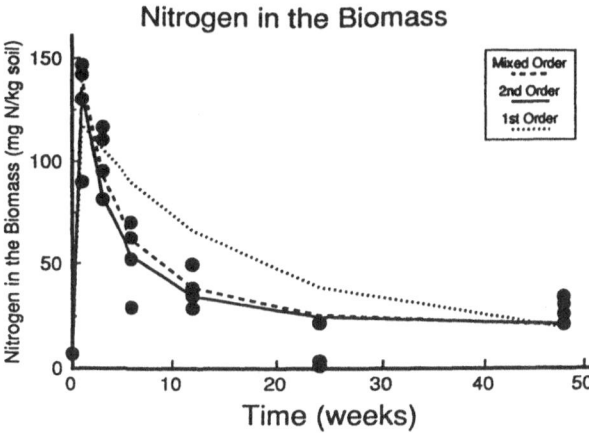

Figure 1. Amounts of nitrogen found in the microbial biomass (●) and simulations of this using: mixed order kinetics (—), second order kinetics (− − −), and first order kinetics (· · ·).

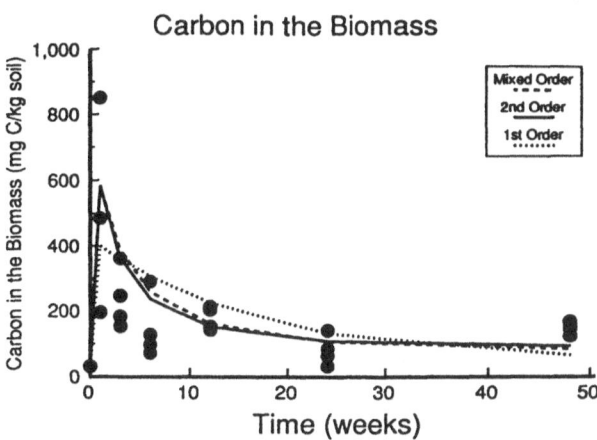

Figure 2. Amounts of carbon found in the microbial biomass (●) and simulations of this using: mixed order kinetics (—), second order kinetics (− − −), and first order kinetics (· · ·).

Table 1. Comparison of the different kinetic laws with the biomass nitrogen data shown in Figure 1

| | Kinetic law | | |
	Mixed	2nd order	1st order
LOFIT	2.82^{NS}	2.77^{NS}	8.44^{***}
Variance accounted for (%)	74.4	74.6	39.9

^{NS} The probability that this difference between model and measurement could have arisen by chance is greater than 1 in 20.
*** The probability that this difference between model and measurement could have arisen by chance is less than 1 in 1000.

N in the biomass in this instance) and new parameter values are sought iteratively until the least residual sum of squares is obtained (Press et al., 1986). Local minima are excluded by re-starting the iteration with the optimum parameters found, searching widely and accepting the minimum only if the program reaches the same parameter values consecutively. Each equation was then assessed against the measured biomass data graphically and with the help of the lack-of-fit statistic (LOFIT) developed by Whitmore (1991) with which the goodness of fit of different models can conveniently be compared against the same dataset. This statistic is a variance ratio and its significance is tested using an F-test. Simply put, the smaller it is the better the fit. These are not validations of the parameterized equations but tests of which equation can best be fitted to the data.

Results and discussion

Figure 1 shows the measurements and simulations of the change in the amounts of nitrogen found in the microbial biomass and Figure 2 the change in the amounts of microbial carbon. The most striking thing about these data is the sharp increase in the amount of biomass from a rather low starting level to a large peak followed by an almost equally sharp decline during the next two or three weeks. Other authors have observed a similar phenomenon (e.g. Nicolardot et al., 1994). This rate of decomposition of biomass itself reduces as time goes by. The first order rate equation is unable

to reproduce both the rapid decline after addition and the persistence of biomass up to 48 weeks later. Alternatively this initial decline can be satisfactorily modelled with first order kinetics (data not shown) but the biomass declines away to almost nothing before 24 weeks. This poor fit is reflected in the LOFIT statistic shown in Table 1 where the predictions from the first order rate law were very significantly different from the measurements. In all of the simulations shown in Figures 1 and 2 allowance was made for the growth of biomass from the decomposition of humic compounds in soil as well as from the decomposing fresh sugar beet residues.

The second order and mixed order equations reproduce the measurements almost equally as well as each other. Figures 1 and 2 show that they each trace the data very well indeed and Table 1 shows that the percentage variance accounted for in the simulations of the biomass N data is almost the same with both models; neither set of simulations differed significantly from the measurements. In fact because the mixed order equation contains two rate parameters and the

172

second order one, the second order rate law is to be preferred with this data. The two parameters k_1 and k_2 in the mixed order equation were also highly correlated with each other indicating that the model was over-parameterized with respect to these data. As good results could be obtained with one parameter less as indeed the second order equation demonstrates.

LOFIT statistics relating to Biomass C are not shown in Table 1 because the measured data on the first sampling date were very variable (Fig. 2); this variability derives from the growth of organisms and not their turnover and it is the ability of the rate laws to simulate the rapid turnover of the microbial biomass which the LOFIT statistic is assessing. The variability in measurements during the initial growth phase reduces the value of any statistical comparisons because it reduces the significance test to a meaninglessly low level; Whitmore (1991) has discussed this point. Nonetheless, the model simulated the mean Biomass C well and accurately reflected the changes in the more precise biomass N, from the same soils, very well.

The large increase in biomass seen in these experiments may be an artifact of the large addition of easily decomposable crop residues. Nonetheless an explanation is needed because it has been observed by others (e.g. Nicolardot et al., 1994; Wu et al., 1993). The use of an average value of the C or N content of the lysed organisms may bias fumigation techniques towards certain species or towards rapidly growing organisms (Bremer and van Kessel, 1990). But the rate of release of carbon dioxide from the residues (data not shown) suggests that the measured microbial growth was representative of the microbial activity in the soil.

Dalenberg and Jager (1981, 1989) and Wu et al. (1993) using ^{14}C-labelling, observed the unexpectedly rapid loss of native soil C on addition of fresh substrate. Both groups concluded that the native biomass died, or to express it another way turned over more rapidly, as a result of this addition. Both second and mixed order rate laws predict an increase in microbial turnover (but of both native and substrate-derived micro-organisms) but still beg the question why. Second order kinetics was derived above by assuming that organisms turnover partly as a result of an interaction among themselves. Increase the number of organisms and the interaction also increases. The observed increase in turnover rate only appears odd from the standpoint of first order kinetics.

Strictly α and β are not constants as has been assumed in Equations (3), (4), (5) and (6). If αB represents predatory protozoa then α is probably in the range 3–7% (e.g. de Ruiter et al., 1993) and $\alpha(1-\alpha)$ then varies between 0.0679 and 0.0651 which is not a very serious variation (4%). The factor β is more problematical however. The second order term varies with β^2 magnifying small changes and an implicit assumption in the discussion of the previous paragraph is that β can vary. Carbon derived from native organisms in soil does appear to turn over more rapidly once substrate is added. This variability in β may account for the failure of the mixed order model to improve on the second order law.

References

Bradbury N J, Whitmore A P, Hart P B S and Jenkinson D S 1993 Modelling the fate of nitrogen in crop and soil in the years following application of ^{15}N-labelled fertilizer to winter wheat. J. Agric. Sci., Cambridge 121, 363–379.

Bremer E and van Kessel C 1990 Extractability of microbial ^{14}C and ^{15}N following addition of variable rates of labelled glucose and $(NH_4)_2SO_4$ to soil. Soil Biol. Biochem. 22, 707–713.

Brookes P C, Landman A, Pruden G and Jenkinson D S 1985 Chloroform fumigation and the release of soil nitrogen: a rapid direct extraction method to measure microbial biomass nitrogen in soil. Soil Biol. Biochem. 17, 837–842.

Dalenberg J W and Jager G 1981 Priming effect of small glucose additions to ^{14}C-labelled soil. Soil Biol. Biochem. 13, 219–223.

Dalenberg J W and Jager G 1989 Priming effects of some organic additions to ^{14}C-labelled soil. Soil Biol. Biochem. 21, 443–448.

De Ruiter P C, van Veen J A, Moore J C, Brussard L and Hunt H W 1993 Calculation of nitrogen mineralization in soil food webs. Plant and Soil 157, 263–274.

Hansen S, Jensen H E, Nielsen N E and Svendsen H 1991 Simulation of nitrogen dynamics and biomass production in winter wheat using the Danish simulation model DAISY. Fert. Res. 27, 245–260.

Jenkinson D S and Powlson D S 1976 The effects of biocidal treatments on metabolism in soil-IV. A method for measuring the soil biomass. Soil Biol. Biochem. 8, 209–213.

Jenkinson D S and Rayner J H 1977 The turnover of soil organic matter in some of the Rothamsted classical experiments. Soil Sci. 123, 298-305.

Jenkinson D S, Hart P B S, Rayner J H and Parry L C 1987 Modelling the turnover of organic matter in long-term experiments at Rothamsted. Intecol Bull. 15, 1-8.

McGill W B, Hunt H W, Woodmansee R G and Reuss J O 1981 Phoenix, a model of the dynamics of carbon and nitrogen in grassland soils. In Terrestrial Nitrogen Cycles. Eds. F E Clark and T Rosswall. Ecol. Bull. 33, 49–115.

Molina J A E, Clapp C E, Shaffer M J, Chichester F W and Larson L W 1983 NCSOIL, a model of nitrogen and carbon transformations in soil: description, calibration and behaviour. Soil Sci. Soc. Am. J. 47, 85–91.

Nicolardot B, Molina J A E and Allard M R 1994 C and N fluxed between pools of soil organic matter: model calibration with long-term incubation data. Soil Biol. Biochem. 26, 235–243.

Press W H, Flannery B P, Teukolsky S A and Vetterling W T 1986 Numerical Recipes, the Art of Scientific Computing, Cambridge University Press, Cambridge, UK. 818 p.

Rutherford P M and Juma N G 1992 Simulation of protozoa-induced mineralization of bacterial carbon and nitrogen. Can. J. Soil Sci. 72, 201–216.

Stol W, Rouse D I, van Kraalingen D W G and Klepper O 1992 FSEOPT a Fortran program for calibration and uncertainty analysis of simulation models. Simulation Report CABO-TT, no 24, AB-DLO, Wageningen, the Netherlands. 24 p.

Vance E D, Brookes P C and Jenkinson D S 1987 An extraction method for measuring soil microbial biomass C. Soil Biol. Biochem. 19, 703–707.

Verberne E L J, Hassink J, de Willigen P, Groot J J R and van Veen J A 1990 Modelling organic matter dynamics in different soils. Neth. J. Agric. Sci. 38, 221–238

Whitmore A P 1991 A method for assessing the goodness of computer simulation of soil processes. J. Soil Sci. 42, 289–299.

Whitmore A P and Groot J J R 1994 The mineralization of N from finely or coarsely chopped crop residues: measurements and modelling. Eur. J. Agron. 3, 367-373

Wu J, Brookes P C and Jenkinson D S 1993 Formation and destruction of microbial biomass during the decomposition of glucose and ryegrass in soil. Soil Biol. Biochem. 25, 1435–1441.

Section editor: R Merckx

O. Van Cleemput et al. (eds.), Progress in Nitrogen Cycling Studies, 465–469, 1996.

The decomposition of wheat and clover residues in soil: measurements and modelling

A.P. Whitmore and F.J. Matus
DLO Research Institute for Agrobiology and Soil Fertility (AB-DLO), Oosterweg 92, P.O. Box 129, NL-9750 AC Haren, The Netherlands

Key words: crop residues, decomposition, modelling

Abstract

A computer model is described that is able to trace the fate of nitrogen in crop residues added to field soils. Clover and wheat residues labelled with ^{15}N were added to a clay and a sand soil and the fate of the label traced over a period of almost 16 months under field conditions. Using a simple function to retard the decomposition of crop residues according to how much fibrous tissue they contain, the model was able to estimate the organic N remaining in soil, and the mineral N and microbial biomass N derived from the crop residues. It proved necessary, however, to postulate the existence of a pool of organic matter derived from crop residues that was more labile than native humus in soil.

Introduction

Computer simulation models of carbon or nitrogen turnover in soil have enjoyed much success in recent years in their ability to estimate the mineralization of C or the mineralization or immobilization of N (e.g. Bradbury et al., 1993, Parton et al., 1987, Verberne et al., 1990). The relative amounts of C and N, or their ratio in organic material or crop residues added to soil, have proved excellent indicators of the speed of release of N or C and as predictors of whether N is immobilized or mineralized. Nonetheless not all the variation in residue decomposition is accounted for with this simple hypothesis and a number of authors have noticed that the residual variation is well correlated with the fibre content of the organic materials (e.g. Matus, 1997; Whitmore and Groot, 1997; Vanlauwe et al., 1993). Here we show that a simple adaptation of a computer simulation model to take account of the lignin, cellulose and hemi-cellulose contents of crop residues is able to produce very good simulations of the amount of nitrogen remaining behind in soil as the residues decompose. The same model satisfactorily estimated the amount of labelled N found in the microbial biomass and in mineral forms in soil.

Materials and methods

Incubation experiments

The shoots (12.7 g) or roots (3.2 g) from each of clover or wheat residues labelled with ^{15}N were dried, ground to pass through a 2 mm sieve and added to either 6100 g of sand soil (containing 5% clay) or 5400 g of clay soil (containing 48% clay) in a series of field incubation experiments (Matus, 1997). All weights expressed on a dry matter basis. The plant materials differed in their fibre contents and these and other relevant chemical properties are given in Table 1. Dried residues were mixed with field moist soil and the mixture confined within PVC cylinders (20 cm in diameter by 30 cm in length) that were kept in the field under ambient conditions for about 500 days beginning in May 1991. The fate of the ^{15}N-label derived from these plant materials has been described in detail by Matus (1997) who has presented the distribution of label with depth in the soil. Since virtually no organic ^{15}N was found below 15 cm the results simulated here are for the top 15 cm of soil in each cylinder only.

Table 1. Chemical properties of the wheat and clover

Residue	C (%)	N (%)	Cellulose + Hemi-cellulose (%)	Lignin (%)
Wheat shoot	35.1	3.6	36.6	3.7
Wheat root	32.8	1.8	59.7	4.0
Clover shoot	33.3	4.4	20.5	3.3
Clover root	32.0	3.5	31.4	6.9

Computer simulations

Whitmore and Groot (1994) have described a simple computer simulation model of the decomposition of crop residues; in this model the decomposition of both crop residues and humified organic matter follows first order kinetics, the only difference being that the turnover times differ. A proportion α of the carbon from the decomposing residues becomes biomass and a proportion β becomes humified (Bradbury et al., 1993) the remainder $(1-\alpha-\beta)$ is lost as CO_2. Whether nitrogen is mineralized or immobilized depends on the C:N ratios of the residues (Z), biomass (X) and humified organic matter (Y). Equation (1) shows the

$$\frac{1}{Z} > (\frac{\alpha}{X} + \frac{\beta}{Y}) \qquad (1)$$

necessary condition for mineralization directly from crop residues. Whitmore (1996) has described a simple adaptation of the kinetic law describing the turnover of the microbial biomass in soil. The amount of biomass B remaining in soil with initial amount B_0 turns over in time t with rate constant k_B as described by Equation (2).

$$B = \frac{B_0}{1 + k_B B_0 t} \qquad (2)$$

A combination of these models was used to estimate the fate of ^{15}N-labelled crop residues in soil. Because the cylinders containing the residues were kept outside in the field, Burn's leaching equation (1976) was used to estimate the loss of mineralized N from the soil cylinders assuming that the mineralization of N was uniform throughout the 15 cm length and assuming that all ammonified N was rapidly nitrified. If P is the percolating excess rainfall (rain minus evaporation) falling in time t, h is the depth of soil (15 cm here) and θ is the volumetric moisture holding capacity of the soil.

$$N_{leached} = (\frac{P}{P + \theta})^{\frac{h}{2}} \qquad (3)$$

The fibre content was assumed to retard the decomposition of the crop residues. As the easily decomposable parts of the residues break down the fibrous parts become relatively more concentrated and retard the decomposition process to a greater and greater extent. The retardation factor f_R is calculated as follows in a manner analagous to that proposed by Parton et al. (1987):

$$f_R = \exp(\frac{-k_R F C_0}{C_t}) \qquad (4)$$

where k_R is a constant and F is the fraction of fibre in crop residues (cellulose plus lignin in these simulations) C_0 the initial mass of the crop residues and C_t the mass at time t. Moisture and temperature are assumed to modify the rates of turnover and organic matter in the same way as described by Bradbury et al. (1993). The factor f_R reduces the rate of residue decomposition along with f_T (effect of temperature) and f_W (effect of moisture) as follows:

$$k^l = f_R * f_W * f_T * k \qquad (5)$$

note that f_T can be greater than unity (e.g. the temperature can exceed the base temperature of 9 °C) and that there is no interaction between these rate-modifying factors.

Daily values of the rainfall and soil temperature were recorded on site; these are given by Matus (1995). Evaporation was not recorded locally and values for the long term mean were taken from Eelde airport 5 km distant. Occasional missing values among the rain and temperature data were also supplied from the mean monthly value at Eelde.

Results and discussion

The residual ^{15}N remaining in soil and the simulations of these data are presented in Figure 1. Here all the organic ^{15}N estimated with the model to be remaining in soil microbial biomass and humus has

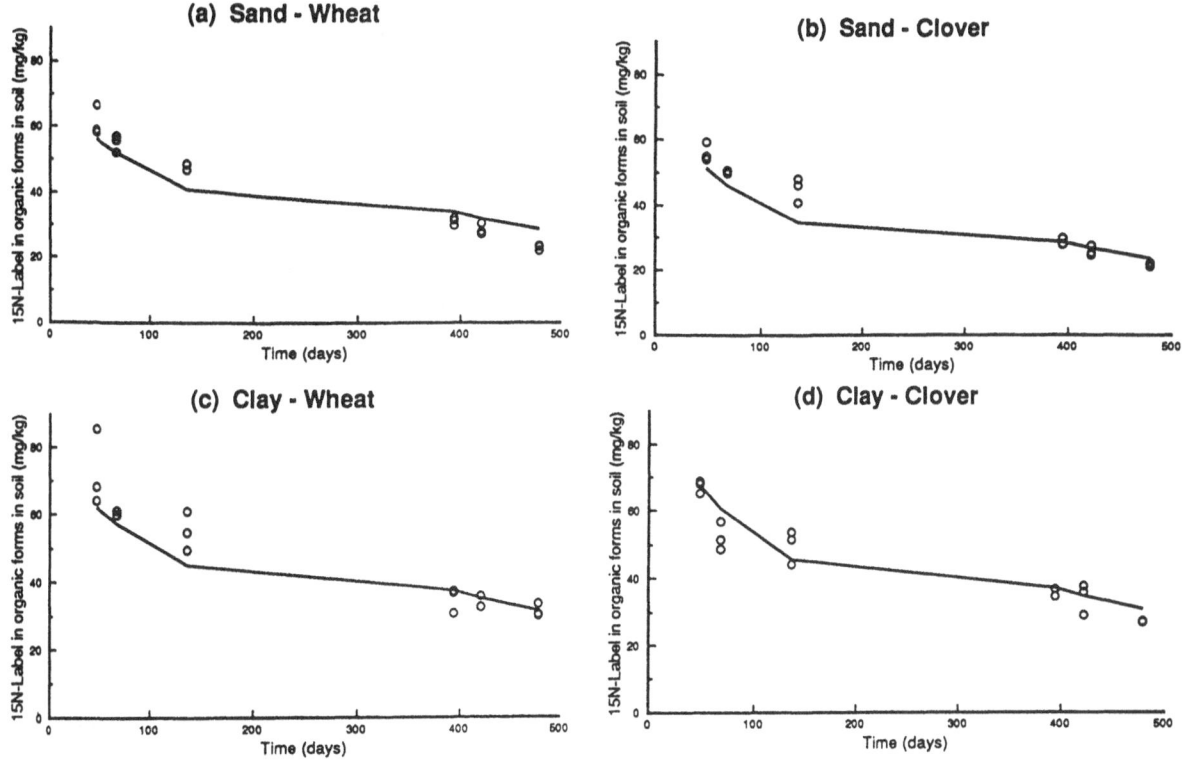

Figure 1. Measured and modelled amounts of ^{15}N-label remaining in soil in organic forms, (**a**) sand soil with wheat, (**b**) sand soil with clover, (**c**) clay soil with wheat, (**d**) clay soil with clover.

been summed. Measurements of total ^{15}N-label in soil have had mineral ^{15}N subtracted. It is clear that the model simulates the retention of ^{15}N-label in organic forms in soil very well indeed. Of particular interest is the fact that results from both the clay and sand soils and with clover and wheat residues can be simulated by the same model without having to make any changes in the parameter values apart from θ which determines the moisture holding capacity of the soil. The behavior of the N derived from crop residues can be described with exactly the same organic matter turnover model.

Figure 2 shows the amounts of ^{15}N-label recovered in the microbial biomass. Relatively large amounts of microbial biomass were found in these soils and the amounts decline throughout the experiment. The model does not predict a sharp increase in the amount of ^{15}N-label found in the biomass but rather a gradual enrichment over the first month or so followed, as shown here, by an equally gradual decline. Matus (1995) reported that the soils used here were taken shortly after harvest and that they already contained large amounts of crop residues. The model predicts that the total biomass N in the control soil declines

over the course of the experiment as indeed was found with the measured values (data not shown). As such it is not surprising therefore that the model predicts no rapid increase in the microbial biomass [15]N.

The ^{15}N-label found in mineral forms in soil is shown in Figure 3 together with the simulations. Here the most striking feature is the rapid decline and increase in mineral N between the first and third sampling times (48 and 135) days. Although these measurements took place in the early summer and autumn a great deal of ^{15}N-label was lost from the top 15 cm of soil. Matus (1995) did not recover a great deal more in the 35 cm of soil below these cylinders. Assuming that all the loss was leaching, Burn's (1976) leaching equation has simulated the pattern of loss very well indeed. Even so some of the loss was undoubtedly denitrification and it is clear in the clay soil, for example (Figs. 2c and d), that the model over-estimated the amount of mineral N in the soil on the first sampling date; more denitrification might reasonably be expected in this soil under these conditions. Careful examination of the data in all figures reveals that much ^{15}N-label is missing. Nitrogen equivalent to 50 mg N per 100

468

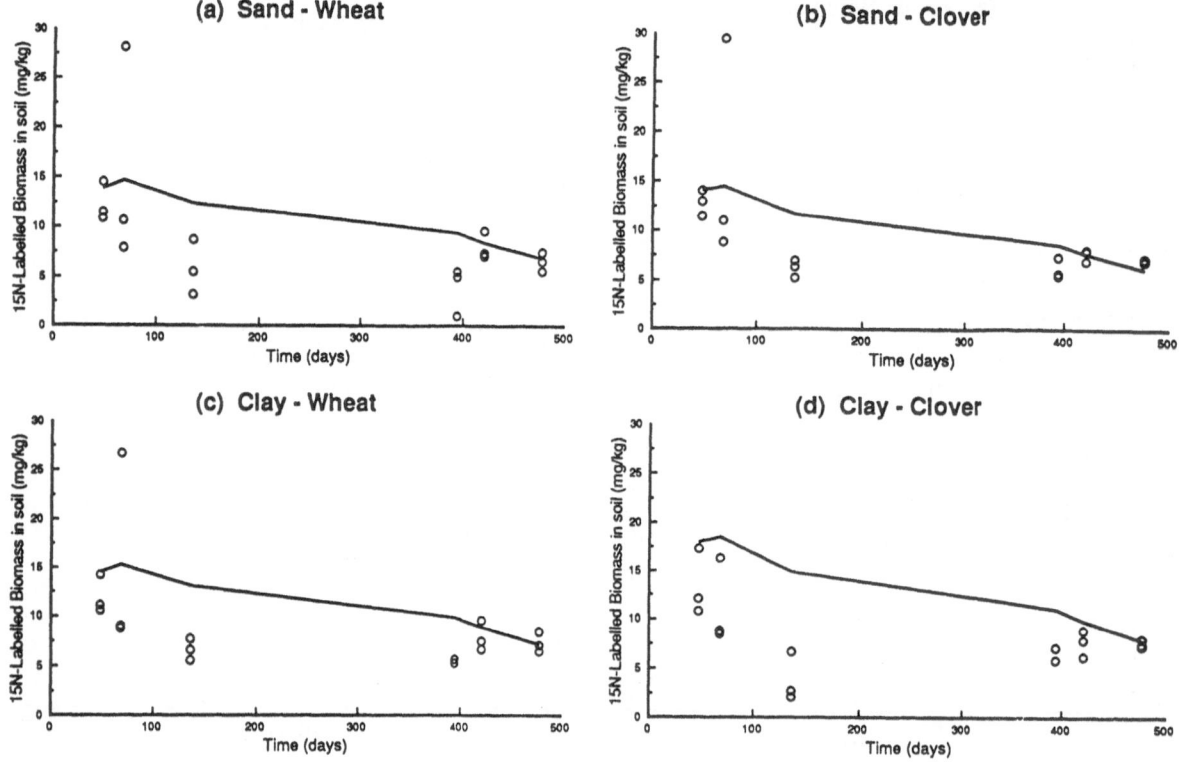

Figure 2. Measured and modelled amounts of ^{15}N-label remaining in soil in the microbial biomass, (**a**) sand soil with wheat, (**b**) sand soil with clover, (**c**) clay soil with wheat, (**d**) clay soil with clover.

g soil was lost during the first 48 days. During this period some 90 mm of rain fell, much of it during thunderstorms. Indeed on this first sampling day 21.9 mm of rain fell which probably accounts for much of the variation found in the measurements made on the clay soil.

Although almost all of the parameters in the organic matter model are as previously described (Bradbury et al., 1993, Whitmore and Groot, 1994), we could not obtain the fits shown in these diagram for the decline in ^{15}N-label in organic forms in soil without incre as-ing the rate of turnover of ^{15}N-labelled humified substances in soil. The value of this rate constant needed to be about three times faster than was calculated for the previous articles where either non-labelled materials were studied or the long-term fate of resistant crop residues was traced (mature wheat roots and stubble). One novel element in these simulations is the use of a 2nd order rate law to describe the decomposition of the microbial biomass. Equally good simulations of the retention of ^{15}N-label in organic forms in soil as those shown in Figure 1 were obtained using a 1st order rate law in which the microbial biomass turned

over much more slowly during the initial months of the experiment; the amounts of microbial biomass N simulated were unreasonably high, however. In effect the biomass was fulfilling the same function of storing ^{15}N-label and releasing it more quickly than humified organic matter.

Either way it seems likely that an active fraction of organic matter was produced from these easily decomposable crop residues and that this itself is relatively rapidly re-mobilized in soil.

Acknowledgements

FJM gratefully acknowledges the support of a Fellowship Programme from the Ministerio de Planificación (MIDEPLAN), Agencia de Cooperación Internacional (AGCI), Santiago, Chile and of Sandwich-Programme Fellowship (P806), from the Agricultural University, Wageningen, The Netherlands.

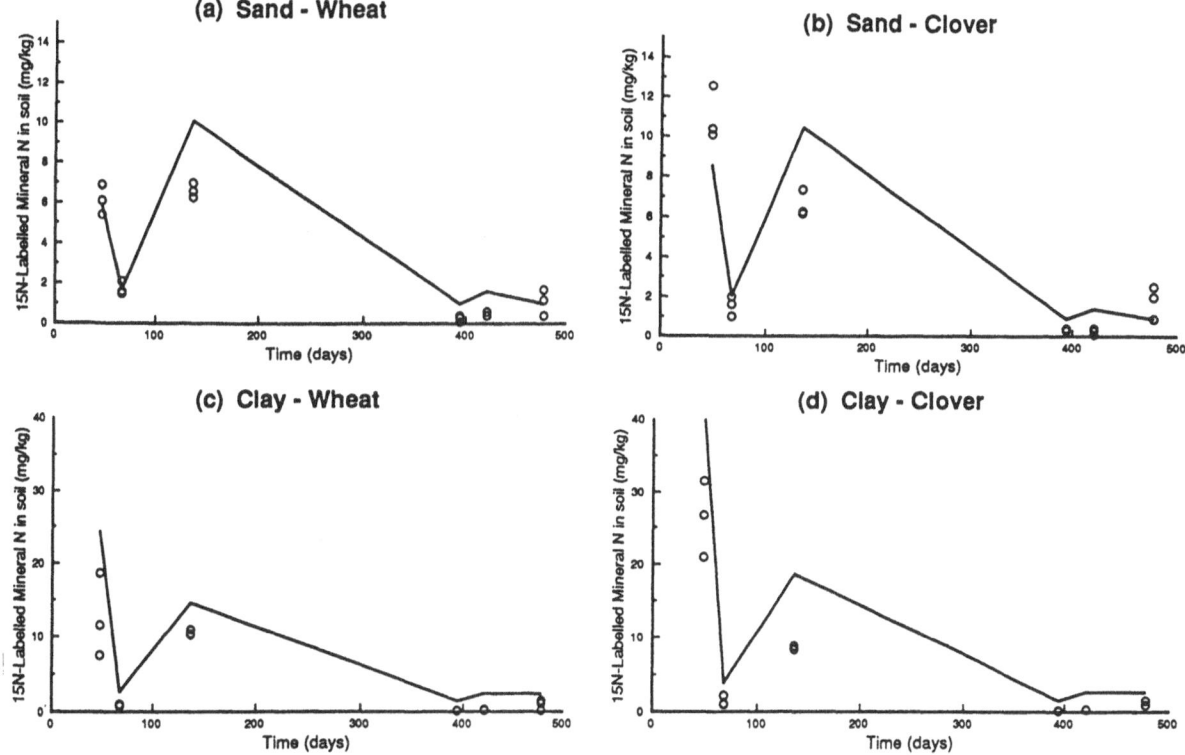

Figure 3. Measured and modelled amounts of ^{15}N-label remaining in soil as mineral N, (**a**) sand soil with wheat, (**b**) sand soil with clover, (**c**) clay soil with wheat, (**d**) clay soil with clover.

References

Bradbury N J, Whitmore A P, Hart P B S and Jenkinson D S 1993 Modelling the fate of nitrogen in crop and soil in the years following application of ^{15}N-labelled fertilizer to winter wheat. J. Agric. Sci., Cambridge 121, 363–379.

Burns I G 1976 Equations to predict the leaching of nitrate uniformly incorporated to a known depth or uniformly distributed throughout a soil profile. J. Agric. Sci., Cambridge 86, 305–313.

Matus F J 1995 Effects of soil texture, soil structure and cropping on the decomposition of crop residues. II: Residual organic N, microbial biomass N and inorganic N derived from decomposing ^{15}N-labelled wheat and clover in a clay soil and sand soil. Soil Biol. Biochem. (*Submitted*).

Parton W J, Schimel D S, Cole C V and Ojima D S 1987 Analysis of factors controlling soil organic matter levels in great plains grasslands. Soil Sci. Soc. Am. J. 51, 1173-1179.

Verberne E L J, Hassink J, de Willigen P, Groot J J R and van Veen J A 1990 Modelling organic matter dynamics in different soils. Neth. J. Agric. Sci. 38, 221–238

Vanlauwe B, Dendooven, L and Merckx R 1994 Residue fractionation and decomposition: The significance of the active fraction. Plant and Soil 158, 263–274.

Whitmore A P and Groot J J R 1994 The mineralization of N from finely or coarsely chopped crop residues: measurements and modelling. Eur. J. Agron. 3, 103-109.

Whitmore A P 1996 Alternative kinetic laws to describe the turnover of the microbial biomass. Plant and Soil 181.

6. ASPECTS RELATED TO AMMONIA VOLATILIZATION

O. Van Cleemput et al. (eds.), Progress in Nitrogen Cycling Studies, 473–477, 1996.

Evaluation of urease inhibitors for the reduction of ammonia volatilization following urea addition to flooded rice

D.G. Keerthisinghe and J.R. Freney
Division of Plant Industry, CSIRO, G.P.O. Box 1600, Canberra, A.C.T. 2601, Australia

Key words: phosphoroamides, urea, urease inhibition, volatilization

Abstract

Ammonia volatilization is one of the main reasons for the poor efficiency of urea as a fertilizer for flooded rice. The rapid hydrolysis of urea, catalyzed by soil urease, and the high pH of the floodwater during daylight hours result in large losses of N by ammonia volatilization. One approach to decrease ammonia losses is to use compounds that retard the hydrolysis of urea, thus allowing the urea to diffuse deeper into the soil. This paper reports some results of laboratory and field experiments conducted to evaluate five compounds, N-(n-butyl) phosphorictriamide (NBPTO), cyclohexylphosphorictriamide (CHPT), phosphoryltriamide (PT), N-(n-butyl)thiophosphorictriamide and thiophosphoryltriamide, as inhibitors of urease activity in flooded soils.

The potential of these compounds to inhibit urease was tested by using jack bean urease in an enzyme assay. CHPT, NBPTO and PT were more effective in inhibiting urease activity than the thio compounds. Studies on these compounds at different concentration ranges showed that NBPTO and CHPT were promising urease inhibitors. Incubation studies on soil under flooded conditions confirmed the above findings. However, in the presence of photosynthetic algae the effectiveness of CHPT and NBPTO were markedly reduced. In field experiments with flooded rice CHPT was more effective than NBPTO; it reduced ammonia loss by 90%. Addition of CHPT in the presence of an algicide significantly increased the recovery of applied N by rice from 8 to 23%.

Introduction

The efficiency of urea applied to flooded rice is low (Fillery et al., 1986), and one of the main reasons for the poor efficiency is that substantial amounts of N are lost to the atmosphere by ammonia volatilization (Simpson and Freney, 1988). This results from the rapid hydrolysis of urea to ammoniacal N by soil urease, the increased growth of algae in the floodwater in response to the added N, and the increased pH during daylight hours due to the increased photosynthetic activity. Field studies have shown that up to 56% of the N applied to flooded rice can be lost in this way (Freney et al., 1990).

Ammonia volatilization from flooded rice can be reduced by application of urease inhibitors. Urease inhibitors delay urea hydrolysis, allowing the urea to move to deeper soil layers before hydrolysis. Ammonia released then would be retained by the cation exchange complex in the soil. A large number of compounds has been tested for their capaci-

ty to inhibit soil urease, but most failed to inhibit this enzyme or did not persist in soil (Austin et al., 1984; Byrnes et al., 1989). Of the compounds tested N-(n-butyl)thiophosphorictriamide (NBPT) seemed to be one of the most promising (Bremner and Chai, 1986). However, laboratory studies with non-flooded soils show that NBPT must be converted to the oxygen analogue N-(n-butyl)phosphorictriamide before it is active (McCarty et al., 1989). A similar study with thiophosphoryltriamide (TPT) found that its inhibitory activity was largely due to the formation of phosphoryltriamide (PT) (McCarty and Bremner, 1989). It appears that the capacity of thiophosphorictriamides to inhibit urease depends on their rate of oxidation to the corresponding oxygen analogue.

This paper reports some results of laboratory and field experiments conducted to investigate the effectiveness of three phosphorictriamides and two thiophosphoric triamides in flooded soils.

Materials and methods

Urease inhibitors

The compounds tested were N-(n-butyl)thiophosphorictriamide, thiophosphoryltriamide, N-(n-butyl)phosphorictriamide, phosphoryltriamide and cyclohexylphosphorictriamide.

Enzyme assays

The enzyme assays using jackbean (*Canavalia ensiformis* (L.)) urease were performed in pH 7.0 TRIS (2-amino-2-hydroxymethyl-1,3-propanediol) buffer prepared according to the directions of McCarty et al. (1990). To 25 mL of TRIS buffer containing 100 mM urea, 50 μL of inhibitor solution was added. The assay was then initiated by the addition of 50 μL of urease solution (4.2 units μL^{-1}). All the assays were conducted at 30 °C in a water bath. The ammoniacal N (NH$_3$+ NH$_4^+$) produced by the enzymatic hydrolysis of urea was measured every 5 minutes for a total assay period of 30 minutes with an ammonia electrode (Orion Model 95–12). The inhibitors were compared at 100 μM and 1 μM final concentration in the assay.

Soil incubation studies

To test the effect of phosphoroamides on inhibition of soil urease an incubation experiment was conducted with two soils used for growing rice (a Typic Pelloxerert from Griffith, N.S.W.; 34°21'S, 146°2'E; and a Typic Haploxerert, SMSS 1983, from Deniliquin, N.S.W. 35°21' S., 144°51'E). The incubation procedure and conditions are described in Cai et al. (1989).

The experiments consisted of factorial combinations of six urease inhibitor treatments (a control and the five triamides), two algal treatments and six sampling dates. All the treatment combinations were replicated twice. Two algal treatments were established by adding the algicide terbutryn (2-(tert-butylamino)-4-(ethylamino)-6-(methylthio)-s-triazine) with the water (0.2 mg active ingredient L^{-1}) to half of the bottles, which were kept in the dark, and water only to the other half, which were kept in the light. The inhibitors were added to the respective treatments at a rate equivalent to 1% of the weight of urea (14 mg of urea per bottle). The soils were incubated for 12 days and sampled every 2 days and analyzed for urea (Mulvaney and Bremner, 1979). Photosynthetic algae were separated by hand from the two flooded soils every 2 days

throughout 12 days and analysed to determine uptake of total N.

Field experiments

To study the effectiveness of CHPT and NBPTO under flooded conditions an experiment was conducted on a clay soil (classification: Fluvic Tropaquept, SMSS, 1983 ; fresh water alluvial soil, Phimai series, local classification) at the Rice Experiment Station, Suphanburi (100°8'E, 14°28'N) in the Central Plain region of Thailand during the 1993 dry season. The inhibitors were tested with or without an algicide treatment. The experiment was carried out in 4 m × 4 m plots with three replicates in a randomized block design. The inhibitors were dissolved in water and sprayed onto the surface of the floodwater at the rate of 1% of the weight of urea immediately before the addition of urea (60 kg N ha^{-1}). As the algicide treatment, copper sulphate (5 kg Cu ha^{-1}) and terbutryn (2-(tert-butylamino)-4-(ethylamino)-6-methylthio)-triazine, 0.2 mg active ingredient L floodwater^{-1}) were applied alternately at 3 day intervals for 12 days. The potential for ammonia loss from the plots was determined from 2-hourly measurements of floodwater pH, temperature and ammoniacal N concentration by a bulk aerodynamic method (Freney et al., 1985).

A mass balance study using ^{15}N labelled urea was conducted in 0.40 m × 0.40 m microplots to determine the effect of the inhibitor treatments on the recovery of applied N.

Results and discussion

Enzyme assay

The effect of phosphoroamides at concentrations of 1 and 100 μM on percentage inhibition after an incubation period of 30 minutes are presented in Table 1. At a concentration of 100 μM, the phosphorictriamides (CHPT, NBPTO, and PT) inhibited urea hydrolysis almost completely (Table 1). In these treatments, the percentage inhibition was above 99% during the whole assay (data not shown). The thiophosphorictriamides reduced urea hydrolysis, but did not completely inhibit urease activity. At the end of the assay, the percentage inhibition in the TPT and NBPT treatments was 36 and 61, respectively. At a 1 μM concentration the percentage inhibition by NBPTO and CHPT was above

Table 1. Effect of phosphoroamides on percent inhibition of jackbean (*Canavalia ensiformis* L.) urease[a]

Phosphoroamide	Concentration of phosphoroamide (μM)	
	1	100
	% inhibition	
Thiophosphorictriamide		
TPT	7.5	35.5
NBPT	11.3	60.9
Phosphorictriamide		
PT	16.7	99.2
NBPTO	76.8	99.9
CHPT	52.3	99.7
l.s.d[b]	5.2	8.9

[a] 50 μL of jackbean urease (4.2 units μL^{-1}) to 25 mL of TRIS containing 100 mM urea at 30 °C. Incubation time: 30 minutes.
[b] Least significant difference at $p=0.05$

Table 2. Effect of NBPTO and CHPT and algicide on ammonia loss, grain yield and recovery of fertilizer N in the grain of flooded rice

Urease inhibitor	Algicide	Ammonia loss (kg N ha^{-1})	Grain yield (t ha^{-1})	N-recovery (% of applied N)
Control (urea only)	-	8.67	3.1	4.3
	+	7.05	3.4	5.2
NBPTO	-	4.58	3.2	5.1
	+	2.09	3.7	8.8
CHPT	-	0.65	3.2	7.5
	+	0.85	3.8	13.4
l.s.d[a]		3.2	0.4	2.6

[a] Least significant difference at $p=0.05$.

50% whereas all the other compounds inhibited only slightly (less than 16.7%, Table 1)

These results show that the phosphorictriamides, NBPTO and CHPT were very effective inhibitors of jackbean urease. As in other enzyme assay studies (McCarty and Bremner, 1989; McCarty et al., 1989) NBPT and TPT were poor inhibitors of jackbean urease.

Soil incubation

As urea hydrolysis in both soils was affected in the same way by the addition of the phosphoroamides, the results of the Griffith soil only are presented. The effect of NBPTO and CHPT in the presence or absence of photosynthetic algae on the inhibition of soil urease is shown in Figure 1a and 1b. The addition of the thiophosphorictriamides TPT and NBPT and the phosphorictriamide PT to flooded soils had an insignificant effect on the rate of hydrolysis of urea and therefore, are not presented. In the control treatment (i.e. without urease inhibitors), most of the urea had disappeared from flooded soil by day 2, and all had gone by day 4. CHPT was the most effective of the inhibitors tested. For example, in the absence of photosynthetic algae, 59% of the urea added to the Griffith soil remained unhydrolyzed after 12 days (Fig. 1b). The corrosponding value for NBPTO was 41% (Fig. 1a).

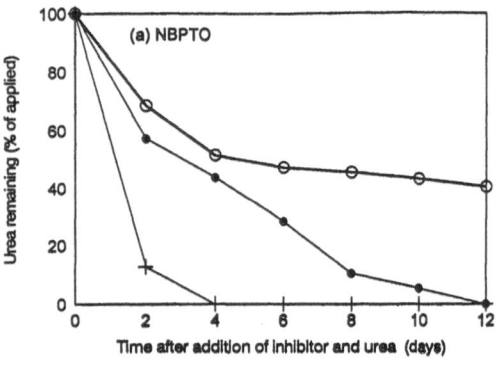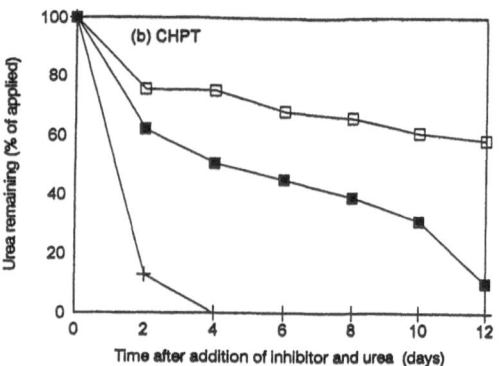

Figure 1. Effect of algae on inhibition of soil urease by (**a**) NBPTO (No inhibitor (+); NBPTO (○); NBPTO + algae (●)), (**b**) CHPT (No inhibitor (+); CHPT (□); CHPT + algae (■))

In the presence of algae the effectiveness of CHPT and NBPTO declined significantly (Fig. 1a and 1b). The apparent reduced effectiveness of CHPT and NBP-TO in prolonging the lifetime of urea in the presence of algae needs to be further explored. It is possible that algal growth has increased the urease activity of the flooded soils, as has been observed by Lindau et al. (1989), and that the amount of CHPT and NBP-TO added was then insufficient to inhibit all of the urease. Vlek et al. (1980) demonstrated that algae grew profusely after addition of urea to the floodwater and that the algae assimilated considerable amounts of urea from the floodwater. However, in our experiment, where N uptake by the algae was measured throughout the study, the uptake amounted to less than 12% of the applied N. This is probably insufficient by itself to account for the observed effect on urea hydrolysis. The algae increased the rate of nitrification of ammoniacal N in both soils (data not shown) and also increased the pH of the flooded soil to > 8.5 thereby increasing ammonia volatilization. The increased removal of ammoniacal N by these two processes may well have resulted in increased urea hydrolysis.

The results of the enzyme assay and the soil incubation studies show that CHPT and NBPTO were effective in inhibiting urease activity. Similar results were reported by Christianson et al. (1990).

Field studies

Ammonia volatilization from flooded rice fields is controlled by floodwater ammoniacal N, pH and temperature, and wind speed, (Freney and Denmead, 1992). The floodwater ammoniacal N concentrations in the control treatment ranged from 0 to 12 g

m^{-3}, temperatures varied between 25 °C and 39 °C, pH values were between 7.2 and 9.2 , and wind speeds were usually between 1 and 3 m s^{-1}. These factors combined to produce a loss of ammonia from the control treatment amounting to 14.4% of the applied N. Application of NBPTO and CHPT, with or without algicide, resulted in much lower ammoniacal N concentrations in the floodwater. For example, in the presence of algicide the ammoniacal N in the CHPT treatment was <2 g N m^{-3}. Addition of NBPTO or CHPT alone, or in combination with algicide, significantly ($p < 0.05$) reduced ammonia loss with the lowest losses (90% reduction) being obtained with CHPT (Table 2).

It is apparent that the soil at this site was not very responsive to applied N. In the plot without added N grain yield was 2.7 t ha^{-1} and this was increased to only 3.1 t ha^{-1} by the addition of 60 kg N ha^{-1}. Application of the urease inhibitors NBPTO and CHPT alone did not result in a significant increase in grain yield (Table 2). However, the addition of NBPTO or CHPT with algicide significantly increased grain yield by about half a tonne per hectare (Table 2).

Recovery of fertilizer N in the grain was significantly increased by application of NBPTO and CHPT (Table 2). In the presence of algicide addition of NBPTO doubled the recovery of applied nitrogen in the grain , while the corresponding CHPT treatments increased the recovery three fold (Table 2).

The results show that fertilizer N can be conserved and grain yield increased by the application of urease, and algal inhibitors.

Acknowledgement

This work was made possible by a grant from the Australian Centre for International Agricultural Research.

References

Austin E R, Bradford T J and Lupin M S 1984 High-performance liquid chromatographic determination and hydrolysis studies of phenyl phosphorodiamidate, a urease inhibitor. J. Agric. Food Chem. 32, 1090–1095.

Bremner J M and Chai H S 1986 Evaluation of N-butyl phosphorothioic triamide for retardation of urea hydrolysis in soil. Commun. Soil Sci. Plant Anal. 17, 337–351.

Byrnes B H, Vilsmeier K, Austin E and Amberger A 1989 Degradation of the urease inhibitor phenyl phosphorodiamidate in solutions and floodwaters. J. Agric. Food Chem. 37, 473–477.

Cai G X, Freney J R, Muirhead W A, Simpson J R, Chen D L and Trevitt A C F 1989 The evaluation of urease inhibitors to improve the efficiency of urea as a N-source for flooded rice. Soil Biol. Biochem. 21, 137–145.

Christianson C B, Byrnes B H and Carmona G 1990 A comparison of the sulfur and oxygen analogs of phosphoric triamide urease inhibitors in reducing urea hydrolysis and ammonia volatilization. Fert. Res. 26, 21–27.

Fillery I R P, De Datta S K and Craswell E T 1986 Effect of phenyl phosphorodiamidate on the fate of urea applied to wetland rice fields. Fert. Res. 9, 251–263.

Freney J R and Denmead O T 1992 Factors controlling ammonia and nitrous oxide emissions from flooded rice fields. Ecol. Bull. 42, 188–194

Freney J R, Leuning R, Simpson J R, Denmead O T and Muirhead W A 1985 Estimating ammonia volatilization from flooded rice fields by simplified techniques. Soil Sci. Soc. Am. J. 49, 1049–1054.

Freney J R, Trevitt A C F, De Datta S K, Obcemea W N and Real J G 1990 The interdependence of ammonia volatilization and denitrification as nitrogen loss processes in flooded rice in the Philppines. Biol. Fertil. Soils 9, 31–36.

Lindau C W, Reddy K R, Lu W, Khind C S, Pardue J A and Patrick W H 1989 Effect of redox potential on urea hydrolysis and nitrification in soil suspensions. Soil Sci. 148, 184–190.

McCarty G W and Bremner J M 1989 Formation of phosphoryl triamide by decomposition of thiophosphoryl triamide in soil. Biol. Fertil. Soils 8, 290–292.

McCarty G W, Bremner J M and Chai H S 1989 Effect of N-(n-butyl) thiophosphoric triamide on hydrolysis of urea by plant, microbial, and soil urease. Biol. Fertil. Soils 8, 123–127.

McCarty G W, Bremner J M and Lee J S 1990 Inhibition of plant and microbial ureases by phosphoroamides. Plant and Soil 127, 269–283.

Mulvaney R L and Bremner J M 1979 A modified diacetyl monoxime method for colorimetric determination of urea in soil extracts. Commun. Soil Sci. Plant Anal. 10, 1163–1170.

Simpson J R and Freney J R 1988 Interacting processes in gaseous nitrogen loss from urea applied to flooded rice fields. In Proceedings of International Symposium on Urea Technology and Utilization. Eds. E Pushparajah, A Husin and A T Bachik. pp 281–290. Malaysian Society of Soil Science, Kuala Lumpur, Malaysia.

SMSS 1983 Keys to Soil Taxonomy. Soil Management Support Services, Technical Monograph No. 6, United States Department of Agriculture. U.S. Government Printing Office, Washington, USA.

Vlek P L G, Stumpe J M and Byrnes B H 1980 Urease activity and inhibition in flooded soils systems. Fert. Res. 1, 191–202.

O. Van Cleemput et al. (eds.), Progress in Nitrogen Cycling Studies, 479–482, 1996.
© 1996 *Kluwer Academic Publishers*.

Cation effects on ammonia volatilization loss from urea applied to a tropical soil

Y.M. Khanif, H. Pancras and C. Daud
Department of Soil Science, Universiti Pertanian Malaysia, UPM 43400, Serdang, Selangor, Malaysia

Key words: ammonia, N losses, urea, volatilization

Abstract

Ammonia volatilization loss from surface-applied urea is an important problem even in tropical soils. The presence of cations at urea microsites during urea hydrolysis can reduce ammonia volatilization loss. Laboratory, and field experiments were carried out to study the effect of cations on ammonia volatilization loss from urea applied to a tropical soil. In the laboratory study urea treated with Ca^{2+}, Mg^{2+} or K^+ at the rate 0, 0.25, 0.5, 1.0 and 2.0 cation/N (chemical equivalent ratio) were applied on the surface of the Bungor top soil (Typic Paleudult). Ammonia volatilization loss was determined daily for a duration of one week by using the force draft technique. In the field study ammonia volatilization loss was determined by using ^{15}N recovery technique. In this experiment ^{15}N labelled urea treated with Ca^{2+}, Mg^{2+} or K^+ at the rate of 0, 0.5 and 1.0 cation/N (chemical equivalent ratio) were surface-applied in micro plots. Ammonia volatilization loss was determined at the end of one week and two weeks. In the laboratory study treatments of urea with Ca^{2+}, Mg^{2+} or K^+ were effective in reducing ammonia volatilization loss. The effectiveness improved with increasing the rate of cations. At the highest rate of the cation added (2.0) the losses were reduced from 36.9% to less than 13.1%. In the field experiment Ca^{2+}, Mg^{2+} and K^+ effectively reduced ammonia volatilization loss. The ammonia volatilization loss was reduced from 27.7% to less than 15.2% in one week. Subsequently more fertilizer N was recovered in the treated soils. Thus we conclude that surface application of urea with Ca^{2+}, Mg^{2+} or K^+ reduced ammonia volatilization

Introduction

Urea is an important source of N fertilizer worldwide. It is, however, subjected to high ammonia volatilization loss. Although substantial ammonia volatilization occurred only at pH > 7.5 (Vlek and Stumpe, 1978), high ammonia volatilization losses were reported in tropical soils with pH< 4.5 (Khanif, 1992). This is because during urea hydrolysis tremendous pH increase occurs at the urea microsites, pH values of > 9.0 in tropical soils had been reported (Khanif and Pancras, 1987). Thus, suppression of soil pH at urea microsites during urea hydrolysis can be an effective approach to reduce ammonia volatilization.

Several earlier studies had shown that addition of cations such as Ca^{2+}, Mg^{2+} and K^+ had effectively reduced ammonia volatilization loss (Fenn et al., 1981a, 1981b, 1982; Rapport and Axley, 1984). Although the mechanism is not fully understood it is speculated that the effect is due to precipitation of

CO_3^{2-} by Ca^{2+} and Mg^{2+}. This prevents the formation of $(NH_4)_2CO_3$ required for ammonia volatilization. The Ca^{2+} and Mg^{2+} originate directly from the added cation or from the soils with high Ca^{2+} and Mg^{2+}, these cations are desorbed when K^+ is added. Another explanation was due to suppression of pH increase at urea micro site during urea hydrolysis.

The demonstration of the effectiveness of cations in reducing ammonia volatilization loss had been limited to calcareous soils. The performance of these cations under tropical conditions has not been studied. Thus, this study was carried out to evaluate the effectiveness Ca^{2+}, Mg^{2+} and K^+ treatments on ammonia volatilization loss from surface - applied urea.

Materials and methods

The study consisted of a laboratory and a field experiments. It was carried out on a Bungor series soil (Typic

Paleudult) collected at the depth of 0–15 cm. The soil has a sandy clay loam texture, with pH 4.5, CEC 8.7 cm$(^+)$ kg^{-1}, 1.72% organic carbon and 0.13% total N.

Laboratory experiment

Measurement of ammonia volatilization loss was carried out by the forced-draft technique as described by Fenn and Kissel (1973). The ammonia evolved was trapped in boric acid and determined by titration with standard HCl. Urea or treated urea was evenly applied on the soil surface in the air exchange chamber at the rate of 400 μg N g^{-1} soil. The amount of soil used for each treatment was 400 g and the water content was maintained at field capacity. The urea were treated with Ca^{2+} (CaCl$_2$), Mg^{2+} (MgSO$_4$) and K$^+$ (KCl) at the rate of 0, 0.25, 0.5, 1.0 and 2.0 cation/N (chemical equivalent ratio). The chemicals were thoroughly mixed with urea before they were evenly applied on the soil surface. The measurement of ammonia volatilization loss was made daily up to seven days. Each treatment was replicated three times.

Field experiment

The treatments consisted of three soluble salts of Ca^{2+} (CaCl$_2$), Mg^{2+} (MgCl$_2$) and K$^+$ (KCl) applied at the rate of 0, 0.5 and 1.0 cation/N (chemical equivalent ratio). Urea labelled with ^{15}N (a.e. 8.8425%) was uniformly mixed with one of the cations and broadcasted on the soil surface in a micro-plot. The rate of urea applied was 150 kg N ha^{-1} (1.47 g plot^{-1}). The micro-plot was made up of a metal cylinder with 24 cm diameter and 15 cm height. The cylinders were pushed into the soil until a margin of 1 cm remained above the soil surface. The experiment was carried out in a randomized complete block design with four replications. Soil samples up to 30 cm depth were collected in each micro-plot at one and two weeks after urea application. The samples were cut at 10 cm intervals and analysed for total N and ^{15}N (Bremner and Mulvaney, 1982; Hauck, 1982). Ammonia volatilization loss was calculated by taking the difference between the applied and the recovered ^{15}N in the soil (leaching below 30 cm and denitrification losses was found to be negligible from an earlier study).

Table 1. Effect of soluble salts on ammonia volatilization loss from urea under field conditions

Treatment	Cation/N	Sampling time	
		1st. week	2nd. week
		Ammonia loss (%)	
Calcium	0.5	12.2$_b$	16.4$_b$
	1.0	12.4$_b$	14.3$_b$
Magnesium	0.5	13.0$_b$	16.1$_b$
	1.0	14.8$_b$	15.6$_b$
Potassium	0.5	12.2$_b$	16.6$_b$
	1.0	15.2$_b$	16.3$_b$
Control	-	27.7$_a$	31.8$_a$

Figures followed by the same letter are not significantly different ($p=0.05$).

Results and discussion

Laboratory experiment

The ammonia volatilization loss from surfaced applied urea was significantly reduced with Ca^{2+}, Mg^{2+} and K$^+$ treatments (Fig. 1). The effectiveness of the cations was similar. The loss was significantly reduced when more than 0.5 cation/N was added to urea. Increase of cation level from 0.5 to 1.0 cation/N did not significantly reduce ammonia volatilization loss. With addition of Ca^{2+} and K$^+$ at 2.0 cation/N, ammonia loss was significantly lower than at 1.0 cation/N. However, the reduction at this level was not significant for Mg-treatment.

Field experiment

The ^{15}N recovery in the soil increased significantly when Ca^{2+} Mg^{2+} and K$^+$ were mixed with urea (Fig. 2, 3). The ^{15}N was leached into the profile; however, it was within the 30 cm sampling depth.

Treatments of urea with Ca^{2+} Mg^{2+} and K$^+$ effectively reduced ammonia volatilization loss (Table 1). The loss however, was not affected by the type of cations, sampling dates and the rate of cations used. Increasing the rate of cations from 0.5 to 1.0 cation/N did not significantly reduce ammonia volatilization loss. The loss also did not increase significantly when the sampling time was increased from one week to two weeks. This was because ammonia volatilization occurred very rapidly and most of the loss occurred during the first week (Khanif and Pancras, 1987).

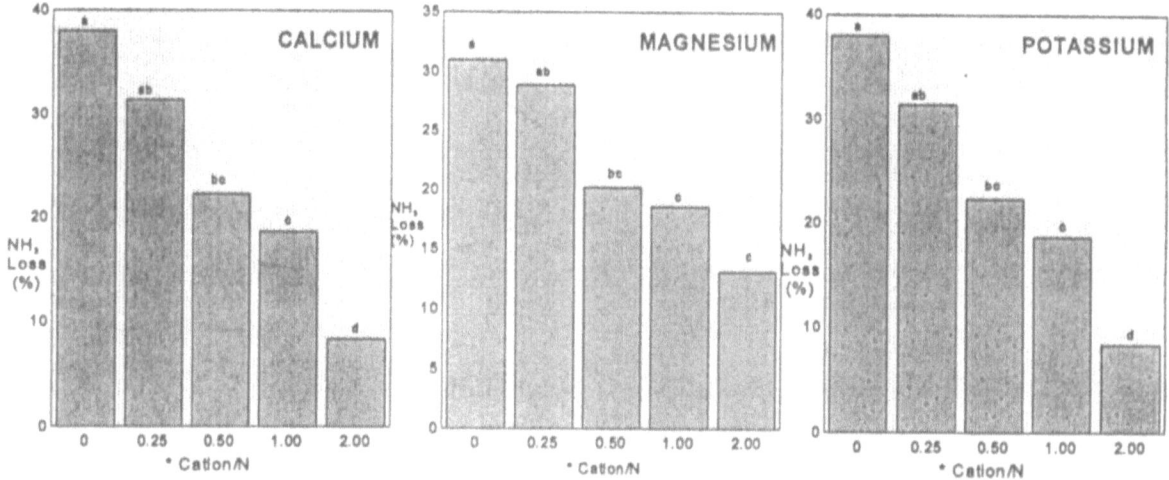

Figure 1. Cumulative ammonia volatilisation loss from urea with addition of different rates of Ca, Mg and K. **Values followed by the different letters in same column are significantly different by Duncan Multiple Range Test (*p*=0.05). *Ratio of chemical equivalent of cation to urea - N.

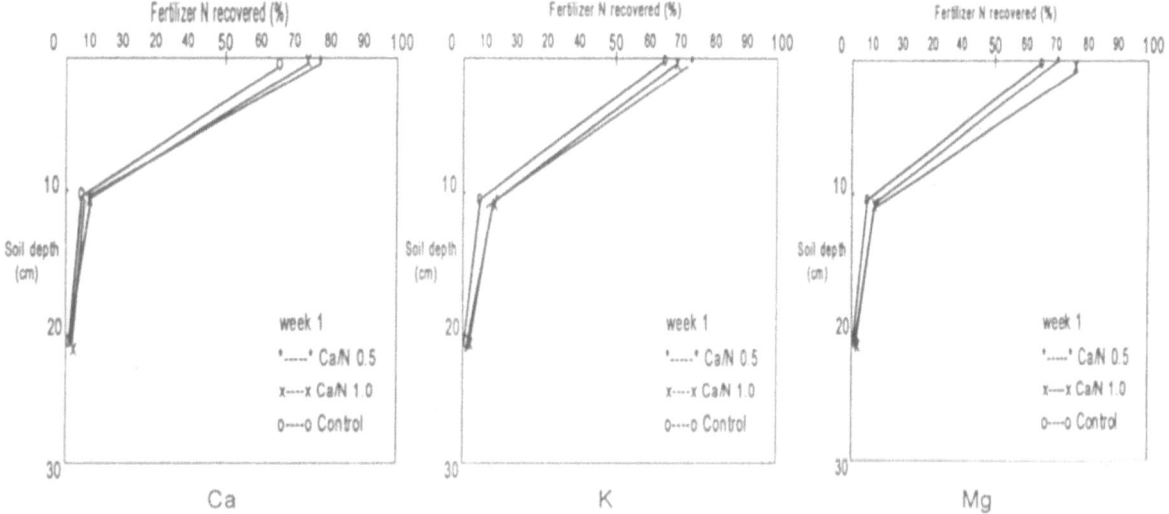

Figure 2. Effects of Ca, K and Mg on Fertilizer N recovery in the soil (week 1).

The effectiveness of the cations in reducing ammonia volatilization loss from urea in calcareous soils was demonstrated earlier by Fenn et al. (1981a, 1982). In calcareous soils, the effect was attributed to precipitation of $CaCO_3$ or $MgCO_3$ which prevented the formation of $(NH_4)_2CO_3$ required for ammonia volatilization (Fenn et al., 1981c). In our study the loss was probably controlled by suppression of pH increase during urea hydrolysis rather than carbonate precipitation. Such precipitation was unlikely to occur in tropical soils due to low soil pH and low exchangeable Ca^{2+} to be released in the case of K^+ treatment (Fenn, 1982). Also in this soil the exchange complex is dominated by Al^{3+} which can be displaced by the cations into the

soil solution. The hydrolysis of Al^{3+} then reduces the micro-site pH, preventing ammonia volatilization loss.

Using soluble salts to reduce ammonia volatilization with urea is very promising, because cations used, are needed as plant nutrients and often added as fertilizer. Thus no extra cost is needed, it only requires modification of the fertilization schedule such that urea is added with the cations.

Conclusion

Addition of Ca^{2+}, Mg^{2+} and K^+ with urea at the rate 0.5 cation/N (chemical equivalent ratio) signif-

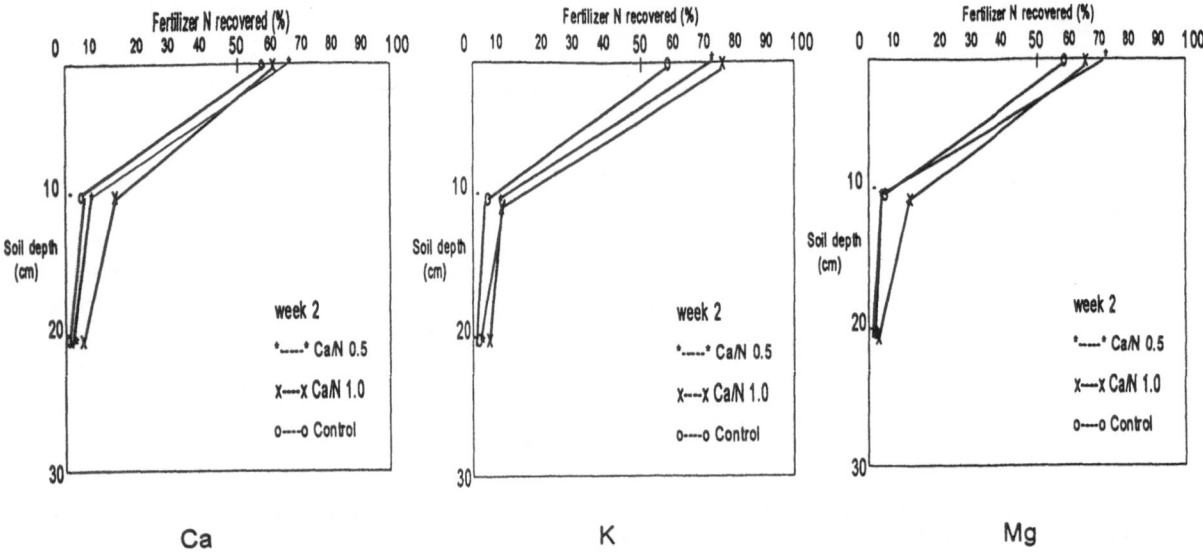

Figure 3. Effects of Ca, K and Mg on Fertilizer N recovery in the soil (week 2).

icantly reduced ammonia volatilization from surface applied urea. The effectiveness of Ca^{2+} Mg^{2+} and K^+ in reducing ammonia volatilization loss from urea was similar. The use of cations in reducing ammonia volatilization loss has practical application. It does not require extra cost since the cations used, are plant nutrients and being applied as fertilizers.

Acknowledgement

The authors wish to acknowledge PETRONAS for providing financial support for this study.

References

Bremner J M and Mulvaney C S 1982 Nitrogen-Total. *In* Methods of Soil Analysis, Part 2. Eds. A L Page, R H Miller and D R Keeney. pp 595–624. American Society of Agronomy, Madison, Winconsin, USA.

Fenn L B and Kissel D E 1973 Ammonia volatilization of ammonia compounds. I. Theory. Soil Sci. Soc. Am. Proc. 37, 855–859.

Fenn L B, Matocha J E and Wu E 1981a A comparison of calcium carbonate precipitation and pH depression on calcium-reduced ammonia loss from surface-applied urea. Soil Sci. Soc. Am. J. 45, 1128–1131.

Fenn L B , Matocha J B and Wu E 1981b Ammonia losses from surface-applied urea and ammonium fertilizers as influenced by rate of soluble calcium. Soil Sci. Soc. Am. J. 46, 771–776.

Fenn L B, Matocha J E and Wu E 1982 Substitution of ammonium and potassium for added in reduction of ammonia loss from surface-applied urea. Soil Sci. Soc. Am. J. 46, 771–776.

Fenn L B, Taylor R M and Matocha J E 1981c Ammonia losses from surface applied nitrogen fertilizer as controlled by soluble calcium and magnesium : General theory. Soil Sci. Soc. Am. J. 45, 777–781.

Hauck R D 1982 Nitrogen-isotope-ratio analysis. *In* Methods of Soil Analysis, Part 2. Eds. A L Page, R H Miller and D R Keeney. pp 735–779. American Society of Agronomy, WI, USA.

Khanif Y M 1992 Ammonia volatilization of loss of surface applied urea on some Malaysian soils. Pertanika 15, 115–120..

Khanif Y M and Pancras H 1987 Urea transformation in Malaysian soils. *In* Proceedings, International Symposium of urea Technology and Utilization. Eds. E Pushparajah, H Alias and A T Bachik. pp 259–266. Malaysian Society of Soil Science, Kuala Lumpur, Malaysia.

Rapport B D and Axley J H 1984. Potassium chloride from improved fertilizer efficiency. Soil Sci. Soc. Am. J. 48, 399–401.

Vlek P L G and Stumpe J M 1978 Effects of solution chemistry and environmental conditions on ammonia volatilization losses from aqueous systems. Soil Sci. Soc. Am. J. 42, 416–421.

O. Van Cleemput et al. (eds.), Progress in Nitrogen Cycling Studies, 483–490, 1996.

Technical possibilities to reduce ammonia emissions from animal husbandry

A review of research in The Netherlands

G.J. Monteny

DLO-Institute for Agricultural and Environmental Engineering (IMAG-DLO), P.O. Box 43, 6700 AA Wageningen, The Netherlands

Key words: acidification, ammonia emission, animal housing, cattle, emission reduction, pigs, poultry

Abstract

By the emission of ammonia, animal production contributes considerably to environmental acidification. In the Netherlands, legislative measures concerning slurry application and covering of slurry storages enforce a substantial reduction of the ammonia release. Research in a model system showed that ammonia volatilization starts almost immediately after urine excretion by cattle and pigs. For poultry, the degradation of uric acid to ammonia is a relatively slow process. Both temperature and air velocity have a marked effect on ammonia release.

Studies in an experimental cubicle dairy cow house with a slatted floor showed that emission is effectively reduced by flushing slats and concrete, sloped floors with water or a diluted formalin solution. Acidification of the slurry under the slats may reduce the total barn emission by 40 percent. Floor systems affect the flow of urine and therefore emission. A sloped solid, concrete floor enabled an emission reduction of 40–60 percent relative to a slatted floor. Comparison between cattle housing systems indicate that tied barns are also favourable in terms of emission. Experiments showed substantial effects of diet composition for dairy cows on emission. For pig houses, frequent removal by scrapers over a sub-floor system showed high emission reduction potential in piglet and sow houses, but not for fattening pigs. Flushing with aerated or acidified liquid, obtained after slurry separation, showed to be very effective as a measure to reduce ammonia emission from fattening pigs housing.

Ammonia emission from hen houses can be substantially reduced by means of forced drying on manure belts and by daily removal of the manure, combined with drying in tunnels outside the hen house.

In conclusion, various technical possibilities exist for reducing the ammonia emission from animal houses.

Introduction

By the emission of ammonia, animal production contributes to environmental acidification. Estimates for the Netherlands show that in the year 1989 ammonia (NH_3) contributed approx. 45% to environmental acidification (Heij and Schneider, 1991). This relatively high figure is directly related to high livestock densities. The emission of sulphur dioxide (SO_2) and nitrogen oxides (NO_X) was responsible for the remaining part of acidification. Concerning ammonia, approx. 95% of the emission sources are related to primary agricultural production; the other sources are related to industrial and household activities. This paper will deal with agricultural sources only. In order to protect the environment in the longer run, the Dutch govern-

ment has enforced legislation, based upon so called critical loads of nitrogen deposition for ecosystems. For ammonia this has been translated in a national average reduction of 70 percent in the year 2005, relative to the level in 1980 (approx. 240 million kg; Oudendag and Wijnands, 1989). For agricultural production units in ecologically vulnerable parts of the Netherlands, higher reductions might be demanded in the future (Anonymus, 1993).

Within the framework of ammonia legislation, the Dutch government and the farmers organizations have raised funding for research into technical solutions for the ammonia emission problem. This started in the early eighties as a part of the Dutch Additional Programme on Environmental Acidification Research. After 1988, the ammonia research activities were integrated with the Dutch Research Programme on Manure, which continued until 1995. Until now, emphasis in ammonia research have changed from slurry application and storage to animal housing systems, animal nutrition and fundamental aspects.

This paper describes major developments in on-farm technology and farm management to reduce ammonia emission. The extent of the ammonia emission problem is shown first, followed by a description of the most important theoretical backgrounds concerning ammonia emission. The major results from research and development activities concerning emission reducing solutions for slurry application, slurry storage and animal housing are discussed, followed by conclusive remarks.

Ammonia emission in the Netherlands

Recent model calculations show that the emission of ammonia in the Netherlands for the year 1990 was approx. 202 million kg. The relative emission from cattle, pigs and poultry was 55, 37 and 8 percent respectively. Manure application was the most important source representing 55 percent of the national emission. Animal housing, including outside slurry storage, and grazing contributed for 35 and 10 percent respectively (Oudendag, 1993). Compared to earlier model calculations performed by Oudendag and Wijnands (1989), the emission of ammonia has slightly decreased during the eighties. This effect is for a major part related to changes in livestock numbers. Because slurry application with low ammonia emission rates and the covering of silo are law-enforced nowadays, the national emission can be expected to decrease much further. However, no figures of the actual situation are present yet.

Theoretical backgrounds

Ammonia (NH_3) in livestock production is formed following the degradation of urea, present in the urine of pigs and cattle. This degradation takes place in an aqueous environment, from which ammonia can volatilize (emission). The degradation and emission can be represented by the following, simplified, formula:

$$\text{urea} \xrightarrow[\text{degradation}]{\text{urease}} CO_{2,\text{gas}} + NH_{4,\text{liq}}^+/NH_{3,\text{liq}} \xrightarrow[\text{emission}]{} NH_{3,\text{gas}}$$

Poultry produce manure containing uric acid, which is microbially converted to urea first before urea degradation and emission can take place. The complex processes relative to the degradation of uric acid are kept outside the scope of this paper.

Urea degradation, mainly taking place inside the animal houses, is catalized by the enzyme urease, produced by micro organisms that are present in faeces. This means that these micro organisms will also be present on floors in animal houses (Ketelaars and Rap, 1994), where faeces are deposited regularly. The urea conversion is taking place directly after urine is deposited on the floor (Elzing et al., 1992).

Whereas the process of urea degradation is mainly dependent on the presence and activity of micro organisms, the ammonia emission process is strongly, influenced by the chemical and physical conditions of surfaces where there is contact with air (Elzing et al., 1992). The emission can be described by the following simplified formula:

$$E = K\,A\,dC$$

where:

E	=	emission of ammonia (kg s^{-1})
A	=	area of emitting surface (m^2)
k	=	mass transfer coefficient (m s^{-1})
dC	=	difference in ammonia concentration at the emitting surface and in the air (kg m^{-3})

The most important parameters in the emission process are temperature of the emitting surface and air velocity at the emitting surface, mainly determining the mass transfer coefficient (k) of ammonia and the area (A) of the emitting surface. Furthermore, urease activity of the emitting surface, nitrogen content of the substrate (ureum) and pH are also of importance, since they determine the actual concentration of volatile ammonia at the emitting surface and therefore the concentration gradient dC.

The above mentioned parameters and processes may be modified by technical measures in order to reduce the ammonia emission under practical conditions. In housing systems both degradation and emission processes can be managed. Because of the characteristics of these processes, emission reduction techniques for pig and cattle housing have to be designed either aiming at the permanent lowering of the urease activity or at interfering instantly in the emission processes. For poultry housing, the slowing down of uric acid degradation processes by the removal of moisture is the most important technique to reduce emission, since drying results in suboptimal conditions for microbial activity. With respect to land spreading and slurry storage, only an intervention in the emission process is possible, since degradation has already taken place inside the animal house.

Techniques for the reduction of ammonia emission

Through research activities in the past years, technical possibilities to reduce ammonia emission after slurry application, drying slurry storage outside the animal houses and from animal housing systems themselves have been developed. The major results are discussed successively. Consequently, the perspectives of emission reduction from grazing and through animal nutrition will highlighted briefly.

Slurry application

As the major source of ammonia emission, methods of slurry application were first investigated in ammonia research. Experiments were carried out under semi-practical conditions, where several circular plots of approx. 0.15 ha each were used. On one plot, slurry was spread in a traditional way (surface spreading; reference plot), while on one or more other plots, slurry was applied using alternative techniques. Air velocity was measured at several heights using anemometer masts in the centre and at the edge (down wind) of the plots; ammonia concentration profiles were measured at the same points using an impinger method with which ammonia was captured in acid filled bottles. The emission rate was calculated by using both air velocity and ammonia concentration profiles; emission reduction for the plots treated with alternative techniques was calculated relative to the emission from the reference plot. The method used is described in detail by Denmead (1983).

Technical possibilities for the reduction of ammonia emission by land spreading of slurry on grassland and arable land and their relative emission reduction are given in the following table. Pig slurry was applied on sandy and clay soils in nearly all experiments.

Not all of the investigated techniques, however, are legally allowed in the Netherlands. Low emission slurry application techniques are only allowed, if they result in 80 percent emission reduction compared to traditional slurry application, in order to achieve a maximal effect on the national emission. Until now, only

Table 1. Relative ammonia emission as compared to surface spreading (in percent) by different techniques for slurry application (after: Mulder and Huijsmans, 1994)

Application techniques	Relative emission
Grassland	
Surface spreading	100
Dilution (1:3) with water	20–80
Irrigation	25–75
Acidification pH < 5	0–30
pH 5–6	20–70
Trailing feet	20—75
Sod injection	< 20
Deep injection	< 5
Arable land	
Direct tillage	5–30
Tillage after application	5–65
Trailing feet	30–65
Sod injection	< 20
Deep injection	< 5

shallow and deep injection on grassland and injection and direct soil tillage following application on arable land are allowed (see Table 1).

Slurry storage

Research into the possibilities of emission reduction from slurry storage facilities was carried out with scale models of silo's, the so called mini-silo's. A Lindvall-box (see Elzing et al., 1992) was used for emission measurements.

In Dutch legislation, only covers that have a reduction potential of 75 percent or more relative to uncovered silo's are allowed. Furthermore, covers must be observable at a distance. This means that only roof constructions made from concrete, wood and tents are allowed. Also underground slurry storage combined with silage storage on the roof and slurry bags are allowed, because of expected high emission reduction. The covering of cattle slurry with a crust, induced by the addition of straw during filling of the silo, is effective, but not allowed as it can hardly be monitored.

Measurements of the ammonia emission from full scale storage facilities have only been carried out in one experiment with an induced straw crust on cattle slurry.

The measurement strategy and method, however, were not satisfactory.

Except for diluted and fresh poultry manure, no regulations are present yet for storage of pre-dried and dried poultry manure. Refering to the described topics under poultry housing, pre-drying of fresh poultry droppings on belts inside the house, followed by frequent removal becomes more and more of interest from manure quality (dry matter content and emission point of view. The manure from those systems, with a dry matter content of 40–60 percent, is usually stored at the farm in an open building or is transported directly to arable farmers. During storage of this pre-dried manure, microbial degradation processes will continue, leading to high emissions of ammonia. In order to prevent emissions, dry matter contents of 70 percent and higher have to be achieved shortly after production of the manure (Kroodsma et al., 1993). Further drying in a tunnel alongside the henhouse, following belt drying in side the house, was found to be effective (Demmers et al., 1992).

Animal housing

Research into ammonia emission reduction from animal housing systems concerns mainly dairy cattle, pigs and poultry. Potential emission reducing techniques are tested in a so called model system under standardized experimental conditions (Elzing et al., 1992). A Lindvall-box is used for emission measurements. The model system is also used to study the effect on the emission of factors that influence the degradation and emission processes both separately and as a whole. In semi-practical, mechanically ventilated units, further investigations are being carried out with small groups of animals. One unit is layed out as a traditional housing system and is regarded as reference unit. One or more other units are equipped with alternative systems. Continuous ammonia emission measurements (ventilation rate × ammonia concentration) are simultaneously carried out per unit. Ventilation rate is measured in ventilation shafts using windtunnel calibrated anemometers; ammonia concentration in ventilation air is measured by sampling exhaust air and using a combination of an ammonia convertor and a NO_X-monitor. The measuring method is described in detail by Bleijenberg and Ploegaert (1994). The most important aim of the experimental set up with semi-practical units is to determine the ammonia emission, relative to the emission from the reference unit. In order to establish absolute emission levels for tech-

Table 2. Investigated manure handling techniques and floor systems to reduce ammonia emission from cow housing systems and average emission reduction relative to cubicle houses with slatted floors

Technique	Relative emission
Cubicle house, slatted floors	100
Additional flushing	
Water	75
Formalin	50
Cubicle house, sloped concrete floor	50
Additional flushing	
Water	20
Formalin	15
Acidification of slurry in pit	60
Additional flushing with acid slurry	40

niques that reduce ammonia emission from housing systems, long term measurements are carried out under practical conditions on experimental and commercial farms. Ammonia emission measurements are carried out in the same way as in semi practical experiments.

In the following paragraphs, the most important developments in techniques that reduce ammonia emission from housings systems for dairy cows, pigs and poultry are discussed.

Dairy cow housing

Dairy cows in the Netherlands are traditionally kept in cubicle houses with slatted floors and slurry storage under slats and cubicles. Voorburg and Kroodsma (1992) reported an ammonia emission level of approx. 1 kg per cow per month, during the period that the cows were kept inside permanently. Slurry handling techniques and floor systems designed for use in cow houses with cubicles and two alternative housing systems have been investigated on their effect on ammonia emission.

Slurry handling and floor systems. In cubicle houses for dairy cows, approx. 40 percent of the ammonia emission originates from the slurry pit under the slatted floor and 60 percent from the floor surface (Elzing et al., 1992). Research has mainly been focussed on the latter, largest source.

The technical solutions that were investigated and their relative ammonia emission reduction are given in Table 2.

In all experiments, manure was removed by scrapers. Flushing with water over the slatted floors appeared to be effective: ammonia emission was reduced by 20 percent or less, depending on the flushing strategy (interval, duration, water pressure). Flushing was carried out once every one or two hours, by means of a pipe system with nozzles alongside the alleys. The amount of water varied between 50–110 L per cow per day (De Boer et al., 1993). Flushing with 45 L per cow per day of a diluted formalin added to the water resulted in a 50 percent emission reduction (Bleijenberg, pers. commun.).

Replacing the slatted floor by sloped concrete floors with a central urine gutter gave a 50 percent lower emission (Swierstra et al., 1994). Additional flushing with water and with a diluted formalin solution resulted in an emission reduction with 80 percent and 85 percent respectively compared to slatted floors. The daily amount of flushing liquid used in both experiments was approx. 40 L per cow (Bleijenberg, pers. commun.).

Improving the runoff of urine from the floor surface in experiments with slatted floors and concrete floors by using an epoxy cement layer did not result in additional emission reduction. Acidification with nitric acid of the slurry in the pit in a cubicle cow house with slatted floors lowered the ammonia emission by 40 percent. The pH-level had to be 4 in order to prevent denitrification to occur, leading to the loss of the nitrate that is present in the slurry following the addition of nitric acid (Voorburg and Kroodsma, 1992). Flushing the slats with the acidified slurry in the pit lowered ammonia emission with 60 percent (Kroodsma, pers. commun.).

Housing systems. Although 80 percent of Dutch cattle are kept in cubicle houses, ammonia emission research has also focussed on other types of housing systems. Tying stalls and housing with straw bedding were investigated in both field studies and in experiments in units. Results show that tying stalls might have a substantial potential for emission reduction compared to cubicles (Groenestein, 1993); straw bedding had hardly any effect on ammonia emission, but appeared to give high methane emissions (Groenestein and Reitsma, 1993).

Discussion
Various slurry handling and floor systems have been tested for their ability to reduce ammonia emission. By

flushing, urine is removed from the floor surface and the already formed ammonium is diluted. The flushing frequency and the amount of water used are determining the emission reduction. So, frequent flushing with much water will give the largest effect. However, this will lead to an unacceptable increase in slurry volume. Farmers who have to invest in additional capacity for slurry storage and land application will find flushing systems there for economically less attractive.

Flushing with water in combination with sloped concrete floors is very effective from the point of view of emission reduction. Problems however may be expected with locomotion of animals, because concrete floors can become slippery when faeces are not completely removed from the surface. Recycling of flushing liquid combined with biological or chemical treatment (aeration, acidification) is technically possible in combination with concrete floors with an urine gutter, but this will be economically not feasable. Running costs for necessary hardware and energy are expected to be high. With respect to flushing with diluted fomalin, the acceptability will have to be investigated; formaldehyde concentrations in the air inside the building have to be kept low. The quality of the surface of floors appears to be of high importance to the emission reduction effect, since all crucial emission related processes are taking place there. Recent experiments showed that floors from various producers developed differences in urease activity and ammonia emission. So, the relationship between floor and floor surface characteristics on one hand and urease activity and emission on the other hand need further study.

Acidification of slurry with nitric acid has been under research in various EU countries. When denitrification is prevented, acidification could be a very effective technique to reduce ammonia emission from the house and the slurry storage silo. In addition, landspreading of acidified slurry gives little ammonia emission, as shown before. This means that acidification in a whole-farm approach could result in substantial reduction of ammonia emission.

The emission reducing effect of a tying stall might be explained from the smaller fouled area per cow (approx. 1 m^2) compared to cubicle housing (3.5 m^2). Recent measurements in experimental units, however, showed a substantial smaller reduction in an experiment where a tying stall and a traditional cubicle house with 10 non lactating cows each were compared directly. The difference in findings is yet to be explained.

Other housing systems in general for cows need to be studied not only, from the perspective of the environment, but in combination with animal welfare, labour and costs.

Pig housing

In traditional pens for slaughter pigs and sows with fully and partly slatted floors and long term slurry storage in the pit, in general approx. 30 percent of the emission originates from the floor and approx. 70 percent from the slurry pit. Reduction of the emission can be achieved by:

- fast and complete removal of slurry from the pit;

- reduction of the emitting area.

Fast and complete removal of slurry from the pit. For slaughter pigs, a system with flushing over a sloped pit floor, flushing in combination with a drive out system and flushing in combination with a drain system were studied. The technical principles of the systems are described in detail by Hoeksma et al. (1993). Ammonia emissions could be reduced by approx. 70 percent relative to the traditional system. Flushing was carried out every three hours, using the aerated or acidified (pH 6) watery fraction of separated slurry.

Slurry removal from the pit is also possible using scrapers. Various systems are already on the market. To comply with the requirement of complete slurry removal, technical specifications for scrapers and pit floors are stringent. In a house for weaned piglets, emission reductions of more than 70 percent relative to the traditional system with storage in pits were measured (Groenestein, 1993). For housing of slaughter pigs and pregnant sows, however, only a small emission reduction was achieved (Groenestein, 1994).

Reduction of emitting area. Although the floor on average only contributes a small part in the total emission, the floor emission can vary greatly depending on the extent to which slats and solid floors are fouled with urine and faeces. This pen fouling can be prevented by controlled climatization to minimize the urination and defecation area, combined with a reduction of the percentage of slatted floors from 50 to 25 percent (Aarnink et al., 1993) and by different floor types (Elzing and Swierstra, 1993). At normal temperatures, pigs will lie on the solid floor and perform urination and defecation on the slatted floor. This behaviour changes at high temperatures, when animals attempt to stay cool by laying on slatted floor and urinate and defecate on the solid floor. The emission reduction by minimizing the area with slats, reduction of pen fouling and different floor systems, is currently being investigated.

The emitting pit area can be reduced by covering the slurry surface (Derikx and Aarnink, 1993). Mineral oil and vegetable oil were used as a covering. Results under laboratory conditions showed that the emission from the pit was nearly eliminated. Emission reduction under practical conditions and in combination with a system for regular slurry removal is currently being studied.

Discussion

Flushing with aerated liquid after separation is very effective from the point of view of ammonia emission reduction. After aeration, the liquid will be free of ammonia so fresh slurry that is collected in the liquid will be diluted resulting in lower ammonia emission. The treatment of flushing liquid, however, is highly energy consuming because not only ammonia will be converted, but also organic matter will be oxydized to some extent. Acidification can be carried out with organic and inorganic acids. Organic acids leave no residues in the slurry, but are expensive. Inorganic acids are relatively cheap, but have the disadvantage that from an environmental point of view unwanted anions (nitrate, phosphate or chloride) are added.

The difference in emission reduction through scraping in pig housing can not yet be easily explained. Except for possible differences in lay out of the system, floor characteristics, composition of the slurry and climatic conditions might have been of influence. The research will be continued.

Poultry housing

Battery cages are for several years traditional housing for layer hens in the Netherlands. Manure is stored for longer periods of time under the cages, is moved by scrapers to a closed pit or is removed from the house twice per week by manure belts. In the mid eighties, systems for drying manure under the cages (so called 'deep pits') or drying on belts were introduced, because of a demand for higher quality poultry manure and reduction of costs of transportation. These systems were found to reduce ammonia emission when drying was combined with frequent removal. Broiler chickens are usually kept on litter, Drying of litter is effective in reducing ammonia emission.

Layer hens. The emission level from layer hen houses with battery cages and inside storage of manure are quite high (84 g per bird per day; Kroodsma et al., 1993) because of degradation of uric acid arid ammonia emission during long term storage inside the house. Removing the manure twice weekly by manure belts reduces the emission by 60 percent or more. The same reduction is achieved by drying manure on belts, combined with weekly removal (Kroodsma et al., 1993). The, relatively slow, degradation process then continues during storage. Possibilities for the reduction of ammonia emission during storage of wet or pre dried manure is discussed earlier in this paper.

Daily removal of the manure has a potential of even further emission reduction, since hardly any degradation takes place inside the house (Demmers et al., 1992).

Broilers. Relative to housing on litter, partly and fully wired floors combined with belt removal of the throughfall manure (including drying; Kroodsma, 1993), as well as a wired floor system combined with forced drying of the litter/manure reduces ammonia emissions up to 90 percent (Groenestein, 1993).

Grazing

During summer season, when cows are kept outside permanently or during day time, grazed pastures emit ammonia. Emission from grazed swards is thought to arise mainly from deposited urine. Losses increase with fertilizer application rate. Bussink (1994) reported emission rates that varied between 0.4 and 0.8 kg NH_3 per cow per month in case of a fertilizer application rate of 250 kg N ha^{-1}. In case of 400 kg N ha^{-1} losses varied between 1.4 and 2.4 kg NH_3 per cow per month. These data refer to dairy cows that grazed outside permanently. If animals are kept inside during nighttime, losses from the pastures will be reduced. However, emission from the cow house will take place. According to Voorburg and Kroodsma (1992), emission from a cubicle house where cows were grazing outside during daytime ran up to more than 1 kg per cow per month in May and June, mainly caused by high inside temperatures.

Animal nutrition

An important way of handling not only the nitrogen surplus problem but also the ammonia emission problem might be to adjust the nitrogen intake of the animals more closely to their needs. For cattle urine, Elzing and Kroodsma (1993) reported a linear rela-

tionship between the concentration urea nitrogen and the ammonia emission measured in a scale model system. Smits et al. (1993) found a 39 percent reduction in ammonia emission from a mechanically ventilated cubicle house for dairy cows by substantially lowering the surplus of rumen degradable protein. The reduction in practice might be lower, because diets might vary less than in the experiments. Furthermore, the effects on milk production and animal growth need further study. Still, the results show that adjusting diets to the animals' need has perspective.

Conclusive remarks

Ammonia emission in the Netherlands needs to be reduced in order to protect ecosystems from acidification.

Emission reduction technology for on-farm use has been developed for all major emission sources. In this development, knowledge of basic processes related to and parameters influencing urea degradation and ammonia emission are used.

New technology in slurry application and covering of outside slurry silo's can reduce ammonia emission substantially compared to traditional application techniques and storage. For cow houses, manure handling techniques and floor systems, both separately used and in combination, have perspective from emission reduction point of view. Possible side effects need attention.

Frequent manure removal from pig houses by flushing or scraper systems can lower the emission substantially, although for some animal types results may vary. For poultry houses, quick drying of manure reduces ammonia emissions drastically.

Through nutrition and fertilization management, emissions from grazing and housing systems can be reduced.

References

Aarnink A J A, Koetsier A C and van den Berg A J 1993 Dunging and lying behaviour of fattening pigs in relation to pen design and ammonia emission. *In* Livestock Environment IV - Fourth International Symposium University of Warwick Coventry, England, 6–9 July 1993. Eds. E Collins and C Boon. pp 1176–1183. ASAE, Michigan, USA.

Anonymus 1993 Notice Dutch Manure and Ammonia Policy. Tweede Kamer, vergaderjaar 1992–1993, 19 882, nr. 34. SDU Publishers, Den Haag, The Netherlands. 55p. (*In Dutch*).

Bleijenberg R and Ploegaert J P M 1994 Manual for the IMAG-DLO measuring method for ammonia emissions from mechanically

490

ventilated animal houses. Report 94-1, IMAG-DLO, Wageningen, The Netherlands. 77p. (*In Dutch with English summary*).

Boer W J de, Keen A and Monteny G J 1994 The effect of flushing on ammonia emission from dairy houses; estimates of treatment effects and accuracies by time series analysis. Report 94-6, IMAG-DL0, Wageningen, The Netherlands. 36p. (*In Dutch with English summary*).

Bussink D W 1994 Relationship between ammonia volatilization and nitrogen fertilizer application rate, intake and excretion of herbage nitrogen by cattle on grazed swards. Fert. Res. 38, 111–121.

Demmers T G M, Hissink M G and Uenk G H 1992 Drying poultry manure in a tunnel and the effect on ammonia emission, Report 92-6, IMAG-DLO, Wageningen, The Netherlands. 19p. (*In Dutch with English summary*).

Denmead O T 1983 Micrometeorological methods for measuring gaseous losses of nitrogen in the field. *In* Developments in Plants and Soil Sciences, Vol. 9. Gaseous loss of nitrogen from plant-soil systems. Ed. J R Freney and J R Simpson. pp 1–29. Martinus Nijhoff, Den Haag, The Netherlands.

Derikx P L J and Aarnink A J A 1993 Reduction of ammonia emission from slurry by application of liquid top layers. *In* Nitrogen Flow in Pig Production and Environmental Consequences – Proceedings of the First International Symposium on Nitrogen Flow in Pig Production and Environmental Consequences, Wageningen (Doorwerth), The Netherlands, 8–11 June 1993 (EAAP Publication No. 69). Eds. M W A Verstegen, L A den Hartog, G J M van Kempen and J H M Metz. pp 344–349. Pudoc Scientific Publishers, Wageningen, The Netherlands.

Elzing A, Kroodsma W, Scholtens R and Uenk G H 1992 Measurements of ammonia emission in a model system of cow houses: Theoretical considerations. Report 92-3, IMAG-DLO, Wageningen, The Netherlands. 25p. (*In Dutch with English summary*).

Elzing A and Kroodsma W 1993 Relationship between ammonia emission and nitrogen concentration in cattle urine. Report 93-3, IMAG-DLO Wageningen, The Netherlands. 22p. (*In Dutch with English summary*).

Elzing A and Swierstra D 1993 Ammonia emission measurements in a model system of a pig house. *In* Nitrogen Flow in Pig Production and Environmental Consequences. Proceedings of the First International Symposium on Nitrogen Flow in Pig Production and Environmental Consequences, Wageningen (Doorwerth), The Netherlands, 8–11 June 1993 (EAAP Publication No. 69). Eds. M W A Verstegen, L A den Hartog, G J M van Kempen and J H M Metz. pp 280–285. Pudoc Scientific Publishers, Wageningen, The Netherlands.

Groenestein C M 1993 Animal waste managment and emission of ammonia from livestock housing systems: Field studies. *In* Livestock Environment IV – Fourth International Symposium University of Warwick Coventry, England, 6–9 July 1993. Eds. E Collins and C Boon. pp 1169–1175. ASAE, Michigan, USA.

Groenestein C M 1994 Ammonia emission from pig houses after frequent removal of slurry with scrapers. *In* Proceedings XII World Congress on Agricultural Engineering, Milano, Italy, August/September 1994 (Vol. 1). pp 543–550. CIGR, Merelbeke, Belgium.

Groenestein C M and Reitsma B 1993 Field measurements of the ammonia emission from animal housing systems, Vol. 10, Loose housing with straw bedding for dairy cows. Report no 93-1005, DLO, Wageningen, The Netherlands. (*In Dutch*).

Heij G J and Schneider T 1991 Final report second phase Dutch additional programme for acidification research, Report no. 200-09, RIVM, Bilthoven, The Netherlands. 250p. (*In Dutch*).

Hoeksma P, Verdoes N and Monteny G J 1993 Two options for manure treatment to reduce ammonia emission from pig housing. *In* Nitrogen Flow in Pig Production and Environmental Consequences, Proceedings of the First International Symposium on Nitrogen Flow in Pig Production and Environmental Consequences, Wageningen (Doorwerth), The Netherlands, 8–11 June 1993 (EAAP Publication No. 69). Eds. M W A Verstegen, L A den Hartog, G J M van Kempen and J H M Metz. pp 301–306. Pudoc Scientific Publishers, Wageningen, The Netherlands.

Ketelaars J J M H and Rap H 1994 Ammonia volatilization from urine applied to the floor of a dairy cow barn. *In* Grassland and Society. Proceedings of the 15th General Meeting of the European Grassland Federation, 6–9 June 1994. Eds. L 't Mannetje and J Frame. pp 413–417. Wageningen Pers, Wageningen, The Netherlands.

Kroodsma W, Huis in 't Veld J W H and Groot Kroerkamp P W G 1993 Technical Possibilities for reduction of ammonia emission from poultry houses by manure handling. *In* Reduction of ammonia emissions from housing systems: state of the art – Proceedings of Study Day organized by Technologisch Instituut. KVIV, Merelbeke, september 1993. Groep Voeding en Landbouwtechnologie, Genootschap voor Landbouwtechniek, Merelbeke, Belgium.

Mulder E M and Huijsmans J F M 1994 Reduction of ammonia emission from slurry application review of field experiments 1990–1993. Research into Manure and Ammonia Problems in Animal Husbandry Vol. 8, DLO, Wageningen, The Netherlands. 77p. (*In Dutch*).

Oudendag D A 1993 Reduction of ammonia emission: technical possibilities and costs of emission reduction on national and regional scale. Research report 102, LEI-DLO, Den Haag, The Netherlands. 22p.(*In Dutch*).

Oudendag, D A and Wijnands J H M 1989 Reduction of ammonia emission from animal wastes: Technical possibilities and costs. Research report 56, LEI-DLO, Den Haag, The Netherlands. 72p. (*In Dutch*).

Smits M C J, Valk H, Elzing A, Huis in 't J W H Veld and Keen A 1993 Perspectives of reduction of ammonia emission from dairy cattle houses by attuning the ration. Report 93-31, IMAG-DLO, Wageningen, The Netherlands. 38p. (*In Dutch with English summary*).

Swierstra D, Huis in 't Veld J W H, Kroodsma W and Smits M C J 1994 Ammonia emission and roughness of slatted floors and concrete floors in cubicle houses for dairy cows. Report 93-12, IMAG-DLO, Wageningen, The Netherlands. 26p. (*In Dutch with English summary*).

Voorburg J H and Kroodsma W 1992 Volatile emissions from cattle housing. Livestock Prod. Sci. 31, 67–70.

Plant and Soil **181**: 123–129, 1996.
© 1996 *Kluwer Academic Publishers.*

Laboratory measurements and simulations of ammonia volatilization from urea applied to calcareous Chinese loess soils

M. Roelcke[1], Y. Han[2], S.X. Li[3] and J. Richter[1]

[1]*Institute of Geography and Geoecology, Technische Universität Carolo-Wilhelmina, Langer Kamp 19 c, D-38106 Braunschweig, Germany*, *[2]*Nanjing Institute of Soil Science, Chinese Academy of Sciences, 210008 Nanjing, P.R. China and *[3]*Department of Soil Science, Northwestern Agricultural University, 712100 Yangling, Shaanxi, P.R. China*

Key words: ammonia volatilization, calcium carbonate, China, loess, simulations, urea

Abstract

Ammonia volatilization is the major pathway for mineral nitrogen loss in the calcareous soils of the Chinese loess plateau, with maximum losses reaching 50% of the fertilizer-N applied. A volatilization-diffusion experiment was carried out in the laboratory using a forced-draft system and soil columns of 15.5 cm depth. Urea was surface applied at rates of 210 kg N ha^{-1} to a soil with 10% $CaCO_3$ and a pH of 7.7. The amount of ammonia volatilized as well as the concentration profiles of ammoniacal-nitrogen and soil pH in the upper 50 mm of the soil columns after 4, 7 and 10 days were measured and subsequently modelled. The mechanistic model of Rachhpal-Singh and Nye, originally developed for neutral, non-calcareous soils, was modified to include the pH-buffering action of the soil carbonates. Model parameters were independently determined or taken from the literature. Measured and predicted cumulative NH_3 losses agreed very well in the first 10 days following fertilizer application. However, in contrast to the simulations, NH_3-volatilization was still proceeding in the experiment even after 13 days, with cumulative losses reaching 60% of the applied N. In addition to the high initial soil pH, the low bulk density and high volumetric air content of the soil columns used for the experiment proved decisive for the high rates of ammonia volatilization, provoking a strong increase in the amount of ammoniacal-N diffusing towards the soil surface as gaseous NH_3. The simulations showed that due to the high soil pH, the buffering action of the soil carbonates played a comparatively smaller role.

Introduction

The region at the southern edge of the loess plateau in Shaanxi Province, China, has a subhumid climate and topsoils with a pH of 7.7 and high carbonate content. A double crop winter wheat (*Triticum aestivum* L.) - summer maize (*Zea mays* L.) rotation is practiced. Ammonium bicarbonate (NH_4HCO_3) and urea are the predominant mineral nitrogen fertilizers, applied at average rates of about 150 (maximum 200) kg N ha^{-1} per crop, usually in a single application. The forms of N-fertilizer and these high application rates lead to ammonia volatilization as main pathway of nitrogen loss. As much as 32% of urea-N applied can be lost by volatilization from calcareous soils in northern China (Zhang et al., 1992), and losses from NH_4HCO_3 applied to summer crops usually exceed

those from urea. Extensive in situ and laboratory measurements of NH_3-volatilization using a forced-draft system have been carried out in earlier experiments (Roelcke, 1994). In the laboratory, total NH_3 losses after 17–21 days amounted to 66% and 51% of the surface applied N for NH_4HCO_3 and urea respectively. In field trials using ^{15}N-labelled urea, the amounts of unaccounted for fertilizer-N ranged from 43% to 62% for summer maize and from 36% to 46% for winter wheat (Rees et al., 1996).

NH_3-volatilization rates are controlled by many different processes, with significant interactions occurring between variables (Kirk and Nye, 1991). The relationships between NH_3-volatilization, soil pH and the buffering action of $CaCO_3$ have been described both theoretically and experimentally (Avnimelech and Laher, 1977; Ferguson et al., 1984; Vlek and Stumpe,

Plant and Soil is the original source of publication of this article. It is recommended that this article is cited as: *Plant and Soil* **181**: 123–129, 1996.

124

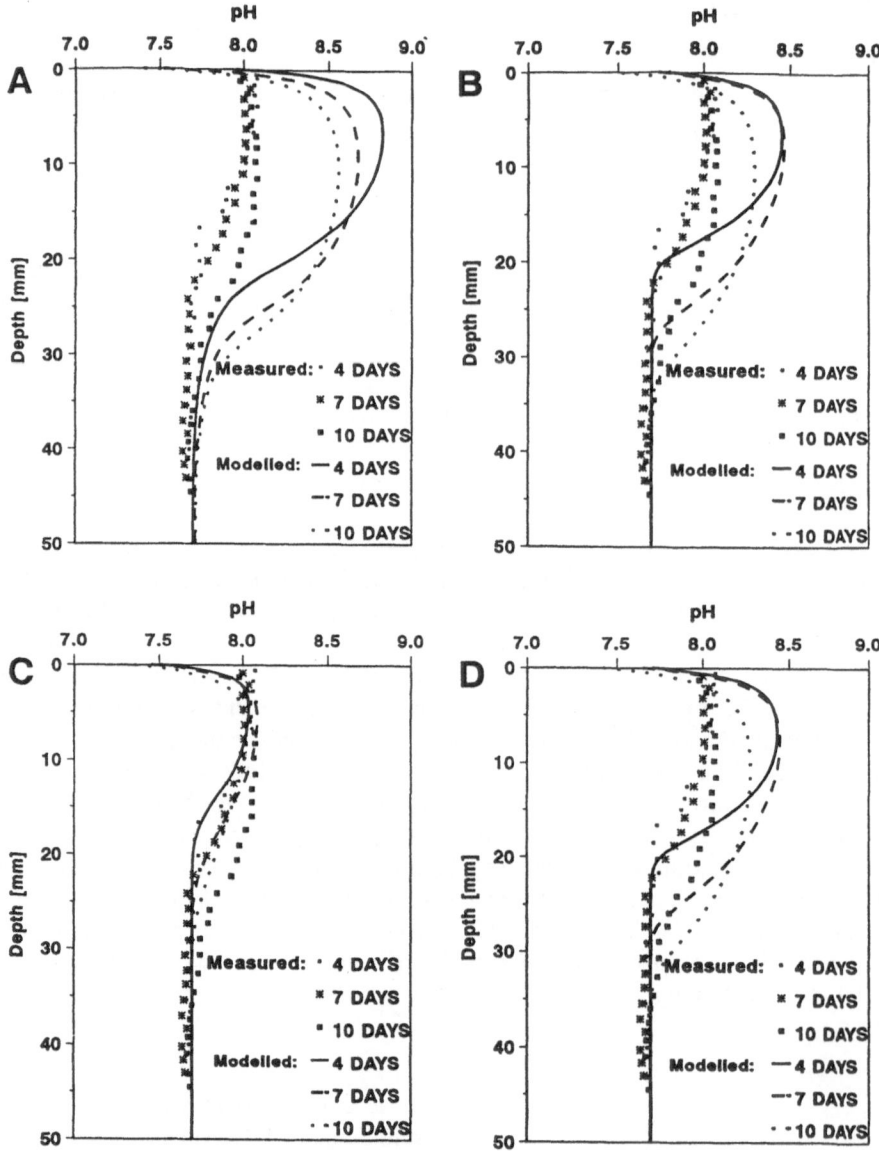

Figure 1. Measured and modelled soil pH profiles in the soil columns at three different diffusion times after surface application of urea (simulations A-D).

1978). For a neutral, carbonate-free soil, the diffusion of HCO_3^--ions to the soil surface, resisting the decrease in surface pH caused by the reaction $NH_4^+ \leftrightarrow NH_3 \uparrow +H^+$, was shown to be the main rate limiting process of NH_3-volatilization in Rachhpal-Singh and Nye's (1986) measurements and simulations. So far, existing simulation models of NH_3-volatilization do not properly account for the effect of high soil $CaCO_3$-contents on changes in soil pH resulting from urea hydrolysis and ammonia volatilization. The purpose of this study was to examine the influence of a high

$CaCO_3$-content and soil pH as well as relatively dry soil conditions on ammonia volatilization following urea application.

Materials and methods

Experiment

A laboratory volatilization-diffusion experiment with forced-draft system, similar to the one described by

Rachhpal-Singh and Nye (1986) was carried out. Four volatilization chambers were running in parallel, each with an area of 6 cm × 12.5 cm, a depth of the soil columns of 15.5 cm and a headspace volume of 255 cm³. The soil used for the experiments was a highly calcareous ($\approx$ 10% $CaCO_3$) Udic Haplustalf (USDA, 1994) developed from loess, with a silt loam texture and a pH (0.01 M $CaCl_2$) of 7.7. Its CEC was 164 meq kg⁻¹ (replacement of all exchangeable cations with a 0.1 M $BaCl_2$ solution buffered at pH 8.1, Thomas, 1982; subsequent replacement of Ba-ions with $MgCl_2$ solution and determination of Ba-ions via AAS). Total N content was 803 μg g⁻¹, NH_4^+-N_{fix} (method A by Silva and Bremner, 1966) was 422 μg g⁻¹, C_{org} (dry combustion after removal of free carbonates; CO_2 measured coulometrically with a Ströhlein Coulomat 701, Balesdent et al., 1988) was 0.65% (w/w). Soil pretreatment was carried out as described in Rachhpal-Singh and Nye (1986), with addition of 50 μg g⁻¹ soil 4-amino-1,2,4-triazole (ATC), a nitrification inhibitor. The soil bulk density ρ in the upper 50 mm of the columns was only 0.95 kg dm⁻³ and the gravimetric water content 15.0%. The corresponding relative porosity ϕ was 0.642, volumetric water content θ was 0.143 and the volumetric air content θ_g 0.499. Laboratory temperature was 18.0 ± 1.1 °C. Urea was surface applied at rates of 210 kg N ha⁻¹, an amount in the upper range of farmers' practice in Shaanxi Province. Precleaned, water-saturated air was passed over the soil in the chambers, with an air flow rate of 16.3 exchange volumes min⁻¹ (equalling a wind speed of 0.034 m s⁻¹). Ammonia was trapped in 2 subsequent flasks, each containing 200 cm³ of 0.0125 N H_3PO_4; samples were analyzed for NH_4^+-N using a CHEM-LAB continuous-flow autoanalyzer. Four, seven and ten days after urea application, the soil column of one of the chambers was sectioned at approximately 0.75 mm increments down to 42.5 mm depth. Alternate slices were analyzed for NH_4^+-N and pH as described in the original experiment. Unhydrolyzed urea was not determined.

Simulations

To predict NH_3-volatilization and concentration profiles of urea-N and ammoniacal-N as well as soil pH profiles, simulations were run with the mechanistic model of Rachhpal-Singh and Nye (1986), which only considers diffusive transport processes. Urease activity of this soil was not determined independently. The Arrhenius equation was used to adapt the rates of ure-

ase activity used in the original model (for 25 °C) to the temperature of 18 °C as described in Sadeghi et al. (1988). Using the proper values for θ and θ_g, the solution diffusion (f_l) and gaseous (f_g) diffusion impedance factors were calculated from empirical relationships taken from the literature (Millington and Quirk, 1961; Nye, 1979). Soil pH buffer capacity was determined by serial titration with 0.05 M NaOH and 0.06 M HCl and an equilibration time of 72 h. A linear approximation gave a value of 0.03 mol OH⁻ kg⁻¹ soil pH⁻¹ in the pH range 7.4–9.4, which is similar to the value in the original model. The original values were used for the Freundlich NH_4^+ adsorption isotherm. Four different simulation runs were compared:

A) The model's standard values were used, with the initial pH set at 7.7: ρ=1.5 kg dm⁻³, θ=0.296, θ_g=0.114, f$_l$=0.29, f$_g$=0.17.
The soil pH buffer capacity was kept unchanged at 0.03 mol OH⁻ kg⁻¹ soil pH⁻¹.

B) The parameters determined as described above were used: ρ=0.95 kg dm⁻³, θ=0.143, θ_g=0.499, f$_l$=0.06, f$_g$=0.24.
The initial pH was 7.7, the soil pH buffer capacity was kept unchanged.

C) The soil pH buffer capacity was increased three-fold, other parameters as in B.

D) The processes of $CaCO_3$ precipitation and dissolution following pH changes were deterministically included in the model. The two ionic species Ca^{2+} and CO_3^{2-}, and the solubility product of $CaCO_3$ (K_{sp}) and dissociation constant of HCO_3^- (K_2) were newly incorporated. Other parameters as described in B, with the pH buffer capacity unchanged. Instead of the Richtmyer-algorithm used in the original model, a block tridiagonal matrix method with *LU* decomposition procedure (Press et al., 1992) was used for computing the numerical solution.

Results

Figure 1 shows the measured and modelled soil pH profiles after 4, 7 and 10 days in the upper 50 mm of the soil columns. Due to urea hydrolysis, the measured profiles showed an increase of 0.4 pH units to a maximum of 8.07 at the soil surface four days after urea application. Although not very pronounced, a pH drop (0.1 pH units) was observed in the upper 6 mm of soil after 10 days, due to the process of NH_3-volatilization. Figure 2 shows the measured and modelled concen-

126

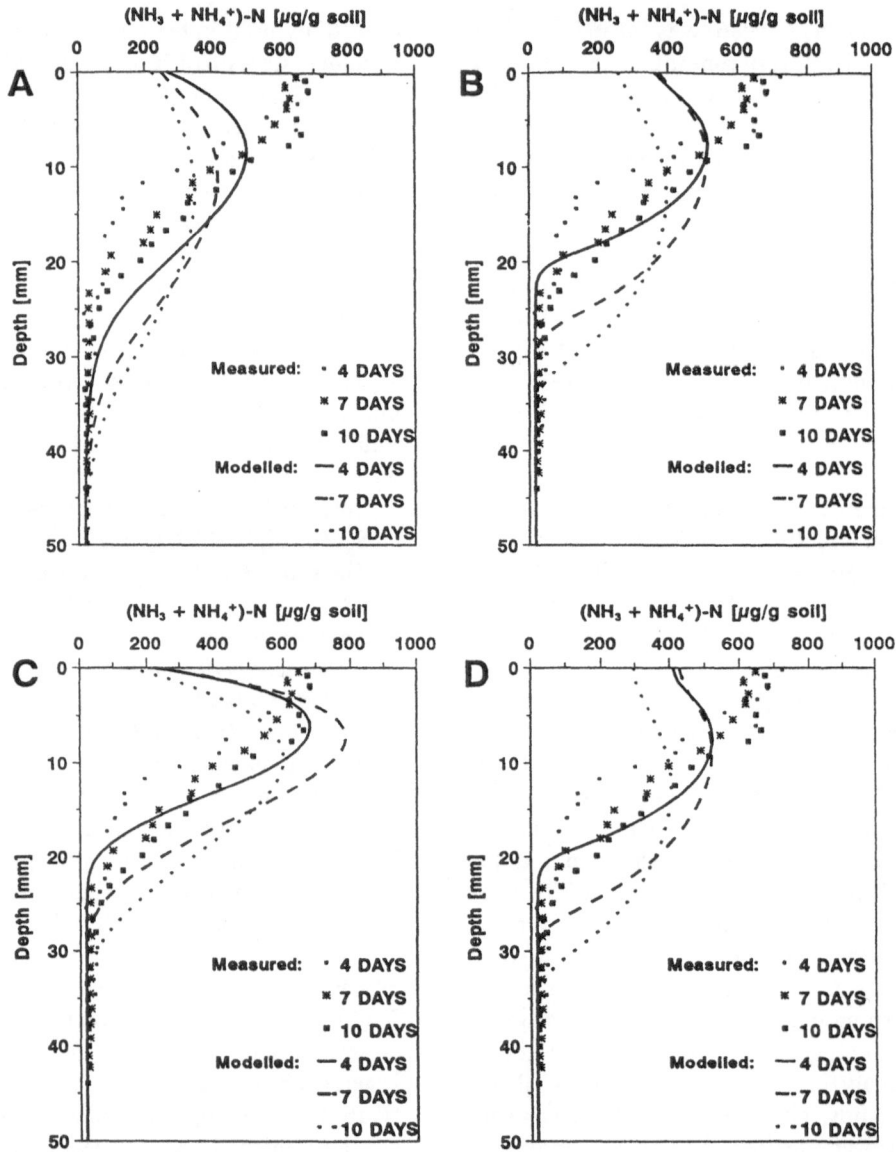

Figure 2. Measured and modelled concentration profiles of ammoniacal-N in the soil columns at three different diffusion times after surface application of urea (simulations A-D).

tration profiles of ammoniacal-N after a diffusion time of 4, 7 and 10 days in the upper 50 mm of the soil columns. The marked concentration increase between the 4th and the 7th day, despite NH_3-volatilization losses, was due to the ongoing urea hydrolysis. Figure 3 shows the modelled concentration profiles of unhydrolyzed urea-N at 2, 4 and 7 days after urea application. At 18 °C, most of the urea was hydrolyzed after 7 days. Since urease activity is not affected by soil pH in this model, the modelled urea profiles were identical in B, C and D. Urea diffused to almost 30 mm depth in

simulation A, but hardly moved below 15 mm in simulations B-D. Figure 4 shows the measured cumulative ammonia losses in the four chambers as well as the modelled cumulative losses for the different simulation runs. Highest NH_3 fluxes were measured after 4–7 days, with total cumulative losses after 13 days reaching 60% of the total applied N. The best agreement between measured and predicted cumulative NH_3 losses was obtained using simulation C. However, while simulated volatilization rates were beginning to decline after 10 days, NH_3-volatilization was still proceeding

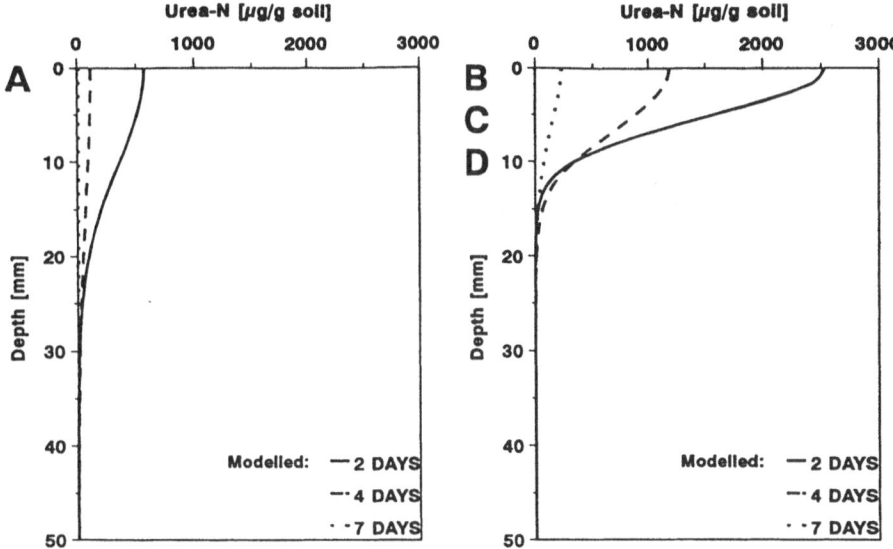

Figure 3. Modelled concentration profiles of urea-N in the soil columns at different diffusion times following surface application of urea (simulations A and B-D).

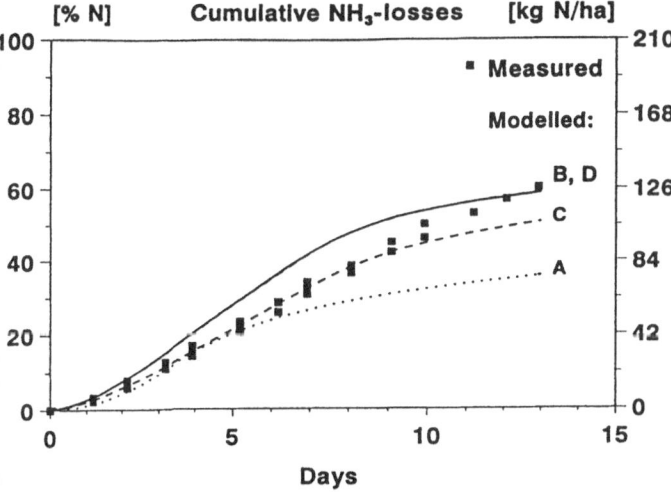

Figure 4. Measured and modelled cumulative NH₃-volatilization losses in the 4 volatilization chambers (simulations A-D).

in the experiment even after 13 days. Correspondingly, modelled concentration profiles of ammoniacal-N (Fig. 2) all showed a marked decrease near the soil surface, while no such depletion of the upper soil layer was measured.

Discussion

Since urea is not adsorbed, all of it is present in the liquid phase (Kirk and Nye, 1991). Due to the lower water content in our experiment, urea diffused less deep into the soil and urea hydrolysis took place nearer to the soil surface in simulations B-D than in simulation A. Correspondingly, the increase in soil pH and higher ammoniacal-N concentration were confined to a more shallow layer at the surface in simulations B-D than in simulation A. The low bulk density and the high soil air content in our experiment led to an increase in gaseous NH₃ diffusion and a reduction of liquid phase diffusion. The product $(\theta f_l) = 0.009$ was ten times smaller and the product $(\theta_g f_g) = 0.120$ was more than six times greater than in the original experiment of Rachhpal-Singh and Nye (1986). Simulations by Rachhpal-Singh (1987) have shown that at pH values of 8 and (θf_l) values < 0.042, gas phase diffusion of NH₃ is the dominating pathway of movement of ammoniacal-N through soil. Since the diffusion coefficient of NH₃ in air is about 4 orders of magnitude higher than that of ammoniacal-N in solution, soil drying leads to a strong increase in NH₃-volatilization (Kirk and Nye, 1991). The effect of the higher volumetric air content is very pronounced and explains the marked increase in NH₃-volatilization losses in simulations B-D as compared to simulation A. The processes of convective transport of urea and ammoniacal-N towards the soil surface caused by soil water evaporation are not included in the simulation model. Situations similar to the one described are frequently occurring in the area under investigation, with dry topsoils and a low soil bulk density (1.25-1.4 kg dm⁻³ in the Ap horizon).

Measured and modelled changes in pH profiles (simulations A, B and D) showed the same tendency, although measured pH changes were much less pronounced. The threefold increase in pH buffer capacity (simulation C) buffered against an increase in pH very well, but did not prevent the drop in soil pH at the soil surface due to NH_3-volatilization, which was even greater than in simulation B. This suggests that the description of the pH buffer capacity in Rachhpal-Singh and Nye's model may be incomplete. Since NH_3 losses are highly sensitive to soil pH, the greater rise in pH in simulations B and D led to the steeper increase in NH_3-volatilization and higher total losses in simulations B and D than in C (Fig. 4).

Simulation D showed hardly any effect of the buffering action of $CaCO_3$ against the increase in pH caused by urea hydrolysis. The high initial soil pH value is close to the upper limit of the buffering range of the $CaCO_3$ - H_2O - $CO_{2(gas)}$ system ($K_{sp} = 10^{-8.3}$). In the pH range between 7.7 and 8.1 and a simulated P_{CO2} between 0.3 and 3 mbar in the upper 50 mm of the soil columns, the concentration of Ca^{2+}-ions in soil solution is already very low ($\approx 10^{-2} M$ - $10^{-4} M$). The amount of protons released in connection with the precipitation of $CaCO_3$ (approximately one mole of H^+ per mole of $CaCO_3$) is insufficient to resist the pH increase following urea hydrolysis. Moreover, the soil pH buffer capacity determined experimentally was smaller at pH values > 7.8 and showed a greater dependency on the equilibration time (24–120 h) than at pH < 7.8, suggesting that precipitation of $CaCO_3$ is a rather slow process. Therefore, the pH and (NH_3 + NH_4^+)-N profiles of simulations B and D at the different diffusion times were almost identical, resulting in no difference in total NH_3 losses.

Conclusions

The effect of the high volumetric air content in our experiment on NH_3-volatilization rates was the most pronounced, provoking a strong increase in the amount of ammoniacal-N diffusing to the soil surface as ammonia. The very high initial soil pH also proved decisive for the high NH_3 losses. By comparison, due to the high soil pH, the buffering action of the soil carbonates only played a smaller role. However, since the equilibration time strongly affected the measured pH buffer capacity of our soil, the kinetics of $CaCO_3$ precipitation and dissolution should be considered in future simulations. While the processes of H_2CO_3 and HCO_3^- dissocia-

tion are treated as instantaneous in this model, it is possible that a kinetic term may have to be included in order to adequately describe $CaCO_3$ precipitation and dissolution for this soil.

Acknowledgements

The authors wish to express thanks to Mr Peter H Nye, Oxford and Dr Rachhpal-Singh, Ludhiana, India, for the provision of a copy of their computer program. The experiments were carried out at the Northwestern Agricultural University in Yangling, Shaanxi Province, China; we thank the university for logistical support. This work was supported by the German Science Foundation (DFG Ri 269/25–1,2), the Volkswagen Foundation (VW I/65 586) and the Chinese Ministry of Agriculture.

References

Avnimelech Y and Laher M 1977 Ammonia volatilization from soils: Equilibrium considerations. Soil Sci. Soc. Am. J. 41, 1080–1084.

Balesdent J, Wagner G H and Mariotti A 1988 Soil organic matter turnover in long term field experiments as revealed by carbon-13 natural abundance. Soil Sci. Soc. Am. J. 52, 118–124.

Ferguson R B, Kissel D E, Koelliker J K and Basel W 1984 Ammonia volatilization from surface-applied urea: Effect of hydrogen ion buffering capacity. Soil Sci. Soc. Am. J. 48, 578–582.

Kirk G J D and Nye P H 1991 A model of ammonia volatilization from applied urea. V. The effects of steady-state drainage and evaporation. VI. The effects of transient-state water evaporation. J. Soil Sci. 42, 103–125.

Millington R J and Quirk J P 1961 Permeability of porous solids. Trans. Faraday Soc. 57, 1200–1207.

Nye P H 1979 Diffusion of ions and uncharged solutes in soils and soil clays. Adv. Agron. 31, 225–272.

Press W H, Teukolsky S A, Vetterling W T and Flannery B P 1992 Numerical recipes in C: The art of scientific computing. 2nd ed. Cambridge University Press, Cambridge, UK. 994 p.

Rachhpal-Singh 1987 Predicting the effect of soil-water-air dynamics on ammonia volatilization from applied urea with a mechanistic model. Fert. Res. 13, 277–285.

Rachhpal-Singh and Nye P H 1986 A model of ammonia volatilization from applied urea. I. Development of the model. II. Experimental testing. III. Sensitivity analysis, mechanisms, and applications. J. Soil Sci. 37, 9–40.

Rees R M, Roelcke M, Li S X, Wang X Q, Li S Q, Stockdale E A, McTaggart I P, Smith K A and Richter J 1996 The effect of fertilizer placement on nitrogen uptake and yield of wheat and maize in Chinese loess soils. Fert. Res. (Submitted).

Roelcke M 1994 Die Ammoniak-Volatilisation nach Ausbringung von Mineraldünger-Stickstoff in carbonatreichen chinesischen Löß-Ackerböden. Ph.D. thesis, Braunschweig Technical University. Göttinger Beiträge zur Land- und Forstwirtschaft in den Tropen und Subtropen, Vol. 92, Erich Goltze, Göttingen, Germany. 194 p.

Sadeghi A M, McInnes K J, Kissel D E, Cabrera M L, Koelliker J K and Kanemasu E T 1988 Mechanistic model for predicting ammonia volatilization from urea. *In* Ammonia Volatilization from Urea Fertilizers. Eds. B R Bock and D E Kissel. pp 67–92. Bulletin Y-206, NFDC, Tennessee Valley Authority, Muscle Shoals, Alabama, USA.

Silva J A and Bremner J M 1966 Determination and isotope-ratio analysis of different forms of nitrogen in soils: 5. Fixed ammonium. Soil Sci. Soc. Am. Proc. 30, 587–594.

Thomas G W 1982 Exchangeable cations. *In* Methods of Soil Analysis, Part 2. Eds. A L Page, R H Miller and D R Keeney. pp 159–165. Agronomy 9, pp 159–165. American Society of Agronomy, Madison, WI, USA.

Vlek P L G and Stumpe J M 1978 Effects of solution chemistry and environmental conditions on ammonia volatilization losses from aqueous systems. Soil Sci. Soc. Am. J. 42, 416–421.

United States Department of Agriculture, Soil Conservation Service (ed.) 1994 Keys to Soil Taxonomy. Sixth Edition.

Zhang S L, Cai G X, Wang X Z, Xu Y H, Zhu Z L and Freney J R 1992 Losses of urea-nitrogen applied to maize grown on calcareous fluvo-aquic soil in North China Plain. Pedosphere 2, 171–178.

Section editor: R Merckx

7. ASPECTS RELATED TO NITRIFICATION AND DENITRIFICATION

Plant and Soil **181**: 7–12, 1996.

Production of N_2O in soil during decomposition of dead yeast cells with different spatial distributions

Per Ambus

*Department of Population Biology, Copenhagen University, DK-2100, Denmark. Present address: Plant Nutrition, Environmental Science and Technology Department, Risø National Laboratory, PO Box 49, DK-4000, Denmark**

Key words: dead yeast cells, denitrification, nitrification, N_2O production, spatial distribution

Abstract

Production and sources of N_2O were determined in soil columns amended with autoclaved yeast cells either mixed into or added as 0.5 cm^3 lumps to the soil in combination with no or 200 μg NO_3^--N g^{-1}. At four occasions over a two-week study period, subsets of cores were measured for N_2O production during 4-hour incubations under atmospheres of ambient air, 10 Pa of C_2H_2, and N_2, respectively. Denitrification enzyme activity (DEA) was assessed in subsamples of cores that had been incubated continuously under air.

Autoclaved yeast provided a C-source readily available for denitrifying bacteria in the soil. Nitrous oxide production was negligible in unamended columns whereas accumulated N_2O losses in the presence of yeast material were substantial, varying between 15 to 49 ng N_2O-N g^{-1} h^{-1}. Mixing yeast into the soil caused the highest production of N_2O followed by the yeast lump and no yeast treatments. Incubation in the presence of 10 Pa C_2H_2 indicated that denitrification was the sole source of N_2O, in accordance with an increase in DEA. Nitrous oxide production and DEA peaked after 4–7 days of incubation, and both were unaffected by additional NO_3^-. Two- to four-fold responses to anaerobiosis and accumulation of NO_3^- and NH_4^+ in proximity of the lumps indicated that N_2O production here was limited by relatively low C-availability. In contrast, 10- to 12-fold responses to anaerobiosis and no accumulation of inorganic N suggested a higher C-availability where yeast was mixed into the soil.

Introduction

Biological nitrification and denitrification are the dominant sources of N_2O production in soil. Nitrifier N_2O production is due to the reduction of NO_2^- under microaerophilic conditions (Goreau et al., 1980; Poth and Focht, 1985); N_2O, along with N_2, is produced by the denitrification process under anaerobic conditions (Tiedje, 1988). Since N_2O originates from reductive processes, the N_2O production is anticipated to increase as O_2 availability decreases. However, while this is true for the denitrification process (Tiedje, 1988) it is not certain whether a lowered O_2 availability translates into enhanced N_2O production by nitrification since the overall process may be limited at low O_2 tensions (Firestone and Davidson, 1989).

Enhanced N_2O production has been observed in response to amendments with plant residues (Aulakh et al., 1991; McKenney et al., 1993; Rolston et al., 1982) and glucose (Bergstrom et al., 1994). However, distinctions between nitrification and denitrification as sources of N_2O production have rarely been attempted. Extremely high denitrification activity in soil has been associated with O_2 depletion during rapid decomposition of particulate organic material (Christensen et al., 1990; Parkin, 1987). Depending on the C/N ratio of the material and the oxygen concentration, nitrification may increase significantly in the microenvironment of decomposing organic matter supplying additional NO_3^- for denitrification (Henriksen and Larsen, 1991; Rice et al., 1988). It is not known if this increase in nitrification will also contribute to the N_2O formation of such organic hot spots. The ratio of N_2O:NO_3^- produced via nitrification is usually below 0.01 (Firestone and Davidson, 1989). Significant nitrification N_2O production may occur under certain circumstances, e.g.

---

* FAX No.: +4546323383

Plant and Soil is the original source of publication of this article. It is recommended that this article is cited as: *Plant and Soil* **181**: 7–12, 1996.

in urea fertilized soils (Duxbury and McConnaughey, 1986).

The aim of my experiment was to determine the total N_2O production, including the contribution of nitrification to total N_2O production, as influenced by different spatial distributions of dead yeast cells in soil columns.

Materials and methods

The soil was a sandy loam (pH(H_2O) of 7.6; 7.5% OM; 72% sand; 11% silt; 17% clay; 26 μg NO_3^--N g^{-1}; 2 μg NH_4^+-N g^{-1}) cropped to winter wheat (*Triticum aestivum* Vill.). Samples were collected randomly at 0–15 cm depth on June 3, 1993, bulked, sieved < 4mm and stored field moist (14.3% H_2O w/w) at 5 °C until incubations began four days later. Two bulk samples were moistened to 60% FC (26% H_2O w/w) by adding water without and with KNO_3-N providing 200 μg N g^{-1} dry soil; two other bulk samples were moistened to 60% FC by mixing with a suspension of autoclaved yeast (47.2% C; 8.2% N) providing 5.6 mg dry matter g^{-1} soil ± 200 μg NO_3^--N g^{-1}, respectively. Thirty-g (dry weight) portions of each of the four bulk samples were subsequently transferred to decapitated 60-mL disposable syringes, sealed at one end by butyl stoppers, giving a 50 mm high soil column at a bulk density of 1.1 g cm^{-3}. A third yeast treatment was made by adding a lump of 0.5 cm^3 yeast paste, providing 5.6 mg dry matter g^{-1} soil, into the center of water amended columns ±NO_3^-. The cylinders were incubated at 20 °C in the dark.

At day 1, 4, 7 and 14, respectively, 3 duplicate sets of cylinders of each treatment were sealed for subsequent incubation under 3 different atmospheres: 1) ambient air, 2) air with 10 Pa of C_2H_2, and 3) N_2. Incubation under C_2H_2 was generated by mixing vigorously 5 mL of 0.1% C_2H_2 with the cylinder gasphase by syringe followed by removal of 5 mL headspace gas. Anaerobiosis was obtained by exchanging the headspace with N_2 3 times. Three 3-mL gas samples were withdrawn at 1, 3, and 5 h after sealing to measure N_2O in the gasphase. Each sampling was preceded by injecting 3 mL of air or N_2 to restore pressure. The gas samples were stored in 2-mL pre-evacuated crimp seal vials until analysis for N_2O which took place on a Shimadzu 8A gas chromatograph equipped with [63]Ni EC-detector and Porapak Q column operated at 340 °C and 35 °C, respectively, and Ar/CH_4 (5/95) carrier at 60 mL min^{-1}. Immediately after the last gas sam-

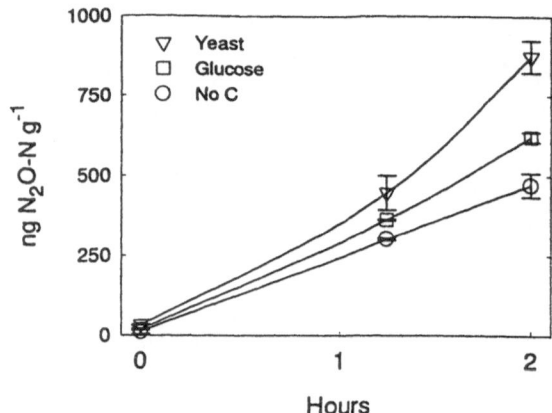

Figure 1. Denitrification in soil slurries amended with NO_3^- and no C, glucose or autoclaved yeast, respectively. Means of duplicate samples; vertical lines indicate ± SD.

pling the soil from the columns incubated under air was removed and bulked for the no yeast and mixed yeast treatments. Yeast lump amended columns were split into 3 sections taken at 0-20, 20-30, and 30-50 mm depth, respectively. The 0–20 mm and 30-50 mm depth sections were bulked and are termed the distal compartment. The 20-30 mm section, termed the proximal compartment, contained the yeast lump. Duplicate subsamples of each soil portion were analyzed for denitrification enzyme activity (DEA; Smith and Tiedje, 1979) using a slurry of 10 g soil mixed with 10 mL of substrate (0.17 g KNO_3-N L^{-1}, 1.8 g glucose-C L^{-1} and 1 g chloramphenicol L^{-1}) under an atmosphere of 90% N_2 and 10% C_2H_2. Nitrate and ammonium was measured in 0.1 M KCl extracts of 10-g soil portions (2:1 vol:w) on a Tecator Aquatec Ion Analyzer; only one sample was included for the analysis of inorganic N in the proximal soil compartment. Gravimetric soil moisture was obtained by oven drying (105 °C, 24 h).

A preliminary experiment was conducted to assess the availability of the autoclaved yeast for denitrification. This was done by measuring denitrification in duplicate soil slurries incubated as for the DEA assay, and amended with no C, 1.8 g glucose-C L^{-1}, and yeast dry matter equivivalent to 2.4 g C L^{-1}, respectively.

Results

The amount of N denitrified over a 2 hour period in the soil slurry containing autoclaved yeast was 79% and 39% higher than denitrified N in soil slurries without C and with glucose-C, respectively (Fig. 1). This indi-

Table 1. Nitrous oxide production in repacked soil columns following short-term incubations under different atmospheres at day 1, 4, 7, and 14 after amendment with autoclaved yeast without or with additional NO_3^-. Data are weighted arithmetic means($\pm$SE) of four duplicate sets of columns (n=8)

Atmosphere	Treatment		
	No C	Yeast lump	Mixed yeast
	($ng\ N_2O\text{-}N\ g^{-1}\ h^{-1}$)		
No NO_3^-			
Air	$\ll 1e^z$	24bc	15cd
10 Pa C_2H_2	$<0e$	27bc	9cde
N_2	$<1de$	92ab	185ab
With 200 $\mu g\ g^{-1}\ NO_3^-$-N			
Air	$\ll 1e$	33abc	49bc
10 Pa C_2H_2	$<0e$	66abc	61abc
N_2	$<1e$	76ab	488a

^{z} Means with the same letter are not significantly different (LSD on LOG-transformed data; $p<0.05$; SAS 1990).

cate that autoclaved yeast provides a C-source readily available for the denitrifying bacteria present in this soil.

Nitrous oxide production throughout the incubation period was negligible in soil columns without yeast (Fig. 2 top and Table 1). The unamended columns only produced appreciable amounts of N_2O when oxygen had been removed, and only within the first 4 days; the mean activity remained below 1 ng N_2O-N g^{-1} h^{-1}. The N_2O production in yeast amended columns peaked after 4–7 days of incubation except when measured in soil with yeast lumps under anaerobiosis in which a high N_2O production persisted for at least two weeks (Figs. 2 mid and bottom). In the presence of O_2 there was almost no detectable N_2O production at day 14.

The amount of N_2O produced in columns exposed to 10 Pa of C_2H_2 did not differ from N_2O produced in columns exposed to air for any treatment (Fig. 2). Removal of oxygen increased N_2O production 2- to 4-fold in soil with yeast lumps and 10- to 12-fold in soil mixed with yeast, respectively (Table 1).

The mean N_2O production increased 2–3 orders of magnitude in response to the two yeast amendments (Table 1). The highest production of N_2O was observed when yeast was mixed into the soil, however, the difference between the two yeast treatments was ambiguous and insignificant. The N_2O production tended to be higher, though not significantly, with added NO_3^-.

The DEA increased, up to 8-fold on day 4, in response to mixing the yeast material into the soil

(Fig. 3). A similar increase of DEA was observed in proximity of the yeast lumps, whereas DEA in distal compartments remained at about 580 ng N_2O-N g^{-1} h^{-1}, comparable to the 450 ng N_2O-N g^{-1} h^{-1} recorded in the control columns. Additions of NO_3^- did generally not influence the DEA except on day 1 and day 7 in proximity of yeast lumps, and on day 4 in soil mixed with yeast; in either case the lowest DEA was observed at elevated NO_3^- ($p<0.05$; LSD of SAS 1990). The DEA in soil mixed with yeast and in proximity of yeast lumps peaked around day 4 (Fig. 3), concomitantly with the peak in N_2O production in columns incubated in air (Figs. 2 mid and bottom); at day 14, however, the DEA in both treatments was still about 3-fold higher than at day 1, in contrast to the negligible N_2O production in soil columns (Fig. 2). In unamended soil as well as in distal compartments, the DEA did not vary throughout the experiment.

Nitrate concentrations increased markedly at day 14 in both proximal and distal compartments of the columns amended with yeast lumps (Fig. 4). This increase in NO_3^- occured concomitantly with a vigourous increase in NH_4^+ in proximity of the yeast lumps. In unamended soil and soil mixed with yeast, the NO_3^- concentration remained rudely at the initial level. This was also the case for NH_4^+ in unamended soil, whereas NH_4^+ in soil mixed with yeast increased temporary within day 1–4 whereafter it declined to the initial level (Fig. 4).

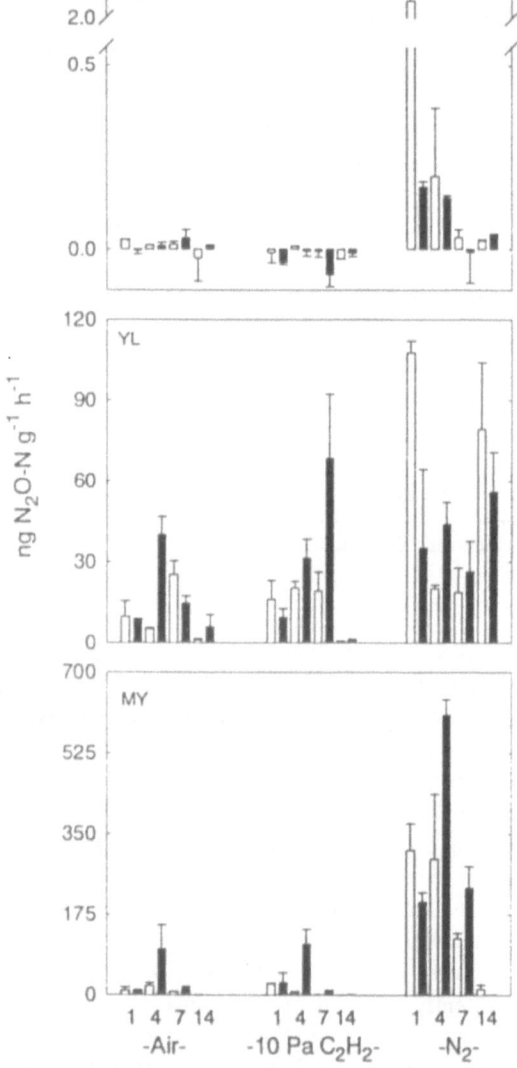

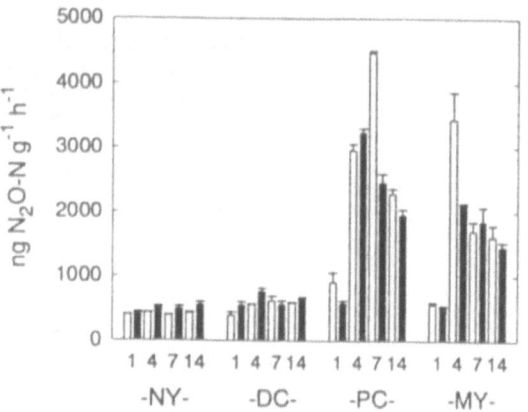

Figure 3. Denitrification enzyme activity in soil columns without yeast (NY), distal (DC) and proximal (PC) compartments of columns amended with yeast lumps, and columns mixed with yeast (MY), respectively, and without (□) and with (■) additional 200 μg NO_3^--N g^{-1}. The DEA was measured four times over a two-week period indicated by day number on the horizontal axis. Means of duplicate samples; vertical lines indicate one SD.

Figure 2. Nitrous oxide production in soil columns without yeast (NY), columns amended with yeast lumps (YL), and columns mixed with yeast (MY), respectively, and without (□) and with (■) additional 200 μg NO_3^--N g^{-1}. The N_2O production was measured under atmospheres of air, 10 Pa C_2H_2, and N_2, respectively, four times over a two-week period, indicated by day number on the horizontal axis. Notice different scales on vertical axis. Means of duplicate samples; vertical lines indicate one SD.

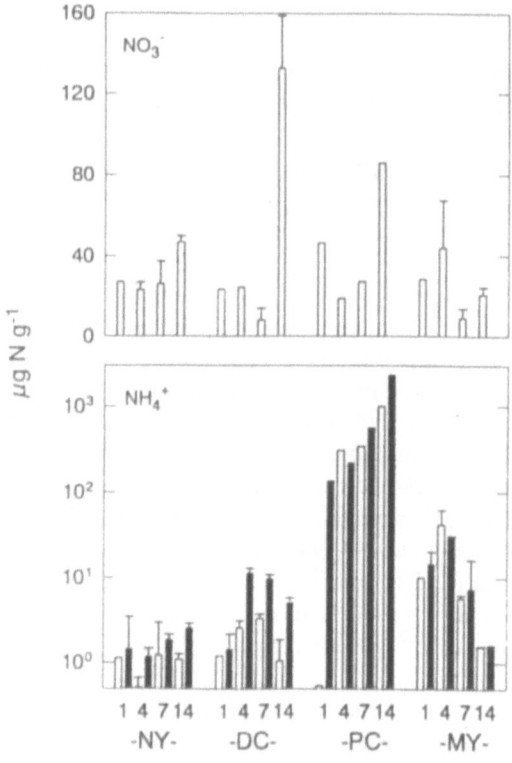

Figure 4. Concentrations of NO_3^- and NH_4^+ in soil columns without yeast (NY), distal (DC) and proximal (PC) compartments of columns amended with yeast lumps, and columns mixed with yeast (MY), respectively, and without (□) and with (■) additional 200 μg NO_3^--N g^{-1}. No NO_3^- data are available for the NO_3^--amended columns. The inorganic N contents were measured four times over a two-week period indicated by day number on the horizontal axis. Notice LOG-scale for NH_4^+ concentrations. Means of duplicate samples; vertical lines indicate one SD.

Discussion

Dead microbial tissue stimulated denitrification similarly to glucose, and in accordance with the finding of Rice et al. (1988). The apparently higher availability of yeast-C than glucose-C could be an artefact due to inactivation of the growth inhibitor or growth of bacteria insensitive to chloramphenicol in the presence of yeast material as suggested by the non-linear N_2O evolution on Figure 1. Inactivation of chloramphenicol in DEA assays in soils has not been reported, for all I know, but this preliminay result suggests that it may occur when particulate organic matter concentrations are high.

Denitrification was seemingly the sole source of N_2O since the production of N_2O remained unaltered in the presence of 10 Pa C_2H_2, an inhibitor of autotrophic nitrification. Accumulation of NO_3^- indicated that nitrification was intense in columns with yeast lumps, whereas no NO_3^- accumulation was found in columns mixed with yeast probably because the better contact between yeast material and soil increased microbial NO_3^- uptake. The observed nitrification was most likely due to autotrophic activity (Rudaz et al., 1991). However, heterotrophic nitrification not sensitive to C_2H_2 can not be ruled out as a source of N_2O (Anderson et al., 1993).

Nitrous oxide losses were substantial upon incorporation of the yeast material. Translated to the field scale the rates under air (Table 1) equals losses of 3–9 kg N ha^{-1} (0-5 cm depth) over the two week period. This is between 8 to 26% of the total denitrification loss observed upon injection of dead *E. coli* material into columns of a loam soil (Rice et al., 1988), suggesting a total N loss of tens of kg ha^{-1} 14 d^{-1} in the present study. However, the applicaton rate used here, approximately 500 kg ha^{-1} biomass derived N, is an extreme dosis 5-fold higher than that recommended for agricultural use of fermentation waste, i.e. dead microbial tissuc (H. Ørtenblad, 1994 pers. comm.). The consistent high NO_3^- content in both yeast treatments makes it likely, that the weak increase in N_2O production in response to NO_3^- (Table 1) was due to an increasing denitrification $N_2O:N_2$ ratio (Firestone et al., 1979) rather than an overall increase in activity.

The initial denitrification enzyme activity compared to that found in other studies (Martin et al., 1988; Myrold and Tiedje, 1985), and the temporal pattern was similar to that observed in proximity of clover leaf bundles (Henriksen and Larsen, 1991). The coincident patterns of N_2O production and DEA supports the dominance of denitrification in N_2O production. Nitrate was not important for the establishment of DEA in this agricultural soil, in agreement with the finding of Myrold and Tiedje (1985). The patterns of DEA indicated that N_2O production in soil columns with yeast lumps took place only in proximity of the lumps. Assuming this, and taking into account a 4:1 ratio of distal vs. proximal compartments, the ratios of N_2O production vs. DEA varied between $6.2 \cdot 10^{-6}$-0.47 and 0.002-0.071 in columns mixed with yeast and columns with yeast lumps, respectively. The ratio was extremely low on day 14 in the mixed columns where N_2O production was negligible; within the first week the $N_2O:DEA$ ratio was on average 4-fold higher in proximity of the yeast lumps than in the mixed columns suggesting conditions favourable for a high denitrification N_2O production in the former environment.

The higher anaerobic N_2O production in columns mixed with yeast than in those with yeast lumps for the first 7 days indicate a higher carbon availability when the yeast material was mixed into the soil (Fig. 2), since NO_3^- was generally high (Fig. 4). At day 14 anaerobic N_2O production was still high in proximity of the lumps whereas O_2 inhibition apparently became increasingly important at air incubations (Fig. 2). In contrast, anaerobic N_2O production was strictly C-limited in the mixed columns at day 14. The different substrate availability depending on physical locations suggested above is supported by the much higher availability of mineral N in proximity of the yeast lumps compared to within the soil mixed with yeast (Fig. 4). A low $C:NO_3^-$ ratio, as at the yeast lumps, suggests a high denitrification $N_2O:N_2$ ratio (Firestone and Davidson, 1989) which may help explain the higher $N_2O:DEA$ ratio in proximity of the lumps than in soil mixed with yeast.

The study demonstrated that denitrification was the main source of N_2O during decomposition of a low C/N ratio organic material, and that N_2O may be produced at similar rates despite disparate substrate accessibility.

Acknowledgements

I wish to thank K Kofoed and H-O Kraglund for skilful technical assistance, S Christensen and ES Jensen for useful suggestions and comments, and the European Commission (STEP program CT90-0028) and the Danish Research Council for financial support.

References

Anderson I C, Poth M, Homstead J and Burdige D 1993 A comparison of NO and N_2O production by the autotrophic nitrifier *Nitrosomonas europaea* and the heterotrophic nitrifier *Alcaligenes faecalis*. Appl. Environ. Microbiol. 59, 3525–3533.

12

Aulakh M S, Doran J W, Walters D T, Mosier A R and Francis D D 1991 Crop residue type and placement effects on denitrification and mineralization. Soil Sci. Soc. Am. J. 55, 1020–1025.

Bergstrom D W, Tenuta M and Beauchamp E G 1994 Increase in nitrous oxide production in soil induced by ammonium and organic carbon. Biol. Fertil. Soils 18, 1–6.

Christensen S, Simkins S and Tiedje J M 1990 Temporal patterns of soil denitrification: their stability and causes. Soil Sci. Soc. Am. J. 54, 1614–1618.

Duxbury J M and McConnaughey P K 1986 Effect of fertilizer source on denitrification and nitrous oxide emissions in a maize-field. Soil Sci. Soc. Am. J. 50, 644–648.

Firestone M K, Smith M S, Firestone B and Tiedje J M 1979 The influence of nitrate, nitrite, and oxygen on the composition of the gaseous products of denitrification in soil. Soil Sci. Soc. Am. J. 43, 1140–1144.

Firestone M K and Davidson E A 1989 Microbiological basis of NO and N_2O production and consumption in soil. In Exchange of trace Gases between Terrestrial Ecosystems and the Atmosphere. Eds. M O Andreae and D S Schimel. pp 7–21. John Wiley and Sons, Chichester, UK.

Goreau T J, Kaplan W A, Wofsy S C, McElroy M B, Valois F W and Watson S W 1980 Production of NO_2^- and N_2O by nitrifying bacteria at reduced concentrations of oxygen. Appl. Environ. Microbiol. 40, 526–532.

Henriksen K and Larsen L 1991 Temporal and spatial variation in nitrification and denitrification activities in millimeter zones in and around organic hot spots. In Denitrification in Forest Soils - Summaries from a Workshop arranged in Copenhagen. Eds. P Ineson et al. Denmark 6–7 November 1989. EC, Brussels, Belgium.

Martin K, Parsons L L, Murray R E and Smith M S 1988 Dynamics of soil denitrifier populations: relationships between enzyme activity, most-probable number counts, and actual N gas loss. Appl. Environ. Microbiol. 54, 2711–2716.

McKenney D J, Wang S W, Drury C F and Findlay W I 1993 Denitrification and mineralization in soil amended with legume, grass, and corn residues. Soil Sci. Soc. Am. J. 57, 1013–1020.

Myrold D D and Tiedje J M 1985 Establishment of denitrification capacity in soil: effects of carbon, nitrate, and moisture. Soil. Biol. Biochem. 17, 819–822.

Parkin T B 1987 Soil microsites as a source of denitrification variability. Soil Sci. Soc. Am. J. 51, 1194–1199.

Poth M and Focht D D 1985 ^{15}N kinetic analysis of N_2O production by Nitrosomonas europaea: an examination of nitrifier denitrification. Appl. Environ. Microbiol. 49, 1134–1141.

Rice C W, Sierzega P E, Tiedje J M and Jacobs L W 1988 Stimulated denitrification in the microenvironment of a biodegradable organic waste injected into soil. Soil Sci. Soc. Am. J. 52, 102–108.

Rolston D E, Sharpley A N, Toy D W and Broadbent F E 1982 Field measurements of denitrification: III. Rates during irrigation cycles. Soil Sci. Soc. Am. J. 46, 289–296.

Rudaz A O, Davidson E A and Firestone M K 1991 Sources of nitrous oxide production following wetting of dry soil. FEMS Microbiol. Ecol. 85, 117–124.

SAS 1990 SAS/STAT User's guide, Version 6, Fourth Edition. SAS Institute, Cary, NC, USA.

Smith M S and Tiedje J M 1979 Phases of denitrification following oxygen depletion in soil. Soil. Biol. Biochem. 11, 261–267.

Tiedje J M 1988 Ecology of denitrification and dissimilatory nitrate reduction to ammonium. In Biology of Anaerobic Microorganisms. Ed. A J B Zehnder. pp 179–244. John Wiley and Sons, New York, USA.

Section editor: R Merckx

O. Van Cleemput et al. (eds.), Progress in Nitrogen Cycling Studies, 507–511, 1996.

An open gas-flow system for investigating the response of nitrous oxide fluxes from soil cores to different oxygen concentrations

L. Anderson, R. Parsons and D.W. Hopkins[1]
Department of Biological Sciences, University of Dundee, Dundee DD1 4HN, UK. [1]*Corresponding author*

Key words: denitrification, gas mixing device, nitrification, nitrous oxide, oxygen, soil

Abstract

The design of an inexpensive device for creating mixtures of gases in different but known proportions and perfusing them through soil cores is described. Soil in cores were perfused with gas mixtures containing 0, 2, 10, 33, 81 and 100% O_2 plus 0.1% CO_2 made up to 100% with He for 48 hours at a rate which ensured that the air-filled pore volume of the soil was replaced once every 3.3 minutes. After this period the N_2O emission from the soil was measured. There was significant N_2O flux from all soils within the first 12 hours, including those which had been exposed to 100% O_2. The amount of N_2O emitted over 324 hours increased with decreasing O_2 concentration in the perfusing gas mixture, although the initial (0 to 108 hours) rate of N_2O emission from the 100% O_2 treatment was significantly greater than that of the 33 and 81% O_2 treatments N_2O emissions from soil exposed to 0, 2 and 10% O_2 were linear with time, whereas the rate of N_2O emission from soil exposed to the greater O_2 concentrations increased significantly after 210 hours.

Introduction

N_2O is a relatively long-lived (approximately 160 years) trace atmospheric gas, the concentration of which has been rising at a rate between 0.2 and 0.3% per year in recent years, and which absorbs long-wave radiation emitted from the earth's surface (IPCC, 1990). Although stable in the troposphere, N_2O contributes to photolysis of stratospheric ozone (Cicerone, 1987).

Emission of biogenic N_2O from soils is a major contributor to global N_2O fluxes (Van Bremmen and Feijtal, 1990), with two functional groups of bacteria, denitrifiers and nitrifiers, being regarded as quantitatively the most important (Robertson and Tiedje, 1987). N_2O production during both denitrification and nitrification is sensitive to oxygen. Denitrification occurs in soils primarily under anaerobic conditions in which N_2O has been identified as a free obligatory intermediate preceding N_2 (Payne, 1981), although the significance of aerobic denitrification (simultaneous use of O_2 and NO_3^- as terminal electron acceptors) should not be overlooked (Lloyd et al., 1987; Robertson and Kuenen, 1984). Nitrifying (ammonium-oxidizing) bacteria may also produce N_2O under a range of O_2 conditions.

Nitrifier-denitrification, the reduction of NO_2^- to N_2O by *Nitrosomonas europaea*, occurs under conditions of O_2 limitation (Poth and Focht, 1985) and N_2O production by non-denitrifying nitrifiers can occur under aerobic conditions (Blackmer et al., 1980).

In soil aerobic and anaerobic sites can simultaneously exist in close proximity with the result that the actual sources and their partitioning is unclear (Davidson et al., 1986). In this contribution we describe a simple system for preparing mixtures of gases with a range of known O_2 concentrations and passing them through soil cores with the intention of determining the N_2O production response to O_2 in the range 0 to 100% by the soil microbial community. The gas mixing device described has the advantage of being relatively cheap to manufacture and the approach can be adapted to provide mixtures containing a variety of gases over wide concentration ranges.

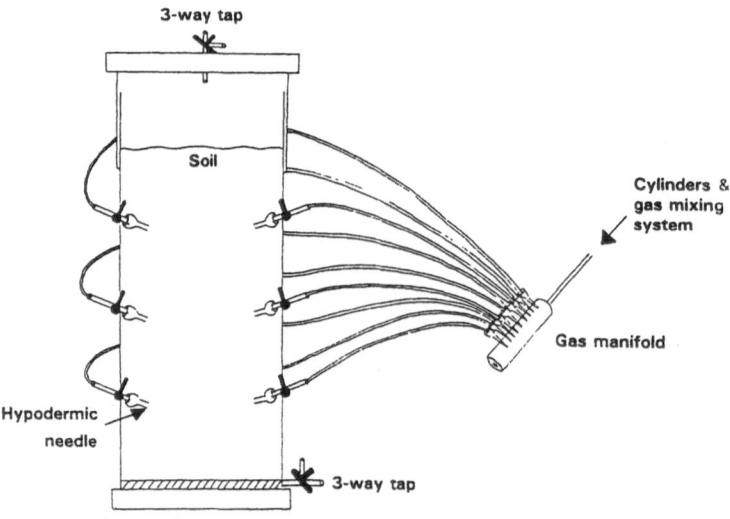

Figure 1. Diagram showing soil core design

Materials and methods

Soil core experiments

The soil used was a brown forest soil of the Carpow series (Laing, 1976) (sandy-loam texture and pH 6.4) collected from the Scottish Crop Research Institute, Invergowrie, Scotland (National Grid reference NO 346296) and sieved (to pass a 4 mm mesh) in the field moist state and stored at 5 ± 2 °C until it was packed into cores made from 2 mm thick PVC-C (Fig. 1). The cores had an internal diameter of 105 mm and were 350 mm high with bonded PVC bases and removable gas-tight lids made from PVC-U. Nine hypodermic needles (30 mm long and 0.6 mm internal diameter) were sealed into the walls of each core with epoxy resin and then fitted with gas-tight taps. The lid and the bottom of the core were fitted with 3-way taps for head-space sampling and draining, respectively. 2.2 kg (dry weight) sieved soil was packed into each core on top of a 1 cm deep bed of washed sand. The head-spaces above the soil were 10 cm deep ($866 \, cm^3$) and the soil bulk density in each core was a $1.06 \, g \, cm^{-3}$. Prior to packing into the cores, the soil was amended with $100 \, \mu g \, NH_4NO_3\text{-N} \, g^{-1}$ soil and deionised water soil was added to the soil bring it to field capacity (matric potential = -5.0 kPa) at which point the water content of the soil was $0.31 \, cm^3 \, H_2O \, g^{-1}$ soil. The soil cores were incubated at 15 ± 2°C for 60 days after which the air-filled pore volume of the soil was calculated to be $0.28 \, cm^3 \, g^{-1}$ soil ($605 \, cm^3$ per core). After this period gas mixtures containing different O_2 concentrations,

0.1% CO_2 and made up to 100% with He were passed through the soil via the hypodermic needles fitted to the cores for 48 hours. The design of the gas mixing device used to provide the mixtures is described below. The total gas flow rate through each core was $3.06 \, cm^3 \, s^{-1}$ (i.e. one core air-filled pore volume every 3.3 minutes or 873 air-filled pore volumes in 48 hours). The O_2 partial pressures in the gas mixtures ranged between 0 and 100 kPa (%). There were triplicate cores for each gas mixture.

After 48 hours, the gas supplies were switched off and the cores were sealed by closing all the valves. After 12 hours, and every 24 hours thereafter up to 324 hours, two $1 \, cm^3$ gas samples were removed from the head-space of each core. N_2O was determined by gas chromatography using a Perkin Elmer $\Sigma 4$ gas chromatograph fitted with a 1.8 m long, 3 mm internal diameter glass column packed with 80–100 mesh Porapak Q and a ^{63}Ni electron capture detector with oxygen-free N_2 as the carrier gas.

Gas mixing device

Mixtures of O_2, CO_2 and He in specified proportions were produced using a device which operates on the same principal as that described by Parsons et al. (1992). The device uses regulator valves (Fig. 2) to regulate pressures of O_2, CO_2 and He, and the relative flow rates of each gas were determined using appropriate lengths of fine bore (0.6 mm diameter) tubing which acts as a resistance. Since gas flow rate is approximately directly proportional to pressure and inversely

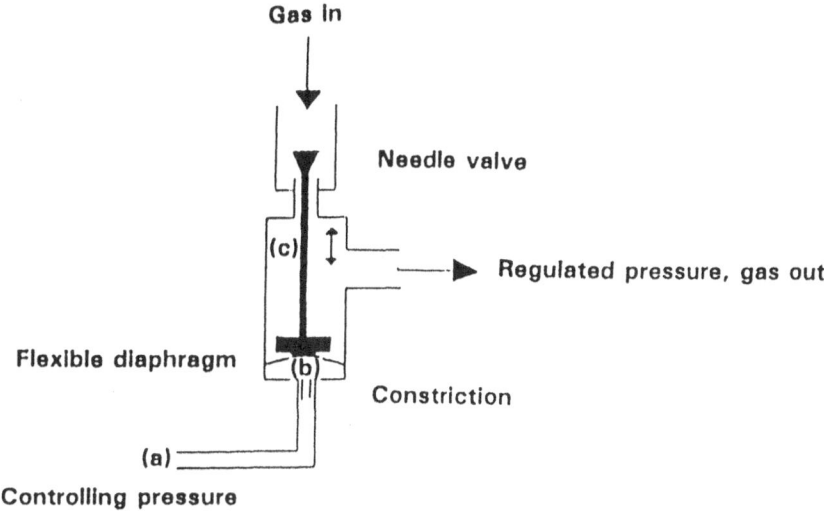

Gas in

Needle valve

(c)

Regulated pressure, gas out

Flexible diaphragm

(b)

Constriction

(a)

Controlling pressure

Figure 2. Design of the pressure regulating valve developed to supply gases at a constant and equal pressure to allow accurate mixing of gases (Parsons et al., 1992).

Table 1. Rates of N_2 emission from soil treated with different concentrations of O_2 over the period 0 to 210 and 210 -324 hours or 0–324 hours. Each value is the mean of three replicates and the standard deviations are shown in brackets

Percent O_2 (kPa)	Rate of N_2O production (nmol N_2O g^{-1} soil hour^{-1})		
	0–210 hours	210–324 hours	0–324 hours
0	-	-	3.52×10^{-2} (1.74×10^{-2})
2	-	-	5.25×10^{-3} (2.99×10^{-3})
10	-	-	9.57×10^{-4} (1.14×10^{-4})
33	2.75×10^{-4} (3.36×10^{-5})	1.25×10^{-3} (1.14×10^{-4})	-
81	2.03×10^{-4} (1.53×10^{-5})	8.40×10^{-4} (5.8×10^{-5})	-
100	2.39×10^{-4} (3.33×10^{-5})	6.36×10^{-4} (7.16×10^{-5})	-

ber (c) until the pressure inside both chambers is equal. The overall design of the device is shown in Figure 3. CO_2 flow was kept constant at 0.1% of the total, O_2 flow was controlled to provide a range between 0 and 100% of the total flow and the remaining balance was made up with He. The output lines from the device were each fed into separate gas manifolds from which 27 lines were taken to supply triplicate cores (9 gas inputs per core), each with a flow rate of 0.34 cm^3 s^{-1}. The proportions of O_2 in the resulting gas mixtures were determined by gas chromatography (Varian 90 GC fitted with a 2.6 m long×3 mm internal diameter stainless steel column packed with molecular sieve A and a thermal conductivity detector). The measured O_2 contents of the gas mixtures were 0, 2, 10, 33, 81 and 100% (kPa).

Results and discussion

Purging the soil with gas mixtures containing different amounts of O_2 led to different amounts of N_2O accumulating in the head-space above the soils (Fig. 4) and significant N_2O was observed from soil treated with 100% O_2. After 324 hours, the amounts of N_2O which had been emitted were significantly different between all treatments and followed the trend of increasing N_2O with decreasing O_2 (Fig. 4). It is likely that at low O_2 concentrations, production of N_2O was by both heterotrophic denitrifying (Robertson and Tiedje, 1987) and denitrifying-nitrifying (Poth

proportional to tube length (Poiseuille's law), the proportions of each gas in the mixture can be determined by adjusting the length of tube through which it must pass. O_2 was used as the controlling gas, with its pressure being set at the cylinder head regulator (a), and the pressures of CO_2 and He were regulated by regulator valves in the mixing device. The O_2 was connected to the valve chamber (b) below the diaphragm and its pressure causes the needle valve to open. The regulated gas, either CO_2 or He, then flows into the upper cham-

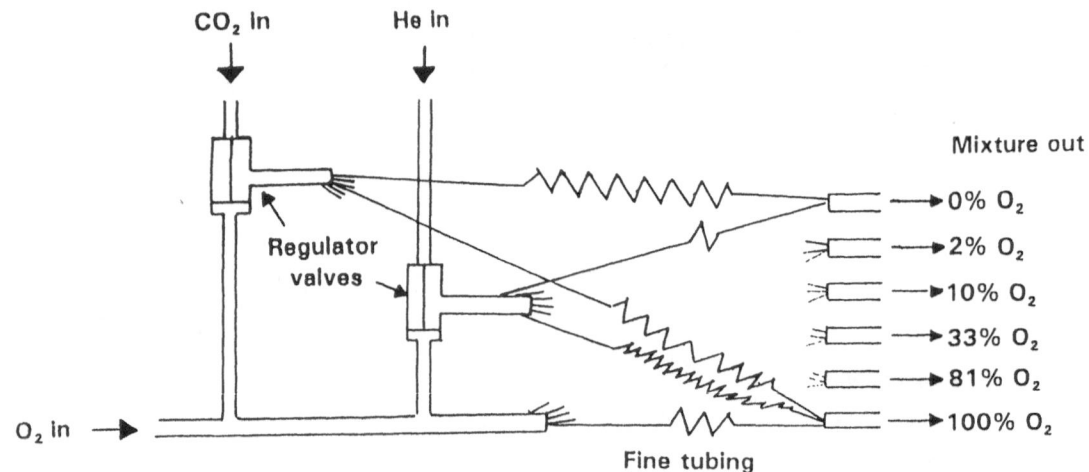

Figure 3. Basic design of gas mixing system. Adapted from Parsons et al. (1992).

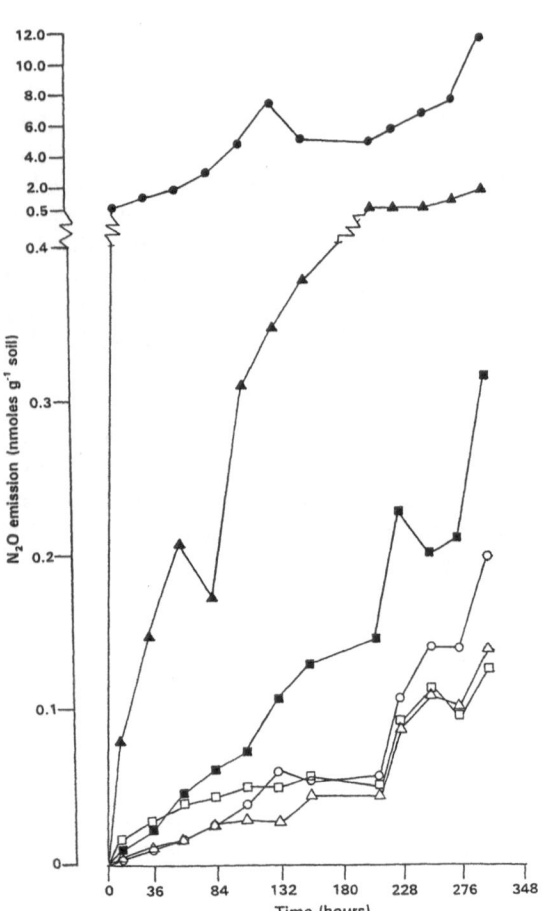

Figure 4. Nitrous oxide emissions over 324 hours from soil cores treated with 0% O_2 (●), 2% O_2 (▲), 10% O_2 (■), 33% O_2 (○), 81% O_2 (△) and 100% O_2 (□). Points represent the mean of 3 replicates.

and Focht, 1985) bacteria and at the greater O_2 concentrations non-denitrifying (Blackmer et al., 1980) and aerobic denitrifying (Anderson and Levine, 1986; Robertson and Kuenen, 1984) bacteria were likely to be responsible for an increasing contribution to N_2O flux. The greater initial rate of N_2O emission (i.e. that between 0 and 108 hours) from the soil treated with 100% O_2 compared with the 33 and 81% O_2 treatments suggests that N_2O production by aerobic denitrifiers or non-denitrifying nitrifiers was stimulated at the greater O_2 concentrations. Such stimulation could be a direct effect on N_2O production or the indirect result of increased NO_3^- production by nitrifying bacteria. N_2O emission was not linear for the soil cores which had been treated with the greater concentrations of O_2 (33, 81 and 100%). For these treatments there was an apparent lag of between approximately 210 hours during which the rate of N_2O emission was significantly less than that between about 210 hours and the end of the experiment (Table 1). It is likely that this lag corresponds to a period of relatively abundant O_2 in the soils, and as O_2 was depleted, the rate of N_2O emission increased. It is also possible that the two phases of N_2O were associated with different functional groups of N_2O producing microorganisms in the soil.

The data presented here show that significant N_2O fluxes can occur in soil under highly aerobic conditions and indicate, therefore, that contributions to N_2O flux from aerobic processes, such as aerobic denitrification and nitrification, need to be estimated in order to achieve a more complete understanding of N_2O metabolism in soil.

Acknowledgements

We wish to acknowledge the UK Natural Environment Research Council for a postgraduate studentship for LA and the support received from the British Society of Soil Science, and the assistance of Dr R E Wheatley, SCRI, Invergowrie.

References

Anderson I C and Levine J S 1986 Relative rates of nitrous oxide production by nitrifiers, denitrifiers and nitrate respirers. Appl. Environ. Microbiol. 51, 938–945.

Blackmer A M, Bremner J M and Schmidt E L 1980 Production of nitrous oxide by ammonia-oxidizing chemoautotrophic microorganisms in soil. Appl. Environ. Microbiol. 40, 1060–1066.

Cicerone R J 1987 Changes in stratospheric. Science 237, 35–42.

Davidson E A, Swank W T and Perry T O 1986 Distinguishing between nitrification and denitrification as sources of gaseous nitrogen production in soil. Appl. Environ. Microbiol. 52, 1280–1286.

IPCC (Intergovernmental Panel on Climate Change) 1990 Greenhouse gases and aerosols. In Climate Change. Eds. J T Houghton, G J Jenkins and J J Ephraums. p 27. Cambridge University Press, Cambridge, UK.

Laing D 1976 Soils of the country round Perth, Arbroath and Dundee. HMSO, Edinburgh, UK.

Lloyd D, Davies K I P and Boddy I 1987 Persistance of bacterial denitrification capacity under aerobic conditions - the rule rather than the exception. FEMS Microbiol. Ecol. 45, 185–190.

Parsons R, Raven J A and Sprent J I 1992 A simple open flow system used to measure acetylene reduction activity of Sesbania rostrata stem and root nodules. J. Exp. Bot. 43, 595–604.

Payne W J 1981 The status of nitric oxide and nitrous oxide as intermediates in denitrification. In Denitrification, Nitrification and Atmospheric Nitrous Oxide. Ed. C C Delwiche. pp 85–103. John Wiley and Sons, Inc., New York, USA.

Poth M and Focht D D 1985 ^{15}N kinetic analysis of N_2O production by Nitrosomonas europaea: an examination of nitrifier denitrification. Appl. Environ. Microbiol. 49, 1134–1141.

Robertson L A and Kuenen J G 1984 Aerobic denitrification: a controversy revived. Arch. Microbiol. 139, 351–354.

Robertson G P and Tiedje J M 1987 Nitrous oxide sources in aerobic soils; nitrification, denitrification and other biological processes. Soil Biol. Biochem. 19, 187–193.

Van Breemen N and Feijtal T C J 1990 Soil processes and properties involved in the production of greenhouse gases, with special relevance to soil taxonomic systems. In Soils and the Greenhouse Effect. Ed. A F Bouwmann. pp 195–223. John Wiley and Sons, Inc., New York, USA.

O. Van Cleemput et al. (eds.), Progress in Nitrogen Cycling Studies, 513–515, 1996.
© 1996 *Kluwer Academic Publishers.*

Correcting for leaks from megachambers

Jonathan Arah[1], Albert Scott and Keith Smith
Soils Department, SAC, School of Agriculture, West Mains Road, Edinburgh EH9 3JG, UK. [1]*Present address:
Institute of Terrestrial Ecology, Bush Estate, Penicuik, Midlothian EH26 0QB, UK*

Key words: flux chamber, leak correction, trace gas

Abstract

Large chambers ("megachambers") used to quantify trace gas exchange inevitably leak. This leakage can be allowed for by fitting a simple equation to measured data. Where such correction is not performed, fluxes are likely to be underestimated by 10–50%, the extent of the underestimate depending on megachamber construction, external windspeed, and closure period.

Introduction

Conventional chamber measurements of trace gas emissions from soil frequently encounter great spatial variability (e.g. for nitrous oxide, N_2O, Arah et al., 1991; Folorunso and Rolston, 1984). This makes it difficult to obtain representative values. One way round the problem is to increase the area enclosed by the chamber. This approach is especially attractive for use on relatively small experimental plots and other areas where micrometeorological methods cannot be applied (Fowler and Duyzer, 1989; Smith et al., 1994a).

However, large chambers (especially portable ones) are practically impossible to seal. Some allowance should be made for this. One possibility is to inject an inert tracer gas and monitor its disappearance from the chamber at the same time as the gas of interest (N_2O in the work reported here) accumulates. Another is to fit an appropriate equation to the N_2O data itself. The first approach requires additional instrumentation (it must be possible to measure the tracer); where this is unavailable, modelling is the only option.

Theory

If the flux from the soil x and the leak rate constant k (the rate at which air is exchanged between the inside and the outside of the chamber) remain uniform during a given measurement period, then:

$$dy/dt = (x/h) - ky \qquad (1)$$

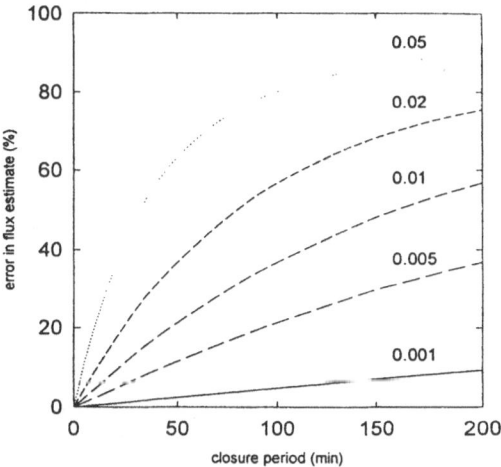

Figure 1. Percentage error (underestimate) in uncorrected flux as a function of leak rate constant and closure period: numbers on lines are leak rate constants k (min^{-1}).

where y is the concentration enhancement in the chamber, h is its effective height (volume divided by area enclosed) and t is the time. This may be integrated to give:

$$y = (x/hk)[1 - \exp(-kt)] \qquad (2)$$

Fitting Equation 2 to measured data gives estimates of the leak rate constant k and the flux x. If no correction is made for leakage (i.e. if the concentration increase is assumed to be linear), the flux x is underestimated. Figure 1 shows how the extent of this error depends on the leak rate constant k and the closure period.

Table 1. Leak rate constants and flux corrections

Date	Chamber material	Average windspeed (m s^{-1})	Leak rate constant (min^{-1})
April 1992	canvas	ca. 2	0.02
August 1993	plastic	2.0	0.005
	plastic	0.5	0.002

Curve-fitting may not be feasible where relatively few measurements are available (e.g. where more than one megachamber is employed simultaneously, and discrete samples are taken for off-line analysis). Leak correction is still possible, however: Equation (2) can be manipulated to give:

$$k = (1/t_1) \ln(y_1/(y_2 - y_1)) \qquad (3)$$

$$x = hky_1^2/(2y_1 - y_2) \qquad (4)$$

where y_1 and y_2 are the concentration enhancements measured at times t_1 and t_2 after closure of the chamber, and $t_2 = 2\,t_1$. Estimating the leak rate constant k and the flux x thus requires a minimum of one background and two properly-timed internal concentration measurements per chamber.

Materials and methods

We used two "megachambers", both approximately 50 m^2 in area, to evaluate the potential of the large-chamber method for measuring N$_2$O flux. The first, used in April 1992 at a grassland site near Stirling, Scotland, was a lightweight tent with an extended removable midsection (Galle et al., 1994; Smith et al., 1994b), the second (August 1993, grain stubble, Lammefjord, Denmark) a sheet of builder's plastic stretched over a series of cane hoops and weighed down with sandbags around its edge (Smith et al., 1994b). Nitrous oxide concentrations were measured using a long-path infrared gas monitor (Partridge, 1991; Smith et al., 1994b) installed within the chamber. A second instrument was used in 1993 to measure concentrations of methane, CH$_4$, in the chamber.

Results and discussion

Nitrous oxide concentrations in the chambers were fitted to Equation 2 and flux rates x and leak rate constants

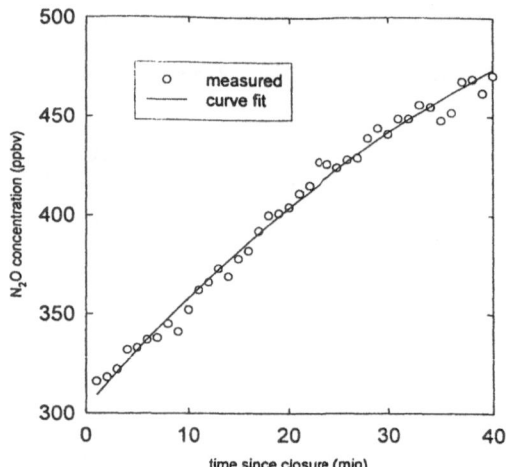

Figure 2. Nitrous oxide concentration increase in canvas megachamber: leak rate constant 0.02 min^{-1}; flux 11.9 μmol m^{-2} h^{-1}.

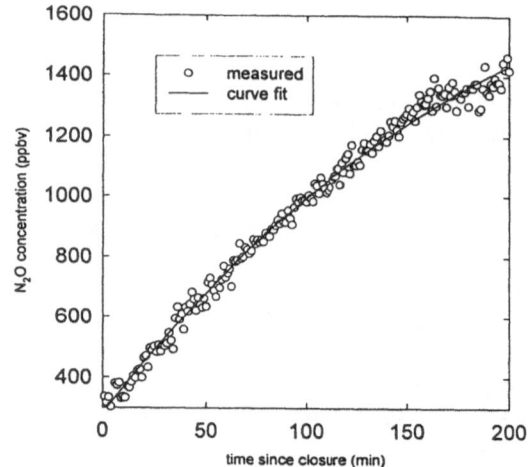

Figure 3. Nitrous oxide concentration increase in plastic megachamber: leak rate constant 0.005 min^{-1}; flux 18.0 μmol m^{-2} h^{-1}.

k were derived. Specimen data sets are illustrated in Figures 2 and 3, and average leak rate constants are presented in Table 1. The plastic megachamber was less leaky than the canvas one: leak rate constants increased with increasing windspeed.

Values of k derived from N$_2$O data compared reasonably well with those estimated using CH$_4$ as a tracer (Fig. 4), but generally exceeded them. This probably reflects the fact that, as defined in Equation 1, our "leak rate constant" k actually incorporates all concentration-dependent factors retarding further concentration enhancement. The mere buildup of concentration in the chamber is itself such a factor (Hutchin-

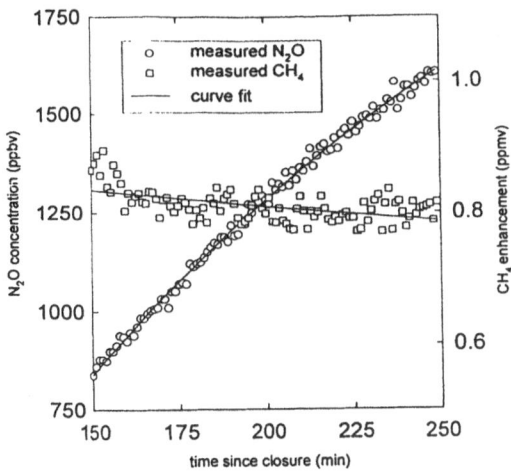

Figure 4. Nitrous oxide accumulation and methane decline in plastic megachamber: nitrous oxide leak rate constant 0.004 min^{-1}; flux 20.6 μmol m^{-2} h^{-1}; methane leak rate constant 0.002 min^{-1}.

son and Mosier, 1981), and one which would not affect an inert tracer.

Conclusion

Where no attempt is made to correct for leakage, the size of the error in the estimated flux depends on the leak rate constant k and the length of the enclosure period. For (mega)chambers with leak rate constants greater than about 0.005 min^{-1} some kind of leak correction should be applied even where the closure period is as short as one hour.

Acknowledgements

We thank the Commission of the European Communities for supporting this work as part of STEP project CT 90–0028. We also acknowledge the financial support provided by the Natural Environment Research Council through its TIGER (Terrestrial Initiative in Global Environmental Research) program, award GST/02/600 (UK).

References

Arah J R M, Smith K A, Crichton I J and Li H S 1991 Nitrous oxide production and denitrification in Scottish arable soils. J. Soil Sci. 42, 351–367.
Folorunso O A and Rolston D E 1984 Spatial variability of field-measured denitrification gas fluxes. Soil Sci. Soc. Am. J. 48, 1214–1219.
Fowler D F and Duyzer J 1989 Meteorological techniques for the measurement of trace gas exchange. *In* Exchange of Trace Gases between Terrestrial Ecosystems and the Atmosphere. Ed. M O Andreae and D S Schimel. pp 189–207. John Wiley and Sons, New York.
Galle B, Klemedtsson L and Griffith D W T 1994 Application of an FTIR system for measurements of nitrous oxide fluxes using micrometeorological methods, an ultra-large chamber system, and conventional field chambers. J. Geophys. Res. 99, 16575-16584.
Hutchinson G L and Mosier A R 1981 Improved soil cover method for field measurement of nitrous oxide fluxes. Soil Sci. Soc. Am. J. 45, 311–316.
Partridge R H 1991 Long-path monitoring of atmospheric pollution. Meas. Control 23, 293-298.
Smith K A, Clayton H, Arah J R M, Christensen S, Ambus P, Fowler D, Hargreaves K J, Skiba U, Harris G W, Wienhold F G, Klemedtsson L and Galle B 1994 Micrometeorological and chamber methods for measurement of nitrous oxide fluxes between soils and the atmosphere: overview and conclusions. J. Geophys. Res. 99, 16541–16548.
Smith K A, Scott A, Galle B and Klemedtsson L 1994 Use of a long-path infrared gas monitor for measurement of nitrous oxide flux from soil. J. Geophys. Res. 99, 16585–16592.

O. Van Cleemput et al. (eds.), Progress in Nitrogen Cycling Studies, 517–522, 1996.
© 1996 *Kluwer Academic Publishers.*

N_2O production in freshwater wetland sediments with and without added nitrate

V.N. Astorga, E. Novella and F.A. Comín
Departament d'Ecologia, Facultat de Biologia, Universitat de Barcelona, Avgda. Diagonal 645, 08028 Barcelona, Spain

Key words: acetylene, denitrification, freshwater wetlands, Mediterranean, nitrogen

Abstract

In-situ N_2O production rates of sediments from a restored freshwater wetland in the Ebro River Delta (NE Spain) were measured with the acetylene inhibition technique. Cores were taken at two sites in the wetland, one near the flood water inlet and the other near the outlet. One series of sediment cores from each site was treated with deionized water saturated with acetylene and another series with a solution of KNO_3 54 μM saturated with acetylene. The cores were incubated in-situ for 4 h.

N_2O production in the cores with added nitrate was 5 to 500 times higher than in the cores without nitrate. Rates in sediments without nitrate added ranged from positive values, in spring and early summer (up to 6.5 μmol N-N_2O $m^{-2}h^{-1}$), to negative values, in winter. N_2O production rates in cores with nitrate were higher at the floodwater inlet (50 to 480 μmol N-N_2O $m^{-2}h^{-1}$) than at the outlet (0.8 to 145 μmol N-N_2O $m^{-2}h^{-1}$) throughout the year.

Results show that N_2O production in these restored wetland is strongly limited by nitrate. Furthermore, when the nitrate limitation is removed, N_2O production shows correlation with temperature.

Introduction

Denitrification is a key process in the nitrogen cycle in wetlands, where the sediment conditions for this process, high organic matter and low oxygen concentrations are optimal (Faulkner and Richardson, 1989). Denitrification as well as nitrification are the main sources of N_2O in soils and sediments, the rates of both processes depending on ammonia, nitrate, organic matter and oxygen availability in the soil or sediment (Chalamet, 1985; Seitzinger, 1988). The rate of these processes in wetland sediments is determined by oxygen penetration in the sediment for nitrification, and nitrate diffusion from nitrification sites or from the overlying water for denitrification. That makes the nitrification rate the "limiting factor for nitrogen removal through denitrification in wetlands" (Hsieh and Coultas, 1989).

There is abundant literature on denitrification in wetlands, mainly in North America, Northern Europe and Australia (Johnston, 1991; Knowles, 1982; Seitzinger, 1988, 1994). However, few studies refer to this process in wetlands in Southern Europe, even though wetlands are key ecosystems for the regions they are located in, socially, economically and ecologically. Some rates have been reported for the freshwater wetlands of the Rhone river delta, the Camargue (El-Habr, 1987; Minzoni et al., 1987), and indirect rates have been calculated for the ricefields of the Ebro river delta (Forés and Christian, 1993).

The Ebro river delta is located in the North East of Spain, at 40°40'N. It has an area of 320 km^2, of which 8% are occupied by natural wetlands. In the last 10 years, rice growing has been discontinued in some ricefields and they have been restored as wetlands by the Ebro Delta Natural Park Authority to add to the few remaining pockets of natural wetland. In a previous study on denitrification in Ebro Delta ricefields sediments we observed rates of N_2O production between 10 and 160 μmol N m^{-2} h^{-1} (Astorga et al., 1994). The present study was designed to measure the production of N_2O in natural conditions and under a higher nitrate load in a restored non-tidal freshwater wetland, which was three years old, and compare it with the production from the ricefields in use.

Material and methods

Sediment samples (36% sand, 63.8% silt, 0.2% clay, 7.9% C, 0.4% N, pH 7.1) were collected on seven occasions, from April to February, at two sites within the wetland, one near the flood water inlet and another near the outlet. Flood water comes from the Ebro River through irrigation channels, during the wet period of the rice growing season (April to October). During the dry period, the water level in the marshes depends mainly on the level of the water table.

Sediment sampling was done with PVC tubes 4.5 cm i.d. and 17 cm long. Eight cores were taken at each site: 3 treatments with two replicates per treatment, and 2 cores for initial nutrient concentration of the sediments. The tubes containing the cores were then transported to the field laboratory, 5 min away, where the following treatments were applied:

- Blanks: deionized water, 50% volume of the interstitial water of the sediments
- Denitrification: deionized water saturated with C_2H_2. Deionized water was added to give a final concentration of 10% C_2H_2 v/v in the sediment.
- "Potential" denitrification: the same as in the denitrification treatment, but with the addition of 54 μM N as nitrate in each core.

All the manipulations were done with the cores remaining inside the tubes, to preserve their structure. Treatments were applied to the cores through their top surfaces with 5 mL syringes fitted with 12 cm long needles. The needles were inserted into the 10 cm long sediment cores from the top to the very bottom, and the corresponding treatment was injected while the needle was being withdrawn from the sediment. This process was repeated 9 times for each core at homogeneously distributed points, in such a way that the treatments diffused through all the volume of sediment, but with the minimum disturbance to its structure. This procedure was adapted to our sediments from Law et al. (1991) and Koike and Sorensen (1988), and was chosen to avoid incomplete inhibition due to limited C_2H_2 diffusion from the headspace into the sediments, which may occur when the acetylene is applied in the gas phase (Tiedje et al., 1989).

Once the treatment had been injected, water from the surface of the core was reduced to 5 mL so that oxygen was able to diffuse to the top of the sediment during the short incubation time. Cores were closed with thick rubber stoppers, and acetylene added to their atmosphere through a lateral rubber septa, in the same proportion as in the sediment. Cores were then incubated in the dark, sunken in the sediment from which they were taken, for 4 h, always between 11 am and 5 pm, hours between which peak temperature was reached in the sediments.

Gas samples were taken from the cores' atmosphere at the start and at the end of the incubation time in the blanks, and at the end of the incubation in the other two treatments. These samples were taken through the lateral septa, and stored in Vacutainer (Beckton Dickinson) and kept at -18 °C until their analysis by gas chromatography (HP gas chromatograph with 1/8"i.d., 10 ft long Porapak Q columns at 70 °C, ECD at 300 °C and injector at 100 °C, N_2 as a carrier gas). Rates of N_2O production were calculated subtracting the concentration of N_2O in the blanks at time zero from the N_2O concentration found in the three treatments at the end of the incubation time.

The two cores for the initial nutrient concentrations at each sampling site were cut into 2 cm slices to a depth of 10 cm. Each slice was then added to 125 mL serum bottles filled with 50 mL of 1 M KCl, shaken for 3 min and frozen at -18 °C until the moment of the extraction and analysis of the supernatant. The supernatant was analyzed for nitrate, nitrite and ammonia with automated colorimetry, following standard methods adapted for the Skalar Autoanalizer. Eh was measured with platinum electrodes (Orion Research), at a depth of 2 cm, and values corrected in reference to the H_2 electrode.

Statistical analysis were performed with Statgraphics Statistical Systems.

Results and discussion

N_2O concentration in the blank treatment results from the production of N_2O by denitrification and nitrification, as well as by the reduction of N_2O to N_2 by denitrifiers. N_2O concentration in the blanks is then the result of those three processes occurring simultaneously. When acetylene is added to the sediment cores, nitrification (Davidson et al., 1986; Sloth et al., 1992) and N_2O reduction are inhibited, thus giving total denitrification rates. N_2O production rates in the blank and denitrification treatments were very low throughout the year (Fig. 1), with rates in the blank treatment being generally lower than in the denitrification treatment, except in August and October. Rates in the blanks ranged between -0.99 and 1.2 μmol N m^{-2} h^{-1} at both sites in the marsh. The denitrification cores gave rates of -0.5 to 6.5 μmol N m^{-2} h^{-1} at the marsh

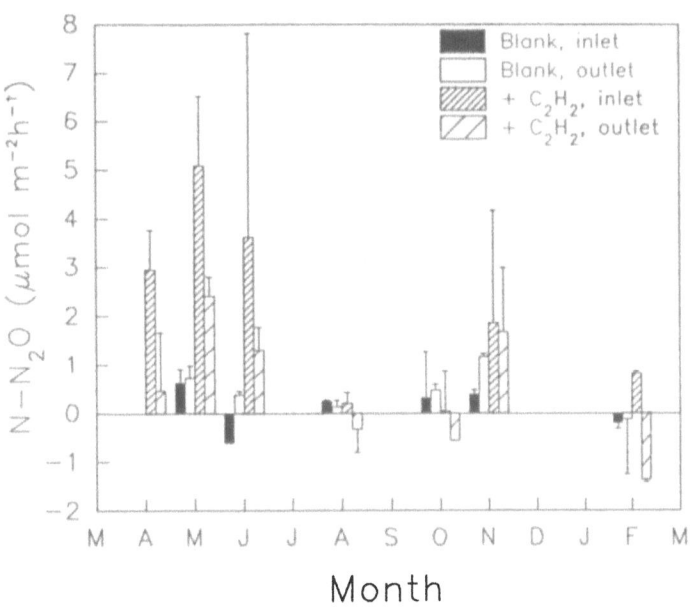

Figure 1. N$_2$O released from marsh sediments with only deionized water added (blanks) or deionized water with C$_2$H$_2$ (denitrification), at two sites within the marsh (inlet and outlet of flood water).

Table 1. Nutrient concentrations in the upper 2 cm of the sediment (nmol N g^{-1} dry weight). Eh values at 2 cm depth (mV). Se = standard error (n=2)

Sampling point	Date	NH$_4$		NO$_2$		NO$_2$+NO$_3$		PO$_4$		Eh	
		Avg	Se	Avg	Se	Avg	Se	Avg	Se	Avg	Se
Inlet	01-Apr	217	-	0.4	0.1	-	-	-	-	-	-
	19-May	721	370	2.3	0.9	6.1	7.7	11.7	3.5	-54.6	-
	30-Jun	522	30	2.5	0.3	22.0	6.3	16.5	8.3	-47.4	22.5
	4-Aug	797	166	43.4	35.2	4.4	6.1	28.0	12.1	-59.9	13.6
	1-Oct	4512	3210	11.4	8.1	36.8	48.4	80.6	56.1	-180.2	13.3
	18-Nov	97	-	8.2	5.6	49.7	68.3	29.2	10.0	102.2	177.0
	16-Feb	1165	258	7.2	2.0	26.3	9.7	35.2	20.2	-	-
Outlet	1-Apr	1115	-	9.1	5.3	-	-	-	-	-	-
	19-May	579	243	2.0	0.7	2.9	0.8	13.0	4.9	35.9	37.0
	30-Jun	679	188	11.6	2.1	19.3	9.4	10.6	3.3	-108.5	34.0
	4-Aug	466	92	1.5	1.2	2.4	1.0	37.1	24.8	-151.9	21.7
	1-Oct	366	2	1.9	0.1	1.3	0.2	8.8	-	-222.5	21.0
	18-Nov	439	213	3.5	0.2	9.7	9.1	114.1	23.5	12.5	42.5
	16-Feb	980	612	5.6	2.6	18.4	17.9	94.0	72.5	-	-

inlet, and -1.4 and 2.6 μmol N m^{-2} h^{-1} at the outlet, with higher rates in spring and early summer. A similar pattern of annual denitrification rates was described by Jensen et al. (1988) for coastal sediments, coinciding with an input of organic matter from phytoplankton. In our study though, organic matter do not appear to be a limiting factor for denitrification, the more so when the major input of organic matter to the sediments occurs

during late summer and autumn as dead plant material (Novella, pers. comm.).

Negative rates of N$_2$O production are due to higher N$_2$O concentrations in the blanks at time zero than in the treatments at the end of the incubations. In the blank treatment these rates mean that N$_2$O reduction to N$_2$ in the sediments is faster than the production of nitrous oxide by denitrification and nitrification, which

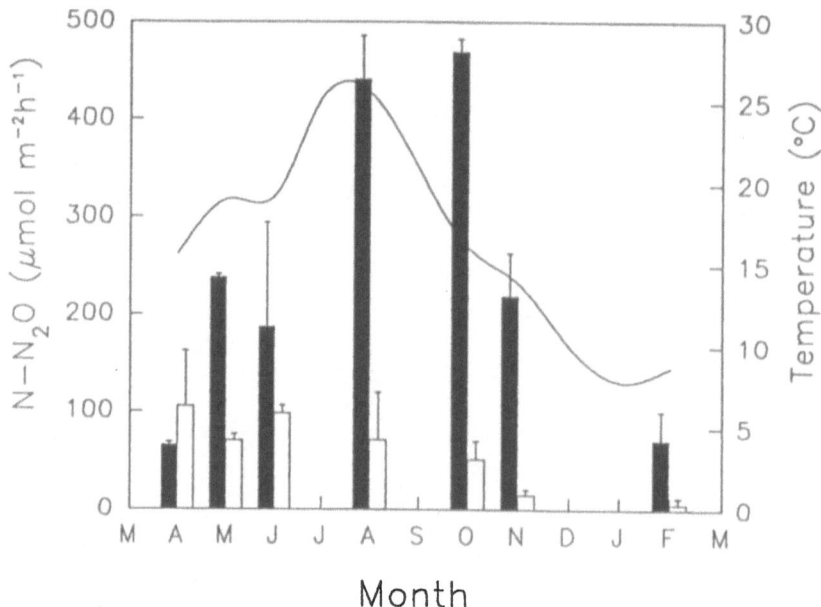

Figure 2. N$_2$O released from marsh sediments with C$_2$H$_2$ and nitrate added, at two sites within the marsh (inlet and outlet of flood water). Superimposed is average monthly temperature.

may indicate low rates of nitrification. This agrees with results we obtained in a short experiment with ATU nitrification inhibitor for the same sediments (rates from -0.4 to 0.2 μmol N m^{-2} h^{-1}). The negative rates of N$_2$O production could also be due to reabsorption into the sediment of the nitrous oxide released when applying the treatments. Most of the time, however, nitrous oxide production rates were positive, although at the low end when compared with literature (Koike and Sorensen, 1988; Seitzinger, 1988, 1994), even considering that the rates we measured may be underestimated by 50% due to the use of the acetylene inhibition technique (Seitzinger et al., 1993).

Differences in N$_2$O production for the denitrification treatment between both sites were significant ($p<0.05$). Differences between the blank and denitrification treatments were statistically significant ($p<0.01$), in spite of the high variability occurring in these sediments, which were highly vegetated with macrophyte roots and rhizomes, and contained pockets of decomposing plant material. This variability has also been encountered in works by other authors for denitrification (Lalisse-Grundman, 1987; Tiedje et al., 1989) and was also to be found in the other variables measured in the sediment in this work (nutrients, Eh).

The sediments of this three-year-old marsh had recovered from rice cultivation so the biomass of macrophytes was high (500 to 4000 g DW m^{-2},

Menéndez, pers. comm). In these conditions, competition for nitrogen between plants and soil microorganisms would be very high (Buresh et al., 1981; Dean and Biesboer, 1985), leading to the low rates of denitrification observed, especially during the peak growing period for macrophytes (July-August).

N$_2$O production rates in the treatment with added nitrates ("potential" denitrification) were between 5 and 500 times higher than the rates without nitrates. "Potential" denitrification at the floodwater inlet showed maximum rates in August and October (Fig. 2). Rates at the outlet were significantly lower ($p<0.000$) for all the year except April, with a maximum in June of 98.6 μmol N m^{-2} h^{-1}. These results could be related to the input of fresh particulate organic matter, which is greater at the inlet than at the outlet (J A Romero, pers. commun.).

"Potential" denitrification was correlated with temperature at the inlet ($r^2=0.53$, $p<0.01$), but not at the outlet (Fig. 2). We were unable to explain this differences with the variables presented in this work. The increase in the production of N$_2$O after the addition of nitrates to the sediments confirms the strong limitation of denitrification due to lack of available nitrates for denitrifiers. Once this nitrate limitation is removed, N$_2$O production shows a clearer seasonal trend. No correlation with temperature was observed in the denitrification treatment (without nitrates added), which as other authors suggest (Koch et al., 1992; review by Knowles, 1982), would be due to denitrification being limited by nitrate availability.

Sediment nutrient concentrations in the upper 2 cm of sediment are presented in Table 1, as are Eh measurements. $NO_2^- + NO_3^-$ was around or below 50 nmol N g^{-1} of sediment (dry weight) at all sampling occasions, except in October and November. There were no significant differences between the inlet and the outlet, nor between sampling dates, the latter due again to the high variability of the two replicates in October and November. Ammonia concentrations were much higher, as was expected, and showed a similar pattern to $NO_2^- + NO_3^-$. No correlation was observed between either NH_4^+ or $NO_2^- + NO_3^-$ and production of N_2O in the blank or in the denitrification treatments. Eh values (Table 1) were negative at all sampling occasions, except in November at the inlet and May and November at the outlet. No correlation with the other variables was observed.

Conclusions

In the marsh studied denitrification was limited by nitrate supply. Even though ammonia concentrations are high, nitrification does not seem to be a major source of nitrates for denitrification. Competition for nitrogen between marsh vegetation and microorganisms would also explain the low rates of denitrification observed. The uncoupling of nitrification-denitrification by the use of acetylene has also to be considered as a cause for low denitrification.

Denitrification rates in this recovered marsh were smaller than denitrification rates in ricefields under exploitation (between 10 and 160 μmol N m^{-2} h^{-1}, Astorga et al., 1994), when no nitrate was added to the marsh sediments. When nitrate was added, the pattern reversed, rates being higher for the marsh sediments. This is probably caused either by a shortage of easily degradable organic matter (Novella et al., 1994) in the ricefield sediments studied, or by a shortage of nitrate or both. These results show that under different management practices, the factors controlling the nitrogen processes in simular sediments are different.

As we stated in the introduction, there are not many studies of the processes of the nitrogen cycle in the mediterranean area. More studies are needed on the nutrient cycles in this area, especially on an annual basis.

Acknowledgements

We would like to thank C J de Groot for his helpful suggestions and corrections on the first draft of this manuscript. This work was funded by the EEC Environmental Programme 1986–90 (EV4V-0132-E-TT), CE 91–0016 and CYCIT NAT89–844-CE. Thanks are also given to the Ebro Delta Natural Park - Generalitat de Catalunya. V N Astorga was in receipt of a FPI grant from the Spanish Ministry of Education and Science and E Novella in receipt of a FPI grant from the Generalitat de Catalunya.

References

Astorga V N, Novella E and Comín F A 1994 Denitrification related to organic matter in ricefield sediments. Verh. Int. Ver. Limnol. 25, 1365–1368.

Buresh R J, Delaune R D and Patrick W H, Jr (1981) Influence of Spartina alterniflora on nitrogen loss from marsh soil. Soil Sci. Am. J. 45, 660–661.

Chalamet A 1985 Effects of environmental factors on denitrification. In Denitrification in the Nitrogen Cycle. Ed. H L Golterman. pp 7–29. Plenum Publishing, New York, USA.

Davidson E A, Swank W T and Perry T O 1986 Distinguishing between nitrification and denitrification as sources of gaseous nitrogen production in soils. Appl. Environ. Microbiol. 52, 1280–1286.

Dean J V and Biesboer D D 1985 Loss and uptake of ^{15}N-ammonium in submerged soils of a cattail marsh. Am. J. Bot. 72, 1197–1203.

El-Habr H 1987 Les elements nutritifs du Rhone; leur devenir dans les canaux d'irrigation et les marais en Camargue. Thèse de Doctorat. Université Claude Bernard, Lyon I, France.

Faulkner S P and Richardson C J 1989 Physical and chemical characteristics of freshwater wetland soils. In Constructed Wetlands for Wastewater Treatment. Ed. R D Hammer. pp 41–72. Lewis Publishers, Chelsea, MI, USA.

Forés E and Christian R R 1993 Network analysis on nitrogen cycling in temperate wetland ricefields. Oikos 67, 299–308.

Hsieh Y P and Coultas C L 1989 Nitrogen removal from freshwater wetlands: nitrification - denitrification coupling potential. In Constructed Wetlands for Wastewater Treatment. Ed. R D Hammer. pp 493–500. Lewis Publishers, Chelsea, MI, USA.

Jensen M H, Andersen T K and Sorensen J 1988 Denitrification in coastal bay sediment: regional and seasonal variation in Aarhus Bight, Denmark. Mar. Ecol. Prog. Ser. 48, 155–162.

Johnston C 1991 Sediment and nutrient retention by freshwater wetlands: effects on surface water quality. Crit. Rev. Environ. Control 21, 491–565.

Knowles R 1982 Denitrification. Microbiol. Rev. 46, 43–70.

Koch M S, Maltby E, Oliver G A and Bakker S A 1992 Factors controlling denitrification rates of tidal mudflats and fringing salt marshes in south-west England. Estuarine Coastal Shelf Sci. 34, 471–485.

Koike I and Sorensen J 1988 Nitrate reduction and denitrification in marine sediments. In Nitrogen Cycling in Coastal Marine Environments. (SCOPE 33). Eds. T H Blackburn and J Sorensen. pp 251–273. John Wiley and Sons, New York, USA.

522

Lalisse-Grundmann G 1987 Dénitrification en sol de culture: definition des conditions imposées par les facteurs de l'environment en vue de sa modelisation. Thèse de Doctorat. Université Claude Bernard, Lyon I, France.

Law C S, Rees A P and Owens N P J 1991 Temporal variability of denitrification in estuarine sediments. Estuarine, Coastal Shelf Sci. 33, 37–56.

Minzoni F, Bonneto C and Golterman H L 1988 The nitrogen cycle in shallow water sediment systems of ricefields. Part I: The denitrification process. Hydrobiologia 159, 189–202.

Novella E, Astorga V N and Forés E 1994 Effects of organic matter mineralization on the respiratory activity in ricefield sediments. Verh. Int. Ver. Limnol. 25, 1361–1364.

Seitzinger S P 1988 Denitrification in freshwater and coastal marine ecosystems: Ecological and geochemical significance. Limnol. Oceanogr. 33, 702–724.

Seitzinger S P, Nielsen L P, Cafrey J and Christensen P B 1993 Denitrification measurements in aquatic sediments: A comparison of three methods. Biogeochemistry 23, 147–167.

Seitzinger S P 1994 Linkages between organic matter mineralization and denitrification in eight riparian wetlands. Biogeochem. 25, 19–39.

Sloth N P, Nielsen L P and Blackburn T H 1992 Nitrification in sediment cores measured with acetylene inhibition. Limnol. Oceanogr. 37, 1108–1112.

Tiedje J M, Simkins S and Groffman P M 1989 Perspectives on measurement of denitrification in the field including recommended protocols for acetylene based methods. Plant and Soil 115, 261–284.

O. Van Cleemput et al. (eds.), Progress in Nitrogen Cycling Studies, 523–525, 1996.

Nitrous oxide emissions from soil incorporation of crop residues

Elizabeth Baggs[1], Robert Rees[2] and Keith Smith[1,2]

[1]IERM, University of Edinburgh/SAC and [2]Soils Department, SAC, School of Agriculture, West Mains Road, Edinburgh, EH9 3JG, UK

Key words: cultivation, grass/clover, green manure, nitrous oxide

Abstract

The practice of incorporating plant residues may significantly contribute to emissions of nitrous oxide (N_2O) from agricultural systems. Field trials were undertaken on two experimental sites in Scotland: a sandy loam and a loamy sand. At the first site, N_2O emissions were compared after incorporation of different varieties of Italian ryegrass, and different mixes of grass/clover varieties. At the second site comparisons of N_2O emissions were made between five different crop residues - white clover, mustard, oats, trefoil and forage peas. These emissions were related to measurements of soil available N, temperature and gravimetric soil moisture content. Short-lived fluxes of N_2O occurred immediately after incorporation of residues at both sites. The greatest emissions, 23 g N_2O-N ha^{-1} d^{-1}, occurred after incorporation of the grass/clover trial, and were positively correlated with a rise in air temperature ($r=0.5$, $p<0.01$). Soil available N increased throughout the experiments, the greatest increases being at the second experimental site.

Introduction

Previous work has shown that emissions of N_2O after incorporation of plant residues may make a significant contribution to emissions from agricultural systems (Aulakh et al., 1983; Ryden et al., 1979). Such addition of degradable organic material enhances microbial activity, leading to oxygen consumption, and the development of anaerobic microsites in the soil. The decomposition of plant residues can also lead to N_2O production as a result of nitrification. Such N_2O emissions are of concern since they have been estimated to contribute up to 5% of greenhouse warming (Van Breemen and Feitjel, 1990). Nitrous oxide also destroys stratospheric ozone through the formation of NO (Sahrawat and Keeney, 1986). Granli and Bockman (1994) state that it is not at present possible to make a precise estimate of the global N_2O production from agricultural land given that there are too few field measurements, with widely variable results, and an incomplete coverage of potentially important areas and sources. This work compares emissions after incorporation of different types of crop residues, including both legumes and non legumes, and considers the importance of various controlling soil environmental variables.

Sites, materials and methods

Sites and treatments

Two experimental sites were used. The first site was a sandy loam at the Edinburgh School of Agriculture's experimental farm on the Bush estate, where two trials were established. The second was a loamy sand near Elgin, north-east Scotland. At Bush comparisons were made following the incorporation of different varieties of Italian ryegrass and grass/clover swards in a blocked experimental design. Augusta, 85/22, and Bab 424 varieties of Italian ryegrass, and Menna, Kent, and Huia varieties of clover were included in the trials, on plots of 5 m by 1.2 m. The grass/clover swards had been cut 6 times throughout the growing season in an attempt to simulate grazing.

At Elgin an overwintering trial of various green manures had been set up which included white clover, mustard, oats, trefoil and forage peas. The residues of these crops were incorporated in the spring on replicated plots (12 × 3.65 m). Bare fallow soil, both with and without weeds, was also included in this comparison.

At both sites crop residues were ploughed into the soil, and followed by rotary tillage. The plots were sown to barley at Bush and oats at Elgin, and rolled. Ploughing took place on 6 April (Julian day 96) and rotary tillage, sowing and rolling on 28 April (Julian day 118) at Bush. At Elgin all cultivation was under-taken on 14 April (Julian day 104).

Nitrous oxide measurements

Circular chambers of 40 cm diameter by 20 cm height were inserted 5 cm into the soil on each plot. Gas samples were taken from these chambers both prior to and periodically following cultivation. The chambers were closed with an airtight lid for a known period of time (1 hour), after which samples were taken. These samples were analysed for their N_2O content by gas chromatography, using an electron capture detector, in the laboratory. Measurements of air temperature were made at the time of gas sampling.

Mineral N

Monthly determinations of available NH_4^+ and NO_3^- were made by soil extraction using 1 M KCl in a 1:5 soil:solution ratio. Extracts were filtered and NO_3^- and NH_4^+ determined by continuous flow analysis. Gravi-metric soil moisture contents were also determined at monthly intervals.

Results and discussion

Nitrous oxide

Short-lived fluxes of N_2O occurred immediately after rotary tillage in both trials at the Bush estate (Fig.1). Ploughing alone appeared to have little or no effect on emissions. Maximum N_2O emissions at Julian days 121 and 124 were greater on the grass/clover trial than the Italian ryegrass trial, reaching values of 23 and 14 g N_2O-N ha^{-1} d^{-1}, respectively. This may have been due to a lower C:N ratio of the clover residues, stimulating microbial activity.

These N_2O fluxes were positively correlated with a rise in air temperature (r=0.5, $p<0.01$ on the grass/clover plots, and r=0.3, $p<0.001$ on the ryegrass plots), but not with soil moisture content. All varieties of Italian ryegrass showed weaker correlations with air temperature than the various grass/clover mixtures. The sudden rise in temperature from 4 to 15 °C on Julian day 120 coincided with the greatest N_2O emis-sions. This created some difficulty in separating the

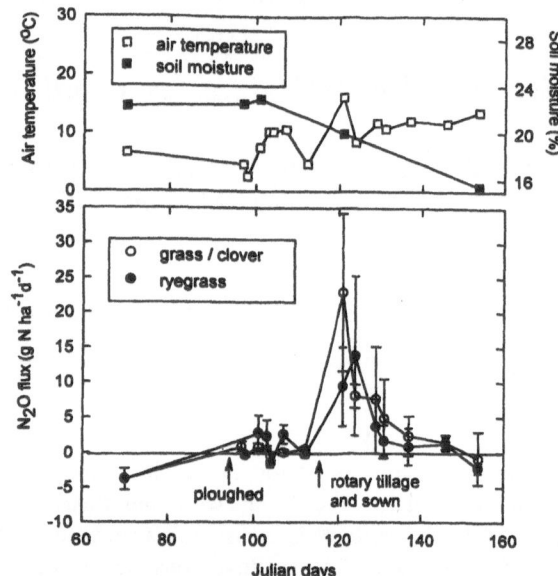

Figure 1. Daily N_2O - N emissions, air temperature and soil moisture following residue incorporation at Bush.

effects of rotary tillage and temperature. The gravi-metric soil moisture contents correlated poorly with N_2O emissions (r=0.06, $p<0.01$ for grass/clover; r=0.1, $p<0.01$ for ryegrass), suggesting that temperature is a more significant variable for gaseous emissions.

Nitrous oxide emissions were less pronounced fol-lowing residue incorporation at the Elgin site (Fig.2), with no significant difference between treatments. A minor flux of N_2O occurred on all treatments imme-diately after incorporation. The greatest emissions of 14 g N_2O-N ha^{-1} d^{-1} were from the fallow - weeds plots on Julian day 116, but this was very short-lived. Cultivation did not appear to have a large effect on emissions, confirming the importance of temperature. The peak N_2O emission of 14 g N_2O-N ha^{-1} d^{-1} on day 140 corresponded with a rise in temperature from 11 to 12 °C at this site. The temperature correlated strongly with N_2O emissions throughout the sampling period (r=0.8, $p<0.01$).

The mean daily N_2O emissions from both sites (Fig. 3) show that the greatest total emissions over the sam-pling period resulted from the incorporation of the Huia variety of clover. Incorporation of var. Augusta result-ed in the lowest emissions from the two trials at the Bush Estate. Nitrous oxide emissions from the various crop residues at Elgin were relatively similar (Fig. 3).

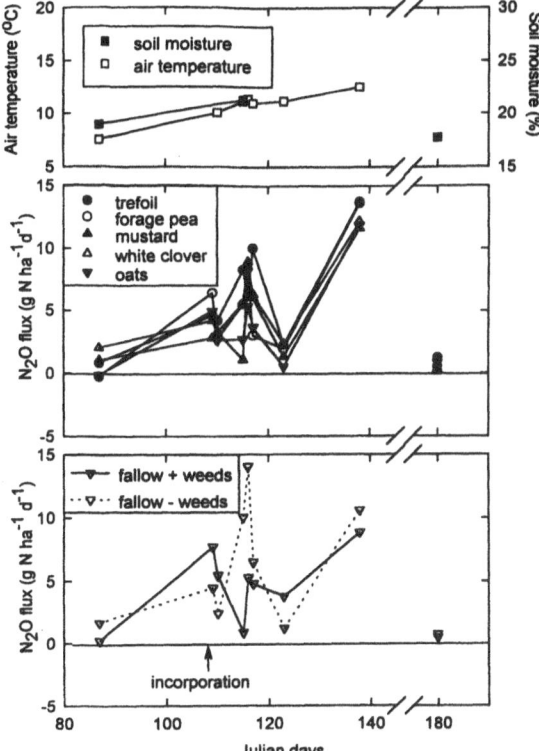

Figure 2. Daily N₂O - N emissions, air temperature and soil moisture following residue incorporation at Elgin.

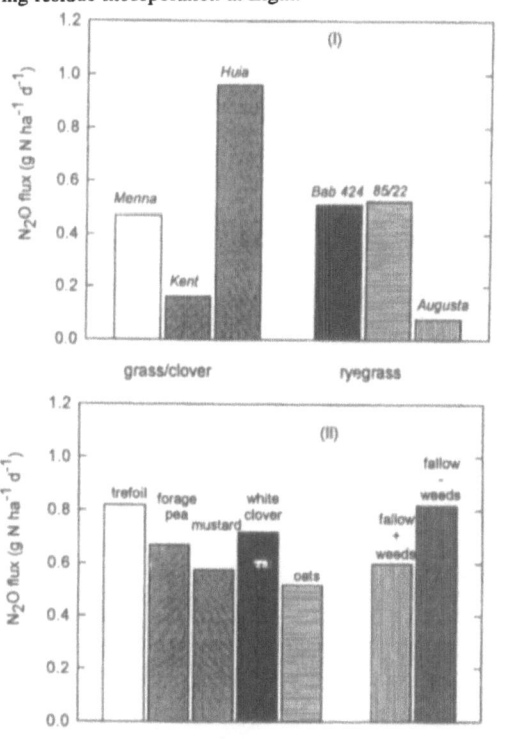

Figure 3. Mean daily N₂O - N emissions over sampling period at (I) Bush, and (II) Elgin.

Mineral N

The available N concentrations of the soils increased in all plots during the sampling period from April until June, with largest increases following the incorporation of crop residues at the Elgin site. The greatest increase in available NO_3^--N, from 3 to 15 μg N g soil^{-1}, occurred on the fallow + weeds treatment, with available NH_4^+-N increasing from 0.5 to 12 μg N g soil^{-1}. Available NH_4^+-N on the trefoil plots increased from 1 to 16 μg N g soil^{-1} over this period. At the Bush estate the largest increases in available N were found on the grass/clover trial, with increases of available NO_3^--N from 1 μg g soil^{-1} to 11 μg N g soil^{-1}, and available NH_4^+-N from 2 μg N g soil^{-1} to 4 μg N g soil^{-1}.

Nitrous oxide may be released from soils as a product of either nitrification or denitrification (Bremner and Blackmer, 1981). The relative contributions of these two processes to the N₂O fluxes in these trials are unquantifiable without the use of selective nitrification inhibitors, or ^{15}N methodology. However, the work presented here suggests the importance of nitrification as a source of N₂O emission, and further process based studies will be undertaken to verify this.

Acknowledgements

Funding for this work was provided by the Ministry of Agriculture, Fisheries and Food, Scottish Natural Heritage, and the Scottish Office Agriculture and Fisheries Department.

References

Aulakh M S, Rennie D A and Paul E A 1983 The effect of various clover management practices on gaseous N losses and mineral N accumulation. Can J. Soil Sci. 63, 593–605.

Bremner J M and Blackmer A M 1981 Terrestrial nitrification as a source of atmospheric nitrous oxide. *In* Denitrification, Nitrification and Atmospheric Nitrous Oxide. Ed. C C Delwiche. pp 151–170. John Wiley and Sons, New York, USA.

Granli T and Bockman O C 1994 Nitrous oxide in agriculture. Norw. J. Agric. Sci. Suppl. 12, 1–128.

Ryden J C, Lund L J, Letey J and Focht D D 1979 Direct measurement of denitrification loss from soils. II. Development and application of field methods. Soil Sci. Soc. Am. J. 43, 110–118.

Sahrawat K L and Keeney D R 1986 Nitrous oxide emissions from soils. Adv. Soil Sci. 4, 103–148.

Van Breemen N and Feijtel T C J 1990 Soil processes involved in the production of greenhouse gases, with special relevance to soil taxonomic systems. *In* Soils and the Greenhouse Effect. Ed. A F Bouwman. pp 195–223. John Wiley and Sons, New York, USA.

O. Van Cleemput et al. (eds.), Progress in Nitrogen Cycling Studies, 527–531, 1996.

Losses and flows of nitrogen in the Norwegian society

Marina Azzaroli Bleken[1] and Lars R. Bakken[2]

[1]Department of Economics and Social Sciences and [2]Departments of Soil and Water Sciences, Agricultural University of Norway, P.O. Box 5033, N 1432 Aas, Norway

Key words: nitrogen budget, nitrous oxyde, pollution

Abstract

The anthropogenic fixation of atmospheric nitrogen by industry and biological fixation in agriculture "pumps" a significant amount of nitrogen into the biosphere. Most research concerning the pollution effects of these inputs has been focused primarily on the agricultural sector and the contamination of surrounding waters, whereas the fate of the flow of nitrogen from arable land through plant products is often ignored in such contexts. We expect that a clue to the global aspect of the nitrogen pollution problem may be found by tracing and quantifying such flows of nitrogen through society. This approach may be particularly useful to show the connection between human activity and the present accumulation of nitrous oxide in the atmosphere. Although the primary N-source for the N_2O accumulation is thought to be the N-inputs to agriculture, the fertilized soil is not necessarily the primary site where the fertilizer-N is transformed to N_2O. Shifting focus from the fertilized soil to the recipients of nitrogen losses may allow a more realistic estimate of present and future anthropogenically driven biological production of N_2O.

We are investigating the nitrogen flows in Norway; preliminary data for the main flows are presented in this paper. These data are based on both official and non-official statistics for imports and exports, and commodities produced and traded within the Norwegian society. Not surprisingly, fertilizer use represents a large primary input of N to the system (26 kg N per capita annually, **pca**), and urea used for industrial purposes is the second largest one (1.4 kg N pca). Less than 1/5 of the fertilizer + fodder-N input is recovered in edible agricultural products. Hence, the agricultural system as a whole operates with a considerable nitrogen surplus (21 kg N pca), which is partly lost to the surroundings as leached nitrate or volatilized ammonium (10–15 kg N pca), as products unaccounted for (wastes + nonedibles), or stored within the system (as soil organic N). The amount of nitrogen in other waste products is comparable to the total amount that enters the sewage system (4 kg N pca). In comparison, the total NO_x production by combustion and industry is 14 kg N pca.

Introduction

Nitrogen pollution, a local/regional or global pollution problem ?

The human impact on the nitrogen cycle has increased rapidly over the last decades, the result being that the scale of the nitrogen pollution has shifted from regional to continental and even to a global level (Heathwaite et al., 1993). Nitrogen pollution was originally perceived as a local problem resulting in eutrophication (aquatic habitats) and health risks (nitrate in drinking water). The shift in scale from local to regional is obvious when considering the documented atmospheric trans-port of nitrogen over large areas, and the penetration of nitrate into continuously larger fraction of the available drinking water sources, fresh water bodies as well as marine environments (ibid). So far, however, the nitrogen accumulation in soil and water bodies has only to a limited extent been recognized and approached as a truly global problem.

Attempts to estimate global rates of fixation have demonstrated that the anthropogenic inputs (fertilizers, biologically fixed nitrogen, and production of NO_x during combustion) greatly exceed historic or prehuman inputs (Delwich, 1977; Granli and Bøckman, 1994). The likely outcome of this increased input is a transient period of net accumulation of biologically available fixed nitrogen (primarily as organic nitrogen). It will be transient, because removal through denitrification (and possibly by "fossilization" in sediments) sooner or later will balance the new level of input. A number of factors will influence the perturbations brought about by the elevated N-inputs to the biosphere. The reactivity of the various nitrogen pools, and the transfer rates between them is critical (Delwiche, 1977). The primary distribution of the extra (anthropogenic) N between systems (topsoils, subsoils, fresh water and marine waters), and its chemical form (mineral-N versus organic-N) is also decisive.

Atmospheric N_2O

The atmospheric concentration of nitrous oxide (N_2O) is currently increasing with about 0.25% per year (Prinn et al., 1990). Analyses of historic atmospheric samples (bubbles entrapped in Antarctic ice) indicate a gradual increase during the last 200 years, but the most dramatic increase has taken place after World War II. This coincides rather closely with growth in industrial activity, combustion of fossil fuels and application of chemically fixed nitrogen in agriculture.

A number of direct chemical sources have been identified and assessed (industry, combustion etc.), but their total contribution explains only a fraction of the observed accumulation of atmospheric N_2O. For this reason, emissions from biological processes must account for most of the increased N_2O emission that has occurred during the last 50 years (Granli and Bøckman, 1994).

Direct measurements of emissions induced by fertilization of agricultural land have shown that this may be a significant source, but contributes little to the overall emission (Robertsson, 1991). Measured emissions from other habitats (terrestrial as well as aquatic) have demonstrated that nitrogen inputs as well as a variety of "disturbances/perturbations" may increase the N_2O emissions (Granli and Bøckmann, 1994). Integration of such data to yield estimates of regional/continental emissions has been difficult due to the almost overwhelming spatial and temporal variability of the fluxes (ibid).

The link between N-dissipation and N_2O

The apparently obvious causal relationship between fertilization and N_2O emissions has led to a massive research on the role of fertilized agricultural land as an N_2O source. It may be argued that this approach is somewhat misleading, not because agricultural land is an insignificant source but because the approach is focusing on the "tip of the iceberg".

Modern agricultural systems have extremely open nitrogen cycles, where most of the imports of nitrogen (as mineral fertilizers and biological fixation) is balanced by equal exports of N as plant products plus emissions of ammonia (from manure) and leached nitrate. This nitrogen flow from the agricultural system is a potential source of N_2O elsewhere in the biosphere, wherever and whenever it ends up as mineral forms of nitrogen susceptible to bacterial oxidation/reduction processes by which N_2O is produced and liberated as a side product. This may happen gradually over many years since much of the nitrogen will be held in an organic form with considerable residence time (as biomass and humic substances).

The turnover time of the various organic nitrogen pools, together with the N_2O/N_2 ratio of the N-losses from the system, will determine the kinetics and total quantities of N_2O emitted as a result of nitrogen inputs to each habitat.

This view calls for new approaches in the study of the human impact on biological production of N_2O. We need more information about the fate of the nitrogen that leaves agriculture in the form of various products. We need to know the ways in which nitrogen is transported and where it ends up. This is what the present project is about. The long range aim is to arrive at more realistic estimates of present and future emissions of N_2O.

Methods

The primary aim was to estimate the amount of nitrogen in the various commodities of the Norwegian society. The Norwegian Official Statistics were used to estimate the amounts of N in more than 1400 commodities. Data for imports and exports were adequate for this purpose but the industry statistics were difficult to use due to lack of information about physical units and inadequate resolution level (pooling of commodities with widely differing N-contents). Thus, direct contact with producers and trade organizations was necessary. The Rubin foundation provided statistics on byproducts from fish. Data are average annual values for the period 1988–1991; in some cases data are based on shorter periods.

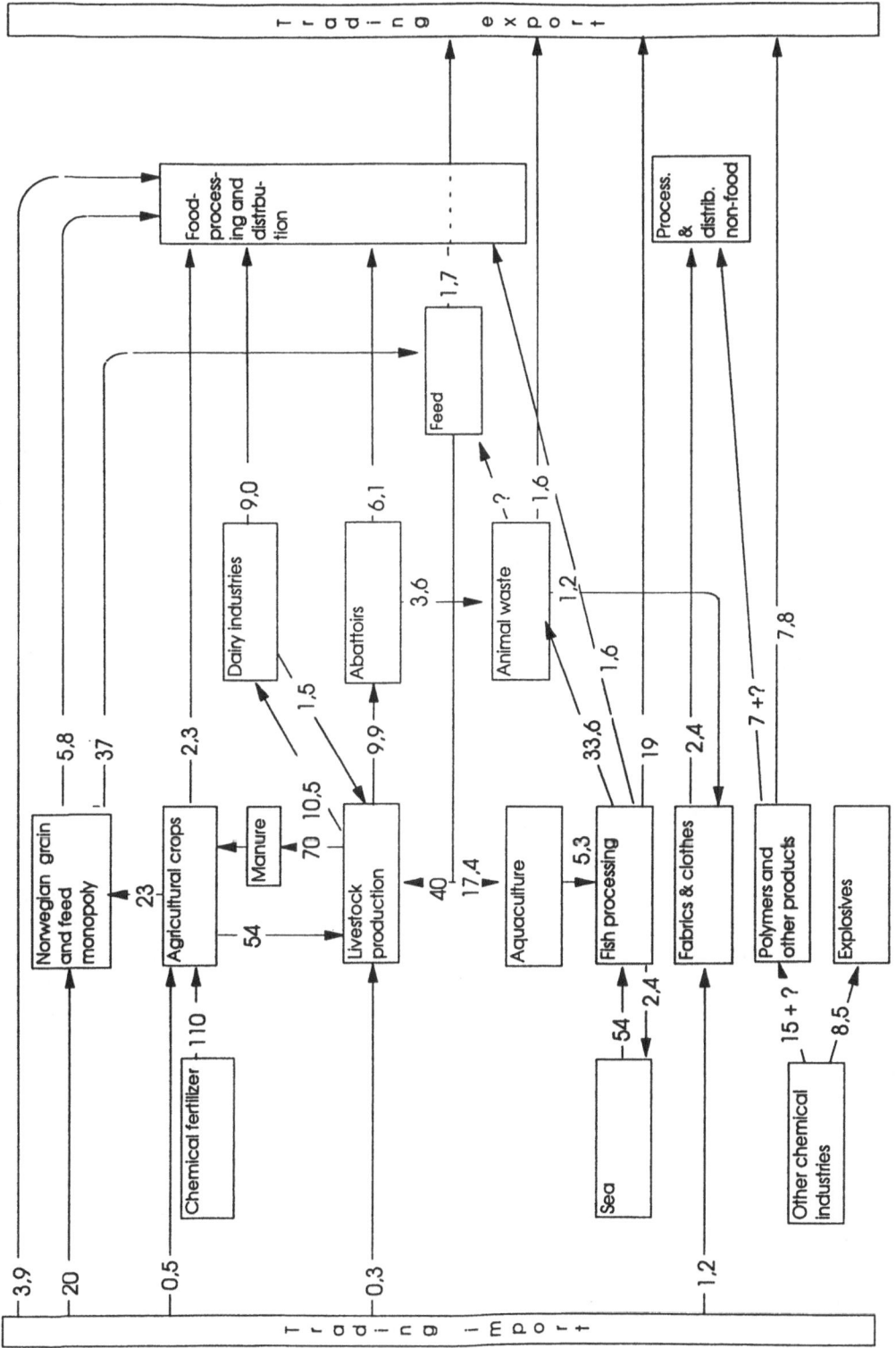

Figure 1. Annual major transport of nitrogen in the Norwegian economy, in Gg. Average values for the period 1988 - 1991, except for waste from fish and for some chemical products which are based on data for 1991. The population of Norway is 4.3 million. Only net trading import and export is shown. We have not been making distinctions between the final consumer and final processing/transport, as in both cases waste is likely to end up in municipal renovation.

530

Results and discussion

We are reporting preliminary and somewhat incomplete estimates of major N-flows. Please contact the corresponding author before citing any specific numbers, as they are under revision.

Figure 1 shows the annual flows of nitrogen (in 10^9 g N, or Gg N) through the different sectors of the Norwegian economy. Trading imports are shown at the extreme left. The flows between major commodities and sectors within Norway is at the center of the figure, from which the products are either consumed in Norway as edible and non-edible commodities (towards right) or exported to other nations (extreme right hand side of the figure).

Agricultural crop production receives 110 Gg N through mineral fertilizers and 70 Gg through manure. The latter estimate is based on excreted N from animals, of which a substantial fraction is likely to be lost through ammonia volatilization during storage and immediately after spreading of the manure (Osthoek et al., 1991). The net export of N through cropped plant material is 23 (cereals, mainly to feed) + 1.8 (other crops human consumption) + 54 (fodder crops) = 79.3 Gg N. Thus, exported N is close to 50% of the total input to this system (higher if the lost ammonia is not included as an input).

Livestock production receives 95.5 Gg N in fodder, and gives off 70 Gg in manure, 10.5 Gg in "dairy products" (eggs included) and 9.9 Gg in meat. This leaves about 5 Gg of N unaccounted for.

As a whole, the agricultural sector receives 150 Gg N (110 in fertilizer and 40+1.5 in feedstuff), and gives off only 23+1.8+10.5+9.9=45.2 Gg N in sold products. The difference between inputs and outputs is thus around 100 Gg N. Ammonia losses, leaching of nitrate and denitrification may account for a substantial fraction of this N-loss, but experiences with single farm analyses indicate that such losses may only account for a fraction of the difference between input and output of N (Jarvis, 1991; Kristensen and Kristensen, 1992).

The fish processing sector receives 54 Gg N from the sea and 5.3 Gg N from aquaculture, in 1991. The major N output from this industry goes to fodder production and to export, and only a small fraction (1.6 Gg) to human consumption in Norway. About 4 Gg fron cut-offs are dumped either directly into the sea or deposited on land. Fish quantity varies greatly with years.

The N content of all edible products for the Norwegian market ("Food processing and distribution" at

Table 1. Nitrogen input and output from Norwegian agriculture, abattoirs, dairy industry and mills, taken as one sector. Averages for the period 1988-1991

Input	Gg N/year^{-1}
Chemical fertilizer	110
Feed non derived from Norwegian agriculture	17
Sum	127
Output	
Cereals for human consumption	2.2
Other crops for human consumption	1.8
Dairy products	9.0
Abattoirs[a]	5.7
Sum	18.7

[a] Meat production, mainly lambs, from natural pastures not included.

right in Fig. 1) is 28.7 Gg. The fate of this nitrogen has not yet been analyzed. Sewage-N may account for about 17 Gg of this N, based on a per capita sewage-N of 4 kg per year (Petter D. Jenssen, Agricultural University of Norway pers. commun.). Only 50% of the sewage-N reaches the sewage treatment plants, the rest is assumed lost through a leakages in the sewage systems (ibid). Thus about 12 Gg. are expected to end up as garbage. We intend to compare these results with the National garbage inventory which is soon to be published.

Nitrogen in non-edible products are apparently moderate, but the analysis of these commodities is not complete. Polymers and other products amount to at least 15 Gg N, of which at least 8,5 Gg is sold to consumers in Norway. The major part of this nitrogen is urea used in combination with aldehydes as hardeners. The further fate of this nitrogen has not yet been investigated, but it may be justifiable to assume a considerable time lag before it becomes available for biological transformations. The annual consumption of explosives amounts to 8.5 Gg N. The use of explosives releases oxidized forms of nitrogen.

The agricultural sector appears to be the main source of N-dissipation to the outer environment. It should be kept in mind, however, that this sector also has a considerable "buffer capacity" in its large reservoirs of soil organic nitrogen. This means that the N-dissipation from the agricultural sector is not necessarily equal to the net balance between inputs and outputs of nitrogen.

The processing of edible raw materials is apparently a minor source of N-dissipation, but the analyses of these processes are still incomplete. The nitrogen in sold food is substantial, and hitherto a very small fraction of this nitrogen is returned to productive agricultural land. A considerable fraction is deposited as garbage (10 Gg), lost through leaky sewage pipes (8 Gg), collected in sewage treatment plants (mainly returned to unproductive areas) or discharged into the sea.

The analysis of other commodities indicates that they represent a small but significant flow of nitrogen, which sooner or later will become available for biological transformations.

Considering agriculture, abattoirs, dairy production and mills as one sector, the input of nitrogen by chemical fertilizers and imported feed is 110 + 17 = 127 Gg N, and the output is about 19 Gg N (Table 1). On a per capita basis (population approximately 4,3 million), the annual input is about 30 kg N and output 4.1 kg. The annual N-surplus for the agricultural sector is 21 kg N per capita. The annual consumption of N in industrial products is 1.4 kg N per capita. It may be worth mentioning that the total emissions of NO_x-N through combustion (stationary and mobile) and industrial processes is around 14 kg N per capita.

These preliminary investigations illustrate that a number of pathways contribute to the total dissipation of nitrogen from society to the environment. Further refinement of the analysis will be conducted, so as to get more precise information about the size of the different N-discharges and the further fate of this nitrogen, emphasizing the production of N_2O.

References

Delwiche C C 1977 Energy relations in the global nitrogen cycle. Ambio 6, 106–111.

Granli T and Bøckman O C 1994 Nitrous oxide from agriculture. Norw. J. Agric. Sci., Suppl. 12, 128.

Heathwaite A L, Burt T P and Trudgill S T 1993 Nitrate, Processes, Patterns and Managements. John Wiley and Sons, Chichester, UK. 444 p.

Jarvis S C 1993 Nitrogen cycling and losses from dairy farms. Soil Use Manage. 9, 99–105.

Kristensen E S and Kristensen I S 1992 Analysis of nitrogen surplus and efficiency on organic and conventional dairy farms. Report no 710 from the National Institute of Animal Science, DK-8830 Tjele, Denmark. pp 1–53.

Osthoek J, Kroodsma W and Hoeksma P 1991 Ammonia emission from dairy and pig housing systems. *In* Odour and Ammonia Emissions from Livestock Farming. Eds. V C Nielsen, J H Voorburg and P L'hermite. pp 31–42. Elsevier Applied Science, London, UK.

Prinn R, Cunnold D, Rasmussen R, Simmonds P, Alyea F, Crawford A, Fraser P and Rosen R 1990 Atmospheric emissions and trends of nitrous oxide deduced from 10 years of ALE-GAGE data. J. Geophys. Res. 95D, 18369-18385.

Robertson K 1991 Emission of N_2O in Sweden - Natural and anthropogenic sources. Ambio 20, 151–155.

O. Van Cleemput et al. (eds.), Progress in Nitrogen Cycling Studies, 533–536, 1996.

Comparison of N_2O flux measurements using gas chromatography and photo-acoustic infra-red spectroscopy

C.A.M. De Klein[1], R. Harrison[1] and E.I. Lord[2]
[1]ADAS, Anstey Hall, Maris Lane, Trumpington, Cambridge CB2 2LF, UK and [2]ADAS, Wergs Road, Wolverhampton WV6 8TQ, UK

Key words: carbon dioxide, gas chromatography, nitrous oxide, photo-acoustic infra-red spectroscopy, soil enclosures

Abstract

The Trace Gas Analyser (TGA) uses photo-acoustic infra-red spectroscopy to measure the N_2O concentration in a gas sample. In association with soil enclosures, this fully portable device allows in-field N_2O flux measurements. The accuracy of photo-acoustic infra-red spectroscopy (PAIRS) for N_2O analysis was tested by comparing PAIRS analysis with GC analysis of both standard N_2O gases and of head-space samples taken from soil enclosures. In addition, the CO_2 interference on N_2O analysis by PAIRS was investigated by analysing N_2O/CO_2 gas mixtures.

Photo-acoustic infra-red spectroscopy accurately measured the N_2O concentration of standard gases. The CO_2 interference on N_2O analysis by PAIRS was linear and the average CO_2 correction factor was 4.2×10^{-4}. There was some discrepancy between flux measurements using GC and PAIRS, which might be due to cross contamination between the soil enclosures.

Abbreviations: PAIRS–photo-acoustic infra-red spectroscopy, TGA–Trace Gas Analyser.

Introduction

N_2O fluxes from soil are commonly measured using soil enclosures and gas chromatography (e.g. Hutchinson and Brams, 1992; Mosier et al., 1993; Skiba et al., 1992). Although accurate, GC analysis of N_2O samples has important drawbacks. In the absence of a field-laboratory, samples taken in the field have to be transported to a remote laboratory for analysis. This increases the risk of leakage. Moreover, since the results are not directly available, flexible adjustments of the sampling strategy are not possible. This could lead to spending valuable time in the field when N_2O fluxes are low or not detectable.

The Trace Gas Analyser (TGA) represents an alternative approach for measuring N_2O fluxes from soil (De Klein et al., 1994). The fully portable and robust analyser allows in-field N_2O flux measurements in association with soil enclosures. The TGA uses a technique known as photo-acoustic infra-red spectroscopy to measure the N_2O concentration in a gas sample. Since CO_2 has its infra-red spectrum in the same range

as N_2O, this gas will interfere in the N_2O measurements and a correction for CO_2 present in the sample should be made.

The objectives of the work presented in this paper were (1) to test the accuracy of the TGA by comparing N_2O measurements using gas chromatography and photo-acoustic infra-red spectroscopy; and (2) to investigate the effect of interference by CO_2 on N_2O analysis by photo-acoustic infra-red spectroscopy.

Materials and methods

TGA accuracy

Teflon gas bags (3 L) were filled with the following range of certified standard gases: 0.96, 3.6, 9, 19, 41.2 and 103 $\mu L\ L^{-1}$ N_2O in N_2. After connecting the gas bags to the Trace Gas Analyser with 1 m Teflon tubing (3 mm diameter), the standards were analysed in duplicate.

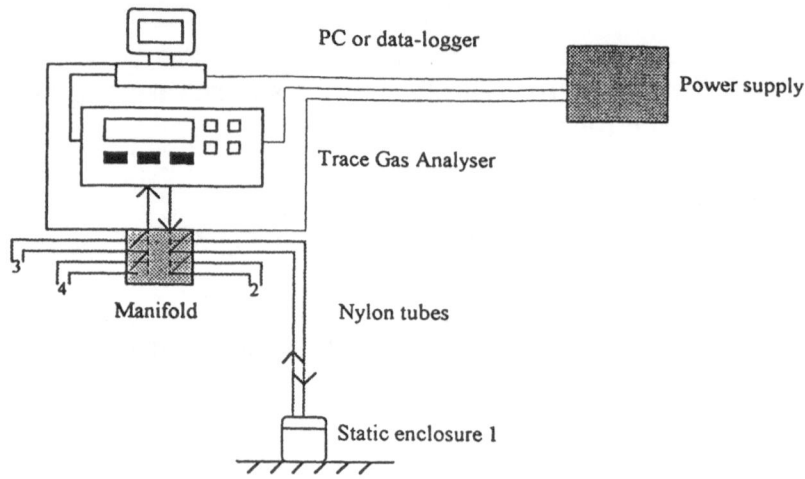

Figure 1. Schematic overview of the system for measuring N_2O emissions in the field.

For field measurements of N_2O fluxes the TGA was used in association with soil enclosures (15 cm diameter and 12 cm high). The enclosures consisted of a PVC cylinder with a screw-on PVC lid. The lid was fitted with four ports: an in- and an outlet port, a vent-port and a port fitted with a three-way valve to connect a syringe. The vent-port contained a short piece of tubing (3.3 mm diameter and 8 cm long) to allow pressure equilibration in and outside the enclosures (Hutchinson and Mosier, 1981). The enclosures were insulated with a thin layer of foam, coated with a reflecting aluminium layer (radiator insulation material).

In the greenhouse, plastic containers (50 cm wide, 30 cm deep and 20 cm high) were filled with sandy loam soil. The soil was irrigated twice (2×10 mm) and fertilised with NH_4NO_3 at a rate of approximately 4 g N m^{-2}. Four enclosures were inserted into the soil to 2 cm depth and connected to the TGA via a manifold using nylon tubing with a diameter of 3.3 mm (Fig. 1). N_2O flux measurements were carried out by sampling each enclosure every 6 minutes during a 2 hour period. During the measurements the soil temperature at 5 cm depth was 19 °C.

In addition, 0, 60, 90 and 120 minutes after closing the lids, head space samples from the enclosures were taken by syringe and analysed on a gas chromatograph (Hewlett Packard 5890 A with ECD detector and Poropack Q column).

CO_2 interference

To investigate the CO_2 interference with photo-acoustic N_2O analysis, N_2O/CO_2 gas mixtures were analysed using both photo-acoustic infra-red spectroscopy and gas chromatography. The mixtures were made using the following certified standards:

    – 0, 0.9, 1.8, 3.6, and 9 $\mu L\ L^{-1}$ N_2O in N_2 and

    – 0, 100, 350, 500, 1750, and 2500 $\mu L\ L^{-1}$ CO_2 in N_2

Using a 1L gas syringe, 30 Teflon gas bags (3L) were filled with 200 mL of the N_2O standards (6 bags for each N_2O standard). Thereafter, 600 mL of one of the CO_2 standards was added so that each of the N_2O standards was mixed with each of the CO_2 standards. This resulted in the following N_2O and CO_2 concentrations in the mixtures:

    N_2O – 0, 0.24, 0.45, 0.9 and 2.25 $\mu L\ L^{-1}$

    CO_2 – 0, 75, 262, 375, 1312, 1875 $\mu L\ L^{-1}$

A 5 mL sample was taken from each bag by syringe and analysed on the GC for N_2O. Thereafter, the gas bags were connected to the TGA and analysed in duplicate.

Results and discussion

TGA accuracy

The analysis of certified N_2O standards gases using photo-acoustic infra-red spectroscopy showed a linear response up to 100 $\mu L\ L^{-1}$ (Fig. 2). The slope of the regression line was close to unity, indicating good agreement with the known concentrations of the standard gases. In contrast, GC analysis in the same range of concentrations showed a non-linear response.

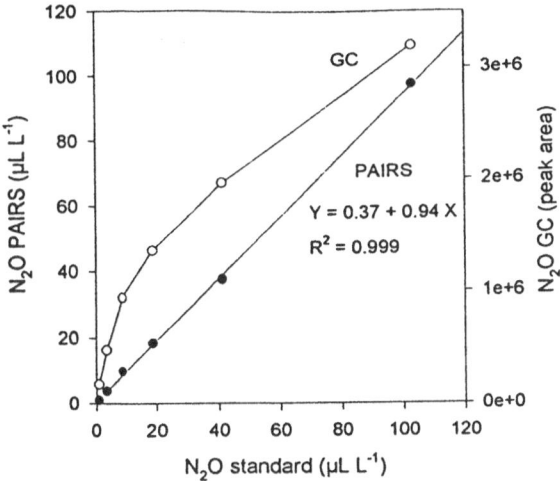

Figure 2. Analysis of certified standard gases (N_2O in N_2) using photo-acoustic infra-red spectroscopy (PAIRS) and gas chromatography (GC).

Table 1. Comparison of N_2O flux measurements from 4 enclosures using gas chromatography (GC) and photo-acoustic infra-red spectroscopy (PAIRS). Mean value is the back log-transformed mean of log-transformed data

| | N_2O flux in g N ha^{-1} day^{-1} | |
	PAIRS	GC
Flux 1	3	nd[a]
Flux 2	5	nd[a]
Flux 3	4	8
Flux 4	9	23
Mean	6	4

[a]nd = not detectable.

Measurements of N_2O fluxes from soil using GC and PAIRS analysis in association with soil enclosures showed some discrepancy (Table 1). In 2 enclosures the flux was about 2.5 times higher when calculated using GC analysis. The fluxes in the other 2 enclosures were not detectable with the GC, whereas PAIRS analysis revealed low fluxes. To calculate a mean value for the 4 enclosures, the data were log-transformed, averaged and back log-transformed. The back log-transformed means were similar for the 2 analysis techniques (Table 1).

The discrepancy between the flux measurements might be due to the current field set-up. At each sampling for PAIRS analysis, the head-space air from an enclosure is diluted with head-space air from the previous enclosure still present in the dead volume between

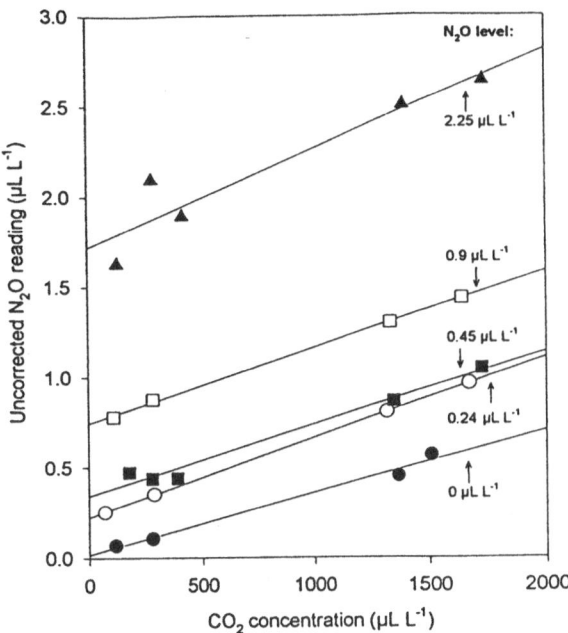

Figure 3. Uncorrected N_2O reading of the Trace Gas Analyser (TGA) in relation to CO_2 concentration at various N_2O levels.

the TGA and the manifold. The dead volume amounts to approximately 25 mL, which is pumped into a head space volume of 1770 mL. When the variation in fluxes between the enclosures is high, this might lead to artificially high or artificially low N_2O concentrations. Problems associated with cross contamination between the enclosures can be solved by increasing the size of the enclosures, or by decreasing the sampling frequency of each enclosure. Further investigations into this are currently in progress.

CO_2 interference

The results from the PAIRS analysis of the N_2O/CO_2 gas mixtures are shown in Figure 3. The N_2O output of the Trace Gas Analyser linearly increased with CO_2 concentration at any given N_2O level.

The results were analysed per N_2O level, so that the CO_2 interference could be determined using the following equation:

$$N_2O(TGA) = a \times CO_2(TGA) + b \qquad (1)$$

where, N_2O (TGA) and CO_2 (TGA) are the N_2O and CO_2 output given by the TGA; a is the correction factor for the CO_2 concentration and b represents the actual N_2O concentration of the sample (i.e. the N_2O concentration given by the GC). Table 2 shows the results

Table 2. Regression equations of CO_2 versus N_2O at various N_2O levels

Target N_2O ($\mu L\ L^{-1}$)	N_2O output =	R^2
0	$0.02 + 3.4 \times 10^{-4}[CO_2]$	0.987
0.24	$0.22 + 4.4 \times 10^{-4}[CO_2]$	0.999
0.45	$0.34 + 4.0 \times 10^{-4}[CO_2]$	0.977
0.9	$0.81 + 3.7 \times 10^{-4}[CO_2]$	0.944
2.25	$1.72 + 5.5 \times 10^{-4}[CO_2]$	0.887

of the regression equations. The slope of the equations (i.e. the correction factor for CO_2 interference) ranged from 3.4×10^{-4} to 5.5×10^{-4}. The average value was 4.2×10^{-4} with a standard deviation of 0.8×10^{-4}.

Conclusions

Photo-acoustic infra-red spectroscopy accurately measured the N_2O concentration of standard gases. The CO_2 interference on N_2O analysis by PAIRS was linear and the average CO_2 correction factor was 4.2×10^{-4}.

There was some discrepancy between flux measurements using GC and PAIRS, which might be due to cross contamination between the soil enclosures, but the mean values were close. Photo-acoustic infra-red spectroscopy potentially represents an important step forward in measuring N_2O fluxes from soil.

Acknowledgements

GC analysis was performed by C Nicholls, ADAS Wolverhampton. This work was part of the Ministry of Agriculture, Fisheries and Food funded ADAS Seed-corn Programme.

References

De Klein C A M, Harrison R and Lord E I 1994 Measurement of N_2O fluxes from soils using photo-acoustic infra-red spectroscopy. 4th Research Conference of the British Grassl. Soc., 26–28 Sept., Reading British Grassl. Soc., Reading, UK (*Accepted*).

Hutchinson G L and Brams E A 1992 NO versus N_2O emissions from an NH_4^+-amended Bermuda grass pasture. J. Geophys. Res. 97, 9889–9896.

Hutchinson G L and Mosier A R 1981 Improved soil cover method for field measurement of nitrous oxide fluxes. Soil Sci. Soc. Am. J. 45, 311–316.

Mosier A R, Klemedtsson L K, Sommerfeld R A and Musselman R C 1993 Methane and nitrous oxide flux in a Wyoming subalpine meadow. Global Biogeochem. Cycles 7, 771–784.

Skiba U, Hargreaves K J, Fowler D and Smith K A 1992 Fluxes of nitric and nitrous oxide from agricultural soils in a cool temperate climate. Atmos. Environ. 26A, 2477–2488.

O. Van Cleemput et al. (eds.), Progress in Nitrogen Cycling Studies, 537–542, 1996.
© *1996 Kluwer Academic Publishers.*

Denitrification as affected by anaerobic and aerobic events

L. Dendooven[1], S. M. Jackson and J. M. Anderson
Department of Biological Sciences, University of Exeter, Prince of Wales Road, Exeter, EX4 4PS, UK. [1] *Present address: Department of Soil Science, 1ACR-Rothamsted, Harpenden, Hertfordshire AL5 2JQ, UK*

Key words: C mineralization, denitrification, N_2O production

Abstract

The effect of an extended period of anaerobiosis and subsequent period of aerobiosis on the link between CO_2 production and denitrification, and the kinetics of NO_3^-, NO_2^-, N_2O and N_2 were studied in laboratory systems using a soil from permanent pasture. Soil was conditioned anaerobically for 7 days and then aerobically for the same period of time. At regular intervals assays for enzymes activity were carried out.

Glucose increased the CO_2 production significantly compared to the unamended soil, but anaerobic conditioning of the soil compared to aerobic conditioning did not affect the CO_2 production in the unamended soil. NO_3^- concentrations in the anaerobic conditioned soil were constant between 2–5 h while they decreased in the aerobic conditioned soil (11.7 mg NO_3^--N kg^{-1}). On average, NO_2^- concentrations in the assays for enzyme activity in the anaerobic conditioned soil were significantly lower (1.43 mg NO_2^--N kg^{-1} after 2 h) than in the aerobic conditioned soil (1.75 mg NO_2^--N kg^{-1} after 2 h). There were no significant differences between the treatments of the unamended soil but the N_2O production was higher in the control soil relative to the C_2H_2 amended soil for the anaerobic conditioned soil. The rates of N_2O production and the N_2O-to-CO_2 ratios were highly variable with time but there was no obvious relation with way or period of conditioning of the soil.

It was concluded that the high variability in N_2O productions as measured after 5 h was presumably related to other factors than those normally considered as main controls over the denitrification process. They were most likely related to denitrifier characteristics and the rate of uptake or release of nitrogenous oxides. As a consequence, denitrification resultant from sudden changes in the O_2 content of the soil, such as heavy rainfall, could be much more variable than expected from factors normally considered as controls of the denitrification process, i.e. NO_3^- concentration, C substrate, previous water regime of the field, etc.

Introduction

It is generally accepted that denitrification and its gaseous products are largely controlled by NO_3^- concentration, water content, pH and C substrate (Firestone, 1982). We have some evidence to suggest that preceding anaerobic events not only affect the gaseous products of the denitrification process but also the link between the C mineralization and the reduction of NO_3^-. This paper describes experiments set up to examine possible effects of anaerobiosis and subsequent aerobiosis on the link between C mineralization and the denitrification process and the kinetics of NO_3^-, NO_2^-, N_2O and N_2. Soils were incubated anaerobically to induce the formation of reduction enzymes followed by aerobic incubation, with periodic denitri-

fication assays to study their persistence. This is part of a programme of research aimed at refining a model of nitrogen dynamics in pasture systems (Scholefield et al., 1991) to accommodate the high denitrification rates which occur during summer rain events on field plots.

Materials and methods

Experimental site

The experiments were carried out using soil from permanent pasture at the Agriculture and Food Research Council Institute for Grassland and Environmental Research at North Wyke, Devon, UK. Characteristics of the experimental site are described in Dendooven and Anderson (1994).

Pre-treatment of soils

Soil samples were collected on 13.10.93 from the 0–10 cm layer of an experimental plot that had been grazed by cattle but had received no N fertiliser for at least ten years. The soil was moist sieved (5 mm) and sub-samples of 300 g soil were added to 3000 mL PVC containers containing a vessel with 50 mL distilled water. The containers were air-tight sealed and stored at 25 °C for 14 days. Every 3 days, the containers were left with the lids removed for 15 min to ensure the maintenance of aerobic conditions in the soils, then resealed and incubation continued at 25 °C. After 14 days, the soil was pooled and mixed. Sub-samples of 300 g soil were added to 1000 mL Kilner jars fitted with rubber septa for gas sampling. The jars were sealed connected to a vacuum line for 10 min and then He was passed through at atmospheric pressure for 10 min. The procedure of vacuum sucking and flushing with He was repeated twice to scrub oxygen from soil pores before the units were stored at 25 °C for 7 days. This way of handling the soil has proven to create complete anaerobiosis.

Assays for enzyme activity during anaerobic incubation

On days 0, 1, 2, 3, 5, 6, 7, one jar was selected at random. The headspace was sampled and analysed for O_2, N_2O and CO_2 using a gas chromatograph (Pye Unicam 4600, UK) fitted with a thermal conductivity detector at 30 °C. The CTR column from Alltech (a Porapak Q column of 2 m in line with a Molecular sieve + Porous Polymer of 1.82 m) with the carrier gas He flowing at a rate of 35 mL min^{-1}, was maintained at 30 °C. Concentrations of O_2 were monitored to ensure that the soils had been incubated under anaerobic conditions. All samples met this criterion.

After analysis of the headspace, the jars were opened, the soil was pooled and mixed. Sub-samples of 7.5 g soil were added to 30, 100 mL Erlenmeyer flasks. The flasks were sealed with an air-tight silicone rubber stopper, connected to a vacuum line for 10 min and then flushed with He for 10 min. Ten mL of a degassed 3.39 mM KNO$_3$ solution was injected in twelve flasks (CON treatment), twelve were injected with a 3.39 mM KNO$_3$ solution with 0.44 mM chloramphenicol (CHL treatment), three were amended with a solution containing 3.39 mM KNO$_3$ and 0.67 mM glucose (GLA treatment) and three were amended with a solution containing 3.39 mM KNO$_3$, 0.44 mM chloramphenicol and 0.67 mM glucose (DEA treatment). This procedure resulted in a concentration of approximately 100 mg NO$_3^-$ N kg^{-1} soil, a concentration of chloramphenicol of ca. 300 mg kg^{-1}, and a concentration of glucose of ca. 100 mg glucose C kg^{-1}. We have shown that this concentration of chloramphenicol inhibits protein synthesis in this soil (Dendooven et al., 1994) and that the concentration of glucose obtained was sufficient enough to overcome possible limitations in C substrate as compared to the nitrate reduction capacity of the soil (Dendooven and Anderson, 1994).

After injection of the different solutions, 10 mL of C_2H_2 (10% v/v) was added to six sub-samples applied with KNO$_3$ (ACE treatment), to six applied with KNO$_3$ + chloramphenicol (CHA treatment) and to the three samples of the GLA and DEA treatment. Sub-samples were then placed on an orbital shaker (rotating at 180 r.p.m.) at room temperature 22 ± 2 °C pending sampling.

After 1 h and 2 h, the headspace of each Erlenmeyer from the GLA and DEA treatment was sampled, analysed for O_2, N_2O and CO_2 and returned to the shaker. The concentrations of O_2, N_2O and CO_2 were corrected for gas dissolved in the water by using published values of the Bunsen absorption coefficient. Concentrations of O_2 were monitored to assure an anaerobic incubation. After 2 and 5 h, three sub-samples were taken from the CON, ACE, CHL and CHA treatment and after 5 h from the GLA and DEA treatment. The headspace of each Erlenmeyer was sampled and analysed for O_2, N_2O and CO_2. After analysis of the headspace each sub-sample was used for assays for NO$_3^-$ and NO$_2^-$. Each sub-sample was extracted for 30 min with 20 mL of distilled water. The extracts were filtered through a GFC Whatman filter paper and nitrate was measured on an ion chromatograph (Dionex, UK). Nitrite in the extracts was measured on a Bemas automatic analyser (Burkhard Instruments, UK) by the sulphanilamide method.

Aerobic incubation

After 7 days of anaerobic incubation, the headspace of each remaining jar was sampled and analysed for O_2, N_2O and CO_2. Concentrations of O_2 were monitored to ensure that the soils had been incubated under anaerobic conditions. All samples met this criterion. The soil in the jars was pooled and mixed. Five sub-samples of 300 g soil were added to 3000 mL PVC containers containing a vessel with 50 mL of distilled water. The jars were sealed and stored at 25 °C for 7 days. Every 3 days, the containers were left with the lids removed for 15 min to ensure the maintenance of aerobic conditions in the soils, then resealed and incubation continued at 25 °C.

Assays for enzyme activity during aerobic incubation
On days 1, 2, 4, 6 and 7, one container was selected at random. The soil in the containers was pooled and mixed. The assays for enzyme activity, as mentioned earlier, were repeated.

Statistical analysis

Regression coefficients were calculated and covariance analysis was carried out with the SAS statistical package (1988) using the general linear model procedure.

Results

CO_2 production

The CO_2 production apparently followed zero-order kinetics in each of the assays for enzyme activity. Rates of CO_2 production are given in Figure 1. The CO_2 production was the same in the CON, ACE, CHL and CHA but significantly higher in the glucose amended soil. The CO_2 was higher in the GLA treatment as compared to the DEA treatment but not significantly different ($p=0.061$). Rates of CO_2 production for the glucose amended soil were significantly higher ($p<0.05$) in the anaerobic conditioned soil as compared to the aerobic conditioned soil but there was no significant ($p>0.05$) difference between the treatments of the unamended soil.

NO_3^- concentrations

Nitrate concentrations were highly variable within each assay for enzyme activity. The decrease in NO_3^- concentrations after 2 h appeared to be greater in the assays for enzyme activity taken from the anaerobic conditioned soil as compared to the aerobic conditioned soil. For the assays taken from the anaerobic conditioned soil, on average NO_3^- concentrations appeared to increase by 0.3 mg N kg^{-1} from 2 to 5 h while in the assays taken from the soil aerobically incu-

bated the NO_3^- concentrations on average decreased by 11.7 mg N kg^{-1}.

NO_2^- concentrations

Nitrite concentrations as measured after 2 h in the assays for enzyme activity were significantly lower ($p<0.05$) for the anaerobic conditioned soil as compared to the aerobically conditioned soil. This was most obvious for assays carried out at the end of the aerobic conditioning (day 13 and 14). Differences between treatments were not significant but higher concentrations were measured in the chloramphenicol amended soil as compared to the control soil. After 5 h, NO_2^- concentrations in the glucose amended soil were significantly higher ($p<0.05$) than in the unamended soil. After 5 h, the effect of addition of chloramphenicol and conditioning of the soil, anaerobic versus aerobic incubation, was comparable with those obtained after 2 h.

N_2O production

The rate of N_2O production was slightly lower for the first 2 h as compared to the next 3 h. Rates of N_2O production were highly variable with no clear pattern (Figure 2). The highest rates of N_2O production were found in the glucose amended soil with a slightly lower production in the DEA treatment as compared to the GLA treatment. Rates of N_2O production in the glucose amended soil were significantly higher ($p<0.05$) than in the unamended soil but there was no difference between the treatments. However, on average the N_2O production in the CON soil was higher than in the ACE treatment and higher in the CHL treatment as compared to the CHA treatment. This was most noticeable for assays carried out with anaerobically incubated soil. The N_2O production was not significantly different ($p>0.05$) in the assays for enzyme activity of the anaerobically conditioned soil amended with glucose as compared to the aerobically conditioned soil. The N_2O production in the anaerobically conditioned unamended soil, however, was significantly different from the aerobic conditioned soil.

N_2O-to-CO_2 ratios were highly variable for each treatment without any relationship to period or type of soil conditioning. In the ACE treatment the N_2O-to-CO_2 ratio was 0.12 at the onset of the experiment and 0.35 after 3 days of conditioning while it was 0.03 after 7 days and 0.34 after 14 days in the CON treatment.

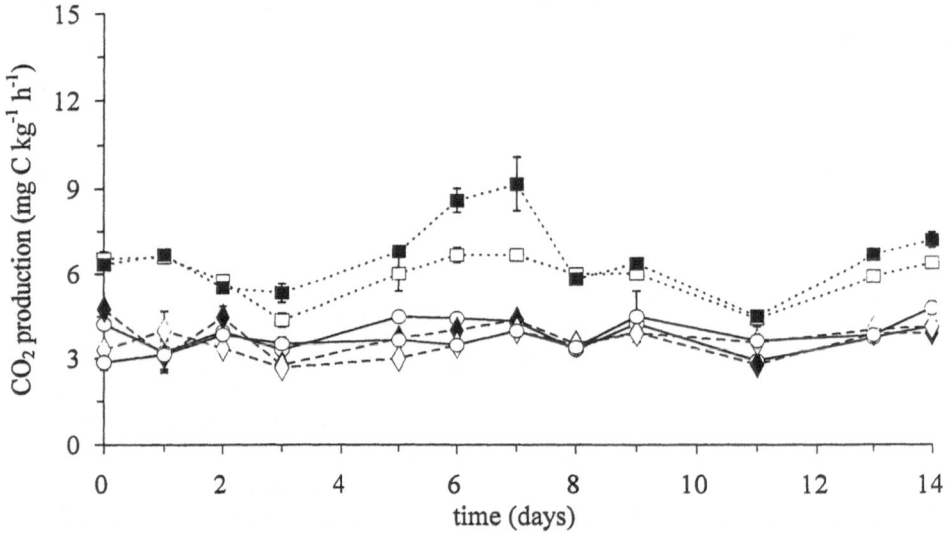

Figure 1. The CO_2 production rates (mg CO_2 C kg^{-1} h^{-1}) in assays for enzyme activity of the Rowden soil conditioned anaerobically for 7 days and then aerobically for 7 days at 25 °C. Soil amended with 100 mg NO_3^- N kg^{-1} and then incubated anaerobically for 5 h at 25 °C. Control (—o—) CON, (—●—) C_2H_2(10% V/V) ACE, (- -◇- -) 300 mg chloramphenicol kg^{-1} CHL, (- -◆- -) 300 mg chloramphenicol kg^{-1} + C_2H_2 (10% V/V) CHA, (- -■- -) 100 mg glucose C kg^{-1} + C_2H_2 (10% V/V) GLA, (- -□- -) 100 mg glucose C kg^{-1} + C_2H_2 (10% V/V) + 300 mg chloramphenicol kg^{-1} DEA. Bars indicate plus and minus standard deviation.

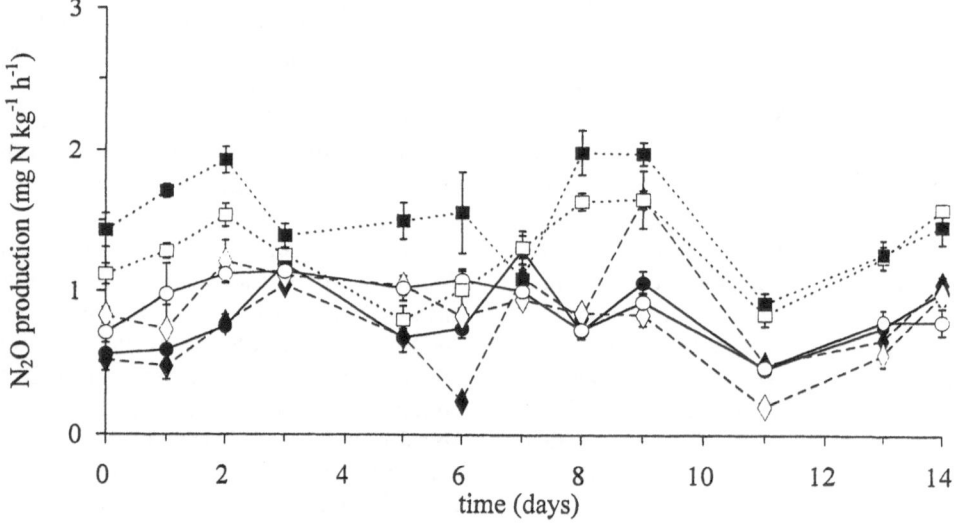

Figure 2. The N_2O production rates (mg N_2O N kg^{-1} h^{-1}) in assays for enzyme activity of the Rowden soil conditioned anaerobically for 7 days and then aerobically for 7 days at 25 °C. Soil amended with 100 mg NO_3^- N kg^{-1} and then incubated anaerobically for 5 h at 25 °C. Legends were the same as in Figure 1. Bars indicate plus and minus standard deviation.

Discussion

Chloramphenicol did not increase the CO_2 production as found in a previous experiment (Dendooven et al., 1994). The time of incubation was presumably to short to detect any significant increase.

In the assays for enzyme activity, NO_3^- concentrations dropped between 2–5 h in the aerobically conditioned soil but slightly increased in the anaerobically conditioned soil. We have some evidence to suggest that the uptake of NO_3^- within the first 2 h under anaerobic conditions can be in excess as compared to the

541

amount reduced depending on the previous conditioning of the soil (unpubl. data).

The anaerobic conditioning of the soil reduced the NO_2^- concentrations as compared to the aerobic conditioned soil. Nitrate reductase had a higher persistence than nitrite reductase and its de novo synthesis occurred earlier (between 1–3 h) compared to nitrite reductase (after 5 h) (Dendooven and Anderson, 1994). Concentrations of NO_2^- were higher in the assays for enzyme activity of soil taken during the aerobic conditioning of the soil as compared to the anaerobic conditioned one. Consequently, the incubation of 5 h was to short to detect any significant effect of chloramphenicol on the de-repression of nitrite reductase.

Rates of N_2O production in the assays for enzyme activity were highly variable with time. It was very difficult to relate the pattern observed with the period and type of conditioning of the soil except for the CON treatment. In the CON treatment, the N_2O production increased sharply from day 0 to day 1 as more nitrite reductase was formed during the anaerobic conditioning of the soil. As a result, more NO_2^- was reduced and thus more N_2O produced. The N_2O production then remained constant in the CON treatment for the next 6 days, i.e. as long as the soil was kept under anaerobic conditions. The rates of N_2O production dropped again when the soil was conditioned aerobically and was more variable than during the anaerobic conditioning of the soil.

Rates of N_2O production in the ACE treatment were lower on average than in CON treatment when the soil was conditioned anaerobically. Changes in pressure due to the injection of C_2H_2 could not account for the differences and the phenomenon was not observed when the soil was conditioned aerobically. It is very difficult to indicate a possible effect of C_2H_2 on the N_2O production as Payne (1984) stated that C_2H_2 is a gas that appears to act as a non-damaging, non-competitive inhibitor whose effects do not linger. It was also difficult to explain the fluctuations in rates of N_2O production with time. Did that reflect a change in denitrifier characteristics or was it a result of manipulating the soil? However, why was that variation absent in the CON treatment when the soil was conditioned anaerobically and why was the variation in N_2O production in each of the treatments more or less the same when the soil was conditioned aerobically?

Chloramphenicol reduced the N_2O production in the glucose amended soil. The application of 'readily' decomposable substrate increased the oxidative capacity of the soil while de novo synthesis of nitrate

reductase, which normally occurred within the first 3 h was inhibited (Dendooven and Anderson, 1994). The relationship between N_2O production in the DEA treatment and the N_2O in the CON treatment was fairly constant except between day 3 and 7 for the anaerobically conditioned soil.

As rates of N_2O production were highly variable so were the N_2O-to-CO_2 ratios. An increase in the N_2O-to-CO_2 ratio as observed after an anaerobic conditioning of the soil for 3 days could not be found: the fluctuations in rates of N_2O production were too high and not in phase with variations in the rates of CO_2 production.

A high variability in N_2O production was observed at the end of a 5 h incubation for soil conditioned anaerobically and aerobically in an experiment under well defined conditions. It appeared that controls other than the ones normally considered to affect denitrification, i.e. NO_3^- concentration, C substrate, etc. could be important on a short time scale. The importance of concentrations of reduction enzymes involved in the denitrification process has been observed previously (Dendooven and Anderson, 1994). The amount of NO_2^- formed, dependent on the concentrations of nitrite reductase available, affected the N_2O production. However, the fluctuations measured in N_2O production were presumably not only affected by changes in reduction enzymes because the amounts of e^- taken up by NO_2^- and N_2O showed a comparable change in time. This would mean that during denitrification in the field, where there is a high variability in C substrate availability and composition, and NO_3^- concentrations, where conditions may rapidly change from anaerobic to aerobic and where the previous water regime of the field will affect the adaptation of the microbial biomass to anaerobiosis, variability in N_2O production could be increased by factors only important on a short time scale, i.e. dynamics of denitrifiers, rate of uptake of nitrate and rate of subsequent release of, if any, NO_2^- and N_2O.

It was concluded that the high variability in N_2O production, as measured after 5 h, was presumably related to factors other than those normally considered as main controls over the denitrification process. They were most likely related to denitrifier characteristics and the rate of uptake or release of nitrogenous oxides. Consequently, denitrification occurring by sudden changes in the O_2 content of the soil, such as heavy rainfall, could be much more variable than expected from factors generally considered as controls of the denitrification process, i.e. NO_3^- concentration, C substrate, previous water regime of the field, etc.

Acknowledgements

We are grateful to Dr D Scholefield for access to the Rowden plots and to S Ellis for comments on an early draft. The research was funded by the Agricultural and Food Research Council.

References

Dendooven L and Anderson J M 1994 Dynamics of reduction enzymes involved in the denitrification process in pasture. Soil Biol. Biochem. 26, 1501–1506.

Dendooven L, Splatt P and Anderson J M 1994 The use of chloramphenicol in the study of the denitrification process: some side-effects. Soil Biol. Biochem. 26, 925–927.

Firestone M K 1982 Biological denitrification. *In* Nitrogen in Agricultural Soils. Ed. F J Stevenson. pp 289–326. Agronomy Monograph, 22, American Society of Agronomy, Madison, WI. John Wiley and sons, New York, USA.

Payne W J 1984 Influence of acetylene on microbial and enzymatic assays. J. Microbiol. Meth. 2, 117–133.

SAS Institute 1988 Statistic Guide for Personal Computers. Version 6.03, ed. SAS Inst., Cary NC. 429 p.

Scholefield D, Lockyer D R, Whitehead D C and Tyson K C 1991 A model to predict transformations and losses of nitrogen in UK pastures grazed by beef cattle. Plant and Soil 132, 165–177.

O. Van Cleemput et al. (eds.), Progress in Nitrogen Cycling Studies, 543–547, 1996.
© 1996 *Kluwer Academic Publishers.*

Denitrification in acid soils, in a leaching tube decomposition study of bean residues

Costas Ehaliotis, Georg Cadisch, Lynn Garraway and Ken E. Giller
Department of Biological Sciences, Wye College, University of London, Wye, Ashford, Kent TN25 5AH, UK

Key words: acetylene inhibition method, denitrification, N mineralization, soil incubation

Abstract

The decomposition of bean residues (*Phaseolus vulgaris*) with small N content (1.38%) was studied in temperate and tropical acid soils (pH 4.6–5.6) of varying texture over 244 days in leaching tubes. Initial net N immobilization was observed in all soils treated with the residues compared to the unamended controls. Thereafter, in four of the residue-treated soils net mineralization occurred. At the end of the incubation the cumulative net mineral N release from these four soils accounted for 19% to 63% of the N initially added in the form of bean residue and was less in the clay soils than the sandy soils.

However, in four of the residue-amended soils the initial net immobilization phase apparently lasted until the end of the experiment (day 244). In some of the leaching tubes with these soils, dark patches were observed and it was suspected that these were sites of denitrification, a process often indicated as a potential problem in leaching tube experiments. Denitrification was then measured using the acetylene inhibition method at the end of the incubation period. There was a high rate of N_2O evolution from the four soils which showed apparent net immobilization and practically no evolution of N_2O from the first four soils. Denitrification rate was not directly related to soil texture nor to the initial residue additions but to total soil organic matter content. The log-transformed estimates of N_2O release correlated significantly with the total carbon content of all eight soils ($r^2=0.66$) and even better with the CO_2 evolved from the controls during the first 96 days of incubation ($r^2=0.76$). Thus the carbon availability as expressed by the CO_2 evolution from soils was a good indicator of their denitrification potential.

Introduction

Denitrification often occurs at significant rates when soil samples are incubated at water contents close to their water holding capacity. The leaching tube methodology, introduced by Standford and Smith (1972) to measure N-mineralization potentials (N_o) in soils aims to achieve maximum organic-N mineralization under aerobic conditions (Standford and Epstein, 1974). Therefore, it involves long-term incubation of soils under high water content regimes. The methodology has been widely used with various modifications to measure not only the N_o of soils but also to give N immobilization-mineralization patterns of plant residues decomposing in soils (Bonde and Rosswall, 1987; Frankenberger and Abdelmagid, 1985; Handayanto et al., 1994). The high moisture level however, may favour the creation of anaerobic microsites which, together with available C and NO_3^- at certain

stages of the incubation, lead to denitrification. The high moisture level may also promote the redistribution of available C and NO_3^- to the denitrifying population of the soil, as well as directly restricting O_2 availability in the soil profile (Sexstone et al., 1988). When plant residues are incorporated into soil, the enhanced microbial activity and consequent O_2 demand may lead to even more severe local O_2 limitations for the microbial biomass of the incubated soils.

A leaching tube experiment was carried out in which the decomposition of low quality bean residues was studied in eight different acid soils selected from temperate and tropical regions. The experiment was primarily designed to examine the effect of soil clay content on the decomposition rate and the nitrogen immobilization-mineralization balance of the residue used. However, during the progress of the experiment, visual observations and immobilization patterns obtained from certain soils suggested denitrification was occurring. This report focuses on the variability of denitrification within the treatments in which it occurred and on its relation to total organic matter and CO_2 evolution in soils incubated in leaching tubes.

Materials and methods

Soils

Acid soils from temperate and tropical climatic regions were selected with a wide range of clay contents. Temperate soils were collected in Kent, UK (soils 1, 2 and 3) and tropical soils were from Brazil (soils 4, 7 and 8), Thailand (5) and Indonesia (6). Characteristics of the soils selected are given in Table 1.

The leaching tube methodology

Net nitrogen mineralization and CO_2 evolution were monitored in all soils throughout the experimental period of eight months (244 days). The leaching tube methodology used was based on the method proposed by Stanford and Smith (1972) but with a modified leaching solution as used by Cassman and Munns (1980). The method was further modified by performing an initial leaching at time zero (t_0) and by adjusting the KH_2PO_4 concentration in the leaching solution to acid soils (Cadisch et al., 1994). Soil samples were mixed with 75 g of sand (1:1 w/w) and 0.451 g of bean residue (leaves and stems of Phaseolus vulgaris) was added to form the +residue treatment while no residue was added to the control treatments. The nitrogen content of the residue added was 1.38% and the amount added corresponded to 83 mg N kg^{-1} soil. The leaching tubes comprised of three parts: the main plexiglass tube, a top stopper fitted with a septum and a funnel shaped bottom stopper (McDonagh, 1993). The soil-sand mixture was poured in between two thin sand layers (0.5 cm), 150 mL of dilute-N nutrient solution (1 mM $CaCl_2$, 1 mM $MgSO_4$, 0.9 mM KCl and 0.1 mM KH_2PO_4) was added to the soil/sand mixture in each tube at t_0 and suction was applied to achieve approx. 85% of the water holding capacity. The leachates collected were stored with a drop of toluene added in each container at 4 °C prior to analysis. Beakers containing 20 mL of 0.25 M KOH were placed on top of the leached soils to trap CO_2 evolving from soils and the tubes were closed. A syringe containing soda-lime was fitted on each top stopper in order to permit replacement of air inside the tube without allowing entry of

CO_2. The leaching treatment was repeated using 100 mL of solution at increasing time intervals (more frequent leaching during the first two weeks of the experiment) when the CO_2 traps were also replaced. The leaching tubes were stored in the dark at 28 ° C. CO_2 evolution data were collected only for the first 96 days of incubation as no significant difference in CO_2 evolution could be detected between treatments and controls of the same soil after this period.

Chemical analysis

$(NO_3^- + NO_2^-)$-N and NH_4^+-N were measured colorimetrically using a Technicon autoanalyzer. The $(NO_3^- + NO_2^-)$-N was measured using hydrazine sulphate as the reducing solution and sulphanilamide as the colour reagent. The NH_4^+-N was measured using alkaline sodium salicylate and sodium hypochlorite to react with ammonium salts and form a bluish-green indophenol colour.

CO_2 trapped in the KOH traps was titrated with 0.25 M HCl after the addition of 1 mL 3 N $BaCl_2$ and absorbed CO_2 was calculated on the basis that 1 mL 0.25 M HCl is equivalent to 5.5 mg CO_2 (Anderson, 1973).

Denitrification rate measurements

On day 244, acetylene was added to the leaching tubes atmosphere to inhibit N_2O reduction, at a concentration of 10% v/v. After 24 h, 1 mL gas samples were taken by syringe from the tubes. The samples were injected and analyzed in a PYE UNICAM gas chromatograph equipped with a Porapak Q column at 50 °C using a ^{63}Ni electron capture detector at 310 °C with N_2 as the carrier gas. The denitrification rates calculated were corrected for N_2O dissolved in the soil solution (Moraghan and Buresh, 1977).

Results and discussion

An initial net immobilization period was observed in all soils treated with the residues compared to the controls (data not presented). This immobilization period lasted 60 days for the high clay Brazilian Oxisol (soil 8), 36 days for the medium clay Brazilian Oxisol (soil 7), and only 18/23 days for the two sandy soils, a Brazilian Oxisol (soil 4) and an English Brown Sand (soil 1), respectively. Thereafter, the release of mineral nitrogen from these residue treated soils exceeded that

Table 1. Characteristics of soils used in the leaching tube experiment

Soil	Location	pH	% C	% N	C/N ratio	% Sand 2–0.06 mm	% Silt 60-2 μm	% Clay < 2μm
1. Brown Sand	S.E. England	5.1	1.33	0.12	11.37	80	8	12
2. Brown Earth	S.E. England	5.0	3.02	0.26	11.71	64	14	22
3. Brown Earth	S.E. England	5.2	2.64	0.20	13.07	37	38	25
4. Oxisol	West Brazil	4.8	0.56	0.04	13.88	91	2	7
5. Ultisol	N.E. Thailand	5.6	0.98	0.06	15.98	77	13	10
6. Ultisol	Sumatra	4.6	2.08	0.17	12.46	70	12	18
7. Oxisol	Central Brazil	5.1	0.97	0.07	13.79	56	13	31
8. Oxisol	Central Brazil	5.0	1.09	0.12	8.79	25	18	57

from the controls. On day 244, the cumulative net mineral nitrogen release from these four soils accounted for 19% to 63% of the nitrogen incorporated as bean residue and was less for the clay soils than the sandy soils (Fig. 1a). This cumulative net nitrogen mineralization effect caused by the residue was estimated by subtracting the cumulative net nitrogen mineralization in the unamended controls from the cumulative net nitrogen mineralization in the +residue treatments. For the remaining four soils (2, 3, 5 and 6) the net immobilization phase apparently lasted throughout the whole incubation period and was at no time succeeded by net mineralization (Fig. 1a).

The denitrification test performed at the end of the incubation period (Fig. 1b), reflected the mineralization/immobilization patterns obtained. It revealed virtually no denitrification in the four soils (8, 7, 4, and 1) in which net N mineralization had occurred within the first two months of the incubation period. However for the other four soils (5, 2, 6 and 3) in which net N mineralization did not occur at any stage throughout the incubation, denitrification rates were significant and very high in some cases. High denitrification rates from the soils corresponded to high CO_2 evolution rates during the first 96 days of incubation (Fig. 1c).

The absence of net mineralization together with the increased variability of the net immobilization data for these last four soils suggested that most of the bean residue N which was mineralized had been lost by denitrification during the early stages of the incubation. However, during these early stages, significant denitrification is unlikely to have occurred since the large carbon availability from the residue induced strong N immobilization in all treated soils, dramatically reducing NO_3^- availability to denitrifiers. Denitrification probably occurred at a later stage (when remineralization of immobilized N occurred) depending on carbon availability from the soil rather than from the residue. At the time of the denitrification measurement, loss of the residue N was probably almost complete as there were no detectable differences in mineral NO_3^- release rates (data not presented) and in denitrification rates between the +residue treatments and the controls (Fig. 2a). Aulakh and Rennie (1987) showed that practically all the denitrification they observed by adding ground straw to nitrate treated soils occurred within the first 8 days of incubation in both treatments and control soils. Similar results were found using soils under different glucose, NO_3^- and water content treatments by Weier et al. (1993) who concluded that available carbon rather than nitrate was the factor limiting denitrification. The denitrification test in this study revealed denitrification also occurring at a late incubation stage, in both residue-treated and control soils.

Burford and Bremner (1975) found a significant correlation between total organic C and denitrification capacities of 17 soils and a close relationship between water soluble carbon and denitrification capacity. Similar relationships have also been reported between denitrification and total organic C and with anthrone reactive C (extracted after boiling soils in 0.01 M $CaCl_2$ solution) in anaerobically incubated soils (Stanford et al., 1975). In our results correlation between total soil C content and N_2O production for the 8 soils after eight months of incubation was poor ($r^2=0.35$) due to increased variability of denitrification at high total C contents (Fig. 2a). The analysis of variance showed increasing variability of measurement and residuals with higher total C contents. Skewed frequency distributions of N_2O values and residuals were also observed. Such data are often obtained when estimations of denitrification are carried out and may be explained by accepting that factors influencing denitrification are combined in a multiplicative man-

546

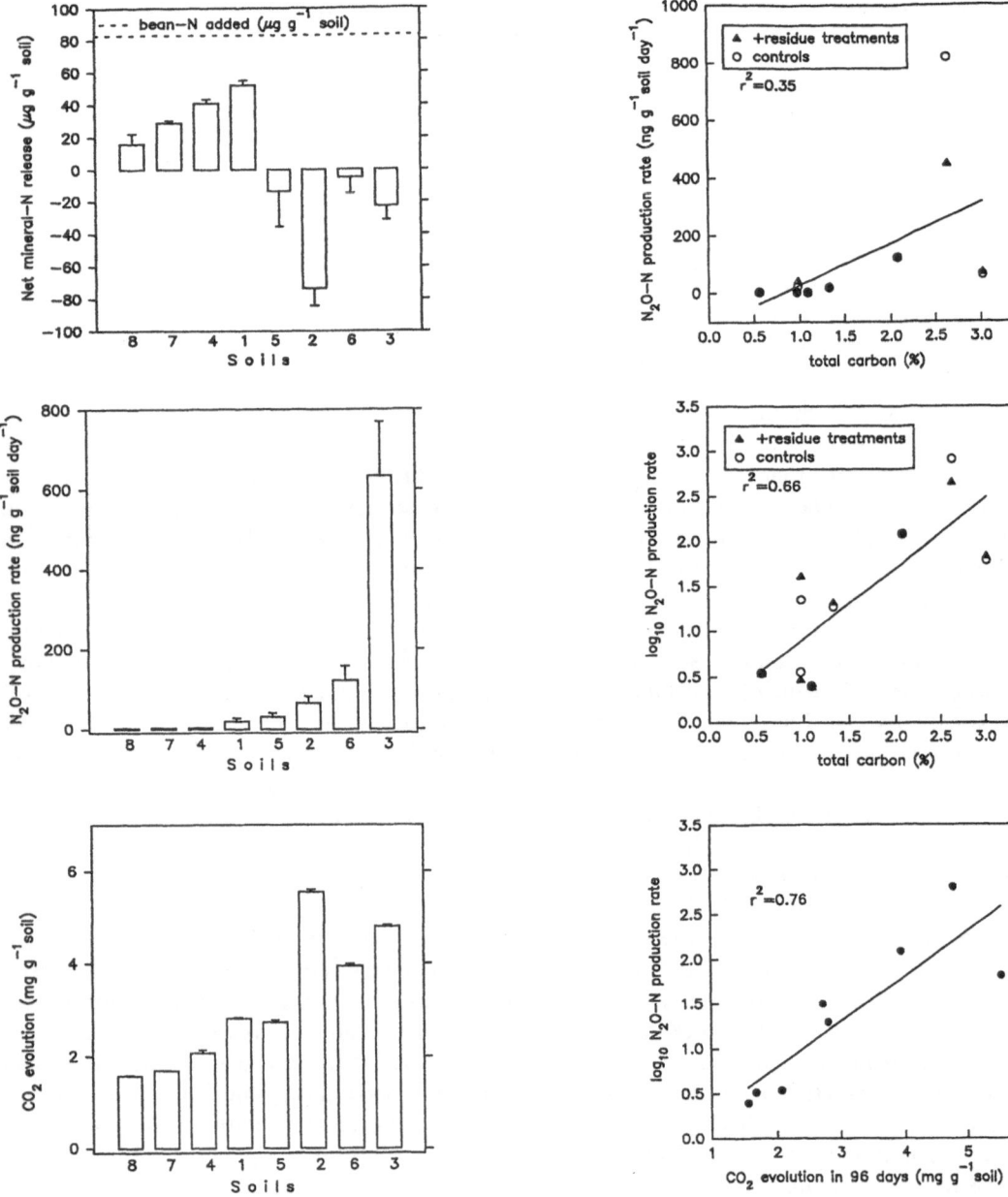

Figure 1. Mineral-N release, denitrification rates and CO_2 evolution from eight acid soils with different organic matter contents amended with bean residues (83 μg bean-N g^{-1} soil) and incubated in leaching tubes together with unamended control soils for eight months. Error bars are standard errors of the means; (a) (upper) Cumulative mineral-N release from the eight residue treated soils during the eight months of incubation, estimated after subtracting cumulative N mineralized in the control soils (n=4); (b) (middle) N_2O-N production rates at the end of the incubation period. Values are means of residue amended and control treatments for each soil (n=8); (c) (bottom) Cumulative CO_2 evolution from the eight unamended control soils during the first 96 days of incubation (n=4).

Figure 2. Relationship between denitrification rates at the end of an eight month incubation period and carbon availability indices of eight acid soils with different organic matter contents amended with bean residues (▲) and incubated together with unamended control soils (○) in leaching tubes. Regression was calculated on the combined data of amended and unamended soils; (a) (upper) Relationship between total carbon content of the soils and the estimates of their denitrification rates; (b) (middle) Relationship between total carbon of the soils and the log-transformed estimates of their denitrification rates; (c) (bottom) Relationship between the cumulative CO_2 evolution from the unamended control soils (●) and the log-transformed estimates of the denitrification rates from all soils.

ner (Parkin, 1987). Plotting total soil C against log-transformed N_2O data (Fig. 2b) improved the correlation ($r^2=0.66$) and revealed the relationship between organic C and denitrification rates. This relationship indicated that carbon availability was driving denitrification at this late decomposition stage, deriving from the soil rather than from the residue. When the CO_2 production from the control soils during the first 96 days of incubation was used as an indicator of the C availability and was plotted against the log transformed denitrification data (Fig. 2c) an even better correlation was obtained ($r^2=0.76$).

Conclusions

Denitrification may often be a major problem in leaching tube decomposition studies even at late incubation stages. Availability of soil C was a major factor controlling the occurrence of significant denitrification during the whole incubation period. Thus even when no residues are added to soils, denitrification tests should be also carried out, especially when working with soils with high organic carbon content and/or high basal respiration. Additionally the use of transparent tubes may help to detect the occurrence of denitrification.

Acknowledgement

C Ehaliotis acknowledges financial support by the Greek State Scholarship Foundation.

References

Anderson J M 1973 Carbon dioxide evolution from two temperate deciduous woodland soils. J. Appl. Ecol. 10, 361–378.

Aulakh M S and Rennie D A 1987 Effect of wheat straw incorporation on denitrification of N under anaerobic and aerobic conditions. Can. J. Soil Sci. 67, 825–834.

Burford J R and Bremner J M 1975 Relationships between the denitrification capacities of soils and total, water-soluble and readily decomposable soil organic matter. Soil Biol. Biochem. 7, 389–394.

Bonde T A and Roswall T 1987 Seasonal variation of potentially mineralizable nitrogen in four cropping systems. Soil Sci. Soc. Am. J. 51, 1508–1517.

Cadisch G, Giller K E, Urquiaga S, Miranda C H B, Boddey R M and Schunke R M 1994 Does phosphorus supply enhance soil-N mineralization in Brazilian pastures? Eur. J. Agron. 3, 339–345.

Cassman K G and Munns D N 1980 Nitrogen mineralization as affected by soil moisture, temperature and depth. Soil Sci. Soc. Am. J. 44, 1233–1237.

Frankenberger W T and Abdelmagid H M 1985 Kinetic parameters of nitrogen mineralisation rates of leguminous crops incorporated into soil. Plant and Soil 87, 257–271.

Handayanto E, Cadisch G and Giller K E 1994 Nitrogen release from prunings of legume hedgerow trees in relation to quality of the prunings and incubation method. Plant and Soil 160, 237–248.

McDonagh J F 1993 Nitrogen benefits from legumes to cropping systems in Northeast Thailand. Ph.D. Thesis, Wye College, University of London, UK.

Moraghan J T and Buresh R 1977 Correction for dissolved nitrous oxide in nitrogen studies. Soil Sci. Soc. Am. J. 41, 1201–1202.

Parkin T 1987 Soil microsites as a source of denitrification variability. Soil Soc. Am. J. 51, 1194–1199.

Stanford G and Smith S J 1972 Nitrogen mineralisation potentials in soils. Soil Sci. Soc. Am. Proc. 36, 465–472.

Stanford G and Epstein E 1974 Nitrogen mineralisation-water relations in soils. Soil Sci. Soc. Am. Proc. 38, 103–106.

Stanford G, Van der Pol R A and Dzienia S 1975 Denitrification rates in relation to total and extractable soil carbon. Soil Sci. Am. Proc. 39, 284–289.

Sexstone A J, Parkin T B and Tiedje J M 1988 Denitrification response to soil wetting in aggregated and unaggregated soil. Soil Biol. Biochem. 20, 767–769.

Weier K L, Doran J W, Power J F and Walters T D 1993 Denitrification and the dinitrogen/nitrous oxide ratio as affected by soil water, available carbon and nitrate. Soil Sci. Soc. Am. J. 57, 66–72.

O. Van Cleemput et al. (eds.), Progress in Nitrogen Cycling Studies, 549–551, 1996.
© 1996 *Kluwer Academic Publishers.*

N_2O emission from soil fertilized with Fymol treated slurry

A.B. Eriksen[1], T. Granli[2] and H. Høyvik[2]
[1]*The Phytotron, P.O.Box 1066, University of Oslo, 0316 Oslo, Norway and* [2]*Norsk Hydro Research Center, Norsk Hydro, 3901 Porsgrunn, Norway*

Key words: Fymol, N-level, nitrate, N_2O emission, soil temperature, soil water content

Abstract

N_2O emission from soil fertilized with FYMOL, a nitrate containing compound, has been examined in the laboratory. Soil fertilized with slurry was enclosed in glass cuvettes (1 L) and the N_2O concentration in the headspace of the cuvette was measured daily by gas chromatograph. The cuvettes were ventilated after gas sampling and the cumulative N_2O emission for 14 days was determined.

No significant difference in the N_2O emission from soil was found when comparing FYMOL treated slurry with untreated slurry. Slurries from two farms were compared and no significant difference was found. However, the N_2O emission was notably enhanced (from 3 to 9 times), when soil temperature, soil water content and N-level were increased from 10 to 20 °C, 60 to 100% and 50 to 150 kg N ha^{-1}, respectively.

Introduction

The production of dihydrogensulfide (H_2S) in stored pig and cattle slurry can be a problem for the farmers, as H_2S give both unpleasant odour and human health problems. Repeated exposures to even low concentration of H_2S (20 mmol mol^{-1}) are considered as a health risk and brain damage can be a long term effect. The production of H_2S in the slurry is stopped and H_2S already present is removed by addition of FYMOL, a nitrate containing compound. The effect is probably a result of autotrophic denitrification (Sublette and Sylvester, 1987). As FYMOL contains nitrate, the substrate for the denitrification process, a negative side effect of the FYMOL treatment may be N_2O emission both from the slurry and the soil fertilized with this slurry.

The aim of the present study was to examine the N_2O emission from soil fertilized with Fymol treated slurry at various soil conditions.

Materials and methods

The pig slurries used in the experiments (Table 1) were collected at two farms, Rustan and Sørli. The slurries were stored at 2 °C before the treatment with FYMOL.

Table 1. Chemical characteristics of the pig slurries before FYMOL treatment. The slurries came from two farms, Sørli and Rustan

Analysis	Sørli	Rustan
Dry matter, g 100 g^{-1} wet	12.1	1.2
Tot. N, g 100 g^{-1} wet	0.34	0.30
NO_3-N, mg 100 g^{-1} wet	0.26	0.35
NH_4-N, mg 100 g^{-1} wet	187	236
Tot. C, g 100 g^{-1} DM	44.1	25.2

Two samples of 1 L were taken from one batch of slurry. 7.5 mL 42% FYMOL was mixed into one of the samples and both samples were placed at 15 °C. The amount of S^{2-} and NO_3^- in the samples were determined by the methylene blue method and the cadmium reduction method, respectively (Greenberg et al., 1985). In the present study the FYMOL treatment was completed and the slurry was ready for application into soil, when the amount of S^{2-} in the slurry was less than 1.0 mg L^{-1}. The amount of nitrate in the slurry ranged from 1.5 to 695 mg NO_3^--N L^{-1}, when the slurry was used.

The soil fertilized with slurry was enclosed in 1 L glass cuvette. The cuvette consisted of two joined glass tubes (diam. 60 mm; height 140 mm and diam. 75 mm; height 140 mm) and was sealed by a glass lid (diam.

Table 2. Chemical and physical characteristics of the Mysen soil used

Analysis	
Clay (%)	2
Silt (%)	16
Sand (%)	82
Tot. N (%)	0.22
Tot. C (%)	4.4
pH (H$_2$O)	4.9
NH$_4$-N, mg kg^{-1}	9.33
NO$_3$-N, mg kg^{-1}	9.39

Table 3. Main effects on the total N$_2$O emission (μg N$_2$O-N) for 14 d from soil ferilized with pig slurry

Variables	Treatment	N$_2$O emission, (μg N$_2$O-N)	
		2^{5-1} design	2^3 design
FYMOL	No	788	
	With	806	
Slurries	Sørli	643	
	Rustan	952	
Temperature	10 oC	160	186
	20 °C	1433	877
Soil water	60% FC	180	234
	100% FC	1398	827
N-level	50 kg N ha^{-1}	215	266
	150 kg N ha^{-1}	1365	801

75 mm; height 20 mm) with a silicon rubber packing and a lock ring.

A loamy sand soil from Mysen, Norway was used in the experiments (Table 2). The soil was sieved to pass 2 mm and 190 g soil (air-dry basis) was filled into each cuvette. Slurry (18,4 and 55.3 μg NH$_4$-N g^{-1} soil) was added to the soil and the application rates were equivalent to 50 and 150 kg N ha^{-1}. Slurry and 10 mL water were mixed into the upper 30 mm of the soil, then water was added and drained into the soil to attain a soil water content of 60 and 100% of the field capacity.

Gas samples from the headspace of the cuvettes were taken daily for N$_2$O analysis by a gas chromatograph (Carlo Erba Series 6000) with ^{63}Ni electron capture detector (ECD) and a widebore Poraplot Q capillary column (25 m × 0.53 mm). The temperature was 50 °C in the injector; 26 °C in the column and 340 °C in the ECD. The flow rate was 2.5 mL min^{-1} for the carrier gas (99.999% He) and 40 mL min^{-1} for the make up gas (99.999% N$_2$). After gas sampling the cuvettes were ventilated for 5 minutes.

A half-fraction factorial experimental design at two levels with five variables (2^{5-1} design) was used to study the effect of FYMOL treatment, two different slurries, soil temperature, soil water content and N-level. In addition a factorial experimental design at two levels with 3 variables, a 2^3 design, was used to study further the effect of soil water content, soil temperature and N-level. In these experiments soil fertilized with untreated slurry from Sørli was used. The N$_2$O emission from the different treatments was given as the total N$_2$O emitted during 14 days. The results were transformed to logarithmic values and T-test of paired two-samples was used to determine significant differences.

Results and discussion

The main object of the present work was to compare the N$_2$O emission from soil fertilized with slurry with and without FYMOL treatment. The effect of FYMOL was studied by using two different slurries and two levels of N-content, soil water content and soil temperature. Table 3 shows the main effects on the total N$_2$O emission during 14 days. No significant difference in the N$_2$O emission was found when soil fertilized with Fymol treated slurry was compared to untreated slurry. In the present study treatment of slurry with a nitrate containing product had no effect on the N$_2$O emission when using the treated slurry as fertilizer.

The two slurries used in the experiments were quite different. Both the dry matter and the total carbon content was highest in the slurry from Rustan. Even though, organic carbon may stimulate the microbial activity in the soil (Drury et al., 1991) and positive correlation between soil organic content and N$_2$O emission has been reported (Iqbal, 1992; Robertson and Tiedje, 1984), we found no significant difference in the N$_2$O emission from the two slurries (Table 3). In agreement with Goodroad and Keeney (1984) we found that N$_2$O emission increased strongly by increasing the temperature from 10 to 20 °C. At a soil water content of 100% of field capacity, where both nitrification and denitrification will proceed, a higher N$_2$O emission rate was found compared to 60% soil water content. The same soil water effect has been reported by Klemedtsson et al. (1988). Application of higher amount of slurry which increased the amount of NH$_4^+$ in the soil, enhanced the N$_2$O emission in agreement with report from Duxbury and McConnaughey (1986).

Conclusion

Production of H_2S in stored slurry causes both an odour and a human health problem. Fymol treatment of the stored slurry eliminates the production of H_2S. However, as Fymol contains nitrate, the treatment may give both an increased N_2O emission from slurry during storage and from soil fertilized with Fymol treated slurry. In the present study, the N_2O emission from soil fertilized with pig slurry was examined. No significant difference in the N_2O emission was found between Fymol treated slurry and untreated slurry. The large effect of soil temperature, soil water content and N-level on the N_2O emission from soil demonstrated that our laboratory experiments with cuvettes were suitable for evaluating the Fymol treatment.

References

Drury C F, McKenney D J and Findlay W J 1991 Relationships between denitrification, microbial biomass and indigenous soil properties. Soil Biol. Biochem. 23, 751–755.

Duxbury J M and McConnaughey P K 1986 Effect of fertilizer source on denitrification and nitrous oxide emissions in a maize field. Soil Sci. Soc. Am. J. 50, 644–648.

Goodroad L L and Keeney D R 1984 Nitrous oxide production in aerobic soils under varying pH, temperature and water content. Soil Biol. Biochem. 16, 39–43.

Greenberg A E, Trussel R R and Clesceri L S 1985 Standard methods for the examination of water and wastwater. 16th ed. American Public Health Association, Washington DC, USA.

Klemedtsson L, Svensson B H and Rosswall T 1988 Relationships between soil moisture content and nitrous oxide production during nitrification and denitrification. Biol Fertil. Soils. 6, 106–111.

Iqbal M 1992 Potential rates of denitrification in 2 field soils in southern England. J. Agric. Asc. 118, 223–227.

Robertson G P and Tiedje J M 1984 Denitrification and nitrous oxide production in successional and old-growth Michigan forests. Soil Sci. Soc. Am. J. 48, 383–389.

Sublette K L and Sylvester N D 1987 Oxidation of hydrogen sulfide by continuous cultures of *Thiobacillus denitrificans*. Biotechnol. Bioeng. 29, 753–758.

O. Van Cleemput et al. (eds.), Progress in Nitrogen Cycling Studies, 553–557, 1996.

Nitric oxide (NO) release in activated sludge plants with nitrification and denitrification

Joost Groeneweg, Ingrid Leuther and Thomas Muckenheim
Institut für Biotechnologie 3, Forschungszentrum Jülich GmbH D-52425 Jülich, Germany

Key words: denitrification, nitric oxide, nitrification, waste water treatment

Abstract

The potential of NO release from waste water treatment plants was investigated, using an intermittently fed and aerated model activated sludge plant.

During denitrification, but only in the presence of nitrite and sparging the activated sludge suspension with N_2, significant amounts of NO were released. When the gas stream was led over (instead of through) the stirred suspension, the NO release was below the detection limit. In the latter situation which is closer to denitrification conditions in sewage treatment plants, NO is obviously further reduced to N_2O or N_2.

When the conditions for complete nitrification are unfavourable (e.g. low pH) and nitrite appears as an intermediate, considerable amounts of nitric oxide are released. Under complete nitrification conditions, maximally 0.1% of the oxidized nitrogen is released as NO.

If one assumes that the total amount of waste water nitrogen produced by the 80 mio. inhabitants of Germany is completely nitrified in treatment plants, the NO emission would be 380 t a^{-1} or only 0.01% of the amount of nitrogen oxides released by combustion processes in Germany.

Introduction

The trace gas nitric oxide (NO) plays an important role in tropospheric chemistry. Ozone may be produced or destroyed depending on the NO mixing ratio (Conrad, 1990; Crutzen, 1979). Besides burning of fossil fuels, the biological processes of nitrification and denitrification in soil are important sources of NO emission (Ehhalt and Drummond, 1982; Williams et al., 1992).

In advanced waste water treatment the microbial processes of nitrification and denitrification are widely used for the elimination of nitrogen compounds. To get an idea about the potential emission of nitric oxide from waste water treatment plants we investigated the release of NO from a lowly loaded model activated sludge plant under a variety of conditions.

Materials and methods

The activated sludge unit consisted of an aeration tank with 24 L working volume and a clarifier with a volume of 15 L (Fig. 1).

Table 1. Composition of the synthetic waste water

Carbon source		N/P source	
$CH_3 COOH$ (99.8%)	8.91 mL L^{-1}	$NaHCO_3 \cdot H_2O$	11.61 g L^{-1}
C_2H_5OH (96%)	6.4 mL L^{-1}	NH_4Cl	0.58 g L^{-1}
NaOH	4.33 g L^{-1}	$K_2HPO_4 \cdot 3H_2O$	1.39 g L^{-1}
Na_2CO_3	4.33 g L^{-1}	$FeSO_4 \cdot 7H_2O$	0.055 g L^{-1}
$MgSO_4 \cdot 7H_2$)	0.72 g L^{-1}	Na_2 EDTA	0.04 g L^{-1}
Trace elements[a]	1 mL L^{-1}		

[a] After Payer and Trültzsch, 1972.

The waste water was added intermittently to prevent bulking sludge and to have an optimal use of the organic carbon content in the waste water for denitrification (Soeder et al., 1993). Every 60 minutes a batch of synthetic waste water (70 mL of both the carbon and the N/P source together with 1920 mL deionized water, see Table 1) was added to the aeration tank within one minute. Shortly before the waste water addition aeration was stopped and the activated sludge suspension was only stirred (and in some experiments sparged with N_2) to allow for denitrification. After 20 min-

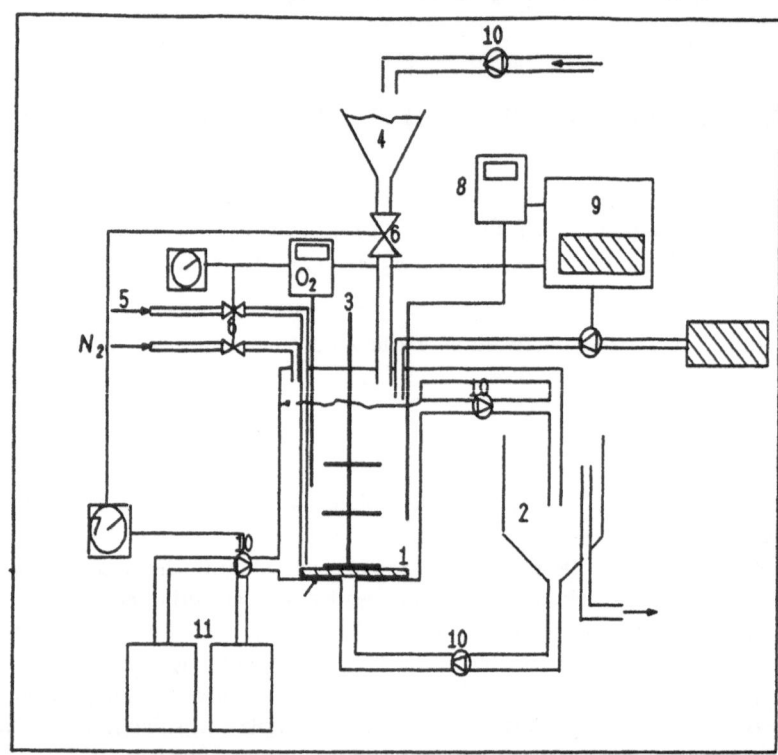

Figure 1. Schematic layout of the model activated sludge unit 1) activated sludge tank (24 L); 2) clarifier (15 L); 3) stirrer; 4) substrate container for pulsed feeding 5) air/N₂ inlet; 6) solenoid valve; 7) timer; 8) pH measurement; 9) hybride recorder; 10) peristaltic pumps 11) a) organic nutrients acetate and ethanol b) inorganic nutrients.

Table 2. Operational parameters of the activated sludge plant

Sludge loading	0.3 g COD g^{-1} SS.d^{-1}
Sludge concentration	3.0 g SS L^{-1}
Sludge age	13 d(ays)
Hydraulic residence time	12 h(ours)
Cycle time	1 h
Anoxic period	20 min
Oxic period (3 mg O_2 L^{-1})	40 min
Temperature	20–22 °C

utes of every one-hour cycle the aeration started again (the oxygen concentration in the suspension was set at 3 mg O_2 L^{-1}) to have a nitrification period of 40 minutes before the cycle was repeated. The suspended solids (SS) concentration was adjusted daily to 3 g L^{-1} of dry matter. From the amount of surplus sludge withdrawn daily and the dry matter content in the effluent, the sludge age was calculated to be 13 days (see operational parameters, Table 2).

A suspensa-free filtrate stream obtained from a continuously operating filtration unit (Tappe, 1992) directly connected to the activated sludge tank ensured the online measurement (one sample per minute) of ammonia, nitrite and nitrate concentrations by flow injection analysis (Aquatec, Perstorp). The activated sludge tank was covered with a lid and the exhaust was pumped into an NO/NO₂ gas analyzer (Nucletron). This apparatus had to be provided with a gas stream of at least 60 L h^{-1}. During the nitrification period the aeration rate was 100 L h^{-1} to maintain a suitable gas flow of compressed air and changed automatically to N₂ if the preset value of 3 mg O_2 L^{-1} was reached. During denitrification, the activated sludge suspension was sparged with N₂ at a rate of 100 L h^{-1}. In some experiments the N₂ gas stream was led over (and not through) the activated sludge suspension during denitrification.

Results

Without pH control of the activated sludge suspension the pH value varied between 6.7 at the end of the nitrification period and 7.4 at the end of the denitrification period. During the oxic phase most of the ammonia was oxidized to nitrite first and the aeration time last-

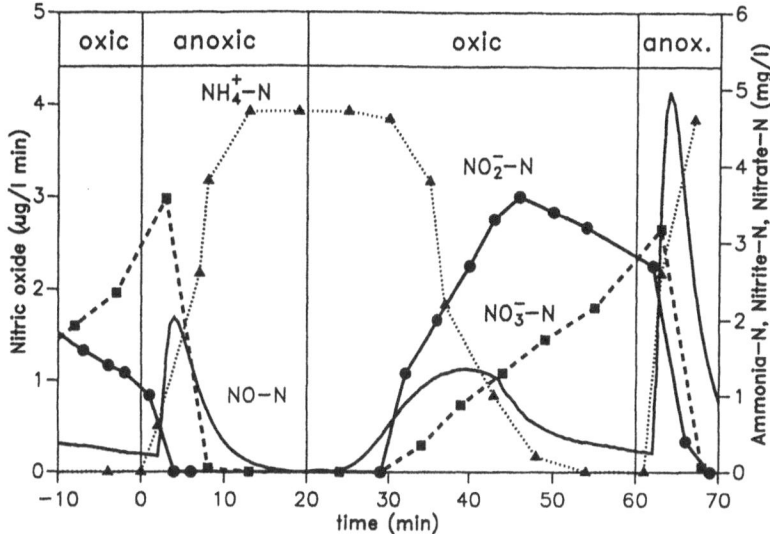

Figure 2. Changes in NH_4^+-N (▲...▲), NO_2^--N (● – ●), NO_3^--N (■ - - ■) concentrations in the activated sludge tank and NO-N (——) during a one-hour cycle at pH 7.0.

ed not long enough to convert the nitrite into nitrate. At the beginning of the anoxic phase both nitrite and nitrate were present. NO was released in considerable amounts during the nitrification as well as during the denitrification phase. One day after the installation of a pH control unit (titration of the suspension by either 1 M NaOH or 1 M H_2SO_4) the changes in time of the concentrations of ammonia, nitrate and nitrate were obtained as shown in Figure 2. The release of NO at pH 7 during the denitrification phase seemed to depend on the amount of nitrite at the start of the denitrification period. At time zero the nitrite concentration was 1.2 mg N L^{-1} and the NO release was much lower than in the next denitrification phase at the start of which the water contained 3.7 mg NO_2^--N L^{-1} (60 minutes, Fig. 2).

During nitrification NO release seems to be coupled with the oxidation of ammonium to nitrite. After about two weeks of operation at pH 7 (approximately one sludge age) nitrite was oxidized much faster to nitrate, resulting in a significantly smaller appearance of nitrite and a complete conversion of ammonia to nitrate during the oxic phase (Fig. 3). Corresponding to the lower nitrite concentrations, the release of NO was also low during nitrification and almost no NO was released during denitrification.

Increasing the pH to constantly 8.0 resulted in nitrite not appearing as an intermediate product during nitrification, and the amount of NO released was only 0.09% of the amount of ammonia oxidized (Fig.

4). The corresponding values for the adapted activated sludge suspension at pH 7 were 0.14% and 0.45% for the unadapted suspension at pH 7.

After the pH value of the activated sludge suspension was reset to pH 6.5, it took approximately two weeks before nitrification almost stopped and the ammonia concentration increased. Under these conditions, NO release was as low as the values obtained at pH 8.

Sparging the activated sludge suspension with N_2 during denitrification is rather unnormal as compared to the situation in waste water treatment plants. For this reason, some experiments were carried out with a flow of N_2 led over the suspension. In this situation, no NO could be detected in the gas stream. Sparging the suspension with N_2 obviously interrupts the denitrification process and the measured NO release is probably an artifact.

Discussion

During denitrification, but only in the presence of nitrite and sparging the activated sludge suspension with N_2, considerable amounts of NO are released from the model activated sludge plant. When the gas stream was led over the stirred suspension, the NO release decreased below detection limit. In the latter situation which is closer to denitrification conditions in technical treatment plants, the intermediately formed NO

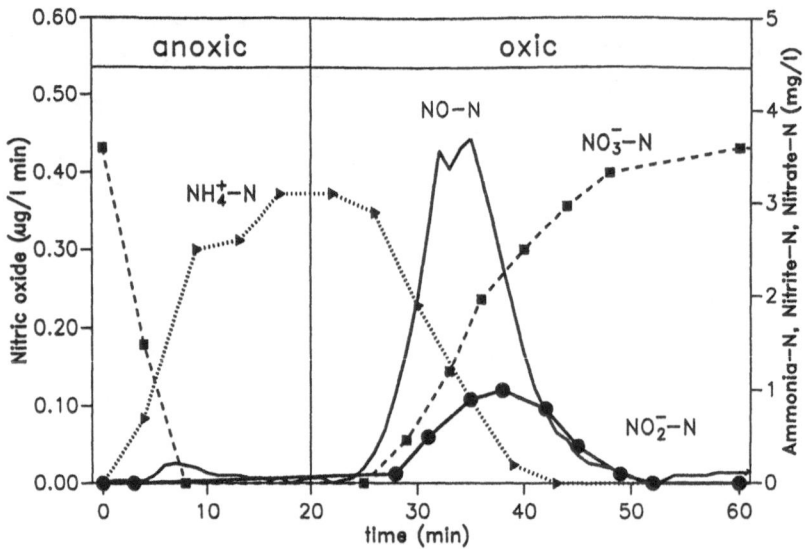

Figure 3. Changes in NH_4^+ -N (▲...▲), NO_2^- -N (● – ●), NO_3^- -N (■ - - ■) concentrations in the activated sludge tank and NO-N (——) during a one-hour cycle adapted to pH 7.0.

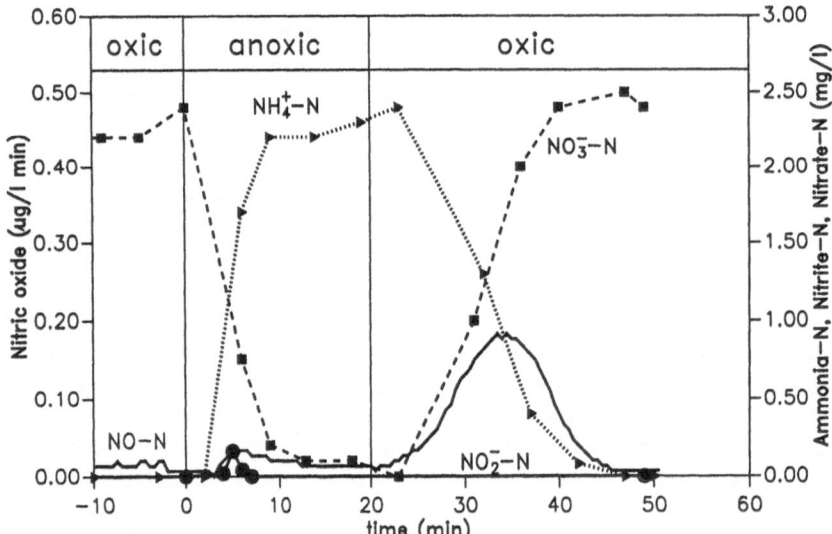

Figure 4. Changes in NH_4^+ -N (▲...▲), NO_2^- -N (● – ●), NO_3^- -N (■ - - ■) concentrations in the activated sludge tank and NO-N (——) during a one-hour cycle at pH 8.0.

(Zumft, 1993) is obviously further reduced to N_2O or N_2.

When the conditions for complete nitrification are unfavourable and nitrite appears as an intermediate or as the end product, considerable amounts of nitric oxide are released. Nitrite accumulation in waste water treatment plants with nitrification occurs only exceptionally. Normally, most of the ammonia is oxidized to nitrate, without measurable appearance of nitrite. After having simulated such conditions experimentally, we

concluded that the release of NO reaches maximally 0.1% of the amount of ammonia nitrogen oxidized.

If one assumes that the 13 g of nitrogen produced per inhabitant and day is completely oxidized to nitrate, while 0.01% are converted into nitric oxide, the annual emission of NO from waste water treatment plants and 80 million inhabitants in Germany would be 380 t. The annual emission in Germany from e.g. power stations and traffic amounts to 2.95 mio t (UBA, 1990). Compared to this value waste water treatment plants with biological nitrogen removal are no significant sources of NO emission.

References

Conrad R 1990 Flux of NO_x between soil and atmosphere: importance and soil microbial metabolism. *In* Denitrification in Soil and Sediments. Eds. P N P Revsbach and J Sorensen. pp 105–128. Plenum Press, New York, USA.

Crutzen P J 1979 The role of NO and NO_2 in the chemistry of the troposphere and stratosphere Ann. Rev. Earth Plant Sci. 7, 443–472.

Ehhalt D M and Drummond J W 1982 The tropospheric cycle of NO_x. *In* Chemistry of the unpolluted and polluted Troposphere. Eds. H W Georgii and W Jaeschke. pp 219–251. Reidel, Dordrecht, the Netherlands.

Payer H D and Trültzsch U 1972 Ein Beitrag zur Versorgung dichter Kulturen von Grünalgen mit Mangan, Vanadium und anderen Spurenelementen. Arch. Mikrobiol. 84, 43–53,

Soeder C J, Zanders E, Muckenheim T and Groeneweg J 1993 Weitgehende Elimination des anorganisch gebundenen Stickstoffs (TIN) durch schubweise Abwasserzugabe bei einstufigen Belebungsanlagen: Modellversuche bei abgestuftem BSB_5:TIN-Verhältnis. Wasser Abwasser gwf 134, 462–467.

Tappe W 1992 Wachstum von Bakterienreinkulturen und definierten Mischkulturen bei kontinuierlicher und diskontinuierlicher Substratzugabe unter besonderer Berücksichtigung der biomassespezifischen Nährstoffzufuhr. Bericht des Forschungszentrums Jülich Nr. 2677, Jülich, Germany.

UBA 1990 Umweltbundesamt Berichte 1990 Luftverschmutzung durch Stickstoffoxide; Ursachen, Wirkungen, Minderungen. E Schmidt Verlag, Berlin, Germany.

Williams E J, Hutchinson G L and Fehsenfeld F C 1992 NO_x and N_2O emissions from soil. Global Biogeochem. Cycles 6, 351–388.

Zumft W G 1993 The biological role of nitric oxide in bacteria. Arch. Microbiol. 160, 253–264.

O. Van Cleemput et al. (eds.), Progress in Nitrogen Cycling Studies, 559–566, 1996.
© 1996 *Kluwer Academic Publishers.*

Estimation of N_2O fluxes under rape for biological fuel production, bare soil and grass fallow

C. Hénault[1], C. Devroe[1], R. Reau[2] and J.C. Germon[1]
[1]*INRA, CMSE, Laboratoire de Microbiologie des Sols, 17 rue Sully, 21000, Dijon, France and* [2] *CETIOM, 174 Av. V. Hugo, 75116 Paris, France*

Key words: bare soil, fertilization, grass fallow, nitrous oxide emissions, rapeseed, static chamber method

Abstract

Nitrous oxide emission from agricultural soils is an important environmental concern as this gas contributes to the partial destruction of the stratospheric ozone layer and to global warming. European farmers are being encouraged to develop crops for producing biofuel. However, the excessive use of nitrogenous fertilizers on these crops could substantially increase agricultural emissions of N_2O and hence eliminate the beneficial effects of recycling CO_2. N_2O fluxes were estimated from soil under four treatments: (1) bare soil, (2) unfertilized grassland fallow, (3) soil cropped to rapeseed and fertilized at rate of 180 kg N ha^{-1} yr^{-1} and (4) as for (3) but fertilized at 280 kg N ha^{-1} yr^{-1}. Measurements were made using the static chamber method throughout spring and summer 1994. No particularly beneficial effect on N_2O production was observed under reasonable fertilized rapeseed. Fluxes in the order of 1 kg N ha^{-1} yr^{-1} were measured under rape fertilized at 180 kg N ha^{-1} yr^{-1}. The amount of fertilizer applied to rape has a very definite effect on the amount of N_2O released. Bare soils constitute a risk in terms of N_2O release as a 'hot spot' was observed after the harvest. Grass fallow seemed to be best adapted to limiting N_2O release.

Introduction

Nitrous oxide is a natural gas and its present concentration in the atmosphere is 310 ppbv. It was 285 ppbv before the industrial era and the estimated annual increase is concurrently 0.8 ppbv. This gas, however, is implicated in the greenhouse effect and destruction of the ozone layer (Knowles, 1982; Smith and Arah, 1990), its role in the greenhouse effect being 200 times greater than that of CO_2, per mole of gas (Rodhe, 1990). Human activities account for about 50% of the N_2O released and agriculture would be responsible for about two thirds of this (Granli and Bockman, 1994).

The following points of relevance have been taken from various studies concerned with N_2O production associated with agricultural activities:

1. The N_2O fluxes measured in arable soils under temperate climates are in the order of 1 to 5 kg N ha^{-1} yr^{-1} in the whole (Eichner, 1990; Germon and Hénault, 1994).

2. Two microbial processes, denitrification and nitrification, are involved. Nitrification takes place mainly in the superficial soil layer under aerobic conditions and more or less continuously. Denitrification, in contrast, occurs much deeper in the soil, and as it occurs during very wet periods, it is extremely transitory (Bouwman, 1990). The proportion of N_2O produced during denitrification is also highly variable, ranging from 0 to 100% of the reaction products, according to Aulakh et al. (1992).

3. The soil regulation of N_2O release is still poorly understood. Generally acknowledged factors include vegetation and the type of crop, soil type, water supply, temperature, and the nature and amount of fertilization (Bouwman, 1990; Eichner, 1990; Sahrawat and Keeney, 1986). However, the results obtained for these various factors have often been contradictory as, for example, in the presence or absence of plant cover. Granli and Bockman (1994) reported some experiments in which the presence of plants on the soil led to an increase in N_2O fluxes (Cribbs and Mills, 1979; Klemedtsson et al., 1977) whereas in others the N_2O fluxes

were greater under bare soil (Aulakh et al., 1982; Duxburry et al., 1982).

A project is currently underway to develop rape for energetic use (diester), the principal argument being the recycling by a crop-plant of CO_2 produced by using fossil fuels. In 1992, however, German ecologists put forward the hypothesis that rape cultivation would encourage the release of soil N_2O, thus cancelling the beneficial effect of using diester (der Spiegel, 1992). We have not yet found in the literature any measurements of N_2O flux, used to compare different crops, that include rape.

The work described here assembles various measurements of N_2O flux obtained from March to July 1994 from a Burgundy soil (France) under different fallow systems (bare; seeded with rye grass; and energy-producing under rape receiving two levels of fertilizer application). This paper includes (1) measurements of the N_2O flux and of agro-climatic parameters generally examined in studies of denitrification regulation and N_2O production, (2) comparison of the different techniques used to estimate N_2O flux and (3) interpretation of the results using a predictive denitrification model (Hénault, 1993).

Materials and methods

Experimental plots

The experiments were carried out on a large scale plot in Burgundy, France. The soil was a deep, calcareous brown soil of clay texture, with a clay content of 45%, a relatively poor organic matter content (1.9%) and alkaline pH; ($pH_{water} = 7.7$). The experiment consisted of the 4 treatments presented in Table 1.

The experimental plot was divided into 4 blocks, each containing a replicate of the 4 treatments. No differentiation was made between treatments T3 and T4, when the different measurements were taken, until April 12th.

Monitoring of the agro-climatic parameters affecting N_2O production

A meteorological unit (ENERCO CE 395–42) was set up on the experimental site for the daily monitoring of rainfall, and temperature at 2 m and −10 cm in 3 plots corresponding to treatments T1, T2 and T3. The moisture per unit weight of samples of soil was determined by oven drying (105 °C – 24 h). Moisture was

measured in the 0–20 cm horizon on January 94 and on the days of N_2O flux measurement, 2 samples being taken from this horizon in each of the sub-plots. The nitrate concentrations of the soil samples were determined by colorimetric assay of the filtrate collected after soil ion extraction in the presence of molar KCl, with a soil/solution ratio of 1/5. The colorimetric analyses were performed with a continuous flux Technicon II analyser, using the procedure developed by Griess-Illosvay. Measurements were made using soil samples taken at the time as those used for the moisture content determination.

In situ measurements of N_2O flux

The technique used corresponded to the so-called 'Chamber system', the principal characteristics of which have been described by Hutchinson and Livingston (1993). The chambers consisted of hollow PVC cylinders, 50 cm in diameter, and 15 cm in height, sunk into the soil to a depth of approximately 8 cm. These chambers were set up on the experimental plot at the end of February, with 2 chambers on each of the T1, T2 and T3 treatment subplots, corresponding to 8 replicates per treatment. During March and April, the vegetation was not very high and so the chambers were not displaced between the different times of assessment. A set of 8 new chambers was installed after the third nitrogen application corresponding to treatment T4. From this date onwards, the chambers were displaced within the rape plots, between each time of assessment, as the vegetation had to be cut back so that the chambers could be closed. The chambers were therefore set up at least one day before the measurements so as to limit the disturbance caused by their installation. Each chamber was sealed with a PVC lid fitted with a ring of foam rubber to ensure adhesion to the chamber, and kept in place with a metal cross placed on the lid and fixed to the chamber. Tiny holes were pierced in the lides to maintain the internal pressure of each chamber at atmospheric pressure. The following protocol was adopted for the field measurements:

1. Closure of the previously installed chambers.
2. Injection of 50 mL of Kr into each chamber. This inert gas was used to check the absence of excessive lateral flux between the inside and outside of the chamber.
3. Removal of gas samples from each chamber. These samples were taken with 1 mL syringes, which were stabbed into rubber corks and brought back to the laboratory. Four samples of gas were taken over

Table 1. Presentation of the different treatments

Treatment	Type of crop	Sowing date	Nitrogen fertilizer		
			Date	Amount (kg N ha^{-1})	Form[a]
T1	Bare soil	–	–	0	
T2	Rye grass	6.09.93	–	0	
T3	Rape	6.09.93	14.02	80	S.N
			9.03	100	L.N.U
T4	Rape	6.09.93	14.02	80	S.N
			9.03	100	L.N.U
			12.04	100	S/N

[a]S.N–Solid (NH_4NO_3), L.N.U–Liquid (NH_4NO_3 + Urea).

a total period of 2.25 h, with one gas sample every 3/4 h, to determine the kinetics of N_2O release. The lids were taken off the chambers after the 4th sampling session.

4. Return to the laboratory of the syringed gas samples for rapid analysis of their N_2O content.

The gas samples were analysed in the laboratory using Gas Phase Chromatography. The N_2O analyses were performed with a Varian 3400 Cx apparatus fitted with an electron capture detector and coupled to the Varian Star Workstation integrated software. The gaseous volume of each chamber was calculated by measuring the average height above the soil surface at the periphery. This volume was used to convert the measured concentrations into the amounts contained inside the chambers. The flux were then calculated in relation to air temperature on the day of assessment. The mean values for each treatment were thus the arithmetic means of 8 replicates, and the kinetics were considered significant when the correlation coefficient r was such that the absolute value was greater than 0.9 ($\alpha < 0.1$). If not, they were considered as equal to 0. Results obtained are expressed in g N ha^{-1} day^{-1}.

Laboratory measurement of natural and potential N_2O production and potential denitrification

These measurements were carried out using a similar procedure to the 'core system with gas recirculation' described by Tiedje et al. (1989). Measurements were made on undisturbed soil cores taken with stainless steel cylinders, 10 cm diameter and 25 cm in height, used to remove soil samples from the 0–20 cm horizon. These measurements were obtained as fallows in the laboratory at 20 °C:

1. Removal of 8 cylinders from each T1, T2 and (T3/T4) treatment on April 13th and incubation of these samples at 20 °C in contact with the ambient atmosphere.

2. Determination of the kinetics of natural N_2O production. The soil cylinders where then tightly closed. Gas samples were taken from each cylinder every hour for three hours. The atmosphere in the chamber was made homogeneous by operating a hermetically sealed aquarium pump during 3 min, at a flow rate of 200 mL min^{-1}, before removal of the gas samples into 3 mL Venoject tubes.

3. Drip application of a nitrate solution (500 mL, 0.1 M N-KNO$_3$) for 24 h to bring the samples to 0.39 moisture (w/w), representing approximately 86% of the total porosity, and to a nitrate content of about 180 mg N kg^{-1} soil, this in theory corresponding to the establishment of zero order kinetics in relation to nitrate concentration (Hénault, 1993). This period of application was followed by resting for 24 h.

4. Determination of the kinetics of potential N_2O production, using the above described procedure (2), and of potential denitrification after enrichment of the gaseous atmosphere in each chamber with acetylene (2%) to inhibit N_2O reductase and terminate the conversion at the N_2O stage (Yoshinari and Knowles, 1977), still using the same procedure (2).

The gas samples were analysed on a Girdel 30 type gas phase chromatograph, fitted with an electron capture detector and coupled to an HP 3390A integrator and an automatic sampling system (Dani 3950). N_2O solubility was taken into consideration in the production rate calculations (Germon, 1980). The final results are expressed as the arithmetic means of the 8 replicates

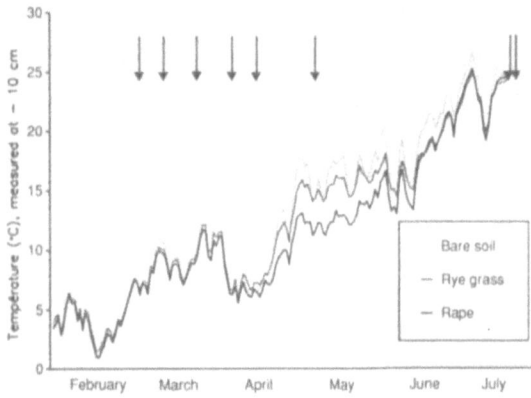

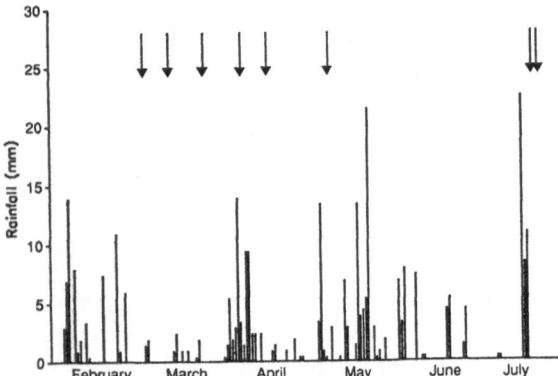

Figure 1. Climatic data measured between 24.01 and 23.07. The arrows indicate the days of N_2O measurement.

Table 2. Overall measurements of N_2O flux

Treatment	Mean N_2O flux ($g N ha^{-1} day^{-1}$)
T1	3.56
T2	0.4
T3	3.88
T4	6.13

Results and discussion

Agro-climatic parameters affecting N_2O production

The recorded temperature are indicated in Figure 1. It should be noted that the curves obtained for the three treatments T1, T2 and T3 can be almost perfectly surimposed up to mid-April. From this date onwards until the end of June, there is divergence of the recorded temperature between treatments, with a higher mean daily temperature under bare soil than under rye grass and rape. This order may be compared with the inverse order of the extent of plant cover suggesting that greater vegetative cover will produce lower temperatures at the depth of -10 cm. This result, in terms of nitrogen losses by denitrification, may be interpreted as follows. The function Ft – empirical function incorporating the limiting effect of soil temperature on denitrification – (Hénault, 1993; Stanford et al., 1975), is applied to the temperature measurements in all three treatments. A mean limiting effect associated with the temperature factor can thereby be calculated. It corresponds to the proportion of potential denitrification measured with the protocol previously described, which may be expressed as a result of the actual temperatures being lower than in the potential denitrification measurement made at 20 °C. This calculation produces the following result. The observed temperature differences, between April 15th and June 15th 1994, in the three treatments lead to an expression of 70% of the potential denitrification under bare soil, 65% under rye grass and 53% under rape. Indirectly, by retarding the temperature increase during the spring, the rape crop helps to create temperature conditions that are less favourable to denitrification than bare soil.

Concerning the rainfall and soil moisture, it should be noted that the beginning of 1994 was relatively wet with a total of 315 mm falling from January 24th to July 23rd (Fig. 1). The individual moisture measurements are presented in Figure 2. It should be noted

in $g N g^{-1}$ soil h^{-1} and in $g N ha^{-1} day^{-1}$. This conversion was performed by considering an apparent soil density of 1.2, estimated from the mean ratio (mass/volume) of the soil cylinders.

Predictive denitrification model

A predictive denitrification model (Hénault, 1993) was used to calculate the estimated nitrogen losses by denitrification and to interpret the level in the soil of the regulatory parameters of this process in terms of limiting effects. This model will be described in detail later but has, in brief, the following structure. The actual denitrification activity of the soil, is a multiplicative function of the potential soil denitrification, measured with the protocol described in the previous paragraph, and empirical dimensionless functions englobing the effects of soil moisture (Grundmann and Rolston, 1987), temperature (Stanford et al., 1975) and nitrate concentration (Hénault, 1993).

563

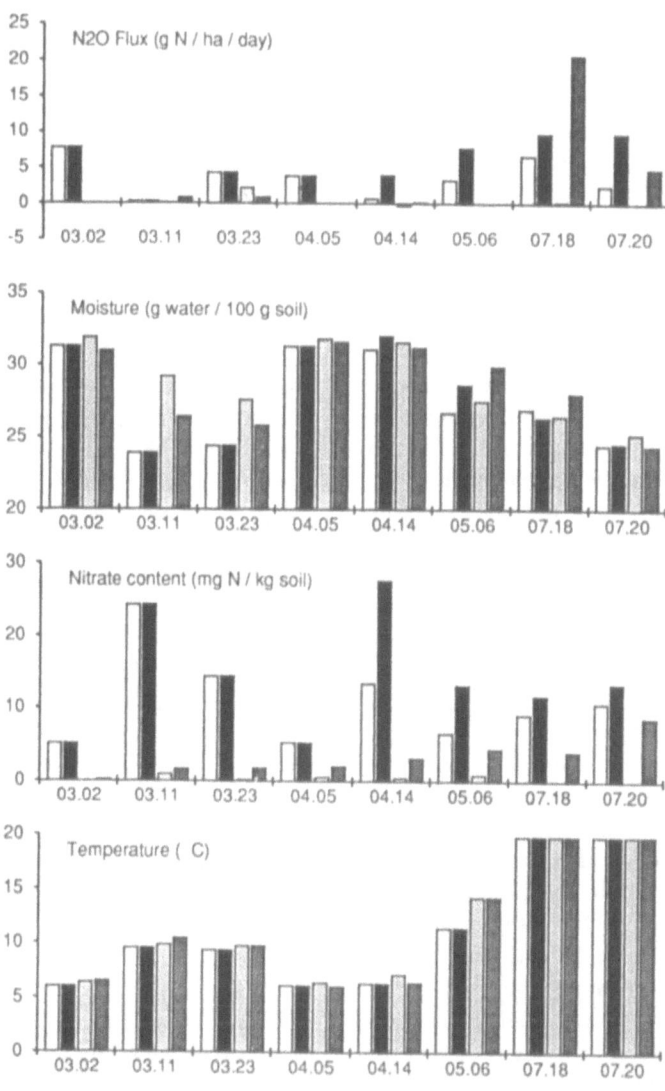

Figure 2. Measurements of N₂O flux and monitoring of regulatory parameters. ☐ Rape (T3), ■ Rape (T4), ▤ Rye grass (T2), ▦ bare soil (T1).

Table 3. Measurements obtained with the 'Core System'

Crop	Natural N$_2$O production	Potential N$_2$O production	Potential denitrification
	(ng N g^{-1} soil h^{-1})		
Rape	1.37 ± 4.02	200 ± 150	257 ± 225
Rye grass	not detected	104 ± 90	137 ± 88
Bare soil	0.4 ± 0.74	140 ± 138	191 ± 243
	(g N ha^{-1} d^{-1})		
Rape	78.91 ± 231.5	11520 ± 8640	14802 ± 12960
Rye grass	not detected	5990 ± 5190	7891 ± 5063
Bare soil	16.99 ± 38.01	8582 ± 7545	11001 ± 13997

that there is no significant difference between the different treatments except for the 23.03 and 6.05 dates when the values observed under rye grass and bare soil, respectively, were higher than the others. It is not therefore clear at the end of this study that the type of fallow is able to indirectly influence N_2O production by inducing differences in soil moisture behaviour.

The nitrate factor, by definition, is not independent of the treatments. The amount of nitrates contained in the 0–20 cm horizon before the first nitrogen application in January, was relatively low in the different treatments (< 1 mg N kg^{-1} soil) but nevertheless significantly higher under bare soil. According to the function F_n (limiting effect of soil nitrate concentration on denitrification) established by Hénault (1993), a nitrate content of 1 mg N kg^{-1} soil permits expression of only 4.5% of the potential soil denitrification, given that all the other factors are favourable. After the first application, the nitrate content under rape becomes significantly higher than under the other treatments. The maximal value measured during this period was 8.6 mg N kg^{-1} soil corresponding to a possible expression of 28% of the potential denitrification. High nitrate concentrations were observed, between the second and third applications, under rape permitting expression of more than 50% of the potential denitrification compared with less than 8% in the other treatments. After the third application and before the rape harvest, 3 groups of nitrate contents could be distinguished, one consisting of the over-fertilized rape with a nitrate content permitting expression of more than 50% of the potential denitrification immediatly after the application, followed by 35% fifteen days later, the second consisting of rape receiving the recommended dose for which the nitrate content corresponded to expression of 38% of the potential followed by 23%, and lastly a group consisting of rye grass and bare soil (less than 15% of the potential expressed). After harvesting the observed nitrate content under bare soil increased to approach that observed under rape. The effect of the third application to rape was still visible after harvesting.

In situ measurements of N_2O production

The values of the measurements were between -0.5 and 22 g N ha^{-1} day^{-1} (Fig. 2) and this corresponds to the generally observed order of magnitude (Eichner, 1990). Taking the eight times of assessment together, the overfertilized rape exhibited the highest nitrogen production and rye grass the lowest (Table 2). The greatest release of nitrogen, except for the measurements on July 18th, was obtained under rape. The third nitrogen application on April 12th had a definite effect on the amount of N_2O release. It should be noted that this effect was still visible after harvesting (July 18th and 20th). The N_2O flux measured in soil sown with rye grass were very weak throughout the study period and did not exceed 2.4 g N ha^{-1} day^{-1}. Very small negative flux (about $- 0.5$ g N ha^{-1} day^{-1}) were even observed on April 14th and May 6th, althought the environmental conditions on those days were not unusual. The N_2O fluxes in bare soil were very weak throughout the spring (< 1 g N ha^{-1} day^{-1}). In contrast relatively large fluxes, exceeding 20 g N ha^{-1} day^{-1} on July 18th falling to approximately 5 g N ha^{-1} day^{-1} two days later, could be observed in the two July assessments. These results show that the study of microbial phenomena should be continued during the inter-crop period and, as a 'hot spot' was observed under bare soil on July 18th on a rainy day, illustrate the highly transient nature of N_2O release.

In conclusion, the overall N_2O flux remained relatively weak. During the period of crop production they were highest under rape and so was the soil nitrate content. A third application of nitrogen fertilizer to rape resulted in a net increase in N_2O release. In view of our detection of a 'hot spot', bare soil seemed to constitute a risk in terms of N_2O release. In contrast grass fallow seemed very well adapted to limiting nitrous oxide production.

Laboratory measurements of natural and potential N_2O production and potential denitrification

The experimental results are summarized in Table 3. The experimental soil exhibits a relatively high potential of N_2O production and denitrification, i.e. approximately 10 kg N ha^{-1} day^{-1}. This value is in accordance with other results obtained in our laboratory (Hénault, 1993). The values obtained by this method are, however, associated with very large confidence intervals leading to coefficients of variation of about 100%, so that between-treatment comparison is difficult. Nevertheless the tendency seems to be greater for the potential denitrification and N_2O production under rape than under bare soil, followed by rye grass. Similarly the actual N_2O activity, measured in the laboratory, is much higher under rape than under bare soil, again followed by rye grass.

Table 4. Potential N$_2$O production to denitrification ratio (r)

Crop	Ratio (r)
Rape	0.76
Rye grass	0.77
Bare soil	0.78

Table 5. Comparison of methods used to estimate natural N$_2$O production

Crop	Natural N$_2$O production (g N ha^{-1} day^{-1})		
	In situ measurement	Laboratory measurement	Estimation with model
Rape	2.05	5.02	13.0
Rye grass	0.30	0	1.76
Bare soil	0.37	2.02	5.43

Finally, it is interesting to note that the ratio (r) between potential N$_2$O production and denitrification is the same for all three treatments (Table 4).

Comparison of measurement techniques

The measurements made between April 12th and 16th were used to compare the following techniques for estimating soil N$_2$O production:

1. Direct measurement 'in situ' using the chamber method.
2. Direct measurement in the laboratory using the 'core system' with application of correction based on the temperature effect. We have used the function F$_t$ determined from the experimental results of Stanford et al. (1975).
3. Application of the predictive denitrification model developed in the Laboratory of Soil Microbiology (Hénault, 1993) to the natural production of N$_2$O. In this case the calculations are based on the potential N$_2$O production rather than the potential denitrification.

The results are given in Table 5. The values obtained in the laboratory or estimated with the model were higher than the 'in situ' measurements. All three methods nevertheless produced the same order of flux intensity, i.e. in increasing order, rye grass, bare soil, and rape. The results can be analysed as follows. It is not ready surprising, in so far as the 'in situ' and laboratory methods are concerned, that higher values are obtained with the 'Core System', in which the chamber gases are homogenized and those dissolved in the soil aqueous phase are taken into account in the calculation. The accordance between the two methods is therefore satisfactory. It should be noted, in the estimation with the predictive model, that this model was designed to simulate total denitrification rather than N$_2$O production. The simple adaptation applied for the purposes of this study is not sufficiently satisfactory to permit its systematic use.

Conclusion

This 6-month study does not enable us to attribute rape with a particularly beneficial effect on N$_2$O production. The values obtained with bare soil and rape receiving 180 kg N ha^{-1} are in fact very similar and if the means obtained for the eight assessment times are extrapolated to one year, flux in the order of 1 kg N ha^{-1} year^{-1}, which are classically observed under our climatic conditions, are obtained with both treatments.

In contrast this study does show that the amount of fertilizer applied to rape has a very definite effect on the amount of N$_2$O released. The third nitrogen application in this crop led to a 57% increase in flux, and the effect was still apparent after harvesting.

These measurements need to be completed with a longer experiment, which, in view of the considerable flux observed under bare soil during July, should include the intercrop period.

Finally, we were able to see that the soil can act as an N$_2$O sink leading to the establishment of very small negative flux, particularly under rye grass. The N$_2$O flux would seem to be weakest under this type of nonfertilized 'fallow' system.

References

Aulakh M S, Doran J W and Mosier A R 1992 Soil denitrification – significance, measurement, and effects of management. Adv. Soil Sci. 18, 1–57.
Aulakh M S, Rennie D A and Paul E A 1982 Gaseous nitrogen slosses from cropped and summer-fallowed soils. Can. J. Soil Sci. 62, 187–196.
Bouwman A F 1990 Analysis of global nitrous oxide emissions from terrestrial natural and agroecosystems. Trans. 14th Cong. Soil Sci. 2, 261–266.
Cribbs W J and Mills H A 1979 Influence of nitrapyrin on the evolution of nitrogen oxide (N$_2$O) from an organic medium with and without plants. Commun. Soil Sci. Plant. Anal. 10, 785–794.
Duxbury J M, Bouldin D R, Terry R E and Tate R L 1982 Emissions of nitrous oxide from soils. Nature 298, 462–464.

Eichner M J 1990 Nitrous oxide emissions from fertilized soils: Summary of available data. J. Environ. Qual. 19, 272–280.

Germon J C and Henault C 1994 Quantifier la dénitrification et la production naturelle de N_2O dans les sols. Synthèse bibliographique. Ministère de l'environnement. Convention n° 90385. 37p.

Germon J C 1980 Etude quantitative de la dénitrification biologique dans le sol à l'aide de l'acétylène. I. Application à différents sols. Ann. Microbiol. (Inst. Pasteur). 131 B, 69–80.

Granli T and Bockman O O 1994 Nitrous oxide from agriculture. Norw. J. Agric. Sci., Suppl. 12, 128.

Grundmann G L and Rolston D E 1987 A water function approximation to degree of anaerobiosis associated with denitrificaton. Soil Sci. 144, 437–441.

Henault C 1993 Quantification de la dénitrification dans les sols à l'échelle de la parcelle cultivée, à l'aide d'un modèle prévisionnel. Thèse ENSA, Montpellier, France. 108p.

Hutchinson G L and Livingston G P 1993 Use of chamber systems to measure trace gas fluxes. *In* Agricultural Ecosystem Effects on Trace Gases and Global Climate Change. Eds. L A Harper, A R Mosier and J M Duxburry. pp 63–78. Am. Soc. Agron., Madison WI, USA.

Klemedtsson L, Svensson B H, Lindberg T and Rosswall T 1977 The use of acetylene inhibition of nitrous oxide reductase in quantifying denitrification in soils. Swed. J. Agric. Res. 7, 179–185.

Knowles R 1982 Denitrification. Microbiol. Rev. 46, 43–70.

Rodhe H 1990 A comparison of the contribution of various gases to the greenhouse effect. Science 248, 1217–1219.

Sahrawat K L and Keeney D R 1986 Nitrous oxide emission from soils. Adv. Soil Sci. 4, 103–148.

Smith K A and Arah J R M 1990 Losses of nitrogen by denitrification and emissions of nitrogen oxides from soils. Paper read before the Fertiliser Society in London on 13 December 1990. Proceedings n° 299. 34p.

Spiegel der 1992 Lachgas vom Acker. Der Spiegel 22, 86.

Stanford G, Dziena S and Vander Pol R A 1975 Effect of temperature on denitrification rate in soil. Soil Sci. Am. Proc. 39, 867–870.

Tiedje J M, Simkins S and Groffman P M 1989 Perspectives on measurement of denitrification in the field included recommended protocols for acetylene based methods. Plant and Soil 115, 261–284.

Yoshinari T and Knowles R 1976 Acetylene inhibition of nitrous oxide reduction by denitrifying bacteria. Biochem. Biophys. Res. Commun. 69, 705–710.

Plant and Soil **181**: 57–63, 1996.

Temporal changes in N_2O-losses from two arable soils

Ernst-August Kaiser and Otto Heinemeyer
Institut für Bodenbiologie, Bundesforschungsanstalt für Landwirtschaft (FAL), Bundesallee 50, 38116 Braunschweig, Germany*

Key words: arable soil, nitrous oxide, soil texture

Abstract

N_2O-loss rates from two soils were measured over a continuous observation period of 2 years. The two soils, differing in texture (sandy loam and silty loam), are frequently used for intensive crop production. Rates were estimated using a closed soil cover box technique. N_2O-losses obtained were scrutinised with physical, chemical and microbiological properties of the soils as well as with climatic data.

Large temporal changes in N_2O-emission rates were found. The data were approximately log-normal distributed. In spring maximal values of 20 g N_2O-N ha^{-1} d^{-1} were observed. According to this observation, two situations associated with high flux rates could be distinguished; 1. N_2O- production by soil at spring thaw and 2. N_2O-production within one week after N-fertilizer application. For both soils equal N_2O-losses were found, which are adequate to 1 kg N_2O-N ha^{-1} per year. From this data was calculated that N_2O-losses ranged from 0.8–1.5% of the applied fertilizer N.

Introduction

The annual increase in the atmospheres overall concentration of nitrous oxide makes N_2O a potent cause of global climate change and stratospheric ozone depletion. Agricultural soils are suspected to be responsible for the increased N_2O-emission into the atmosphere. According to large temporal fluctuations, precise informations about representative mean annual N_2O-losses from different soils and management practices are rare. Most field studies concerning seasonal fluctuations are based on a limited period within the year only. These included the influence of irrigation (Mosier et al., 1986), soil management (Arah et al., 1991; Hansen et al., 1993), spring thaw (Christensen and Tiedje, 1990; Goodroad and Keeny, 1984) or crop growth (Vinther, 1984). Few data exists on annual losses based on an investigation period of more than 1 year (Aulakh et al., 1983; Cates and Keeny, 1987; Weir et al., 1991).

This study was conducted to provide information from intensive monitoring concerning mean annual N_2O-losses. An attempt was made to investigate the factors controlling the temporal changes in N_2O-emission and to distinguish between the N_2O pro-

duction processes (denitrification/ nitrification). Two selected soils/management systems, which represent approximately 40% of the arable land area in Germany, were selected.

Material and methods

Field sites

Two field sites were established in the surrounding of Braunschweig, Lower Saxony, Germany (52°17'35 North, 10°26'55 East). The soils are frequently used for intensive crop production (sugar beet–winter wheat–winter barley) and are under cultivation for more than 100 years. In 1993, winter barley was substituted by winter wheat on site 1, in contrast to the previous years (Kaiser and Heinemeyer, 1993). Site 1 (Timmerlah) is a silty loam (luvisol from loess) and site 2 (FAL) is a sandy loam (alfisol, Kaiser et al., 1992). Due to a low water-holding capacity and a gravel layer (70 cm depth), water is a limiting factor for plant growth on site 2. In spite of this possible limitation no irrigation was applied throughout the experimental period. In accordance with conventional soil manage-

---

* FAX No.: +49531596375
Plant and Soil is the original source of publication of this article. It is recommended that this article is cited as: *Plant and Soil* **181**: 57–63, 1996.

ment, plant residues (chopped straw) were incorporated by mouldboard ploughing (30 cm deep) after chiselling (10 cm deep) the stubble. A disc drill was used for sowing. In 1992, the crops were fertilised (calcium ammonium nitrate) according to conventional agricultural practice. In 1993, only one application (60 kg N ha^{-1} as KNO_3) was applied on the experimental fields.

Table 1. Microbial biomass-C (C_{mic}) (μg C g^{-1}), metabolic quotient (qCO_2) (mg C g^{-1} h^{-1}) and C_{mic}:C_{org} ratio (mg C g^{-1} C) of the two soils (0–30 cm, Nov. 1991, 1992, 1993)

Soil	C_{mic}	qCO_2	C_{mic}:C_{org}
Sandy loam	246.2[a]	1.27[a]	28.9[a]
Loamy silt	319.8[b]	1.20[a]	31.0[a]

N_2O flux measurements in field

To monitor the N_2O-flux from soil into the atmosphere the closed soil cover box technique was used. Nine plastic tubes (19 cm in diameter) were driven into the ground to a depth of 40 cm on each site after seeding. At least twice a week, the tubes were sealed by a lid equipped with a rubber stopper. Gas samples were taken at the beginning and the end of the covering period. A covering period of 3 h was sufficient for flux rate measurements at both soils. A linear increase in the N_2O concentration was found throughout this period. This was checked by additional gas samples (once an hour). Following plant growth, the tube height was increased by extension tubes. The tubes were removed before tillage, and reinstalled afterwards.

Incubation of soil cores

The acetylene inhibition technique was applied to soil cores to distinguish between the N_2O production processes (denitrification/ nitrification). Acetylene concentrations higher than 1% in soil inhibit the reduction of N_2O to N_2 by denitrification, and nitrification is suppressed. Therefore, denitrification is indicated by an increase in the N_2O production after acetylene fumigation. Since May 1992, intact soil cores (5.3 cm in diameter) were taken from the soil surface (0–10 cm) of each field using 20 cm plastic tubes. The tubes were sealed with rubber stoppers. 25 mL acetylene was injected at the bottom of the 10 tubes, to reach an acetylene concentration of 10%. All cores were incubated at 15 °C (mean soil temperature during the vegetation period). The cumulative N_2O- production within 24 h of all samples was measured after 6 h of pre incubation. At that time the acetylene concentration was higher than 1% in the headspace of the fumigated tubes. Flux rates were calculated for the surface of cores. Bulk density and water-filled pore space (Weir et al., 1993) were estimated after the incubation.

Analyses

Evacuated Vacutainers^{TM} (10 mL, Bector and Dickinson, Rutherford, USA) were used for gas sampling, after a pretreatment described by Heinemeyer and Kaiser (1996). Gas samples were analysed using a ^{63}Ni-electron capture detector equipped with an automatic sample injector system (Heinemeyer and Kaiser, 1996).

Soil samples were analysed for mineral N. The triplicated filtered extracts (2 N KCl, 5:1) were automatically analysed for nitrate and ammonium (Technicon AA II, Bad Vibel, Germany). Sieved moisture adjusted soil samples were analysed for soil microbial biomass (SIR-method, Anderson and Domsch, 1978) and the metabolic quotient (qCO_2). Soil organic carbon (C_{org}) was determined by dry combustion. Details of the analyses and calculations were described by Kaiser and Heinemeyer (1993).

Statistical analysis

Analysis of variance was used to determine variances attributable to the soils and treatments. Logarithmic (base 10) transformations were done to normalize the N_2O-data. Means were compared using t-test. For each parameter, values not marked with the same letter are significantly different ($p \leq 5\%$).

Results and discussion

The physical and microbial characteristics of the 2 soils are presented in Table 1 and 2. Although rainfall was comparable on both sites, a lower mean soil moisture content was found in the sandy loam during the experimental period. This was also indicated by the water-filled pore space (WFPS), although the absolute differences were reduced due to different total pore spaces (Table 2). Because of seasonal variations in the microbial biomass-C (data not presented) on both sites,

Table 2. Total and water-filled pore space, soil moisture (%) for the surface layer (0–10 cm) of the two soils and rainfall (mm a^{-1})

Soil	Year	Rainfall	Moisture	Water-filled pore space	Total pore space
Sandy loam	1992	611	9.77[a]	33.79[a]	41.5[a]
	1993	686	12.22[b]	46.34[b]	46.0[b]
Loamy silt	1992	599	15.43[c]	45.73[b]	47.9[b]
	1993	677	16.41[c]	47.26[c]	49.8[c]

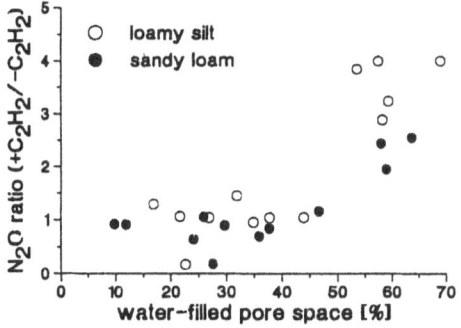

Figure 1. Ratio between the N$_2$O-release from the acetylene fumigated (+C$_2$H$_2$) and unfumigated (−C$_2$H$_2$) soil cores as affected by the water-filled pore space (n=10).

microbial characteristics are given for the soil sampling in November (Kaiser and Heinemeyer, 1993). Higher microbial biomass-C was found in the loamy silt soil, but the C$_{mic}$:C$_{org}$ ratios were similar. These values were representative for soils under continuous crop rotation (Anderson and Domsch, 1989; Kaiser and Heinemeyer, 1993).

N$_2$O-release from soil cores

No correlation was found between the N$_2$O rates calculated from the soil core incubations and the corresponding field estimations, but the results were used to identify soil conditions where denitrification took place. Therefore the ratio between the N$_2$O-release from the fumigated and unfumigated soil cores was calculated, and related to the WFPS (Fig. 1). Because of a great variability in the N$_2$O-release (coefficient of variation =1.9), no significant difference between the fumigated and unfumigated cores were found for ratios between 0–1.5. Only ratios higher than 1.5 indicated a N$_2$ production from denitrification. On both soils denitrification was detected where WFPS exceeded 50%. This corresponded to data calculated from a ^{15}N field

experiment at the loamy silt site (Walenzik and Heinemeyer, 1989).

Significant differences in the N$_2$O-release from the fumigated and unfumigated soil cores were found between both soils ($p \leq 5\%$). More N$_2$O was released from the loamy silt (data not presented), which could be attributed to a higher microbial biomass (Kaiser, 1994), a higher WFPS, and a finer soil texture (Weir et al., 1993) (Table 1, Fig. 1).

Temporal changes in N$_2$O-losses

Temporal changes in N$_2$O-losses from two arable soils and the climatic data are presented in Figures 2 and 3. The total rate of precipitation in 1992 was representative for this area (Deutscher Wetterdienst, 1992). Information about the vegetation period of each crop and soil tillage is given on the time scale. The missing values in Figures 2 and 3 are caused by an interruption of the measurements due to tillage operations.

A large seasonal variability of N$_2$O-emission rates was observed. Small N$_2$O-flux rates (<2 g N$_2$O-N ha^{-1} d^{-1}) were detectable every day. On the other hand, in spring 20 g N$_2$O-N ha^{-1} d^{-1} were estimated as a maximum value for N$_2$O-flux. These high flux rates were found after rainfall, when the WFPS were above 50%, and thus probably resulted from denitrification (Fig. 1). Two situations associated with high flux rates (Figs. 2 and 3) could be distinguished. (1) In January high N$_2$O-flux rates were found on both soils, when daily freezing and thawing cycles occurred (Figures 2 and 3). The observation of a brief and vigorous N$_2$O-production from soil at spring thaw was also described by Christensen and Tiedje (1990). They attributed this increase in N$_2$O-production to an increased microbial activity induced by additional available carbon from killed microorganisms by freezing or detritus that became available by freezing and thawing processes. In addition, a physical N$_2$O-release after thawing may have occurred (Goodroad and Keeny, 1984). (2) High N$_2$O-release rates (20 g N$_2$O-N ha^{-1} d^{-1}) were also found

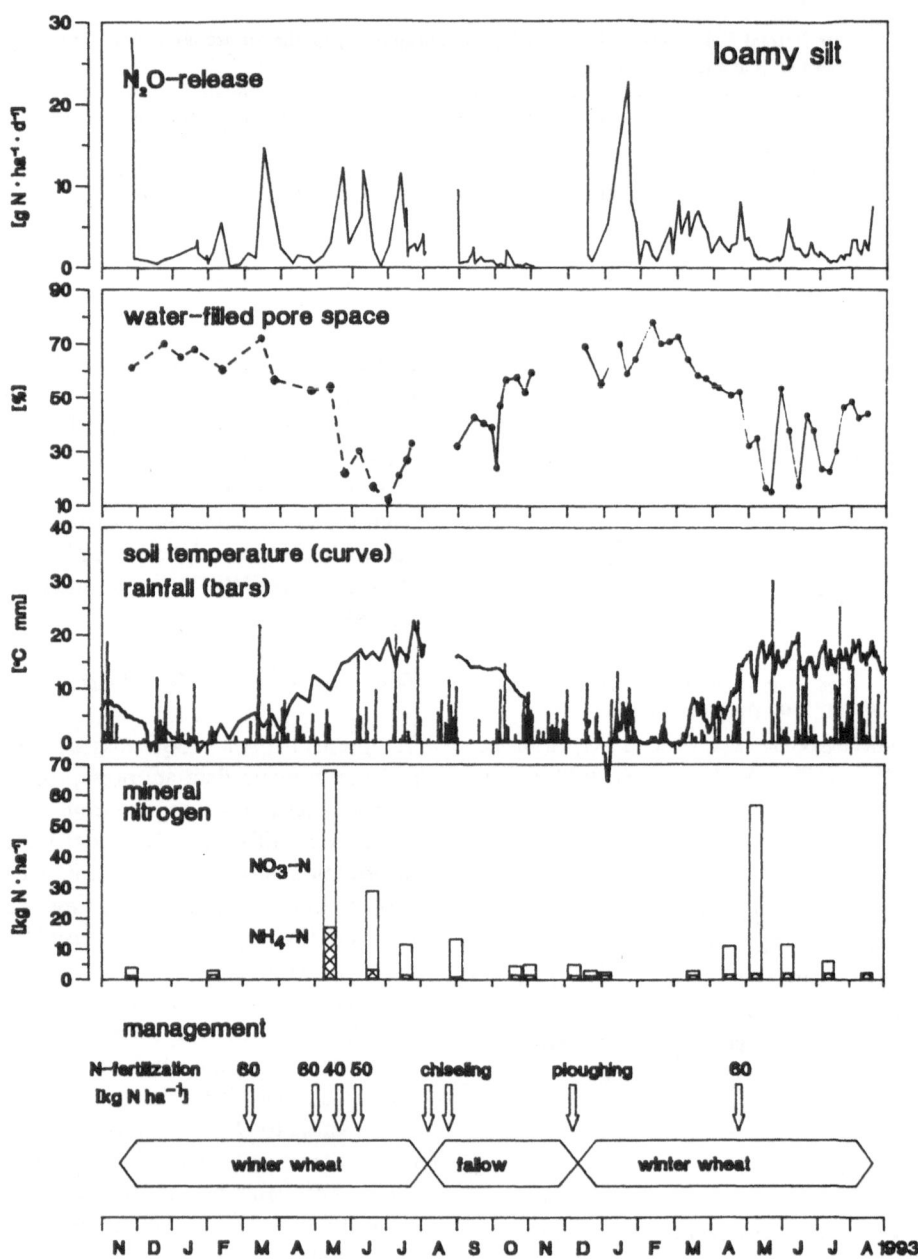

Figure 2. N$_2$O-release, water-filled pore space, rainfall, soil temperature, mineral nitrogen and soil from the loamy silt (site 1, Timmerlah) during the investigation period.

within one week after N-fertilization. 10–15% of the total N$_2$O-production on both soils occurred during these periods (Table 3). This intensive N$_2$O- production could also be attributed to denitrification, induced by oxygen depletion in soil and the high nitrate supply. The oxygen depletion probably resulted from root growth after fertilization and increased microbial activity (Kaiser, 1994). However, large variability between the years and the soils exist (Table 3).

Influence of the two arable soils on N$_2$O-losses

The mean N$_2$O-losses throughout the experimental period were similar for both soils. This observation was in contrast to the other results (Table 1, Fig. 1). A

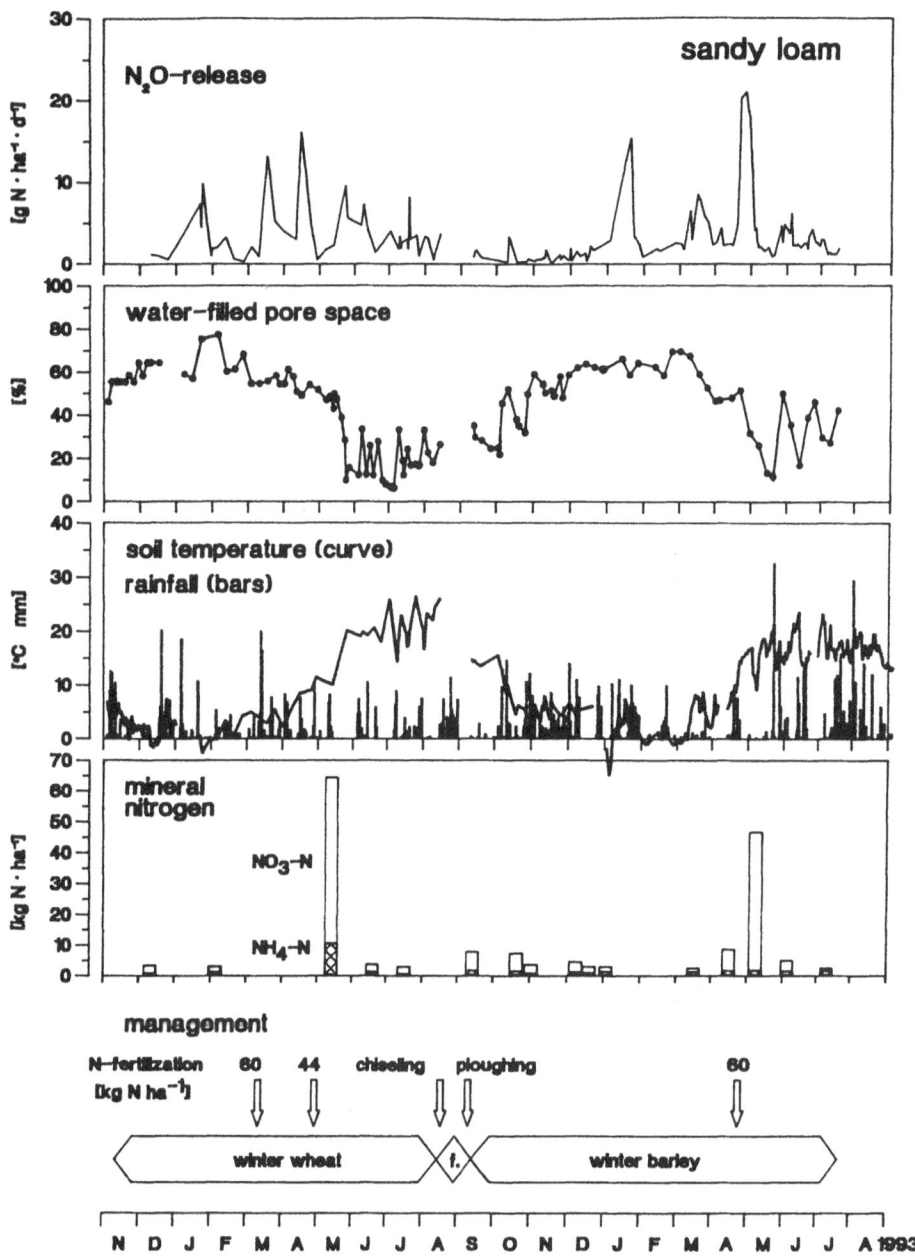

Figure 3. N$_2$O-release, water-filled pore space, rainfall, soil temperature and mineral nitrogen from the sandy loam (site 2, FAL) during the investigation period.

higher N$_2$O production in the loamy silt compared to the sandy loam was expected from their different soil properties.

Arah et al. (1991) observed a discrepancy between the production of N$_2$O within the soil profile and the corresponding N$_2$O-release at the soil surface. N$_2$O produced at depth was reduced to dinitrogen on its way up to the soil profile, especially on heavier-textured soils where diffusion was slow. Reference to comparable processes in the loamy silt were found during the experiment (Fig. 2). High flux rates were found in the loamy silt within 24 h after the installation of cover boxes. These tubes are reaching 40 cm into the soil. Therefore it is likely that the observed high N$_2$O-fluxes originated from N$_2$O-production of deeper soil layers. This N$_2$O escaped along cracks created by the installa-

62

Table 3. N₂O-losses and fertilizer nitrogen for the two soils

Soil		Nitrogen (kg N ha^{-1})			Ratio (%)	
		Fertilizer	N₂O-losses			
		Year^{-1}	Year^{-1}	1 week after fertilization		
		A	B	C	B/A	C/B
Sandy loam	1992	108	0.970[a]	0.104	0.9	10.7
	1993	60	1.621[b]	0.142	2.7	8.8
	total	168	2.591	0.246	1.5	9.5
Loamy silt	1992	210	1.182[c]	0.279	0.6	23.6
	1993	60	1.090[c]	0.057	1.8	5.2
	total	270	2.272	0.336	0.8	14.8

tion process. No comparable observations were recognised for the coarse-textured sandy soil. Consequently, a detectable influence of the increased precipitation in 1993 was only observed at the sandy loam, indicated by an increase in WPFS and N₂O-losses (Tables 2 and 3). The moisture controlled the coexistence of N₂O-production and reduction processes within the soil.

With respect to the different fertilizer management, the percentage of fertilizer N released as N₂O was calculated (Table 3). Over the 2 years an average of 1.5% N of the applied fertilizer N was lost from the sandy loam, while the corresponding figure from the loamy silt was only 0.8%. Both ratios are low compared with other studies (Granli and Bøckman, 1994; Heinemeyer and Kaiser, 1993).

Nevertheless, the risk of N₂O-losses is clearly higher from the sandy loam, because of its deficit in buffer capacity for water compared with the fine-textured soil. Problems will increase, because irrigation is often necessary on this soils. The unexpected result of almost similar amounts of N₂O (1 kg N₂O-N ha^{-1} a^{-1}) evolved during this experiment and the given explanation for this phenomenon indicates that the role of soil water buffer capacity needs further clarification. The soil water buffer capacity influences not only N₂O production, but also transport across soil layers and possible reduction of N₂O to N₂.

With respect to the contribution of N₂O to global change this study may highlight the need for further accurate (based on an intensive frequent monitoring) and longterm studies covering the effects of both soil properties and agricultural management practices.

Acknowledgements

We would like to thank Petra Mitschke, Andrea Oehns-Rittgerodt, Sabine Schintzel, Michael Schön and Bernd Volkmar for expert technical assistance and Donal Murphy for language editing. This work was funded by the European Community "STEP NEMIS".

References

Anderson J P E and Domsch K H 1978 A physiological method for the quantitative measurement of microbial biomass in soils. Soil Biol. Biochem. 10, 215–221.

Anderson T-H and Domsch K H 1989 Ratio of microbial biomass to total organic carbon in arable soils. Soil Biol. Biochem. 21, 471–479.

Arah J R M, Smith K A, Crichton I J and Li H S 1991 Nitrous oxide production and denitrification in Scottish soils. J. Soil Sci. 42, 351–367.

Aulakh M S, Rennie D A and Paul E A 1983 Field studies on nitrogen losses under continuous wheat versus a wheat-fallow rotation. Plant and Soil 73, 15–27.

Cates R L and Keeny D R 1987 Nitrous oxide production throughout the year from fertilized and manured maize fields. J. Environ. Qual. 16, 443–447.

Christensen S and Tiedje J M 1990 Brief and vigorous N₂O production by soil at spring thaw. J. Soil Sci. 41, 1–4.

Deutscher Wetterdienst 1992 Jahresbericht des Deutschen Wetterdienstes, Zentrale agrarmeteorologischer Forschungsstelle Braunschweig, Germany.

Goodroad L L and Keeny D R 1984 Nitrous oxide emissions from soils during thawing. Can. J. Soil Sci. 64, 187–194.

Granli T and Bøckman O C 1994 Nitrous oxide from agriculture. Norw. J. Agric. 12, 7–128.

Hansen S, Mæhlum J E and Bakken L R 1993 N₂O, CO₂ and O₂ fluxes in soil influenced by fertilization and tractor traffic. Soil Biol. Biochem. 25, 621–630.

Heinemeyer O and Kaiser E-A 1993 Landwirtschaftliche Bodennutzung und N₂O-Emissionen sowie CH₄-Umsetzungen im Boden. In Anthropogene N₂O- und CH₄-Emissionen in der Bundesrepublik Deutschland- Phase 1: Emissionsbilanz, Identifikation von Forschungs- und Handlungsbedarf sowie Erarbeitung von Handlungsempfehlungen. UBA- Berichte. pp 143–162. Erich Schmidt Verlag, Berlin, Germany.

Heinemeyer O and Kaiser E-A 1996 Automated gas injector system for gas chromatography - atmospheric nitrous oxide analysis. Soil Sci. Soc. Am. J. (*In press*).

Kaiser E-A 1994 Significance of microbial biomass for carbon and nitrogen mineralisation in soil. Z. Pflanzenernähr. Bodenkd. 157, 271–278.

Kaiser E-A and Heinemeyer O 1993 Seasonal changes in microbial biomass carbon within the plough layer. Soil Biol. Biochem. 25, 1649–1655.

Kaiser E-A, Mueller T, Joergensen R G, Insam H and Heinemeyer O 1992 Evaluation of methods to estimate the soil microbial biomass and the relationship with soil texture and organic matter. Soil Biol. Biochem. 24, 675–683.

Mosier A R, Guenzi W D and Schweizer E E 1986 Soil losses dinitrogen and nitrous oxide from irrigated crops in Northeastern Colorado. Soil Sci. Soc. Am. J. 50, 344–348.

Vinther F P 1984 Total denitrification and the ratio between N_2O and N_2 during the growth of spring barley. Plant and Soil 76, 227–232.

Walenzik G and Heinemeyer O 1989 Direkte Messung der N_2 und N_2O Abgabe aus mechanisch belastetem Boden im Freiland. Mitt. Dtsch. Bodenk. Ges. 59, 629–634.

Weir K L, Macrae I C and Myers R J K 1991 Seasonal variations in denitrification in a clay soil under a cultivated crop and permanent pasture. Soil Biol. Biochem. 23, 629–635.

Weir K L, Doran J W, Power J F and Walters D T 1993 Denitrification and the denitrogen/nitrous oxide ratio as affected by soil water, available carbon, and nitrate. Soil Sci. Soc. Am. J. 57, 66–72.

Section editor: R Merckx

O. Van Cleemput et al. (eds.), Progress in Nitrogen Cycling Studies, 575–579, 1996.

Factors affecting nitrous oxide emission from a sandy grassland soil under controlled field conditions

C.A. Langeveld, P.A. Leffelaar and J. Goudriaan
Department of Theoretical Production Ecology, Wageningen Agricultural University, Wageningen, The Netherlands

Key words: diffusion, moisture content, nitrate fertiliser, nitrogen, nitrous oxide

Abstract

The atmospheric concentration of nitrous oxide (N_2O) currently rises by about 0.25% per year. N_2O contributes to the depletion of the stratospheric ozone layer and the enhanced greenhouse effect. Soils are the most important N_2O sources. Many factors affect N_2O emission from soils, via their influence on transport and the microbiological processes nitrification and denitrification. This study reports results of research on the processes and factors determining N_2O emission from a sandy grassland soil in the Rhizolab, Wageningen, The Netherlands, during the second half of 1993. Four grassland plots on sand (1.25 m × 1.25 m × 1 m) were intensively monitored with respect to fluxes from the soil and belowground profiles of temperature, soil water potential, soil moisture content, several gases (N_2O, CO_2 and O_2) and several dissolved nutrients. Treatments were: height of groundwater table and amount of nitrate-N fertiliser. The first results indicate that the relation between the mean N_2O fluxes from soil and the nitrate-N fertilisation rate exhibits a threshold phenomenon. Fluxes at 0 and 250 kg N ha^{-1} (in 5 equal dressings, 3–5 weeks apart) were about equal. An application of 500 kg N ha^{-1} (in 5 equal dressings, 3–5 weeks apart), however, increased fluxes by a factor 4. No relation was found between N_2O fluxes and belowground N_2O profiles in the upper soil layer. From the surface to the groundwater, the N_2O concentrations first increased and then decreased with depth. Surprisingly, soil moisture content showed a similar profile.

Above-mentioned results will be used to develop, calibrate and validate a mechanistic simulation model.

Introduction

The concentration of atmospheric nitrous oxide (N_2O) is increasing. The concentration of this gas, having an atmospheric residence time of 100–200 years (Bouwman, 1990a), was about 310 parts per billion by volume (ppbv) in 1990 and rises by about 0.25% per year (Bouwman et al., 1993). N_2O contributes to two environmentally harmful effects: the depletion of the ozone layer (no estimation of the contribution available) and the enhanced greenhouse effect (estimated contribution 5%).

Soils are estimated to be responsible for more than 50% of the global emission of N_2O (Bouwman, 1990b). Soil emission data show a large variation, depending on land use, soil type and local conditions. This variation results from the influence of various factors on the processes transport, nitrification and denitrification that determine N_2O emission from soils. These factors are temperature, aeration (related to moisture content), nitrogen mineralisation rate, amount of fertiliser added and content of readily decomposable carbohydrates. The current research is focussed on quantifying and modelling the influence of these factors under relatively controlled conditions, using the Wageningen Rhizolab (Van de Geijn et al., 1994), with the final purpose to propose measures to suppress N_2O emissions.

Materials and methods

In May 1993, 4 compartments in the Wageningen Rhizolab were packed with sandy soil from a non-fertilised lawn in Ede, The Netherlands. The soil was installed within watertight Aquatex 1.0 mm covers upon a filter bed layer and separated from this layer by a gauze (Fig. 1).

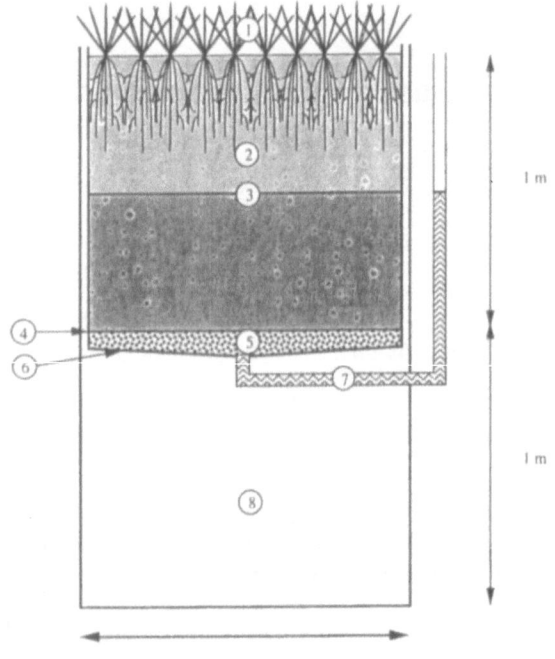

Figure 1. Scheme of a compartment in the Wageningen Rhizolab.
1. Perennial ryegrass; 2. 1 m sandy soil; 3. Groundwater table; 4.
Gauze; 5. Filter bed layer; 6. Watertight cover; 7. Tube to impose
the groundwater table; 8. Base soil layer.

The imposed groundwater tables in 1993 were
about 50 cm below ground level ('high' groundwater
table) in three of the compartments and about 100 cm
below ground level ('low' groundwater table) in the
remaining compartment. During rain the Rhizolab is
sheltered (Van de Geijn et al., 1994). From May to
October, water was applied regularly at the surface
with a watering can or via the drip irrigation system
to mimic a rain regime. The intervals (1–9 days) and
applications (2–12 L per compartment per day) were
adapted to (a) apply water directly after fertiliser application (see below), and (b) maintain relatively constant moisture contents in the upper soil. Total applications during this period corresponded to 230 mm rain.
After October 1993 no irrigation took place anymore.
Within the soil various sensors, samplers and (horizontal) root observation tubes were installed at various depths. The sensors included capacity sensors to
measure soil moisture content, temperature and electric conductivity, tensiometers to measure soil water
potential, and thermocouples to measure temperature.
Hourly averages (from 9–12 measuring points) of the
sensor data were determined and collected with the
data-acquisition/control unit of the Rhizolab. The samplers included ceramic suction cups (for point mea-

surements) and 50 cm long microporous tubings (for
spatially more integrated measurements) to extract soil
solution, and gas exchange cells to sample the soil air.
The soil solution was analysed for nitrate, ammonium, total nitrogen, phosphorus, sodium, potassium and
total water soluble carbon (results not reported here);
gas samples were analysed for N_2O, carbon dioxide
and oxygen. For the positioning of the sensors and
samplers, the scheme presented in Van de Geijn et al.
(1994) was slightly adapted to comply with the specific
needs of this experiment.

On 26 May 1993, perennial ryegrass (*Lolium
perenne* L.), var. Preference, was sown. In the period July–October 1993 the grass was cut and harvested
5 times, at intervals of 3–5 weeks (stubble height 3–4
cm). On 18 May 1993, and also directly after cuts 1,
2, 3 and 4 fertiliser was applied. In the applications,
phosphorus and potassium were amply supplied to all
compartments. Furthermore, two of the compartments
with the high groundwater table (50 cm) also received
nitrogen as $Ca(NO_3)_2$, a simple, commonly used N
fertiliser, in amounts corresponding to 50 and 100 kg
N ha^{-1} per application, respectively. Four days before
and three days after the cuts N_2O fluxes were determined with the equipment described in Velthof and
Oenema (1993 and 1994). An extra observation was
done on 13 December 1993. For this purpose closed
flux chambers (inner diameter 20 cm, height 15 cm)
were installed at two fixed spots on each compartment
during the measurements.

Gas samples were collected from the gas exchange
cells, at the same times as the flux measurements.
Carbon dioxide and oxygen profiles were determined
with a gas chromatograph with Thermal Conductivity Detector as described by Leffelaar (1986). Nitrous
oxide profiles were determined with the aid of a
gas chromatograph with an Electron Capture Detector
(Philips Unicam PU 4400, equipped with a backflush
system; detector temperature 350 °C; carrier gas N_2;
specified minimally detectable N_2O concentration 0.1
ppmv (parts per million by volume)).

Gas samples from 0, 5 and 10 cm depth (in twofold)
were used to determine N_2O gradients in the soil at
the ground level. The least-squares method as implemented by Microsoft Excel 4.0 for Apple Macintosh
was applied to fit a parabola to the obtained depth-
concentration pairs, according to:

$$c = a_1 z^2 + a_2 z + a_3 \qquad (1)$$

where c represents the N_2O concentration, z the depth
and a_1, a_2 and a_3 are coefficients.

Table 1. For the four compartments, the compartment number, GWT: the groundwater table (m below the surface), N_{tot}: the total amount of N applied via $Ca(NO_3)_2$ in 1993 (g N m^{-2}), N_2O efflux: the N_2O efflux averaged over all measuring dates ± standard deviation (g N m^{-2} h^{-1}). Means were calculated from 16 flux values

Compartment number	GWT (m)	N_{tot} (g N m^{-2})	N_2O efflux (g N m^{-2} hr^{-1})
11	1	0	$(0\pm20)\cdot10^{-6}$
13	0.5	0	$(15\pm8)\cdot10^{-6}$
9	0.5	25	$(16\pm11)\cdot10^{-6}$
15	0.5	50	$(70\pm50)\cdot10^{-6}$

Differentiation of Equation (1) with respect to z and taking the limit z $\longrightarrow$ 0 shows that the N_2O gradient in the soil at the ground level equals the coefficient of the linear term:

$$(dc/dz)_0 = a_2 \qquad (2)$$

Results and discussion

The results of the flux measurements are represented in Table 1. On the basis of the available data, no conclusions about expected changes of the fluxes in time (systematic differences between the values before and after the cuts and possibly a trend during the observation period) could be drawn. Therefore, only mean fluxes are given.

The N_2O fluxes (Table 1) are of similar size as those described by Webster and Dowdell (1982) and the data for 1979/1980 in Eggington and Smith (1986) for comparable grassland soils. For the compartment with the lowest groundwater table, going with relatively low moisture contents at all depths, the lowest flux was observed. This finding supports the suggestion of several other authors that there is a (cor)relation between the wetness of soils and the N_2O fluxes (Sahrawat and Keeney, 1986). An explanation is that wet conditions go with low partial oxygen pressures in the soil that favour N_2O production in denitrification. The data for the compartments with the groundwater table at 50 cm show a 'threshold phenomenon': no clear flux differences between the treatments without or with an intermediate N application (250 kg N ha^{-1}) were observed, while application of 500 kg N ha^{-1} resulted in a flux increase by a factor 4. This phenomenon was also reported by Watson et al. (1992) for grazed perennial ryegrass swards fertilised with calcium ammonium

nitrate (split application) in Northern Ireland. The standard deviation of the N_2O fluxes is large. Such a large variability has been found in several other studies, e.g. Velthof and Oenema (1994). Similarly to these authors we found that temporal variations of the fluxes increase if N fertilisation is applied.

In Figure 2 the mean N_2O effluxes per measuring date are plotted against the N_2O gradients at the ground level as calculated from the upper three gas concentration measurements for the 3 compartments with the high groundwater table. The mean effluxes and the gradients appear to have no relation. Points combining small gradients with large effluxes as well as points combining large gradients with small effluxes were observed. These points deviate strongly from the line. This line represents the relationship that was to be expected under the hypothesis that the upper soil layer can be described as a homogeneous system in which (a) N_2O transport only occurs in the gas-filled space according to Fick's first law, (b) N_2O transport is not influenced by geometrical restrictions by the remaining space and (c) no net N_2O production or consumption occurs:

$$J = D_{eff}(dc/dz)_0 \qquad (3)$$

$$D_{eff} = \varepsilon D_o = (\phi - \theta_v) D_o, \qquad (4)$$

where J is the N_2O efflux, D_{eff} the effective diffusion coefficient, $(dc/dz)_0$ the N_2O gradient in the soil at the surface (z=0, z increasing with depth; from Equation 2), ε the gas-filled porosity, ϕ the porosity, θ_v the volumetric moisture content and D_o the diffusion coefficient of N_2O in air.

The values $\phi = 0.5$, $\theta_v = 0.25$ and $D_0 = 0.15 \cdot 10^{-4}$ m^2 s^{-1} were used to determine the slope in Figure 2. These values were, respectively, the mean porosity of the soil at installation, the mean volumetric moisture content at 5 cm and the diffusion coefficient of CO_2, which has the same molecular weight as N_2O, in air at 15 °C (calculated from Leffelaar, 1987).

The deviations from the theoretical line indicate that our results do not support the hypothesis underlying Equations (3) and (4). A part of the explanation is possibly of statistical nature. The mean fluxes per compartment were determined by linear regression on the observed N_2O concentration change in time in two flux chambers. Often the two obtained values differed strongly, probably reflecting spatial variability of the system. The vertical error bars reflect the difference *between* these two values. However, they do not

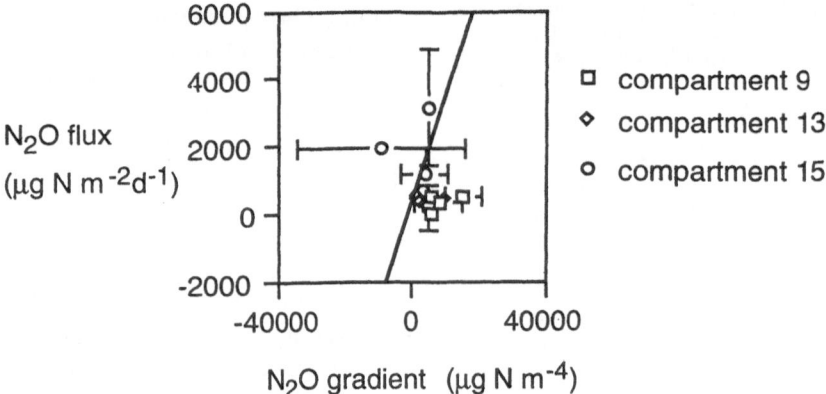

Figure 2. Mean N_2O fluxes versus subsurface N_2O gradients for the containers with the high groundwater table (0.5 m depth). Each point stands for one measurement date. Horizontal error bars indicate standard errors obtained from the least-squares method. Vertical error bars represent the difference between the two measured values and the mean. The line represents diffusion according to Equations (3) and (4) with $D_{eff} = 0.0375 \cdot 10^{-4}$ m^2 s^{-1}.

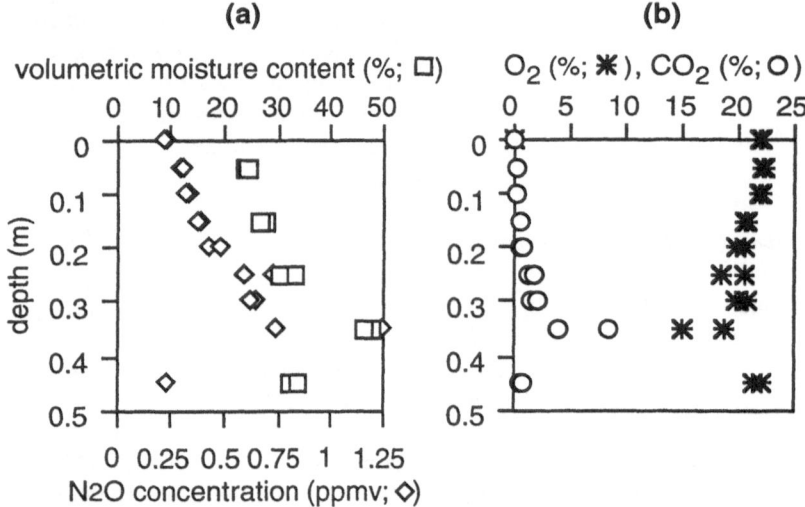

Figure 3. Profiles of (**a**) volumetric moisture content (%), N_2O (ppmv), (**b**) O_2 (%) and CO_2 (%). Data for compartment 13 (groundwater table at 0.5 m depth, porosity 50%) on October 25, 1993.

account for uncertainties *in* the respective values connected with non-steady concentration increases in the flux chambers. Thus, the vertical error bars underestimate the total uncertainty in the flux. Another part of the explanation may be that diffusion according to Equations (3) and (4) does not give a reasonable description of N_2O emission in our system. A refined description including N_2O sources in the upper layer and a diffusion coefficient depending on variables, such as moisture content and root density, might be necessary. This is part of the modelling action to be performed.

In Figure 3 an example is shown of the simultaneously measured, time-dependent profiles of the volumetric moisture content, N_2O (Fig. 3a), O_2 and CO_2 (Fig. 3b). The observation that the steepest gradients did not occur at the surface was not expected a priori as well as the behaviour just above the groundwater table (0.5 m depth). The profiles should be explained by the model to be developed as the result of transport processes of water and gases and (biological) transformations of nutrients, organic substances and gases.

Acknowledgements

The support of J E Hofman, M H van den Bergh and the staff of the Wageningen Rhizolab (J Groenwold, G

Versteeg and H Wessels) is gratefully acknowledged. Measurement of N_2O fluxes and CH_4 and N_2O concentrations was made possible through cooperation with the Nutrient Management Institute (NMI), Wageningen (O Oenema, G L Velthof, A B Brader, A Van Dasselaar and J G Koops). A M van Dam and R Segers are thanked for their critical comments on earlier versions of this paper and logistic support. The research is part of the integrated N_2O grassland research project in which also participate the NMI, Wageningen, and the Research Institute for Agrobiology and Soil Fertility (AB-DLO), Wageningen and Haren, The Netherlands. This Project is financially supported by the Dutch National Research Program on Global Air Pollution and Climate Change (Project 852074).

References

Bouwman A F 1990a Chapter 2. Introduction. *In* Soils and the Greenhouse Effect. Ed. A F Bouwman. pp 25–32. John Wiley and Sons, Chichester, UK.

Bouwman A F 1990b Chapter 4. Exchange of greenhouse gases between terrestrial ecosystems and the atmosphere. *In* Soils and the Greenhouse Effect. Ed. A F Bouwman. pp 61–128. John Wiley and Sons, Chichester, UK.

Bouwman A F, Fung I, Matthews E and John J 1993 Global analysis of the potential for N_2O production in natural soils. Global Biogeochem. Cycles 7, 557–597.

Eggington G M and Smith K A 1986 Nitrous oxide emission from a grassland soil fertilized with slurry and calcium nitrate. J. Soil Sci. 37, 59–67.

Leffelaar P A 1986 Dynamics of partial anaerobiosis, denitrification, and water in a soil aggregate: experimental. Soil Sci. 142, 352–366.

Leffelaar P A 1987 Dynamic simulation of multinary diffusion problems related to soil. Soil Sci. 143, 79–91.

Sahrawat K L and Keeney D R 1986 Nitrous oxide emission from soils. Adv. Soil Sci. 4, 103–148.

Van de Geijn S C, Vos J, Groenwold J, Goudriaan J and Leffelaar P A 1994 The Wageningen Rhizolab – a facility to study soil-root-shoot-atmosphere interactions in crops. I. Description of main functions. Plant and Soil 161, 275–287.

Velthof G L and Oenema O 1993 Nitrous oxide flux from nitric-acid-treated cattle slurry applied to grassland under semi-controlled conditions. Neth. J. Agric. Sci. 41, 81–93.

Velthof G L and Oenema O 1994 Nitrous oxide emission from grasslands on sand, clay and peat soils in the Netherlands. *In* Non-CO_2 Greenhouse Gases: Why and How to Control? Eds. J van Ham, L J H M Janssen and R J Swart. pp 439–444. Kluwer Acad. Publ., Dordrecht, the Netherlands.

Watson C J, Taggart P J, Jordan C and Steen R W J 1992 Denitrification losses from grazed grassland. *In* Proceedings of the Second Congress of the European Society for Agronomy. Ed. A Scaife. pp 436–437. ESA, Warwick. UK.

Webster C P and Dowdell R J 1982 Nitrous oxide emission from permanent grass swards. J. Sci. Food Agric. 33, 227–230.

O. Van Cleemput et al. (eds.), Progress in Nitrogen Cycling Studies, 581–584, 1996.

Emission of nitrous oxide and denitrification from Danish soils amended with slurry and fertilizer

M. Maag, A.M. Lind and F. Eiland
Danish Institute for Plant and Soil Science, Department of Soil Science, Research Centre Foulum, Postbox 23, DK-8830 Tjele, Denmark

Key words: crops, denitrification, nitrification, N-emission, N-fertilizer, nitrous oxide, slurry

Abstract

Interactions between crops and soil factors in relation to N_2O emission and denitrification from soil are not well understood. This study was conducted to determine N-emission from arable soil under various crops in Denmark as a part of a larger European investigation. Denitrification losses from a coarse sandy loam cropped with spring barley, fodder beets or spring rape, and from a fine sandy loam under spring barley or winter wheat were measured during a 2-years period. In the measuring years 1992 and 1993, the N_2O-emission factors varied from 0.2 to 5 kg N ha^{-1} yr^{-1} between the four different crops and the two soil types. Due to the high spatial variability, the 95% confidence interval for the mean varied between 0.7–40 kg N ha^{-1} yr^{-1} for the highest emission and between 0.05–0.7 kg N ha^{-1} yr^{-1} for the lowest emission. The *highest emission* occurred from the fine sandy loam soil cropped with spring barley and fertilized with cattle slurry supplemented with inorganic fertilizer. The *lowest emission* occurred during a dry year (1993) on the coarse sandy soil cropped with spring barley and fertilized with inorganic nitrogen. A similar low emission was measured from a soil cropped with winter wheat despite of a larger amount of nitrogen applied. From this study it became evident that N_2O-emission during periods with low soil moisture was produced by nitrification and that the emission was dependent on agricultural management and weather, and that no single emission factor can be used for a given soil type.

Introduction

Emission of nitrous oxide (N_2O) from agricultural land is an important contribution to the greenhouse effect as well as to the loss of soil nitrogen. The soil processes emitting N_2O (nitrification and denitrification) are enhanced by nitrogen fertilization. Therefore strategies for N-fertilization and field management are important tools to minimize the emission and for assessment of emission factors for modelling.

N$_2$O-emission measurements from a joint European project called 'NEMIS' with participants from 6 European countries (Denmark, Belgium, France, Germany, United Kingdom and Spain) have been performed. The purpose of the project was comprehensive studies of N_2O-emissions in the field as well as in the laboratory. Different European cropping systems with typical management strategies have been studied. A further purpose was to set up a database comprising data of N_2O-emission, denitrification, climatic parameters together with dynamic and static soil parameters. This database can be used for modelling of N-transport and transformation in the rootzone, and as input to atmospheric N-modelling. In this paper results from the Danish project are reported.

Materials and methods

Characteristics of field sites

The two sites (Foulum and Odum) are situated in the northwestern part of Denmark.

Foulum. Longitude 9° 35$'$ E; latitude 56° 30$'$ N; elevation above sea 48 m; field area: 25 ha; plot area: 75 m^2; soil profile: dark-brown clayey sand (0–35 cm), yellowish-brown silty sand (35–100 cm), weak pseudogley at 70 cm depth; soil texture (0–35 cm): 7.5% clay, 9.5% silt and 80.0% sand.

Odum. Longitude 10° 08′ E; latitude 56° 18′ N; elevation above sea 62 m; field area: 16.5 ha; plot area: 75 m^2; soil profile: dark-brown sandy loam (0–50 cm), yellowish-brown sandy loam (50–80 cm), yellowish-brown silty loam (80–100 cm), weak pseudogley at 55 cm depth; soil texture (0–35 cm): 9.2% clay, 14.4% silt and 73.8% sand.

Climate

The studied area is located in a temperate coastal climatic zone: Annual precipitation: Foulum 704 mm and Odum 714 mm. Annual average temperature: Foulum and Odum 7.3 °C.

The experimental areas were equipped with automatic meteorological stations with hourly registration of air temperature in two heights, accumulated daily precipitation, wind speed, mean relative humidity and global radiation.

Due to some very special weather conditions in 1992, which influenced the crop and the soil processes, a brief summary of the weather situation is given. The spring was cold, windy and unsteady, and caused severe problems at ploughing and sowing time, and for the establishment of the crops. In June the weather changed completely and this month was the driest ever recorded in Denmark since 1874. The drought continued until mid July. The water balance from May until mid July showed a deficit of 210 mm water while the temperature sum (with basis 5 °C) was 960 °C as compared to 795 °C (10 years normal). This severely hampered the development of the crops. The rest of the year was quite normal.

Also in 1993 the weather behaved atypically. The spring and early summer was warmer and drier than normal; the precipitation in this period was 100 mm less than normal. The small precipitation had the effect that spring applied inorganic fertilizer remained undissolved on top of the soil. In the summer month, the weather changed and became cool and rainy.

Land use, history and field treatments

The sites have been used for long-term research in farming systems since 1986. The basic field experiments at the two locations are set up with an ecological, an integrated and a technological system. In the fields where measurements of N-emission were performed various levels of slurry, N-fertilizers and organic matter (catch crops, incorporation of straw) were used. After surface application, the slurry was incorporated into the soil by rotavation within 24 hours. Field treatments and crops are given in Table 1.

Sampling and incubation of soils for field measurements

Static soil core method (Christensen, 1985) was used for measuring N$_2$O emission as well as denitrification by acetylene inhibition.

Soil sampling. The soil cores (0–10 cm) taken for measurements of nitrous oxide emission and denitrification were taken in PVC cylinders (diam. 3.5 cm, length 25.0 cm). From each plot 16 soil cores were taken, 8 cores were incubated with and without acetylene, respectively. Three field replications were used. It resulted in 48 samples from each N-treatment. The sampling frequency was approximately one to two times a month from early spring to late autumn. During very dry periods, samples were not taken. The incubation experiments were initiated at the sampling date.

Incubation experiments. The soil cores were incubated in the cylinders which were sealed with a rubber stopper in the bottom and a silicone stopper in the top for gas sampling from the head space. Acetylene (10% v/v) was injected into the headspace of half of the cylinders from each plot and all the samples were preincubated for 24 hours at approximately the average actual field temperature in 10 cm depth. The N$_2$O concentration was measured 2 times with one hour interval after it has been verified that acetylene has reached the bottom of the soil cores. The gas samples (4 mL) taken from the headspace were injected in 3.5 mL Venoject® for later automatically sampling for GC analyses.

Calculation of denitrification rates. The amount of N$_2$O in each soil core was calculated using the concentration of N$_2$O in the headspace, Bunsen's coefficient of N$_2$O solubility in water (Tiedje, 1982), the soil weight, the soil water content, and the total core volume. Accumulated denitrification losses were calculated using lognormal statistics (Finney's methods) as described by Parkin et al. (1988).

Results and discussion

Calculated yearly N$_2$O-emissions from all treatments (1992 and 1993) varied between 0.2 and 5 kg N ha^{-1}

Table 1. N$_2$O-emission and denitrification from two Danish sites with different crops and types and levels of N application

	Mineral N added	Organic N added[a]	Measuring period	N$_2$O emission[b]	95% conf. interval	N$_2$O + N$_2$ emission[2]	95% conf. interval
	(kg N ha^{-1} yr^{-1})		(days)	(kg N ha^{-1})			
Spring barley 1992							
Odum	60	0	350	2	0.1–46	1	0.5–3
Odum	120	25	350	5	0.7–40	2	0.5–8
Foulum	60	0	349	4	0.8–16	4	2.5–6.6
Foulum	60	36	349	3	0.6–15	3	1.4–6
Foulum	120	36	349	5	1.2–22	8.3	3–13
Spring barley 1993							
Foulum	90	0	188	0.2	0.05–0.7	0.5	0.06–3
Foulum	90	36	188	0.5	0.08–2.6	0.4	0.08–2.3
Fodder beets 1992							
Foulum	113	30	346	5	1–27	9	1.9–42
Foulum	188	45	346	3	0.8–13	4	1.1–15
Foulum	57 + 56[c]	15 + 15[c]	346	4	0.8–19	5	1.1–19
Foulum	94 + 94[c]	23 + 23[c]	346	4	0.8–16	5	1.3–21
Spring rape 1993							
Foulum	90	0	185	1	0.06–8	1	0.5–1.5
Foulum	90	36	242	2	0.2–19	0.8	0.4–1.4
Winter wheat 1993							
Odum	90	0	287	0.8	0.3–2	0.6	0.2–2
Odum	90	54	287	1	0.3–3	0.6	0.2–2
Odum	180	54	287	0.7	0.2–3	1	0.3–3

[a] Organic N added as animal slurry.

[b] N$_2$O-emission and confidence intervals are calculated using lognormal statistics.

[c] Split application, $\frac{1}{2}$ at the sowing time and $\frac{1}{2}$ at the growing season.

(Table 1). The N$_2$O-emissions from soils cropped with spring barley were generally 10 times higher in 1992 than in 1993. At both sites (1992 and 1993), there was no effect of different levels of mineral N, but a slight increase in N$_2$O-emission and denitrification were found after addition of a combination of inorganic N and organic material. It should be emphasized that the soil only had been treated with organic amendment (slurry, straw and catch crops) for a few years (since 1989). Thus, a long-term effect of the organic matter cannot be expected. Split-application for fodder beets had not a marked effect on the N$_2$O-emission because the crop on this stage of development does not use very much nitrogen.

In 1992, the long and wet spring might have stimulated N$_2$O from denitrification, and the extremely dry early summer (May to July) might have stimulated N$_2$O-formation by nitrification. There was no significant difference between the N$_2$O-emissions from spring barley and fodder beets in 1992. The highest N$_2$O-emissions were found in spring barley soils during 1992 (2–5 kg N ha^{-1}). The denitrification losses were in the same order of magnitude as found in other studies (Mosier and Hutchinson, 1981; Mosier et al., 1982).

In 1993 N$_2$O-emissions from the investigated soils (cropped with spring barley, spring rape and winter wheat) ranged from 0.2 to 2.0 kg N ha^{-1} and N$_2$O-emission and denitrification were slightly higher from the site with spring rape than from spring barley and winter wheat. Probably, this was caused by the climatic conditions this year, because the spring was dryer than

584

normal and the summer more cold and rainy. These conditions resulted in lower levels of N_2O-emission than in 1992.

N_2O-emission and denitrification were calculated for periods based upon time of sowing, fertilization and weather. In 1992 and 1993 the mean daily N_2O-emission from all treatments and crops varied between 0.001 and 0.16 kg N_2O-N ha^{-1} day^{-1}, while the mean daily denitrification rate varied between 0.001 and 0.097 kg N ha^{-1} day^{-1}. The general picture seen in this study was a low daily N_2O-emission in spring and autumn, and higher emission in late spring–summer. Denitrification ($N_2O + N_2$) was larger in spring and autumn than N_2O-emission (1–3 times). However, during late spring – summer denitrification was lower or equal to N_2O-emission as a result of favourable conditions for nitrification. This in accordance with other studies, where no denitrification was found below 60–65% water-filled porespace (Klemedtson et al., 1991; Kroekel and Stolp, 1986; Linn and Doran, 1984; Myrold and Tiedje, 1985).

Nitrification is also a source of N_2O and this process is inhibited by the amounts of acetylene normally used in denitrification studies (> 3% v/v). Denitrification losses were in most cases not significantly different from N_2O emissions. In some periods, N_2O was the most dominant gas emitted from the soil indicating a loss during nitrification. Nitrification is normally at optimum at water contents between 30% and 60% waterholding capacity. This implies that during the very dry periods in 1992 and 1993 (late spring-early summer), N_2O was mainly emitted from nitrification.

Lowest N_2O-emission was found from coarse sandy soil cropped with spring barley and fertilized with inorganic N which resulted in an emission factor of 0.05–1 kg N ha^{-1} yr^{-1}. The highest N_2O-emission was found from fine sandy loam cropped with spring barley and added cattle slurry supplemented with inorganic fertilizer resulting in an the emission factor of 0.7–40 kg N ha^{-1} yr^{-1}.

Acknowledgements

We thank our colleagues of the STEP-NEMIS group for the fruitful cooperation during the three years project period. Also the skilful technical assistance of Rita Bundgaard, Mette Hansen, Bodil Møllnitz and Jørgen Wohlfahrt are very much appreciated. The financial support from EU, contract no STEP-CT900029, and from The Danish Environmental Research Programme are gratefully acknowledged.

References

Christensen S 1985 Denitrification in a sandy loam as influenced by climatic and soil conditions. Danish J. Plant Soil Sci. 89, 351–365.

Klemedtsson L, Simkins S, Svensson B H, Johnsson H and Rosswall T 1991 Soil denitrification in three cropping systems characterized by differences in nitrogen and carbon supply. II. Water and NO_3 effects on the denitrification process. Plant and Soil 138, 273–286.

Kroekel L and Stolp H 1986 Influence of the water regime on denitrification and aerobic respiration in soil. Biol. Fert. Soils 2, 15–21.

Linn D M and Doran J W 1984 Effect of water-filled pore space on carbon dioxide and nitrous oxide production in tilled and notilled soil. Soil Sci. Soc. Am. J. 48, 1267–1272.

Mosier A R and Hutchinson G L 1981 Nitrous oxide emissions from cropped fields. J. Environ. Qual. 10, 169–173.

Mosier A R, Hutchinson G L, Sabey B R and Baxter J 1982 Nitrous oxide emissions from barley plots treated with ammonium nitrate or sewage sludge. J. Environ. Qual. 11, 78–81.

Myrold D D and Tiedje J M 1985 Establishment of denitrification capacity in soil: effect of carbon, nitrate and moisture. Soil Biol. Biochem. 17, 819–822.

Parkin T B, Meisinger J J, Chester S T, Starr J L and Robinson J A 1988 Evaluation of statistical methods for lognormally distributed variables. Soil Sci. Soc. Am. J. 52, 323–329.

Tiedje J M 1982 Denitrification. In Methods of Soil Analysis, part 2. Chemical and Microbiological Properties. Agronomy Monographs no. 9. Eds. A L Page, R H Miller and D R Keeney. pp 1011–1026. Am. Soc. Agron., Madison, WI, USA.

O. Van Cleemput et al. (eds.), Progress in Nitrogen Cycling Studies, 585–588, 1996.
© 1996 *Kluwer Academic Publishers.*

N$_2$O-emissions of forest soils in northern Germany: seasonal variability and influencing parameters

B. Mogge[1], O. Heinemeyer[2], E.-A. Kaiser[2] and J.Ch. Munch[1]

[1]*Institute for Soil Ecology, GSF-Research Center for Environment and Health Neuherberg, 85764 Oberschleißheim, Germany and* [2]*Institute for Soil Biology, Federal Research Centre (FAL), Bundesallee 50, 38116 Braunschweig, Germany*

Key words: alder, beech, nitrous oxide, seasonal variability

Abstract

In situ-collection of gas samples showed that the N$_2$O-emissions of an alder forest were significantly higher than of a beech forest. This is demonstrated in a study which is part of a long term ecosystem research program in the Bornhöved Lakes region, Northern Germany. Six soil covers according to Hutchinson and Mosier were installed for the measurement. The in situ-collection of gas samples occurred in 1993 usually once a week. Soil samples were also collected every 14 days and investigated on following parameters: carbon-, nitrate-, water content and soil temperature. The results show that N$_2$O-emissions on both sites were high in August. In autumn emissions decreased. But quantitatively much higher emissions were found from the study site in the alder forest. This effect was propably due to generally higher water and nitrate contents in the alder forest and therefore better conditions for denitrifying bacteria. The best direct predictor for N$_2$O-emissions of the alder forest in 1993 was the soil temperature (r=0.747, $p<0.005$).

Introduction

Nitrous oxide (N$_2$O) is an important chemically-reactive greenhouse gas and involved in stratospheric ozone destruction. In soils the production and release of N$_2$O is mainly linked on two microbial processes: the 'Denitrification' (reduction of nitrate to gaseous nitrogen compounds under oxygen limited conditions) and the 'Nitrification' (aerobic oxidation of ammonia to nitrate). Until now very little is known about the N$_2$O-release from forest soils, especially from alder forest soils. The aim of the presented study is to identify the main N$_2$O-sources in a forest-catena. Parameters which are known to influence the N$_2$O-production by microorganisms were also recorded in order to explain the regulation of these processes.

Materials and methods

Study site and soils

The study site was located in Bornhöved, 30 km south of Kiel in Northern Germany. A cross section through the research area is shown in Figure 1.

Two different soil types were investigated. The beech forest soil (FAO: cambic Arenosol) dominated by *Fagus sylvatica*, about 90 years old. The alder forest soil (FAO: fibric Histosol) was dominated by *Alnus glutinosa* and occurs in an area adjacent to 'Lake Belau'. Some properties of the investigated soils are given in Table 1.

Soil samples and climatic parameters

Soil samples were taken from the upper horizons (beech forest: 0–5 cm; alder forest: 0–20 cm). The soils were sieved to 2 mm in the laboratory immediately after sampling, placed in Polyethylene bags and stored at 4 °C. Subsamples for water content were oven dried at 105 °C for 24 h. The nitrate content

Table 1. Properties of the investigated soils

	Horizons depth (cm)	Bulk dens. (g cm^{-3})	C_{org} (% ww)	N_t (% ww)	C/N	pH (CaCl$_2$)
Beech forest	0–5	1.1	3.4	0.2	17	3.0
Alder forest	0-20	0.6	42.2	2.4	17	2.9

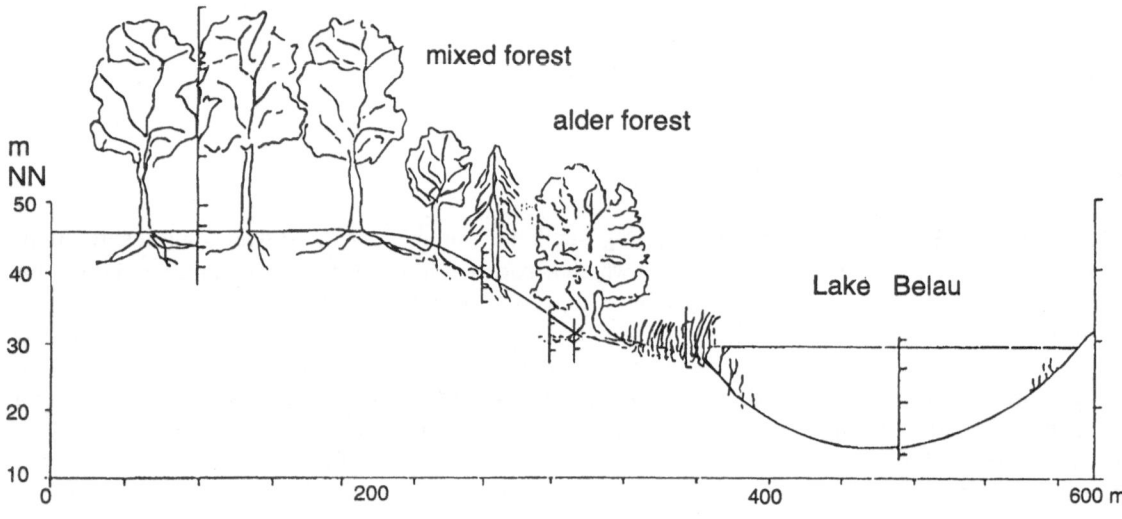

Figure 1. Cross section through the research area.

was determined after extracting the wet soil with 1:4 slurries in 2 M KCl (10 g in 40 mL), shaking 1.5 h and filtering. Measurements were done according to Scharpf and Wehrmann (1976, modified) via UV-spectroscopy at 210 nm after reducing NO_3^- to NO_2^- in a control vessel with cuppered zinc. Watersoluble carbon content was analysed according to Burford and Bremner (1975, modified). Carbon was extracted after adding 40 mL destilled water to 20 g wet soil, shaking over head for 15 min, centrifugating and filtering (0.45 μm). Measurements of total carbon were done with an infrared gas analyzer (TOC-Analyzer, Shimadzu, oven temperature: 680 °C). The soil temperature was measured in 10 cm depth at the times of gas sampling. The precipitation was recorded automatically by rainfall collectors.

Gas samples

A 'closed soil cover-system' according to Hutchinson and Mosier (1981, modified) was used for the collection of the N_2O-emissions. The material was PVC and the system had a volume of about 15 L. The covered

area was 0.07 m^2. One box consisted of two parts: a 'PVC-ring' that was installed permanently and the main soil cover. For gassampling both parts were put together and installed for no longer than 1.5 h. The in situ-collection of gas samples occured in 1993 usually once a week (n=12). However, for better comparison, the 14-day-rhythm is shown in Figures. Gas samples were analysed on N_2O according to Heinemeyer et al. (1991) with a Chrompack Gas Chromatograph (model: CP9000; carrier gas: N_2; make-up gas: Ar with 5% CH_4; detector temperature: 400 °C, oven and column temperature: 50°C).

Statistical analysis

The dependence between N_2O-emissions and soil and climatic parameters was checked via calculation of correlation coefficients ('Spearman-rank-order-method'). Significant differences of all variables between the two soil types were computed with the 'Mann-Whitney-Test'.

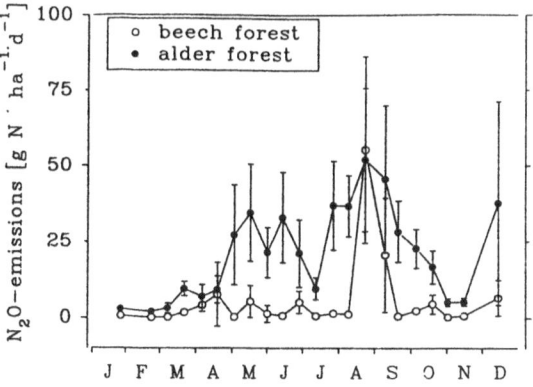

Figure 2. N₂O-emissions from soils of the study site in 1993.

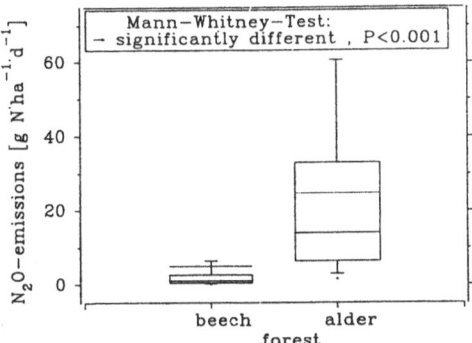

Figure 3. Box Plots of N₂O-emissions from a beech and alder forest (n=264 and n=319, respectively; dot line represents the mean).

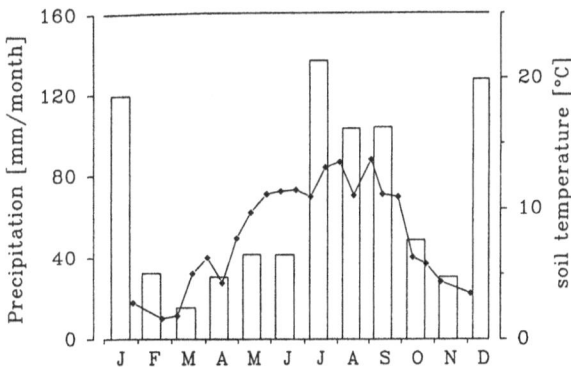

Figure 4. Precipitation (bars) and soil temperature in 10 cm depth (line) in the study site in 1993.

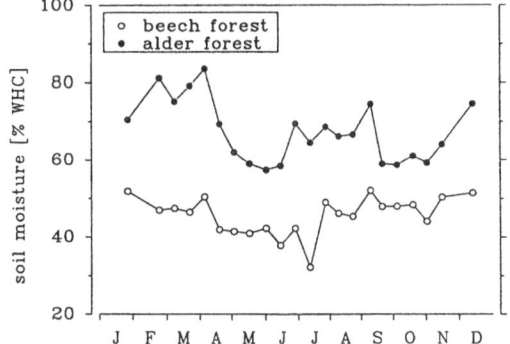

Figure 5. Soil moisture (% water holding capacity) from soils of the study site in 1993.

Results

N₂O-emissions

In Figure 2, N₂O-losses from both forest soils are shown. Each symbol presents the mean of 12 measurements that were conducted on two following days (n=6 boxes per day). The coefficient of variation showed a high spatial variabilty (sometimes > 150%).

The N₂O-emissions from both soils were high from August to September and decreased in autumn. However, the beech forest soil showed over the year a marked lack of significant N₂O-emissions. About 62% of all N₂O-losses occurred in 2 weeks and after this events no further significant emissions from this soil were detected. On the contrary the N₂O-emissions from the alder forest appeared in 1993 mainly from the beginning of May to the end of September with a short depression in July. About 80% of the emitted N₂O fell in this time. The 'Mann-Whitney-Test' shows that N₂O-emissions between the soils were significantly different ($p < 0.001$, Fig. 3).

Influencing parameters

Climatic parameters in 1993 showed a marked tendency: high water contents occurred in winter and from August to September in both soils after strong precipitation events. The soil temperature in 10 cm depth

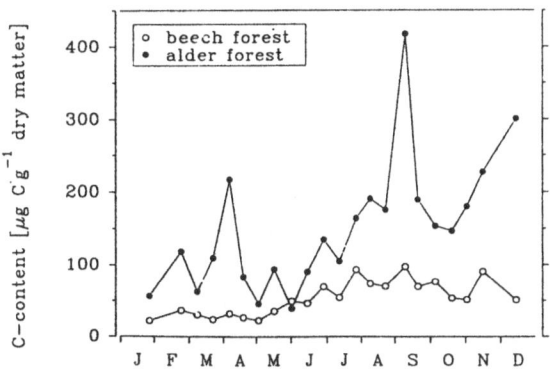

Figure 6. Watersoluble carbon content from soils of the study site in 1993.

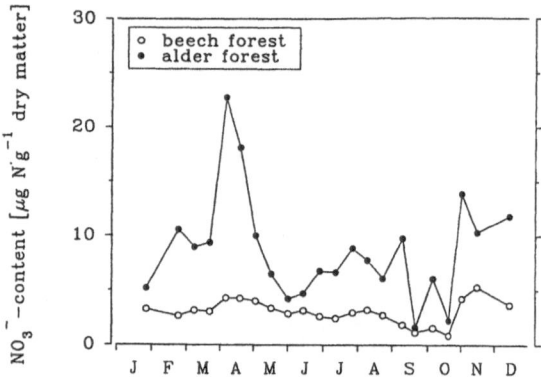

Figure 7. Nitrate content from soils of the study site in 1993.

Table 2. Average N_2O-emissions, nitrate-, carbon, and water contents from forest soils in the bornhöved lakes region (n=22). Computing the 'Mann-Whitney-Test' with raw data gave significant differences in each case ($p<0.001$)

	Beech forest	Alder forest
N_2O-emissions (g N ha^{-1} d^{-1})	4.93	24.65
Nitrate (μg N g^{-1} dry matter)	3.01	8.74
Carbon (μg C g^{-1} dry matter)	53.46	150.05
Water (% WHC)	45.67	67.39

reached maximum values of 13 °C from end of July to middle of September (see Figs. 4 and 5).

The nitrate content was high in spring and in summer rather low values predominated. The carbon content increased in both soils to maximum values in September. However, between alder and beech forest soils occurred obvious differences in the contents of these parameters (3-fold higher nitrate and 5-fold higher carbon values per gram dry soil in the alder in comparison to the beech forest soil).

Discussion

By comparing both soils distinct differences appeared. Much higher N_2O-emissions from the alder forest soil were propably due to significant higher water, nitrate and carbon contents in comparison to the beech soil (Mann-Whitney-Test, $p<0.001$, n=22 in each case, see Table 2).

The interpolation of the data to the whole year shows that from the beech forest soil only 1.9 kg N_2O-N ha^{-1} yr^{-1} were emitted. On the contrary the alder forest represents an area that is mainly influenced through near 'Lake Belau'. High water contents of that

Table 3. Spearman rank order correlations of the investigated parameters (* statistically significant, $p< 0.005$)

	T	NO_3^-	Cws	H_2O	pH
			r		
Alder forest					
N_2O	0.747*	−0.290	0.336	−0.284	−0.128
Beech forest					
N_2O	0.349	−0.115	0.220	0.186	−0.100

soil, caused by trespassing water, generally affect the microbial turnover (Dilly and Munch, 1994) and especially the N_2O-production by microorganisms (8.9 kg N_2O-N ha^{-1} yr^{-1}).

The seasonal fluctuations of the soil temperature show almost the same pattern as the N_2O-emissions from the alder forest, but not from the beech forest soil.

Table 3 presents the correlation coefficients between N_2O-emissions and edaphic parameters:

Only one direct dependence between N_2O-emissions and soil temperature demonstrates that the interaction of several variables must be known to predict these emissions.

Acknowledgements

I am grateful to Elke Erlebach for her technical assistance. These studies were supported by the German Ministry of Research and Technology (BMFT).

References

Burford J R and Bremner J M 1975 Relationships between the denitrification capacities of soils and total, water-soluble and readily decomposable soil organic matter. Soil Biol. Biochem. 7, 389–394.

Dilly O and Munch J C 1995 Microbial biomass and activities in partly hydromorphic agricultural and forest soils in the Bornhöved Lake Region of Northern Germany. Biol. Fertil. Soils 19, 343–347.

Heinemeyer O, Walenzik G and Kaiser E A 1991 Zur Methodik der Bestimmung gasförmiger N-Abgaben in Freilandexperimenten. Mitt. Dtsch. Bodenkd. Gesellsch. 66, 499–502.

Hutchinson G L and Mosier A R 1981 Improved soil cover method for field measurement of nitrous oxide fluxes. Soil Sci. Soc. Am. J. 45, 311–316.

Scharpf H C and Wehrmann J 1976 Die Bedeutung des Mineralstickstoffvorrats des Bodens zu Vegetationsbeginn für die Bemessung der N-Düngung zu Winterweizen. Landw. Forsch. 32, 100–114.

Plant and Soil **181**: 95–108, 1996.
© 1996 *Kluwer Academic Publishers.*

Nitrous oxide emissions from agricultural fields: Assessment, measurement and mitigation

A.R. Mosier[1], J.M. Duxbury[2], J.R. Freney[3], O. Heinemeyer[4] and K. Minami[5]
[1]*USDA/ARS, Fort Collins, CO 80522, USA*,* [2]*Cornell University, Ithaca, NY, USA,* [3]*CSIRO, Canberra, ACT, Australia,* [4]*BFAL, Braunschweig, Germany and* [5]*JIRCAS, Tsukuba, Japan*

Key words: dentrification, nitrification, nitrification inhibitors, N_2O

Abstract

In this paper we discuss three topics concerning N_2O emissions from agricultural systems. First, we present an appraisal of N_2O emissions from agricultural soils (**Assessment**). Secondly, we discuss some recent efforts to improve N_2O flux estimates in agricultural fields (**Measurement**), and finally, we relate recent studies which use nitrification inhibitors to decrease N_2O emissions from N-fertilized fields (**Mitigation**).

To assess the global emission of N_2O from agricultural soils, the total flux should represent N_2O from all possible sources; native soil N, N from recent atmospheric deposition, past years fertilization, N from crop residues, N_2O from subsurface aquifers below the study area, and current N fertilization. Of these N sources only synthetic fertilizer and animal manures and the area of fields cropped with legumes have sufficient global data to estimate their input for N_2O production. The assessment of direct and indirect N_2O emissions we present was made by multiplying the amount of fertilizer N applied to agricultural lands by 2% and the area of land cropped to legumes by 4 kg N_2O-N ha^{-1}. No regard to method of N application, type of N, crop, climate or soil was given in these calculations, because the data are not available to include these variables in large scale assessments. Improved assessments should include these variables and should be used to drive process models for field, area, region and global scales.

Several N_2O flux measurement techniques have been used in recent field studies which utilize small and ultra-large chambers and micrometeorological along with new analytical techniques to measure N_2O fluxes. These studies reveal that it is not the measurement technique that is providing much of the uncertainty in N_2O flux values found in the literature but rather the diverse combinations of physical and biological factors which control gas fluxes. A careful comparison of published literature narrows the range of observed fluxes as noted in the section on assessment. An array of careful field studies which compare a series of crops, fertilizer sources, and management techniques in controlled parallel experiments throughout the calendar year are needed to improve flux estimates and decrease uncertainty in prediction capability.

There are a variety of management techniques which should conserve N and decrease the amount of N application needed to grow crops and to limit N_2O emissions. Using nitrification inhibitors is an option for decreasing fertilizer N use and additionally directly mitigating N_2O emissions. Case studies are presented which demonstrate the potential for using nitrification inhibitors to limit N_2O emissions from agricultural soils. Inhibitors may be selected for climatic conditions and type of cropping system as well as the type of nitrogen (solid mineral N, mineral N in solution, or organic waste materials) and applied with the fertilizers.

* FAX No.: +19704908213
Plant and Soil is the original source of publication of this article. It is recommended that this article is cited as: *Plant and Soil* **181**: 95–108, 1996.

Introduction

Nitrous oxide is an important atmospheric constituent because it is a long-lived greenhouse gas and it is also the major source of stratospheric NO (Cicerone, 1989). The atmospheric concentration of N_2O is about 310 ppbv, it is increasing at a rate of 0.6–0.9 ppbv yr^{-1}, and its lifetime is 166 ± 16 years (Prinn et al., 1990). The global warming potential (GWP) of each molecule of N_2O is about 250 times greater than each molecule of CO_2. Nitrous oxide currently accounts for 2–4% of total GWP (Watson et al., 1992) and could contribute as much as 10% of GWP in the future (Cicerone, 1989). The only known significant removal mechanism for atmospheric N_2O is transport into the stratosphere where it is photolytically oxidized to NO which reacts with stratospheric ozone and absorbs harmful solar ultra violet radiation (Crutzen, 1981). It has been estimated that doubling the concentration of N_2O in the atmosphere would result in a 10% decrease in the ozone layer and this would increase the ultraviolet radiation reaching the earth by 20% (Crutzen and Ehhalt, 1977). This could result in increased skin cancer and other health problems (Lijinsky, 1977).

The concentration of N_2O in the atmosphere is increasing, because of biotic and anthropogenic activities (Andreae and Schimel, 1989; Rodhe, 1990). Previously it was estimated that about 1.5 Tg of N is injected directly into the atmosphere each year as N_2O as a result of fertilizer applications to agricultural ecosystems (CAST, 1992; Watson et al., 1992). This represented about 44% of the anthropogenic input and about 13% of the total input of N_2O to the atmosphere annually. These estimates (Watson et al., 1990, 1992) did not include N_2O production from other major N input sources; animal manures and biological N fixation. As we discuss later, these are important inputs to be considered in assessing global N_2O emission budgets and we include them in revised emission estimates presented below.

It is likely that N_2O production resulting from fertilizer and increased use of biological nitrogen fixation is underestimated because the effect of a N input is usually only partially traced through the environment (Duxbury et al., 1993). A complete accounting of fertilizer N, biologically fixed N, and N mineralized from soil organic matter is difficult to achieve, but is needed if we are to accurately assess the impact of increased use of N in agricultural ecosystems on terrestrial N_2O emissions. Since only about half of the fertilizer N added to a field is taken up by a crop and agricultur-

al soils are tending to lose N through time (CAST, 1992), about half of the N is eventually returned to the atmosphere through denitrification. Loss vectors from the field include nitrate leaching, erosion and gaseous emissions. The N removed from the field in the crop is fed to animals which generate manure. The manure is returned to cropland to fertilize a second crop and, in the process, about half of the N in the manure is volatilized as NH_3. Ammonia volatilization from agricultural systems is globally important (Isermann, 1993), but its impact on N_2O emissions has not been directly quantified.

Since about 70% of the N_2O emitted from the biosphere into the atmosphere is derived from soils (Bouwman, 1990) we assume that changes in N cycling in soil systems have influenced the increases in atmospheric N_2O during the past century and will help dictate future changes. In the past, land use conversion has caused major perturbations in soil N processes. Conversions of forests and grasslands to croplands accelerated C and N cycling and increased N_2O emissions from soil. Globally, land use conversion is important now only in tropical areas. Most of the conversions of forests and grasslands in the northern hemisphere occurred 50 to 200 years ago (Hammond, 1990). Continued growth in the atmospheric concentration of N_2O can be partly attributed to the increase in N input into the soil system. The increased input comes from atmospheric deposition [which ranges from about 0.5 g N $m^{-2} yr^{-1}$ in the central US and other relatively pristine areas of the globe, to 6 g N $m^{-2} yr^{-1}$ in western Europe (Andreae and Schimel, 1989)], N fertilization with mineral N sources or animal manures and biological N fixation. Nitrogen fertilizer use and biological N-fixation are projected to continue to increase during the next 100 years (Hammond, 1990). Much of this increase in N use is necessary to continue global food production to meet the needs of the rapidly expanding population, since crop production can be directly related to N-fertilizer use (World Development Report, 1992).

Soil production processes

Research during the past several decades provides an understanding of how N_2O is produced, factors that control its production, source/sink relationships and gas movement processes. However, even with this large amount of knowledge, we are not yet able to reliably predict the fate of a unit of N that is applied or deposited on a specific agricultural field. Studies of emissions of N_2O from presumably 'similar' agricultural systems show highly variable results in both time and space. It is the complex interaction of the physical and biological processes involved that must be understood before appropriate predictive capability can be developed (Mosier, 1993).

N_2O is produced primarily from the microbial processes, nitrification and denitrification in soil. In well aerated, yet moist conditions, N_2O emissions from nitrification of ammonium based fertilizers can be substantial (Bremner and Blackmer, 1978; Duxbury and McConnaughey, 1986). Other work suggests that N_2O is a byproduct of nitrification (Yoshida and Alexander, 1970) through denitrification of nitrite by nitrifying organisms which denitrify nitrite under oxygen stress (Poth and Focht, 1985). In wet soils, where aeration is restricted, denitrification is generally the source of N_2O (Smith, 1990). Under these conditions both the rate of denitrification and the $N_2O/(N_2 + N_2O)$ ratio must be known to evaluate N_2O emissions through denitrification. According to Smith (1990), soil structure and water content, affecting the balance between diffusive escape of N_2O and its further reduction to N_2 are important in determining the proportions of the two gases.

Purpose of paper

In this paper we discuss three topics concerning N_2O emissions from agricultural systems. First we present a current appraisal of N_2O emissions from agricultural soils (**Assessment**). Secondly we discuss some recent efforts to improve N_2O flux estimates in agricultural fields (**Measurement**), and finally we relate some recent field studies which use nitrification inhibitors to decrease N_2O emissions from N-fertilized fields (**Mitigation**).

Assessment

The increase in N_2O emissions from agricultural soils during the past several decades has been related to the increase in N application to these soils (McElroy et al., 1977; Granli and Bockman, 1994). To assess N_2O emissions during a calendar year it is necessary to know the amount of N used as fertilizer on a global basis. Estimates of the total amount of N applied to agricultural soils as mineral N, animal waste (manure + urine) N and the area of harvested N-fixing crops are given in Table 1. The amount of mineral N con-

sumed in each of the seven regions is tabulated from information provided by FAO (1990a). The land area of harvested pulses and soybeans is also obtained from FAO (1990b). The amount of N from animal waste that is used to fertilize agricultural fields is not well known. We derived the estimates in Table 1 from calculations that were adapted from Safley et al. (1992) and Bouwman (1994a). We then estimated an amount of N for each region that may be used as fertilizer N. These numbers were based on estimates of animal distribution and management systems for each region and we do not consider the N estimates to be very reliable because of the lack of information.

Galbally et al. (1992) indicate that legumes may contribute to N_2O emission in a number of ways. Atmospheric N_2 fixed by the legumes can be nitrified and denitrified in the same way as fertilizer N, thus providing a source of N_2O. In addition, symbiotically living Rhizobia in root nodules are able to denitrify and produce N_2O (O'Hara and Daniel, 1985). Galbally et al. (1992) suggested an emission rate of 4 kg N ha^{-1} y^{-1} for improved pastures, and Duxbury et al. (1982) suggested that legumes can increase N_2O emissions by a factor of 2 to 3 compared to unfertilized fields.

In addition to including animal manure and N from biological N fixation with synthetic fertilizer N as total agricultural N sources for N_2O, we present below a revised method for calculating the contribution of N_2O from agricultural systems. Earlier estimates (OECD/OCDE, 1991) were generally based upon assessments derived from reviews of published N_2O emissions data (Bouwman, 1990; Eichner, 1990). More recently, Bouwman (1994a, b) reviewed the literature again and presented a more comprehensive assessment of N_2O emissions. He notes that loss of N_2O from agricultural soils may be presented in three ways: (1) the total loss of N_2O during the period covered by the measurements; (2) the difference between fertilized and control plots which is referred to as 'fertilizer induced N_2O loss' and (3) the total loss of N_2O calculated as a percentage of fertilizer N applied.

The total N_2O flux should represent N_2O from all possible sources; native soil N, N from recent atmospheric deposition, past years fertilization, N from crop residues (depending upon the C/N ratio of the residue this N could either be mineralized or immobilized), N_2O from subsurface aquifers below the study area, and current N fertilization. Because unfertilized control plots mostly represent the fertilization and cropping history of previous years, it is difficult to assess a proper control value for the recent fertilizer effects on

Table 1. Estimated nitrogen derived from synthetic fertilizers and animal wastes applied to agricultural lands and land area cropped with pulses and soybeans

Region	Fertilizer N applied (Tg)	Manure N produced (Tg)	N used as fertilizer (% of total)	Manure N used as fertilizer (Tg)	Harvested area of pulses + soybeans (ha $\times 10^6$)
Africa	2.1	20.9	50	10.5	12.8
Am. N and C	13.1	7.8	70	5.5	28.1
Am. S	1.7	21.9	50	11.0	23.5
Asia	37.3	37.4	70	26.2	48.5
Europe	13.6	12.3	90	11.1	4.0
Oceania	0.9	0.5	30	1.5	1.4
USSR[a]	8.7	10.1	90	9.1	6.6
Total	77.4	115.3		74.9	124.9

[a] Former USSR.

N_2O loss. Even though we recognize that subtracting a control may underestimate the amount of N_2O directly derived from added fertilizer we use method (2) in our calculations to estimate the contribution of fertilization to N_2O emissions from agricultural soils.

We have discarded the concept suggested by Eichner (1990) of different amounts of N_2O evolving from fertilized soils according to N source. As noted in Mosier (1993), we consider soil management and cropping systems to impact N_2O emissions more than mineral N source. Limited data does, however, indicate that organic N sources such as animal manures and sewage sludge do induce larger N_2O emissions per unit of N added to the soil than does mineral N (Bouwman, 1990, 1994a, b). Because of the lack of adequate parallel experiments that cover the range of possibilities of mineral and organic N applications, we use a single conversion coefficient for all sources.

Bouwman (1994a) estimated the total emissions of N_2O from a regression equation: total annual direct field N_2O (kg N ha^{-1}) loss = 1 + 0.0125 $\times$ N-application. The value of 1 kg N_2O-N ha^{-1} represents the background emission while the 0.0125 factor accounts for the contribution from fertilization. This estimate includes N sources from a variety of mineral and organic N-fertilizers and was based on long term data sets.

Conducting other analyses with the data tabulated by Bouwman (1994a) we arrived at similar N_2O conversion estimates. From the data listed, we selected data satisfying the following criteria: soils were cropped, measurements were made for more than 80

days, and 50 to < 500 kg fertilizer N^{-1}ha^{-1} was added. Seventy of these studies included a nonfertilized control. The average 'fertilized induced N_2O flux' was 0.0124 kg N_2O-N kg^{-1} N applied, with a standard error of 0.0016. The range of these data is 0 to 0.063 and the median is 0.009. Of these studies 1 found no 'fertilizer induced N_2O flux' while 28 studies found 0.001 to 0.004 kg N kg^{-1} N, 10 cases found 0.004 to 0.008, 21 found 0.008 to 0.02, 8 found 0.02 to 0.036 and 2 were greater than 0.036.

Although these emission estimates are variable, the variability is lower than suggested in the OECD/OCDE (1991) calculation methodology. Experience in conducting field flux experiments tells us that much narrower constraints can be placed on the N_2O flux predictions. We expect that the Bouwman (1994a) estimate of 1.25% $\pm$ 1.0% of the applied N will encompass approximately 90% of the direct contributions of fertilization to N_2O emissions.

The indirect contribution of fertilizer N additions to N_2O emissions, apart from the fields where fertilizer N is applied, must also be considered in Agriculture's contribution to atmospheric N_2O. Based upon the discussions of Duxbury et al. (1993) and Mosier (1993) and the large amounts of N_2O frequently found in subsurface aquifers (Bowden and Bormann, 1986; Minami and Ohsawa, 1990; Ronen et al., 1988) we estimate that an additional 0.75% of N applications will eventually be evolved to the atmosphere as N_2O resulting from N leaching, runoff and NO_x and NH_3 volatilization.

Table 2. Estimates of direct and indirect emissions of N_2O produced by application of synthetic fertilizer N or animal waste N to agricultural soils or by growing legumes

Region	Estimated N_2O emitted from			Total estimate	Range of estimates
	Fertilizer N	Manure	N-Fixation		
	(Tg N_2O-N)				
Africa	0.04	0.21	0.05	0.30	0.15–0.45
Am. N and C	0.26	0.11	0.11	0.48	0.24–0.72
Am. S	0.03	0.22	0.09	0.34	0.17–0.51
Asia	0.75	0.52	0.19	1.46	0.73–2.19
Europe	0.27	0.22	0.02	0.51	0.26–0.77
Oceania	0.01	0.03	0.01	0.05	0.03–0.08
USSR[a]	0.17	0.18	0.03	0.3	0.19–0.57
Total	1.53	1.49	0.50	3.5	1.8–5.3

[a] Former USSR.

The total direct and indirect N_2O-N emissions from application of mineral or organic N total approximately $2 \pm 1\%$ annually. This estimate is expected to encompass more than 90% of field situations. Nitrous oxide from biological N-fixation is calculated by multiplying the area of land used for growing pulses plus soybeans in each region by 4 kg N ha^{-1} (Duxbury et al., 1982; Galbally et al., 1992) (Table 2).

The N_2O emission estimates from Table 2 suggest that N_2O from fertilizer application to croplands is about equally proportioned between synthetic fertilizer and manure fertilizer sources. About 40% of the estimated N_2O production is from North and Central America, Europe and the former Soviet Union where about 20% of the world human population reside. Asian countries which hold about 55% of the global human population (World Development Report, 1992) contribute about 40% of the estimated annual N_2O production.

Improvements in local, regional and national N_2O emission assessments are needed to be more certain that these global emission estimates are realistic. Improving assessment methodologies may evolve in a series of steps that culminates in development of process based models which are coupled to land use, crop and soil data bases to provide geographically related information. With these models, if a relatively simple set of input information can be supplied on a country or region basis, then detailed emission calculations may be made. Because of the inherent spatial and temporal variability associated with N_2O production and emissions from soils, it appears that very simple approaches

may not provide realistic emission estimates (Mosier and Bouwman, 1993).

Measurement

Reducing uncertainty

A multitude of reviews on N_2O flux measurement methods have been published during the past 15 years (e.g. Baldocchi et al., 1988; Denmead, 1983, 1994; Desjardins and MacPherson, 1989; Fowler and Duyzer, 1989; Mosier, 1989). The reader is referred to some of the current reviews to gain historical and theoretical insight to gas flux measurement problems.

Nitrous oxide emissions from N-fertilized agricultural fields have been found to vary between 0.001 and 6.8% of the N applied to the field (Bouwman, 1990; Eichner, 1990). A portion of this variability in N_2O estimates, relative to the amount of fertilizer applied, has been attributed to spatial and temporal variability of the processes which produce N_2O in soil (Denmead, 1994). Since small chamber systems were used to collect most of these data, the data are assumed to be inherently flawed (Denmead, 1994). It is also assumed that by using flux measurement techniques which integrate over larger areas that the average N_2O flux value for a field will be more accurate and subject to less uncertainty. This assumption has recently driven the development of new techniques and funding for new research in many areas of the world. There is still little data from comparative field sites which discern whether or not the average N_2O flux made with chambers in a field has any true value or if the variability noted in the literature reviews of Eichner (1990) and Bouwman (1990) is primarily from field factors such as crop, cultivation practices, former history of the field being studied, weather, irrigation practices or a host of other variables that regulate gas flux.

During the past two decades the need for detection systems which are rapid, sensitive and accurate has been expressed (Andreae and Schimel, 1989) and several N_2O measurement techniques have been proposed which permit using micrometeorological techniques to provide field scale spatial integration. Denmead (1983), Desjardins and MacPherson (1989), Fowler and Duyzer (1989) and Mosier (1990) discuss these possibilities and a variety of aspects as they relate to N_2O flux measurements. These techniques include N_2O detection by tunable diode laser (TDL) which permits eddy correlation or flux gradient micrometeorological flux measurements, Fourier transformed infrared (FTIR) spectrometry which can be used for flux gradient techniques or ultra-large chambers, and long path infrared analyzers for use in ultra-large chambers. Measurement techniques are now becoming available to permit making interfield comparisons without using small chamber techniques. These techniques are now being used in field studies and are presented as case studies to demonstrate some recent advances in N_2O measurement technology and to demonstrate recent efforts to compare the capabilities of flux techniques.

Case studies

Case 1. Stirling, Scotland, UK

Measurement of N_2O flux from a fertilized grassland using three different analytical techniques with the flux gradient micrometeorological technique (Hargreaves et al., 1994).

A level timothy grass (*Phleum pratense*) pasture about 0.08 m high on a clay soil with poor drainage was fertilized with ammonium nitrate (185 kg N ha^{-1}) on 3 April 1992 and N_2O emission rates were measured over the following three weeks. During this time the soil moisture content remained in the range of 40–50% on a dry weight basis and total soil mineral N content declined from about 200 to 80 mg N kg^{-1} (Clayton et al., 1994).

Gas chromatography. Gas samples were collected from 0.06, 0.23, 0.51, 1.05, and 1.99 m above the

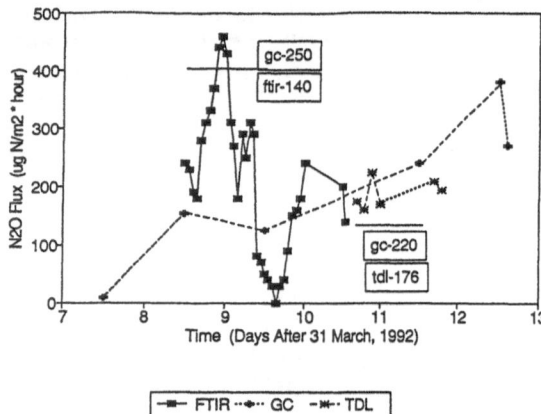

Figure 1. Nitrous oxide flux using three different N_2O detectors and the flux gradient micrometeorological technique at Stirling, Scotland, April 1992 (adapted from Hargreaves et al., 1994).

grass canopy during a 30-min period by filling 25-L Tedlar bags at the rate of 0.7 L min^{-1}. Aliquots of these samples were analyzed 10 to 15 times each by GC, using electron capture detection. Repeated analyses of the same sample increased analytical precision to 1.0 ppbv so that fluxes of > 100 g ha^{-1} d^{-1} could be detected (Arah et al., 1994).

Tunable diode laser. The N_2O concentration gradient, 0.06 and 1.05 m above the grass canopy, was measured using a TDL (Hargreaves et al., 1994). This TDL employed a lead salt laser which rapidly scanned (1 kHz) across a single rotational-vibrational absorption line centered on 4.472 μm. Ambient air was pumped through a multiple reflection white cell which provided an optical absorption path length of 38.7 m. Precision of 1 ppm in 1-min measurement periods were attained and 15 min averages were used for flux comparisons.

Fourier transformed infrared spectrometry. Galle et al. (1994) also used a FTIR to measure N_2O concentration differences at 0.06 and 1.05 m above the pasture canopy. Air was drawn from each height and pumped alternatively into a 25-L multiple reflection cell with an effective path length of 139 m. A differential spectrum was obtained by ratioing spectra from the two heights and 12 min averages of concentration differences between heights were obtained.

The three techniques provided generally similar estimates for field scale measurement of N_2O fluxes (Fig. 1). Hargreaves et al. (1994) concluded that 'the magnitude of the uncertainty of N_2O fluxes measured with the three techniques is fairly large because of the small difference required to be measured against a large background concentration.'

Case 2. Lammefjorden, Denmark

Short-term comparison of N_2O fluxes measured by TDL eddy correlation, ultra-large chambers, and small chambers.

During 19–28 August 1993 another field comparison of analytical techniques was conducted in a high organic matter, silty clay soil where wheat (*Triticum aestivum* L.) had recently been harvested (Fowler et al., 1994; Smith et al., 1995).

Eddy correlation. Fowler and Duyzer (1989) describe the eddy correlation micrometeorological method for measuring gas fluxes. Eddy correlation measures the vertical transport of a gas past a point in the atmosphere. It is obtained by correlating the instantaneous vertical wind speed at a point with the instantaneous concentration of the gas. In the natural environment the eddies which are important in the transport process occur with frequencies of up to 5 or 10 Hz. The new TDL used by the ITE Edinburgh research group (Fowler et al., 1994) provides a fast response detector having sufficient sensitivity to perform eddy correlation measurements. Measurements were made almost continuously between 19 and 28 August 1993.

Ultra-large chamber and long path IR (Hawk) detector. A large chamber was constructed from a series of plastic hoops covered with either tent fabric or polyethylene sheet to form a hemi-cylindrical chamber which was 2 m wide, about 30 m long and covered an area of about 60 m^2. The accumulation of N_2O released from the soil surface under the cover was measured with a long-path infrared absorption spectrometer tuned to an N_2O absorption band. The 'Hawk' gas monitor manufactured by Siemens Plessy[1] was used in these studies (Smith et al., 1994, 1995). This instrument can detect a N_2O concentration change of about 25 ppbv.

Small chambers. Small, closed chambers (IAEA, 1992; Smith et al., 1995) were arrayed along the assumed fetch of the eddy correlation technique and near the ultra-large chamber. The small chambers anchors were installed before the study was initiated

Table 3. Comparison of N_2O flux using small chambers (SC), ultra-large chamber with long path IR (ULC), and eddy correlation (EC) in a recently harvested wheat field in Lammefjorden, Denmark (Smith et al., 1995)

Date	Method		
	SC	ULC	EC
	N_2O Flux (μg N m^{-2} h^{-1})		
22 August	175	367	359
23/24 August	209	254	367
26/27 August	254	309	225

and each chamber covered 0.16 m^2. The concentration buildup within the chambers was measured by periodically withdrawing a gas sample from within the chamber and determining the quantity of N_2O by gas chromatography using an electron capture detector (Arah et al., 1994).

Table 3 shows the results of three comparisons on three occasions. The small chamber, ultra-large chamber and the eddy correlation technique provided flux measurements that were of similar variability to the three measurement techniques in Case 1 (Fig. 1). Smith et al. (1995) note that their previous experience indicates that similar flux measurements should be expected, provided that the area sampled by the chambers is representative of the micrometeorological fetch.

Case 3. Guelph, Ontario, Canada

Flux gradient technique using TDL. A sensitive, fast response N_2O analyzer, based on TDL absorption (IAEA, 1992) was used to make flux measurements using micrometeorological techniques. The TDL technique is based on infrared (IR) absorption spectrometry, whereby the extent of absorption depends upon path length, line strength and absorber concentration. The N_2O instrument has a total system noise of 1 ppb (v/v) based on a 30 minute sampling period, a resolution of 8 ppt (v/v) can be achieved. The maximum sampling rate is 10 Hz and data point averaging time is 0.1 s. This system is now commercially available from Campbell Scientific, Logan, Utah, USA[1].

The TDL system was used to measure N_2O fluxes utilizing the gradient-transport micrometeorological method (Wagner-Riddle et al., 1994) within four one ha adjacent fields at the Elora Research Station near Guelph, Ontario, Canada. The plots contained the following treatments: (1) bare fallow; (2) bare fallow fertilized with liquid cattle manure; (3) alfalfa (*Me-*

[1] Name of specific supplier of instrumentation does not constitute an endorsement by USDA/ARS or imply that other comparable instruments produced elsewhere are not suitable.

dicago sativa L.) established in 1992 and (4) a mature stand of Kentucky bluegrass (*Poa pratensis* L.). Air samples were taken from two heights above each plot for one hour, six times each day. Flux measurements were begun before snow melt in March and continued through the remainder of 1993. Hourly flux measurements were averaged to provide a daily flux value for each site.

N_2O emissions from March to December totaled 3.5, 4.3, 3.2 and 0 kg N ha^{-1} for bare fallow, manured bare fallow, alfalfa and bluegrass fields, respectively. Fluxes for the year were highest from the noncropped fields immediately following snow melt, when soil surface thawing occurred. This initial burst in emissions was not observed for the cropped plots, although the average fluxes were higher from March through April than the remainder of the year. Plowing the alfalfa field at the end of September resulted in large N_2O emissions, increasing the monthly average from 10.3 to 59.4 ng m^{-2} s^{-1} (or 0.17 kg N ha^{-1} month^{-1} and 1 kg N ha^{-1} month^{-1}) (Wagner-Riddle et al., 1994).

These studies demonstrate that TDL can now provide micrometerological measurement of N_2O fluxes on a routine basis. Particularly important in this demonstration is that the technology can be used to continuously monitor gas fluxes throughout the year. The studies also point out that year-round measurements are necessary because maximum fluxes are observed in different treatments during different times of the year. Such systems should be ideal to compare N_2O emissions from different management systems at the same time. Simultaneous measurement of different treatments on the field scale are required to determine if the uncertainty in N_2O emissions from agricultural fields under different management systems is as large as generally perceived (Eichner, 1990).

Methods comparisons

The comparisons of flux measurement techniques presented above in Cases 1 and 2 (Galle et al., 1994; Hargreaves et al., 1994; Smith et al., 1994, 1995) indicate that with appropriate experimental design and analyses the different methods provide essentially the same information concerning N_2O flux. According to Smith et al. (1995) the small chambers require a high labor requirement and the need for a large number of replicate measurements to overcome spatial variability. The ultra-large chambers average fluxes over areas of 2–3 orders of magnitude larger than conventional chambers and are useful where site character-istics rule out micrometeorological methods. These ultra-large chambers are, however, cumbersome to set up and cannot be used in windy conditions. Overall, small chambers have been invaluable in establishing the effects of variables such as temperature, the supply of mineral N and soil water content on N_2O fluxes (Smith et al., 1995) and will probably remain the major flux measurement technique for many types of studies. Micrometerological methods which supply 'real time' spatial integration of N_2O fluxes over areas of 0.1 to 1 km^2 should also provide the ability to interpret N_2O emissions in terms of short term environmental variables (Hargreaves et al., 1994).

Further interpretation of the field N_2O flux measurements discussed above indicates that the data collected thus far, primarily using chamber techniques, may after all, be reflective of the N_2O fluxes that occurred in the field. Utilization of techniques which provide spatial integration more readily may not necessarily change the interpretation of the numbers acquired. From this one may conclude that it is not the analytical technique that is providing the major uncertainty in N_2O flux values found in the literature but rather the diverse combinations of physical and biological factors which control gas fluxes. A careful comparison of published literature narrows the range of observed fluxes as noted in the section on assessment. An array of careful field studies which compare a series of crops, fertilizer sources, and management techniques in controlled parallel experiments are needed to improve flux estimates and decrease uncertainty in prediction capability using techniques which best fit the experimental design. Another point which is demonstrated clearly by the Guelph, Canada case study, is that studies must be conducted throughout the year and not just during the cropping period.

Mitigation

As noted in the above assessment section about 3 Tg of N_2O-N is emitted from agricultural soils that are fertilized with synthetic and manure N each year. This represents a small loss of the about 150 Tg of N (inorganic and organic) used as fertilizer world wide (Table 1), but does amount to approximately 20% of the annual global efflux of N_2O to the atmosphere.

The low efficiency of fertilizer N use in agricultural systems is primarily caused by the large losses of N from those systems in gaseous forms (Freney and Simpson 1983; Hauck, 1984; Peoples et al., 1994) and

N_2O emission is directly linked to the loss processes. It is axiomatic then that any strategy which increases the efficiency of fertilizer N use will reduce emissions of N_2O, and this has been directly demonstrated for a number of strategies (Aulakh et al., 1984a, b; Bremner et al., 1981; Bremner and Blackmer, 1979; Bronson et al., 1992; Magalhaes et al., 1984; Minami et al., 1990). In general, N_2O emissions from mineral and organic N can be decreased by management practices which optimize the crop's natural ability to compete with processes whereby plant available N is lost from the soil-plant system (i.e. NH_3 volatilization, denitrification and leaching), and directly lowering the rate and duration of the loss processes (Doerge et al., 1991).

There are indirect contributions to N_2O emission through volatilization of NH_3 and emission of NO_x into the atmosphere, and its redistribution over the landscape through wet and dry deposition. Strategies to increase the overall efficiency of N are therefore necessary and some practices are noted in Table 4.

Although it is widely recognized that NH_3 losses from grazed pastures are a major source of atmospheric NH_3 (Freney et al., 1983; Schlesinger and Hartley, 1992), and loss by denitrification can be substantial from heavily fertilized grasslands in wet climates (e.g. up to 79 kg N ha^{-1} yr^{-1}, Jordan, 1989), most of the research effort has been devoted to reducing gaseous emissions from cropped soils rather than from pastures. Therefore, many of the strategies described in Table 4 which potentially can minimize gaseous emissions, deal primarily with cropping systems. Although most of the practices listed are assumed to decrease N_2O emissions there have been relatively few systematic studies which compare a variety of farming practices as to their ability to conserve N and limit N_2O emissions. A number of field studies have been conducted with nitrification inhibitors which demonstrate a high likelihood of decreased N_2O emissions when used.

Nitrification inhibitors

Since NH_3 or ammonium producing compounds are the main sources of fertilizer nitrogen (FAO, 1990a), maintenance of the applied nitrogen in the ammonium form should result in lowered emission of N_2O from cultivated soils. One mechanism of maintaining added N as ammonium is to add a nitrification inhibitor with the fertilizer (Broadbent et al., 1957; Bundy and Bremner, 1973; Sahrawat et al., 1987).

Numerous substances including pyridines, pyrimidines, mercapto-compounds, succinamides, acetylenes, thiazoles, triazoles, triazines and carbon disulfide have been tested for their ability to inhibit nitrification and several have been patented (Hauck, 1984). Unfortunately most of these compounds have limitations to their usefulness (Keeney, 1983). For example, the most commonly used nitrification inhibitor, nitrapyrin (2-chloro-6 (trichloromethyl)pyridine), is seldom effective because of sorption on soil colloids, hydrolysis to 6-chloropicolinic acid and loss by volatilization (Hoeft, 1984).

It has been established in laboratory studies that acetylene is a potent inhibitor of nitrification (Bremner and Blackmer, 1979; Walter et al., 1979), but because it is a gas there are problems in introducing it into the soil in the field and maintaining it during the growing period at the concentration required to limit nitrification. This problem may be overcome by the use of calcium carbide coated with layers of wax and shellac to provide a slow-release source of acetylene (Banerjee and Mosier, 1989). Addition of wax-coated calcium carbide to the fertilized soil has reduced nitrification and increased yield, or recovery of N, in irrigated wheat (Freney et al., 1992), maize (Bronson et al., 1992) and cotton (Chen et al., 1994; Freney et al., 1993), and flooded rice (Banerjee et al., 1990; Bronson et al., 1994; Keerthisinghe et al., 1993).

Another way of overcoming the problem of applying gaseous acetylene is to use substituted acetylenes such as 2-ethynylpyridine or phenylacetylene which are liquids at ambient temperatures. These two compounds have proved to be effective inhibitors in laboratory studies (Crawford and Chalk, 1992; McCarty and Bremner, 1986, 1990) and the use of 2-ethynylpyridine in irrigated cotton has resulted in greatly increased recovery of applied N (Freney et al., 1993).

Using nitrification inhibitors does not always result in increased crop yields (Scharf and Alley, 1988), but a number of field studies indicate that nitrification inhibitors do limit N_2O emissions from ammonium based fertilizers (Aulakh et al., 1984b; Bremner and Blackmer, 1978; Bremner et al., 1981; Bronson et al., 1992; Magalhaes et al., 1984; Minami and Ohsawa, 1990). Several recent field tests continue to show that utilization of a variety of nitrification inhibitors does significantly limit N_2O emissions from the application of ammonium-based fertilizers. The following are three case studies from recent publications which illustrate this point.

Table 4. List of practices to improve fertilizer and manure N use efficiency in agriculture

1– Match N supply with crop demand
    a) Use soil/plant testing to determine fertilizer N needs
    b) Minimize fallow periods to limit mineral N accumulation
    c) Optimize split application schemes
    d) Match N application to reduced production goals in regions of crop over production

2– Close N flow cycles
    a) Integrate animal and crop production systems in terms of manure reuse and plant production
    b) Maintain plant residue N on the production site

3– Use advanced fertilization techniques
    a) Controlled release fertilizers
    b) Place fertilizers below the soil surface
    c) Foliar application of fertilizers
    d) Use nitrification inhibitors
    e) Match fertilizer amount and type to seasonal precipitation

4– Optimize tillage, irrigation and drainage

Table 5. Effect of nitrification inhibitors nitrapyrin and DCD on the emission of N_2O from ryegrass plots fertilized with ammonium sulfate, ammonium nitrate, or urea in an imperfectly drained clay loam soil (McTaggart et al., 1994) during the 1992 and 1993 growing seasons

Treatment	N_2O-N Emitted (g ha^{-1})	
	1992	1993
AS[a]	1100	500
AS + DCD[b]	500	250
AS + NP[c]	NA	250
Urea[d]	5000	2700
Urea + DCD[e]	2000	1000
Urea + NP[f]	NA	1500
AN[g]	3900	1500
AN + DCD[h]	2700	NA
Urea/AN/AN[i]	2900	NA

On 2 April, 9 June and 4 August 120 kg N ha^{-1} of the respective fertilizers were applied, nitrification inhibitors were applied with the first fertilization and a second application of dicyandiamide (DCD) was made with the third fertilization.
[a]Ammonium sulfate, [b]Ammonium sulfate + 12.5 kg DCD ha^{-1}, [c]Ammonium sulfate + 7.5 kg ha^{-1} nitrapyrin, [d]Urea, [e]Urea + dicyandiamide, [f]Urea + nitrapyrin, [g]Ammonium nitrate, [h]Ammonium nitrate + DCD, [i]First fertilization of urea and second and third with ammonium nitrate.

Table 6. Effect of ECC, DCD and nitrapyrin on N_2O emissions from irrigated wheat and maize

Treatment	Wheat	Maize
	(g N_2O-N ha^{-1})	
Urea[z]	930[a][y]	1650[a]
Urea + ECC[z]	510[b]	480[c]
Urea + DCD[z]	440[b]	NA
No N Control	440[b]	110[c]
Urea + nitrapyrin[z]	NA	980[b]

[z]Prilled urea was banded between plant rows at the rate of 150 kg N ha^{-1} at planting for winter wheat and 218 kg N ha^{-1} for maize one month after planting. Nitrification inhibitors were applied with the urea, ECC (20 kg ha^{-1} as calcium carbide) as solid material, DCD (15 kg ha^{-1}) in aqueous solution and nitrapyrin as an aqueous emulsion (0.5 L ha^{-1}).
[y]Numbers within each column followed by the same letter are not significantly different at $p=0.05$.

Case studies

Case 1. Scotland, UK

The effect of various N sources and nitrification inhibitors were quantified in a ryegrass (*Lolium* sp.) field located near Edinburgh, Scotland in 1992 and 1993 (McTaggart et al., 1994). Nitrapyrin and DCD decreased N_2O emissions by 45 to 64% in plots fertilized with urea or ammonium sulfate (Table 5).

Case 2. Colorado, US

The effect of acetylene applied as encapsulated calcium carbide (ECC) (Banerjee and Mosier, 1989),

Table 7. Effect of ECC and nitrapyrin on soil nitrate content (0-30 cm soil depth) in a urea-fertilized maize field (Bronson and Mosier, 1993)

Time after fertilizing	Treatment		
	Urea	Urea + ECC	Urea + NP
(weeks)	Soil nitrate content (mg N kg^{-1})		
0	2	2	2
2	38	1	16
4	32	3	28
7	10	6	18

See Table 6 for treatment details.

Table 8. Effect of split application of ammonium sulfate and DCS on N$_2$O emission and nitrate leached from a carrot field (Minami, 1994)

Treatment	N$_2$O Emission (g N ha^{-1})	Nitrate leached (kg N ha^{-1})
AS (150 + 50)[a]	630	9.5
AS (200)[b]	360	11
AS (200 + DCS)[c]	260	6.8
No fertilizer control	200	5.2

[a] A basal application of 150 kg N ha^{-1} was made on 25 August and a 50 kg N ha^{-1} top dressing was applied on 2 October, N$_2$O flux and nitrate leaching measurements were conducted until harvest on 21 December.
[b] A basal application of 200 kg ammonium sulfate-N was made at planting.
[c] In addition to the basal application of ammonium sulfate 10 kg ha^{-1} of DCS was applied.

DCD and nitrapyrin on N$_2$O emissions were tested in two field studies in Northeastern Colorado in irrigated maize (*Zea mays* L.) and winter wheat (Table 6) grown on clay loam and clay soils, respectively. Experimental details are described in Bronson and Mosier (1993). The amount of nitrate detected early in the maize growing season reflects the relative effects of ECC and nitrapyrin on nitrification and N$_2$O production (Table 7).

Case 3. Tsukuba, Japan
A study to quantify the effect of the nitrification inhibitor DCS (N-2,5-dichlorophenyl succinamic acid) and split applications of ammonium sulfate on N$_2$O emissions was conducted in field lysimeters using carrot (*Daucus carota* L.) as a test crop (Minami, 1994). The soil at this site was a clay loam having a pH of 6.5. In this study either 200 or 150 kg N ha^{-1} of ammonium sulfate was applied at planting. With plots fertilized with 200 kg N ha^{-1}, 10 kg ha^{-1} of DCS was added with the fertilizer. The addition of DCS markedly reduced N$_2$O emission and leaching of nitrate (Table 8).

These case studies clearly demonstrate the potential for using nitrification inhibitors to limit N$_2$O emissions from agricultural soils. Inhibitors may be selected for climatic conditions and type of cropping system as well as the type of nitrogen (solid mineral N, mineral N in solution, or organic waste materials) and applied with the fertilizers. Crop yield response is not always improved with nitrification inhibitors but a clear reduction in N$_2$O emissions is generally achieved. Case 1 and 3 also show that timing of fertilizer application and type of fertilizer can affect N$_2$O emissions. Where split fertilizer applications are possible application of ammonium based materials decrease N$_2$O emissions when denitrification seems the likely N$_2$O producing process and application of nitrate based fertilizers decrease N$_2$O emissions during the time when nitrification causes N$_2$O production (McTaggart et al., 1994).

Acknowledgements

We thank the following research groups for providing technical reports an and preprints of manuscripts to provide the information for the case studies discussed above: Institute of Terrestrial Ecology at Midlothian, Scotland, UK under the direction of Dr David Fowler with specific help from Dr Ute Skiba; the Environmental Research Program in the Department of Land Resource Science in the University of Guelph, Guelph, Ontario, Canada under the direction of Dr George Thurtell with specific help from Dr Claudia Wagner-Riddle; and Soil Science Department, National College for Food, Land and Environmental Studies, Edinburgh, Scotland, UK, under the direction of Dr Keith Smith.

References

Andreae M O and Schimel D S 1989 Exchange of Trace Gases Between Terrestrial Ecosystems and the Atmosphere. John Wiley and Sons, Chichester, UK.

Arah J R M, Crichton I J, Smith K A, Clayton H and Skiba U 1994 Automated gas chromatographic analysis system for micrometeorological measurements of trace gas fluxes. J. Geophys. Res. 99, 16593–16598.

Aulakh M S, Rennie D A and Paul E A 1984a Gaseous nitrogen losses from soils under zero-till as compared with conventional-till management systems. J. Environ. Qual. 13, 130–136.

Aulakh M S, Rennie D A and Paul E A 1984b Acetylene and N-serve effects upon N_2O emissions from NH_4^+ and NO_3^- treated soils under aerobic and anaerobic conditions. Soil Biol. Biochem. 16, 351–356.

Baldocchi E D, Hicks B B and Meyers T P 1988 Measuring biosphere-atmosphere exchanges of biologically related gases with micrometerological methods. Ecology 69, 1331–1340.

Banerjee N K and Mosier A R 1989 Coated calcium carbide as a nitrification inhibitor in upland and flooded soils. J. Indian Soc. Soil Sci. 37, 306–313.

Banerjee N K, Mosier A R, Uppal K A and Goswami N N 1990 Use of encapsulated calcium carbide to reduce denitrification losses from urea-fertilized flooded rice. Mitt. Dtsch. Bodenk. Ges. 60, 245–248.

Bouwman A F 1990 Exchange of greenhouse gases between terrestrial ecosystems and the atmosphere. In Soils and the Greenhouse Effect. Ed. A F Bouwman. pp 61–127. John Wiley and Sons, New York, USA.

Bouwman A F 1994a Direct emission of nitrous oxide from agricultural soils. RIVM report No. 773004004. RIVM, Bilthoven, the Netherlands.

Bouwman A F 1994b Estimated global source distribution of nitrous oxide. In NIAES Series 2 - CH_4 and N_2O Global Emissions and Controls from Rice Fields and Other Agricultural and Industrial Sources. Ed. K Minami, A Mosier and R Sass. pp 147–159. Yokendo Publishers, Tokyo, Japan.

Bowden W B and Borman F H 1986 Transport and loss of nitrous oxide in soil water after forest clear-cutting. Science 233, 867–869.

Bremner J M and Blackmer A M 1978 Nitrous oxide: Emissions from soils during nitrification of fertilizer nitrogen. Science 199, 295–296.

Bremner J M and Blackmer A M 1979 Effects of acetylene and soil water content on emissions of nitrous oxide from soils. Nature (London) 280, 380–381.

Bremner J M, Breitenbeck G A and Blackmer A M 1981 Effect of nitrapyrin on emission of nitrous oxide from soil fertilized with anhydrous ammonia. Geophys. Res. Lett. 8, 353–356.

Broadbent F E, Tyler K B and Hill G N 1957 Nitrification of ammoniacal fertilizers in some California soils. Hilgardia 27, 247–267.

Bronson K F, Mosier A R and Bishnoi S R 1992 Nitrous oxide emissions in irrigated corn as affected by encapsulated calcium carbide and nitrapyrin. Soil Sci. Soc. Am. J. 56, 161–165.

Bronson K F and Mosier A R 1993 Nitrous oxide emissions and methane consumption in wheat and corn-cropped systems. In Agricultural Ecosystem Effects on Trace Gases and Global Climate Change. Eds. L A Harper, A R Mosier, J M Duxbury and D E Rolston. pp 133–144. ASA Special Pub. No. 55. Am. Soc. Agron., Madison, WI, USA.

Bronson K F, Mosier A R, Bollich P K and Lindau C W 1994 Grain yield and ^{15}N uptake of drill-seeded rice as affected by coated calcium carbide. IRRN 19, 22.

Bundy L G and Bremner J M 1973 Inhibition of nitrification in soils. Soil Sci. Soc. Am. Proc. 37, 396–398.

CAST 1992 Preparing US Agriculture for Global Climate Change. Task Force Report. No. 119. P.E. Waggoner, Chair. Council for Agricultural Science and Technology. Ames, IA, USA. 96 p.

Chen D L, Freney J R, Mosier A R and Chalk P M 1994 Reducing denitrification loss with nitrification inhibitors following presowing applications of urea to a cottonfield. Aust. J. Exp. Agric. 34, 75–83.

Cicerone R J 1989 Analysis of sources and sinks of atmospheric nitrous oxide (N_2O). J. Geophys. Res. 94, 18265–18271.

Clayton H, Arah J R M and Smith K A 1994 Measurement of nitrous oxide emissions from fertilized grassland using closed chambers. J. Geophys. Res. 99, 16599–16607.

Crawford D M and Chalk P M 1992 Mineralization and immobilization of soil and fertilizer nitrogen with nitrification inhibitors and solvents. Soil Biol. Biochem. 24, 559–568.

Crutzen P J 1981 Atmospheric chemical processes of the oxides of nitrogen including nitrous oxide. In Denitrification, Nitrification and Atmospheric Nitrous Oxide. Ed. C C Delwiche. pp 17–44. John Wiley and Sons, New York, USA.

Crutzen P J and Ehhalt D H 1977 Effects of nitrogen fertilizers and combustion on the stratospheric ozone layer. Ambio 6, 112–117.

Denmead O T 1983 Micrometerological methods for measuring gaseous losses of nitrogen in the field. In Gaseous Loss of Nitrogen from Plant-Soil Systems. Eds. J R Freney and J R Simpson. pp 133–157. Martinus Nijhoff/Dr W Junk Publishers, The Hague, Netherlands.

Denmead O T 1994 Measuring fluxes of CH_4 and N_2O between agricultural systems and the atmosphere. In CH_4 and N_2O Global Emissions and Controls from Rice Fields and Other Agricultural and Industrial Sources. Eds. K Minami, A Mosier and R Sass. pp 209–234. NIAES Series 2. Yokendo Publishers, Tokyo, Japan.

Desjardins R L and MacPherson J I 1989 Aircraft-based measurements of trace gas fluxes. In Exchange of Trace Gases Between Terrestrial Ecosystems and the Atmosphere. Eds. M O Andreae and D S Schimel. pp 135-154. John Wiley and Sons, Chichester, UK.

Doerge T A, Roth R L and Gardner B R 1991 Nitrogen Fertilizer Management in Arizona. College of Agriculture, University of Arizona, Tucson, USA.

Duxbury J M, Bouldin D R, Terry R E and Tate R L III 1982 Emissions of nitrous oxide from soils. Nature 298, 462–464.

Duxbury J M and McConnaughey P K 1986 Effect of fertilizer source on denitrification and nitrous oxide emissions in a maize field. Soil Sci. Soc. Am. J. 50, 644–648.

Duxbury J M et al. 1993 Contributions of agroecosystems to global climate change. In Agricultural Ecosystem Effects on Trace Gases and Global Climate Change. Eds. L A Harper, A Mosier, J M Duxbury and D E Rolston. pp 1–18. ASA Special Pub. No. 55, Am. Soc. Agron. Inc., Madison, WI, USA.

Eichner M J 1990 Nitrous oxide emissions from fertilized soils: Summary of available data. J. Environ. Qual. 19, 272–280.

FAO 1990a Fertilizer Yearbook Volume 39. FAO statistics series No. 95. FAO, Rome, Italy.

FAO 1990b Production Yearbook Volume 43. FAO statistics series no. 94. FAO, Rome, Italy.

Fowler D, Skiba U, Hargreaves K J, Sheppard L J and Cape J N 1994 N_2O flux measurements from agricultural grasslands and natural soils using micrometeorological and chamber techniques. TIGER TO 3064H6.

Fowler D and Duyzer J H 1989 Micrometeorological techniques for the measurement of trace gas exchange. In Exchange of Trace Gases Between Terrestrial Ecosystems and the Atmosphere. Eds. M O Andreae and D S Schimel. pp 189–208. John Wiley and Sons, Chichester, UK.

Freney J R and Simpson J R 1983 Gaseous Loss of Nitrogen from Plant-Soil Systems. Martinus Nijhoff/Dr W Junk Publishers, The Hague, the Netherlands.

Freney J R, Simpson J R and Denmead O T 1983 Volatilization of ammonia. In Gaseous Loss of Nitrogen from Plant-Soil Systems. Eds. J R Freney and J R Simpson. pp 1–32. Martinus Nijhoff/Dr W. Junk Publishers, The Hague, the Netherlands.

Freney J R, Smith C J and Mosier A R 1992 Effect of a new nitrification inhibitor (wax coated calcium carbide) on transformations and recovery of fertilizer nitrogen by irrigated wheat. Fert. Res. 32, 1–11.

Freney J R, Chen D L, Mosier A R, Rochester I J, Constable G A and Chalk P M 1993 Use of nitrification inhibitors to increase fertilizer nitrogen recovery and lint yield in irrigated cotton. Fert. Res. 34, 37–44.

Galbally I E 1992 Biosphere-atmosphere exchange of trace gases over Australia. *In* Australia's Renewable Resources: Sustainability and Global Change. Eds. R M Gifford and M M Barson. pp 117–149. Bureau of Rural Resources, Canberra, Australia.

Galle B, Klemedtsson L and Griffith D W T 1994 Application of an micrometeorological method, an ultra-large chamber system, and conventional field chambers. J. Geophys. Res. 99, 16599–16608.

Granli, T and Bockman O C 1994 Nitrous Oxide from Agriculture. Norw. J. Agric. Sci. Suppl. No. 12. 128 p.

Hammond A L 1990 World Resources 1990-91. A report by the World Resources Institute. Oxford Univ. Press, Oxford, UK. 383 p.

Hargreaves K J, Skiba U, Dowler D, Arah J, Wienhold F G, Klemedtsson L and Galle B 1994 Measurement of nitrous oxide emission from fertilized grassland using micrometeorological techniques. J. Geophys. Res. 99, 16569–16574.

Hauck R D 1984 Technological approaches to improving the efficiency of nitrogen fertilizer use by crop plants. *In* Nitrogen in Crop Production. Ed. R D Hauck. pp 551–560. Am. Soc. Agron., Madison, WI, USA.

Hoeft R G 1984 Current status of nitrification inhibitor use in US agriculture. *In* Nitrogen in Crop Production. Ed. R D Hauck. pp 561–570. Am. Soc. Agron., Madison, WI, USA.

IAEA 1992 Manual on measurement of methane and nitrous oxide emissions from agriculture. International Atomic Energy Agency. IAEA-TECHDOC-674. IAEA, Vienna, Austria. 91 p.

Isermann K 1993 Territorial, Continental and Global Aspects of C, N, P and S Emissions from Agricultural Ecosystems. *In* NATO Advanced Research Workshop (ARW) on Interactions of C, N, P and S Biochemical Cycles ASI Series. pp 79–121. Springer-Verlag, Heidelberg, Germany.

Jordan C 1989 The effect of fertilizer type and application rate on denitrification losses from cut grassland in Northern Ireland. Fert. Res. 19, 45–55.

Keeney D R 1983 Factors affecting the persistence and bioactivity of nitrification inhibitors. *In* Nitrification Inhibitors- Potentials and Limitations. Eds. J J Meisinger, G W Randall and M L Vitosh. pp 33–46. Am. Soc. Agron., Madison, WI, USA.

Keerthisinghe D G, Freney J R and Mosier A R 1993 Effect of wax-coated calcium carbide and nitrapyrin on nitrogen loss and methane emission from dry-seeded flooded rice. Biol. Fertil. Soils 16, 71–75.

Lijinsky W 1977 How nitrosamines cause cancer. New Sci. 27, 216–217.

Magalhaes A M T, Chalk P M and Strong W M 1984 Effect of nitrapyrin on nitrous oxide emission from fallow soils fertilized with anhydrous ammonia. Fert. Res. 5, 411–421.

McCarty G W and Bremner J M 1986 Inhibition of nitrification in soil by acetylenic compounds. Soil Sci. Soc. Am. J. 50, 1198–1201.

McCarty G W and Bremner J M 1990 Evaluation of 2-ethynylpyridine as a soil nitrification inhibitor. Soil Sci. Soc. Am. J. 54, 1017–1021.

McElroy M P, Wofsy S C and Yung Y L 1977 The nitrogen cycle: Perturbations due to man and their impact on atmospheric N_2O and O_3. Philos. Trans. R. Soc. (London) B. 277, 159–181.

McTaggart I, Clayton H and Smith K 1994 Nitrous oxide flux from fertilized grassland: Strategies for reducing emissions. *In* Proceedings of the Symposium on Non-CO_2 Greenhouse Gases.

Maastricht, December 1993. pp 421–426. Kluwer, Dordrecht, the Netherlands.

Minami K et al. 1990 Effect of nitrification inhibitors on emission of nitrous oxide from soils. Trans. 14th Int. Congress Soil Sci. 2, 267–272.

Minami K and Ohsawa A 1990 Emission of nitrous oxide dissolved in drainage water from agricultural land. *In* Soils and the Greenhouse Effect. Ed. A F Bouwman. pp 503–509. John Wiley and Sons, New York, USA.

Minami K 1994 Effect of nitrification inhibitors and slow-release fertilizer on emission of nitrous oxide from fertilized soils. *In* CH_4 and N_2O Global Emissions and Controls from Rice Fields and Other Agricultural and Industrial Sources. Eds. K Minami, A Mosier and R Sass. pp 187–196. NIAES Series 2. Yokendo Publishers, Tokyo, Japan.

Mosier A R 1989 Chamber and isotope techniques. *In* Exchange of Trace Gases Between Terrestrial Ecosystems and the Atmosphere. Eds. M O Andreae and D S Schimel. pp 175–188. John Wiley and Sons, Chichester, UK.

Mosier A R 1990 Gas flux measurement techniques with special reference to techniques suitable for meas urements over large ecologically uniform areas. *In* Soils and the Greenhouse Effect. Ed. A F Bouwman. pp 289–302. John Wiley and Sons, Chichester, UK.

Mosier A R 1993 Nitrous oxide emissions from agricultural soils. *In* Methane and Nitrous Oxide: Methods in National Emission Inventories and Options for Control Proceedings. Ed. A R van Amstel. pp 273– 285. National Institute of Public Health and Environmental Protection, Bilthoven, The Netherlands.

Mosier A R and Bouwman A F 1993 Working group report: Nitrous oxide emissions from agricultural soils. *In* Methane and Nitrous Oxide: Methods in National Emission Inventories and Options for Control Proceedings. Ed. A R van Amstel. pp 343–346. National Institute of Public Health and Environmental Protection, Bilthoven, The Netherlands.

OECD/OCDE 1991 Estimation of Greenhouse Gas Emissions and Sinks. Final report from the OECD Experts Meeting. 18–21 Feb., 1991. Prepared for Intergovernmental Panel on Climate Change. Revised August, 1991. OECD, Paris, France.

O'Hara G W and Daniel R M 1985 Rhizobial denitrification: a review. Soil Biol. Biochem. 17, 1–9.

Peoples M B, Mosier A R and Freney J R 1994 Minimizing gaseous loss of nitrogen. *In* Nitrogen Fertilization in the Environment. Ed. P E Bacon. pp 565–602. Marcel Dekker Inc., New York.

Poth M and Focht D D 1985 ^{15}N kinetic analysis of N_2O production by Nitrosamonas europae: An examination of nitrifier denitrification. Appl. Environ. Microbiol. 49, 1134–1141.

Prinn R D, Cunnold R, Rasmussen R, Simmonds P, Alyea F, Crawford A, Fraser P and Rosen R 1990 Atmospheric emissions and trends of nitrous oxide deduced from 10 years of ALE-GAGE data. J. Geophys. Res. 95, 18369–18385.

Rodhe H 1990 A comparison of the contribution of various gases to the greenhouse effect. Science (Washington) 248, 1217–1219.

Ronen D, Magaritz M and Almon E 1988 Contaminated aquifers are a forgotten component in the global N_2O budget. Nature 335, 57–59.

Safley L M, Casada M E, Woodbury J N and Roos K F 1992 Global methane emissions from livestock and poultry manure. USEPA report 400/1-91/048. Office of Air and Radiation, Washington, DC, USA.

Sahrawat K L, Keeney D R and Adams S S 1987 Ability of nitrapyrin, dicyandiamide and acetylene to retard nitrification in a mineral and an organic soil. Plant and Soil 101, 179–182.

108

Scharf P C and Alley M M 1988 Nitrogen loss pathways and nitrogen loss inhibitors: a review. Fert. Res. 5, 109–125.

Schlesinger W H and Hartley A E 1992 A global budget for atmospheric NH_3. Biogeochem. 15, 191–211.

Smith K A 1990 Greenhouse gas fluxes between land surfaces and the atmosphere. Progr. Phys. Geography 14, 349–372.

Smith K A, Scott A, Galle B and Klemedtsson L 1994 Use of a long-path infrared gas monitor for measurement of nitrous oxide flux from soil. J. Geophys. Res. 99, 16585–16592.

Smith K A, Clayton H, McTaggart I P, Thomson P E, Arah J R M and Scott A 1995 The measurement of nitrous oxide emissions from soil using chambers. Phil. Trans. R. Soc. London A. (*In Press*).

Wagner-Riddle C, Thurtell G W, Kidd G E, King K M, Sweetman R, Beauchamp E G and Bergstrom D 1994 Efflux of trace greenhouse gases from agricultural sites into the atmosphere. Annual Report to the Ontario Ministry of Environment and Energy, Toronto, Canada.

Walter H M, Keeney D E and Fillery I R 1979 Inhibition of nitrification by acetylene. Soil Sci. Soc. Am. J. 43, 195–196.

Watson R T et al. 1990 Greenhouse gases and aerosols. *In* Climate Change, the IPCC Scientific Assessment. Ed. J T Houghton. pp 1–40. Cambridge Univ. Press, Cambridge, UK.

Watson R T et al. 1992 Climate change 1992, The supplementary reports to the IPCC scientific assessment. *In* Greenhouse Gases: Sources and Sinks. Eds. J T Houghton, B A Callander and S K Varney. pp 25–46. Cambridge Univ. Press, Cambridge, UK.

World Development Report 1992 Development and the Environment. Oxford Univ. Press, New York, NY, USA. 308 p.

Yoshida T and Alexander M 1970 Nitrous oxide formation by *Nitrosomonas europea* and heterotrophic microorganisms. Soil Sci. Soc. Am. Proc. 34, 880–882.

Section editor: R Merckx

O. Van Cleemput et al. (eds.), Progress in Nitrogen Cycling Studies, 603–606, 1996.
© 1996 Kluwer Academic Publishers.

Nitric oxide (NO) emission by higher plants

P. Rockel[1], A. Rockel[1], J. Wildt[2] and H.-J. Segschneider[3]

[1]Institut für Biotechnologie 3, [2]Institut für Chemie und Dynamik der Geosphäre and [3]Institut für Radioagronomie Forschungszentrum Jülich GmbH D - 52425 Jülich, Germany

Key words: ammonium, atmosphere, emission, higher plants, nitrate, nitric oxide

Abstract

In experiments conducted on trace gas emission from plants in a continuously stirred tank reactor we found that several nitrate nourished plant species (sunflower, soybean, corn, spinach, tobacco, sugar cane) emitted NO during the photoperiod of up to $1.1 \cdot 10^{-14}$ mol cm^{-2} s^{-1} (one sided leaf area). Emission rates were independent of nitrate concentrations between 0.07–3.5 mM nitrate in the nutrient solution but decreased when plants were transfered to N-free or ammonium nutrient solution. NO emission was dependent on light flux, increasing with higher light intensities. During the photoperiod similar ratios for the rate of NO emission and the CO_2 uptake were obtained for all plants studied. During darkness all plant species studied except sugar cane evolved small to negligible amounts of NO. Sugar cane evolved NO in high quantities every night. NO emission during darkness could be induced in all plants by adding nitrate to the nutrient solution. In the dark period following application emission rates of up to $3.9 \cdot 10^{-13}$ mol cm^{-2} s^{-1} were measured. Our results imply that NO emission is a physiological characteristic common to all plants and that plants may contribute significantly to the NO-budget of the continental background atmosphere.

Introduction

Gaseous nitrogen oxides (NO_x = NO + NO_2) such as nitric oxide (NO) and nitrogen dioxide (NO_2) play an important role in the chemistry of the atmosphere since they lead to the photochemical production of oxidants such as ozone, influence the radical balance and are precursors of nitric acid (HNO_3) (Williams et al., 1992 and references cited therein). Among the five natural sources of NO in the lower atmosphere (lightning, intrusion from the stratosphere, oxidation of atmospheric ammonia, photolytic processes in oceans, microbial activities in soils, (Williams et al., 1992) that have been identified until now microbial activities are believed to be the most important. Although vegetation has been included in factors affecting the exchange across the soil-atmosphere boundary (Williams et al., 1992) plants have not been taken into account as potential sources of NO in the lower atmosphere. However, during experiments conducted on trace gas emission from plants, we found that several nitrate nourished plant species emitted NO.

Materials and methods

Sunflower (*Helianthus annuus* L. var. giganteus), corn (*Zea mays* L. var. dento), spinach (*Spinacia oleracea* L. var. Monnopa) and soybean (*Glycine max* (L.) Merr.) were germinated in moist quartz sand, tobacco (*Nicotiana tabacum* L. var. Bel W3) in soil at a 14 h photoperiod (280 μE m^{-2} s^{-1}) with a 28/23 °C (day/night) temperature regime and 90/80% RH. Sugar cane (*Saccharum officinarum*) was propagated by cutting pieces from the stem of a soil grown plant and placing them into a nitrate nutrient solution for rooting. After 1–2 weeks plants were transferred to a hydroponic system and grown at 20 °C and 60% RH during a 12 h photoperiod (350 μE m^{-2} s^{-1}) and 17 °C and 80% RH during darkness, respectively. The aerated nutrient solution was automatically adjusted to pH 6.0 by addition of 0.1 N NaOH (NH_4^+-nutrient solution) or 0.1 N H_2SO_4 (NO_3^--nutrient solution). Nutrient concentrations were kept nearly constant by automated addition of 10-fold stock solution. The nutrient solution for nitrate grown plants contained 0.2 mM KNO$_3$, 0.25 mM Ca(NO$_3$)$_2$, 0.05 mM NaNO$_3$, 0.05 mM

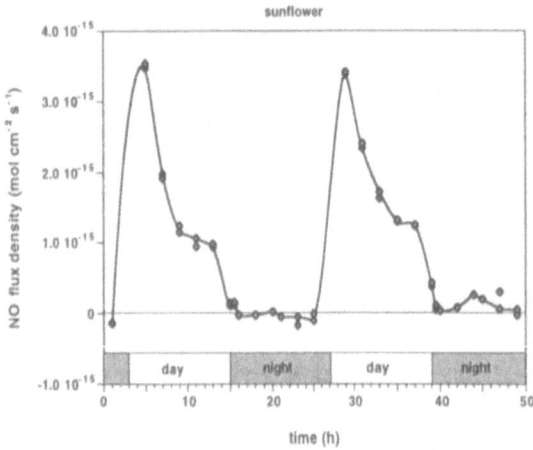

Figure 1. NO emission rates of a nitrate nourished sunflower.

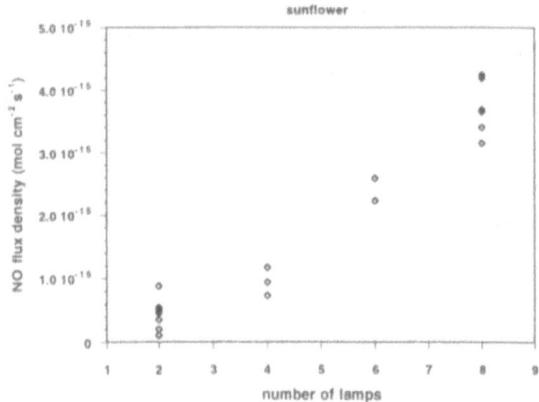

Figure 2. NO emission rates of a nitrate nourished sunflower as a function of light intensities (number of lamps).

KH$_2$PO$_4$, 0.1 mM MgSO$_4$, the NH$_4^+$-nutrient solution 0.05 mM (NH$_4$)$_2$HPO$_4$, 0.35 mM (NH$_4$)$_2$SO$_4$, 0.125 mM K$_2$SO$_4$, 0.1 mM MgSO$_4$, 0.25 mM CaCl$_2$, N-free nutrient solution 0.05 mM KH$_2$PO$_4$, 0.125 mM K$_2$SO$_4$, 0.1 mM MgSO$_4$, 0.25 mM CaCl$_2$. Micronutrient concentrations in all solutions were 10 μM KCl, 6,3 μM H$_3$BO$_3$, 4,0 μM FeH$_2$-EDTA, 0.4 μM ZnSO$_4$·7H$_2$O, 0.4 μM MnSO$_4$·H$_2$O, 0.1 μM Na$_2$MoO$_4$·2H$_2$O, 0.1 μM CuSO$_4$·5H$_2$O.

After plants had reached a leaf area of about 500–1000 cm^2 they were transferred to the experimental setup for NO measurements. The upper part of the plant was placed into a continuously stirred tank reactor, a glass vessel of 164 l volume through which purified air was passed with a constant flow of 40 L/min. A teflon plate sealed the glass vessel from the environment and held the plant in position. Background concentrations of ozone, NO and NO$_2$ of the inflowing air were less than 100 ppt. Concentrations of NO, NO$_2$, water vapor, CO$_2$ and ozone of the inflowing and outflowing air were measured alternately at 1 h intervals as described in Neubert et al. (1993). NO emission fluxes were calculated from the concentration differences between the inlet and outlet of the chamber.

The roots of the plant hung in a hydroponic system especially designed for this setup, in which O$_2$, pH and nutrient concentrations, i.e. nitrate or ammonium concentration were continuously controlled and regulated.

Results

NO emission during the photoperiod

At constant light fluxes during the photoperiod (820 μE m^{-2} s^{-1}) and ambient NO concentrations below 1 ppb all nitrate nourished plants emitted NO up to $1.1 \cdot 10^{-14}$ mol cm^{-2} s^{-1} (one sided leaf area). Plants which had solely received ammonium or were held under N deficient conditions did not evolve NO. The NO emission rates showed daily fluctuations with higher rates shortly after the light had been switched on and decreased during the course of the day (Fig. 1).

Emission rates in the light were independent of nitrate concentrations between 0.07–3.5 mM nitrate in the nutrient solution but decreased when plants adapted to 0.75 mM nitrate in the nutrient solution were transferred to a N-free or ammonium nutrient solution. In this case NO emission during the photoperiod declined from 100% to values of about 40% and 15% in the following days. Three days after this transfer no measurable quantities of NO were detected. On resupply of nitrate (0.75 mM) NO emission during the photoperiod was detectable again after 4–5 h.

Dependence on light flux and ambient CO$_2$ concentration

NO emission was dependent on light intensity: rising light flux densities resulted in increased NO emission which superimposed the emission pattern found at constant light flux densities (Fig. 2).

At constant light intensity during the photoperiod similar ratios for the rate of NO emission and CO$_2$ uptake were obtained for all plants. At a light intensity

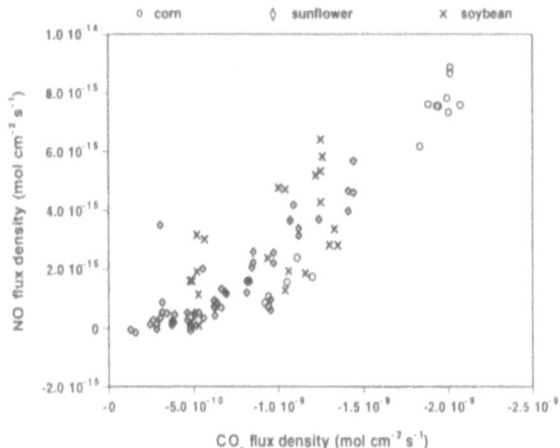

Figure 3. NO flux density as a function of CO_2 uptake for different plants (corn, sunflower, soybean).

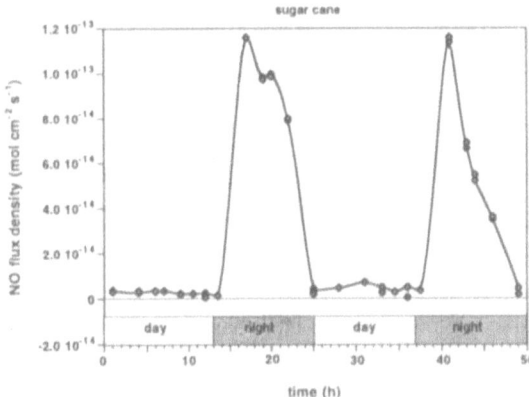

Figure 4. NO emission rates of a nitrate nourished sugar cane.

of 820 μE cm^{-2} s^{-1} the mean value for the ratio of NO flux and CO_2 uptake was $(3.1 \pm 0.6) \cdot 10^{-6}$. Thus, it was assumed that CO_2 uptake and NO emission are related. This was confirmed by changing the rate of net photosynthesis by varying light intensity or CO_2 concentrations. Enhancement or reduction of the rate of net photosynthesis correspondingly changed NO emission (Fig. 3).

NO emission during darkness

During darkness all plants except sugar cane emitted small to negligible amounts of NO. In the case of sugar cane, emission was observed every night with rates more than tenfold higher than during the photoperiod (Fig. 4).

However, it was possible to induce NO production during darkness in the other plants by increasing the nitrate supply. Then emission rates of up to $3.9 \cdot 10^{-13}$

mol cm^{-2} s^{-1} were measured in the dark period following application. A transfer of plants from N-free or ammonium nutrition to nitrate nutrition resulted in even more pronounced emission.

NO emission from soil grown plants

During the phototperiod, NO flux densities of soil grown plants had the same order of magnitude as those of hydroponically grown plants. In addition, NO emission during darkness could also be induced when soil grown plants were wetted with nitrate nutrient solution.

Discussion

NO emission by plants under normal conditions has not been observed as such until now. Klepper was able to evoke NO$_x$ emission by treating soybean plants with phototsynthetic inhibitor herbicides (Klepper, 1974, 1978, 1979) or certain chemicals (Klepper, 1990, 1991) as well as under dark anaerobic conditions (Klepper 1987, 1990). He suggested that this emission was due to chemical reactions of accumulated nitrite with plant metabolites such as salicylate derivates or the chemical decomposition of HNO$_2$.

However, NO was also observed to be the predominant compound evolved during a purged in vivo assay with soybean nitrate reductase derived from accumulated nitrite (Harper, 1981). Results obtained with boiled leaflets (Harper, 1981) and a mutant soybean line (nr$_1$) (Dean and Harper, 1986, 1988; Nelson et al., 1983; Ryan et al., 1983) further indicated that the enzymic reaction of a constitutive NAD(P)H:nitrate reductase is responsible for evolution of NO$_x$. Experiments with ^{15}N-labelled nitrate as substrate for nitrate reduction showed that NO$_x$ is produced from ^{15}N-NO$_3$$^-$ (Dean and Harper, 1986).

On the other hand, Siegel and Wilkerson (1989 and references cited therein) reported that the predominant enzymic species during the turnover of spinach nitrite reductase with nitrite and reductants is an Fe^{II}-sirohaem-NO-complex. This complex persists in solutions when the turnover occurs in the presence of reductants and excess nitrite. It might therefore also be possible that a nitrite reductase isoenzyme with properties similar to that of the bacterial nitrite reductase exists which is able to reduce nitrite to NO.

Our results that only nitrate nourished plants evolve NO are in accordance with Klepper's (1978) observation that only nitrate nourished herbicide treated but not ammonium nourished plants emitted NO. This and the possibility to diminish or increase and to induce NO emission by nitrate withdrawal or supply supports Dean and Harper's (1986) results that NO is produced during nitrate reduction. However, contrary to Dean and Harper's (1988) postulation that the ability to evolve NO is restricted to the *Phaseoleae* tribe of the familiy *Leguminosae* due to the unique occurrence of this isoenzyme, our results imply that NO emission is a physiological characteristic common to all plants.

The release of NO from natural sources is mainly attributed to the activity of soil microorganisms. However, NO emission rates from plants are comparable to the highest emission rates found for recently fertilized soils (up to 2.4×10^{-12} mol $cm^{-2}s^{-1}$, Williams et al., 1992) considering that the leaf areas of plants usually exceed the area of soil they cover and increases in nitrate supply cause emission rates from plants of up to 3.9×10^{-13} mol $cm^{-2} \, s^{-1}$. Therefore, NO emission by plants may contribute significantly to the NO budget of the continental background atmosphere, although less than 0.1% of the N taken up by plants is volatilized as NO.

References

Dean J V and Harper J E 1986 Nitric oxide and nitrous oxide production by soybean and winged bean during the in vivo nitrate reductase assay. Plant Physiol. 82, 718–723.

Dean J V and Harper J E 1988 The conversion of nitrite to nitrogen oxide(s) by the constitutive NAD(P)H-nitrate reductase enzyme from soybean. Plant Physiol. 88, 389–395.

Harper J E 1981 Evolution of nitrogen oxide(s) during in vivo nitrate reductase assay of soybean leaves. Plant Physiol. 68, 1488–1493.

Klepper L 1974 A mode of action of herbicides: inhibition of nitrite reduction. Nebraska Exp. Sta. Res. Bull. 259, 1–42.

Klepper L 1978 Nitric oxide (NO) evolution from herbicide-treated soybean plants. Plant Physiol. Sup. 61, 65.

Klepper L 1979 Nitric oxide (NO) and nitrogen dioxide (NO_2) emissions from herbicide-treated soybean plants. Atmos. Environ. 13, 537–542.

Klepper L A 1987 Nitric oxide emissions from soybean leaves during in vivo nitrate reductase assays. Plant Physiol. 85, 96–99.

Klepper L 1990 Comparison between NO_x evolution mechanisms of wild-type and nr_1 mutant soybean leaves. Plant Physiol. 93, 26–32.

Klepper L 1991 NO_x evolution by soybean leaves treated with salicylic acid and selected derivatives. Pestic. Biochem. Physiol. 39, 43–48.

Nelson R S, Ryan S A and Harper J E 1983 Soybean mutants lacking a constitutive nitrate reductase activity. I. Selection and initial plant characterization. Plant Physiol. 72, 503–509.

Neubert A, Kley D, Wildt J, Segschneider H J and Förstel H 1993 Uptake of NO, NO_2 and O_3 by sunflower (*Helianthus annuus* L.) and tobacco plants (*Nicotiana tabacum* L.): Dependence on stomatal conductivity. Atmos. Environ. 27A, 2137–2145.

Ryan S A, Nelson R S and Harper J E 1983 Soybean mutants lacking constitutive nitrate reductase activity. II. Nitrogen assimilation, chlorate resistance, and inheritance. Plant Physiol. 72, 510–514.

Siegel L M and Wilkerson J O 1989 Structure and function of spinach ferredoxin-nitrite reductase. *In* Molecular and Genetic Aspects of Nitrate Assimilation. Eds. J L Wray and J R Kinghorn. pp 263–283. Oxford Science Publications, Oxford, New York, Tokyo.

Williams E J, Hutchinson G L and Fehsenfeld F C 1992 NO_x and N_2O emissions from soil. Global Biogeochem. Cycles 6, 351–388.

O. Van Cleemput et al. (eds.), Progress in Nitrogen Cycling Studies, 607–611, 1996.
© 1996 *Kluwer Academic Publishers*.

Nitrous oxide formation in black earth soils depending on the soil water content

R. Russow and M. Körschens
UFZ - Centre for Environmental Research Leipzig-Halle, Department of Soil Research, Leipzig, Germany

Key words: black earth, denitrification, nitrification, nitrogen-15, nitrous oxide, water content

Abstract

Normally nitrous oxide (N_2O) is formed by biological denitrification under anaerobic conditions in soils, but recent results have indicated that nitrous oxide can also be formed as a byproduct of the microbial nitrification. This is important particularly for the black earth soil in Central Germany because of the absence of typical denitrification conditions under semi-arid climate conditions. Measurements of the N_2O liberation by this soil (Haplic Phaeozem) in relation to the soil water content showed in general the following results: - below 80% of the water holding capacity (WHC) (aerobic conditions) the N_2O emission is on a nearly constant low level (≤ 0.4 μg N kg^{-1} soil h^{-1}); - above 80% WHC (anaerobic conditions) the N_2O emission increases exponentially and reaches up to 28 μg N kg^{-1} soil h^{-1} at 90–95% WHC. These findings indicate that there are probably two different paths for the N_2O formation in soils. First studies on Haplic Phaeozem using [^{15}N]ammonium and [^{15}N]nitrate and a GC-MS aided incubation system confirmed the assumption given above as follows:- under anaerobic conditions ($\geq$90% WHC) N_2O originates mainly from the nitrate pool by denitrification; - under aerobic conditions ($\leq$ 80% WHC) the N_2O formation is lower, but the gas originates directly from the ammonium and not from the nitrate pool, probably as a byproduct of the nitrification process.

Introduction

Soils, especially those which are fertilized with high rates of nitrogen, contribute to a considerable extent to the emission of nitrous oxide (N_2O) an environmentally important trace gas in the atmosphere (green house effect, destruction of the ozone layer). Furthermore the N_2O emission (together with the N_2 release) can lead to a remarkable loss of fertilizer N. This loss is estimated of up to 3% ($\approx$ 3 mill. tons N a^{-1}) of the fertilizer N world-wide applicated (Isermann, 1993; Ottow, 1991; Umarov, 1989).

Under anaerobic conditions, biological denitrification is well known as a process of N_2O formation in soils. But recent results have indicated that nitrous oxide can also be formed as a byproduct of microbial nitrification (Davidson, 1992; Firestone and Davidson, 1989) (Fig. 1). This is important particularly for the black earth soil in Central Germany because of the absence of typical denitrification conditions in a semi-arid climate (e.g. long-standing average annual precipitation 480 mm). There is little information about

Table 1. Preparation of the soil samples incubated

Parameters	Anaerobic A	Aerobic E
Soil mass (g)	100	100
Incubat.-temp. (°C)	25	25
relative WHC (%)	95–100	75–80
^{15}N-labelling	KNO$_3$	KNO$_3$
15 mg N[a] (95 at.-% per vessel)	(NH$_4$)$_2$SO$_4$	(NH$_4$)$_2$SO$_4$

[a] Corresponds to about 300 kg N ha^{-1} with a soil bulk density = 1,35 g cm^{-3} and a soil layer = 15 cm

denitrification, especially N_2O formation in this black earth.

Materials and methods

To investigate the influence of the soil water content on the amount and the path of N_2O-formation, laborato-

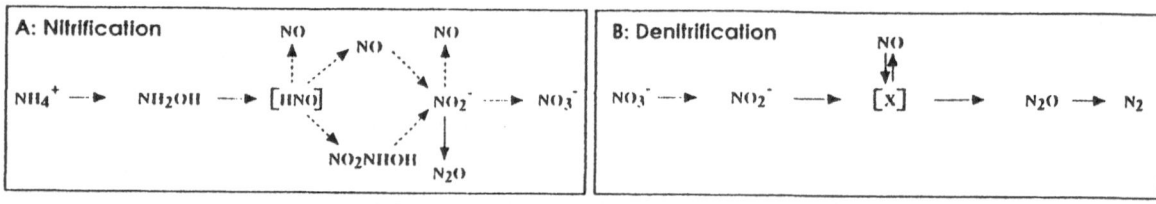

Figure 1. Mechanism of microbiological nitrification and denitrification (according Davidson, 1992; Fireston and Davidson, 1989)

N2O, μg N/kg soil·h

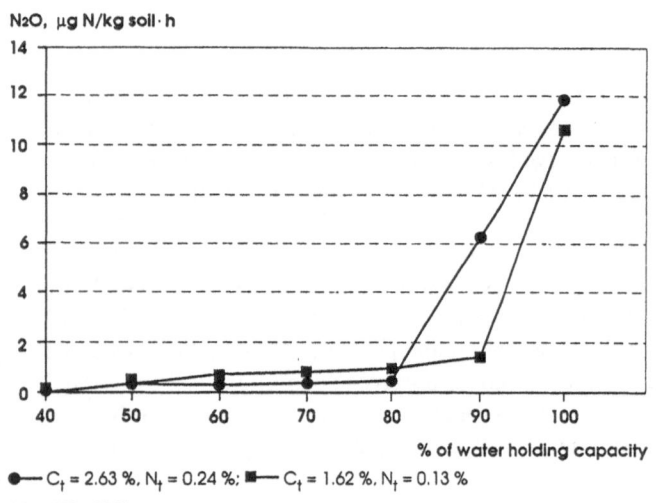

● — C_t = 2.63 %, N_t = 0.24 %; ■ — C_t = 1.62 %, N_t = 0.13 %

SD = ±30 - 80 %

Figure 2. Nitrous oxide emission from soil vs. soil water content C_t - total carbon, N_t - total nitrogen.

ry incubation experiments with [¹⁵N]ammonium- and [¹⁵N]nitrate-labelled soils were conducted (Table 1):

 — Soil type: Haplic Phaeozem
 — Soil form: well textured loess-black earth
 — Total C (C_t): 1.6; 2.6 %
 — Total N (N_t): 0.13; 0.24 %

The experiments were carried out with a special analytical installation where the incubation vessels are directly connected via a gas injection device with a combined gas chromatograph quadrupole mass spectrometer system (GCQMS, Shimadzu QP 2000) (see details Russow et al., 1993, 1994). This system allows the simultaneous determination of the N_2O concentration and its ¹⁵N abundance in the incubation atmosphere. Before and after incubation the inorganic N forms, ammonium and nitrate, as well as their ¹⁵N abundance (Emission Spectrometer NOI-6 PC, FAN, Germany) were measured in the soil sample.

Results and discussion

Relation between nitrous oxide and soil water content

In general the results concerning the influence of the water content on the N_2O formation in black soils show that, when a water saturation of 80–90% of the maximum water holding capacity (max. WHC) has been reached, a drastic increase in the N_2O emission occurs (Fig. 2) and extreme values of up to 12 μg N kg⁻¹ h⁻¹ (corresponds to ≈ 210 kg N ha⁻¹ y) can be attained. Below this threshold value (≤ 80% max. WHC), only a negligible base emission of 0.2–0.4 μg N kg⁻¹ h⁻¹ (corresponds to ≈ 3.5–7 kg N ha⁻¹ y) appears. This indicates two different formation paths, depending on the water content. In the following, more detailed information about possible formation mechanisms should be derived from the ¹⁵N tracer experiments conducted.

The paths of the N_2O formation

The use of ¹⁵N labelled ammonium and nitrate as tracers permits the investigation of the fate of the ammonium and nitrate nitrogen in the soil. The results of this experiments are shown in Table 2 and in Figure 3. They should be discussed on the basis of an principal reaction scheme (Fig. 4) according to the kinetic isotope method introduced in 1943 by Zilversmit et al., (Rev. Nejman and Gal, 1971).

Because there are no temporal intermediate values for the inorganic soil nitrogen (Table 2), only a qualitative estimate of the proportions of the individual reaction paths is possible. This estimate can be made by comparison the ¹⁵N abundance of the participating compounds ammonium (A), nitrate (N) and nitrous oxide (O) (Table 3, see details Nejman and Gal, 1971 and Russow et al., 1994).

Table 2. Inorganic nitrogen at the beginning (B) and the end (E) of incubation

Incubation		Ammonium-N		Nitrate-N	
		(mg kg^{-1} soil)	(at.–%)	(mg kg^{-1} soil)	(at. –%)
Anaerobic	B	156	92.7	40	0.37
(^{15}N) Amm.	E	50	61.5	165	45.0
Anaerobic	B	5.7	0.37	190	75.3
(^{15}N) Nitr.	E	23.5	12.0	140	65.0
Aerobic	B	155	93.0	40.7	0.37
(^{15}N) Amm.	E	6.8	29.0	235	53.1
Aerobic	B	5.7	0.37	187	75.8
(^{15}N) Nitr.	E	7.7	20.0	251	55.3

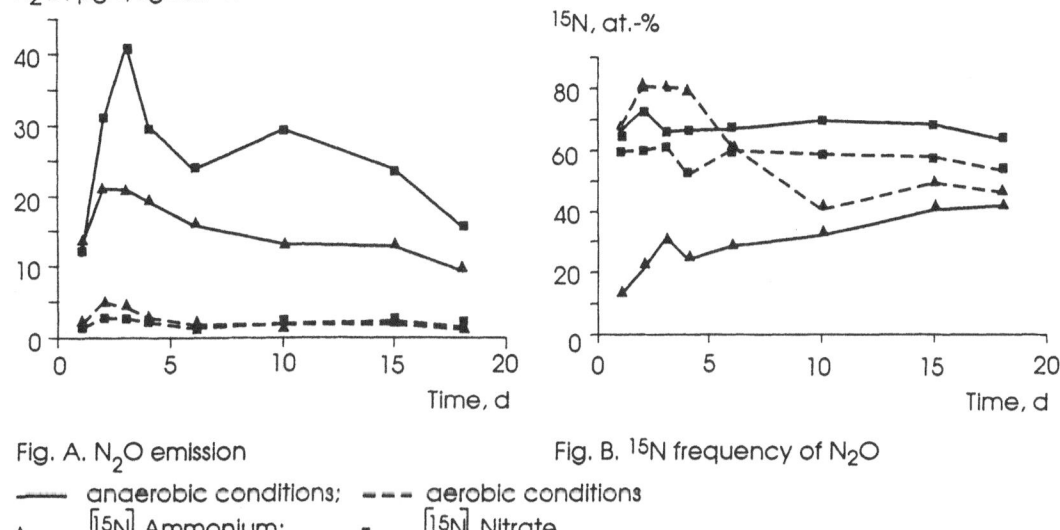

Fig. A. N$_2$O emission
Fig. B. ^{15}N frequency of N$_2$O

—— anaerobic conditions; – – – aerobic conditions
▲ $[^{15}$N$]$ Ammonium; ■ $[^{15}$N$]$ Nitrate

Figure 3. N$_2$O production (**A**) and ^{15}N abundance (**B**) from soil (N$_t$ = 0.24% w/w) for two forms of N fertilizer (150 mg N kg^{-1} soil) vs. incubation duration.

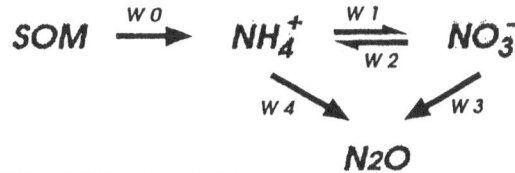

SOM - Soil Organic Matter
W 0 - Mineralization
W 1 - Nitrification
W 2 - Nitrate Ammonification
W 3 - Denitrification
W 4 - Ammonium - [NO] - Path

Figure 4. Principal scheme of possible paths for N transformation in soils related to N$_2$O formation

In summary this qualitative analysis allows the following statements:

* **Mineralization (w 0)**: During incubation a considerable mineralization of active soil organic matter to ammonium (W 0) has taken place, which was strongly promoted by aerobic conditions (dilution of the initial ammonium ^{15}N abundance).

* **Nitrification (w 1)**: Nitrification is clearly recognizable by the increase in nitrate concentration (Table 2) and the decrease or increase of the ^{15}N abundance in nitrate if [^{15}N]nitrate or ammonium has been applied, respectively (Table 3). It is surprising that even under anaerobic conditions (water saturation of the soil) clearly measurable nitrification occurs.

Table 3. Comparison of the ^{15}N abundance of ammonium (A), nitrate (N) and nitrous oxide (X) measured

Condition	(^{15}N) Ammonium (at.– %)			(^{15}N) Nitrate (at. – %)	
Anaerobic	$A_B = 93$			$N_B = 75$	
	$A_E = 62$	$\frac{\Delta A}{\Delta t} < 0$		$N_E = 65$	$\frac{\Delta N}{\Delta t} < 0$
	$N_E = 45$			$A_E = 12$	
	$X = 20-42$			$X = 70 - 60$	
	$A > N > X$			$X \approx N >> A$	
Aerobic	$A_B = 93$			$N_B = 76$	
	$A_E = 29$	$\frac{\Delta A}{\Delta t} < 0$		$N_E = 55$	$\frac{\Delta N}{\Delta t} < 0$
	$N_E = 53$			$A_E = 20$	
	$X = 80 - 55$			$X = 65 - 53$	
	$X > N > A$			$X \approx N > A$	
Comparison anaerobic aerobic	$X^e > X^a$			$X^a > X^e$	

B – Beginning of incubation; E – End of incubation; a – anaerobic, e – aerobic.

* **Nitrate-Ammonification (w 2)**: The relatively high ^{15}N abundance of ammonium if labelled with [^{15}N]nitrate indicates a considerable contribution of this microbial process to nitrate decomposition. This observation is in contrast to the literature which reports that dissimilatory nitrate ammonification only exists in waterlogged sedimentary areas or soils with high available carbon content (e.g. Fazzolari et al., 1990; Stanford et al., 1975).

* **Path of N$_2$O formation (w 1-w3/4)**: Anaerobic Condition ($\geq$ 90% max. WHC): The comparison of the ^{15}N abundances of the relevant compounds under anaerobic conditions supports the reaction path 0–1-3 (Fig. 4), because the ^{15}N abundance of the nitrous oxide is similar to that of nitrate with [^{15}N]nitrate-labelling but significantly smaller in the case of [^{15}N]ammonium-labelling. The high N$_2$O formation under anaerobic conditions occurs via the nitrate pool of the soil, i.e. by denitrification. As expected, the ^{15}N abundance of nitrous oxide produced with nitrate-labelling is, therefore, greater here than under aerobic conditions.

– Aerobic condition ($\leq$ 80% max. WHC): The ^{15}N abundance of nitrous oxide produced with ammonium-labelling is greater than that of nitrate and also greater in comparison to that obtained under anaerobic conditions. These findings are consistent with reaction path 0–4. That means that the low but constantly N$_2$O base emission under aerobic conditions does not occur preferentially via the nitrate pool of the soil. It is probably a by-product of nitrification. Aerobic conditions permits a high nitrate production, but the release of the small amounts of N$_2$O must take place before nitrate formation via the nitrification chain.

– One result can not be placed into the findings summarized above: The ^{15}N abundance of the nitrous oxide in the case of [^{15}N]nitrate-labelling under aerobic conditions is in contrast to the expectations equal to that of nitrate and greater than that of the ammonium (Falsification by nitrate ammonification and/or remineralization?).

* **Total denitrification (N$_2$ + N$_2$O)**: To quantify the total gaseous N losses an additional pot experiment with extremely high fertilizer rates (double ^{15}N marked ammonium nitrate) up to 500 mg N kg^{-1} loess-black earth soil at 60% WHC was carried out. After a four months vegetation period (green oats: 27.04.-25.06.; cruciferous species: 26.06.-27.08.) results were calculated for ^{15}N and total N (N$_t$) balance, respectively.

Although it is not possible to differentiate the nitrogen into distinctive N fractions from the results of the pot experiment a considerable extent of gaseous

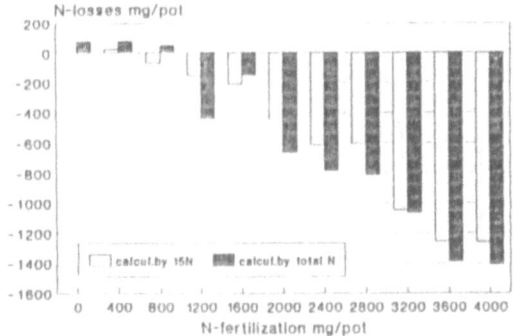

Figure 5. Gaseous N-losses determined in a pot experiment with black earth using ^{15}N in comparison to total N balance; Vegetation period 27.04.-27.08.

losses are obvious (Fig. 5). The positive values (N input) calculated by N_t for the first three treatments can be ascribed to an additional N input by "other sources" (e.g. air born N deposition). Losses amount up to 35% (N_t) and 32% (^{15}N) of the fertilizer N, respectively. E.g. for a dose of 1200 mg N/pot ($\approx$ 440 kg N ha^{-1} at 5.5 kg soil/pot) the loss is in average 300 mg N, that means about 110 kg N ha^{-1}. The results show that under aerobic conditions gaseous N losses can be high if large amounts of inorganic N and easily convertible carbon are available. The latter were supplied in the pot experiment by the accumulation of roots and roots exudates.

References

Davidson E A 1992 Sources of nitric and nitrous oxide following wetting of dry soil. Soil Sci. Soc. Am. J. 56, 95–102.

Fazzolari E, Mariotti A and Germon J C 1990 Dissimilatory ammonium production vs. denitrification in vitro and in inoculated soil samples. Can. J. Microbiol. 36, 786–793.

Firestone M K and Davidson E A 1989 Microbiological basis of NO and N$_2$O production and con- sumption in soil. *In* Exchange of Trace Gases between Terrestrial Ecosystems and the Atmosphere, Dahlem Konferenz. Eds. M O Andreae and D S Schimel. pp 7–21. John Wiley and Sons, Chichester, UK.

Isermann K 1993 Territorial, continental and global aspects of C, N, P,and S emission from agricultural ecosystems. NATO ASI Ser. 14, 79–121.

Nejman M B and Gal D 1971 The kinetic isotope method and its application. Akademiai Kiado, Budapest, Hungary.

Ottow J C G 1991 Denitrifikation, eine kalkulierbare Größe der Stickstoffbilanz? Ergebnisse land- wirtschaftlicher Forschung an der Justus-Liebig-Universität, Volumen XX, Gießen, Germany.

Russow R and Förstel H 1993 Use of GC-QMS for stable isotope analysis of environmentally relevant main and trace gases in the air. Isotopenpraxis Environ. Health Stud. 29, 327–334.

Russow R, Höfer M and Faust H 1994 ^{15}N tracer studies on nitrous oxide formation in soils: On the mechanism of the N$_2$O formation. Isotopenpraxis Environ. Health Stud. 30, 157–164.

Stanford G, Legg J O, Dezienia S and Simpson E C 1975 Denitrification and associated nitrogen transformations in soils. Soil Sci. 120, 147–152.

Umarov K 1989 Biotic sources of nitrous oxide in the context of the global budget of nitrous oxide. *In* Soils and the Greenhouse Effect. Ed. A F Bouwman. pp 263–268. John Wiley and Sons, Chichester, UK.

O. Van Cleemput et al. (eds.), Progress in Nitrogen Cycling Studies, 613–619, 1996.

N$_2$O emission in selected representative soils of the North-Eastern Iberian Peninsula

E. Saguer and M.A. Gispert
Universitat de Girona Department E.Q.A.T.A., Crop and Soil Sciences Section, Avda. Lluís Santaló s/n., 17003 Girona, Spain

Key words: C/N, nitrate, nitrous oxide losses, organic matter, soil water content

Abstract

Soils of sixteen different sampling locations of the north-east part of the Iberian Peninsula have been chosen according to their diversity in parent material and plant cover in order to study the most important parameters governing soil N$_2$O emissions. Five samples were collected in each sampling location. Soil water content (SWC), pH, organic matter (OM), total nitrogen, C/N ratio and nitrate content were analysed to establish their relationships with N$_2$O emissions. Significant correlations were observed between N$_2$O losses and nitrates ($p<0.01$), soil water content ($p<0.01$) and pH ($0.01<p<0.05$). A negative correlation was observed with particles size fraction >2mm ($0.01<p<0.05$). A regression model was developed and showed that the parameters more closely related with N$_2$O losses were NO$_3$ content, OM content and, in minor extent C/N ratio.

Introduction

N$_2$O is a well known gas because of its participation to the depletion of stratospheric ozone. Bouwman (1990) suggested that soils contribute with over 50% in the global N$_2$O emission. The edaphic processes which participate to the production of these emissions are either biotics or abiotics (Firestone and Davidson, 1989). The main biological processes of N$_2$O formation are the nitrification and denitrification. It was reported that in both processes, the N$_2$O production in the field is mainly controlled by the oxygen status in the soil (Yoshida and Alexander, 1970). However, the production of N$_2$O may also be favoured by chemical reactions which take place under fully aerobic conditions and, in this case, the process is generally called chemodenitrification.

Several authors have shown that SWC, OM, C/N, pH and nitrate content represent some of the edaphic parameters that may affect the above mentioned processes (Goodroad and Keeney, 1984; Hutchinson et al., 1993).

The objective of the present work is the assessment of the parameters that may be involved in the production of N$_2$O losses from the soil. In order to have a sufficiently wide range for each studied parameter (else-where in this work called variable) we have chosen 16 soil types differing in parent material and vegetation. The physico-chemical characterization data were statistically elaborated in order to find correlations with N$_2$O losses from the soil.

Material and methods

In order to get information about N$_2$O emission capacities of different soil types located in the north-east part of the Iberian Peninsula, we have chosen sixteen different sampling locations according to their diversity in parent material, vegetation and microclimatological conditions (Table 1).

Five undisturbed soil cores were collected in each sampling site by using a steel cylinder of 12 cm long and 6 cm wide so that a total of 80 samples was obtained. The cores were incubated for 24 hours at 20 °C in a close-shut glass jar. After incubation a gas sample was extracted and analysed by ECD Gas-Chromatography to quantify N$_2$O losses. Experimental conditions during gas-chromatography analyses were as follows: temperature of injector 25 °C, temperature of the oven 40 °C, temperature of the detector 350 °C, carrier Ar (95%) + CH$_4$ (5%) 40 mL min $^{-1}$. A soil

Table 1. Principal characteristics of the studied areas

Soil	Zone	Parent material	Geomorphology	Vegetation
1	Albera	Gneiss	Slope: 15%	Beech forest
2	Estanys de Capmany	Granite	Plane	Rushlike
3	St. Climent Sasebes	Granodiorite	Plane	Vineyard
4	Aiguamolls de Vilaüt	Alluvial deposits	Plane	Uncultivated
5	Crosa de St. Dalmai	Volcanic material (Piroclast)	Slope: 10%	Typic evergreen oak
6	St. Hilari Sacalm	Granite	Slope: 5%	Cork oak grove
7	Massanet de la Selva	Volcanic material (Basalt)	Plane (Slope: 1%)	Mixed forest
8	St. Grau d'Ardenyà	Granite	Slope: 5%	Cork oak grove
9	Beget	Marls	Plane (Slope: 1%)	Meadow
10	La Pastura	Gneiss	Plane	Meadow
11	Setcases	Sandstone	Slope: 5%	Meadow
12	Coll de Merolla	Marls	Plane (Slope: 1%)	Meadow
13	Adri	Volcanic material (Basanite)	Plane (Slope: 1%)	Evergreen oak
14	Serra del Corb	Volcanic material (Basanite)	Plane	Beech forest
15	Fageda d'en Jordá	Volcanic material (Basanite)	Plane	Beech forest
16	St. Jaume de Llierca	Alluvial deposits	Plane	Uncultivated

Table 2. Physico-chemical characteristics of the different soil types

Soil	Part.>2mm (%)	pH	OM (%)	SWC (%)	Nt (%)	C/N	NO$_3$ (mg kg^{-1} dw)
1	20.8 ± 9.6	4.0 ± 0.2	14.3 ± 7.8	25.6 ± 10.7	0.34 ± 0.09	17.4 ± 2.6	3.0 ± 3.5
2	0.0 ± 0.0	5.4 ± 0.9	8.5 ± 6.2	64.3 ± 60.3	0.36 ± 0.33	12.7 ± 5.8	18.1 ± 63.3
3	22.3 ± 5.6	5.7 ± 1.9	1.4 ± 0.4	3.2 ± 0.8	0.11 ± 0.01	7.0 ± 2.2	26.2 ± 13.9
4	0.0 ± 0.0	6.9 ± 0.1	11.2 ± 6.0	74.8 ± 10.7	0.66 ± 0.15	8.8 ± 2.7	138.0 ± 137.9
5	40.7 ± 13.7	5.7 ± 0.5	8.0 ± 3.7	14.6 ± 5.1	0.33 ± 0.12	13.4 ± 1.9	12.5 ± 21.0
6	30.6 ± 10.6	5.1 ± 0.5	5.9 ± 5.1	6.3 ± 3.3	0.20 ± 0.07	15.3 ± 6.2	0.8 ± 1.0
7	18.2 ± 16.2	5.0 ± 0.9	14.4 ± 11.6	23.9 ± 11.2	0.43 ± 0.20	16.8 ± 5.3	1.7 ± 1.6
8	25.8 ± 11.6	5.5 ± 0.7	4.5 ± 1.6	6.6 ± 4.8	0.20 ± 0.09	13.0 ± 1.9	1.3 ± 2.6
9	33.1 ± 13.9	6.9 ± 0.2	5.9 ± 3.3	13.4 ± 1.4	0.38 ± 0.09	8.8 ± 3.4	7.4 ± 7.0
10	31.4 ± 25.4	4.9 ± 0.3	18.7 ± 4.5	29.3 ± 14.4	0.98 ± 0.30	9.6 ± 1.1	38.8 ± 58.7
11	23.1 ± 11.6	5.1 ± 0.1	14.0 ± 7.1	27.5 ± 14.4	0.73 ± 0.44	10.1 ± 2.1	43.4 ± 63.4
12	5.4 ± 13.5	7.2 ± 0.5	7.8 ± 6.9	18.6 ± 9.0	0.45 ± 0.34	9.6 ± 0.4	20.5 ± 35.1
13	26.7 ± 18.7	5.5 ± 1.7	7.9 ± 3.4	12.9 ± 4.4	0.30 ± 0.15	14.9 ± 6.4	0.0 ± 0.0
14	25.1 ± 12.8	5.3 ± 1.4	20.2 ± 4.9	37.6 ± 21.8	0.80 ± 0.08	12.6 ± 2.1	27.8 ± 36.9
15	33.8 ± 34.1	5.4 ± 0.6	19.8 ± 8.6	34.8 ± 24.1	0.64 ± 0.47	17.6 ± 9.1	6.5 ± 10.4
16	7.9 ± 29.1	5.5 ± 0.2	1.5 ± 0.2	6.8 ± 1.3	0.14 ± 0.02	6.6 ± 2.1	0.0 ± 0.0

sample was collected in the outer part of each sampling site (just close to the respective soil core hole) in order to determine physico-chemical characteristics of each area of sampling. According to Methods of Soil Analysis (1982), pH, particle size fraction >2 mm, soil water content (SWC), organic matter (OM) and total nitrogen (Nt) were determined whilst nitrate content was determined colorimetrically by using the Aplkem FIA autoanalyzer.

Table 3. Descriptive of the soil parameters studied. Normality of distributions was tested according to Shapiro-Wilk test

Parameter	Mean	St. Error	CV	Median	Mode	W:Normal	Prob<W
Part.>2mm (%)	21.78	1.73	69.91	23.18	0.0	0.9358	0.0012
pH	5.57	0.1	16.34	5.4	5.1	0.9260	0.0002
SWC (%)	24.65	2.41	86.57	16.11	2.62	0.8238	0.0001
OM (%)	10.44	0.78	65.45	8.36	23.0	0.9069	0.0001
Nt (%)	0.45	0.03	61.71	0.37	0.11	0.9039	0.0001
C/N	12.16	0.47	34.3	11.97	6.64	0.9552	0.0260
NO_3 (mg kg^{-1} dw)	21.81	4.82	196.28	35.0	0.0	0.5739	0.0000
N_2O (g ha^{-1} day^{-1})	42.44	4.79	99.71	32.62	9.89	0.5583	0.0000

Table 4. Values of the Shapiro-Wilk test over the transformed variables

Parameter	W:Normal	Prob<W
log SWC	0.9593	0.0468
log N_2O	0.9612	0.0624
log (NO_3+1)	0.8645	0.0001
log OM	0.9049	0.0001
log Nt	0.9294	0.0003
log (C/N)	0.9793	0.5696
1/OM	0.6695	0.0000
1/Nt	0.8343	0.0001

Table 5. Summary of mean separation test for N_2O losses

Soil	Zona	Mean[a]
4	Aiguamolls de l'Empordà	2.078 a
10	La Pastura	1.664 b
8	St. Grau d'Ardenyà	1.654 b
5	Crosa de St. Dalmai	1.654 b
12	Coll de Merolla	1.594 b
11	Setcases	1.570 b
9	Beget	1.560 b
6	St. Hilari Sacalm	1.552 b
2	Estanys de Capmany	1.504 b
1	Albera	1.452 b
3	St. Climent Sasebes	1.432 b
16	St. Jaume de Llierca	1.394 b
7	Massanet de la Selva	1.390 b
15	Fageda d'en Jordà	1.372 b
14	Serra del Corb	1.276 b
13	Adri	1.227 b

[a] Means followed by the same letter are not significantly different (p<0,05) according to the REG-WM test

Statistical analysis

Statistical analysis was carried out with the Statistical Analytical System (SAS,1990). Frequency distributions were assessed for each variable studied by using the Shapiro-Wilk (W) test (α=0.05) for normality. The same test has been applied, in same instances, on the log-transformed data. Likewise, log-transformed data of N_2O emissions were used to realize an ANOVA for non-compensated models for the different studied areas. The Ryan-Einot-Gabriel-Welchs (REGWF) multiple range test (α=0.05) has been used for multiple comparison of mean values.

Results

Soil characteristics

Physico-chemical characteristics of the different soil types are reported in Table 2. Their range of variation is considerably high, as expected: pH (3.9 - 7.4);% particles > 2 mm (0 - 57.44);% soil water content (2.62 - 88.79);% organic matter content (1.14 - 24.5);% total

nitrogen content (0.11 - 1.22); C/N (5.07 - 27.15); nitrate content (0 - 222.5 mg kg^{-1} d.w.)

Descriptive statistics and the values of the Shapiro-Wilk test of the different variables studied are reported in Table 3. In agreement with the W values and the associated probability (p<W) obtained, none of the variables studied presents a normal distribution. We have considered normal distributions when W>0.90 and the existence of two different populations of data is not detected. By using this criterium only the C/N ratio values present a normal distribution. In Table 3, values of W and p<W from log-transformed parameters are reported. With regard to coarse particle size fraction > 2 mm, pH, and NO_3, the existence of a bimodal distribution of data does not allow their normal distribution. Data of organic matter and total nitrogen did not allow neither a log-normal distribution nor a normal distribution through a reciprocal transformation.

Relationships between soil parameters and N_2O losses

ANOVA of N_2O losses performed with log-transformed data (Table 5) shows that there are highly significant differences ($R^2=0.51$, $p=0.0001$) among sampling areas. In Table 5 it may be observed that in the Aiguamolls of Empordà (S4) the mean is significatively higher than other areas that are all included in another group, despite the variations presented. If we eliminate the S4, not log-transformed data of N_2O losses are almost normally distributed (W=0.9484; Prob<W=0.0104). Despite ANOVA shows that there are significant differences ($0.01<p<0.05$) among sampling areas, the REGWM test does not allow their separation into different groups. A positive correlation between N_2O losses and nitrates ($p<0.01$), soil water content ($p<0.01$) and pH ($0.01< p<0.05$) has been observed (Table 6). A negative correlation between N_2O losses and coarse particle size fraction ($0.01< p<0.05$) was also observed. Log-transformation of data did not produced an increase in correlation coefficients.

A regression model relating N_2O losses and mean values of global data which allowed to identify the parameters more closely related with N_2O losses and their respective specific weight. The equation obtained was the following:

$$y = -1.93x + 2.84z + 1.03t - 5.29 \qquad (1)$$

where y is the rate of N_2O losses in g ha^{-1}day^{-1}, x is the organic matter content in %, z is the C/N ratio, and t is NO_3content in mg kg^{-1} dw.

The coefficient of determination (R^2) was 0.901 ($p=0.0001$). Residuals were normally distributed according to a Shapiro-Wilk test (W=0.9357; Prob<W=0.2977). Regression equation obtained from predicted versus observed N_2O loss rates had a slope of 0.999 ($p=0.0001$) and an intercept that was not different from 0 (Fig. 1). According to this regression model, the N_2O losses are a function of NO_3 content, organic matter content and, to a minor extent, of the C/N ratio (Table 7).

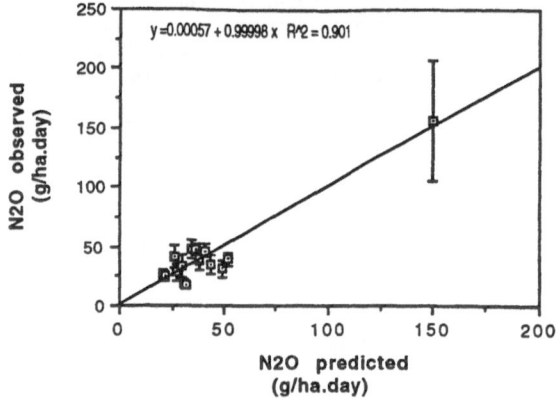

Figure 1. Regression between the observed N_2O losses and the predicted for the model from Equation 1.

Discussion

It has been argued that the parent material or vegetation may be considered as factors of soil formation which determine its characteristics. Soils used in this work represent a wide range of physico-chemical characteristics (Table 2). Several authors have observed that, when working at field scale, many of the physical, chemical and biological soil parameters follow a log-normal distribution (Groffman and Tiedje, 1989; Parkin et al., 1985). This was not our case (Table 4) because of the rather limited number of sampled points. Log-normal distributions only were observed for N_2O emission and soil water content.

In S4, N_2O losses were significantly higher than in the remainder soils according to ANOVA and mean separation test applied to the log-transformed N_2O emission data (Table 5). The high pH of this soil suggests that N_2O production was not due to chemical decomposition of HNO_2 (Scott Smith and Zimmerman, 1981). This is a soil which presents an hydric regime close to saturation. According to Davidson (1992), denitrification is the predominant source of N_2O when the soil water content is higher than field capacity. The presence of large amounts of organic matter together with the high nitrate content also may favour this process. Christensen et al., (1990) showed that the high content of organic matter may affect the denitrification rate due mainly to an increase in the anoxic conditions rather than to an increase of the organic substrate for the process. Similarly, the high amounts of nitrates found in this soil were in agreement with the results obtained by Groffman and Tiedje (1989) with poorly drained soils. We have found a

Table 6. Correlation matrix between the parameters from pooled data of different soils

	N_2O	Part.>2mm	pH	SWC	OM	Nt	C/N	NO_3
N_2O	1	-0.23025	0.29327	0.43182	-0.01247	0.16108	-0.18065	0.60377
	0	**0.0484	**0.0107	***0.0001	0.9148	0.1645	0.1184	***0.0001
Part.>2mm		1	-0.24024	-0.35158	0.17843	0.06670	0.18466	-0.26768
		0	**0.0421	***0.0021	0.1310	0.5750	0.1178	** 0.0202
pH			1	0.13067	-0.33782	-0.08516	-0.44379	0.29681
			0	0.2605	***0.0030	0.4676	***0.0001	***0.0092
SWC				1	0.46797	0.55008	-0.05178	0.53277
				0	***0.0001	***0.0001	0.6547	***0.0001
OM					1	0.85635	0.37352	0.18020
					0	***0.0001	***0.0008	0.1168
Nt						1	-0.07135	0.37099
						0	0.5375	***0.0009
C/N							1	-0.24641
							0	**0.0308
NO_3								1
								0

Part.>2mm = Soil Particles >2mm SWC–Soil Water Content(%), OM–Organic Matter(%), Nt–total nitrogen (%), C/N–C/N ratio, NO_3 = Nitrates Content (mg kg^{-1} dw), N_2O = N_2O Emission (g ha^{-1} day $^{-1}$).
ns $p>0.1$; * $0.05 < p<0.01$; ** $0.01 <p <0.05$; *** $p<0.01$.

positive correlation (Table 6) between NO_3 content and water content of soil despite the fact that generally the NO_3 content decreases when soil water content increases. This is probably due to the peculiar dynamics of this soil; it is well known that during strong anaerobic conditions N_2 emissions prevail on N_2O emision (Focht, 1974) and that the pH close to neutrality favours denitrifying activity by stimulating the reduction of NO_3 to N_2 (Koshinen and Keeney, 1982). Nevertheless, the high NO_3 content may result in the inhibition of N_2O reductase activity (Firestone et al., 1979; Weier et al., 1993). The relationship obtained in our study between NO_3 content and N_2O losses agreed with the findings of the above mentioned authors. Several authors (Van Screven, 1967; Sørensen, 1974) have reported that wetting-drying cycles stimulate decomposition of organic matter affecting inorganic N transformations. It is therefore reasonable to find high levels of N losses in waterlogged soils ('Aiguamolls de l'Empordà') where this kind of cycles are fairly frequent. Generally in waterlogged soils the denitrification rates show coefficients of variation and skewness lower than in no waterlogged soils. Nevertheless, the N_2O losses in the soil of 'Aiguamolls of Empordà' presented a coefficient of variation which was higher (approximately a 70%) than inthe other areas studied.

Significant differences between areas were observed after removing S4 and applying the ANOVA on the non-transformed N_2O losses data; not withstanding, it is not possible to separate them into significantly different groups. The pH of these soils, except for 'Beget' soil and 'Coll de Merolla 'soil, was acid. It has been shown that even very little variation of pH may affect denitrifying enzymes (Knowles, 1981). Therefore, the fact that we found N_2O losses where the presence of nitrate was not detected (Capmany, Adri, St. Jaume), could be explained by: a) the high affinity for NO_3 utilization in this systems with Michaelis-Menten constants around $<10 M$ (Murray et al., 1989); b) N_2O reduction to N_2 is negligible at acid pH (Christensen, 1985; Christensen et al. 1990); and c) chemo-denitrification of nitrite may also play an important role in the gaseous N-products evolution under acid conditions. In fact, we have found that there was a significant positive correlation between the pH and N_2O losses. The lack of nitrate, at least at a detectable level, may be due to the fact that nitrification is one of soil processes more affected by the acid pH. In this study we reported a significant positive correlation between pH and nitrate content (Table 6). Likewise, we have observed that a significant positive correlation exists between the soil water content and the N_2O losses. The soil of 'Capmany' area presents a high soil water

Table 7. Significance of regression model parameters for Equation 1

| | Parameter estimate | | | |
	Intercept	OM	C/N	NO$_3$
Std error	12.24	0.60	1.09	0.10
Prob>F	0.6735	0.0075	0.0232	0.0001

content and organic matter. However, its N$_2$O losses were lower when compared to the soil of 'Aiguamolls of Empordà' probably due to the lack of nitrates.

We have also observed that, except for the soil of 'Crosa de Sant Dalmai', soils originated on volcanic material present low N$_2$O losses which decreased, particularly when the soil was formed on basanites ('Adri', 'Serra del Corb' and 'Fageda d'en Jordà'). In contrast, within this gradient, the soils that support a meadow vegetation were in the high level range.

The model obtained by multiple regression (Table 7) suggested that organic matter content, C/N ratio and nitrate content explained the variability in N$_2$O losses from our soils. The lack of soil water content as one of the main regulating factors was striking. The model indicates also that not only the organic matter content but its quality has a significant effect for N$_2$O losses. Several authors have shown that the denitrification rate is significantly correlated with easily decomposable organic matter content (Burford and Bremner, 1975; Stanford et al., 1975). In this study, this parameter was not determined but we used the C/N ratio because of its primary importance in regulating the magnitude of mineralization. In spite of our model was highly significant (R^2=0.90) it should be considered that the 'Aiguamolls de l'Empordà' area alone basically governs the regression obtained.

Conclusions

We have shown that the main soil physico-chemical characteristics affecting the N$_2$O losses in the soils studied are the organic matter, C/N ratio and nitrate content. However, a large number of areas should be studied in order to know the importance of other factors such as parent material and vegetation on N$_2$O losses.

Acknowledgements

This work was supported in part by the STEP programme (CT90-0029) of the European Community.

References

Bouwman A F 1990 Exchange of greenhouse gases between terrestrial ecosystems and the atmosphere. *In* Soils and the Greenhouse Effect. Eds. A F Bouwman. pp 61–127. John Wiley and Sons, New York, USA.

Burford J R and Bremner J M 1975 Relationships between the denitrification capacities of soils and total, water soluble and readily decomposable soil organic matter. Soil Biol. Biochem. 7, 389–394.

Christensen S 1985 Denitrification in an acid soil: Effects of slurry and potassium nitrate on the evolution of nitrous oxide and on nitrate-reducing bacteria. Soil Biol. Biochem. 17, 757–764.

Christensen S, Simkins S and Tiedje J M 1990 Spatial variation in denitrification: Dependency of activity centers on the soil environment. Soil Sci. Soc. Am. J. 54, 1608–1613.

Davidson E A 1992 Sources of nitric oxide and nitrous oxide following wetting of dry soil. Soil Sci. Soc. Am. J. 56, 95–102.

Firestone M K and Davidson E A 1989 Microbiological basis of NO and N$_2$O production and consumption in soil. *In* Exchange of Trace Gases between Terrestrial Ecosystems and the Atmosphere. Eds. M O Andreae and D S Schimel. pp. 7–21. John Wiley and Sons, New York, USA.

Firestone M K, Smith M S, Firestone R B and Tiedje J M 1979 The influence of nitrate, nitrite and oxygen on the composition of the gaseous products of denitrification in soil. Soil Sci. Soc. Am. J. 43, 1140–1144.

Focht D D 1974 The effect of temperature, pH and aeration on the production of nitrous oxide and gaseous nitrogen - a zero-order kinetic model. Soil Sci. 118, 173–179.

Goodroad L L and Keeney D R 1984 Nitrous oxide emissions from forest, marsh and prairie ecosystems. J. Environ. Qual. 13, 448–452.

Groffman P M and Tiedje J M 1989 Denitrification in north temperate forest soils: Spatial and temporal patterns at the landscape and seasonal scales. Soil Biol. Biochem. 21, 613–620.

Hutchinson G L, Guenzi W D and Livingston G P 1993 Soil water controls on aerobic soil emission of gaseous nitrogen oxides. Soil Biol. Biochem. 25, 1–9.

Knowles R 1981 Denitrification. *In* Soil Biochemistry, Vol. 5. Eds. E A Paul and J N Ladd, Marcel Dekker, New York, USA.

Koskinen W C and Keeney D R 1982 Effect of pH on the rate of gaseous products of denitrification in a silt loam soil. Soil Sci. Soc. Am. J. 46, 1165–1167.

Murray R E, Parsons L L and Smith M S 1989 Kinetics of nitrate utilization by mixed populations of denitrifying bacteria. Appl. Environ. Microbiol. 55, 717–721.

Parkin T B, Sexstone A J and Tiedje J M 1985 Comparison of field denitrification rates determined by acetylene-based soil core and nitrogen-15 methods. Soil Sci. Soc. Am. J. 49, 95–99.

SAS Institute 1990 SAS/STAT Guide for personal computers. Ver 6 ed. SAS Institute, Cary, NC. USA. 1028 p.

Scott Smith M and Zimmerman K 1981 Nitrous oxide production by nondenitrifying soil nitrate reducers. Soil Sci. Soc. Am. J. 45, 865–871.

Sørensen L N 1974 Rate of decomposition of organic matter in soil as influenced by repeated air drying-rewetting and repeated additons of organic material. Soil Biol. Biochem. 6, 287–292.

Stanford G, Vander Pol R A and Dzienia S 1975 Denitrification rates in relation to total and extractable soil carbon. Soil Sci. Soc. Am. Proc. 39, 284–289.

Van Screven D A 1967 The effect of intermittent drying and wetting of a calcareous soil on carbon and nitrogen mineralization. Plant and Soil 26, 14–32.

Weier K L, Doran J W, Power J F and Walters D T 1993 Denitrification and the dinitrogen/nitrous oxide ratio as affected by soil water, available carbon, and nitrate. Soil Sci. Soc. Am. J. 57, 66–72.

Yoshida T and Alexander M 1970 Nitrous oxide formation by *Nitrosomonas europaea* and heterotrophic microorganisms. Soil Sci. Soc. Am. Proc. 34, 880–882.

O. Van Cleemput et al. (eds.), Progress in Nitrogen Cycling Studies, 621–626, 1996.
© 1996 *Kluwer Academic Publishers.*

Nitrite accumulation in soils upon urea application

A.H. Samater and O. Van Cleemput
Laboratory of Applied Physical Chemistry, Faculty of Agricultural and Applied Biological Sciences, University of Ghent, Coupure links 653, B–9000 Gent, Belgium

Key words: hydroquinone, nitrite, N-(n-butyl) phosphorothioic triamide, urea, urease inhibitors

Abstract

Nine soils with different physico-chemical characteristics from different areas in Belgium were studied under laboratory conditions for their ability to nitrite (NO_2^--N) accumulation. Most low pH soils (<7) did not show any considerable nitrite accumulation during the ten days of incubation, even when treated with 100 mg urea-N kg^{-1}. The soils with a high pH (>7) responded strongly to the N-fertilizer and nitrite accumulation was as high as 67 mg NO_2^--N kg^{-1} when applied with 100 mg urea-N kg^{-1}.

For a more detailed study, some high pH soils were amended with a higher fertilization rate and longer periods of incubations. They showed a higher NH_4^+-N and NO_2^--N accumulation and consequently a higher NO_3^--N formation due to the hydrolysis of the N-fertilizer added. The NO_2^--N accumulation as well as its persistence was increased with the increase of urea fertilizer application rate for all three soils.

Only the pH and clay content were found to have a significant correlation with the NO_2^--N production in this specific experiment.

In order to limit or retard urea hydrolysis, which may contribute to nitrite accumulation, and consequently lower the efficiency of urea, two urease inhibitors, NBPT (N-(n-butyl) phosphorothioic triamide) and HQ (Hydroquininone), were used and compared for their effectiveness at the rate of 1% on urea weight basis. For this experiment, NBPT was more effective than HQ in retarding urea hydrolysis and nitrite accumulation for two soils, while in the third soil, neither NBPT nor HQ has an effect on NH_4^+-N and NO_2^--N accumulation. However, they had no influence the final NO_3^--N accumulation. The effectiveness of inhibitors in this condition was limited to less than a week.

Introduction

Nitrite (NO_2^-) is an intermediary compound when ammonium is oxidized to nitrate. Its accumulation in soils has been reported during nitrification of hydrolysed urea (Chapman and Liebig, 1952; Court et al., 1962). The nitrite ion is to some extent stable under alkaline conditions and accumulates in some basic soils (Reuss and Smith, 1965; Tyler and Broadbent, 1960). Its disappearance is accompanied by a corresponding increase in nitrate. Nitrite may not accumulate in acid soils, because it is chemically unstable under these conditions (Smith and Clark, 1960; Tyler and Broadbent, 1960). Furthermore, it has been shown that nitrite can decompose into NO and NO_3^- (Van Cleemput and Baert, 1976). It is an important compound in chemodenitrification (Van Cleemput and Baert, 1984). Nitrite

can react with soil organic matter and become fixed and at the same time nitrogenous gases can be involved (Smith and Chalk, 1980).

Nitrite toxicity to plants has been reported by several authors. Gould et al. (1986) noted that accumulation of nitrite in the soil may be toxic to the plant and may result in losses of gaseous nitrogen by chemical denitrification. Also Court et al. (1962) showed that nitrite accumulation was accompanied by marked phytotoxicity and reduction in plant yield. Chapman and Liebig (1952) on the other hand, suggested that under neutral and alkaline conditions nitrite is not likely to be very toxic to plants.

It was also found that the accumulation of nitrite was higher upon hydrolysis and nitrification of urea-N than upon nitrification of ammonium sulphate and DAP fertilizers (Bezdicek et al., 1971).

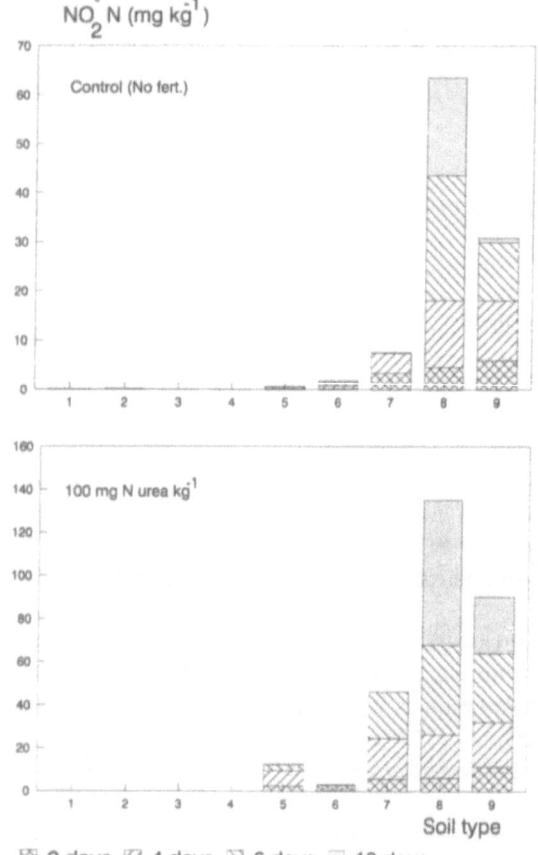

NO$_2$ N (mg kg^{-1})

Control (No fert.)

100 mg N urea kg^{-1}

Soil type

⊠ 2 days ▨ 4 days ◨ 6 days ▢ 10 days

Figure 1. Nitrite accumulation pattern in different soils

To reduce the hydrolysis rate of urea and the accompanying pH rise and consequently to lower the loss of ammonia by volatilization, urease inhibitors can be used. N-(n-butyl)phosphorothioic triamide (NBPT) as well as hydroquinone (HQ) have been shown to be effective urease inhibitors (Van Cleemput and Wang, 1991).

In this study, we present data on the influence of different levels of urea mixed with and without NBPT and HQ on nitrite accumulation in different soils. It is our aim to present data on the (1) accumulation pattern of nitrite in different soils, (2) effect of urea application rates on nitrite accumulation, (3) effect of urease inhibitors on nitrite accumulation.

Materials and methods

Nine soils having different physico-chemical and biological properties (Table 1) were collected from different areas in Belgium. The selection was based on

having a different range of pH,% organic carbon and texture. These soils were air dried at room temperature (22 °C), grounded to pass through a 2 mm sieve, mixed throughly and stored in tightly closed containers.

In each study of these soils, a sample of thirty gram were placed in 150 ml plastic jars with lid. Water was added to a moisture content of 3/4 field capacity. The nitrogen (N) used in these experiments was urea and was applied to the soil as a solution with the water. After mixing well, to homogenise the urea and the water throughout the soil, they were incubated at 25°C and opened every other day for a few minutes for aeration.

Nitrite accumulation pattern in different soils

This experiment was carried out to measure the capability of different soils to accumulate nitrite under normal conditions (no addition of fertilizer) and upon application of urea. To the soil samples, 100 mg urea-N kg^{-1} was applied. Also untreated soil samples (without addition of fertilizer) were compared. They were incubated in duplicate for 0, 2, 4, 6 and 10 days before analysis.

Nitrite accumulation at different urea doses for high pH soils

This study was intended to verify the influence of higher doses of urea on the pattern of nitrite accumulation and its stability in these concerned soils. Three soils (5, 7 and 8) of the preceding series showing a high capability of nitrite accumulation were chosen for further studies on their responses to different urea doses (0, 50, 100, 150 and 200 mg urea-N kg^{-1}) and longer incubation periods (0, 2, 4, 6, 10, 15 and 20 days).

Nitrite accumulation in the presence of urease inhibitors

This experiment was conducted to compare the effectiveness of two urease inhibitors for reducing nitrite accumulation under aerobic conditions. Two urease inhibitors, hydroquinone (HQ) and N-(n-butyl) phosphorothioic triamide (NBPT), were used with the same three soils (5, 7 and 8) with incubation periods of 0, 2, 4, 6, 10, 15 and 20 days. One hundred mg urea-N kg^{-1} was used with 1% urease inhibitor on urea weight basis.

Table 1. Some physico-chemical characteristics of the used soils

No.	Texture[a]	pH	CaCO$_3$(%)	Clay(%)	O.M(%)	Total-N(%)	C/N
1	Sand	4.69	0.20	1.89	2.28	0.07	17.00
2	Loam	5.82	0.30	10.32	1.68	0.10	8.45
3	Sand loam	6.05	0.30	9.38	2.32	0.12	10.10
4	Silt loam	7.01	0.30	15.60	5.04	0.25	10.08
5	Sand loam	7.12	0.39	13.61	1.72	0.11	8.00
6	Clay loam	7.36	0.98	36.90	5.44	0.26	10.62
7	Silt loam	7.70	0.65	16.04	2.08	0.10	10.43
8	Silt clay loam	7.72	15.65	28.56	4.80	0.20	12.06
9	Clay	7.80	16.83	58.41	2.64	0.18	7.17

[a]USDA soil textural classification

Table 2. Correlation between soil factors and nitrite production at 2 days with and without fertilizer by using simple regression model

Soil factor	Treatment	Correlation (r^2)(%)	Equation model
pH	Without fertilizer(-)	56.00*	$NO_2^- \text{-N} = Exp(-17.74+2.32(pH))$
	With fertilizer(+)	57.76*	$NO_2^- \text{-N} = Exp(-198.08+2.59(pH))$
	(100 mg N urea kg^{-1})		
Organic matter	(-)	3.99[ns]	
(%)	(+)	3.99[ns]	
Clay (%)	(-)	51.56*	$NO_2^- \text{-N} = -0.43+0.10(Clay)$
	(+)	56.74*	$NO_2^- \text{-N} = -0.75+0.18(Clay)$

* Significant at $p < 0.05$.
[ns] Not significant.

Analytical methods

The exchangeable ammonium (NH_4^+), nitrite (NO_2^-) and nitrate (NO_3^-) was extracted from the soil by shaking the soil samples for one hour with 1 N KCl solution. The NH_4^+-N and NO_3^--N was determined by distillation method according to Keeney and Nelson (1982). NO_2^--N was determined by the modified Griess-Ilosvay colorimetric method (Barnes and Folkard, 1951; Bremner, 1965).

Results and discussion

Nitrite accumulation pattern in different soils

In Figure 1 the evolution of NO_2^--N accumulation is given for the different soils. In both treatments, with and without urea application, the soils 5–9 showed to accumulate a higher amount of nitrite than soils 1–4. Even without urea application, in soil 8, up to 25 mg NO_2^--N kg^{-1} was found in the 6th day of incubation. In this high NO_2^--N accumulating soil, the maximum

accumulation was also later than in the soils with less NO_2^--N. The same observation was made in the series to which urea was added: the higher the nitrite accumulation, the longer it takes before the maximum is attained. In the urea amended soils, all samples accumulated some NO_2^--N. One soil (nr. 8) showed almost 70 mg NO_2^--N kg^{-1}. In comparison to the accumulation in the control, this corresponded to almost half of the added urea-N. Taking into account the determined soil characteristics, no clear soil property could be indicated as being solely responsible for the nitrite accumulation. A possible reason for nitrite accumulation is a lower activity of **Nitrobacter** in alkaline conditions, created by urea hydrolysis. This was not the case for soils nr. 1, 2 and 3 who had initially a low pH. The soil with both the highest% clay content and pH definitely showed the highest accumulation of nitrite. Moreover, comparing the correlation between soil factors and nitrite production it was found that only the pH and the clay have a significant correlation with NO_2^--N production in this specific experiment (Table 2). At this stage, it could only be stated that an

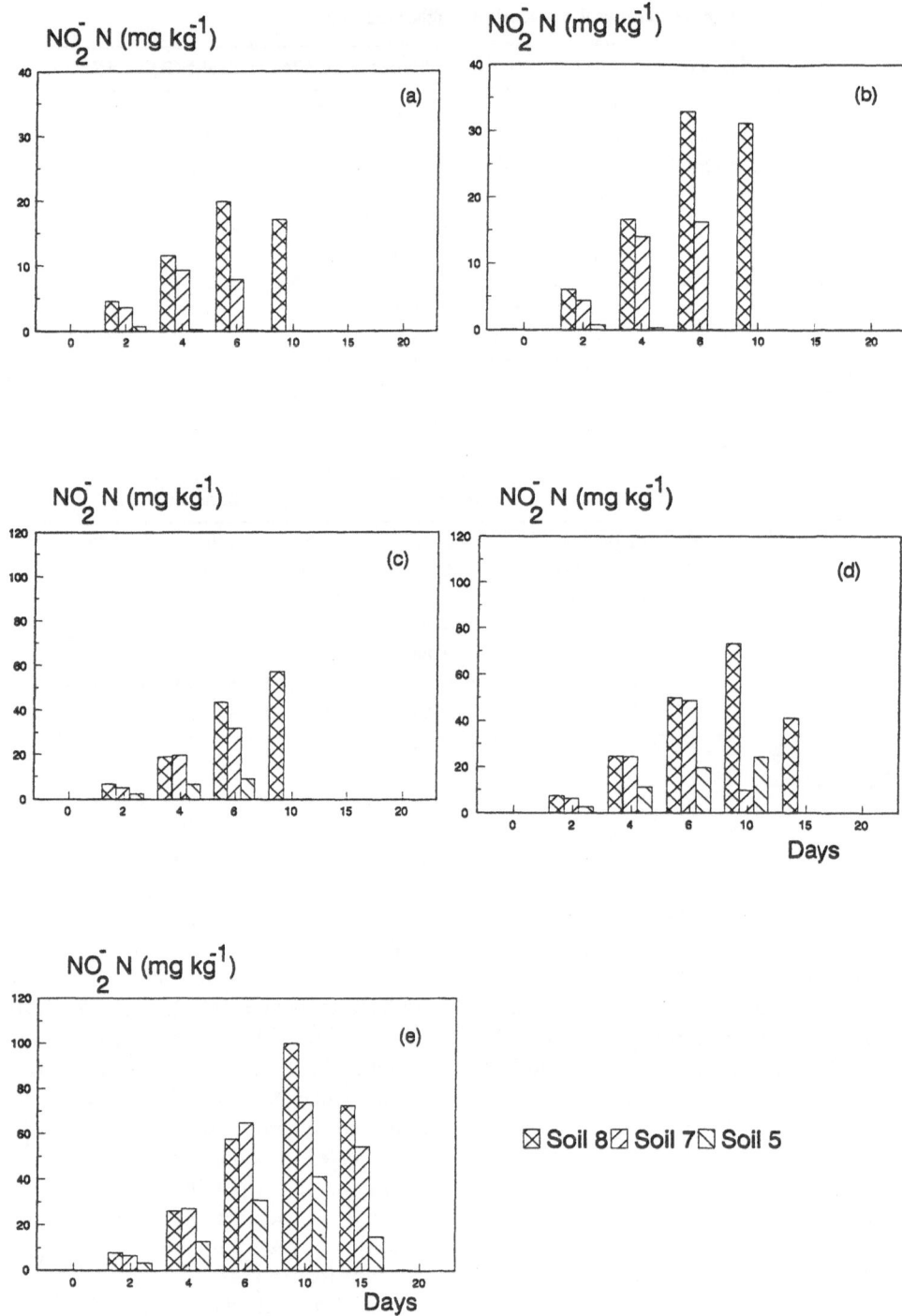

Figure 2. Influence of urea rates on nitrite accumulation in three high pH soils incubated at 3/4 field capacity and 25 °C (**a**) No fertilizer, (**b**) 50 mg N kg^{-1} soil, (**c**) 100 mg N kg^{-1} soil, (**d**) 150 mg N kg^{-1} soil, (**e**) 200 mg N kg^{-1} soil.

alkaline pH alone does not necessarily lead to a high NO_2^--N accumulation.

Nitrite accumulation at different urea doses

It is clear from Figure 2 that the maximum NO_2^--N accumulation was reached at a later stage the more

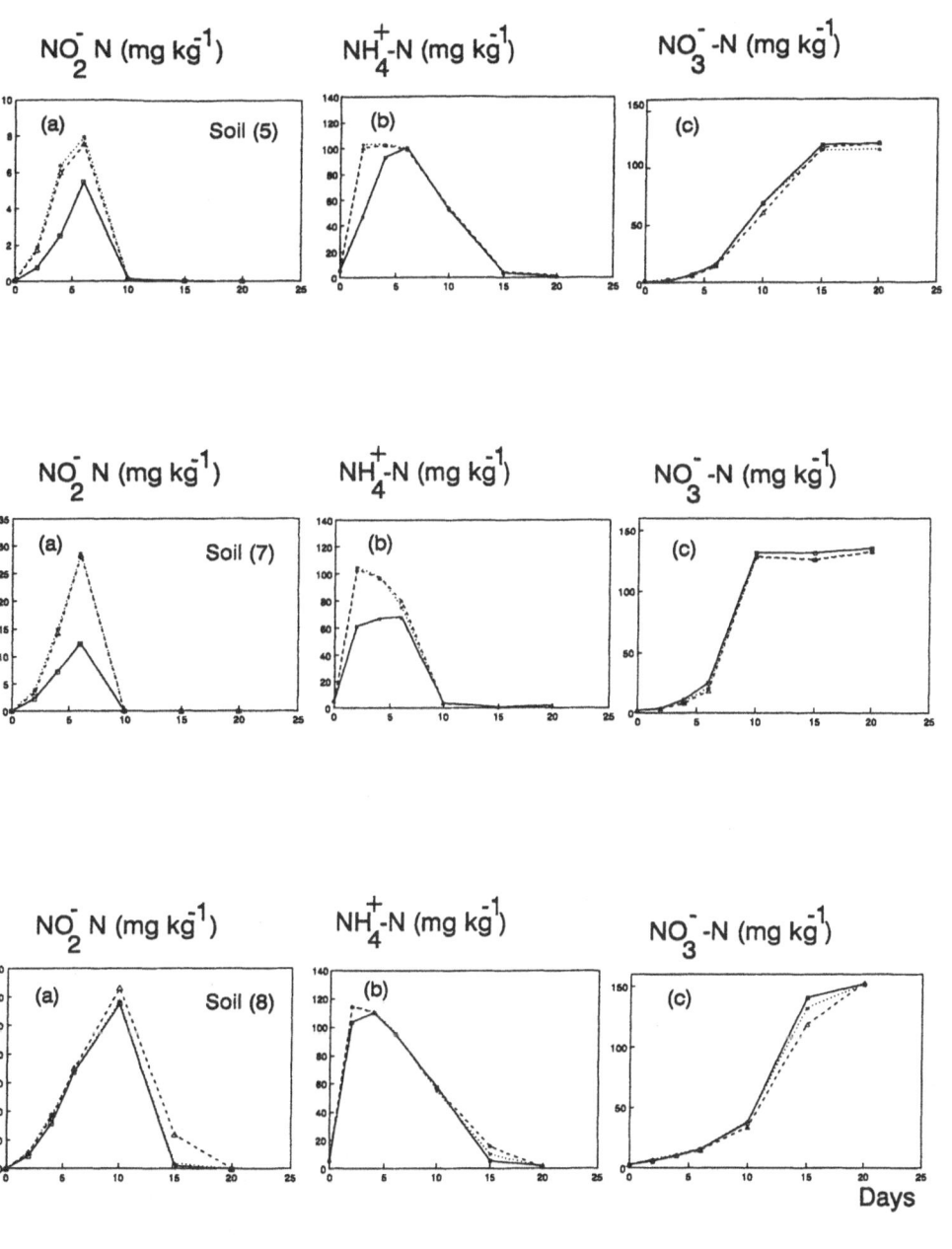

Figure 3. Influence of urease inhibitors (1%) on mineral-N evolution in soils treated with 100 mg N urea kg^{-1} at 3/4 F.C. and 25 °C

urea-N was added. It is also clear that the NO_2^--N accumulated only for about 10 days except for the high doses. Apparently, the nitrite reductase activity becomes very active upon some period of NO_2^--N accumulation. At the highest urea concentration NO_2^--N persisted for all soils at least up to 15 days. During this period the accumulated NO_2^--N might be toxic to plants. However, to demonstrate a clear toxic effect a specific experiment for this purpose should be conducted. The same sequence of importance of NO_2^--N accumulation among the three soils was found as in the preceding experiment. It was also found that the NH_4^+-N level attained a maximum at days 2 to 4, indicating a rapid urea hydrolysis. The sequence of nitrogen transformations is as follows: urea hydrolysis, NH_4^+-N formation, NO_2^--N accumulation and disappearance and finally NO_3^--N accumulation. Looking into the other mineral nitrogen fractions at 50, 100, 150 and 200 mg urea-N kg^{-1}, it can be seen that in all soils, the NO_2^--N was present as long as there was NH_4^+-N. An important point is that as long as NO_2^--N is present, the NO_3^--N did not appreciably increased. Comparing the mineral-N at zero time plus the added urea-N and the mineral-N after 20 days, it was found that the amount of non-accounted for N was about 10 to 25%. Immobilization and gaseous N loss might be the reason for it.

Nitrite accumulation in the presence of urease inhibitors

In Figure 3a it is shown that in two soils (5 and 7) NBPT was clearly more effective in lowering the NO_2^--N accumulation than HQ. This result confirms the findings of Bremner (1990), who stated that NBPT is the most effective compound currently available for retarding hydrolysis of urea and for decreasing ammonia volatilization and NO_2^--N accumulation.   In all three soils the NO_2^--N accumulation was almost the same in the non-inhibited as in the HQ-inhibited treatment. In soil 5 and 7, NBPT drastically reduced the NO_2^--N accumulation. Apparently in soil 8 both inhibitors had no strong effect on NO_2^--N accumulation. It is suggested that the effect of NBPT on pH rise is limited due to the already high native pH of this soil. Because the maximum accumulation was reached at about the same time, it indicates that the nitrite reduction activity was not influenced. The different behaviour between HQ and NBPT is also seen in Figure 3b. The NH_4^+-N accumulation is almost similar between the HQ-inhibition and non-inhibition treatment for all treatments. In soil 5 and 7, a lower NH_4^+-N content was found with NBPT. The inhibitors do not seem to be active for a period longer than 6 days. At that time, the NBPT treatment resumes the other curves. The consequence of this is a lower NO_2^--N accumulation. The curves (Fig. 3c) showing the NO_3^--N evolution, indicate that the final NO_3^--N accumulation is not influenced by the NO_2^--N due to the inhibitors.

Conclusion

Out of the different treatments it was seen that NO_2^--N can drastically accumulate upon urea application. Especially high pH soils were susceptible for it. The high NO_2^- accumulation also retarded the NO_3^--N formation. Although the urea hydrolysed within 2 days, the NO_2^--N accumulation maximum was observed much later. This is the result of pH rise which holds on for several days. Only the pH and clay content were found to have a significant correlation with NO_2^--N production in this specific experiment. Addition of HQ or NBPT inhibitors decreased the NO_2^--N content, but did not influence the final NO_3^--N accumulation.

References

Barnes H and Folkard A R 1951 The determination of nitrites. Analyst (London) 76, 599–603.
Bezdicek D F, Mac Gregor J M and Martin W P 1971 The influence of soil fertilizer geometry on Nitrification and nitrite accumulation. Soil Sci. Soc. Am. Proc. 35, 997- 1002.
Bremner J M 1965 Inorganic forms of nitrogen *In* Methods of Soil Analysis, part 2. Ed. C A Black et al. Agron. 9, 1179–1237. ASA, Madison, WI, USA.
Bremner J M 1990 Problems in the use of urea as a nitrogen fertilizer. Soil Use Manage. 6,70–71.
Chapman H D and Liebig G F 1952 Field and laboratory studies of nitrite accumulation in soil. Soil Sci. Soc. Am. Proc. 16, 276–282.
Court M N, Steven R C and Waid J S 1962 Nitrite toxicity arising from the use of urea as fertilizer. Nature 194, 1263–1265.
Gould W D, Hagedorn C and McCready R G L 1986 Urea transformations and fertilizer efficiency in soil. Adv. Agron. 40, 209–238.
Keeney D R and Nelson D W 1982 Nitrogen-Inorganic forms. *In* Methods of Soil Analysis, Part 2. 2nd ed. Eds. A L Page et al., Agron. 9, 643–698. ASA and SSSA, Madison, WI, USA.
Reuss J O and Smith R L 1965 Chemical reactions of nitrites in acid soils. Soil Sci. Soc. Am. Proc. 29, 267–270.
Smith C J and Chalk P M 1980 Fixation and loss of nitrogen during transformations of nitrite in soils. Soil Sci. Soc. Am. J. 44, 288–291.
Smith D H and Clark F E 1960 Volatile losses from acid or neutral soils containing nitrite and ammonium ions. Soil Sci. 90, 86–92.
Tyler K B and Broadbent F E 1960 Nitrite transformations in California soils. Soil Sci. Soc. Am. Proc. 24, 279–282.
Van Cleemput O and Baert L 1976 Theoretical considerations on nitrite self-decomposition reactions in soils. Soil Sci. Soc. Am. J. 40, 322–323.
Van Cleemput O and Baert L 1984 Nitrite: a key compound in nitrogen loss processes under acid conditions. Plant and Soil 76, 233–241.
Van Cleemput O and Wang Zhengping 1991 Urea transformations and urease inhibitors. Trends Soil Sci. 1, 45–52.

Plant and Soil **181**: 139–144, 1996.
© 1996 *Kluwer Academic Publishers.*

Measurement of field scale N_2O emission fluxes from a wheat crop using micrometeorological techniques

U. Skiba[1], K.J. Hargreaves[1], I.J. Beverland[2], D.H. ONeill[2], D. Fowler[2] and J.B. Moncrieff[2]
[1] *Institute of Terrestrial Ecology, Bush Estate, Penicuik, Midlothian, EH26 0QB, Scotland* * and [2] *Institute of Ecology and Resource Management, University of Edinburgh, Mayfield Road, Edinburgh, EH9 3JU, Scotland*

Key words: eddy covariance, eddy accumulation, micrometeorology, nitrous oxide

Abstract

Measurements of N_2O emission fluxes from a 3 ha field of winter wheat were measured using eddy covariance and relaxed eddy accumulation continuously over 10 days during April 1994. The measurements averaged fluxes over approximately 10^5 m^2 of the field, which was fertilised with NH_4NO_3 at a rate of 43 kg N ha^{-1} at the beginning of the measurements. The emission fluxes became detectable after the first heavy rainfall, which occured 4 days after fertiliser application. Emissions of N_2O increased rapidly during the day following the rain to a maximum of 280 ng N m^{-2}s^{-1} and declined over the following week. During the period of significant emission fluxes, a clear diurnal cycle in N_2O emission was observed, with the daytime maximum coinciding with the soil temperature maximum at 12 cm depth. The temperature dependence of the N_2O emission was equivalent to an activation energy for N_2O production of 108 kJ mol^{-1}. The N_2O fluxes measured using relaxed eddy accumulation, averaged over 30 to 270 min, were in agreement with those of the eddy covariance system within 60%. The total emission of N_2O over the period of continuous measurement (10 days) was equivalent to about 10 kg N_2O-N, or 0.77% of the N fertiliser applied.

Introduction

Emissions of N_2O from soil dominates the global N_2O budget, but for all regions the estimated fluxes are uncertain and this is partly due to the difficulty of extrapolating from the conventional emission measurements by chambers at the 0.1 to 1 m^2 scale to the field scale (Folorunso and Rolston, 1984; Smith et al., 1994). Spatial variability problems for scales smaller than 1 km^2 can be overcome by micrometeorological techniques, which average the fluxes typically over 100 m^2 to 1 km^2. The most robust micrometeorological method, being free of much of the empiricism of other techniques is eddy covariance. The flux F is provided by the product of w', the instantaneous deviation of vertical wind speed from the mean (m s^{-1}) and χ', the instantaneous deviation of the gas concentration from the mean (μg m^{-3}) (Fowler and Duyzer, 1989):

$$F_\chi = W'_{\chi'}$$

However, eddy covariance measurements require instrumentation capable of sampling at rates of typically 10 Hz, and in order to adequately sample the spectrum of flux carrying eddies, it is also necessary to detect concentration differences of the order of 0.1 to 1 ppbv. Ultrasonic anemometers have been the instrument of choice for some time to measure the three components of the turbulence, but few instruments are available for rapid measurements of trace gas concentrations other than CO_2 and water vapour. For N_2O these conditions were only very recently met, mainly by the development of a tunable diode laser spectrometer (TDL) (Zahniser et al., 1995).

An alternate approach, not involving the use of fast response sensors, is provided by the 'relaxed' eddy accumulation or conditional sampling technique (Businger and Oncley, 1990). The vertical wind velocity sensed using an ultrasonic anemometer is used to switch continuously sampled air into one of two collecting resevoirs, i.e. one to hold air from updrafts and one to hold air from downdrafts. The flux is given by

---

* FAX No.: +4411314453943
Plant and Soil is the original source of publication of this article. It is recommended that this article is cited as: *Plant and Soil* **181**: 139–144, 1996.

$$F = \beta(L)\sigma_w\rho_a(\chi_u - \chi_d) \qquad (1)$$

where β is a dimensionless empirical constant, σ_w is the standard deviation of the vertical windspeed (m s^{-1}), ρ_a is the mean dry air density over the half hour integration period (g m^{-3}) and χ_u-χ_d are the mixing ratios of the trace gas (ppbv). The conditional sampling system requires a vertical wind sensor and sampling valve assembly with a time resolution of at least 0.1 s and gas sensors capable of detecting the small concentration differences between updraft and downdraft air.

Here we report measurements of N_2O fluxes over a field of winter wheat by eddy covariance and relaxed eddy accumulation, using a TDL, which provided hourly average fluxes for an area of $\sim 10^5 m^2$ for a 10 day period after the application of N fertiliser.

Methods

Site description

Measurements of N_2O were made over a winter wheat crop at Howmuir Farm in SE Scotland (National Grid Reference NT 613767) in April 1994, starting immediately after the second application of NH_4NO_3 (43 kg N ha^{-1}). The soil was a sandy clay loam of the Winton series. The site had a gentle slope of 15 m, running perpendicular to the prevailing wind direction with a fetch of over 0.5 km to the South and West. Simultaneous flux measurements by chambers were made by colleagues from the School of Agriculture, Edinburgh, these results will not be discussed here.

Micrometeorological measurements

Eddy covariance measurements of N_2O were made on a near continuous 24 hour basis from 20th to 29th April 1994, using a TDL. Air was sampled from the top of a 4 m high mast and was pumped through nylon tubing (3/16" internal diameter, 25 m long) at a flowrate of 10 L min^{-1} into the TDL, which was housed in a mobile laboratory at the edge of the field. Power was provided by a generator located about 200 m to the NE, generally downwind of the mast. The three components of turbulence were measured using an ultrasonic anemometer (Gill Instruments), which was mounted on top of the 4 m mast. The gas inlet for the eddy covariance measurements was attached as close to the measuring point of the ultrasonic anemometer as possible, without obstructing the path of the anemometer. The instrumentation for eddy covariance and the TDL are described in detail by Hargreaves et al. (1996) and Zahniser et al. (1995), respectively. Eddy covariance fluxes of N_2O were calculated using the software Eddysol, providing raw data and mean fluxes over 10 min sample periods (Moncrieff et al., 1995). For conditional sampling two miniature solenoid valves with a response time of approximately 10 ms (Clippard Minimatic) were mounted onto the mast and were located 10 cm from the centre point of the anemometer path. Computers were used to monitor the N_2O concentration, the vertical wind speed signal and to control the switching of the conditional sampling valves. Updraft and downdraft air was collected into Tedlar bags (5 or 20 L) housed in an aluminium case 25 m from the base of the mast. The bags were filled over 30–270 minute periods and were analysed by TDL spectroscopy at a later stage. Updraught and downdraught pairs of bags were analysed together by switching between each at 30 second intervals. The average of 1–2 updraught (or downdraught) measurement blocks and the average of the surrounding 3–5 downdraught (or updraught) blocks were calculated in order to determine a concentration difference between the pair. This 'chopping' technique was employed to remove any systematic bias in the gas analysis on the assumption that this bias was linear over the relatively short measurement period, e.g. drift caused by relatively linear temperature change.

Corrections to the fluxes of N_2O for the effect of latent heat on air density were made as described by Hargreaves et al. (1994). The energy balance for this site were provided by a Bowen - Ratio system (Campbell Scientific Ltd.).

Results and discussion

The eddy covariance measurements of N_2O for the first 10 days after the second fertiliser application are shown in Figure 1. Although measurements were made almost continuously, there are gaps in the data set, caused by unsuitable wind directions from the NE,N or NW, especially in the first 3 days of the experiment, and by the daily disruptions, e.g. refilling the TDL with liquid N_2. Eddy covariance fluxes were corrected for the effect of simultaneous latent heat fluxes as described by Webb et al. (1980). Corrections for the air density effects of the sensible heat were not required, as the sampled air was

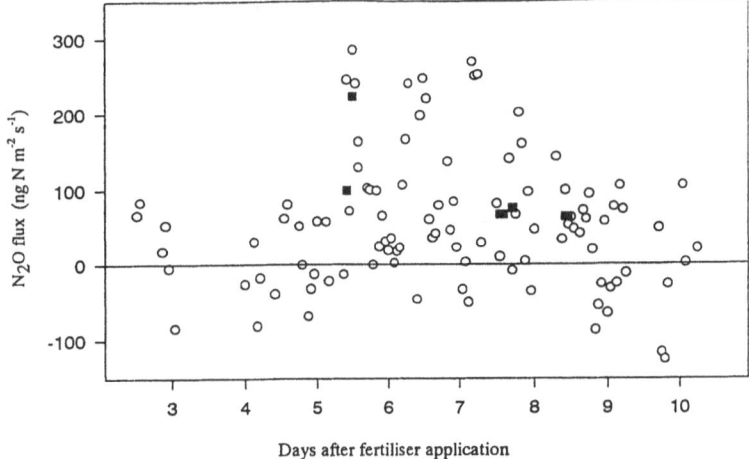

Figure 1. Nitrous oxide emissions measured by Eddy covariance (hourly averages) (o) and conditonal sampling (30 to 270 min periods) (■) from a winter wheat crop at Howmuir farm, starting on 20 April 1994, 1 day after fertiliser application.

brought to a uniform temperature before analysis. The very small fluxes imply uptake of N_2O. However soils have never been shown to be a sink for N_2O at the rates suggested by the negative fluxes in Figure 1. The small fluxes were within the noise of the measuring system, which in these conditions had a detection threshold of about 10 ng N_2O-N m^{-2}s^{-1}. It is also possible that the apparent N_2O uptake rates were an artefact caused by the storage and advection problems outlined by Fowler et al. (1995). The horizontal and vertical flux divergence during periods when N_2O concentrations are changing systematically during the measurement appears to be a greater problem for very patchy sites and was not observed in a previous experiment of N_2O measurements by eddy covariance over a very homogeneous agricultural site in an area in Denmark reclaimed from the sea (Hargreaves et al., 1996). There is evidence, that at the Howmuir site spatial variability was present, as simultaneous chamber measurements by the School of Agriculture, Edinburgh (SAC), within the same field, but not within the prevalent fetch of the tower showed much smaller fluxes, with a maximum emission rate of 30 ng N_2O-N m^{-2} s^{-1} (I Mc Taggart, pers. comm.).

With the conditional sampling technique, a total of 12 pairs of bag samples were collected on days when the eddy covariance and static chamber (I Mc Taggart, pers. comm.) N_2O flux measurements suggested that the expected differences in the up- and downdraft air should be measureable on the TDL. Of these, the analysis of 5 pairs of sample bags provided significant differences between up- and downdraft and the N_2O fluxes obtained were in good agreement, within 60% of fluxes

measured simultaneously by eddy covariance (Fig. 1). Instrument drift on the TDL and insufficent sample for the repeated analysis necessary for statistical evaluation made it impossible to analyse the remaining 7 sample pairs. The validity of the conditional sampling method was confirmed by the comparison of the sensible heat flux measured by eddy covariance and a conditional sampling simulation (Fig. 2). Both calculations were made using the virtual temperature output by the sonic anemometer at 20.5 Hz resolution. The conditional sampling flux was calculated using a simple binning technique depending on the sign of the vertical velocity and Businger and Oncley's original estimate of β =0.6 (Businger and Oncley, 1990).

The fertiliser induced response in N_2O emission was not detected until rainfall 3.5 days after fertiliser application (7.6 mm of rain in 22 hours) had wetted the soil sufficiently for the fertiliser to be transported into the soil and to increase the number of anaerobic microsites necessary for N_2O production (Arah et al., 1988) (Fig. 3). The maximum N_2O emission (280 ng N_2O-N m^{-2} s^{-1}, hourly mean) was measured 1 day after the first heavy rainfall since fertiliser application (5 days after fertiliser application) (Fig. 1). From then on the daily average N_2O emission declined steadily to a value of 15 ng N_2O-N m^{-2} s^{-1}, 10 days after fertiliser application. Diurnal variations in the N_2O emission were observed for several days, with emission rates peaking around noon at 284 ng N_2O-N m^{-2} s^{-1} and with a minimum of about 25 ng N_2O-N m^{-2} s^{-1} during the night (5 days after fertiliser application) (Fig. 4). The diurnal cycle appears to be closely linked with soil temperature, and appeared linear over the

142

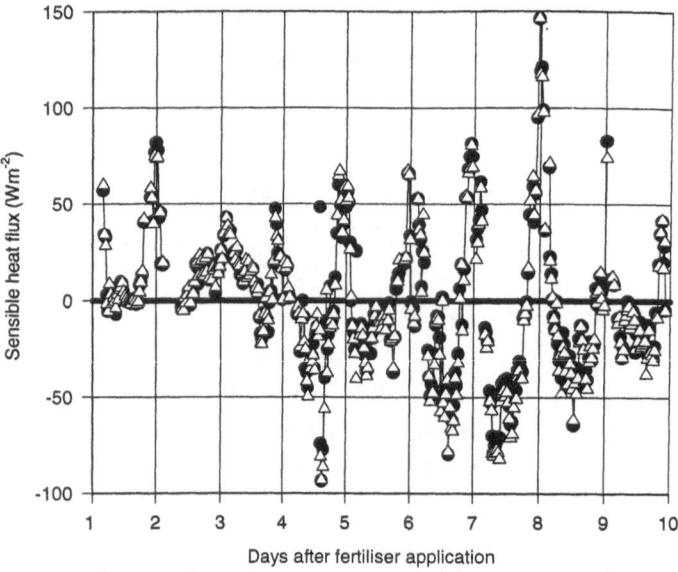

Figure 2. A comparison of the sensible heat flux measured by eddy covariance (Δ) and conditional sampling (●).

small temperature range measured (r^2= 0.406, d.f.=28, p< 0.001). An activation energy of 108 kJ mol^{-1} was calculated from the Arrhenius plot in Figure 5. These data are consistent with the range of activation energies observed for microbial N_2O evolution by denitrification and nitrification (28 to 166 kJ mol^{-1}) (Conrad et al., 1983).

The delayed response in N_2O flux to fertiliser application caused by lack of rainfall and the response to changes in soil temperature are typical of the observations for N_2O emission measurements using chambers (Goodroad and Keeney, 1984; Skiba et al., 1992), but in the case of the micrometeorological measurements the averaging area is larger by many orders of magnitude. The measurements and analysis show, that eddy covariance measurements of N_2O can be used to study processes as well as obtaining fluxes. Not enough conditional sampling measurements were made to observe the response in fluxes to rainfall or diurnal temperature variation. However, conditional sampling measurements were larger 5 days after fertiliser application (161 ng N_2O-N m^{-2} s^{-1}), when maximum N_2O emissions by eddy correlation were measured than 7 and 8 days after fertiliser application (71 and 32 ng N_2O-N m^{-2} s^{-1}, respectively) (Fig. 1).

Enhanced emissions of N_2O are usually only observed for 2 to 3 week periods following the application of N fertiliser, and although micrometeorological measurements of N_2O were stopped 10 days after fertiliser application, measurements of N_2O by cham-

bers (SAC), showed, that in the following 2 weeks the N_2O emissons continued to decline (I McTaggart, pers. comm.). The N_2O emissions measured by 12 chambers (0.13 m^2) were much smaller than those measured by eddy covariance and conditional sampling. The peak chamber measurement made late in the morning, also 5 days after fertiliser application, was around 30 ng N_2O-N m^{-2} s^{-1}, about an order of magnitude smaller than the 280 ng N_2O-N m^{-2} s^{-1} measured by eddy covariance and 220 ng N_2O-N m^{-2} s^{-1} measured by conditional sampling. The discrepency between chamber and micrometeorological measurements was also observed for a grassland site in Central Scotland (Smith et al., 1994). Such differences are not surprising, considering the spatial variabilities generally observed for N_2O emissions and that the 12 chambers were integrating a flux over a total area of less than 2 m^2, in a part of the field, which was not sampled by the eddy covariance and conditional sampling systems, and that the eddy covariance and conditional sampling techniques were integrating fluxes over an area of 0.3 km^2.

The total loss of N_2O during the first 10 days after fertiliser application averaged over the field was 10 kg N_2O-N, and accounted for 0.77% of the total N fertiliser added. Very similar figures for the percentage of NH_4NO_3-N fertiliser lost as N_2O from Scottish soils have been reported by Mc Taggart et al. (1994).

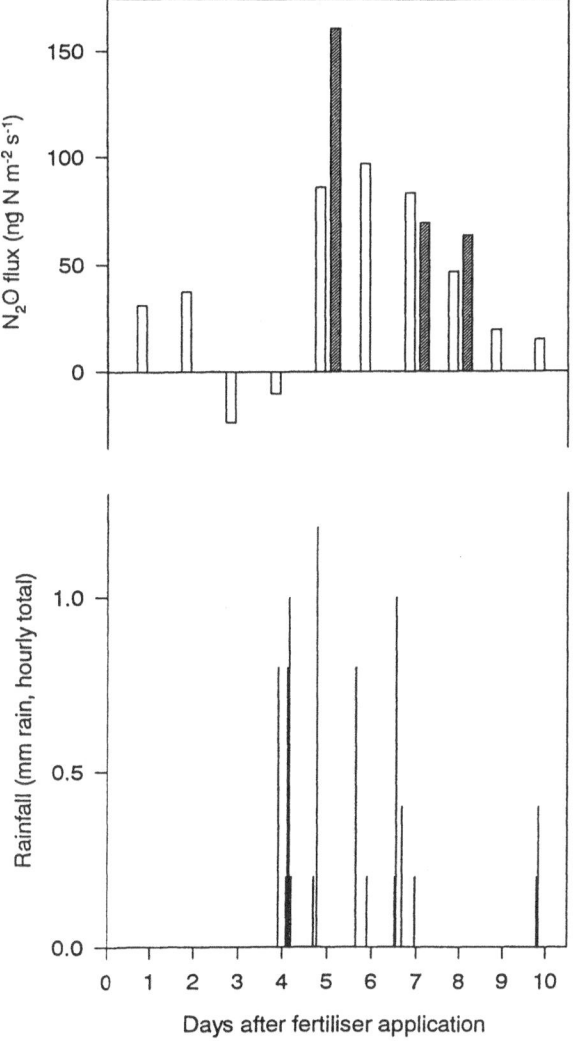

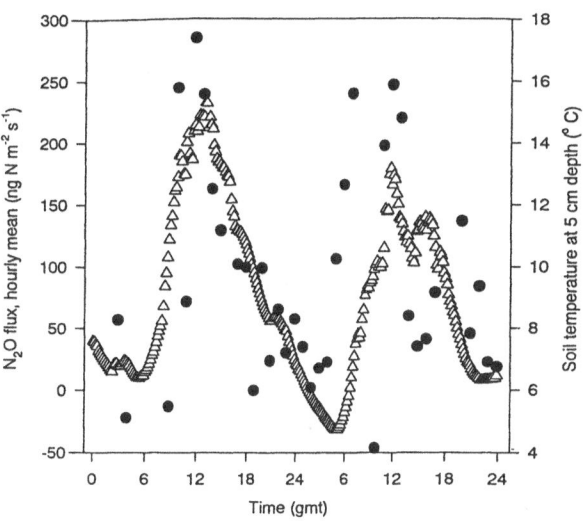

Figure 4. Diurnal variations of the N_2O flux (mean hourly flux) (●) 5 and 6 days after fertiliser application. The soil temperature was measured at a depth of 5 cm (Δ).

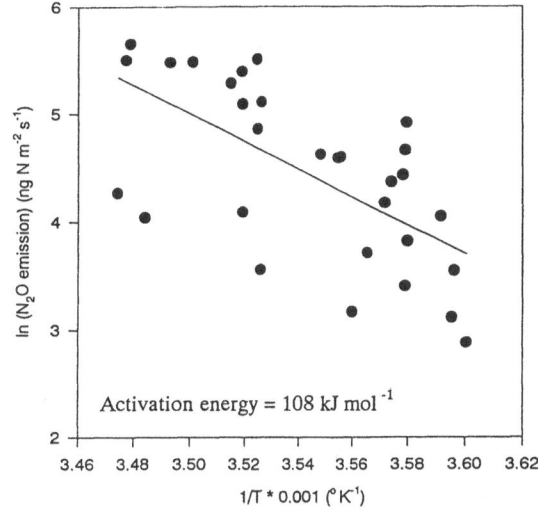

Figure 5. Temperature dependence of the N_2O emission 5 and 6 days after fertiliser application.

Figure 3. Daily mean N_2O fluxes measured by eddy covariance (open bar) and rainfall (hourly total) (solid bar). For comparison the short period conditional sampling fluxes are included (shaded bar).

Acknowledgements

The authors wish to thank the Natural Environment Research Council, Tiger programme, for financial support and Mr C G Spence for providing the field site.

References

Arah J R M 1988 Modelling denitrification in aggregated and structureless soils. *In* Nitrogen Efficiency in Agricultural Soils. Eds. D S Jenkinson and K A Smith. pp 433–444. Elsevier, London, UK.

Businger J A and Oncley S P 1990 Flux measurement with conditional sampling. J. Atmos. Ocean. Technol. 7, 349–352.

Folorunso O A and Rolston D E 1984 Spatial variability of field-measured denitrification gas fluxes. Soil Sci. Soc. Am. J. 48, 1214–1219.

Conrad R, Seiler W and Bunse G 1983 Factors influencing the loss of fertilizer - nitrogen into the atmosphere as N_2O. J. Geophys. Res. 88, 6709–6718.

Fowler D and Duyzer J H 1989 Micrometeorological techniques for the measurement of trace gas exchange. *In* Exchange of Trace Gases between Terrestrial Ecosystems and the Atmosphere. Eds. M O Andreae and D S Schimel. pp 189–207. John Wiley and Sons, London, UK.

Fowler D, Hargreaves K J, Skiba U 1995 Measurements of CH_4 and N_2O at the landscape scale using micrometeorological methods. Proc. R. Soc. A 351, 339–356.

144

Goodroad L L and Keeney D R 1984 Nitrous oxide production in aerobic soils under varying pH, temperature and water content. Soil Biol. Biochem. 16, 39–43.

Hargreaves K J, Skiba U, Fowler D 1994 Measurements of nitrous oxide emission from fertilised grassland using micrometeorological techniques. J. Geophys. Res. 99, 16569-16574.

Hargreaves K J, Wienhold F and Klemedtsson L 1996 Measurements of nitrous oxide from agricultural land using micrometeorological methods. Atmos. Environ. (*In press*).

Mc Taggart I P, Clayton H and Smith K A 1994 Nitrous oxide flux from fertilised grassland: strategies for reducing emissions. *In* Non-CO$_2$_ Greenhouse Gases. Eds. J van Ham, L J H M Janssen and R J Swart. pp 421–426. Kluwer Academic Publishers, Dordrecht, the Netherlands.

Moncrieff J B, Massheder J M and DeBruin H 1995 A system to measure surface fluxes of momentum, sensible heatflux, water vapour and carbon dioxide J. Hydrol. (*In press*).

Skiba U, Hargreaves K J, Smith K A and Fowler D 1992 Fluxes of nitric and nitrous oxides from agricultural soils in a cool temperate climates. Atmos. Environ. 26A, 2477-2488.

Smith K A, Clayton H, Arah J R M 1994 Comparison of methods for measurement of nitrous oxide fluxes between soils and the atmosphere, Stirling, Scotland, 1992: overview and conclusions. J. Geophys. Res. 99, D8, 16541–16548.

Webb E K, Pearman G I and Leuning R 1980 Correction of flux measurements for density effects due to heat and water vapour transfer. Quat. J. Meteorol. Soc. 106, 85–100.

Zahniser M S, Nelson D D and McManus J B 1995 Measurement of trace gas fluxes using tunable diode laser spectroscopy. Proc. R. Soc. A 351, 371-382.

Section editor: R Merckx

Plant and Soil **181**: 145–151, 1996.

Influence of carbon availability on the production of NO, N_2O, N_2 and CO_2 by soil cores during anaerobic incubation.

M. Swerts, R. Merckx and K. Vlassak
*Laboratory of Soil Fertility and Soil Biology, K.U.Leuven, Kardinaal Mercierlaan 92, B 3001 Heverlee, Belgium**

Key words: aerobic preincubation, available carbon, CO_2 production, denitrification, N_2O/N_2 ratio, water soluble carbon

Abstract

Net productions of permanent soil atmosphere gases (N_2, CO_2, O_2) and temporary gases (N_2O, NO) were monitored in soil cores using a non-interfering, fully automated measuring technique allowing highly time resolved measurements over prolonged periods. The influence of changes in available organic carbon on CO_2, N_2O, NO and N_2 production was studied by changing the soil carbon content through aerobic preincubations of different length, up to 21 days.

The aerobic preincubation caused an increase in NO_3^- concentration and a decrease in available carbon content. Available carbon content dominated both CO_2 and total N gas (N_2+N_2O+NO) production during anaerobiosis. Both CO_2 and total N gas production rates decreased with increasing length of the previous aerobic preincubation, this in spite of the higher initial NO_3^- concentration.

Total denitrification rates were closely related to the anaerobic CO_2 production rates. No relation was found between water soluble carbon content and total denitrification. The N_2O/N_2 ratio could be explained by an interaction of carbon availability, NO_3^- concentration and enzyme status. Net N_2O consumption was monitored. The balance between cumulative total N gas production and NO_3^- consumption varied according to the different treatments. Cumulative N_2O production exceeded cumulative N_2 production for 0 up to 5 days.

Introduction

Denitrification is a major source of N_2O, especially in intensively cultivated nitrogen rich soils (Bouwman, 1990). N_2O is implicated in the destruction of the stratospheric ozone layer (Crutzen, 1981) and N_2O is a long-lived infrared-absorbing greenhouse gas. Apart from its effects on the atmosphere, denitrification results in a decrease of available soil-N to plants. Both N_2 and N_2O result, in varying amounts, from denitrification. The major factors influencing the total denitrification rate are O_2-concentration, C-availability and NO_3^- concentration. The amounts of N_2O produced can vary from 0 to the majority of N reduced. The impact of several factors on the N_2O/N_2 ratio can largely be reduced to their influence on the relative availability of oxidant vs. reductant (Hutchinson and Davidson, 1993). Among the parameters explaining N_2O/N_2 ratios are N-oxide concentration, O_2 availabil-

ity, organic C availability, ratio's of enzyme activity, time since initiation of denitrification etc. .

Most studies on denitrification use the acetylene inhibition method (Yoshinari et al., 1977), often with measurement intervals of one to several days. The acetylene inhibition method has however several well known disadvantages.

We studied denitrification under a He atmosphere to avoid the use of acetylene. The aim of the research was (1) to study the dynamics of the denitrification process, and (2) to monitor the influence of NO_3^- and C-availability on the N_2O/N_2 ratio of the denitrification products.

* FAX No.: +32-16-321997
Plant and Soil is the original source of publication of this article. It is recommended that this article is cited as: *Plant and Soil* **181**: 145–151, 1996.

146

Materials and methods

Soil

The soil used for all experiments was sampled from the upper 0–10 cm of a clayey silt loam soil (Mal, Belgium) with following characteristics : texture : 0–2 μm : 17%, 2–50 μm : 69%, >50 μm : 14%; pH_{H2O} : 6.7 (1:2.5 soil:water); pH_{KCl} : 5.6 (1:2.5 soil:KCl 1N); total organic carbon : 12.3 mg C g^{-1} dry soil. The bulked samples were air dried to a moisture content of 180 mg water g^{-1} dry soil, sieved to pass a 2 mm sieve, mixed and stored at 4 °C.

Measurement set-up

A fully automated set-up was used allowing measurement of net gas production from nine independent soil cores during prolonged periods under an artificial atmosphere. The nine circuits, each containing one soil core, are connected through a selection valve to an injection valve with two sample loops for simultaneous injection in two gaschromatographs (GC), one with electron capture detector (ECD) for the analysis of O_2, NO, CO_2 and N_2O, and one with thermal conductivity detector (TCD) allowing the analysis of O_2, N_2, CO_2 and N_2O. Gas is continuously recirculated through the soil cores with membrane pumps to avoid diffusional problems. The artificial atmosphere allows direct measurement of N_2-production. A complete description of the set-up, the analysis conditions and the accuracy of the method is given by Swerts et al. (1995).

Chemical analysis

Mineral N was analysed in a soil extract (25 g soil : 50 mL KCl 1 N). Nitrate, NO_2^- and NH_4^+ concentrations were determined colorimetrically on a Skalar-autoanalyser. Water soluble carbon was determined on a soil extract (35 g soil : 100 mL H_2O_{dest}) using a persulfate oxidation method according to McCardell and Fuhrmann (1992).

Experimental

Bulk soil was aerobically preincubated at 25 °C for 7, 13 or 21 days (experiments AER7, AER13, and AER21, respectively). Moisture content after incubation was 173.5 mg water g^{-1} dry soil. For experiment ADDNO3, 6.9 mg NO_3^--N kg^{-1} dry soil was added to the soil (40.5 mL kg^{-1} moist soil of a solution of

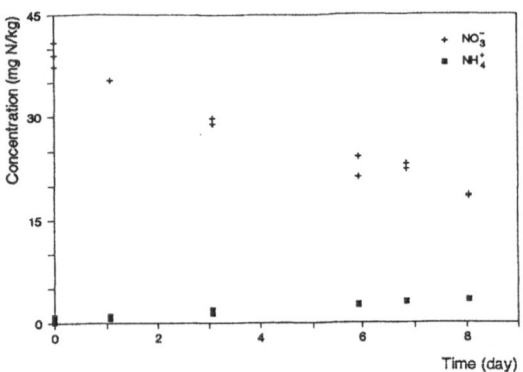

Figure 1. Time course of mineral N concentrations (mg N kg^{-1} soil) for experiment AER21.

104 mg KNO_3 100 mL^{-1} H_2O) to reach an initial NO_3^- concentration comparable to AER13, but without aerobic preincubation. Moisture content after addition was 214.3 mg water g^{-1} dry soil. For experiment ADDH2O, 40.5 mL H_2O kg^{-1} moist soil was added to the soil. The achieved moisture content was 215.5 mg H_2O g^{-1} dry soil.

For each experiment, 9 core holders (length 20 cm, diam. 6 cm) were packed with 450 g, for AER7, AER13 and AER21, or 468 g, for ADDNO3 and ADDH2O, carefully mixed moist soil and incubated at 25 +/- 0.1°C. Anaerobicity was obtained by flushing the soil cores for 10 min with He. The end of the He flushing was set as time 0 for the experiments.

Results

The different treatments resulted in initial NO_3^- concentrations of 31.3 +/- 1.3, 35.6 +/- 1.0, 39.0 +/- 1.0, 36.4 +/- 1.9 and 34.7 +/- 0.7 mg N kg^{-1} dry soil and initial water soluble carbon contents of 65 +/- 2, 57 +/- 6, 78 +/- 13, 64 +/- 8, and 76 +/- 16 mg C kg^{-1} dry soil for experiment AER7, AER13, AER21, ADDNO3 and ADDH2O, respectively. Nitrate concentrations declined linearly with time for all experiments. Changing NO_3^- and NH_4^+ concentrations are shown for experiment AER21 (Fig. 1). Nitrate concentrations reached zero at approximately day 9 for ADDNO3 and day 7 for ADDH2O. For all other experiments NO_3^- remained available throughout the experiment.

Patterns in water soluble carbon content changes were less obvious. In general an apparent initial decrease in soluble carbon content was followed by

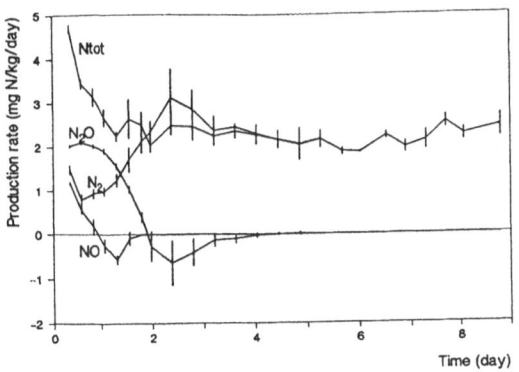

Figure 2. Time course of the N_2, N_2O, NO, and Ntot (N_2+N_2O+NO) production rates (average +/- standard deviation in mg N kg^{-1} day^{-1}) for experiment AER13.

a slight increase after day 5 (data not shown). Data for AER7 were extremely scattered.

The dynamic changes in CO_2 production rate and in evolution of several N gases were measured very accurately. As an example, Figure 2 shows the avg +/- std for production rates of the different N-gases against time for AER13. The average coefficient of variation for total N gas (N_2+N_2O+NO) production rates against time was 9.1% (0.9% to 36.1%) and 7.7% for CO_2 (2.2% to 20.2%).

The overall denitrification rates (total N gas production rates) of the different experiments are compared in Figure 3a. Total N gas production rates for AER13 and AER21 are clearly lower. The initially high rates for ADDNO3 and ADDH2O decline rapidly to reach a fairly constant level. At NO_3^- exhaustion (+/- day 7 for ADDH2O, +/- day 9 for ADDNO3) the production rates drop. For the N_2O production rates (Fig. 3b) AER7 and AER13 show a clear minimum with net consumption between day 2 and 3. AER21 has the same initial N_2O production rate but consumption is delayed. ADDNO3 and ADDH2O initially have higher N_2O production rates, after day 2 however there is no significant difference between AER21, ADDNO3 and ADDH2O. N_2 production rates (Fig. 3c) for AER7 and AER13 show a clear maximum at the time of maximal N_2O consumption. N_2 production rates for ADDNO3 and ADDH2O are quite constant until NO_3^- is exhausted. ADDNO3 and ADDH2O show a very fast reduction of the initially produced NO (Fig. 3d), for AER7, AER13, and AER21 consumption occurs later. CO_2 production rates (Fig. 3e) for AER13 and AER21 are significantly lower than for AER7, ADDNO3 and ADDH2O.

For experiments AER7, AER13, and AER21 there is a clear change in the balance between cumulative total N gas production and cumulative NO_3^- consumption (Fig. 4). For AER7 total amounts of NO_3^- consumed balance N gas production. For AER13 cumulative NH_4^+ production has to be subtracted from the cumulative NO_3^- consumption. In AER21 NO_3^- consumption clearly exceeds total N gas production. For ADDNO3 total N gas production is higher then NO_3^- consumption. In ADDH2O total N gas production is balanced by NO_3^- consumption minus NH_4^+ production.

Discussion

The main impact of aerobic preincubations of various length is expected to be an increase in NO_3^- concentration and a decrease in available carbon content with time. For NO_3^- this assumption is confirmed by the differences in initial NO_3^- concentration of AER7, AER13, and AER21. Although water soluble carbon content is considered to be a good indicator of available carbon content (Burford and Bremner, 1975; Katz et al., 1985), the measured water soluble carbon contents do not confirm the expected tendency in this study.

CO_2 production rates are considerably higher for AER7 compared to AER13. Production rates for AER13 are slightly higher than for AER21. Thus CO_2 production rates do confirm the decreasing C availability after aerobic preincubation. Perhaps the method to determine water soluble carbon content is not accurate enough to detect small differences at low C concentrations. Another possible explanation may be the extraction of carbon components which are in situ not readily available.

The higher availability of carbon in experiments AER7, ADDNO3, and ADDH2O compared to AER13 and AER21 is clearly reflected in the CO_2 production rates of Figure 3e. The easily available carbon is consumed throughout the experiment and CO_2 production rates decline accordingly, to approach about the initial CO_2 production rates of AER13 and AER21 by the end of the experiment.

Total denitrification rates (taken as the total N gas production or the sum of N_2O, N_2, and NO) (Fig. 3a) show, throughout the experiment a pattern parallel to CO_2 production rates. In spite of the increase in initial NO_3^- concentrations from AER7 over AER13 to AER21, their respective total N gas production rates decline. This points to available carbon as the most

148

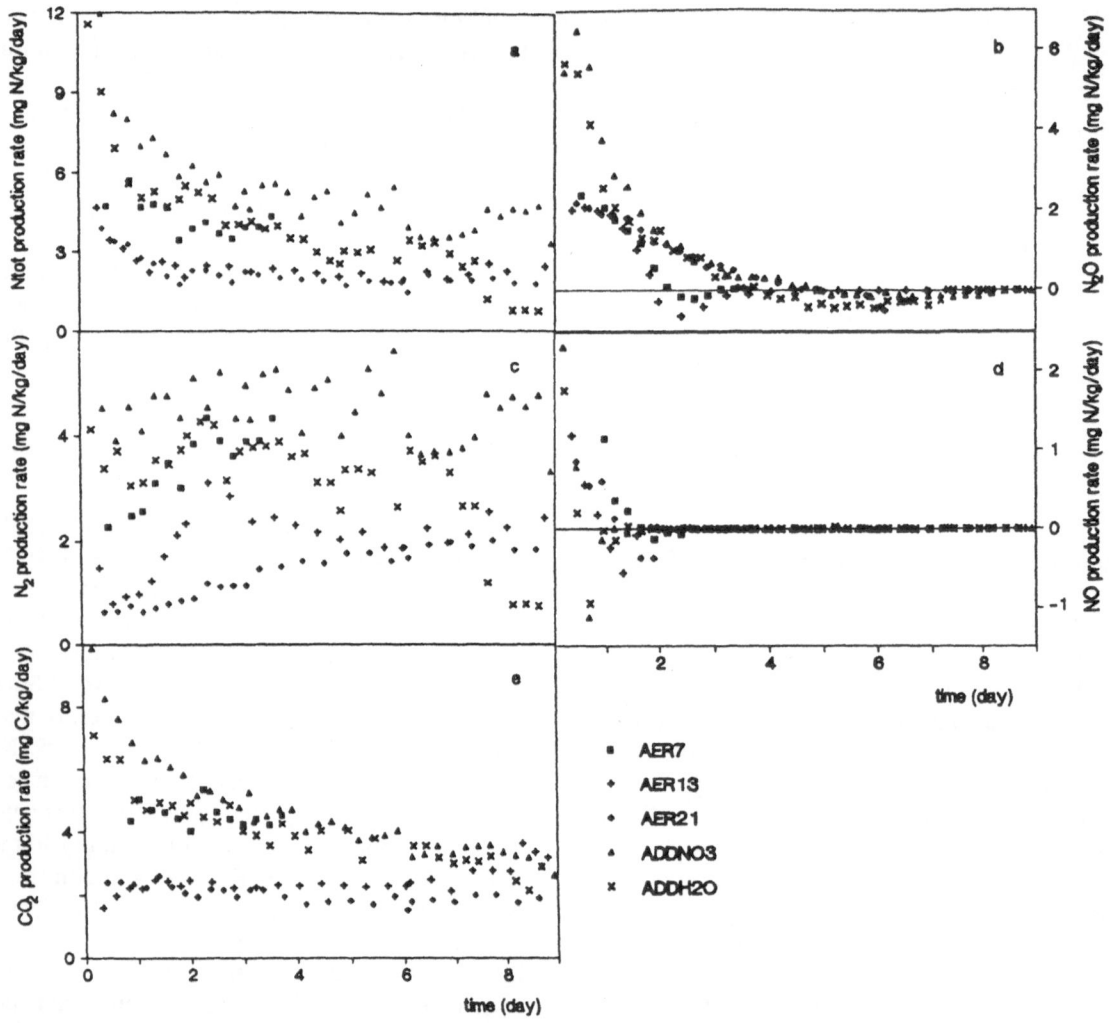

Figure 3. Time course of the average production rates of (a) Ntot (N_2+N_2O+NO), (b) N_2O, (c) N_2, (d) NO in mg N kg^{-1} day^{-1}, and (e) CO_2 in mg C kg^{-1} day^{-1} for the experiments AER7, AER13, AER21, ADDNO3, and ADDH2O.

influencing factor for the total denitrification rate under the given experimental conditions. This is conform to the findings of several other authors who concluded that carbon availability is the most limiting factor for denitrification under anaerobic conditions (Lalisse-Grundmann et al., 1988; Paul and Beauchamp, 1989; Weier et al., 1993). Most of those data, however, were obtained either through addition of carbon sources or through comparison of soil types differing in native carbon content. Our data, using one soil and aerobic preincubations of varying length, confirm the importance of available carbon.

In our study anaerobic CO_2 production was closely related to total denitrification. Water soluble carbon content however was a poor indicator. This is in accor-

dance with Paul and Beauchamp (1989) and Bijay-Singh et al. (1988) as well as with data in the review papers of Sahrawat and Keeney (1986) and Aulakh et al. (1992).

The conversion of net N_2O production to N_2 production is generally attributed to differences in reaction rate for the reduction of NO_3^-, NO_2^-, NO and N_2O (Comfort et al., 1990; Tiedje, 1982). Because of the preferential acceptance of electrons by NO_3^- over N_2O, N_2O accumulates whenever NO_3^- supply exceeds the reducing demand of the denitrifying organisms. We do not expect NO_3^- to inhibit N_2O reductase as the maximum NO_3^- concentrations were low (less then 40 mg NO_3^--N kg^{-1}).

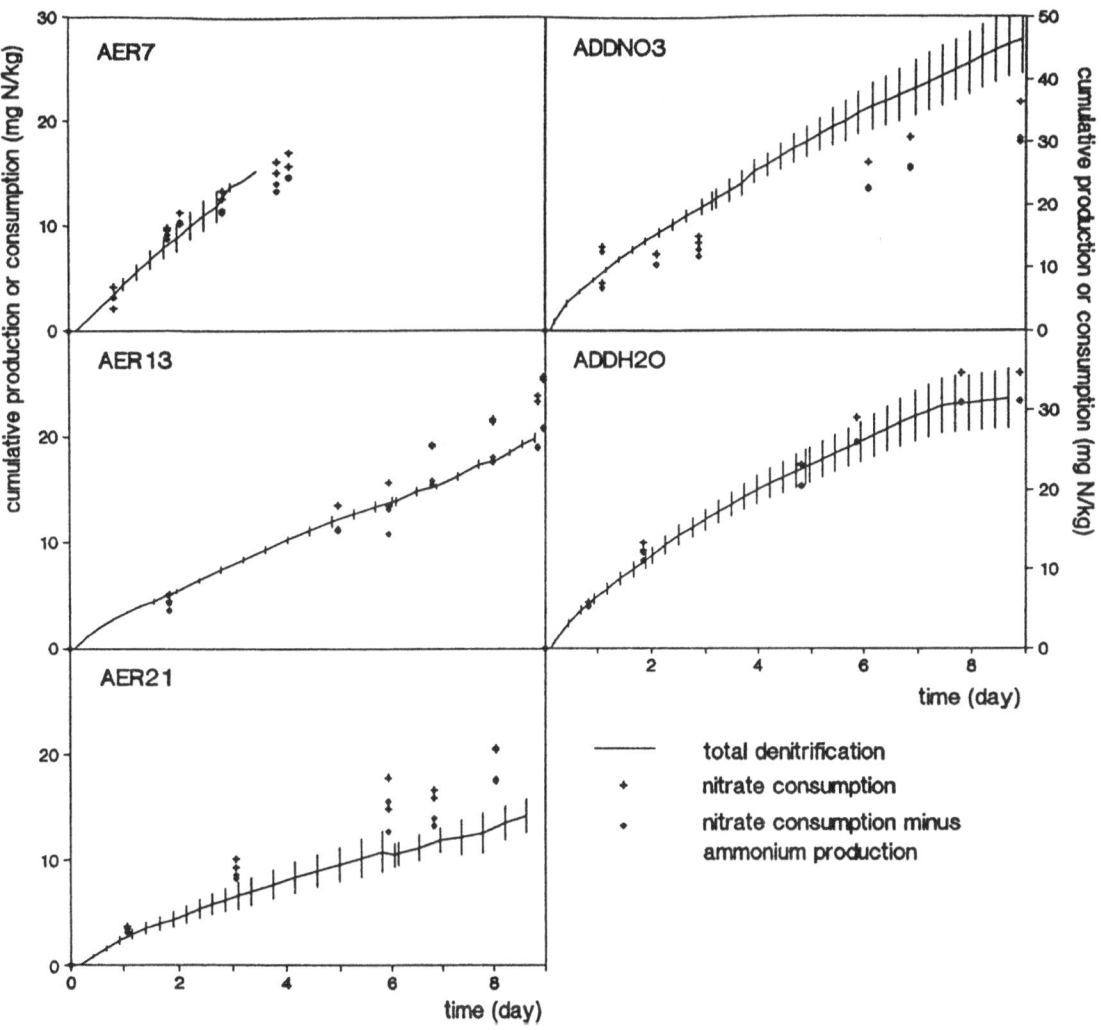

Figure 4. Time course of the cumulative denitrification (N$_2$+N$_2$O+NO) (average +/- standard deviation in mg N kg^{-1}) against cumulative nitrate consumption (mg N kg^{-1}) and cumulative nitrate consumption minus ammonium production (mg N kg^{-1}).

NO is reduced very quickly, in accordance with its high affinity for electron acceptance.

The sequence of increasing NO$_3^-$ concentration against decreasing available carbon content of AER7 over AER13 to AER21 has a pronounced influence on the relative amounts of N$_2$O produced compared to N$_2$. This is a clear example of the influence of the relative availability of oxidant versus reductant. Hutchinson and Davidson (1993) state that when the available oxidant largely exceeds the reductant the substrate N-oxide will be incompletely reduced resulting in a larger N$_2$O/N$_2$ ratio. Analysis based on Michaelis-Menten enzyme kinetics predicts that the proportion of N$_2$O should increase whenever any of the other controllers slows the overall rate of reduction below the maxi-

mum that can be supported by existing enzyme. This is in accordance with Sahrawat and Keeney (1986) who state that an increase in NO$_3^-$ concentration will cause an increase in N$_2$O/N$_2$ ratio while an increase in available carbon or in enzyme status will cause a decrease in the N$_2$O/N$_2$ ratio.

For AER7 and AER13 N$_2$ production rates increase to reach a maximum at day 2, at the time of maximum N$_2$O consumption (Fig. 3b,c). For ADDNO3 and ADDH2O, more C is available, but the N$_2$ production rates remain constant throughout the experiment until NO$_3^-$ becomes limiting. Net N$_2$O production rates reach net consumption only between day four and seven. This apparent slower reduction of N$_2$O to N$_2$ under higher C availability conditions might be due to differences in initial enzyme concentrations. The denitrification rates during the first phase of denitrification are determined by the preexisting enzymatic capacity to denitrify, rather than by nitrate or carbon availability. Although the enzymes involved in denitrification are considered to be rather stable over prolonged periods, the aerobic preincubations at 25°C may have caused some decay, resulting in lower initial N_2O and N_2 production rates. Subsequent increases in reductase concentrations could explain the faster reduction of N_2O to N_2 for AER7 and AER13 compared to ADDH2O and ADDNO3. The slower N_2O reduction for AER21 could be due both, to a higher nitrate to available carbon ratio or, to a slower enzyme synthesis activity caused by the lower C-availability.

Cumulative N_2O production exceeds the cumulative N_2 production for 2 days in AER13, for 5 days in AER21, for 1 day in ADDNO3, and for 1.5 day in ADDH2O (data not shown). In AER7 the N_2O production does not exceed the N_2 production.

Cumulative total N gas production during denitrification is balanced by NO_3^- consumption in AER7 (Fig. 4). For AER13 to AER21 NO_3^- consumption increasingly exceeds total N gas production. Due to the decreasing availability of carbon an increasing part of the oxidant becomes available for incorporation in the biomass.

In ADDNO3 the total N gas production exceeds the NO_3^- consumption. This is often attributed to an unknown amount of NO_3^- originating from nitrification (Paul and Beauchamp, 1989; de Catanzaro et al., 1987). Indeed some trace amounts of O_2 were detected during this experiment, probably due to a leak in the system. Together with the O_2, N_2 will have intruded into the system as well, resulting in an overestimation of N-gases produced through denitrification. On the other hand, another N oxide source may have attributed to total N gas production. Addition of 6.9 mg NO_3^--N kg^{-1} soil only resulted in an increase of the initial NO_3^- concentration of +/- 2 mg NO_3^--N kg^{-1} soil. Part of the missing NO_3^- might have been available to denitrification.

For ADDH2O total N production is well balanced by NO_3^- consumption.

From the presented data, it can be concluded that available carbon was the most influencing parameter on total denitrification. Anaerobic CO_2 production was closely related to total denitrification. Water soluble carbon content, however, could not be correlated with the denitrification rates. The N_2O/N_2 ratio was influenced by the balance between NO_3^- concentration and available carbon. Hypothetical differences in enzyme status had to be taken into account to further explain N_2O/N_2 ratios. Cumulative N_2O production exceeded cumulative N_2 production for several days during the anaerobic incubations at 25°C. At lower temperatures in the field N_2O is likely to be the predominant gas produced during anaerobicity. For most experiments total N gas loss was well balanced by NO_3^- consumption.

Acknowledgements

This research was conducted in the frame of the Impulse Programme "Global Change" supported by the Belgian State - Prime Minister's Service - Federal Office for Scientific, Technical and Cultural Affairs. We thank Griet Uytterhoeven for excellent technical support.

References

Aulakh M S, Doran J W and Mosier A R 1992 Soil denitrification - significance, measurement, and effects of management. Adv. Soil Sci. 18, 1–57.

Bijay-Singh, Ryden J C and Whitehead D C 1988 Some relationships between denitrification potential and fractions of organic carbon in air-dried and field-moist soils. Soil Biol. Biochem. 5, 737–741.

Bouwman A F 1990 Nitrous Oxide. In Soils and the Greenhouse Effect. Ed. A F Bouwman. pp 100–120. John Wiley and Sons, Chichester.

Burford J R and Bremner J M 1975 Relationship between the denitrification capacities of soils and total, water-soluble and readily decomposable soil organic matter. Soil Biol. Biochem. 7, 389–394.

Comfort S D, Kelling K A, Keeney D R and Converse J C 1990 Nitrous oxide production from injected liquid dairy manure. Soil Sci. Soc. Am. J. 54, 421–427.

Crutzen P J 1981 Atmospheric chemical processes of the oxides of nitrogen, including nitrous oxide. In Denitrification, Nitrification, and Atmospheric Nitrous Oxide. Ed. C C Delwiche. pp 17–44. John Wiley and Sons, New York, USA.

de Catanzaro J B, Beauchamp E G and Drury C F 1987 Denitrification vs. dissimilatory nitrate reduction in soil with alfalfa, straw, glucose, and sulfide treatments. Soil Biol. Biochem. 19, 583–587.

Hutchinson G L and Davidson E A 1993 Processes for production and consumption of gaseous nitrogen oxides in soil. In Agricultural Ecosystem Effects on Trace Gases and Global Climate Change. Eds. L A Harper, A R Mosier, J M Duxbury and D E Rolston. pp 79–93. ASA Special Publication 55, Madison, WI, USA.

Katz R, Hagin J and Kurts L T 1985 Participation of soluble and oxidizable soil organic compounds in denitrification. Biol. Fertil. Soils 1, 209–213.

Lalisse-Grundmann G, Brunel B and Chalamet A 1988 Denitrification in a cultivated soil : optimal glucose and nitrate concentrations. Soil Biol. Biochem. 20, 839–844.

McCardell A and Fuhrmann J J 1992 Determination of persulfate-oxidizable carbon by gas chromatography. Soil Biol. Biochem. 24, 615–616.

Paul J W and Beauchamp E G 1989 Effect of carbon constituents in manure on denitrification in soil. Can. J. Soil Sci. 69, 49–61.

Sahrawat K L and Keeney D R 1986 Nitrous oxide emission from soils. Adv. Soil Sci. 4, 103–148.

Swerts M, Uytterhoeven G, Merckx R and Vlassak K 1995 Semi-continuous measurement of soil atmosphere gases with a gas-flow soil core method. Soil Sci. Soc. Am. J. 59, 1336–1342.

Tiedje J M 1982 Denitrification. *In* Methods of Soil Analysis, part 2. Eds. A L Page, R H Miller and D R Keeney. pp 1011–1026. Agronomy Monograph 9, ASA-SSSA, Madison, USA.

Weier K L, Doran J W, Power J F and Walters D T 1993 Denitrification and the dinitrogen/nitrous oxide ratio as affected by soil water, available carbon, and nitrate. Soil Sci. Soc. Am. J. 57, 66–72.

Yoshinari T, Hynes R and Knowles R 1977 Acetylene inhibition of nitrous oxide reduction and measurement of denitrification and nitrogen fixation in soil. Soil Biol. Biochem. 9, 177–183.

Section editor: R Merckx

O. Van Cleemput et al. (eds.), Progress in Nitrogen Cycling Studies, 641–643, 1996.

Nitric oxide production in *Nitrosomonas europaea* under aerobic and anoxic conditions

C. Tomaschewski, I. Leuther, J. Groeneweg and W. Tappe

Forschungszentrum Jülich GmbH Institut für Biotechnologie 3 D-52428 Jülich, Germany

Key words: aerobic/anoxic conditions, ammonia, *N. europaea*, nitric oxide

Abstract

The trace gas nitric oxide (NO) substantially influences the chemistry of the troposphere. Depending on the mixing ratio, NO contributes to both, the net formation or degradation of ozone. Besides denitrification, nitrification plays a substantial role regarding the NO release from soils.

In this study, experiments are presented with the ammonia oxidizing bacterium *Nitrosomonas europaea* in pure cultures. Batch tests with this organism give an indication of different mechanisms regarding the formation of NO. Under aerobic conditions, the NO release was linearly correlated with the ammonia concentration. About 1.3% of the oxidized ammonia was released as NO. Under anoxic conditions, nitrite and additionally a biochemical reductant (e.g. acetate) were required for NO production.

We especially refer to the increased NO formation (doubling) after ammonia is added under anoxic conditions.

Abbreviation: $NH_x = NH_3 + NH_4^+$

Introduction

The trace gas nitric oxide (NO) has an essential influence on the chemistry of the troposphere. Depending on its concentration,, NO drives either the destruction or the formation of ozone. Among the natural origin of NO, nitrification and denitrification are significant, especially in soils. Under what conditions and how NO is produced by ammonia oxidizers like *Nitrosomonas europaea*, is not sufficiently clear yet. It is still an open question whether NO is a direct product of the hydroxylamine oxidoreductase-reaction (Hooper and Terry, 1979) or denitrification is the most abundant process for NO and N_2O production (Poth and Focht, 1985).

Some results on the influence of different oxygen partial pressures and substrates (ammonia, nitrite and acetate) on the production of NO are presented.

Materials and methods

Nitrosomonas europaea (Strain Nm35 from the culture collection of the University of Hamburg) was continuously cultivated at 25 °C in fermenters with 1 L net volume with biomass retention (dilution rate of biomass:0.03 h^{-1}, dilution rate of substrate: 0.1 h^{-1}). The inorganic medium contained per litre: KCl (16 mg), $MgSO_4 \times 7 H_2O$ (10 mg), $CaCl_2 \times 2 H_2O$ (3 mg), Na_2HPO_4 (100 mg), $FeCl_2 \times 4 H_2O$ (0.71 mg), Titriplex III (1.3 mg), $NaHCO_3$ (1129 mg), $MnCl_2 \times 4 H_2O$ (0.099 mg), $ZnSO_4 \times H_2O$ (0.0063 mg), $NiSO_4 \times 6 H_2O$ (0.0263 mg), $CoSO_4 \times 7 H_2O$ (0.0028 mg), $CuSO_4 \times 5 H_2O$ (0.005 mg), $(NH_4)_6Mo_7O_{24} \times 4 H_2O$ (0.008 mg), H_3BO_3 (0.305 mg), pH 7.8 (with 0.25 M H_2SO_4).

As substrates were added: for continuous culture: 80 mg NH_x-N L^{-1}(NH_4Cl); for batch experiments: 5, 10 or 12 mg nitrite- ($NaNO_2$) or ammonia-N (according to conditions, see figures).

In one experiment sodium acetate (100 mg C L^{-1}) was added as an organic carbon source.

Because the biomass-suspension contained a lot of nitrate, the cells were harvested by centrifugation (for 20 min. at 20000 g Sorvall) and washed in a medium without ammonia or nitrite. After washing, the cells were resuspended in 1 L of the same medium to a cell-concentration of 10^8 mL^{-1}. This culture suspension was put in a glass vessel with water jacket (25 °C) and

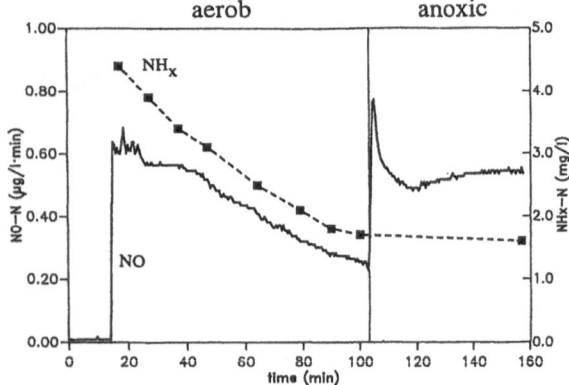

Figure 1. NO release under aerobic and anoxic conditions after addition of 5 mg NH$_x$-N L^{-1}.

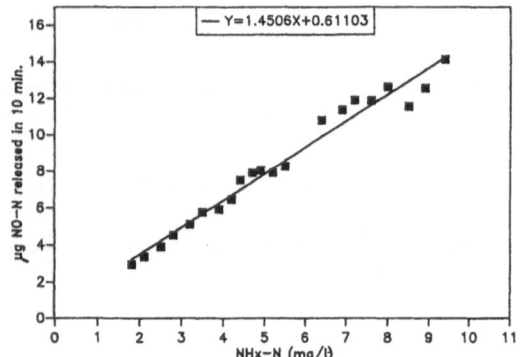

Figure 2. Correlation between the ammonia concentration and. NO produced (integrated over 10 minutes) during oxidation of ammonia.

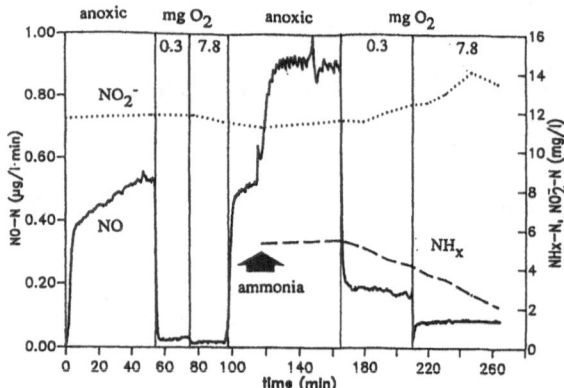

Figure 3. NO release at different oxygen partial pressures (0; 0.3; 7.8 mg O$_2$ L^{-1}). Starting conditions of the experiment: 12 mg nitrite-N L^{-1} and 100 mg acetate-C L^{-1}. After 2 hours, 5 mg NH$_x$-L^{-1} were added.

darkened to prevent photoinhibition. The suspension was gassed (with N$_2$ or compressed-air) with 70 L h^{-1} at 25 °C. The exhaust was fed into a NO chemoluminescence gas analyzer (Fa. Nucletron CLD 3.2-H) The pH value was kept constant at 7.8 by titration with 0.25 M H$_2$SO$_4$. The different oxygen partial pressures were measured and controlled by means of a Clark electrode (Syland) and a valve in the gas inlet.

Results

Under aerobic as well as anoxic conditions, *Nitrosomonas europaea* produced NO (Figs. 1 and 3). Figure 2 concludes the results from two experiments with biomass of the same age and pretreatment. In one experiment 5 mg L^{-1} NH$_x$-N (Fig. 1) were added in the other 10 mg L^{-1} NH$_x$N. The amounts of NO produced, integrated over ten minutes, were linearely correlated with the ammonia concentrations at the beginning of the ten minutes period. The aerobic release of NO

was linearly correlated with the ammonia concentration between 2–10 mg NH$_x$-N L^{-1} (Fig. 2), and about 1.3% of the oxidized ammonia was released as NO (Fig. 1).

Figure 3 shows the NO release at different partial pressures of oxygen (anoxic, 0.3 mg O$_2$ L^{-1} and 7.8 mg O$_2$ L^{-1}. Under anoxic conditions between 0.4–0.55 μg NO-N L^{-1} min^{-1} was released in the presence of nitrite and acetate.

The NO formation seemed to depend on nitrite concentration (data not shown) but only in the presence of an electron-donor. Immediately after aeration, the NO release dropped significantly. Restoring the anoxic conditions resulted in the same amount of NO released at the beginning of the experiment. Surprisingly subsequent addition of ammonia almost doubled the amount of the NO produced. Under renewed aerobic conditions (0.3 mg O$_2$ L^{-1}), but in the presence of ammonia the NO release was higher than under previous conditions in the absence of ammonia. Under oxygen limitation (0.3 mg O$_2$ L^{-1}) it seems that both an oxidation of ammonia and a reduction of nitrite took place as already Poth and Focht (1985) postulated.

Discussion

NO was released under aerobic as well as under anoxic conditions. There are probably different mechanisms responsible for the NO-production. The linear correlation between ammonia-concentration and the produced NO-amounts shows that an oxidative process causes the NO-formation. A possible explanation for this is the suggestion of Hooper and Terry (1979), who postulated a mixed functional hydroxylation, where NO is released during the oxidation of HNO.

Stüven et al. (1992) postulated that the oxidation of organic compounds (in addition to the ammonia oxidation) would generate additional electrons. They would feed the ammonia monooxygenase and lead to an increased hydroxylamine production. Furthermore hydroxylamine would reduce nitrite to NO or N_2O by chemodenitrification.

Under oxygen deprivation a nitrite reductase, which shows similarities with the nitrite reductase of denitrifiers, should be responsible for production of NO (Hooper, 1968). In the experiments of Hooper (1968) nitrite was reduced by hydroxylamine. Poth and Focht (1985) and Poth (1986) described a N_2O-release due to nitrite reduction under oxygen deprivation, during the ammoniaoxidation ammonia served as energy source. From the experimental set up of Poth and Focht (1985) and Poth (1986) it is not clear, whether real anoxic conditions were realized during the experiments.

Under true anoxic conditions (Fig. 3) the NO-release was dependent on the presence of nitrite *and* an electron-donor. Neither nitrite, nor acetate or ammonia alone caused NO-release (data not shown). However the release of NO was strongly increased by addition of ammonia. Abeliovich and Vonshak (1992) found that nitrite reduction in the presence of ammonia and pyru-

vate was stronger than if the components were given separately. The observed results could raise the question if ammonia could serve as a reduction equivalent during the nitrite reduction. Broda (1977) showed that this reaction is thermodynamically possible. Another question is, if there is a gain of energy by this reaction at least to satisfy maintenance demand of *N. europaea*. Perhaps ammonia oxidizers have a better chance to survive under anoxic conditions.

References

Abeliovich A and Vonshak A 1992 Anaerobic metabolism of *Nitrosomonas europaea*. Arch. Microbiol. 158, 267–270.

Broda E 1977 Two kinds of lithotrophs missing in nature. Z. Allg. Microbiol. 17, 491–493.

Hooper A B 1968 A nitrite reducing enzyme from *Nitrosomonas europaea*. Preliminary characterization with hydroxylamine as electron donor. Biochim. Biophys. Acta. 162, 49–65.

Hooper A B and Terry K R 1979 Hydroxylamine oxidoreductase of *Nitrosomonas* production of nitric oxide from hydroxylamine. Biochim. Biophys. Acta. 571, 12–20.

Poth M and Focht D D 1985 [15]N Kinetic analysis of N_2O production by *Nitrosomonas europaea*: An examination of nitrifier denitrification. Appl. Environ. Microbiol. 49, 1134–1141.

Poth M 1986 Dinitrogen production from nitrite by a *Nitrosomonas* isolate. Appl. Environ. Microbiol. 52, 957–959.

Stüven R, Vollmer M and Bock E 1992 The impact of organic matter on nitric oxide formation by *Nitrosomonas europaea*. Arch. Microbiol. 158, 439–443.

Plant and Soil **181**: 153–162, 1996.
© 1996 *Kluwer Academic Publishers.*

Effect of ammonium and nitrate application on the NO and N_2O emission out of different soils

Annick Vermoesen, Cornelis-Jan de Groot, Lode Nollet, Pascal Boeckx and
Oswald van Cleemput
*Faculty Agricultural and Applied Biological Sciences, University of Ghent, Coupure 653, B–9000 Ghent,
Belgium**

Key words: denitrification, greenhouse gas, nitric oxide, nitrification, nitrous oxide

Abstract

The effect of nitrate and ammonium application (0, 50, 100 and 150 mg N kg^{-1} soil) was studied in an incubation experiment. Four Belgian soils, selected for different soil characteristics, were used. The application of both nitrate and ammonium caused an increase of the NO and N_2O emission. The NO production from nitrate and ammonium was found to be of the same order of magnitude. At low pH the NO production was found to be highest from nitrate, at higher pH values the production was found to be higher from ammonium. This seems to be the result of the negative effect of low pH on nitrification.

The ANOVA analysis was carried out to separate the effect of the form of nitrogen, quantity of N applied and soil characteristics. The total production of NO was found to depend for 97% on the soil characteristics and for 3% on the quantity of N added. The total N_2O production depended for 100% on the soil characteristics.

Stepwise regression analysis showed that the total NO production was best predicted by a combination of the factors $CaCO_3$ content and NH_4^+ concentration in the soil. Total N_2O production was best described by a combination of $CaCO_3$, water soluble carbon (WSC) and sand-content.

The N_2O/NO ratio was found to be highly variable, indicating that their productions react differently to changes in conditions, or are partly independent.

It may be concluded that to NO and N_2O from soils both nitrification and denitrification may be equally important, their relative importance depending on local conditions such as substrate availability, water content of the soil etc. However, the NO production seems to be more nitrification dependent than the N_2O production.

Introduction

Nitric oxide plays an important role in the chemistry of the atmosphere by catalysing both the generation of O_3 and the oxidation of CO (Crutzen, 1979). Furthermore, it is a component of acid deposition. The contribution of NO_x (NO and NO_2) to total acidic deposition has been shown to increase rapidly (Logan, 1983). Although large differences exist between the estimates of the contribution of different sources, the NO-emission from soils seems to contribute significantly to global emission (Conrad, 1990).

Nitrous oxide is a greenhouse gas whose concentration has been increasing over the last decades. Its increase in concentration has been estimated at 50% as compared to pre-industrial times (Houghton et al., 1990). Although its absolute concentration is low compared to CO_2, it may significantly contribute to the greenhouse effect due to its high IR absorbing capacity and its long residence time in the atmosphere. In the stratosphere, N_2O may be converted into NO and thus contribute to the destruction of stratospheric ozone (Crutzen, 1981). Microbial processes in soils are the most important sources of N_2O (Granli and Bøckman, 1994).

Nitrification and denitrification are the main microbial processes producing N_2O and NO. Other biochemical oxidation or reduction reactions, like N_2-fixation and dissimilatory nitrate reduction may yield some traces of N_2O and NO as well. Abiotic production

* FAX No.: +3292646242

Plant and Soil is the original source of publication of this article. It is recommended that this article is cited as: *Plant and Soil* **181**: 153–162, 1996.

may occur through chemodenitrification (Van Cleemput and Baert, 1984).

Agricultural soils are important to the emission of nitrogen oxides due to the relatively high availability of inorganic nitrogen compounds and organic carbon. Nitrogen availability is high due to two sources of nitrogen: application of N-fertilizer and the mineralization of organic nitrogen, harvest residues and soil organic matter. The use of organic fertilizer may constitute a third source of organic carbon and of nitrogen.

In grassland N_2O losses normally account for less than 2% of the fertilizer applied (Granli and Bϕckman, 1994; Van Cleemput et al., 1994; Velthof and Oenema, 1994).

The production of N_2O and NO is generally considered to be positively related to the inorganic N availability. Mosier et al. (1983) described the production of $N_2 + N_2O$ during denitrification using Michaelis-Menten kinetics. They suggested that the N_2O production increased with increasing NO_3^- concentration, whereas the N_2 production had an optimum. The NO_3^- concentration giving the maximum denitrification rate varied between different studies (Granli and Bϕckman, 1994).

The aim of this study was to investigate the effect of inorganic N fertilizer concentration (NO_3^- and NH_4^+) on the emission of NO and N_2O from 4 soils with different characteristics. Furthermore, it was tried to identify the factors determining total NO and N_2O emissions and the responsible processes.

Material and methods

Soils were collected at the same time during the winter. Directly after collection, the samples were air-dried, ground, sieved (2 mm mesh size), and stored until use. Soil analyses were carried out according to Black et al. (1965). Inorganic soil nitrogen was extracted by shaking for 2 h with 1 M KCl (soil (g d.w.)/ KCl (mL) = 1/2). NH_4^+-N was determined by titration with HCl after steam-distillation. NO_3^--N was determined as NH_4^+-N after reduction with devarda alloy. To determine field capacity, the dry soil was saturated with water. Field capacity was defined as the water content after letting the soil drain freely for 24 hours. Table 1 shows the characteristics of the soils.

NO was measured using a Thermo Environmental Instruments Chemiluminescence NO-NO2-NOx analyzer Model 42, like described by De Groot et al. (1994a). Calibration was carried out using 10 ppm

NO standard gas. Using a gas blender, different concentrations were made for the calibration curve. Air, which was sucked through different filters (removing NO_x, N_2O and O_3) was used as blank.

N_2O was determined using a Chrompack 437A gas chromatograph with a ^{63}Ni ECD detector and a stainless steel Chromosorb 102 column of 4.88 m length and 3.175 mm diameter under the following conditions: injector temperature 90 °C, oven temperature 90 °C and detector temperature 300 °C. A detailed description is given by De Groot et al. (1994b).

To determine the emission of NO and N_2O from the 4 Belgian soils, 200 grams of dry soil was put in one litre jars. Water was added to reach the desired pre-incubation water content (75% of field capacity). The soil was left to equilibrate for 1 week to avoid artifacts due to the wetting effect. At t=0, fertilizer nitrogen was added. Then, water was added until field capacity was reached. The nitrogen fertilizers studied were NH_4Cl and KNO_3. Both fertilizers were applied in the following quantities: 0, 50, 100, and 150 mg N kg^{-1} of dry weight.

The soil NH_4^+ and NO_3^- content was determined before and after the experiment. The emissions of NO and N_2O were measured after 1, 2, 4, 7, 10, 14, 17 and 22 days. In case the NO and N_2O emissions became too low, measurements were stopped after 14 days. This had almost no influence on the total N_2O or NO production in comparison to longer incubation periods. First, a gas sample was taken for the analysis of N_2O, after which NO was measured. Between the measuring days, the recipients remained closed. But at every measuring day, the headspace of the jars was replaced by fresh air that was led through the filters. This means that no real anaerobic conditions were created.

Linear regression analyses, stepwise regression analyses, ANOVA and comparison of treatments were carried out with the Statgraphics package.

Results and discussion

The addition of fertilizer-N resulted in an increase in NO and N_2O in all soils (Figs. 1-4). This may indicate that nitrogen was limiting both the nitrification and denitrification rates. The results of the soil analyses generally confirm this idea.

In soil I (pH = 4.8) the addition of NO_3^- provoked a higher increase than the addition of NH_4^+. The pH = 4.8 of soil I is far from optimal for autotrophic nitrification (Hadas et al., 1986; Granli and Bϕckman, 1994) and,

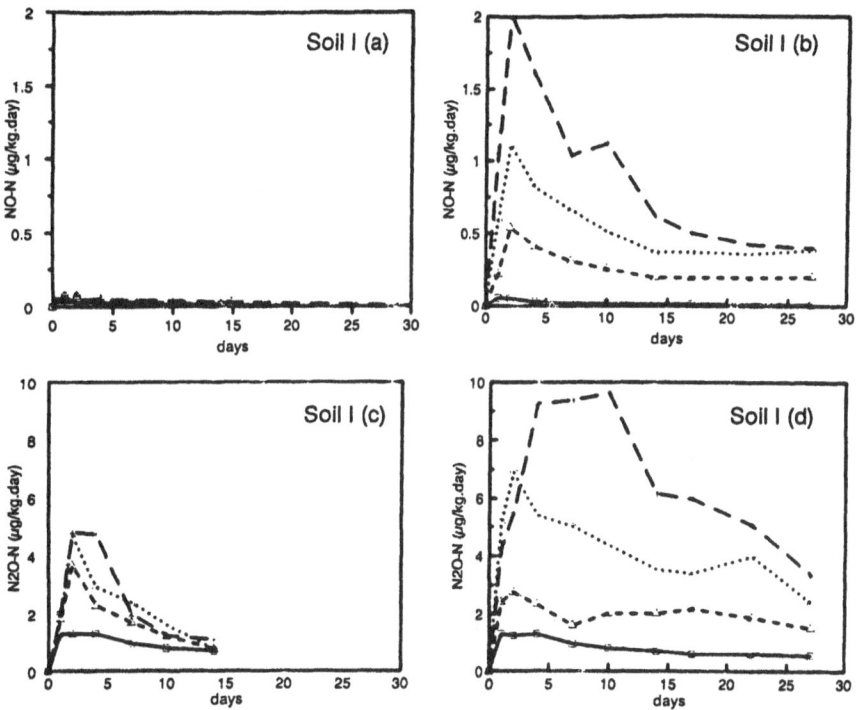

Figure 1. The emission of NO and N_2O from soil I upon the application of 0 (□), 50 (△), 100 (○) and 150 (*) mg N kg^{-1} of NH_4Cl and KNO_3: the emission of NO-N ($\mu g\ kg^{-1}\ day^{-1}$) upon application of (**a**) NH_4Cl and (**b**) KNO_3, the emission of N_2O-N ($\mu g\ kg^{-1}\ day^{-1}$) upon application of (**c**) NH_4Cl and (**d**) KNO_3.

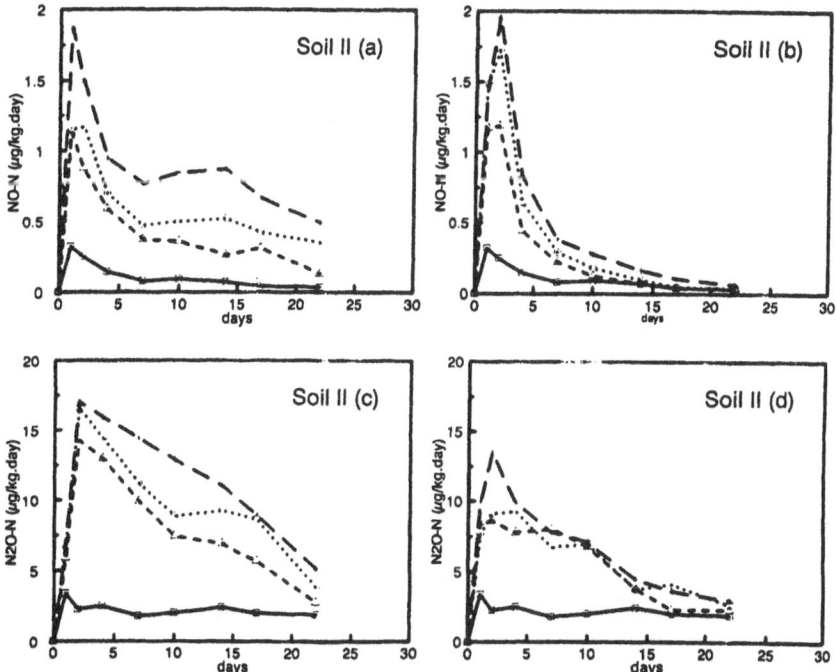

Figure 2. The emission of NO and N_2O from soil II upon the application of 0 (□), 50 (△), 100 (○) and 150 (*) mg N kg^{-1} of NH_4Cl and KNO_3: the emission of NO-N ($\mu g\ kg^{-1}\ day^{-1}$) upon application of (**a**) NH_4Cl and (**b**) KNO_3, the emission of N_2O-N ($\mu g\ kg^{-1}\ day^{-1}$) upon application of (**c**) NH_4Cl and (**d**) KNO_3.

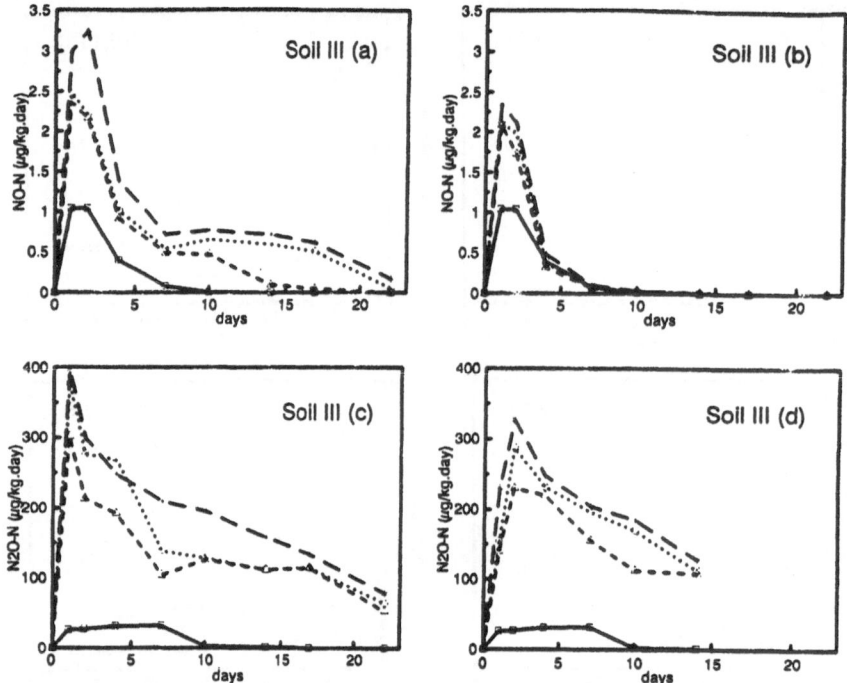

Figure 3. The emission of NO and N$_2$O from soil III upon the application of 0 (□), 50 (△), 100 (○) and 150 (*) mg N kg^{-1} of NH$_4$Cl and KNO$_3$: the emission of NO-N (μg kg^{-1} day^{-1}) upon application of (a) NH$_4$Cl and (b) KNO$_3$, the emission of N$_2$O-N (μg kg^{-1} day^{-1}) upon application of (c) NH$_4$Cl and (d) KNO$_3$.

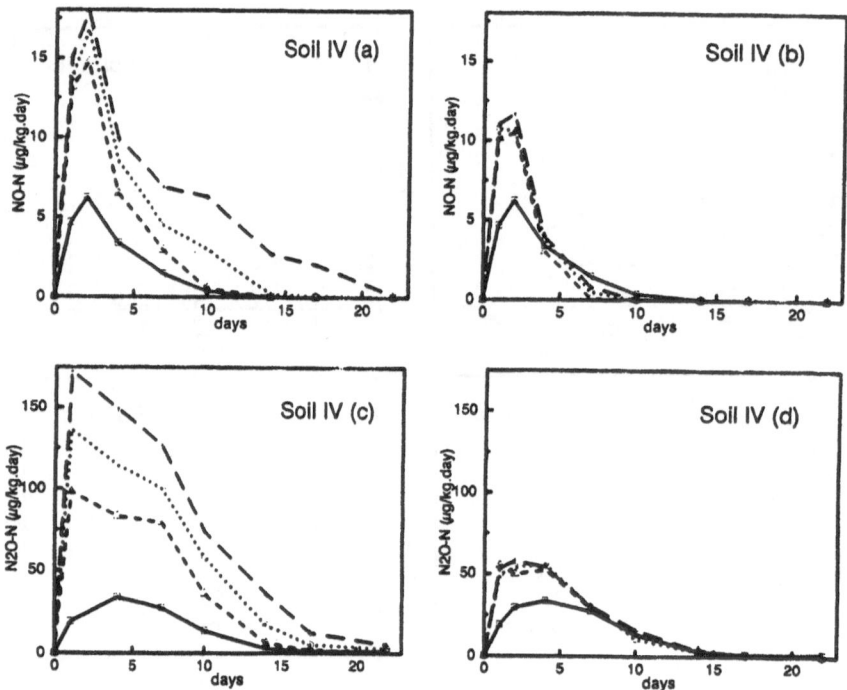

Figure 4. The emission of NO and N$_2$O from soil IV upon the application of 0 (□), 50 (△), 100 (○) and 150 (*) mg N kg^{-1} of NH$_4$Cl and KNO$_3$: the emission of NO-N (μg kg^{-1} day^{-1}) upon application of (a) NH$_4$Cl and (b) KNO$_3$, the emission of N$_2$O-N (μg kg^{-1} day^{-1}) upon application of (c) NH$_4$Cl and (d) KNO$_3$.

Table 1. Characteristics of studied soils (organic matter is abbreviated O.M., water soluble carbon WSC)

	Soil I	Soil II	Soil III	Soil IV
Soil type	Sandy	Sandy-loamy	Clayey	Heavy clay
Land-use	Heath	Arable land	Arable land	Arable land
pH (H_2O)	4.8	5.9	7.1	7.8
Sand (%)	97.5	47.2	65.4	19.0
Silt (%)	0.6	42.5	21.2	39.7
Clay (%)	1.9	10.3	13.6	58.4
CaCO3 (%)	0.30	0.30	0.39	16.8
O.M. (%)	2.52	1.58	2.28	2.82
WSC (mg C kg^{-1})	2.0	2.5	1.2	1.2
NH_4^+ t=0 (mg N kg^{-1})	11.2	25.2	16.5	8.0
NO_3^- t=0 (mg N kg^{-1})	0	3.0	12.0	2.9

therefore, nitrification rates were low. After incubation, the NH_4^+ concentration in the 0 mg kg^{-1} treatment was higher than at the beginning of the experiment (11.2 mg N kg^{-1}), indicating that NH_4^+ production by mineralization was higher than the consumption by nitrification (Table 3). At pH < 5 the oxidation of NO_2^- is often inhibited by HNO_2 toxicity (Granli and Bφckman, 1994). Under acidic conditions, NO_2^- accumulation is likely to occur. The denitrification rate is considerably lower at low pH, but this is compensated for by an increase in N_2O/N_2 ratio (Bremner and Blackmer, 1978; Weier and Gilliam, 1986), and, furthermore, chemo-denitrification may occur under acidic conditions (Van Cleemput and Baert, 1984). The decomposition of HNO_2 yields NO and HNO_3, or NO_2. The relatively low N_2O/NO ratio in this soil I, may be an indication of the occurrence of chemo-denitrification.

In soil IV (pH = 7.7) the addition of NH_4^+ gave the highest increase in NO and N_2O emission rather than NO_3^- addition. pH 7.7 is optimal for both nitrification and denitrification. As far as denitrification is concerned, the increase in process rate is generally counteracted by a decrease in N_2O/N_2 ratio (Granli and Bφckman, 1994; Weier and Gilliam, 1986). At pH 7.7 nitrification seems to have been the main process determining NO and N_2O production. This is in agreement with the results of many studies in which nitrification was identified as the major N_2O source (Bremner and Blackmer, 1978; Davidson et al., 1996; De Groot et al., 1994a,b; Skiba et al., 1994; Vermoesen et al., 1992) and NO producing process (De Groot et al., 1994a; Skiba et al., 1994). In Soil II (pH = 5.9) and III (pH = 7.1) the increase in the emission of NO

and N_2O was found to be of the same order of magnitude for the addition of NH_4^+ and of NO_3^-, indicating a simultaneous NO and N_2O formation by nitrification and denitrification.

Soil III and IV gave total N_2O emissions an order of magnitude higher than soil I and II. The NO emission (NO_3^- treatment) of the soils I, II and III was of the same order of magnitude. The NH_4^+ treatment of soil I resulted in total NO emissions an order of magnitude lower than the NO emission in soils II and III, whereas the NO emission of soil IV was considerably higher (Table 2). This conflicts with the conclusions of Remde et al. (1989), who found the highest NO production in an acid soil (pH = 4.7) as well as with the results of Nägele and Conrad (1990), who found a negative relation between NO and N_2O production and pH, in laboratory experiments adapting the pH of an acid (pH = 4.0) and an alkaline soil (pH = 7.8). The results obtained in their experiments, carried out under flushed conditions, can not directly be compared to ours, due to important differences in experimental set-up. Krämer and Conrad (1991) found that incubation in closed bottles resulted in much higher N_2O/NO ratios, than incubation under flushed conditions, probably due to an overestimation of NO production under flushed conditions. In a closed system, NO can still be reduced afterwards. Baumgärtner and Conrad (1992) found no significant correlation between soil pH and NO-production in a series of oxic soils. Out of the stepwise regression analysis on these data, it seemed that pH indeed has no influence on the N_2O and NO production.

The relative fertilizer effect (Table 2) gives an indication of the response of the NO and N_2O emission to

Table 2. Total emission of NO and N$_2$O (μg N kg^{-1}), N$_2$O/NO ratio and relative fertilizer effect (% of the increase caused by the lowest fertilizer dosis)

Soil	Fertilizer	Dose (mg N kg^{-1})	Total NO (μg N kg^{-1})	Total N$_2$O (μg N kg^{-1})	N$_2$O/NO N-ratio	Rel. fertil. effect NO	Rel. fertil. effect N$_2$O
I	NH$_4^+$	0	0.3	14	46.7	0	0
		50	0.4	22	55.0	100	100
		100	0.5	27	54.0	260	177
		150	0.6	30	50.0	460	216
	NO$_3^-$	0	0.4	18	45.0	0	0
		50	6.6	52	7.9	100	100
		100	13.0	112	8.6	206	277
		150	22.5	170	7.5	361	449
II	NH$_4^+$	0	2.1	98	46.7	0	0
		50	8.0	157	19.6	100	100
		100	11.7	194	16.6	163	164
		150	18.0	232	12.9	270	229
	NO$_3^-$	0	2.1	98	46.7	0	0
		50	5.0	128	25.6	100	100
		100	6.8	134	19.7	163	122
		150	8.6	143	16.6	223	154
III	NH$_4^+$	0	3.3	227	68.8	0	0
		50	9.7	2735	281.9	100	100
		100	14.4	3088	214.4	173	114
		150	19.0	3819	201.0	246	143
	NO$_3^-$	0	3.3	227	68.8	0	0
		50	4.8	2248	468.3	100	100
		100	5.1	2452	480.8	108	110
		150	5.8	2731	470.9	123	124
IV	NH$_4^+$	0	23.2	253	10.9	0	0
		50	51.4	738	14.4	100	100
		100	70.0	1059	15.1	166	166
		150	110.0	1451	13.2	308	247
	NO$_3^-$	0	23.2	253	10.9	0	0
		50	28.3	353	12.5	100	100
		100	30.3	350	11.5	141	97
		150	33.1	373	11.3	196	120

different doses of fertilizer. In all soils an approximately linear response of the total NO emission to increasing NH$_4^+$ addition was observed (Table 2), suggesting first order kinetics. For the emission of N$_2$O, this was only the case in soil IV. The addition of NO$_3^-$ caused an approximately linear increase in total emission of both NO and N$_2$O in soil I and of total NO emission in soil II. In all other cases increasing the fertilizer concentration did not lead to a corresponding increase in emission. Since this was not the case for N$_2$O emission, it seems as if the production of the NO and N$_2$O are not necessarily linked, or at least respond differently to changes in conditions. This is in agreement with the

fact that for certain soils (II: NH$_4^+$ and III: NO$_3^-$) a very poor relation was obtained between N$_2$O emission and NO emission (Table 4). For most soil-fertilizer combinations an excellent relation was obtained between N$_2$O and NO emission rates (Table 4). The variation in N$_2$O/NO ratio was considerable (Table 2) and seems to depend both on the soil characteristics and form of fertilizer applied. The variation in N$_2$O/NO ratio was higher than in the study of De Groot et al. (1994a).

The soil NH$_4^+$ and NO$_3^-$ content after incubation is given in Table 3. Both N-compounds depend on a producing and a consuming process. In the case of NH$_4^+$ there is a production via mineralization and a consump-

Table 3. Soil NH_4^+ and NO_3^- content after incubation

Soil	Fertilizer	Measuring period (days)	Dose (mg N kg^{-1})	NH_4^+ end (mg N kg^{-1})	NO_3^- end (mg N kg^{-1})
I	NH_4^+	14	0	17.6	0
			50	nd	nd
			100	nd	nd
			150	nd	nd
	NO_3^-	27	0	17.6	0
			50	29.4	34.0
			100	30.7	66.8
			150	30.6	97.1
II	NH_4^+	22	0	1.4	32.0
			50	35.1	31.5
			100	93.0	19.7
			150	150.2	14.2
	NO_3^-	27	0	1.4	32.0
			50	23.8	78.9
			100	26.6	118.1
			150	27.8	163.8
III	NH_4^+	22	0	0.0	39.2
			50	9.1	40.3
			100	29.6	60.7
			150	40.1	84.6
	NO_3^-	14	0	0.0	38.2
			50	0.0	85.3
			100	0.0	127.3
			150	0.0	166.0
IV	NH_4^+	22	0	0.8	44.9
			50	23.1	49.8
			100	43.3	68.9
			150	72.9	89.9
	NO_3^-	14	0	0.8	44.8
			50	0.0	97.6
			100	0.0	146.5
			150	0.0	190.5

Table 4. Results of linear regression analyses of NO (depend variable) against N_2O for the individual soils (* $p<0.01$, ** $p<0.0001$, *** $p<0.00001$)

Fertilizer	Soil	x-Coefficient	Constant	R^2	F	Degrees of freedom
NH_4Cl	I	0.0134	0.0145	0.525**	24.36	22
	II	0.0523	0.237	0.264*	10.75	30
	III	0.00706	-0.123	0.732***	82.09	30
	IV	0.0784	-0.0176	0.800***	108.4	27
KNO_3	I	0.140	-0.0315	0.635***	59.06	34
	II	0.139	-0.326	0.663***	66.74	34
	III	0.00392	0.103	0.177	4.737	22
	IV	0.177	1.60	0.631***	37.59	22

Table 5. Nested loop design ANOVA analyses of total NO emission and total N$_2$O emission

Source of var.	Sum of sq.	DF	Mean sq.	Var. comp.	%
Total NO					
Fertilizer	675.3	1	675.3	20.12	3.37
Dosis	2119.8	6	353.3	0.000	0.00
Soil	13848.4	24	577.0	577.0	96.63
Total N$_2$O					
Fertilizer	578619	1	578619	0	0
Dosis	4.5e+07	6	757889	0	0
Soil	3.2e+08	24	1318491	100,000	100

tion by nitrification; in the case of NO$_3^-$: production via nitrification and consumption by denitrification.

In soil I, without addition of fertilizer, there is formation of NH$_4^+$ by mineralization. Although the amount of available NH$_4^+$, almost no NO or N$_2$O were formed. Also upon addition of NH$_4^+$ there was only a slight increase in the N$_2$O production. Upon addition of NO$_3^-$, it is consumed and this results in higher NO and N$_2$O emissions. So it is clear that in soil I the NO and N$_2$O production were due to denitrification rather than to nitrification. This is in agreement with the observation that, for soil I, the emissions of NO and N$_2$O were higher after the addition of NO$_3^-$ than after NH$_4^+$. Krämer and Conrad (1991) only found denitrifying bacteria and no nitrifying bacteria in an acidic forest soil (pH=4.0), indicating that the observed NO and N$_2$O production were mainly denitrification linked, like soil I in our study. In soil I, the addition of NO$_3^-$ led to an increase in NH$_4^+$ end (Table 3). This suggests that the addition of NO$_3^-$ had a stimulating effect on the mineralization of organic N. The addition of inorganic nitrogen compounds is known to speed up mineralization in systems with a high C/N ratio, in which N-availability limits mineralization (Russell, 1961).

In soils II, III and IV without fertilizer addition, there was a consumption of NH$_4^+$ (almost 100%) and a production of NO$_3^-$. This means that under these conditions, nitrification is higher than denitrification. Also upon the addition of fertilizers, the consumption of NH$_4^+$ and production of NO$_3^-$ remained, so that in these soils the NO and N$_2$O production were mainly due to nitrification rather than denitrification.

The ANOVA (Table 5) showed that of the 3 factors investigated in this study (Soil, Fertilizer and Dose) the factor Soil was the most important determining both the total NO (96.63% of variation) and the total N$_2$O emission (100% of variation). The remaining 3.37% of variation in total NO emission was accounted for by the factor Fertilizer. This is in agreement with Groffman (1993) who stated that on landscape scale, soil type (together with plant community) was the most important factor controlling both nitrification and denitrification.

To investigate which of the soil characteristics determined the variation in NO and N$_2$O emission, stepwise regression analyses were carried out, confronting total NO and N$_2$O emission with soil characteristics. For total NO emission, the best fit was obtained with a model, containing the variables CaCO$_3$ and NH$_4^+$ t=0 (R^2 adj. 0.7328). The NO emission was positively correlated with both (Table 6a). For total N$_2$O emission a model was obtained, containing pH, sand content and NH$_4^+$ t=0 (R^2 adj. 0.6750). The total N$_2$O emission was positively correlated with all 3 variables (Table 6b). In this model pH was the most significant factor, explaining most of the variation. This is in agreement with the results of Bandibas et al. (1994) who showed that pH was the most important factor predicting total N$_2$O emission from 16 Belgian soils. Soil pH and CaCO$_3$ content are related due to the buffering capacity of CaCO$_3$ and a positive relation exists between the two. However CaCO$_3$ also has an effect on the texture of the soil by gluing particles together (Browning and Millam, 1944; Czeratzki, 1957) and may thus influence the diffusion characteristics of the soil as well. Bandibas et al. (1994) found that CaCO$_3$ was the most significant factor predicting the maximum N$_2$O concentration under saturated conditions, comparing 16 soils. They suggested that its effect on the soil texture may be the reason why CaCO$_3$ sometimes better explained variation in N$_2$O emission

Table 6. Results of stepwise regression analyses confronting total NO (**a**) and N_2O (**b**) emission with soil characteristics

Independent variables	Coefficient	Std. err.	t-Value	R^2-adj.	Sig. level
a.					
$CaCO_3$	2.337	0.3089	7.5665	0.7328	0.0000
NH_4^+ $t=0$	0.206	0.0433	4.7403		0.0001
Constant	-2.494	3.4117	-0.7311		0.4712
Source	Sum of sq.	DF	Mean sq.	F	P-value
Model	11848.3	2	5924.15	39.3862	0.0000
Error	3910.51	26	150.404		
Total(corr)	15758.8	28			
b.					
pH	11.66	174.3	6.6931	0.6750	0.0000
Sand	42.67	7.083	6.0241		0.0000
NH_4^+ $t=0$	9.025	2.290	3.9330		0.0061
Constant	-9521	1464	-6.5001		0.0000
Source	Sum of sq.	DF	Mean sq.	F	P-Value
Model	24871040	3	8290347	20.3852	0.0000
Error	10167123	25	406685		
Total(corr)	34038164	28			

linked parameters than did pH. NH_4^+ $t=0$ appears in the total NO as well as in the total N_2O model. This is probably due to its direct effect on nitrification on one hand and the stimulating effect on denitrification on the other hand. Baumgärtner and Conrad (1992) also found a significantly positive correlation between NH_4^+ and NO production rates in a series of soils.

Conclusions

The emission of N_2O and NO in the 4 soils studied, was found to increase both upon the application of NH_4^+ and NO_3^-, indicating that both nitrification linked and denitrification linked production occurred simultaneously. At low pH (pH = 4.8), denitrification was the dominant process. The low N_2O/NO ratio observed in the NO_3^- treatment may indicate the occurrence of chemo-denitrification. At high pH (pH = 7.7) nitrification was dominant. A stimulating effect of nitrification on denitrification was observed in all soils except soil I (pH = 4.8). This seems to be caused by nitrification linked O_2 consumption creating conditions more favourable to denitrification.

ANOVA showed that the variation in total NO (96.63%) and N_2O emission (100%) was almost entirely explained by the factor Soil. The factor Fertilizer explaining the remaining 3.37% of variation in total NO emission.

Stepwise regression analysis was used to determine the soil characteristics explaining the highly explicative value of the factor Soil in the ANOVA. For total NO emission the best description was obtained with the variables $CaCO_3$ content and NH_4^+ $t=0$ (R^2 adj. 0.7328), whereas for total N_2O a model was selected containing pH, sand content and NH_4^+ $t=0$ (R^2 adj. 0.6750), which seemed to give a more realistic description than a model with $CaCO_3$ content, WSC and NH_4^+ $t=0$ (R^2 adj. 0.6915).

For six of the eight soil-fertilizer combinations a good to excellent correlation was obtained between N_2O emission rates and NO emission rates. The N_2O/NO ratio was found to be highly variable, indicating that the processes responsible for the emission do not depend in the same way on different conditions. The lack of correlation between the N_2O emission rates and NO emission rates in two soil-fertilizer combinations and the fact that total NO emission did not respond in the same way to increasing doses of fer-

162

tilizer as did the N_2O emission, indicates that NO production and N_2O production are not necessarily linked.

Acknowledgements

Financial support was provided by the IWONL (Institute for Encouraging Scientific Research in the Industry and Agriculture, Brussels), the Commission of the European Community DGXII (the "Human Capital and Mobility" Programme and the "STEP" programme) and by the "Ministerie voor Wetenschapsbeleid" (Project EUREKA, EUROTRAC, BIATEX EU7/03).

References

Bandibas J, Vermoesen A, De Groot C J and Van Cleemput O 1994 The effect of different moisture regimes and soil characteristics on nitrous oxide emission and consumption by different soils. Soil Sci. 158, 106–114.

Baumgärtner M and Conrad R 1992 Effects of soil variables and season on the production and consumption of nitric oxide in oxic soils. Biol. Fertil. Soils 14, 166–174.

Black C A, Evans D D, White J L, Ensminger L E and Clark F E 1965 Methods of soil analysis. Am. Soc. Agron., Madison, WI, USA. 1572 p.

Bremner J M and Blackmer A M 1978 Nitrous oxide: emission from soils during nitrification of fertilizer nitrogen. Science 199, 295–296.

Browning G M and Millam F M 1944 Effect of different types of organic materials and lime on soil aggregation. Soil Sci. 57, 91–106.

Conrad R 1990 Flux of NO_x between soil and atmosphere: Importance of microbial metabolism. In Denitrification in Soil and Sediment. Eds. N P Revsbech and J Sorensen. pp 105–128. Plenum press, New York.

Czeratzki W 1957 Untersuchungen über Krümelstabilität an einem Kalkversuch. Pflanzenernähr. Düng. Bodenkd. 78, 121–135.

Crutzen P J 1979 The role of NO and NO_2 in the chemistry of the troposphere and the stratosphere. Ann. Rev. Earth Planet. Sci. 7, 443–472.

Crutzen P J 1981 Atmospheric chemical processes of the oxides of nitrogen, including nitrous oxide. In Denitrification, Nitrification and Atmospheric N_2O. Ed. C C Delwiche. pp 17–44. John Wiley and Sons, Chichester, UK.

Davidson E A, Swank W T and Perry T O 1986 Distinguishing between nitrification and denitrification as sources of gaseous nitrogen production in soil. Appl. Environ. Microbiol. 52, 1280–1286.

De Groot C J, Vermoesen A and Van Cleemput O 1994a Laboratory study of the emission of NO and N_2O from some Belgian soils. Environ. Monit. Assess. 31, 183–189.

De Groot C J, Vermoesen A and Van Cleemput O 1994b Laboratory study of the emission of N_2O and CH_4 from a calcareous soil. Soil Sci. 158, 355–364.

Granli T and Bϕckman O 1994 Nitrous oxide from agriculture. Norw. Agric. Sci. Suppl. 12. 128 p.

Groffman P M 1993 Ecology of nitrification and denitrification in soil evaluated at scales relevant to atmospheric chemistry. In Microbiol Production and Consumption of Greenhouse Gases: Methane, Nitrogen oxides and Halomethane. Eds. J E Rogers and W B Whitman. pp 201–217. American Society for Microbiology, Washington DC, USA.

Hadas A, Feigenbaum S, Feigin A and Portnoy R 1986 Nitrification rates in profiles of differently managed soil types. Soil Sci. Soc. Am. J. 50, 633–639.

Houghton J T, Jenkins G J and Ephraums J J 1990 IPCC (Intergovernmental Panel on Climate Changes) Climate changes. pp 36–37. Cambridge University Press, Cambridge, UK.

Krämer M and Conrad R 1991 Influence of oxygen on production and consumption of nitric oxide in soil. Biol. Fertil. Soils 11, 38–42.

Logan J 1983 Nitrogen oxides in the troposphere: Global and regional budgets. J. Geophys. Res. 88, 10785–10807.

Mosier A R, Parton W J and Hutchinson G L 1983 Modelling nitrous oxide evolution from cropped and native soils. Environ. Biogeochem. Ecol. Bull. 35, 229–241.

Nägele W and Conrad R 1990 Influence of pH on the release of NO and N_2O from fertilized and unfertilized soil. Biol. Fertil. Soils 10, 139–144.

Remde A, Slemr F and Conrad R 1989 Microbiol production and uptake of nitric oxide in soil. FEMS Microbiol. Ecol. 62, 221–230.

Russell E W 1961 Soil Conditions and Plant Growth. Ed. G Longmans. Green, London.

Skiba U and Fowler D 1994 Emissions of NO and N_2O from soils. Environ. Monit. Assess. 31, 153–158.

Van Cleemput O and Baert L 1984 Nitrite: a key compound in N loss processes under acid conditions. Plant and Soil 76, 233–241.

Van Cleemput O, Vermoesen A, de Groot C J and Van Ryckeghem K 1994 Nitrous oxide emission out of grassland. Environ. Monit. Assess. 31, 145–152.

Velthof G L and Oenema O 1994 Nitrous oxide emission from grasslands on sand, clay and peat soils in the Netherlands. In Non-CO_2 Greenhouse Gases. Why and how to Control? Proceedings of an International Symposium, Maastricht, The Netherlands, 13–15 December 1993. Eds. J Van Ham, L J H M Janssen and R J Swart. pp 439–444. Kluwer Academic Publishers, Dordrecht, The Netherlands.

Vermoesen A, Corre M and Van Cleemput O 1992 Nitrous oxide emission during nitrification. In Nitrogen cycling and leaching in cool and wet regions of Europe, COST 814, Commission of the European Communities. Eds. K François, K Pitman and N Bartiaux-Thill. pp 33–34. CEC, Brussels, Belgium.

Weier K L and Gilliam J W 1986 Effect of acidity on denitrification and nitrous oxide evolution from atlantic coastal-plain soils. Soil Sci. Soc. Am. J. 50, 1202–1205.

Section editor : R Merckx

O. Van Cleemput et al. (eds.), *Progress in Nitrogen Cycling Studies*, 655–658, 1996.

The effect of soil pH on nitrification in coarse sandy soil

Finn P. Vinther and Finn Eiland
Danish Institute of Plant and Soil Science, Department of Soil Science, Research Centre Foulum, Postbox 23, DK–8830 Tjele, Denmark

Key words: field incubations, laboratory incubations, nitrification, N-mineralization, soil pH

Abstract

The effect of soil pH on the nitrification at low temperatures was studied in a coarse sandy soil. Field incubations were carried out during two winter periods in a field trial with soil pH-values from 4.2 to 6.2. In the laboratory the effect of soil pH on N-mineralization and nitrification was measured after application of liquid pig manure and potential nitrification was also measured.

Both potential nitrification, field and laboratory incubations showed that the nitrification rates increased with increasing soil pH, whereas the N-mineralization after application of liquid pig manure was not affected.

Introduction

Use of animal manure in agriculture is an important source of nitrogen supply to the crops, but may also be an environmental problem due to leaching of nitrate. To reduce N-leaching from arable land, many fields are covered by catch crops or wintercrops during the winter periods. However, there is still a need for supplementary approaches.

It has been shown in several investigations, e.g. Nyborg and Hoyt (1978), Lyngstad (1979), Terry et al., (1981), Weier and Gilliam (1986) and Killham (1987) that the nitrification rates increase with increasing soil pH. All of these investigations have been carried out at relative high temperatures (20–25 °C). Vilsmeier and Amberger (1988) found in laboratory incubation experiments that nitrification after decomposition of liquid manure in two different soils with different pH was influenced by soil pH also at low temperature. The nitrification was slower at pH 5.7 than at 7.2, when the soil temperature was below 5 °C. These observations indicated that there could be a possibility of reducing the nitrate leaching from arable land by a reduction of chalk additions, i.e lowering of soil pH.

To elucidate this hypothesis, we have performed studies both in the field and in the laboratory with one type of soil from a Danish field trial (fertilizer/chalk experiments), where various amounts of chalk have been added during several years resulting in plots with

Table 1. Physical and chemical characteristics of the soils

Characteristics	Soil pH			
	4.2	5.0	5.9	6.2
Texture, %				
clay (< 2 μm)	4.6	5.1	4.9	5.0
silt (2–20 μm)	3.8	3.4	3.6	3.4
sand (20–200 μm)	15.8	15.9	15.9	15.9
sand (> 200 μm)	73.7	73.4	73.4	73.5
Humus, %	2.1	2.3	2.2	2.2
Total N, %	0.086	0.095	0.093	0.095
Al^{3+}, (meqv)	0.7	2.4	3.3	3.9
Ca^{2+}, (meqv)	0.15	0.72	2.45	2.89

different soil pH in the range from 4.2 to 6.2 in the same soil.

Materials and methods

The soil used in the present investigation originated from a field trial at Jyndevad Experimental Station, where various amounts of chalk have been added since 1946. The various liming intensities have resulted in four different soil pH-values; pH 4.2, 5.0, 5.9, and 6.2, respectively. During the last 10 years spring barley has been grown and it has annually been fertilized with 90 kg N ha^{-1} and 15 kg P ha^{-1}. Physical and chemical

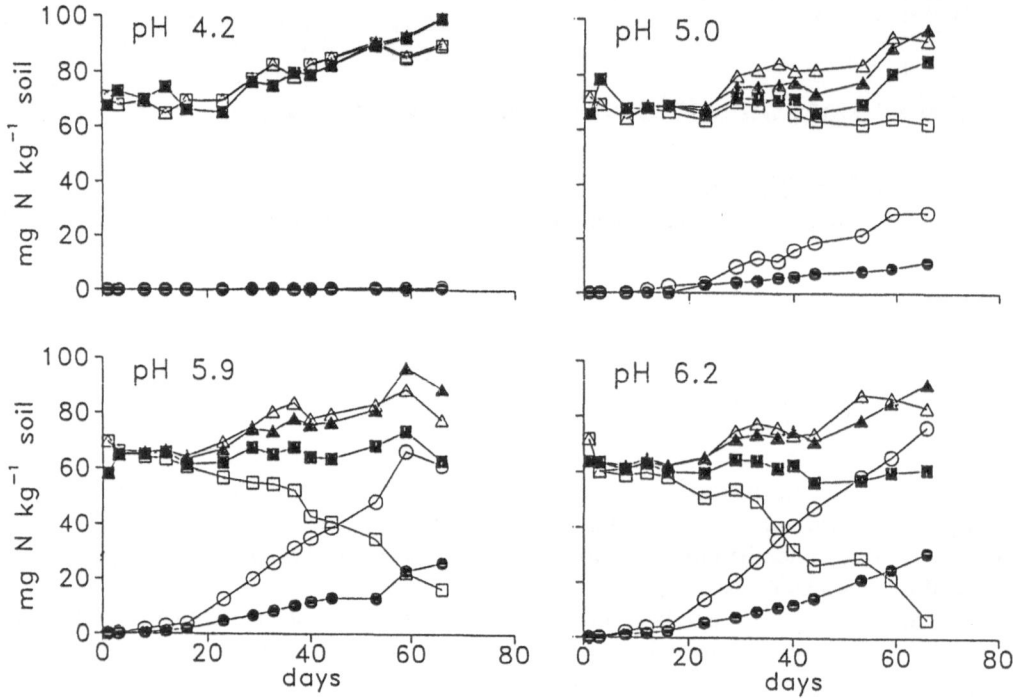

Figure 1. Content of NO$_3^-$-N (circles), NH$_4^+$-N (squares) and inorganic N (triangles) in soil ammended with liquid pig manure and incubated at 2 °C (closed symbols) and 5 °C (open symbols).

characteristics of the coarse sandy soil is shown in Table 1.

Potential nitrification was measured according to Staley et al. (1990). Portions of fresh soil (15 g) were weight out in 200 mL flasks. Then 50 mL 1 mM K$_2$HPO$_4$ buffer (pH 7.2) containing 0.5 mM (NH$_4$)$_2$SO$_4$ and 1 mL of 1.0 M NaClO$_3$ was added to each flask. The flasks (4 replicates) were incubated at 20 °C for 24 h on a rotary shaker (180 rpm). Then the suspension was centrifuged and NO$_2$-N in the supernatant was determined on a Flow Injection Analyzer (FIAstar®, Tecator, Sweden).

Net production of nitrate, in the present investigation referred to as net nitrification, was determined in the field by an in situ technique (Debosz and Vinther, 1989; Raison et al., 1987). At the start of each incubation period, soil samples were collected and steel tubes with a diameter of 35 mm were inserted 20 cm into the soil and equipped with a lid to prevent leaching of nitrogen during the incubation period. At the end of the incubation period, which lasted from one to two months depending on the climatic conditions, the soil from the tubes were collected, and a second incubation was started. The number of replicates per treatment were 10. The content of inorganic nitrogen in the bulk

soil as well as in the incubated soil was determined as described below.

Net N-mineralization and net nitrification after application of liquid pig manure was measured in a laboratory incubation experiment. In 300 mL flasks 40 g of fresh soil (sieved through a 5 mm sieve) was added and 1 mL of homogenized (blended in a Waring blender for 1 min.) liquid manure (3% dry matter, 0.52% total N, 0.4% NH$_4^+$-N) was spread on the soil surface, whereafter 10 g of soil was added to cover the manure. The flasks were incubated at 2 and 5 °C. During the incubation period of 66 days, flasks in triplicates were removed at intervals of 3–7 days and the content of inorganic nitrogen was determined as described below.

Inorganic nitrogen in the soil samples was extracted by shaking for 1 h with 2 M KCl (soil:liquid ratio 1:2). Concentrations of extracted NH$_4^+$ and NO$_3^-$ were determined colorimetrically (FIAstar®, Tecator, Sweden). The content of nitrogen was calculated on dry soil basis. Moisture content was determined gravimetrically by drying at 105 °C for 24 h.

Table 2. Accumulated net nitrification in the field (0–20 cm) during two winter periods. Period 1: 11/11 1992 to 10/3 1993. Peiod 2: 9/11 1993 to 15/3 1994

Soil pH	kg NO$_3^-$-N ha^{-1}	
	Period 1	Period 2
4.2	2.4	4.8
5.0	n.m.	9.3
5.9	7.1	12.6
6.2	9.0	15.3

n.m. = not measured

Results and discussion

The results of the potential nitrification measurements showed a linear correlation ($r^2 = 0.99$) between nitrification and soil pH, with an increase in the nitrification rate of 1.1 mg NO$_3^-$-N kg^{-1} soil per unit pH (not shown). This is similar as previously found e.g. by Lyngstad (1979).

The in situ measurements resulted in nitrification rates between 0.011 and 0.099 mg NO$_3^-$-N kg^{-1} soil. In all the incubation periods the rates at pH 5.9 and 6.2 were higher than at pH 4.2 and 5.0. The accumulated nitrification during the two winter periods (Table 2) showed the same trends as the potential nitrification; 2 to 3 times higher nitrification at pH 6.2 and 5.9 as compared to pH 4.2. The accumulated nitrification was at the two highest pH-values (5.9 and 6.2) between 7.1 and 15.3 kg NO$_3^-$-N ha^{-1}, which is in the same order of magnitude as previously reported by Vinther (1994), who in the same coarse sandy soil (pH 6.4) measured accumulated nitrification at 12, 13, and 11 kg NO$_3^-$-N ha^{-1} during the winter periods of 1988–89, 1989–90, and 1990–91, respectively.

One of the factors which is known to affect the nitrification activity in the field is the soil moisture (e.g. Focht and Verstraete, 1977; Malhi and McGill, 1982; White and Gosz, 1987). By incubating the soil in tubes with a lid to prevent leaching of nitrate, it could be expected that this would affect the moisture conditions inside the tubes. However, measurements of the soil moisture showed that there was no significant difference between the water content inside and outside the tubes.

The accumulated nitrification which occured during the two winter periods resulted in relative small values. The difference between the nitrification in the two years at pH 6.2 and 4.2 was 6.6 and 10.5 kg NO$_3^-$-N,

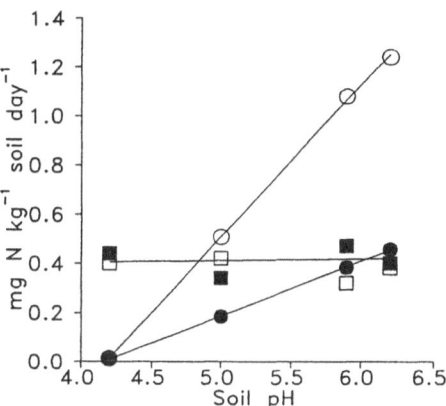

Figure 2. Relationship between soil pH and net N-mineralization (mg inorganic N kg^{-1} dry soil day^{-1}) and net nitrification (mg NO$_3^-$-N kg^{-1} dry soil day^{-1}). ■ = N-mineralization at 2 °C, □ = N-mineralization at 5 °C, ● = nitrification at 2 °C, and ○ = nitrification at 5 °C.

respectively. This means that a consequence of lowering pH from 6 to 4 only would reduce the nitrate leaching by maximum 10 kg N ha^{-1}. An effect of environmental importance can only be expected in situations with high nitrification rates, e.g. after application of farmyard manure or liquid manure.

Results from the incubation experiment with liquid pig manure are shown in Figure 1. After a lag period of 1–2 weeks with immobilization, the content of inorganic nitrogen (NO$_3^-$-N + NH$_4^+$-N) increased linearly during the incubation period at all soil pH values and at both temperatures. The increase in inorganic nitrogen (N-mineralization), was in the same order of magnitude (0.32 to 0.47 mg N kg^{-1} dry soil day^{-1}) at the four soil pH-values indicating that the N-mineralization was not affected by soil pH in the range from 4.2 to 6.2 (Fig. 2). This is in good agreement with results of e.g. Terry et al. (1981), Fu et al. (1987) and Müller (1988), who measured the mineralization after incorporation of different types of plant material or organic waste. In these experiments the mineralization rate was not affected by soil pH in the range from 4 to 8. There was no difference in the N-mineralization rates at 2 and 5 °C.

The nitrification was significantly affected by soil pH. At 2 and 5 °C the nitrification rates at soil pH 4.2 were 0.012 and 0.014 mg NO$_3^-$-N kg^{-1} dry soil day^{-1}, respectively. The corresponding values at pH 6.2 were 0.46 and 1.24 mg NO$_3^-$-N kg^{-1} dry soil day^{-1}. At both 2 and 5 °C the nitrification rates increased linearly with increasing pH (Fig. 2). The increase in nitrification rates was per unit pH 0.22 mg NO$_3^-$-N kg^{-1} dry soil day^{-1} at 2 °C and 0.62 mg NO$_3^-$-N kg^{-1} dry soil day^{-1} at 5 °C, or approximately 3 times higher at 5 °C than at 2 °C. Reid and Waring (1979), who used a fine loamy soil at pH 4.2 adjusted to different pH-values in the range 4.2 to 6, also found a linear relationship between soil pH and nitrification rates. But their results were best described by one relationship below pH 5 and another above pH 5 with approximateley the double nitrification rates per unit pH.

From the relationship between soil pH and nitrification rates (Fig. 2) a pH-value can be calculated at which nitrification will be zero. These values are pH 4.15 at 2 °C and pH 4.18 at 5 °C, which agrees with results of Reid and Waring (1979), who found the lowest limit for nitrification at pH 4.05 in a fine sandy loam. Also Weber and Gainey (1962), who investigated four soils of different types, found that the nitrification stopped at soil pH 3.9–4.1.

The results of the present field and laboratory incubation experiments indicate that it may be possible to reduce the nitrate production and subsequently the leaching of nitrate by lowering the soil pH in a coarse sandy soil. How much the soil pH can be lowered will depend upon the effects on the crop yield and on other factors influenced by soil pH such as solubility of aluminium and heavy metals. The crop yield was measured in the field plots in 1993, and at soil pH 4.2 no harvest was possible, whereas the yield from field plots with soil pH 5.0, 5.9 and 6.2 was 2.04, 2.16 and 2.06 Mg grain ha^{-1}, respectively. This indicate that the soil pH can be lowered to about 5 without affecting the yield of spring barley. By lowering soil pH from about 6, which is normal in this soil type, to about 5, a reduction in nitrification and subsequent nitrate leaching to 45–50% could possibly be expected.

Acknowledgements

We wish to thank Helle Nielsen for technical assistance and The Danish Enviromental Research Progamme for financial support.

References

Debosz K K and Vinther F P 1989 An in situ technique for simultaneous measurements of mineralization, leaching and plant uptake of nitrogen applied to agricultural soils. *In* Nitrogen in organic Wastes applied to Soils. Eds. K Henriksen and J Aa Hansen. pp 3–10. Academic Press, London, UK.

Focht D D and Verstraete W 1977 Biochemical ecology of nitrification and denitrification. Adv. Microbiol. Ecol. 1, 135-214.

Fu M H, Xu X C and Tabatabai M A 1987 Effect of pH on nitrogen mineralization in crop-residue-treated soils. Biol. Fertil. Soils 5, 115–119.

Killham K 1987 A new perfusion system for measurement and characterization of potential rates of soil nitrification. Plant and Soil 97, 267–272.

Lyngstad I 1979 The effects of some physical and chemical factors on ammonification and nitrification in soil. Jord og Myr 3, 52–59 (*In Norwegian*).

Malhi S S and McGill W B 1982 Nitrification in three Alberta soils: Effect of temperature, moisture and substrate concentration. Soil Biol. Biochem. 14, 393–399.

Müller M M 1988 The fate of clover-derived nitrogen (^{15}N) during decomposition under field conditions: Effects of liming and fertilization. Plant and Soil 111, 121–126.

Nyborg M and Hoyt P B 1978 Effects of soil acidity and liming on mineralization of soil nitrogen. Can. J. Soil Sci. 58, 331-338.

Raison R J, Connell M J and Khanna P K 1987 Methodology for studying fluxes of soil mineral-N in situ. Soil Biol. Biochem. 19, 521–530.

Reid R E and Waring S A 1979 Nitrogen transformations in a soil of the Lower Burdekin, Queensland 2. Laboratory incubation studies. Aust. J. Exp. Agric. Anim. Husb. 19, 739–745.

Staley T E, Caskey W H and Boyer D G 1990 Soil denitrification and nitrification potentials during the growing season relative to tillage. Soil Sci. Soc. Am. J. 54, 1602–1608.

Terry R E, Nelson D W and Sommers L E 1981 Nitrogen transformations in sewage sludge - Ammended soils as affected by soil environmental factors. Soil Sci. Soc. Am. J. 45, 506–513.

Vilsmeier K and Amberger A 1988 Nitrifikation von unbehandelter Gülle und "Kalkgülle" in Abhangigkeit von Temperatur und pH-wert des Bodens. Bayer. Landwirtsch. Jahr. 6, 764–766.

Vinther F P 1994 Nitrogen fluxes in a cropped sandy and a loamy soil measured by sequential coring. Eur. J. Agron. 3, 311–316.

Weber D F and Gainey P L 1962 Relative sensitivity of nitrifying organisms to hydrogen ions in soils and in solutions. Soil Sci. 94, 138–145.

Weier K L and Gilliam J W 1986 Effect of acicidy on nitrogen mineralization and nitrification in Atlantic coastal plain soils. Soil Sci. Soc. Am. J. 50, 1210–1214.

White C S and Gosz J R 1987 Factors controlling nitrogen mineralization and nitrification in forest ecosystems in New Mexico. Biol. Fertil. Soils 5, 195–202.

O. Van Cleemput et al. (eds.), Progress in Nitrogen Cycling Studies, 659–662, 1996.
© *1996 Kluwer Academic Publishers.*

Nitrous oxide production from different forms of fertiliser nitrogen

W.J. Wang and R.M. Rees[1]
Soil Science Department, SAC, School of Agriculture, West Mains Road, Edinburgh EH9 3JG, UK.
[1] *Corresponding author*

Key words: denitrification, nitrification, N-fertiliser, nitrous oxide, NBPT

Abstract

Laboratory incubations with different forms of fertilizer-nitrogen were carried out in order to determine the effects of fertiliser N additions on N_2O production by soils. In the first experiment, ten different soils were incubated with no N (control), 60 μg NO_3^--N g^{-1} or 60 μg urea-N g^{-1}. The N_2O produced by the ten urea treated soils (on average 60.1 μg N_2O-N kg^{-1} over 14 days) was consistently greater than that produced by the control and nitrate treated soils (8.4 and 11.6 μg N_2O-N kg^{-1} over 14 days respectively). In the second experiment a single soil was incubated with a wider range of forms of nitrogen (urea with and without NBPT, $Ca(NO)_3$, and $(NH_4)_2SO_4$). In this experiment the highest rate of N_2O production (35 μg N_2O kg soil^{-1} day^{-1}) was in the urea treatment, however the largest accumulation of N_2O was in the $(NH_4)_2SO_4$ treatment (215 μg kg^{-1}) over 16 days. A regression equation using NH_4^+ concentrations and pH accounted for 92% of the variability in N_2O production.

Introduction

The form in which fertiliser N is applied to soil can have a significant effect on the amount of N_2O produced (Breitenbeck and Bremner, 1986). Observations from field trials carried out near Edinburgh have shown that emission of N_2O following the application of urea can exceed those from comparable amounts of other nitrogen fertilizers (McTaggart and Smith, 1994). The purpose of this study was to examine the mechanism that controls N_2O emissions from different forms of fertiliser N.

Materials and methods

In the first experiment, ten arable soils from different parts of SE Scotland with contrasting soil properties, were incubated at a constant moisture potential (500 mm) for 14 days. Three treatments were applied in triplicate to each soil. These were: control, calcium nitrate and urea. Solutions were applied using a pipette, and each fertiliser addition supplied 60 μg N g^{-1}.

A second incubation experiment was carried out using a clay loam collected in June 1994 from the Bush Estate near Edinburgh. The soil had a pH (in water) of

6.9 and an organic-carbon concentration of 50.5 mg kg^{-1}. Immediately after sampling, the soil was passed through a 4 mm sieve and stored at 4 °C. For use in the incubation experiment, the water content was adjusted to 25.3%, equivalent to a water potential of 500 mm water.

Five treatments were applied:

(1) Control: 11 mL distilled water
(2) Ammonium sulphate dissolved in 11 mL distilled water
(3) Calcium nitrate dissolved in 11 mL distilled water
(4) Urea dissolved in 11 mL distilled water
(5) Urea plus 144 μg NBPT

Each fertiliser treatment supplied 0.2 mg N g^{-1} dry soil. The solutions were applied to soils in 500 mL erlenmyer flasks using a pipette to distribute the solution evenly across the soil surface. Each treatment was replicated 15 times to allow analysis of triplicate samples on each sampling occasion throughout the experiment.

Analyses

In both experiments, the flasks were sealed with subaseals at the start of the experiment, and triplicate gas samples were taken at regular intervals from each treat-

660

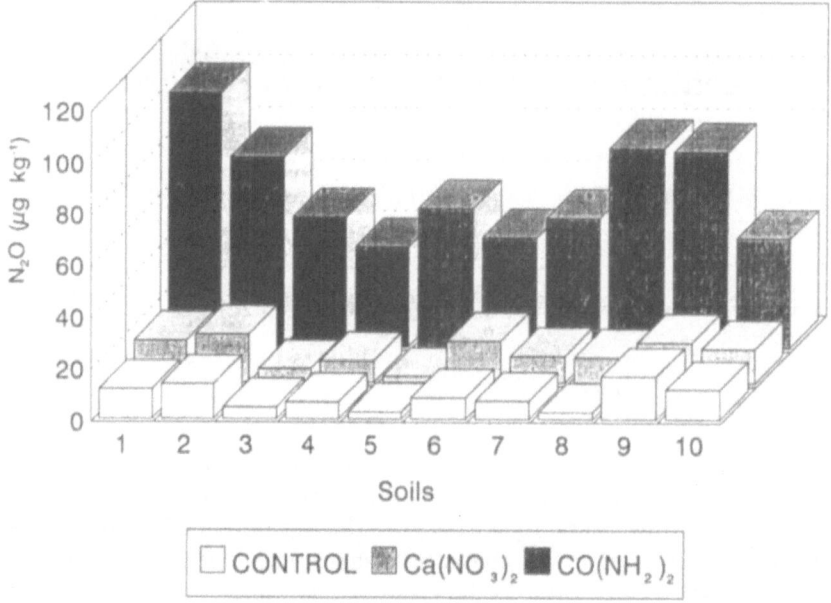

Figure 1. The amounts of N_2O produced within two weeks by soils treated with different forms of fertiliser N.

ment to allow analysis for N_2O (using an HP 5890 gas chromatograph equipped with an ECD). Following the collection of each gas sample, flasks were flushed with air, and resealed. Available N was measured by taking 15 g of fresh soil from triplicate flasks in each treatment. A subsample was used to determine the soils' moisture content, and the remainder was shaken with 1 M KCl for 30 min on an orbital shaker. The sample was filtered through Whatman No 42 filter papers, and the NH_4^+ and NO_3^- contents of the sample determined by continuous flow analysis. Samples of soil from the same flasks were used to measure pH after mixing with water (1:2.5) using a glass calomel electrode.

Results and discussion

In experiment one, production of N_2O by the urea treated soils (60.1 μg N_2O kg^{-1} over 14 days) was always greater than that by the controls or calcium nitrate treatments (8.4 and 11.6 μg kg^{-1} over 14 days respectively; Fig. 1). Given the relative constancy with which this observation was made, fertiliser incubations were carried out using just one soil type in the second experiment.

In experiment two, the production of N_2O by the urea and ammonium sulphate treated soils was significantly greater ($p<0.001$) than that from the control and calcium nitrate treatments (Fig. 2). Nitrous oxide

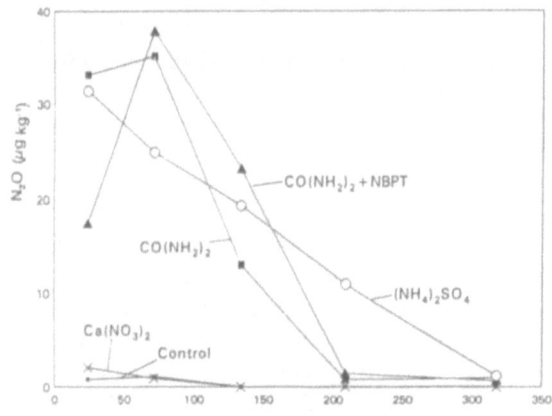

Figure 2. The rate of emission of N_2O from soils treated with different forms of fertiliser N.

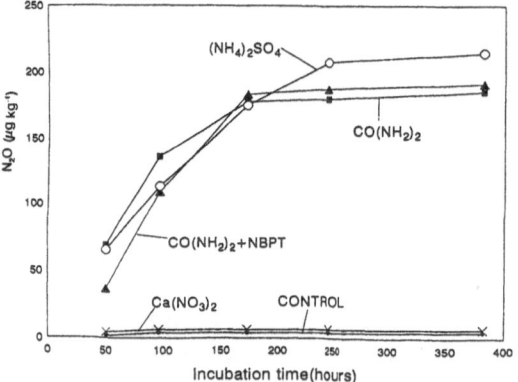

Figure 3. The amounts of N_2O produced by soils treated with different forms of fertiliser N.

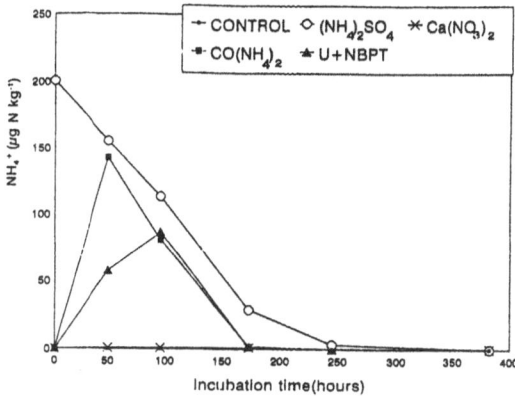

Figure 4. The dynamics of NH_4^+ in soils treated with different forms of fertiliser N.

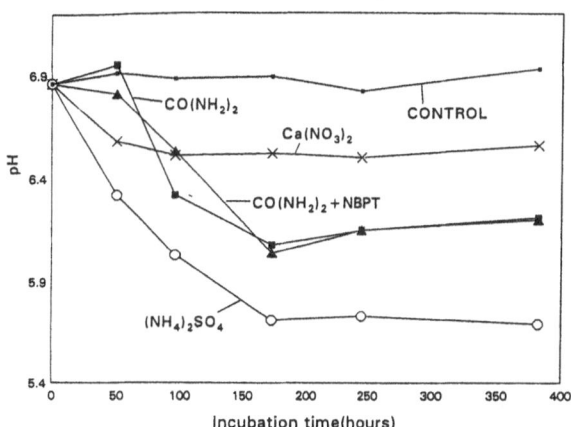

Figure 5. The dynamics of pH in soils treated with different forms of fertiliser N.

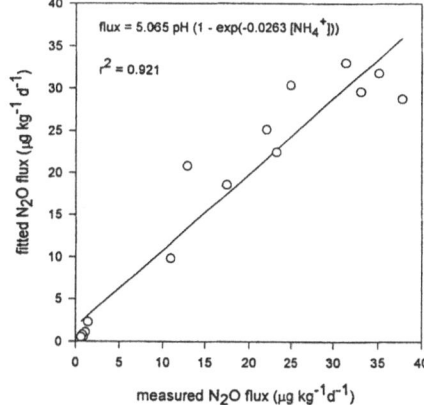

Figure 6. The prediction of N_2O production from soils using soil pH and NH_4^+-N concentrations.

production reached a peak in the urea treated soils of around 35 μg N_2O kg soil^{-1} day^{-1} from day 2 to day 4. After nine days it declined to around 0, but the rate of production of N_2O by the ammonium sulphate treated soils was still significantly higher ($p<0.001$) than that from all other treatments. At the end of incubation there were no differences between any of the treatments, with emissions close to zero; the total amount of N_2O produced by the different treatments differed significantly (Fig. 3). The largest quantity of N_2O (215 μg kg^{-1}) was produced by the ammonium sulphate treatment with least produced by the calcium nitrate and control (6.3 and 4.2 μg kg^{-1} respectively). It is probable that much of the N_2O produced during this experiment resulted from losses during nitrification, since very litttle emission was observed from the control or nitrate treated soils. Largely aerobic conditions were maintained throughout the experiment which may responsible for the apparently low rates of denitrification. Skiba et al. (1993) found that in relatively dry soils, most N_2O was produced by nitrification. By contrast when the soils were watered (increasing the water content from 18 to 20.4%) denitrification was the main N_2O source.

The concentration of NH_4^+-N in the ammonium sulphate treatment declined steadily from 200 μg g^{-1} at day 0 to around 1 μg g^{-1} at day 11 (Fig. 4). In the urea treated soils the peak in NH_4^+-N (150 μg g^{-1}) occurred after 50 hours. When urea plus NBPT were applied in combination, the peak in ammonium production was lower (100 μg g^{-1}) and occurred later after 100 hours. These results indicate that ammonium concentrations alone are unable to predict N_2O production.

The application of different nitrogen fertilisers had a significant effect on soil pH. The pH of the control

soils remained at around 6.8 throughout the incubation. However, the pH of the calcium nitrate, urea alone, urea and NBPT and the ammonium sulphate treatments were 6.53, 6.20, 6.18 and 5.71 respectively after 16 days (Fig. 5). The combination of pH and NH_4^+-N concentrations was very effective at predicting N_2O production in this experiment. The regression equation describing the N_2O flux was:

$$\text{flux} = 5.07\text{pH}(1 - \exp(-0.026[NH_4^+]))$$
$$r^2 = 0.92 (p < 0.001; \text{Fig. 6})$$

Conclusion

The results of this study indicate that relatively large differences in N_2O production can result from the

662

manipulation of the form of fertiliser N applied to soils, the soil pH and water content. This information is likely to be valuable in designing strategies to reduce emissions of N_2O from arable soils.

Acknowledgements

The authors wish to thank the Sino British Friendship Scholarship Scheme and the Scottish Office Agriculture and Fisheries Department for financial support. We would also like to express gratitude to J R M Arah for discussion of the results and preparation of Figure 6 and to other to colleagues in the Soil Science Department, SAC for their valuable help.

References

Breitenbeck G A and Bremner J M 1986 Effects of various nitrogen fertilisers on emission of nitrous oxide from soils. Biol. Fertil. Soils 2, 195–199.

McTaggart I P and Smith K A 1994 Nitrous oxide flux from fertilised grassland: strategies for reducing emissions. *In* Proceedings of the Symposium on Non-CO_2 Greenhouse Gases, Maastricht, 1993. Eds. J van Ham, L J Janssen and R J Swart. pp 421–426. Kluwer Acad. Publ., Dordrecht, the Netherlands.

Skiba U, Smith K A and Fowler D 1993 Nitrification and denitrification as sources of nitric and nitrous oxide in a sandy loam soil. Soil Biol. Biochem. 25, 1527–1536.

8. ASPECTS RELATED TO LEACHING

Plant and Soil **181**: 1–6, 1996.

Measuring and modelling nitrogen leaching: parallel problems

T.M. Addiscott
IACR-Rothamsted, Harpenden, Herts AL5 2JQ, UK*

Key words: leaching, nitrate, nitrogen

Abstract

In studies of nitrate leaching both experimenters and modellers experience problems arising from soil variability. Because of the small-scale heterogeneity that gives rise to mobile and immobile categories of water, both measurements and modelling are easiest in homogeneous sandy soils and most difficult in strongly structured clay soils. There are also parallels at plot and field scale in the problems caused to experimenters by log-normal distributions of nitrate concentrations and those caused to modellers by non-linearity in models. All researchers need to be aware that a reliable estimate of the mean from a set of measurements or a model may necessitate considerations of variances as well as means.

Introduction

Measurements and models of nitrate leaching are both influenced by the physical and statistical properties of the soil. There are parallels between the problems these properties cause to experimenters and those they cause to modellers. Researchers in both categories have to take account of the heterogeneous nature of the soil, and the uncertainties that arise from measured values that are not normally distributed have something in common with those that occur because models are not linear with respect to their parameters. This paper examines these parallel problems with particular reference to their impact at various scales.

Problems associated with small-scale soil heterogeneity

Measurements of nitrate leaching

The problems of measuring nitrate leaching are essentially those of capturing the water that is about to move beyond the rooting zone of the soil. The determination of the nitrate concentration in this water is a matter of routine chemical analysis.

Measurements of nitrate leaching can be made most readily in the most homogeneous soils, which usual-ly means sandy soils. These soils permit the use of porous ceramic cups, which are usually the easiest and the cheapest technique for measuring nitrate concentrations in soil water. Webster et al. (1993) give an up-to-date account of the use of porous cups, together with a note of precautions that need to be taken and results from two different sites. Their advice on installing porous cups is particularly important; incorrect installation can lead to misleading results as discussed in earlier reviews (Addiscott, 1990; Addiscott et al., 1991).

As the percentages of clay and silt in the soil increase, the soil becomes more structured and less homogeneous, with clearly-defined mobile and immobile categories of water in the soil matrix and, particularly in the most clayey soils, cracks and other channels by-passing the whole soil matrix. At the same time porous ceramic cups become less and less useful for collecting the water likely to pass from the soil. This was illustrated by a study made by Barbee and Brown (1986) who installed porous cups in three types of soil adjacent to collecting pans inserted beneath the soil at the same depth. Chloride was applied to the soil surface and allowed to leach under natural rainfall. Weekly samplings showed no significant differences between the chloride concentrations measured by the two systems in a sandy soil, and the peak concentration occurred after about 100 days in both. In a moderately-structured silt soil, however, there were

---
* FAX No.: +44 1582 760981
Plant and Soil is the original source of publication of this article. It is recommended that this article is cited as: *Plant and Soil* **181**: 1–6, 1996.

clear differences in the patterns found, with the porous cups recording the peak concentration after 8 days and the horizontal collecting vessels after 42 days. In a clay soil, the porous cups produced a large enough water sample for analysis on only one occasion and seemed to be by-passed by the water flow on all other occasions.

Soil heterogeneity is a problem for users of porous cups because of the scale of the measurement. In sandy soils, the volume of the soil from which the cup extracts water is large compared with the scale of the variability in soil properties. In heavier soils, the volume from which the cup extracts water is probably smaller than in sandy soils while the scale of the variability is larger. This can lead to the problems found by Barbee and Brown (1986). Thus porous cups are most effective in sandy soils, but it does not prevent their use in heavier soils, provided they are used with a clear idea of what they measure. Goulding and Webster (1992) placed porous cups in plots of the Broadbalk Experiment at Rothamsted that had field drains. They concluded that in this silty clay loam soil the porous cups sampled the immobile water while the field drains collected the mobile water. As a result, the nitrate concentrations sampled by the porous cups were larger than those in the drainage before fertilizer applications, because most of the nitrate was produced by microbes in the soil, but smaller after fertilizer N had been applied and was at risk of being carried down in the mobile water.

The results of Goulding and Webster (1992) described above make it clear that porous cups are not suitable for estimating losses of nitrate from heavier soils without supplementary measurements. Lysimeters provide a better, but more labour-intensive, approach. There are two basic categories, Ebermeyer lysimeters and monolith lysimeters. The construction and use of these devices are reviewed elsewhere (Addiscott, 1990; Addiscott et al., 1991), and Belford (1979) gives a useful account of the setting-up of monolith lysimeters. We concentrate here on the problems that may be encountered in their use.

Ebermeyer lysimeters are constructed by inserting a collecting vessel into a horizontal aperture in the soil, usually from a trench. A monolith lysimeter is made by driving a suitable casing, often a fibreglass pipe, into the soil and cutting the soil away around it, so that the column of soil it contains can be severed at the base. The monolith can then be installed either where it was taken, so that it can be subjected to the same agronomic practice as the soil around it, or in a bank of lysimeters elsewhere. In either event a suitable collecting vessel must be attached to the base.

Both types of lysimeter can be constructed with relatively little disturbance to the soil within them, and the scale of the lysimeter is sufficiently large to accommodate the heterogeneity that leads to the presence of mobile and immobile water. The Ebermeyer lysimeter has a degree of uncertainty that arises because the water that is collected may have moved laterally as well as vertically and may not have come from directly above the collector. The casing of the monolith lysimeter avoids this uncertainty, but there can be problems of preferential water flow if the soil shrinks away from the casing (Belford, 1979).

There is, however, one problem that affects both types of lysimeter, and that is the air-water interface at the base of the soil above the collector. Because many of the pores that conduct water through the soil are continuous, there is a 'hanging column' of water that extends to a reasonable depth in the soil. Cutting this hanging column has the effect that the surface tension that arises from the air-water interface holds back the water from draining until the soil becomes saturated, even in a freely-draining soil (Richards et al., 1939). Webster et al. (1993) found clear evidence of this effect in an experiment on a sandy soil in which the movement of a pulse of chloride was followed using porous cups and lysimeters. The peak chloride concentration was delayed in the lysimeters and the experimenters attributed the delay to the air-water interface. Only by applying suction can the effect of this interface be overcome; Coleman (1946) showed that suction controlled both the rate of drainage and the amount of water in the base of the soil. The suction applied to the base of a lysimeter should be as close as possible to that corresponding to the hanging column of water. Morton et al. (1988) achieved this by adjusting the suction with reference to tensiometers inserted in nearby soil to the depth of the lysimeter's base. This question of appropriate suction is not just an academic one. Haines et al. (1982) found that subjecting the collector in an Ebermeyer lysimeter to a 1 m hanging column of water doubled the average water flow and changed its nitrate concentration by a factor of three.

The heaviest clay soils often have subsoils that are so nearly impermeable to water that they need artificial drainage. This is a situation that can be turned to the experimenter's advantage, because most of the nitrate that is leached from the soil is carried through the drainage pipes and can be measured. The Brimstone Experiment near Wantage in the UK is an example of such an experiment (Cannell et al., 1984). This experiment has plots that are large enough to permit fairly normal agricultural practice, separated from each other by vertical barriers of heavy-gauge polythene sheet. Three categories of water can be collected and analysed for nitrate; surface run-off, interflow (flow at the base of the plough layer) and water carried by the drainage system. Goss et al. (1993) recently reported nitrate losses in several years from plots subjected to various treatments.

The plots in this experiment give few physical problems because the conditions at the interface between air and water are the same as in fields elsewhere subjected to the same standard drainage practice.

Models of nitrate leaching

Just as measuring nitrate leaching is simplest in the most homogeneous soils, so is modelling nitrate leaching. It is in sandy and other homogeneous soils that the classic mechanistic modelling approach to leaching, the Richards equation used with the convection-dispersion equation (Wagenet, 1983), can be applied most relevantly. Soils of this kind also permit the use of several simple models for leaching, such as the early 'piston-flow' model of Rouselle (1913) and the leaching model and equation developed by Burns (1974, 1975). Both categories of model have been investigated and used reasonably widely.

All the models cited above presume the soil to be homogeneous, with water and solute moving equally freely in all parts of the soil. Like porous cups, therefore, they become less relevant as the soil becomes more structured and less homogeneous and the water becomes divided into mobile and immobile categories, with by-pass flow becoming a possibility. For soils such as these models that take account of mobile and immobile water become necessary. Probably the earliest 'mobile-immobile' model for leaching in the soil was that of Van Genuchten and Wierenga (1976). This was an adaptation of the classical mechanistic approach that included a category of 'stagnant' water, the movement of solute into which was governed by a transfer coefficient. This model was applied mainly to columns of soil in the laboratory, but Barraclough (1989) used an adaptation of it at the field scale. A simpler 'mobile-immobile' model developed by Addiscott (1977) was intended for use at field plot or field scale. Despite its simplicity this model gave a good simulation of the intricate nature of the 'break-through curves' obtained when chloride was applied to the Drain Gauges (lysimeters) at Rothamsted (Addiscott et al., 1978).

The simple mobile/immobile model, and the Burns leaching model and equation, are classified as 'capacity' models because their main parameters derive from the volumetric moisture contents, θ, of the soil at various suctions. The classical approach has as its main water parameter the hydraulic conductivity and is therefore classified as a 'rate' model. One model, the SLIM model of Addiscott and Whitmore (1991) has both a capacity parameter and a simplified rate parameter. A recent development of this approach is the SLM model of Hall (1993) which takes account of water moving at various rates.

A rather different approach has evolved for heavy, cracking clay soils in the form of the CRACK model of Jarvis and Leeds-Harrison (1987). This model takes account of the cracking and swelling of clay soils and the resulting changes in water flow pathways. This was originally a water-flow model but it has recently been adapted to simulate nitrate leaching (P B Leeds-Harrison, pers. commun.). Jarvis et al. (1991) recently developed the MACRO model, which is also intended for heavy soils.

Problems arising from variability at plot and field scale

Measurements of nitrate leaching

Statistically-related problems in both soil measurements and soil modelling arise from the variability of the soil. In measurements of nitrate leaching we are concerned mainly with the variability of nitrate concentrations. These can arise in part from the physical properties of the soil, but some of the variability in nitrate concentrations arises from non-uniform excretion by organisms of various types at widely ranging scales. The nitrate concentrations measured with porous cups in arable soils by Webster et al. (1993) were not particularly variable, with coefficients of variation in the range 14–20 percent. There are probably two reasons for this relatively small variability. One is tillage, which tends to smooth out variability by mixing the soil. Another is the fact that the organisms excreting ammonium or nitrate are mainly microbes and the scale of even a porous cup measurement is so much larger than that of the variability of the excretion that the latter is not detected by the measurement. The factor most likely to increase variability in nitrate is the irregular distribution of crop residues at harvest.

Grassland that is cut for hay or silage but not grazed would probably have nitrate concentrations showing as little variability as arable land, or possibly less because the grass would tend to even out the variability by taking up most nitrate where most was available. Most grassland, however, is grazed by farm animals, usually cattle or sheep. These animals excrete at a scale that is very readily picked up by porous cups or soil sampling. A cow, for example, may deliver 2 L of urine on an area of about 0.5 m^2, giving a localized application of about 500 kg ha^{-1} of N, far more than the grass can use. The concentrations of nitrate found by Cuttle et al. (1992) using porous cups in grassland grazed by sheep ranged over four orders of magnitude and were distributed log-normally. Similar degrees of variability were found by White et al. (1987) in soil samples for nitrate in grassland.

When measurements deliver a skewed population of concentrations that includes same very large values, some care is needed in choosing a suitable statistical estimator to represent the distribution. The type of estimator needed will depend on the precise nature of the information required. If concern is centred on the concentration per se, because of the EC nitrate limit, for example, the estimator needed is one that will not be influenced excessively by a few large values among mainly smaller ones, so the mean of the log distribution is likely to be the most reliable estimator; this is equivalent to taking the geometric mean. If, however, our main concern is with the overall loss of solute from the area of land, we need to take full account of those few large values because they contribute so much to the loss. This means that we need the arithmetic mean of the concentrations, but if calculated directly this can be an inefficient estimator for skewed distributions, because the skew gives a large error to the estimate. To obtain an efficient and unbiased estimate it may be better to back-transform from the log distribution. There are two methods for doing so.

The simpler method is that of Aitchison and Brown (1957) which estimates the mean μ and variance σ^2 of a population represented by a log-normal distribution from the sample mean m and variance s^2:

$$\mu = \exp\left(m + \frac{1}{2}s^2\right) \quad (1)$$

$$\sigma^2 = \mu^2[\exp(s^2) - 1] \quad (2)$$

The other method developed independently by Finney (1941) and Sichel (1952) estimates μ and σ^2 through a power series:

$$\mu = \exp(m)\Psi\left(\frac{1}{2}s^2\right) \quad (3)$$

$$\sigma^2 = \exp(2m)\left\{\Psi(2s^2) - \Psi\left[\frac{(n-2)}{(n-1)}s^2\right]\right\} \quad (4)$$

where the power series Ψ is given by:

$$\Psi(t) = 1 + \frac{t(n-1)}{n} + \frac{t^2(n-1)^3}{n^2(n+1)2!}$$
$$+ \frac{t^3(n-1)^5}{n^3(n+1)(n+3)3!}$$
$$+ \frac{t^4(n-1)^7}{n^4(n+1)(n+3)(n+5)4!} + \cdots \quad (5)$$

where n is the sample size.

The key point to note in both sets of equations is that the estimate of the population mean has to take account of the sample variance as well as the sample mean. Similarly the estimate of the population variance has to take account of the sample mean as well as the sample variance.

These equations were evaluated by Parkin et al. (1988) who wished to find out (a) whether it was necessary to use these equations rather than the simple untransformed mean and variance of the sample, and (b) whether there was any benefit from using the more complex Finney-Sichel equations. Their conclusions can be summarized as follows:

1 Use the standard untransformed estimator to estimate the mean of a slightly-skewed population (skew about 1.625 or less).

2 Use the Finney-Sichel estimator to estimate the means of moderately- to markedly-skewed distributions (skew about 4.0 or about 16.0 respectively). For large samples and moderate skews the Aitchison-Brown estimator may be used.

3 Use the Finney-Sichel estimator for the variance with all combinations of skew and sample size, except when small samples (4–20) are taken from a slightly skewed population, when the standard untransformed estimator should be used. The Aitchison-Brown estimator may be used for sample sizes greater than 40 from slightly-skewed populations.

4 Use the Finney-Sichel estimator for the coefficient of variation for all combinations of skew and sample size. The Aitchison-Brown estimator may be used when more than 40 samples were taken or when more than 20 were taken and the skew was slight.

Parkin et al. (1988) studied populations that were generated mathematically and whose distributions were clearly defined, but the results cited above that

Cuttle et al. (1992) and White et al. (1987) obtained in the field generally support their conclusions.

Models of nitrate leaching

The previous section was concerned mainly with the problems caused by the variability in nitrate concentrations. Here we are concerned with the variability in the parameters of models. The underlying problem can be seen in some equations presented by Rao et al. (1977) that relate the mean of a function f(x,y) to the means and variances of x and y, μ_x, μ_y, σ_x^2 and σ_y^2 where x and y are distributed normally:

$$\mu_{f(x,y)} = f(\mu_x, \mu_y) + C \qquad (6)$$

That is to say, the mean of the function is not necessarily the function of the means. They are the same only when C is zero, and C is given by:

$$C = \left(\frac{\partial^2 f(x,y)}{\partial x^2} \right) \frac{\sigma_x^2}{2}$$
$$+ \left(\frac{\partial^2 f(x,y)}{\partial y^2} \right) \frac{\sigma_y^2}{2}$$
$$+ \rho \left(\frac{\partial^2 f(x,y)}{\partial x \partial y} \right) \frac{\sigma_x \sigma_y}{2} \qquad (7)$$

where ρ is the product moment correlation of x and y. Thus C is zero and $\mu_{f(x,y)}$ is equal to f(μ_x, μ_y) only if the second partial differentials are all zero or if the variances are all zero (or both).

If we take f(x,y) as a very simple representative of models we can see another parallel between problems in measurements and problems in modelling. Just as the estimate of the population mean obtained by back-transforming a log-normal distribution depended on the variance of the log-normal distribution as well as its mean, so the mean obtained from the function or model depends on the variance of the parameter as well as its mean if the function or model is non-linear with respect to the parameter. The non-linearity shows itself in the fact that the second partial differentials are not zero.

This parallel is not, however, a complete one. In the back-transformation of the log-normal distribution, not only did the population mean depend on the variance of the distribution, but the population variance depended on the mean of the distribution. The equation given by Rao et al. (1977) for the variance of f(x,y) shows it to depend only on the variances of x and y and not on

Table 1. Effects of including or omitting the variances of the rate parameter, α, and the capacity parameter, W_r, in the simulations of the downward movement of surface-applied nitrate with the SLIM leaching model. (From Addiscott and Bland, 1988)

Treatment of variance	LOF Mean Square[a]
All variances included	11
All variances omitted	47
Variance of α omitted	26
Variance of W_r (topsoil) omitted	12
Variance of W_r (subsoil) omitted	11

[a] Lack of fit mean square as defined by Whitmore (1991). The smaller the value, the better the simulation.

their means:

$$\sigma_{f(x,y)}^2 = \left[\frac{\partial f(x,y)}{\partial x} \right]^2 \sigma_x^2 + \left[\frac{\partial f(x,y)}{\partial y} \right]^2 \sigma_y^2 \qquad (8)$$

The problems caused by the term C in Equation 6 vary greatly between different types of model. The volumetric moisture content of the soil usually varies relatively little, so capacity parameters such as those used in the models of Burns (1974) and Addiscott (1977) usually have coefficients of variation of the order of 10 percent. These models are also more or less linear with respect to their parameters, so the term C does not cause any problems. By contrast the classical approach to modelling leaching, the combination of the Richards equation and the convection-dispersion equation, has problems of both kinds. The main parameters of these equations, the hydraulic conductivity and the dispersivity, are highly variable and the models are not linear with respect to these parameters.

The difference in behaviour between capacity parameters and rate parameters such as the hydraulic conductivity is illustrated by a study made by Addiscott and Bland (1988) on the SLIM model (Addiscott and Whitmore, 1991), which has both a capacity parameter and a simplified rate parameter. The model was run with or without allowance for the variances of these parameters. Ignoring the variance of the capacity parameter had little or no effect on the ability of the model to simulate the proportions of a pulse of applied nitrate found at various depths down to 1 m (Table 1), but ignoring the variance of the rate parameter clearly made the simulation less satisfactory. The latter effect was not very large because of the 'stabilizing' influence of the capacity parameter.

6

Discussion

Both experimenters and modellers seek true representations of processes occurring in the soil. Both should therefore be expected to find the same types of problems if they are effective in their depictions of what is happening in the soil. This seems to happen. The heterogeneity at small scales that results in mobile/immobile water phenomena gives broadly similar physical problems in both measurements and models of nitrate leaching, such that both are easiest in homogeneous sandy soils and most difficult in strongly-structured clay soils. The variability problems met by experimenters and modellers also have parallels in the problems caused to the former by log-normal distributions of concentrations and to the latter by nonlinearity in models. The possibility that a reliable estimate of the mean from either a set of measurements or a model may necessitate the consideration of variances as well as means is one that both experimenters and modellers ignore at their peril.

References

Addiscott T M 1977 A simple computer model for leaching in structured soils. J. Soil Sci. 28, 554–563.

Addiscott T M 1990 Measurement of nitrate leaching: a review of methods. *In* Nitrate, Agriculture, Water:Problems and Challenges. Ed. R Calvert. pp 157–168. INRA, Paris, France.

Addiscott T M and Bland G J 1988 Nitrate leaching models and soil heterogeneity. *In* Nitrogen Efficiency in Agricultural Soils. Eds. D J Jenkinson and K A Smith. pp 394–408. Elsevier Applied Science, Barking, UK.

Addiscott T M and Whitmore A P 1991 Simulation of solute leaching in soils of differing permeabilities. Soil Use Manage. 7, 94–102.

Addiscott T M, Rose D A and Bolton J 1978 Chloride leaching in the Rothamsted Drain Gauges: Influence of rainfall pattern and soil structure. J. Soil Sci. 29, 305–314.

Addiscott T M, Whitmore A P and Powlson D S 1991 Farming, fertilizers and the nitrate problem. CAB International, Wallingford, UK. 176 p.

Aitchison J and Brown J A C 1957 The Lognormal Distribution. Cambridge University Press, Cambridge.

Barbee G C and Brown K W 1986 Comparison between suction and free-drainage soil solution samplers. Soil Sci. 141, 149–154.

Barraclough D 1989 A usable mechanistic model of nitrate leaching. I. The model. J. Soil Sci. 40, 543–554.

Belford R K 1979 Collection and evaluation of large-scale monoliths for soil and crop studies. J. Soil Sci. 30, 363–373.

Burns I G 1974 A model for predicting the redistribution of salts applied to shallow soils after excess rainfall or evaporation. J. Soil Sci. 25, 165–178.

Burns I G 1975 An equation to predict the leaching of surface applied nitrate. J. Agric. Sci. 85, 443–454.

Cannell R Q, Goss M J, Harris G L, Jarvis M G, Douglas J T, Howse K R and Le Grice S 1984 A study of mole-drainage with simplified cultivation for autumn-sown crops on a clay soil. Background, experiment and site details, drainage systems, measurement of drainflow and summary of results. J. Agric. Sci. 102, 539–559.

Coleman E A 1946 A laboratory study of lysimeter drainage under controlled soil moisture tension. Soil Sci. 62, 365–382.

Cuttle S P, Hallard M, Daniel G and Scurlock R V 1992 Nitrate leaching from sheep-grazed grass/clover and fertilized grass pastures. J. Agric. Sci., Cambridge 119, 335–343.

Finney D J 1941 On the distribution of a variate whose logarithm is not normally distributed. R. Statist. Soc. London J. Suppl. 7, 155–161.

Goss M J, Howse K R, Lane P W, Christian D G and Harris G L 1993 Losses of nitrate-nitrogen in water draining from under autumn-sown crops established by direct drilling or mouldboard ploughing. J. Soil Sci. 44, 35–48.

Goulding K W T and Webster C P 1992 Methods for measuring nitrate leaching. Asp. Appl. Biol. 30, 63–70.

Haines B L, Waide J B and Todd R L 1982 Soil solution nutrient concentrations sampled with tension and non-tension lysimeters: Report of discrepancies. Soil Sci. Soc. Am. J. 46, 658–661.

Hall D G M 1993 An amended functional leaching model applicable to structured soils. I. Model description. J. Soil Sci. 44, 574–588.

Jarvis N J and Leeds-Harrison P B 1987 Modelling water movement in a drained clay soil. Description of model, sample output and sensitivity analysis. J. Soil Sci. 38, 487–498.

Jarvis N J, Janssen P-E, Dik J P and Messing I 1991 Modelling water and solute transport in macroporous soil. I. Model description and sensitivity analysis. J. Soil Sci. 42, 59–70.

Morton T G, Gold A J and Sullivan W M 1988 Influence of over-watering and fertilization on nitrogen losses from home lawns. J. Environ. Qual. 17, 124–130.

Parkin T B, Meisinger J J, Chester S T, Starr J L and Robinson J A 1988 Evaluation of statistical estimation methods for lognormally distributed variables. Soil Sci. Soc. Am. J. 52, 323–329.

Rao P S C, Rao P V and Davidson J M 1977 Estimation of the spatial variability of the soil-water flux. Soil Sci. Soc. Am. Proc. 41, 1208–1209.

Richards L A, Neal O R and Russell M B 1939 Observations on moisture conditions in lysimeters, II. Soil Sci. Soc. Am. Proc. 4, 55–59.

Rouselle V 1913 Le mouvement des nitrates dans le sol et ses consequences relative a l'emploi du nitrate de soude. Ann. Sci. Agron. 4th Ser. pp 97–115.

Sichel H S 1952 New methods in the statistical evaluation of mine sampling. Trans Inst. Mining Metall. (London) B61, 261–288.

Van Genuchten M Th and Wierenga P J 1976 Mass transfer studies in porous sorbing media. I. Analytical solutions. Soil Sci. Soc. Am. J. 40, 473–480.

Wagenet R J 1983 Principles of salt movement in soil. *In* Chemical Mobility and Reactivity in Soil Systems. Eds. D W Nelson et al. pp 123–140. Special Publication No 11. Soil Science Society of America, Madison, WI, USA.

Webster C P, Shepherd M A, Goulding K W T and Lord E I 1993 Comparisons of methods for measuring the leaching of mineral nitrogen from arable land. J. Soil Sci. 44, 49–62.

White R E, Haigh R A and MacDuff J H 1987 Frequency distributions and spatially-dependent variability of ammonium and nitrate concentrations in soil under grazed and ungrazed grassland. Fertil. Res. 11, 193–208.

Whitmore A P 1991 A method of assessing the goodness of simulation of soil processes. J. Soil Sci. 42, 289–299.

Section editor: R Merckx

O. Van Cleemput et al. (eds.), Progress in Nitrogen Cycling Studies, 671–674, 1996.
© 1996 Kluwer Academic Publishers.

Near real-time determination of nitrate leaching as an integrated component of a large soil block experiment for fine resolution investigation of preferential flow

N.M. Holden[1], D. Scholefield[2], A.G. Williams[1] and J.F. Dowd[3]
Department of Geographical Sciences, University of Plymouth, Plymouth, PL4 8AA, UK, [2]Institute of Grassland and Environmental Research, North Wyke, Okehampton, Devon, EX20 2SB, UK and [3]Department of Geology, University of Georgia, Athens, GA 30602, USA

Key words: nitrate test-strip, preferential flow, resolution, spatial, temporal

Abstract

A large soil block was isolated in situ at North Wyke, Devon, UK. It was designed as the basis for fine spatial and temporal resolution investigations of water and chemical movement by preferential flow through a well structured soil. Three simultaneous tracer experiments were conducted within the block using a chloride/nitrate/deuterium cocktail. This paper consideres the nitrate experiment. The block was densely instrumented with time domain reflectometry waveguides (to measure volumetric water content), tensiometers (to measure matric potential) and suction samplers (to extract soil water and tracer). All instrumentation was automated using computers, and chloride concentration was determined using in situ flow-injection systems. To determine nitrate concentrations, samples were obtained from the samplers (54) once an hour. In the day-time some samples were analysed using test-strip technology and some stored for analysis the next day. In a 24 hour period a total of 1296 samples were available for analysis. The temporal resolution obtained, while not as fine as that for chloride, was greatly in excess of anything previously attempted in the field. The fine resolution was possible, partially because a single operator could perform analysis as sample became available without total reliance on a busy laboratory facility.

Abbreviations: TDR–time domain reflectometry.

Introduction

Nitrate leaching is of great importance with regard to ground water pollution (Addiscott et al., 1991). EC directives on levels of nitrate in water have raised concerns about agricultural practice, and demanded greater understanding of the processes involved in leaching and increased the need for reliable models of nitrate leaching for planning and agricultural development purposes. A major component of chemical transport in soils is preferential flow, which is a rapid movement of water through larger pores in which capillary tension is only a minor factor.

To improve understanding of nitrate movement through soil, fine temporal and spatial resolution sampling is required in order to observe rapid changes in concentrations, and the variability in concentration of nitrate leaching from the soil.

This paper describes a large soil block designed to permit fine spatial and temporal resolution investigation of preferential flow. It will focus on the near real-time measurement of nitrate concentrations within the soil block during a controlled tracer experiment. The methodology described illustrates the utility of using test-strip technology (Scholefield and Titchen, 1995) which negates the need for laboratory analysis of nitrate.

Materials and methods

The North Wyke soil block

A large block (5.4 m × 3.4 m × 1.2 m) of undisturbed, well structured, sandy-loam soil (Dystric Eutrocrept) under pasture was isolated in situ by excavating

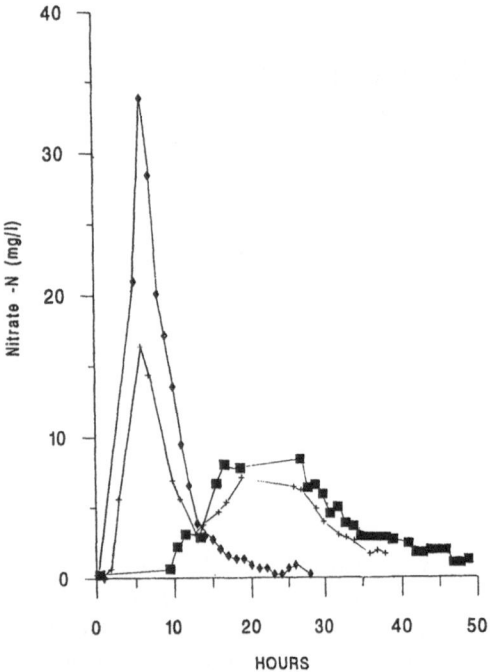

Figure 1. Nitrate breakthrough curves from soil water extracted at 0.4 m in the soil block under a constant surface flux of 5.5 mm hr^{-1}. ($\blacklozenge$ = A, $\blacksquare$ = B, + = C).

a trench 1.2 m deep around the volume of interest. The vertical faces of the block were sealed by rendering with puddling clay, the sides of the trench and block were then supported using plywood and the cavity between clay and wood filled with watertight expanding foam. The block was instrumented with 54 tensiometers, 54 TDR waveguides and 54 suction lysimeters. These were located using a stratified random strategy, 6 of each per 0.1 m layer, starting 0.1 m below the soil surface. All probes were installed horizontally into the soil volume to prevent any possibility of creating artificial preferential flow paths through the block.

Automatic data acquisition

The tensiometers were equipped with calibrated pressure transducers (Dowd and Williams, 1989) that were multiplexed to a Campbell 21X datalogger which recorded matric potential every 15 minutes for each tensiometer. The TDR waveguides (for measuring volumetric water content every 15 minutes) were connected to a Tektronix 1502B cable-tester via RF relays, and the whole system controlled by a PC. The suction lysimeters were connected to sample traps that formed

part of an in situ flow-injection system. There were six such systems, each connected to nine samplers. Further details of all the systems can be found in Holden et al. (1994). The flow-injection system was designed to be used for either chloride or nitrate determination. In the experimental situation described here the system was used for chloride analysis. To facilitate nitrate analysis, samples were manually removed by replacing sample traps once every hour.

Test-strip analysis for nitrate

A "Merck RQflex" reflectometer (E. Merck, 64271 Darmstadt, Germany) was used with "Reflectoquant" nitrate-test strips (E. Merck, 64271 Darmstadt, Germany) for the analysis of soil water nitrate collected in the sample traps. Each batch of test strips was supplied with a barcode which contained information for wave length correction and a batch specific calibration curve. The bar code was read by the meter to initialise it. The meter was laboratory tested before field use with 6 nitrate standards (5, 10, 20, 80, 100 and 120 mg L^{-1}). It was found that a single test-strip produced a reliable result that was reproducible. An older system described by Titchen and Scholefield 1995 required at least 5 replicates to provide a reliable figure. The RQflex produced better results because of its batch specific information and each strip had two reactive pads which the machine measured to produce a mean value. To analyse a sample, the meter's clock was started at the same time as a strip was dipped into a sample. The clock counted down 60 seconds and at 5 seconds, the strip was placed in the meter and a nitrate concentration value displayed. As many as 6 strips could be read consecutively at ten second intervals, and thus the meter was able to analyse a theoretical 180 samples an hour. The experiment yielded a maximum of 54 samples an hour so all samples could be analysed during the sample interval of 1 hour. Samples taken at night were stored and analyzed in the field with the RQflex reflectometer the following morning. After analysis the sample vials were capped and stored for selective laboratory analysis if confirmation of chloride or nitrate results was required.

The tracer experiment

The large soil block was designed with the intension of conducting a number of leaching experiments under various hydrological conditions. A single experimental run and its results are described here.

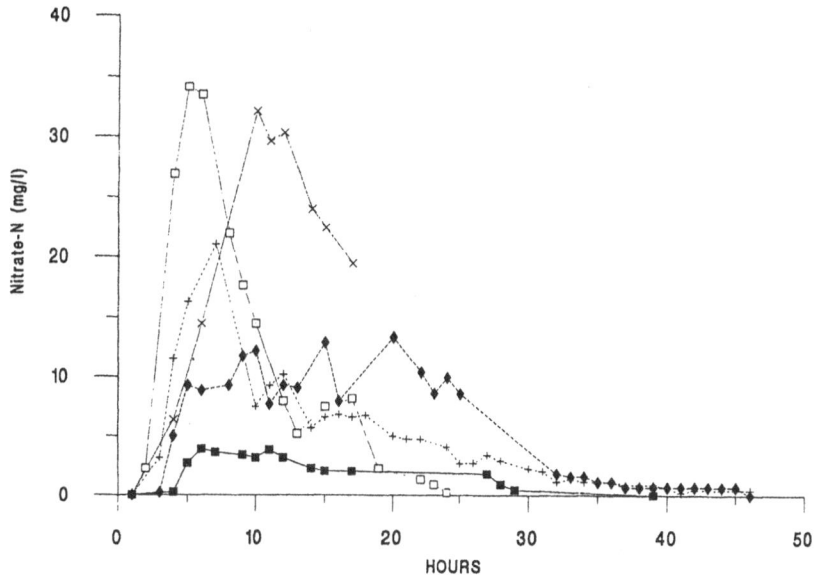

Figure 2. Nitrate breakthrough curves from soil water extracted at 0.5 m ($\times$ = C1T6, + = C1T9, $\square$ = C2T4, $\blacklozenge$ = C3T5 and $\blacksquare$ = C3T9) in the soil block under a constant surface flux of 5.5 mm hr^{-1}.

A spray-rig consisting of a solenoid control valve, 15 nozzles (each spraying 1 m^2) and 5 pressure regulators set at 3hPa (to ensure even spray over the block) was placed 0.3 m above the soil surface with the nozzles pointing upwards. Water was applied to the soil surface at a realistic rate in a sequence of 5 second pulses each seperated by 20 seconds. This simulated rainfall of 5.5 mm hr^{-1}. A poly-tunnel was constructed over the block to prevent natural rain and wind interference and the spray was on for 10 days prior to tracer application to allow the block to reach equilibrium soil water conditions for a 5.5 mm hr^{-1} surface flux as indicated by TDR and tensiometer data. The flow-injection system was then started to analyse background chloride concentrations, and some samples were taken for background nitrate determinations. 415 L of tracer were mixed using 449 g KNO$_3$ and 218g KCl. This produced tracer concentrations of 250 mg L^{-1}-Cl$^-$ and 150 mg L^{-1}-NO$_3^-$-N. The tracer was applied to the soil for 4 hours through the spray-rig using a pump, and samples for nitrate analysis were taken after 30 minutes and then every hour until the nitrate concentration returned to background. Chloride was analysed automatically every 30 minutes (these results are not shown or discussed here).

Results

Figures 1 illustrates the types of breakthrough curves characteristic of the soil block at equilibrium with a surface water flux of 5.5 mm hr^{-1}. The case illustrated by curve A had a rapid peak and decline indicating that the nitrate moved through the soil in large pores with high flow rates which permitted little interaction with existing soil water in the matrix. The breakthrough shown in curve B had a peak that occured quickly but was more rounded and long-lived suggesting an increased interaction of applied water with existing soil water. Curve C represents a response with early breakthrough followed by a decline and later an increase in concentration before tailing off. This appears to be a combination of the responses illustrated incurves A and B. The variation in nitrate breakthrough was found at all depths through the soil block.

Figure 2 represents breakthrough curves for 5 locations at 0.5 m depth. The data demonstrates the degree of spatial variability of peak tracer concentration found in the block. Peak concentrations ranged from 4–32 mg L^{-1} NO$_3^-$-N while time of peak concentration was fairly consistent. It can be seen that some curves have a second smaller peak about 36 hours into the experiment.

Discussion

The large soil block and associated equipment permitted controlled experiments to investigate leaching processes linked to preferential flow. The main advantage of the suite of data acquisition systems is that many locations within the soil block could be sampled and analysed contemporaneously. The analysis of samples in the field is particularly useful because there is then little need for laboratory facilities. Due to the main chemical analysis systems (field flow-injection equipment) being used for chloride determination, an alternative technology for nitrate had to be found. The RQflex reflectometer was an ideal tool with many advantageous features: it was cheap to purchase compared to other analysis systems; it required little operator training; analysis was rapid; no sample preparation was required; samples did not need storing so did not degrade and the samples once analysed were still available for any further analysis required. Real-time data acquisition is useful because it permits more complex experimental manipulations and sampling controlled by the soil condition rather than by arbitrary schemes. Examples of this are selective gas sampling, facilities for which were installed in the soil block, and complex tracer experiments with more than one tracer pulse. Having soil water status data and tracer data available simultaneously permits trouble shooting to be undertaken and enables clear understanding of response to environmental variables.

Demand on laboratory time and equipment use means that it is not possible for most experimenters to analyse thousands of samples rapidly; it would take 50 working hours to analyse 4000 samples in the laboratory with the need to prevent sample degradation.

The results from the single experimental run for conditions described illustrate that there is great variability in the nature of nitrate leaching in time and space. Peak concentration values show a wide range (ca.80 mg L^{-1}) and the shape of curves show that a number of interacting transport mechanisms are occurring simultaneously. This information could not be easily found using smaller cores due to the lack of three-dimensional interactions, while field scale experiments yield little useful data about the mechanisms of solute transport. Fine resolution data sets as obtained are particularly useful for model testing and development. It is intended that these data should be used to improve our understanding of variability in transport processes and to assess existing predictive models in order to highlight their capabilities and limitations.

In conclusion, the RQflex reflectometer in conjunction with the large soil block, permited many samples to be analysed in near real-time. This facilitated fine temporal and spatial resolution experimentation which is necessary to improve understanding of mechanisms of transport and their modelling.

Acknowledgements

The project was funded by Natural Environment Research Council grant GR38556. We are greteful to Ms Kirsty Phillips for the nitrate analysis.

References

Addiscott T M, Whitmore A P and Powlson D S 1991 Farming, fertilizers and the nitrate problem. CAB International, Wallingford, UK. 170 p.

Dowd J F and Williams A G 1989 Calibration and use of pressure transducers in soil hydrology. Hydro. Process. 3, 43–49.

Holden N M, Scholefield D, Williams A G and Dowd J F 1994 Design of a large soil block for fine temporal and spatial resolution investigations of preferential flow. *In* Transactions of the 15th World Congress of Soil Science, Acapulco, Mexico. Volume 2b, Ed. J D Etchevers. pp 169–170. Int. Soc. of Soil Sci./Mex. Soc. of Soil Sci., Mexico.

Scholefield D and Titchen N M 1995 Development of a rapid field test for soil mineral nitrogen and its application to grazed grassland. Soil Use Manage. 11, 33–43.

O. Van Cleemput et al. (eds.), Progress in Nitrogen Cycling Studies, 675–680, 1996.

N leaching in manured and ammonium nitrate fertilized lysimeters in Zimbabwe

W. Kamukondiwa[1], L. Bergström[2], B.M. Campbell[1], P.G.H. Frost[1] and M.J. Swift[3]

[1] The University of Zimbabwe, Department of Biological Sciences, P.O. Box MP 167, Mount PLeasant, Harare, Zimbabwe. [2] The Swedish University of Agricultural Sciences, Division of Water Management, P.O. Box 7072, S-750 07 Uppsala, Sweden and [3] UNESCO ROSTA, UN Complex, Gigiri, P.O. Box 30592, Nairobi, Kenya.

Key words: fertilizer-N, leachate, maize, manure, N-load, wheat

Abstract

Leaching of N was measured in eight lysimeters with maize (summer 1991/92) and wheat (winter 1992). In 1991/92, 100kg N ha^{-1} was applied as manure or ammonium nitrate in each set of four lysimeters and rainfall amounted to 391 mm. Wheat received 100 kg Nha^{-1} ammonium nitrate in combination with either 772 mm or 927 mm water. Although wheat received more water than maize, it lost less N through leaching (0.2–2.7 and 0.1–1.7 kg ha^{-1}) and had greater aboveground-N (88–116 and 93–141 kg ha^{-1}) in the low and high water treatments than the manured and fertilized maize (52–84 and 55–91 kg ha^{-1}) which lost more N (1.4–4.2 and 4.7–8.6 kg N ha^{-1}) through leaching. Leaching of ammonium-N never exceeded 0.5 kg ha^{-1}.

Introduction

Application of fertilizer-N (Baker and Johnson, 1981) or dumps of manure (Addiscott et al., 1992) increase the risk of nitrate leaching. In Zimbabwe, use of manure is common among communal area farmers (Tanner and Mugwira, 1984). Chemical fertilizers are increasingly being used by communal farmers since the last fourteen years. The risk of N leaching may also have increased.

Apart from the transportive effect of water, climate also affects N leaching by influencing the amounts and quality of soil organic matter, a major source of leached N. Rainfall received in preceding years may also affect leaching in subsequent years through its previous influence on the distribution of N with depth.

We investigated the impact of manure and fertilizer-N on mineral-N leaching in a soil derived from granite and resembling those found in most communal areas of Zimbabwe. Each lysimeter had been subjected to either a low (W1) or a high (W2) water treatment in combination with manure or fertilizer-N over the three preceding seasons (Kamukondiwa and Bergström, 1994a) and it was hypothesized that N leaching is influenced by the history of fertilization and irrigation.

Materials and methods

The study was conducted in Marondera (18^0 11'S, 31^0 28'E), Zimbabwe. Annual rainfall is about 900 mm (November - April) with about 30% coefficient of variation. Lysimeters (1 m diameter and 1.2 m deep) were installed in September 1989 by reconstructing disturbed profiles of soils comparable to kanhaplic Haplustalfs of the USDA taxonomy. Paired lysimeters were buried in four field plots each measuring 88 m^2 (Kamukondiwa and Bergström, 1994a).

Two maize seeds were sown in each lysimeter on 12 December 1991. Rainfall amounting to 101 mm had been received over the 30 days that preceded sowing. At sowing, lysimeters 1, 2, 7 and 8 received 100 kg N ha^{-1} in cattle manure whilst lysimeters 3–6 received 60 kg ha^{-1} each of P_2O_5 and K_2O, and 6 t ha^{-1} agricultural lime. Lysimeters 3–6 also received 100 kg N ha^{-1} in ammonium nitrate on 16 January 1992. Summer 1991/92 was abnormally dry and the low level of water in the reservoir forced irrigation to be discontinued after only 80 mm had been applied. Rainfall received between sowing and harvest (23/04/92) amounted to 289 mm only. The field plots received identical treatments to the lysimeters buried in them.

On 12 May 1992, wheat was sown in all lysimeters after applying 60 kg ha^{-1} each of P_2O_5 and K_2O. Only about 30 m^2 immediately surrounding and adjoining the two lysimeters in each plot was sown to wheat because of a limited water supply. Wheat received 100 kg N ha^{-1} in ammonium nitrate on 1 June and rainfall amounting to 17 mm only. The wheat received one of two watering regimes that differed in amounts and frequency of irrigation which was simulated using perforated plastic containers. Rates of water application were designed to avoid ponding, mostly at 25 mm per irrigation which was done about twice weekly during May - September. Lysimeters previously on the W1 treatment received 772 mm and those previously on the W2 treatment received 927 mm. The rather liberal irrigation was necessary to offset the high evaporative losses expected when a small wet area is surrounded by a large dry area. Wheat was harvested in the first week of October 1992.

Leachate was collected by free drainage in 10 L plastic buckets from which 100 mL samples were taken for mineral-N analysis after every 5–7 mm flow. Ammonium and nitrate were determined colorimetrically by methods based on the indophenol blue complex and the cadmium reduction of nitrate (Keeney and Nelson, 1982). Samples of maize and wheat were analyzed for total N by mass spectrometry (ANCA-MS Europa) along with ^{15}N that was applied in December 1990 (see Kamukondiwa and Bergström, 1994b).

Results

Leachate volumes, ammonium and nitrate concentrations

During 1991/92, leaching largely occurred in December - January (Table 1) and breakthrough occurred before sowing. Leachate volumes from the manured lysimeters were insignificantly smaller than those from the fertilized lysimeters (Table 2) but inclusion of the previous water treatments in a two way ANOVA brought significance ($p < 0.05$, F = 8.85 and d.f. = 1,4). Although the 1991/92 maize received less water than the 1992 wheat, volumes of leachate were 21–36% and 1.4–5.7% of the water received in each season respectively (Tables 2 and 3). Leaching ceased at the onset of active wheat growth (July - August).

Nitrate concentrations were generally higher in leachate from fertilized than in that from manured lysimeters (Fig. 1). This was also the case even

before applying fertilizer-N. In 1991/92, amounts of ammonium-N in leachate were smaller (0–0.6 mg L^{-1}) than in winter 1992 (Fig. 2). Maximum amounts of nitrate were 6 and 10 mg N L^{-1} for leachates from the previously manured and the previously fertilized lysimeters. No differences of N concentration existed between water treatments.

N loads and aboveground-N

More N was leached from the fertilized than from the manured lysimeters (Table 2). A two way ANOVA based on previous treatments, showed that N-loads from the previously W2 lysimeters were larger ($p < 0.05$, F = 9.22 and d.f. = 1, 4) than those from the previously W1 lysimeters. A similar and more pronounced trend ($p < 0.01$, F = 34.34 and d.f. = 1, 4) was obtained for loads of leached ammonium-N. There were no differences in aboveground-N between the two N sources but a two way ANOVA based on previous treatments showed greater aboveground-N in the previously W1 lysimeters ($p < 0.05$, F = 8.38 and d.f. = 1, 4). Grain yields were 1.2±0.72 and 0.8±0.65 t ha^{-1} in fertilized and manured lysimeters and represented 26 and 15% of aboveground dry matter.

Loads of leached N for winter 1992 (Table 3) also include ammonium and nitrate but amounts of the former were negligible. Leaching of N was less than in summer 1991/92 and never exceeded 3 kg N ha^{-1} whereas aboveground-N was larger and mostly exceeded 100 kg N ha^{-1}, particularly in the W2 treatment. Lysimeters that had previously received manure had smaller N leaching loads than those that had always received fertilizer-N.

In 1992, aboveground dry masses were insignificantly larger in the W2 lysimeters (Table 3). However, these differences became significant when the previous N treatments were also considered ($p < 0.05$, F = 15.50 and d.f. = 1, 4).

Discussion

Leachate volumes, ammonium and nitrate concentrations

The 101 mm rainfall received before sowing must have favoured early leaching which occurred in December 1991 before germination. Transpirational losses were then small.

Table 1. Distribution of rainfall, irrigation and leachate from lysimeters 1–8 in the 1991/92 maize season. *W1* and *W2* represent the previous low and high water treatments

	Nov	Dec	Jan	Feb	Mar	Apr
			(mm)			
Rainfall	91.1	73.6	67.2	52.2	75.4	32.8
Irrigation		18.3	54.1	7.9	–	–
			Leachate			
Fertilized						
3^{W1}	–	63.2	42.2	1.4	–	–
4^{W1}	–	88.0	43.1	1.0	–	–
5^{W2}	–	72.4	91.5	4.6	–	–
6^{W2}	–	53.4	90.4	3.3	–	–
Manured						
1^{W1}	–	77.1	39.5	0.8	–	–
2^{W1}	–	62.2	34.4	0.7	–	–
7^{W2}	–	62.4	35.3	2.1	–	–
8^{W2}	–	73.5	40.8	1.7	–	–

Table 2. Leachate volumes, N loads, aboveground dry mass and N for the 1991/92 maize in lysimeters 1–8. *W1* and *W2* represent the previous low and high water treatments

Lysimeter	Leachate (mm)	Ammonium (kg N ha^{-1})	Nitrate (kg N ha^{-1})	N-load (kg ha^{-1})	Dry mass (t ha^{-1})	Above-ground-N (kg ha^{-1})
Fertilized						
3^{W1}	107	0.02	4.69	4.7	5.3	91
4^{W1}	132	0.07	5.44	5.5	5.2	86
5^{W2}	169	0.50	8.12	8.6	4.1	60
6^{W2}	147	0.27	5.83	6.1	3.7	55
Mean	138.8			6.23	4.58	73.0
S.E.	13.03			0.729	0.399	9.06
Manured						
1^{W1}	117	0.03	1.70	1.7	5.8	84
2^{W1}	97	0.03	1.35	1.4	4.8	72
7^{W2}	100	0.37	2.76	3.1	4.2	52
8^{W2}	116	0.42	4.19	4.2	6.3	79
Mean	107.5			2.60	5.28	71.8
S.E.	5.235			0.562	0.475	7.03
t	2.229			3.942**	1.129	0.105

** Significant at $p < 0.01$

The preferential demand for ammonium-N by the heterotrophic soil biomass (Jansson et al., 1955) presumably led to small concentrations of ammonium-N in leachate. Furthermore, leaching of ammonium is generally low because ammonium-N is adsorbed onto the negatively charged surfaces of clays. Higher nitrate concentrations in leachate from the fertilized than in

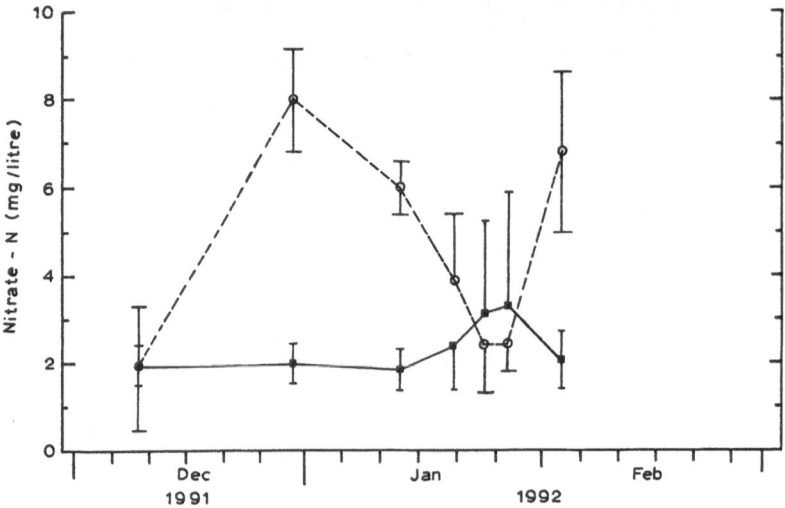

Figure 1. Nitrate in leachates from manured (boxes) and fertilized (circles) lysimeters during the 1991/92 season (bars represent ± 1 s.d.).

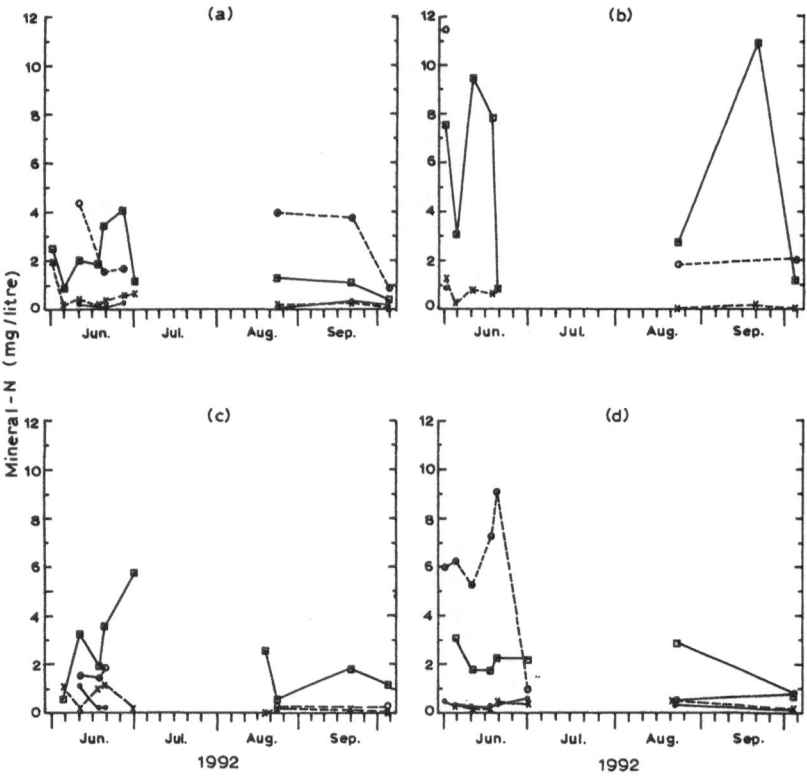

Figure 2. Ammonium (crosses and dots) and nitrate (boxes and circles) in leachates from the previously manured (**a** and **c**) and the previously fertilized (**b** and **d**) lysimeters on the low (a and b) and high (c and d) water treatments in winter 1992.

those from manured lysimeters reflects greater availability of the N in inorganic fertilizers.

Increases of nitrate concentration in leachate from the manured lysimeters presumably resulted from nitrification of manure. Increases in leachates from the fertilized lysimeters two weeks before application of fertilizer-N would similarly result from nitrification of soil-N. Also, residual fertilizer-N from a previous season may remain in the inorganic form until the second season from the time of application (Broad-

Table 3. Aboveground dry masses, N, leachate volumes and N loads for the 1992 wheat in lysimeters 1–8

Lysimeter	Leachate (mm)	N-load (kg N ha^{-1})	Dry mass (t ha^{-1})	Above-ground-N (kg N ha^{-1})	Previous treatment
Low moisture (772 mm)					
1	26	0.8	10.8	116	Manure
2	30	0.4	11.1	108	Manure
3	11	0.2	10.9	113	Fertilizer
4	45	2.7	10.2	88	Fertilizer
Mean	28.0	1.03	10.75	106.3	
S.E.	6.99	0.572	0.194	6.31	
High moisture (927 mm)					
5	44	1.7	14.0	130	Fertilizer
6	34	0.8	15.8	141	Fertilizer
7	27	0.1	12.2	120	Manure
8	29	0.4	10.7	93	Manure
Mean	33.5	0.75	13.18	121.0	
S.E.	3.80	0.348	1.105	10.27	
t	0.691	0.418	2.166	1.220	

bent and Carlton, 1978). An increase in the concentration of nitrate at the end of January was attributed to fertilizer-N (Baker and Johnson, 1981) although the actual amount of fertilizer-N leached in the short-term is small (Kamukondiwa and Bergström, 1994b). Less nitrate in leachate from the previously manured lysimeters during winter 1992 presumably reflects that manure still provided a source of energy for the N-immobilizing microbes.

N loads and aboveground-N

Loads of leached N were smaller in manured lysimeters partly because of the small proportion (< 2%) of manure-N that was initially in mineral form. Because the history of fertility management influenced N leaching, N availability in the manured lysimeters may have been limited by the accumulation of lignin, a stable constituent of humified organic substances, in fine silt and coarse clay fractions (Schulten and Leinweber, 1991).

Less N leaching and greater aboveground-N in the previously W1 lysimeters can be explained by increased crop growth under greater availability of N. These lysimeters must have had less mineral-N at greater depths and at risk of leaching. Over the preceding three seasons, the W2 lysimeters had on average lost 13 kg N ha^{-1} more nitrate than the W1 lysimeters (Kamukondiwa and Bergström, 1994a). Also, the previously W1 lysimeters must have experienced more intense drying leading to a greater flush of mineral-N in soil after the rains (Birch, 1960). This would increase early growth and reduce leaching.

A smaller proportion of dry matter made up by grain in the manured (15%) than in the fertilized (26%) lysimeters that had no differences of aboveground-N or dry masses (Table 2), suggests delayed anthesis and cob development in manured lysimeters. N-immobilization can reduce early growth of manured plants (Tanner and Mugwira, 1984) thereby extending vegetative growth and decreasing grain yield.

Wheat experienced less N leaching than maize although the former received more water. This is because most irrigation water was applied when the wheat was well established and actively transpiring so that additions of water increased crop growth and N-uptake, thereby reducing N leaching.

Significant effects of previous water and N treatments in dry matter production and N leaching indicated that previous applications of manure still influenced the turnover and availability of N (Schulten and Leinweber, 1991). Because moisture influences mineralization of manure (Murwira, 1990), previous water treatments on manure presumably influenced humification of manure and subsequent mineralization hence leaching.

Acknowledgements

We thank the Swedish Agency for Research Cooperation with Developing Countries for financing this study and Mr Kamukondiwa's travel to Belgium. Technical support was also received from the National Environment Research Council (UK).

References

Addiscott T M, Whitmore A P and Powlson D S 1992 Farming, Fertililizers and the Nitrate Problem. CAB International, Wallingford, Oxon, UK. 170 p.

Baker J L and Johnson H P 1981 Nitrate-nitrogen in tile drainage as affected by fertilizer. J. Environ. Qual. 10, 519–522.

Birch H F 1960 Nitrification in soils after different periods of dryness. Plant and Soil 12, 81–96.

Broadbent F E and Carlton A B 1978 Field trials with isotopically labeled nitrogen fertilizer. *In* Nitrogen in the Environment. I. Nitrogen behaviour in Field Soil. Eds. D R Nielsen and J G MacDonald. pp 1–41. Academic Press, New York, USA.

Jansson S L, Hallam M J and Bartholomew W V 1955 Preferential utilization of ammonium over nitrate by micro-organisms in the decomposition of oat straw. Plant and Soil 6, 382–390.

Kamukondiwa W and Bergström L 1994a Nitrate leaching in field lysimeters at an agricultural site in Zimbabwe. Soil Use Manage. 10, 118–124.

Kamukondiwa W and Bergström L 1994b Leaching and crop recovery of ^{15}N from ammonium sulphate and labelled maize (*Zea mays*) material in lysimeters at a site in Zimbabwe. Plant and Soil 162, 193–201.

Keeney D R and Nelson D W 1982 Nitrogen-Inorganic forms. *In* Methods of Soil Analysis, Part II. 2nd ed. Eds. A L Page et al. pp 643–698. ASA and SSSA, Madison, WI, USA.

Murwira H K, Kirchmann H and Swift M J 1990 The effect of moisture on the decomposition of cattle manure. Plant and Soil 122, 197–199.

Schulten H-R and Leinweber P 1991 Influence of long-term fertilization with farmyard manure on soil organic matter: Characteristics of particle-size fractions. Biol. Fertil. Soils 12, 81–88.

Tanner P D and Mugwira L 1984 Effectiveness of communal area manures as sources of nutrients for young maize plants. Zim. Agric. J. 81, 31–35.

O. Van Cleemput et al. (eds.), Progress in Nitrogen Cycling Studies, 681–687, 1996.
© 1996 *Kluwer Academic Publishers.*

Agricultural nitrogen and phosphorus losses into Belgian surface waters

P. Nyssen, P. de Cooman and P. Scokart
Ministry of Agriculture, Administration of Agricultural Research, Institute for Chemical Research, Leuvensesteenweg 17, 3080 Tervuren, Belgium

Key words: agriculture, nitrogen, nutrient losses, phosphorus, surface waters

Abstract

In Belgium, nutrient losses to surface waters, resulting from agricultural activities, can be calculated by a pragmatic and empirical model developed at the IRC/ISO. In a first step, a theoretical maximum for seven categories of nutrient losses was determined using general statistical data. Then, more specific conditions were taken into account, such as local agricultural practices, geographical conditions or, when available, data from detail studies. The losses were calculated for each of the 586 municipalities and summarized by hydrological catchment area. In order to validate these theoretical results, a field study has been conducted in three basins. The experimental measurements available until now indicate that the theoretical estimations are only slightly underestimated. The model seems thus to be a good tool to evaluate nutrient losses for Belgium.

Introduction

Nutrients, as naturally occurring compounds necessary for life, can not be classified as toxificants sensu stricto. However, their concentration in surface waters can reach toxic levels causing important eutrophication problems and disturbing the biological and chemical balances in aquatic ecosystems. During the Meeting of Den Haag, the North-Sea-countries agreed to reduce by 50 percent their nutrient losses to the North Sea with reference to the year 1985, and several measures leading to direct action were implemented.

In order to quantify the nutrient losses from agriculture and to evaluate the impact of the measures taken, a pragmatic and empirical methodology was developed at ISO/IRC (De Cooman et al., 1993). The nutrient losses have been calculated for each of the 586 Belgian municipalities and were totalled by hydrological catchment area. In order to validate these theoretical results, a field study has been started in three representative catchment areas.

Methodology

The model

In the proposed model, a distinction is made between seven sources of nutrient losses into surface waters :

- *Atmospheric losses* : input of nutrients via atmospheric deposition from agriculture
- *Direct losses* :input of nutrients without intermediate transport, divided into:
- Direct losses of mineral fertilizers during fertilizer storage, transport and application
- Direct losses of organic fertilizers during grazing
- Direct losses of organic fertilizers during stabling
- Direct losses of nutrient from manure and silo juices
- *Run-off losses* : direct input of nutrients after spreading of slurry and run-off on country-roads
- *Drain losses* : input of nutrients via drain water with normal agricultural practices
- *Ground water losses* : input of nutrients via transfer of ground water to surface waters with normal agricultural practices
- *Erosion losses* : input of nutrients via soil erosion and run-off during agressive rainfall

Catchment area :

1. Beneden Schelde
2. Boven Schelde
3. Demer
4. Dender
5. Dijle (Fl)
6. IJzer
7. Lele
8. Maas
9. Nete
10. Polders en Gentse Kanalen

11. Amblève
12. Chiers
13. Dendre
14. Dyle (W)
15. Escaut
16. Geer
17. Gette
18. Gueule
19. Haine
20. Hermeton
21. Inde
22. Lesse
23. Lys
24. Méhaigne
25. Meuse-Nord
26. Meuse-Sud

27. Olse
28. Our
29. Ourthe
30. Roer
31. Sambre
32. Semols
33. Senne (W)
34. Sûre
35. Vesdre
36. Viroin

1-10 : Basins defined by V.M.M.
11-36 : Basins defined by "Service
de la politique générale de l'eau"

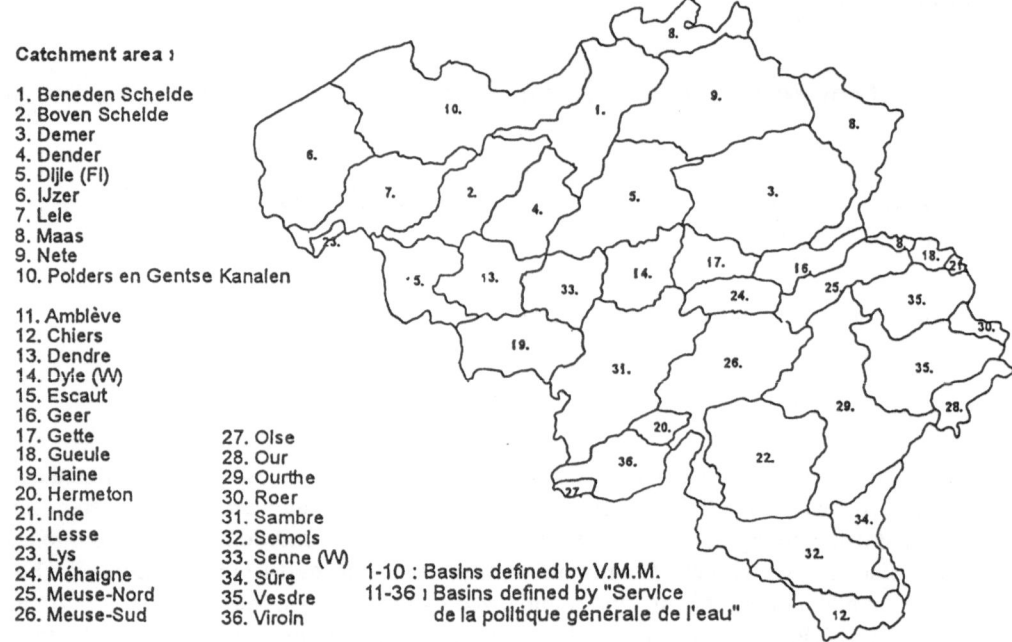

Figure 1. Belgian hydrological catchment area.

— *Excess losses* : input of nutrients via drain water, ground water, run-off,... due to application of live-stock effluents

The "nutrient stream" from agriculture into surface waters was determined for each hydrological catchment area (Fig. 1). In a first calculation, a theoretical loss was computed for each of the seven sources of nutrient losses. These calculations were based on available statistical data describing mineral fertilizer application rates, the use of farmland, live-stock densities... When possible the data were specified to the smallest geographical calculation unit : the municipality. Sometimes it was necessary to work with more global figures (averages for agricultural districts, provinces...). Then, starting from this potential loss, an actual nutrient loss was estimated taking into account the local agricultural practices, geographical conditions or, when available, extrapolated data and parameters from detail studies.

Statistical data, and agricultural practices were obtained from the "Institut Economique Agricole of the Ministry of Agriculture" and climatological data from the "Institut Royal Météorologique de Belgique". Organic nutrients produced per animal and per year in manure were calculated using coefficients defined by the "Mestdekreet".

Field study

A field study has been started in three catchment areas in Belgium to evaluate the theoretical results. The choice of the basin was based on soil type, agricultural practices and differences in calculated nutrient losses. Three basins were selected : the Senne, the Zwalm (Boven Schelde) and the Mark (Maas).

Every three weeks, about thirty water samples were taken in basin, and nitrates, organic nitrogen and phosphorus concentrations were determined. The flow rate at each sampling point was estimated by distributing the flow measured at the exutory proportionally to the surface drained at each point.

Multiplying concentration by flow rate gives the total nutrient losses at each point (TL).

Five sources of nutrient losses are assumed to compose the total nutrient losses :

— *Natural losses (NL)* : estimated respectively at 5; 6.5 and 8 kg N ha^{-1} for the permeable surfaces of the Senne, Zwalm and Mark, and at 0.2 kg P ha^{-1} for the permeable surfaces of the three basins accounting to the soil texture (De Becker, 1986; Hanbuckers, 1993).

— *Atmospherical losses (AL)* : 31 kg N ha^{-1} and 0 kg P ha^{-1} for the impermeable surfaces (Oslo and Paris Commissions)

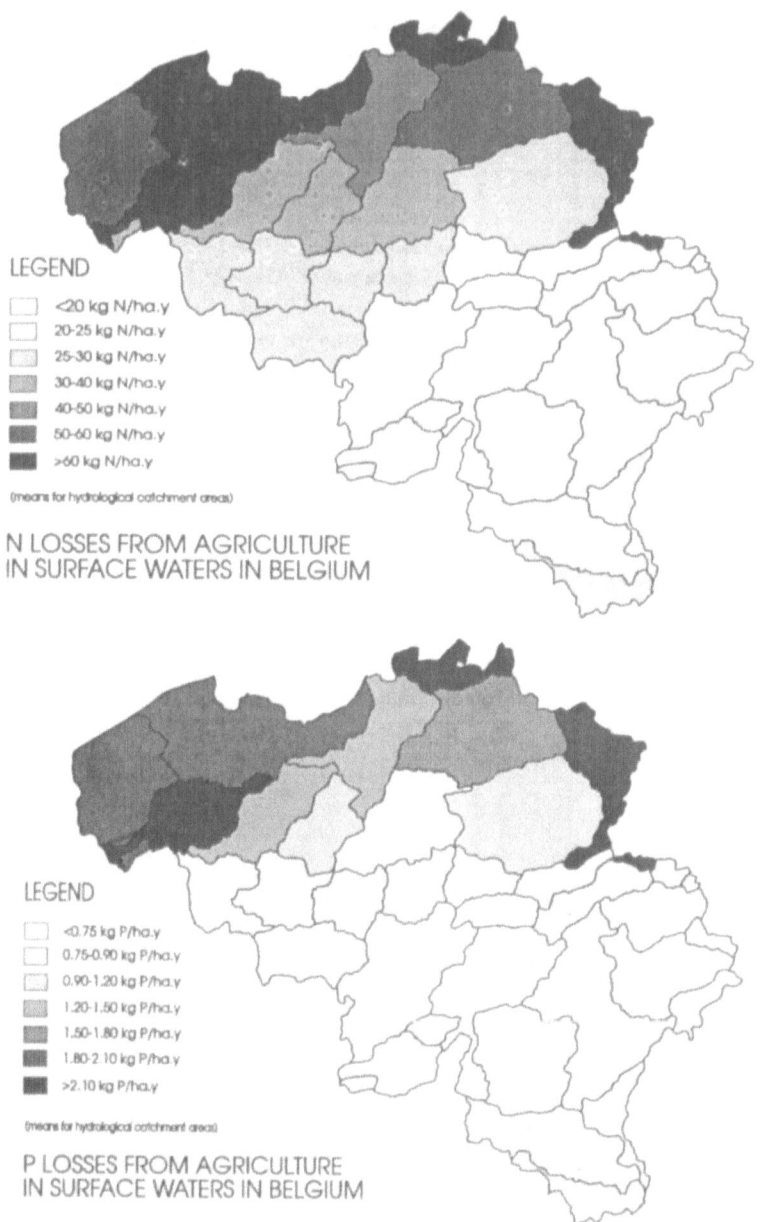

Figure 2. Theoretical N en P losses from agriculture into Belgian surface waters (1993–94).

— *Domestical losses (DL)* : 14.3 g N day^{-1} person^{-1} (Vander Borght, 1980) and 1.4 g P day^{-1} person^{-1} (Durand and Golicheff 1978)

— *Industrial losses (IL)* : data have been given by the "Division de l'eau" for the Senne and by the "Vlaamse Milieu Maatschappij" for the Zwalm and the Mark

— *Agricultural losses (AGL)* : calculated by the difference between total losses and the four other sources of losses AGL = TL - NL - AL - DL - IL

The permeable and impermeable surfaces are composed respectively of woods, gardens, parks, arable lands, and of roads, roofs, parking...

Table 1. Agricultural N and P losses into surface waters per basin (1993–94)

Basin	Region	kg N ha^{-1}	kg P ha^{-1}
Beneden Schelde	Vlaanderen	45.3	1.37
Boven Schelde	Vlaanderen	37.7	1.25
Demer	Vlaanderen	29.9	1.12
Dender	Vlaanderen	32.1	1
Diujle	Vlaanderen	32.9	0.87
Ijzer	Vlaanderen	54.4	1.91
Leie	Vlaanderen	67.0	2.3
Maas	Vlaanderen	69.5	2.1
Nete	Vlaanderen	54.2	1.71
Polders en Gentse Kanalen	Vlaanderen	64.4	1.91
Ambléve	Wallonie	21.9	0.68
Chiers	Wallonie	25.1	0.71
Dendre	Wallonie	26.9	0.86
Dijle	Wallonie	27.8	0.74
Escaut	Wallonie	25.0	0.85
Geer	Wallonie	19.8	0.8
Gette	Wallonie	19.7	0.72
Gueule	Wallonie	23.0	0.79
Haine	Wallonie	25.4	0.75
Hermeton	Wallonie	22.8	0.71
Inde	Wallonie	21.5	0.68
Lesse	Wallonie	23.1	0.72
Lys	Wallonie	38.0	1.81
Méhaigne	Wallonie	19.4	0.73
Meuse-Nord	Wallonie	22.5	0.78
Meuse-Sud	Wallonie	21.1	0.74
Oise	Wallonie	24.8	0.74
Our	Wallonie	21.4	0.68
Outhe	Wallonie	23.2	0.73
Roer	Wallonie	21.3	0.66
Sambre	Wallonie	20.9	0.72
Semois	Wallonie	24.1	0.71
Senne	Wallonie	26.0	0.82
Sûre	Wallonie	23.6	0.74
Vesdre	Wallonie	22.5	0.73
Viroin	Wallonie	23.0	0.69

Results

Theoretical model

The estimation given by the model are presented in Figure 2 for the agricultural N and P losses. The losses do range for nitrogen from 19.4 kg ha^{-1} for the catchment areas of the Mehaigne to 69.5 kg ha^{-1} for the Maas, and, for Phosphorus, from 0.66 kg ha^{-1} for the Roer to 2.3 kg ha^{-1} for the Leie.

If we compare Wallonie to Vlaanderen, one can notice that only one walloon catchment area (the Lys) out of the 26 basins has nutrient losses greater than 30 kg N ha^{-1} and 1.2 kg P ha^{-1}, whereas 9 Flemish catchment areas out of 10 are in this situation!

Figure 3 clearly shows the differences between the two regions. This figure shows the contribution of the seven nutrient sources to the total agricultural nutrient losses.

The average nitrogen losses amount to 52.89 kg N ha^{-1} in Vlaanderen, versus only 23.15 kg N ha^{-1}

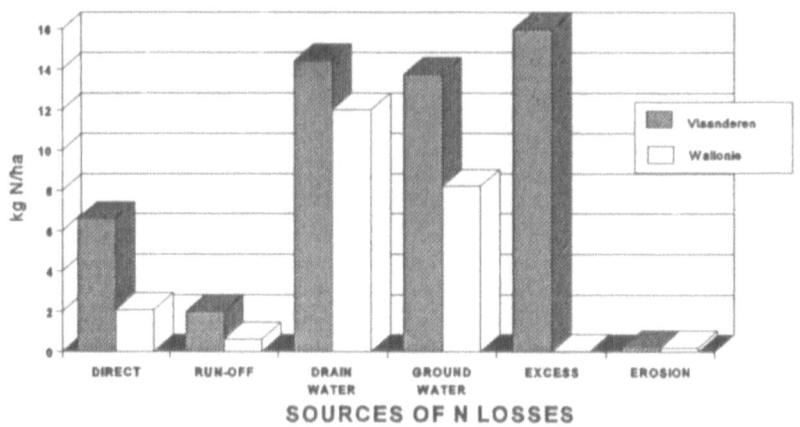

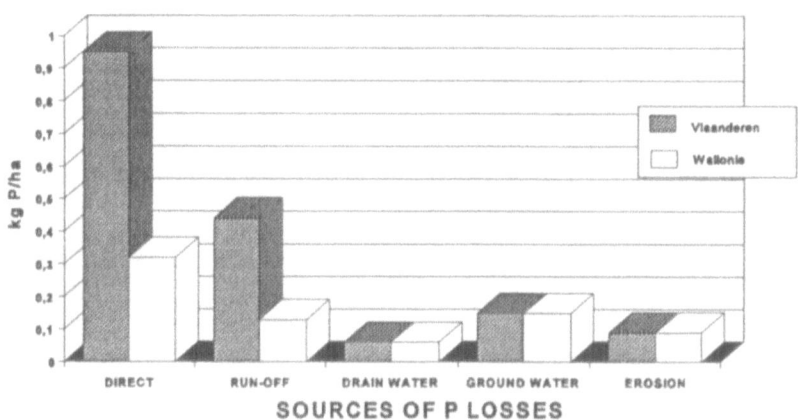

Figure 3. Agricultural N and P losses into surface waters for Vlaanderen (black) and Wallonie (white) during a one year period (1993-94).

in Wallonie. Most of this difference is due to excess fertilization in Vlaanderen.

The average phosphorus losses amount to 1.69 kg P/ha in Vlaanderen, versus 0.75 kg P ha^{-1} in Wallonie. The major source in both regions are direct losses, followed by run-off losses.

Field study

The evolution of the agricultural N and P losses for the three basins is showed in Figure 4. Three nitrogen loss peaks are observed in January, March and June. The biggest losses occur in the Mark basin, the smallest losses in the Senne basin. The only important phosphorus losses were observed in the Mark basin, in March and June.

Discussion

Regional differences

The evolution of the N and P losses according to the axis south-east north-west can be explained by two parameters : the soil type and the agricultural practices bound to the exploitation structures.

A lot of studies have demonstrated that the leaching of nitrogen, and more particulary of nitrates, is influenced by type of soil, which is not the case for phosphorus. The sandy soil type of Vlaanderen is more favourable to leaching than the loamy soil type of Wallonie and this can partially explain the greater losses for Vlaanderen.

However, the fundamental explanation has to be found in differences in agricultural practices and in the structures of the farms. Indeed, most of the farms off-soil (especially pig farms) are situated in Vlaanderen. These structures have led to a slurry production above

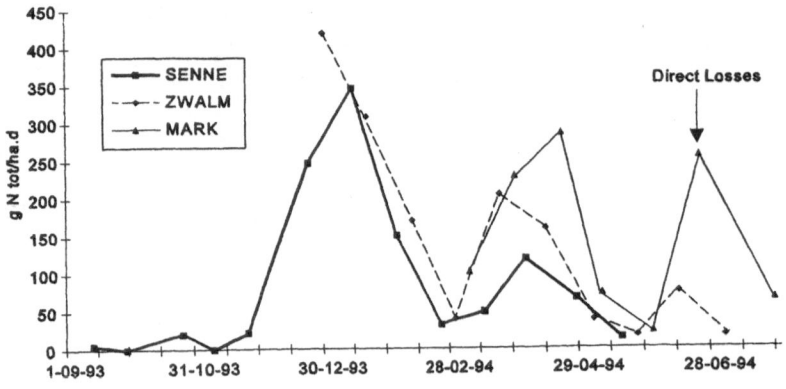

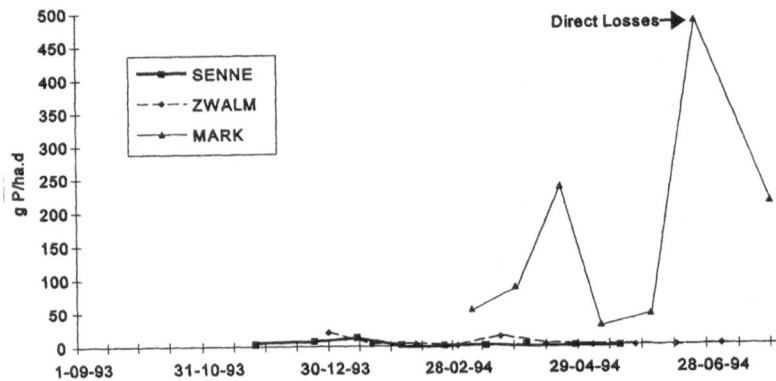

Figure 4. Evolution of the agricultural N and P losses during the field study.

Table 2. Comparison between the results of the theoretical model and the field study results (1993–94)

Basin	Model results				Field study			
	Period	(kg N ha^{-1})	Period	(kg P ha^{-1})	Period	(kg N ha^{-1})	Period	(kg P ha^{-1})
Senne	1/7/93	24.52	1/7/93	0.97	1/9/93	25.19	9/11/93	0.64
	30/6/94		30/6/94		19/5/94		19/5/94	
Zwalm	1/7/93	34.07	1/7/93	0.86	1/12/93	33.86	1/12/93	0.91
	30/6/94		30/6/94		5/7/94		5/67/94	

REM : The shortness of the studied period for the Mark didn't allow us to compare the field results with the model results.

the needs of the crops, which leads to excess in fertilization. These quantities of excess nutrients aren't absorbed by plants and are free to join the water surfaces through run-off, drain or ground water. One consequence of this slurry overproduction is direct losses into surface waters. This can be seen in Figure 4. Indeed, the heavy N and P losses in June in the Mark basin must be due to (illegal) direct losses, since the water flow rate was very low at that period.

Validation of theoretical model

Before attempting a comparison between the results from the model and those from the field study, we must underline that the period during which the experimental study was conducted, is not the same for the three basins. For the Senne, we have results for 261 days (1/9/93 to 19/5/94) for nitrogen, and only 191 days (9/9/93 to 19/5/94) for phosphorus. For the Zwalm, we have results for 217 days (1/12/93 to 5/7/94) for both elements. The results for the Mark are not included in this table because of the shortness of the studied period.

The estimations based on the theoretical model are only slightly underestimated relative to the experimental results (Table 2). Correction for the period (1 year for the theoretical results) should be small in the case of the Senne, since only small losses occur in the summer months. In the Zwalm basin, the losses over the whole year will probably reach 38 to 40 kg N ha^{-1}, and 1 to 1.15 kg P ha^{-1}.

The model, after minor adaptation, seems thus to be a good tool for nutrient management at the catchment level.

Acknowledgements

The authors wish to thank the Laboratory of Hydrology and Water Management - University of Ghent, the Royal Meteorological Institute - Department of Hydrology and the Direction Générale des Voies hydrauliques - SETHY for their practical help measuring the flow rates of the Zwalm, Mark and Senne.

References

De Cooman P, Scokart P and De Borger R 1993 Evaluation par bassin hydrographique des pertes en nutriments dans les eaux de surfaces provoquées par l'activité agricole en Belgique. Water 74, 26–32.

De Becker E 1986 Apports, transferts et transformations de l'azote dans les réseaux hydrographiques : développement d'une méthologie générale et applications au réseau belge. Thèse de doctorat, Faculté des Sciences, Université Libre de Bruxelles, Belgium.

Durand A and Golicheff A 1978 Etude des composants des eaux usées domestiques. Tech. Eau Assainissement 383, 27–43.

Oslo and Paris Commissions 1993 Nutrients in the Convention Area. OSPARCOM, London, UK. 38p.

Vander Borght P 1980 Etude physico-chimique des eaux de la Semois en vue d'une gestion qualitative. Thèse de doctorat, Fondation Universitaire Luxembourgeoise, Arlon, Belgium.

O. Van Cleemput et al. (eds.), Progress in Nitrogen Cycling Studies, 689–693, 1996.

Influence of long-term fertilisation on nitrogen leaching

D.A. Stoichev, D.I. Stoicheva, G.K. Angelov and K.S. Dimitrova
N. Poushkarov Institute of Soil Science and Agroecology, 7 Chaussee Bankya, Sofia 1080, Bulgaria

Key words: ground water, lysimetric water, mineral fertilisation, nitrogen migration, soil solution

Abstract

The influence of mineral fertilisation on the content of nitrogen in the soil profile, the soil solution, the lysimetric and underground waters has been studied in the conditions of a 20-year-long field experiment with maize (monoculture). Results show that, under fertilisation rates estimated to cover 100–125% of the nitrogen uptake with the plant production, a nitrogen accumulation in the entire profile is observed.

The long-term nitrogen fertilisation with rates over 210 kg per ha leads to an increased amount of nitrate nitrogen in the soil solution and the lysimetric water, which is the precondition for its increased migration outside the root zone and a subsequent nitrate pollution of the ground water.

Introduction

A significant part of the experiments conducted during the last years has been connected with the problems of nitrate migration in the soil profile (Addiscott and Bland, 1988; Bashkin, 1987; Sinkevitch, 1989; Snakin, 1989; Stoichev et al., 1983) and the search for technological decisions for the maintenance of their level under the permissible concentrations (Bizik, 1993; Collins and Johnson, 1988; Power, and Schepers, 1989, Stoichev et al., 1986; Varvel and Peterson, 1990).

Parameters have been obtained caracterizing the liquid phase of the soil as a dynamic component reflecting the influence of anthropogenic factors (Madison and Brunett, 1985; Snakin, 1989; Stoichev and Stoicheva, 1987).

The relationship between the nitrogen fertilisation rate and the nitrate content in the lysimetric waters has been proved under various soil-climatic conditions (Atanasov et al., 1981; Hergert, 1986; Stoichev et al., 1985). Different methods and means have been tested for a maximum inclusion of nitrogen in the biological cycle of nutrients and a decrease of its migration throughout the soil profile (Bizik, 1993; Stoichev et al., 1990).

The purpose of the present work is to evaluate the influence of long-term fertilisation on the mineral nitrogen content along the soil profile, in the soil solution, and in the lysimetric and underground waters.

Materials and methods

The area of study is located in South Bulgaria (24° 35′ E, 42° 14′ N, and 180 m above the sea-level), near the town of Plovdiv. This region is characterized by the following mean annual climatic data: rainfall - 548 mm, a 212-day period of temperature above 10 °C, a total temperature sum of 3886 °C, and 298 mm rainfall for the same period; air humidity for the growing period varies from 41% to 59%; the mean temperature of the arable soil layer is 13.1°C and the average annual air temperature is 12.0 °C.

The experiment was carried out on an alluvial-meadow soil. The arable horizon has the following mean characteristics: bulk density at MHC - 1.66 g/cm^3, moisture holding capacity - 14.99%; particles<0.01 mm - 27.3%; CEC - 7.92 meq $100\ g^{-1}$; pH_{H2O} - 6.00; humus content - 0.70% and total nitrogen - 0.052%. Calcium carbonate is leached under one meter depth. The level of the shallow ground water in the experimental field varies from 3.0 to 4.0 m below the soil surface. The water level is monitored in six drill wells situated in different parts of the experimental station.

The field experiment started in 1972 with maize grown as a monoculture under irrigation. The experimental design includes five variants with nitrogen and phosphorus, with no potassium application. The rates for the period of 1990–1992 were: 1 - N_0P_0; 2 - N_{120} P_{90}; 3 - $N_{170} P_{140}$; 4 - $N_{230} P_{190}$; 5 - $N_{290} P_{240}$; 6 - fallow. To estimate the fertiliser after-effect, in 1992 the wheat was grown with no fertiliser.

The field plots were equipped with Ebermeir lysimeters (Stoichev, 1974), collecting water from a depth of 50 and 100 cm from the surface, as well as with drill wells to study the relationship between the nitrogen rate fertilisation and the nutrient's concentration in the liquid phase. The chemical composition of the lysimetric and ground water was determined by monthly sampling throughout the year.

In order to study the changes in the mineral nitrogen content under long-term fertiliser application, soil samples were taken from the soil profile (0–100 cm) in a 20 cm step at the end of the 20 years after the beginning of the field trial. The mineral N was extracted by 1 N KCl (1:2.5) and measured by ADM-300 analyser. The soil solution of the samples taken monthly from the 0–30 and 30–60 cm layers during the growing season has been displaced by ethyl alcohol 96° proof (Stoicheva, 1979).

The NO_3-N analyses in soil solution, lysimetric and ground waters were carried out by direct distillation with 10% Fe_2SO_4 and 0.5% Ag_2SO_4 reducing agents.

Results and discussion

Mineral nitrogen content in the soil profile

The concentration of mineral nitrogen in the one-meter soil layer is of a significant ecological importance. The data obtained for the alluvial meadow soil (Fig. 1) show that the lowest residual nitrogen is found not in the non-fertilised plots, but in the soil profile, where the fertilisation rate covers only 50% (variant 2) of the uptake with the yields. When the fertilisation covers 100% (variant 4) and 125% (variant 5) of the uptake a significant nitrogen accumulation in all soil profiles is observed.

The mineral nitrogen's distribution curve in a fallow land profile (maintained for many years) clearly shows that neither the fertilisation exclusion, nor the lack of bioproduction are a perspective for a reduction of the mineral nitrogen in the soil as a potential source of ground water pollution. As shown in Figure 1 the

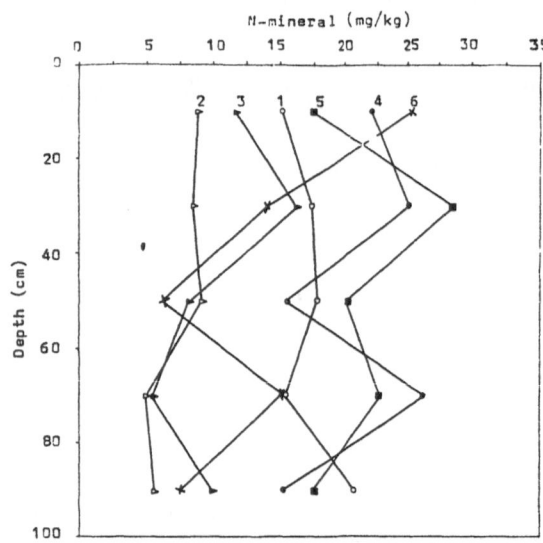

Figure 1. Mineral nitrogen content in the one-meter soil layer after 20-year maize cultivation at different rates of fertilisation. Treatments: 1- $N_0 P_0$; 2-$N_{120} P_{90}$; 3-$N_{170} P_{140}$; 4-$N_{230} P_{190}$; 5-$N_{290} P_{240}$; 6- fallow.

content of mineral nitrogen at the fallow land profile during the spring of 1992 is significantly higher than that of the low rate fertilisation variants ($N_{120} P_{90}$ and $N_{170} P_{140}$), and it gets closer to the full uptake compensated variants. It's all too obvious that the sustaining of a fallow land leads to almost the same risk of accumulation of remaining nitrogen, as the one from a fertilisation for higher yields.

Nitrogen concentration in the soil solution

The obtained analytical data (Fig. 2) show that under the conditions of the maize (as a monoculture) and the wheat (aftereffect) experiments, the fertilisation with different nitrogen rates have led to a 1.5 to 4.0 times increase in the nitrate nitrogen concentration in the soil solution. The highest nitrogen concentrations are observed in solutions of the maximum fertilised (N_{290} P_{240}), and the optimum fertilised ($N_{230} P_{190}$) variants. There is a direct correlation between the nitrogen rates applied in the experiment and the nitrate nitrogen found in the soil solution (r=0.835). The obtained analytical data for the nitrate nitrogen content in the soil solution from fallow land (Fig. 2a) show that in those types of lands the use of higher nitrogen concentration is observed when compared to the non-fertilised control. Comparison of data for the nitrogen concentration in the soil solution from the fallow and the non-fertilised variant, shows that the fallow land produces significant

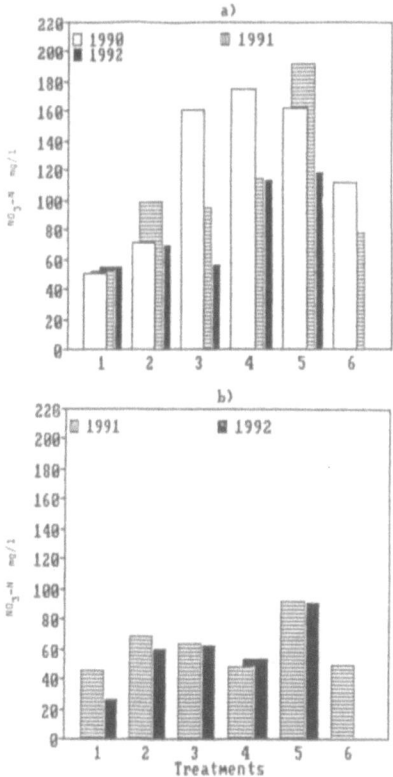

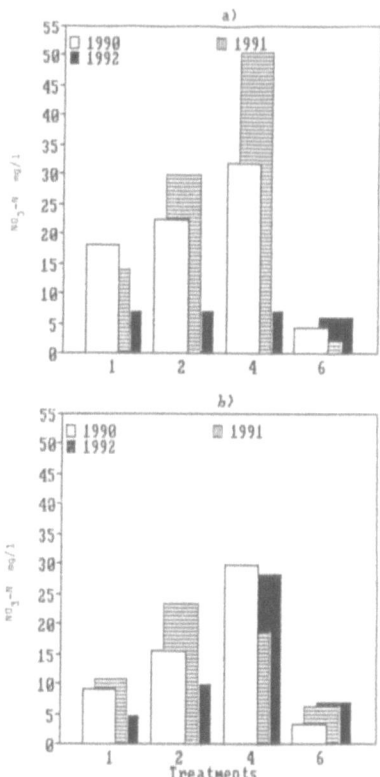

Figure 2. Nitrate nitrogen concentration in the soil solution (average for the whole growing season). Treatments: 1-N_0 P_0; 2-N_{120} P_{90}, 3-N_{170} P_{140}; 4-N_{230} P_{190}; 5-N_{290} P_{240}; 6- fallow. a) 0–30 cm layer; b) 30–60 cm layer.

Figure 3. Nitrate nitrogen content in the lysimetric water (average for the whole year). Treatments: 1-N_0 P_0; 2-N_{120} P_{90}; 4-N_{230} P_{190}; 6- fallow. a) 0–50 cm layer b) 0–100 cm layer.

quantities of nitrogen. The soil cultivation improves aeration, facilitates moisture holding and intensifies nitrification. The cultivated crops (variant 1) use a significant part of the soluble forms of nitrogen and its concentration in the soil solution therefore is greatly reduced. Due to a long-term application of fertiliser rates (over 170 kg ha^{-1} N) in alluvial meadow soil, there is a high level of N-NO_3 in the soil solution, which is a precondition for a higher migration of nitrates outside the root layer.

Nitrate nitrogen concentration in the lysimetric water

Comparison between Figure 2 and Figure 3 shows that the lysimetric water does not completely reflect the soil solution composition, because, when using lysimeters we collect only the rapidly filtrated water which is not in equilibrium with the solid phase of the soil. However, this is the only way to estimate the accumulative effects of the N-treatment in conditions similar to the natural ones, and to make a complete assessment of

that part of the nitrogen geochemical cycle which contributes to the ground water nitrate pollution.

The long term investigations show that despite the application rate the NO_3-N content of the lysimetric water ranges widely more markedly in the 0–50 cm layer.

When maize is grown (1990–1991) even without fertilisation (variant 1) lysimetric water is formed in the 0–50 cm layer with NO_3-N concentration exceeding the standard level of drinking water (Fig. 3a). In the same conditions, wheat growing contributes to the significant purification of drainage water. The lysimetric water from the N_{120} P_{90} variants has a higher or closer to the N_0P_0 (control) nitrogen concentration. Treatment with N_{230} P_{190} (variant 4) has resulted in a significant increase in the NO_3-N content of the lysimetric water from both depths.

The nitrate nitrogen content in the lysimetric water (Fig. 3) from a fallow land (variant 6) is the lowest, despite the above mentioned (Fig. 2) high NO_3-N concentration in the soil solution, displaced from fallow land samples. This is a direct evidence for the different

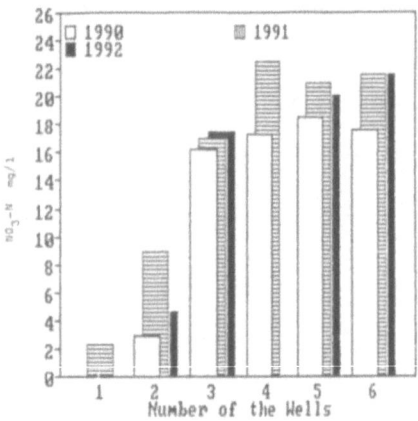

Figure 4. Nitrate nitrogen concentration in the ground water (average for the whole year). 1, 2, 3, 4, 5, 6 - Number of the monitored drill wells. Standard level - 11.3 mg L^{-1} NO_3-N.

mechanisms determining the mineral nitrogen releases in the soil solution and lysimetric water.

The wheat cultivated in 1992 (aftereffect) leads to a fast equalisation in the nitrogen concentration for the 0–50 cm layer in all studied variants of fertilisation (Fig. 3a). The difference between the variants in the 0-100 cm layer however is still existing (Fig. 3b). As shown in the variants with an optimum treatment (N_{230} P_{190}) of the previous crop, the NO_3-N concentration is still 2.5 times higher than the standard limitation for drinking water.

Nitrate nitrogen concentration in the ground water

The nitrate content in almost all of the cases (Fig. 4) is over the maximum permissible level for drinking water. An exception are wells No. 1 and 2. It is noteworthy that these two wells were giving nitrate containing water in the 1985–1990 period that were exceeding the concentrations in the rest of the wells. This may be explained by the fact that they give information basically for a part of the experimental station where there have been no fertiliser experiments during the last several years, as well as the fact that there is a reduction in the fertilisation practices for the farmer fields over the wells. The remaining four wells (No. 3, 4, 5, 6) are situated in the lower part of the field and two of them are used for irrigation. During the period of study (1990-1992), the water in these wells is characterised by a concentration of nitrate nitrogen that exceeds the standard level for drinking water. The scheme location of the wells enables the control over the quality of the underground water as a result of the

experiment. There is an obvious connection between the fertilisation and the concentration of nitrates in the shallow ground water in the experimental field.

Conclusions

Long-term field experimental data show that nitrogen fertilisation, especially with rates exceeding the uptake by the yield, leads to a significant mineral nitrogen accumulation in the 0–100 cm layer. Under these conditions, the nitrate content in the soil solution and the lysimetric water increases, which is a precondition for nitrogen leaching and shallow ground water pollution.

It was found that neither fertiliser exclusion nor fallow land sustension are a perspective for the prevention of ground water pollution. Wheat growing without fertilisation is a suitable way for an effective inclusion of residual nitrogen in the biological cycle of nutrients and soil remediation.

References

Atanasov I, Stoicheva D and Stoichev D 1981 Comparative characterization of the composition of soil solution and lysimetric waters. Soil Sci. Agrochem. 16, 62–67.

Addiscott T M and Bland G J 1988 Nitrate leaching models and soil heterogeneity. *In* Nitrogen Efficiency in Agricultural Soils. Eds. K E Smith and D S Jenkenson. pp 394–408. Elsevier Applied Science, Barking, UK.

Bashkin V N 1987 Agrogeochemistry of Nitrogen. SCBR, Puschino USSR. 270p.

Bizik J 1993 Accumulation of some forms of nitorgen in soils by two models of farming systems. *In* Proc. of Integrated arable Farming Systems. Eds. K Kovač and B Proházka. pp 178–188. DTZS, VTS, Nitza, Slovakia.

Collins A G and Johnson A I (ed) 1988 Ground water contamination: field methods. ASTM, Baltimore, USA. 491p.

Hergert G W 1986 Nitrate leaching through sandy soils as affected by sprinkler irrigation method. J. Environ. Qual. 15, 272–278.

Madison R J and Brunett J O 1985 Overview of the occurrences of nitrates in ground water of the United States. US Geological Survey Water Supply Paper 2275. pp 93–105.

Power J F and Schepers J S 1989 Nitrate contamination of ground water in North America 26. *In* Agriculture, Ecosystems and Environmental. Eds. H Bouwers and R S Bowman. pp 165–187. Elsevier, Amsterdam, the Netherlands.

Sinkevitch Z A 1989 Modern processes in the chernozem of Moldavia. Shtiintsa, Kishinev. 214 p.

Snakin V V 1989 Chemical composition of soil liquid phase. Nauka, Moscow. 118 p.

Stoicheva D I 1979 A study on soil solution substitution by a liquid. Soil Sci. Agrochem. 14, 27–35.

Stoichev D A 1974 A device to obtain lysimetric water. Soil Sci. Agrochem. 9, 13–18.

Stoichev D, Stoicheva, Mateva K and Akhchiisky P 1983 Distribution of nitrates in the profile of a leached smolnitza and their relationship to underground water. Soil Sci. Agrochem. 18, 30–40.

Stoichev D, Atanasov I and Qawasmi W 1985 Effect of fertilising of the chemical composition of lysimetric water from leached cinnamonic forest soil. Soil Sci. Agrochem. Plant Proc. 20, 42–47.

Stoichev D, Petkova D, Stoicheva D, Donov D and Stoicheva M 1986 Evaluation of some intensive cropping factors for their residual effect on a leached smolnitza. Soil Sci. Agrochem. Plant Proc. 21, 27–33.

Stoichev D and Stoicheva D 1987 Nitrate content of liquid phase of soil under intensive agriculture. *In* Proc. 5th International Symposium of CIEC, Balatonfured, Hungary. Eds. L Vermes, A Mihálefy and T Nemeth. pp 284–291 MTESZ, Balatonfured, Hungary.

Stoichev D, Angelov G and Stoicheva D 1990 Effect of 1-carbamoyl-3(5) methylpyrazol inhibitor of nitrification on the fate of nitrogen in paddy fields. *In* Proc. 14th International Congress of Soil Science, Kyoto. pp 754–755. ISSS, Kyoto, Japan.

Varvel G E and Peterson T A 1990 Residual soil N as affected by continuous, two-year and four-year crop rotation systems. Agron. J. 82, 958–962.

O. Van Cleemput et al. (eds.), Progress in Nitrogen Cycling Studies, 695–701, 1996.
© *1996 Kluwer Academic Publishers.*

Nitrate attenuation under fertiliser–free grass strips

S.K. White, H.F. Cook and J.L. Garraway
Departments of Environment and Biological Sciences, Wye College, University of London, Wye, Ashford, Kent, TN25 5AH, UK

Key words: buffer strips, drainage, nitrate, water quality

Abstract

One approach in the attempt to reduce diffuse agrochemical pollution is to use buffer zones between the surface waters and the polluted source. Current research on Walland Marsh, East Sussex, UK is concerned with the evaluation of nitrate concentration reduction beneath field margins with and without 5–m wide grass strips, overlying strongly layered soils developed in mineral alluvium with a predominantly loamy texture. Preliminary findings have demonstrated that strips located over these relatively rapidly draining soils are effective in nitrate attenuation.

Introduction

Conversion to arable production, including deep drainage, of former grazing marsh during the last fifty years has threatened dyke water quality and the characteristic aquatic ecosystems associated with permanent grassland. Buffer zone research in the UK and elsewhere has demonstrated that agro-chemical free grassland established in riparian situations is effective in reducing nitrate pollution through subsurface runoff (Muscutt et al., 1993). Research at Wye College, University of London since 1989 on soils of reclaimed alluvial marshland with a drainage density in the order of 6–9 $km.km^{-2}$, typical of the East of England (Cook and Moorby, 1993), has shown grass strips as narrow as 5–m may well be effective for nitrate attenuation. Moorby and Cook (1992) found the nitrate concentration of water samples from dipwells was substantially reduced beneath 5–m wide strips adjacent to dykes in arable land on Walland Marsh, East Sussex, UK. Soils were stratified and mapped as the Newchurch-Walland complex (Green, 1968) which have predominantly silty clay or silty clay loam profiles with the occasional sandy horizon below 1.0 m. On account of the great increases in horizontal permeability with stratification (Childs et al., 1957), the current study was located on sites over soils of the Agney and Gueldford series which are predominantly loamy, but contain intrusions of sandy subsoil horizons. The objective was to examine the observation of Haycock and Burt (1993) that there is a substantial reduction in nitrate concentration 8–m into buffer zones, particularly in soils with a potential for rapid horizontal permeability, and hence pollutant transport.

Methods

The investigation was based upon two treatments, an arable site with a 5–m wide fertiliser-free grass strip, (UK grid reference 59481205) and an arable site without a grass strip, (grid reference 59361251). Both arable fields were cropped with winter wheat and received the same amounts of fertiliser (140 kg N ha^{-1} $annum^{-1}$). Samples were collected weekly from December to May and thereafter every two weekly until the end of July. Sampling was replicated three times at each location, but the sites were not replicated. Both sites were located parallel to the dykes where there were no tile drain outfalls, thus permitting direct movement of pollutants to the dyke (Fig. 1a and Fig. 1b).

Nitrate leaching into the ground water was measured using dipwells installed to 1.9 m below the surface, collecting water indiscriminately from the unsaturated and saturated zones. Samples were also obtained from piezometers at depths between 0.7–0.9 m from the arable site with the grass strip and 1.0–1.2 m from

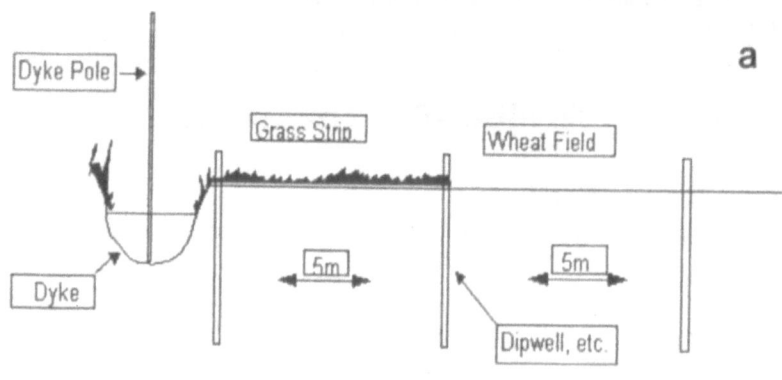

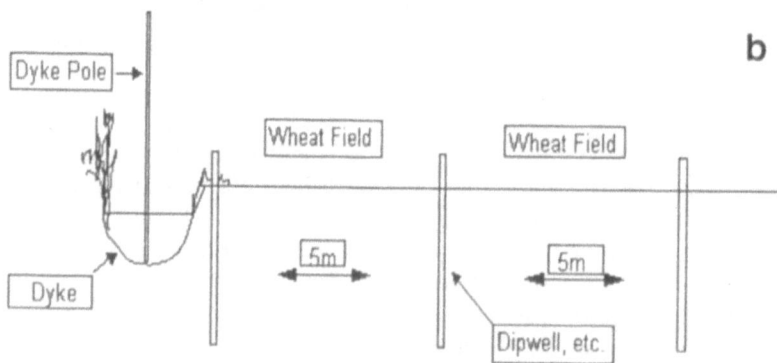

Figure 1. Cross section of arable site with (**a**) and without (**b**) 5–m strip showing positions of the dyke/ strip/ field situation (a) and the dyke/field situation (b). Only dipwells shown as other instruments are perpendicular to the dipwells.

Table 1. The average monthly rainfall (mm) for Walland Marsh, East Sussex for the period of 1993–1994 and 1941 -1970 (Smith and Trafford, 1976). The period 1941–1970 is given as monthly averages (mm)

Date	Oct.	Nov.	Dec.	Jan.	Feb.	March	April	May
1993/94	117	50	138	93	35	43	79	79
1941/70	66	79	68	64	49	45	44	51

the arable site without the grass strip and vacuum (0.7 bar) sampling ceramic cups located at depths of 0.6 m. Instrument sets were placed at the dyke edge, at 5–m (strip/field interface) and 10m into the field (Fig. 1a). At the arable site without a grass strip the edge instruments were located 0.15 m into the narrow grass verge (Fig. 1b). Additional piezometric samplers were put in to a depth of 1.7 m - 1.9 m at the edge position to sample for water that could upwell by deep percolation into the dykes. Data thus collected permits investigation of both the downward and lateral flow of water and nitrate into the saturated zone and dyke, and in turn the assessment of nitrate attenuation under the fertiliser-free strips to be monitored. Each site had two rain gauges which were monitored weekly. The site average rainfall for the period October 1993 - June 1994 was 79.3 mm (Table 1). October, December, January, April and May being generally wetter than the corresponding months for the period of 1941 -1970; November and February were also drier than those from the period of 1941 -1970 (Smith and Trafford, 1976).

The agroclimatic region has an annual rainfall of 683 mm with a potential evapotranspiration of 563 mm for the period 1941–1970. The actual evapotranspiration derived from the UK Met. office MORECS model, for the monthly means from October 1993 to June 1994 were 33.8 mm and 37.4 mm respectively.

All water samples collected were stored at -4 °C on return from the field and analysed usually within 48 hours of collection. Nitrate-nitrogen was determined using an automated colourimetry technique, more specifically, based on alkaline copper (II) - hydrazine reduction using a Burkard Scientific SFA 2 autoanalyser.

Results and discussion

Water flow direction

The results from the dipwells (Fig. 2a) indicates that the water movement was predominantly from the field into the dyke at the arable site with the 5–m strip, from late November until mid-February and again in late April and late May; at the arable site without the strip (Fig. 2b), water movement towards the dyke was evident from mid-December onwards. Fluctuations of the water level in the dykes were also recorded this being 0.28 m at the arable site with the strip and 0.25 m at the arable site without the strip over the period of the investigation. Comparison of Figure 2a and Table 1 indicates that the groundwater movement with respect to the dyke is somewhat sensitive to rainfall, although during drier periods (especially February until April) there is no net water movement.

Attenuation of nitrate

Figures 3a and 3b show the mean nitrate concentration of 3 replicate samples from ceramic cups at 0.6 m in the unsaturated zone. However during mid-December to late-January sampling at the arable site without the strip tended to be from the saturated zone due to the raised watertable. Results from the instrument sets, which were located close to the edge, at 5 m and 10 m, at the arable site with the 5–m strip (Fig. 3a), showed nitrate concentrations appreciably lower than those at the arable site without the strip (Fig. 3b). Differences are also apparent at the 10 m sampling points at both sites even though they received the same fertiliser

application during spring 1994, probably reflecting differences in past management. In both cases the nitrate concentrations recovered from ceramic cups along the edge were appreciably lower than in the field over the sampling period. For the arable site with the strip (Fig. 3a), at 10m the concentrations ranged 27.98 mg L^{-1} (22.12.93) to 0.01 mg L^{-1} (1.6.94), at 5 m concentrations ranged from 1.03 mg L^{-1} (15.12.93) to 0.00 mg L^{-1} (27.7.94), and at the edge from 0.32 mg L^{-1} (15.12.93) to 0.00mg L^{-1} (1.6.94). In contrast, at the arable site without the strip (Fig. 3b) the concentrations were wider ranging; at 10 m concentrations ranged 57.20 mg L^{-1} (22.12.93) to 1.53 mg L^{-1} (29.6.94), at 5 m from 82.95 mg L^{-1} (15.12.93) to 0.09 mg L^{-1} (1.6.94), and at the edge from 6.62 mg L^{-1} (26.1.94) to 0.00 mg L^{-1} (27.7.94). After mid-May the concentration of nitrate in the unsaturated zone at both sites fell, despite fertiliser being applied during March to April by the farmer.

Figures 4a and 4b show the mean nitrate concentration of replicate samples from the dipwells. At the arable site with the strip (Fig. 4a), the measured concentrations were considerably lower than those at the site without the strip, the greatest nitrate-nitrogen concentration measured being only 44.6% of the highest concentration found at the arable site without the strip (Fig. 4b). The concentration range varied from 16.29 mg L^{-1} (19.1.94) to 0.12 mg L^{-1} (27.7.94) for 10 m, 1.78 mg L^{-1} (22.12.93) to 0.04 mg L^{-1} (26.6.94) for 5 m and 0.85 mg L^{-1} (15.12.93) to 0.02 mg L^{-1} (13.4.94) for the edge. It is apparent from Figure 2a, that during drier periods (February to April) the direction of groundwater flow is reversed under the site with the grass strip, i.e. movement is from dyke to the field; however this appears to have little affect on the pattern of nitrate levels in the samples from either the ceramic cups (Fig. 3a) or the dipwells (Fig. 4a), due mainly to the low levels of nitrate in the dyke.

At the arable site without the strip (Fig. 4b), the nitrate content of the dipwell samples was high in both dipwells at 10 m and 5 m (i.e., in the crop). The concentration range was found to vary from 27.19 mg L^{-1} (26.1.94) to 0.82 mg L^{-1} (27.7.94) for 10–m, 36.51 mg L^{-1} (26.1.94) to 0.78 mg L^{-1} (27.7.94) for 5–m and 5.15 mg L^{-1} (16.2.94) to 0.17 mg L^{-1} (13.7.94) for 0.15 m. The concentration of nitrate in the dyke at this site was more variable on account of its function as a main drain receiving water from a variety of sources. The concentration of nitrate decreased from late March onwards particularly with considerable aquatic plant growth in the dyke.

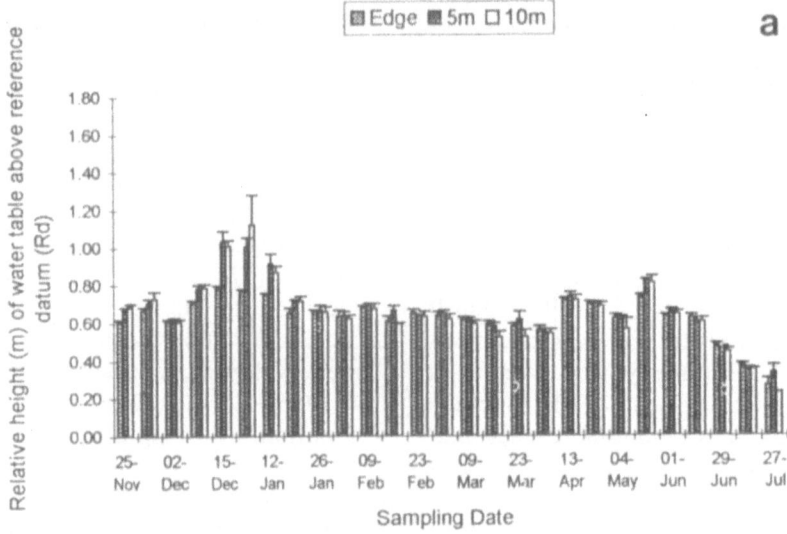

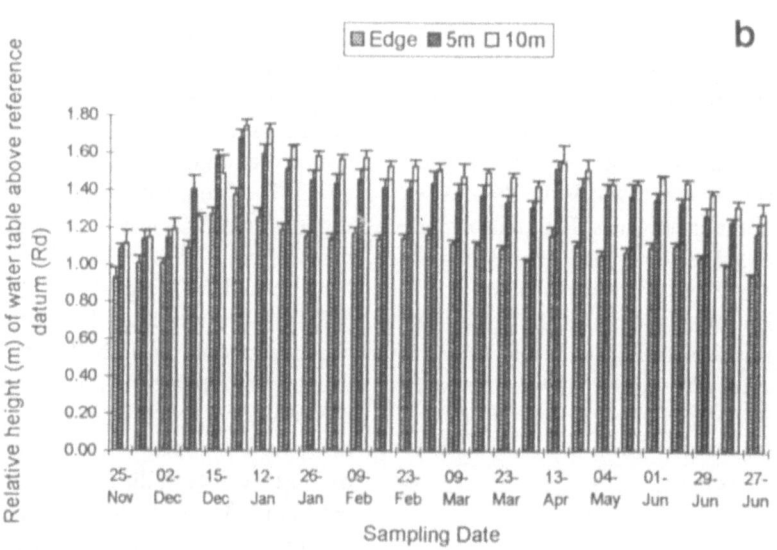

Figure 2. The relative height (m Rd) of the water table in the dipwells; **a** Arable site with 5–m strip. **b** Arable site without 5–m strip. Error bars show standard error of the mean.

The piezometric sampler data (not displayed graphically), while being referenced to a specific depth, was discontinuous on account of watertable fluctuations. There was a general trend towards high concentrations of nitrate at 10 m at both sites ranging, for the site with the 5–m strip, from 19.95-mg L^{-1} to 6.25 mg L^{-1} and for the other site from, 30.86 mg L^{-1} to 3.70 mg L^{-1}, nitrate concentrations being higher from March onwards. The piezometric samplers from the edge and

at 2.5m at the site with the 5–m grass strip, suggested progressive attenuation of nitrate under the strip with ranges from 3.07 mg L^{-1} to 0.33 mg L^{-1}. At both sites the nitrate concentrations measured in samples collected from the deep piezometric sampler (1.7m - 1.9m); ranged from zero to 2.20 mg L^{-1} for the site with the 5–m grass strip and from 2.07 mg L^{-1} to 0.22 mg L^{-1} for the site without the grass strip. These values were comparable to the concentrations found in the shallow-

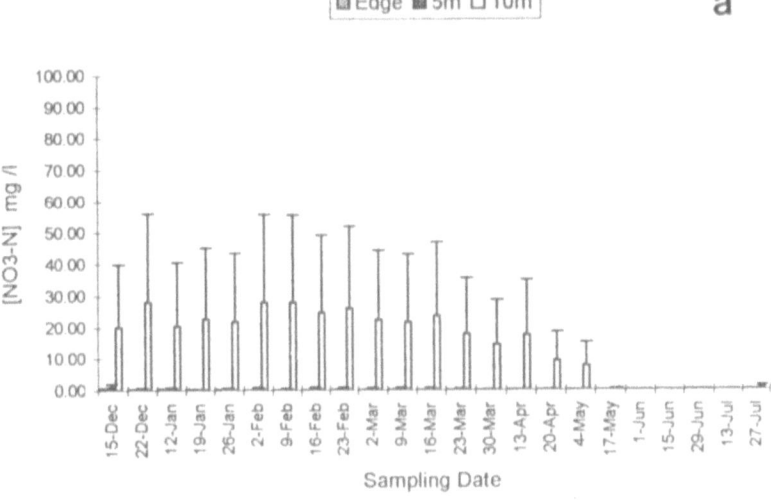

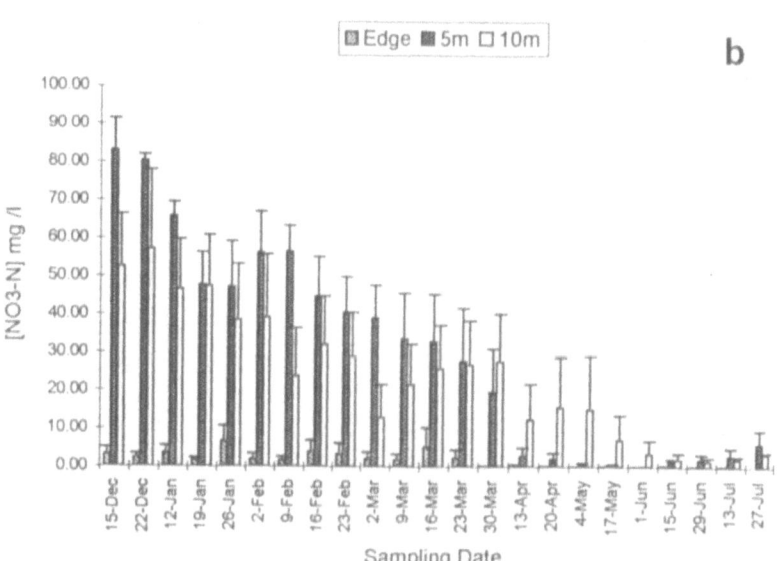

Figure 3. Nitrate-nitrogen concentration in ceramic cups at 0.6 m; **a** Arable site with 5–m strip. **b** Arable site without 5–m strip. Error bars show standard error of the mean.

er edge piezometers (0.7–0.9 m and 1.0–1.2 m); which ranged from 1.22 mg L^{-1} to 0.36 mg L^{-1} and 1.16 mg L^{-1} to 0.16 mg L^{-1} for the sites with and without strips respectively, suggesting low nitrate concentration in any deep percolating water before May 1994. The slightly elevated values were from the samples collected after fertiliser application.

Conclusion

An investigation of nitrate attenuation under narrow (5–m) fertiliser-free grass strips established on sites with relatively permeable subsoil has revealed encouraging results, confirming previous research demonstrating that agrochemical free buffer zones are effective in reducing nitrate pollution provided artificial

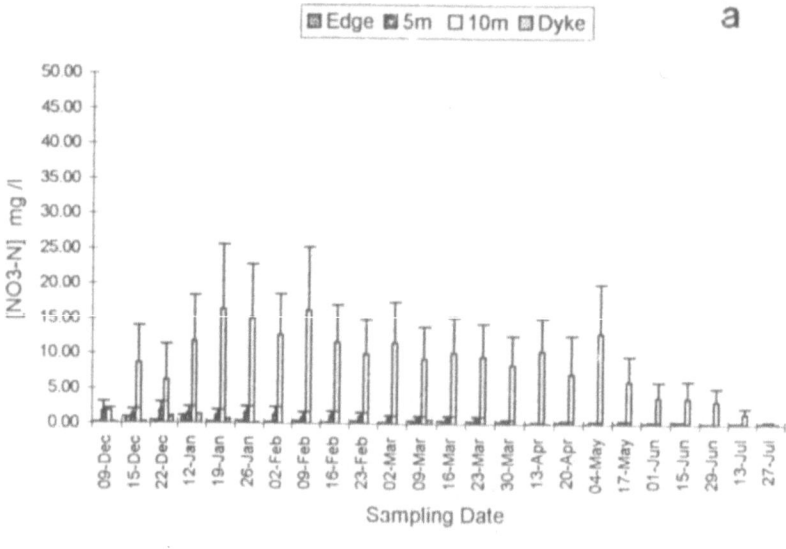

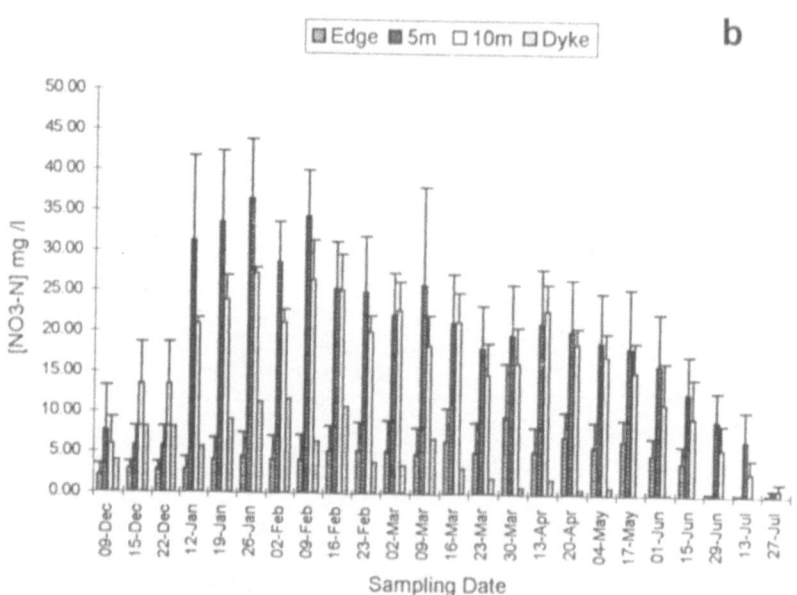

Figure 4. Nitrate-nitrogen concentration in the dipwell; **a** Arable site with 5–m strip. **b** Arable site without 5–m strip. Error bars show standard error of the mean.

underdrainage is absent. In appropriate situations, established grass strips of 5–m compared to the more conventional 15–80 m wide strips, are effective in nitrate attenuation of groundwater. This work is continuing and will be monitoring nitrate-nitrogen concentrations in both saturated and unsaturated zones, as well as investigation of the mechanism of nitrate loss and in particularly in nitrous oxide production.

Acknowledgements

This research is funded by a Ministry of Agriculture, Fisheries and Food studentship. We also acknowledge the assistance of Mr L Cooke, the farmer, English Nature, the National Rivers Authority who supplied the climatic data, Dr C P Burnham for identifying the soil series, Mr K Hart and Mr K Roger for their help in both the field and laboratory.

References

Childs E C, Collis-George N and Holmes J W 1957 Permeability measurements in the field as an assessment of anisotrophy and structural development. J. Soil Sci. 8, 27–41.

Cook H F and Moorby H 1993 English Marshlands reclaimed for grazing: a review of the physical environment. J. Environ. Man. 38, 55–72.

Green R D 1968 Soils of Romney Marsh. Soil Survey of England and Wales, Bulletin 14, Harpenden, UK.

Haycock N E and Burt T P 1993 The sensitivity of rivers to nitrate leaching: The effectiveness of near-stream land as a nutrient retention zone. *In* Landscape Sensitivity. Eds. R J Allison and D S G Thomas. pp 261–272. John Wiley and Sons, London, UK.

Moorby H and Cook H F 1992 The use of fertiliser - free grass strips to protect dyke water from nitrate pollution. Aspects Appl. Biol. 30, 231–234.

Muscutt A D, Harris G L, Bailey S W and Davis D B 1993 Buffer zones to improve water quality: a review of their potential use in UK agriculture. Agric. Ecosyst. Environ. 45, 59–77.

Smith L P and Trafford B D 1976 Climate and Drainage. MAFF Technical Bulletin 34. HMSO, London, UK.

9. ASPECTS RELATED TO METHODOLOGY

O. Van Cleemput et al. (eds.), Progress in Nitrogen Cycling Studies, 705–708, 1996.
© 1996 Kluwer Academic Publishers.

Is deep-freezing a safe method for storing soil samples for inorganic nitrogen determination?

M.J. Esala
Agricultural Research Centre of Finland, FIN-31600 Jokioinen, Finland

Key words: ammonium nitrogen, deep-freezing, inorganic nitrogen, nitrate nitrogen, soil.

Abstract

Soil samples for inorganic nitrogen determination are usually stored deep-frozen to avoid microbial transformations of nitrogen and then thawed before analysis. In some studies, no changes have been observed in the ammonium or nitrate nitrogen content after freezing and thawing of samples, whereas in others considerable changes have been found in one or both.

In this work, three laboratory experiments were performed to study the effect of deep-freezing and thawing on the ammonium and nitrate nitrogen contents of a sandy soil, a clay soil, and a peat soil. A special mill for grinding the frozen samples to minimize these changes, was tested. One experiment was done to study whether the time of extraction of the samples could be extended to 20 hours.

Thawing of the samples increased the concentration of nitrate in the extracts in one experiment and that of ammonium in another experiment. Grinding increased the concentration of ammonium in both of the experiments. In the third experiment, there was no change in either form of nitrogen during freezing of the soil nor up to 4 hours of thawing. After that an increase occurred, especially in the content of nitrate nitrogen. This increase occurred few hours after a rise in temperature above zero degrees. In the fourth experiment, extending the time of extraction from 0.5 or 1 hour to 20 hours increased the concentration of ammonium in the extracts. Nitrate content was also increased slightly.

It is concluded that the divergence in the results of various studies on the subject are a consequence of mineralization of soil nitrogen during the thawing period. During this period, the temperature of the soil should not rise above zero for more than the minimum time. The period of extraction should not exceed two hours.

Introduction

Determination of inorganic nitrogen is essential if we want to know plant available or leacheable nitrogen in soil. The microbial transformations of nitrogen in soil during pre-treatment and storing can cause a large error in the result. The best method would be to analyse the soil samples as soon as possible after sampling. However, in routine work, when a large number of samples have to be analysed, this is often not possible. Therefore, the samples have to be stored the meantime between sampling and analysing.

Deep-freezing is the method most commonly applied when soil samples are stored for inorganic nitrogen determination (Keeney and Nelson, 1982). Other methods applied are air-drying or treatment with certain chemicals. These two latter methods, however,

have been shown to be unsatisfactory in storing the soil samples without any changes in ammonium or nitrate nitrogen contents.

Usually the samples are thawed before analysis to allow homogenization and weighing. The results of several studies on the subject are nevertheless inconsistent. In some studies, no changes have been observed in the ammonium or nitrate nitrogen content after freezing and thawing of samples (Gasser, 1958; Westfall et al., 1978), whereas in others considerable changes have been found in one or both (Allen and Grimshaw, 1962; Nelson and Bremner, 1972; Robinson, 1967). To avoid the microbial changes when thawing the samples, a mill for grinding the frozen soil samples is available from a Swedish manufacturer.

Usually, the soil samples are extracted for two hours or less (Keeney and Nelson, 1982). Extraction

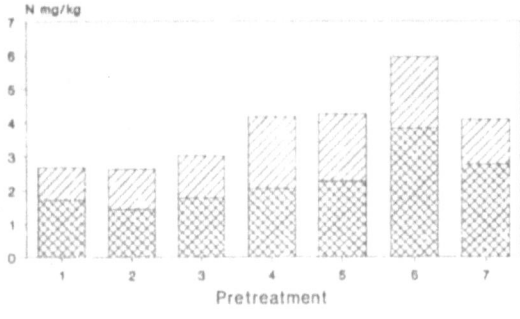

Figure 1. Effect of pretreatment on ammonium and nitrate nitrogen determinations. 1) Fresh soil, KCl added in the field, 2) fresh soil, KCl added in the laboratory, 3) freezing , KCl added to frozen soil, 4 and 5) freezing, thawing, 6) freezing, grinding, thawing, 7) freezing, grinding.

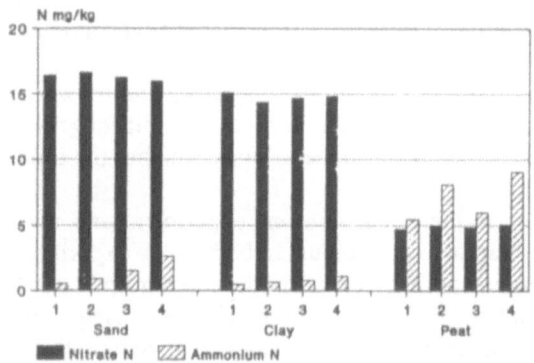

Figure 2. Effect of pretreatment of soil samples on ammonium and nitrate nitrogen contents in soil. 1) Fresh Soil, 2) frozen and thawed soil, 3) frozen and ground soil, 4) frozen, ground, and thawed soil.

overnight has been suggested for clay soils to improve the dispersion of the clay aggregates (Linden, 1981). It would also be more convenient in routine work.

Materials and methods

Four laboratory experiments were performed to study the effect of deep-freezing, thawing and time of extraction on the ammonium and nitrate nitrogen contents of three soils (Esala, 1994 and 1995). In the three first experiments, the soils were 1) a sandy soil (total N=0.140%, organic carbon=1.659%), 2) a heavy clay soil (total N=0.351%, organic carbon=4.379%), and 3) a peat soil (total N=1.581%, organic carbon=27.980%). In the fourth experiment, soil no 1 as above and a heavy clay topsoil and subsoil were studied. A special mill for grinding the frozen samples to minimize the microbial changes, was tested.

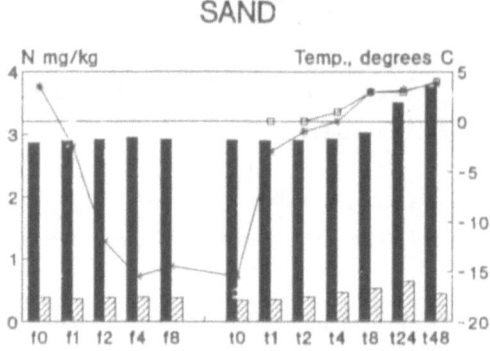

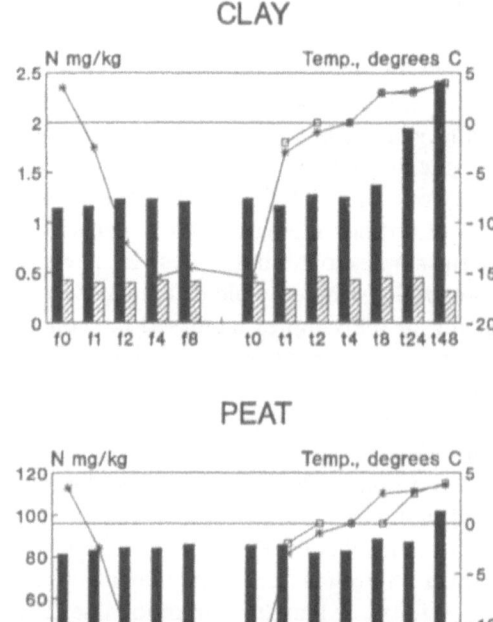

Figure 3. Changes in temperature and in ammonium and nitrate nitrogen contents in soil samples during freezing and thawing. Numbers refer to hours of freezing (f) and thawing (t).

The soils were sieved through a 6 mm sieve and frozen at -18 °C. After pretreatment the soil samples were extracted with $2M$ KCl (100 g soil, 250 mL KCl). The soils were extracted overnight in the first experiment and for two hours in the second and third exper-

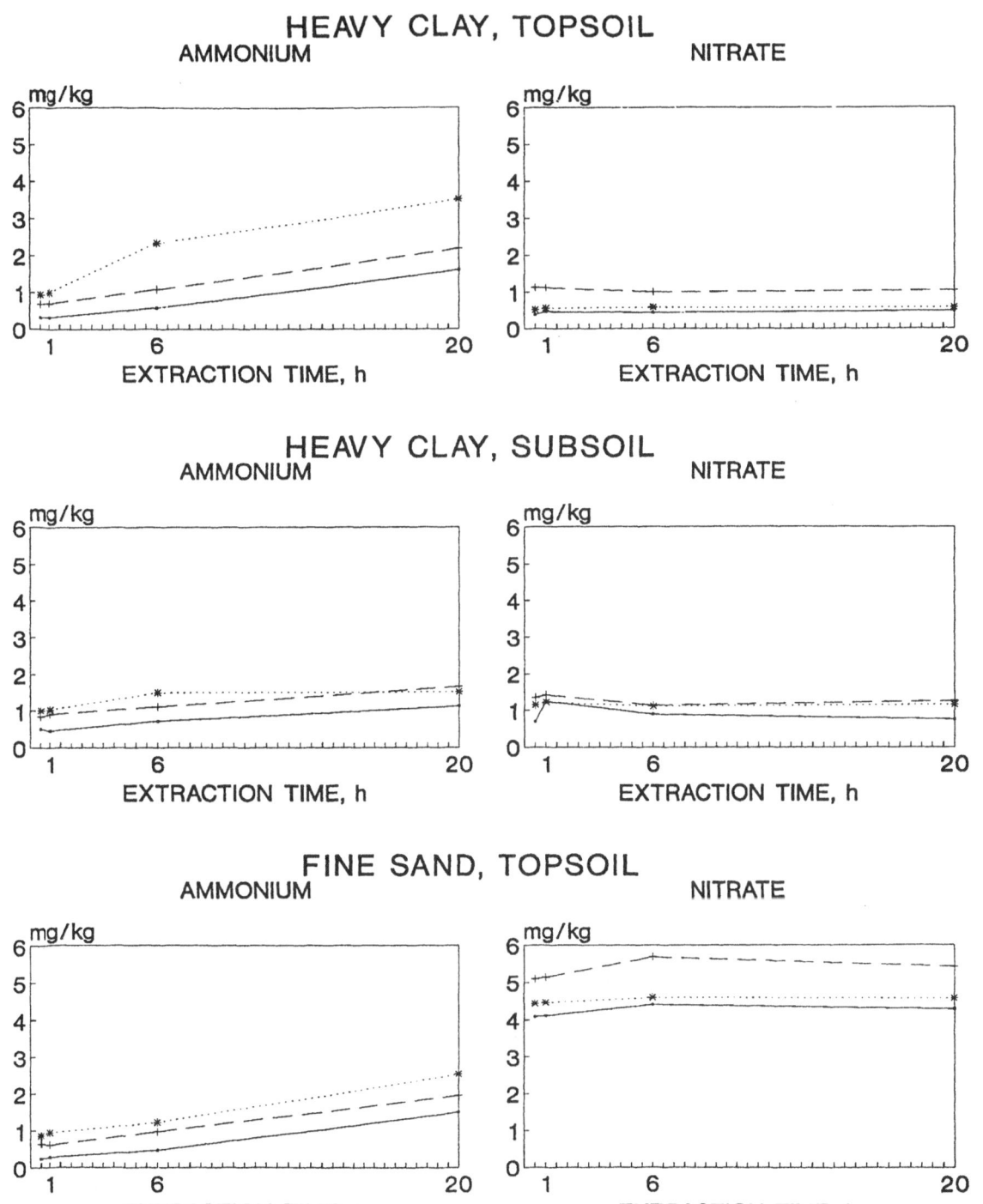

Figure 4. Effect of pretreatment and extraction time on ammonium and nitrate nitrogen determination. (—■—) fresh soil, (– + –) freezing, thawing (– * –) freezing, grinding.

iment. The ammonium and nitrate nitrogen content in the extracts was determined by a Skalar autoanalyser.

Results

The contents of ammonium or nitrate nitrogen did not change when the soil was taken from the field to the

708

laboratory and frozen (Fig. 1 and 2). Thawing of the samples increased the concentration of nitrate in the extracts in one experiment and that of ammonium in another experiment. In both of the experiments, grinding of the soil increased the content of ammonium nitrogen.

In the third experiment, there was no change in either form of nitrogen during freezing of the soil nor up to 4 hours of thawing (Fig. 3). After that an increase occurred, especially in the content of nitrate nitrogen. This increase occurred few hours after a rise in temperature above zero degrees.

In the fourth experiment, extending the time of extraction from 0.5 or 1 hour to 20 hours increased the concentration of ammonium in the extracts (Fig. 4). Nitrate content was also increased slightly.

Conclusions

The divergence in the results of various studies on deep-freezing and thawing is a consequence of mineralization of soil nitrogen during the thawing period. This period should not be extended longer than is necessary to render the soil loose enough for homogenizing and weighing. The temperature of the soil should not rise above zero for more than the minimum time. The period of extraction should not exceed two hours.

Grinding of the soils did not prove to be a safe method of avoiding the changes in ammonium nitrogen content of the soils. Apparently, to avoid these changes, the mill would need some modifications to make the ground soil coarser.

In principle, deep-freezing and thawing of samples for inorganic nitrogen determination is a reliable method providing care is taken in thawing the samples. The safest procedures for pretreatment would be to extract the soil samples fresh. If the samples have to be frozen, they should be thawed just enough to make the soil loose. There should still be some ice crystals in the soil at the end of the thawing period. If time allows, the fresh soil samples could also be weighed in plastic bags for freezing and then extracted frozen later.

References

Allen S E and Grimshaw H M 1962 Effect of low-temperature storage on the extractable nutrient ions in soils. J. Sci. Food Agric. 13, 525–529.

Esala M J 1994 Deep-freezing pretreatment and time of extraction of soil samples for inorganic nitrogen determination. Commun. Soil Sci. Plant Anal. 25, 651–662.

Esala M J 1995 Changes in the extractable ammonium- and nitrate-nitrogen contents of soil samples during freezing and thawing. Commun. Soil Sci. Plant Anal. 26, 61–68.

Gasser J K R 1958 Use of deep-freezing in the preservation and preparation of fresh soil samples. Nature 181, 1334–1335.

Keeney D R and Nelson D W 1982 Nitrogen - inorganic forms. In Methods of Soil Analysis. Part 2. Chemical and microbiological Properties. Agronomy No 9. Eds. A L Page et al. pp 643–698. American Society of Agronomy, Madison, WI, USA.

Linden B 1981 Ammonium- och nitratkvävets rörelser och fördelning i marken. II. Metoder för mineralkväveprovtagning och analys. Swedish Univ. Agric. Sci., Dept. Soil Sci., Rep. 137.

Nelson D W and Bremner J M 1972 Preservation of soil samples for inorganic nitrogen analysis. Agron. J. 64, 196–199.

Robinson J B D 1967 The preservation unaltered of mineral nitrogen in tropical soils and soil extracts. Plant and Soil 27, 53–80.

Westfall D G, Henson M A and Evans E P 1978 The effect of soil sample handling between collection and drying on nitrate concentration. Commun. Soil Sci. Plant Anal. 9, 169-185.

O. Van Cleemput et al. (eds.), Progress in Nitrogen Cycling Studies, 709–712, 1996.

Continuous cultivation of *Nitrosomonas europaea* with complete biomass retention

W. Tappe, S. Rittershaus, C. Tomaschewski and J. Groeneweg
Forschungszentrum Jülich GmbH, Institut für Biotechnologie 3, P.O. Box 1913, D-52425 Jülich, Germany

Key words: biomass retention, cell volume, maintenance, *Nitrosomonas europaea*, retentostat

Abstract

A bioreactor (360 mL working volume) was developed for continuous axenic cultivation of bacteria with internal retention of biomass. *Nitrosomonas europaea* was cultivated in this system at a constant ammonia input of 0.57 mmol NH_4Cl L^{-1} h^{-1}. Since only cell-free filtrate left the reactor at a rate equivalent to the input rate, *Nitrosomonas* biomass was accumulating. After four weeks of operation, the biomass concentration reached a stable maximum of 2.7×10^9 cells mL^{-1} (equivalent to 400 mg dry matter per liter) and the growth rate approached zero. In this state, the ammonia input met the energy required for maintenance. The calculated specific maintenance energy demand was 0.02 mg NH_3-N mg DM^{-1} h^{-1}. Cell volumes decreased from an arithmetic mean volume of 0.6 μm^3 at μ_{max} to 0.25 μm^3 at zero growth. Nutrient additions to the non growing culture resulted in a shift of the whole population to bigger cell volumes thus indicating homogeneity of the population at zero growth.

Introduction

Autotrophic nitrifiers like *Nitrosomonas* are ecologically important, since they catalyze the first step in the oxidative pathway of the nitrogen turnover. Their maximum specific growth rates are low with doubling times in the range of 10–25 h and the requirement for reverse electron flow for synthesis of reducing power results in low growth yields (Keen and Prosser, 1987). In addition, they are inhibited by self induced conditions such as low pH and high NO_2^- (HNO_2) concentrations (Hunik et al., 1992). These unfavourable factors have to be considered when pure culture studies are carried out with relatively highly concentrated nutrient solutions (Groeneweg et al., 1994). Since maintenance energy demand plays a major role as a size-limiting factor for natural populations (Belser, 1979), there is more need for studies on pure cultures involving maintenance, starvation and reactivity under various growth-limiting conditions. The aim of this study was, therefore, to grow a pure culture of *Nitrosomonas europaea* on a continuous feed supply completely retaining the biomass until zero net increase thus allowing the direct determination of maintenance energy demand.

Materials and methods

Bacterial strains, medium and growth conditions

Nitrosomonas europaea (strain Nm 35, obtained from H -P Koops, University of Hamburg) was grown in inorganic medium contained per liter: NH_4Cl, 0.306 g; Na_2HPO_4, $12H_2O$, 0.1 g; $MgSO_4$. $7H_2O$, 0.05 g; $CaCl_2$. $2H_2O$, 0.015 g; $NaHCO_3$, 1,13 g; Fe-EDTA, 1 mL of the stock solution (Triplex III, 9.3 g L^{-1} and $FeSO_4$, 6.95 g L^{-1}); trace minerals, 1 mL of stock solution.

The medium was supplied continuously at a rate of 36 mL h^{-1} (Abimed HP4 pump) to give a dilution rate of 0.1 h^{-1}, and the filtrate was removed at the same rate. The inoculum consisted of 360 mL suspension with a concentration of $1.5 \cdot 10^7$ cells mL^{-1} from a continuous culture. During biomass accumulation the pH remained stable at 7.8, and there was no need for pH adjustment. The bioreactor was aerated with 15 L air h^{-1}. All experiments were carried out at 25 ± 1 °C. The cultures were regularly checked for heterotrophic contamination by incubating samples in tryptic soy broth (Merck) for two weeks at 30 °C.

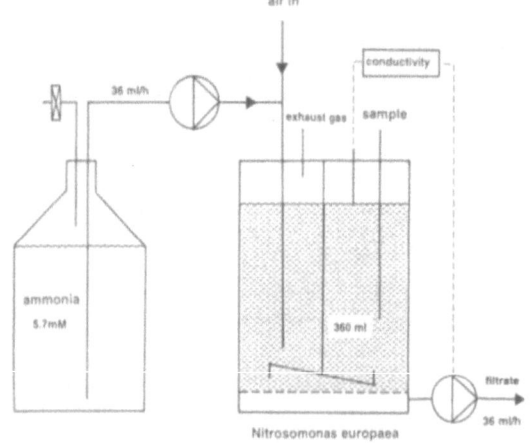

Figure 1. Schematic diagram of the bioreactor with internal biomass retention (retentostat).

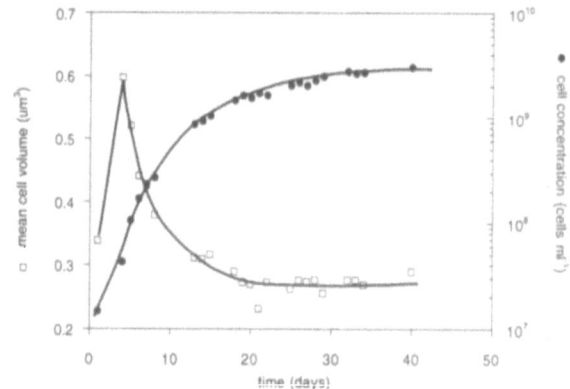

Figure 2. Cell concentrations and mean cell volumes during the run with complete biomass retention.

Bioreactor with internal biomass retention (retentostat)

The bioreactor consisted of a glass cylinder pressed together by wing bolts between a bottom and a cover made from stainless steel (Fig. 1). A membrane filter (Millipore, PVDF, 0.22 μm pore size) was placed at the bottom of the reactor upon a support membrane and a stainless steel grid so that a cell-free filtrate could be drawn off from underneath the grid. A teflon-covered magnetic stirring bar positioned above the membrane produced a cross-flow effect and prevented the bacteria from clogging. The nutrient solution was dosed by a peristaltic pump, then dropped through a glass sphere which was also gassed with filtered air and pressed into the reaction vessel close to the stirrer. This effectively prevented bacterial growth to the feed supply and ensured thorough mixing. A constant culture volume of 360 mL was maintained by means of a conductivity probe that triggered the tube pump for filtrate removal.

Analytical methods

Ammonia and nitrite were measured with a segmented-flow autoanalyzer (Skalar). Cell counts, cell volume distribution and mean cell volumes were determined with an electronic particle counter (Elzone PC 280; 18 μm orefice). In earlier experiments, a direct correlation between biovolume and POC was found. Thus, only small samples (max. 3 mL per day) could be taken for electronic particle counting to minimize disturbances in growth kinetics and POC and dry matter were calculated from the measured biovolume.

Results

A pure culture of *Nitrosomonas europaea* was continuously grown in a retentostat for 6 weeks with complete biomass retention. An ammonia input concentration of 5.7 mM ensured energy-limited growth and excluded product inhibition (NO_2^-/HNO_2) as it occurred in former chemostat experiments at higher input concentrations (30 mM NH$_4$Cl, Groeneweg et al., 1994). Despite the low concentration of 5.7 mM NH$_4$Cl, the retentostat allowed higher amounts of NH$_4$Cl to be fed per unit time as compared to a chemostat, because the biomass retention time is uncoupled from the hydraulic retention time. The substrate could be provided at a 'dilution rate' of 0.1 h^{-1} exceeding the maximum growth rate of *N. europaea* of 0.05 h^{-1} at 25 °C.

Nitrosomonas cells from a continuously grown preculture were inoculated to give a concentration of 1.5×10^7 cells mL^{-1} at the beginning of the experiment. During the following 5 days growth was not limited by ammonia and *Nitrosomonas* reached a μ_{max} of 0.05 h^{-1}. This phase was characterized by a drastic increase in mean cell volumes from initially 0.32 μm^3 up to 0.6 μm^3.

Subsequently, ammonia became growth-limiting and the pH remained constant at 7.8. During the accumulation of biomass, the ratio of ammonia per cell and unit of time decreased continuously. The decrease in biomass-specific ammonia availability resulted in decreasing growth rates accompanied by a diminution in mean cell volume (Fig. 2).

Finally, the culture reached a stable maximum cell concentration of 2.7 . 10^9 cells mL^{-1} equivalent to 180 mg L^{-1} POC or 400 mg L^{-1} of dry matter. At this stage, energy generated from ammonia oxidation must

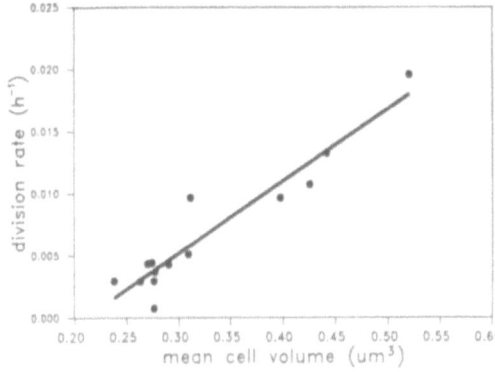

Figure 3. Mean cell volumes versus division rates (obtained from the derivative of increase in cell concentration versus time shown in Figure 2).

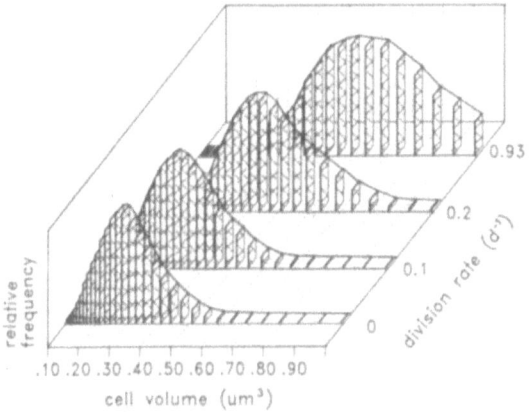

Figure 4. Cell volume distribution at four selected division rates.

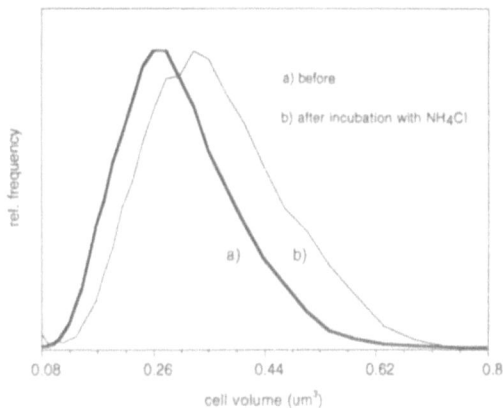

Figure 5. Cell volume distribution before (**a**) and after (**b**) incubation with 0.7 mM NH$_4$Cl for 12 hours.

somonas cells with a mean volume of 0.25 μm^3 already indicates a homogeneous population. Faster growing subpopulations would lead to an uneven distribution in cell volume class (Fig. 4). The reactivity of the non-growing cells was tested by incubating diluted samples (10^7 cells mL^{-1}) with 0.7 mM NH$_4$Cl for 12 hours to permit a growth response. From the almost homogeneous shift in mean cell volume from 0.25 to 0.31 μm^3 after substrate addition, we conclude that really the whole population was in an active state (Fig. 5).

Discussion

Although nutrient limited growth and starvation are quite common situations for natural microbial populations (Morgan and Dow, 1986; Poindexter, 1981; Schlegel, 1992), the prerequisites for the experimental simulation of such growth conditions with pure bacterial cultures are insufficient.

Chemostats, often proclaimed as appropriate tools for experimental microbial studies (Gottschal and Dijkhuizen, 1988; Jannasch, 1967), are unsuitable for the investigation of bacterial physiology at very poor or zero growth. This is mainly based on the fact that inhomogeneities in time and space arise with decreasing dilution rates (Hansford and Humphrey, 1966).

Furthermore, zero growth conditions are principally impossible to obtain in chemostats, because any addition of substrate results in a volumetric adequate loss of biomass suspension that is compensated by growth again. In contrast the retentostat enables the cultivation without any loss or with controlled loss of bacteria despite continuous feeding. With complete biomass retention, *Nitrosomonas* run through all

have been used completely for meeting the maintenance energy demand. A specific maintenance energy demand of 0.02 mg NH$_x$-N mg DM^{-1} h^{-1} was calculated as the amount of ammonia consumed per mg dry matter and hour at zero growth.

In order to verify that growth was limited by the specific substrate availability, the ammonia input was shifted up by doubling the dilution rate from 0.1 h^{-1} to 0.2 h^{-1}. In this manner, the amount of ammonia provided per unit of time was doubled without changing the input concentration. An increase in cell concentration indicated that ammonia was the growth-limiting factor (not shown in Fig. 2).

The division rate obtained from the derivative of the increase of cell concentration with time correlated linearly with the mean cell volume (Fig. 3). Therefore, the mean cell volume could serve as a parameter for the actual growth rate. The cell volume distribution should reflect the homogeneity of the population. The narrow log-normal distribution of non-growing *Nitro-*

states of growth from μ_{max} at the beginning until zero growth when the maximum biomass concentration was reached and all energy was consumed for maintenance. The specific energy demand for maintenance (m) of *N. europaea* in this study was about one order of magnitude lower, when measured directly, compared with the scarce literature reports (Helder and de Vries, 1983; Keen and Prosser, 1987), where m was calculated indirectly with the method of Pirt (1965). Usually, the method suggested by Pirt (1965) is employed for the calculation of maintenance energy demand from chemostat cultures, but a growth rate independent m as formulated by Pirt is not in accordance with our findings and not so with many other authors' results (Arbige and Chesbro, 1982; Bulthuis et al., 1989; Neijssel and Tempest, 1976). The evident deviations of growth yields below a 'critical' dilution rate from the calculated values based on Pirt's original maintenance energy model gave rise to the so-called 'dormant cell model' (Pirt, 1987). Herein, the formation of dormant cells with zero maintenance energy demand is postulated to account for the observed decreasing overall m with very low dilution rates.

Our experiments with *N. europaea* gave no hint on a dormant or non-viable fraction of the population, not even when growth ceased. Since the homogeneity and reactivity of the population was tested indirectly as a substrate-induced shift in mean cell volume, a further proof for homogeneity would be the combination of this procedure with methods for the direct visualisation of active and inactive cells like e.g. the CTC method described in Rodriguez et al. (1992). The importance of maintenance energy demand and reactivity for population size and competition of bacteria in nature requires further research on bacterial behaviour at very slow growth, zero growth and decay. For all these purposes, the retentostat is a suitable tool.

References

Arbige M and Chesbro W R 1982 Very slow growth of *Bacillus polymyxa*: Stringent response and maintenance energy. Arch. Microbiol. 132. 338–344.

Belser L W 1979 Population ecology of nitrifying bacteria. Ann. Rev. Microbiol. 33, 309–333.

Bulthuis B A, Koningstein G M, Stouthamer A H and Verseveld H W 1989 A comparison between aerobic growth of *Bacillus licheniformis* in continuous culture and partial-recycling fermenter, with contributions to the discussion on maintenance energy demand. Arch. Microbiol. 152, 499–507.

Gottschal J C and Dijkhuizen L 1988 The place of the continuous culture in ecological research. *In* Handbook of Laboratory Model Systems for Microbial Ecosystems, Vol. I. Ed. J W T Wimpenny. pp 19–43. CRP Press, Boca Raton, FL. USA.

Groeneweg J, Sellner B and Tappe W 1994 Ammonia oxidation in *Nitrosomonas* at NH_3 concentrations near k_m: Effects of pH and temperature. Wat. Res. 28, 2561–2566.

Hansford G S and Humphrey A E 1966 The effect of equipment scale and degree of mixing on continuous fermentation yield at low dilution rates. Biotechnol. Bioeng. 8, 85–96.

Helder W and De Vries R T P 1983 Estuarine nitrite maxima and nitrifying bacteria (Ems-Dollard estuary). Neth. J. Sea Res. 17 (1), 1–18.

Hunik J H, Meijer J H G and Tramper J 1992 Kinetics of *Nitrosomonas europaea* extreme substrate, product and salt concentrations. Appl. Microbiol. Biotechnol. 37, 802–807.

Jannasch H W 1967 Enrichments of aquatic bacteria in continuous culture. Arch. Microbiol. 59, 165–173.

Keen G A and Prosser J I 1987 Steady-state and transient growth of autotrophic nitrifying bacteria. Arch. Microbiol. 147, 73–79.

Morgan P and Dow C S 1986 Bacterial adaptations for growth in low nutrient environments. *In* Microbes in extreme Environments. Ed. R A Herbert. pp 187–192. Academic Press, London, UK.

Neijssel O M and Tempest D W 1976 Bioenergetic aspects of aerobic growth of *Klebsiella aerogenes* NCTC 418 in carbon-limited and carbon-sufficient chemostat culture. Arch. Microbiol. 107, 215–221.

Pirt S J 1965 The maintenance energy of bacteria in growing cultures. Proc. R. Soc. London Ser. B 163, 224–231.

Pirt S J 1987 The energetics of microbes at slow growth rates: Maintenance: energies and dormant organisms. J. Ferment. Technol. 65 J, 173–177.

Poindexter 1981 Oligotrophy: Fast and famine existence. Adv. Microb. Ecol. 5, 63–89.

Rodriguez G G, Phipps D, Ishiguro K and Ridgway H F 1992 Use of a fluorescent redox probe for direct visualization of actively respiring bacteria. Appl. Environ. Microbiol. 58, 1801–1808b.

Schlegel H G 1992 Allgemeine Mikrobiologie 7, Aufl. pp 562–563. Georg Thieme-Verlag, Stuttgart, Germany.

Index

Developments in Plant and Soil Sciences

Developments in Plant and Soil Sciences

Developments in Plant and Soil Sciences

Developments in Plant and Soil Sciences

70. M. Rahman, A. Kumar Podder, C. van Hove, Z.N. Tahmida Begum, T. Heulin and A. Hartmann (eds.): *Biological Nitrogen Fixation Associated with Rice Production.* 1996
ISBN 0-7923-4197-X

Kluwer Academic Publishers – Dordrecht / Boston / London